Chemistry and Life

An Introduction to General, Organic, and Biological Chemistry

Sixth Edition

John W. Hill

University of Wisconsin–River Falls

Stuart J. Baum

SUNY Distinguished Teaching Professor,
State University of New York, Plattsburgh

Rhonda J. Scott-Ennis

Southern Adventist University

PRENTICE HALL
Upper Saddle River, NJ 07458

Library of Congress Cataloging-in-Publication Data

Hill, John William.
Chemistry and life: an introduction to general, organic, and biological chemistry.—6th ed. / John W. Hill, Stuart J. Baum, Rhonda J. Scott-Ennis.
p. cm.
Includes index.
ISBN 0-13-082181-0
1. Chemistry. I. Baum, Stuart J. II. Scott-Ennis, Rhonda J.

QD31.2 H56 2000 99-048049
540-dc21

Executive Editor: *John Challice*
Media Editor: *Paul Draper*
Development Editor: *Mary Ginsburg*
Production Editor: *Debra A. Wechsler*
Editor in Chief: *Paul F. Corey*
Associate Editor in Chief, Development: *Carol Trueheart*
Executive Managing Editor: *Kathleen Schiaparelli*
Assistant Managing Editor: *Lisa Kinne*
Assistant Vice President of Production and Manufacturing: *David W. Riccardi*
Marketing Manager: *Steve Sartori*
Director of Marketing, ESM: *John Tweeddale*
Manufacturing Buyer: *Michael Bell*
Manufacturing Manager: *Trudy Pisciotti*
Editorial Assistants: *Gillian Buonanno; Amanda K. Griffith*
Art Director: *Joseph Sengotta*
Associate Creative Director: *Amy Rosen*
Director of Creative Services: *Paul Belfanti*
Art Manager: *Gus Vibal*
Art Editor: *Karen Branson*
Interior Designer: *Judy Matz-Coniglio*
Cover Designer: *John Christiana*
Photo Researcher: *Stuart Kenter Associates*
Photo Research Administrator: *Melinda Reo*
Art Studios: *Academy Artworks, Inc.; selected figures © Kenneth Eward/Biografx*
Cover Photo: *Alan and Linda Detrick/ALD Photo, Inc.*

Printed in the United States of America

10 9 8 7 6 5 4 3 2 1

ISBN 0-13-082181-0

Prentice-Hall International (UK) Limited, *London*
Prentice-Hall of Australia Pty. Limited, *Sydney*
Prentice-Hall Canada Inc., *Toronto*
Prentice-Hall Hispanoamericana, S. A., *Mexico*
Prentice-Hall of India Private Limited, *New Delhi*
Prentice-Hall of Japan, Inc., *Tokyo*
Pearson Education Asia Pte. Ltd.
Editora Prentice-Hall do Brasil, Ltda., *Rio de Janeiro*

Brief Contents

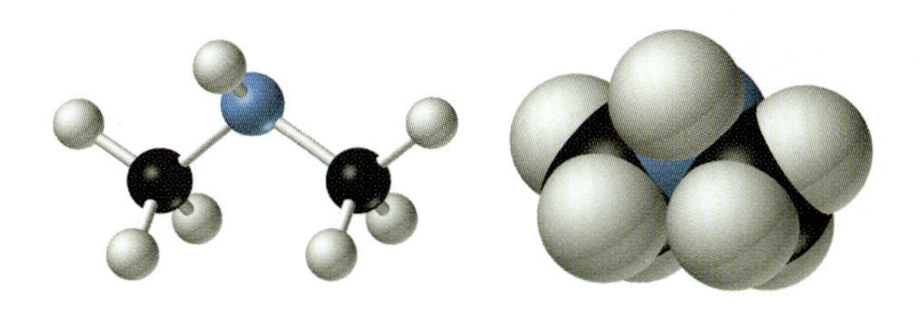

Application Boxes

Contents

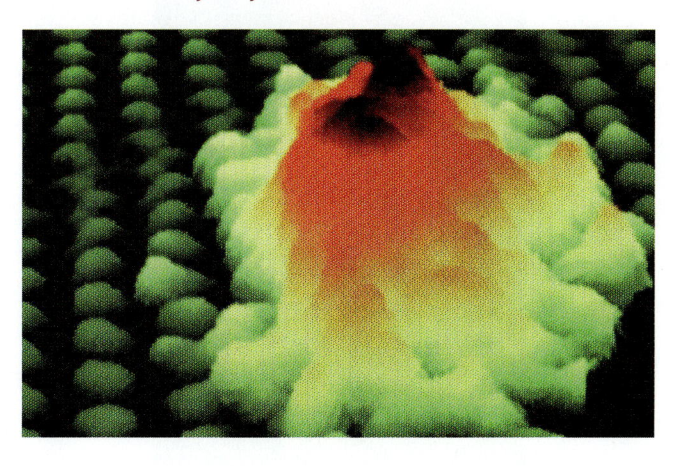

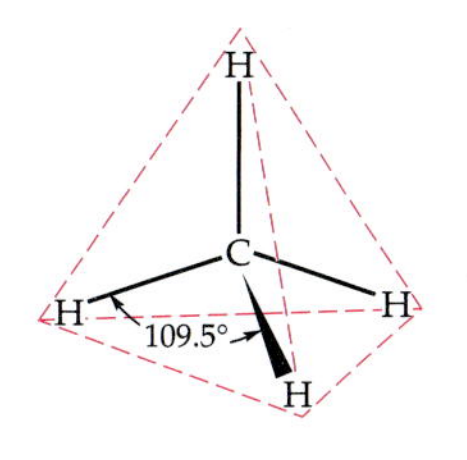

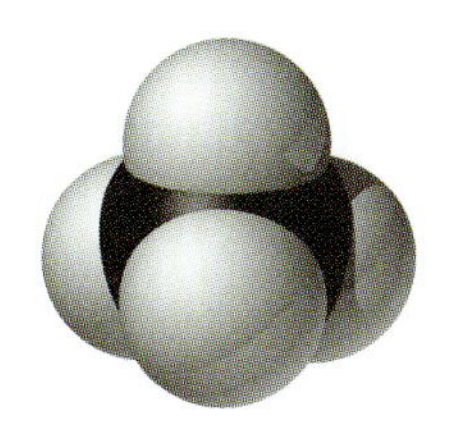

11 Electrolytes 271

12 The Atomic Nucleus 305

13 Hydrocarbons 328

14 Alcohols, Phenols, and Ethers 367

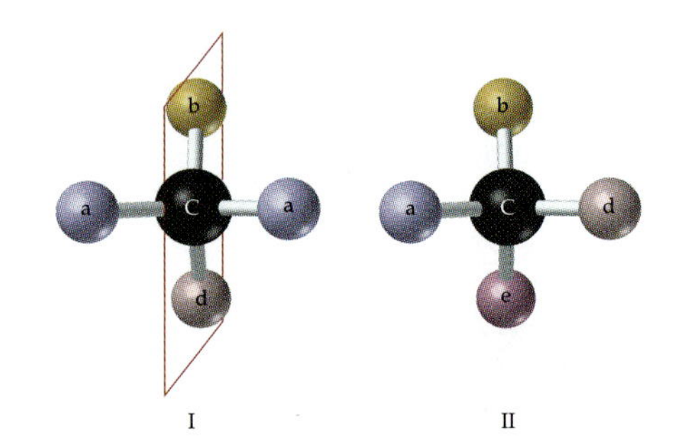

b
a
C
a
d
b
a
C
d
e
I
II

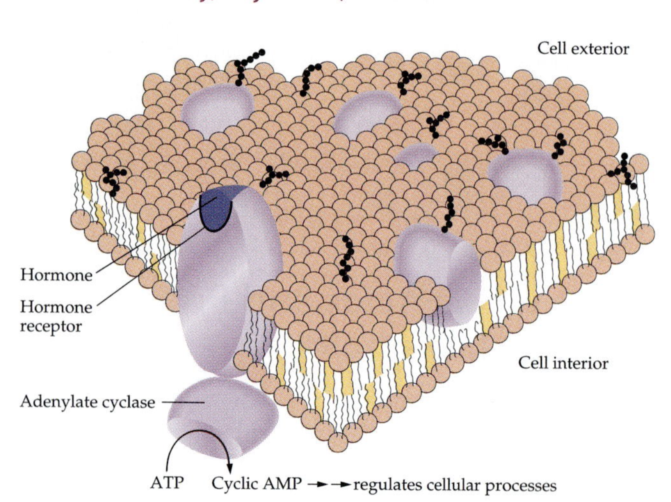

Cell exterior
Hormone
Hormone receptor
Cell interior
Adenylate cyclase
ATP
Cyclic AMP → → regulates cellular processes

Preface

Our world has been transformed by science and technology. The impact of science on the quality of human life is profound. To beginning students, the scientific disciplines that daily influence their lives often seem mysterious and incomprehensible. Those of us who enjoy the study of science, however, find it a fascinating and rewarding experience precisely because it can provide reasonable explanations for seemingly mysterious phenomena.

Chemistry and Life has been written in that spirit. We help explain apparently obscure phenomena in an informal, readable style. We assume that the student has little or no chemistry background, so we clearly explain each new concept as it is introduced. Chemical principles and biological applications are carefully integrated throughout the text, with liberal use of drawings, diagrams, and photographs.

For this new edition, the entire text has been updated to reflect the latest scientific knowledge. In addition, we have responded to suggestions of users and reviewers of the fifth edition and used our own writing and teaching experience to make some important improvements.

Effective, Flexible Organization

Our selection of topics and choice of examples make the text especially appropriate for students in health and life sciences, but it is also suitable for anyone seeking to become a better-informed citizen of our technological society. The text provides ample material for a full-year course. We consciously increase the sophistication of chemical understanding as the student progresses through the chapters.

Selected Topics Offer Flexibility to the Instructor

We have included in this edition, as in past editions, a number of Selected Topics that cover key optional material in additional detail. These are introduced at the appropriate times (for example, the Selected Topic on Vitamins, which discusses key coenzymes, follows immediately after the chapter on Enzymes), and each includes its own end-of-topic problems. These Selected Topics offer instructors maximal flexibility; they may be omitted or assigned as outside reading without loss of continuity.

New to this edition:

In this new edition, unit conversions and significant figures are now in Chapter 1. VSEPR theory and the shapes of molecules are in Chapter 3, with our discussion of chemical bonding. Nuclear chemistry is now Chapter 12, following the general chemistry part of the text and just before the organic chapters. The chapters dealing with metabolism (24–27) have been extensively reorganized and rewritten and include a more complete discussion of anabolic pathways. Chapter 24 is now an overview of metabolism, with a particular emphasis on digestion and energy production (Krebs cycle and cellular respiration). Chapter 25 is concerned with the metabolic pathways unique to the metabolism of carbohydrates; Chapter 26 discusses the unique metabolism of lipids; and Chapter 27 presents protein metabolism.

Many sections have undergone extensive rewriting, especially the Selected Topics and sections dealing with molecular biology (Chapter 23) and body fluids (Chapter 28).

Rich in Applications

Capturing students' attention and curiosity is critical in teaching this course. To aid in this effort, we have created a text rich in applied chemistry. We offer applications in three places:

- In a series of special boxed essays within each chapter (you can find a list of these on page iv)
- In marginal notes located throughout the text
- In the prose itself (where even those students who tend to skip boxes and marginal notes, thinking they "won't be on the exam," can see the importance of chemistry to their lives and future careers).

New to this edition:

Most of the health-related topics from the fifth edition have been retained, and in some cases expanded. For example, the essay on "Aspartame" in Chapter 19 has been expanded to include other artificial sweeteners. We have added several new essays, including Body Temperature, Hypothermia, and Hyperthermia; Sizes and Masses of Objects: Powers of Ten; Oxidation-Reduction: Bleaches and Stain Removal; Reducing Fat Intake; Prions; Human Genome Project; Polymerase Chain Reaction; Creatine Phosphate; Cyanide Poisoning; Obesity Genes; and Genetic Diseases of Amino Acid Catabolism.

Pedagogy to Help Students

Each chapter has a list of Key Terms and a chapter Summary. The Key Terms are boldfaced when they are introduced in the text, and all are defined in the Glossary (Appendix II).

At the end of each chapter we offer **two classes of end-of-chapter exercises:**

- **Problems arranged by topic** test mastery of the material and—where pertinent—of problem-solving techniques introduced in the chapter. These problems are usually arranged in matched pairs.
- The **Additional Problems** are not grouped by type. Some are intended to be a bit more challenging; they often require a synthesis of ideas from more than one chapter. Others, however, are not any more difficult than those arranged by topic. Rather, they pursue an idea further than is done in the text, or they introduce new ideas.

New to this edition:

New to this edition are **Learning Objectives/Study Questions,** given at the beginning of each chapter. These are in the form of questions that students should be able to answer after completing the chapter.

Most sections of each chapter are followed by new **Review Questions** intended to provide an immediate assessment of the student's understanding of the section's material. Many worked-out Examples and Practice Exercises are also interspersed in the body of each chapter. Where appropriate, we provide two Exercises, labeled **A** and **B,** after a worked Example. The A exercise is much like the Example it follows; the B exercise often requires incorporation of knowledge acquired previously. Many of the worked-out Examples have been revised to improve the pedagogy.

Supplements for the Student

- **Student Study Guide and Solutions Manual,** by Marvin L. Hackert of the University of Texas at Austin, Roger K. Sandwick of the State University of New York at Plattsburgh, Michael Pelter of Purdue University–Calumet, and Libbie Pelter. This student-friendly manual contains chapter summaries, additional examples and problems, and numerous self-tests (with answers). Solutions correspond to the odd-numbered problems in the text. (ISBN 0-13-085385-2)
- **Chemistry and Life Companion Website: http://www.prenhall.com/hill.** This student-oriented website features computer-graded quizzes with detailed, book-specific feedback, pre-built molecular models for students to view using Chime, downloadable animations, and up-to-date links to chemistry and career-oriented websites.
- **Chemistry on the Internet,** by Thomas Gardner of Tennessee State University. This brief review of the Internet is perfect for students using the Internet and World Wide Web for the first time. It focuses on using the Internet to study chemistry. Available free with new copies of the text. Ask your Prentice Hall representative for details.
- **Chemistry and Life in the Laboratory: Experiments in General, Organic, and Biological Chemistry,** by Victor L. Heasley and Val J. Christensen of Point Loma Nazarene College, and Gene E. Heasley of Southern Nazarene University. This Manual contains 36 experiments that cover the same general topics as the text. Laboratory instructions are clear and thorough and the experiments are well-written and imaginative. This revision includes expanded information on issues of safety and disposal. All experiments have been thoroughly class tested. (ISBN 0-13-085376-3)
- **Allied Health Chemistry: A Companion,** by Tim Smith and Diane Vukovich, both of the University of Akron. This student companion teaches students how to apply the basic mathematics needed for this course. The book features study tips, examples, and careful explanations. Chapters cover metric conversions, unit conversions, simple algebra, temperature conversions, mole conversions, and stoichiometry. (ISBN 0-13-470460-6)
- **Prentice Hall/*The New York Times* Themes of Times.** Through this unique program, adopters of *Chemistry and Life* are eligible to receive our *New York Times* supplement for their students. This newspaper-format resource uses current chemistry-related articles to emphasize the importance and relevance of chemistry in everyday life. (Free in quantity to qualified adopters through your local Prentice Hall representative.)

Supplements for the Instructor

- **Instructor's Solutions Manual with Test Bank,** by Sandwick, Pelter, Pelter, and Aninna Carter of Adirondack Community College. The Instructor's Manual contains solutions to all the problems in the text. The extensively reviewed Test Bank contains over 1100 multiple-choice questions. (ISBN 0-13-085377-1)
- **PH Custom Test for Windows** (ISBN 0-13-085379-8) and **PH Custom Test for Macintosh** (ISBN 0-13-085378-X). These electronic versions of the *Chemistry and Life* Test Bank allow you to customize tests and questions.
- **Transparencies:** 137 full-color transparency acetates selected by the text authors. (ISBN 0-13-085381-X)
- **GOB Presentation Manager** is designed for instructors who use a computer to present material in-class. This CD-ROM features most of the art from the text

as well as animations relevant to general, organic, and biological chemistry. All images can be shown using the Presentation Manager program on the CD-ROM or can be downloaded into other presentation programs (such as PowerPoint). (ISBN 0-13-0853836)

- **Instructor's Manual to the Laboratory Manual,** by Heasley, Christensen, and Heasley. This Manual features equipment lists, chemical lists, teaching suggestions, and precautions for instructors using the Lab Manual. It also includes answers to the pre- and post-lab questions posed in the Lab Manual. (ISBN 0-13-085370-4)

Acknowledgments

We especially want to acknowledge the many magnificent contributions of Dorothy M. Feigl of Saint Mary's College, Notre Dame, Indiana, to earlier editions. Her love of learning and the joy she shares in teaching live on in this sixth edition.

JWH would like to thank his colleagues at the University of Wisconsin–River Falls for so many ideas that made their way into his other texts—and some of which appear in this one. He is especially indebted to Ina Hill and Cynthia Hill for library research, typing, and unfailing support throughout the several editions of this book.

SJB thanks his wife, Sharon, and children, Derek and Kym, for their love, support, and encouragement in the preparation of this edition and all previous texts.

RJSE would like to thank her colleagues at Southern Adventist University and her family and friends for their help and encouragement during her work on this sixth edition. She is especially indebted to her husband Paul and sons Michael and Christopher for their patience and support during a rather hectic time.

We greatly appreciate the substantial support and guidance from many creative people at Prentice Hall: John Challice, our chemistry editor, for imaginative guidance throughout the project; Amanda Griffith and Gillian Buonanno, editorial assistants, for diligence and patience in managing reviews and other correspondence; and the production staff for their care and forbearance in bringing all the parts together to yield a finished work. We owe a special debt of gratitude to Mary Ginsburg, our development editor, for her many creative contributions, and to Debra Wechsler, production editor, for her unswerving diligence and unending patience in guiding us through this process.

All three of us would like to thank our students, who have challenged us to be better teachers, and the users and reviewers of this book, who have challenged us to be better writers.

Hani Y. Awadallah
Montclair State University
Richard F. Drushel
Case Western Reserve University
Blaise Frost
West Chester University
Grace Gagliardi
Bucks County Community College
Mark Hemric
Oklahoma Baptist University
Jon R. Iverson
Western Iowa Technical Community College
Raifah M. Kabbani
Pace University
Glen Lawrence
Long Island University
Lauren E. H. McMills
Ohio University
Helen E. Mertwoy
Bucks County Community College
Carl E. Minnier
Essex Community College
Ruth Ann Murphy
University of Mary Hardin-Baylor
John H. Nickles
Hudson Valley Community College, New York
Mary O'Sullivan
Indiana State University

Sara Selfe
University of Washington
Michael Serra
Youngstown State University

Ronald Swisher
Oregon Institute of Technology
Donald H. Williams
Hope College

No book—or other educational device—can replace a good teacher; thus we have designed this book as an aid to the classroom teacher. The only valid test of this or any text is in a classroom. We would greatly appreciate receiving comments and suggestions based on your experience with this book.

JOHN W. HILL
jwhill@pressenter.com

STUART J. BAUM

RHONDA J. SCOTT-ENNIS
rscottenn@southern.edu

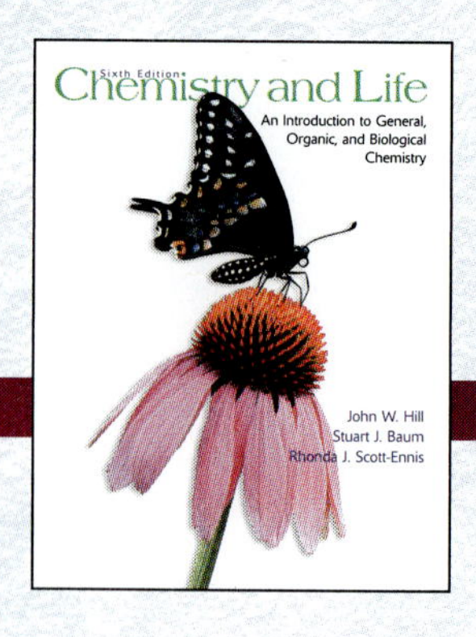

A Guide to Using this Text

What is chemistry? Chemistry is such a broad, all-encompassing area of study that people almost despair in trying to define it. Indeed, some have taken a cop-out approach by defining chemistry as "what chemists do." But that won't do; it's much too narrow a view.

Chemistry is what we all do. We bathe, clean, and cook. We put chemicals on our faces, hands, and hair. Collectively, we use tens of thousands of consumer chemical products in our homes. Professionals in the health and life sciences use thousands of additional chemicals as drugs, antiseptics, or reagents for diagnostic tests.

Your body itself is a remarkable chemical factory. You eat and breathe, taking in raw materials for the factory. You convert these supplies into an unbelievable array of products, some incredibly complex. This chemical factory—your body—also generates its own energy. It detects its own malfunctions and can regenerate and repair some of its component parts. It senses changes in its environment and adapts to these changes. With the aid of a neighboring facility, this fabulous factory can create other factories much like itself.

Everything you do involves chemistry. As you read this sentence, light energy is converted to chemical energy. As you think, protein molecules are synthesized and stored in your brain. All of us do chemistry.

Chemistry affects society as well as individuals. Chemistry is the language—and the principal tool—of the biological sciences, the health sciences, and the agricultural and earth sciences.

Chemistry has illuminated all the natural world, from the tiny atomic nucleus to the immense cosmos. We believe that a knowledge of chemistry can help you. We have written this book in the firm belief that from the beginning, chemistry is related to problems and opportunities in the life and health sciences. And we believe that this can make the study of chemistry interesting and exciting, especially to nonchemists.

For example, an "ion" is more than a chemical abstraction. Enough mercury ions in the wrong place can kill you, but the right number of calcium ions in the right place can keep you from bleeding to death. "$PV = nRT$" is an equation, but it is also the basis for the respiratory therapy that has saved untold lives in hospitals. "Hydrogen bonding" is a chemical phenomenon, but it also helps to account for the fact that a dog has puppies while a cat has kittens and a human has human babies. There are hundreds of similar fundamental and interesting applications of chemistry to life.

A knowledge of chemistry has already had a profound effect on the quality of life. Its impact on the future will be even more dramatic. At present we can control diabetes, cure some forms of cancer, and prevent some forms of mental retardation because of our understanding of the chemistry of the body. We can't *cure* diabetes or cure *all* forms of cancer or *all* mental retardation, because our knowledge is still limited. So learn as much as you can. Your work will be enhanced and your life enriched by your greater understanding.

Be prepared. Something good might happen to you—and to others because of you.

You and your classmates come to this course with a variety of backgrounds and interests. Most of you plan to be professionals in a biological or allied health field. Knowledge of chemistry is essential to a true understanding of everything from DNA replication to drug discovery to nutrition. Indeed, the chemical properties and principles you learn in this course will pervade almost every aspect of your private and professional lives. In this text, we provide you with both the principles and applications of chemistry that will help you in your professional practice and enrich your everyday life as well.

This text is rich in pedagogical aids, both within and at the ends of the chapters. We present this "user's guide" to the text to help you get the most out of this book and your course.

Applications

78 Chapter 3 Chemical Bonds

▲ Some familiar foods with a high Na^+ content.

Some manufacturers enrich their cereals with very fine specks of iron filings. These iron filings dissolve in the acidic environment of the stomach, producing iron(II) ions.

What Is a Low-Sodium Diet?

People with high blood pressure are usually advised to follow a low-sodium diet. Just what does this mean? Surely they are not being advised to reduce their consumption of sodium metal. Sodium is an extremely reactive metal that reacts violently with moisture; eating it wouldn't be safe. The concern is really with sodium *ions*, Na^+. Most people in the United States consume much more Na^+ than they need, most of it from sodium chloride, common table salt. It is not uncommon for some individuals to eat 6 or 7 g of sodium chloride a day, most of it in prepared foods. Many snack foods, such as potato chips, pretzels, and corn chips, are especially high in salt. The American Heart Association recommends that adults limit their salt intake to about 3 g per day.

Note the important difference between ions and the atoms from which they are made. A metal atom and its cation are as different as a whole peach (atom) and a peach pit (ion). The names and symbols look a lot alike, but the species themselves are quite different. Unfortunately, the situation is confused because people talk about needing "iron" to perk up "tired blood" and "calcium" for healthy teeth and bones. What they really mean is iron(II) *ions* (Fe^{2+}) and calcium *ions* (Ca^{2+}). You wouldn't think of eating iron nails to get "iron." Nor would you eat highly reactive calcium metal. Many people do not always make careful distinctions, but we will try to use precise terminology here.

Margin notes highlight the intriguing ways in which you can put your knowledge of chemistry to work. We touch on fields as diverse as medicine, engineering, agriculture, and consumer products. ▶

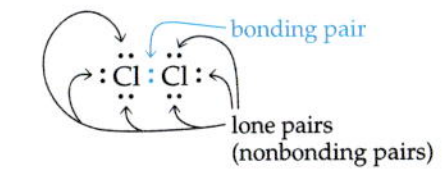

licity, the hydrogen molecule is often represented as H_2 and the chlorine as Cl_2. In each case, the covalent bond between the atoms is understood. es the covalent bond is indicated by a dash, H—H and Cl—Cl. Nonbond- of electrons often are not shown.

Reducing Fat Intake

We often hear of the need for a low fat diet to reduce the risk of cancer, heart disease, and other problems associated with obesity. But fats provide texture, flavor, and a creamy "mouthfeel" to foods. Artificial or high-intensity sweeteners have been around for over a century (see boxed essay in Chapter 19), but the same is not true for fat replacers. The first fat replacers, introduced in the 1960s, used carbohydrates as the primary ingredient. Carrageenan, a seaweed derivative approved for use in food in 1961, was initially used as an emulsifier, stabilizer, and thickener in food. In the early 1990s it was used as a fat replacer. Protein-based fat substitutes came along in the early 1990s. Unlike many of the initial carbohydrate-based products that were first used for other purposes, these were specifically designed to replace fat in foods. Microparticulated proteins, such as Simplesse®, are made from whey protein or milk and egg protein. The carbohydrate- and protein-based fat replacers can give the "mouthfeel," bulk, and texture of fats, but cannot be used for frying.

In January 1996 the first true fat-based fat replacer was approved—olestra (marketed as Olean® by Procter & Gamble). Olestra is composed of a sucrose core with six to eight fatty acids attached. Because olestra is a much larger molecule than a triglyceride, it is too large to be hydrolyzed by lipases and digested or absorbed by the body or metabolized by the microorganisms in the intestinal tract. Thus it adds no fat or calories to foods. Olestra is not broken down or degraded when it is exposed to high temperatures; thus it can be used for frying foods.

Olestra does have some drawbacks to its use. Clinical studies have shown that it may cause intestinal cramps and loose stools in some individuals. It also reduces the absorption of fat-soluble nutrients, such as vitamins A, D, E, and K and carotenoids, from foods eaten at the same time. Because of these concerns, the Food and Drug Administration (FDA) requires that foods containing olestra be fortified with vitamins A, D, E, and K and that the following statement appear on the package: "This product contains olestra. Olestra may cause abdominal cramping and loose stools. Olestra inhibits the absorption of some vitamins and other nutrients. Vitamins A, D, E, and K have been added."

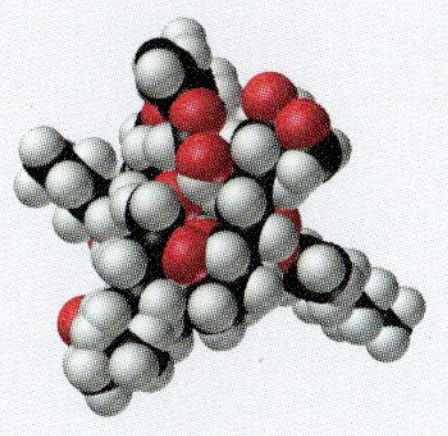

Olestra

◀ Molecular models of olestra and a trizglyceride. Olestra's core is sucrose with six to eight fatty acids attached.

◀ There are many boxes in this text that focus on how we apply our chemical knowledge to solve real-life problems, and on historical topics of interest. These readings will help you see how chemistry affects everyday life and how we arrived at our current understanding of chemistry.

Learning Tools

At the start of each chapter you'll find a set of learning objectives in the form of study questions. Read these objectives before you start each chapter; they will help you identify key points in each section. They will also help you make sure you can answer each question when you finish the chapter. This is an excellent way to test your understanding of the material.

Solutions

Fish, plants, and other marine organisms reside within a water solution that contains all the necessary ingredients to support life.

Learning Objectives/Study Questions

1. What is a solution? What are the different types of solutions?
2. What is the difference between a soluble substance and a miscible substance?
3. What is the difference between a saturated solution and a supersaturated solution?
4. What factors determine the solubility of ionic compounds? Of covalent compounds?
5. How do temperature and pressure affect the solubility of gases in water?
6. What ... conce... used?
7. What is a colligative property? How do colligative properties explain the use of antifreeze in radiators or of salt to melt snow and ice?
8. What are osmosis and osmotic pressure? What is their relevance to human biology and to medicine?
9. What are colloids? How do the properties of a colloid compare to those of a solution or suspension? Give some examples of colloids.
10. What is dialysis? What is its relevance to

Numerous worked Examples help you build your problem-solving skills by showing you how to solve various types of problems. Study the Examples carefully to make sure you understand the model solution. Then start to master the problem-solving process by working the Practice Exercises that follow.

Example 4.4

What volume of oxygen is required to burn 0.556 L of propane, if both gases are measured at the same temperature and pressure?

$$C_3H_8(g) + 5\,O_2(g) \longrightarrow 3\,CO_2(g) + 4\,H_2O(g)$$

Solution

The equation indicates that *five* volumes of $O_2(g)$ are required for every volume of $C_3H_8(g)$. Thus, we use $5\text{ L }O_2(g)/\,1\text{ L }C_3H_8(g)$ as the ratio to find the volume of oxygen required.

$$?\text{ L }O_2(g) = 0.556\text{ L }C_3H_8(g) \times \frac{5\text{ L }O_2(g)}{1\text{ L }C_3H_8(g)} = 2.78\text{ L }O_2(g)$$

Practice Exercises

A. Using the equation in Example 4.4, calculate the volume of $CO_2(g)$ that is produced when 0.492 L of propane is burned if the two gases are compared at the same temperature and pressure.

B. Calculate the volume of $CO_2(g)$ that is produced when 5.42 L of butane, $C_4H_{10}(g)$, is burned if the two gases are compared at the same temperature and pressure. (Hint: First write a balanced chemical equation.)

Often, the Practice Exercises are labelled A and B. Exercise A asks a question similar to that in the Example. Exercise B usually asks you to extend your understanding a bit further. For example, you may need to apply a problem-solving technique you learned earlier. These Exercises help prepare you for solving more complex problems, such as those you might face on an exam.

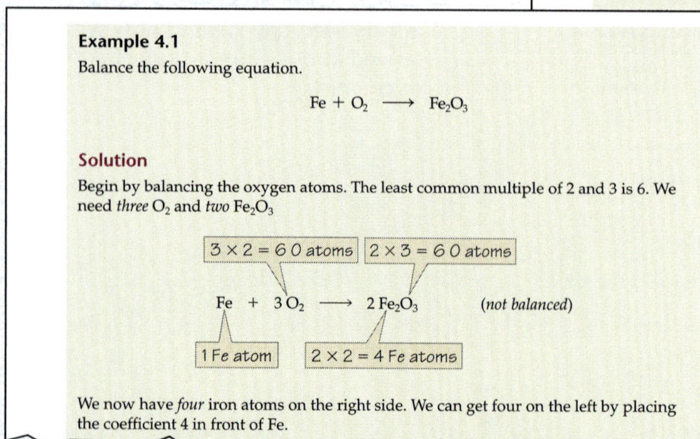

Example 4.1

Balance the following equation.

$$Fe + O_2 \longrightarrow Fe_2O_3$$

Solution

Begin by balancing the oxygen atoms. The least common multiple of 2 and 3 is 6. We need *three* O_2 and *two* Fe_2O_3

$$Fe + 3\,O_2 \longrightarrow 2\,Fe_2O_3 \quad \textit{(not balanced)}$$

We now have *four* iron atoms on the right side. We can get four on the left by placing the coefficient 4 in front of Fe.

Voice balloons help you understand each step in the solution to a problem. Make sure you understand where each answer comes from; don't just memorize them.

Review questions conclude each section. Make sure you can answer these questions before you proceed. You can check your answers to the questions numbered in red at the back of the book in Appendix III.

…velop like charges. The brush develops an opposite charge, and the hairs are attracted to it.

Review Questions

1.11 What is a force? How do charged particles exert a force? What sort of force is exerted by Earth on an object at its surface?

1.12 Describe what happens to two particles with like charges when they are brought close together. What happens to particles with unlike charges when they are brought close together?

End-of-chapter problems test your mastery of the problem-solving techniques and material presented in the chapter. They are arranged by topic and come in matched pairs (each odd- and even-numbered pair test the same concept). Problems may emphasize estimation skills or conceptual understanding. Work many of the Problems to develop the strong problem-solving skills that will help you succeed. Problems numbered in red are answered at the back of the book.

Problems

Some General Considerations

1. How do liquids and solids differ from gases in their compressibility, spacing of molecules, and intermolecular forces?
2. List four types of interactions between particles in the liquid and solid states. Give an example of each type.
3. Why does it take longer to boil an egg at high altitude than at sea level? Does it take longer to fry an egg at high altitude? Explain.
4. Why does steam at 100 °C cause more severe burns than liquid water at the same temperature?
5. In which process is energy absorbed by the material undergoing the change of state?
   a. melting or freezing
   b. condensation or vaporization
6. Label each arrow with the term that correctly identifies the process described.

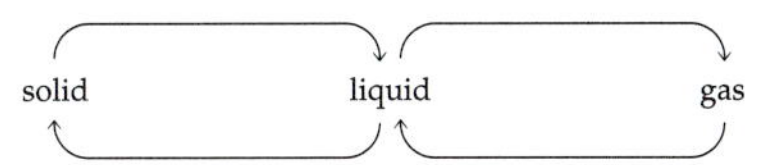

Intermolecular Forces

7. Which of the following would you expect to have the lower boiling point: carbon disulfide (CS_2) or carbon tetrachloride (CCl_4)? Why?
8. Which of the following would you expect to have the lower boiling point: phosphine (PH_3) or arsine (AsH_3)? Why?
9. Which of the following would you expect to have the higher boiling point: propane (C_3H_8) or ethanol (C_2H_5OH)? Why?
10. Which of the following would you expect to have the higher boiling point: water (H_2O) or carbon monoxide (CO)? Why?
11. Arrange the following substances in the expected order of increasing boiling point: H_2S, H_2Se, H_2Te. Give the reasons for your ranking.
12. Arrange the following substances in the expected order of increasing boiling point: H_2O, HCl, CH_4. Give the reasons for your ranking.

Additional Problems

61. The speed limit in rural Ontario is 90 km/hr. What is this speed in miles per hour?
62. The speed of light is 186,000 mi/sec. What is this speed in meters per second?
63. If your heart beats at a rate of 72 times per minute and your lifetime will be 70 years, how many times will your heart beat during your lifetime?
64. How many 325-milligram aspirin tablets ca… from 875 grams of aspirin?
65. A doctor puts you on a diet of 5500 kilojoul… day. How many calories is that? How many kilocalories (food Calories)?
66. Milk costs \$3.89 per gallon or \$1.15 per liter. Which is cheaper?
67. What is the thickness of a 45.4 cm × 104.6 cm rectangular piece of cast iron ($d = 7.76\ g/cm^3$) that has a mass of …

Some of the Additional Problems are more challenging than the Problems, requiring a synthesis of concepts from multiple chapters. Others will help you to attain a stronger mastery of key concepts in this chapter. Additional Problems may also emphasize estimation skills or conceptual understanding.

…to 35.0 °C by the same quantity of heat that is capable of raising the temperature of 145 g H_2O from 22.5 °C to 35.0 °C?

71. Arrange the following in order of increasing length (shortest first): (1) a 1.21-m chain, (2) a 75-in. board, (3) a 3 ft 5-in. rattlesnake, (4) a yardstick.
72. Arrange the following in order of increasing mass (lightest first): (1) a 5-lb bag of potatoes, (2) a 1.65-kg cabbage, (3) 2500 g sugar.
73. One of the women pictured below has a mass of 38.5 kg and a height of 1.51 m. Which one is it likely to be?

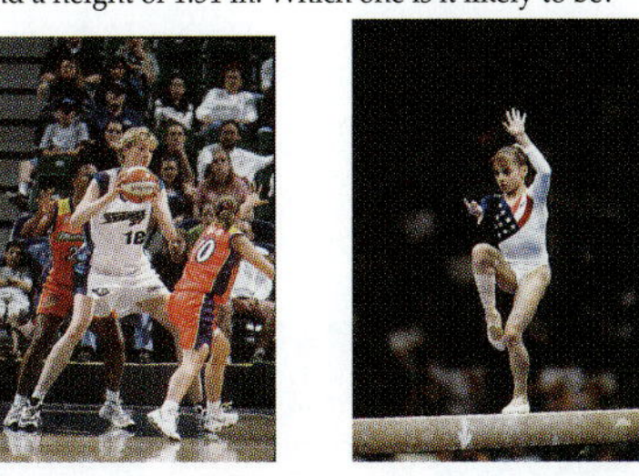

…includes the air between the pieces of polystyrene foam.)

75. Which of the following items would be most difficult to lift into the bed of a pickup truck: (1) a 100-lb bag of potatoes, (2) a 15-gal plastic bottle filled with water, (3) a 3.0-L flask filled with mercury ($d = 13.6\ g/cm^3$)?
76. A rectangular block of gold-colored material measures 3.00 cm × 1.25 cm × 1.50 cm and has a mass of 28.12 g. Can the material be gold?
77. An experiment calls for 8.65 grams of carbon tetrachloride ($d = 1.59\ g/mL$). What is the volume of such a mass?
78. Adult male Hooker's sea lions are 250 to 350 cm long and weigh 300 to 450 kg. Convert these measurements to inches and pounds.
79. Pediatric drug dosages are usually based on infant weight and are expressed in units such as milligrams of drug per kilogram of body weight (mg/kg). If the recommended dosage of a particular drug is 5.0 mg/kg, what is the proper mass of drug for a 17-lb baby?
80. Each Tylenol chewable tablet contains 80 mg of acetaminophen. The recommended dosage for children is 10 mg/kg of body weight. How many tablets constitute the proper dosage for a 55 lb child?
81. In its nonstop, round-the-world trip, the aircraft Voyager traveled 25,102 mi in 9 days, 3 min, and 44 s. Calculate the average speed of Voyager in miles per hour.

Selected Topics

SELECTED TOPIC E

Hormones

Humans and all other multicellular organisms must have a way for cells to communicate with each other—intercellular communication. Informational signals must be sent from one cell to adjacent cells or to cells or tissues at a greater distance. Table E.1 outlines the ways that cells and tissues communicate with each other. Selected Topic C discussed neurotransmitters, needed for synaptic communication. In this special topic we consider the molecules needed for paracrine and endocrine communication. A distinguishing characteristic of these compounds is their production of dramatic effects at very low concentrations. In **paracrine communication** the chemical messengers, known as **paracrine factors,** move from one cell to an-

Table E.1 Mechanisms of Intercellular Communication

Mechanism	Transmission	Chemical Mediators	Distribution of Effects
Direct communication	Through gap junctions from cytoplasm to cytoplasm	Ions, small solutes, lipid-soluble materials	Limited to adjacent cells that are directly interconnected by interlocking membrane proteins
Paracrine communication	Through extracellular fluid	Paracrine factors	Primarily limited to local area, where concentrations are relatively high; target cells must have appropriate receptors
Endocrine communication	Through the circulatory system	Hormones	Target cells are primarily in other tissues and organs and must have appropriate receptors
Synaptic communication	Across synaptic clefts	Neurotransmitters	Limited to very specific area; target cells must have appropriate receptors

557

Interspersed between chapters are seven Selected Topics. These mini-chapters treat selected subjects in more detail. Your instructor may assign these, or you may choose to read them on your own. In either case, each has end-of-topic problems you can use to test your understanding of the material.

Illustrations

Study the illustrations and graphics carefully. Chemistry is a visual science, and the art will help you to visualize atoms, molecules, and chemical processes that cannot be seen with the unaided eye.

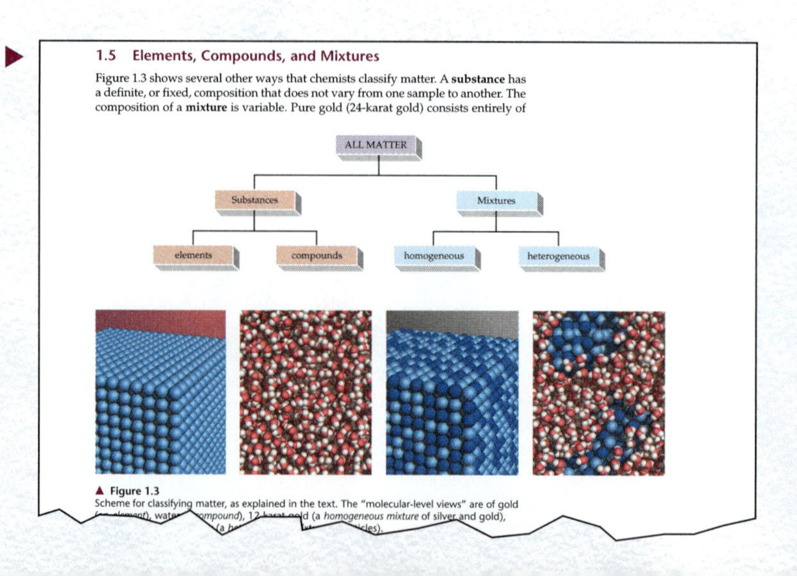

1.5 Elements, Compounds, and Mixtures

Figure 1.3 shows several other ways that chemists classify matter. A **substance** has a definite, or fixed, composition that does not vary from one sample to another. The composition of a **mixture** is variable. Pure gold (24-karat gold) consists entirely of

▲ **Figure 1.3**
Scheme for classifying matter, as explained in the text. The "molecular-level views" are of gold ... water ... 12-karat gold (a *homogeneous mixture* of silver and gold), ...

Matter and Measurement

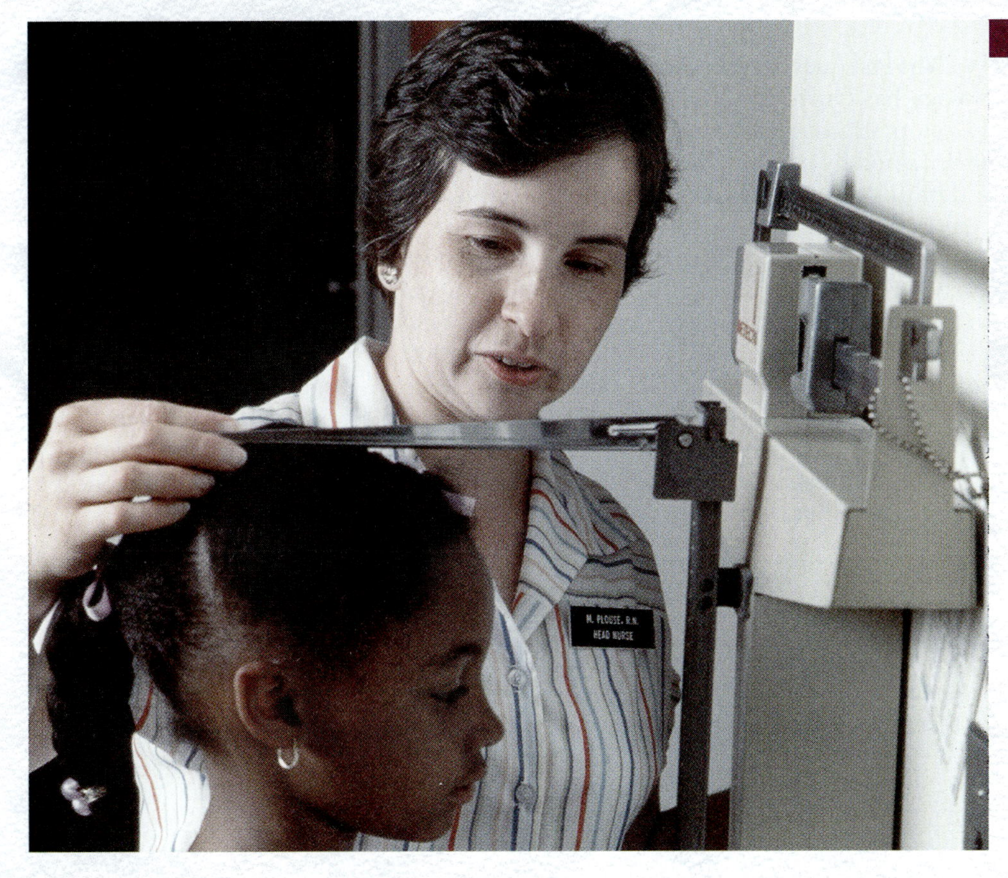

Accurate measurements are important in science and in medicine. Measurement of height and weight are meaningful first steps in any physical examination. Other measurements, often more sophisticated than these, are also essential in many medical procedures.

Learning Objectives/Study Questions

1. How do scientists obtain new scientific knowledge? Do views or concepts in science change?
2. What is chemistry?
3. How does mass differ from weight?
4. How do physical properties of matter differ from chemical properties?
5. What is the difference between a mixture and a compound?
6. What is energy? What is potential energy? What is kinetic energy?
7. What is the metric system of measurement? How does one convert between metric units and the units commonly used in the United States?
8. What is the difference between precision and accuracy in measurements?
9. Why is an understanding of significant figures important in chemistry? How do we determine the number of significant figures to report?
10. What is density?
11. What is the difference between temperature and heat?
12. How are the different temperature scales related to one another?

This book is called *Chemistry and Life.* Our choice of title might well raise three questions:

- What does chemistry have to do with life?
- What is chemistry?
- For that matter, what is life?

The last question is more than rhetorical. Progress in medical science and technology has blurred the distinction between life and death. Is a person whose heart has stopped beating necessarily dead? Is a person whose vital functions are being maintained by machine truly alive? In this book, we won't even attempt to supply a definitive answer to the question "What is life?" We will simply note that the question is a critical one for our society.

Now consider the first question, "What does chemistry have to do with life?" A chemist would answer, "Just about everything." The human body, for example, is the most extraordinarily complicated, most elegantly designed, and most efficiently operated chemical laboratory there is. Most of this text is devoted to our attempts to answer this first question.

That leaves us with the middle question, the subject of this first chapter, "What is chemistry?" We shall see how science in general and chemistry in particular have developed from earlier human endeavors. We will then develop some basic concepts necessary to our study of chemistry and its relationship to life.

Science

We will begin with a look at the nature of science. Our study will include a consideration of the methods of science and the manner of its progress. We will also look at some of the problems that arise from the applications of science.

1.1 Science and the Human Condition

People generally have three basic needs: food, clothing, and shelter. Certainly those three things—if adequate in quantity and quality—are enough to keep us alive. Most of us, however, would agree that there are two more requirements for a *good* life: reasonably good health and some chance for happiness.

In early human societies, people put nearly all their effort into the hunting and gathering of food, the making of clothing, and the provision of shelter. They had no knowledge of the biological and chemical basis of illness. Other than a few herbal medicines, they could do little about their health except pray and make sacrifices to their gods. With the coming of civilization, some people began to have enough leisure time to turn their thoughts to the human condition and to the natural world around them. Over the centuries, what we now call science grew out of their speculations. As this scientific study of the material universe progressed, the responsibility for adding to the growing body of knowledge was divided among various disciplines, and one of these disciplines was chemistry.

The roots of modern chemistry are planted in alchemy, a kind of mystical chemistry that flourished in Europe during the Middle Ages (about 500 to 1500 C.E.). Modern chemists inherited from the alchemists an abiding interest in human health and the quality of life. Alchemists not only searched for a philosopher's stone that would turn cheaper metals into gold but also sought an elixir that would confer immortality on those exposed to it. Alchemists never achieved these goals, but they

◀ *The Alchemist,* a painting done by the Dutch artist Cornelis Bega around 1660, depicts a laboratory of the seventeenth century.

▲ A Swiss physician, Theophrastus Bombastus von Hohenheim (1493–1541), who preferred the self-chosen name Paracelsus, helped alchemists to turn away from their attempts to make gold and to undertake humanitarian tasks.

discovered many new chemical substances and perfected techniques, such as distillation and extraction, that are still used today.

In the early sixteenth century Paracelsus, a Swiss physician, urged alchemists to turn away from their attempts to make gold and to seek instead medicines with which to treat disease. Possessed of a monstrous ego, Paracelsus alienated many of his contemporaries. His followers, however, were numerous enough to ally forever the science of chemistry with the art of medicine.

By the seventeenth century astronomers, physicists, physiologists, and philosophers had adopted a new attitude, characterized by a reliance on experimentation. This change in orientation signaled the emergence of chemistry from alchemy. The English philosopher Francis Bacon (1561–1626) had visions of these new scientific methods endowing human life with new inventions and wealth.

By the middle of the twentieth century science and its application in technology appeared to have made the dreams of Bacon and Paracelsus a reality. Many once-dreaded diseases such as smallpox, polio, and plague had been almost eliminated. Fertilizers, pesticides, and scientific animal breeding had increased and enriched our food supply. Transportation was swift, and communication was nearly instantaneous. Nuclear energy seemed to promise an unlimited source of power for our every need. New materials—plastics, fibers, metals, and ceramics—were developed to improve our clothing and shelter.

Indeed, it seemed that, despite its sometimes less than honorable intentions, science could do no wrong. For example, during World War I the German armies' supply of nitrates (which they needed to make explosives) was cut off by a naval blockade. However, Fritz Haber (1868–1934) invented a chemical process to make ammonia, which could be converted to ammonium nitrate. This alternative supply of explosives probably lengthened the war, but Haber's work is far more significant for its influence on modern agriculture. Ammonia and nitrates are the stuff of which fertilizers are made, and fertilizers are essential to modern high-yield farming. In fact, most of the ammonia made by the Haber process today goes into fertilizers.

Ammonium nitrate in contact with organic materials has considerable explosive potential, as has been dramatically demonstrated. An enormous explosion aboard a ship carrying ammonium nitrate in Texas City, Texas, in 1947, killed nearly 600 people and injured several thousand. Explosive mixtures of ammonium nitrate fertilizer and fuel oil have been used in many terrorist attacks on buildings around the world.

Much of modern technology has grown out of scientific discoveries, and technological developments are used by scientists as tools for even more discoveries. These developments in science and technology are, to a considerable extent, the cornerstone of what we mean by the "modern" world.

(a)

(b)

▲ (a) Synthetic pesticides and fertilizers have greatly increased the production of food and fiber on our farms. (b) Unfortunately, pesticides also kill beneficial insects such as honeybees, essential to the pollination of many crops.

1.2 Problems in Paradise

If during the first half of the twentieth century science was viewed as humankind's savior, during the latter half it was sometimes viewed as quite the opposite. Many people's views of science changed dramatically in the latter decades of the twentieth century. They saw science not as savior, but as quite the opposite. Fertilizers and pesticides have dramatically increased the food supply, but fertilizer runoff from farms has polluted streams, and pesticides have had a devastating effect on some wildlife. Chlorine that has made water supplies safe from disease-carrying bacteria has also produced cancer-causing substances in the water. Anesthetics that make surgery painless for the patient have caused miscarriages in female operating-room personnel. Wastes from the manufacture of many of the products we need and want contaminate the water, land, and air.

We can't solve these problems by throwing out science. Do we really wish to return to surgery without anesthetics and to water supplies carrying cholera and other deadly diseases? Most of us don't. Rather, we should use science to develop safer anesthetics, to provide approaches to increased agricultural production more compatible with the natural environment, and to develop analytical techniques that will ensure a healthful environment for us all.

Chemistry and its products, both good and bad, are so intimately involved in determining the quality of life that to ignore the subject is to court disaster. It will take an educated, informed society to ensure that science is used for the betterment of the human condition.

1.3 The Way Science Works—Sometimes

Science is not simply a "body of knowledge," nor is it a finished work. Rather, science is an ever changing approach to learning. Science is organized into concepts. For example, even though we will often speak of the structure of an atom as if it were readily observable, atomic structure is just a convenience that successfully describes many observable facts in a metaphorical way. It is not the body of facts that characterizes science but the *organization* given to those facts and the theories that we use to explain them.

The most distinguishing characteristic of science is its use of processes or methods. These intellectual processes usually begin with the making of observations and the gathering of data. Often data is summarized in a **scientific law.** These laws are frequently stated in mathematical form. For example, Robert Boyle (1627–1691) conducted many experiments to determine how the volume of a gas varies with the pressure acting on the gas. In words, his experiments are summarized in the statement that the volume (V) of a gas is inversely proportional to the pressure (P) acting on it. Mathematically, we write Boyle's law as $V = a/P$, where a is a constant. (We will discuss Boyle's law in Chapter 6; we mention it here only as an example of a scientific law.)

Scientists must be able to make careful measurements, but they must also be able to grasp the central theme of these observations. They must recognize the variables and be able to note the effect of changing **variables.**

A *variable* is a data item that changes over the course of an experiment. Historically, scientists have tried to hold most factors constant and observe changes as they vary only one quantity. For example, to see how the volume of a gas varies with the pressure acting on the gas, the temperature and the amount of gas are held constant. With computers to do the complicated mathematics, scientists nowadays can vary several items at once and still determine the effect of each variable.

Basic to science is the ability to formulate testable **hypotheses**—guesses about what will happen in an experiment. Even an educated guess is of little value to scientists unless an experiment can be devised to test its validity. If a hypothesis cannot be tested, the question is generally considered to lie outside the realm of science. If a hypothesis stands up to continued experimental testing, it may be expanded into a **theory,** a more comprehensive explanation of a set of related experimental data.

To be useful, scientific theories must have *predictive* value. If the atomic theory is to be useful, it should enable scientists to predict how matter will behave.

Science is not totally different from other disciplines. For instance, creativity is central to the arts and humanities as well as to science. Science does not involve cold logic to the exclusion of other human characteristics. Albert Einstein recognized that there was no *logical* path to some of the laws that he formulated. Even he relied on intuition based on experience and understanding.

It is important to realize that there is no single "scientific method" that, when followed, produces guaranteed results. Scientists observe, gather facts, and make hypotheses, but somewhere along the way they test their hunches and their organization of facts by *experimentation.* Scientists, like other human beings, use intuition and may generalize from a limited number of facts. Sometimes they are wrong. A great strength of science is that results of experiments are published in scientific journals, and these results are read and checked by other scientists around the world. To become an accepted part of the "body of knowledge," the results must be *reproducible.* Scientists also extend each other's work, sometimes to the point that we see a bandwagon effect. One breakthrough sometimes results in the unleashing of vast quantities of new data and leads to the development of new concepts. For example, early in the nineteenth century it was thought that certain chemical substances, called *organic* compounds, could be produced only by living tissues, such as someone's liver or the leaf of a plant. These substances were in contrast to other materials, labeled *inorganic,* which could be prepared by a chemist in a laboratory. In 1828 Friedrich Wöhler (1800–1882), a German chemist, succeeded in making an organic compound from an inorganic one in the laboratory. The belief that such a compound could not be prepared in this manner was so strong that Wöhler did the same thing over and over again to assure himself that he had really done the "impossible." When he finally published his work, other chemists quickly repeated it and then proceeded to make hundreds of thousands of organic compounds. That bandwagon is still rolling today, with chemists making hormones, vitamins, and even genes.

Thus, contrary to an often expressed popular notion, scientific knowledge is not absolute. Science is cumulative, and the "body of knowledge" is dynamic and constantly changing. Old concepts are discarded as new tools, new questions, and new techniques reveal new data or generate new concepts. To truly understand what science is, one has to observe what the entire worldwide community of scientists has done over several years rather than look over the shoulder of a single scientist for a few days.

Review Questions

1.1 What is a hypothesis? How are hypotheses tested?

1.2 What are some of the distinguishing characteristics of science?

Chemistry

Chemistry is called the central science not only because it is important in itself, but also because a knowledge of chemistry is essential to the life sciences and related medical fields, to materials sciences and much of engineering, to geology and related planetary sciences, and to many other human endeavors. Chemistry provides answers to questions about the universe in which we live, from the materials that make up distant galaxies to those that compose the cells of our bodies.

1.4 Some Fundamental Concepts

Chemistry is the study of the composition, structure, and properties of matter and of changes that occur in matter. What is matter? It is the material of which things are made. One way to think about matter is that particles or objects of **matter** occupy space, and that no two objects can occupy the same space at the same time. Wood, sand, people, water, and air are all examples of matter. Light is not matter; it is a form of energy.

Chemists are concerned with the tiny, microscopic building blocks of matter known as atoms. **Atoms** are the smallest units that we associate with the chemical behavior of matter. Ultimately, samples of matter are what they are because of the atoms that form them.

Mass is a measure of the quantity of matter that an object contains. The greater the mass of a thing, the more difficult it is to change its velocity. You can easily deflect a tennis ball coming toward you at 30 meters per second (m/s), but you would have difficulty stopping a cannonball of the same size moving at the same speed. A cannonball has more mass than a tennis ball of equal size. The mass of an object does not vary with location. An astronaut has the same mass on the moon as on Earth (Figure 1.1). **Weight,** in contrast, measures a force. On Earth it measures the force of attraction between our planet and the mass in question. On the moon, where gravity is one-sixth that on Earth, an astronaut weighs only one-sixth as much as on Earth. Weight varies with gravity; mass does not.

▲ **Figure 1.1**
Astronaut John W. Young leaps from the lunar surface in 1972, where the force of gravity is only one-sixth that on Earth.

Example 1.1

On the surface of Mars, gravity is one-third that on the surface of Earth.

a. What would be the mass on Mars of a person who has a mass of 55 kilograms (kg) on Earth?

b. What would be the weight on Mars of a person who weighs 150 lb on Earth?

Solution

a. The person's mass would be the same (55 kg) as on Earth; the quantity of matter does not depend on location.

b. The person would weigh only 50 lb; the force of attraction between planet and person is only one-third that on Earth.

Practice Exercise

At the surface of Planet X, the force of gravity is 2.4 times that on Earth's surface.

a. What would be the mass of a standard 1.00-kg object on Planet X?

b. A man who weighs 198 lb on Earth would weigh how much on Planet X?

Physical and Chemical Properties

To distinguish between samples of matter, we can compare their *properties* (see Figure 1.2). The **physical properties** of a substance are its physical characteristics and behavior, such as color, odor, or hardness. Its **chemical properties** describe how it reacts with other types of matter—how its atomic building blocks can change.

Chemical properties are inherent in a substance; they are made manifest through chemical change.

A **physical change** is one that does not entail any change in chemical composition: solid iron melting in a blast furnace or an ice cube changing to liquid water in a glass. The process of melting is a physical change, and the temperature at which it occurs—the melting point—is a physical property. A **chemical change** involves a change in chemical composition. In exhibiting a chemical property, matter undergoes a chemical change: The original substance is replaced by one or more new substances. Iron metal reacts with oxygen from the air to form rust (iron oxide); when sulfur burns in air, sulfur, which is made up of one type of atom, and oxygen from air, which is made up of another type of atom, combine to form sulfur dioxide, which is comprised of molecules that have sulfur and oxygen atoms in the ratio 1:2. A **molecule** is an electrically neutral group of atoms bound together in a recognizable unit (see Section 1.5).

It is difficult at times to decide whether a change is physical or chemical, but we can decide on the basis of what happens to the composition or structure of the matter involved. *Composition* refers to the types of atoms that are present and their relative proportions, and *structure* to the arrangement of those atoms in particular

◄ **Figure 1.2**
Copper (left) and ethyl alcohol (right) are easily distinguished by their properties. Copper is a solid; ethyl alcohol is a liquid. Copper is opaque and has a red-brown color. Ethyl alcohol is transparent and colorless. Also, ethyl alcohol burns whereas copper does not. (Which of these are physical properties, and which are chemical?)

assemblages or in space. A chemical change results in a change in composition or structure, whereas a physical change does not. Table 1.1 lists some examples of physical and chemical properties.

Example 1.2

Which of the following events involve chemical changes and which involve physical changes?

a. You trim your fingernails.
b. Lemon juice converts milk to curds and whey.
c. Water boils.
d. Water is broken down into hydrogen gas and oxygen gas.

Solution

a. Physical change: the composition of the nail is not changed by clipping.
b. Chemical change: the compositions of curds and whey are different from the composition of the milk.
c. Physical change: liquid water and invisible water vapor formed when liquid water boils have the same composition; the water merely changes from a liquid to a gas.
d. Chemical change: new substances, hydrogen and oxygen, are formed.

Practice Exercise

Which of the following events involve chemical changes and which involve physical changes?

a. Liquid alcohol vaporizes from an open container.
b. A piece of lithium metal burns in air to form a white powder called lithium oxide.
c. A dull saw is sharpened with a file.

Table 1.1 Some Examples of Physical and Chemical Properties

Physical Properties	
Property	**Example**
Temperature	Water for a bath is at 40 °C.
Mass	A nickel has a mass of 5 grams.
Color	Sulfur is yellow.
Odor	Hydrogen sulfide stinks.
Boiling point	Water boils at 100 °C.
Solubility	Table salt dissolves in water.
Heat capacity	Water has a high heat capacity.
Hardness	Diamond is exceptionally hard.
Electrical conductance	Copper conducts electricity.
Density	Water has a density of 1.00 grams per milliliter.
Chemical Properties	
Substance	**Typical Chemical Property**
Iron	Will rust (combine with oxygen to form iron oxide)
Carbon	Will undergo combustion (will combine with oxygen to form carbon dioxide)
Silver	Will tarnish (combine with sulfur to form silver sulfide)
Sodium	Will react violently with water to form hydrogen gas and a solution of sodium hydroxide
Nitroglycerin	Will explode (decompose, when detonated, to a mixture of gases)

States of Matter

There are three familiar *states of matter:* solid, liquid, and gas. A **solid** object ordinarily maintains its shape and volume regardless of its location. A **liquid** occupies a definite volume, but assumes the shape of the occupied portion of its container. If you have a 12-oz soft drink, you have 12 oz of liquid whether it is in a can, in a bottle, or, through some slight mishap, on the floor—which demonstrates another property of liquids. Unlike solids, liquids flow readily. A **gas** maintains neither shape nor volume. It expands to fill completely whatever container or space it occupies. Gases can be easily compressed. For example, enough air for many minutes of breathing can be compressed into a steel tank for underwater diving. We shall take up the topic of the states of matter in more detail in Chapters 6 and 7.

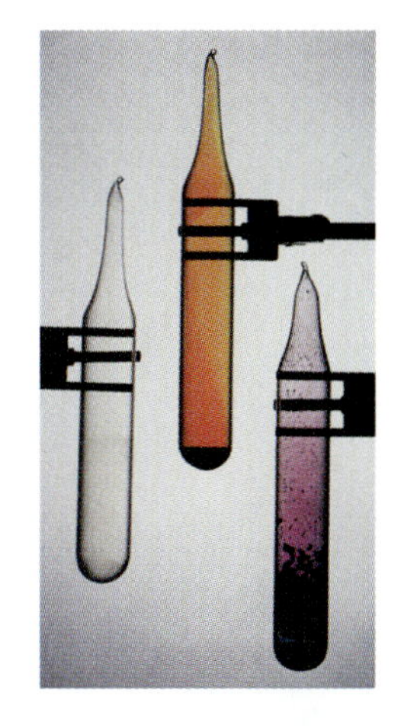

▲ The elements chlorine, bromine, and iodine illustrate the three states of matter. Chlorine (left) is a pale yellow-green gas at room temperature. Bromine (middle) is a reddish brown liquid, but gaseous bromine is clearly visible above the liquid in the bottom of the tube. Iodine (right) is dark as a solid, but purple iodine vapor (gaseous iodine) is visible above the solid in the bottom of the tube.

✓ Review Questions

1.3 Define each of the following terms.

a. chemistry b. matter
c. mass d. weight

1.4 How do physical and chemical properties differ?

1.5 How do gases, liquids, and solids differ in their properties?

1.5 Elements, Compounds, and Mixtures

Figure 1.3 shows several other ways that chemists classify matter. A **substance** has a definite, or fixed, composition that does not vary from one sample to another. The composition of a **mixture** is variable. Pure gold (24-karat gold) consists entirely of

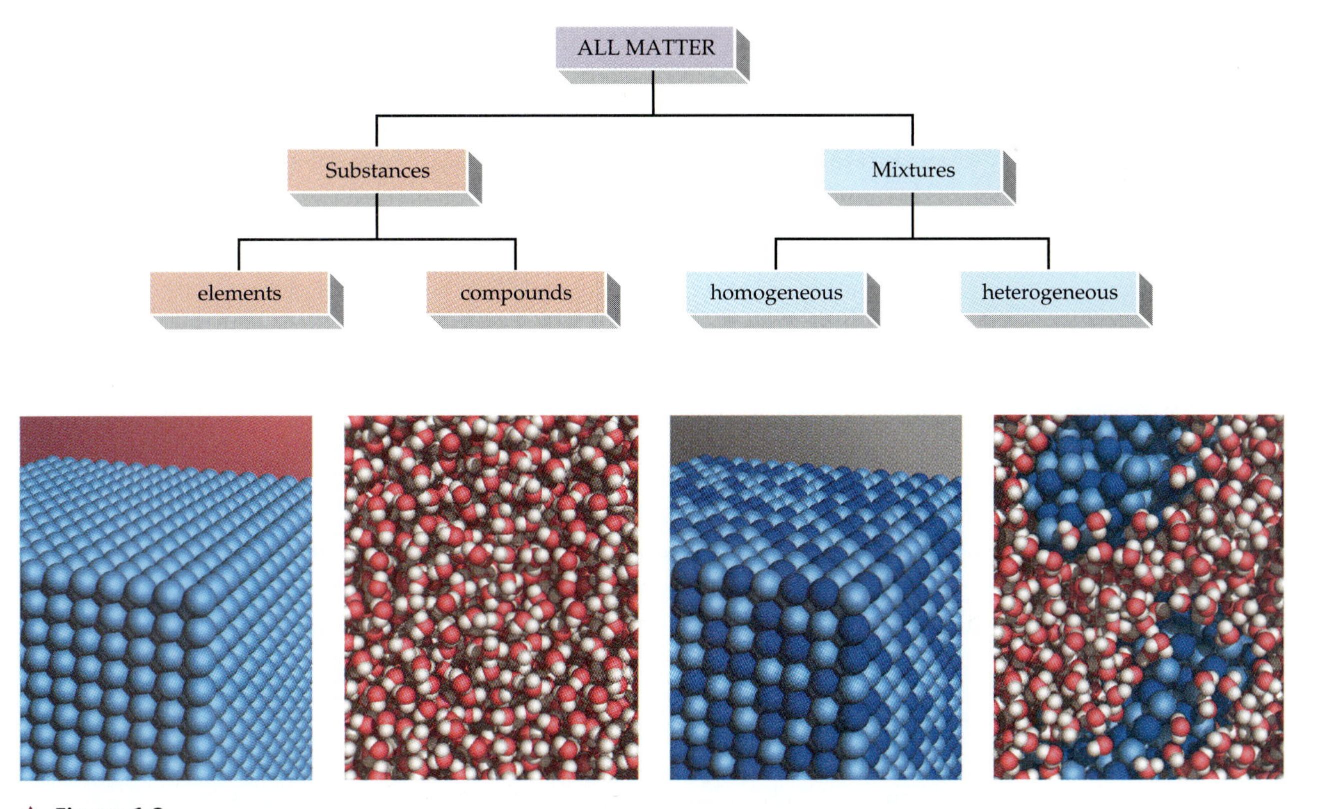

▲ **Figure 1.3**
Scheme for classifying matter, as explained in the text. The "molecular-level views" are of gold (an *element*), water (a *compound*), 12-karat gold (a *homogeneous mixture* of silver and gold), and 12-karat gold in water (a *heterogeneous mixture* of particles).

one type of atom; it is a substance. All samples of water are comprised of molecules consisting of two hydrogen atoms and one oxygen atom; water is a substance. On the other hand, a saline solution—a solution of salt in water—is a mixture. The proportions of salt and water can vary from one saline solution to another.

Any given saline solution is a **homogeneous mixture:** it has the same composition and properties—the same "saltiness"—throughout the solution. By contrast, a **heterogeneous mixture** varies in composition and/or properties from one part of the mixture to another. A mixture of ice and water is heterogeneous. Although the composition of ice and liquid water is the same, the physical properties of the two are different. In a sand-water mixture, both the composition and properties vary.

The names and symbols of all the elements are listed on the inside front cover of this book.

All *substances* are either elements or compounds. An **element** is composed entirely of a single type of atom. Because bulk matter cannot be made exclusively from any particles smaller than atoms, elements are the simplest of all substances. At present, 115 elements are known, but many are quite rare. Among the familiar elements are oxygen, nitrogen, carbon, sulfur, iron, copper, aluminum, silver, and gold.

The modern definition of an element is based on its *atomic number.* This definition is given and explained in Chapter 2.

A **compound** is a substance made up of atoms of two or more elements, with the different kinds of atoms joined in fixed ratios. In contrast to the limited number of elements, the number of possible compounds is essentially limitless. Currently, more than 20 million compounds are known. Water, carbon dioxide, sodium chloride (table salt), sucrose (cane or beet sugar), and iron oxide (rust) are all compounds.

Many compounds are made up of small distinguishable units called molecules. A *molecule* has distinctive characteristics that may not be related in any obvious way to the properties of the atoms that make up the molecule.

Because elements and compounds are so fundamental to the study of chemistry, we find it useful to refer to them by symbols and formulas. A **chemical symbol** is a one- or two-letter designation derived from the name of an element. Most symbols are based on English names; a few are based on the Latin name of the element or one of its compounds (see Table 1.2). The first letter of a symbol is capitalized and the second is always lowercase. (It makes a difference. For example, Co is the symbol for the element cobalt, whereas CO represents the compound carbon monoxide.) Compounds are designated by combinations of chemical symbols called *formulas.*

A chemical symbol in a formula stands for one atom of the element. If more than one atom is to be indicated in a formula, a subscripted number is used after the symbol. For example, the formula Cl_2 represents two atoms of chlorine in a chlorine molecule, and the formula CCl_4 stands for one atom of carbon and four atoms of chlorine in a carbon tetrachloride molecule. The formula $C_{12}H_{22}O_{11}$ indicates that there are 12 carbon atoms, 22 hydrogen atoms, and 11 oxygen atoms in each molecule of sucrose (table sugar). We will consider the writing of chemical formulas in some detail in Chapter 4.

✓ Review Questions

1.6 What is the difference between a substance and a mixture?

1.7 All samples of glucose, a simple sugar, consist of eight parts (by mass) oxygen, six parts carbon, and one part hydrogen. Is glucose a substance or a mixture?

1.8 Distinguish between an element and a chemical compound.

1.6 Energy and Energy Conversion

Nearly all chemical processes are accompanied by changes in energy. **Energy** is the capacity for doing work. **Work** must be done to make something happen that wouldn't happen by itself. Getting out of bed, building a house, and mining coal all require energy. Eating requires energy; that forkful of spaghetti would never make it to your mouth by itself. Energy is the basis for change in the material world.

Table 1.2 Names, Symbols, and Physical Characteristics of Some Common Elements

Name (Latin name)	Symbol	Selected Properties
Aluminum	Al	Light, silvery metal
Argon	Ar	Colorless gas
Arsenic	As	Grayish white solid
Barium	Ba	Silvery white metal
Beryllium	Be	Steel gray, hard, light solid
Boron	B	Black or brown powder; several crystal forms
Bromine	Br	Reddish brown liquid (Br_2)[a]
Calcium	Ca	Silvery white metal
Carbon	C	Soft black solid (graphite) or hard brilliant crystal (diamond)
Chlorine	Cl	Greenish yellow gas (Cl_2)[a]
Copper (Cuprum)	Cu	Light reddish brown metal
Fluorine	F	Pale yellow gas (F_2)[a]
Gold (Aurum)	Au	Yellow malleable metal
Helium	He	Colorless gas
Hydrogen	H	Colorless gas (H_2)[a]
Iodine	I	Lustrous purple-black solid (I_2)[a]
Iron (Ferrum)	Fe	Silvery white, ductile, malleable metal
Lead (Plumbum)	Pb	Bluish white, soft, heavy metal
Lithium	Li	Silvery white, soft, light metal
Magnesium	Mg	Silvery white, ductile, light metal
Mercury (Hydrargyrum)	Hg	Silvery white, liquid, heavy metal
Neon	Ne	Colorless gas
Nickel	Ni	Silvery white, ductile, malleable metal
Nitrogen	N	Colorless gas (N_2)[a]
Oxygen	O	Colorless gas (O_2)[a]
Phosphorus	P	Yellowish white waxy solid or red powder (P_4)[a]
Plutonium	Pu	Silvery white, radioactive metal
Potassium (Kalium)	K	Soft, silvery white metal
Silicon	Si	Lustrous gray solid
Silver (Argentum)	Ag	Silvery white metal
Sodium (Natrium)	Na	Silvery white, soft metal
Sulfur	S	Yellow solid (S_8)[a]
Tin (Stannum)	Sn	Silvery white, soft metal
Uranium	U	Silvery, dense, radioactive metal
Zinc	Zn	Bluish white metal

[a]The formula of the substance described is provided for possible later reference.

Energy is involved any time an object moves or breaks or cools or shines or grows or decays.

Energy exists in two basic forms: potential energy and kinetic energy. **Potential energy** is energy of position or arrangement; it is the energy associated with forces of attraction or repulsion between objects. **Kinetic energy** is energy of motion. Of

▲ **Figure 1.4**
The chemical potential energy stored in the sugars in this orange is converted to the kinetic energy needed to move this velodrome racer along the race course.

particular interest to us is *chemical potential energy,* such as that stored in sugar molecules (Figure 1.4). We convert the potential energy stored in sugar and other molecules to kinetic energy for all our activities, from breathing to vigorous exercise. Most chemical reactions absorb or release heat energy (see Section 1.12).

A hydroelectric dam provides another familiar illustration of potential energy and kinetic energy. Water at the top of a dam has potential energy by virtue of its gravitational attraction to Earth's center. The water has the capacity to do work, but as long as it remains behind the dam it does none. When the water is allowed to flow through a pipe to a lower level, some of its potential energy is converted to kinetic energy. Water rushing through the pipe can be made to turn the blades of a turbine (a water wheel), which in turn can rotate a coil of wire in an electrical generator, producing electricity. The net result is that some of the potential energy originally stored in the water is converted to electrical work.

Kinetic energy depends on mass and velocity. The bigger an object is and the faster it is moving, the more kinetic energy it has and the more work it can do. Mathematically, kinetic energy equals one-half the mass times the square of the velocity.

$$KE = \frac{1}{2}mv^2$$

Example 1.3

What kind of energy—potential or kinetic—does each of the following possess?

a. a thrown softball
b. a softball resting at the edge of a table
c. gasoline

Solution

a. A thrown softball has kinetic energy; it is moving through the air.
b. A softball at the edge of a table has potential energy by virtue of its position; it could release energy by falling.
c. Gasoline contains chemical energy, a form of potential energy; the energy is released when the gasoline burns.

Practice Exercise

What kind of energy—potential or kinetic—does each of the following possess?

a. a stick of dynamite
b. a bird in flight toward a plate-glass window
c. a hammer poised at the top of an arm-swing

Energy may be classified in other ways, often based on some characteristic such as the source of the energy. For example, we speak of *nuclear energy,* but—like chemical energy—it is a form of potential energy. Both nuclear energy and chemical energy originate in forces of attraction or repulsion between particles.

Example 1.4

What energy changes occur during each of the following events?

a. Fuel oil is burned.
b. A softball falls off the edge of a table.

Solution

a. Potential (chemical) energy is converted to kinetic (heat or thermal) energy.
b. Potential energy is converted to kinetic energy.

Review Questions

1.9 Define each of the following terms.

a. energy **b.** work

1.10 What is the difference between kinetic energy and potential energy?

1.7 Electric Forces

An electrostatic generator causes an electric charge on this woman's hair. The hairs all have like charges and repel one another.

To deal with energy transformations, chemistry often borrows fundamental concepts from its neighboring discipline, physics. One such concept is force. A force is a push or a pull that sets an object in motion, or stops a moving object, or holds an object in place. *Gravity* is a force. Objects are held to the surface of the Earth by gravity, the attraction of Earth's mass for other masses. The weight of an object is the force of gravity that exists between the object and the Earth.

If you have ever encountered static electricity, you know that matter can be electrified. Electric forces are extremely important in chemistry. Some particles of matter are neutral, and some bear an electric charge, either positive (+) or negative (−). A particle with a charge can exert a force—that is, can push or pull another particle that also has a charge. The particles do not have to be touching to attract or repel one another; they do so even at a distance. For this reason, we say that charged particles have *force fields* around them. These forces get weaker as the particles get farther apart. Particles with *like* charges (both positive or both negative) repel one another. Those with *unlike* charges (one positive and one negative) attract one another (Figure 1.5).

You are no doubt familiar with some forms of static electricity. Clothes tumbled in an automatic dryer pick up electrical charges as they rub against one another. Some fabrics tend to pick up positive charges and others, negative charges. When taken from the dryer, the pieces of clothing stick to one another by what commercials call static cling. The "cling" results from the attraction of unlike charges. If, on the other hand, you brush your hair vigorously on a cold day, your hair may become "unmanageable." Instead of lying flat against your head, the hair sticks out in all directions. Each strand repels all the other strands because all are alike and develop like charges. The brush develops an opposite charge, and the hairs are attracted to it.

Review Questions

1.11 What is a force? How do charged particles exert a force? What sort of force is exerted by Earth on an object at its surface?

1.12 Describe what happens to two particles with like charges when they are brought close together. What happens to particles with unlike charges when they are brought close together?

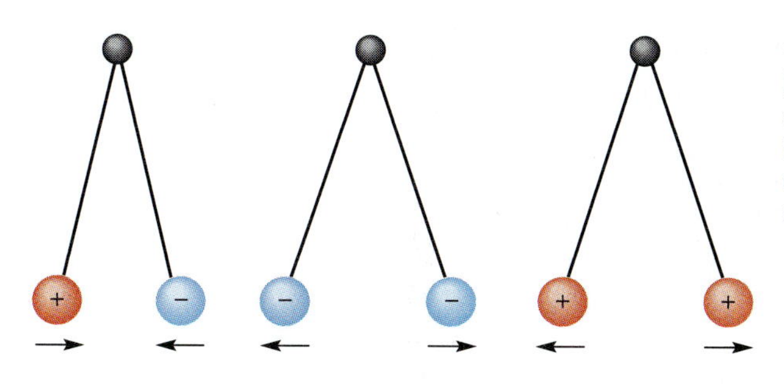

Figure 1.5
Diagram illustrating (a) the mutual attraction of particles with unlike charges and (b) the mutual repulsion of particles with like charges.

Scientific Measurements

Scientists gather data by making measurements. To be accepted, the measured data must be verified by other scientists. Verification is easier if all scientists use a common system of measurement. Measurements in medicine are no less important. Medical workers routinely measure body temperature and blood pressure. Modern medical diagnosis depends on many types of measurements, including careful chemical analyses of blood and urine.

1.8 The Modern Metric System

The measurement system agreed upon since 1960 is the **International System of Units (SI)** (from the French *Système International*), a modernized version of the metric system established in France in 1791. Most countries use metric measures in everyday life. In the United States, SI is used mainly in science laboratories. However, metric measures are increasingly used in commerce, especially in businesses with an international component. The contents of most bottled beverages are now given in metric units, and metric measurements are also common in sporting events. Figure 1.6 compares some customary and metric units.

Because SI is based on the decimal system, it is easy to convert from one unit to another. All measured quantities can be expressed in terms of the seven base units listed in Table 1.3. We will use the first six in this text. An essential part of the SI system is the use of exponential ("powers of ten") notation for numbers. If you are not already familiar with this notation, you will find a discussion of this topic in Appendix I.

A disadvantage of the basic SI units is that they are often of awkward magnitude. Thus we use prefixes (see Table 1.4) to indicate units larger and smaller than the base unit.

Length, Area, and Volume

Today the meter is defined precisely as the distance that light travels in a vacuum during 1/299,792,458 of a second.

The SI base unit of length is the **meter (m),** a unit about 10% longer than the yard. To measure distances along a highway, we often use the kilometer (km) (see Table 1.5). In the laboratory, we usually find lengths smaller than the meter most convenient. We use the centimeter (cm)—about 0.4 inch—and the millimeter (mm)—about the

Figure 1.6 ▶
A comparison of metric and customary units of measure. The beaker at the left contains 1 kg of candy and the one next to it contains 1 lb (1 kg = 2.2046 lb). The green ribbon is 1 cm wide around a stick 1 m long (1 m = 1.0936 yd and 1 cm = 0.3937 in.). The yellow ribbon is 1 in wide around a stick 1 yd long. The flasks each hold 1 L when filled to the mark. The flask on the right and the bottle behind it each contain 1 L of orange juice. The flask on the left and the carton behind it each contain 1 qt of milk (1 L = 1.057 qt).

Table 1.3 The Seven SI Base Units

Physical Quantity	Name of Unit	Symbol of Unit
Length	meter[a]	m
Mass	kilogram	kg
Time	second	s
Temperature	kelvin	K
Amount of substance	mole	mol
Electric current	ampere	A
Luminous intensity	candela	cd

[a]Spelled *metre* in most countries other than the United States.

Table 1.4 Some Approved Numerical Prefixes[a]

Exponential Expression	Decimal Equivalent	Prefix	Pronounced	Symbol
10^{12}	1,000,000,000,000	tera-	TER-uh	T
10^{9}	1,000,000,000	giga-	GIG-guh	G
10^{6}	1,000,000	mega-	MEG-uh	M
10^{3}	1,000	kilo-	KIL-oh	k
10^{2}	100	hecto-	HEK-toe	h
10	10	deka-	DEK-uh	da
10^{-1}	0.1	deci-	DES-ee	d
10^{-2}	0.01	centi-	SEN-tee	c
10^{-3}	0.001	milli-	MIL-ee	m
10^{-6}	0.000,001	micro-	MY-kro	μ
10^{-9}	0.000,000,001	nano-	NAN-oh	n
10^{-12}	0.000,000,000,001	pico-	PEE-koh	p
10^{-15}	0.000, 000,000,000,001	femto	FEM-toe	f

[a]The most commonly used prefixes are shown in color.

Table 1.5 Some Metric Units of Length

1 kilometer (km)	= 1000 meters (m) = 10^3 m
1 meter (m)	= 100 centimeters (cm)
1 centimeter (cm)	= 10 millimeters (mm) = 0.01 m = 10^{-2} m
1 millimeter (mm)	= 1000 micrometers (μm) = 0.001 m = 10^{-3} m
1 micrometer (μm)	= 1000 nanometers (nm) = 10^{-6} m
1 nanometer (nm)	= 1000 picometers (pm) = 10^{-9} m
1 picometer (pm)	= 1000 femtometers (fm) = 10^{-12} m

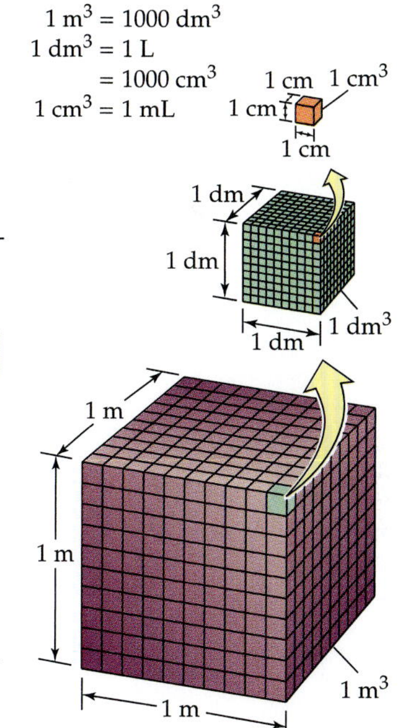

▲ A cube one meter on a side is one cubic meter (1 m³). A cube one decimeter on a side is one cubic decimeter (1 dm³). (1 m³ = 1000 dm³.) A cube one centimeter on a side is one cubic centimeter (1 cm³). (1 dm³ = 1000 cm³.)

thickness of the cardboard backing in a note pad. For measurements at the atomic and molecular level, we use the micrometer (μm), the nanometer (nm), and the picometer (pm). For example, a chlorophyll molecule is about 0.1 μm or 100 nm long, and the diameter of a sodium atom is 372 pm.

The units for area and volume are derived from the base unit of length. The SI unit of area is the square meter (m^2), but we often find square centimeters (cm^2) or square millimeters (mm^2) more convenient for laboratory work.

$$1\ cm^2 = (10^{-2}\ m)^2 = 10^{-4}\ m^2 \qquad 1\ mm^2 = (10^{-3}\ m)^2 = 10^{-6}\ m^2$$

Similarly, the SI unit of volume is the cubic meter (m^3), but two units more likely to be used in the laboratory are the cubic centimeter (cm^3)—often called a "cc" in medical laboratories—and the cubic decimeter (dm^3). A cubic centimeter is about the volume of a sugar cube, and the cubic decimeter is slightly larger than one quart.

$$1\ cm^3 = (10^{-2}\ m)^3 = 10^{-6}\ m^3$$

$$1\ dm^3 = (10^{-1}\ m)^3 = 10^{-3}\ m^3$$

The units liter, deciliter, and milliliter are commonly used for volumes of liquids and gases; the units cubic centimeter and cubic decimeter are more often used for solids.

The cubic decimeter is commonly called a liter. A **liter (L)** is one cubic decimeter or 1000 cubic centimeters.

$$1\ L = 1\ dm^3 = 1000\ cm^3$$

The milliliter (mL) (cubic centimeter) and the deciliter (dL) are frequently used in medical laboratories:

$$1\ mL = 1\ cm^3$$

$$1\ dL = 100 mL = 0.1\ L$$

Mass

Today the kilogram is based on a standard platinum-iridium bar kept at the International Bureau of Weights and Measures.

The SI base quantity of mass is the **kilogram (kg),** about 2.2 pounds. This base quantity is unusual in that it already has a prefix. A more convenient mass unit for most laboratory work is the gram (g).

$$1\ kg = 10^3\ g = 1000\ g$$

The milligram (mg) is a suitable unit for small quantities of materials, such as some drug dosages.

$$1\ mg = 10^{-3}\ g$$

Chemists can now detect masses in the microgram (μg), the nanogram (ng), and even the picogram (pg) range.

Time

The SI base unit for measuring intervals of time is the second (s). Extremely short time periods are expressed through the usual SI prefixes: milliseconds, microseconds, nanoseconds, and picoseconds. Long time intervals, in contrast, are usually expressed in traditional, non-SI units: minute (min), hour (h), day (d), and year (y).

Example 1.5

Convert each of the following measurements to a unit that replaces the power of ten by a prefix.

a. 2.89×10^{-3} g **b.** 4.30×10^3 m

Solution

Our goal is to replace each power of ten with the appropriate prefix from Table 1.4. For example, $10^{-3} = 0.001$, leading to: milli(unit).

a. 10^{-3} corresponds to the prefix milli; 2.89 mg

b. 10^3 corresponds to the prefix kilo; 4.30 km

Practice Exercise

Convert each of the following measurements to a unit that replaces the power of ten by a prefix.

a. 7.24×10^{-3} g **b.** 5.14×10^{-6} m

c. 1.91×10^{-9} s **d.** 5.58×10^3 m

Example 1.6

Use exponential notation to express each of the following measurements in terms of an SI base unit.

a. 4.12 cm **b.** 947 μs

Solution

a. Our goal is to find the power of ten that relates the given unit to the SI base unit. That is, centi(base unit) = $10^{-2} \times$ (base unit)

$$4.12 \text{ centimeter} = 4.12 \times 10^{-2} \text{ m}$$

b. To change microsecond to the base unit second, we need to replace the prefix micro by 10^{-6}. To get our answer in the conventional exponential form, we also need to replace the coefficient 947 by 9.47×10^2. The result of these two changes is

$$947 \text{ μs} = 947 \times 10^{-6} \text{ s} = 9.47 \times 10^2 \times 10^{-6} \text{ s} = 9.47 \times 10^{-4} \text{ s}$$

Practice Exercise

Use exponential notation to express each of the following measurements in terms of an SI base unit.

a. 745 nm **b.** 525 ns
c. 1415 km **d.** 2.06×10^6 g

Review Questions

1.13 What are the names and symbols of the SI base units for length, mass, and temperature?

1.14 What is the basic unit of length in the SI system? What are the derived units for area and volume?

1.9 Precision and Accuracy in Measurements

Counting usually can give exact numbers. We can count exactly 24 students in a room. Measurements, on the other hand, are subject to error. One source of error is in the measuring instruments themselves. For example, an incorrectly calibrated thermometer may consistently yield a result that is 0.2 °C too low. Other errors may result from the experimenter's lack of skill or care in using measuring instruments.

We will consider the conversion between U.S. customary units and metric units in Section 1.10.

Suppose you were one of five students asked to measure a person's height, using a meter stick marked off in millimeters. The five measurements are recorded in Table 1.6. The **precision** of a set of measurements refers to how closely individual measurements agree with one another. We say there is good precision if each of the measurements is close to the average of the series. The precision is poor if there is a wide deviation from the average value. How would you describe the precision of the data in Table 1.6? Examine the individual data, note the average value, and determine how much the individual data differ from the average. You will find that the maximum deviation from the average value is 0.003 m. The precision is therefore quite good.

Table 1.6 Five Measurements of a Person's Height

Student	Height, m
1	1.827
2	1.824
3	1.826
4	1.828
5	1.829
Average:	1.827

The **accuracy** of a set of measurements refers to the closeness of the average of the set to the "correct" or most probable value. Measurements of high precision are more likely to be accurate than are those of poor precision, but even highly precise measurements are sometimes inaccurate. For instance, what if the meter stick used to obtain the data in Table 1.6 were actually 1005 mm long, but still carried 1000 millimeter markings? The accuracy of the measurements would be rather poor, even

Sizes and Masses of Objects: Powers of Ten

Scientists deal with objects smaller than atoms and as large as the universe. We usually use exponential numbers (Appendix I) to describe the sizes of such objects. Subatomic particles have dimensions of about 1×10^{-15} m and masses in the range of 1×10^{-30} kg (electrons) to 1×10^{-27} kg (protons and neutrons). At the other extreme, a galaxy typically measures about 1×10^{23} m across and has a mass of about 10^{41} kg. It is difficult even to imagine numbers so small or large. The following data offer some perspectives on size.

Object	Typical Diameter (m)	Typical Mass (kg)
Atom (a)	10^{-10}	10^{-25}
Molecule (hemoglobin)	10^{-9}	6×10^{-22}
Virus (tobacco mosaic) (b)	10^{-7}	4×10^{-19}
Red blood cell (c)	6×10^{-6}	6×10^{-9}
Grain of sand	10^{-3}	2×10^{-6}
Baseball (d)	8×10^{-2}	1.5×10^{-1}
Earth (e)	6×10^{6}	6×10^{24}
Sun (f)	1.4×10^{9}	2×10^{3}

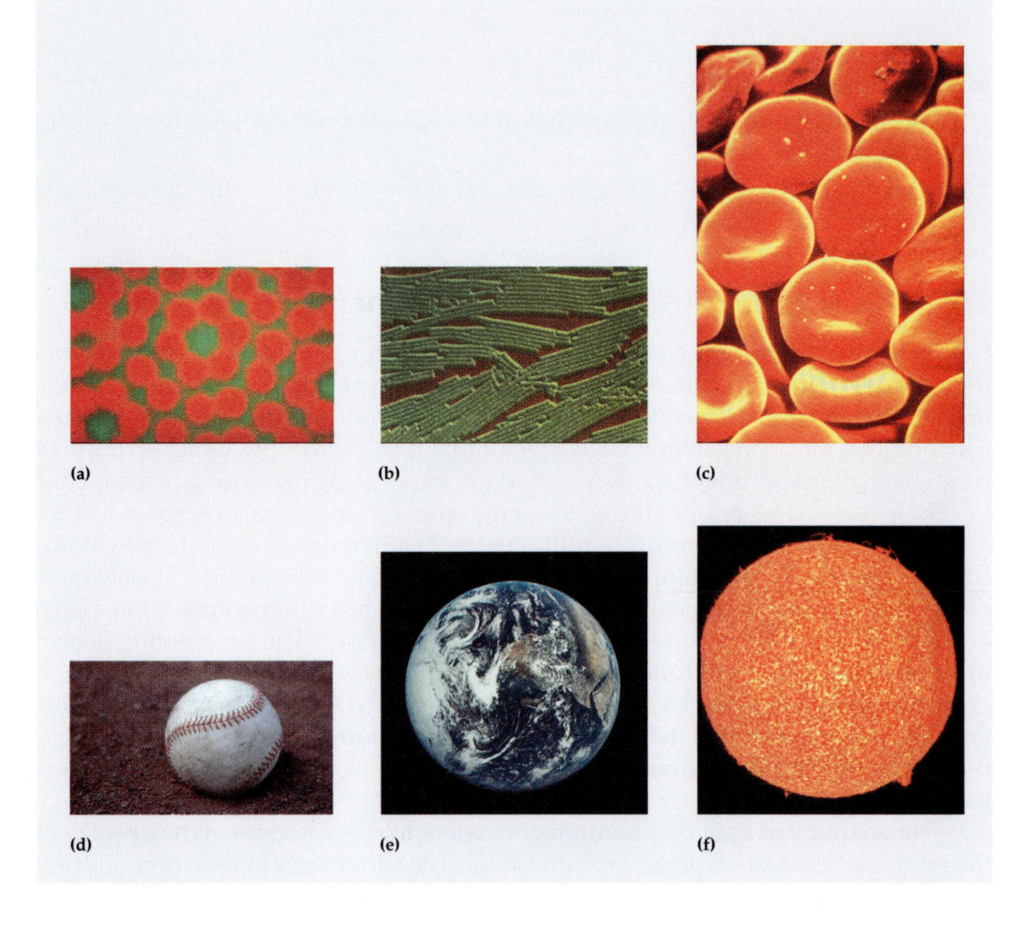

(a) (b) (c) (d) (e) (f)

though the precision would remain good. Accuracy and precision are compared in another manner in Figure 1.7.

Sampling Errors

No matter how accurate an analysis, it will not mean much unless it was performed on valid, representative samples. Consider determining the level of glucose in the blood of a patient. High glucose levels are associated with diabetes, a debilitating disease. Results vary, depending on several factors such as the time of day and what and when the person last ate. Glucose levels are much higher soon after a meal high in sugars. They also depend on other factors, such as stress. Medical doctors usually take blood for analysis after a night of fasting. These fasting glucose levels tend to be more reliable, but physicians often repeat high or otherwise suspicious results. There is therefore no one true value for the level of glucose in a person's blood. Repeated samplings provide an average level. Similar results from several measurements give much more confidence in the findings.

Diagnosis of diabetes is usually based on two consecutive fasting blood glucose levels above 120 mg/dL.

Significant Figures

Look again at Table 1.6. Notice that the five measurements of height agree in the first three digits (1.82); they differ only in the fourth digit. We say that the fourth digit is uncertain. All digits known with certainty, plus the first uncertain one, are called **significant figures.** The precision of a measurement is reflected in the number of significant figures—the more significant figures, the more precise the measurement. The measurements in Table 1.6 have four significant figures. In other words, we are quite sure that the person's height is between 1.82 m and 1.83 m. Our best estimate of the average value, including the uncertain digit, is 1.827 m.

The number 1.827 has four digits; we say it has four significant figures. In any properly reported measurement, all nonzero digits are significant. Zeros, however, may or may not be significant because they can be used in two ways: as a part of the measured value or to position a decimal point.

- Zeros between two other significant digits are significant. Examples: 4807 (four significant figures); 70.004 (five).
- We use a lone zero preceding a decimal point for aesthetic purposes; it is not significant. Example: 0.352 (three significant figures).
- Zeros that precede the first nonzero digit are also not significant. Examples: 0.000819 (three significant figures); 0.03307 (four).
- Zeros at the end of a number are significant if they are to the right of the decimal point. Examples: 0.2000 (four significant figures); 0.050120 (five).

We can summarize these four situations with a general rule: When we read a number from left to right all the digits starting with the first nonzero digit are significant. Numbers without a decimal point that end in zeros are a special case, however.

- Zeros at the end of a number may or may not be significant if the number is written without a decimal point. Example: 700. We do not know whether the number 700 was measured to the nearest unit, ten, or hundred. To avoid this confusion, we can use exponential notation (see Appendix I). In exponential notation, 700 would be recorded as 7×10^2 or 7.0×10^2 or 7.00×10^2 to indicate one, two, or three significant figures, respectively. The only significant digits are those in the coefficient, not in the power of ten.

We use significant figures only with measurements—quantities subject to error. The concept does not apply to a quantity that is (a) inherently an integer, such as

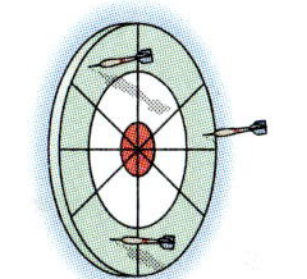
(a) Low accuracy Low precision

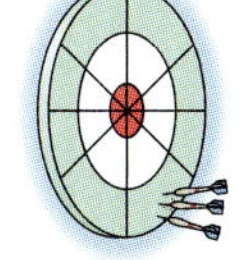
(b) Low accuracy High precision

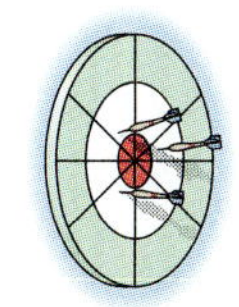
(c) High accuracy Low precision

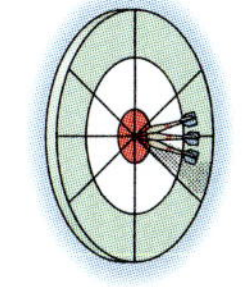
(d) High accuracy High precision

▲ **Figure 1.7**
Comparing precision and accuracy: a dart board analogy. (a) The darts are both scattered (low precision) and off-center (low accuracy). (b) The darts are in a tight cluster (high precision) but still off-center (low accuracy). (c) The darts are somewhat scattered (low precision) but evenly distributed about the center (high accuracy). (d) The darts are in a tight cluster (high precision) and well centered (high accuracy).

three sides to a triangle or 12 items in a dozen, (b) inherently a fraction, such as the radius of a circle = one-half the diameter, or (c) obtained by an accurate count, such as 18 students in a class. It also does not apply to defined quantities, such as 1 km = 1000 m. In these contexts, the numbers 3, 12, 1/2, 18, and 1000 can have as many significant figures as we want. More properly, we say that each is an exact value.

Significant Figures in Calculations: Multiplication and Division

If we measure a sheet of notepaper and find it to be 14.5 cm wide and 21.7 cm long, we can find the area of the paper by multiplying the two quantities. A calculator gives the answer as 314.65. Can we conclude that the area is 314.65 cm^2? That is, can we know the area to the nearest hundredth of a square centimeter when we know the width and length only to the nearest tenth of a centimeter? It just doesn't seem reasonable—and it isn't.

A calculated quantity can be no more precise than the data used in the calculation, and the reported result should reflect this fact.

A strict application of this principle involves a fairly complicated statistical analysis that we will not attempt here, but we can do a pretty good job through a practical rule involving significant figures:

- *In multiplication and division, the reported result should have no more significant figures than the factor with the fewest significant figures.*

In other words, a calculation is only as precise as the least precise measurement that enters into the calculation.

To give a numerical answer with the proper number of significant figures often requires that we round off numbers. In rounding, we drop all digits that are not significant and, if necessary, adjust the last reported digit. We will follow these rules in rounding:

- If the leftmost digit to be dropped is less than 5, leave the final digit unchanged. Example: If we need four significant figures, 69.744 rounds to 69.74, and to 69.7 if we need three significant figures.
- If the leftmost digit to be dropped is greater than 5, increase the final digit by one. Example: 538.76 rounds to 538.8 if we need four significant figures. Similarly, 74.397 rounds to 74.40 if we need four significant figures and 74.4 if we need three.
- If the leftmost to be dropped is exactly 5, we round up if the preceding digit is odd and down if the preceding digit is even. Example: If we need three significant figures in each case, 4.735 rounds to 4.74 and 5.625 rounds to 5.62.

Example 1.7

What is the area, in square centimeters, of a rectangular gauze bandage that is 2.54 cm wide and 12.42 cm long? Use the correct number of significant figures in your answer.

Solution

The area of a rectangle is the product of its length and width. In the result, we can show only as many significant figures as there are in the least precisely stated dimension, the width, which has three significant figures.

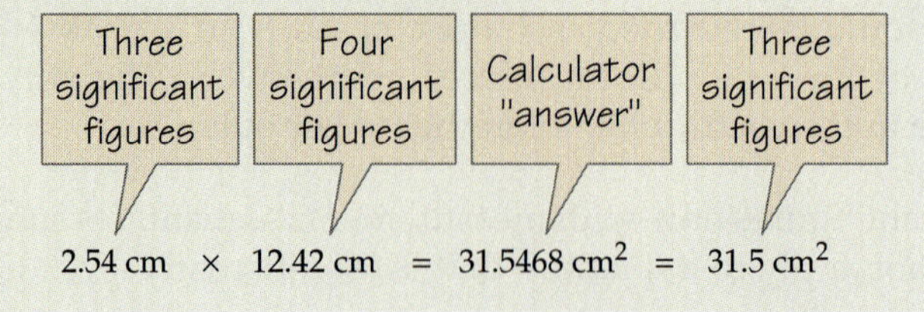

We use the rules on rounding off numbers as the basis for dropping the digits *468*.

Practice Exercise

Calculate the volume, in cubic meters, of a rectangular block of foamed plastic that is 1.827 m long, 1.04 m wide, and 0.064 m thick. Use the correct number of significant figures.

Example 1.8

For a laboratory experiment, a teacher wants to divide all of a 226.8 g sample of glucose equally among the 18 members of her class. How many grams of glucose should each student receive?

Solution

Here we need to recognize that the number "18" is a counted number; that is, an exact number that is not subject to significant figure rules. The answer should carry four significant figures, the same as in 226.8 g.

$$\frac{226.8\ \text{g}}{18} = 12.60\ \text{g}$$

In this calculation a calculator displays the result "12.6." We add the digit "0" to emphasize that the result is precise to four significant figures.

Practice Exercise

A dozen eggs has a mass of 681 g. What is the average mass of one of the eggs, expressed with the appropriate number of significant figures?

Addition and Subtraction

In addition or subtraction, we are concerned not with the number of significant figures but with digits to the right of the decimal point. If we add or subtract quantities with varying numbers of digits to the right of the decimal point, we need to note the one with the fewest such digits. The result should contain the same number of digits to the right of its decimal point. For example, if you are adding several masses, and one of them is measured only to the nearest gram, the total mass cannot be stated to the nearest milligram, no matter how precise the other measurements are.

We apply this idea in Example 1.9. Note that in a calculation involving several steps, we need round off only the final result.

Example 1.9

Perform the following calculation and round off the answer to the correct number of significant figures.

$$2.146\ \text{g} + 72.1\ \text{g} - 9.1434\ \text{g} = ?$$

Solution

In this calculation, we add two numbers and subtract a third from the sum of the first two.

$$\begin{array}{rl} 2.146\ \text{g} & \text{Three decimal places} \\ \underline{+72.1}\ \text{g} & \text{One decimal place} \\ 74.246\ \text{g} & \\ \underline{-9.1434}\ \text{g} & \text{Four decimal places} \\ 65.1026\ \text{g} = 65.1\ \text{g} & \text{One decimal place} \end{array}$$

Note that we do not round off the intermediate result (74.246). Using a calculator, we generally don't need to write down an intermediate result.

Practice Exercise

Perform the indicated operations and give answers with the proper number of significant figures. Note that in addition and subtraction all terms must be expressed in the same unit.

a. 48.2 m + 3.82 m + 48.4394 m
b. 148 g + 2.39 g + 0.0124 g
c. 15.436 L + 5.3 L − 6.24 L − 8.177 L
d. (51.5 m + 2.67 m) × (33.42 m − 0.124 m)
e. $\dfrac{(125.1\text{ g} - 1.22\text{ g})}{(52.5\text{ mL} + 0.63\text{ mL})}$
f. $\dfrac{(0.307\text{ g} - 14.2\text{ mg} - 3.52\text{ mg})}{(1.22\text{ cm} - 0.28\text{ mm}) \times 0.752\text{ cm} \times 0.51\text{ cm}}$

Problem Solving

The ability to solve problems is important in scientific work—and in everyday life. You will find many problems in this text. To solve them, you generally will need to use the given information and apply some basic principles to a new situation. Many chemistry problems require calculations. In the following section we will describe a useful method for doing many such calculations.

1.10 Unit Conversions

We live in a world in which almost all other countries use metric measurements. Suppose that an American applies for a job in another country, and the job application asks for her height in centimeters. She knows her height in customary units is 66.0 in. Her actual height is the same, whether we express it in inches, feet, centimeters, or millimeters. Thus when we measure something in one unit and then convert it to another one, we must not change the measured quantity in any fundamental way.

In mathematics, multiplying a quantity by "1" does not change its value. We can therefore use a factor equivalent to "1" to convert between inches and centimeters. We find our factor in the definition of the inch in Table 1.7.

$$1\text{ in.} = 2.54\text{ cm}$$

Notice that if we divide both sides of this equation by 1 in., we obtain a ratio of quantities that is equal to 1.

$$1 = \frac{1\text{ in.}}{1\text{ in.}} = \frac{2.54\text{ cm}}{1\text{ in.}}$$

If we divide both sides of the equation by 2.54 cm, we also obtain a ratio of quantities that is equal to 1.

$$\frac{1\text{ in.}}{2.54\text{ cm}} = \frac{2.54\text{ cm}}{2.54\text{ cm}} = 1$$

The two ratios, one printed in red and the other in blue, are conversion factors. A **conversion factor** is a ratio of terms, equivalent to the number one, used to change the unit in which a quantity is expressed. We call this process the **unit-conversion method** of problem solving. Because we set up problems by examining the *dimensions*

Table 1.7 Some Conversions Between Common (U.S.) and Metric Units

Metric		Common
Mass		
1 kg	=	2.205 lb
453.6 g	=	1 lb
28.35 g	=	1 ounce (oz)
Length		
1 m	=	39.37 in.
1 km	=	0.6214 mi
2.54 cm	=	1 in. [a]
Volume		
1 L	=	1.057 qt
3.785 L	=	1 gal
29.57 mL	=	1 fluid ounce (fl oz)

[a]The U.S. inch is defined as exactly 2.54 cm. The other equivalencies are rounded off.

associated with the given quantity, those associated with the desired result, and those needed in the conversion factors, the process is sometimes called *dimensional analysis*.

What happens when we multiply a known quantity by a conversion factor? The original unit cancels out and is replaced by the desired unit. Thus our general approach to using conversion factors is

$$\text{Desired quantity and unit} = \text{given quantity and unit} \times \text{conversion factors}$$

Now let's return to the question about the woman's height, a measured quantity of 66.0 in. To get an answer in centimeters, we must use the appropriate conversion factor (in red). Note that the desired unit (cm) is in the numerator, and the unit to be replaced (in.) is in the denominator. Thus we can cancel the unit in. so that only the unit cm remains.

$$66.0\ \cancel{\text{in.}} \times \frac{2.54\ \text{cm}}{1\ \cancel{\text{in.}}} = 168\ \text{cm} \quad \text{(rounded off from 167.64)}$$

You can see why the other conversion factor (blue) won't work. It gives a nonsensical unit.

$$66.0\ \text{in.} \times \frac{1\ \text{in.}}{2.54\ \text{cm}} = 26.0\ \frac{\text{in.}^2}{\text{cm}}$$

This method of problem solving is best learned by practice.

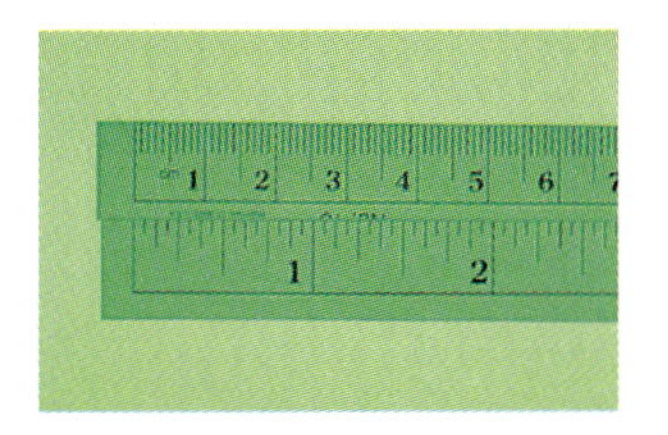

▲ Customary and metric measures of length. 1 in. = 2.54 cm (exactly).

▲ A person's height is the same whether it is measured in meters or in feet; it is simply expressed differently.

Example 1.10

a. Convert 0.742 kg to grams.
b. Convert 0.615 lb to ounces.
c. Convert 135 lb to kilograms.

Solution

a. Use a knowledge of prefixes to convert from kilograms to grams.

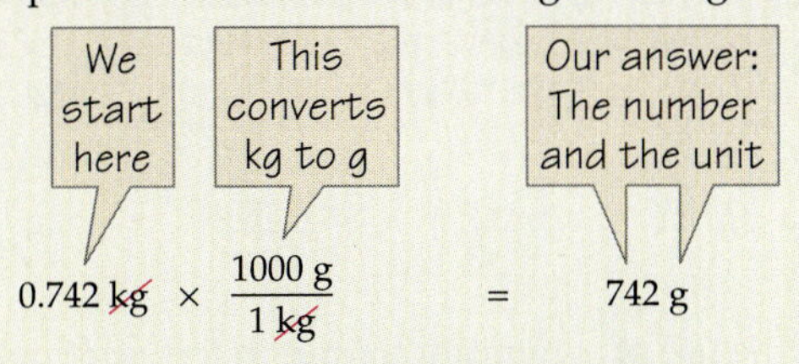

$$0.742\ \cancel{\text{kg}} \times \frac{1000\ \text{g}}{1\ \cancel{\text{kg}}} = 742\ \text{g}$$

b. Use the fact that 1 lb = 16 oz to convert from pounds to ounces. Then proceed as in part (a), here arranging the conversion factor to cancel the unit lb.

$$0.615\ \cancel{\text{lb}} \times \frac{16\ \text{oz}}{1\ \cancel{\text{lb}}} = 9.84\ \text{oz}$$

c. Use data from Table 1.7 to convert from pounds to kilograms. Then proceed as in part (b), arranging the conversion factor to cancel the unit lb.

$$135\ \cancel{\text{lb}} \times \frac{1\ \text{kg}}{2.205\ \cancel{\text{lb}}} = 61.2\ \text{kg}$$

Now let's look at the use of significant figures in this problem. In part (a), the measured quantity, 0.742 kg, is given with three significant figures. The relationship 1000 g = 1 kg is exact; it *defines* the way we relate grams and kilograms. Our answer should have three significant figures.

In part (b), the measured quantity, 0.615 lb, is given with three significant figures. The relationship 16 oz = 1 lb is exact, and this will not affect the precision of our calculation. Our answer should have three significant figures.

In part (c), the measured quantity, 135 lb, is given with three significant figures. The relationship 1 kg = 2.205 lb is stated to four significant figures; there is no need to write 1 kg as 1.000 kg. However, our answer should have only three significant figures.

Practice Exercise

Carry out the following conversions.

a. Convert 16.3 mg to grams.
b. Convert 24.5 oz to pounds.
c. Convert 12.5 fl oz to milliliters.

Quite often we must use more than one conversion factor to get the desired unit. We can do so by arranging all the necessary conversion factors in a single setup that yields the final answer in the desired unit.

Example 1.11

What is the length, in millimeters, of a 3.25-ft piece of tubing?

Solution

No relationship between feet and millimeters is given in Table 1.7, so we need more than one conversion factor. We can think of the problem as a series of three conversions:

1. Use the fact that 1 ft = 12 in. to convert from feet to inches.
2. Use data from Table 1.7 to convert from inches to centimeters.
3. Use a knowledge of prefixes to convert from centimeters to millimeters.

We could solve this problem in three distinct steps by making one conversion in each step, but it is just as easy to combine three conversion factors into a single setup. Then we proceed as sketched below.

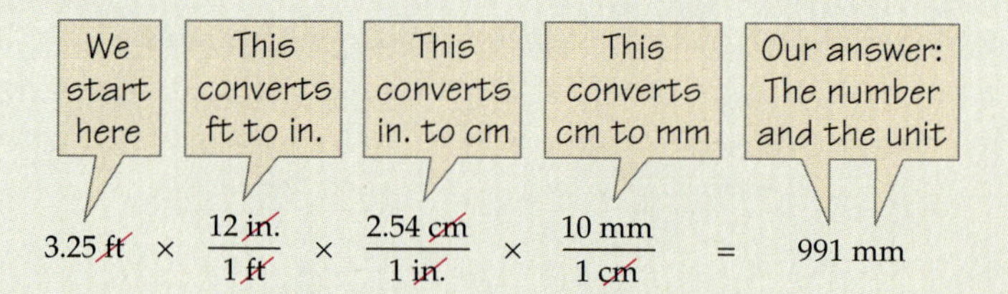

$$3.25\ \text{ft} \times \frac{12\ \text{in.}}{1\ \text{ft}} \times \frac{2.54\ \text{cm}}{1\ \text{in.}} \times \frac{10\ \text{mm}}{1\ \text{cm}} = 991\ \text{mm}$$

Now let's look at the use of significant figures in this problem. The measured quantity, 3.25 ft, has three significant figures. The relationship 12 in. = 1 ft is exact, and will thus not affect the precision of our calculation. The relationship 1 m = 39.37 in. is stated to four significant figures. Finally, the relationship 100 cm = 1 m is exact. Our answer should have three significant figures.

Practice Exercise

Carry out the following conversions.

a. 90.3 mm to meters
b. 729.9 ft to kilometers
c. 1.17 gal to fluid ounces

Sometimes we need to convert two or more units in the measured quantity. We can do this too by arranging all the necessary conversion factors in a single setup so that the starting units cancel and the conversions yield the final answer in the desired units.

Example 1.12

A saline solution has 1.00 lb of salt in 1.00 gal of solution. Calculate the quantities in grams per liter of solution.

Solution

First let's identify the measured quantities. It is a ratio of a mass of salt in pounds to a volume in gallons.

$$\frac{1\text{ lb (salt)}}{1\text{ gal (solution)}}$$

We want to convert this ratio to one expressed in grams per liter. We must convert from pounds to grams in the numerator and from gallons to liters in the denominator. We need this set of equivalent values to formulate conversion factors:

$$1\text{ lb} = 453.6\text{ g} \qquad 1\text{ gal} = 3.785\text{ L}$$

We start here.

To convert lb to g in numerator

To convert gal to L in denominator

Our answer: The number and the unit

$$\frac{1.00\ \cancel{\text{lb}}}{1.00\ \cancel{\text{gal}}} \times \frac{453.6\text{ g}}{1\ \cancel{\text{lb}}} \times \frac{1\ \cancel{\text{gal}}}{3.785\text{ L}} = 120\text{ g/L}$$

Note that we could also do the conversions in the numerator and denominator separately and then divide the numerator by the denominator.

Numerator: $1.00\ \cancel{\text{lb}} \times \dfrac{453.6\text{ g}}{1\ \cancel{\text{lb}}} = 454\text{ g}$

Denominator: $1.00\ \cancel{\text{gal}} \times \dfrac{3.785\text{ L}}{1\ \cancel{\text{gal}}} = 3.785\text{ L}$

Division: $\dfrac{454\text{ g}}{3.785\text{ L}} = 120\text{ g/L}$

Note that both methods give the same answer to three significant figures.

Practice Exercise

Carry out the following conversions.

a. 88.0 km/h to meters per second
b. 1.22 ft/s to kilometers per hour
c. 4.07 g/L to ounces per quart

Because a conversion factor has an intrinsic value of 1—the numerator and denominator are equivalent—we can raise the factor to a power and it still has an intrinsic value of 1. For example,

$$\frac{2.54\text{ cm}}{1\text{ in.}} = 1 \text{ and } \left(\frac{2.54\text{ cm}}{1\text{ in.}}\right)^2 = 1^2 = 1 \text{ and } \left(\frac{2.54\text{ cm}}{1\text{ in.}}\right)^3 = 1^3 = 1$$

Example 1.13

A piece of cardboard has an area of 976 cm^2. What is its area in square inches?

Solution

We can base this conversion on a single conversion factor, but the factor (1 in./2.54 cm) must be squared (raised to the second power).

$$976\ \text{cm}^2 \times \left(\frac{1\ \text{in.}}{2.54\ \text{cm}}\right)^2 = 976\ \cancel{\text{cm}^2} \times \frac{(1)^2\ \text{in.}^2}{(2.54)^2\ \cancel{\text{cm}^2}} = 151\ \text{in.}^2$$

Practice Exercise

Carry out the following conversions.

a. 1.56×10^4 in.^{3} to cubic meters

b. 14.7 lb/in.^{2} to kilograms per square meter

Review Questions

1.15 What is a conversion factor? How must the numerator and denominator of a conversion factor be related?

1.16 How would you describe a young man who is 160 cm tall and has a mass of 94 kg?

1.11 Density

An important property of matter, particularly in scientific work, is density. We define **density** as the quantity of mass per unit of volume:

$$d = \frac{m}{V}$$

For substances that don't mix, such as oil and water, the concept of density allows us to predict which one will float on the other. It isn't just the mass of the materials, but the *mass per unit volume*—the density—that determines the result. Density is one of the properties that helps us to know that in an oil-and-vinegar salad dressing, oil (lower density) floats on water (higher density). Figure 1.8 gives more elaborate examples.

We can rearrange the equation for density to give

$$m = d \times V \quad \text{and} \quad V = \frac{m}{d}$$

These equations are useful for calculations. Densities (see Table 1.8) are usually reported in grams per milliliter (g/mL) for liquids or grams per cubic centimeter (g/cm^3) for solids.

▲ The brass weight and the pillow have the same mass. Which has the greater density?

Example 1.14

What is the density of iron if 156 g of iron occupies a volume of 20.0 cm^3?

Solution

The given quantities are

$$m = 156\ \text{g} \quad \text{and} \quad V = 20.0\ \text{cm}^3$$

We can use the equation that defines density.

$$d = \frac{m}{V} = \frac{156\ \text{g}}{20.0\ \text{cm}^3} = 7.80\ \text{g/cm}^3$$

Table 1.8 Densities of Some Common Substances at Specified Temperatures

Substance[a]	Density	Temperature
Solids		
Copper (Cu)	8.94 g/cm^3	25 °C
Gold (Au)	19.3 g/cm^3	25 °C
Magnesium (Mg)	1.738 g/cm^3	20 °C
Water (ice) (H_2O)	0.917 g/cm^3	0 °C
Liquids		
Blood plasma (a mixture)	1.027 g/mL	25 °C
Ethyl alcohol (C_2H_5OH)	0.789 g/mL	20 °C
Hexane (C_6H_{14})	0.660 g/mL	20 °C
Mercury (Hg)	13.534 g/mL	25 °C
Urine (a mixture)	1.003–1.030 g/mL	25 °C
Water (H_2O)	0.997 g/mL	0 °C
Water (H_2O)	1.0000 g/mL	3.98 °C
Gases[b]		
Air (a mixture)	1.185 g/L	25 °C
Carbon dioxide (CO_2)	1.976 g/L	0 °C
Helium (He)	0.194 g/L	25 °C
Oxygen (O_2)	1.429 g/L	0 °C

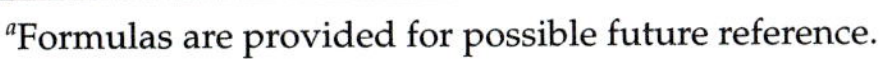

[a]Formulas are provided for possible future reference.

[b]Densities at 1 atmosphere pressure.

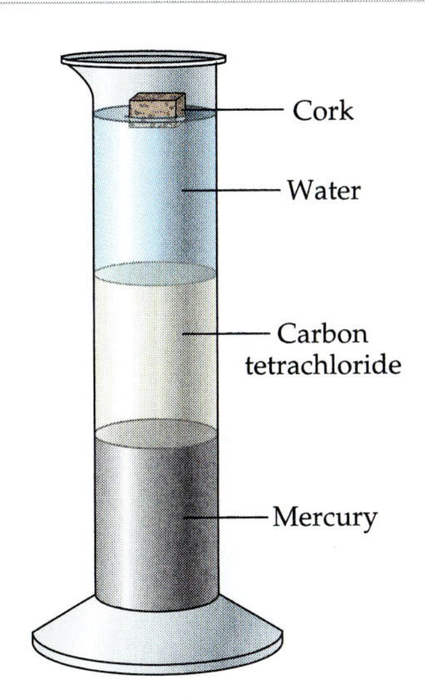

▲ **Figure 1.8**
Liquid carbon tetrachloride (d = 1.54 g/mL) floats on liquid mercury (d = 13.6 g/mL). Water (d = 1.00 g/mL), which does not mix with carbon tetrachloride, floats on it. Finally, a cork (d = 0.22 g/cm^3) floats on the water.

Practice Exercise

What is the density of a metal alloy if a cube that measures 2.00 cm on an edge has a mass of 74.4 g?

Example 1.15

What is the mass of 1.00 L of gasoline if its density is 0.660 g/mL?

Solution

We can express density as a ratio, 0.660 g/1.00 mL, and use it as a conversion factor. We also need the factor 1000 mL/1L to convert liters to milliliters.

$$m = d \times V = \frac{0.660\ \text{g}}{1.00\ \cancel{\text{mL}}} \times 1.00\ \cancel{\text{L}} \times \frac{1000\ \cancel{\text{mL}}}{1\ \cancel{\text{L}}} = 600\ \text{g}$$

Practice Exercise

What is the mass of 50.0 mL of glycerol, which has a density of 1.264 g/mL at 20 °C?

Example 1.16

What volume is occupied by 461 g of mercury? (The density of mercury is given in Table 1.8.)

Solution

Here we can express the inverse of density as a ratio, 1.00 mL/13.534 g, and use it as a conversion factor.

$$V = 461\ \text{g} \times \frac{1\ \text{mL}}{13.534\ \text{g}} = 34.1\ \text{mL}$$

Practice Exercise

What volume would be occupied by a 10.0-g piece of aluminum? (The density of aluminum is 2.70 g/mL.)

Mass does not vary with temperature, but volume does, and so does density. We can assume the density of water to be 1.00 g/mL near room temperature.

Ethylene glycol (the principal ingredient in many antifreeze solutions), water, and ethanol are all colorless liquids. They can be distinguished by measuring their densities. At 20 °C, the density of ethylene glycol is 1.114 g/mL, that of water is 0.998 g/mL, and that of ethanol is 0.789 g/mL.

A term related to density is **specific gravity,** the ratio of the mass of any substance to the mass of an equal volume of water. The specific gravity of water itself, therefore, is 1.00. Mercury (the silvery liquid in thermometers) has a specific gravity of 13.5. That means it has a density 13.5 times as great as that of water. The specific gravity of methyl alcohol is 0.791; thus, methyl alcohol is less dense than water. Because it is the ratio of two values, specific gravity is a number without units, whereas density is reported in units of mass per volume. Because the density of water is 1.00 g/mL, the specific gravity of a substance is numerically the same as its density.

$$\text{Specific gravity of a substance} = \frac{\text{density of the substance}}{\text{density of water}}$$

Example 1.17

The density of chloroform is 1.48 g/mL at 20 °C. What is its specific gravity?

Solution

$$\text{Specific gravity} = \frac{1.48\ \cancel{\text{g/mL}}}{1.00\ \cancel{\text{g/mL}}} = 1.48$$

Practice Exercise

The specific gravity of a sample of battery acid is 1.39. What is its density?

The specific gravity of a liquid is frequently measured by a device called a hydrometer. The hydrometer is placed in the liquid. How far it dips down into the liquid is determined by the density of the liquid (Figure 1.9). The stem of the hydrometer is calibrated in such a way that the specific gravity can be read directly at the surface of the liquid. Hydrometers can be used to measure parameters of solutions that are related to specific gravity, such as the strength of the "battery acid" in your car, the percentage of antifreeze in the radiator, sugar content in maple syrup, dissolved solids in urine, and the alcohol content of wine.

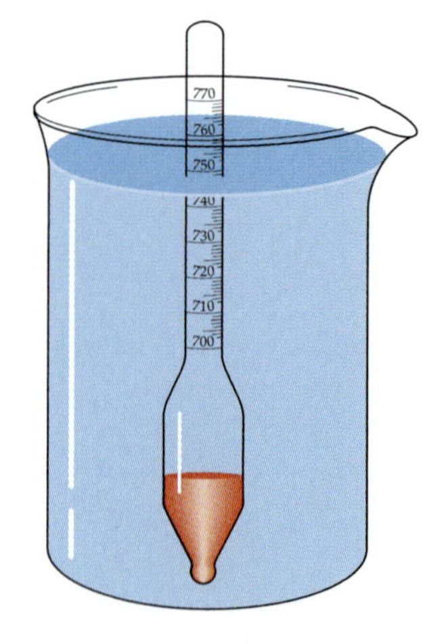

▲ **Figure 1.9**
The hydrometer shown here measures specific gravities over a range of 0.700 to 0.770.

Review Questions

1.17 Define each of the following terms.

a. density **b.** specific gravity

1.18 What is a hydrometer? How does it work and what is it used for?

1.12 Energy: Temperature and Heat

Two important concepts in science are the related ones of temperature and heat. It is difficult to define either in simple terms,[1] but for now we can use the following ideas. We can think about what happens if two objects at different temperatures are brought together: **Heat** flows from the warmer to the colder object. The temperature of the warmer object drops and that of the colder object increases, until finally the two objects are at the same temperature. **Temperature** is therefore a property that tells us in what direction heat will flow. For example, if you touch a hot test tube, heat will flow from the tube to your hand. If the tube is hot enough, your hand will be burned.

Temperature

The SI base unit of temperature is the **kelvin** (K). For routine laboratory work, we often use the more familiar **Celsius** temperature scale. On this scale, the freezing point of water is 0 degrees Celsius (°C) and the boiling point is 100 °C. The interval between these two reference points is divided into 100 equal parts, each a degree Celsius. The Kelvin scale is called an absolute scale because its zero point is the coldest temperature possible, or absolute zero. (This fact was determined by theoretical considerations and has been confirmed by experiment, as we will see in Chapter 6.) The zero point on the Kelvin scale, 0 K, is equal to −273.15 °C. A kelvin is the same size as a degree Celsius, so the freezing point of water on the Kelvin scale is 273.15 K. The Kelvin scale has no negative temperatures, and we don't use a degree sign with the K. To convert from degrees Celsius to the Kelvin scale, simply add 273.15 to the Celsius temperature.

$$\text{K} = {}^\circ\text{C} + 273.15$$

Example 1.18

The boiling point of water is 100.00 °C. What is the boiling point of water on the Kelvin scale?

Solution

$$100.00\ {}^\circ\text{C} + 273.15 = 373.15\ \text{K}$$

Practice Exercise

Express a temperature of −78 °C in kelvins.

The **Fahrenheit** temperature scale is widely used in the United States. Figure 1.10 compares the three temperature scales. Note that on the Fahrenheit scale, the freezing point of water is 32 °F and the boiling point is 212 °F, so that a 10-degree temperature interval on the Celsius scale equals an 18-degree interval on the Fahrenheit scale. From these facts we can derive two equations that relate temperatures on the two scales. One of these requires multiplying the degrees of Celsius temperature by the factor 1.8 (that is, 18/10) to obtain degrees of Fahrenheit temperature, followed by adding 32 to account for the fact that 0 °C = 32 °F. In the other equation, we subtract 32 from the Fahrenheit temperature to get the number of degrees Fahrenheit above the freezing point of water. Then this quantity is divided by 1.8.

[1]More formal definitions are as follows: **Heat** is energy in transit into or out of an object caused by a difference in temperature between the object and its surroundings. **Temperature** is a physical property related to the kinetic energies of the atoms or molecules in a substance. It indicates the direction of heat flow. Kinetic energy, as heat, is transferred from more energetic (higher temperature) to less energetic (lower temperature) atoms or molecules.

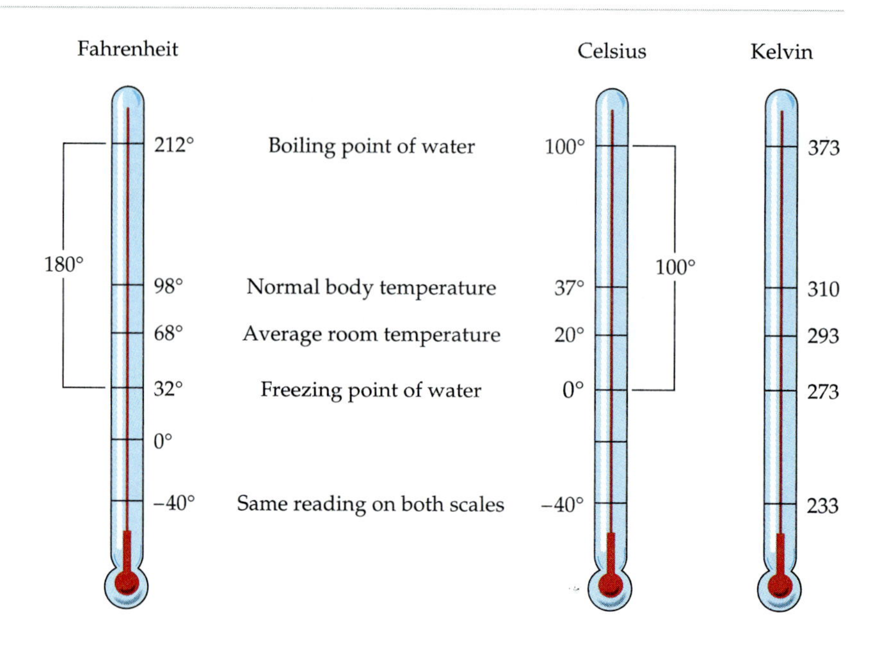

Figure 1.10 ▶
A comparison of the Fahrenheit, Celsius, and Kelvin temperature scales.

$$°F = 1.8\,(°C) + 32 \qquad\qquad °C = \frac{(°F - 32)}{1.8}$$

The following are alternate forms of the equations to interconvert Fahrenheit and Celsius.

$$°F = \frac{9}{5}(°C) + 32$$

$$°C = (°F - 32)\,\frac{5}{9}$$

Example 1.19 illustrates a practical situation where conversion between Celsius and Fahrenheit temperatures would be necessary.

Example 1.19

At home you keep your room thermostat set at 68 °F. While traveling, you have a room with a thermostat that uses the Celsius scale. What Celsius temperature would give you the same temperature as at home?

Solution

$$°C = \frac{(°F - 32)}{1.8}$$

$$= \frac{(68 - 32)}{1.8} = 20\ °C$$

Practice Exercise

Carry out the following conversions.

a. Convert 85.0 °C to degrees Fahrenheit.
b. Convert −12.2 °C to degrees Fahrenheit.
c. Convert 355 °F to degrees Celsius.
d. Convert −20.8 °F to degrees Celsius.

▲ The lowest temperature ever recorded on Earth was −128.6 °F at Vostok Station, Antarctica. What is that temperature in degrees Celsius and in kelvins?

Heat

Scientists often need to measure quantities of heat energy. The SI unit of heat is the joule. A **joule** (J) is the work done by a force of 1 newton[2] acting over a distance of 1 meter. We will also use the more familiar *calorie.*

[2] A *newton* (N) is the basic unit of force in SI. It is the force required to give a 1 kg mass an acceleration of 1 m/s^2. That is, 1 N = 1 $kg{\cdot}m/s^2$. Therefore 1 J = 1 N·m = 1 $kg{\cdot}m^2/s^2$.

Body Temperature, Hypothermia, and Hyperthermia (Fever)

Carl Wunderlich, a German physician, first recognized fever as a symptom of disease. During the 1840s and 1850s he averaged a large number of human temperature measurements and rounded the value to 37 °C. When this value was converted to the Fahrenheit scale, it somehow (improperly) acquired an extra significant figure, and 98.6 °F became widely (and incorrectly) known as the *normal body temperature.* In recent years millions of measurements have revealed that *average* normal body temperature is actually 98.2 °F, and that body temperatures in healthy people range from 97.7 °F to 99.5 °F.

The healthy human body maintains a fairly constant temperature. Heat is a by-product of metabolism, and we must constantly get rid of some of it. Because heat always flows spontaneously from a hot object to a cold one, we can get rid of excess heat by simple conduction only if the surrounding temperature is less than 37 °C. Above that temperature, the body depends on air movement, blood vessel dilation, and perspiration to keep its temperature normal. If the outside temperature is above 40 °C, the National Weather Service usually issues a heat advisory because the body can *gain* heat from the environment. The body temperature then rises above the normal range, a condition known as *hyperthermia.* Certain diseases also cause abnormally high body temperatures or *fevers.* Mild fevers (up to 102 °F) are not dangerous and may even help the body fight off an infection. Prolonged high fevers (104 °F or more in adults) can be fatal.

When exposed to prolonged cold, the body can lose too much heat to the environment. The body temperature drops below the normal range, a condition known as *hypothermia.* A drop of only 2 °F or 3 °F leads to shivering, a condition in which muscles contract in an attempt to generate more heat. Prolonged or severe (body temperature below 93 °F) hypothermia can lead to unconsciousness and death.

"Homeostasis"

$$1 \text{ cal} = 4.184 \text{ J}$$

$$1 \text{ kcal} = 1000 \text{ cal} = 4184 \text{ J}$$

A **calorie** (cal) is the amount of heat required to raise the temperature of 1 g of water 1 °C. (This quantity varies slightly with temperature; a calorie is defined more precisely as the amount of heat required to raise the temperature of 1 g of water from 14.5 to 15.5 °C).

Some substances gain or lose heat more readily than others (see Table 1.9). The **specific heat** of a substance is the amount of heat required to raise the temperature

The specific heat of a substance varies somewhat with temperature, and some of the calculations we do here yield only approximate answers. The most important specific heat value that we use in these calculations is that of water. At about room temperature the specific heat of water is 1.00 cal/(g·°C), and over the temperature range from 0 °C to 100 °C it remains within 1% of this value.

Table 1.9 Specific Heat Values of Some Common Substances at 25 °C

Substance[a]	Specific Heat cal/(g·°C)	J/(g·°C)
Aluminum (Al)	0.216	0.902
Copper (Cu)	0.0921	0.385
Ethyl alcohol (C_2H_5OH)	0.588	2.46
Iron (Fe)	0.106	0.443
Ethylene glycol (HOC_2H_4OH)	0.561	2.35
Magnesium (Mg)	0.245	1.025
Mercury (Hg)	0.0332	0.139
Sulfur (S_8)	0.169	0.706
Water (H_2O)	1.000	4.182

[a]Formulas are provided for possible future reference.

of 1 g of the substance by 1 °C. The definition of the calorie indicates that the specific heat of water is 1.00 cal/(g·°C). In SI, the specific heat of water is 4.182 J/(g·°C) at 25°C. Water is a poor conductor of heat. Metals are good heat conductors. A sample of water must absorb much more heat to raise its temperature than a metal sample of similar mass. The metal sample may become red hot after absorbing a quantity of heat that will make the water sample only lukewarm.

We can use the following equation, in which ΔT is the change in temperature (in either °C or kelvins), to calculate the quantity of heat absorbed or released by a system.

$$\text{Heat absorbed or released} = \text{mass} \times \text{specific heat} \times \Delta T$$

Example 1.20

How much heat, in calories, kilocalories, and kilojoules, does it take to raise the temperature of 225 g of water from 25.0 °C to 100.0 °C?

Solution

Let's list the quantities that we need for the calculations.

$$\text{mass of water} = 225\ \text{g}$$

$$\text{specific heat of water} = 1.00\ \text{cal/(g}\cdot{}^\circ\text{C)} = 4.182\ \text{J/(g}\cdot{}^\circ\text{C)}$$

$$\Delta T = (100.0 - 25.0)\ {}^\circ\text{C} = 75.0\ {}^\circ\text{C}$$

Now, using the equation

$$\text{Heat absorbed} = \text{mass} \times \text{specific heat} \times \Delta T$$

$$= 225\ \cancel{\text{g}} \times 1.00\ \text{cal/}(\cancel{\text{g}\cdot{}^\circ\text{C}}) \times 75.0\ \cancel{{}^\circ\text{C}}$$

$$= 16{,}900\ \text{cal}$$

We can then convert the unit cal to kcal and kJ

$$16.900\ \cancel{\text{cal}} \times \frac{1\ \text{kcal}}{1000\ \cancel{\text{cal}}} = 16.9\ \text{kcal}$$

$$16.9\ \cancel{\text{kcal}} \times \frac{4.182\ \text{kJ}}{1\ \cancel{\text{kcal}}} = 70.7\ \text{kJ}$$

Practice Exercise

How much heat, in calories, kilocalories, and kilojoules, is released by 975 g of water as it cools from 100.0 °C to 18.0 °C?

Example 1.21

What mass of water, in kilograms, can be heated from 5.5 °C to 55.0 °C by 878 kJ of heat?

Solution

Let's list the quantities that we need for the calculations.

$$\text{Heat absorbed} = 878\ \text{kJ}$$

$$\text{mass of water} = ?\ \text{kg}$$

$$\text{specific heat of water} = 4.182\ \text{J/(g}\cdot{}^\circ\text{C)}$$

$$\Delta T = (55.0 - 5.5)\ {}^\circ\text{C} = 49.5\ {}^\circ\text{C}$$

Now, solving the equation for mass H_2O, we get

$$\text{Heat absorbed} = \text{mass} \times \text{specific heat} \times \Delta T$$

$$\text{mass } H_2O = \frac{\text{heat absorbed}}{\text{specific heat} \times \Delta T}$$

$$\text{mass } H_2O = \frac{878\ \text{kJ}}{4.182\ \text{J/(g}\ {}^\circ\text{C)} \times 49.5\ {}^\circ\text{C}} \times \frac{1000\ \text{J}}{1\ \text{kJ}} \times \frac{1\ \text{kg}}{1000\ \text{g}} = 4.24\ \text{kg}$$

Practice Exercise

What temperature change occurs when a 475-g sample of water absorbs 2.44 kJ of heat?

The energy content of foods is usually measured in kilocalories (kcal), but in everyday life most people just call the unit a "calorie." Sometimes the food "calorie" is called a *large Calorie* and written with a capital C. A dieter might be aware that a banana split contains 1500 Calories. If the same dieter realized that this really means 1,500,000 calories, giving up the banana split might be easier.

Review Questions

1.19 Define each of the following terms.

a. heat **b.** calorie

c. temperature **d.** specific heat

1.20 What is the difference between the food Calorie and the scientific calorie?

Summary

Science is an ever changing collection of information. Scientists observe events and catalog data, often summarizing large sets of data in **scientific laws** that may be in the form of mathematical equations. Scientists use their observations and data, along with intellectual creativity and intuition, to formulate testable **hypotheses.** A hypothesis, verified through numerous experiments and expanded to explain additional data, may become a **theory.** Scientists use theories to explain phenomena and to predict the results of new experiments. A theory is discarded if proven wrong.

Matter is anything that has mass and occupies space, and **chemistry** is the study of all the changes that take place in matter. **Mass** is a measure of how much matter an object contains, whereas **weight** is a measure of a gravitational force on the object.

Every type of matter has characteristic **chemical properties,** which describe how that particular matter reacts with other types of matter, and characteristic **physical properties,** which describe its physical appearance and behavior. A **chemical change** involves a change in chemical composition. A **physical change** does not entail any change in chemical composition.

Matter exists in three states: **solid** (fixed volume and shape), **liquid** (fixed volume, variable shape), or **gas** (variable volume and shape).

A pure **substance** is one that has a definite, unchanging chemical composition. A **mixture** can have various compositions. Mixtures can be either **homogeneous,** with the same properties throughout, or **heterogeneous,** in which case the components are not evenly distributed

An **element** is a fundamental substance that cannot be broken down into simpler substances and retains the chemical and physical properties of the original substance. A **compound** is composed of two or more elements chemically combined in fixed proportions. An **atom** is the smallest unit that we associate with the chemical behavior of matter. A **molecule** is an electrically neutral group of atoms bound together in a recognizable unit.

Potential energy is the energy an object has because of its position or its configuration; **kinetic energy** is the energy an object has whenever it is moving. Potential energy can be converted to kinetic energy, and the kinetic energy of an object is related to its mass and velocity by the equation

$$KE = 1/2\ mv^2$$

A force is a push or pull on any object. The force of gravity is the pull exerted by Earth on all objects at or near its surface. Electric forces, which are forces that two electrically charged bodies exert on each other, can be positive or negative. Two like electric forces (+/+ or −/−) repel each other; two unlike electric forces (+/−) attract each other.

Scientists use the **International System of Measurement (SI)** to report measured quantities. In SI the **meter** is the basic unit of length and the **kilogram** is the basic unit of mass. The **liter** is a unit of volume, and the **kelvin** is the SI unit of temperature. In the United States, nonscientific measurements are often made in non-SI units, such as feet instead of meters and gallons instead of liters.

To indicate its **precision,** a measured quantity must be expressed with the proper number of significant figures. Calculations themselves frequently can be done by the **unit-conversion method.**

The physical property of **density** also serves as an important conversion factor. The density of a material is its mass per unit volume: $d = m/V$. The **specific gravity** of any substance is the ratio of the density of the substance to that of water at the same temperature.

Temperature is a measure of how much internal energy, or thermal energy (heat), a body has. The three temperature scales in use today are the Kelvin scale, the Celsius scale, and the Fahrenheit scale. Some relationships among the three scales are

$$K = °C + 273, \quad °F = 1.80(°C) + 32, \quad °C = 0.555(°F - 32)$$

Heat is a measure of the transit of thermal energy from one body to another. The SI unit of heat is the **joule,** and one non-SI heat unit is the **calorie,** defined as the amount of heat needed to raise the temperature of 1 g of water by 1 °C. These two units are related to each other by the equations

$$1 \text{ cal} = 4.184 \text{ J and } 1000 \text{ cal} = 1 \text{ kcal} = 4184 \text{ kJ}$$

The **specific heat** of a substance is the heat required to raise the temperature of 1 g of the substance by 1 °C (or 1 K).

Key Terms

(See Glossary for definitions of these terms.)

accuracy (1.9)
atom (1.4)
calorie (cal) (1.12)
Celsius scale (1.12)
chemical change (1.4)
chemical property (1.4)
chemical symbol (1.5)
chemistry (1.4)
compound (1.5)
conversion factor (1.10)
density (d) (1.11)
element (1.5)
energy (1.6)
Fahrenheit scale (1.12)
gas (1.4)
heat (1.12)
heterogeneous mixture (1.5)
homogeneous mixture (1.5)
hypothesis (1.3)
International System of Units (SI) (1.8)
joule (J) (1.12)
kelvin (K) (1.12)
kilogram (kg) (1.8)
kinetic energy (1.6)
liquid (1.4)
liter (L) (1.8)
mass (1.4)
matter (1.4)
meter (m) (1.8)
mixture (1.5)
molecule (1.4)
physical change (1.4)
physical property (1.4)
potential energy (1.6)
precision (1.9)
scientific law (1.3)
significant figures (1.9)
solid (1.4)
specific gravity (1.11)
specific heat (1.12)
substance (1.5)
temperature (1.12)
theory (1.3)
unit-conversion method (1.10)
variables (1.3)
weight (1.4)
work (1.6)

Problems

(A word of advice: You cannot learn to work problems by reading them or watching your teacher work them any more than you could become a basketball player solely by reading about basketball skills or attending a basketball game. Plan to work through most of these practice problems.)

Some Fundamental Concepts

1. Which of the following are examples of matter?
   a. air **b.** anger **c.** the human body
2. Which of the following are examples of matter?
   a. gasoline **b.** prayer **c.** an idea
3. Two samples are weighed under identical conditions in a laboratory. Sample A weighs 143 g and sample B weighs 286 g. Does sample B have twice the mass of sample A? Explain your answer.
4. Sample A, which is on the moon, has the same mass as sample B on Earth. Do the two samples weigh the same? Explain your answer.
5. Which of the following describes a chemical change, and which describes a physical change?
   a. Sheep are sheared and the wool is spun into yarn.
   b. A lawn grows thicker after being fertilized and watered.
   c. Milk turns sour when left out of the refrigerator for many hours.
6. Which of the following describes a chemical change, and which describes a physical change?
   a. Silkworms feed on mulberry leaves and produce silk.
   b. An overgrown lawn is manicured by mowing it with a lawn mower.
   c. Molten lava from an erupting volcano flows down the side of a mountain and solidifies.
7. Which of the following is a physical property and which is a chemical property?
   a. Methanol boils at 64 °C.
   b. White phosphorus bursts into flame when exposed to air.
   c. Lead is a dull gray metal.
   d. Sodium bicarbonate (baking soda) dissolves in water.
8. Which of the following is a physical property and which is a chemical property?

a. Methanol burns in air with a blue flame.
b. Sodium bicarbonate (baking soda) reacts with acids in baking powder, forming carbon dioxide gas that causes bread to rise.
c. Jasmone, the chemical responsible for the odor of jasmine, has a sweet, floral odor.
d. Hexane boils at 69 °C.

9. Which has the greater kinetic energy: (a) a sprinter or a long-distance runner, assuming that the two weigh the same? (b) a bicyclist traveling at 15 mi/hr or an automobile traveling at 40 mi/hr?
10. Which has the greater kinetic energy: (a) a cannonball or a bullet, both fired at the same speed? (b) a 110-kg football tackle moving slowly across the field or an 80-kg halfback racing quickly down the field?
11. Which has the greater potential energy: a diver on the 1-m board or the same diver on the 10-m platform?
12. Which has the greater potential energy: an elevator stopped at the twentieth floor or one stopped at the twelfth floor?

Elements, Compounds, and Mixtures

13. Which of the following represent elements and which do not? Explain.
a. Hf b. HF c. Cl
14. Which of the following represent elements and which do not? Explain.
a. $CaCl_2$ b. Na c. KI
15. Which of the following are substances and which are mixtures?
a. Helium (the gas used to fill a balloon)
b. Maple syrup (made from the sap of a maple tree)
c. Smog formed in the early morning air
16. Which of the following are substances and which are mixtures?
a. Vinegar made from wine
b. Salt used to de-ice roads
c. Mercury used in a thermometer
17. Which of the following mixtures are homogeneous and which are heterogeneous?
a. high-octane gasoline b. iced tea
18. Which of the following mixtures are homogeneous and which are heterogeneous?
a. Italian salad dressing b. white wine
19. Without consulting Table 1.2, name the elements with the following symbols.
a. He b. N c. F
d. K e. Fe f. Cu
20. Without consulting Table 1.2, name the elements with the following symbols.
a. Mg b. Si c. S
d. Br e. P f. Sn
21. Without consulting Table 1.2, give the symbols for the following elements.
a. hydrogen b. carbon c. oxygen
d. zinc e. iodine f. mercury
22. Without consulting Table 1.2, give the symbols for the following elements.
a. aluminum b. phosphorus c. sodium
d. chlorine e. calcium f. cobalt

Metric Measurement

23. Change each of the following measurements to one in which the unit has an appropriate SI prefix.
a. 4.54×10^{-3} g b. 3.76×10^{-2} m c. 6.34×10^{-6} g
24. Change each of the following measurements to one in which the unit has an appropriate SI prefix.
a. 1.09×10^{-6} L b. 9.01×10^{-3} s c. 7.77×10^{3} m
25. For each of the following, indicate which is the larger unit.
a. mm or cm b. kg or g c. dL or μL d. lb or g
26. For each of the following, indicate which is the larger unit.
a. L or cm^3 b. cm^3 or mL c. in. or m d. L or gal

Significant Figures

27. How many significant figures are there in each of the following measured quantities?
a. 7007 m b. 0.00235 s c. 0.043300 g
d. 6.012×10^{5} m e. 1.200×10^{8} s f. 0.050 °C
28. How many significant figures are there in each of the following measured quantities?
a. 2021 m b. 0.0169 s c. 0.00430 g
d. 5.00×10^{12} m e. 1.60×10^{-9} s f. 0.0150 °C
29. Perform the indicated operations and give answers with the proper number of significant figures.
a. 48.2 m + 3.82 m + 48.4394 m
b. 151 g + 2.39 g + 0.0124 g
c. 100.53 cm − 46.1 cm
d. 451 g − 15.46 g
e. 15.44 mL − 9.1 mL + 105 mL
f. 12.52 cm + 5.1 cm − 3.18 cm − 12.02 cm
30. Perform the indicated operations and give answers in the indicated unit and with the proper number of significant figures. (Hint: You must convert all quantities to a common unit.)
a. 13.25 cm + 26 mm − 7.8 cm + 0.186 m (in cm)
b. 48.834 g + 717 mg − 0.166 g + 1.0251 kg (in kg)
c. 73.0 cm × 1.340 cm × 0.41 cm
d. 265.02 mm × 0.000581 m × 12.18 cm (in mm)

Unit Conversions

(You may refer to Table 1.7 when necessary.)

31. Carry out the following conversions.
a. 50.0 km to meters b. 546 mm to meters
c. 97.5 kg to grams d. 47.9 mL to liters
e. 577 μg to milligrams f. 237 mm to centimeters
32. Carry out the following conversions.
a. 87.6 μg to kilograms
b. 1.00 m to micrometers
c. 0.0962 km/min to meters per second
d. 55 mi/h to kilometers per minute
33. Carry out the following conversions.
a. 413 in. to yards b. 86.2 oz to pounds
c. 64.0 fl oz to quarts d. 12.6 ft/s to miles per hour

34. Carry out the following conversions.
 a. 4.53 ft to inches b. 8.08 lb to ounces
 c. 6.13 qt to fl oz d. 70.06 mi/hr to ft/s
35. Carry out the following conversions.
 a. 16.4 in. to centimeters b. 4.17 qt to liters
 c. 1.61 kg to pounds d. 9.34 g to ounces
36. Carry out the following conversions.
 a. 2.05 fl oz to milliliters b. 105 lb to kilograms
 c. 143 cm to feet d. 775 mL to quarts
37. Convert 90.0 kilometers per hour to meters per second.
38. A sprinter runs a 100.0-meter dash in 11.00 seconds. What is her speed in kilometers per hour?

Density and Specific Gravity

(You may need data from Table 1.8 for some of these problems.)

39. What is the density, in grams per milliliter, of (**a**) a salt solution if 75.0 mL has a mass of 87.5 g? (**b**) 2.75 L of the liquid glycerol, which has a mass of 3.46 kg?
40. What is the density of (**a**) a sulfuric acid solution if 5.00 mL has a mass of 7.52 g (in grams per milliliter)? (**b**) a 10.0-cm^3 block of plastic which has a mass of 9.23 g (in grams per cubic centimeter)?
41. What is the mass, in grams, of (**a**) 125 mL of castor oil, a laxative, which has a density of 0.962 g/mL? (**b**) a 33.0-mL sample of the liquid hexane (C_6H_{14}), a solvent used to extract oil from soybeans?
42. What is the mass, in grams, of (**a**) 30.0 mL of the liquid propylene glycol, a moisturizing agent for foods, which has a density of 1.036 g/mL at 25 °C? (**b**) 30.0 mL of grenadine, which has a density of 1.32 g/mL?
43. What is the volume of (**a**) a 475-g piece of copper (in cubic centimeters)? (**b**) a 253-g sample of mercury (in milliliters)?
44. What is the volume of (**a**) 227 g of hexane (in milliliters)? (**b**) a 454-g block of ice (in cubic centimeters)?
45. Some metal chips having a volume of 3.29 cm^3 are placed on a piece of paper and weighed. The combined mass is found to be 18.43 g. The paper itself weighs 1.21 g. Calculate the density of the metal.
46. A glass container weighs 48.462 g. A sample of 4.00 mL of antifreeze solution is added, and the container plus the antifreeze weigh 54.51 g. Calculate the density of the antifreeze solution.
47. The density of a normal urine sample is 1.02 g/mL. What is its specific gravity?
48. The specific gravity of an antifreeze solution is 1.1044. What is its density in grams per milliliter?

Energy: Temperature and Heat

(You may need data from Table 1.9 for some of these problems.)

49. For each of the following, indicate which is the larger unit.
 a. °C or °F b. cal or Cal
50. Order the temperatures from coldest to hottest: 0 K, 0 °C, 0 °F.
51. Carry out the following conversions.
 a. 37.0 °C to degrees Fahrenheit
 b. 5.5 °F to degrees Celsius
 c. 273 °C to degrees Fahrenheit
52. Carry out the following conversions.
 a. 98.2 °F to degrees Celsius
 b. 2175 °C to degrees Fahrenheit
 c. 25.0 °F to degrees Celsius
53. Carry out the following conversions.
 a. 2.75 kcal to calories b. 0.741 Cal to joules
 c. 8.63 kJ to calories
54. Carry out the following conversions.
 a. 1.36 kcal to kilojoules b. 345 cal to joules
 c. 873 kJ to kilocalories
55. How many calories are required to raise the temperature of 50.0 g of water from 20.0 °C to 50.0 °C?
56. How many kilojoules are required to raise the temperature of 131 g of water from 15.0 °C to 95.0 °C?
57. How much heat, in kilojoules, would be released by 2.00 kg of water cooling from 90.0 °C to 20.0 °C?
58. How much water, in grams, can be heated from 20.0 °C to 50.0 °C by 836 kJ of energy?
59. A 454-g block of iron is at an initial temperature of 22.5 °C. What will be the temperature of the iron after it absorbs 17.6 kcal of heat from its surroundings?
60. What will be the final temperature if a 5.00-g block of aluminum at 37.0 °C gives off 25.0 J of heat to its surroundings?

Additional Problems

61. The speed limit in rural Ontario is 90 km/hr. What is this speed in miles per hour?
62. The speed of light is 186,000 mi/sec. What is this speed in meters per second?
63. If your heart beats at a rate of 72 times per minute and your lifetime will be 70 years, how many times will your heart beat during your lifetime?
64. How many 325-milligram aspirin tablets can be made from 875 grams of aspirin?
65. A doctor puts you on a diet of 5500 kilojoules (kJ) per day. How many calories is that? How many kilocalories (food Calories)?
66. Milk costs $3.89 per gallon or $1.15 per liter. Which is cheaper?
67. What is the thickness of a 45.4 cm × 104.6 cm rectangular piece of cast iron ($d = 7.76\ g/cm^3$) that has a mass of 89.8 kg?
68. What are the dimensions of a cubic piece of silver that has a mass of 5.79 g? The density of silver is 10.5 g/cm^3.
69. What is the volume of 5.79 mg of gold ($d = 19.3\ g/cm^3$)?

If the gold is hammered into gold leaf of uniform thickness with an area of 44.6 cm^2, what is the thickness of the gold leaf?

70. How many grams of copper can be heated from 22.5 °C to 35.0 °C by the same quantity of heat that is capable of raising the temperature of 145 g H_2O from 22.5 °C to 35.0 °C?

71. Arrange the following in order of increasing length (shortest first): (1) a 1.21-m chain, (2) a 75-in. board, (3) a 3 ft 5-in. rattlesnake, (4) a yardstick.

72. Arrange the following in order of increasing mass (lightest first): (1) a 5-lb bag of potatoes, (2) a 1.65-kg cabbage, (3) 2500 g sugar.

73. One of the women pictured below has a mass of 38.5 kg and a height of 1.51 m. Which one is it likely to be?

(a)

(b)

(c)

74. A box with a base 0.80 m on a side and a height of 1.20 m is filled with 3.2 kg of expanded polystyrene packing material. What is the bulk density, in grams per cubic centimeter, of the packing material? (The bulk density includes the air between the pieces of polystyrene foam.)

75. Which of the following items would be most difficult to lift into the bed of a pickup truck: (1) a 100-lb bag of potatoes, (2) a 15-gal plastic bottle filled with water, (3) a 3.0-L flask filled with mercury ($d = 13.6\ g/cm^3$)?

76. A rectangular block of gold-colored material measures 3.00 cm × 1.25 cm × 1.50 cm and has a mass of 28.12 g. Can the material be gold?

77. An experiment calls for 8.65 grams of carbon tetrachloride ($d = 1.59\ g/mL$). What is the volume of such a mass?

78. Adult male Hooker's sea lions are 250 to 350 cm long and weigh 300 to 450 kg. Convert these measurements to inches and pounds.

79. Pediatric drug dosages are usually based on infant weight and are expressed in units such as milligrams of drug per kilogram of body weight (mg/kg). If the recommended dosage of a particular drug is 5.0 mg/kg, what is the proper mass of drug for a 17-lb baby?

80. Each Tylenol chewable tablet contains 80 mg of acetaminophen. The recommended dosage for children is 10 mg/kg of body weight. How many tablets constitute the proper dosage for a 55 lb child?

81. In its nonstop, round-the-world trip, the aircraft Voyager traveled 25,102 mi in 9 days, 3 min, and 44 s. Calculate the average speed of Voyager in miles per hour.

82. The furlong is a unit used in horse racing, and the units chain and link are used in surveying land. There are 8 furlongs in 1 mi, 10 chains in one furlong, and 100 links in one chain. Calculate the length of one link, in inches. (1 mi = 5280 ft; 1 ft = 12 in.)

83. In the United States land area is commonly measured in acres: 640 acre = 1 mi^2. In most of the rest of the world land area is measured in hectares: 1 hectare = 1 hm^2. [1 hectometer (hm) = 100 m]. Which is the larger area, the acre or the hectare? Write a conversion factor that relates the acre and the hectare. (1 mi = 5280 ft)

84. In scientific work, densities usually are expressed in g/cm^3. In engineering work, the unit lb/ft^3 is often used. The density of water at 20 °C is 0.998 g/cm^3. What is its density in pounds per cubic foot?

85. A square of aluminum foil ($d = 2.70\ g/cm^3$) is 5.10 cm on a side and has a mass of 1.762 g. Calculate the thickness of the foil, in millimeters.

CHAPTER 2 Atoms

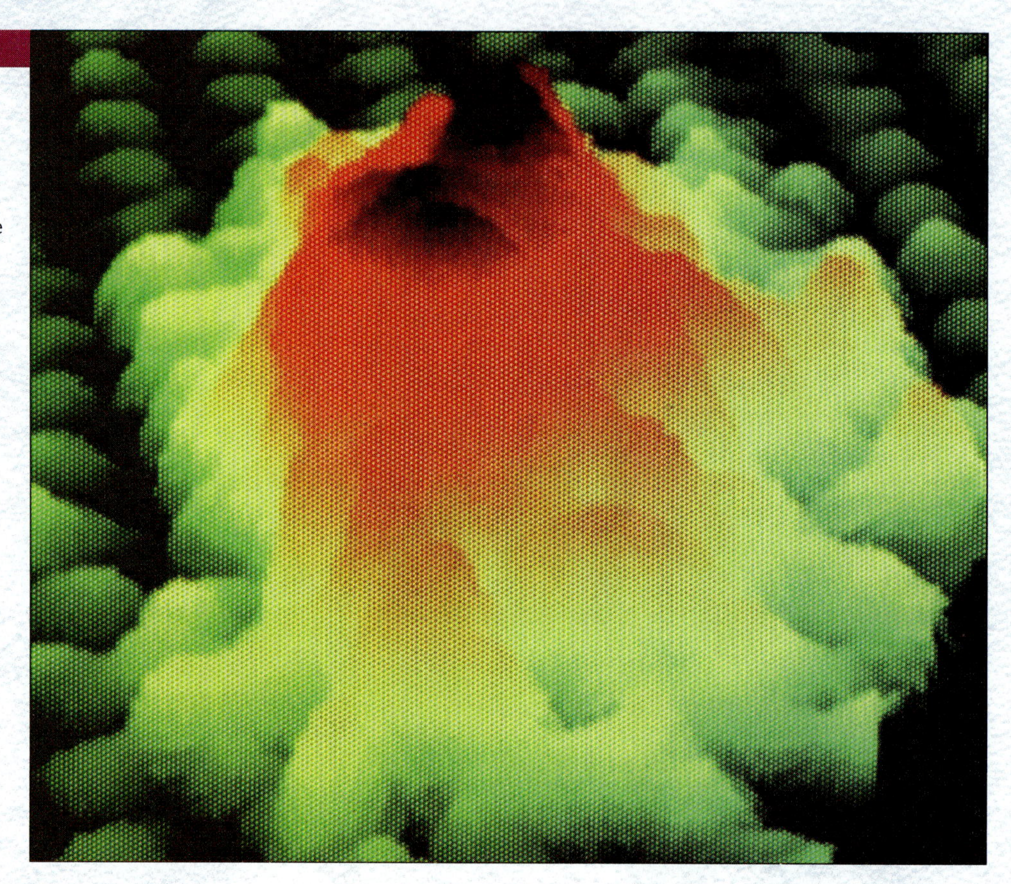

The concept of atoms has been around for more than two millennia, but it became an important part of chemistry only about two centuries ago. And only in the latter years of the twentieth century were scientists able to obtain images of atoms. This photograph, made by a scanning tunneling microscope, shows gold atoms (yellow, orange, and red) on a background of carbon atoms (green). The image has been enhanced and false color added by a computer.

Learning Objectives/Study Questions

1. What are the main points of Dalton's atomic theory? What modifications have been made in Dalton's theory? Why did the theory have to be modified?
2. Describe the nuclear model of the atom.
3. What are the three main subatomic particles that make up an atom? What are the approximate mass and electric charge on each of the particles?
4. What are (a) the atomic number and (b) the mass number of an element?
5. Describe the Bohr model of the atom.
6. What is meant by the electronic configuration of an atom? What is meant by principal shell and subshell?
7. How are electronic configurations related to the periodic table?
8. What is the valence electronic configuration of (a) noble gas atoms, (b) alkali metal atoms, (c) alkaline earth metal atoms, and (d) halogen atoms?
9. Describe periodic trends in (a) atomic size, (b) ionization energy, and (c) electron affinities.

In Chapter 1 we noted that an atom is the smallest characteristic particle of an element. The concept of the atom has been around for thousands of years, but for most of that time it was not widely accepted. Chemistry developed as a science primarily after scientists accepted the existence of atoms and then worked to define the properties of the atom more precisely.

The concept of the atom as originally proposed by both ancient Hindu and Greek scholars was as philosophical as it was practical. Both groups spoke of earth, water, fire, and air as elements composed of eternal and unchanging atoms in perpetual motion. The atomic view of matter fell into disfavor when Aristotle, the greatest of the Greek philosophers, sided with those who believed that matter was continuous, that is, infinitely divisible. According to Aristotle and his followers, there was no particle of matter so small that it could not be subdivided into still smaller pieces. It is a tribute to Aristotle that his view prevailed for 2000 years, even though we now know it was wrong.

Atomic Theories

Data gradually accumulated that could not be explained within Aristotle's philosophy, and—starting about 1500 C.E.—experimental science finally forced a reevaluation of his views. This was a major victory for the experimentalists. Scientific theories had to be accepted or rejected on the basis of their ability to explain experimental data. They could not be accepted solely on the basis of their elegance or even their appeal to common sense.

Why do we care about the structure of atoms? It is the arrangement of the parts of atoms that determines the properties of different kinds of matter. Only by understanding atomic structure can we know how atoms combine to make the many different substances in nature and determine how to modify materials to meet our needs. A knowledge of atomic structure is even essential to your health. Many medical diagnoses are based on the analysis of body fluids such as blood and urine, and several such analyses depend on knowledge of how the structure of atoms is changed when energy is absorbed.

2.1 Dalton's Atomic Theory

In the early nineteenth century John Dalton, an English schoolteacher, developed an atomic theory. His theory explained the best available experimental data. Following are the important points of Dalton's atomic theory with some modern modifications that we will consider later.

▲ John Dalton (1766–1844) was not a good experimentalist—perhaps because he was color blind. However, he skillfully used the results of others' experiments in formulating his atomic theory.

Dalton's Theory	Modern Modifications
1. All matter is composed of extremely small particles called atoms.	**1.** Dalton assumed atoms to be indivisible. This isn't quite so, as we will see with subatomic particles (Section 2.2) and radioactivity (Chapter 12).
2. All atoms of a given element are alike, but atoms of one element differ from the atoms of any other element.	**2.** Dalton assumed that all the atoms of a given element were identical in all respects, including mass. This is now known to be incorrect, as we will see in the next section.

3. Compounds are formed when atoms of different elements combine in fixed proportions.

3. Unmodified: The numbers of each kind of atom in a compound usually form a simple ratio. For example, the ratio of carbon atoms to oxygen atoms is 1:1 in carbon monoxide and 1:2 in carbon dioxide.

4. A chemical reaction involves a *rearrangement* of atoms. No atoms are created or destroyed or broken apart in a chemical reaction.

4. Unmodified for *chemical* reactions. Atoms are broken apart in *nuclear* reactions.

Dalton said that all atoms of a given element are the same and would thus have a unique mass. Atoms are incredibly tiny, and in Dalton's time scientists could not determine the actual masses of atoms. Indirect measurements, however, could indicate their relative masses. Dalton set up a table of relative masses for the elements, based on hydrogen, the lightest element. As we might expect, the rather crude equipment available at the time gave quite a few imprecise masses. We will consider the modern basis for relative masses in Section 2.3. For now we will simply say relative atomic masses are expressed in terms of the *atomic mass unit,* sometimes abbreviated *amu,* but designated by the symbol u. (For example, we say the atomic mass of carbon is 12 atomic mass units, 12 amu, or 12 u.)

Although Dalton was wrong in some details of his atomic theory, his theory led other scientists to think about matter as made of atoms. Over the years scientists adopted Dalton's ideas and modified them. As we noted in Chapter 1, science is not simply an accumulation of correct bits of information. Rather, it is the gradual development of models that help us to understand the workings of nature. Some ideas have to be corrected or modified as our understanding increases. This situation is not unique to chemistry. An accepted medical treatment of 50 (to say nothing of 100) years ago is likely to be unacceptable today. Yet a now outdated treatment did once keep people alive and helped medical researchers develop better, safer, more effective procedures.

✓ Review Questions

2.1 What is the distinction between the atomic view and the continuous view of matter?

2.2 If we applied the ideas of divisible ("atomic") or continuous to foods at the macroscopic level, which designation would you use for each of the following?

a. hard-boiled eggs **b.** hot dogs **c.** mashed potatoes
d. milk **e.** peas **f.** scrambled eggs

2.3 Outline the main points of Dalton's atomic theory.

2.2 The Nuclear Atom

If we begin with the premise that matter is made of atoms, we can then devise experiments to find out more about these atoms. In the century following Dalton's work many such experiments were conducted. As data continued to accumulate, scientists found that many of the results could not be explained by Dalton's atomic theory without modification. Scientists around the world discovered that the atom was not indivisible. It could still be regarded as the smallest characteristic unit of an *element,* but it certainly was not the smallest particle of matter.

The studies of a German physicist, Eugen Goldstein, and an English physicist, William Crookes, demonstrated that atoms could be torn apart. In an apparatus

▲ **Figure 2.1**
Cathode rays are deflected in a magnetic field in a discharge tube. Although these rays are invisible, we can detect them because they produce a green fluorescence as they strike a screen coated with zinc sulfide. The beam of cathode rays begins at the cathode on the left. It is deflected by the field of a magnet that is located slightly behind the anode. The deflection we observe is that expected of negatively charged particles.

called a discharge tube (Figure 2.1), electric energy was used to kick out small pieces from atoms of the metal making up one electrode. The tiny subatomic particles were shown to be negatively charged and could be detected as a beam of particles (called a **cathode ray**) that crossed from one electrode in the apparatus to the other. No matter what element the electrode was made of, identical negatively charged subatomic particles, now called **electrons** , were kicked out to form the cathode ray.

Cathode rays travel in straight lines in the absence of an external applied field.

An electron is a subatomic particle with one unit of negative charge (1−) and essentially no mass.

When a small amount of gas was admitted into the tube, the cathode-ray particles struck the atoms of the gas, knocking more electrons loose from these atoms (Figure 2.2). Those atoms that lost negatively charged electrons were left with a positive charge. Although all the negatively charged particles were identical, the positively charged particles differed depending on what gas was in the tube. For example, helium produced more massive positively charged particles than did hydrogen. The lightest positive particles found were those from hydrogen and were

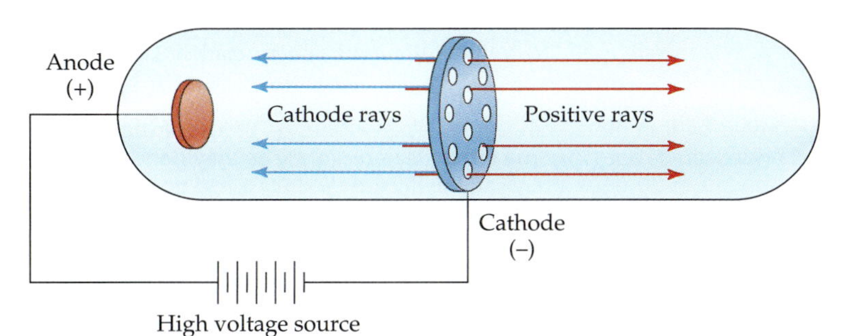

◀ **Figure 2.2**
Positive rays are also produced in cathode-ray tubes that contain some residual gas. Cathode rays (blue) stream toward the anode (+). They collide with gas atoms and knock electrons from the atoms to produce positively charged ions. These ions are attracted to the cathode (−), but some pass through the holes in the cathode and appear as a stream of positive particles (red) on the other side.

A proton is a subatomic particle with one unit of positive charge (1+) and a mass of approximately 1 atomic mass unit (1 u). A proton is 1837 times as massive as an electron.

called **protons.** The charges on the proton and the electron are equal in magnitude but opposite in sign. However, the proton is 1837 times as massive as the electron. Thus the smallest positive particle isolated was many times heavier than the negatively charged electron.

A French physicist, Antoine Henri Becquerel, discovered that atoms of uranium emitted some type of "rays." Scientists soon found that uranium and some other elements emitted three basic types of "rays" and labeled them alpha (α), beta (β), and gamma (γ) after the first three letters of the Greek alphabet. Two of these later were shown to be particles a fraction of the size of an atom. Alpha particles were found to have a mass of about 4 u and a charge of 2+. Beta particles were found to be identical to the electrons formed in a cathode-ray tube; that is, beta particles have a charge of 1− and very little mass. The third type of radiation, gamma rays, were shown to be a form of energy similar to X-rays. These phenomena were given the name **radioactivity** by Polish-born Marie Curie (1867–1934). We will consider them in more detail in Chapter 12.

Certain kinds of atoms are radioactive; they naturally decay with the release of subatomic particles (alpha or beta particles) and energy (gamma rays) from the nucleus.

A British scientist, Ernest Rutherford, used radioactive atoms as "atomic guns," with the emitted alpha particles playing the role of bullets (Figure 2.3). By

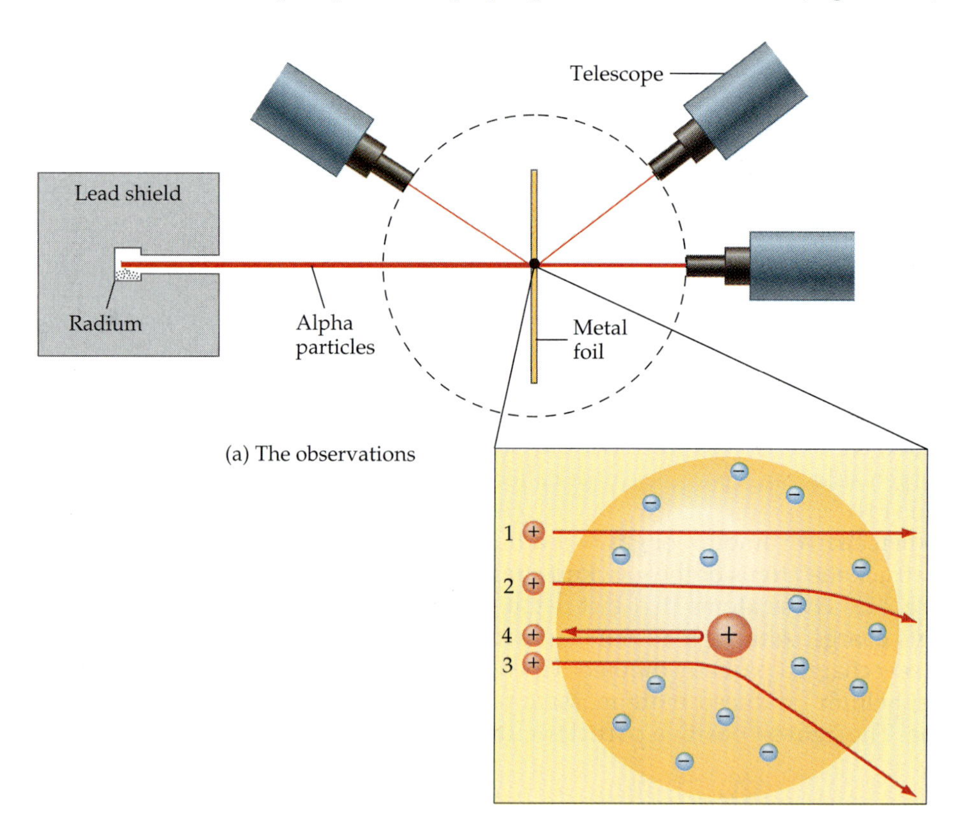

(a) The observations

(b) Rutherford's interpretation

▲ **Figure 2.3**
Alpha (α) particles are scattered by thin metal foil. (a) The observations. (1) Most of the α particles pass through the foil undeflected, but (2) some are deflected slightly as they penetrate the foil. (3) A few α particles (about 1 in 20,000) are greatly deflected, and (4) a few do not penetrate the foil at all but are reflected back toward the source. (b) Rutherford's interpretation. If the atoms of the foil have a massive, positively charged nucleus and light electrons outside the nucleus, we can see that: (1) most α particles can pass through the atom undeflected, (2) but an α particle is deflected slightly as it passes near an electron; (3) an α particle is deflected significantly as it passes close to the atomic nucleus; and (4) an α particle bounces back toward the source when it approaches the nucleus head on.

▲ Ernest Rutherford (1871–1937) was born in New Zealand but worked in Canada and England. He developed the theory of the nuclear atom.

Table 2.1 Subatomic Particles

Particle	Symbol[a]	Approximate Mass (u)	Charge	Location in Atom
Proton	p^+	1	1+	Nucleus
Neutron	n	1	0	Nucleus
Electron	e^-	0	1−	Outside nucleus

[a]The superscripts are often omitted. We will keep them to remind us of the charge of each particle.

▲ If an atom were as large as this stadium, the nucleus would be the size of a pea at the center of the stadium.

shooting such "bullets" at other atoms, he made two important observations and came to two crucial conclusions: (1) Most of the bullets passed right through the target atoms, suggesting that the atoms were mostly empty space. (2) Some of the bullets were deflected, from which observation he concluded that there was a concentrated bit of positively charged "solid" material at the center of the atom.

A student of Rutherford, James Chadwick, discovered a subatomic particle unlike those previously found. This particle, called a **neutron,** has about the same mass as the proton, but it is neutral—neither positively nor negatively charged.

Like protons, neutrons have a mass of approximately 1 u, but unlike protons, neutrons have no charge.

Early in the twentieth century scientists developed a new model of the atom to account for the new data. According to this model there are three main types of subatomic particles: protons, neutrons, and electrons. The relative masses of a proton and a neutron are 1.007276 u and 1.008665 u, respectively. The differences are so small that they often can be ignored. For many purposes we can assume that the masses of the proton and the neutron are 1 atomic mass unit (1 amu) each. The proton has a charge equal in magnitude but opposite in sign to that of an electron. This charge is regarded as the basic unit of charge. We write the charge on a proton as 1+ and that on an electron as 1−. In atomic mass units, the relative mass of an electron is only 0.000549 u (5.49×10^{-4} u). The electrons in an atom contribute so little to its total mass that their mass is often disregarded.[1] These properties of the subatomic particles are summarized in Table 2.1.

When combined to form an atom, the more massive particles (protons and neutrons) are crowded into the **nucleus,** a tiny volume of space at the center of the atom (Figure 2.4). The electrons are scattered throughout the remaining volume of the atom. To visualize how "empty" an atom is, picture a balloon that stands 10 stories high. If the balloon were an atom, its nucleus would be the size of the letter O as printed here. The rest of the space in the balloon-atom is the domain of the electrons, but the total mass of the electrons is less than that of the nucleus by a factor of thousands.

The **atomic number** of an element is simply the number of protons in the nucleus of an atom of the element. An *element* is a substance that cannot be broken down into simpler substances by chemical reactions. All atoms of a given element have the same atomic number, which determines the identity of the element. Recall that Dalton said that the mass of an atom determines the element. We now know that it is not the mass but the number of protons that determines the element. For example, an atom with atomic number 26 has 26 protons and is an atom of iron (Fe). An atom with 92 protons is an atom of uranium (U).

Atomic number = number of protons. All the atoms of a given element have the same number of protons. For neutral atoms, the atomic number also gives the number of electrons. The positively and negatively charged particles balance, and the atom as a whole is neutral.

[1]Subatomic particles are exceedingly small. Their masses in grams are: proton = 1.673×10^{-24} g (0.000000000000000000000001673 g); neutron = 1.675×10^{-24} g; electron = 9.107×10^{-28} g. (We review exponential notation in Appendix I.)

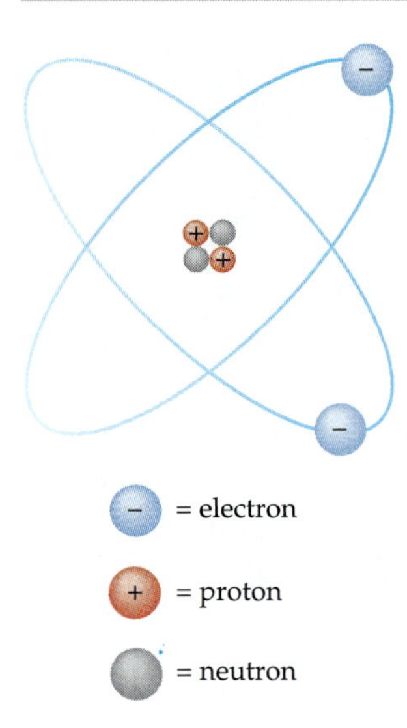

▲ **Figure 2.4**
The nuclear model of an atom—illustrated here by helium—shows that an electrically neutral atom has the same number of electrons outside the nucleus as it has protons in the nucleus. This drawing shows the electrons much closer to the nucleus than they really are. The actual case is more accurately represented by imagining the entire atom to be a room measuring 5 m × 5 m × 5 m and the nucleus to be the period at the end of this sentence.

The isotopes of a given element have the same number of protons but different numbers of neutrons.

✓ Review Questions

2.4 Give the distinguishing characteristics of the proton, the neutron, and the electron.

2.5 What is the atomic nucleus? What subatomic particles are found in the nucleus?

2.6 What subatomic particles are found outside the nucleus?

2.3 Isotopes, Atomic Masses, and Nuclear Symbols

The atoms of a given element can have different numbers of neutrons in their nuclei. For example, all hydrogen atoms have one proton in their nucleus. Most hydrogen atoms have a nucleus consisting of just that proton. Having no neutrons, they have a mass of 1 atomic mass unit (1 u). However, about one hydrogen atom in 6700 has a neutron as well as a proton in the nucleus. This heavier hydrogen is called *deuterium* and has a mass of 2 u. Both are hydrogen atoms because both have only one proton. Atoms that have the same number of protons but different numbers of neutrons are **isotopes** (Figure 2.5). A third, rare isotope of hydrogen is *tritium*, which has one proton—as do all hydrogen atoms—but two neutrons in the nucleus. Tritium has a mass of 3 u.

Most elements exist in nature as a mixture of isotopes, and this requires a modification of Dalton's original theory. He said that all atoms of the same element have the same mass. Because most elements have several isotopes—that is, atoms with different numbers of neutrons—an element can have atoms with different masses.

Naturally occurring carbon consists of a *mixture* of two main isotopes. Scientists now use the much more abundant pure isotope carbon-12, which is assigned a mass of exactly 12 atomic mass units, as the atomic mass standard. Based on this standard, we can now define an **atomic mass unit (u)** as exactly 1/12 the mass of a carbon-12 atom. The other principal carbon isotope is carbon-13, with a mass of 13.00335 u. When carbon atoms participate in a chemical reaction, even though a mixture of isotopes is involved, it's more convenient to think in terms of a hypothetical, "average" atom. This "average" carbon atom would have a mass somewhere between 12.00000 u and 13.00335 u. However, this mass must be "weighted" toward the mass of the more abundant carbon-12 isotope. The weighted average of the masses of the naturally occurring carbon atoms is 12.011 u. The **atomic mass**[2] of an element, then, is the weighted average of the masses of the naturally occurring isotopes

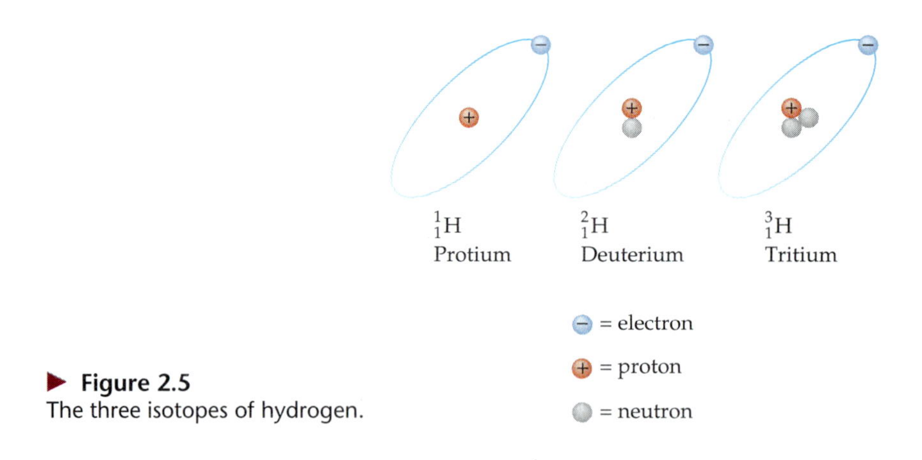

▶ **Figure 2.5**
The three isotopes of hydrogen.

[2]Many chemists and others use the term "atomic weight" rather than "atomic mass." However, "atomic mass" is more appropriate.

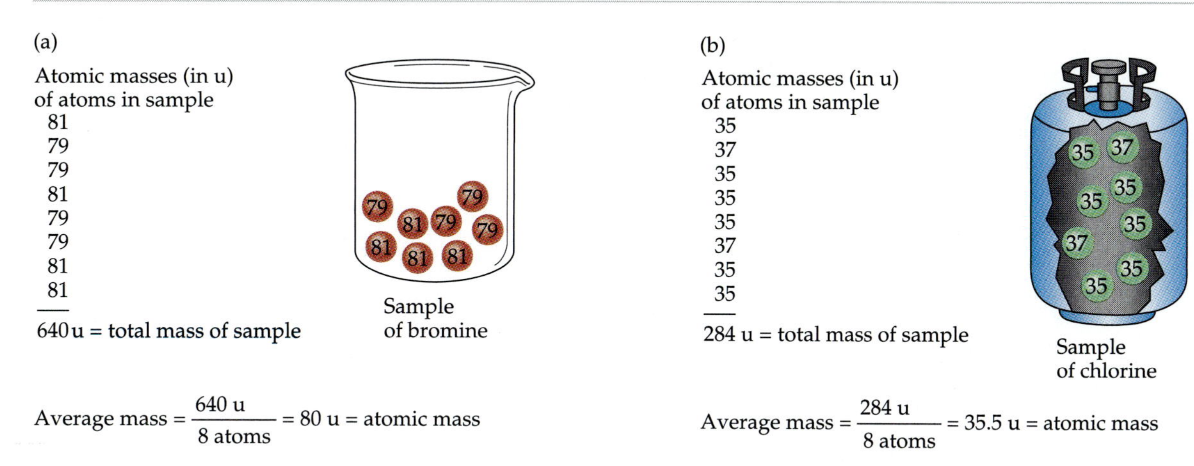

(a)
Atomic masses (in u)
of atoms in sample
81
79
79
81
79
79
81
81
640 u = total mass of sample
Sample
of bromine
Average mass = 640 u / 8 atoms = 80 u = atomic mass
(b)
Atomic masses (in u)
of atoms in sample
35
37
35
35
35
37
35
35
284 u = total mass of sample
Sample
of chlorine
Average mass = 284 u / 8 atoms = 35.5 u = atomic mass

▲ **Figure 2.6**
Atomic masses of elements are averages of the masses of the atoms in representative samples of the elements. In a sample of bromine (a), about half the atoms have a mass of 79 u and half a mass of 81 u; the atomic mass of bromine is therefore about 80 u. (The actual value listed in the periodic table, 79.9 u, indicates that there is slightly more ^{79}Br than ^{81}Br.) In a sample of chlorine (b), about three-fourths of the atoms have a mass of 35 u and one-fourth a mass of 37 u; the atomic mass of chlorine is therefore about 35.5 u. The average mass is much closer to that of the ^{35}Cl than ^{37}Cl because there is more of that isotope in the sample.

of that element. Figure 2.6 illustrates an approximation of how the atomic masses of chlorine and bromine are derived from the masses of their individual atoms.

We need two quantities to calculate the weighted-average atomic mass of carbon: (1) the atomic masses of carbon-12 and carbon-13; and (2) the naturally occurring fractional abundances of the two isotopes. These values are determined experimentally by mass spectrometry. We need not discuss the details of the process. Instead, let's just describe the results.

The masses of the two carbon isotopes are 12.00000 u for carbon-12 (by definition) and 13.00335 u for carbon-13 (by measurement). Fractional abundances refer to the fraction of all the naturally occurring carbon atoms that are carbon-12 and the fraction that are carbon-13. The two fractions must total to 1. These abundances are often expressed as percentages. The percent natural abundances of the carbon isotopes are 98.892% carbon-12 and 1.108% carbon-13. To obtain the weighted average in Example 2.1, we express the contribution of each isotope to the weighted average as the product of its atomic mass and fractional abundance.

Example 2.1

Use the data above to complete the determination of the weighted average atomic mass of carbon.

Solution

To get any weighted average, we multiply each of the values to be averaged by a "weighting factor." In this case the weighting factor is the fractional abundance. Because percentages are fractions expressed on a "per hundred" basis, that is, percent = fraction × 100%, fractional abundances are percent abundances divided by 100%.

$$\text{Fraction} = \text{percent}/100\%$$

Thus 98.892% carbon-12 is the same as a fractional abundance of 0.98892, and 1.108% carbon-13 represents a fractional abundance of 0.01108.

The contribution that each isotope makes to a weighted-average atomic mass is given by the product:

$$\text{Contribution of isotope} = \text{fractional abundance} \times \text{mass of isotope}$$

Thus, the contributions of carbon-12 and carbon-13 are

$$\text{Contribution of carbon–12} = 0.98892 \times 12.00000\ \text{u} = 11.867\ \text{u}$$

$$\text{Contribution of carbon–13} = 0.01108 \times 13.00335\ \text{u} = 0.1441\ \text{u}$$

The weighted average is the sum of these two contributions.

$$\text{Atomic mass of carbon} = 11.867\ \text{u} + 0.1441\ \text{u} = 12.011\ \text{u}$$

This is the value listed in a table of atomic masses.

This is quite similar to a *grade-point average,* a weighted average in which letter grades are converted to points (A = 4.0, B = 3.0, . . .) and the weighting factor for each course is the fraction of the total number of credit units of enrollment assigned to that course.

Practice Exercises

A. The three naturally occurring isotopes of neon, their percent abundances, and their atomic masses are: neon-20, 90.51%, 19.99244 u; neon-21, 0.27%, 20.99395 u; neon-22, 9.22%, 21.99138 u. Calculate the weighted-average atomic mass of neon.

B. The two naturally occurring isotopes of copper are copper-63, with a mass of 62.9298 u, and copper-65, with a mass of 64.9278 u. If the tabulated atomic mass of copper is 63.546 u, what must be the percent natural abundances of the two isotopes? (Hint: The fractional abundances are unknowns. If one of them is x the other must be $1.000 - x$.)

To represent the different isotopes of the elements, symbols with subscripted and superscripted numbers are used. Consider the following generalized symbol, in which the subscript Z represents the atomic number and the superscript A represents the **mass number,** which is defined as the number of protons plus the number of neutrons.[3]

Atomic number Z = number of protons

Mass number A = number of protons + number of neutrons

The mass number (no. of protons + no. of neutrons) → $^{A}_{Z}E$
The atomic number (no. of protons) → $^{A}_{Z}E$

For example, the symbol $^{35}_{17}Cl$ tells us that there are 17 protons and $35 - 17 = 18$ neutrons in a chlorine atom.

A particular isotope is identified by its nuclear symbol—for example, $^{35}_{17}Cl$ or $^{37}_{17}Cl$. Because the atomic number of an element is implicit in its chemical symbol (the element with the symbol Cl always has the atomic number 17), simplified forms such as ^{35}Cl and ^{37}Cl are sometimes used. More often, isotopes are identified by the name of the element followed by the mass number, such as chlorine-35 and chlorine-37.

Example 2.2

Write the symbol for an isotope with a mass number of 58 and an atomic number of 27. Identify the element.

Solution

In the general symbol $^{A}_{Z}E$, A is the mass number, which is 58 in this case, and Z is the atomic number, or 27.

$$^{58}_{27}E$$

[3]The mass number is sometimes called the *nucleon number.* (Protons and neutrons, collectively are called nucleons.)

From the list of elements (inside front cover), we see that the element with atomic number 27 is cobalt.

$$^{58}_{27}\text{Co}$$

Practice Exercises

A. Write the symbol for an isotope with a mass number of 90 and an atomic number of 42.
B. Write the symbol for an isotope of calcium that has 25 neutrons in its nucleus.

Example 2.3

Indicate the number of protons, neutrons, and electrons in a neutral atom of the isotope $^{235}_{92}\text{U}$.

Solution

The atomic number gives the number of protons and electrons in a neutral atom of the isotope.

$$\text{Atomic number} = \text{no. of protons} = \text{no. of electrons} = 92$$

We obtain the number of neutrons by subtracting the atomic number from the mass number.

$$\text{Number of neutrons} = \text{mass number} - \text{atomic number}$$
$$= 235 - 92 = 143$$

Practice Exercises

A. Indicate the number of protons, neutrons, and electrons in a neutral atom of the isotope $^{90}_{38}\text{Sr}$.
B. Write the nuclear symbol for the isotope that has 18 protons and 22 neutrons in its nucleus.

Example 2.4

Which of the following (a) are isotopes of the same element, (b) have identical mass numbers, and (c) have the same number of neutrons? We are using the letter E as the symbol for all elements so that the symbol does not identify the element.

$$^{16}_{8}\text{E} \qquad ^{16}_{7}\text{E} \qquad ^{14}_{7}\text{E} \qquad ^{14}_{6}\text{E} \qquad ^{12}_{6}\text{E}$$

Solution

a. $^{16}_{7}\text{E}$ and $^{14}_{7}\text{E}$ are isotopes of the element nitrogen (N). $^{14}_{6}\text{E}$ and $^{12}_{6}\text{E}$ are isotopes of the element carbon (C).
b. $^{16}_{8}\text{E}$ and $^{16}_{7}\text{E}$ have the same mass number. The first is an isotope of oxygen, and the second is an isotope of nitrogen. $^{14}_{7}\text{E}$ and $^{14}_{6}\text{E}$ also have the same mass number. The first is an isotope of nitrogen, and the second is an isotope of carbon.
c. $^{16}_{8}\text{E}$ ($16 - 8 = 8$ neutrons) and $^{14}_{6}\text{E}$ ($14 - 6 = 8$ neutrons) have the same number of neutrons.

Practice Exercises

A. Which of the following are isotopes of the same element? Is there more than one pair of isotopes?

$$^{90}_{37}\text{E} \qquad ^{90}_{38}\text{E} \qquad ^{88}_{37}\text{E} \qquad ^{88}_{36}\text{E} \qquad ^{93}_{38}\text{E}$$

B. Write the nuclear symbol for each of the following isotopes.
a. uranium-238 **b.** nitrogen-15 **c.** sodium-24

Review Questions

2.7 The nucleus of an atom with a mass number of 23 has 11 protons.

a. How many electrons are there in the neutral atom?

b. How many neutrons are there in the nucleus of the atom?

2.8 What are isotopes, and what is meant by the mass number of an isotope?

2.9 Explain what tabulated atomic mass values, such as those found on the inside front cover, represent.

Electron Structure of Atoms

Have you ever watched colored fireworks or put chemicals into a fireplace to color the flames? If you have, then you have seen a phenomenon that long puzzled scientists and triggered another modification of our concept of the atom. Light is pure energy. Light of different color is light of different energy. For example, blue light packs more energy than red light. If you throw compounds containing lithium into a fire, you see the flames colored red. With sodium compounds (such as ordinary table salt), the light is yellow-orange; with strontium, red; with potassium, lavender (Figure 2.7). Why? That is what early twentieth-century scientists asked themselves.

▲ The colors of fireworks result from changes in electron energy levels in atoms.

2.4 The Bohr Model of the Atom

In an analogy to the motion of planets, Danish physicist Niels Bohr proposed that electrons cannot be located just anywhere outside the nucleus. Instead, they must move around only in well-defined paths or orbits (Figure 2.8), much as the planets orbit the sun. The orbits represent *energy levels.* An electron in one of these **energy levels** has a characteristic amount of energy: a certain amount of kinetic energy because of its motion around the orbit and a certain amount of potential energy because of its distance from the nucleus. If an electron changes from one energy level to another, its energy also changes. If it moves from a higher energy level to a lower energy level, it releases energy, and that energy appears as light. The color of the light depends on the difference in energy between the two levels.

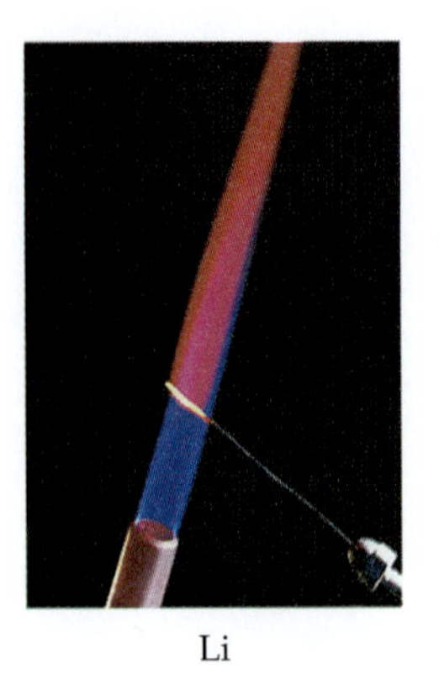

Li

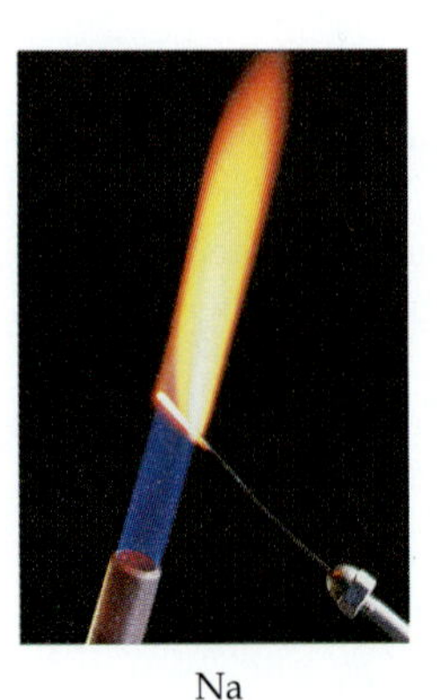

Na

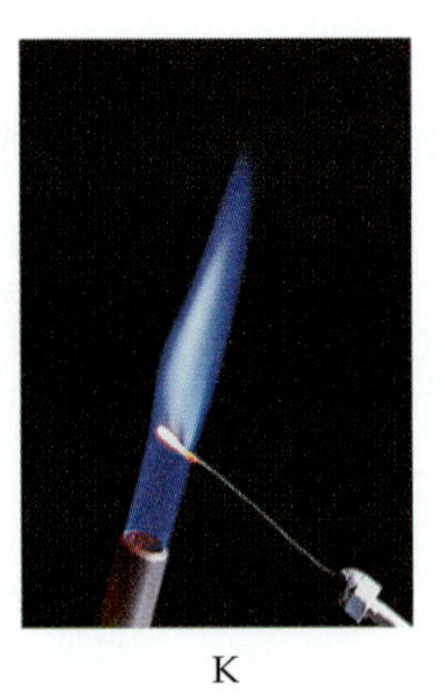

K

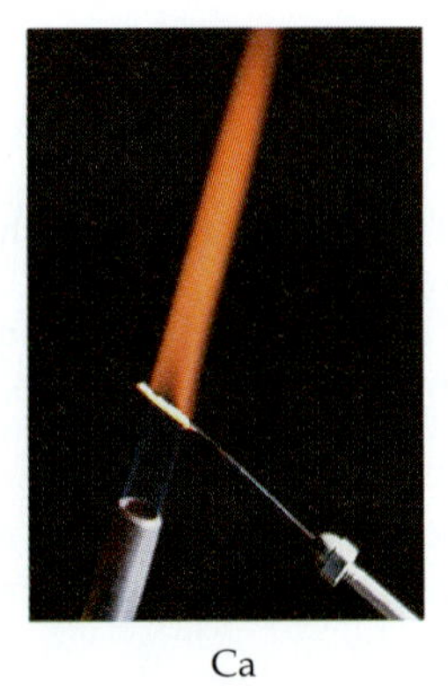

Ca

Sr

▲ **Figure 2.7**
Certain chemical elements can be identified by the characteristic colors that their compounds impart to flames. Five examples are shown.

Atoms in which all electrons are in their lowest possible energy levels are said to be in the *ground state.* If one or more electrons occupy higher energy levels than the ground state, the atom is in an *excited state.*

Bohr's theory also held that a given energy level within an atom can hold only a certain number of electrons (Figure 2.9). These energy levels are often called *shells,* and Bohr said that the maximum number of electrons in a given energy level or shell is given by the formula $2n^2$, where n is the shell being considered. For the first shell—the one closest to the nucleus—$n = 1$ and the maximum population is $2(1^2)$, or 2. For the second shell, $n = 2$ and the maximum number of electrons is $2(2^2)$, or 8. For the third shell, the maximum number is $2(3^2)$, or 18.

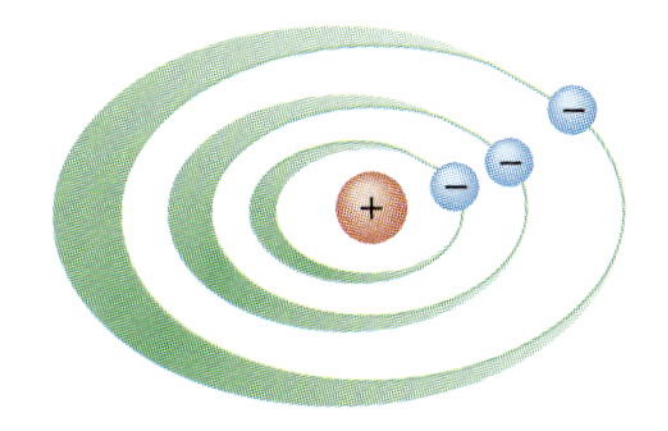

▲ **Figure 2.8**
Bohr visualized the atom as planetary electrons circling a nuclear sun.

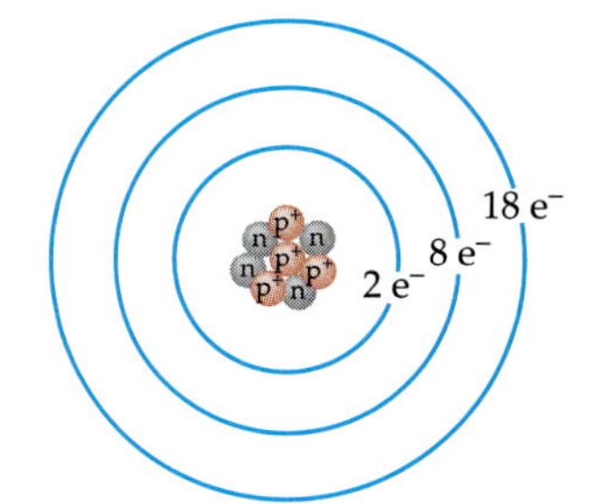

▲ **Figure 2.9**
A Bohr diagram for atoms of an element pictures specified numbers of electrons in distinct energy levels.

Example 2.5

What is the maximum number of electrons in the fifth shell?

Solution

For the fifth shell, $n = 5$, so we have

$$2n^2 = 2 \times 5^2 = 2 \times 25 = 50$$

Practice Exercises

A. What is the maximum number of electrons in the fourth shell?

B. The largest known atoms have electrons in seven shells. According to Bohr's theory, what would be the maximum number of electrons in that shell?

Electronic shells are regions outside the atomic nucleus in which electrons have different energies.

For an atom in the ground state, electrons enter the lowest energy level that has a vacancy. After Bohr put forth his planetary model, chemists used circles to represent the various shells. We now know that electrons don't travel in circular orbits. (We will see why in the next section.) We can still indicate the arrangement of electrons in shells by using a modified Bohr diagram. For example, sulfur ($Z = 16$) has two electrons in the first shell, eight in the second, and the remaining six in the third shell.

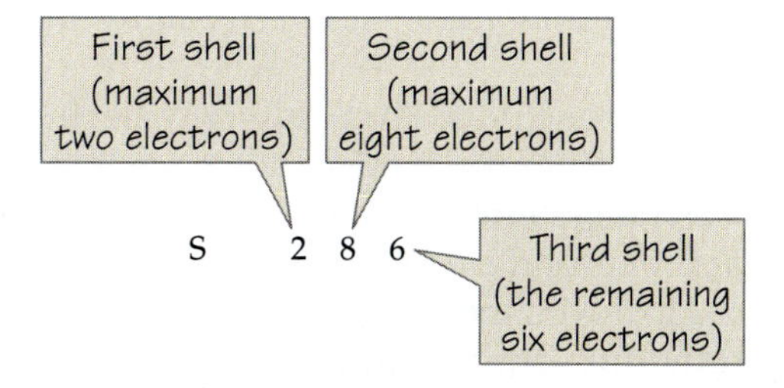

Bohr diagrams have limited uses, and for many purposes we need a better model.

✓ Review Questions

2.10 According to Bohr's theory, what is the maximum number of electrons in the third shell? If the third shell contains two electrons, what is the total number of electrons in the atom?

2.11 According to Bohr's theory, what is the number of electrons in each shell for each of the following elements?

a. silicon **b.** nitrogen **c.** sulfur

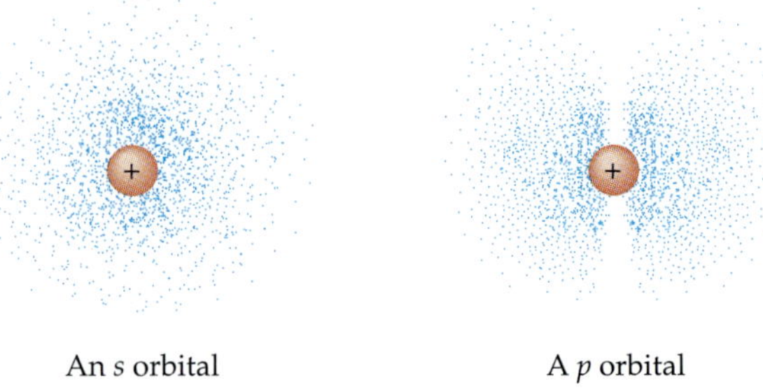

▶ **Figure 2.10**
Charge-cloud representations of atomic orbitals.

2.5 Electronic Configurations

Scientists have replaced the simple but poetic planetary model of the atom by more sophisticated models that explain more data in greater detail than does Bohr's model. These newer models involve sophisticated mathematics and are therefore more difficult for many of us to understand. Fortunately, we don't need to understand the mathematics that generates these models to make use of some of the results.

In the late 1920s Austrian physicist Erwin Schrödinger used quantum mechanics to develop elaborate equations that describe the properties of electrons in atoms. The results provide a measure of the probability of finding an electron in a given volume of space, replacing the definite planetary orbits of the Bohr model by shaped volumes of space in which electrons move. This description of the location of electrons is called an **orbital** (replacing Bohr's *orbits*).

An orbital is the defined volume of an atom where an electron exists. An orbital can hold no more than two electrons.

Suppose you had a camera that could photograph electrons (there is no such thing, but we are just supposing) and you left the shutter open while an electron zipped about the nucleus. When you developed the picture, you would have a record of where the electron had been. Doing the same thing with an electric fan that was turned on would give you a picture in which the blades of the fan would look like a disk of material. The blades move so rapidly that their photographic image is blurred. Similarly, electrons in the first shell of an atom would appear in our imaginary photograph as a fuzzy ball (Figure 2.10). The fuzzy ball (frequently referred to as a *charge cloud* or *electron cloud*) is the rough equivalent of an orbital.

Like Bohr, Schrödinger concluded that only two electrons could occupy the first shell in an atom. In quantum mechanical notation, this energy level is designated the 1*s* orbital (which is spherical—that is, shaped like a ball). Also like Bohr, Schrödinger stated that the second electron shell of an atom could contain a maximum of eight electrons. However, these eight electrons must be located in four different orbitals. Each individual orbital can contain a maximum of two electrons.[4] One of the orbitals of the second energy level is spherical in shape and is designated the 2*s* orbital. The remaining three orbitals of the second level have identical dumbbell shapes (see again Figure 2.10). They differ in their orientation in space (that is, in the direction in which they point; see Figure 2.11). These are called the 2*p* orbitals, and collectively they make up the 2*p* subshell. To distinguish the 2*p* orbitals from one another they are individually referred to as the $2p_x$, $2p_y$, and $2p_z$ orbitals. Electrons in these three orbitals all lie at the same energy level and possess

[4]The two electrons in an orbital must have opposite *spin*. An electron seems to act like a tiny top; it can spin either clockwise or counterclockwise. That the maximum capacity of an orbital is two electrons and that they must have opposite spin is called the *Pauli Exclusion Principle*.

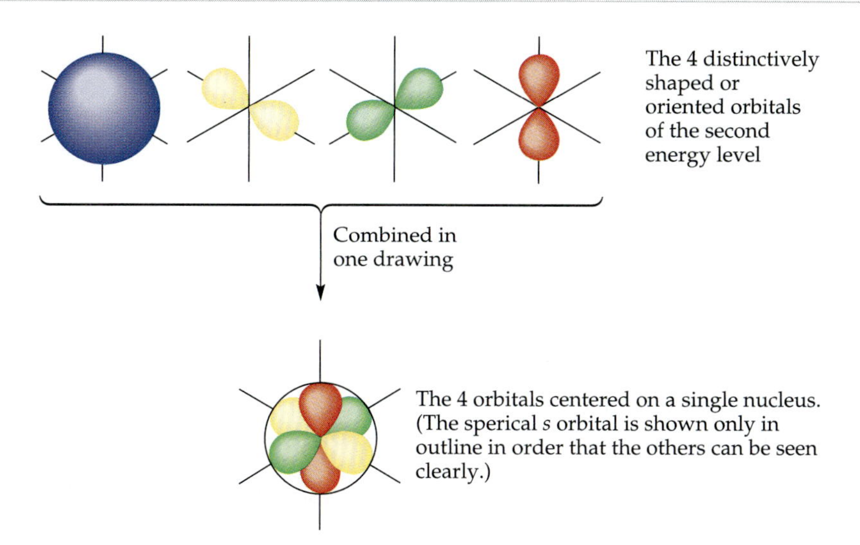

▲ **Figure 2.11**
In these drawings of electron orbitals, the nucleus is located at the intersection of the axes. The eight electrons of the second energy level are distributed among these four orbitals, with two electrons in each orbital.

Table 2.2 Energy Levels (Shells) and Orbitals[a]

Energy Level (Shell)	Orbital(s)
1	*s*
2	*s*, *p*
3	*s*, *p*, *d*
4	*s*, *p*, *d*, *f*

[a]We do not consider the *f* orbitals in this text.

slightly more energy than the electrons in the 2*s* orbital. The quantum mechanical model of the atom thus distinguishes between principal shells (1, 2, 3, and so on) and subshells (2*s* and 2*p*, for example). See Tables 2.2 and 2.3.

The third shell, both in Bohr's model and in Schrödinger's, can hold 18 electrons. In the quantum mechanical atom, however, the third principal shell is divided into three subshells totaling nine orbitals: one 3*s* orbital, three 3*p* orbitals, and five 3*d* orbitals. Each higher principal shell adds another subshell. The orbitals in each new subshell have their own special shapes.

The energy-level diagram for nitrogen ($Z = 7$) is shown in Figure 2.12. The quantum mechanical description of the nitrogen atom focuses on a more detailed description of its **electronic configuration** (electron arrangement). We indicate the electronic configuration of the nitrogen atom as follows, where the superscripts indicate the total number of electrons contained in a particular energy subshell.[5]

The electronic configuration of an atom is the specific ordering of electrons in the atom.

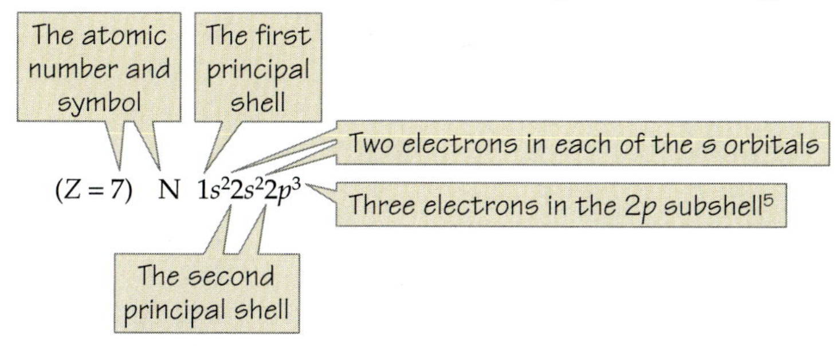

Table 2.4 lists the electronic configurations of the first 20 elements. Notice that the orbitals fill in order of increasing energy: first the 1*s*, then the 2*s*, 2*p*, 3*s*, and so on. The *d* orbitals introduce a minor complication because the 3*d* orbitals turn out to be at a slightly higher energy level than the 4*s* orbitals. Figure 2.13 illustrates the order in which orbitals are filled.

[5]Each of the three *p* orbitals is half-filled with one electron before any orbital is filled with two electrons. This statement, called *Hund's rule,* means that the electronic configuration of nitrogen is $1s^22s^22p_x^{\ 1}2p_y^{\ 1}2p_z^{\ 1}$.

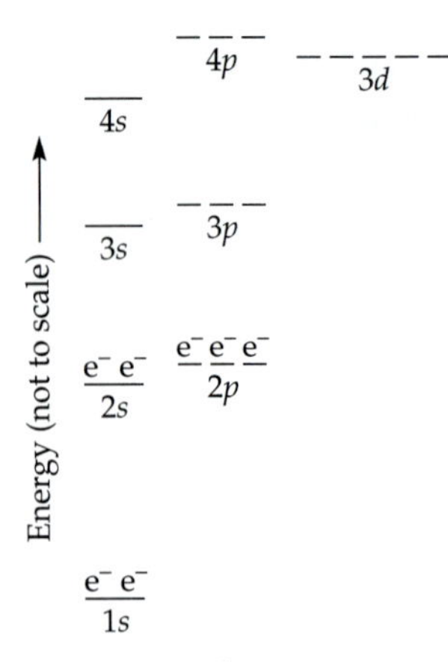

▲ **Figure 2.12**
Energy-level diagram for a nitrogen atom.

Table 2.3 Electrons in Orbitals[a]

Orbital Type	Orientation	Number of Orbitals	Electrons per Orbital	Total Number of Electrons
s	Spherical	1	2	2
p	Perpendicular	3	2	6
d	Perpendicular	5	2	10

[a] We do not consider the *f* orbitals in this text.

The Aufbau Principle

Imagine building up atoms by adding one electron to the proper energy level as each proton is added to the nucleus. This idea is known as the **aufbau principle.** (*Aufbau* is a German word that means "building up.") This principle is a hypothetical process (we can't actually do this) in which we pretend to build up each atom from the one that precedes it in atomic number. We imagine adding a proton (and some neutrons) to the nucleus and one electron to an atomic orbital. That is, we think about building a helium atom from a hydrogen atom; a lithium atom from a helium atom; and so on. We focus on the particular atomic orbital into which the added electron goes to produce a ground-state atom, an atom in its lowest energy state. Electrons will go to the lowest energy subshell available.

Hydrogen: Nucleus of only one proton.

Single electron goes into the first principal shell, which has only one orbital, the 1*s*.

(Z = 1) H $1s^1$

Table 2.4 Electronic Configurations for Atoms of the First 20 Elements

Name	Atomic Number	Electron Structure
Hydrogen	1	$1s^1$
Helium	2	$1s^2$
Lithium	3	$1s^2 2s^1$
Beryllium	4	$1s^2 2s^2$
Boron	5	$1s^2 2s^2 2p^1$
Carbon	6	$1s^2 2s^2 2p^2$
Nitrogen	7	$1s^2 2s^2 2p^3$
Oxygen	8	$1s^2 2s^2 2p^4$
Fluorine	9	$1s^2 2s^2 2p^5$
Neon	10	$1s^2 2s^2 2p^6$
Sodium	11	$1s^2 2s^2 2p^6 3s^1$
Magnesium	12	$1s^2 2s^2 2p^6 3s^2$
Aluminum	13	$1s^2 2s^2 2p^6 3s^2 3p^1$
Silicon	14	$1s^2 2s^2 2p^6 3s^2 3p^2$
Phosphorus	15	$1s^2 2s^2 2p^6 3s^2 3p^3$
Sulfur	16	$1s^2 2s^2 2p^6 3s^2 3p^4$
Chlorine	17	$1s^2 2s^2 2p^6 3s^2 3p^5$
Argon	18	$1s^2 2s^2 2p^6 3s^2 3p^6$
Potassium	19	$1s^2 2s^2 2p^6 3s^2 3p^6 4s^1$
Calcium	20	$1s^2 2s^2 2p^6 3s^2 3p^6 4s^2$

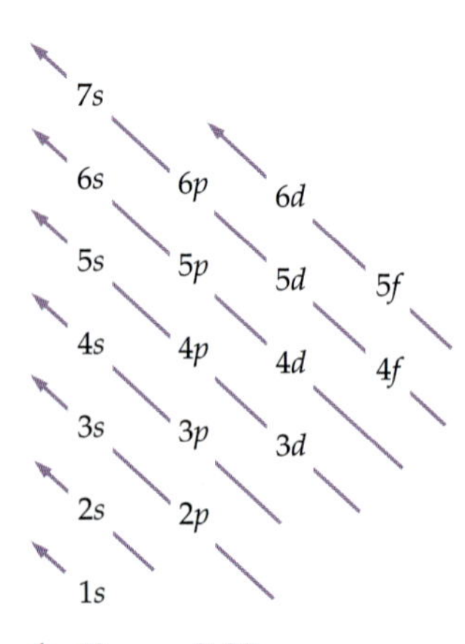

▲ **Figure 2.13**
An order-of-filling chart, using the aufbau principle, for determining the electron configurations of atoms. The first energy level has only *s* orbitals, the second has *s* and *p* orbitals, the third has *s*, *p*, and *d* orbitals, and so on.

Helium: Nucleus of two protons.

The electron we add in "building up" from hydrogen also goes into the 1*s* orbital.

(Z = 2) He $1s^2$

There is a maximum of two electrons in any orbital and only one orbital in the first principal shell. In the aufbau process, then, the first shell is filled with the two electrons of the helium atom. Helium is the first of a series of especially unreactive elements called *noble gases* (Section 2.7).

Lithium: Nucleus of three protons.

The first two electrons are in the 1*s* orbital, but according to the $2n^2$ rule, the third electron must go into the 2*s* orbital, the vacant orbital next lowest in energy. With lithium, the first principal shell is filled and the second begins to fill.

(Z = 3) Li $1s^2 2s^1$

Beryllium: Nucleus of four protons.

The added electron also goes into the 2*s* orbital.

(Z = 4) Be $1s^2 2s^2$

Boron: Nucleus of five protons.

The added electron must enter a 2*p* orbital.

(Z = 5) B $1s^2 2s^2 2p^1$

The 2*p* subshell fills as the atomic number increases from Z = 6 to Z = 10 in the series: carbon, nitrogen, oxygen, fluorine, and neon.

Neon: Nucleus of ten protons.

(Z = 10) Ne $1s^2 2s^2 2p^6$

Like helium, neon has a filled outer shell. Also like helium, neon is especially unreactive, and it too is a noble gas.

We can abbreviate electronic configurations by using the idea of a noble gas core. In a *noble-gas-core abbreviated electronic configuration*, we replace the portion that corresponds to the electronic configuration of a noble gas by a bracketed chemical symbol. Thus, [He] replaces the configuration $1s^2$; so the electronic configuration of lithium, $1s^2 2s^1$, becomes [He]$2s^1$. Similarly, [He] $2s^2 2p^3$ is equivalent to $1s^2 2s^2 2p^3$, the electronic configuration of nitrogen.

The electronic configuration of neon has both the first and second principal shells filled to their maximum capacities (recall Table 2.3). In extending the aufbau process from neon to sodium, the added electron goes into the available orbital of lowest energy, the 3*s*.

$$(Z = 11)\ \text{Na}\ 1s^2 2s^2 2p^6 3s^1 \text{ or } (Z = 11)\ \text{Na}\ [\text{Ne}]3s^1$$

We demonstrate how to write electronic configurations of other atoms with electrons entering the third shell in Example 2.6

Example 2.6

Write out the full electronic configuration and the noble-gas-core abbreviated electronic configuration for sulfur.

Solution

The sulfur atom, with atomic number 16, has 16 electrons. We must place them into the lowest-energy subshells available. Two go into the 1*s* subshell, two into the 2*s* subshell, six into the 2*p* subshell, and two into the 3*s* subshell. That leaves four electrons to be placed in the 3*p* subshell.

$$(Z = 16)\ \text{S}\ 1s^2 2s^2 2p^6 3s^2 3p^4$$

In the noble-gas-core abbreviated electronic configuration, we substitute the symbol [Ne] for $1s^2 2s^2 2p^6$, the portion of the electronic configuration that corresponds to that of neon.

$$(Z = 16)\ \text{S}\ [\text{Ne}]3s^2 3p^4$$

Practice Exercise

Write out the full electronic configuration and the noble-gas-core abbreviated electronic configuration for (**a**) phosphorus and (**b**) chlorine.

The electronic configurations predicted by the aufbau principle for the rest of the third period elements follow a regular pattern through argon, in which the 3*p* subshell fills.

$$(Z = 18)\ \text{Ar}\ 1s^2 2s^2 2p^6 3s^2 3p^6 \quad \text{or} \quad (Z = 18)\ \text{Ar}\ [\text{Ne}]3s^2 3p^6$$

Now we need to be careful. In proceeding to potassium (Z = 19), we must use the order of filling of subshells (see Figure 2.13), rather than the $2n^2$ rule, which states that the maximum capacity of the third principal shell is $2n^2 = 2\ (3)^2 = 18$. The order of filling indicates that the 19th electron goes into the 4*s* subshell, not the 3*d*.

$$(Z = 19)\ \text{K}\ 1s^2 2s^2 2p^6 3s^2 3p^6 4s^1 \quad \text{or} \quad (Z = 19)\ \text{K}\ [\text{Ar}]4s^1$$

In calcium, the 20th electron pairs up with the 19th electron in the 4*s* orbital.

$$(Z = 20)\ \text{Ca}\ 1s^2 2s^2 2p^6 3s^2 3p^6 4s^2 \quad \text{or} \quad (Z = 20)\ \text{Ca}\ [\text{Ar}]4s^2$$

In the series scandium (Z = 21) through zinc (Z = 30), electrons enter the 3*d* subshell, and that subshell fills before the 4*p*. The 4*p* subshell fills in proceeding from gallium (Z = 31) through krypton (Z = 36).

The filling of the 3*d* subshell is somewhat irregular, and we will not concern ourselves further with that matter here.

Quantum mechanics offers detailed descriptions of the electronic structure of atoms. However, we shall make most use of the picture it paints of electrons as clouds of negative charge. Sometimes the shape of the cloud (that is, the orbital) is presented simply in outline. (Figure 2.11 uses shaded drawings to present the combined orbitals of the second energy level.) Later we shall see how the shape and orientation of the electron cloud can determine the shapes of molecules.

✓ Review Questions

2.12 Consider the quantum mechanical notation $2p^6$.

- **a.** How many electrons does this notation represent?
- **b.** What is the general shape of the orbitals described?
- **c.** How many orbitals are included?

2.13 State in words what is meant by the notation $1s^2 2s^2 2p^6 3s^2 3p^5$. What element has this configuration?

The Periodic Table

As our picture of the atom becomes more detailed, we find ourselves in a dilemma. With 115 elements to deal with, how can we keep all this information straight? One way is by using the periodic table of the elements. (See the inside front cover of this book.) The periodic table neatly tabulates information about atoms. It records the atomic number, which tells us how many protons and electrons the atoms of a particular element contain. It even stores information about how electrons are arranged in the atoms of each element.

▲ Dmitri Ivanovich Mendeleev (1834–1907). Mendeleev invented the periodic table while trying to systematize the properties of the elements for presentation in a chemistry textbook. His highly influential text lasted for 13 editions—five of them after his death.

2.6 Mendeleev's Periodic Table

Not long after Dalton presented his model for the atom (an indivisible particle whose mass determined its identity), chemists began listing the elements according to their atomic masses. They observed patterns among the elements. In 1869, Dmitri Ivanovich Mendeleev, a Russian chemist, published a periodic table of the elements that he had prepared by taking into account both the atomic masses and the periodicity of certain properties of the elements. For example, elements that occurred at specific intervals shared similarities in certain properties. Among the approximately 60 elements known at that time, the second and ninth showed similar properties, as did the third and tenth, the fourth and eleventh, the fifth and twelfth, and so on.

Scientists had no idea of either electrons or atomic number at that time, so Mendeleev arranged the elements primarily in order of increasing atomic mass. However, in a few cases, he placed a slightly heavier element before a lighter one. He did this only when it was necessary in order to keep elements with similar chemical properties in the same column. For example, he placed tellurium (atomic mass 128) ahead of iodine (atomic mass 127) because tellurium resembles sulfur and selenium in its properties, whereas iodine is similar to chlorine and bromine.

Mendeleev left several gaps in his table. Instead of looking upon those blank spaces as defects, he boldly predicted the existence of elements as yet undiscovered. Furthermore, he even predicted the properties of some of these missing elements. In succeeding years, many of the gaps were filled by the discovery of new elements, with properties that were often quite close to those predicted by Mendeleev. This predictive value of Mendeleev's table led to its wide acceptance.

2.7 Electronic Structure and the Modern Periodic Table

The modern **periodic table of the elements** has vertical columns called **groups** (or families) of elements. The elements in a given group have similar properties. The horizontal rows of elements are called **periods.** We can summarize periodic relationships by a general statement called the periodic law. In its modern form, the **periodic law** states that certain sets of physical and chemical properties recur at regular intervals (periodically) when the elements are arranged according to increasing atomic number.

Columns in the periodic table are called groups or families.

Rows in the periodic table are called periods.

Valence-Shell Electrons

The properties of an element depend mainly on the configuration of electrons in the outermost shell of the ground-state atoms of the element. This shell is called the **valence shell** of an atom. (The nucleus and the inner shells of electrons are called the *core* of the atom.) With a few exceptions, all the elements in a vertical group of the periodic table have the same number of electrons in the valence shell of their atoms. This generalization is summarized in the following statement and pictured in Figure 2.14.

1 1A	2 2A	3 3B	4 4B	5 5B	6 6B	7 7B	8	9 8B	10	11 1B	12 2B	13 3A	14 4A	15 5A	16 6A	17 7A	18 8A
1 **H** $1s^1$																	2 **He** $1s^2$
3 **Li** $2s^1$	4 **Be** $2s^2$											5 **B** $2s^22p^1$	6 **C** $2s^22p^2$	7 **N** $2s^22p^3$	8 **O** $2s^22p^4$	9 **F** $2s^22p^5$	10 **Ne** $2s^22p^6$
11 **Na** $3s^1$	12 **Mg** $3s^2$											13 **Al** $3s^23p^1$	14 **Si** $3s^23p^2$	15 **P** $3s^23p^3$	16 **S** $3s^23p^4$	17 **Cl** $3s^23p^5$	18 **Ar** $3s^23p^6$
19 **K** $4s^1$	20 **Ca** $4s^2$	21 **Sc** $3d^14s^2$	22 **Ti** $3d^24s^2$	23 **V** $3d^34s^2$	24 **Cr** $3d^54s^1$	25 **Mn** $3d^54s^2$	26 **Fe** $3d^64s^2$	27 **Co** $3d^74s^2$	28 **Ni** $3d^84s^2$	29 **Cu** $3d^{10}4s^1$	30 **Zn** $3d^{10}4s^2$	31 **Ga** $4s^24p^1$	32 **Ge** $4s^24p^2$	33 **As** $4s^24p^3$	34 **Se** $4s^24p^4$	35 **Br** $4s^24p^5$	36 **Kr** $4s^24p^6$
37 **Rb** $5s^1$	38 **Sr** $5s^2$	39 **Y** $4d^15s^2$	40 **Zr** $4d^25s^2$	41 **Nb** $4d^45s^1$	42 **Mo** $4d^55s^1$	43 **Tc** $4d^55s^2$	44 **Ru** $4d^75s^1$	45 **Rh** $4d^85s^1$	46 **Pd** $4d^{10}$	47 **Ag** $4d^{10}5s^1$	48 **Cd** $4d^{10}5s^2$	49 **In** $5s^25p^1$	50 **Sn** $5s^25p^2$	51 **Sb** $5s^25p^3$	52 **Te** $5s^25p^4$	53 **I** $5s^25p^5$	54 **Xe** $5s^25p^6$
55 **Cs** $6s^1$	56 **Ba** $6s^2$	57 ***La** $5d^16s^2$	72 **Hf** $5d^26s^2$	73 **Ta** $5d^36s^2$	74 **W** $5d^46s^2$	75 **Re** $5d^56s^2$	76 **Os** $5d^66s^2$	77 **Ir** $5d^76s^2$	78 **Pt** $5d^96s^1$	79 **Au** $5d^{10}6s^1$	80 **Hg** $5d^{10}6s^2$	81 **Tl** $6s^26p^1$	82 **Pb** $6s^26p^2$	83 **Bi** $6s^26p^3$	84 **Po** $6s^26p^4$	85 **At** $6s^26p^5$	86 **Rn** $6s^26p^6$
87 **Fr** $7s^1$	88 **Ra** $7s^2$	89 **†Ac** $6d^17s^2$	104 **Rf** $6d^27s^2$	105 **Db** $6d^37s^2$	106 **Sg** $6d^47s^2$	107 **Bh**	108 **Hs**	109 **Mt**	110	111	112						

s-block elements

d-block elements

p-block elements

f-block elements

*	58 **Ce** $4f^26s^2$	59 **Pr** $4f^36s^2$	60 **Nd** $4f^46s^2$	61 **Pm** $4f^56s^2$	62 **Sm** $4f^66s^2$	63 **Eu** $4f^76s^2$	64 **Gd** $4f^75d^16s^2$	65 **Tb** $4f^96s^2$	66 **Dy** $4f^{10}6s^2$	67 **Ho** $4f^{11}6s^2$	68 **Er** $4f^{12}6s^2$	69 **Tm** $4f^{13}6s^2$	70 **Yb** $4f^{14}6s^2$	71 **Lu** $4f^{14}5d^16s^2$
†	90 **Th** $6d^27s^2$	91 **Pa** $5f^26d^17s^2$	92 **U** $5f^36d^17s^2$	93 **Np** $5f^46d^17s^2$	94 **Pu** $5f^67s^2$	95 **Am** $5f^77s^2$	96 **Cm** $5f^76d^17s^2$	97 **Bk** $5f^97s^2$	98 **Cf** $5f^{10}7s^2$	99 **Es** $5f^{11}7s^2$	100 **Fm** $5f^{12}7s^2$	101 **Md** $5f^{13}7s^2$	102 **No** $5f^{14}7s^2$	103 **Lr** $5f^{14}6d^17s^2$

▲ **Figure 2.14**
Valence electronic configurations and the periodic table.

*For A-group elements (also called **main-group elements**), the number of valence-shell electrons is the same as the periodic table group number.*

Thus, except for helium, which has only two electrons, all the **noble gases** (Group 8A) have eight valence-shell electrons. These electrons have the configuration ns^2np^6, where n is the principal quantum number for the outer shell of electrons (that is, $2s^22p^6$ for neon, $3s^23p^6$ for argon, $4s^24p^6$ for krypton, and so on).

The B-group elements are generally known as the **transition elements.** The correlation between an A-group number and the number of valence-shell electrons is useful, but it does not extend to the B-group elements.

Each element in Group 1A (the **alkali metals**)[6] has a single electron in an s orbital of the valence shell: $2s^1$ for Li, $3s^1$ for Na, $4s^1$ for K, or, in general, ns^1. Similarly, the Group 2A (the **alkaline earth metals**) elements have the valence-shell electronic configuration ns^2, that is, with the two outermost electrons in an s orbital. As a further example, the seven valence-shell electrons of the Group 7A elements (the **halogens**) are in the configuration ns^2np^5.

Look again at Figure 2.14 and at the periodic table inside the front cover and note the correlation between the valence-electronic configuration of an element in an A group and the location of the element in the periodic table.

The period number is the same as the number of the principal shell of the valence-shell electrons.

All elements in the fourth period, for example, have one or more electrons with $n = 4$ and none with a higher value of n. The period begins with K, which has the electronic configuration $[Ar]4s^1$, and ends with Kr ($[Ar]3d^{10}4s^24p^6$).

[6]Although it is a member of Group 1A, hydrogen is NOT an alkali metal; it isn't even a metal. It exists as H_2 molecules and is a gas at room temperature.

Unsettled Issues Concerning the Periodic Table

The periodic table has been around for well over a century, and in that time many variations have been proposed. Yet there still is no single table accepted by all chemists. A particular state of confusion results from differing use of the letters A and B. In the United States, A is used to designate the main-group (nontransition) elements and B the transition elements. In Europe, the groups to the left of Fe/Ru/Os have typically been designated as A groups, and those to the right of Ni/Pd/Pt as B groups. In order to eliminate the confusion over the use of A and B, the International Union of Pure and Applied Chemistry (IUPAC; see Section 13.5) has recommended that the groups simply be numbered from 1 to 18. These numbers are shown above the A/B group designations in the inside front cover. There are advantages to each system. The A/B system is used in this text. It is helpful, for example, in that the A-group numbers are equal to the numbers of valence-shell electrons in atoms of the main-group elements. (By the IUPAC system, the last digit of the group number gives the number of valence-shell electrons.) Elements 110 and above have not yet been named. These synthetic elements are rarely encountered and will not be considered further here. Although chemists disagree over which form of the periodic table is most useful, there is no disagreement over the great utility of the table.

Each period of the periodic table begins with a Group 1A element and ends with a noble gas element. The periods differ in length because the number of subshells that must fill to get from the electronic configuration ns^1 to ns^2np^6 differs. In the two-member first period, we go directly from $1s^1$ to $1s^2$ (there is no p subshell in the first principal shell). In the eight-member second and third periods the subshells that must fill are the $2s$ and $2p$ and the $3s$ and $3p$, respectively. The fourth and fifth periods have 18 members because the subshells that must fill are the $4s$, $3d$, and $4p$ and the $5s$, $4d$, and $5p$, respectively. Finally, the 32-member sixth and seventh periods require the filling of the $6s$, $4f$, $5d$, and $6p$ and the $7s$, $5f$, $6d$, and $7p$, respectively.

Example 2.7

Write out the subshell notation for the valence-shell electrons of strontium and arsenic.

Solution

Strontium is in Group 2A and the fifth period of the periodic table. Its valence-shell electronic configuration is $5s^2$. Arsenic is in Group 5A and the fourth period. Its five valence-shell electrons have the configuration $4s^24p^3$.

Practice Exercise

Write out the subshell notation for the valence-shell electrons of gallium and tellurium.

Metals, Nonmetals, and Metalloids

In the periodic table on the inside front cover of this book, elements are also divided into two main classes by a bold, stepped, diagonal line. Except for hydrogen in blue, those shown in tan, to the left of the line are **metals,** elements that have a characteristic luster and are generally good conductors of heat and electricity. Most metals are *malleable,* which means that they can be hammered into thin sheet or foil. Also, most metals are *ductile;* they can be drawn into wire. Except for mercury, which is a liquid, all the metals are solid at room temperature.

▲ The metals copper (top) and gold (bottom) are malleable and ductile.

Elements to the right of the stepped line are **nonmetals** (blue), elements that lack metallic properties. For example, nonmetals generally are poor conductors of heat and electricity. At room temperature, several nonmetals, such as oxygen, nitrogen, fluorine, and chlorine, are gases. Others, such as carbon, sulfur, phosphorus, and iodine, are brittle solids. Bromine is the only nonmetal that is a liquid at room temperature.

Most of the elements along the stepped line that divides metals and nonmetals have the physical appearance of metals but some nonmetallic properties as well; these "borderline" elements are sometimes called **metalloids** (green).

Looking at the positions of metals (tan), metalloids (green), and nonmetals (blue) in the periodic table (inside front cover), we see that, generally speaking, metal atoms have few (usually one to three) electrons in their valence shells. Metallic character is closely related to atomic size and the ease with which electrons can be removed from the valence shell of an atom. The easier the removal of these electrons, the more metallic is the element. We can relate metallic character to location of the element in the periodic table. Except for hydrogen, all elements in Groups 1A and 2A are metals. All B group elements are metals. A few of the elements in Groups 3A through 5A also are metals.

Nonmetal atoms generally have more (usually five to seven) electrons in their valence shell than do metals. Except for the special case of hydrogen, the nonmetals are all in Groups 3A through 5A.

The elements in the column at the extreme right of the periodic table (Group 8A) are all gases at room temperature. They are especially resistant to chemical reactions and are often singled out as a special group, the noble gases (purple). Each period of the periodic table ends with a noble gas element. Figure 2.15, which is in the format of the periodic table, summarizes this information.

Metallic character increases from top to bottom in a group and decreases from left to right in a period of the periodic table.

Nonmetallic character decreases from top to bottom in a group and increases from left to right in a period of the periodic table.

Metals | Metalloids | Nonmetals | Noble gases

Period	1 1A	2 2A	3 3B	4 4B	5 5B	6 6B	7 7B	8	9 8B	10	11 1B	12 2B	13 3A	14 4A	15 5A	16 6A	17 7A	18 8A
1	H																	He
2	Li	Be											B	C	N	O	F	Ne
3	Na	Mg											Al	Si	P	S	Cl	Ar
4	K	Ca	Sc	Ti	V	Cr	Mn	Fe	Co	Ni	Cu	Zn	Ga	Ge	As	Se	Br	Kr
5	Rb	Sr	Y	Zr	Nb	Mo	Tc	Ru	Rh	Pd	Ag	Cd	In	Sn	Sb	Te	I	Xe
6	Cs	Ba	La*	Hf	Ta	W	Re	Os	Ir	Pt	Au	Hg	Tl	Pb	Bi	Po	At	Rn
7	Fr	Ra	Act	Rf	Db	Sg	Bh	Hs	Mt	**	**	**						

*Lanthanide series	Ce	Pr	Nd	Pm	Sm	Eu	Gd	Tb	Dy	Ho	Er	Tm	Yb	Lu
†Actinide series	Th	Pa	U	Np	Pu	Am	Cm	Bk	Cf	Es	Fm	Md	No	Lr

** Not yet named

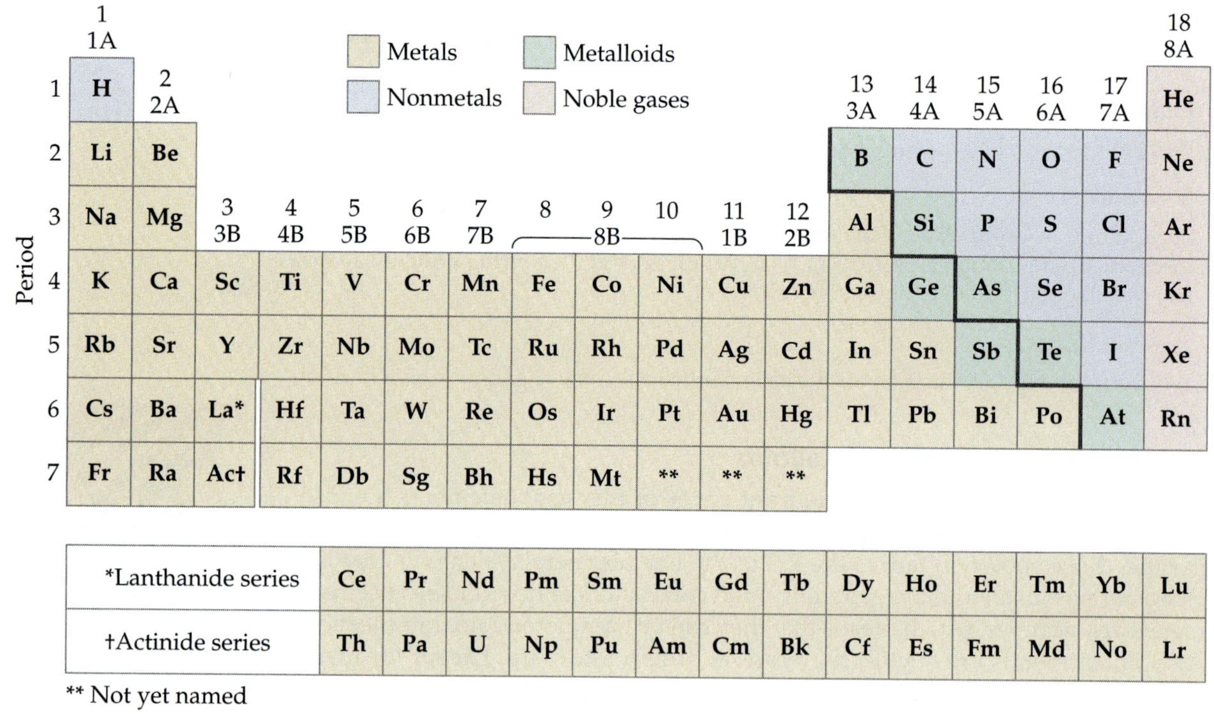

▲ **Figure 2.15**
Metals, nonmetals, metalloids, and noble gases and their location in the periodic table.

From the first generalization we can identify the alkali metals (Group 1A) as highly metallic, and from the second generalization we know that the halogens (Group 7A) are highly nonmetallic. In the middle of the periodic table we see intermediate behavior. Group 4A has carbon, a nonmetal, at the top and two metals, tin and lead, at the bottom. In between are two metalloids, silicon and germanium.

Selected Topic A (which follows Chapter 11) discusses the chemistries of some of the members of these various groups in more detail.

More metallic character

More metallic character

Example 2.8

In each set, indicate which is the more metallic element.

a. Ba, Ca **b.** Sb, Sn **c.** Ge, S

Solution

a. Ba is below Ca in Group 2A, and therefore the Ba atom is larger than the Ca atom. It is easier to remove a valence electron from the larger atom. As a result, Ba is more metallic than Ca.

b. Sn is to the left of Sb in the fifth period and is therefore more metallic than is Sb.

c. Ge is to the left of S and in the following period. Because of its larger atoms, we expect Ge to be more metallic than S. (Actually, Ge is a metalloid and S is a nonmetal.)

Practice Exercise

In each set, indicate which is the more nonmetallic element.

a. O, P **b.** As, S **c.** P, F

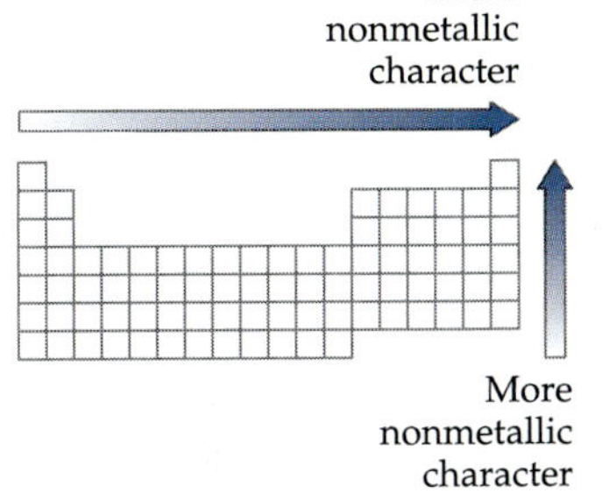

2.8 Periodic Atomic Properties

Let's take a brief look at some *atomic properties,* those associated with individual atoms. We have already discussed one such property, electronic configuration. Here we examine three others: atomic radii, ionization energies and electron affinities.

Atomic Radii

We cannot measure the exact size of an isolated atom because its outermost electrons have a chance of being found relatively far from the nucleus. However, we can measure the distance between the nuclei of two atoms and derive an **atomic radius** from this distance. These internuclear distances depend on the particular environment in which the atoms are found and are therefore not unique. Thus there can be more than one value for the "atomic" radius of an element. Let's consider two specific cases.

The *covalent radius* of an atom is one-half the distance between the nuclei of two like atoms joined into a molecule. Consider, for example, the I_2 molecule pictured in Figure 2.16. The distance between the two iodine nuclei is 266 pm, and the covalent radius of iodine is one-half of that (133 pm). The *metallic radius* is one-half the distance between the nuclei of adjacent atoms in a solid metal. In this text, we will use the *covalent* radius for *nonmetals* and *metallic* radius for *metals.*

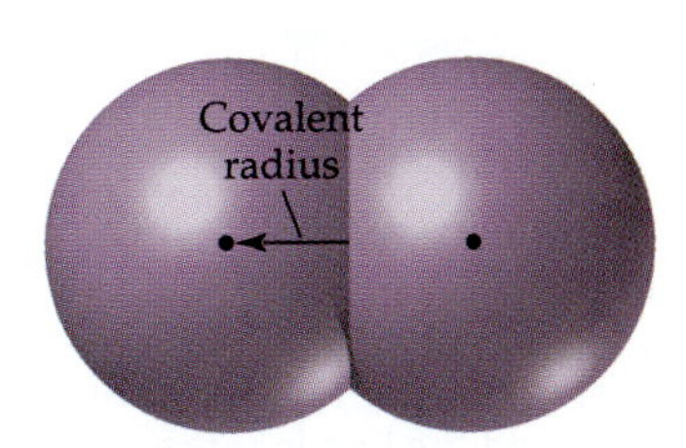

▲ **Figure 2.16**
We can determine an atomic radius for an iodine atom from the internuclear distance in the I_2 molecule. The covalent radius is one-half the distance between the centers of the two I atoms in the I_2 molecule.

Figure 2.17, a graph of atomic radius versus atomic number, is an example of a periodic atomic property of the elements. The first member of each period, a Group 1A metal, has the largest atomic radius for the period (Li for period 2, Na for period 3, and so on). The radii then generally are smaller as we proceed through the period, reaching the lowest value with the Group 7A nonmetal at the end of the period. To explain these trends, let's think of an atomic radius as roughly equal to the distance from the nucleus to the outer-shell (valence) electrons. Any factor that causes this distance to increase makes for a larger atomic radius.

Within a vertical *group* of the periodic table, each succeeding member has one more principal shell with electrons than its immediate predecessor. Thus the sodium

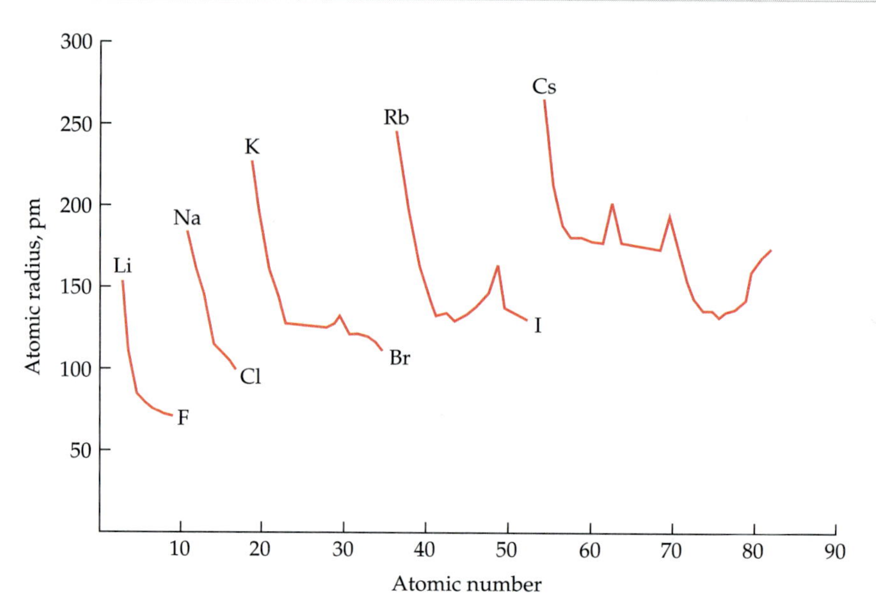

▲ **Figure 2.17**
A graph of atomic radii versus atomic number of the elements. The values, in picometers (pm), are metallic radii for metals and covalent radii for nonmetals. Data are not included for the noble gases because it is difficult to assess their covalent radii (only Kr and Xe compounds are known).

atom has electrons in the principal shells, $n = 1, 2$, and 3, whereas the lithium atom has electrons only in $n = 1$ and 2. The potassium atom has electrons in the principal shells, $n = 1, 2, 3$, and 4; and so on. The electrons in the outer shell are farther from the nucleus as n increases.

Atomic radii increase from top to bottom within a group of the periodic table.

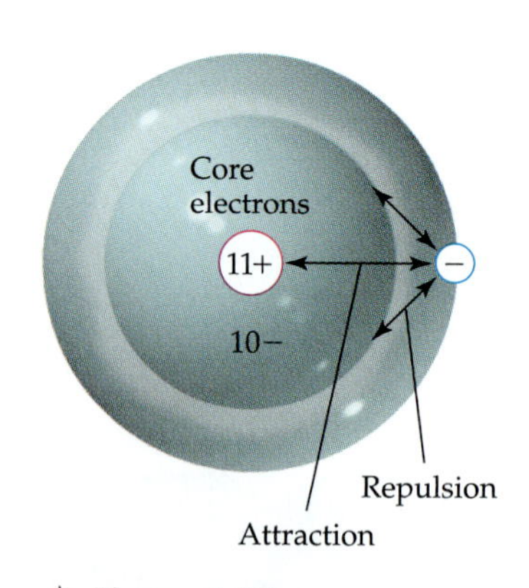

▲ **Figure 2.18**
The effective nuclear charge for a sodium atom. The charge on the nucleus is +11. If the nucleus and the two inner electron shells (contained within the inner circle) acted as a unit, the effective nuclear charge would be +1. Shielding by the inner electrons is not perfect, however, and the effective nuclear charge is greater than +1.

To explain the trend in atomic radii within a *period* of the periodic table, we consider the apparent nuclear charge acting on an electron. The *effective nuclear charge* is the actual nuclear charge less the screening effect of other electrons in the atom.

First, as an oversimplification, consider the sodium atom pictured in Figure 2.18. *If* the 3*s* valence electron were at all times completely outside the region in which the ten electrons of the neon core ($1s^2 2s^2 2p^6$) are found, the 3*s* electron would be perfectly screened from the positively charged nucleus; it would experience an attraction to a net positive charge of only $+11 - 10 = +1$. Similarly, for a magnesium atom there would be two 3*s* electrons outside the neon core and a net positive charge of $+12 - 10 = +2$ acting on each 3*s* electron. Proceeding in this fashion, we would find effective nuclear charge increases from left to right in a period of main-group elements. Because the effective nuclear charge increases, valence electrons are pulled in toward the nucleus and held more tightly.

Atomic radii of the A-group elements decrease from left to right in a period of the periodic table.

(In a series of B-group elements, electrons enter an *inner* electron shell, not the valence shell. The effective nuclear charge therefore remains essentially constant instead of increasing.)

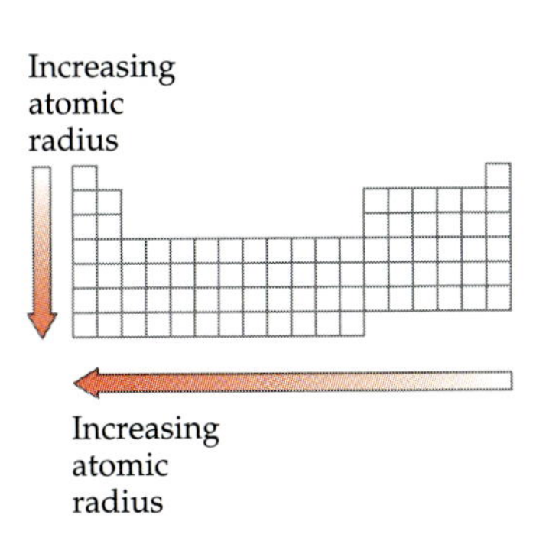

Example 2.9

Arrange each set of elements in order of increasing atomic radius, that is, from smallest to largest.

a. Mg, Si, S **b.** As, N, P

Solution

a. All three elements are A-group elements in the same period (third). Atomic radii within a period decrease from left to right. The order of *increasing* radius is:

$$\text{S (smallest)} < \text{Si} < \text{Mg (largest)}.$$

b. All three elements are in the same group (5A). Atomic radii increase from top to bottom. The order of *increasing* radius is:

$$\text{N (smallest)} < \text{P} < \text{As (largest)}.$$

Practice Exercise

Arrange each set of elements in order of increasing atomic radius.

a. Be, F, N **b.** Ba, Be, Ca **c.** Cl, F, S **d.** Ca, K, Mg

Ions: Noble Gas Configurations

When an atom gains or loses electrons, it becomes charged and is called an **ion.** Thus, if an atom of lithium loses an electron, its electronic configuration becomes $1s^2$, the same as that of the helium atom. It would then be Li^+. In the same way, if an atom of fluorine were to gain an electron, it would become F^- with the electronic configuration $1s^22s^22p^6$. Many simple ions are formed by an atom giving up or gaining electrons to achieve a noble gas configuration.

Some ions are simply electrically charged atoms. Other ions consist of an electrically charged group of atoms that act as a unit (see Section 3.14).

Ionization Energy

In chemical reactions, metal atoms tend to give up valence electrons. A measure of this tendency is the **ionization energy,** the energy required to remove an electron from a ground-state atom (or ion) in the gaseous state. Atoms with more than one electron can ionize in successive steps, but the second step requires considerably more energy because it involves removing an electron from a positive ion whereas the first electron is removed from a neutral atom. Also, in general, it is much easier to remove an electron from the valence shell than from an inner shell. This is consistent with the observation that only valence electrons are associated with the chemical reactivity of the main-group elements.

Figure 2.19 is a graph of first ionization energy versus atomic number. From the graph we can infer the following useful generalizations.

- Ionization energies *decrease* down a group in the periodic table, that is, from lower to higher atomic numbers. Note the gradual decrease in the minimum point of the graph of Figure 2.19 each time a minimum recurs.
- In general, ionization energies *increase* in going from left to right through a period in the periodic table. In Figure 2.19 this is seen as a rise in ionization energies (with some exceptions) from the minima for Group 1A (alkali metal) elements to the maxima for Group 8A (noble gas) elements.

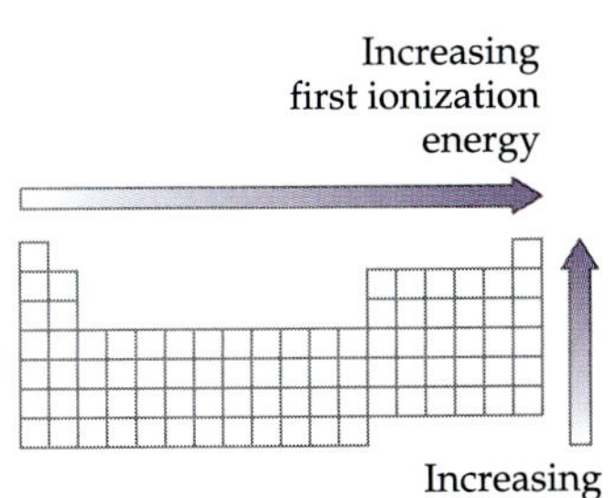

We can explain these general trends in terms of atomic sizes. The greater the distance between the atomic nucleus and the electron to be removed, the less tightly that electron is held to the nucleus, and the more readily ionization occurs. As we

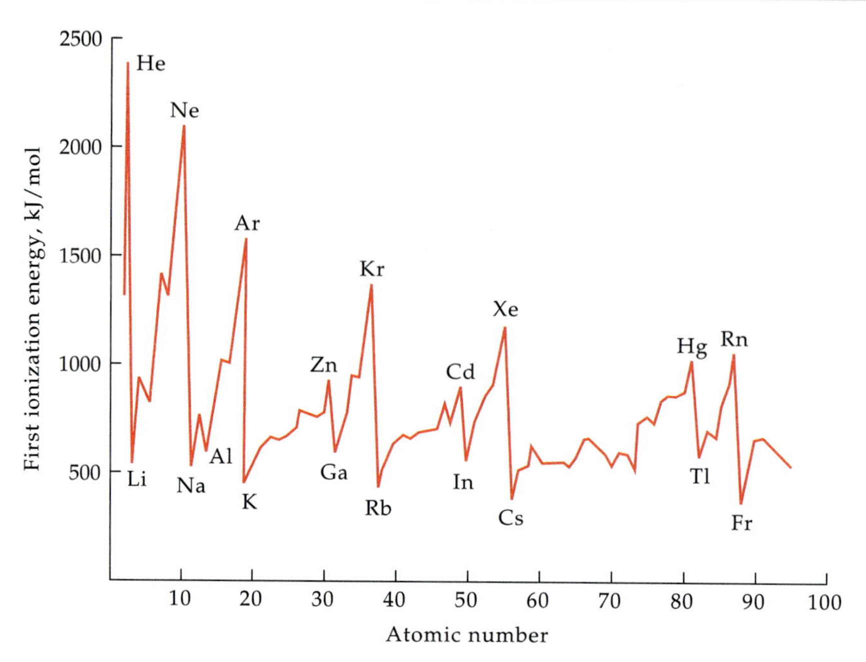

▲ **Figure 2.19**
A graph of the first ionization energy as a function of atomic number of the elements.

saw (page 60), the general trends in atomic radii are an increase down a group and a decrease across a period of the periodic table.

We will not concern ourselves with the irregularities in the trends across periods.

Example 2.10

Without reference to Figure 2.19, arrange each set of elements in order of increasing first ionization energy.

a. Mg, S, Si **b.** As, N, P

Solution

a. All three elements are in the same period. Within a period the ionization energy *increases* from left to right as the atoms become smaller. The order of increasing ionization energy is:

$$\text{Mg (lowest)} < \text{Si} < \text{S (highest)}$$

b. All three elements are in the same group. Within a group the ionization energy *decreases* from top to bottom as atoms become larger. The order of increasing ionization energy is:

$$\text{As (lowest)} < \text{P} < \text{N (highest)}$$

Practice Exercise

Arrange each set of elements in order of increasing first ionization energy.

a. Be, F, N **b.** Ba, Be, Ca

Electron Affinity

Ionization energy is a measure of the energy required to *remove* an electron from an atom, forming a *positive* ion. The corresponding energy change that occurs when an electron is *added* to a gaseous atom to form a gaseous *negative* ion is **electron affinity.**

When an electron approaches a neutral atom, it is attracted to the positively charged nucleus but repelled by electrons already present in the atom. In most cases, however, the incoming electron is absorbed by the atom and energy is evolved.

There are fewer clear-cut periodic trends in electron affinities than in ionization energies. In general, the smaller an atom, the greater its affinity for an electron. The closer the added electron can approach the atomic nucleus, the more strongly it is attracted to the nucleus. This trend seems to be the case for the Group 1A elements, for Group 6A from S to Po, and for Group 7A from Cl to At. The first row elements present some problems, though. The electron affinity of O is not as negative as that of S, nor is that of F as negative as that of Cl. It may be that the electron repulsions in small compact atoms keep the added electron from being as tightly bound as we might expect.

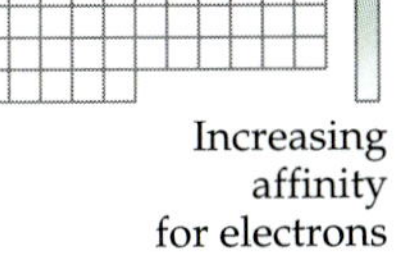

In most cases, the added electron goes into an energy sublevel that is already partly filled. For the Group 2A and 8A atoms, however, the added electron would be required to enter a significantly higher energy level, the *np* level for the Group 2A atoms and the *s* level of the next principal level for the Group 8A atoms. Atoms in these groups do not form a stable negative ion.

Atoms in some groups can add more than one electron. Generally, an atom adds one (Group 7A), two (Group 6A), or three (Group 5A) electrons—enough to complete a valence shell or subshell.

✓ Review Questions

2.14 State the periodic law in its modern form.

2.15 Explain why the several periods of the periodic table do not all have the same number of members.

2.16 Explain why the sizes of atoms do not simply increase uniformly with increasing atomic number.

2.17 What is the general trend in first ionization energies within a period? Within a group? Explain each trend.

Summary

Dalton's atomic theory states that (1) elements are made up of small, indivisible atoms; (2) all atoms of a given element are identical, (3) atoms of different elements combine to form compounds, and (4) chemical reactions involve change in the composition of compounds, not change in the atoms making up those compounds.

We now know that atoms are not indivisible but rather are made up of three kinds of subatomic particles: protons, electrons, and neutrons. Each **proton** carries a positive electric charge of 1+ and has a mass of approximately 1 u. Each **neutron** is electrically neutral and has a mass of approximately 1 u. Each **electron** carries a negative electric charge of 1− and has a mass of 5.49×10^{-4} u, a mass so small relative to the mass of the proton and the neutron that we often ignore it. In any atom the number of protons equals the number of electrons, and so the atom is electrically neutral.

The protons and neutrons are clustered tightly together in a relatively small volume at the center of an atom called the **nucleus.** The electrons are scattered outside the nucleus in all the remaining volume of the atom.

The **atomic number** *Z* of an element tells us the number of protons in the nucleus of each atom of the element. The **mass number A** of an atom is the number of protons plus neutrons in the atom. The number of electrons in an atom is the same as the number of protons.

All atoms of a given element have the same number of protons, but the number of neutrons can vary from one atom to another of the same element. Atoms having the same number of protons but different numbers of neutrons are **isotopes.** Isotopes are represented by the notation $^{A}_{Z}E$, where E stands for the symbol of the element. The two isotopes of carbon are represented by the symbols $^{12}_{6}C$ and $^{14}_{6}C$.

The relative masses of atoms are given in units called **atomic mass units (u).** The carbon-12 isotope, $^{12}_{6}C$, is arbitrarily assigned a mass of exactly 12 u, and the masses of all other elements are measured relative to this standard mass.

Bohr's model of the atom pictures electrons circling the nucleus much like planets circling a central sun. Each orbit represents a certain **energy level** or shell of electrons. In any given orbit an electron has a certain (constant) amount of energy. Whenever an electron jumps to another orbit, the energy of the electron changes.

Normally an atom has all its electrons in the lowest possible energy levels, an arrangement called the **ground state** of the atom. When an atom has absorbed energy to boost one or more electrons to a higher shell, the atom is in an **excited state.** An electron gains energy as it is promoted to a higher shell; it releases energy as it returns to a lower shell. Each main energy level (principal shell) can hold only a certain number of electrons. That number is given by the formula $2n^2$ where n is the number of the shell.

In the quantum-mechanical model of the atom, the well-defined orbits of the Bohr model are replaced by fuzzy "volumes of space," called **orbitals,** occupied by the electrons. Each orbital can hold at most two electrons. Further, the principal shells are divided into subshells, designated by the letters *s*, *p*, *d*, and *f*. The first shell has only one orbital, designated 1*s*, where the 1 designates the first shell and the *s* describes a type of orbital that is spherical in shape. The second shell has two subshells, the 2*s* with one spherical orbital and the 2*p* with three dumbbell-shaped orbitals oriented perpendicular to each other.

As with the Bohr model, each shell in the quantum-mechanical model can hold a maximum of $2n^2$ electrons. The way these electrons are distributed in any atom is called the **electronic configuration** for that atom.

An atom that loses or gains one or more electrons acquires a net electric charge (because now the number of negative electrons is no longer balanced by the number of positive protons). Such a charged species is called an **ion.**

The **periodic table of the elements** organizes all the known elements into rows and columns. The table is periodic because it exhibits a pattern that repeats at regular intervals. The columns are called **groups,** and all the atoms of elements in a given A group element have the same number of electrons in the valence shell. All members of a group therefore have similar chemical properties. The rows of the periodic table are called **periods,** and the beginning of each new period indicates an additional shell in the atoms.

When plotted versus atomic number, certain atomic properties recur periodically. In this chapter, we discussed trends in **atomic radius, ionization energy,** and **electron affinity.** Values of these atomic properties strongly influence physical and chemical properties of the elements.

Metallic character decreases from left to right in the periodic table. It increases from top to bottom with increasing atomic size. The **alkali metals** are all quite metallic, and the **halogens** are all distinctly nonmetallic. In the middle of the table we see intermediate behavior in **metalloids.**

Key Terms

alkali metal (2.7)
alkaline earth metal (2.7)
atomic mass (2.3)
atomic mass unit (u) (2.3)
atomic number (*Z*) (2.2)
atomic radius (2.8)
aufbau principle (2.5)
cathode ray (2.2)
electron (2.2)
electron affinity (2.8)
electronic configuration (2.5)
energy level (2.4)
group (2.7)
halogen (2.7)
ion (2.8)
ionization energy (2.8)
isotope (2.3)
main-group elements (2.7)
mass number (*A*) (2.3)
metal (2.7)
metalloid (2.7)
neutron (2.2)
noble gas (2.7)
nonmetal (2.7)
nucleus (2.2)
orbital (2.5)
period (2.7)
periodic law (2.7)
periodic table of the elements (2.7)
proton (2.2)
radioactivity (2.2)
transition element (2.7)
valence shell (2.7)

Problems

Dalton's Atomic Theory

1. How did the discovery of radioactivity contradict Dalton's atomic theory?
2. Compare Dalton's model of the atom with the nuclear model of the atom.
3. To the nearest atomic mass unit, an atom of calcium has a mass of 40 u and an atom of vanadium has a mass of 50 u. Do these findings contradict Dalton's atomic theory? Explain.
4. To the nearest atomic mass unit, an atom of calcium has a mass of 40 u and an atom of potassium has a mass of 40 u. Do these findings contradict Dalton's atomic theory? Explain.
5. To the nearest atomic mass unit, one atom of calcium has a mass of 40 u and another calcium atom has a mass of 44 u. Do these findings contradict Dalton's atomic theory? Explain.
6. An atom of uranium-235 is struck by a neutron and splits into two smaller atoms. Do these findings contradict Dalton's atomic theory? Explain.

The Nuclear Atom

7. Based on electrical charges, should a proton and an electron attract or repel one another?
8. Based on electrical charges, should a neutron and a proton attract or repel one another?
9. Indicate how many electrons and how many protons there are in a neutral atom of each of the following elements. (You may use the periodic table.)
   a. calcium **b.** sodium **c.** fluorine **d.** argon

10. Indicate how many electrons and how many protons there are in a neutral atom of each of the following elements. (You may use the periodic table.)
    a. beryllium **b.** nitrogen **c.** iron **d.** uranium
11. Indicate the number of protons and the number of neutrons in atoms of the following isotopes.
    a. ^{62}Zn **b.** ^{241}Pu **c.** ^{99}Tc **d.** ^{99}Mo
12. Indicate the number of protons and the number of neutrons in atoms of the following isotopes.
    a. ^{11}B **b.** ^{154}Sm **c.** ^{81}Kr **d.** ^{121}Te
13. Which of the following pairs of symbols represent isotopes?
    a. $^{70}_{33}E$ and $^{70}_{34}E$ **b.** $^{57}_{28}E$ and $^{66}_{28}E$ **c.** $^{186}_{74}E$ and $^{186}_{74}E$
    d. $^{7}_{3}E$ and $^{3}_{2}E$ **e.** $^{22}_{11}E$ and $^{11}_{6}E$
14. Use the symbolism $^{A}_{Z}E$ to represent each of the following isotopes. (You may refer to the periodic table.)
    a. boron-8 **b.** carbon-14
    c. uranium-235 **d.** cobalt-60

Atomic Masses

15. The two principal isotopes of lithium are lithium-6 and lithium-7. The atomic mass of lithium is 6.9 u. Which is the predominant isotope of lithium?
16. Out of every five atoms of boron, one has a mass of 10 u and four have a mass of 11 u. What is the atomic mass of boron? Use the periodic table only to check your answer.
17. There are two naturally occurring isotopes of silver, existing in approximately equal proportions. One of these is ^{107}Ag. Which of the following must be the other: ^{106}Ag, ^{108}Ag, ^{109}Ag? Explain.
18. Gallium in nature consists of two isotopes, gallium-69, with a mass of 68.926 u and a fractional abundance of 0.601, and gallium-71, with a mass of 70.925 u and a fractional abundance of 0.399. Calculate the weighted-average atomic mass of gallium.

Electronic Configurations

19. Give the electronic configurations for each of the following elements.
    a. silicon **b.** nitrogen **c.** sulfur
20. Give the electronic configurations for each of the following elements.
    a. helium **b.** chlorine **c.** magnesium
21. Which elements have the following electronic configurations?
    a. $1s^2 2s^2$ **b.** $1s^2 2s^2 2p^3$ **c.** $1s^2 2s^2 2p^6 3s^2 3p^1$
22. None of the following electronic configurations is reasonable. Explain why in each case.
    a. $1s^2 2s^2 3s^2$ **b.** $1s^2 2s^2 2p^2 3s^1$ **c.** $1s^2 2s^2 2p^6 2d^5$
23. If three electrons were added to the valence shell of a phosphorus atom, the new electronic configuration would resemble that of what element?
24. If two electrons were removed from the valence shell of a magnesium atom, the new electronic configuration would resemble that of what element?
25. How many electrons are in the valence shell of a Group 7A atom?
26. How many electrons are in the valence shell of a Group 2A atom?
27. What is meant by the ground state of an atom?
28. What is meant by the excited state of an atom?
29. When light is emitted by an atom, what change has occurred within the atom?
30. Which has absorbed more energy, an atom in which an electron has moved from the second shell to the third shell, or an otherwise identical atom in which an electron has moved from the first shell to the third?
31. What is the difference between the electronic configurations of oxygen (O) and fluorine (F)?
32. What is the difference between the electronic configurations of fluorine (F) and sulfur (S)?

The Periodic Table

(You may use the periodic table unless directed otherwise.)

33. List some characteristic properties of metals, and indicate where metallic elements are located in the periodic table.
34. List some characteristic properties of nonmetals, and indicate where nonmetallic elements are located in the periodic table. What is a metalloid?
35. Indicate the group and period in which each of the following elements is found. Classify each as a metal or nonmetal.
    a. C **b.** Ca **c.** Cd **d.** Cl
    e. B **f.** Ba **g.** Bi **h.** Br
36. Indicate the group and period in which each of the following elements is found. Classify each as a metal or nonmetal.
    a. S **b.** Sn **c.** Sm **d.** Sr
    e. Ta **f.** Tc **g.** Ti **h.** Tl
37. Identify the element in each of the following descriptions.
    a. Group 3A, Period 4 **b.** Group 1B, Period 4
    c. Period 5, halogen
38. Identify the element in each of the following descriptions.
    a. Group 4A, nonmetal **b.** Group 7B, Period 5
    c. Period 2, alkali metal
39. Which of the following elements are halogens?
    a. Ag **b.** At **c.** As **d.** Hf **e.** F
40. Which of the following elements are alkali metals?
    a. K **b.** Y **c.** W **d.** Al **e.** Rb
41. Which of the following elements are noble gases?
    a. Fe **b.** Ne **c.** Ge **d.** He **e.** Xe
42. Which of the following elements are transition metals?
    a. Ti **b.** Tc **c.** Te **d.** Co **e.** Cs

Periodic Properties

43. By consulting only the periodic table, determine which member of the following pairs has the *larger* radius? Explain.
    a. Cl or S **b.** Al or Mg
44. By consulting only the periodic table, determine which member of the following pairs has the *smaller* radius? Explain.
    a. Ca or Rb **b.** N or S

45. By consulting only the periodic table, arrange each set of the following elements in order of increasing atomic radius and explain the basis for this order.
    a. Al, Mg, Na **b.** Ca, Mg, Sr
46. By consulting only the periodic table, arrange each set of the following elements in order of increasing atomic radius and explain the basis for this order.
    a. Ca, Rb, Sr **b.** Al, C, Si
47. Arrange each set of the following elements in order of increasing first ionization energy and explain the basis for this order.
    a. Ca, Mg, Ba **b.** P, Cl, Al
48. Arrange each set of the following elements in order of increasing first ionization energy and explain the basis for this order.
    a. Ca, Na, As **b.** S, As, Sn
49. Which group of elements has the most affinity for electrons? Explain.
50. Which of the main groups of elements do not form stable negative ions? Use electronic configurations to explain this behavior.

Additional Problems

51. In what location of the periodic table would you expect to find the two or three elements having the largest atoms? Explain.
52. Fill in the blanks for the six elements listed.

	Atomic No.	No. of Protons	No. of Electrons	No. of Neutrons	Mass No.
Pb	____	____	____	126	208
Sr	____	38	____	50	____
N	7	____	____	____	14
Cr	____	____	24	____	52
Ag	47	____	____	60	____
As	____	____	33	42	____

53. Indicate the group number or numbers of the following families.
    a. alkali metals **b.** transition metals
    c. halogens **d.** alkaline earth metals
54. Which atom of each pair is larger?
    a. K or Rb **b.** Ne or Na **c.** Zn or Zr
55. Which atom will ionize more readily? That is, which has the lower ionization energy?
    a. Li or Cs **b.** N or Sb **c.** Ar or Ac
56. Elements are defined on a theoretical basis as being composed of atoms that share the same atomic number. On the basis of this theory, would you think it possible for someone to discover a new element that would fit between magnesium (atomic number 12) and aluminum (atomic number 13)?
57. Referring only to the periodic table, indicate what similarity in electronic structure is shared by fluorine (F) and chlorine (Cl). What is the difference between their electronic structures?

Chemical Bonds

CHAPTER 3

These cubic crystals of sodium chloride—common table salt—reflect the arrangement of charged particles called ions. The particles are held in place by ionic bonds, one of the two major types of chemical bonds discussed in this chapter. Bonds hold atoms and ions in place in molecules and crystals, giving structure to matter.

Learning Objectives/Study Questions

1. What electronic configuration appears to be the most stable? How is an understanding of this important in determining the most common ion(s) formed from a given atom?
2. What are Lewis structures?
3. What is an ionic bond? How does it differ from a covalent bond?
4. How are the chemical formulas of ionic compounds determined? How are ionic compounds named?
5. What determines if a covalent bond is polar or nonpolar? How does electronegativity influence whether a bond is polar or nonpolar?
6. How are the chemical formulas of simple molecules determined? How are these molecules named?
7. What is a polyatomic ion?
8. What steps are followed in determining the Lewis structure for a given molecule or ion?
9. What is the octet rule? What are some exceptions to the octet rule?
10. How is VSEPR theory used to predict the shapes of molecules and ions?
11. How is the shape of a molecule important in determining whether the molecule is polar or nonpolar?

There are 115 known chemical elements. There are about 20 million known chemical compounds. To form these compounds, atoms of different elements must combine in specific groupings. The forces maintaining these arrangements are called **chemical bonds.** Chemical bonding also plays a major role in determining the state of matter. At room temperature water is a liquid, carbon dioxide is a gas, and table salt is a solid—all because of differences in their chemical bonds. Bonds even determine the physical shape and flexibility of molecules—that is, whether they are spherical or flat, rigid or wobbly.

As scientists developed an understanding of the nature of chemical bonding and how atoms are arranged in molecules, they gained the ability to manipulate the structure of compounds. Dynamite, birth control pills, synthetic fibers, and thousands of other products fashioned in chemical laboratories have dramatically changed our lives. We are now entering an era that promises even greater change.

By understanding the bonding in DNA and the joining of two DNA molecules in a helical structure, we can now explain the chemical basis of heredity (Chapter 23). Whether an organism is fish, fowl, hippopotamus, or human is determined by the structure of DNA. Scientists have also learned how to manipulate the structure of DNA and thus have gained the ability to exert limited control over the structure of living matter. Such genetic engineering promises cures for some diseases and even enables scientists to custom-tailor certain organisms.

3.1 Stable Electronic Configurations

As we noted in our discussion of the atom and its structure (Chapter 2), the electronic structure of the noble gas (Group 8A) atoms appears to be especially stable; the noble gases undergo few chemical reactions. In contrast, the alkali metals (Group 1A) and the halogens (Group 7A) are quite reactive. Is there a useful concept that we can deduce from these facts? There is, and although there are other factors and important exceptions, we can use the concept as follows:

1. Fact: The noble gases, such as helium, neon, and argon, are inert (that is, they undergo few, if any, chemical reactions).
2. Theory: The inertness of the noble gases is due to their electron structures.
3. Deduction: If other elements could alter their electron structures to become more like the noble gases, they would become less reactive.

To illustrate, let's look at an atom of the element sodium (Na, Z = 11), which has the electronic configuration $1s^22s^22p^63s^1$. That is, the Na atom has 11 electrons—two in the first shell, eight in the second, and one in the third. If the Na atom gave up an electron, it would have the same electronic configuration as an atom of the noble gas neon (Ne, Z = 10).

$$(Z = 11)\ \text{Na}\ (1s^22s^22p^63s^1) \longrightarrow \text{Na}^+\ (1s^22s^22p^6) + 1\,e^-$$

Recall that neon has the electronic structure

$$(Z = 10)\ \text{Ne}\ (1s^22s^22p^6)$$

Similarly, if a chlorine atom (Cl) gained an electron, it would have the same electronic configuration as argon (Ar).

$$(Z = 17)\ \text{Cl}\ (1s^22s^22p^63s^23p^5) + 1\,\text{e}^- \longrightarrow \text{Cl}^-\ (1s^22s^22p^63s^23p^6)$$

The electronic configuration of the argon atom is

$$(Z = 18)\ \text{Ar}\ (1s^2 2s^2 2p^6 3s^2 3p^6)$$

The sodium atom, having lost an electron, becomes positively charged. It has 11 protons (11+) and only 10 electrons (10−). Its symbol is written Na^+ and it is called a *sodium ion.* The chlorine atom, having gained an electron, becomes negatively charged. It has 17 protons (17+) and 18 electrons (18−). Its symbol is written Cl^- and it is called a *chloride ion.* Note that a *positive* charge, as in Na^+, indicates that one electron has been *lost* per atom. Similarly, a *negative* charge, as in Cl^-, indicates that one electron has been *gained* per atom. **Ions** are charged units—that is, units in which the number of electrons is *not* equal to the number of protons.

All the noble gases except helium have a stable *octet* of electrons in the highest main energy level (ns^2np^6). When the atoms of other elements react to gain an inert gas configuration, they are obeying what is called the octet rule: Atoms seek an arrangement that will surround them with eight electrons in the outer shell.

By gaining an electron, a chlorine atom achieves the same electronic configuration as argon, but it does not *become* an argon atom. The chloride ion has 17 protons in its nucleus and a net charge of 1−. The argon atom has 18 protons in its nucleus and no net charge; it is electrically neutral. Nor does a sodium atom become neon by giving up an electron. The sodium ion has the same electronic configuration as neon, but the nuclei differ in the number of protons and are therefore different elements.

Some *ions* are simply single atoms with an electrical charge (monatomic ions; Table 3.2). Other ions consist of a group of atoms with an electrical charge that act as a unit (polyatomic ions; Table 3.4). Positively charged ions are called *cations* and negatively charged ions are called *anions* (see Section 12.1).

Example 3.1

Aluminum atoms form ions by giving up three electrons. What is the charge on an aluminum ion?

Solution

The neutral aluminum atom (Z = 13) has 13 electrons and 13 protons. The aluminum ion thus has 13 protons (13+) but only 10 electrons (10−). The net charge on the aluminum ion is 3+. (Its symbol is Al^{3+}.)

Practice Exercises

A. Sulfur atoms form ions by gaining two electrons. What is the charge on the ion?

B. Iron atoms form two kinds of ions by giving up either two or three electrons. What are the charges on the two ions?

✓ Review Questions

3.1 Which group of elements in the periodic table is characterized by especially stable electronic configurations?

3.2 What is an ion? How is an ion formed from an atom?

3.2 Lewis Structures

The core electronic configurations of sodium and chlorine atoms do not change when they form ions. For convenience, we can let the chemical symbol represent the core of an atom—that is, the nucleus plus the inner electrons. The **valence electrons**—those in the outer shell—are represented by dots placed around the symbol. For example, we can write the process

$$\text{Na}\ [1s^2 2s^2 2p^6]3s^1 \longrightarrow \text{Na}^+\ [1s^2 2s^2 2p^6] + e^-$$

more simply as

$$\text{Na}\cdot \longrightarrow \text{Na}^+ + e^-$$

In this representation, the symbol Na stands for the core of the sodium atom and the dot stands for the single valence electron.

▲ Both the octet rule and the Lewis structure method of representation were formulated by the University of California chemist Gilbert Newton Lewis (1875–1946).

A useful practice in placing dots around a chemical symbol is to place lone dots on each of the four sides of the symbol before pairing any of the dots.

Similarly, we can represent the process

$$\text{Cl } (1s^2 2s^2 2p^6 3s^2 3p^5) + 1\text{ e}^- \longrightarrow \text{Cl}^- \ (1s^2 2s^2 2p^6 3s^2 3p^6)$$

as

$$:\ddot{\underset{..}{\text{Cl}}}\cdot + \text{e}^- \longrightarrow :\ddot{\underset{..}{\text{Cl}}}:^-$$

Representations in which the symbol of an element stands for the core of the atom and dots stand for its valence electrons are called **Lewis structures** or (sometimes) *electron dot structures.*

It is especially easy to write Lewis structures for most main-group elements. As we can see from Table 3.1, the number of valence electrons for each of these elements is equal to the group number.

Symbolism is a convenient, shorthand way to convey a lot of information in compact form. It is the chemist's most efficient and economical form of communication. Learning this symbolism is much like learning a foreign language. Once you master a certain basic "vocabulary," the rest is easier.

Example 3.2

Without referring to Table 3.1, give Lewis structures for calcium, oxygen, and phosphorus.

Solution

Calcium is in Group 2A, oxygen is in Group 6A, and phosphorus is in Group 5A. The Lewis structures are therefore

$$\cdot\text{Ca}\cdot \qquad :\ddot{\underset{\cdot}{\text{O}}}\cdot \qquad :\dot{\underset{\cdot}{\text{P}}}\cdot$$

Practice Exercises

A. Without referring to Table 3.1, give Lewis structures for each of the following elements.
a. Ar **b.** Sr **c.** F **d.** N **e.** K **f.** S

B. Without referring to Table 3.1, give Lewis structures for each of the following elements.
a. Rb **b.** Xe **c.** I **d.** As **e.** Ra **f.** Se

Table 3.1 Lewis Structures of Selected Main-Group Elements

1A	2A	3A	4A	5A	6A	7A	Noble Gases
H·							He:
Li·	·Be·	·Ḃ·	·Ċ· (dot below)	:Ṅ· (dot below)	:Ö· (dot below)	:F̈· (pair below)	:N̈e: (pair below)
Na·	·Mg·	·Ȧl·	·Ṡi· (dot below)	:Ṗ· (dot below)	:S̈· (dot below)	:C̈l· (pair below)	:Är: (pair below)
K·	·Ca·	·Ġa·	·Ġe· (dot below)	:Ȧs· (dot below)	:S̈e· (dot below)	:B̈r· (pair below)	:K̈r: (pair below)
Rb·	·Sr·				:T̈e· (dot below)	:Ï· (pair below)	:Ẍe: (pair below)
Cs·	·Ba·						:R̈n: (pair below)

Review Questions

3.3 How does a sodium atom differ in structure from a sodium ion? How does a sodium ion differ in structure from a neon atom? How are the sodium ion and neon atom similar?

3.4 How does a chlorine atom differ in structure from a chloride ion? How do the two species differ in chemical reactivity?

Ionic Bonds and Ionic Compounds

There are two major categories of chemical compounds based on the type of bonds between the atoms. Compounds composed of ions are held together by mutual attraction of the oppositely charged ions and are called *ionic compounds.* Compounds composed of molecules in which the atoms are held together by *shared* electrons are called *molecular* compounds, and the shared electrons are called *covalent bonds.*

We will first consider ionic bonding, starting with one of the most familiar ionic compounds, sodium chloride—common table salt. Then, in later sections, we will turn our attention to covalent bonding.

3.3 The Sodium-Chlorine Reaction

Sodium is a highly reactive metal. It is so soft that it can be cut with a knife. When freshly cut, it is bright and silvery, but it dulls rapidly because it reacts with oxygen in the air. In fact, it reacts so readily in air that it is usually stored under oil or kerosene. Sodium reacts violently with water also, producing so much heat that it melts. A small piece will form a spherical bead after melting and race around on the surface of the water as it reacts.

Chlorine is a greenish yellow gas. It is familiar as a disinfectant for swimming pools and city water supplies. (The actual substance added may be a compound that reacts with water to form chlorine.) Chlorine is extremely irritating to the respiratory tract. In fact, chlorine was used as a poison gas in World War I.

Macroscopic View

Let's first look at what we see at the macroscopic (apparent to the unaided eye) level. When a piece of sodium is dropped into a flask containing chlorine gas, a violent reaction ensues. A white solid that is quite unreactive is formed. It is a familiar compound—sodium chloride (table salt) (Figure 3.1).

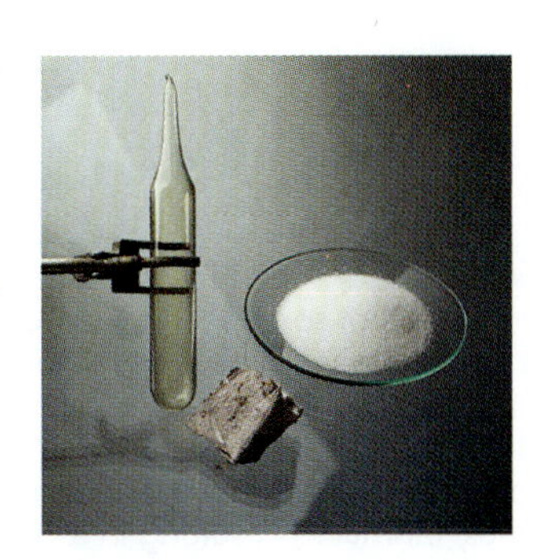

▲ Sodium is a soft, silvery, reactive metal. Chlorine is a poisonous greenish-yellow gas. Sodium chloride is a crystalline white solid used to enhance the flavor of food.

Microscopic View: Theory

The sodium–chlorine reaction illustrates a tendency common in nature. More reactive (more energetic) substances tend to become less reactive (less energetic) substances, releasing energy in the process. A sodium atom becomes less reactive by *losing* an electron. A chlorine atom becomes less reactive by *gaining* an electron. (Both acquire an outer-shell octet.) At the microscopic (atomic) level, we can imagine what happens when sodium atoms come into contact with chlorine atoms: A chlorine atom gets an electron from a sodium atom. Using Lewis structures, the equation for this reaction is written

$$\mathrm{Na}\cdot + :\ddot{\underset{..}{\mathrm{Cl}}}\cdot \longrightarrow \mathrm{Na}^{+} + :\ddot{\underset{..}{\mathrm{Cl}}}:^{-}$$

Actually, chlorine gas is composed of Cl_2 molecules. Each atom of the molecule receives an electron from a sodium atom. Two sodium ions and two chloride ions are formed.

$$\mathrm{Cl_2} + 2\,\mathrm{Na} \rightarrow 2\,\mathrm{Cl^-} + 2\,\mathrm{Na^+}$$

The two ions formed from sodium and chlorine atoms have opposite charges, strongly attracted to one another. Keep in mind, however, that in even a tiny 1.0 mg grain of salt, there are about 10^{19} of each kind of ion, which arrange themselves in

▲ **Figure 3.1**
The reaction of sodium and chlorine. Sodium metal and chlorine gas provide striking visual evidence of their reaction to produce the ionic substance sodium chloride.

an orderly fashion. The arrangements are repeated many times in all directions—front and back, left and right, top and bottom—to make a *crystal* of sodium chloride (Figure 3.2). (A **crystal** is a solid substance with a regular arrangement of its constituent particles. The solid as a whole has a well-defined, regular shape.) Each sodium ion attracts (and is attracted by) six chloride ions (the ones to its front and back, its top and bottom, and its two sides). Each chloride ion attracts (and is attracted by) six sodium ions. The forces holding the crystal together (the attractive forces between positive and negative ions) are called **ionic bonds.**

Scientists sometimes use different models to represent the same system. The model employed in Figure 3.2 is a space-filling model showing the relative sizes of the sodium and chloride ions. Sometimes a ball-and-stick model is employed to better show the geometry of the crystal (Figure 3.3). From this model it is easy to see the cubic arrangement of the ions. In the crystal as a whole, for each sodium ion there is one chloride ion; thus the ratio of ions is 1:1, and the simplest formula for the compound is NaCl.

3.4 Ionic Bonds: Some General Considerations

As we might expect, potassium, a metal in the same family as sodium and with the same valence electron configuration, also reacts with chlorine. The reaction yields potassium chloride (KCl).

$$\text{K}\cdot \; + \; \cdot\ddot{\underset{..}{\text{Cl}}}\!: \longrightarrow \text{K}^+ \; + \; :\ddot{\underset{..}{\text{Cl}}}\!:^-$$

And potassium reacts with bromine, a reddish brown liquid in the same family as chlorine, to form a stable white crystalline solid called potassium bromide (KBr).

$$\text{K}\cdot \; + \; \cdot\ddot{\underset{..}{\text{Br}}}\!: \longrightarrow \text{K}^+ \; + \; :\ddot{\underset{..}{\text{Br}}}\!:^-$$

Sodium also reacts with bromine to form sodium bromide.

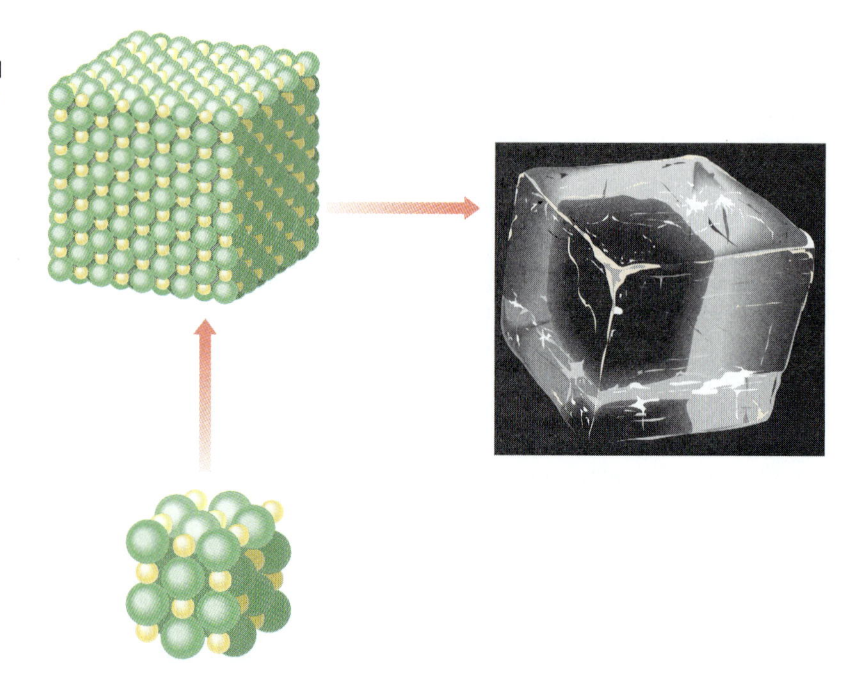

Figure 3.2 ▶
In a crystal of sodium chloride, ions are arranged in a regular fashion. Each Na^+ ion (small sphere) is surrounded by six Cl^- ions (large spheres). In turn, each Cl^- ion is surrounded by six Na^+ ions. This arrangement repeats itself many, many times, ultimately resulting in a crystal of sodium chloride (right).

Example 3.3

Use Lewis structures to show the transfer of electrons from sodium atoms to bromine atoms to form ions with noble gas configurations.

Solution

Sodium has one valence electron, and bromine has seven. Transfer of the single electron from sodium to bromine leaves each with a noble gas configuration.

$$\mathrm{Na}\cdot + \cdot\ddot{\underset{..}{\mathrm{Br}}}: \longrightarrow \mathrm{Na}^{+} + :\ddot{\underset{..}{\mathrm{Br}}}:^{-}$$

Practice Exercises

A Use Lewis structures to show the transfer of electrons from lithium atoms to fluorine atoms to form ions with noble gas configurations.

B Use Lewis structures to show the transfer of electrons from potassium atoms to iodine atoms to form ions with noble gas configurations.

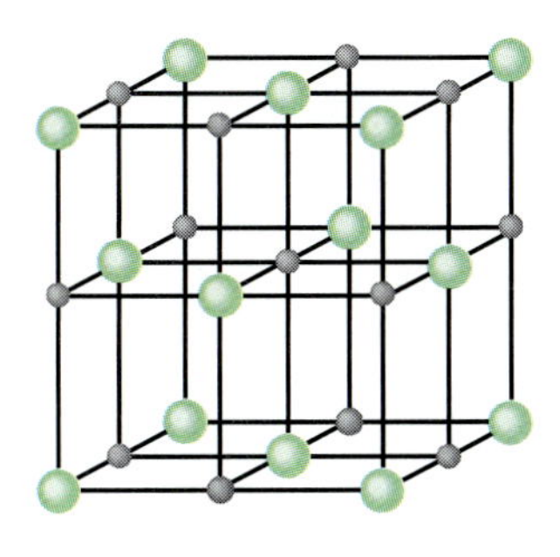

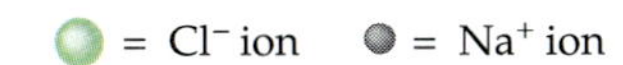

▲ **Figure 3.3**
A ball-and-stick model of a sodium chloride crystal.

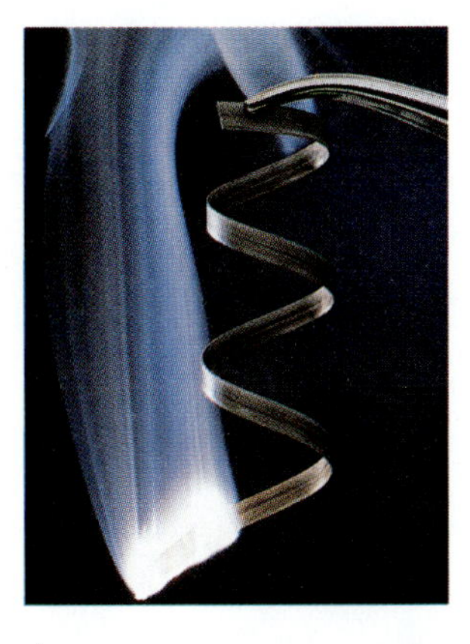

▲ Magnesium metal burns in air by combining with oxygen gas to produce a brilliant white light and a "smoke" of solid, white magnesium oxide.

Magnesium, a Group 2A metal, is harder and less reactive than sodium. Magnesium reacts with oxygen, a Group 6A element (a colorless gas), to form another stable white crystalline solid called magnesium oxide (MgO).

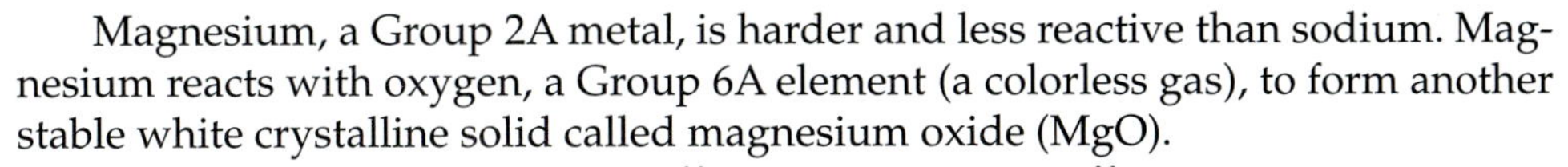

$$\cdot\mathrm{Mg}\cdot + \cdot\ddot{\underset{\cdot}{\mathrm{O}}}: \longrightarrow \mathrm{Mg}^{2+} + :\ddot{\underset{..}{\mathrm{O}}}:^{2-}$$

Magnesium must give up two electrons and oxygen must gain two electrons for each to have the same configuration as the noble gas neon with an outer-shell octet.

An atom such as oxygen, which needs two electrons to complete a noble gas configuration, may react with potassium atoms, each of which has only one electron to give. In this case, two atoms of potassium are needed for each oxygen atom. The product is potassium oxide (K_2O). By this process, each potassium atom achieves the argon configuration and the oxygen atom assumes the neon configuration.

Example 3.4

Use Lewis structures to show the transfer of electrons from magnesium atoms to nitrogen atoms to form ions with noble gas configurations.

Solution

$$\begin{matrix} \cdot\mathrm{Mg}\cdot & & & \mathrm{Mg}^{2+} & \\ & \cdot\ddot{\underset{\cdot}{\mathrm{N}}}\cdot & & & :\ddot{\underset{..}{\mathrm{N}}}:^{3-} \\ \cdot\mathrm{Mg}\cdot\ + & & \longrightarrow & \mathrm{Mg}^{2+}\ + & \\ & \cdot\ddot{\underset{\cdot}{\mathrm{N}}}\cdot & & & :\ddot{\underset{..}{\mathrm{N}}}:^{3-} \\ \cdot\mathrm{Mg}\cdot & & & \mathrm{Mg}^{2+} & \end{matrix}$$

Each of the three magnesium atoms gives up two electrons (a total of six), and each of the two nitrogen atoms acquires three electrons (a total of six). Notice that the total positive and total negative charges on the products are equal (6+ and 6−). Magnesium reacts with nitrogen to produce magnesium nitride (Mg_3N_2).

Practice Exercises

A. Use Lewis structures to show the transfer of electrons from aluminum atoms to oxygen atoms to form ions with noble gas configurations.

B. Use Lewis structures to show the transfer of electrons from calcium atoms to phosphorus atoms to form ions with noble gas configurations.

Generally speaking, metallic elements in Groups 1A, 2A, and 3A of the periodic table react with nonmetallic elements in Groups 5A, 6A, and 7A to form stable crystalline ionic solids. The metals tend to give up electrons to the nonmetals. The ions so formed have noble gas electronic configurations. The crystalline ionic solids are held together by the attraction of oppositely charged ions—that is, by ionic bonds.

3.5 Names of Simple Ions and Ionic Compounds

Names of simple monatomic (single-atom) *positive* ions (*cations*) are derived from those of their parent elements by addition of the word *ion*. A sodium atom (Na), upon losing an electron, becomes a *sodium ion* (Na^{+}). A magnesium atom (Mg), upon losing two electrons, becomes a *magnesium ion* (Mg^{2+}). The name of a simple monatomic *negative* ion *(anion)* is derived from that of the parent element by changing the usual ending to *-ide* and adding the word *ion*. A chlor*ine* atom (Cl), upon gaining an electron, becomes a chlor*ide ion* (Cl^{-}). A sulf*ur* atom (S), upon gaining two electrons, becomes a sulf*ide ion* (S^{2-}).

Names and symbols for several important monatomic ions are given in Table 3.2. Note that the charge on an ion of a Group 1A element is 1+ (usually written simply as +). The charge on an ion of a Group 2A element is 2+, and that on an ion of a Group 3A element is 3+. You can calculate the charges on the negative ions in Table 3.2 by subtracting 8 from the group number. For example, the charge on an oxide ion (oxygen is in Group 6A) is $6 - 8 = 2-$. The charge on a nitride ion (nitrogen is in Group 5A) is $5 - 8 = 3-$. The periodic relationship of these monatomic ions is shown in Figure 3.4.

There is no simple way to determine the most likely charge on ions formed from elements in B subgroups. Indeed, you may have noticed that these elements can form ions with different charges. In such cases, chemists use Roman numerals with the names, to indicate the charge. Thus, *iron(II) ion* means Fe^{2+}, and *iron(III) ion* means Fe^{3+}.

Table 3.2 Symbols and Names for Some Monatomic Ions

Group	Element	Name	Symbol for Ion
1A	Hydrogen	Hydrogen ion[a]	H^{+}
	Lithium	Lithium ion	Li^{+}
	Sodium	Sodium ion	Na^{+}
	Potassium	Potassium ion	K^{+}
2A	Magnesium	Magnesium ion	Mg^{2+}
	Calcium	Calcium ion	Ca^{2+}
3A	Aluminum	Aluminum ion	Al^{3+}
5A	Nitrogen	Nitride ion	N^{3-}
6A	Oxygen	Oxide ion	O^{2-}
	Sulfur	Sulfide ion	S^{2-}
7A	Fluorine	Fluoride ion	F^{-}
	Chlorine	Chloride ion	Cl^{-}
	Bromine	Bromide ion	Br^{-}
	Iodine	Iodide ion	I^{-}
1B	Copper	Copper(I) ion (cuprous ion)	Cu^{+}
		Copper(II) ion (cupric ion)	Cu^{2+}
	Silver	Silver ion	Ag^{+}
2B	Zinc	Zinc ion	Zn^{2+}
8B	Iron	Iron(II) ion (ferrous ion)	Fe^{2+}
		Iron(III) ion (ferric ion)	Fe^{3+}

[a]Does not exist independently in aqueous solution.

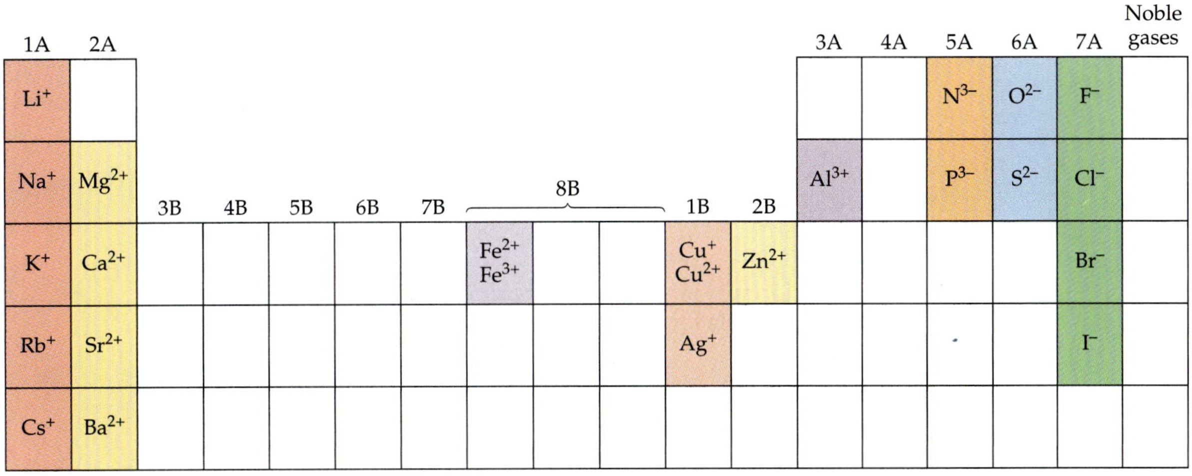

1A	2A	3B	4B	5B	6B	7B	8B	8B	8B	1B	2B	3A	4A	5A	6A	7A	Noble gases
Li^+														N^{3-}	O^{2-}	F^-	
Na^+	Mg^{2+}											Al^{3+}		P^{3-}	S^{2-}	Cl^-	
K^+	Ca^{2+}						Fe^{2+} Fe^{3+}			Cu^+ Cu^{2+}	Zn^{2+}					Br^-	
Rb^+	Sr^{2+}									Ag^+						I^-	
Cs^+	Ba^{2+}																

▲ **Figure 3.4**
The periodic relationships of some monatomic ions.

Formulas for Ionic Compounds

Compounds such as sodium chloride, potassium bromide, magnesium oxide, and potassium oxide are called **ionic compounds.** The constituent units of these compounds are charged particles—ions—but the compound as a whole is electrically neutral. We can use this principle of electrical neutrality to determine the combining ratio of ions. Potassium ions (K^+) combine with bromide ions (Br^-) in a ratio of 1:1. The formula KBr expresses this ratio and represents the compound potassium bromide. The combining ratio (1:1) and the ionic charges are understood.

Let's try another example. One calcium ion (Ca^{2+}) combines with *two* chloride ions (Cl^-). This ratio is expressed in the formula $CaCl_2$. In this formula, the ionic charges are understood. As with a coefficient of 1 in algebra, a subscript of 1 is understood, but higher numbers are written explicitly. Thus, $CaCl_2$ not only gives us the combining ratio (1:2) but also stands for the ionic compound calcium chloride. It is a shorthand way of writing 1 Ca^{2+} and 2 Cl^-.

You can use the charges on the ions in Table 3.2 to determine formulas for compounds of these elements. You can predict the charges on ions formed from elements in the A groups by using the periodic table.

Example 3.5

What is the formula for sodium sulfide?

Solution

First, write the symbols for the ions (positive ion first). Sodium is in Group 1A; the charge on the sodium ion is 1+. Sulfur is in Group 6A, and the charge on the sulfide ion is $6 - 8 = 2-$. The symbols are Na^+ and S^{2-}. The smallest number into which both charges can be evenly divided—that is, the *least common multiple* (LCM)—is two. The least common multiple simply indicates the smallest number of electrons that can be evenly exchanged between the two elements. The subscript for each symbol can be determined by division of its charge (without the plus or minus) into the least common multiple. This step determines how many atoms of each element are needed to supply or accept the smallest common number of electrons. For Na^+,

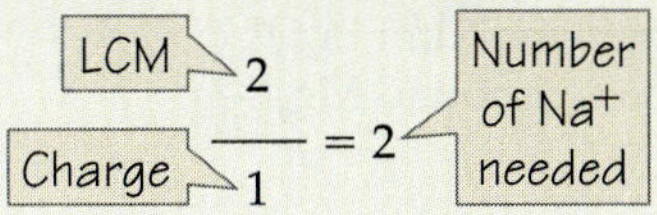

$$\frac{2 \text{ (LCM)}}{1 \text{ (Charge)}} = 2 \text{ (Number of } Na^+ \text{ needed)}$$

For S^{2-},

$$\frac{2 \text{ (LCM)}}{2 \text{ (Charge)}} = 1 \text{ (Number of } S^{2-} \text{ needed)}$$

Thus, we have the formula Na_2S_1 (2 Na^+ and 1 S^{2-}), or Na_2S.

Practice Exercises

A. What is the formula for magnesium bromide?
B. What is the formula for sodium nitride?

Example 3.6

What is the formula for aluminum oxide?

Solution

The symbols are Al^{3+} and O^{2-} (Al is in Group 3A, and O is in Group 6A). The LCM is 6. The number of Al^{3+} needed is $6 \div 3 = 2$, and the number of O^{2-} needed is $6 \div 2 = 3$. The formula is therefore Al_2O_3 (2 Al^{3+} and 3 O^{2-}).

Practice Exercises

A. What is the formula for calcium nitride?
B. What is the formula for aluminum phosphide?

We can arrive at the same result by using the crossover method: the charge number for one ion becomes the subscript for the other. Thus, in Example 3.6, the charge on the aluminum ion is 3, which becomes the subscript for oxygen in aluminum oxide; and the charge on oxygen is 2, which becomes the subscript for the aluminum ion.

Note that the crossover method works because it is based on the transfer of electrons and the conservation of charge. Two Al atoms lose three electrons each (that's six electrons lost), and three O atoms gain two electrons each (that's six electrons gained). Electrons lost equal electrons gained, and all is well.

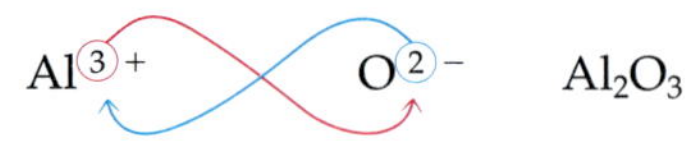

Thus, two aluminum ions have six positive charges, and three oxide ions have six negative charges. The resulting compound, Al_2O_3, is therefore neutral—as all compounds are.

Names for Binary Ionic Compounds

Naming these binary (two-element) ionic compounds is simple. First write the name of the positive ion and then the name of the negative ion. (The word ion is not used; it is understood in each case.)

Example 3.7

What is the name of the compound Na_2S?

Solution

Find the constituent ions in Table 3.2. They are sodium ion (Na^+) and sulfide ion (S^{2-}). The compound is sodium sulfide.

Practice Exercises

A. What is the name of the compound Li_2O?
B. What is the name of the compound Mg_3P_2?

Example 3.8

a. What is the name of the compound FeS?
b. What is the name of the compound $FeCl_3$?

Solution

a. There are two kinds of iron ions. Because sulfur exists as the S^{2-} ion and one iron ion is combined with it, the iron ion in this compound must be Fe^{2+}. The name of FeS is iron(II) sulfide.
b. Because the charge on the chloride ion is 1− and three of these ions are combined with one iron ion, the iron ion must be Fe^{3+}. The name of $FeCl_3$ is iron(III) chloride.

Practice Exercises

A. What is the name of the compound $CuBr_2$?
B. What is the name of the compound Fe_2O_3?

An older system of naming ions is occasionally used, based on the Latin stems and using "ous" and "ic" endings. The "ous" indicates the lower of two possible charges and "ic" denotes the higher charge: Fe^{2+} is *ferrous* ion and Fe^{3+} is *ferric* ion. See similar names for the two copper ions in Table 3.2.

Ionic compounds generally exist as crystalline solids. However, many of them are soluble in water. Ions are found dissolved in all natural waters—including the water in the cells of our bodies, where they are involved in such critical functions as the transmission of nerve impulses.

3.6 Covalent Bonds: Shared Electron Pairs

One might expect a hydrogen atom, with its one electron, to acquire another electron and assume the helium configuration. Indeed, hydrogen atoms do just that in the presence of atoms of a reactive metal such as lithium—that is, a metal that readily gives up an electron.

$$\text{Li}\cdot + \cdot\text{H} \longrightarrow \text{Li}^+ + :\text{H}^-$$

But what if there are no other kinds of atoms around? What if there are only hydrogen atoms? One hydrogen atom can't gain an electron from another because hydrogen atoms all have an equal attraction for electrons. They do combine, however, by *sharing a pair* of electrons.

$$\text{H}\cdot + \cdot\text{H} \rightarrow \text{H:H}$$

Recall (Chapter 2) that a **molecule** is a group of atoms that are chemically bonded together. Molecules are represented by chemical formulas. The symbol H represents an atom of hydrogen; the formula H_2 represents a *molecule* of hydrogen, which is composed of two hydrogen atoms.

By sharing electrons, the two hydrogen atoms form a hydrogen molecule. The bond formed by a shared pair of electrons is called a **covalent bond.**

Consider next the case of chlorine. A chlorine atom will pick up an extra electron from an atom of a reactive metal such as sodium. But, again, if the only atom around is another chlorine atom, chlorine atoms too can attain a more stable arrangement by sharing a pair of electrons in a covalent bond. The two shared electrons are called a **bonding pair.** The other electrons that stay on one atom and are not shared are called **lone pairs** or *nonbonding pairs.*

Chlorine atoms — Chlorine molecule

$$:\ddot{\underset{\cdot\cdot}{\text{Cl}}}\cdot + \cdot\ddot{\underset{\cdot\cdot}{\text{Cl}}}: \longrightarrow :\ddot{\underset{\cdot\cdot}{\text{Cl}}}:\ddot{\underset{\cdot\cdot}{\text{Cl}}}:$$

Each chlorine atom in the chlorine *molecule* has eight electrons around it, an arrangement like that of the noble gas argon. Most of the covalently bonded atoms that we shall consider, except hydrogen, follow the **octet rule.**

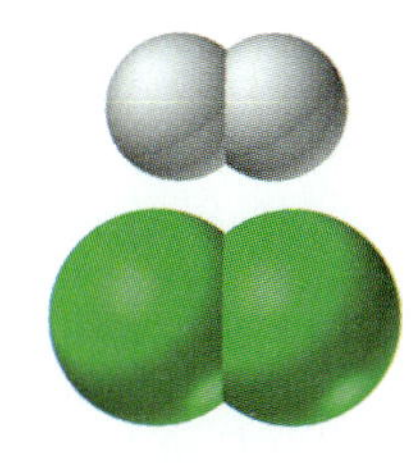

▲ Molecular models of H_2 (top) and Cl_2 (bottom).

▲ Some familiar foods with a high Na^+ content.

Some manufacturers enrich their cereals with very fine specks of iron filings. These iron filings dissolve in the acidic environment of the stomach, producing iron(II) ions.

What Is a Low-Sodium Diet?

People with high blood pressure are usually advised to follow a low-sodium diet. Just what does this mean? Surely they are not being advised to reduce their consumption of sodium metal. Sodium is an extremely reactive metal that reacts violently with moisture; eating it wouldn't be safe. The concern is really with sodium *ions*, Na^+. Most people in the United States consume much more Na^+ than they need, most of it from sodium chloride, common table salt. It is not uncommon for some individuals to eat 6 or 7 g of sodium chloride a day, most of it in prepared foods. Many snack foods, such as potato chips, pretzels, and corn chips, are especially high in salt. The American Heart Association recommends that adults limit their salt intake to about 3 g per day.

Note the important difference between ions and the atoms from which they are made. A metal atom and its cation are as different as a whole peach (atom) and a peach pit (ion). The names and symbols look a lot alike, but the species themselves are quite different. Unfortunately, the situation is confused because people talk about needing "iron" to perk up "tired blood" and "calcium" for healthy teeth and bones. What they really mean is iron(II) *ions* (Fe^{2+}) and calcium *ions* (Ca^{2+}). You wouldn't think of eating iron nails to get "iron." Nor would you eat highly reactive calcium metal. Many people do not always make careful distinctions, but we will try to use precise terminology here.

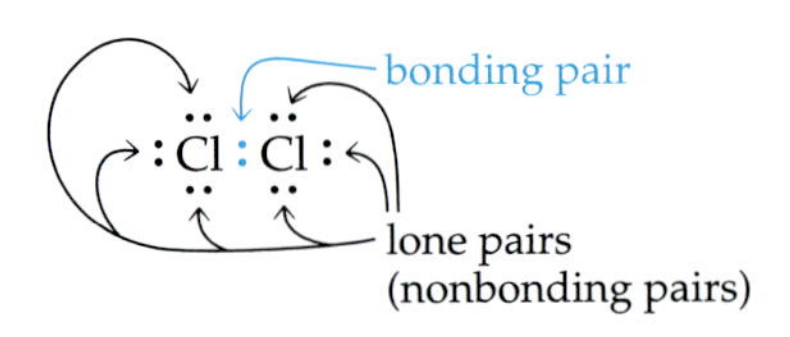

For simplicity, the hydrogen molecule is often represented as H_2 and the chlorine molecule as Cl_2. In each case, the covalent bond between the atoms is understood. Sometimes the covalent bond is indicated by a dash, H—H and Cl—Cl. Nonbonding pairs of electrons often are not shown.

Review Questions

3.5 Use Lewis structures to show the sharing of electrons between two iodine atoms to form an iodine molecule. Label all electron pairs as bonding or lone pairs.

3.6 How does a chlorine molecule differ in structure from a chloride ion? How do the two species differ in chemical reactivity?

3.7 Multiple Covalent Bonds

A shared single pair of electrons is called a **single bond.** Some atoms can share more than one pair. In carbon dioxide, for example, the carbon atom shares two pairs of electrons with each of the two oxygen atoms.

Double bond

:Ö::C::Ö: (or O=C=O)

Note that each atom has an octet of electrons about it as a result of this sharing. These atoms are joined by a **double bond,** a covalent linkage in which the two atoms share two pairs of electrons.

To obtain an octet of electrons, an oxygen atom must share electrons with *two* hydrogen atoms, a nitrogen atom must share electrons with *three* hydrogen atoms, and a carbon atom must share electrons with *four* hydrogen atoms. In general, many nonmetals often form a number of covalent bonds equal to eight minus the group number.

- Oxygen, which is in Group 6A, forms 8 − 6 = 2 covalent bonds in most compounds.
- Nitrogen, in Group 5A, forms 8 − 5 = 3 covalent bonds in most of its compounds.
- Carbon, in Group 4A, forms 8 − 4 = 4 covalent bonds in most carbon compounds, including the millions of organic compounds.

▲ Molecular models of CO_2 (top) and N_2 (bottom).

Atoms also can share three pairs of electrons. Nitrogen can share a pair of electrons with another nitrogen atom to look like this:

$$:\dot{\underset{\cdot}{N}}\cdot \; + \; \cdot\dot{\underset{\cdot}{N}}: \longrightarrow \; :\dot{\underset{\cdot}{N}}:\dot{\underset{\cdot}{N}}: \qquad \textit{(incorrect structure)}$$

To satisfy the octet rule, each nitrogen atom shares *three* pairs of electrons with the other.

Triple bond

$$:N:::N: \quad (\text{or } :N\vdots\vdots N: \text{ or } N{\equiv}N)$$

The atoms are joined by a **triple bond,** a covalent linkage in which two atoms share three pairs of electrons. Each nitrogen atom also has a lone pair of electrons. Note that we could have drawn the lone pair above or below the atomic symbol. Such a drawing would represent the same molecule.

Carbon atoms form a total of four bonds. These can be (1) four single bonds, as in methane (page 89) (2) two double bonds, as in carbon dioxide (below); (3) two single and one double bond, as in formaldehyde (Chapter 15); or (4) one triple and one single bond, as in acetylene (Chapter 13). Similarly, nitrogen's three bonds may be (1) three single bonds, as in ammonia (page 89); (2) one single and one double bond, as in purine (Chapter 17); or (3) one triple bond, as in the N_2 molecule (below). The two bonds of an oxygen atom may be (1) two single bonds; as in water (page 89); or (2) one double bond, as in carbon dioxide.

Covalent bonds often are represented as dashes. We can therefore show the three kinds of bonds as follows.

Cl—Cl O=C=O N≡N

✓ Review Questions

3.7 How many covalent bonds do each of the following usually form? You may refer to the periodic table.

a. H **b.** C **c.** O **d.** F **e.** N **f.** Br

3.8 Names for Covalent Compounds

Many covalent compounds have common and widely used names. Examples are water (H_2O), methane (CH_4), and ammonia (NH_3). For other compounds, the prefixes *mono-, di-, tri-,* and so on are used to indicate the number of atoms of each element in the molecule. A list of these prefixes for up to 10 atoms is given in Table 3.3. For example, the compound N_2O_4 is called *dinitrogen tetroxide.* (The *a* often is dropped from tetra- and other prefixes when it precedes another vowel.) We often leave off the mono- when it is a prefix of the first word (NO_2 is nitrogen dioxide), but do include it with the second word to distinguish between two compounds of the same pair of elements (CO is carbon monoxide; CO_2 is carbon dioxide).

Table 3.3 Prefixes That Indicate the Number of Atoms of an Element in a Compound

Prefix	Number of Atoms
Mono-	1
Di-	2
Tri-	3
Tetra-	4
Penta-	5
Hexa-	6
Hepta-	7
Octa-	8
Nona-	9
Deca-	10

Example 3.9

What is the name of SCl_2? Of SF_6?

Solution

With one sulfur atom and two chlorine atoms, SCl_2 is sulfur dichloride. With one sulfur atom and six fluorine atoms, SF_6 is sulfur hexafluoride.

Practice Exercise

What is the name of BrF_3? Of BrF_5?

Example 3.10

Give the formula for (**a**) carbon tetrachloride and (**b**) tetraphosphorus hexoxide.

Solution

a. The name indicates one carbon atom and four chlorine atoms. The formula is CCl_4.
b. The name indicates four phosphorus atoms and six oxygen atoms. The formula is P_4O_6.

Practice Exercises

A. Give the formula for dinitrogen pentoxide.
B. Give the formula for tetraphosphorus triselenide. (The symbol for selenium is Se.)

3.9 Unequal Sharing: Polar Covalent Bonds

So far, we have seen that atoms combine in two different ways. Some that are quite different in electronic structure (from opposite ends of the periodic table) react by the complete transfer of an electron from one atom to another to form an ionic bond. Atoms that are identical combine by sharing one or more pairs of electrons to form covalent bonds. Now let's consider bond formation between atoms that are different, but not different enough to form ionic bonds.

Hydrogen and chlorine react to form a colorless gas called hydrogen chloride. Both hydrogen and chlorine need an electron to achieve a noble gas configuration, so they share a pair and form a covalent bond.

Both hydrogen and chlorine consist of diatomic molecules; the reaction is more accurately represented as

$$\text{H:H} + \text{:}\ddot{\underset{..}{\text{Cl}}}\text{:}\ddot{\underset{..}{\text{Cl}}}\text{:} \longrightarrow 2\,\text{H:}\ddot{\underset{..}{\text{Cl}}}\text{:}$$

$$\text{H}\cdot + \cdot\ddot{\underset{..}{\text{Cl}}}\text{:} \longrightarrow \text{H:}\ddot{\underset{..}{\text{Cl}}}\text{:}$$

Example 3.11

Use Lewis structures to show the formation of a covalent bond between (**a**) two fluorine atoms and (**b**) a fluorine atom and a hydrogen atom.

Solution

a. $\text{:}\ddot{\underset{..}{\text{F}}}\cdot + \cdot\ddot{\underset{..}{\text{F}}}\text{:} \longrightarrow \text{:}\ddot{\underset{..}{\text{F}}}\text{:}\ddot{\underset{..}{\text{F}}}\text{:}$

b. $\text{:}\ddot{\underset{..}{\text{F}}}\cdot + \cdot\text{H} \longrightarrow \text{H:}\ddot{\underset{..}{\text{F}}}\text{:}$

Practice Exercises

A. Use Lewis structures to show the formation of a covalent bond between (**a**) two bromine atoms, and (**b**) a hydrogen atom and a bromine atom.
B. Use Lewis structures to show the formation of a covalent bond between an iodine atom and a chlorine atom.

Why should the hydrogen molecule and the chlorine molecule react at all? Molecules form to provide a more stable arrangement of electrons. But there is a broad range of stabilities. The chlorine molecule represents a more stable arrangement than two separate chlorine atoms. However, by forming a bond with a hydrogen atom, a still more stable arrangement is achieved.

For convenience and simplicity, we can represent the reaction of a hydrogen molecule and a chlorine molecule to form a hydrogen chloride molecule as

$$H_2 + Cl_2 \longrightarrow 2\,HCl$$

Molecules of hydrogen chloride consist of one atom of hydrogen and one atom of chlorine. These unlike atoms share a pair of electrons. Sharing does not mean sharing equally, though. Chlorine atoms have a greater attraction for a shared pair of electrons than do hydrogen atoms; they are more *electronegative* than hydrogen. Thus the chlorine atom holds the shared electrons more tightly, and this results in the chlorine end of the molecule being more negative than the hydrogen end (see Section 3.10).

In a covalent bond, two atoms with equal electronegativities share the electron pair equally, and the electrons are no closer to one atom than to the other. A bond of this type is a **nonpolar covalent bond.** The H—H and Cl—Cl bonds are nonpolar. Atoms of different electronegativities share the electron pair in a covalent bond unequally, and the electrons are drawn closer to the atom of higher electronegativity. Such a bond is a **polar covalent bond.** The H—Cl bond is a polar bond. The polar covalent bond is *not* an ionic bond[1]. In an ionic bond, there is a complete transfer of an electron from the metal to the nonmetal. In a polar covalent bond, the atom at the positive end of the bond (H in HCl) still has some share in the bonding pair of electrons.

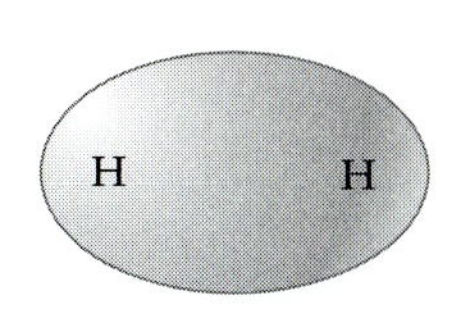

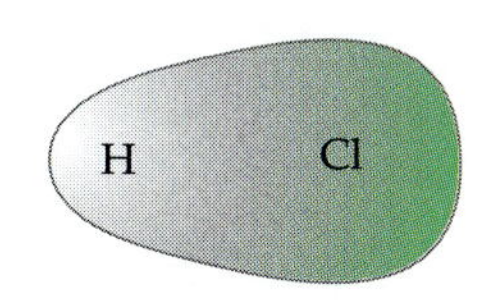

▲ **Figure 3.5**
Nonpolar and polar covalent bonds. In H—H, there is an even distribution of electronic charge density between the atoms. In H—Cl, (not to scale), electron charge density is displaced toward the Cl atom.

Let's picture the electron-pair bond between two atoms as a cloud of negative electric charge that encompasses both atoms. (Modern quantum theory permits us to do this, just as we described atomic orbitals as electron clouds in Section 2.5.) In a *nonpolar* bond such as H—H, the electron cloud density is greatest between the two hydrogen nuclei but is otherwise uniformly distributed (Figure 3.5). Although the greatest electron density is still found between the two nuclei in a *polar* bond such as H—Cl, the cloud is strongly displaced toward the more electronegative chlorine.

To indicate the polar nature of a bond, we use the representation

$$\overset{\delta+}{\text{H}}\text{—}\overset{\delta-}{\text{Cl}}$$

The δ+ and δ− (read "delta plus" and "delta minus") indicate that one end (H) is partially positive and one end (Cl) partially negative. The term *partial charge* signifies something less than the full charges of the ions that would result from complete electron transfer. This unequal sharing of electrons has a marked effect on the properties of a compound.

Hydrogen chloride is a gas that dissolves readily in water. The aqueous solution formed is called hydrochloric acid (sometimes muriatic acid). This acid is used for, among other things, cleaning toilet bowls and removing excess mortar from new brick buildings. Hydrochloric acid is also the well-known "stomach acid." Acids are defined and further discussed in Chapter 9.

✓ Review Questions

3.8 List the differences between nonpolar covalent bonds and polar covalent bonds.

3.9 List the differences between polar covalent bonds and ionic bonds.

3.10 Use Lewis structures to show the sharing of electrons between a hydrogen atom and a fluorine atom. Label the ends of the molecule with symbols that indicate polarity.

3.10 Electronegativity

Let's take a closer look at the concept of *electronegativity.* The **electronegativity** of an atom (element) is a measure of its tendency to attract electrons in a covalent bond to itself.

[1]There is really a continuum, rather than a sharp division between ionic and covalent bonds. Chemists use a *difference* in electronegativity values (see Figure 3.6) to classify bond types. If the difference is 1.8 or more, the bonds are said to be ionic, and if the difference is 1.7 or less, covalent. A difference of ~0.0 means the bonds are nonpolar covalent.

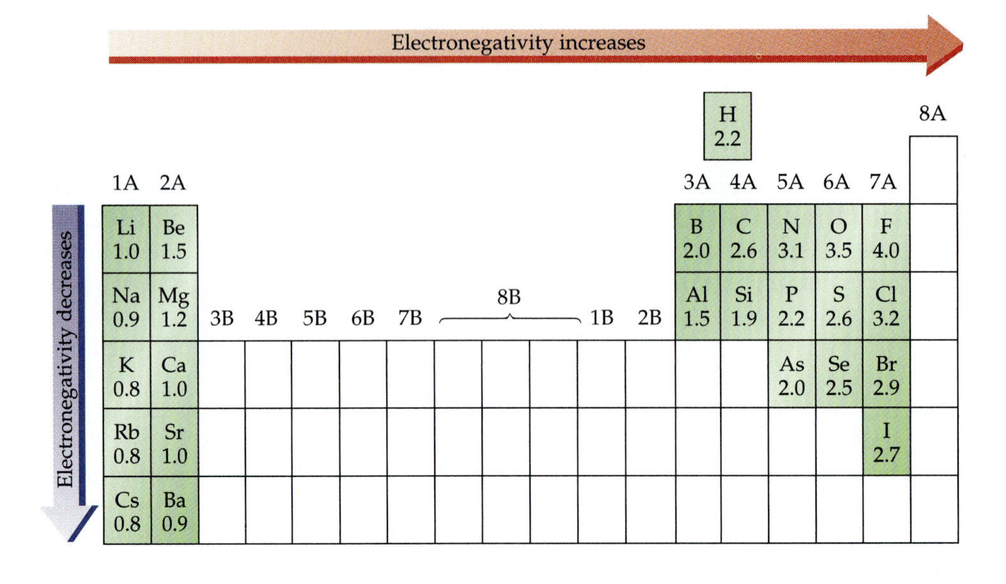

Figure 3.6 ▶ Electronegativities of some common elements.

The greater the electronegativity of an atom in a molecule, the more strongly the atom attracts the electrons in a covalent bond.

The nonmetal elements at the upper right in the periodic table have the greatest electronegativities. The metal elements on the left have the smallest electronegativities. We can use the periodic table to predict trends in electronegativities.

Within a period, elements generally become more electronegative from left to right. Within a group, electronegativity decreases from top to bottom.

Thus chlorine is less electronegative than fluorine, and sulfur is less electronegative than oxygen. Hydrogen, a nonmetal, does not fit well into this scheme. Although usually placed in Group 1A, it is not much like the alkali metals. Its electronegativity is between that of boron and that of carbon, and thus we place it above these two elements in Figure 3.6.

3.11 Rules for Writing Lewis Structures

Recall that electrons are transferred (Section 3.4) or shared (Section 3.6) in ways that leave most atoms with an octet of electrons in the valence shell. In this section we describe the procedure we can follow in writing Lewis structures for molecules. First we must put the atoms of the molecules in their proper places.

The *skeletal structure* of a molecule tells us the order in which the atoms are attached to one another. In the absence of experimental evidence, the following rules help us to devise likely skeletal structures.

1. Hydrogen atoms form only one bond; they are shown at the end of a sequence of atoms. Hydrogen often is bonded to carbon, nitrogen, or oxygen.

```
    H                                      H
    |                                      |
H—C—H      H—N—H      H—O      H—C—O—H
    |           |          |         |
    H           H          H         H
```

2. Polyatomic molecules and ions often consist of a central atom surrounded by more electronegative atoms. (Hydrogen is an exception; it is always on the outside, even when bonded to a more electronegative element.)

```
                      O              Cl
                      |              |
O—C—O              O—S—O         Cl—C—Cl      (incomplete structures)
                                     |
                                     Cl
```

After we choose a skeletal formula for a polyatomic molecule or ion, we can use the following steps to write a Lewis formula.

Step 1: Calculate the total number of valence electrons. The total for a molecule is the sum of the valence electrons for all the atoms. Example: N_2O_4 has $(2 \times 5) + (4 \times 6) = 34$ valence electrons. For a polyatomic anion, add the number of negative charges. Example: NO_3^- has $5 + (3 \times 6) + 1 = 24$ valence electrons. For a polyatomic cation, subtract the number of positive charges. Example: NH_4^+ has $5 + (4 \times 1) - 1 = 5 + 4 - 1 = 8$ valence electrons.

Step 2: Write the skeletal structure, and connect bonded pairs of atoms by a dash (one electron pair).

Step 3: Place electrons about outer atoms so that each (except hydrogen) has an octet. (See also Section 3.15.)

Step 4: Subtract the number of electrons assigned so far from the total calculated in Step 1. Assign any remaining electrons in pairs to the central atom(s).

Step 5: If a central atom has fewer than eight electrons after Step 4, a multiple bond is likely. Move one or more lone pairs from an outer atom to the space between the atoms to form a double or triple bond. A deficiency of two electrons suggests a double bond, and a shortage of four electrons indicates a triple bond or two double bonds to the central atom.

Example 3.12

Give the Lewis structure for (**a**) methanol, CH_4O, and (**b**) nitrogen trifluoride, NF_3.

Solution

a. *Step 1:* The total number of valence electrons is $4 + (4 \times 1) + 6 = 14$.

Step 2: The skeletal structure in which all the hydrogen atoms are on the outside and the least electronegative atom, carbon, is most central is

```
    H
    |
H—C—O—H
    |
    H
```

Step 3: Now, counting each bond as two electrons gives ten electrons. The four remaining electrons are placed (as two lone pairs) on the oxygen atom.

```
    H
    |   ..
H—C—O—H
    |   ..
    H
```

(The remaining steps are not necessary; both carbon and oxygen have an octet of electrons.)

b. *Step 1:* There are 5 + (3 × 7) = 26 valence electrons.
Step 2: The skeletal structure is

```
F—N—F
  |
  F
```

Step 3: Place three lone pairs on each fluorine atom.

```
 ..     ..
:F—N—F:
 ¨  |   ¨
   :F:
    ¨
```

Step 4: We have assigned 24 electrons. Place the remaining two as a lone pair on the nitrogen atom.

```
 ..  ..  ..
:F—N—F:
 ¨   |   ¨
    :F:
     ¨
```

(Each atom has an octet; Step 5 is not needed.)

Practice Exercises

A Give the Lewis structure for ethyl chloride, C_2H_5Cl.
B Give the Lewis structure for oxygen difluoride, OF_2.

Example 3.13

Give the Lewis structure for (**a**) the BF_4^- ion and (**b**) carbon dioxide, CO_2.

Solution

a. There are 3 + (4 × 7) + 1 = 32 electrons in the BF_4^- ion.
Step 2: The skeletal structure is

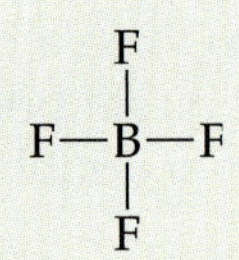

Step 3: Place three lone pairs on each fluorine atom.

```
     ..
    :F:
 ..  |   ..
:F—B—F:
 ¨   |   ¨
    :F:
     ¨
```

Step 4: We have assigned all 32 electrons. (Step 5 is not needed.)

b. *Step 1:* There are 4 + (2 × 6) = 16 valence electrons in CO_2.
Step 2: The skeletal structure is simply O—C—O.
Step 3: Place three lone pairs on each oxygen atom.

```
 ..     ..
:O—C—O:
 ¨      ¨
```

Step 4: We have assigned all 16 electrons.
Step 5: The carbon atom has only four electrons. It needs to form two double bonds in order to have an octet. (There is no reason to expect that carbon would form a triple bond to one of the oxygen atoms.) Move a lone pair from each oxygen atom to the space between the atoms to form a double bond on each side of the carbon atom.

```
 ..     ..
:O=C=O:
```

Practice Exercises

A. Give the Lewis structure for the PH_4^+ ion.
B. Give the Lewis structure for nitryl fluoride, NO_2F.

3.12 The VSEPR Theory

Although molecules are often represented in two dimensions on paper, they have three-dimensional shapes. Shape can determine polarity and other properties. The shapes of biologically active molecules (Chapters 19 to 23) are of utmost importance. They can carry out their vital functions only if they have the right groups of atoms in the right places. The shapes that we consider in this book are shown in Figure 3.7.

We can predict the shapes of many molecules by a simple procedure called the **valence-shell electron-pair repulsion (VSEPR)** theory. The basis of the VSEPR theory is that electron pairs will arrange themselves about a central atom in a way that minimizes repulsion between the like-charged particles. This means that they will get as far apart as possible. Table 3.4 uses ball-and-stick molecular models to show the geometric shapes associated with the arrangement of two, three, or four electron groups about a central atom.

The farthest apart two substituent atoms can get is the distance between the opposite sides of the central atom at an angle of 180°. Three groups assume a triangular arrangement about the central atom, forming angles of separation of 120°. Four groups form a tetrahedral array around the central atom, with a separation of about 109°.

You can determine the shapes of many molecules and polyatomic ions by following these simple rules.

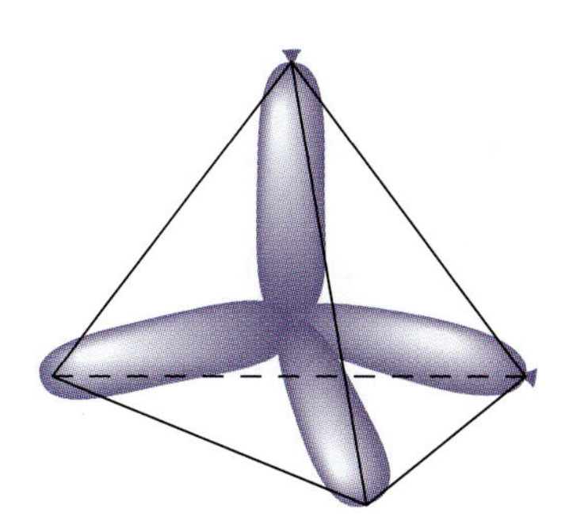

▲ Valence-shell electron-pair repulsion can be pictured using balloons. The lobes of the balloons represent electron pairs. Like electron pairs, the lobes are directed toward the corners of a tetrahedron.

Step 1: Draw a Lewis structure. In the structure, indicate a shared electron pair (**bonding pair, BP**) by a line. Use dots to indicate any **lone pairs (LPs)** of electrons.

Step 2: To determine the shape, we use the steric (spatial arrangement) number. Steric number = no. of atoms bonded to the central atom + no. of lone pairs on the central atom
Consider the following examples.

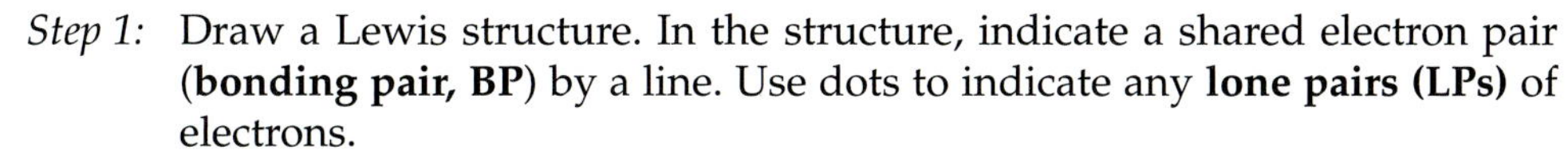

Steric number:	2	2	3	3

Step 3: Determine the steric number and draw a shape *as if* all were bonding pairs.

Step 4: Sketch that shape, placing the electron pairs as far apart as possible (see Table 3.4). If there is *no* lone pair, that is the shape of the molecule. If there

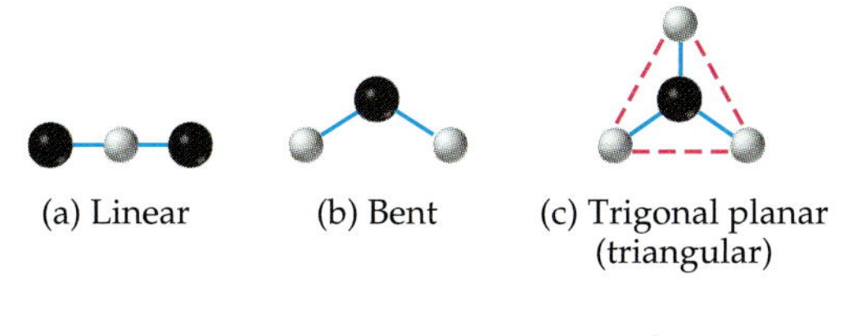

(a) Linear (b) Bent (c) Trigonal planar (triangular)

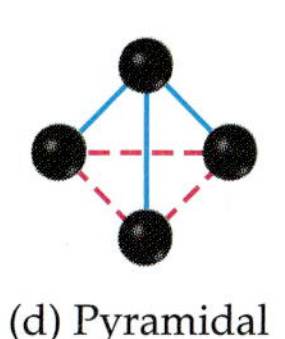

(d) Pyramidal

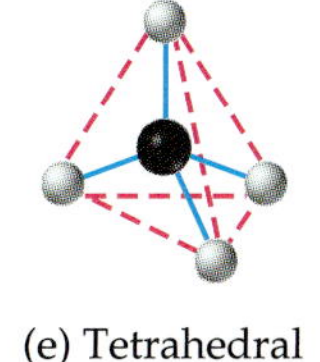

(e) Tetrahedral

◄ **Figure 3.7**
Shapes of molecules. In a *linear* triatomic molecule (a), all the atoms are along a line; the bond angle is 180°. The *bent* or angular molecule (b) has an angle less than 180°. Connecting the three outer atoms of the *triangular* molecule (c) with imaginary lines forms a triangle with an atom at the center, an arrangement called *trigonal planar.* Imaginary lines connecting all four atoms of a *pyramidal* molecule (d) form a three-sided pyramid. Connecting the four outer atoms of a *tetrahedral* molecule (e) with imaginary lines produces a tetrahedron with an atom at the center. (A tetrahedron is a four-sided figure in which each side is a triangle.)

Table 3.4 Bonding and the Shapes of Molecules

Number of Bonded Atoms	Number of Lone Pairs	Steric Number	Molecular Shape	Examples			
2	0	2	Linear	$BeCl_2$	$HgCl_2$	CO_2	$BeCl_2$
3	0	3	Trigonal planar (triangular)	BF_3	$AlBr_3$	CH_2O	BF_3
4	0	4	Tetrahedral	CH_4	CBr_4	$SiCl_4$	CH_4
3	1	4	Pyramidal	NH_3	PCl_3		NH_3
2	2	4	Bent (angular)	H_2O	H_2S	SCl_2	H_2O
2	1	3	Bent (angular)	SO_2	O_3		SO_2

are lone pairs, ignore them, leaving the bonding pairs exactly as they were. (This may seem strange, but it stems from the fact that both bonding electrons and lone pairs *determine* the shape, but that shape of the molecule is only the arrangement of bonded atoms.)

Example 3.14

What is the shape of (**a**) the BH_3 molecule and (**b**) the SCl_2 molecule?

Solution

a. *Step 1:* The Lewis structure is

```
 H
 |
 B—H
 |
 H
```

Step 2: The steric number is *three.* The three bonding pairs get as far apart as possible, forming a triangular arrangement of the sets.

H
B—H
H 120°

Step 3: There are no lone pairs; the molecular shape is trigonal planar, the same as the arrangement of the electrons.

b. *Step 1:* The Lewis structure is

:S̤̈—C̤̈l:
|
:C̤̈l:

Step 2: The steric number is *four,* made up of two chlorine atoms and two lone pairs. The four electron pairs get as far apart as possible, forming a tetrahedral arrangement of the electron pairs about the central atom.

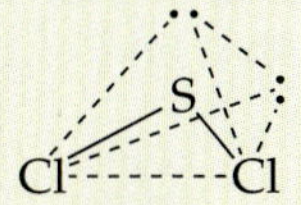

Step 3: Ignore the lone pairs; the molecular shape is bent or angular, with a bond angle of 109.5°.

Practice Exercises

A. What is the shape of the BeH_2 molecule?
B. What is the shape of the PH_3 molecule?

3.13 Polar and Nonpolar Molecules

Diatomic molecules are polar if their bonds are polar and nonpolar if their bonds are nonpolar.

$$\overset{\delta+}{\text{H}}\text{—}\overset{\delta-}{\text{Cl}} \qquad \text{Cl—Cl}$$

Polar Nonpolar

For molecules with three or more atoms, we also must consider the orientation of the bonds to determine whether or not the molecule as a whole is polar.

Water is one of the most familiar chemical substances. Electrolysis experiments (Section 11.4) and other ample evidence indicate that its molecular formula is H_2O. This arrangement completes the valence-shell octet of the oxygen atom, giving it the neon electronic configuration. It also completes the valence shell of the hydrogen atoms, each of which now has the helium electronic configuration.

We should expect the *bonds* in water to be polar because oxygen is more electronegative (Section 3.10) than hydrogen. Just containing polar bonds, however, does not mean that the molecule as a whole is polar. If the atoms in the water molecule were in a linear arrangement, the two polar bonds would cancel one another.

$$\overset{\delta+}{\text{H}}\text{—}\overset{\delta-}{\text{O}}\text{—}\overset{\delta+}{\text{H}} \qquad (\textit{incorrect structure})$$

Instead of one end of the molecule being positive and the other end negative, the electrons would be pulled toward the right in one bond and toward the left in the other. Overall there would be no net dipole. A molecule is a **dipole** if it has a positive end and a negative end.

Dipoles often are represented by an arrow with a plus at the tail end.

$$\overset{+\!\!\longrightarrow}{\text{H—Cl}}$$

The positive end of the arrow is obvious; the head of the arrow indicates the negative end of the dipole. Using these arrows, we see that carbon dioxide is nonpolar despite its polar bonds.

$$\overset{\longleftarrow\!\!+\;+\!\!\longrightarrow}{\text{O=C=O}}$$

The two dipoles cancel one another out.

We show that a water molecule *is* a dipole by placing a sample of water between two electrically charged plates. As shown in Figure 3.8, the water molecules align themselves, one end attracted toward the positive plate and the other end toward the negative plate. To be a dipole, the water molecule must be bent so that the bonds do not cancel one another out. Both dipoles point toward the oxygen, resulting in a net dipole toward that end of the molecule.

$\delta+\text{H}:\ddot{\text{O}}:\delta-$ / $\text{H}\delta+$ or $\delta+\text{H}\text{—O}\delta-$ / $\text{H}\delta+$; H–O–H (bent, dipole arrows toward O) or H–O–H (net dipole arrow)

This shape of the water molecule can be accounted for by a modification of the VSEPR theory. According to VSEPR, the two bonds and two lone pairs should form a tetrahedral arrangement (Figure 3.9). Ignoring the lone pairs, the molecular shape would have the atoms in a bent arrangement, with a bond angle of 109.5° (the tetrahedral angle).

The predicted bond angle of 109.5° for water is a bit larger than the measured angle of 104.5°. The difference is explained by the fact that the lone pairs occupy a greater volume than do the bonding pairs (BPs). These larger orbitals push the smaller BPs closer together.

Ammonia (NH_3) is a gas at room temperature. It is made in vast quantities; most of it is compressed into tanks and used as fertilizer. Ammonia dissolves readily in water, forming a basic solution (Chapter 9). Such aqueous ammonia solutions are familiar household cleaning products.

A nitrogen atom has five electrons in its valence shell. It can assume the neon configuration by sharing electrons with *three* hydrogen atoms. The result is the compound ammonia. In ammonia there are three BPs and one lone pair about the nitrogen atom. The VSEPR theory predicts a tetrahedral arrangement with bond angles of 109.5°. The actual bond angles are 107°, close to the theoretical value. Presumably, the lone pair of electrons occupies a greater volume than does a bonding pair, pushing the latter slightly closer together. The arrangement is therefore a tripod with a hydrogen atom at the end of each leg and the nitrogen atom with its lone pair sitting at the top (Figure 3.10). Each nitrogen–hydrogen bond is somewhat polar, making the ammonia molecule polar..

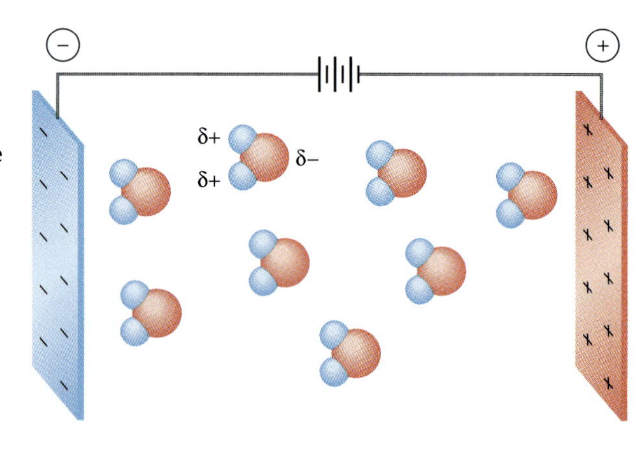

Figure 3.8 ► Polar molecules are aligned in an electric field, with the positive end of the molecule preferentially pointing toward the negative plate and the negative end of the molecule directed toward the positive plate.

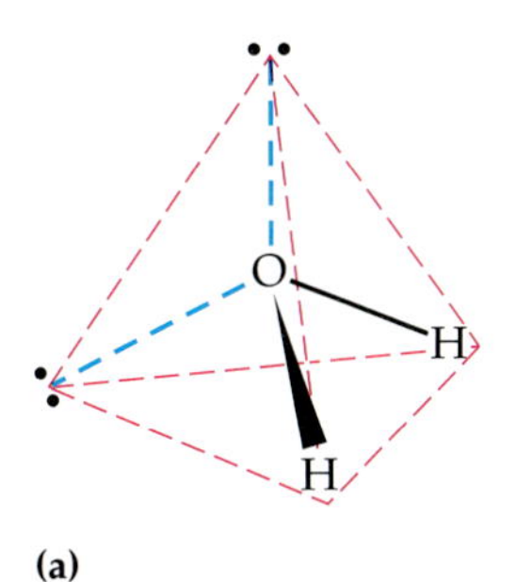

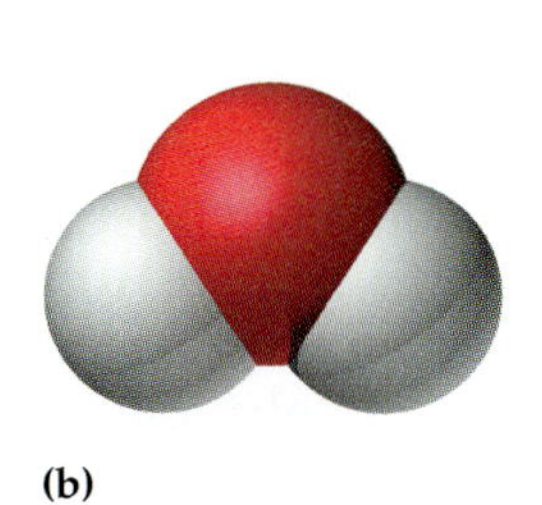

(a) (b)

▲ **Figure 3.9**
The water molecule. In the drawing (a), solid lines indicate covalent bonds, and the dashed lines outline the tetrahedron. The lone pairs of electrons are ignored in determining the *bent* shape of the molecule. The photograph (b) shows a space-filling model of a water molecule.

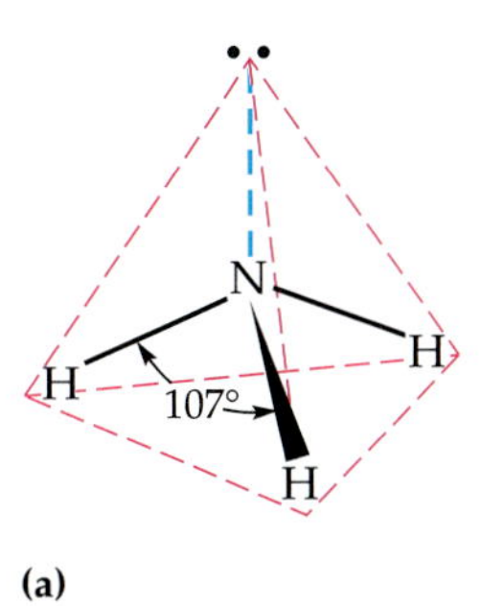

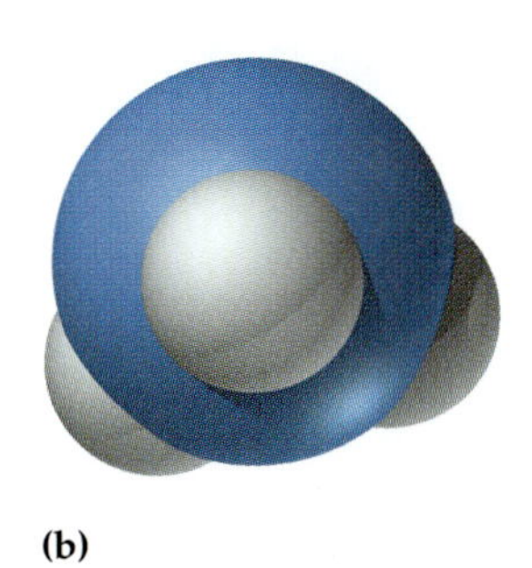

(a) (b)

▲ **Figure 3.10**
The ammonia molecule. In the drawing (a), solid lines indicate covalent bonds, and the dashed lines outline the tetrahedron. The lone pair of electrons is ignored in determining the *pyramidal* shape of the molecule. The photograph (b) shows a space-filling model of an ammonia molecule.

A carbon atom has four electrons in its valence shell. It can assume the neon configuration by sharing electrons with four hydrogen atoms, forming the compound methane.

A methane molecule has four pairs of electrons on the central carbon atom. Using the VSEPR theory, we would expect a tetrahedral arrangement and bond angles of 109.5°. The actual bond angles are 109.5°, in perfect agreement with theory (Figure 3.11). Each of the four electron pairs is shared with a hydrogen atom, and they occupy identical volumes. Each carbon–hydrogen bond is slightly polar, but the methane molecule as a whole is symmetrical. The slight bond polarities cancel out, leaving the methane molecule nonpolar.

Methane (CH_4) is the simplest of the hydrocarbons, a group of organic compounds discussed in detail in Chapter 13. It is the principal component of natural gas.

Many of the properties of compounds—such as melting point, boiling point, and solubility—depend on the polarity of the molecules of the compound.

Bonding: Some Additional Concepts

Now that we have explored the two major types of bonding, ionic and covalent, we can extend our study to somewhat more complex substances and develop some additional concepts that are essential to an understanding of the chemistry that lies ahead.

3.14 Polyatomic Ions

Many compounds contain both ionic and covalent bonds. Sodium hydroxide, commonly known as lye, consists of sodium ions (Na^+) and hydroxide ions (OH^-). The hydroxide ion contains an oxygen atom covalently bonded to a hydrogen atom,

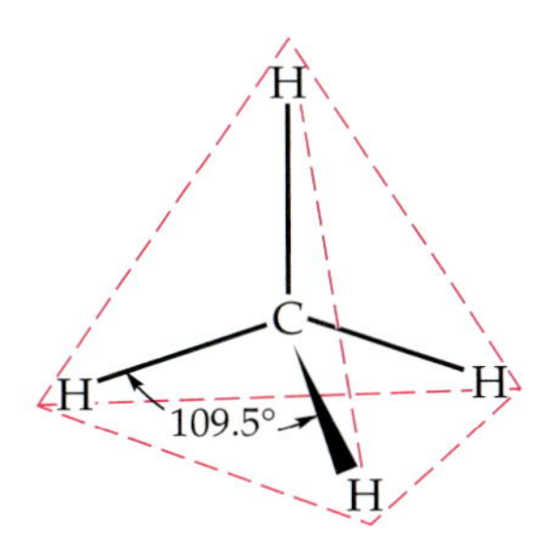

(a) (b)

◄ **Figure 3.11**
The methane molecule. In the drawing (a), solid lines indicate covalent bonds and the dashed lines outline the tetrahedron. All bond angles are 109.5°. The photograph (b) shows a space-filling model of a methane molecule.

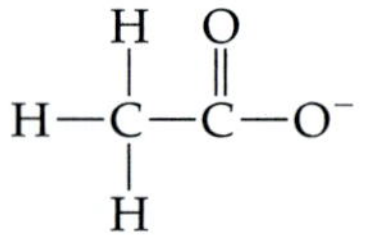

plus an "extra" electron. Whereas the sodium atom becomes a cation by giving up an electron, the hydroxide group becomes an anion by gaining an electron.

$$e^- + \cdot\ddot{\underset{\cdot\cdot}{O}}\cdot + \cdot H \longrightarrow [:\ddot{\underset{\cdot\cdot}{O}}:H]^-$$

The formula for sodium hydroxide is NaOH; for each sodium ion there is one hydroxide ion.

There are many groups of atoms that (like the hydroxide ion) remain together through most chemical reactions. A **polyatomic ion** is a charged particle containing two or more covalently bonded atoms (Figure 3.12). A list of common polyatomic ions is given in Table 3.5. You can use these ions, in combination with the monatomic ions in Table 3.2, to determine formulas for compounds that contain polyatomic ions.

Acetate ion

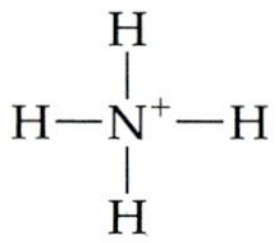

Ammonium ion

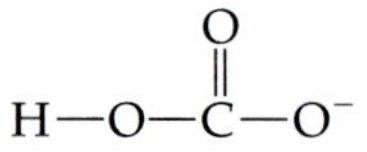

Hydrogen carbonate ion (Bicarbonate ion)

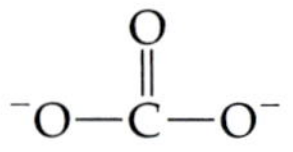

Carbonate ion

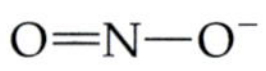

Nitrite ion

▲ **Figure 3.12**
Polyatomic ions have both covalent bonds (dashes) that hold the atoms together as a group and ionic charges (+ or −) that allow them to interact with other ions to form compounds.

Example 3.15

What are the formulas for (**a**) sodium sulfate and (**b**) ammonium sulfide?

Solution

a. First, write the formula for the ions.

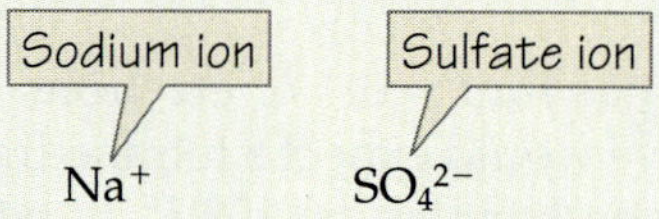

Sodium ion: Na^+ Sulfate ion: SO_4^{2-}

Using the crossover method,

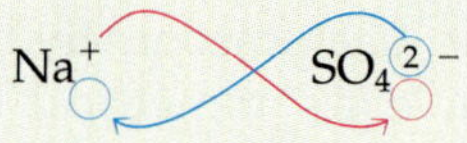

Na^+ SO_4^{2-}

we get

$$Na_2^+(SO_4)_1^{2-}$$

Table 3.5 Some Common Polyatomic Ions

Charge	Name	Formula
1+	Ammonium ion	NH_4^+
	Hydronium ion	H_3O^+
1−	Hydrogen carbonate (bicarbonate) ion	HCO_3^-
	Hydrogen sulfate (bisulfate) ion	HSO_4^-
	Acetate ion	$CH_3CO_2^-$ (or $C_2H_3O_2^-$)
	Nitrite ion	NO_2^-
	Nitrate ion	NO_3^-
	Cyanide ion	CN^-
	Hydroxide ion	OH^-
	Dihydrogen phosphate ion	$H_2PO_4^-$
	Permanganate ion	MnO_4^-
2−	Carbonate ion	CO_3^{2-}
	Sulfate ion	SO_4^{2-}
	(Mono)hydrogen phosphate ion	HPO_4^{2-}
	Oxalate ion	$C_2O_4^{2-}$
	Dichromate ion	$Cr_2O_7^{2-}$
3−	Phosphate ion	PO_4^{3-}

Then, dropping the charges, we have $Na_2(SO_4)_1$ or simply Na_2SO_4

b. The ions are

Ammonium ion | Sulfide ion

NH_4^+ S^{2-}

Crossing over, we get

$$(NH_4^+)_2S_1^{2-}$$

Dropping the charges gives

$$(NH_4)_2S$$

The parentheses with a subscript 2 indicate that the entire ammonium unit is taken twice; there are two nitrogen atoms and $4 \times 2 = 8$ hydrogen atoms.

Practice Exercises

A. What is the formula for potassium phosphate?
B. What is the formula for calcium acetate?

Example 3.16

What are the names of (**a**) the compound NaCN and (**b**) KH_2PO_4?

Solution

a. The ions are Na^+ (sodium ion) and CN^- (cyanide ion). The name of the compound is sodium cyanide.
b. The ions are K^+ (potassium ion) and $H_2PO_4^-$ (dihydrogen phosphate ion). The name is potassium dihydrogen phosphate.

Practice Exercises

A. What is the name of the compound $CaCO_3$?
B. What is the name of $K_2Cr_2O_7$?

3.15 Exceptions to the Octet Rule

Many molecules made of atoms of the main-group elements have electron structures that follow the octet rule. There are many exceptions, however, and they fall into three main groups. Each type is readily identified by some structural characteristic.

Molecules with Odd Numbers of Valence Electrons

Molecules with odd numbers of valence electrons obviously cannot satisfy the octet rule. Examples include nitrogen monoxide (NO, also called nitric oxide), with $5 + 6 = 11$ valence electrons; nitrogen dioxide (NO_2), with 17 valence electrons; and chlorine dioxide (ClO_2), which has 19 outer electrons. One of the atoms in each of these molecules will have an odd number of electrons and therefore cannot have an octet.

Atoms and molecules with unpaired electrons are called **free radicals.** Most free radicals are highly reactive and have only transitory existence as intermediates in chemical reactions. An example is the chlorine atom that is formed from the breakdown of chlorofluorocarbons in the stratosphere and that leads to the depletion of the ozone shield. Some free radicals are quite stable, however. The nitrogen oxides are major components of smog. Chlorine dioxide is made in vast quantities and is used for bleaching flour, paper, and other materials.

Any atom or molecule with an odd number of electrons *must* have one unpaired electron. Filled shells and sublevels have all their electrons paired, with two electrons in each orbital (Section 2.5). We need only consider valence electrons to determine whether or not an atom or molecule is a free radical. Lewis formulas for NO, NO_2, and ClO_2 show that one atom of each has an unpaired electron; that atom obviously does not have an octet of electrons in its valence shell.

:N::O: :O:N::O:

:O:Cl:O:

Molecules with Incomplete Octets

Boron atoms have three valence electrons; fluorine atoms have seven. When boron reacts with fluorine, it shares those electrons with three fluorine atoms to form boron trifluoride.

```
  ..
 :F:
  |  ..
  B—F:
  |  ..
 :F:
  ..
```

The bond between boron and nitrogen, in which the nitrogen atom furnishes both the electrons, is called a *coordinate covalent bond.* Once formed, a coordinate covalent bond is exactly like any other bond; it differs only in the way it is formed.

This structure uses all 24 of the valence electrons; there are none left to put on the central boron atom. Experimental evidence indicates that this structure is consistent with the reactivity of BF_3 toward molecules with lone pairs. For instance, BF_3 reacts readily with ammonia to form BF_3NH_3, a compound in which all atoms (except hydrogen) have octets of electrons.

```
     ..                         ..
    :F:     H                  :F: H
 ..  |      |               ..  |  |
:F—B   +  :N—H    ⟶       :F—B—N—H
 ..  |      |               ..  |  |
    :F:     H                  :F: H
     ..                         ..
```

The second-period elements carbon, nitrogen, oxygen, and fluorine nearly always obey the octet rule. (The odd-electron compounds are obvious exceptions.) The valence electron level of the second-period elements holds a maximum of eight electrons ($2s^22p^6$).

Expanded Valence Shells

The third main shell can hold up to 18 electrons ($3s^23p^63d^{10}$). Third-period elements therefore can violate the octet rule by having more than eight electrons in the valence level. These so-called **expanded valence shells** are evident in the following compounds.

```
      ..                     ..  ..  ..
    :Cl:  ..               :F  :F:  F:
      |  /Cl:               ..\ | /..
 ..   | /  ..                   S
:Cl—P      ..               ../ | \..
 ..   | \Cl:               :F  :F:  F:
    :Cl:  ..                ..  ..  ..
      ..
```

Phosphorus pentachloride   Sulfur hexafluoride

Generally we use the octet rule except in cases where it obviously doesn't apply: when there is an odd number of electrons, when there are too few electrons to make an octet, and (third period and beyond) where there obviously must be more than eight electrons in the valence shell.

Elements in the third period and beyond, then, are not limited to an octet. Yet many of their compounds still follow the octet rule.

✓ Review Question

3.11 List three main types of compounds that are exceptions to the octet rule. Give an example of each.

Summary

Chemical bonds are the forces that hold atoms together in molecules. The electrons in the outermost shell (valence shell) of an atom are called **valence electrons.** An atom that *gains* one or more valence electrons becomes an *anion* and carries a net negative charge. One that *loses* one or more valence electrons is a *cation* and carries a net positive charge.

Valence electrons are shown as dots in **Lewis structures.** For main-group elements, the number of valence electrons is equal to the group number shown on the periodic table.

The noble gas elements (Group 8A) are generally unreactive. Each has a valence electronic configuration ns^2np^6 (except helium; $1s^2$), which seems to be especially unreactive. The **octet rule** states that an atom participating in a chemical reaction tends to acquire a set of eight electrons in its valence shell.

A sodium atom has one valence electron and tends to give it up easily. A chlorine atom has seven valence electrons and readily accepts the lone sodium valence electron. The result is a sodium cation, Na^+, with a neon electronic configuration, and a chloride anion, Cl^-, that has an argon configuration. The positive charge and negative charge on these ions attract each other, and, as a result, Na^+ and Cl^- ions become arranged in a crystal. The forces holding these positive and negative species together in the crystal are called **ionic bonds,** and any compound held together by ionic bonds is an **ionic compound.**

A **covalent bond** is formed when two or more atoms share electrons, and the resulting compound is a *covalent compound.* The electrons participating in the covalent bond are a **bonding pair,** and nonbonding valence-electron pairs are **lone pairs.** As with ionic bonding, covalent bond formation follows the octet rule. A **single bond** has one bonding pair of electrons; a **double bond,** two bonding pairs; and a **triple bond,** three bonding pairs.

Electronegativity is a measure of a bonded atom's tendency to attract electrons to itself. Generally electronegativity increases from left to right across a row in the periodic table and decreases from top to bottom down a column.

When the atoms involved in a covalent bond have identical electronegativity values, the electrons are shared equally by the two atoms and the bond is **nonpolar.** When a covalent bond is formed from two atoms having different electronegativity values, the electrons are pulled closer to the more electronegative atom and the bond is **polar.**

A nonmetal atom generally forms a number of covalent bonds equal to eight minus the group number of the element. Covalently bonded species that carry a net charge are called **polyatomic ions.** Such ions can form ionic compounds with an ion carrying an opposite charge.

We considered three exceptions to the octet rule. (1) Molecules in which the total number of valence electrons is an odd number. (Any molecule containing such an unpaired electron is called a **free radical.**) (2) Molecules in which the total number of valence electrons is too low to allow all atoms to have a filled valence shell. (3) Molecules involving elements from the third and higher periods can have an **expanded valence shell.**

We can use the VSEPR theory to predict the shapes of many molecules. Table 3.4 summarizes the geometric shapes associated with the arrangement of two, three, or four electron groups about a central atom.

Key Terms

bonding pair (BP) (3.6)
chemical bond (introduction)
covalent bond (3.6)
crystal (3.3)
dipole (3.13)
double bond (3.7)
electronegativity (3.10)
expanded valence shell (3.15)
free radical (3.15)
ion (3.1)
ionic bond (3.3)
ionic compound (3.5)
Lewis structure (3.2)
lone pair (LP) (3.6)
molecule (3.6)
nonpolar covalent bond (3.9)
octet rule (3.6)
polar covalent bond (3.9)
polyatomic ion (3.14)
single bond (3.7)
triple bond (3.7)
valence electron (3.2)
valence-shell electron-pair repulsion (VSEPR) theory (3.12)

Problems

You may use a periodic table unless otherwise instructed.

Lewis Structures: Elements

1. Write Lewis structures for each of the following elements.
   a. sodium **b.** oxygen
   c. fluorine **d.** aluminum

2. Write Lewis structures for each of the following elements.
   a. carbon **b.** potassium
   c. magnesium **d.** chlorine

Monatomic Ions

3. Show the formation of an ion from (**a**) barium and (**b**) bromine.

4. Show the formation of an ion from (a) aluminum and (b) sulfur.
5. What is the charge on monatomic ions formed from (a) atoms of Group 2A elements and (b) atoms of Group 7A elements?
6. What is the charge on monatomic ions formed from (a) atoms of Group 1A elements and (b) atoms of Group 6A elements?

Lewis Structures: Ionic Compounds

7. Show the transfer of electrons (a) from calcium to bromine atoms and (b) from magnesium to sulfur atoms. In each case, form ions with noble gas electronic configurations.
8. Show the transfer of electrons (a) from aluminum to sulfur atoms and (b) from magnesium to phosphorus atoms. In each case, form ions with noble gas electronic configurations.
9. Draw Lewis structures for each of the following.
   a. Ca and Ca^{2+} b. S and S^{2-}
   c. Rb and Rb^{+} d. P and P^{3-}
10. Draw Lewis structures for each of the following.
   a. Al and Al^{3+} b. Br and Br^{-}
   c. Mg and Mg^{2+} d. O^{2-} and Ne
11. Give Lewis structures for each of the following.
   a. sodium fluoride b. potassium chloride
   c. sodium oxide d. calcium chloride
   e. magnesium bromide
12. Give Lewis structures for each of the following.
   a. potassium fluoride b. magnesium iodide
   c. potassium sulfide d. sodium nitride
   e. aluminum oxide

Naming Ions and Ionic Compounds

13. Name each of the following.
   a. Na^{+} b. Mg^{2+} c. Al^{3+}
   d. Cl^{-} e. O^{2-} f. N^{3-}
14. Name each of the following.
   a. K^{+} b. Ca^{2+} c. Zn^{2+}
   d. Br^{-} e. Li^{+} f. S^{2-}
15. Name each of the following.
   a. Fe^{3+} b. Cu^{2+} c. Ag^{+}
16. Name each of the following.
   a. Fe^{2+} b. Cu^{+} c. I^{-}
17. Give symbols for each of the following.
   a. bromide ion b. calcium ion
   c. potassium ion d. iron(II) ion
18. Give symbols for each of the following.
   a. sodium ion b. aluminum ion
   c. oxide ion d. copper(II) ion
19. Name each of the following.
   a. NaBr b. $CaCl_2$ c. $FeCl_2$
   d. LiI e. K_2S f. CuBr
20. Name each of the following.
   a. KCl b. $MgBr_2$ c. CuI_2
   d. CaS e. $FeCl_3$ f. Al_2O_3
21. Name each of the following.
   a. CO_3^{2-} b. HPO_4^{2-} c. MnO_4^{-} d. OH^{-}
22. Name each of the following.
   a. NO_3^{-} b. SO_4^{2-} c. $H_2PO_4^{-}$ d. HCO_3^{-}
23. Give formulas for each of the following.
   a. ammonium ion b. hydrogen sulfate ion
   c. cyanide ion d. nitrite ion
24. Give formulas for each of the following.
   a. phosphate ion b. hydrogen carbonate ion
   c. dichromate ion d. oxalate ion
25. Give formulas for each of the following.
   a. magnesium sulfate
   b. sodium hydrogen carbonate
   c. potassium nitrate
   d. calcium monohydrogen phosphate
26. Give formulas for each of the following.
   a. calcium carbonate
   b. potassium dihydrogen phosphate
   c. magnesium cyanide
   d. lithium hydrogen sulfate
27. Give formulas for each of the following.
   a. iron(II) phosphate b. potassium dichromate
   c. copper(I) iodide d. ammonium nitrite
28. Give formulas for each of the following.
   a. iron(III) oxalate
   b. sodium permanganate
   c. copper(II) bromide
   d. zinc monohydrogen phosphate

Covalent Bonds and Molecules

29. Use Lewis structures to show the sharing of electrons between the indicated atoms to form a molecule in which each atom (except hydrogen) has an octet of valence electrons.
   a. P and H b. C and F
30. Use Lewis structures to show the sharing of electrons between the indicated atoms to form a molecule in which each atom (except hydrogen) has an octet of valence electrons.
   a. Si and H b. N and Cl
31. Give formulas for each of the following.
   a. dinitrogen monoxide
   b. tetraphosphorus trisulfide
   c. phosphorus pentachloride
   d. sulfur hexafluoride
32. Give formulas for each of the following.
   a. oxygen difluoride
   b. dinitrogen pentoxide
   c. phosphorus tribromide
   d. tetrasulfur tetranitride
33. Name each of the following.
   a. CS_2 b. N_2S_4 c. PF_5 d. S_2F_{10}
34. Name each of the following.
   a. CBr_4 b. Cl_2O_7 c. P_4S_{10} d. I_2O_5

Electronegativity

35. If atoms of the two elements in each set below are joined by a covalent bond, which atom will more strongly attract the electrons in the bond?
    a. N and S b. B and Cl c As and F
36. Using only the periodic table (inside front cover), indicate which element in each set is more electronegative.
    a. Br or F b. Br or Se c. Cl or As
37. Without referring to figures or tables in the text, arrange each of the following sets of atoms in order of increasing electronegativity.
    a. B, F, N b. As, Br, Ca c. C, O, Ga
38. Without referring to figures or tables in the text, arrange each of the following sets of atoms in order of increasing electronegativity.
    a. I, Rb, Sb b. Cs, Li, Na c. Cl, P, Sb
39. Classify the following bonds as ionic or covalent. For those bonds that are covalent, indicate whether they are polar or nonpolar.
    a. KF b. IBr c. MgS d. F_2
40. Classify the following bonds as ionic or covalent. For those bonds that are covalent, indicate whether they are polar or nonpolar.
    a. NO b. CaO c. NaBr d. Br_2

Lewis Structures

41. Give Lewis structures that follow the octet rule for each of the following.
    a. CH_4O b. CH_2O c. NOH_3
    d. N_2H_4 e. COF_2 f. PCl_3
42. Give Lewis structures that follow the octet rule for each of the following.
    a. NF_3 b. C_2H_2 c. C_2H_4
    d. CH_5N e. H_2SiO_3 f. HCN
43. Give a Lewis structure for each of the following.
    a. NO b. BeI_2 c. PCl_5
44. Give a Lewis structure for each of the following.
    a. BCl_3 b. PF_5 c. $AlBr_3$

Polyatomic Ions and Their Compounds

45. Give Lewis structures that follow the octet rule for each of the following.
    a. ClO^- b. ClO_2^- c. HPO_4^{2-} d. BrO_3^-
46. Give Lewis structures that follow the octet rule for each of the following.
    a. CN^- b. IO_4^- c. PO_3^{3-} d. HSO_4^-
47. Name each of the following.
    a. KNO_2 b. LiCN c. NH_4I
    d. $NaNO_3$ e. $KMnO_4$ f. $CaSO_4$
48. Name each of the following.
    a. $NaHSO_4$ b. $Al(OH)_3$ c. Na_2CO_3
    d. $KHCO_3$ e. NH_4NO_2 f. $Ca(HSO_4)_2$
49. Name each of the following.
    a. Na_2HPO_4 b. $(NH_4)_3PO_4$
    c. $Al(NO_3)_3$ d. NH_4NO_3
50. Name each of the following.
    a. Li_2CO_3 b. $Na_2Cr_2O_7$
    c. $Ca(H_2PO_4)_2$ d. $(NH_4)_2C_2O_4$

VSEPR Theory

51. Use the VSEPR theory to predict the shape of each molecule.
    a. beryllium chloride ($BeCl_2$)
    b. boron chloride (BCl_3)
    c. oxygen difluoride (OF_2)
    d. silicon tetrachloride ($SiCl_4$)
52. Use the VSEPR theory to predict the shape of each molecule.
    a. arsine (AsH_3)
    b. carbon tetrafluoride (CF_4)
    c. silane (SiH_4)
    d. hydrogen selenide (H_2Se)
53. Use the VSEPR theory to predict the shape of each molecule.
    a. nitrogen trichloride (NCl_3)
    b. sulfur dichloride (SCl_2)
54. Use the VSEPR theory to predict the shape of each molecule.
    a. phosphorus trifluoride (PF_3)
    b. dichlorodifluoromethane (CCl_2F_2)

Polar and Nonpolar Molecules

55. Classify the covalent bonds as polar or nonpolar.
    a. H—O b. N—Cl c. C—C
56. Classify the covalent bonds as polar or nonpolar.
    a. Si—Cl b. Cl—Cl c. P—Cl
57. Use the symbols $\delta+$ and $\delta-$ to indicate partial charges, if any, on the bonds in Problem 55.
58. Use the symbols $\delta+$ and $\delta-$ to indicate partial charges, if any, on the bonds in Problem 56.
59. The molecule BeF_2 is linear. Is it polar or nonpolar? Explain.
60. The molecule SF_2 is bent. Is it polar or nonpolar? Explain.
61. Use differences in electronegativity values to arrange each of the following sets of bonds in order of increasing polarity.
    a. Cl—F, F—F, H—F b. H—Br, H—F, H—H
62. Use differences in electronegativity values to arrange each of the following sets of bonds in order of increasing polarity.
    a. H—C, H—F, H—H, H—N, H—O
    b. C—Br, C—C, C—Cl, C—F, C—I
63. Use the symbols $\delta+$ and $\delta-$ to indicate partial charges, if any, on the bonds in Problem 61.
64. Use the symbols $\delta+$ and $\delta-$ to indicate partial charges, if any, on the bonds in Problem 62.

Additional Problems

65. Fill in this table, assuming that elements W, X, Y, and Z are all main-group elements. Use the first column (W) as an example.

Element	W	X	Y	Z
Group Number	7A	1A	___	___
Lewis Symbol	:Ẅ:	___	·Y·	___
Charge on Ion	1−	___	___	2−

66. Consider the hypothetical elements X, Y, and Z with the Lewis symbols:

:Ẍ· :Ÿ· :Ż·

a. To which group in the periodic table would each belong?

b. Write the Lewis structure for the simplest compound of each element with hydrogen.

c. Write Lewis structures for the ions formed when X and Y react with sodium.

67. Explain why SO_2 is a polar molecule whereas SO_3 is not.

68. Draw a charge-cloud picture for the HF molecule. Use the symbols δ+ and δ− to indicate the polarity of the molecule.

69. Predict whether or not each of the following species is probable. For any species that seems improbable, tell why it is so.

a. a linear water molecule, H_2O

b. a planar molecule, SO_3

c. a planar molecule, PH_3

d. a tetrahedral molecule, $GeCl_4$

e. a bent molecule, HCN

70. Predict the molecular shape of each of the following.

a. PCl_3 **b.** ClO_4^- **c.** OCN^-
d. PH_4^+ **e.** NI_3 **f.** Cl_2CO

71. Consider the following statement: The greater the electronegativity difference between the atoms in a molecule, the greater the resultant dipole moment of that molecule. Is this a valid statement? Explain.

72. Potassium is a soft silvery metal that reacts violently with water and ignites spontaneously in air. Your doctor recommends you take a potassium supplement. Why would it be a mistake to take potassium metal? What would you take?

73. Explain what is wrong with the statement "A crystal of ordinary salt is comprised of an enormous number of NaCl molecules in a highly ordered three-dimensional network."

74. Draw acceptable Lewis structures for (**a**) two different molecules having the formula C_2H_6O and (**b**) for two substances having the formula S_2F_2.

75. Give the formula for (**a**) chlorine dioxide, which is used to bleach flour and (**b**) tetraphosphorus trisulfide, which is used in the tips of "strike anywhere" matches.

76. The gas phosphine (PH_3) is used as a fumigant to protect stored grain and other durable produce from pests. Phosphine is generated in the storage area by adding water to aluminum phosphide or magnesium phosphide. Give formulas for the two phosphides.

Chemical Reactions

The seeming "magic" of chemistry is demonstrated in this reaction in which two colorless liquids combine to produce a brilliant yellow solid. (An aqueous solution of potassium chromate poured into aqueous lead nitrate gives solid lead chromate as the product.)

Learning Objectives/Study Questions

1. Why is a balanced chemical equation important? What strategies are used in balancing a chemical equation?
2. What volume relationships are found in gaseous chemical reactions? How did Avogadro explain the law of combining volumes?
3. What is the difference between molecular mass, formula mass, and molar mass?
4. What is the mole and how is it related to Avogadro's number?
5. How are mole and mass relationships used in chemical formulas and equations?
6. What is the difference between an endothermic and an exothermic reaction?
7. What is the importance of knowing the activation energy for a reaction?
8. How do changes in concentration of reactants, temperature changes, and the presence of a catalyst influence the rates of chemical reactions?
9. What is meant by chemical equilibrium?
10. How is Le Châtelier's principle used to predict changes in equilibrium systems?

The complex chemical processes in the living cell involve changes in chemical composition and changes in energy. Some reactions provide the energy that keeps the cell alive and well. Other reactions, vital to life processes, require an input of energy.

Chemical reactions proceed at rates ranging from explosively fast to exceedingly slow. Rates are affected by a number of factors. For living organisms, perhaps the most important factors are complex molecules called enzymes that accelerate reaction rates enormously. Yet, strange as it may seem, the enzymes are still unchanged *after* doing their jobs (see Chapter 22).

Chemical reactions also proceed to different extents. In some, the reactants are converted entirely to products; these reactions are said to go to *completion.* In others, the products react, re-forming the original starting materials. Outside the cell, such reactions come to equilibrium. In a living cell, equilibrium would be deadly. The cellular processes must go to completion—or nearly so. Products can become reactants in the body, but not under equilibrium conditions.

In this chapter, we will examine changes in composition, energy, reaction rates, and equilibria. For the most part, we will deal with simple nonliving systems. The principles developed, however, will be exceedingly important in later chapters, where we will deal with the more complex chemistry of living cells.

4.1 Balancing Chemical Equations

Chemistry is a study of matter and the changes it undergoes and of the energy that brings about those changes or is released when those changes occur. So far we have discussed the symbols and formulas that are used to represent elements and compounds. Now that we have learned the letters (symbols) and words (formulas) of our chemical language, we are ready to write sentences (chemical equations). A **chemical equation** is a shorthand way of describing chemical change, using symbols and formulas to represent the elements and compounds that are involved in the change.

We can describe a chemical reaction in words. For example:

"Carbon reacts with oxygen to form carbon dioxide."

We can also describe the same reaction in chemical shorthand.

$$C + O_2 \longrightarrow CO_2$$

The plus sign (+) indicates the addition of carbon to oxygen (or vice versa) or a combining of the two in some manner. The arrow ($\longrightarrow$) is often read "yields." Substances on the left of the arrow are **reactants,** or *starting materials.* Those on the right are the **products** of the reaction.

Chemists usually work at the *macroscopic* level, using quantities of materials that are visible to the naked eye. They interpret their work at the *microscopic* level, thinking of the chemical processes in terms of atoms, molecules, and ions. They represent both levels *symbolically,* using atomic symbols, formulas, and equations.

At the microscopic (atomic and molecular) level, the chemical equation

$$C + O_2 \longrightarrow CO_2$$

means that one atom of carbon (C) reacts with one molecule of oxygen (O_2) to produce one molecule of carbon dioxide (CO_2).

Sometimes the physical states of the reactants and products are indicated. The initial letter of the state is written immediately following the formula. Thus (g) indicates a gaseous substance, (l) a liquid, and (s) a solid. The label (aq) indicates an aqueous solution—that is, a water solution. Using these labels, our equation becomes

$$C(s) + O_2(g) \longrightarrow CO_2(g)$$

We can't represent all chemical reactions as simply as we can the reaction between carbon and oxygen to form carbon dioxide. For the reaction of hydrogen and oxygen to form water, we might first write

$$H_2(g) + O_2(g) \longrightarrow H_2O(l) \quad (\textit{not balanced})$$

However, this representation is not consistent with the law of conservation of mass. Two oxygen atoms are shown among the reactants, as O_2, but only one appears among the products, in H_2O. For the equation to represent the chemical event correctly, it must be *balanced.* To balance oxygen atoms we first place the coefficient 2 in front of the formula for water.

$$H_2(g) + O_2(g) \longrightarrow 2\,H_2O(l) \quad (\textit{oxygen balanced, hydrogen not balanced})$$

A coefficient preceding a formula multiplies everything in the formula, and a coefficient of 1 is understood when no other number appears. The coefficient 2 means that two molecules of water, and therefore two oxygen atoms, are involved. However, the coefficient 2 not only increases the number of oxygen atoms but also increases the number of hydrogen atoms to four. We have balanced O atoms at the expense of unbalancing H atoms. To balance the numbers of H atoms, we place the coefficient 2 in front of H_2 on the left.

$$2\,H_2(g) + O_2(g) \longrightarrow 2\,H_2O(l) \quad (\textit{balanced})$$

Now there are four H atoms and two O atoms on each side of the equation. The law of conservation of mass is obeyed. Figure 4.1 illustrates two common pitfalls in the process of balancing equations, as well as the correct method.

Although simple equations can be balanced by trial and error, a couple of strategies often help. (1) If an element occurs in just one compound on each side of the equation, try balancing that element *first.* (2) Balance any reactants or products that exist as the *free* element *last.* Perhaps the most important step in any strategy is to check an equation to ensure that it is indeed balanced. Remember that for each element, the same number of atoms of the element must appear on each side of the equation; atoms are conserved in chemical reactions.

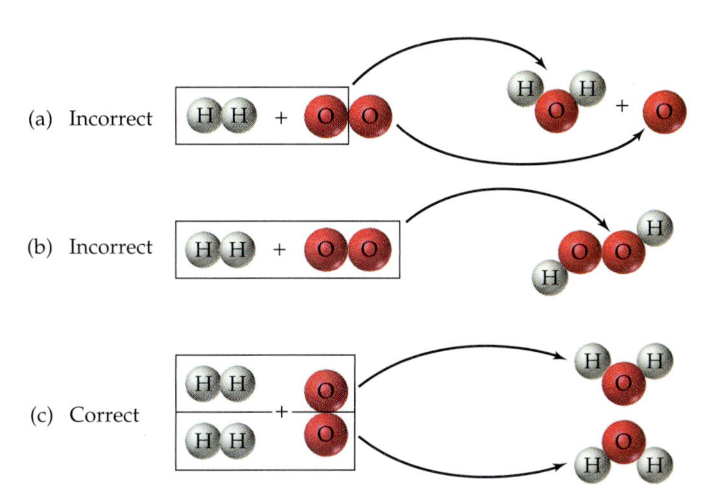

Figure 4.1 Balancing the chemical equation for the reaction between hydrogen and oxygen to form water. (a) *Incorrect.* There is no atomic oxygen (O) as a product. *Extraneous products cannot be introduced simply to balance an equation.* (b) *Incorrect.* The product of the reaction is water (H_2O), not hydrogen peroxide (H_2O_2). *A formula can't be changed simply to balance an equation.* (c) Correct. An equation can be balanced only through the use of *correct formulas* and *coefficients.*

Example 4.1

Balance the following equation.

$$Fe + O_2 \longrightarrow Fe_2O_3$$

Solution

Begin by balancing the oxygen atoms. The least common multiple of 2 and 3 is 6. We need *three* O_2 and *two* Fe_2O_3

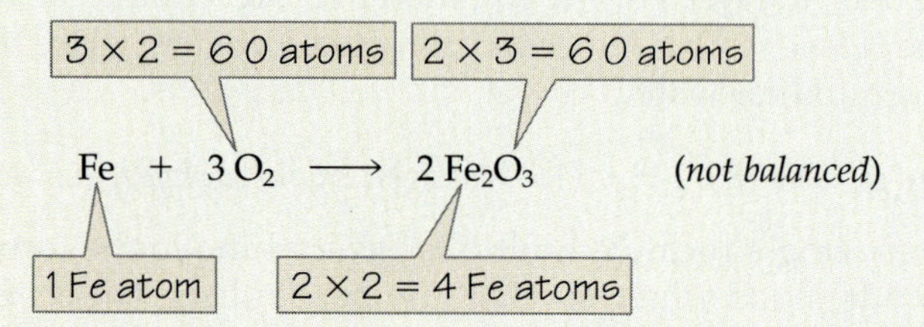

We now have *four* iron atoms on the right side. We can get four on the left by placing the coefficient 4 in front of Fe.

$$4\,Fe + 3\,O_2 \longrightarrow 2\,Fe_2O_3 \quad (\textit{balanced})$$

(Checking that the equation is balanced, we count four Fe atoms and six O atoms on each side.)

Practice Exercise

Balance the following equation.

$$P_4 + H_2 \longrightarrow PH_3$$

Example 4.2

Balance the following equation.

$$CH_4 + O_2 \longrightarrow CO_2 + H_2O$$

Solution

Oxygen appears as the free element (O_2) on the left and in both CO_2 and H_2O on the right. Let's leave O for last and balance C and H first. Carbon is already balanced; there is one C atom on each side. To balance H, we need the coefficient 2 before H_2O.

$$CH_4 + O_2 \longrightarrow CO_2 + 2\,H_2O \quad (\textit{not balanced})$$

Now, if we count O atoms, we find two on the left and four on the right. If we place a 2 in front of O_2, the O atoms balance.

$$CH_4 + 2\,O_2 \longrightarrow CO_2 + 2\,H_2O \quad (\textit{balanced})$$

(Checking that the equation is balanced, we count one C atom, four H atoms, and four O atoms on each side.)

Practice Exercise

Balance the following equations.

a. $Mg + B_2O_3 \longrightarrow B + MgO$

b. $NO_2 + H_2O \longrightarrow HNO_3 + NO$

c. $H_2 + Fe_2O_3 \longrightarrow Fe + H_2O$

Example 4.3

Balance the following equation.

$$H_3PO_4 + NaCN \longrightarrow HCN + Na_3PO_4$$

Solution

Notice that the PO_4 and CN groups remain unchanged in the reaction. In balancing the equation, we can treat each group as a whole rather than considering its constituent atoms separately. To balance hydrogen atoms, we place a 3 before HCN.

$$H_3PO_4 + NaCN \longrightarrow 3\,HCN + Na_3PO_4 \quad (\textit{not balanced})$$

To balance the sodium atoms, we put a 3 in front of the NaCN.

$$H_3PO_4 + 3\,NaCN \longrightarrow 3\,HCN + Na_3PO_4 \quad (\textit{balanced})$$

To check, we note that in the final balanced equation there are three H atoms, three Na atoms, one PO_4 group, and three CN groups on each side of the equation.

Practice Exercise

Balance the following equations.

a. $H_3PO_4 + Ca(OH)_2 \longrightarrow Ca_3(PO_4)_2 + H_2O$
b. $CaO + P_4O_{10} \longrightarrow Ca_3(PO_4)_2$
c. $Al(OH)_3 + H_2SO_4 \longrightarrow Al_2(SO_4)_3 + H_2O$

We have made the task of balancing equations deceptively easy by considering simple reactions, but it is more important at this point for you to understand the principle than to be able to balance complicated equations. You should know what is meant by a balanced equation and be able to handle simple systems.

Review Questions

4.1 What is the purpose of balancing a chemical equation?

4.2 Translate the following chemical equation into words.

$$2\,H_2(g) + O_2(g) \longrightarrow 2\,H_2O(g)$$

4.3 Indicate whether the following equations are balanced. (You need not balance the equation; just determine whether it is balanced as written.)
a. $Mg + H_2O \longrightarrow MgO + H_2$
b. $FeCl_2 + Cl_2 \longrightarrow FeCl_3$
c. $F_2 + H_2O \longrightarrow 2\,HF + O_2$

4.2 Volume Relationships in Chemical Equations

Chemists generally cannot work with individual atoms and molecules. Even the tiniest speck of visible matter contains billions of atoms. John Dalton postulated that atoms of different elements had different masses, and equal masses of different elements would therefore contain different numbers of atoms. Consider the analogous situation of golf balls and Ping-Pong balls. A kilogram of golf balls contains a smaller number of balls than a kilogram of Ping-Pong balls. One could determine the number of balls in each case simply by counting them. For atoms, however, we have no such straightforward method. It was in the experiments of a French chemist and the mind of an Italian scientist that approaches to the problem of numbering atoms were found.

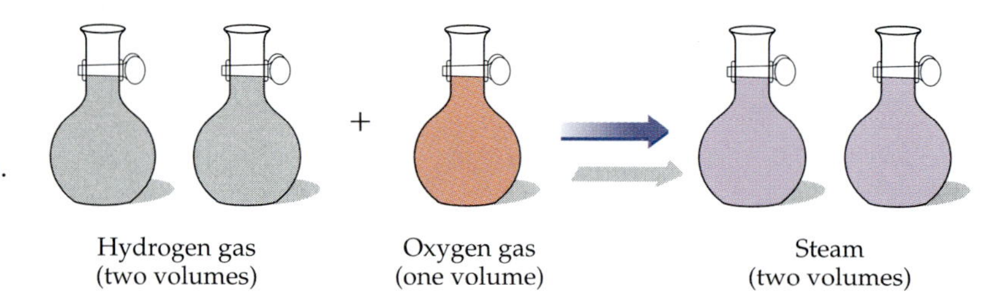

Figure 4.2 ▶
Gay-Lussac's law of combining volumes. When measured at the same temperature and pressure, two volumes of hydrogen gas react with one volume of oxygen gas to yield two volumes of steam.

In 1809 the French scientist Joseph Gay-Lussac (1778–1850) published experimental results summarized in a relationship known as the **law of combining volumes.** The law says that *when gases measured at the same temperature and pressure are allowed to react, the volumes of gaseous reactants and products are in small whole-number ratios.* For example, at 100 °C, two volumes of hydrogen unite with one volume of oxygen to produce two volumes of steam (water vapor), as suggested by Figure 4.2. The combining ratio is 2:1:2.

Figure 4.3 provides another illustration of Gay-Lussac's law. If hydrogen reacts with nitrogen to form ammonia, the combining volumes are three of hydrogen with one of nitrogen to give two of ammonia (3:1:2).

Gay-Lussac thought there must be some relationship between the *numbers* of molecules and the *volumes* of gaseous reactants and products. But it was Amadeo Avogadro who first explained the law of combining volumes. **Avogadro's hypothesis,** based on shrewd interpretation of experimental facts, was that *equal volumes of all gases (at the same temperature and pressure) contain the same number of molecules* (Figure 4.4).

▲ Amadeo Avogadro (1776–1856) did not live to see his ideas accepted by the scientific community. Acceptance finally came in 1860 at a scientific conference at which Stanislao Cannizzaro (1826–1910) effectively communicated Avogadro's ideas from half a century earlier.

The equation for the combination of hydrogen and oxygen to form water (steam) is

$$2\,H_2(g) + O_2(g) \longrightarrow 2\,H_2O(g)$$

The coefficients of the molecules are the same as the combining ratio of the gas volumes, 2:1:2 (see again Figure 4.2). Similarly, the formation of ammonia is described in the equation

$$3\,H_2(g) + N_2(g) \longrightarrow 2\,NH_3(g)$$

The coefficients are identical to the factors of the combining ratio (see again Figure 4.3). The equation says that a nitrogen molecule reacts with three hydrogen molecules to produce two ammonia molecules. It also indicates that if you had 1 million nitrogen molecules, you would need 3 million hydrogen molecules to produce 2 million ammonia molecules. The equation provides the combining ratios. If identical volumes of gases contain identical numbers of molecules, then, according to the equation, one volume of nitrogen reacts with three volumes of hydrogen to produce two volumes of ammonia.

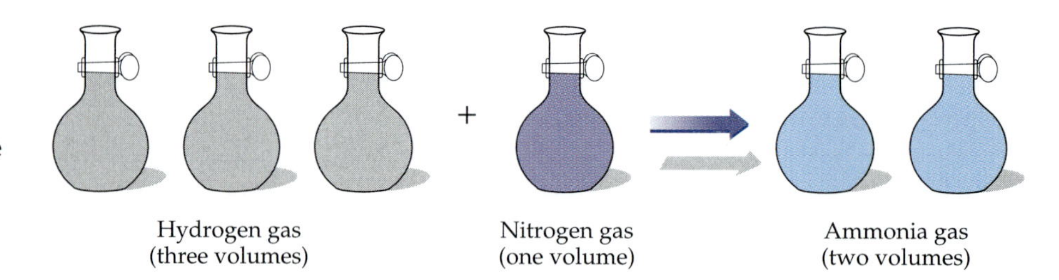

▲ **Figure 4.3**
Gay-Lussac's law of combining volumes. When measured at the same temperature and pressure, three volumes of hydrogen gas react with one volume of nitrogen gas to yield two volumes of ammonia gas.

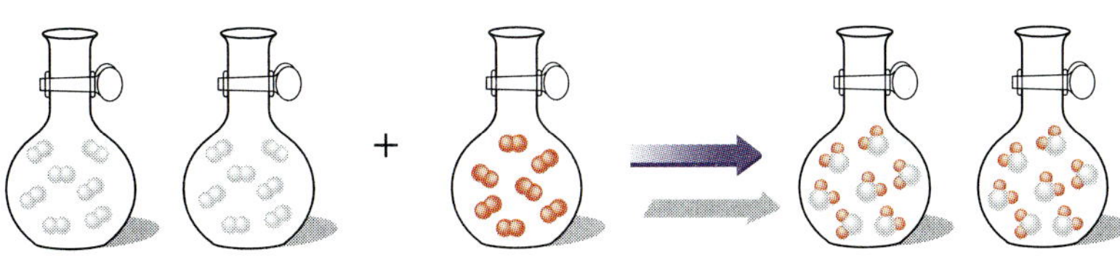

◀ **Figure 4.4**
Avogadro's explanation of Gay-Lussac's law of combining volumes. Equal volumes of each of the gases contain the same number of molecules.

Example 4.4

What volume of oxygen is required to burn 0.556 L of propane, if both gases are measured at the same temperature and pressure?

$$C_3H_8(g) + 5\,O_2(g) \longrightarrow 3\,CO_2(g) + 4\,H_2O(g)$$

Solution

The equation indicates that *five* volumes of $O_2(g)$ are required for every volume of $C_3H_8(g)$. Thus, we use $5\text{ L } O_2(g)/1\text{ L } C_3H_8(g)$ as the ratio to find the volume of oxygen required.

$$?\text{ L } O_2(g) = 0.556\text{ L } C_3H_8(g) \times \frac{5\text{ L } O_2(g)}{1\text{ L } C_3H_8(g)} = 2.78\text{ L } O_2(g)$$

Practice Exercises

A. Using the equation in Example 4.4, calculate the volume of $CO_2(g)$ that is produced when 0.492 L of propane is burned if the two gases are compared at the same temperature and pressure.

B. Calculate the volume of $CO_2(g)$ that is produced when 5.42 L of butane, $C_4H_{10}(g)$, is burned if the two gases are compared at the same temperature and pressure. (Hint: First write a balanced chemical equation.)

Avogadro was also the first to suggest that certain elements such as hydrogen, oxygen, and nitrogen were made up of diatomic molecules. If these substance were monatomic, the equations would be

$$2\,H(g) + O(g) \longrightarrow H_2O(g)$$
(*incorrect*)

$$3\,H(g) + N(g) \longrightarrow NH_3(g)$$
(*incorrect*)

These equations give the wrong ratios of combining volumes. The first, for example, shows a ratio of 2 : 1 : 1, rather than the observed volume ratio of 2 : 1 : 2. On the other hand, if hydrogen, oxygen, and nitrogen molecules are diatomic, we get the observed ratios.

✓ Review Question

4.4 What is Avogadro's hypothesis? How did it explain Gay-Lussac's law of combining volumes?

4.3 Avogadro's Number: 6.02×10^{23}

The periodic table (inside front cover) gives *relative* atomic masses for the various elements. It isn't possible to weigh individual atoms, but we can weigh *equal numbers* of atoms or molecules of different substances and use the ratio of masses as relative masses. Avogadro's hypothesis gives us one way to do that by measuring the masses of equal volumes of gases.

Once the relative masses are known, it is possible to plan reactions such that no materials are wasted. If we could (we can't) weigh out 12 u of carbon and 16 u of oxygen, we would have one atom of each, and we could make one molecule of carbon monoxide (CO). If we weigh out 12 g of carbon and 16 g of oxygen, we still have the proper *ratio* of atoms, and gram quantities are easily weighed. From these amounts of reactants, we could make 28 g of CO, with none of the reactants left over. Similarly, if we wished to make carbon dioxide (CO_2), we could weigh out 12 g of carbon and 32 g of oxygen to make 44 g of CO_2 with no leftover reactants.

Avogadro had no way of knowing how many molecules there were in a given volume of gas. Scientists since his time have determined the number of atoms in various weighed samples of substances. The numbers are enormously large, even for tiny samples. In defining atomic masses (Section 2.3), the mass of a carbon-12 atom is defined as exactly 12 u. The number of carbon-12 atoms in a 12-g sample of carbon-12 is called **Avogadro's number.** The value of Avogadro's number has been determined experimentally to be 6.0221367×10^{23}. For many purposes the number is rounded to 6.02×10^{23}.

Avogadro's number is such a large number that it staggers the imagination. If you had 6×10^{23} dollars, for example, you could spend a billion dollars every second for as long as you lived, and you would still have more than 99.999% of your money left to pass on to your heirs. Or if 6×10^{23} snowflakes fell evenly all across the United States, the blanket of snow would completely cover up every building in the country, including the tallest skyscrapers. See Figure 4.5 for a further illustration of just how gigantic Avogadro's number is.

✓ Review Question

4.5 How do the law of combining volumes and Avogadro's hypothesis indicate that hydrogen gas is composed of diatomic molecules rather than individual atoms?

4.4 Molecular Masses and Formula Masses

We noted in Chapter 2 that each element has a characteristic atomic mass. Because chemical compounds are made up of two or more elements, the masses that we associate with compounds are combinations of atomic masses. For a molecular substance, the **molecular mass** is the average mass of a molecule of a substance relative to that of a carbon-12 atom. More simply, it is the sum of the masses of the atoms represented in a molecular formula. For example, because the formula O_2 specifies two O atoms per molecule of oxygen, the molecular mass of oxygen (O_2) is twice the atomic mass of oxygen.

We use the term "average" when speaking of the mass of an individual molecule because molecules of a compound may have different isotopes of one or more of their constituent elements.

$$2 \times \text{atomic mass of O} = 2 \times 15.9994\ \text{u} = 31.9988\ \text{u}$$

The molecular mass of carbon dioxide (CO_2) is the sum of the atomic mass of carbon and twice the atomic mass of oxygen.

$$\begin{aligned} 1 \times \text{atomic mass of C} &= 1 \times 12.011\ \text{u} &= 12.011\ \text{u} \\ 2 \times \text{atomic mass of O} &= 2 \times 15.9994\ \text{u} &= \underline{31.99988\ \text{u}} \\ \text{Molecular mass of } CO_2 & &= 44.011\ \text{u} \end{aligned}$$

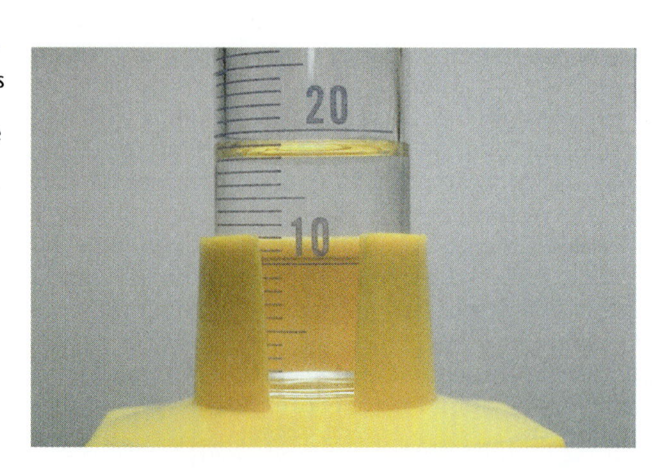

Figure 4.5 ▶ How big is Avogadro's number? There are 6.02×10^{23} molecules of water in 18 mL (about 360 drops) of H_2O, but it would take 100,000 years for 6.02×10^{23} drops of water to pass over Niagara Falls. One tablespoon of water is 15 mL; 18 mL is about one mouthful.

Example 4.5

Calculate the molecular mass of sulfur dioxide, SO_2, an irritating gas formed when sulfur is burned.

Solution

We think about the problem in the following way. Add the atomic mass of sulfur to twice the atomic mass of oxygen. However, if we use a calculator, we need only write down the final answer, 64.065 u. That is, we have no need to record the numbers 32.066 and 15.9994.

$$1 \times \text{atomic mass of S} = 1 \times 32.066\text{ u} = 32.066\text{ u}$$

$$2 \times \text{atomic mass of O} = 2 \times 15.9994\text{ u} = \underline{31.9988\text{ u}}$$

$$\text{Molecular mass of } SO_2 = 64.065\text{ u}$$

Practice Exercise

Calculate the molecular mass of **(a)** $C_6H_4Cl_2$, **(b)** $C_2H_4Cl_2$, and **(c)** H_3PO_4.

The term "molecular mass" is not appropriate for ionic compounds, such as NaCl and K_2O, in which individual molecules do not exist. For ionic compounds, we use the term *formula unit.* **Formula mass** is the average mass of a formula unit relative to that of a carbon-12 atom. In short, the formula mass is the sum of the masses of the atoms or ions represented by the formula.

A **formula unit** is simply the atoms or ions specified by the formula. A formula unit of Al_2O_3 is two aluminum atoms and three oxygen atoms.

Example 4.6

Calculate the formula mass of ammonium sulfate, $(NH_4)_2SO_4$, a fertilizer commonly used by home gardeners.

Solution

To determine a formula mass, we add the atomic masses of the constituent elements. In the summation below, remember that everything within the parentheses must be multiplied by 2.

$$2 \times \text{atomic mass of N} = 2 \times 14.0067\text{ u} = 28.0134\text{ u}$$

$$8 \times \text{atomic mass of H} = 8 \times 1.00794\text{ u} = 8.06352\text{ u}$$

$$1 \times \text{atomic mass of S} = 1 \times 32.066\text{ u} = 32.066\text{ u}$$

$$4 \times \text{atomic mass of O} = 4 \times 15.9994\text{ u} = \underline{63.9976\text{ u}}$$

$$\text{Formula mass of } (NH_4)_2SO_4 = 132.141\text{ u}$$

Practice Exercise

Calculate the formula mass of **(a)** K_2CO_3, **(b)** $K_2Cr_2O_7$, and **(c)** $NaB(C_6H_5)_4$.

✓ Review Question

4.6 Explain the difference between the *atomic mass* of nitrogen and the *molecular mass* of nitrogen. Explain how each is determined from data in the periodic table.

4.5 Chemical Arithmetic and the Mole

We buy socks by the pair (2 socks), eggs by the dozen (12 eggs), pencils by the gross (144 pencils), and paper by the ream (500 sheets). A dozen is the same *number* whether we are counting a dozen melons or a dozen oranges. But a dozen oranges

Even a sample of carbon as small as a pencil-mark period at the end of a sentence contains about 10^{18} C atoms—that is, about 1,000,000,000,000,000,000 C atoms.

and a dozen melons do not *weigh* the same. If a melon weighs three times as much as an orange, then a dozen melons will weigh three times as much as a dozen oranges.

Chemists count atoms and molecules by the mole. (A single carbon atom is much too small to see, but a *mole* of carbon atoms will fill a tablespoon.) A mole of carbon and a mole of magnesium both contain the same number of atoms. But a magnesium atom weighs twice as much as a carbon atom, so a mole of magnesium will weigh twice as much as a mole of carbon.

According to the SI definition, a **mole** (abbreviated *mol*) is an amount of substance that contains the same number of elementary units as there are atoms in 12 g of carbon-12. That number is 6.02×10^{23} (Avogadro's number). The elementary units may be atoms (such as S or Ca), molecules (such as O_2 or CO_2), ions (such as K^+ or SO_4^{2-}), or any other kind of formula unit. A mole of NaCl, for example, contains 6.02×10^{23} NaCl formula units, which means that it contains 6.02×10^{23} Na^+ ions and 6.02×10^{23} Cl^- ions. Figure 4.6 is a photograph of one mole of each of several chemical substances.

The **molar mass** of a substance is the mass of 1 mol of that substance. The molar mass is numerically equal to the atomic mass, molecular mass, or formula mass, but it is expressed in the unit *g/mol*. The atomic mass of sodium is 22.99 u; its molar mass is 22.99 g/mol. The molecular mass of carbon dioxide is 44.01 u; its molar mass is 44.01 g/mol. The formula mass of magnesium chloride is 95.21 u; its molar mass is 95.21 g/mol. We can use these facts, together with the basic definition of the number of elementary units in a mole, to write the following relationships.

$$1 \text{ mol Na} = 22.99 \text{ g Na}$$
$$1 \text{ mol } CO_2 = 44.01 \text{ g } CO_2$$
$$1 \text{ mol } MgCl_2 = 95.21 \text{ g } MgCl_2$$

In turn, these relationships supply the conversion factors that we need to make conversions between mass in grams and amount in moles, as illustrated in the following examples.

Example 4.7

How many grams of Na are there in 0.250 mol of Na?

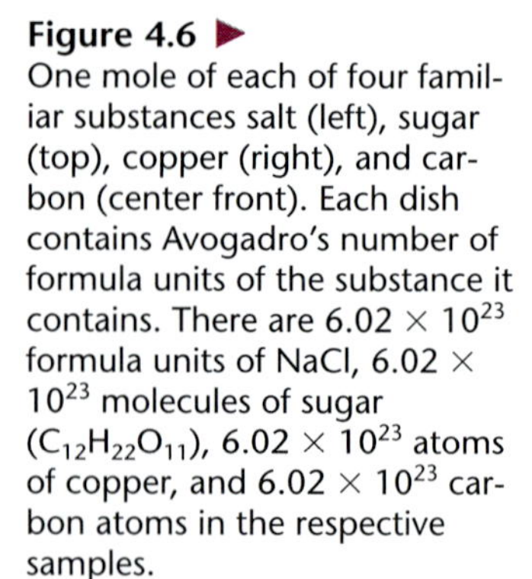

Figure 4.6 ▶
One mole of each of four familiar substances salt (left), sugar (top), copper (right), and carbon (center front). Each dish contains Avogadro's number of formula units of the substance it contains. There are 6.02×10^{23} formula units of NaCl, 6.02×10^{23} molecules of sugar ($C_{12}H_{22}O_{11}$), 6.02×10^{23} atoms of copper, and 6.02×10^{23} carbon atoms in the respective samples.

Solution

Sodium has an atomic mass of 22.99 u and a molar mass of 22.99 g Na/mol Na. Therefore:

$$? \text{ g Na} = 0.250 \text{ mol Na} \times \frac{22.99 \text{ g Na}}{1 \text{ mol Na}} = 5.75 \text{ g Na}$$

Practice Exercise

Calculate the mass, in grams, of (**a**) 55.5 mol H_2O, (**b**) 0.0102 mol $C_4H_{10}O$, and (**c**) 2.45 mol C_2H_6.

Example 4.8

Calculate the number of moles of CO_2 present in a 225-g sample of the gas.

Solution

In this case we need the molar mass of CO_2. We determined the *molecular* mass of CO_2 to be 44.011 (page 104). Its molar mass is therefore 44.01 g CO_2/mol CO_2. To convert from a mass in grams to an amount in moles, we must use the *inverse* of the molar mass to get the proper cancellation of units.

$$? \text{ mol } CO_2 = 225 \text{ g } CO_2 \times \frac{1 \text{ mol } CO_2}{44.01 \text{ g } CO_2} = 5.11 \text{ mol } CO_2$$

Practice Exercise

Calculate the amount, in moles, of (**a**) 3.71 g Fe, (**b**) 76.0 g phosphoric acid (H_3PO_4), and (**c**) 165 g C_4H_{10}.

Review Questions

4.7 What is Avogadro's number and how is it related to the quantity called one mole?

4.8 What are the molecular mass and the molar mass of carbon dioxide? Explain how each is determined from the formula, CO_2.

4.9 How many oxygen molecules are there in 1.00 mol of O_2? How many oxygen atoms are there in 1.00 mol of O_2?

4.10 How many calcium ions and how many chloride ions are there in 1.00 mol of $CaCl_2$?

4.6 Mole and Mass Relationships in Chemical Equations

Chemical equations not only represent ratios of atoms and molecules but also give us mole ratios. The equation

$$C + O_2 \longrightarrow CO_2$$

tells us that one atom of carbon reacts with one molecule (two atoms) of oxygen to form one molecule of carbon dioxide. The equation also indicates that 1 mol (6.02×10^{23} atoms) of carbon reacts with 1 mol (6.02×10^{23} molecules) of oxygen to yield 1 mol (6.02×10^{23} molecules) of carbon dioxide. Because the molar mass in g/mol of a substance is numerically equal to the formula mass of the substance in atomic mass units, the equation also tells us (indirectly) that 12.0 g (1 mol) of carbon reacts with 32.0 g (1 mol) of oxygen to yield 44.0 g (1 mol) of CO_2 (Figure 4.7).

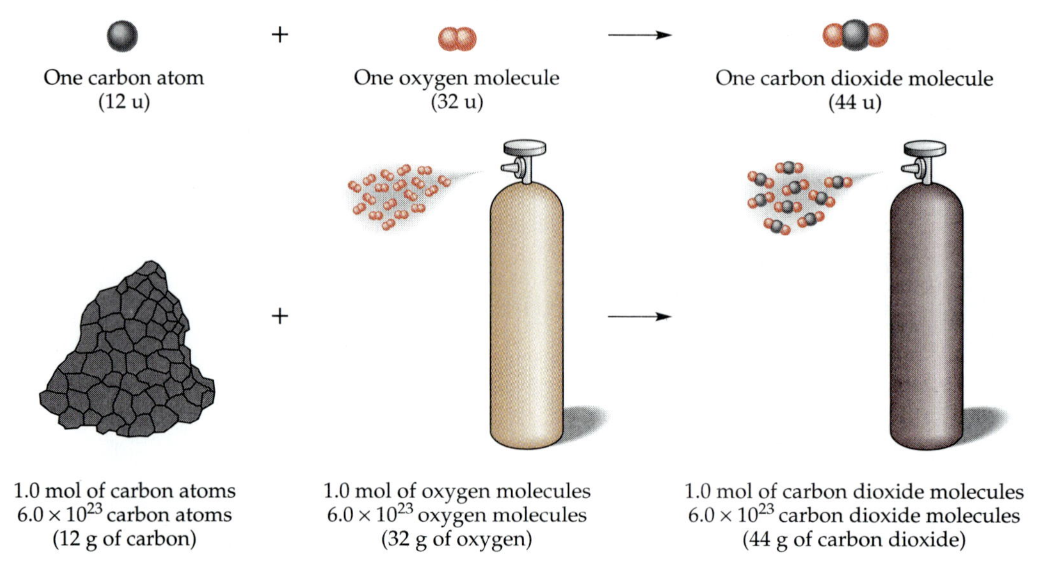

▲ **Figure 4.7**
We cannot weigh single C atoms or O_2 or CO_2 molecules, but we can weigh large numbers of these entities.

Example 4.9

Nitrogen monoxide (nitric oxide) combines with oxygen to form nitrogen dioxide according to the equation

$$2\,NO + O_2 \longrightarrow 2\,NO_2$$

State the molecular, molar, and mass relationships indicated by this equation.

Solution

The coefficients give us the molecular and molar ratios directly. *Molecular:* 2 molecules of NO react with 1 molecule of O_2 to form 2 molecules of NO_2. *Molar:* 2 mol of NO react with 1 mol of O_2 to form 2 mol of NO_2. For the mass ratios, we must first calculate the molar mass of each substance and multiply it by its coefficient.

Coefficient of NO | Atomic mass N and O

$$[2 \times (14.0 + 16.0)] = 60.0\text{ g NO}$$

$$[1 \times 2(16.0)] = 32.0\text{ g } O_2$$

$$\{2 \times [14.0 + (2 \times 16.0)]\} = 92.0\text{ g } NO_2$$

Mass: 60.0 g of NO react with 32.0 g of O_2 to form 92.0 g of NO_2.

Practice Exercise

Hydrogen sulfide burns in air to produce sulfur dioxide and water according to the equation

$$2\,H_2S + 3\,O_2 \longrightarrow 2\,SO_2 + 2\,H_2O$$

State the molecular, molar, and mass relationships indicated by this equation.

Now consider the combustion of propane in air to form carbon dioxide and water.

$$C_3H_8 + 5\,O_2 \longrightarrow 3\,CO_2 + 4\,H_2O$$

The coefficients allow us to make statements such as "1 mol of C_3H_8 reacts with 5 mol of O_2," "3 mol of CO_2 is produced for every 1 mol of C_3H_8 that reacts," and "4 mol of H_2O is produced for every 3 mol of CO_2 produced." Moreover, we can turn these statements into conversion factors known as stoichiometric factors. A **stoichiometric factor** relates the amounts of any two substances involved in a chemical reaction, on a *mole* basis. In the examples that follow, stoichiometric factors are shown in color.

Example 4.10

When 0.105 mol of propane is burned in a plentiful supply of oxygen, how many moles of oxygen are consumed?

$$C_3H_8 + 5\,O_2 \longrightarrow 3\,CO_2 + 4\,H_2O$$

Solution

The equation tells us that 5 mol of O_2 is required to burn 1 mol of C_3H_8. This gives us the stoichiometric factor 5 mol O_2/1 mol C_3H_8 to use as a conversion factor.

$$?\ \text{mol } O_2 = 0.105\ \cancel{\text{mol } C_3H_8} \times \frac{5\ \text{mol } O_2}{1\ \cancel{\text{mol } C_3H_8}} = 0.525\ \text{mol } O_2$$

Practice Exercise

For the combustion of propane in Example 4.10: (**a**) How many moles of carbon dioxide are formed when 0.529 mol of C_3H_8 is burned? (**b**) How many moles of water are produced when 76.2 mol of C_3H_8 is burned? (**c**) How many moles of carbon dioxide are produced when 1.010 mol of O_2 is consumed?

Although the mole is essential in calculations based on chemical equations, we cannot weigh out molar amounts directly. We have to relate them to quantities that we can measure. We usually express mass in grams, in which case we can follow the five-step approach outlined below.

Step 1. Write a balanced equation for the reaction.
Step 2. Determine the molar masses of the substances involved in the calculation.
Step 3. Use the molar mass of the substance for which information is given to convert grams of the *given* substance to moles of that substance.
Step 4. Obtain a stoichiometric factor from the balanced equation to convert from moles of the given substance to moles of the substance about which information is sought—the *desired* substance.
Step 5. Use the molar mass of the desired substance to convert moles to grams of that substance.

We illustrate this five-step approach in Figure 4.8 and Example 4.11, where we also show how the steps easily can be combined into the preferred single setup method.

Example 4.11

Calculate the mass of oxygen needed to react with 10.0 g of carbon in the reaction that forms carbon dioxide.

Figure 4.8 ▶
A chemical equation relates *moles* of reactants and products in a chemical reaction. For *mass* relationships, we must convert the mass of A to moles of A, use a stoichiometric factor to relate moles of A to moles of B, then convert moles of B to a mass of B.

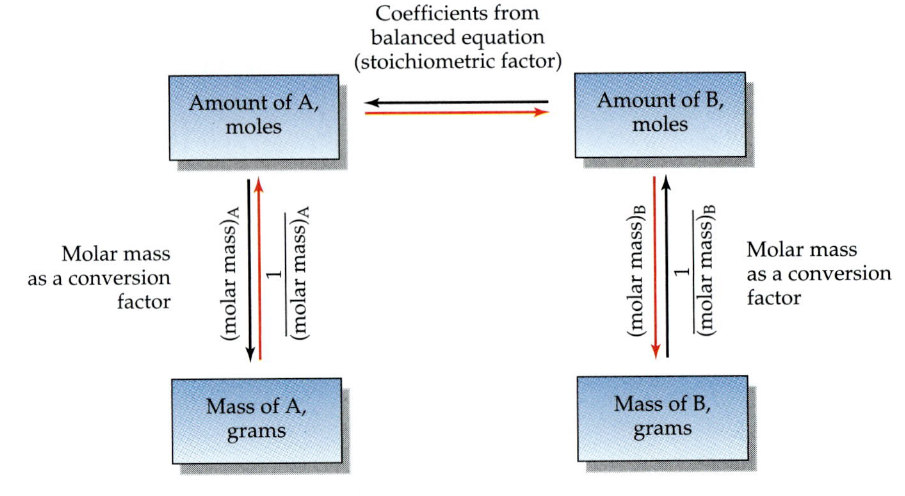

Solution

Step 1. The balanced equation is

$$C + O_2 \longrightarrow CO_2$$

Step 2. The molar masses are $2 \times 15.9994 = 31.9988$ g/mol for O_2 and 12.011 g/mol for C.

Step 3. We convert the mass of the given substance, carbon, to an amount in moles.

$$? \text{ mol C} = 10.0 \text{ g C} \times \frac{1 \text{ mol C}}{12.011 \text{ g C}} = 0.833 \text{ mol C}$$

Step 4. We use coefficients from the balanced equation to establish the stoichiometric factor that relates the amount of oxygen to that of carbon.

$$0.833 \text{ mol C} \times \frac{1 \text{ mol } O_2}{1 \text{ mol C}} = 0.833 \text{ mol } O_2$$

Step 5. We convert from moles of oxygen to grams of oxygen.

$$0.833 \text{ mol } O_2 \times \frac{31.9988 \text{ g } O_2}{1 \text{ mol } O_2} = 26.7 \text{ g } O_2$$

We can also combine all of the steps outlined above into a single setup.

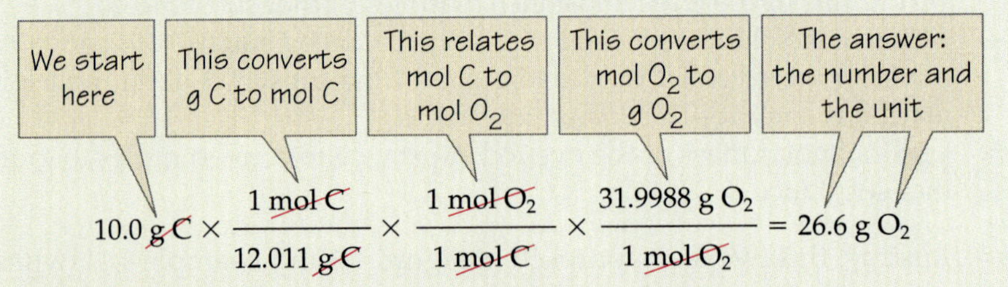

$$10.0 \text{ g C} \times \frac{1 \text{ mol C}}{12.011 \text{ g C}} \times \frac{1 \text{ mol } O_2}{1 \text{ mol C}} \times \frac{31.9988 \text{ g } O_2}{1 \text{ mol } O_2} = 26.6 \text{ g } O_2$$

(The slightly different answers are due to rounding in the intermediate steps.)

Practice Exercises

A. Calculate the mass of oxygen (O_2) needed to react with 0.334 g of nitrogen (N_2) in the reaction that forms nitrogen dioxide.

B. Calculate the mass of carbon dioxide formed by burning 775 g of each of (**a**) methane (CH_4) and (**b**) butane (C_4H_{10}).

Example 4.12

Ammonia, NH_3, a common fertilizer, is made by causing hydrogen and nitrogen to react at a high temperature and pressure. What mass of ammonia, in grams, can be made from 60.0 g of hydrogen?

Solution

Step 1. We start by writing a chemical equation.

$$N_2 + H_2 \longrightarrow NH_3 \quad (\textit{not balanced})$$

Then we balance it.

$$N_2 + 3\,H_2 \longrightarrow 2\,NH_3 \quad (\textit{balanced})$$

Step 2. The molar masses are $2 \times 1.008 = 2.016$ g/mol for H_2 and $14.01 + (3 \times 1.008) = 17.03$ g/mol for NH_3.

Step 3. We convert the mass of the given substance, hydrogen, to an amount in moles.

$$60.0\ \cancel{\text{g H}_2} \times \frac{1\ \text{mol H}_2}{2.016\ \cancel{\text{g H}_2}} = 29.8\ \text{mol H}_2$$

Step 4. We use coefficients from the balanced equation to establish the stoichiometric factor that relates the amount of ammonia to that of hydrogen.

$$29.8\ \cancel{\text{mol H}_2} \times \frac{2\ \text{mol NH}_3}{3\ \cancel{\text{mol H}_2}} = 19.9\ \text{mol NH}_3$$

Step 5. We convert from moles of ammonia to grams of ammonia.

$$19.9\ \cancel{\text{mol NH}_3} \times \frac{17.03\ \text{g NH}_3}{1\ \cancel{\text{mol NH}_3}} = 339\ \text{g NH}_3$$

As is usually the case, all of the steps outlined above can be combined into a single setup.

$$60.0\ \cancel{\text{g H}_2} \times \frac{1\ \cancel{\text{mol H}_2}}{2.016\ \cancel{\text{g H}_2}} \times \frac{2\ \cancel{\text{mol NH}_3}}{3\ \cancel{\text{mol H}_2}} \times \frac{17.03\ \text{g NH}_3}{1\ \cancel{\text{mol NH}_3}} = 338\ \text{g NH}_3$$

Practice Exercises

A. Ammonia reacts with phosphoric acid, H_3PO_4, to form ammonium phosphate, $(NH_4)_3PO_4$. How many grams of ammonia are needed to react completely with 74.8 g of phosphoric acid?

B. The decomposition of potassium chlorate ($KClO_3$) produces potassium chloride (KCl) and O_2 gas. How many grams of oxygen can be made from 2.47 g of potassium chlorate?

✓ Review Questions

4.11 Explain the meaning of the equation

$$CH_4(g) + 2\,O_2(g) \longrightarrow CO_2(g) + 2\,H_2O(g)$$

at the molecular level. Interpret the equation in terms of moles. State the mass relationships conveyed by the equation.

4.12 Translate the following chemical equations into words.

a. $2\,KClO_3(s) \longrightarrow 2\,KCl(s) + 3\,O_2(g)$

b. $2\,Al(s) + 6\,HCl(aq) \longrightarrow 2\,AlCl_3(aq) + 3\,H_2(g)$

Yields of Chemical Reactions

The calculated quantity of product in a reaction is called the *theoretical yield* of the reaction. In Example 4.12, the calculated quantity of product is 339 g; that is, 339 g NH_3 is the theoretical yield of ammonia. The *measured* mass of ammonia formed in the reaction—the *actual yield*—might well be *less* than the theoretical yield. Actual yields of chemical reactions are often less than the theoretical yields for several reasons: (1) The starting materials may not be pure, meaning that the actual quantities of reagents are less than what is weighed out. (In this reaction, for example, if there was some oxygen present in the nitrogen gas, we would have less N_2 than we thought.) (2) Some of the product may be left behind during the process of separating it from excess reactants. (3) Side reactions may occur in addition to the main reaction, converting some of the original reactants into products other than the desired one.

In organic syntheses, such as those employed in the synthesis of drugs, it is difficult to achieve yields above 80–85% even under the best of conditions. Chemists often have to settle for 50%—and sometimes even less than that.

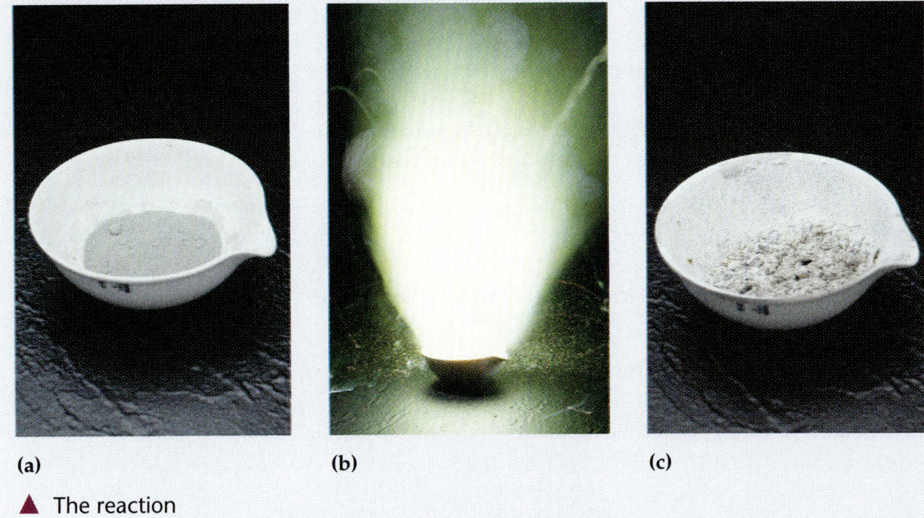

(a) (b) (c)

▲ The reaction

$$8\ Zn(s) + S_8(s) \rightarrow 8\ ZnS(s)$$

gives an actual yield (c) of ZnS(s) that is less than that calculated for the starting mixture (a) for several reasons, including:(1) Neither the powdered zinc nor the powdered sulfur is pure. (2) Zn(s) can combine with O_2(g) in air to produce ZnO(s) and some of the sulfur burns in air to produce SO_2(g). (3) As suggested in (b), some of the product escapes from the reaction mixture as small lumps and as a fine dust.

4.7 Structure, Stability, and Spontaneity

We noted in Chapter 3 that some electron configurations are more stable than others. Sodium *ions* and chloride *ions,* arranged in a crystal lattice, are less reactive than sodium *atoms* and chlorine *molecules.* When sodium metal and chlorine gas are mixed, a vigorous reaction ensues.

$$2\ Na + Cl_2 \longrightarrow 2\ NaCl$$

A great deal of energy is produced as heat and light during this reaction (see again Figure 3.1, page 72). Sometimes this energy is listed as one of the products in the chemical equation.

When heat energy is released during a chemical reaction, we often list heat as a product in the chemical equation. When energy is supplied to keep a chemical reaction going, we can list heat as a reactant in the equation.

$$2\ Na + Cl_2 \longrightarrow 2\ NaCl + \text{energy (heat and light)}$$

The heat released or absorbed during a chemical reaction is called the **heat of reaction.** When the reaction is carried out under constant pressure—as are most of the reactions we consider, which are done in vessels open to the atmosphere—the heat of reaction is called the **enthalpy change** of the reaction and is designated **ΔH.** We generally will use ΔH to specify a given quantity of heat energy and simply write heat as a reactant or product when the exact quantity is not important.

Exothermic and Endothermic Reactions

Chemical reactions that result in the release of heat are said to be **exothermic.** The burning of methane is an exothermic reaction, as is the burning of gasoline or coal. A negative ΔH value indicates an exothermic reaction.

$$CH_4(g) + 2\,O_2(g) \longrightarrow CO_2(g) + 2\,H_2O(l) \quad \Delta H = -890.3\text{ kJ}$$

This equation indicates that 890.3 kJ of heat is liberated when 1 mol $CH_4(g)$ is burned to form $CO_2(g)$ and $H_2O(l)$.

In an exothermic reaction, chemical energy is converted into heat energy. There are other reactions, such as the decomposition of mercury(II) oxide, for which energy must be supplied. If the energy is supplied as heat, such reactions are said to be **endothermic.** A positive ΔH value indicates an endothermic reaction.

$$2\,HgO(s) \longrightarrow 2\,Hg(l) + O_2(g) \quad \Delta H = +181.66\text{ kJ}$$

This equation indicates that 181.66 kJ of heat are absorbed when 2 mol HgO is decomposed to Hg(l) and $O_2(g)$.

The energy released in an exothermic reaction is related to the *chemical energy* present in the reactants. This energy is a form of *potential energy.* Potential energy is released (as heat or light, for example) when reactants are brought together in a way that allows them to achieve more stable electron arrangements.

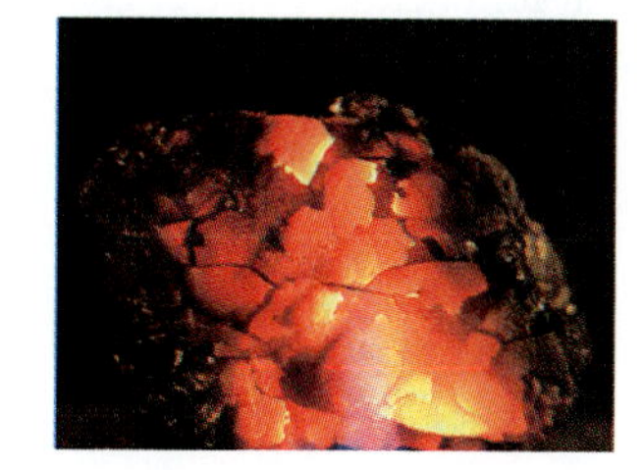

▲ Coal burns in a highly exothermic reaction. The heat released can be used to convert water to steam that can turn a turbine to produce electricity.

To say that a reaction is exothermic does not necessarily mean that it is instantaneous. For example, coal (carbon) has a lot of chemical energy. Carbon reacts with oxygen to form carbon dioxide with the release of considerable heat. But coal does not react very rapidly with oxygen at ordinary temperatures. In fact, one can store a pile of coal in air indefinitely without perceptible change. Before coal will react with oxygen to release its stored energy, it must be heated to a temperature of several hundred degrees. The coal must be supplied with a certain amount of energy (called the activation energy; Section 4.8) before it will begin to burn steadily. Once this activation energy is supplied, the heat evolved in the reaction will keep the coal burning brightly, and the energy eventually produced will exceed the amount of energy required to start the reaction. Overall, you get more energy out than you put in. The reaction is exothermic. In an endothermic reaction, more energy goes in than comes out.

We will return to the burning of coal in a moment, but first let's consider an analogy involving a more familiar situation. If you were in Browning, Montana (elevation 1300 m) and wished to travel to Kalispell (elevation 900 m), you could choose to drive the scenic Going-to-the-Sun Highway through Glacier National Park, which crosses the continental divide at Logan Pass (elevation 2000 m). First, you would have to climb 700 m, but then it would be downhill the rest of the way (Figure 4.9).

The reaction of coal (carbon) with oxygen can be explained in much the same way. The reactants, carbon and oxygen, lie in a rather high potential energy valley (Figure 4.10). A certain amount of energy has to be put into the system for it to get over the top of the "mountain." Once the reaction is under way, the heat released more than compensates for the energy needed to get the reaction going in the first place.

✓ Review Questions

4.13 Define or illustrate each of the following terms.

a. endothermic **b.** exothermic

c. activation energy **d.** catalyst

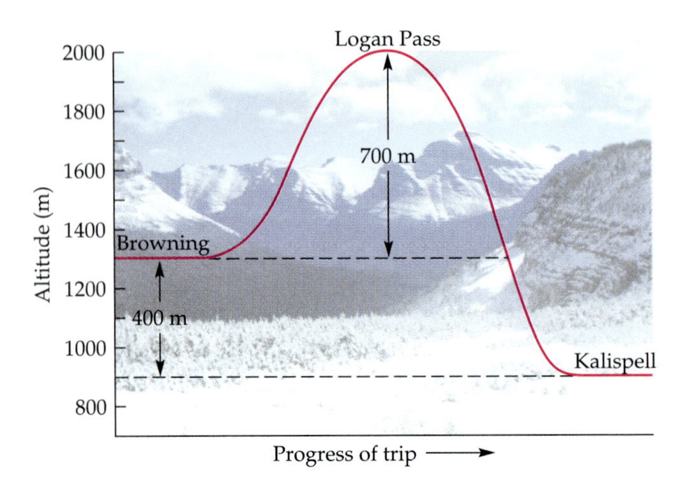

▲ **Figure 4.9**
An analogy to the profile of a reaction and of activation energy. If you were in Browning, Montana (the reactants), and wanted to go to Kalispell (the products), you could drive the scenic Going-to-the-Sun Highway (the reaction profile) through Glacier National Park. First you have to climb 700 m (the activation energy) to the continental divide at Logan Pass, but then little energy is used going downhill.

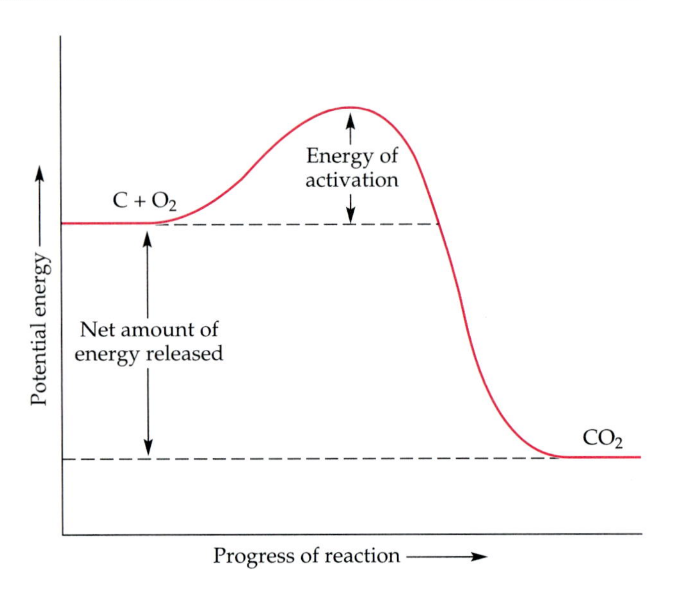

▲ **Figure 4.10**
To get from reactants (carbon and oxygen) to product, we must put some energy (the energy of activation) into the system. (The actual reaction is much more complicated than is indicated by this simplified reaction profile.)

4.14 Refer to the following reaction diagram.

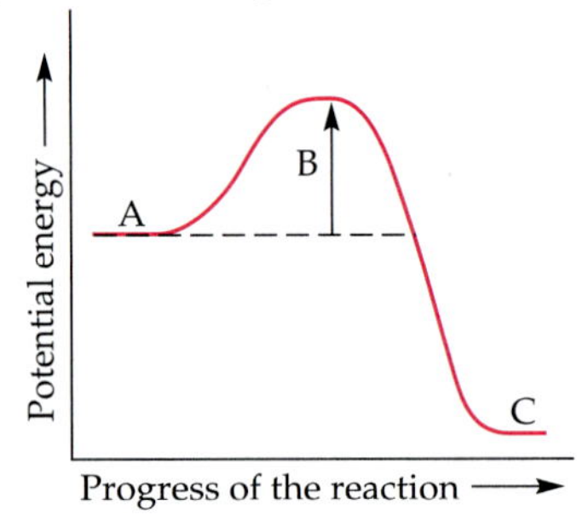

a. Which letter in the diagram refers to the products?
b. Which letter refers to the activation energy?
c. Which letter refers to the reactants?
d. Is the reaction endothermic or exothermic? Explain.

4.8 Forward and Reverse Reactions

As we shall see in Chapter 24, living cells take in oxygen and "burn" glucose to obtain energy, a process called *respiration.*

$$\underset{\text{Glucose}}{C_6H_{12}O_6} + 6\,O_2 \longrightarrow 6\,CO_2 + 6\,H_2O + \text{energy}$$

Green plants carry out the reaction in the opposite direction (an endothermic reaction called **photosynthesis**), using energy from sunlight to convert carbon dioxide and water to glucose and oxygen.

$$6\,CO_2 + 6\,H_2O + \text{energy} \longrightarrow C_6H_{12}O_6 + 6\,O_2$$

Nearly all life on our planet is based on these opposing reactions. But what makes it go one way in one case and the opposite way in another? One obvious answer is energy. Because energy is released when glucose is oxidized, the reactants (glucose

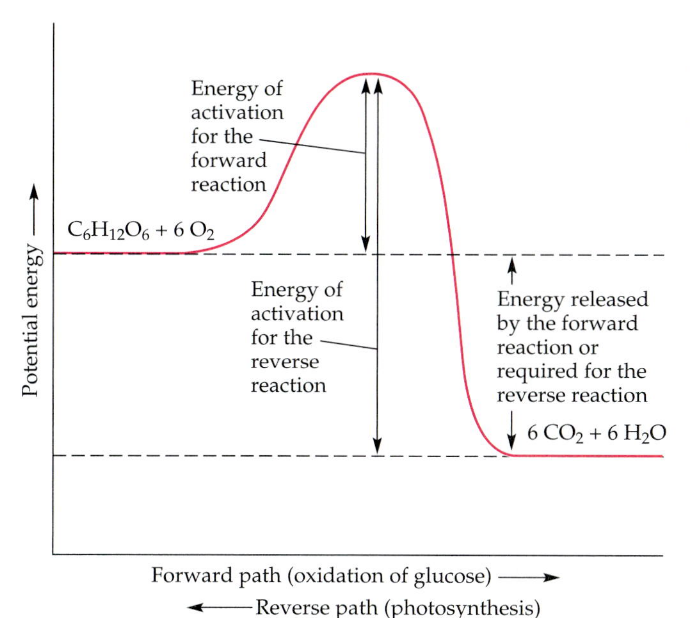

Figure 4.11
A reaction profile for respiration (the forward reaction) and for photosynthesis (the reverse reaction).

and oxygen) must be at a higher energy level than the products (carbon dioxide and water). This does not mean that the path is straight downhill from reactants to products. Indeed, if glucose and oxygen are mixed at ordinary temperatures outside a cell, no perceptible reaction occurs. As with the reaction of coal and oxygen, a certain amount of energy, the activation energy, must be supplied before the reaction takes place. The potential energy diagram for this reaction (Figure 4.11) strongly resembles that for the coal and oxygen reaction (see Figure 4.10).

For reactions that can proceed in either the forward or reverse direction, the energy hill, or the barrier to reaction, may be approached from either side. Although the details of the routes may differ, we can read the diagram in Figure 4.11 from left to right or from right to left. With the burning of glucose, we are considering an exothermic reaction in which higher-energy reactants are converted to lower-energy products. When read in the reverse direction, the diagram presents the net energy changes that occur in the endothermic photosynthesis reaction. The climb to the top of the barrier is longer from one side than from the other. The **activation energy** (**E_a**; also called *energy of activation*) for a chemical reaction is the minimum energy needed to get the reaction started. It is the difference in energy between the level of the reactants and the top of the energy hill. When the reactants are in the valley at the left (as in the forward reaction), the climb to the top is not so long. When the reactants are in the valley to the right (as in the reverse reaction), the climb to the top is much longer. The potential energy diagram shown in Figure 4.11 is much simplified. Just as one seldom crosses a mountain by going straight up one side and straight down the other, so reactions seldom proceed by a smooth, one-hump potential energy change.

Review Question

4.15 If the reaction diagrammed in Review Question 4.14 (page 114) were reversible, would the reverse reaction be endothermic or exothermic? Explain.

4.9 Reaction Rates: Collisions and Orientation

Reaction rates are often profoundly affected by temperature, by catalysts, and by the concentration of reactants. Before atoms, molecules, or ions can react, they must

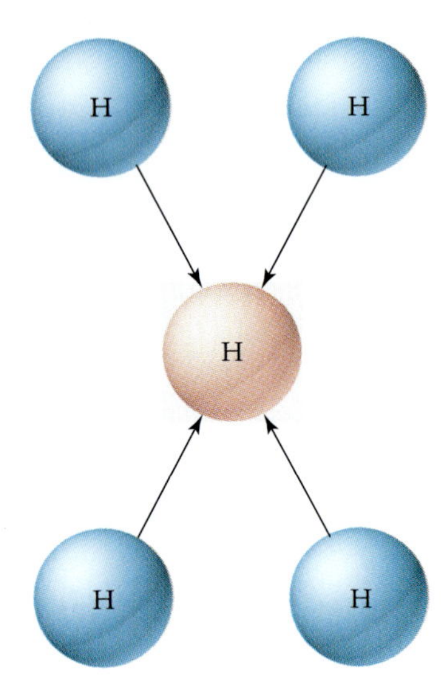

▲ **Figure 4.12**
In the reaction between two hydrogen atoms to form a hydrogen molecule, orientation of the two atoms is unimportant. No matter from which direction the oncoming hydrogen atom (blue) approaches the "target" hydrogen atom (red), its "view" of the impending collision is the same. There is no preferred orientation for the collision.

first collide. Second, for all except the simplest particles, they must come together in the proper orientation. Third, the collision must provide a certain minimum energy, the activation energy.

The frequency of collision is influenced by two factors: temperature and concentration. Molecules move faster at higher temperatures; thus, an increase in temperature increases the frequency of collision. The more concentrated the reactants, the more frequently the particles will collide, simply because there are more of them in a given volume. We will discuss the effects of temperature and concentration on reaction rates in more detail below.

It might not be obvious that orientation is important. Consider a situation in which you want to knock someone down. You can tackle from the front, back, or side and still accomplish your objective. If, however, a kiss is the objective, then an orientation in which the participants are face-to-face works best. For chemical reactions, there are also a few instances in which orientation is not important. If two hydrogen atoms are to react to form a hydrogen molecule,

$$2\,H \longrightarrow H_2$$

then orientation is unimportant (Figure 4.12). Hydrogen atoms have symmetrical (spherical) electron clouds. Front, back, top, and bottom all look the same, and all orientations of two hydrogen atoms are therefore identical. If the atoms come into contact, they can react.

In most cases, however, orientation of the colliding molecules is a factor. Consider the reaction of an iodide ion with methyl bromide (CH_3Br) to produce methyl iodide (CH_3I) and a bromide ion.

$$I^- + CH_3Br \longrightarrow CH_3I + Br^-$$

As shown in Figure 4.13, a collision of an I^- ion with the C atom of the CH_3Br can be effective in leading to a reaction. A collision of an I^- ion with the Br atom of CH_3Br cannot.

The third factor, the requirement for a minimum energy of collision, is more subtle. Temperature can contribute to reaching this energy, but certain substances, called catalysts, may substantially lower the activation energy, thus increasing the reaction rate.

✓ Review Question

4.16 Explain how the orientation of reactant molecules affects the rate of a reaction.

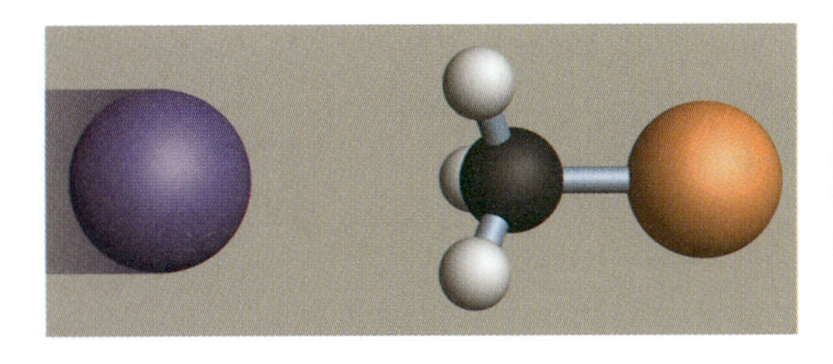

(a)

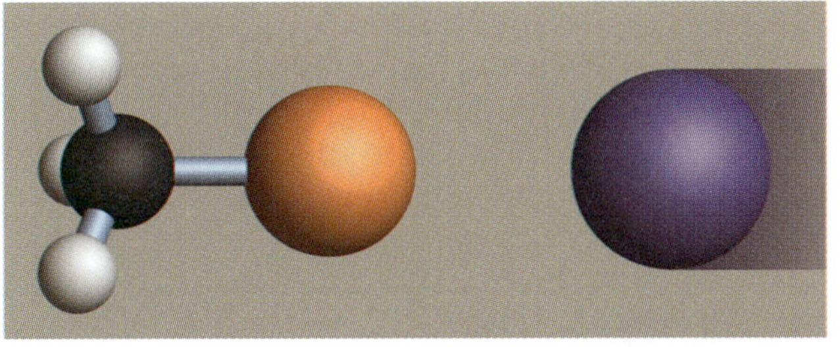

(b)

▲ **Figure 4.13**
The importance of orientation of colliding molecules is illustrated in the reaction of an iodide ion with methyl bromide to produce methyl iodide and a bromide ion. The I^- ion in (a) collides with the C atom of CH_3Br, a favorable collision for reaction to occur. The I^- ion in (b) collides with the Br atom of CH_3Br, an unfavorable collision for chemical reaction.

Temperature

Reactions generally take place faster at higher temperatures. For example, coal (carbon) reacts so slowly with oxygen (from the air) at room temperature that the change is imperceptible. However, when coal is heated to several hundred degrees, it reacts at a much more rapid rate. The heat evolved in the reaction keeps the coal burning smoothly. The effect of temperature on the rates of chemical reactions is explained by the kinetic-molecular theory. We will consider this theory in more detail in Chapter 6. For the moment we will use one of its postulates: Molecules move more rapidly at high temperatures. They collide more frequently, providing a greater chance for reaction, and the rapidly moving molecules also strike one another harder. These harder collisions are more likely to supply the activation energy needed to break chemical bonds and get the reaction going.

Consider the reaction that takes place between hydrogen gas and chlorine gas. In order for hydrogen and chlorine to react, bonds between hydrogen atoms and between chlorine atoms must be broken. In general, energy is absorbed when chemical bonds are broken, and energy is released when chemical bonds are formed.

$$H{-}H + Cl{-}Cl \longrightarrow 2\,H{-}Cl + \text{heat}$$

This reaction is exothermic. Once it has started, the energy released by the formation of hydrogen-chlorine bonds more than compensates for that required to break H—H and Cl—Cl bonds. There is a net conversion of chemical energy to heat energy.

In endothermic reactions, the energy released by bond formation is less than that required to break the necessary bonds. For these reactions energy generally must be supplied continuously from an external source or the process will stop. A typical endothermic reaction is the decomposition of the salt potassium chlorate ($KClO_3$) to give oxygen and another salt, potassium chloride (KCl).

$$2\,KClO_3(s) + \text{heat} \longrightarrow 2\,KCl(s) + O_2(g)$$

The potassium chlorate must be heated continuously. If the source of heat is removed, the reaction quickly subsides.

Every day, we use our knowledge of the effect of temperature on chemical reactions. For example, we freeze foods to retard the chemical reactions that lead to spoilage. On the other hand, if we want to cook our food more rapidly, we turn up the heat. The chemical reactions that occur in our bodies generally do so at a constant temperature of 37 °C. A few degrees' rise in temperature (fever) leads to an increase in respiration, pulse rate, and other physiological reactions. A drop in body temperature of a few degrees slows these same processes considerably, as is exemplified by the slowed metabolism of hibernating animals. This phenomenon is used in some surgical procedures. In some cases of heart surgery the body temperature of the patient is lowered to about 15 °C (60 °F). Ordinarily, the brain is permanently damaged when its oxygen supply is interrupted for more than five minutes. But at the lower temperature metabolic processes slow considerably and the brain can survive much longer periods of oxygen deprivation. The surgeon can stop the heartbeat, perform an hours-long surgical procedure on the heart, and then restart the heart and bring the patient's temperature back to normal (Figure 4.14).

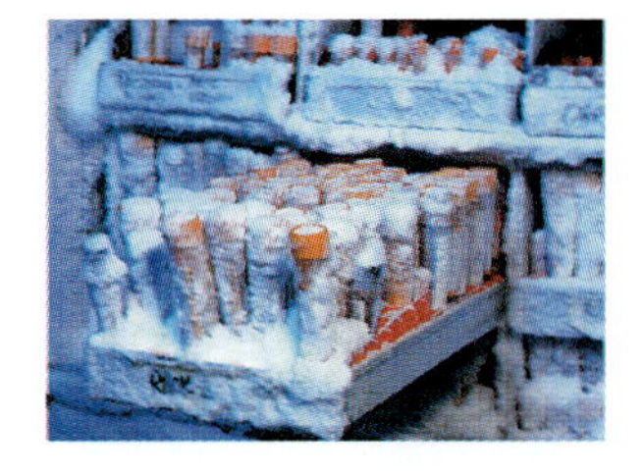

▲ Blood plasma is usually frozen to slow its decomposition and extend its shelf life. Organs harvested for transplant are transported in ice-packed "coolers" for the same reasons.

Despite these examples, the manipulation of reaction rates in living systems through changes in temperature is severely restricted. For living cells there is often a rather narrow range of optimum temperatures. Both higher and lower temperatures can be disabling, if not deadly. Increasing temperature, for example, may deactivate (render inactive) enzymes (page 118) that mediate cellular chemistry and are essential to life. We kill germs by heat sterilization in autoclaves.

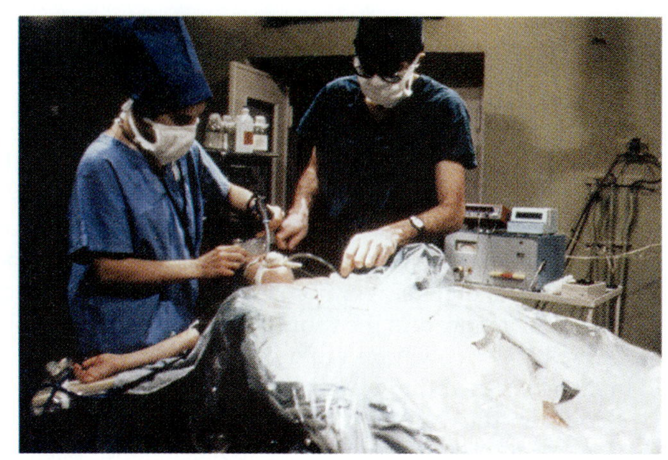

Figure 4.14 ▶ Heart surgery can be performed with the patient's temperature lowered to slow metabolic processes and thus minimize the possibility of brain damage. Instead of packing the patient in ice, as shown here, modern procedures usually use an external heart pump that circulates the blood from an artery through an ice pack and back into a vein. Body temperature is lowered to about 15 °C in this way.

Catalysis

In order for the decomposition of potassium chlorate (page 117) to produce oxygen at a useful rate, the potassium chlorate must be heated to over 400 °C. However, if we add a small amount of manganese dioxide (MnO_2), we can get the same rate of oxygen evolution at only 250 °C. Further, after the reaction is complete, the manganese dioxide can be completely recovered, unchanged. A substance that, like manganese dioxide, changes the rate of a chemical reaction without itself being changed is called a **catalyst.** In general, catalysts act by lowering the activation energy (Figure 4.15). If activation energy is lower, then the collisions of more slowly moving molecules will be sufficient to supply that energy. The lower temperature required for a catalyzed reaction reflects this. A catalyst lowers the activation energy by changing the *path* of the reaction. To return to our analogy of the trip from Browning to Kalispell, it is possible to take an alternate route, U.S. Highway 2, which crosses the continental divide through Marias Pass. This route involves a climb of only 300 m, compared with 700 m via Logan Pass (Figure 4.16). This alternate route is analogous to that provided by a catalyst in a chemical reaction.

Catalysts are of great importance in the chemical industry. The proper catalyst can make an otherwise impractically slow reaction proceed at a reasonable rate. Catalysts are even more important in living organisms, where raising the temperature by 100 °C is not a feasible way to increase the rate of critical reactions. If we raised our body temperature by 100 °C, we'd boil our blood, among other things. Biological catalysts, called **enzymes,** mediate nearly all the chemical reactions that take place in living systems. But these catalysts themselves may suffer irreversible damage when living cells are subjected to excessive temperature. Once a catalyst is deactivated, the reactions that require it no longer proceed at a rate that maintains

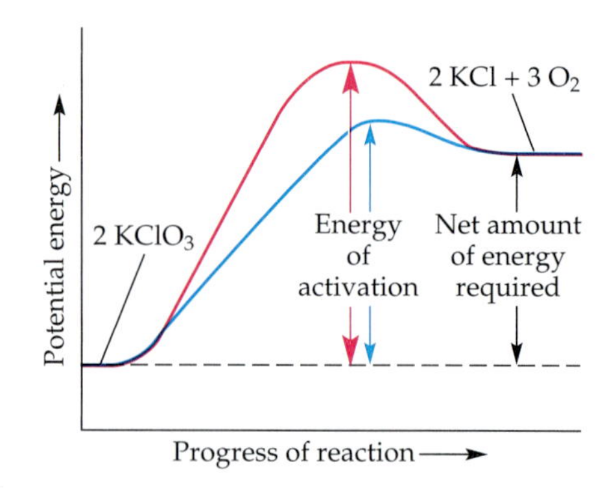

Figure 4.15 ▶ In the decomposition of potassium chlorate, a catalyst acts to lower the energy of activation (blue arrow and curve) as compared with that for the uncatalyzed reaction (red arrow and curve).

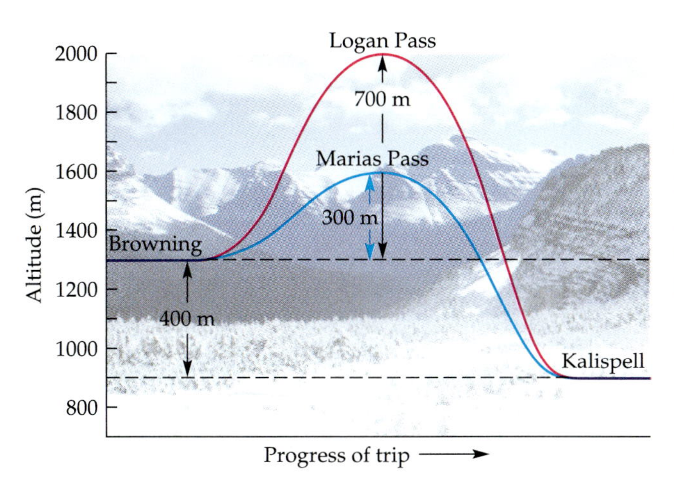

Figure 4.16
An analogy to the profile of a reaction and of activation energy for a catalyzed reaction. In our analogy of the trip from Browning to Kalispell (Figure 4.9), you could take an alternate route, U.S. Highway 2, crossing the continental divide at Marias Pass (catalyzed reaction). Then the climb is only 300 m (lower activation energy), compared with 700 m via Logan Pass (uncatalyzed reaction). This alternate route is analogous to the alternate pathway provided by a catalyst in a chemical reaction.

life. In addition to serving as examples of catalysts, enzymes offer a striking illustration of the importance of orientation in chemical reactions. Each enzyme is a huge protein molecule, with what is called an active site. The reactants must come into contact with this active site if the enzyme is to catalyze the reaction. If the reactants do not collide at the active site, no reaction occurs. So important are enzymes to life that we devote Chapter 22 to them.

Concentration

Another factor affecting the rate of a chemical reaction is the concentration of reactants. The more reactant molecules there are in a volume of space, the more collisions will occur. The more collisions there are, the more reactions will occur. For example, if you light a wood splint and then blow out the flame, the splint will continue to glow as the wood reacts slowly with the oxygen in the air. If the glowing splint is placed in pure oxygen, the splint will burst into flame, indicating a much more rapid reaction. This more rapid reaction can be interpreted in terms of the concentration of oxygen. Air is about one-fifth oxygen. The concentration of O_2 molecules in pure oxygen is therefore about five times as great as in air. The caution against smoking in a hospital room where a patient is in an oxygen tent is not merely a concession to the sensitivity of nonsmokers. It is meant to prevent disaster (Figure 4.17).

For reactions in solution, the concentration of a reactant can be increased by dissolving more of it. One of the first studies of reaction rate was done by Ludwig Wilhelmy in 1850. He studied the rate of reaction of sucrose (cane or beet sugar) with water. The products are two simpler sugars, glucose and fructose (see Chapter 19).

$$\text{Sucrose} + H_2O \xrightarrow{\text{HCl}} \text{Glucose} + \text{Fructose}$$

(We will not be concerned with the chemical formulas for these sugars at this time. The hydrochloric acid serves as a catalyst for the reaction. Chemists generally write formulas or names of catalysts above the arrow, because the catalyst is recovered unchanged.) Wilhelmy found that the rate of the reaction was proportional to the concentration of sucrose. If he dissolved 0.001 mol of sucrose in 1 L of water, the sucrose reacted at a certain rate. If he doubled the concentration of sucrose (for example, if he dissolved 0.002 mol of sucrose in 1 L of water), the reaction rate was doubled.

In general, when the temperature is constant and the catalyst (if any) is present in a fixed amount, the rate of the reaction can be related quantitatively to the

Figure 4.17
The NO SMOKING sign isn't simply an attempt to avoid exposing the patient to the hazards of second-hand smoke. It is a warning about the increased danger of fire in an oxygen-rich atmosphere.

amounts of reacting substances. The relationship is not necessarily a simple one, however.

Consider again the photosynthesis reaction.

$$6\,CO_2 + 6\,H_2O + \text{energy} \longrightarrow C_6H_{12}O_6 + 6\,O_2$$

For this reaction to occur in one step, six carbon dioxide molecules and six water molecules would all have to come together at the same instant in an effective collision. Such an event is highly unlikely. In fact, even three-body collisions are rare. Most reactions proceed through a series of steps involving collisions of only two molecules each. Some steps may involve only the breaking of a bond in a single molecule. The photosynthesis reaction involves an extraordinarily complex web of intermediate steps, but each individual step is fairly simple. Virtually all of these steps require enzymes as catalysts.

The step-by-step process, or **mechanism,** by which a reaction takes place is determined by a study of the rate of the reaction as various factors, such as temperature, concentrations, and catalysts, are changed. Such studies are important in chemistry, for chemists want to get as much product as possible with a minimum expenditure of materials and energy. Investigations of mechanisms are also important in the chemistry of living cells. Some types of cancer are thought to be induced by chemicals. A lot of research is under way to work out the mechanism by which the chemicals act—or are acted upon—in the induction of cancer. A knowledge of the mechanisms of various metabolic reactions has enabled scientists to determine how certain poisons work, and this knowledge has led to effective treatments in many instances. Certain diseases of genetic origin involve the disruption of a single step in the mechanism of a reaction essential to health. Again, this information has permitted physicians to deal successfully with what might otherwise have been a fatal defect.

✓ Review Question

4.17 If all other conditions are held constant, what effect would each of the following have on the rate of a reaction: (**a**) increasing the temperature? (**b**) increasing the concentration of a reactant? (**c**) adding a catalyst?

4.10 Equilibrium in Chemical Reactions

For reactions that proceed in both a forward and a backward direction at the same time, the concept of rate requires further consideration. In such reactions, the reverse reaction cannot occur until some product molecules have been formed. The reverse reaction will be slow at first, because of the low concentration of product molecules. Eventually the forward reaction will produce sufficient product molecules to increase the chance of their re-forming the original reactants. Similarly, as the reactants change to products, there will be fewer collisions between reactant molecules, because their number becomes depleted. In isolated systems (ones that are cut off from outside influences), the rates of the forward and reverse reactions eventually become equal, and a condition called **equilibrium** is established (Figure 4.18).

At equilibrium there *appears* to be no further reaction. In this case appearances are deceiving. Reactants are still changing to products, and products are still changing back to reactants. It is just that these changes are occurring at precisely the same rate. For every reactant molecule lost through the forward reaction, one is gained through the reverse reaction. Once equilibrium is established, we can measure no change in the concentrations of reactants or products. Because molecules are still reacting, even though their concentrations don't change, we say that equilibrium is a dynamic situation.

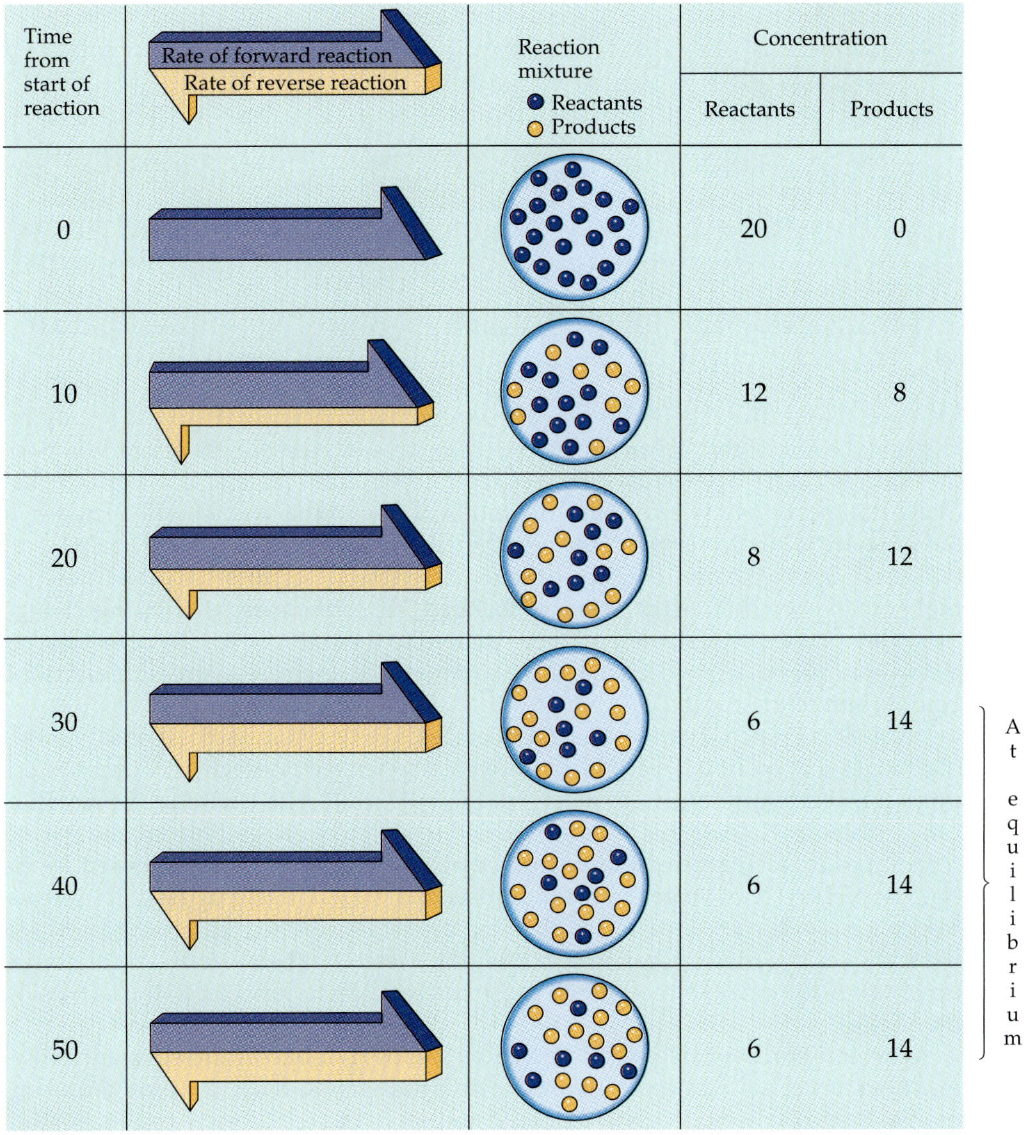

Time from start of reaction	Rate of forward reaction / Rate of reverse reaction	Reaction mixture (● Reactants, ○ Products)	Concentration	
			Reactants	Products
0			20	0
10			12	8
20			8	12
30			6	14
40			6	14
50			6	14

At equilibrium (times 30–50)

▲ **Figure 4.18**
Progress of a reaction toward equilibrium.

Equilibrium is established for reversible reactions in isolated systems, those for which external conditions such as temperature and pressure do not change. What happens if those external conditions do change? To answer this question, we shall once again look at the reaction of nitrogen and hydrogen to form ammonia.

$$N_2(g) + 3\,H_2(g) \rightleftharpoons 2\,NH_3(g) + \text{heat}$$

The double arrow indicates that this is a *reversible* reaction. All of the compounds involved in this reaction are gases. From the work of Gay-Lussac and Avogadro we know that the coefficients given in the equation reflect the combining volumes of the gases. The reactants occupy four volumes and the products occupy two. In Chapter 6, we study the effects of temperature and pressure on the volume of a gas. For the moment let us just say that increased pressure tends to reduce the volume of a gas. If the pressure of this equilibrium system is increased, the equilibrium will be disrupted, and the rate at which N_2 reacts with H_2 to form NH_3 will

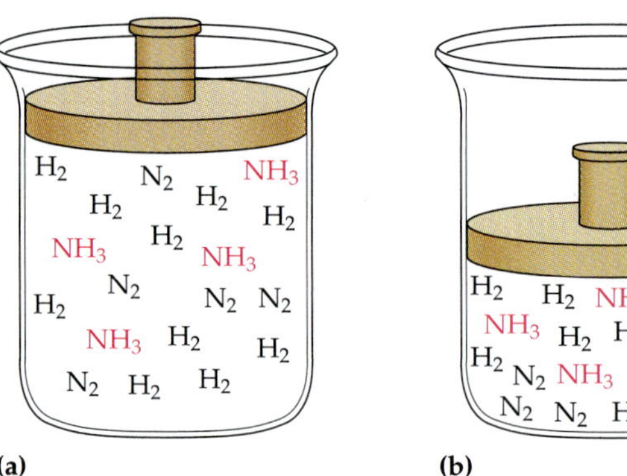

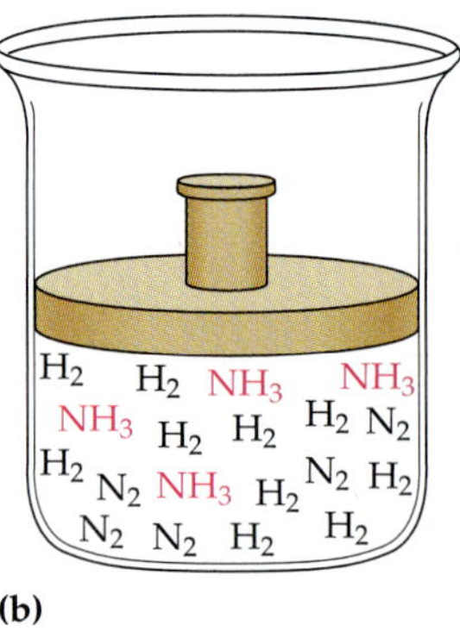

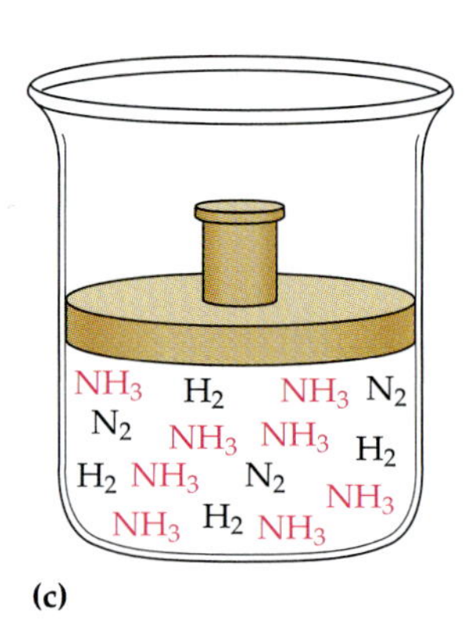

Figure 4.19 ▶ Le Châtelier's principle illustrated. (a) System at equilibrium with 10 H_2, 5 N_2, and 4 NH_3, a total of 19 molecules. (b) The same molecules forced into a smaller volume, creating a stress on the system. (c) Six H_2 and 2 N_2 have been converted to 4 NH_3. A new equilibrium has been established with 4 H_2, 3 N_2, and 8 NH_3, a total of 15 molecules. The stress imposed by the smaller volume is partially relieved by the reduction in the total number of molecules.

increase; that is, the rate of the forward reaction will increase (Figure 4.19). By increasing the rate of the forward reaction, the increased pressure, therefore, will have the effect of changing reactants that occupy four volumes (H_2 and N_2) to product(s) (NH_3) that occupy only two. Thus the total volume of the system will decrease. If the pressure is held constant at the new, higher value, equilibrium will again be established. But in the new equilibrium the concentration of ammonia (NH_3) will be higher than it was before the pressure changed. We sometimes describe this change by saying that the equilibrium has been shifted to the right. So for the reversible reaction of nitrogen and hydrogen to form ammonia, an increase in pressure shifts the equilibrium to the right.

In 1884 a French chemist, Henri Louis Le Châtelier, summarized observations of changes that occur in an equilibrium system when factors such as concentration, pressure, and temperature are changed. His rule, called **Le Châtelier's principle,** can be stated as follows: If a stress is applied to a system in equilibrium, the system rearranges in such a direction as to minimize the stress. If heat is added to the $N_2/H_2/NH_3$ system, the reaction will proceed to the left, using up heat. If more nitrogen gas is added to the system, the reaction will go to the right, using up $N_2(g)$. If additional hydrogen gas is introduced, the system will shift to the right, to use up $H_2(g)$. Additional ammonia will cause the reaction to proceed to the left, using up $NH_3(g)$.

The equilibrium can also be shifted by removal of one of the substances. Removal of ammonia will cause hydrogen and nitrogen to react, forming more ammonia. Removal of hydrogen will cause ammonia to break down and form more hydrogen (and, incidentally, more nitrogen). It should be noted, however, that a catalyst will not shift an equilibrium system. Catalysts change the rate of both the forward and reverse reactions. They do not change the position of the equilibrium, and the equilibrium concentrations of reactants and products are not altered.

▲ Henri Le Châtelier (1850–1936) was one of the first chemists to understand fully the difference between a reaction that goes to completion and one that only reaches an equilibrium condition.

Example 4.13

What effect, if any, will each of the following changes have on the equilibrium in the reaction

$$2\,CO(g) + O_2(g) \rightleftharpoons 2\,CO_2(g) + \text{heat}$$

a. adding CO **b.** removing O_2 **c.** cooling the reaction mixture
d. increasing the pressure **e.** adding a catalyst

Solution

a. The reaction shifts to the right to use up the added CO.
b. The reaction shifts to the left to replace the O_2 that is removed.
c. The reaction shifts to the right to replace the lost heat.

d. The reaction shifts to the right to relieve the pressure by converting three molecules ($2\ CO + 1\ O_2$) into two molecules ($2\ CO_2$).
e. A catalyst has no effect on the position of equilibrium.

Practice Exercises

A. How will the addition of $Cl_2(g)$ to the system below (isolated, constant temperature, and at equilibrium) affect the equilibrium concentration of CO?

$$CO(g) + Cl_2(g) \rightleftharpoons COCl_2(g)$$

B. If gaseous dinitrogen tetroxide is warmed in a closed vessel, it begins to dissociate into nitrogen dioxide gas. At any given temperature an equilibrium will be established. Write the equation for the reaction, showing heat as a reactant or product. How will the addition of $NO_2(g)$ to the system (isolated, constant, and at equilibrium) affect the equilibrium concentration of N_2O_4?

The concept of equilibrium is quite important to our study of the chemistry of life. In Chapter 7, we discuss equilibriums between liquid and solid phases and between liquid and vapor phases. In Chapter 8, we encounter equilibriums involving solutions, and in Chapter 10 we illustrate some equilibrium calculations.

Review Questions

4.18 What is meant by the term dynamic equilibrium?

4.19 State Le Châtelier's principle.

Summary

A **chemical equation** is a shorthand way of using symbols and formulas to show the changes that occur in a chemical reaction. Parenthetical letters are used to indicate physical state: (g) = gaseous state, (l) = liquid state, (s) = solid state.

In a **balanced** chemical equation, for every element shown, the number of atoms on the left equals the number on the right. An equation is balanced by placing numbers, called **coefficients,** before the formulas of the substances. The coefficient 1 is implied when no other number is shown.

Gay-Lussac's **law of combining volumes** states that, for gases measured at the same temperature and pressure, the volumes of the gases are always in a small whole-number ratio. Although he could not *count* the molecules, Avogadro *reasoned* that Gay-Lussac's law is true because gas volume is a macroscopic indicator of number of gas molecules present. **Avogadro's hypothesis** states that at the same temperature and pressure, equal volumes of any two gases contain equal numbers of gas molecules. Scientists since Avogadro's time have determined experimentally that 1 mol of any element contains 6.02×10^{23} atoms of the element. This number 6.02×10^{23} is called **Avogadro's number.**

The **mole** is the basic SI unit for amount of substance. It is defined as the amount of substance containing the number of basic units equal to the number of atoms in exactly 12 g of carbon-12. Because there are 6.02×10^{23} atoms in 12 g of carbon-12, there are 6.02×10^{23} basic units in 1 mol of any substance.

The **molar mass** of any substance is the mass in grams of 1 mol of that substance. The molar mass of any substance is calculated from the weighted-average atomic masses of the atoms in its formula.

The coefficients in chemical equations represent moles as well as atoms and molecules. Knowing the numbers of moles, we can calculate amounts of reactants and products.

Chemical reactions that give off heat are **exothermic reactions;** those that require an input of heat in order to proceed are **endothermic reactions.** The **activation energy** of any chemical reaction is the amount of energy needed to get the reaction started. A reaction can be exothermic and still require some initial input of activation energy, as exemplified by the burning of coal.

The rate at which a chemical reaction proceeds depends on (1) frequency of collisions between reactants, (2) how colliding reactants are aligned with each other, and (3) how much energy each collision produces. Because collision frequency is directly proportional to reactant concentrations and to temperature, reaction rate is also directly proportional to these two variables.

A **catalyst** is any substance that increases the rate of a chemical reaction without itself being changed by the reaction. A catalyst works by lowering the activation energy needed to get a reaction going.

A **reversible reaction** is one that can proceed in either direction. The reaction reactants $\longrightarrow$ products is called the forward direction, and the reaction products $\longrightarrow$ reactants is called the reverse direction. At first, a reversible reaction is mostly forward because the concentration of reactants is high.

As more reactant is used up and more product formed, the forward reaction slows down and the reverse reaction speeds up. **Equilibrium** is established at the point where $rate_{forward}$ equals $rate_{reverse}$.

Once equilibrium is established, it continues until some external agent changes the reaction conditions. We can use **Le Châtelier's principle** to predict how any such change affects equilibrium: Any stress put on an equilibrium system causes the system to react in such a way as to reduce the stress. Here "stress" means a change in product or reactant concentration, a change in temperature, or a change in pressure.

Key Terms

activation energy (4.8)
Avogadro's hypothesis (4.2)
Avogadro's number (4.3)
catalyst (4.9)
chemical equation (4.1)
endothermic (4.7)
enthalpy change (ΔH) (4.7)
enzyme (4.9)
equilibrium (4.10)
exothermic (4.7)
formula mass (4.4)
formula unit (4.4)
heat of reaction (4.7)
law of combining volumes (4.2)
Le Châtelier's principle (4.10)
mechanism (4.9)
molar mass (4.5)
mole (4.5)
molecular mass (4.4)
photosynthesis (4.8)
product (4.1)
reactant (4.1)
stoichiometric factor (4.6)

Problems

Balancing Chemical Equations

1. Indicate whether the following equations are balanced. (You need not balance the equation; just determine whether it is balanced as written.)
   a. $Ca + 2\,H_2O \longrightarrow Ca(OH)_2 + H_2$
   b. $2\,LiOH + CO_2 \longrightarrow Li_2CO_3 + H_2O$
   c. $4\,LiH + AlCl_3 \longrightarrow 2\,LiAlH_4 + 2\,LiCl$
   d. $2\,Sn + 2\,H_2SO_4 \longrightarrow 2\,SnSO_4 + SO_2 + 2\,H_2O$

2. Indicate whether the following equations are balanced as written.
   a. $2\,KNO_3 + 10\,K \longrightarrow 6\,K_2O + N_2$
   b. $2\,NH_3 + O_2 \longrightarrow N_2 + 3\,H_2O$
   c. $SF_4 + 3\,H_2O \longrightarrow H_2SO_3 + 4\,HF$
   d. $4\,BF_3 + 3\,H_2O \longrightarrow H_3BO_3 + 3\,HBF_4$
   e. $3\,Cl_2 + 6\,NaOH \longrightarrow 5\,NaCl + NaClO_3 + 3\,H_2O$

3. Balance the following equations.
   a. $Cl_2O_5 + H_2O \longrightarrow HClO_3$
   b. $V_2O_5 + H_2 \longrightarrow V_2O_3 + H_2O$
   c. $Al + O_2 \longrightarrow Al_2O_3$
   d. $Sn + NaOH \longrightarrow Na_2SnO_2 + H_2$
   e. $PCl_5 + H_2O \longrightarrow H_3PO_4 + HCl$
   f. $Na_3P + H_2O \longrightarrow NaOH + PH_3$
   g. $Cl_2O + H_2O \longrightarrow HClO$
   h. $CH_3OH + O_2 \longrightarrow CO_2 + H_2O$
   i. $Zn(OH)_2 + H_3PO_4 \longrightarrow Zn_3(PO_4)_2 + H_2O$
   j. $C_3H_8 + O_2 \longrightarrow CO_2 + H_2O$

4. Balance the following equations.
   a. $TiCl_4 + H_2O \longrightarrow TiO_2 + HCl$
   b. $C_4H_{10} + O_2 \longrightarrow CO_2 + H_2O$
   c. $WO_3 + H_2 \longrightarrow W + H_2O$
   d. $Al_4C_3 + H_2O \longrightarrow Al(OH)_3 + CH_4$
   e. $Al_2(SO_4)_3 + NaOH \longrightarrow Al(OH)_3 + Na_2SO_4$
   f. $Ca_3P_2 + H_2O \longrightarrow Ca(OH)_2 + PH_3$
   g. $Cl_2O_7 + H_2O \longrightarrow HClO_4$
   h. $MnO_2 + HCl \longrightarrow MnCl_2 + Cl_2 + H_2O$
   i. $Fe + O_2 \longrightarrow Fe_3O_4$
   j. $C_5H_{12} + O_2 \longrightarrow CO_2 + H_2O$

Volume Relationships in Chemical Equations

5. Calculate the volume of methane that must decompose to produce 10.0 liters of hydrogen in the following reaction, if the two gases are compared at the same temperature and pressure.

$$CH_4(g) \longrightarrow C(s) + 2\,H_2(g)$$

6. Using the equation in Example 4.4 (page 103), calculate the volume of $O_2(g)$ that must react to form 10.0 liters of steam, if the two gases are compared at the same temperature and pressure.

7. Consider the following equation.

$$2\,C_4H_{10}(g) + 13\,O_2(g) \rightarrow 8\,CO_2(g) + 10\,H_2O(g)$$

   a. How many liters of $H_2O(g)$ are formed when 0.529 L of $C_4H_{10}(g)$ is burned? Assume both gases are measured under the same conditions.
   b. How many liters of $O_2(g)$ are required to burn 16.1 L of $C_4H_{10}(g)$? Assume both gases are measured under the same conditions.

8. Consider the following equation.

$$C_2H_4(g) + 3\,O_2(g) \rightarrow 2\,CO_2(g) + 2\,H_2O(g)$$

   a. How many liters of $CO_2(g)$ are formed when 2.93 L of $C_2H_4(g)$ is burned? Assume both gases are measured under the same conditions.
   b. How many liters of $O_2(g)$ are required to form 0.370 L of $CO_2(g)$? Assume both gases are measured under the same conditions.

Molecular Formulas and Formula Units

9. How many hydrogen atoms are indicated in one formula unit of (a) NH_4NO_3? (b) CH_3OH? (c) $CH_3CH_2CH_3$? (d) C_6H_5COOH?
10. How many hydrogen atoms are indicated in one formula unit of (a) $(NH_4)_2HPO_4$? (b) $Al(C_2H_3O_2)_3$?
11. How many oxygen atoms are indicated in one formula unit of (a) $Al_2(C_2O_4)_3$? (b) $Ca_3(PO_4)_2$?
12. How many atoms of each kind (Al, C, H, and O) are indicated by the notation 2 $Al(C_2H_3O_2)_3$?

Molecular Masses, Formula Masses, and Molar Masses

13. Calculate the molecular mass or formula mass of each of the following.
    a. C_6H_5Br b. H_3PO_4 c. $K_2Cr_2O_7$
14. Calculate the molecular mass or formula mass of each of the following.
    a. $C_2H_5NO_2$ b. $Na_2S_2O_3$ c. $(NH_4)_3PO_4$
15. Calculate the mass, in grams, of each of the following.
    a. 0.00500 mol MnO_2
    b. 1.12 mol CaH_2
    c. 0.250 mol $C_6H_{12}O_6$
16. Calculate the mass, in grams, of each of the following.
    a. 4.61 mol $AlCl_3$ b. 0.615 mol Cr_2O_3
    c. 0.158 mol IF_5
17. Calculate the amount, in moles, of each of the following.
    a. 98.6 g HNO_3 b. 9.45 g CBr_4
    c. 9.11 g $FeSO_4$ d. 11.8 g $Pb(NO_3)_2$
18. Calculate the amount, in moles, of each of the following.
    a. 16.3 g SF_6 b. 25.4 g $Pb(C_2H_3O_2)_2$
    c. 35.6 g $FeCl_3$ d. 75.3 g $Co(ClO_3)_2$

Mole and Mass Relationships in Chemical Equations

19. Consider the reaction for the combustion of octane.

$$2\,C_8H_{18} + 25\,O_2 \longrightarrow 16\,CO_2 + 18\,H_2O$$

    a. How many moles of CO_2 are produced when 2.09 mol of octane is burned?
    b. How many moles of oxygen are required to burn 4.47 mol of octane?
20. Consider the reaction for the combustion of octane (Problem 19).
    a. How many moles of H_2O are produced when 2.81 mol of octane is burned?
    b. How many moles of CO_2 are produced when 4.06 mol of oxygen is consumed?
21. What mass of ammonia, in grams, can be made from 440 g of H_2?

$$N_2 + H_2 \longrightarrow NH_3 \quad (\textit{not balanced})$$

22. What mass of hydrogen, in grams, is needed to react completely with 892 g of N_2 (see Problem 21)?
23. What mass of oxygen, in grams, can be prepared from 24.0 g of H_2O_2?

$$H_2O_2 \rightarrow H_2O + O_2 \quad (\textit{not balanced})$$

24. Toluene and nitric acid are used in the production of trinitrotoluene (TNT), an explosive.

$$C_7H_8 + HNO_3 \rightarrow C_7H_5N_3O_6 + H_2O \quad (\textit{not balanced})$$

Toluene TNT

What mass of nitric acid, in grams, is required to react with 454 g of C_7H_8?

25. Use the equation in Problem 24 to calculate the mass of TNT that can be made from 829 g of C_7H_8.
26. What mass of quicklime (calcium oxide), in kilograms, can be made when 4.72×10^9 g of limestone (calcium carbonate) is decomposed by heating?

$$CaCO_3(s) \longrightarrow CaO(s) + CO_2(g)$$

27. What mass of nitric acid, in grams, can be made from 971 g of ammonia?

$$NH_3 + O_2 \longrightarrow HNO_3 + H_2O \ (\textit{not balanced})$$

28. In an oxyacetylene welding torch, acetylene (C_2H_2) burns in pure oxygen with a very hot flame.

$$C_2H_2 + O_2 \longrightarrow CO_2 + H_2O \quad (\textit{not balanced})$$

What mass of oxygen, in grams, is required to react with 52.0 g of C_2H_2?

Structure, Stability, and Spontaneity

29. A reaction has a heat of reaction (ΔH) of +36.3 kJ. Is the reaction endothermic or exothermic?
30. What is the ΔH value for the reverse of the reaction in Problem 29? Is the reverse reaction endothermic or exothermic?
31. What is meant by the *mechanism* of a reaction?
32. Why does a catalyst affect the rate of a reaction?

Le Châtelier's Principle

33. According to Le Châtelier's principle, what effect will increasing the temperature have on the following equilibria?
    a. $H_2(g) + Cl_2(g) \rightleftharpoons 2\,HCl(g) + \text{heat}$
    b. $2\,CO_2(g) + \text{heat} \rightleftharpoons 2\,CO(g) + O_2(g)$
    c. $3\,O_2(g) + \text{heat} \rightleftharpoons 2\,O_3(g)$
34. According to Le Châtelier's principle, what effect will decreasing the concentration of $O_2(g)$ have on the following equilibria?
    a. $3\,O_2(g) \rightleftharpoons 2\,O_3(g)$
    b. $2\,CO_2(g) \rightleftharpoons 2\,CO(g) + O_2(g)$
    c. $2\,NO(g) + O_2(g) \rightleftharpoons 2\,NO_2(g)$

Additional Problems

35. Propane burns in air to form carbon dioxide and water. The equation is

$$C_3H_8 + 5\,O_2 \longrightarrow 3\,CO_2 + 4\,H_2O$$

State the molecular, molar, and mass relationships indicated by this equation.

36. The poison gas phosgene reacts with water in the lungs to form hydrogen chloride and carbon dioxide. The equation is

$$COCl_2 + H_2O \longrightarrow 2\,HCl + CO_2$$

State the molecular, molar, and mass relationships indicated by this equation.

37. Write a balanced equation to represent (**a**) the reaction of the gases carbon monoxide and nitrogen monoxide to form the gases carbon dioxide and nitrogen; (**b**) the reaction of the gases methane and water to form the gases carbon monoxide and hydrogen; and (**c**) the reaction of solid magnesium nitride and liquid water to form solid magnesium hydroxide and gaseous ammonia.

38. Write a balanced equation to represent (**a**) the decomposition of solid potassium chlorate ($KClO_3$) upon heating to form solid potassium chloride and oxygen gas as products; (**b**) the combustion of liquid methanol (CH_3OH) in oxygen to produce gaseous carbon dioxide and liquid water; and (**c**) the reaction of the gases ammonia and oxygen to produce the gases nitrogen monoxide and water.

39. Aluminum chloride, used as a topical astringent (a substance that causes contraction of pores and thus retards perspiration), can be made by the reaction of aluminum metal with hydrogen chloride gas. The equation (not balanced) is

$$Al(s) + HCl(g) \longrightarrow AlCl_3(s) + H_2(g)$$

How much $AlCl_3$ can be made from an aluminum beverage can that has a mass of 3.51 g?

40. Calculate the molecular mass of fensulfothion, an insecticide, with the condensed structural formula $(CH_3CH_2O)_2PSOC_6H_4SOOCH_3$.

41. Calculate the molecular mass of trimethobenzamide, used to suppress nausea and vomiting, which has the condensed structural formula

$$(CH_3)_2NCH_2CH_2OC_6H_4CH_2NHCOC_6H_2(OCH_3)_3$$

42. At temperatures above 300 °C, silver oxide, Ag_2O, decomposes to produce metallic silver [Ag(s)] and oxygen gas. How much oxygen, in grams, is produced from a 2.95-g sample of silver oxide?

43. Joseph Priestley discovered oxygen in 1774 by heating "red calx of mercury," mercury(II) oxide. The calx decomposed to the elements. The equation (not balanced) is

$$HgO \rightarrow Hg + O_2$$

How much oxygen, in grams, is produced by the decomposition of 10.8 g of HgO?

44. How much iron can be converted to the magnetic oxide of iron (Fe_3O_4) by 8.80 g of pure oxygen? The equation (not balanced) is

$$Fe + O_2 \rightarrow Fe_3O_4$$

45. Laughing gas (dinitrogen monoxide, N_2O, also called nitrous oxide) can be made by heating ammonium nitrate with great care. The equation (not balanced) is

$$NH_4NO_3 \rightarrow N_2O + H_2O$$

How much N_2O can be made from 4.00 g of ammonium nitrate?

46. Small amounts of hydrogen gas often are made by the reaction of calcium metal with water. The equation (not balanced) is

$$Ca + H_2O \rightarrow Ca(OH)_2 + H_2$$

How many grams of hydrogen are formed by the reaction of water with 0.413 g of calcium?

47. For many years the noble gases were called the "inert gases" because it was thought that they formed no chemical compounds. Neil Bartlett made the first noble gas compound in 1962. Xenon hexafluoride is made according to the equation (not balanced)

$$Xe + F_2 \rightarrow XeF_6$$

How many grams of fluorine are required to make 0.112 g of XeF_6?

48. A piece of aluminum foil that measures 12.3 cm × 14.3 cm × 2.2 mm reacts with an excess of hydrochloric acid. What mass of hydrogen is produced? The density of aluminum is 2.70 g/cm^3.

$$Al(s) + HCl(aq) \rightarrow AlCl_3(aq) + H_2(g) \quad (\textit{not balanced})$$

49. A coal-burning power plant burns 228 trainloads of western subbituminous coal per year. Each train is comprised of 115 cars, and each car carries 90.5 metric tons of coal (1 metric ton = 1000 kg). If the coal is 64.3% carbon, what mass of carbon dioxide, in metric tons, is produced by the plant each year?

50. The following is a side reaction in the manufacture of rayon fibers from wood pulp.

$$3\,CS_2 + 6\,NaOH \rightarrow 2\,Na_2CS_3 + Na_2CO_3 + 3\,H_2O$$

How many grams of Na_2CS_3 is produced in the reaction of 88.0 mL of liquid CS_2 (d = 1.26 g/mL)?

51. Aspartame is an artificial sweetener about 160 times as sweet as sucrose. Aspartame breaks down in acidic solution to produce (among other products) methanol, which is fairly toxic. The equation is

$$C_{14}H_{18}N_2O_5 + 2\,H_2O \xrightarrow{H^+}$$

Aspartame

$$C_4H_7NO_4 + C_9H_{11}NO_2 + CH_4O$$

Methanol

a. A typical can of soda contains about 40 g of sucrose. How much aspartame is required to obtain the same level of sweetness?

b. How much methanol is formed by the complete breakdown of that much aspartame?

c. How many cans of soda with decomposed aspartame would you have to drink to get the approximate lethal dose of 25 g of methanol?

52. Phosphine gas (used as a fumigant to protect stored grain) is generated by the reaction of water with magnesium phosphide. The equation (not balanced) is

$$Mg_3P_2 + H_2O \longrightarrow PH_3 + Mg(OH)_2$$

How much magnesium phosphide is needed to produce 134 g of PH_3?

53. Kerosene is a mixture of hydrocarbons used in domestic heating and as a jet fuel. Assume that kerosene can be represented as $C_{14}H_{30}$, and that it has a density of 0.763 g/mL.

$$C_{14}H_{30}(l) + O_2(g) \rightarrow CO_2(g) + H_2O(l) \quad (\textit{not balanced})$$

How many grams of CO_2 are produced by the combustion of 1.00 gal (3.785 L) of kerosene?

54. Acetaldehyde, CH_3CHO, ($d = 0.788$ g/mL), a liquid used in the manufacture of perfumes, flavors, dyes, and plastics, can be produced by the reaction of ethanol with oxygen.

$$CH_3CH_2OH + O_2 \rightarrow CH_3CHO + H_2O \quad (\textit{not balanced})$$

How many liters of liquid ethanol ($d = 0.789$ g/mL) must be consumed to produce 25.0 L of acetaldehyde?

55. Write a balanced chemical equation for the reaction pictured in the chapter opening photograph.

56. Use the data in the caption to Figure 4.5 to calculate the approximate number of molecules in one drop of water.

CHAPTER 5

Oxidation and Reduction

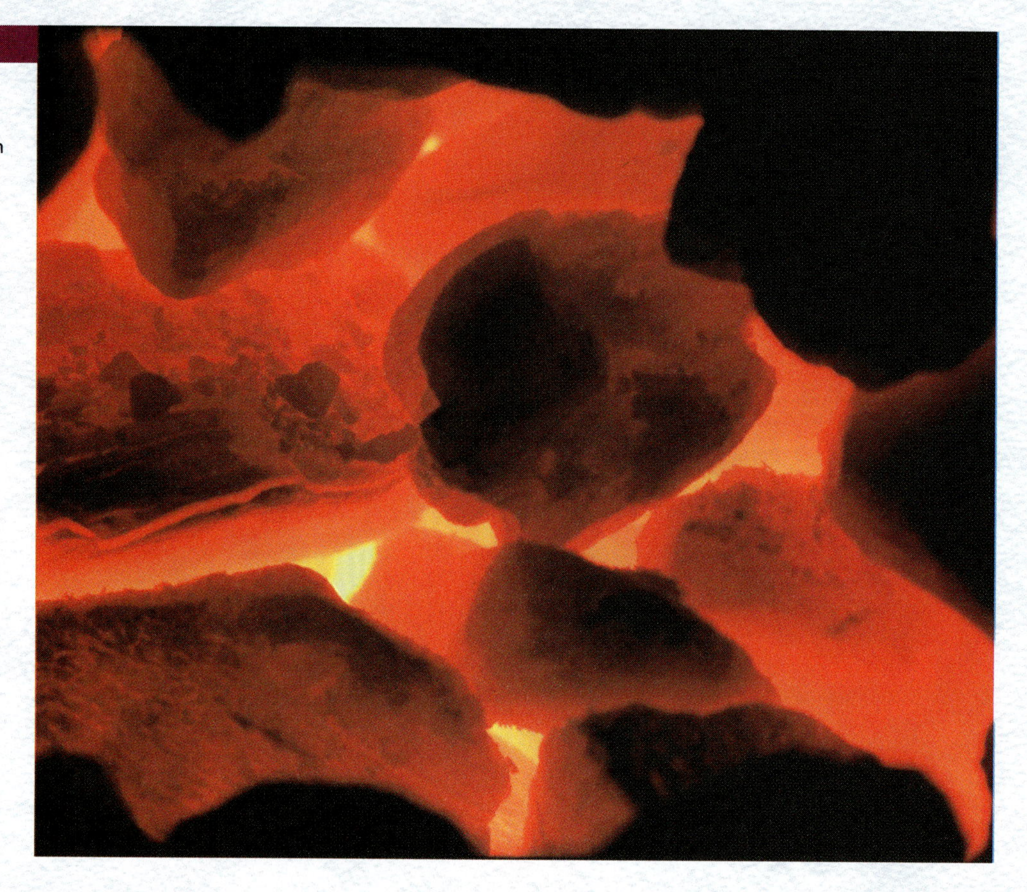

When coal burns, releasing its chemical energy as heat, the carbon in the coal is oxidized to carbon dioxide as oxygen from the air is reduced to water.

Learning Objectives/Study Questions

1. What are some important properties of oxygen?
2. What are some important properties of hydrogen?
3. List some simple ways of describing oxidation. What is its most fundamental definition?
4. List some simple ways of describing reduction. What is its most fundamental definition?
5. How do you identify an oxidation-reduction reaction? How do you use the definitions of oxidation and reduction to identify which substance is oxidized or reduced and which is the oxidizing agent or reducing agent?
6. What are some common oxidizing agents and their uses?
7. What are some common reducing agents and their uses?
8. What are some oxidation-reduction reactions that are important for living organisms?

Chemical reactions can be classified in several ways. In this chapter, we consider an important group of chemical reactions classified as *oxidation-reduction (redox[1]) reactions.* Later on, we will look at acid-base reactions (Chapters 9 and 10), inorganic (Selected Topic A) and organic (Chapters 13–17) reactions, and biochemical reactions (Chapters 19–27). These categories overlap. For example, many inorganic, organic, and biochemical reactions are also redox or acid-base reactions. We will look at some fairly simple redox reactions here, and then will use the concepts often in later chapters.

Our cells obtain energy by oxidizing foods. Green plants, using energy from sunlight, produce food by the reduction of carbon dioxide (photosynthesis). We win metals from their ores by reduction, then lose them again to corrosion as they are oxidized. We maintain our technological civilization by oxidizing fossil fuels (coal, natural gas, and petroleum) to obtain the chemical energy that was stored in these materials eons ago by green plants.

Reduced forms of matter—food, coal, gasoline—are high in energy. Oxidized forms—carbon dioxide and water—are low in energy. Let's examine the processes of oxidation and reduction in some detail, in order to better understand the chemical reactions that keep us alive and enable us to maintain our civilization.

Oxidation and reduction always occur together. You can't have one without the other. When one substance is oxidized, another is reduced. For convenience, however, we may choose to talk about only a part of the process—the oxidation part or the reduction part.

5.1 Oxygen and Hydrogen: An Overview of Oxidation and Reduction

Oxidation-reduction reactions include the extraction of metals from their ores, all combustion processes, the manufacture of countless chemicals, many of the reactions occurring in our natural environment, and most metabolic reactions in living organisms. Before beginning a systematic study, we will look at two important elements—hydrogen and oxygen—that are intimately involved in oxidation and reduction processes and take a brief overview of the subject.

Oxygen: Abundant and Essential

The word "oxidation" originally was used to describe reactions in which a substance combined with oxygen. The word "reduction" described the opposite

(a)

(b)

(c)

▲ Some reduced forms of matter. The energy in food (a) and fossil fuels (b) and (c) is released when these materials are oxidized.

▲ Oxidation and reduction always occur together. In this photograph, a copper penny reacts with nitric acid. The copper metal is oxidized to copper(II) ions (Cu^{2+}), producing a green-blue solution. The nitric acid, HNO_3, is reduced to NO_2 gas, as indicated by the red-brown fumes.

[1]Chemists often refer to oxidation-reduction reactions by turning the words around and abbreviating the term as "redox" reactions; redox is a little easier to say than "oxred."

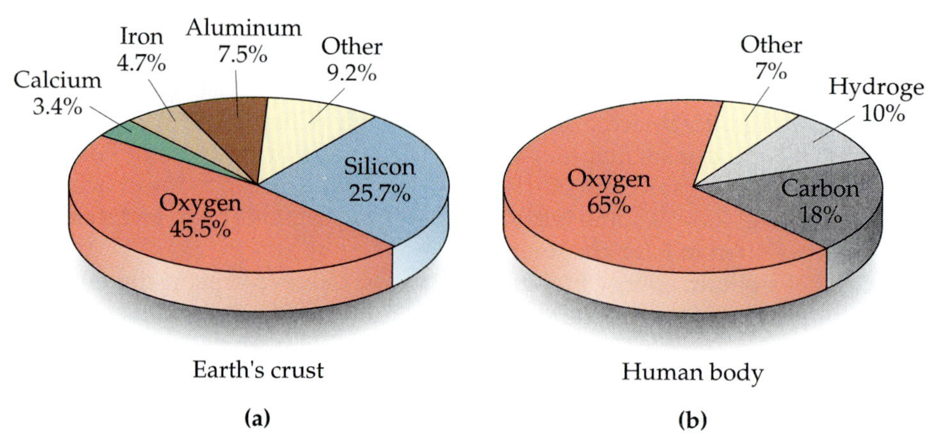

Figure 5.1 ▶
Percent by mass of elements in (a) Earth's crust (including the atmosphere and the oceans, lakes, and streams) and (b) the human body.

process, the removal of oxygen. The most abundant element on Earth, oxygen is one of the two dozen or so elements essential to life. Oxygen occurs in the atmosphere mainly in the form of O_2 molecules. Overall, O_2 makes up about 21% by volume of the air we breathe. (The rest is mostly rather unreactive nitrogen [N_2].) Oxygen also forms compounds with nearly all the other elements. Combined with hydrogen in the remarkable compound water, oxygen makes up about 89% by mass of the oceans, seas, rivers, and lakes that cover three-quarters of Earth's surface. Oxygen constitutes 45.5% by mass of Earth's solid crust, where it is combined with silicon (sand is mainly SiO_2) and many other elements.

Oxygen is found in most of the compounds that are important to living organisms. Foodstuffs—carbohydrates, fats, and proteins—all are oxygen-containing molecules. The human body is approximately 65% water by mass. Because water is 89% oxygen by mass, and many other compounds in the human body also contain oxygen, about two-thirds of our body mass is oxygen (Figure 5.1).

We take O_2 from the air into our lungs, where it passes into our bloodstream and is carried to body tissues. There it combines with the food we eat. This process, called *respiration,* provides us with all our energy. Fuels such as wood, gasoline, and coal also need oxygen to burn and release their stored energy (Figure 5.2).

▲ **Figure 5.2**
Cooking, breathing, and transportation fuel all involve oxidation.

Combustion of such fossil fuels currently supplies about 86% of the energy that turns the wheels of civilization.

Not all that atmospheric oxygen does is immediately desirable. Oxygen causes iron to rust and copper to corrode. It causes food to spoil, and it aids in the decay of wood. All these and many other chemical processes are called oxidation.

Pure oxygen is obtained by liquefying air and then letting the nitrogen and argon boil off. (Nitrogen boils at −196 °C, argon at −186 °C, and oxygen at −183 °C.) About 50 billion pounds of oxygen is produced annually in the United States, but most of it is used directly by industry, much of it by steel plants. About 1% is compressed into tanks for use in welding, hospital respirators, and other purposes.

▲ Corrosion of iron, illustrated by these rusted chains, causes great economic loss, estimated at $70 billion a year in the United States.

✓ Review Questions

5.1 Where does oxygen occur on Earth (**a**) as O_2 and (**b**) in compounds?

5.2 How is most pure oxygen prepared?

Chemical Properties of Oxygen: Oxidation When iron rusts, it combines with oxygen from the air to form a reddish brown powder. We can write the reaction, in a simplified form, as follows.

$$4\,Fe(s) + 3\,O_2(g) \longrightarrow 2\,Fe_2O_3(s)$$

The chemical name for (Fe_2O_3) is iron(III) oxide, but sometimes it is called by an older name, ferric oxide. Many metals react with oxygen to form metal oxides.

Most nonmetals also react with oxygen to form oxides. For example, carbon and sulfur burn in oxygen to form carbon dioxide and sulfur dioxide, respectively.

$$C(s) + O_2(g) \longrightarrow CO_2(g)$$
$$S(s) + O_2(g) \longrightarrow SO_2(g)$$

Example 5.1

Write balanced equations for each of the following reactions.

a. Solid white phosphorous (P_4) reacts spontaneously with oxygen in air to form solid tetraphosphorus decoxide.

b. When ignited in air, magnesium metal combines readily with oxygen to form magnesium oxide.

Solution

a. We are given the formula for phosphorous, we know that oxygen occurs as diatomic molecules (O_2), and the name tetraphosphorus decoxide tells us that the product has the formula P_4O_{10}. We can write the equation.

$$P_4(s) + O_2(g) \longrightarrow P_4O_{10}(s) \quad (\textit{not balanced})$$

To balance the equation, we need 5 O_2 on the left.

$$P_4(s) + 5\,O_2(g) \longrightarrow P_4O_{10}(s) \quad (\textit{balanced})$$

We now have 4 P and 10 O on each side; the equation is balanced.

b. Magnesium metal has the symbol Mg. Oxygen occurs as diatomic molecules (O_2). From Chapter 3, we know that magnesium oxide is ionic, and that magnesium ion is

Mg^{2+} and oxide ion is O^{2-}. The formula for magnesium oxide is therefore MgO. We can write the equation.

$$\text{Mg(s)} + \text{O}_2\text{(g)} \longrightarrow \text{MgO(s)} \quad (\textit{not balanced})$$

To balance it, we need two oxygen atoms (2 MgO) on the right and then 2 Mg on the left.

$$2\,\text{Mg(s)} + \text{O}_2\text{(g)} \longrightarrow 2\,\text{MgO(s)} \quad (\textit{balanced})$$

Practice Exercises

A. When ignited, zinc burns in air with a bluish green flame to form zinc oxide (ZnO). Write the equation for the reaction.

B. When ignited with a hot flame, aluminum metal burns, producing aluminum oxide. Write the equation for the reaction.

Oxygen also reacts with many *compounds*. Methane, the principal ingredient in natural gas and the simplest of a vast number of hydrocarbons (carbon-hydrogen compounds), burns in air to produce carbon dioxide and water. (The other hydrocarbons [Chapter 13] also burn.)

$$\text{CH}_4\text{(g)} + 2\,\text{O}_2\text{(g)} \longrightarrow \text{CO}_2\text{(g)} + 2\,\text{H}_2\text{O}$$

Hydrogen sulfide, a gaseous compound with a rotten-egg odor, burns, producing water and sulfur dioxide.

$$2\,\text{H}_2\text{S(g)} + 3\,\text{O}_2\text{(g)} \longrightarrow 2\,\text{H}_2\text{O} + 2\,\text{SO}_2\text{(g)}$$

In both examples, oxygen combines with both elements of the compound to form their oxides.

The foregoing are just a few of the many reactions in which a substance combines with oxygen. The term "oxidation" originated to describe these reactions. We will examine broader definitions in Section 5.4.

Example 5.2

Carbon disulfide is highly flammable; both elements combine with oxygen. What products are formed? Write the equation.

Solution

Carbon disulfide is CS_2. The products are carbon dioxide and sulfur dioxide. The balanced equation is

$$\text{CS}_2\text{(l)} + 3\,\text{O}_2\text{(g)} \longrightarrow \text{CO}_2\text{(g)} + 2\,\text{SO}_2\text{(g)}$$

Practice Exercises

A. Silane [SiH_4(g)] reacts spontaneously with oxygen in air to form silicon dioxide and water. Write a balanced equation for this reaction.

B. When heated in air, lead sulfide (PbS) combines with oxygen to form lead(II) oxide (PbO) and sulfur dioxide. Write a balanced equation for this reaction.

Review Questions

5.3 Write an equation for the reaction of oxygen with a: **(a)** metal; **(b)** nonmetal. For both reactions, name the type of products formed.

5.4 Write an equation for the reaction of oxygen with: **(a)** propane (C_3H_8); **(b)** hydrogen sulfide. For both reactions, name the type of products formed.

▲ The clouds of gas in interstellar space are mostly hydrogen, by far the most abundant element in the universe.

Hydrogen: A Reactive Lightweight Element

Hydrogen is another crucial element. It is combined with oxygen in the vital compound water. By mass, hydrogen makes up only about 0.9% of Earth's crust (including the atmosphere and oceans). However, because hydrogen is the lightest of all the elements, hydrogen *atoms* are quite abundant. In numbers of atoms in Earth's crust, hydrogen ranks third (15.1%), after oxygen (53.3%) and silicon (15.9%). If we look beyond our home planet, it has been estimated that hydrogen atoms make up 89% of the atoms of the sun, and that 85 to 95% of the atoms in the atmospheres of the outer planets (Jupiter, Uranus, Saturn, and Neptune) are hydrogen atoms. In the universe as a whole, about 90% of all atoms are hydrogen—and most of the rest are helium.

Unlike oxygen, hydrogen is seldom found in a free uncombined state on Earth. Most of it is combined with oxygen in water. Some is combined with carbon in petroleum and natural gas, which are mixtures of *hydrocarbons.* Nearly all compounds derived from plants and animals contain combined hydrogen.

Small amounts of elemental hydrogen can be made for laboratory use by reacting zinc with hydrochloric acid.

$$Zn(s) + 2\,HCl(aq) \longrightarrow ZnCl_2(aq) + H_2(g)$$

Because hydrogen does not dissolve in water, it can be collected by water displacement (Figure 5.3). Commercial quantities of hydrogen are obtained as by-products of petroleum refining or by reaction of natural gas with steam. Each year the United States produces about 200 million kg of hydrogen, at least two-thirds of it being used to make ammonia.

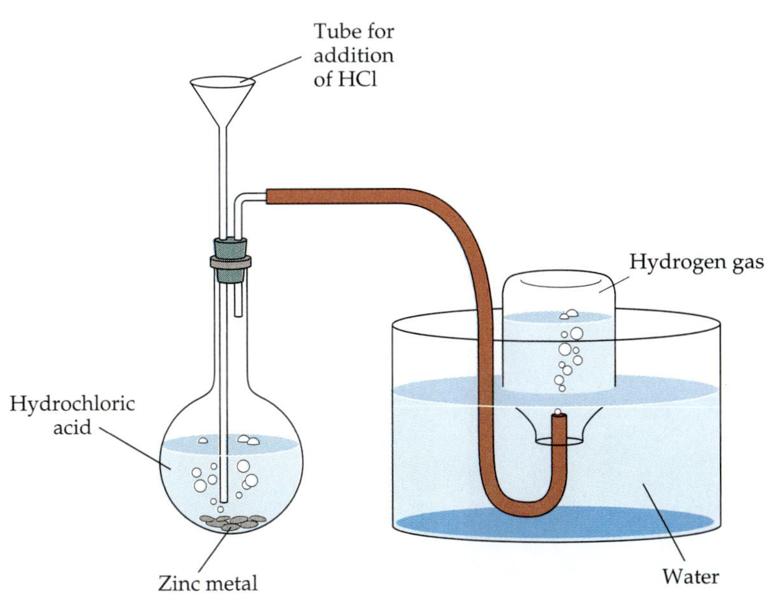

◀ **Figure 5.3**
Hydrogen gas is prepared in the laboratory by the reaction of zinc with hydrochloric acid by water displacement. The gas bubbles from the reaction flask and is trapped in the inverted bottle. When the reaction starts, the bottle is filled with water, but the hydrogen gas pushes the water out as it collects in the bottle.

▲ **Figure 5.4**
Hydrogen is the most buoyant gas, but it is highly flammable. The disastrous fire in the German zeppelin *Hindenburg* led to the replacement of hydrogen by nonflammable helium, which buoys the Fuji blimp.

Hot air balloons, as their name implies, are filled with hot air, which is less dense than the ambient air.

Hydrogen is a colorless, odorless gas and the lightest of all substances. Its density is only one-fourteenth that of air, and for this reason it was once used to fill lighter-than-air craft. Unfortunately, hydrogen can be ignited by a spark, which is what occurred in 1937 when the German airship *Hindenburg* was destroyed in a disastrous fire and explosion as it was landing in New Jersey. The use of hydrogen in airships was discontinued after that, and the dirigible industry never recovered. Today the few airships that are still in service are filled with nonflammable helium; such airships are used mainly in advertising (Figure 5.4).

A stream of pure hydrogen will burn quietly in air with an almost colorless flame. However, when a mixture of hydrogen and oxygen is ignited by a spark or a flame, a violent reaction occurs (Figure 5.5). The product in both cases is water.

$$2\,H_2(g) + O_2(g) \longrightarrow 2\,H_2O$$

Hydrogen has such strong attraction for oxygen that it can remove oxygen atoms from many metal oxides to yield the free metal. For example, when hydrogen is passed over heated copper oxide, metallic copper and water are formed.

$$CuO(s) + H_2(g) \longrightarrow Cu(s) + H_2O$$

Processes similar to this were long used to obtain metals from their ores, a process called "reduction." We will examine broader definitions of reduction in Section 5.2.

Figure 5.5 ▶
(a) A candle is held near a balloon filled with a mixture of hydrogen gas and oxygen gas. (b) The $H_2(g)$ ignites violently, combining with $O_2(g)$ to form water.

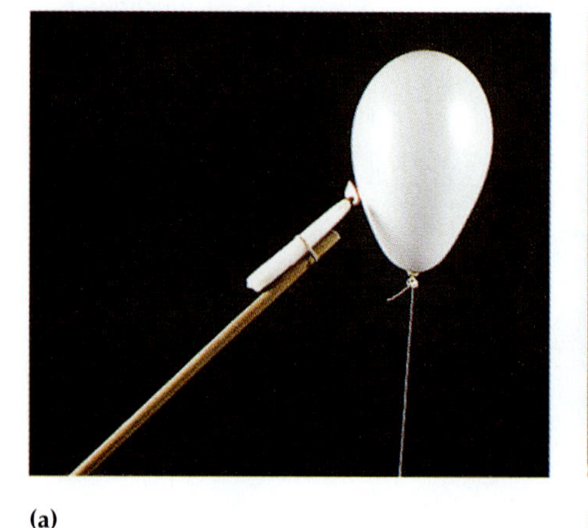

(a)

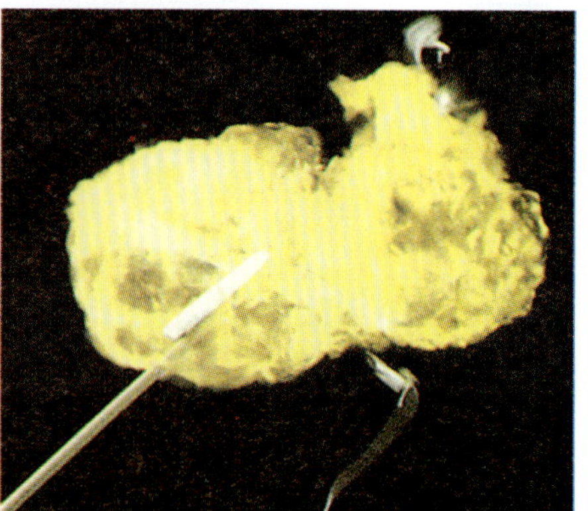

(b)

Certain metals, such as platinum and palladium, have unusual affinity for hydrogen, being able to *absorb* large volumes of the gas. Palladium can absorb up to 900 times its own volume of hydrogen! This absorbed hydrogen is generally much more reactive than $H_2(g)$. Hydrogen and oxygen can be mixed at room temperature with no perceptible reaction. But if a piece of platinum gauze is added, the gases react violently at room temperature. The platinum acts as a catalyst; it lowers the activation energy for the reaction (Section 4.8). The heat from the initial reaction heats up the platinum, making it glow; it then ignites the hydrogen-oxygen mixture, causing an explosion.

Platinum, palladium, and nickel are often used as catalysts for reactions involving hydrogen. The metals have greatest catalytic activity when they are finely divided and have lots of active surface area.

Review Questions

5.5 Describe the abundance of hydrogen on Earth. Is it generally found free or combined? List some important hydrogen compounds. Describe the abundance of hydrogen in the universe.

5.6 What role do metals such as platinum, palladium, and nickel play in hydrogen chemistry?

Some Simple Views of Oxidation and Reduction

As we have noted, "oxidation" originally meant a reaction in which a substance combines with oxygen, and "reduction" described the opposite process, the removal of oxygen from a substance. We find it convenient, especially in organic chemistry and biochemistry (Chapter 13 and following), to use three simple ways of looking at oxidation and reduction (Figure 5.6) to recognize whether a given substance is oxidized or reduced in a chemical reaction. Additional Problem 30 provides a good example of how these ideas are helpful.

1. Oxidation is a gain of oxygen atoms; reduction is a loss of oxygen atoms.

Metals are oxidized when they corrode. Iron combines with oxygen to form iron(III) oxide, the familiar iron rust.

$$4\,Fe(s) + 3\,O_2(g) \longrightarrow 2\,Fe_2O_3(s)$$

The iron atoms have no associated oxygen atoms in the reactant. Two Fe atoms are associated with three O atoms in the product. Iron gains oxygen atoms; it is oxidized.

When lead dioxide is heated at high temperatures, it decomposes as follows.

$$2\,PbO_2(s) \longrightarrow 2\,PbO(s) + O_2(g)$$

The lead dioxide loses oxygen, so it is reduced.

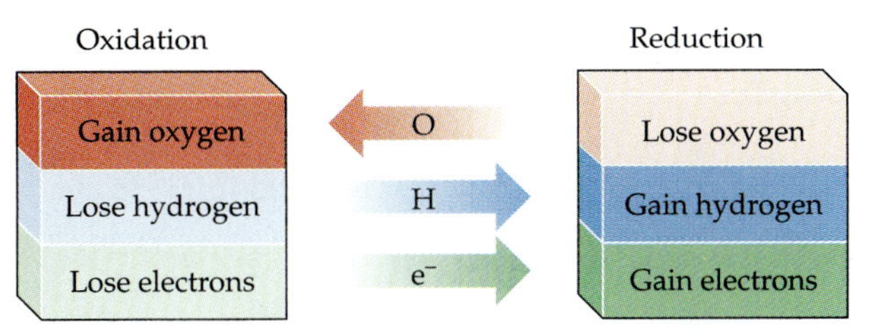

Figure 5.6 Three ways of looking at oxidation and reduction.

Example 5.3

In each of the following, is the reactant undergoing oxidation or reduction? (These are not complete chemical equations.)

a. $Pb \rightarrow PbO_2$ **b.** $SnO_2 \rightarrow SnO$
c. $KClO_3 \rightarrow KCl$ **d.** $C_6H_5CHO \rightarrow C_6H_5COOH$

Solution

a. Pb *gains* oxygen atoms (it has none on the left and two on the right); it is oxidized.
b. Sn *loses* an oxygen atom (it has two on the left and only one on the right); it is reduced.
c. $KClO_3$ has three associated oxygen atoms on the left. KCl has none on the right. The $KClO_3$ *loses* oxygen; it is reduced.
d. Each compound has 7 C atoms and 6 H atoms. On the left, these 7 C and 6 H are associated with 1 O atom. On the right, these 7 C and 6 H are associated with 2 O atoms. The C_6H_5CHO has *gained* oxygen; it is oxidized.

Practice Exercise

In each of the following, is the reactant undergoing oxidation or reduction? (These are not complete chemical equations.)

a. $Fe \rightarrow Fe_3O_4$ **b.** $NO \rightarrow NO_2$
c. $Cr_2O_3 \rightarrow CrO_3$ **d.** $C_3H_6O \rightarrow C_3H_6O_2$

2. Oxidation is a loss of hydrogen atoms; reduction is a gain of hydrogen atoms.

Methanol, when passed over hot copper gauze, forms formaldehyde and hydrogen gas.

$$\underset{\text{Methanol}}{CH_3OH} \longrightarrow \underset{\text{Formaldehyde}}{CH_2O} + H_2$$

Because the methanol loses hydrogen, it is oxidized in this reaction.

Methanol can be made by reaction of carbon monoxide with hydrogen.

$$CO + 2\,H_2 \longrightarrow CH_3OH$$

Because the carbon monoxide has gained hydrogen atoms, it has been reduced.

Example 5.4

In each of the following, is the reactant undergoing oxidation or reduction? (These are not complete chemical equations.)

a. $C_2H_6O \rightarrow C_2H_4O$ **b.** $C_2H_2 \rightarrow C_2H_6$

Solution

a. There are six hydrogen atoms in the compound on the left and only four in the one on the right. The C_2H_6O *loses* hydrogen atoms; it is oxidized.
b. There are two hydrogen atoms in the compound on the left and six in the one on the right. The C_2H_2 *gains* hydrogen atoms; it is reduced.

Practice Exercise

In each of the following, is the reactant undergoing oxidation or reduction? (These are not complete chemical equations.)

a. $C_6H_6 \rightarrow C_6H_{12}$ **b.** $C_3H_6O \rightarrow C_3H_4O$

5.2 Oxidation Numbers and Oxidation-Reduction (Redox) Equations

A third and more fundamental way to look at oxidation and reduction is based on the loss or gain of electrons.

3. Oxidation is a loss of electrons; reduction is a gain of electrons.

When magnesium metal reacts with chlorine, magnesium ions and chloride ions are formed.

$$Mg + Cl_2 \longrightarrow Mg^{2+} + 2\,Cl^-$$

The magnesium atom has two valence electrons; the Mg^{2+} ion has none. Mg loses electrons; it is oxidized. On the left, each Cl atom has seven valence electrons. On the right, the two Cl^- ions have eight electrons each. Because the chlorine atoms gain electrons, Cl_2 is reduced.

By considering the loss or gain of electrons, we can come to a much broader definition of oxidation and reduction. First, however, we introduce a concept that assists us in formulating this comprehensive definition.

Just remember Leo the lion.
LEO says *GER*
*L*oss of
*E*lectrons is
*O*xidation.
*G*ain of
*E*lectrons is
*R*eduction.

Oxidation Numbers

In a general way, the **oxidation number** of a substance is a count of the electrons transferred or shared in the formation or breaking of chemical bonds. For example, in the formation of sodium chloride, Na atoms lose electrons and Cl atoms gain them, forming Na^+ and Cl^- ions. In sodium chloride, the oxidation number of Na^+ is +1 and that of Cl^- is −1. In the compound $CaCl_2$, the oxidation number of the Cl^- ion is −1, and that of the Ca^{2+} ion is +2. The total of the oxidation numbers of the atoms (ions) in a formula unit of $CaCl_2$ is $+2 -1 -1 = 0$.

In the formation of a molecule, electrons are *shared* rather than actually transferred. We can, however, *arbitrarily* assign oxidation numbers. In a binary molecule we assign a negative oxidation number to the more electronegative element. This means that in H_2O the oxidation number of H is positive and that of O is negative. We assign H the oxidation number +1. We also require that the total of the oxidation numbers of the three atoms in H_2O be *zero*. This means that the oxidation number of O must be −2, that is, $+1 +1 -2 = 0$. In the H_2 molecule the H atoms are identical and must have the same oxidation number. The sum of these oxidation numbers must be *zero*, so the oxidation number of each H atom must also be 0.

We can use the following rules to assign oxidation numbers in most compounds. The rules are listed *by priority*, with the highest priority listed first. If two rules are in conflict, use the rule with the higher priority. This generally takes care of exceptions, a few of which are listed in the margin. Some examples are listed below for each rule, and all the rules together are applied in Example 5.5.

1. *The total of the oxidation numbers of all the atoms in a neutral molecule, an isolated atom, or a formula unit is 0.*
   [Examples: The oxidation number of the Fe atom is 0. The sum of the oxidation numbers of all the atoms in each of the molecules Cl_2, S_8, and $C_6H_{12}O_6$ is 0, and the sum of the oxidation numbers of the ions in $MgBr_2$ is 0.]

2. *In their compounds, the Group 1A metals all have an oxidation number of +1, and the Group 2A metals have an oxidation number of +2.*
   [Examples: The oxidation number of Na in Na_2SO_4 is +1, and that of Ca in $Ca_3(PO_4)_2$ is +2.]

The principal exception to rule 3 is when H is bonded to an element that is less electronegative than itself, as in metal hydrides.

The principal exception to rule 4 is when oxygen is bonded to itself, as in peroxides (for example, H_2O_2).

3. *In its compounds, hydrogen has an oxidation number of +1.*
   [Examples: The oxidation number of H is +1 in HCl, H_2O, NH_3, and CH_4.]
4. *In its compounds, oxygen has an oxidation number of −2.*
   [Examples: The oxidation number of O is −2 in CO, CH_3OH, and $C_6H_{12}O_6$.]
5. *In their binary (two-element) compounds with metals, Group 7A elements have an oxidation number of −1, Group 6A elements have an oxidation number of −2, and Group 5A elements have an oxidation number of −3.*
   [Examples: The oxidation number of Br is −1 in $CaBr_2$, that of S is −2 in Na_2S, and that of N is −3 in Mg_3N_2.]

Example 5.5

What is the oxidation number of each element in the following?

a. I_2 **b.** Cr_2O_3 **c.** $AlCl_3$ **d.** Na_2SO_4 **e.** CaH_2

Solution

a. The sum of the oxidation numbers for the two I atoms is 0 (rule 1). The two I atoms are identical, so the oxidation number of each I atom must be 0.
b. The oxidation number of O is −2 (rule 4), and the total for *three* O atoms is −6. The total of the oxidation number for all atoms must be 0 (rule 1). This means that the total oxidation numbers of *two* Cr atoms is +6, and that of one Cr atom is +3.
c. The oxidation number of Cl is −1 (rule 5), and the total for *three* Cl atoms is −3. The total of the oxidation numbers for the molecule must be 0 (rule 1), and therefore the *one* Al atom must have an oxidation number of +3.
d. The oxidation number of Na is +1 (rule 2), and for the *two* Na atoms, +2. The oxidation number of O is −2 (rule 4), and the total for *four* O atoms is −8. The total for the formula unit must be 0 (rule 1). The oxidation number of S must therefore be +6.
e. The oxidation number of Ca is +2 (rule 2). The total for the formula unit must be 0 (rule 1), and so the total oxidation number for the *two* H atoms must be −2. Even though the oxidation number of H is usually +1 (rule 3), here it must be −1. Rule 2 takes priority over rule 3.

Practice Exercise

What is the oxidation number of each element in the following?

a. Al_2O_3 **b.** P_4 **c.** $NaMnO_4$
d. H_2O_2 **e.** CH_3F **f.** $CHCl_3$

(The assignment of oxidation numbers in CH_3F and $CHCl_3$ demonstrates the variability of the oxidation number of carbon in organic compounds.)

Identifying Redox Reactions

We can use the concept of oxidation numbers to help us identify oxidation-reduction reactions. For example, the spectacular reaction pictured in Figure 5.7, called the *thermite* reaction, is used to produce liquid iron for welding large iron objects.

$$2\,Al(s) + Fe_2O_3(s) \longrightarrow 2\,Fe(l) + Al_2O_3(s)$$

▲ **Figure 5.7**
The thermite reaction is an exceptionally vigorous redox reaction.

Even in the original sense, we can see that this an *oxidation-reduction* reaction. Al is *oxidized* to Al_2O_3; aluminum atoms take on or *gain* oxygen atoms. Fe_2O_3 is *reduced* to Fe; iron(III) oxide *loses* oxygen atoms.

Now let's assign oxidation numbers to the elements involved in the thermite reaction. These are the numbers above the chemical symbols.

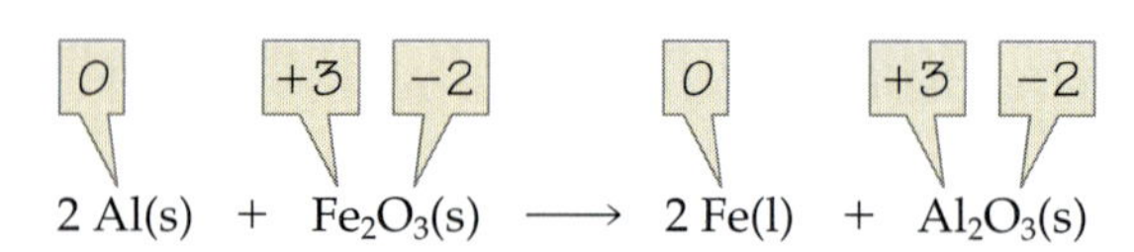

In this reaction, the oxidation number of Al atoms *increases* from 0 to +3, and the oxidation number of Fe atoms *decreases* from +3 to 0.

> *In an oxidation-reduction reaction, the oxidation number of one or more elements increases—an* **oxidation** *process—and the oxidation number of one or more elements decreases—a* **reduction** *process. A redox reaction must always have both an oxidation and a reduction.*

The reaction pictured in Figure 5.8 is strikingly different in appearance from the thermite reaction, but it is also a redox reaction. Oxidation numbers are noted in the following equation.

$$\overset{0}{Mg(s)} + \overset{+2}{Cu^{2+}(aq)} \longrightarrow \overset{+2}{Mg^{2+}(aq)} + \overset{0}{Cu(s)}$$

▲ **Figure 5.8**
The reaction between magnesium and copper(II) ions is a redox reaction. In the top photograph, a coil of Mg(s) is added to a solution of $CuSO_4(aq)$. After several hours, the solution is no longer blue because all the $Cu^{2+}(aq)$ has been displaced from solution, leaving a deposit of red-brown copper metal, some unreacted magnesium, and a clear, colorless solution of $MgSO_4(aq)$.

Half-Reactions

Sometimes it is helpful to visualize a redox reaction as two **half-reactions** that occur simultaneously. For the reaction above, Mg atoms *lose* two electrons and are *oxidized* to Mg^{2+} ions in the oxidation half-reaction.

$$\textit{Oxidation}: Mg(s) \longrightarrow Mg^{2+}(aq) + 2\,e^-$$

In the reduction half-reaction, Cu^{2+} ions *gain* two electrons and are *reduced* to Cu atoms.

$$\textit{Reduction}: Cu^{2+}(aq) + 2\,e^- \longrightarrow Cu(s)$$

Example 5.6

The following are supposed half-reactions (not balanced). In each case, state whether the reactant is undergoing oxidation, reduction, or neither.

a. $Zn(s) \rightarrow Zn^{2+}(aq)$

b. $Fe^{3+}(aq) \rightarrow Fe^{2+}(aq)$

c. $CaCO_3(s) \rightarrow CaO(s) + CO_2(g)$

d. $AgNO_3(aq) \rightarrow Ag(s)$

Solution

a. To form a 2+ ion, zinc *loses* two electrons

$$Zn(s) \longrightarrow Zn^{2+}(aq) + 2\,e^-$$

The oxidation number of zinc increases from 0 to +2. Zinc is *oxidized.*

b. To go from a 3+ ion to a 2+ ion, iron must *gain* an electron.

$$Fe^{3+}(aq) + e^- \longrightarrow Fe^{2+}(aq)$$

The oxidation number of iron decreases from +3 to +2. Iron is *reduced.*

c. Let's start by assigning oxidation numbers to each of the elements.

$$\overset{+2\;+4\;-2}{CaCO_3(s)} \longrightarrow \overset{+2\;-2}{CaO(s)} + \overset{+4\;-2}{CO_2(g)}$$

None of the atoms change in oxidation number; neither oxidation nor reduction is involved.

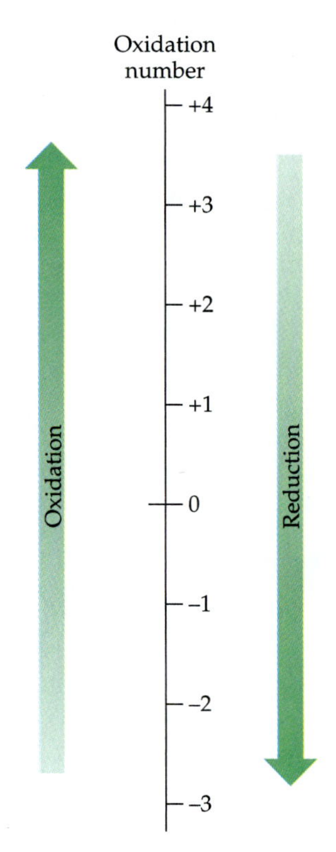

▲ An increase in oxidation number means a loss of electrons and is therefore oxidation. A decrease in oxidation number means a gain of electrons and is therefore reduction.

d. In this case we should recognize that $AgNO_3(aq)$ is an ionic compound with $Ag^+(aq)$ and $NO_3^-(aq)$ ions. In going from $Ag^+(aq)$ to Ag(s), silver *gains* an electron.

$$Ag^+(aq) + e^- \longrightarrow Ag(s)$$

Silver *decreases* in oxidation number from +1 to 0. Silver is *reduced.*

Practice Exercise

The following are supposed half-reactions (not balanced). In each case, state whether the reactant is undergoing oxidation, reduction, or neither.

a. $Cu^{2+}(aq) \rightarrow Cu(s)$
b. $Sn^{2+}(aq) \rightarrow Sn^{4+}(aq)$
c. $CuO(s) \rightarrow Cu^{2+}(aq) + H_2O$
d. $Cu(s) \rightarrow CuSO_4(aq)$

Example 5.7

Does the following equation represent an oxidation-reduction reaction?

$$Mn^{2+}(aq) + O_2(g) + H^+(aq) \longrightarrow MnO_2(s) + H_2O \quad (\textit{not balanced})$$

Solution

We start by designating oxidation numbers.

$$\overset{+2}{Mn^{2+}}(aq) + \overset{0}{O_2}(g) + \overset{+1}{H^+}(aq) \longrightarrow \overset{+4}{Mn}\overset{-2}{O_2}(s) + \overset{+1}{H_2}\overset{-2}{O} \; (\textit{not balanced})$$

From these, we see that the oxidation number of Mn increases from +2 to +4; that is, Mn^{2+} is *oxidized* to MnO_2. The oxidation number of O decreases from 0 to −2; O_2 is *reduced* to H_2O. The equation *does* represent an oxidation-reduction reaction.

Practice Exercise

Does the following equation represent an oxidation-reduction reaction?

$$I_2(s) + 5\,Cl_2(g) + 6\,H_2O \longrightarrow 2\,IO_3^-(aq) + 12\,H^+(aq) + 10\,Cl^-(aq)$$

✓ Review Questions

5.7 How can we recognize that a substance has been oxidized by noting (**a**) oxygen atoms gained, (**b**) hydrogen atoms lost, and (**c**) electrons lost? Give examples of each.

5.8 How can we recognize that a substance has been reduced by noting (**a**) oxygen atoms lost, (**b**) hydrogen atoms gained, and (**c**) electrons gained? Give examples of each.

5.9 Define oxidation and reduction in terms of a change in oxidation number. Give examples.

5.3 Oxidizing and Reducing Agents

When one substance is oxidized, another is reduced. For example, in the reaction

$$CuO(s) + H_2(g) \longrightarrow Cu(s) + H_2O(g)$$

the copper goes from oxidation number +2 to oxidation number 0; the copper(II) oxide is *reduced.* At the same time, the hydrogen goes from oxidation number 0 to oxidation number +1; the H_2 is *oxidized.* The substance that is *oxidized* (H_2, in this case) is called a **reducing agent** because it causes some other substance (CuO) to be reduced. Similarly, the substance that is *reduced* (CuO, in this case) is called an **oxidizing agent** because it causes another substance (H_2) to be oxidized.

Even though the changes in oxidation number occur in Cu and H atoms, we do not refer to the *atoms* as the oxidizing or reducing agents. Rather, the *substances* in which these atoms are found—that is, CuO and H_2, respectively—are given these labels.

Reduction:
copper oxide is being reduced;
CuO is the oxidizing agent.

$$CuO + H_2 \longrightarrow Cu + H_2O$$

Oxidation:
hydrogen is being oxidized;
H_2 is the reducing agent.

Example 5.8

Identify the oxidizing agents and reducing agents in each of the following reactions.

a. $2\,C(s) + O_2(g) \rightarrow 2\,CO(g)$ **b.** $N_2(g) + 3\,H_2(g) \rightarrow 2\,NH_3(g)$
c. $SnO(s) + H_2(g) \rightarrow Sn(s) + H_2O(g)$ **d.** $Mg(s) + Cl_2(g) \rightarrow MgCl_2(s)$

Solution

a. Carbon goes from an oxidation number of 0 in C to +2 in CO. Carbon is *oxidized;* C is the *reducing agent.* Oxygen goes from oxidation number 0 in O_2 to oxidation number −2 in CO; O_2 is *reduced* and is therefore the *oxidizing agent.*
b. Nitrogen goes from an oxidation number of 0 in N_2 to −3 in NH_3. Nitrogen is *reduced;* N_2 is the *oxidizing agent.* Hydrogen goes from oxidation number 0 in H_2 to oxidation number +1 in NH_3; H_2 is *oxidized* and is therefore the *reducing agent.*
c. Tin goes from oxidation number +2 in SnO to 0 in Sn. SnO is *reduced;* it is the *oxidizing agent.* Hydrogen goes from oxidation number 0 in H_2 to +1 in H_2O. H_2 is *oxidized;* it is the *reducing agent.*
d. Magnesium goes from oxidation number 0 in Mg to +2 in $MgCl_2$. Mg is *oxidized;* it is the *reducing agent.* Chlorine goes from oxidation number 0 in Cl_2 to oxidation number −1 in $MgCl_2$. Cl_2 is *reduced;* it is the *oxidizing agent.*

Practice Exercises

A. Identify the oxidizing agents and reducing agents in each of the following reactions.
a. $Se + O_2 \rightarrow SeO_2$ **b.** $2\,K + Br_2 \rightarrow 2\,K^+ + 2\,Br^-$
B. Identify the oxidizing agents and reducing agents in each of the following reactions.
a. $V_2O_5 + 2\,H_2 \rightarrow V_2O_3 + 2\,H_2O$
b. $CH_3C{\equiv}N + 2\,H_2 \rightarrow CH_3CH_2NH_2$

Some Common Oxidizing Agents

We live in an oxidizing atmosphere. Oxygen in the air oxidizes coal in electric power plants, gasoline in our automobiles, and the wood in our campfires. It even "burns" the food we eat to give us the energy to move and to think. Oxygen is also the most common commercial oxidizing agent. Huge quantities of it go into steel furnaces. Purified oxygen is also used in hospital respirators.

Another common oxidizing agent is hydrogen peroxide (H_2O_2). Pure hydrogen peroxide is a syrupy liquid. It is available (in the laboratory) as a dangerous 30% solution that has powerful oxidizing power or as a 3% solution sold in stores for various uses around the home. It has the advantage of being converted to water, an innocuous by-product, in most reactions.

Potassium dichromate ($K_2Cr_2O_7$) is a laboratory oxidizing agent. It can oxidize alcohols to *aldehydes* and *ketones* (Chapter 15). For the oxidation of ethanol (found in alcoholic beverages) to acetaldehyde, the reaction is

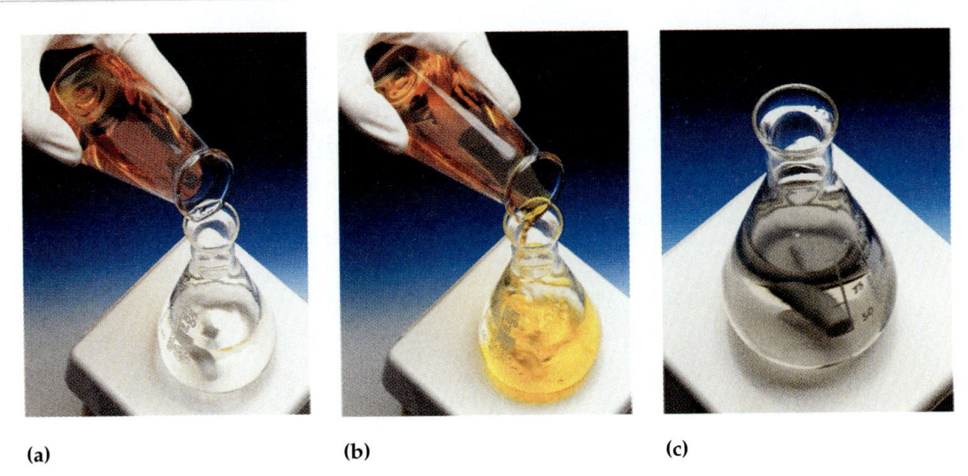
(a) (b) (c)

▶ The oxidation of ethanol by dichromate ion in acidic solution (a) Orange $K_2Cr_2O_7(aq)$ about to be added to colorless $CH_3CH_2OH(aq)$ that has been acidified with H_2SO_4. (b) The ethanol solution becomes colored due to the $Cr_2O_7^{2-}(aq)$. (c) After the reaction, the solution becomes a (barely detectable) pale violet color, signifying that the $Cr_2O_7^{2-}$ is gone; $Cr^{3+}(aq)$ is now present.

$$8\,H^+(aq) + Cr_2O_7^{2-}(aq) + 3\,C_2H_5OH(aq) \longrightarrow 2\,Cr^{3+}(aq) + 3\,C_2H_4O(aq) + 7\,H_2O$$

Potassium permanganate ($KMnO_4$) is a black, shiny, crystalline solid. It dissolves in water to form deep purple solutions. This purple color disappears as the permanganate is reduced, making $KMnO_4$ useful as a test reagent for oxidizable substances. For example, it is used to oxidize iron from Fe^{2+} to Fe^{3+}. The amount of iron(II) ion in a sample can be determined by its reaction with permanganate (Figure 5.9).

One can add the deep purple permanganate solution to a sample of iron(II) ion. The permanganate entering the iron(II) solution is reduced, becoming manganese(II) ion, and no purple color appears in the receiving flask—except momentarily where the permanganate stream enters the solution—until all the iron(II) ion has been oxidized. Further addition of permanganate will cause a purple color to appear and persist in the receiving flask because no iron(II) ion is left to reduce the permanganate. Thus one can measure just how much iron(II) ion there is in the sample by keeping track of how much permanganate is reduced—that is, by measuring how much permanganate is added until the purple color persists. The equation for this reaction is

$$MnO_4^-(aq) + 5\,Fe^{2+}(aq) + 8\,H^+(aq) \longrightarrow Mn^{2+}(aq) + 5\,Fe^{3+}(aq) + 4\,H_2O$$

Permanganate ion (purple)

Manganese(II) ion (pale pink)

Permanganate solutions can also be used to oxidize oxalic acid (a poisonous compound found in rhubarb leaves), sulfur dioxide (SO_2), and many other compounds.

Other common oxidizing agents are the halogens—fluorine (F_2), chlorine (Cl_2), bromine (Br_2), and iodine (I_2). Bromine, for example, oxidizes phosphorus to form phosphorus tribromide.

$$P_4(s) + 6\,Br_2(l) \longrightarrow 4\,PBr_3(l)$$

A *tincture* is a solution made up in alcohol.

Many antiseptics are mild oxidizing agents. (An *antiseptic* is a substance applied to living tissue to kill microorganisms or prevent their growth.) For example, a 3% solution of hydrogen peroxide is often used to treat minor cuts, and tincture of iodine has long been a household antiseptic. Ointments for treating acne often con-

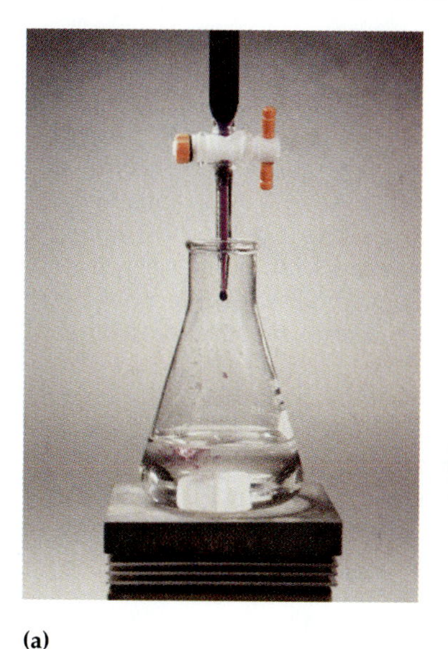

(a)

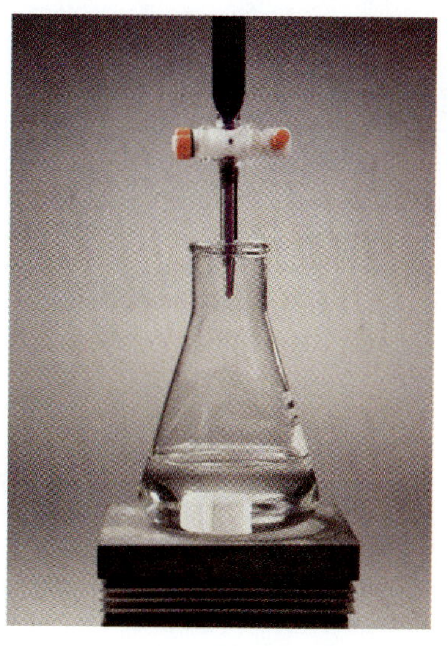

(b)

◀ **Figure 5.9**
Purple permanganate ions (MnO_4^-) are reduced to Mn^{2+} ions by iron(II) ions (Fe^{2+}). (a) The buret contains $MnO_4^-(aq)$ and the flask contains $Fe^{2+}(aq)$. (b) As the permanganate solution is added to the solution of iron(II) ions, the purple color disappears.

tain 5 to 10% of benzoyl peroxide, a powerful oxidizing agent that acts as an antiseptic and as a skin irritant. It causes old skin to slough off and be replaced by new, fresher-looking skin. When used on areas exposed to sunlight, however, benzoyl peroxide may promote skin cancer.

Oxidizing agents are also used as disinfectants. (A *disinfectant* is a substance that is applied to nonliving tissue to kill microorganisms.) A good example is chlorine, which is used to kill disease-causing microorganisms in drinking water. Swimming pools are often "chlorinated" with calcium hypochlorite [$Ca(OCl)_2$], an oxidizing agent.

Swimming pool chemistry also involves acids and bases. Calcium hypochlorite forms a basic (alkaline) solution, raising the pH of the water. (When a pool becomes too alkaline, the pH is lowered by adding hydrochloric acid. Swimming pools are usually maintained at pH 7.2 to 7.8, see Section 10.3.)

Some Reducing Agents of Interest In every redox reaction, the oxidizing agent is reduced, and the substance that is oxidized acts as a reducing agent. Let us consider reactions in which the purpose is reduction of some substance.

Most metals occur in nature as compounds. In order to prepare the free metals, the compounds must be reduced. Metals are often freed from their ores with coal or coke (elemental carbon obtained by heating coal to drive off volatile matter). Tin oxide is one of the many ores that can be reduced with coal or coke.

$$SnO_2(s) + C(s) \longrightarrow Sn(s) + CO_2(g)$$

Sometimes a metal can be obtained by heating its ore with a more active metal. Chromium oxide, for example, can be reduced by heating it with aluminum.

$$Cr_2O_3(s) + 2\,Al(s) \longrightarrow Al_2O_3(s) + 2\,Cr(s)$$

Hydrogen is an excellent reducing agent that can free many metals from their ores, but it is generally used to produce more expensive metals, such as tungsten.

$$WO_3(s) + 3\,H_2(g) \longrightarrow W(s) + 3\,H_2O$$

Hydrogen can be used to reduce many kinds of chemical compounds. Ethylene, for example, can be reduced to ethane.

$$C_2H_4(g) + H_2(g) \xrightarrow{Ni} C_2H_6(g)$$

Oxidation-Reduction in Bleaching and Stain Removal

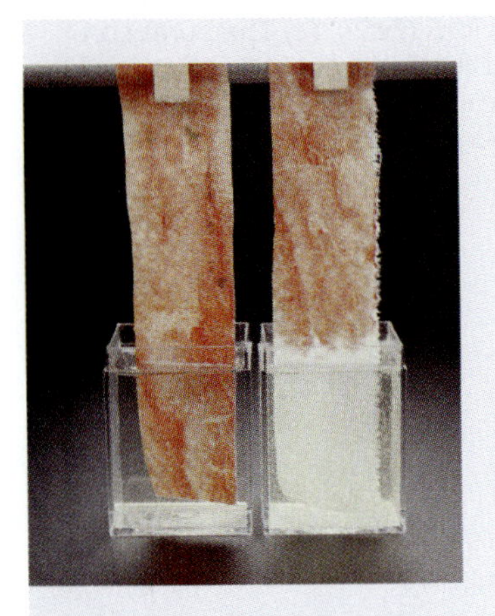

▲ Pure water (left) has little ability to remove a dried tomato sauce stain. Sodium hypochlorite, NaOCl(aq) (right), easily bleaches the stain away by oxidizing the colored pigments of the sauce to colorless products.

Bleaches are oxidizing agents, too. *Bleaches* remove unwanted color from fabrics or other material. Laundry bleaches are usually sodium hypochlorite (NaOCl) as an aqueous solution (in products such as Purex and Clorox) or calcium hypochlorite [$Ca(OCl)_2$], known as bleaching powder. The powder is usually preferred for large industrial operations, such as the whitening of paper or fabrics. It is also used in hospitals to disinfect bedding and clothes. Nonchlorine bleaches often contain sodium perborate (a loose combination of $NaBO_2$ and H_2O_2).

For lightening hair color, the bleaches are usually 6% or 12% solutions of hydrogen peroxide, which oxidizes the dark pigment (melanin) in the hair to colorless products. Hydrogen peroxide can also be used to lighten certain paints by oxidizing sulfides (S^{2-}) to sulfates (SO_4^{2-}). When lead-based paints are exposed to air containing hydrogen sulfide (H_2S), they turn black because of the formation of lead sulfide (PbS). Hydrogen peroxide oxidizes the black lead sulfide to white lead sulfate.

$$\underset{\text{black}}{PbS(s)} + 4\,H_2O_2(aq) \longrightarrow \underset{\text{white}}{PbSO_4(s)} + 4\,H_2($$

This reaction can also be used to restore the once-white areas of old paintings that have darkened by the reaction of white lead compounds (paint pigments) with sulfur compounds. The darkened pigments (black PbS) are converted to white $PbSO_4$ by the hydrogen peroxide.

Stain removal is more complicated than bleaching. A few stain removers are oxidizing agents, but others are reducing agents, solvents, detergents, or compounds with other actions. Some stains require specific stain removers.

Hydrogen peroxide in cold water removes blood stains from cotton and linen fabrics. Potassium permanganate can be used to remove most stains from white fabrics (except rayon). The purple permanganate stain then can be removed in a redox reaction with oxalic acid.

$$5\,H_2C_2O_4(aq) + 2\,MnO_4^-(aq) + 6\,H^+(aq) \longrightarrow 2\,Mn^{2+}(aq) + 8\,H_2O + 10\,CO_2(g)$$

Iodine, used as a disinfectant, often stains clothing in contact with the treated area. The iodine stain is readily removed in a redox reaction with sodium thiosulfate.

$$I_2(s) + 2\,S_2O_3^{2-}(aq) \longrightarrow 2\,I^-(aq) + S_4O_6^{2-}(aq)$$

(Nickel is used as a catalyst in this reaction.) Hydrogen also reduces nitrogen, from the air, in the industrial production of ammonia.

$$N_2(g) + 3\,H_2(g) \xrightarrow{\text{Fe}} 2\,NH_3(g)$$

(A catalyst such as iron is needed in this case.)

Perhaps a more familiar reducing agent, by use if not by name, is the developer used in black-and-white photography. Photographic film is coated with a silver *salt*, usually AgBr(s). Silver ions that have been exposed to light react with the developer, a reducing agent (such as the organic compound hydroquinone), to form metallic silver.

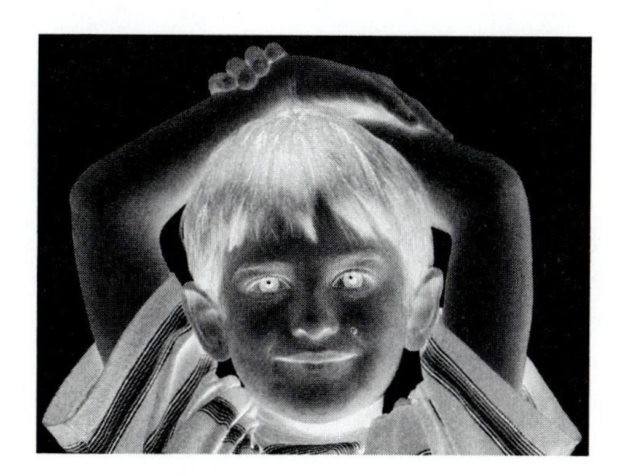

◀ **Figure 5.10**
A photographic negative (left) and a positive print (right).

$$C_6H_4(OH)_2(aq) + 2\,Ag^+(aq) \longrightarrow C_6H_4O_2(aq) + 2\,Ag(s) + 2\,H^+(aq)$$

Hydroquinone (label on $C_6H_4(OH)_2$); Silver metal (label on Ag)

Silver ions not exposed to light are not reduced by the developer. The film is then treated with "hypo," a solution of sodium thiosulfate ($Na_2S_2O_3$), which washes out unexposed silver bromide to form the negative. This leaves the negative dark where the metallic silver has been deposited (where it was originally exposed to light) and transparent where light did not strike it. A similar process produces the positive print by shining light through the negative onto light-sensitive paper containing AgBr in a gelatin base. Figure 5.10 shows positive and negative prints.

In food chemistry, reducing agent are often called **antioxidants.** Ascorbic acid (vitamin C) can prevent the browning of fruit (such as sliced apples or pears) by inhibiting air oxidation. Whereas vitamin C is water soluble, tocopherol (vitamin E) is a fat-soluble antioxidant. Both of these vitamins are believed to retard various oxidation reactions that are potentially damaging to vital components of living cells. (See Selected Topic F.)

✓ Review Questions

5.10 List four common oxidizing agents. Give one use for each.

5.11 Name some oxidizing agents used as antiseptics and disinfectants.

5.12 List several bleaching agents and give a use for each of them.

5.13 List several reducing agents and give a use of each. Write a balanced chemical equation where appropriate.

5.4 Oxidation, Reduction, and Living Things

Perhaps the most important oxidation-reduction processes are photosynthesis and respiration. Bread, cereals, and pasta are largely made up of carbohydrates (Chapter 19). If we represent carbohydrates with the simple example glucose ($C_6H_{12}O_6$), we can write the overall equation for their metabolism as follows.

$$C_6H_{12}O_6 + 6\,O_2 \longrightarrow 6\,CO_2 + 6\,H_2O + \text{energy}$$

The carbohydrate is oxidized in the process; carbon goes from oxidation number 0 in $C_6H_{12}O_6$ to +4 in CO_2.

▲ **Figure 5.11**
Photosynthesis occurs in green plants such as the wheat in this field in eastern Washington. The chlorophyll pigments that catalyze the photosynthetic process provide the green color.

Meanwhile, plants use carbon dioxide and water to produce carbohydrates. The energy needed comes from the sun and the process is called photosynthesis (Figure 5.11). The chemical equation is

$$6\,CO_2 + 6\,H_2O + \text{energy} \longrightarrow C_6H_{12}O_6 + 6\,O_2$$

During photosynthesis, carbon dioxide is reduced to glucose, a carbohydrate.

The carbohydrates produced by photosynthesis are the ultimate source of all our food, since fish, fowl, and other animals either eat plants or eat other animals that eat plants (Figure 5.12). Note that the photosynthetic process not only makes carbohydrates but also yields free elemental oxygen, O_2. In other words, photosynthesis not only provides all the food we eat; it also furnishes all the oxygen we breathe.

We can see that green plants carry out the redox reaction that makes possible nearly all life on Earth. Animals can only oxidize the foods that plants provide. Crop farming is therefore a process of reduction of carbon dioxide to glucose. Energy captured in cultivated plants, whether the plants are used directly or are fed to animals, is the basis for human life.

✓ Review Question

5.14 Relate the chemistry of photosynthesis to the chemistry that provides the energy for your heartbeat.

▲ **Figure 5.12**
The food we eat is oxidized to provide energy for our activities. That energy, which comes ultimately from the sun, is trapped by green plants through photosynthetic reactions that reduce carbon dioxide to carbohydrates.

Summary

We can think of oxidation in several ways, such as (1) the gain of oxygen atoms, (2) the loss of hydrogen atoms, or (3) the loss of electrons. Similarly, we can think of reduction as (1) the loss of oxygen atoms, (2) the gain of hydrogen atoms, or (3) the gain of electrons.

The most fundamental definition of oxidation is based on oxidation numbers. The **oxidation number** is related to the number of electrons that the atom transfers or shares in forming bonds. **Oxidation** entails an *increase* in oxidation number and **reduction,** a *decrease* in oxidation number.

Whenever one reactant in a chemical reaction is oxidized, some other reactant in the reaction is reduced. The overall reaction is often called a *redox reaction.*

An **oxidizing agent** is any substance that causes some other substance to be oxidized; an oxidizing agent is always *reduced* in a redox reaction. A **reducing agent** is any substance that causes some other substance to be reduced; a reducing agent is always *oxidized* in a redox reaction.

Oxygen is the most frequently encountered oxidizing agent. Others are potassium dichromate ($K_2Cr_2O_7$), hydrogen peroxide (H_2O_2), potassium permanganate ($KMnO_4$), and the halogens (F_2, Cl_2, Br_2, I_2). Two frequently used reducing agents are hydrogen gas and carbon.

Antiseptics, compounds used to control microorganism growth on living tissue, are oxidizing agents. *Bleaches,* used to remove color from cloth, hair and other substances, are also oxidizing agents, usually solutions of sodium hypochlorite (NaOCl).

A redox reaction most important to living organisms is *photosynthesis*, the process by which green plants convert solar energy to chemical energy locked up in the chemical bonds of sugar molecules.

Key Terms

antioxidant (5.3)
half-reaction (5.2)
oxidation (5.2)
oxidation number (5.2)
oxidizing agent (5.3)
reducing agent (5.3)
reduction (5.2)

Problems

Reactions of Oxygen

1. Complete and balance the following equations.
   a. $C + O_2 \rightarrow$ b. $C_2H_6 + O_2 \rightarrow$
   c. $N_2 + O_2 \rightarrow$ d. $C_3H_8 + O_2 \rightarrow$
2. Complete and balance the following equations.
   a. $S + O_2 \rightarrow$ b. $CS_2 + O_2 \rightarrow$
   c. $H_2 + O_2 \rightarrow$ d. $C_6H_{12}O_6 + O_2 \rightarrow$

Oxidation Numbers

3. Indicate the oxidation number of the underlined element in each of the following.
   a. $\underline{Cr}$ b. $\underline{Cl}O_2$ c. $K_2\underline{Se}$ d. $\underline{Te}F_6$
4. Indicate the oxidation number of the underlined element in each of the following.
   a. $Ca\underline{Ru}O_3$ b. $\underline{N}H_2OH$ c. $\underline{C}_2H_6$ d. $H\underline{C}OOH$
5. What is the usual oxidation number of hydrogen in its compounds? What is the usual oxidation number of oxygen in its compounds? What are some exceptions?
6. What happens to the oxidation number of one of its elements when a compound is oxidized and when it is reduced?

Recognizing Redox Reactions

7. In the following supposed half-reactions (not balanced), indicate whether oxidation or reduction, or neither, is involved.
   a. $ClO_2(g) \rightarrow HClO_3(aq)$ b. $Mn^{2+}(aq) \rightarrow MnO_2(s)$
   c. $HOBr(aq) \rightarrow Br_2(l)$ d. $SbH_3(g) \rightarrow Sb(s)$
8. In the following supposed half-reactions (not balanced), indicate whether oxidation or reduction, or neither, is involved.
   a. $V_2O_5(aq) \rightarrow V(s)$
   b. $P_4(s) \rightarrow H_3PO_4(aq)$
   c. $CrO_3(s) \rightarrow H_2CrO_4(aq)$
   d. $CH_3CH_2OH(aq) \rightarrow CO_2(g)$
9. In the following reaction, is the $H_2SO_4(aq)$ oxidized, reduced, or neither? Explain.

$$Cu(s) + 2\,H_2SO_4(aq) \longrightarrow CuSO_4(aq) + 2\,H_2O + SO_2(g)$$

10. Is the following reaction a redox reaction? Explain.

$$5\,H_2O_2(aq) + 2\,MnO_4^-(aq) + 6\,H^+(aq) \longrightarrow 2\,Mn^{2+}(aq) + 8\,H_2O + 5\,O_2(g)$$

Oxidizing Agents and Reducing Agents

11. Identify the oxidizing agent and the reducing agent in these reactions.
   a. $4\,Al + 3\,O_2 \rightarrow 2\,Al_2O_3$
   b. $2\,SO_2 + O_2 \rightarrow 2\,SO_3$
   c. $Fe + 2\,HCl \rightarrow FeCl_2 + H_2$
   d. $CS_2 + 3\,O_2 \rightarrow CO_2 + 2\,SO_2$
12. Identify the oxidizing agent and the reducing agent in these reactions.
   a. $Cl_2 + 2\,KBr \rightarrow 2\,KCl + Br_2$
   b. $C_2H_4 + H_2 \rightarrow C_2H_6$
   c. $CuCl_2 + Fe \rightarrow FeCl_2 + Cu$
   d. $2\,AgNO_3 + Cu \rightarrow Cu(NO_3)_2 + 2\,Ag$

13. In the following reactions, which substance is oxidized and which is reduced?
    a. $2\,HNO_3 + SO_2 \rightarrow H_2SO_4 + 2\,NO_2$
    b. $2\,CrO_3 + 6\,HI \rightarrow Cr_2O_3 + 3\,I_2 + 3\,H_2O$
14. In the following reactions, which substance is oxidized and which is the oxidizing agent?
    a. $H_2CO + H_2O_2 \rightarrow H_2CO_2 + H_2O$
    b. $5\,C_2H_6O + 4\,MnO_4^- + 12\,H^+ \rightarrow 5\,C_2H_4O_2 + 4\,Mn^{2+} + 11\,H_2O$
15. To test for an iodide ion (for example, in iodized salt), a solution is treated with chlorine to liberate iodine. The reaction is

$$2\,I^- + Cl_2 \longrightarrow I_2 + 2\,Cl^-$$

Which substance is oxidized? Which is reduced?

16. Molybdenum metal, used in special kinds of steel, can be manufactured by the reaction of its oxide with hydrogen. The reaction is

$$MoO_3 + 3\,H_2 \longrightarrow Mo + 3\,H_2O$$

Which substance is reduced? Which is the reducing agent?

17. Green grapes are exceptionally sour due to a high concentration of tartaric acid. As the grapes ripen, this compound is converted to glucose.

$$\underset{\text{Tartaric acid}}{C_4H_6O_6} \longrightarrow \underset{\text{Glucose}}{C_6H_{12}O_6}$$

Is the tartaric acid being oxidized or reduced?

18. The dye indigo (used to color blue jeans) is formed from indoxyl by exposing it to air.

$$\underset{\text{Indoxyl}}{C_8H_7ON} + O_2 \longrightarrow \underset{\text{Indigo}}{C_{16}H_{10}N_2O_2} + 2\,H_2O$$

What substance is oxidized? What is the oxidizing agent?

19. Acetylene (C_2H_2) reacts with hydrogen to form ethane (C_2H_6). Is the acetylene oxidized or reduced? Explain your answer.
20. Unsaturated vegetable oils react with hydrogen to form saturated fats. A typical reaction is

$$C_{57}H_{104}O_6 + 3\,H_2 \longrightarrow C_{57}H_{110}O_6$$

Is the unsaturated oil oxidized or reduced? Explain.

21. Vitamin C (ascorbic acid) is thought to protect our stomachs from the carcinogenic effect of nitrite ions by converting the ions to NO gas.

$$NO_2^- \longrightarrow NO$$

Is the nitrite ion oxidized or reduced? Is ascorbic acid an oxidizing agent or a reducing agent?

22. In the preceding reaction (Problem 21), ascorbic acid is converted to dehydroascorbic acid.

$$C_6H_8O_6 \longrightarrow C_6H_6O_6$$

Is ascorbic acid oxidized or reduced in this reaction?

Additional Problems

23. Are there any circumstances under which an oxidation half-reaction can occur unaccompanied by a reduction half-reaction? Explain.
24. When the water pump failed in the nuclear reactor at Three Mile Island in 1979, zirconium metal reacted with the very hot water to produce hydrogen gas.

$$Zr + 2\,H_2O \longrightarrow ZrO_2 + 2\,H_2$$

What substance was oxidized in this reaction? What was the oxidizing agent?

25. Calculate the oxidation number of S in sodium peroxodisulfate, $Na_2S_2O_8$. Actually, *two* of the oxygen atoms are involved in a peroxide linkage (—O—O—; that is what the prefix "peroxo" signifies) and have an oxidation number of -1. Make a new calculation to show that the oxidation number of S is really $+6$.
26. Calculate the oxidation number of O in CsO_2. (This is another exception to rule 4 that occurs in compounds called superoxides.)
27. When phosphorus, P_4, is heated with water it forms both phosphine, PH_3, and phosphoric acid, H_3PO_4, in a type of reaction called a disproportionation. Calculate oxidation numbers for the three substances and propose a definition for the word *disproportionation*.
28. Cyanide wastes can be detoxified by the addition of chlorine to a basic solution.

$$NaCN(aq) + NaOH(aq) + Cl_2(g) \longrightarrow NaOCN(aq) + NaCl(aq) + H_2O$$

Following the addition of acid to make the solution less basic, a further reaction occurs.

$$NaOCN(aq) + NaOH(aq) + Cl_2(g) \longrightarrow NaCl(aq) + H_2O + NaHCO_3(aq) + N_2(g)$$

Balance the two equations and identify the oxidizing agent and reducing agent in each equation.

29. Incineration of a chlorine-containing toxic waste such as a polychlorinated biphenyl (PCB) produces CO_2 and HCl.

$$C_{12}H_4Cl_6 + O_2 + H_2O \longrightarrow CO_2 + HCl$$

Balance the equation for this combustion reaction. Comment on the advantages and disadvantages of incineration as a method of disposal of such wastes.

30. In the citric acid cycle, which is important for cellular energy generation, malate (ion) reacts with nicotinamide adenine dinucleotide, NAD^+, to form oxaloacetate (ion) and NADH.

$$\underset{\text{Malate}}{C_4H_4O_5^{2-}} + NAD^+ \longrightarrow \underset{\text{Oxaloacetate}}{C_4H_2O_5^{2-}} + NADH + H^+$$

Which substance is oxidized and which is reduced? Identify the oxidizing agent and the reducing agent. (Hint: Use one or more of the three ways of looking at oxidation and reduction illustrated in Figure 5.6.)

Gases

CHAPTER 6

The properties of gases make it possible for humans to fly in hot-air balloons.

Learning Objectives/Study Questions

1. What gases are found in air?
2. What is the kinetic theory of gases? How does it explain the relationship between the pressure and volume of a gas (Boyle's law)? How does it explain the relationship between the temperature and volume of a gas (Charles's law)? How does it explain the relationship between the number of moles of a gas and its volume (Avogadro's law)?
3. How does a barometer measure atmospheric pressure?
4. How does Boyle's law help us understand human respiration?
5. What is Avogadro's law in words? In mathematical symbols?
6. What are the standard conditions of temperature and pressure for gases?
7. What is the combined gas law? How is it used?
8. What is the ideal gas law?
9. How does pressure influence the solubility of a gas at constant temperature?
10. What is Dalton's law of partial pressures? How is it applied to gases collected over water?
11. Why is the partial pressure of $CO_2(g)$ higher in cellular fluid than in the air we breathe? Why is the partial pressure of O_2 lower in cellular fluid than in the air we breathe?

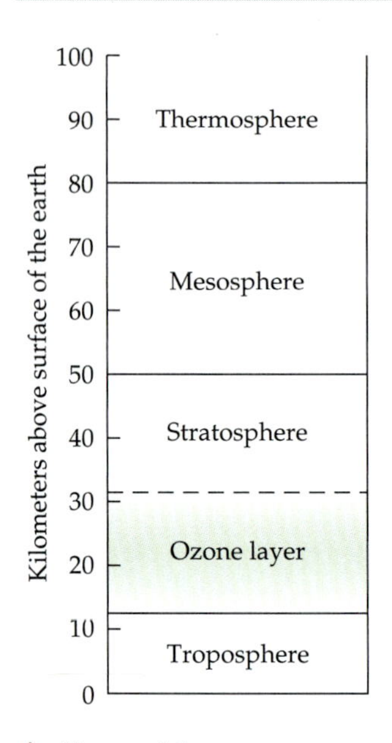

▲ **Figure 6.1**
The altitudes of the several layers of the atmosphere are only approximate. For example, the height of the troposphere varies from about 8 km at the poles to 16 km at the equator. The ozone layer is shown in green.

Humans have walked on the surface of Earth's barren, airless moon. Spacecraft have photographed the dusty desolation of Mercury, the planet nearest the sun, from a few kilometers away. Robotic probes have dropped through clouds of sulfuric acid and a thick blanket of carbon dioxide to land on the hot, inhospitable surface of Venus. Other space probes have descended through the thin, dusty atmosphere of Mars in a vain search for life on its surface. A probe has been dropped through the crushing, turbulent atmosphere of Jupiter, and spacecraft have examined the atmospheres of Saturn, Uranus, and Neptune. Pluto has also been studied, but only from vast distances.

Simple forms of life have been found in many seemingly inhospitable places on Earth. Some scientists hold out hope that primitive life forms might be found on Mars or one of Jupiter's moons, but it now seems clear that our home planet, a small island of green and blue in the vastness of space, is uniquely equipped to serve the needs of higher forms of life.

The life-support system of Spaceship Earth consists in part of a thin blanket of gases called the atmosphere. Although other planets in our solar system have atmospheres, Earth's atmosphere appears to be unique in its ability to support life. The atmosphere is composed of about 5.2×10^{15} metric tons of air spread over a surface area of 5.0×10^{8} km^2.

It is difficult to measure just how deep the atmosphere is. It does not end abruptly, but gradually fades as the distance from the surface of the Earth increases. We do know, however, that 99% of the atmosphere lies within 30 km of the surface of the Earth—a thin layer of air indeed (like the peel of an apple, only relatively thinner).

Earth's atmosphere is divided into layers (Figure 6.1). The layer nearest Earth, the troposphere, harbors nearly all living things and nearly all human activity. The next region, the stratosphere, contains the ozone layer that shields living creatures from deadly ultraviolet radiation.

Air is so familiar, and yet so nebulous, that it is difficult to regard it as matter. But it is matter—matter in the gaseous state. All gases, air included, have mass and occupy space. Like other forms of matter, gases obey certain physical laws. In this chapter, we will examine some of those laws and see how they are related to certain vital processes—such as breathing.

Table 6.1 Composition of Dry Air (Near Sea Level)

Component	Percent by Volume
Nitrogen (N_2)	78.084
Oxygen (O_2)	20.946
Argon (Ar)	0.934
Carbon dioxide (CO_2)	0.036
Plus traces of neon (Ne), helium (He), methane (CH_4), krypton (Kr), hydrogen (H_2), dinitrogen monoxide (N_2O), xenon (Xe), ozone (O_3), sulfur dioxide (SO_2), nitrogen dioxide (NO_2), ammonia (NH_3), carbon monoxide (CO), iodine (I_2)	

6.1 Air: A Mixture of Gases

Air is a mixture of gases. The composition of dry air is summarized in Table 6.1. Water vapor can make up as much as 4% of humid air (Section 6.10). The most important of the several minor constituents of air is carbon dioxide. The concentration of CO_2 in air has increased from about 275 parts per million (ppm) in 1900 to its present value of about 360 ppm. It most likely will continue to rise as more and more fossil fuels (coal, oil, and natural gas) are burned.

Air is a mixture of gases, but what are gases? Perhaps they are best understood in terms of the kinetic-molecular theory.

✓ Review Questions

6.1 Which layer of the atmosphere lies nearest the surface of the Earth? Which contains the ozone layer?

6.2 List the three major components of dry air and give the approximate (nearest whole number) volume percent of each.

6.2 The Kinetic-Molecular Theory

As we noted in Chapter 1, the states of matter—solid, liquid, and gas—differ in obvious ways from one another. Chemists use models to explain these differences. The model used to explain the behavior of gases is the **kinetic-molecular theory.** The basic postulates of this theory are:

1. All matter is composed of tiny, discrete particles called molecules[1].
2. These molecules are in rapid, constant motion and move in straight lines.
3. The molecules of a gas are small compared with the distances between them.
4. There is little attraction between molecules of a gas.
5. Molecules collide with one another, and energy is conserved in these collisions—although one molecule can gain energy at the expense of another.
6. Temperature is a measure of the *average* kinetic energy of the gas molecules.

The kinetic-molecular theory treats gases as collections of individual particles in rapid motion (*kinetic* derives from the Greek word for motion). The particles of nitrogen gas, for example, are molecules (N_2); those of argon gas (Ar) are atoms. The distances between the particles are quite large compared with the dimensions of the particles themselves. Therefore, unlike solids and liquids, gases can be readily compressed. (According to the theory, the individual particles of a solid or liquid are in contact with one another. They can't be pushed closer together because they are already touching.)

The particles of a gas are in such rapid motion, and gases have such low densities, that gravity seems to have little effect on them. They move up and down and sideways with ease and will not fall to the bottom of a vessel (as a liquid will). A flask or cylinder of gas is completely filled. By *filled* we do not mean that gas molecules are packed tightly, but rather that the gas is distributed throughout the container's entire volume. A particle moves along a straight path until it strikes something (another particle or the walls of the container). Then it may bounce off at an angle and travel from the point of collision along a straight path until it hits something else. These collisions are *perfectly elastic;* there is no tendency for the collection of particles to slow down and eventually stop. Two particles that are about to collide have a certain combined kinetic energy. After the collision, the sum of their kinetic energies has not changed. One of the particles may have been slowed down by the collision, but the other will have been speeded up just enough to compensate (Figure 6.2).

Although gas molecules do not settle noticeably under the force of gravity in containers of ordinary size, they do so in that largest of containers—the atmosphere. The density of air decreases with increasing altitude above Earth's surface.

The kinetic-molecular theory explains what we are measuring when we measure temperature. According to the theory, temperature is just a reflection of the kinetic energy of the gas particles. The higher the kinetic energy—that is, the faster the particles are moving—the higher the temperature of the sample. In any single sample, some particles are moving faster than others. Temperature reflects the *average* kinetic energy of the particles. On average, the particles of a cold sample move more slowly than the particles of a hot sample.

As a particle strikes a wall of its container, it gives the wall a little push. (If you were hit with a baseball or a brick, you would feel a push.) The sum of all these tiny pushes over a given area of the wall is what we call *pressure.*

[1]The kinetic-molecular theory was developed to explain the behavior of gases. Expanded to include atoms and ions as well as molecules, some aspects of the theory—particles, the attraction between them, and their motion—can also be used to explain the behavior of liquids and solids (Chapter 7).

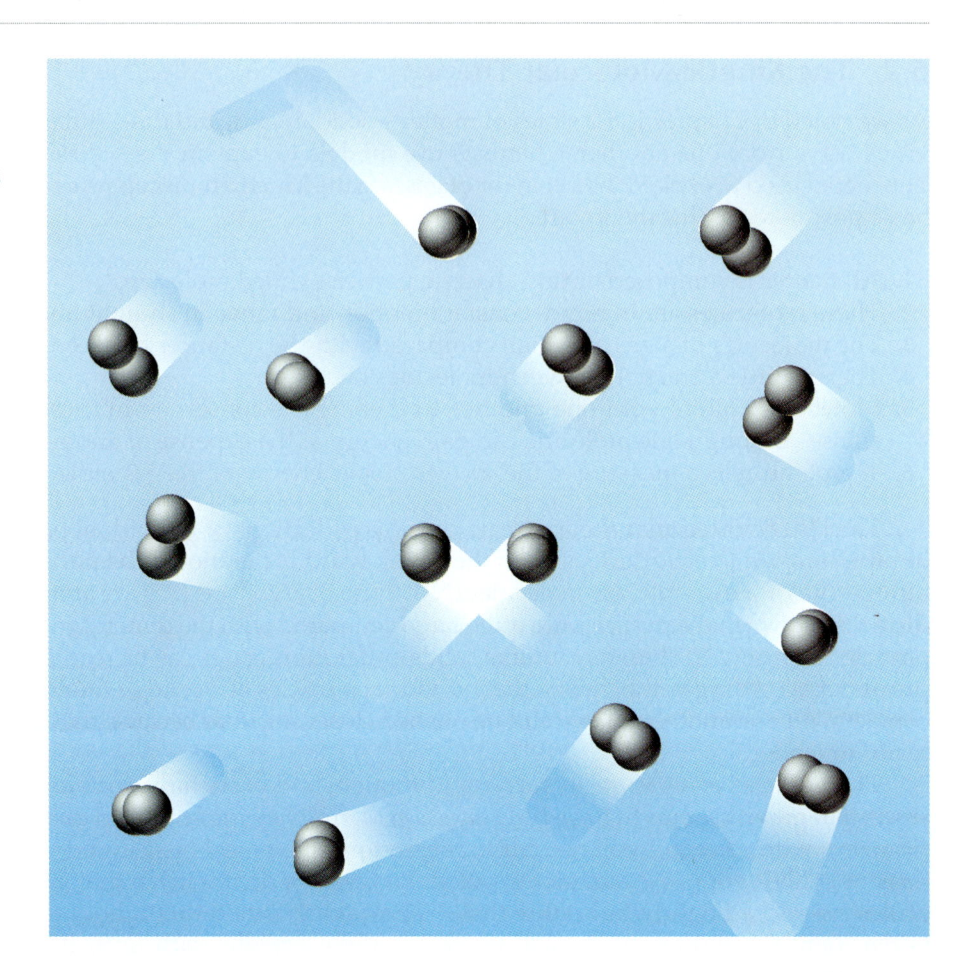

Figure 6.2 ▶ According to the kinetic-molecular theory, molecules of a gas are in constant random motion. They move in straight lines and undergo collisions with each other and with the walls of the container.

✓ Review Questions

6.3 Give a molecular-level description of a gas.

6.4 Give a kinetic-molecular explanation of the origin of gas pressure.

6.3 Atmospheric Pressure

Molecules in the air are constantly bouncing off our skin. They are so tiny, however, that we don't feel their individual impacts. However, when we go up in altitude rapidly—by driving up a mountain or riding an express elevator to the top of a tall building—our ears may "pop." This popping sensation is caused by an unequal air pressure on the two sides of our ear drums. Because the density of air decreases at the higher altitudes, there are fewer molecules of air outside our ear drums pushing in than on the inside pushing out. The popping stops as soon as the pressure inside the ear decreases and becomes equal to that at the higher altitude.

Pressure is force per unit area—that is, force divided by the area over which the force is exerted.

$$\text{Pressure} = \frac{\text{Force}}{\text{Area}} = \frac{F}{A}$$

In the SI system, force is expressed in *newtons* (N) and area in square meters (m^2). (A newton is a force that will give a 1-kg mass an acceleration of 1 meter per second, or m/s.) The derived SI unit for pressure is therefore the newton per square meter, commonly called a **pascal (Pa).**

$$1 \text{ Pa} = 1 \text{ N/m}^2$$

The pascal is such a small unit that the kilopascal (kPa) is often used instead.

The pressure of the atmosphere is measured by a device called a **barometer.** A simple type, known as a mercury barometer, was invented in 1643 by the Italian scientist Evangelista Torricelli (1608–1647). The mercury barometer consists of a long glass tube, closed at one end, filled with mercury, and inverted in a shallow dish that also contains mercury (Figure 6.3). Suppose the tube is 1 m long. Some of the mercury in the tube drains into the dish, but *not all* of it. The mercury drains out only until the pressure exerted by the mercury remaining in the tube exactly balances the pressure exerted by the atmosphere on the surface of the mercury in the dish. The mercury column tends to flow out under the influence of gravity, and the atmospheric pressure tends to push the mercury back into the tube. At some point these two opposing tendencies reach a stalemate.

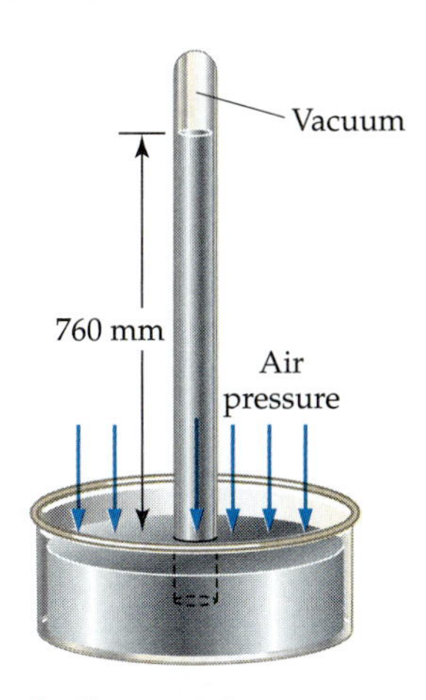

▲ **Figure 6.3**
Measurement of air pressure with a mercury barometer. A column of mercury 760 mm high is maintained in a *closed-end* tube. The space above the mercury is devoid of air and contains only a trace of mercury vapor.

Mercury is a dense liquid. On average, at sea level, a column of mercury about 760 mm high will be supported by air pressure. The pressure that is exerted by a column of mercury 760 mm high is called 1 **atmosphere (atm).** The pressure unit, 1 **millimeter of mercury (mmHg),** is often called a *Torr* (after Torricelli).

$$1 \text{ atm} = 760 \text{ mmHg} = 760 \text{ Torr}$$

The relationship of the atmosphere to SI units is

$$1 \text{ atm} = 101{,}325 \text{ Pa} = 101.325 \text{ kPa}$$

For approximate work, it is helpful to remember that 1 atm is about 100 kPa.

Several other pressure units are widely used. Weather reports in the United States often include atmospheric pressure in *inches of mercury (in.Hg).*

$$1 \text{ atm} = 29.921 \text{ in.Hg}$$

Engineers generally use *pounds per square inch (lb/in.^{2} or psi)* for practical applications such as steam pressure in boilers and turbines.

$$1 \text{ atm} = 14.696 \text{ lb/in.}^2$$

✓ Review Questions

6.5 Why is atmospheric pressure greater at sea level than at the top of a high mountain?

6.6 What does a mercury barometer measure? How does it work? Why is mercury (rather than water or another liquid) used as the fluid in barometers?

▲ Robert Boyle (1627–1691) published *The Sceptical Chymist* in 1661 in which he argued that theories are no better than the experiments on which they are based. Gradually this point of view was accepted, and Boyle's text marked a turning point for the importance of experimentation. His experiments on air helped to found modern science.

6.4 Boyle's Law: The Pressure-Volume Relationship

A simple gas law, discovered by the Irish chemist Robert Boyle in 1662, describes the relationship between the pressure and volume of a gas. **Boyle's law** states that *for a given amount of gas at a constant temperature, the volume of the gas varies* inversely *with its pressure.* That is, in a closed container of gas, when the pressure increases, the volume decreases; when the pressure decreases, the volume increases.

Think of gases as pictured in the kinetic-molecular theory (Figure 6.4). A gas exerts a particular pressure because the gas molecules bounce against the container walls with a certain frequency and speed. If the volume of the container is expanded while the amount of gas remains fixed, the number of molecules per unit volume of gas decreases. The frequency with which molecules strike a unit area of the container walls decreases, and the gas pressure decreases. Thus, as the volume of a gas is increased, its pressure decreases.

Mathematically, for a given amount of gas at a constant temperature, Boyle's law is written

$$V \propto \frac{1}{P}$$

where the symbol $\propto$ means "is proportional to." This may be changed to an equation by inserting a proportionality constant, a.

$$V = \frac{a}{P}$$

Multiplying both sides of the equation by P, we get

$$PV = a \text{ (with constant temperature and mass)}$$

Another way to state Boyle's law, then, is that for a given amount of gas at a constant temperature, the product of the pressure and volume is a constant. This is an elegant and precise, if somewhat abstract, way of summarizing a lot of experimental

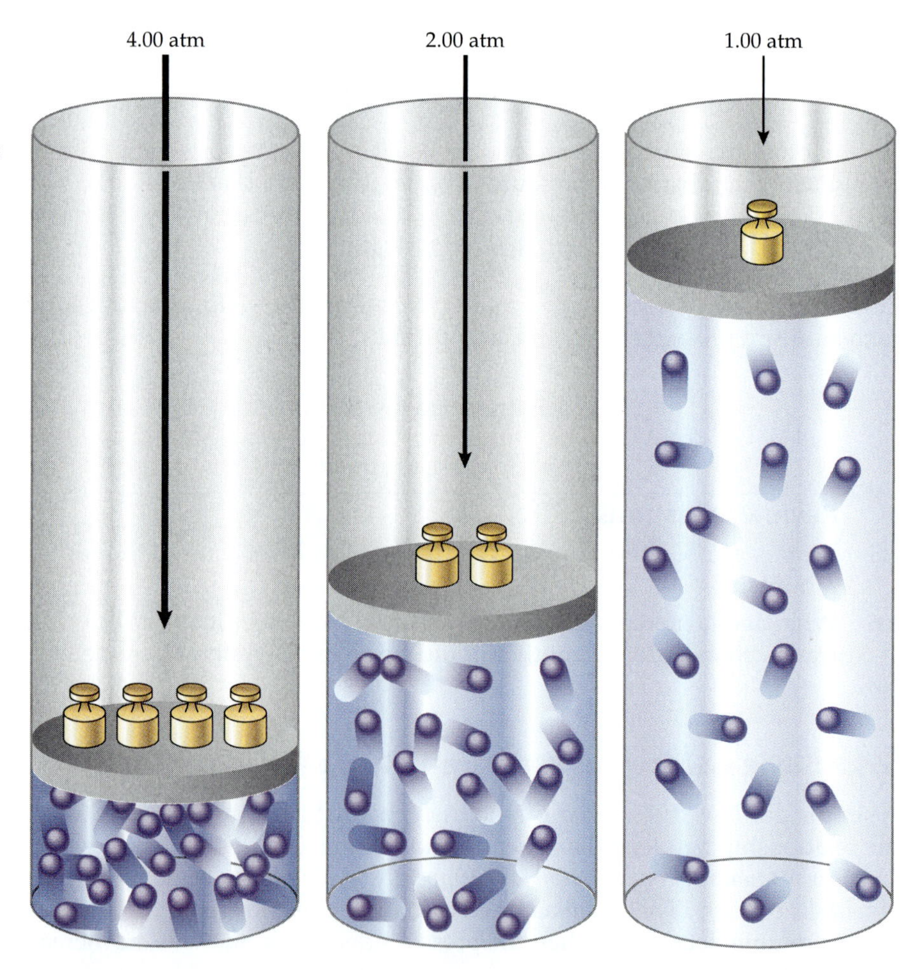

Figure 6.4 ▶ A kinetic-molecular theory view of Boyle's law. As the pressure is reduced from 4.00 atm to 2.00 atm and then to 1.00 atm, the volume of the gas doubles and then doubles again.

data. If the product $P \times V$ is to be constant, then if V increases P must decrease, and vice versa. This relationship is demonstrated in Figure 6.5 by a pressure-volume graph.

Boyle's law has a number of practical applications perhaps best illustrated by some examples. In Example 6.1, we see how to *estimate* an answer. Sometimes an estimation is all we need. Even when we want a quantitative answer, however, the estimate helps us to determine whether or not our answer is reasonable.

Example 6.1

A gas is enclosed in a cylinder fitted with a piston. The volume of the gas is 2.00 L at 0.524 atm. The piston is moved to increase the gas pressure to 5.15 atm. Which of the following is a reasonable value for the volume of the gas at the greater pressure?

0.200 L   0.400 L   1.00 L   16.0 L

Solution

The pressure increase is almost tenfold. The volume should drop to about one-tenth of the initial value. We estimate a volume of 0.20 L. (The calculated value is 0.203 L.)

Practice Exercise

A gas is enclosed in a 10.2-L tank at 1208 mmHg. Which of the following is a reasonable value for the pressure when the gas is transferred to a 30.0-L tank?

25 lb/in.^{2} 300 mmHg   400 mmHg   3600 mmHg

Gases are usually stored under high pressure, even though they will be used at atmospheric pressure. This allows a large quantity of gas to be stored in a small volume.

Examples 6.2 and 6.3 illustrates quantitative calculations using Boyle's law. Note that in these applications any units can be used for pressure and volume, as long as the same units are used throughout a calculation. As long as we use the *same* sample of a confined gas at a constant temperature, the product of the initial volume (V_1) times the initial pressure (P_1) is equal to the product of the final volume (V_2) times the final pressure (P_2). Thus, the following useful equation representing Boyle's law can be written.

$$V_1P_1 = V_2P_2$$

Example 6.2

A cylinder of oxygen has a volume of 2.25 L. The pressure of the gas is 1470 psi at 20 °C. What volume will the oxygen occupy at standard atmospheric pressure (14.7 psi), assuming no temperature change?

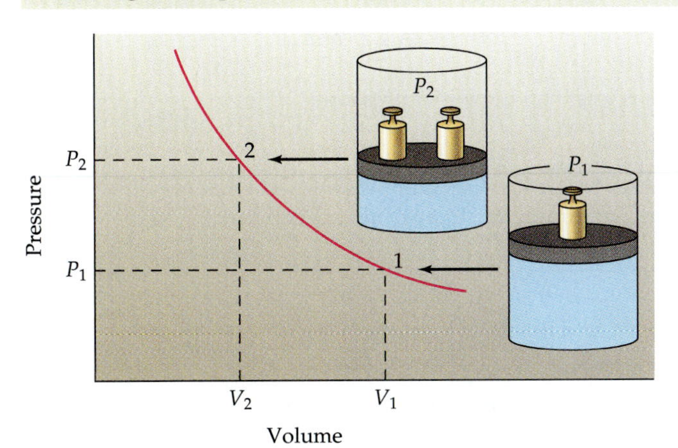

Figure 6.5 A graphical representation of Boyle's law. As the pressure on the gas is increased, its volume decreases. When the pressure is doubled ($P_2 = 2 \times P_1$), the volume of the gas decreases to one half its original value ($V_2 = 1/2 \times V_1$). The pressure-volume product is a constant ($PV = a$)

Helpful hint: Regardless of which variable is to be determined, it is recommended that you solve the equation for the unknown *before* substituting values into the equation. Rearranging a few letters takes less time than rearranging and rewriting complex terms.

Because the final pressure in Example 6.2 is *less than* the initial pressure, we expect the final volume to be *larger than* the original volume and it is.

Solution

It is most helpful to first separate the initial from the final condition.

Initial	**Final**	**Change**
$P_1 = 1470$ psi	$P_2 = 14.7$ psi	↓
$V_1 = 2.25$ L	$V_2 = ?$	↑

The pressure goes down, therefore the volume goes up.

Then use the equation $V_1P_1 = V_2P_2$ and solve for the desired volume or pressure. In this case, we solve for V_2.

$$V_2 = \frac{V_1P_1}{P_2}$$

$$V_2 = \frac{2.25\text{ L} \times 1470\text{ psi}}{14.7\text{ psi}} = 225\text{ L}$$

Practice Exercise

A sample of air occupies 73.3 mL at 98.7 atm and 0 °C. What volume will the air occupy at 4.02 atm and 0 °C?

Example 6.3

A space capsule is equipped with a tank of air that has a volume of 0.125 m^3. The air is under a pressure of 112 atm. After a space walk, during which the cabin pressure drops to zero, the cabin is closed and filled with the air from the tank. What will be the final pressure if the volume of the capsule is 11.0 m^3?

Solution

As in Example 6.2, we first separate the initial from the final condition.

Initial	**Final**	**Change**
$V_1 = 0.125\text{ m}^3$	$V_2 = 11.0\text{ m}^3$	↑
$P_1 = 112$ atm	$P_2 = ?$	↓

The volume goes up, therefore the pressure goes down.

We then solve for P_2.

$$P_2 = \frac{V_1P_1}{V_2}$$

$$P_2 = \frac{0.125\text{ m}^3 \times 112\text{ atm}}{11.0\text{ m}^3} = 1.27\text{ atm}$$

As expected, the final pressure is less than the initial pressure.

Practice Exercise

A sample of nitrogen gas occupies 80.0 mL at 1.00 atm pressure. At what pressure will the nitrogen gas occupy 60.0 mL, assuming the temperature is constant?

Review Questions

6.7 State Boyle's law in words and as a mathematical equation.

6.8 Use the kinetic-molecular theory to explain Boyle's law.

6.9 What is the advantage of storing gases under high pressure—for example, oxygen used in respiratory therapy?

Boyle's Law and Breathing

We can use the pressure-volume relationship to help explain the mechanics of breathing. To induce *inspiration* (breathing in), the diaphragm is lowered and the chest wall is expanded, increasing the volume of the chest cavity (Figure 6.6). According to Boyle's law, the pressure inside the cavity must decrease. Outside air enters the lungs because it is at a higher pressure than the air in the chest cavity. To induce *expiration* (breathing out), the diaphragm rises and the chest wall contracts, decreasing the volume of the chest cavity. The pressure is increased, and some air is forced out.

During normal inspiration the pressure inside the lungs drops about 3 mmHg below atmospheric pressure. During expiration, the internal pressure is about 3 mmHg above atmospheric pressure. About half a liter of air is moved in and out of the lungs in this process, and this normal breathing volume is called the *tidal volume.* The *vital capacity* is the maximum volume of air that can be forced from the lungs and ranges from 3 to 7 L, depending on the individual. A pressure inside the lungs 100 mmHg greater than the external pressure is not unusual during such a maximum expiration.

The lungs are never emptied completely, however. The space around the lungs is maintained at a slightly lower pressure than are the lungs themselves, causing the lungs to be kept partially inflated by the higher pressure within them. If a lung, the diaphragm, or the chest wall is punctured, allowing the two pressures to equalize, the lung will collapse. Sometimes a damaged lung is collapsed intentionally to give it time to heal. The lung reinflates after the opening is closed.

People were breathing long before Boyle formulated his law, but it is satisfying to understand how the process works. We get more than just satisfaction from science, however. An understanding of the pressure-volume relationship has enabled us to keep people alive. Persons unable to breathe due to paralysis can be kept alive by artificial respirators. The iron lung, which kept many polio victims alive during the 1940s and 1950s, is a sealed chamber connected to a compressor and bellows (Figure 6.7). The pressure in the chamber is varied rhythmically. When the bellows moves out, the pressure in the chamber is reduced. The air pressure around the nose and mouth (outside the tank) is greater than the pressure on the chest (inside the tank), so air flows in and fills the lungs. When the bellows moves in, pressure in the tank increases, and air is expelled from the lungs.

The iron lung, designed to enclose the patient completely (except for the head), is cumbersome and uncomfortable. It has been replaced by respirators that enclose the chest only. In fact, the whole area of respiratory therapy has become far more sophisticated in recent years. Specialists in this area are an indispensable part of the medical team.

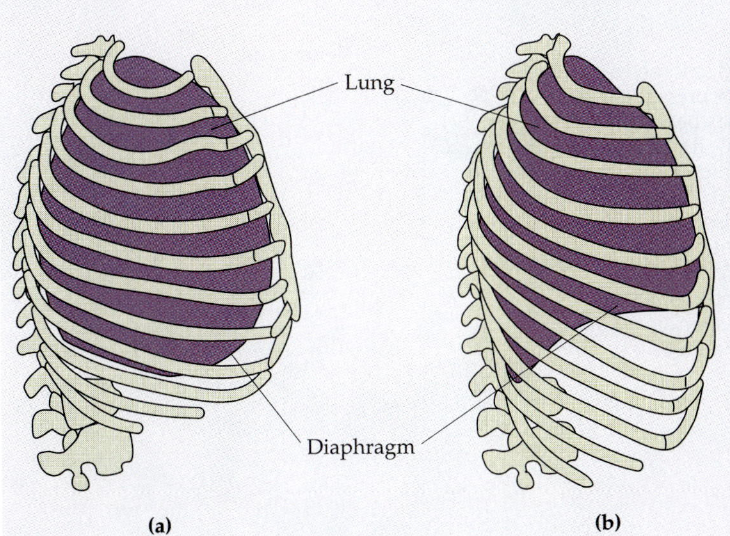

Figure 6.6 The mechanics of breathing.
(a) Inspiration. The diaphragm is pulled down, and the rib cage is lifted up and out, increasing the volume of the chest cavity.
(b) Expiration. The diaphragm is relaxed, the rib cage is down, and the volume of the chest cavity decreases.

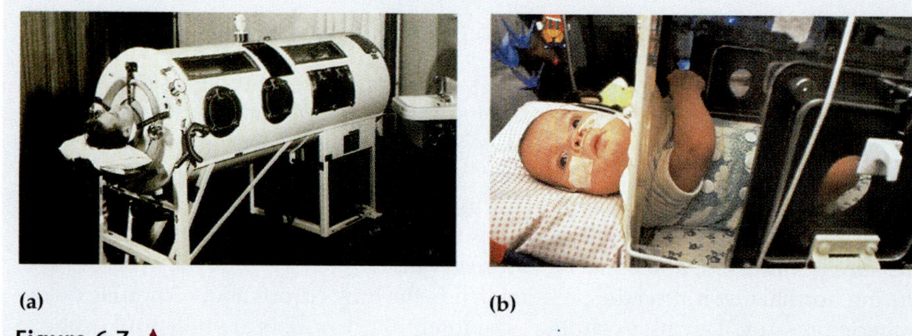

(a) (b)

Figure 6.7 ▲
An iron lung uses changes in pressure to force air into and out of the lungs by enclosing the entire body except for the head. (a) An older version used to treat polio victims in the 1950s. (b) Baby in a modern iron lung.

6.5 Charles's Law: The Temperature-Volume Relationship

In 1787 the French physicist Jacques Charles (1746–1823) studied the relationship between volume and temperature of gases. He found that when a fixed quantity of gas is cooled at constant pressure, its volume decreases. When the gas is heated, its volume increases. Temperature and volume vary directly; that is, they rise or fall together. But this law requires a bit more thought. If a quantity of gas that occupies 1.00 L is heated from 100 °C to 200 °C at constant pressure, the volume does not double but only increases to about 1.27 L. The relationship between temperature and volume is not as tidy as it may seem on first impression.

Zero pressure or zero volume really means zero—no pressure or volume to be measured. Zero degrees Celsius (0 °C) is only the freezing point of water, arbitrarily set, much like mean sea level, which is the arbitrary zero for measurement of altitudes on Earth. Temperatures below 0 °C are often encountered, as are altitudes below sea level.

Charles noted that for each Celsius degree rise in temperature, the volume of a gas expands by 1/273 of its volume at 0 °C. For each Celsius degree drop in temperature, the volume of a gas decreases by 1/273 of its value at 0 °C. If we plot volume against temperature we get a straight line (Figure 6.8). We can *extrapolate* the

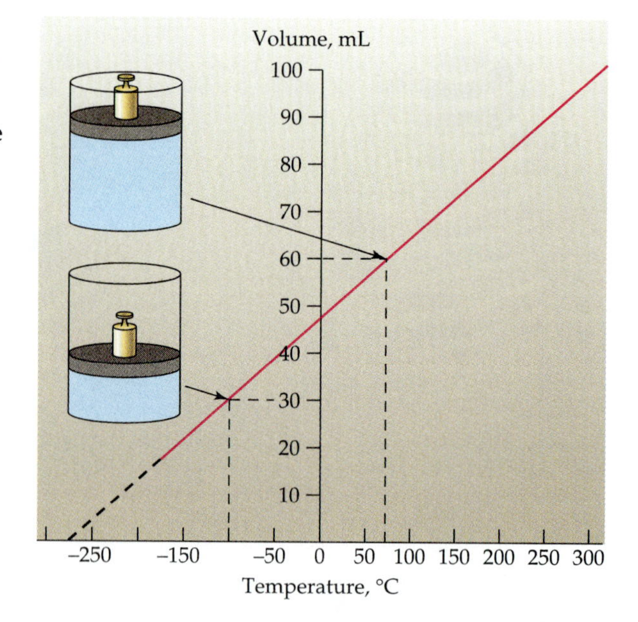

Figure 6.8 ▶
Charles's law relates gas volume to temperature at constant pressure. When the gas shown has been cooled to about −70 °C, its volume is 60 mL. In the temperature interval from about −70 °C to −100 °C, the volume drops to 30 mL. The volume continues to fall as the temperature is lowered. The extrapolated line intersects the temperature axis (corresponding to a volume of zero) at about −270 °C.

line beyond the range of measured temperatures to the temperature at which the volume of the gas would become zero. This temperature is −273.15 °C. In 1848 William Thomson (Lord Kelvin) made this temperature the zero point on an *absolute* temperature scale, which is now called the **Kelvin scale.** The unit of temperature on this scale is the kelvin (K).

Before a gas ever reaches this temperature, however, it liquefies—and then the liquid freezes—so this is an exercise for the imagination.

A modern statement of **Charles's law,** then, is that *the volume of a fixed amount of a gas at a constant pressure is directly proportional to its Kelvin temperature.* Mathematically, this relationship is expressed as

$$V \propto T$$

In the form of an equation, this becomes

$$V = bT \quad \text{or} \quad \frac{V}{T} = b \quad \text{(with constant pressure and mass)}$$

where b is a constant. To keep V/T equal to a constant value, when the temperature increases, the volume must also increase. When the temperature decreases, the volume must also decrease (Figure 6.9).

As long as we are using the *same* sample of trapped gas at a constant pressure, the initial volume (V_1) divided by the initial *absolute* temperature (T_1) is equal to the final volume (V_2) divided by the final *absolute* temperature (T_2). The following equation is useful in solving Charles's law problems.

For all gas law calculations involving temperature, absolute temperatures in kelvins must be used (not °C or °F).

$$\frac{V_1}{T_1} = \frac{V_2}{T_2}$$

The kinetic-molecular model explains the relationship between gas volume and temperature. When we heat a gas, we supply the gas molecules with energy and they begin to move faster. These speedier molecules strike the walls of their container harder and more often. For the pressure to stay the same, the volume of the container must increase so that the increased molecular motion will be distributed over a greater space. In this way, the pressure exerted by the faster molecules in the larger volume (high temperature) is the same as that of the slower-moving molecules in the smaller volume (low temperature).

(a)

(b)

◀ **Figure 6.9**
A dramatic illustration of Charles's law. (a) Liquid nitrogen (boiling point, −196 °C) cools the balloon and its contents to a temperature far below room temperature. (b) As the balloon warms to room temperature, the volume of air increases proportionately (about fourfold).

Example 6.4

A balloon indoors, where the temperature is 27 °C, has a volume of 2.00 L. What would its volume be (a) outdoors, where the temperature is −23 °C, and (b) in a hot room where the temperature is 47 °C? (Assume no change in pressure in either case.)

Solution

First, convert all temperatures to the Kelvin scale

$$T(\text{K}) = T(^\circ\text{C}) + 273$$

The initial temperature is (27 + 273) = 300 K, and the final temperatures are (a) (−23 + 273) = 250 K and (b) (47 + 273) = 320 K.

a. We start by separating the initial from the final condition.

Initial	Final	Change
$t_1 = 27\ ^\circ\text{C}$	$t_2 = -23\ ^\circ\text{C}$	⇓
$T_1 = 300\ \text{K}$	$T_2 = 250\ \text{K}$	⇓
$V_1 = 2.00\ \text{L}$	$V_2 = ?$	⇓

Solving the equation

$$\frac{V_1}{T_1} = \frac{V_2}{T_2}$$

for V_2, we have

$$V_2 = \frac{V_1 T_2}{T_1}$$

$$V_2 = \frac{2.00\ \text{L} \times 250\ \cancel{\text{K}}}{300\ \cancel{\text{K}}} = 1.67\ \text{L}$$

As we expected, the volume decreased because the temperature decreased.

b. We have the same initial conditions as in (a) but different final conditions.

Initial	Final	Change
$t_1 = 27\ ^\circ\text{C}$	$t_2 = 47\ ^\circ\text{C}$	⇑
$T_1 = 300\ \text{K}$	$T_2 = 320\ \text{K}$	⇑
$V_1 = 2.00\ \text{L}$	$V_2 = ?$	⇑

In this case, the temperature increases, the volume must also increase.

$$V_2 = \frac{V_1 T_2}{T_1}$$

$$V_2 = \frac{2.00\ \text{L} \times 320\ \cancel{\text{K}}}{300\ \cancel{\text{K}}} = 2.13\ \text{L}$$

Practice Exercises

A. A sample of oxygen gas occupies a volume of 2.10 L at 25 °C. What volume will this sample occupy at 150 °C? (Assume no change in pressure.)

B. At what Celsius temperature will the initial volume of oxygen in Practice Exercise A occupy 0.750 L? (Assume no change in pressure.)

Review Questions

6.10 State Charles's law in words and in the form of a mathematical equation.

6.11 Use the kinetic-molecular theory to explain Charles's law.

6.12 How is the Kelvin scale related to the Celsius scale? Why must an absolute temperature scale rather than the Celsius scale be used for Charles's law calculations?

6.6 Avogadro's Law: The Molar Volume of a Gas

Avogadro's hypothesis (Section 4.2), explains Gay-Lussac's law of combining volumes. The hypothesis is that equal numbers of molecules of different gases compared at the same temperature and pressure occupy equal volumes. Let's now restate Avogadro's hypothesis in the form generally called **Avogadro's law:** *At a fixed temperature and pressure, the volume of a gas is directly proportional to the number of molecules of gas or to the number of moles of gas, n.* If we double the number of moles of gas at a fixed T and P, the volume of the gas doubles. Because the mass of a gas is proportional to the number of moles, doubling the *mass* of a gas also doubles its volume. Mathematically, we can state Avogadro's law as

$$V \propto n \quad \text{or} \quad V = cn \text{ (where } c \text{ is a constant)}$$

When we use Avogadro's hypothesis to compare different gases, the gases must be at the same temperature and pressure. A convenient temperature/pressure combination for such comparisons is 0 °C (273 K) and 1 atm (760 Torr), known as **standard conditions of temperature and pressure (STP).**

Standard conditions of temperature and pressure (STP): $T = 273$ K and $P = 1$ atm.

Suppose that in comparing different gases we use STP as the fixed temperature and pressure and Avogadro's number as the number of molecules present. Avogadro's hypothesis states that under these conditions, 1 mol (6.022×10^{23} molecules) of *all* gases should occupy the same volume. By experiment, this **molar volume of a gas at** *STP* is found to be 22.428 L of H_2, 22.404 L of N_2, 22.394 L of O_2, 22.360 L CH_4, and so on. To *three* significant figures, we can state that

$$1 \text{ mol gas} = 22.4 \text{ L gas (at STP)}$$

Figure 6.10 pictures a volume of 22.4 L and relates it to some familiar objects. The 22.4-L container would hold 28.0 g of N_2, 32.0 g of O_2, or 44.0 g of CO_2.

Example 6.5

Calculate the volume occupied by 4.11 g of methane (CH_4) gas at STP.

Solution

We must convert the mass of gas to an amount in moles, and then use the molar volume relationship as a conversion factor to go from the amount of gas to its volume at STP. We can do all of this in a single setup.

$$4.11 \cancel{\text{g CH}_4} \times \frac{1 \cancel{\text{mol CH}_4}}{16.04 \cancel{\text{g CH}_4}} \times \frac{22.4 \text{ L CH}_4}{1 \cancel{\text{mol CH}_4}} = 5.74 \text{ L CH}_4$$

Practice Exercises

A. Calculate the volume occupied by 11.2 g of sulfur hexafluoride (SF_6) gas at STP.

B. Solid carbon dioxide, called "dry ice," is useful in maintaining frozen foods because it vaporizes to $CO_2(g)$ rather than melting to a liquid. How many grams of dry ice can be produced from 5.00 L of $CO_2(g)$ measured at STP?

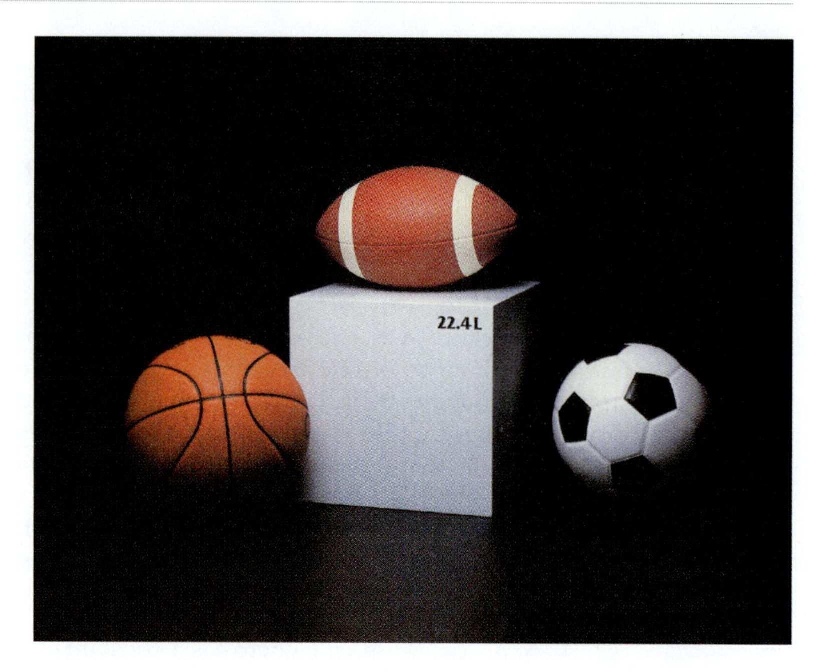

Figure 6.10 ▶
The molar volume of a gas visualized. The wooden cube has the same volume as 1 mol of gas at STP: 22.4 L. By contrast, the basketball has a volume of 6.5 L; the soccer ball, 6.0 L; and the football, 4.4 L.

✓ Review Questions

6.13 What are the standard conditions of temperature and pressure for gases? Why is it useful to define such conditions?

6.14 What is meant by the *molar volume* of a gas? What is the value of the molar volume of a gas at STP?

6.7 The Combined Gas Law

From the three simple gas laws, it seems reasonable that the volume of a gas (V) should be directly proportional to the Kelvin temperature (T) and to the amount of gas (n), and inversely proportional to the pressure (P). That is,

$$V \propto \frac{nT}{P}$$

Or, expressed as an equation rather than a proportionality,

$$\frac{PV}{nT} = \text{constant}$$

This **combined gas law** is most useful in situations where a fixed quantity of gas is described under an initial set of conditions, the gas is subjected to a change, and a question is asked about the final set of conditions. In these cases we write

$$\frac{V_1P_1}{T_1} = \frac{V_2P_2}{T_2}$$

Example 6.6

A balloon is partially filled with helium on the ground at 27 °C and 740 mmHg pressure. Its volume is 10.0 m^3. What would the volume be at a higher altitude where the pressure is 370 mmHg and the temperature is −23 °C?

Solution

The temperature decreases from 27 °C to −23 °C, or from (27 + 273) = 300 K to (−23 + 273) = 250 K.

Initial	Final	Change
$t_1 = 27$ °C	$t_2 = -23$ °C	⇓
$T_1 = 300$ K	$T_2 = 250$ K	⇓
$P_1 = 740$ mmHg	$P_2 = 370$ mmHg	⇓
$V_1 = 10.0\ m^3$	$V_2 = ?$	

Solving the combined gas law equation for V_2, we have

$$V_2 = \frac{V_1 P_1 T_2}{P_2 T_1}$$

Substituting the given values (and using a calculator) yields

$$V_2 = \frac{10.0\ m^3 \times 740\ \text{mmHg} \times 250\ \text{K}}{370\ \text{mmHg} \times 300\ \text{K}} = 16.7\ m^3$$

Practice Exercises

A. A sample of helium occupies 38.4 mL at 40 °C and 680 mmHg. What volume will the helium occupy at 80 °C and 720 mmHg?
B. At what pressure must a sample of carbon dioxide be confined in a 3.75-L flask at 30 °C if its volume at STP is 10.0 L?

Densities of gases are usually reported in the literature in units of grams per liter at STP. We can use the molar mass of a gas and its density at STP to calculate its molar volume. For example, the molar mass of $N_2(g)$ is 28.0 g/mol, and its density at STP is 1.25 g/L. Dividing, we get

$$\frac{28.0\ \text{g/mol}}{1.25\ \text{g/L}} = 22.4\ \text{L/mol}$$

The molar mass of $O_2(g)$ is 32.0 g/mol, and its density at STP is 1.43 g/L. Dividing, we again get a molar volume of 22.4 L/mol. Of course, these are just the values that we expect for the molar volume of a gas at STP. Because we know the molar volume, we can calculate the density of a gas at STP, as illustrated in Example 6.7.

Example 6.7

Calculate the density of $CO_2(g)$ at STP.

Solution

The molar mass of CO_2 is 44.0 g/mol. Dividing by the molar volume gives the density.

$$\frac{44.0\ \text{g/mol}}{22.4\ \text{L/mol}} = 1.96\ \text{g/L}$$

(The experimental value is 1.98 g/L.)

Practice Exercise

Calculate the density of (a) $H_2(g)$ and (b) $C_2H_6(g)$ at STP.

Review Questions

6.15 What effect will the following changes have on the volume of a fixed amount of gas?

a. an increase in pressure at constant temperature
b. a decrease in temperature at constant pressure
c. a decrease in pressure coupled with an increase in temperature

6.16 What effect will the following changes have on the pressure of a fixed amount of gas?

a. an increase in temperature at constant volume
b. a decrease in volume at constant temperature
c. an increase in temperature coupled with a decrease in volume

6.8 The Ideal Gas Law

So far we have done calculations in which the quantity of a gas does not change. As we saw above (in our discussion of molar volume), equal volumes of gases at the same temperature and pressure contain equal numbers of moles. Thus we can write a gas law that takes into account varying quantities of gas. This relationship is called the *ideal gas equation* or **ideal gas law.**

$$PV = nRT$$

In this equation, the pressure is in atmospheres, the volume in liters, and the temperature in kelvins. The number of moles of the gas is given by n. The constant R, which has a value of

We read this R value as 0.0821 liter-atmosphere per mole-kelvin.

$$0.0821 \frac{\text{L·atm}}{\text{mol·K}}$$

is called the **universal gas constant.**

The ideal gas law can be used to calculate any of the four quantities—P, V, n, or T—if the other three are known.

Example 6.8

Use the ideal gas law to calculate (a) the volume occupied by 1.00 mol of nitrogen gas at 244 K and 1.00 atm pressure, and (b) the pressure exerted by 0.500 mol of oxygen in a 15.0-L container at 303 K.

Solution

a. We start by solving the ideal gas equation for V.

$$V = \frac{nRT}{P}$$

$$V = \frac{1.00 \text{ mol}}{1.00 \text{ atm}} \times \frac{0.0821 \text{ L·atm}}{\text{mol·K}} \times 244 \text{ K} = 20.0 \text{ L}$$

b. Here we solve the ideal gas equation for P.

$$P = \frac{nRT}{V}$$

$$P = \frac{0.500 \text{ mol}}{15.0 \text{ L}} \times \frac{0.0821 \text{ L·atm}}{\text{mol·K}} \times 303 \text{ K} = 0.83 \text{ atm}$$

Practice Exercises

A. Determine (**a**) the pressure exerted by 0.0330 mol of oxygen in an 18.0-L container at 40 °C and (**b**) the volume occupied by 0.200 mol of nitrogen gas at 25 °C and 0.980 atm.

B. At what temperature will 1.25 mol of helium gas exert a pressure of 5.00 atm in a 2.00-L container?

Review Question

6.17 State the ideal gas law in words and in the form of a mathematical equation.

6.9 Henry's Law: The Pressure-Solubility Relationship

In the 1760s Joseph Priestley invented soda water by dissolving carbon dioxide gas in water. When we open a bottle of soda pop, we hear a hissing sound and see bubbles form. Carbon dioxide is dissolved in the liquid, and the bottle is capped under pressure.

William Henry, a friend of John Dalton, studied the solubility of gases in liquids. In 1801 he summarized his extensive findings in **Henry's law:** The solubility of a gas in a liquid at a given temperature is directly proportional to the pressure of the gas at the surface of the liquid. For a bottle of soda pop, when the bottle is capped under pressure, a certain amount of carbon dioxide is dissolved in the water. When the bottle is opened, the pressure is *reduced* (the hissing sound is gas escaping), and the solubility of the carbon dioxide is *reduced.* (The bubbles of gas are carbon dioxide escaping from solution).

Henry's law: Solubility of a gas is directly proportional to the pressure of the gas at the surface of the liquid.

The pressure-solubility relationship is also used in therapy. In cases of carbon monoxide poisoning (see Section 28.2), the victim is placed in a hyperbaric (high-pressure) chamber, a device that supplies oxygen at pressures of 3 or 4 atm. More oxygen is forced into the tissues at these pressures to compensate for the lack of oxygen that accompanies carbon monoxide poisoning.

Hyperbaric chambers are also used to treat infections by anaerobic bacteria (bacteria that live without oxygen). Gangrene is one such disease. The organisms that cause gangrene cannot survive in an oxygenated atmosphere. If sufficient oxygen can be forced into the diseased tissues, the infection can be arrested.

Review Questions

6.18 When the cap is removed from a bottle of soda pop, carbon dioxide gas escapes. Explain why this occurs.

6.19 Would it be a good idea to sterilize the water to be used in a fish bowl by boiling? Explain.

6.10 Dalton's Law of Partial Pressures

John Dalton is most renowned for his atomic theory (Section 2.1), but he had wide-ranging interests, including meteorology. In trying to understand the weather, he experimented on water vapor in the air. In one experiment he found that if he added water vapor at a certain pressure to dry air, the pressure exerted by the air would increase by an amount equal to the pressure of the water vapor. Based on this and other experiments, Dalton concluded that each of the gases in a mixture behaves independently of the other gases, exerting its own pressure. The total pressure of the mixture is equal to the sum of the *partial pressures* exerted by the separate gases (Figure 6.11).

Deep-Sea Diving: Applications of Henry's Law

▲ Underwater divers who surface too quickly may experience the painful and dangerous condition known as the "bends."

Divers who go deeply into the sea must take their own supply of air. We breathe about 800 L of air per hour. To be portable, enough air for an hour or so of underwater exploring must be compressed to a much smaller volume. The compressed air is also needed to keep the lungs inflated at the much higher pressures under several meters of water. Henry's law tells us that this high-pressure air is much more soluble in blood and other body fluids than is air at normal pressures. One of the dissolved gases, nitrogen (N_2), causes two kinds of problems: nitrogen narcosis and decompression sickness.

Nitrogen narcosis occurs in people descending to 30 m or more. Ordinarily, the nitrogen we breathe in air has no physiological effect, but at the high pressures at these depths, nitrogen acts as a narcotic. It produces pleasurable sensations similar to those of narcotics such as morphine, accounting for the name "rapture of the deep" sometimes used for the affliction. Nitrogen narcosis impairs judgment and thus sometimes results in serious diving accidents. The divers can minimize the risk by using a helium-oxygen mixture as a substitute for compressed air. Helium is considerably less soluble in blood than is nitrogen and it has no narcotic effect. Excess oxygen presents no problem because it can be consumed in metabolism.

Decompression sickness, caused by bubbles of nitrogen gas in the blood and other tissues, occurs when a person experiences a sudden drop in atmospheric pressure. Tiny bubbles block blood flow in capillaries and impair nerve transmission. Severe joint and muscle pains often cause the victim to curl up in agony, a reaction that explains the common name for the ailment, the "bends." Divers must be careful to return to the surface slowly or to spend considerable time in a decompression chamber where the pressure is gradually lowered. If they don't, excess $N_2(g)$ comes out of solution rapidly, decompression sickness sets in, and fainting, deafness, paralysis, or even death may occur. People who work in deep mines or tunnels, where air pressure is increased to keep water from infiltrating, have similar problems, as do passengers in an airplane at high altitude that suddenly loses pressure.

If the return to normal pressure is slow enough, the excess gases leave the blood gradually. Excess $O_2(g)$ is used in metabolism, and excess $N_2(g)$ is removed to the lungs and expelled by normal breathing. For each atmosphere of pressure above normal that the diver experiences, about 20 minutes of slow decompression is needed.

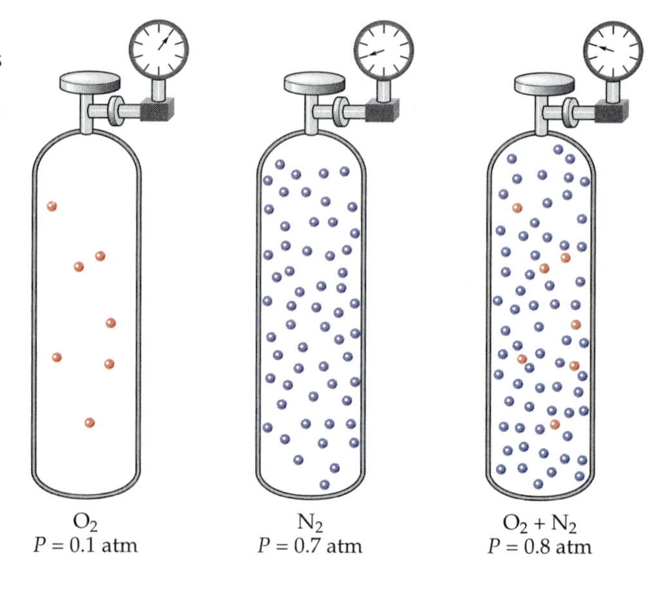

Figure 6.11 ► Dalton's law of partial pressures states that the pressure of a mixture of gases is equal to the sum of the partial pressures of the individual gases.

Mathematically, **Dalton's law of partial pressures** is stated as

$$P_{\text{total}} = P_1 + P_2 + P_3 + \ldots$$

where the terms on the right side refer to the partial pressures of gases 1, 2, 3, and so on.

Gases such as oxygen, nitrogen, and hydrogen are nonpolar. They are only slightly soluble in water and are usually collected over water by displacement (Figure 5.3). Such gases always contain water vapor, and the total pressure in the collection vessel is that of the gas plus that of water vapor. The vapor pressure of water depends on the temperature of the water. (The **vapor pressure** of a substance is the partial pressure exerted by its molecules in the gas phase above the liquid phase of the substance.) The hotter the water, the higher the vapor pressure. If a gas is collected over water, we can make use of vapor pressure tables (Table 6.2) to calculate the pressure due to the gas alone. We need only subtract the vapor pressure of the water, as determined from the table, from the value for the total pressure within the collection vessel.

Table 6.2 Water Vapor Pressure at Various Temperatures

Temperature (°C)	Water Vapor Pressure (mmHg)
0	5
10	9
20	18
30	32
40	55
50	93
60	149
70	234
80	355
90	526
100	760

Example 6.9

Oxygen is collected over water at 20 °C. The total pressure inside the collection jar is 740 mmHg. What is the pressure due to the oxygen alone?

Solution

From Table 6.2 we find that the vapor pressure of water at 20 °C is 18 mmHg. Because the total pressure is equal to 740 mmHg, we have

$$P_{\text{total}} = P_{O_2} + P_{H_2O}$$

$$P_{O_2} = P_{\text{total}} - P_{H_2O}$$

$$P_{O_2} = 740 \text{ mmHg} - 18 \text{ mmHg} = 722 \text{ mmHg}$$

Practice Exercise

Hydrogen gas is collected over water at 20 °C. The total pressure inside the collection jar is set at the barometric pressure of 738 mmHg. If the volume of the gas is 246 mL, what mass of hydrogen is collected?

Humidity is a measure of the amount of water vapor in the air. **Relative humidity** compares the actual amount of water vapor in the air with the maximum amount the air could hold at the same temperature. If the temperature is 20 °C and the vapor pressure of water in the atmosphere is 12 mmHg, the relative humidity is

$$\text{Relative humidity} = \frac{12 \text{ mmHg}}{18 \text{ mmHg}} \times 100 = 67\%$$

Measured vapor pressure of water in the atmosphere. (numerator)

Partial pressure of water at 20 °C (obtained from Table 6.2) (denominator)

When the relative humidity is 100%, the air is saturated with water vapor. (Even at 100% relative humidity, water constitutes only about two or three molecules in every 100 molecules of air.)

▲ A crude measure of relative humidity can be made with strips of filter paper impregnated with an aqueous solution of cobalt(II) chloride and allowed to dry. In dry air the strip is blue, the color of *anhydrous* $CoCl_2$ (right). In more humid air the strip acquires the red color of the *hexahydrate*, $CoCl_2{\cdot}6H_2O$ (left). (Hydrates and anhydrous salts are discussed in Section 8.4).

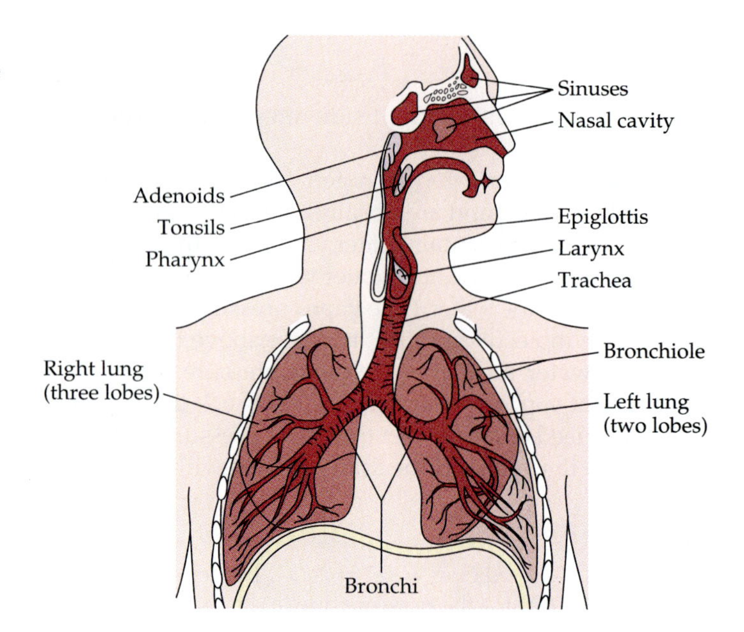

Figure 6.12 ▶
The respiratory system, showing the route of air through the nose, pharynx (throat), larynx (voice box), and trachea (windpipe), into the bronchi and bronchioles (bronchial tubes) and ending in the alveoli (air sacs).

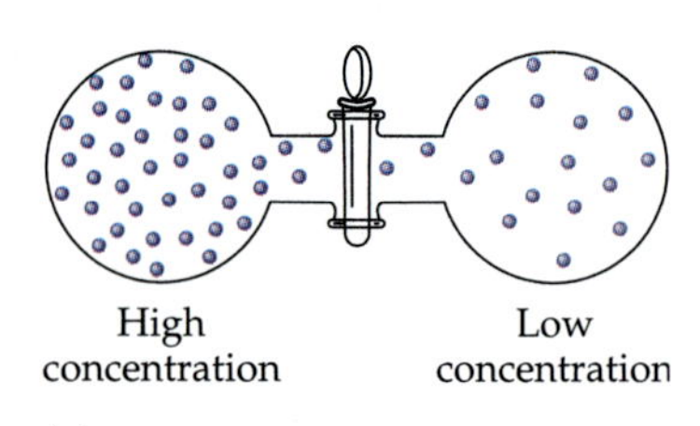

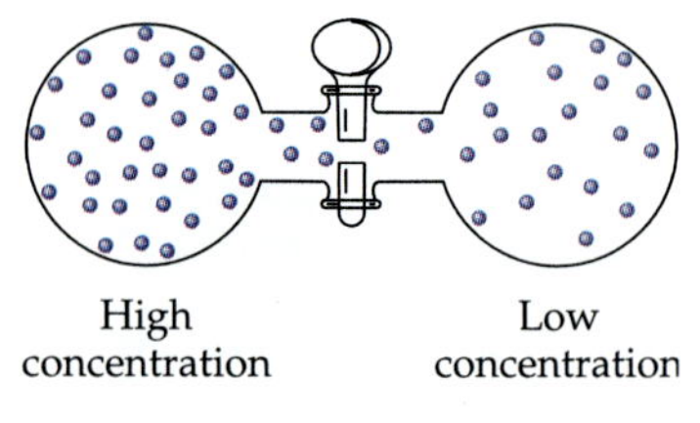

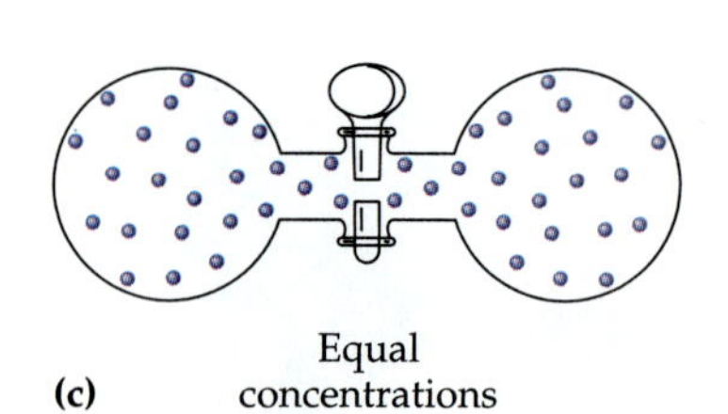

▲ **Figure 6.13**
A gas (air in the lungs, for example) flows from an area of high concentration to an area of low concentration. (a) With the stopcock closed, no flow is possible. (b) With the stopcock open, there is a net flow of gases from the area of higher concentration on the left to the area of lower concentration on the right. (c) When the two concentrations become equal, net flow stops.

Cool air can hold less water vapor than warm air. As the temperature falls during the night, the atmosphere may become saturated. Water vapor condenses from the air as *dew.*

The **heat index** relates temperature to relative humidity. Higher humidity makes you feel hotter by inhibiting evaporation of sweat, which normally cools your body.

Respiratory therapists must concern themselves with the humidity of the gases they administer to patients. Normally, as we breathe, the inspired air is saturated with moisture as it passes through the nose and respiratory passages. Oxygen coming from a tank is quite dry. If oxygen is administered over a long period of time, it must be humidified to prevent it from irritating the mucous linings of the nasal passages and the lungs. If the oxygen or mixture of gases is conducted through the nose, the therapist may humidify the inspired gases to about 30% humidity. If the breathing mixture is conducted directly to the trachea (bypassing the nose), the therapist saturates the gas mixture with water vapor.

✓ Review Questions

6.20 State Dalton's law of partial pressures in words and in the form of a mathematical equation. Describe how Dalton's law of partial pressures is applied to gases collected over water.

6.21 What is meant by the relative humidity of an air sample?

6.11 Partial Pressures and Respiration

When we breathe in, the inspired air becomes moistened and warmed to our body temperature of 37 °C. The air is drawn into the lungs, where it enters a highly branched system of tubes that end in tiny air sacs called alveoli (Figure 6.12). These thin-walled pouches are surrounded by blood vessels that are part of a circulatory system serving every cell in the body.

Inspired air is rich in oxygen (P_{O_2} = 150 mmHg) and poor in carbon dioxide (P_{CO_2} = 0.2 mmHg). Cellular fluid is poor in oxygen (P_{O_2} = 6 mmHg) and rich in carbon dioxide (P_{CO_2} = 50 mmHg). Cells use up oxygen in metabolic reactions that produce energy. Carbon dioxide accumulates in the cells as a waste product of these reactions. To maintain life, we must transfer the oxygen in the inspired air to our cells. At the same time, the carbon dioxide waste in the cells must be transferred to the lungs and then exhaled. Both gases are transferred by the process of **diffusion.** In diffusion, gases flow from regions of higher concentration to regions of lower concentration (Figure 6.13). In our bodies, oxygen makes its way to the cells through a pressure gradient—that is, in a series of steps in which oxygen diffuses from areas where its concentration is higher into areas where its concentration is lower. By the same method, carbon dioxide moves in the opposite direction. It makes its way from the cells, where its partial pressure is high, to the atmosphere, where its partial pressure is low. Figure 6.14 shows the steps in the gradient for both gases. Thus

Diffusion—the flow of a substance from a region of higher concentration to a region of lower concentration.

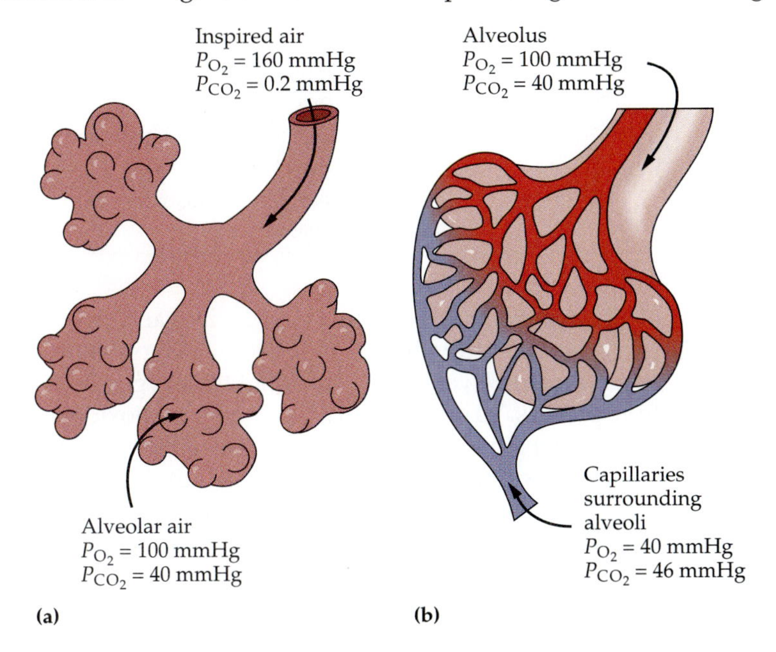

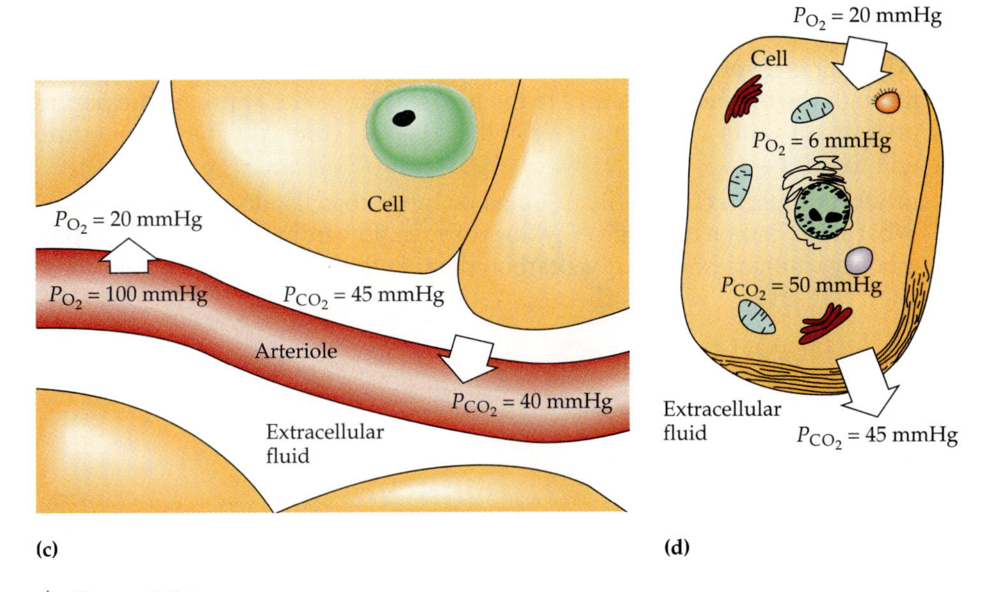

▲ **Figure 6.14**
(a) $O_2(g)$ flows from the inspired air into the alveolar air, and $CO_2(g)$ flows in the opposite direction. (b) $O_2(g)$ diffuses from an arteriole into the extracellular fluid; $CO_2(g)$ flows in the opposite direction. (c) $O_2(g)$ diffuses from an arteriole into the extracellular fluid; $CO_2(g)$ flows in the opposite direction. (d) $O_2(g)$ diffuses into a cell from the extracellular fluid; $CO_2(g)$ moves in the opposite direction. In each case, the flow is from an area of high partial pressure to a region of low partial pressure.

Table 6.3 Nine Compressed Gases Used in Medicine

Gas	Chemical Formula	Use
Air	N_2 and O_2 (mixture)	Life support
Carbon dioxide	CO_2	Laboratory tests, lung function tests
Carbon dioxide-oxygen mixtures	CO_2 and O_2	Diagnosis, inhalation therapy
Helium	He	Laboratory analyses
Helium-oxygen mixtures	He and O_2	Inhalation therapy, diagnostic tests
Nitrogen	N_2	Diagnostic testing, inhalation therapy
Nitrous oxide	N_2O	Anesthetic
Oxygen	O_2	Life support, medical emergencies, adjunct to anesthetics
Oxygen-nitrogen mixtures	O_2 and N_2	Treatment of obstructive lung diseases

once the chest and diaphragm act mechanically to get air into and out of the lungs (Section 6.4), gases go "downhill" from higher to lower partial pressure.

Normally the carbon dioxide level in the blood (not the oxygen level) acts as a trigger for the breathing process. To oversimplify, when carbon dioxide levels build up, we take a breath; if they get too low, we don't. Thus one of the concerns of a therapist is the partial pressure of carbon dioxide in the blood. Under certain unusual conditions, it is possible for the level of carbon dioxide to fall so low that it fails to trigger the breathing mechanism. Even with access to a plentiful supply of oxygen, the person suffocates because he or she simply stops breathing.

Several commercially supplied gases are used in medicine, the majority in respiratory therapy. Table 6.3 lists these gases and some of their uses.

Review Questions

6.22 When air is inspired it becomes fully saturated with water vapor as it passes through the trachea on its way to the lungs. What is the approximate partial pressure of water vapor in the air in the alveoli? (Hint: At what temperature will the air be?)

6.23 In which net direction, cells to lungs or lungs to cells, does oxygen move? What about carbon dioxide? Why does oxygen flow from the alveoli to the pulmonary capillaries? Why does carbon dioxide flow in the reverse direction?

Summary

The air making up Earth's atmosphere is a gaseous mixture of nitrogen, oxygen, argon, trace constituents, and water vapor.

The kinetic-molecular theory of gases pictures any gas as a collection of fast-moving particles constantly bouncing off one another and off the walls of their container. The faster the particles move, the higher the *temperature* of the gas. The *pressure* of the gas is the combined effect of all the forces exerted on the walls of the container by all the moving gas particles.

The **atmosphere** is a unit of air pressure equivalent to 760 mmHg. Standard pressure at sea level is 760 mmHg = 1 atm. Pressure is force per unit area. The SI unit of pressure, the **pascal,** is a force of 1 newton exerted on an area of 1 square meter: 1 Pa = 1 N/m^2.

The various gas laws are summarized as follows.

Gas law	Equation	Variables	Constant(s)
Boyle's law	$P_1V_1 = P_2V_2$	P, V	n, T
Charles's law	$V_1/T_1 = V_2/T_2$	V, T	n, P
Avogadro's law	$V_1/n_1 = V_2/n_2$	V, n	P, T
Combined gas law	$P_1V_1/T_1 = P_2V_2/T_2$	P, V, T	n
Ideal gas law	$PV = nRT$	P, V, n, T	

The **combined gas law** incorporates **Boyle's law, Charles's law,** and **Avogadro's law** in one expression.

A gas is said to be at **standard temperature and pressure (STP)** when its pressure is 1 atm and its temperature is 0 °C. The **molar volume of a gas** is the volume occupied at STP by 1 mol of any gas and is equal to 22.4 L.

The **ideal gas law,** $PV = nRT$, employs the **universal gas constant (*R*)**. It can be solved for any of the four variables if the other three are known.

Henry's law states that, at constant temperature, the solubility of a gas in a liquid is directly proportional to the gas pressure at the liquid surface.

Dalton's law of partial pressures states that when a number of gases are confined to the same container, each gas behaves as if it were alone in the container. The total pressure of the mixture is the sum of the partial pressures exerted by the separate gases.

Some of the surface molecules in any liquid escape to the volume above the surface and thereby enter the gaseous state. The pressure exerted by these gas molecules is the **vapor pressure** of the liquid.

Key Terms

atmosphere (atm) (6.3)
Avogadro's law (6.6)
barometer (6.3)
Boyle's law (6.4)
Charles's law (6.5)
combined gas law (6.7)
Dalton's law of partial pressures (6.10)
diffusion (6.11)
heat index (6.10)
Henry's law (6.9)
ideal gas law (6.8)
Kelvin scale (6.5)
kinetic-molecular theory (6.2)
millimeters of mercury (mmHg) (6.3)
molar volume of a gas (6.6)
pascal (6.3)
pressure (6.3)
relative humidity (6.10)
standard conditions of temperature and pressure (STP) (6.6)
universal gas constant (*R*) (6.8)
vapor pressure (6.10)

Problems

Kinetic-Molecular Theory

1. According to the kinetic-molecular theory, (**a**) what change in temperature is occurring if the molecules of a gas begin to move more slowly, on average? (**b**) what change in pressure results when the walls of the container are being struck less often by molecules of the gas?
2. Container A has twice the volume but holds twice as many gas molecules as container B at the same temperature. Use the kinetic-molecular theory to compare the pressures in the two containers.
3. For each of the following, indicate in which container the gas would be expected to have the higher density.
   a. Containers A and B have the same volume and are at the same temperature, but the gas in A is at a higher pressure.
   b. Containers A and B are at the same pressure and temperature, but the volume of A is greater than that of B.
   c. Containers A and B are at the same pressure and volume, but the gas in A is at a higher temperature.
4. Interpret what we mean by temperature in terms of the kinetic-molecular theory.

Pressure

5. Carry out the following conversions between pressure units.
   a. 0.985 atm to mmHg **b.** 849 mmHg to atm
   c. 721 mmHg to atm
6. Carry out the following conversions between pressure units.
   a. 4.00 atm to mmHg **b.** 642 mmHg to kPa
   c. 105.7 kPa to mmHg
7. What is the height of a mercury column that exerts a pressure of 213 mmHg?
8. Calculate the height of a mercury column that exerts a pressure of 4.36 atm.

Boyle's Law

9. A sample of helium occupies 521 mL at 1572 mmHg. Assume that the temperature is held constant and deter-

mine (**a**) the volume of the helium at 752 mmHg and (**b**) the pressure, in mmHg, if the volume is changed to 315 mL.

10. A decompression chamber used by deep-sea divers has a volume of 10.3 m^3 and operates at an internal pressure of 4.50 atm. What volume, in m^3, would the air in the chamber occupy if it were at 1.00 atm pressure, assuming no temperature change?

11. Oxygen used in respiratory therapy is stored at room temperature under a pressure of 150 atm in gas cylinders with a volume of 60.0 L.

a. What volume would the gas occupy at a pressure of 750.0 mmHg? Assume no temperature change.

b. If the oxygen flow to the patient is adjusted to 8.00 L per minute, at room temperature and 750.0 mmHg, how long will the tank of gas last?

12. The pressure within a 2.25-L balloon is 1.10 atm. If the volume of the balloon increases to 7.05 L, what will be the final pressure within the balloon, if the temperature does not change?

Charles's Law

13. A gas at a temperature of 100 °C occupies a volume of 154 mL. What will the volume be at a temperature of 10 °C, assuming no change in pressure?

14. A balloon is filled with helium. Its volume is 5.90 L at 26 °C. What will be its volume at 278 °C, assuming no pressure change?

15. A 567-mL sample of a gas at 305 °C and 1.20 atm is cooled at constant pressure until its volume becomes 425 mL. What is the new gas temperature?

16. A sample of gas at STP is to be heated at constant pressure until its volume triples. What is the new gas temperature?

Avogadro's Law and Molar Volume

17. Which of the following gas samples at STP contains the greatest number of molecules?

a. 5.0 g of H_2

b. 50 L of SF_6

c. 1.0×10^{24} molecules of CO_2

18. How many molecules are present in 475 mL of $CO_2(g)$ at STP?

19. What is the volume occupied by 0.837 g of xenon gas at STP?

20. What is the mass of 498 L of neon gas at STP?

The Combined Gas Law

21. A sealed can with an internal pressure of 721 mmHg at 25 °C is thrown into an incinerator operating at 755 °C. What will be the pressure inside the heated can, assuming the container remains intact during incineration?

22. A fixed amount of He exerts a pressure of 775 mmHg in a 1.05-L container at 26 °C. To what value must the temperature be changed to change the gas pressure to 725 mmHg? Assume the volume of gas remains constant.

23. If a fixed amount of gas occupies 2.53 m^3 at a temperature of −15 °C and 191 mmHg, what volume will it occupy at 25 °C and 1142 mmHg?

24. What volume will 575 mL of gas, measured at 23 °C and 725 mmHg, occupy at STP?

25. If the gas present in 4.65 L at STP is changed to a temperature of 15 °C and a pressure of 756 mmHg, what will be the new volume?

26. What volume will 498 mL of a fixed amount of gas, measured at 27 °C and 722 mmHg, occupy at STP?

The Ideal Gas Law

27. Calculate the volume, in liters, of 1.12 mol of $H_2S(g)$ at 62 °C and 1.38 atm.

28. Calculate the volume, in liters, of 0.00600 mol of a gas at 31 °C and 661 mmHg.

29. Calculate the pressure, in atmospheres, of 4.64 mol of CO(g) in a 3.96-L tank at 29 °C.

30. Calculate the pressure, in mmHg, of 0.0108 mol of $CH_4(g)$ in a 0.265-L flask at 37 °C.

31. How many moles of Kr(g) are there in 2.22 L of the gas at 698 mmHg and 45 °C?

32. How many grams of CO(g) are there in 745 mL of the gas at 784 mmHg and 36 °C?

Gas Densities

33. Calculate the density, in grams per liter, for each of the following gases at STP.

a. CO **b.** AsH_3 **c.** Ar **d.** N_2

34. What is the molar mass of a gas that has a density of 2.57 g/L at STP?

Dalton's Law of Partial Pressures

35. Oxygen is collected over water at 30 °C and a barometric pressure of 742 mmHg. What is the partial pressure of $O_2(g)$ in the container?

36. A container holds oxygen at a partial pressure of 0.25 atm, nitrogen at a partial pressure of 0.50 atm, and helium at a partial pressure of 0.20 atm. What is the pressure inside the container?

37. A container is filled with equal numbers of nitrogen, oxygen, and carbon dioxide molecules. The total pressure in the container is 750 mmHg. What is the partial pressure of nitrogen in the container?

38. The pressure of the atmosphere on the surface of Venus is about 100 atm. Carbon dioxide makes up about 97% by volume of the atmospheric gases. What is the partial pressure of carbon dioxide in the atmosphere of Venus?

39. Atmospheric pressure on the surface of Mars is about 6.0 mmHg. The partial pressure of carbon dioxide is 5.7 mmHg. What percent by volume of the Martian atmosphere is carbon dioxide?

Additional Problems

40. A sample of hydrogen gas has a volume of 1.10 L at −40 °C and 0.520 atm. What will be its volume at STP?
41. During inhalation, does the chest cavity expand or contract? Is the pressure inside the lungs decreased or increased? What happens during exhalation?
42. Define or explain the following terms.
    a. combined gas law **b.** Henry's law
    c. mmHg **d.** tidal volume
    e. vital capacity **f.** vapor pressure
    g. diffusion
43. The interior volume of the Hubert H. Humphrey Metrodome in Minneapolis is 1.70×10^{10} L. The Teflon-coated fiberglass roof is supported by air pressure provided by 20 huge electric fans. How many moles of air are present in the dome if the pressure is 1.02 atm at 18 °C?
44. Use the ideal gas equation to calculate the pressure exerted by 1.00 mol of $CO_2(g)$ when it is confined to a volume of 2.50 L at 298 K.
45. Calculate the temperature in °C of 0.78 mol of oxygen if its volume is 68 L and its pressure is 0.37 atm.
46. A sample of intestinal gas was collected and found on analysis to consist of 44% CO_2, 38% H_2, 17% N_2, 1.3% O_2, and 0.003% CH_4 by volume. (The percentages do not add to exactly 100% because of rounding.) What is the partial pressure of each gas if the total pressure in the intestine is 820 mmHg?
47. If the P_{H_2O} in air is 12 mmHg on a day when the temperature is 20 °C, what is the relative humidity?
48. Two flasks are connected. Flask A contains only oxygen at a pressure of 460 mmHg. Flask B has oxygen at a partial pressure of 320 mmHg and nitrogen at a partial pressure of 240 mmHg.
    a. Which direction will the net flow of oxygen take?
    b. Which direction will the net flow of nitrogen take?
49. A person at rest breathes about 80 mL of air per second. How long does it take to breathe 22.4 L of air?
50. A hyperbaric chamber is an enclosure containing oxygen at higher-than-normal pressures used in the treatment of certain heart and circulatory conditions. What volume of $O_2(g)$, from a cylinder at 25 °C and 151 atm, is required to fill a 4.20×10^3-L hyperbaric chamber to a pressure of 3.00 atm at 17 °C?
51. Typically, when a person coughs, he or she first inhales about 2.0 L of air at 1.0 atm and 25 °C. The epiglottis and the vocal chords then shut, trapping the air in the lungs, where it is warmed to 37 °C and compressed to a volume of about 1.7 L by the action of the diaphragm and chest muscles. The sudden opening of the epiglottis and vocal chords releases this air explosively. Just prior to this release, what is the approximate pressure of the gas inside the lungs?
52. *Eleodea* is a green plant that carries out photosynthesis under water.

$$6\,CO_2(g) + 6\,H_2O(l) \longrightarrow C_6H_{12}O_6(aq) + 6\,O_2(g)$$

    In an experiment, some *Eleodea* produce 122 mL of $O_2(g)$, collected over water at 743 mmHg and 21 °C. What mass of oxygen is produced? What mass of glucose ($C_6H_{12}O_6$) is produced concurrently?
53. In an attempt to verify Avogadro's hypothesis, small quantities of several different gases were weighed in 100.0-mL syringes. Masses were determined to the nearest 0.1 mg on an analytical balance. The following masses were obtained: 0.0080 g of H_2, 0.1112 g of N_2, 0.1281 g of O_2, 0.1770 g of CO_2, 0.2320 g of C_4H_{10}, and 0.4824 g of CCl_2F_2. Within experimental error, are these results consistent with Avogadro's hypothesis? Explain.
54. A novel energy storage system involves storing air under high pressure. (Energy is released when the air is allowed to expand.) How many cubic feet of air, measured at standard atmospheric pressure (14.7 psi), can be compressed into a 19-million-ft^3 underground cavern at a pressure of 1070 pounds per square inch (psi)?

CHAPTER 7

Liquids and Solids

Water is the only substance on Earth to exist in large amounts in all three physical states: solid (ice and snow); liquid (rivers, lakes, and oceans); and gas (water vapor in the atmosphere). Giant icebergs, like those shown here, are composed of water in the solid state. If they could be easily transported, icebergs could be melted to provide abundant liquid water anywhere on Earth.

Learning Objectives/Study Questions

1. How are liquids and solids similar? How are they different?
2. What are the major types of intermolecular forces? What are the unique features of each?
3. How are intermolecular forces important in determining the physical properties of substances, such as melting point, boiling point, viscosity, and surface tension?
4. What happens when a liquid vaporizes? Why is vaporization important in maintaining body temperature?
5. How is distillation used to purify a liquid?
6. How are the structures of crystalline solids described?
7. What happens when a solid melts?
8. What are the unique properties of water?

In our discussion of bonding in Chapter 3, our primary concern was how atoms combine to form molecules and why elements react to form compounds. Now we're going to expand our consideration of bonding, and this time we will seek an answer to this question: What makes oxygen a gas at room temperature and atmospheric pressure, while water is a liquid and sugar a solid? Why does liquid water become a gas at higher temperatures and a solid at lower temperatures? Some force of attraction holds the particles of solids and liquids in contact with one another. The particles of a gas fly about at random; those of liquids or solids cling together. Gas particles interact so little with one another that a collection of them retains neither a specific volume nor a specific shape. But particles of a liquid are held together with sufficient force that a collection of them has a specific volume. And particles of a solid are so rigidly held together that not only the volume but also the shape of a given sample is fixed (Figure 7.1). Are these forces of attraction important? They are if appearances are important to you. You are built of solids and liquids. If you think you've got it all together, you should thank the special forces of attraction that give shape and volume to the condensed forms of matter: liquids and solids.

▲ Intermolecular and intramolecular forces compared. (*Inter* is Latin for "between" and *intra* is Latin for "within.") In the hypothetical case pictured here, intramolecular forces—chemical bonds—are shown as solid green lines. Forces between molecules—intermolecular forces—are indicated by broken red lines.

In this chapter, we also consider changes in the state of matter—what happens when a solid is converted to a liquid or a liquid to a gas. Is this important? Well, consider perspiration. From the amount of advertising directed against this lowly liquid, you would think it was an unnecessary annoyance. However, were it not for the conversion of this liquid to a vapor on skin surfaces, we would find it difficult just to survive in the temperate and tropical zones of our planet, let alone carry out vigorous physical activity.

Intermolecular Forces

In Chapter 3 we focused on the forces, called *chemical bonds,* that bind atoms to one another *within* molecules—*intramolecular forces.* These bonds determine the geometric shapes and resulting chemical properties of molecules. The attractive forces that exist *between* molecules—**intermolecular forces**—are also quite important. Intermolecular forces determine the *physical* properties of liquids and solids. In fact, if there were no intermolecular forces, there would be no liquids or solids—everything would be a gas.

Intermolecular forces pertain to the forces of attraction between a molecule and neighboring molecules. Intramolecular forces pertain to those within a molecule due to chemical bonding.

In our kinetic-molecular model of gases (Chapter 6), we visualize speedy, energetic molecules undergoing frequent collisions and never coming to rest or clumping together. Intermolecular forces are relatively unimportant in gases, because the molecules are far apart. The ideal gas law (Section 6.8) actually assumes that they do not exist at all. If intermolecular forces are sufficiently strong, however, molecules cluster together into the liquid or solid state. In our discussion of liquids and solids

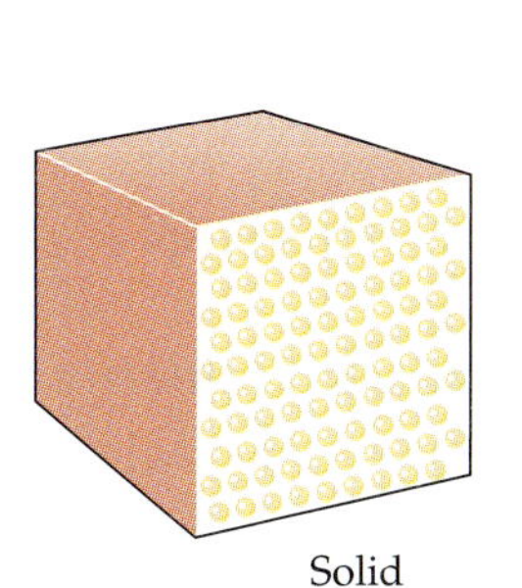

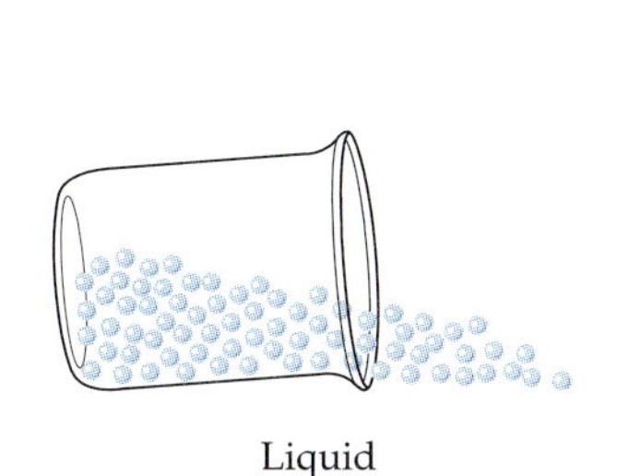

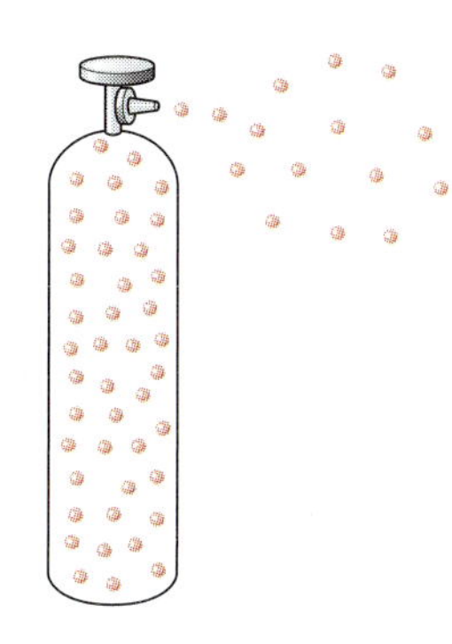

◄ **Figure 7.1**
Solids, liquids, and gases compared. (a) In solids, the particles (atoms, molecules, or ions) are close together and fixed in place; solids retain both shape and volume. (b) In liquids, the particles are close together but free to move over and around each other; liquids retain volume but not shape. (c) In gases, the particles are far apart and moving at random; gases maintain neither shape nor volume.

in subsequent sections, we will consider the different kinds of intermolecular forces and their relative strengths. However, let's first consider some fundamental phenomena involving liquids and solids.

7.1 Intermolecular Forces: Some General Considerations

Intermolecular forces determine the physical properties of substances. As we noted in Chapter 6, *molecules have a tendency to remain apart from each other.* Intermolecular forces of *attraction* are most effective in overcoming this tendency at *low* temperatures, at which the molecules have low energies, and at *high* pressures, where molecules are close together. A substance is likely to exist as a gas at high temperatures, at which the molecules are quite energetic, and at low pressures, where molecules are far apart. The solid state exists at low temperatures and at moderate to high pressures, where molecules are most closely packed. The liquid state—a sort of in-between state—exists at intermediate temperatures and moderate to high pressures.

The adjectives "high" and "low" are relative terms. A temperature of 1000 °C is far above the boiling point of water (100 °C), but is rather low in relation to the melting point of iron (1530 °C).

Before studying intermolecular forces in detail, we can make some generalizations. First, almost all ionic compounds are solids at room temperature. It is possible to obtain them as liquids (by melting them), but generally only at high temperatures. Second, nearly all metals (mercury is a notable exception) are solids at room temperature. They, too, can be melted—some at fairly low temperatures, but others only at high temperatures.

There is no third generalization to cover molecular substances; they exist in all three physical states at room temperature. Nitrogen (N_2) and carbon dioxide (CO_2) are gases, water (H_2O) and octane (C_8H_{18}) are liquids, and sulfur (S_8) and glucose ($C_6H_{12}O_6$) are solids. The physical state of a molecular substance is determined both by molecular mass and by the type of force between molecules. If one of these two variables can be eliminated (or held constant), simple generalizations are possible. For example, the halogens (Group 7A) all exist as nonpolar diatomic molecules. Intermolecular forces are similar; thus any variation in properties can be attributed to variations in molecular mass. And we do find such a trend. Fluorine (F_2), with a molecular mass of 38 u, and chlorine (Cl_2), with a molecular mass of 71 u, are gases at room temperature. Bromine (Br_2), with a molecular mass of 160 u, is a liquid, and iodine (I_2), with a molecular mass of 254 u, is a solid. A similar trend can be found for compounds that are subject to similar intermolecular forces, as is the case for the following compounds of carbon and the halogens.

Compound	CF_4	CCl_4	CBr_4	CI_4
Molecular mass (u)	88	154	332	520
Physical state at 20 °C	Gas	Liquid	Solid	Solid

As we shall see, the types of intermolecular forces are usually of overriding importance if we compare molecules that are subject to dissimilar forces.

Review Questions

7.1 In Chapter 6 we used the kinetic-molecular theory to explain the properties of gases. How is this theory applied to liquids and solids?

7.2 In what ways are liquids and solids similar? In what ways are they different?

7.3 What is the difference between an intramolecular force and an intermolecular force?

7.2 Ionic Bonds as Forces Between Particles

There really is no such thing as a "molecule" of a solid ionic compound. So there are really no *intermolecular* forces in ionic compounds. As we saw in Chapter 3, there are simply *interionic attractions* in which each ion is simultaneously attracted to several ions of the opposite charge. These interionic attractions—ionic bonds—extend throughout an ionic crystal.

We can use the following generalization to compare the strengths of various kinds of interionic attractions: *The attractive force between a pair of oppositely charged ions increases as the charges on the ions increase and as ionic size decreases.* Ionic solids melt if enough thermal energy is supplied to break down the crystalline lattice.

Example 7.1

Which member of each of the following pairs of ionic solids has the higher melting point?

a. MgO or NaCl **b.** $CaBr_2$ or $CaCl_2$

Solution

a. Mg^{2+} and O^{2-} ions have higher charges than do Na^+ and Cl^-. The forces between the doubly charged ions of MgO are greater than those between the singly charged ions of NaCl. By the generalization stated above, MgO should have the higher melting point.

b. In comparing $CaBr_2$ and $CaCl_2$, the cation (Ca^{2+}) is the same in both. Also, both anions carry a charge of 1−. However, the Br^- ion is a larger ion than the Cl^- ion. Consequently, we expect the interionic attractions in $CaBr_2$ to be somewhat weaker than those in $CaCl_2$. We expect $CaBr_2$ to have a lower melting point than $CaCl_2$.

Practice Exercises

A. Which has the higher melting point, KCl or KI? Explain.

B. Which has the higher melting point, HgS or AgCl? Explain.

Review Question

7.4 Describe the general way in which ionic charges and sizes affect the melting point of an ionic solid.

7.3 Dipole Forces

The attractive forces between small molecules are not nearly as great as those between oppositely charged ions. Hydrogen chloride is a gas at room temperature. It melts at −112 °C and boils at −85 °C. We know that in the HCl molecule, the hydrogen atom and chlorine atom are joined by a covalent bond (Chapter 3), but what sort of force holds one HCl molecule to another in the liquid or solid state?

Recall that the HCl molecule is a dipole with a positive end and a negative end. HCl is a polar substance. Two dipoles brought close together will attract one another. The dipoles attempt to align themselves with the positive end of one dipole directed toward the negative ends of neighboring dipoles and vice versa. These **dipole forces** exist throughout the liquid or solid (Figure 7.2). Even if attractive forces between permanent dipoles are fairly weak, they are greater than those in a nonpolar substance of about the same molecular mass. We see this effect in the series: nitrogen, nitrogen monoxide, and oxygen. The NO molecule is polar and has the highest boiling point.

Attractions between polar molecules result from forces between centers of partial charge, whereas the much stronger attraction between ions involves fully charged particles.

	N_2	NO	O_2
Molecular mass	28.0 u	30.0 u	32.0 u
Boiling point	−196 °C	−152 °C	−183 °C

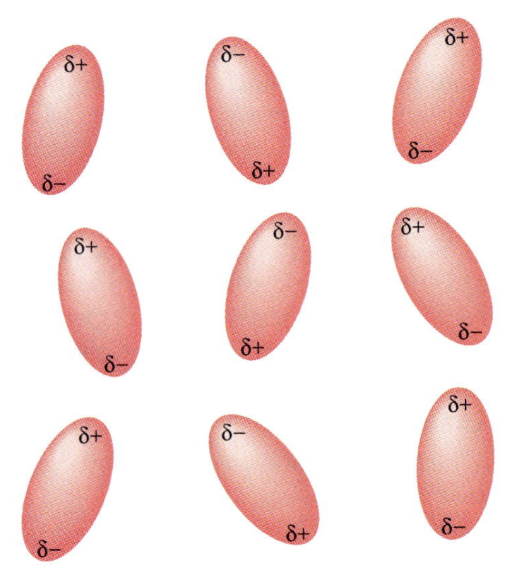

Figure 7.2 ▶
Dipole forces illustrated. Molecular motion prevents a perfect alignment of the dipoles. Nevertheless, the dipoles do maintain a general arrangement leading to the attractions δ+ ... δ−.

Example 7.2

Which member of each of the following pairs of substances has the higher boiling point?

a. Hydrogen bromide, HBr (molecular mass 80.92 u); or krypton, Kr (molecular mass 83.30 u)

b. Silane, SiH_4 (molecular mass 32.09 u); or phosphine, PH_3 (molecular mass 34.00 u)

Solution

a. HBr is polar. Krypton, a monatomic gas, is nonpolar. The molecular masses are comparable; that of Kr is only slightly higher than that of HBr. We expect the polar substance to have greater intermolecular forces than the nonpolar substance. HBr has the higher boiling point.

b. Although the Si—H bonds are slightly polar, the tetrahedral SiH_4 molecule as a whole is nonpolar (compare methane, Section 3.13). P—H bonds are polar and the pyramidal PH_3 molecule is polar. The molecular masses are comparable; that of PH_3 is only slightly higher than that of SiH_4. We expect PH_3 to have the higher boiling point.

Practice Exercises

A. Which has the higher boiling point, hydrogen chloride or fluorine (F_2)? Explain.

B. Which has the higher boiling point, silane (SiH_4), or hydrogen sulfide (H_2S)? Explain.

✓ Review Question

7.5 Why does a polar liquid generally have a higher normal boiling point than a nonpolar liquid of the same molecular mass?

7.4 Hydrogen Bonds

Suppose we knew the boiling points of hydrogen sulfide (H_2S), hydrogen selenide (H_2Se), and hydrogen telluride (H_2Te) and were asked to predict the boiling point of water (H_2O).

	H_2O	H_2S	H_2Se	H_2Te
Molecular mass	18.02 u	34.08 u	80.98 u	129.63 u
Boiling point	?	−60.33 °C	−41.3 °C	−2 °C

All are bent, polar molecules, but molecular mass increases in a regular fashion. We might therefore expect the boiling points to vary in a similar manner. If we plot boiling point versus molecular mass, we could well predict a boiling point of about −68 °C for water (Figure 7.3). However, our prediction would be wrong. Water has a much higher boiling point, 100 °C. Our incorrect prediction suggests that there is an additional kind of intermolecular force in water that is not found in the other molecules. There is indeed such a force, and it is known as a hydrogen bond.

A **hydrogen bond** between molecules is an intermolecular force in which a hydrogen atom covalently bonded to a nonmetal atom in one molecule is simultaneously attracted to a nonmetal atom of a neighboring molecule. The strongest hydrogen bonds are formed if the nonmetal atoms are small and highly electronegative. Only nitrogen, oxygen, and fluorine routinely fit this requirement.

Think of a hydrogen bond in this way: In a covalent bond an electron cloud joins a hydrogen atom to another atom—oxygen, for example. The electron cloud is much denser at the oxygen end of the bond. The bond is polar, with a partial positive charge (δ+) on the H atom and a partial negative charge (δ−) on the O atom. This leaves the hydrogen nucleus somewhat exposed, and an oxygen atom of a neighboring molecule can approach the exposed hydrogen nucleus rather closely and share some of its "electron wealth" with the hydrogen atom. Hydrogen bonding in water is illustrated in Figure 7.4, where we follow the customary convention of representing hydrogen bonds by dotted lines.

Water has both an unusually high boiling point and an unusually high melting point for a substance with such a low molecular mass. Both these unusual values are a result of intermolecular hydrogen bonds between water molecules in the liquid and solid states.

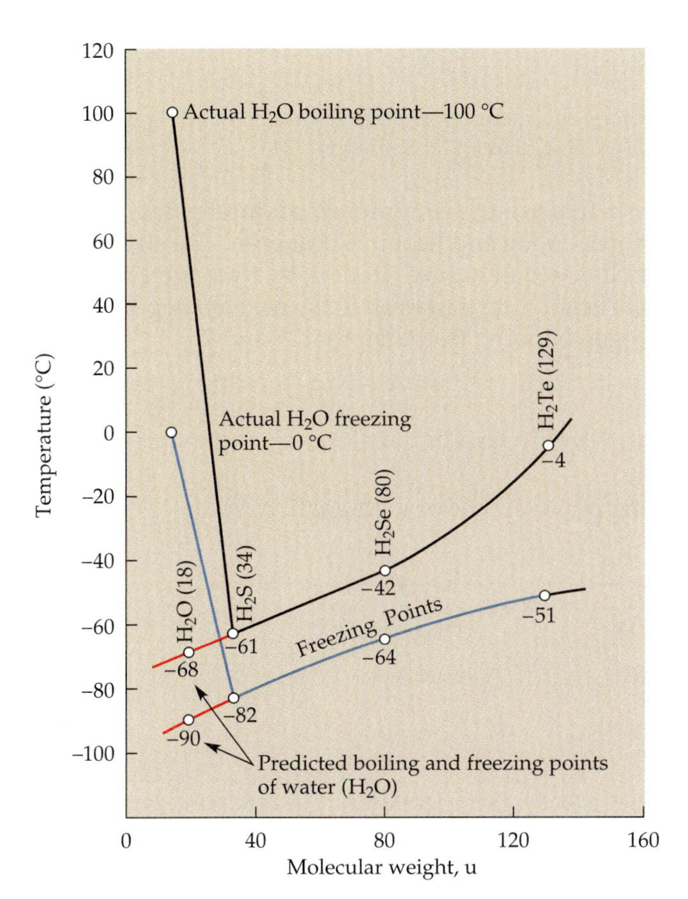

◀ **Figure 7.3**
Boiling points (black lines) and freezing points (blue lines) as a function of molecular masses. In order for ice to melt or liquid water to boil, hydrogen bonds must be broken in addition to overcoming the dipolar forces to achieve a change of state. The boiling point and freezing point of water are therefore much higher than we would otherwise predict (red lines) for a substance composed of such small molecules.

Figure 7.4 ▶
Hydrogen bonds in water. As suggested through (a) Lewis structures and (b) ball-and-stick models, each water molecule is linked to four others through hydrogen bonds. Each H atom lies along a line that joins two O atoms. The shorter distances (100 picometers) correspond to O—H covalent bonds, and the longer distances (180 pm) to the hydrogen bonds.

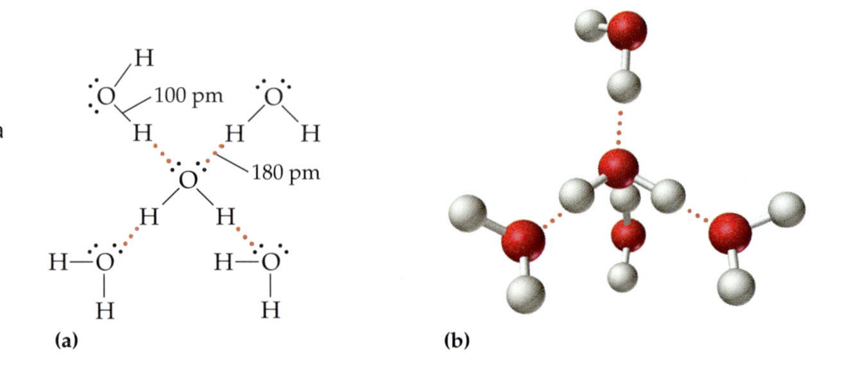

Example 7.3

For each of the following substances, comment on whether hydrogen bonding is an important intermolecular force: N_2, HI, HF, $(CH_3)_2O$, CH_3OH.

Solution

N_2: No. N is a highly electronegative atom, but it can't form hydrogen bonds without H atoms.

HI: No. H atoms are present, but they must be bonded to small, highly electronegative nonmetal atoms. Iodine atoms are quite large.

HF: Yes. Both of the requirements stated above are met: H is bonded to a small, highly electronegative nonmetal atom (fluorine).

$(CH_3)_2O$: No. Both H and a small, highly electronegative nonmetal (oxygen) are present, but the H atoms are bonded to carbon, not oxygen.

CH_3OH: Yes. Both H and highly electronegative O are present, and one of the H atoms is bonded to O.

Practice Exercise

For each of the following substances, comment on whether hydrogen bonding is an important intermolecular force: NH_3, CH_4, C_6H_5OH, H_2S, H_2O_2.

The hydrogen bond may, at this point, seem merely an interesting piece of chemical theory, but its importance to life and health is immense. The structure of proteins, chemicals essential to life, is determined, in part, by hydrogen bonding, and the hereditary traits that one generation passes on to the next are dependent on an elegant application of hydrogen bonding (see Chapters 21 and 23).

✓ Review Questions

7.6 State the principal reasons why H_2O is a liquid at room temperatures whereas CH_4 is a gas.

7.7 Which has a higher boiling point, ethane or methanol? Why?

```
   H  H           H
   |  |           |
H—C—C—H      H—C—O—H
   |  |           |
   H  H           H

  Ethane       Methanol
```

7.5 Dispersion Forces

Because unlike charges attract, it is easy to understand interionic forces and intermolecular forces between dipolar molecules. But how can we explain the fact that nonpolar substances can also exist in liquid and solid states? Helium atoms are nonpolar, yet helium condenses to a liquid at temperatures below about 5 K. Some kind of intermolecular force must hold the He atoms close enough together to keep the helium in the liquid state. What is the source of this force?

To answer this question, we first must realize that the electron cloud pictures we used in Chapter 3 represent average situations only. On average, the electron charge density associated with helium's two 1*s* electrons is evenly distributed about the nucleus. However, at any given instant, purely by chance, the electron charge density may become uneven. The normally nonpolar atom becomes momentarily polar; an instantaneous dipole is formed. This transitory dipole, in turn, can displace electrons in a neighboring helium atom, also producing a dipole. One dipole induces the other, and the newly formed dipole is therefore called an induced dipole. Taken together, these two events lead to an intermolecular force of attraction (Figure 7.5). The force of attraction between the instantaneous dipole and the induced dipole is known as a **dispersion force** (also called a London force, named for Fritz London, a professor of chemical physics at Duke University who offered a theoretical explanation of these forces in 1928.)

Dispersion forces are weak for small, nonpolar molecules such as H_2, N_2, and CH_4 but can be substantial between large molecules. The properties of polymers such as polyethylene (Section 13.10) are determined to a large degree by dispersion forces between long chains of repeating $\text{-(}CH_2CH_2\text{)}_n\text{-}$ units. (In this formula, *n* is several hundred or even several thousand.)

Dispersion forces, to a large extent, determine the physical properties of nonpolar compounds. Recall that at room temperature fluorine (F_2) and chlorine (Cl_2) are gases, bromine (Br_2) is a liquid, and iodine (I_2) is a solid. Dispersion forces are greater for larger molecules than for smaller ones. Larger atoms have larger electron clouds. Their valence electrons are farther from the nucleus than those of smaller atoms. These remote electrons are more loosely bound and can shift toward another atom more readily than the tightly bound electrons in a smaller atom. This makes molecules with larger atoms more polarizable than those with small ones. Iodine molecules are attracted to one another more strongly than bromine molecules are attracted to one another. Bromine molecules have greater dispersion forces than chlorine molecules, and chlorine molecules, in turn, have greater dispersion forces than fluorine molecules. If you look at the periodic table, you will see that this is the order in which these elements, the halogens, appear in Group 7A, with iodine the largest and fluorine the smallest.

Dispersion forces are important even when other types of forces are present. We might think of such forces as individually weaker than dipolar attractions or ionic bonds, but in substances composed of large molecules the cumulative effect of dispersion forces can be considerable. For large ions, such as silver (Ag^+) and iodide (I^-), dispersion forces play a significant role, even though interionic forces also exist. Dispersion forces play a major role in the presence of some dipolar forces. In hydrogen chloride (HCl), for example, dipolar forces may contribute as little as 15% to the intermolecular attraction; the rest is due to dispersion forces.

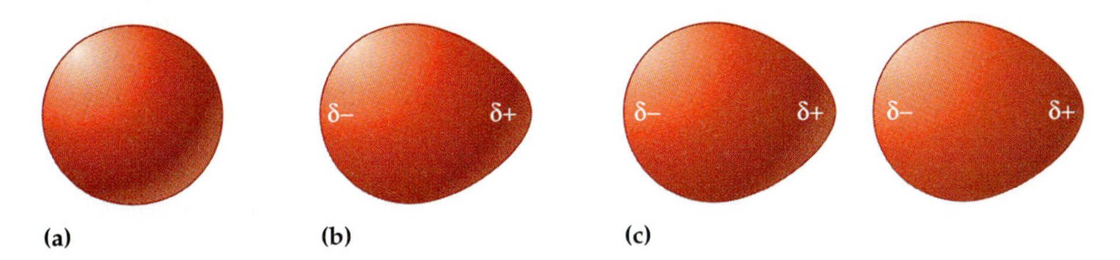

▲ **Figure 7.5**
Dispersion forces explained. (a) *Unpolarized molecule.* The electron charge distribution is symmetrical. (b) *Instantaneous dipole.* A momentary displacement of electron charge density (to the left) produces an instantaneous dipole. (c) *Induced dipole.* The instantaneous dipole on the left induces charge separation in the molecule on the right, making it also a dipole. The attraction between the two dipoles is an intermolecular force called a dispersion force.

▲ The viscosity of motor oils is indicated by a number assigned by the Society of Automotive Engineers (SAE). SAE 40 motor oil (left) is more viscous than SAE 10 oil (right).

Review Questions

7.8 Of each pair of substances, which has the higher boiling point? Explain.
a. Ne or Xe **b.** methane (CH_4) or ethane (C_2H_6)

7.9 What is the difference between an instantaneous and an induced dipole? What is a dispersion force?

7.10 Explain how oxygen can be liquefied if the temperature is lowered sufficiently, even though O_2 molecules are nonpolar.

Liquids, Solids, and Changes of State

Now that we have noted the major types of intermolecular forces, we can move to a more detailed discussion of liquids and solids, and we can examine the changes that occur in going from one physical state to another.

7.6 The Liquid State

Molecules of a liquid are much closer together than those of a gas. Consequently, liquids can be compressed only slightly. The molecules are in constant motion, but their movements are greatly restricted by neighboring molecules. One liquid can diffuse into another, but such diffusion is much slower than in gases because of the restricted molecular motion of liquids.

Viscosity—a measure of the resistance of a liquid to flow; the higher the viscosity, the slower the liquid's rate of flow.

The shape of the molecules that make up a liquid has an effect on an important property of the liquid—its **viscosity,** or resistance to flow. Liquids with low viscosity—"thin" liquids—generally consist of small, symmetrical molecules with weak intermolecular forces. Viscous liquids, on the other hand, are generally made up of large or unsymmetrical molecules with fairly strong intermolecular forces (Figure 7.6). Viscosity generally decreases with increasing temperature. Increased kinetic energy partially overcomes the intermolecular forces. Cooking oil, for example, as it pours from the bottle is thick and "oily." After it's been heated in a frying pan, it becomes thinner and more watery—that is, more like water in its consistency.

▲ The surface tension of the water keeps the water strider from breaking through the surface. Notice how the water surface is indented but not penetrated by the insect's feet.

Another property of liquids is **surface tension.** A clean glass can be slightly overfilled with water before it spills over. A small needle, carefully placed, can be made to float horizontally on the surface of water—even though steel is several times denser than water. A variety of insects can walk or skate across the surface of a pond with ease. These phenomena indicate something quite unusual about the surface of a liquid. There is a special force or tension there that resists disruption by the needle or water bug.

These surface forces can be explained by intermolecular forces. A molecule in the center of a liquid is pulled equally in all directions by the molecules surrounding it. A molecule on the surface, however, is attracted only by molecules at its sides and below it (Figure 7.7). There is no corresponding upward attraction. These unequal forces tend to pull inward at the surface of the liquid and cause it to contract, much as a stretched sheet of rubber would tend to contract. A small amount of liquid will "bead" to minimize its surface area, and a drop will be spherical for the same reason. The smaller the surface area, the smaller the number of molecules subjected

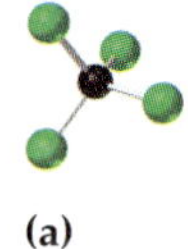

(a)

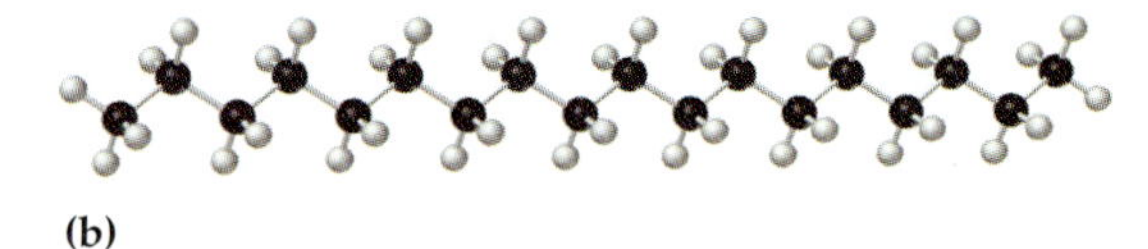

(b)

Figure 7.6 ▶
(a) Carbon tetrachloride (CCl_4) consists of relatively small symmetrical molecules with fairly weak intermolecular forces. It has a low viscosity. (b) Octadecane ($C_{18}H_{38}$) is made up of long-chain molecules with comparatively strong intermolecular forces. It has a higher viscosity than carbon tetrachloride.

to the unequal pull. Soaps and other detergents act in part by lowering surface tension, enabling water to spread out and wet a solid surface (see Section 20.3).

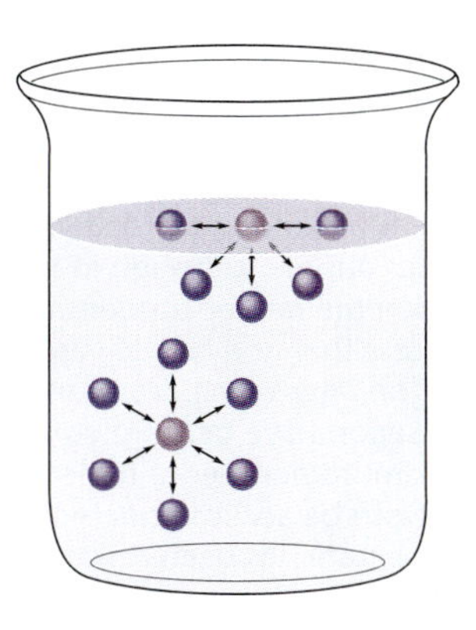

▲ **Figure 7.7**
Surface tension explained. Molecules in the body of a liquid are attracted equally in all directions. Those at the surface, however, are pulled downward and sideways but not upward.

✓ Review Questions

7.11 Use intermolecular forces to explain the phenomena of surface tension.

7.12 What do we mean when we say that a liquid is viscous? How does temperature affect the viscosity of a liquid? Explain.

7.7 From Liquid to Gas: Vaporization

The molecules of a liquid are in constant motion, some moving fast, some more slowly. Occasionally one of these molecules has enough kinetic energy to escape from the liquid's surface and become a molecule of vapor. If a liquid, such as water, is placed in an open container, it will soon disappear through evaporation. The vapor molecules disperse throughout the atmosphere, and eventually all the liquid molecules escape as they enter the vapor state. On the other hand, if the liquid is placed in a closed container, it remains at nearly the same volume. Some of the liquid is converted to vapor, but the vapor molecules are trapped within the container. Eventually the air above the liquid becomes saturated, and evaporation seems to stop. This vapor exerts a partial pressure (the vapor pressure—Section 6.10) that is constant at a given temperature.

The rate of evaporation of a particular liquid depends on the temperature of the liquid and the amount of the exposed surface area.

It may well appear that nothing further is happening, but molecular motion has not ceased. Some molecules of liquid are still escaping into the vapor state. Vapor molecules in the space above the liquid occasionally strike the liquid's surface, are captured, and thus return to the liquid state. At first there are lots of liquid molecules and no vapor molecules. So **vaporization** (conversion of liquid to vapor) is taking place, but **condensation** (conversion of vapor to liquid) is not. As more molecules pass into the vapor state, the rate of condensation increases. Eventually the rate of condensation equals the rate of vaporization. This equilibrium condition appears static but is, in fact, dynamic; two opposing processes are taking place at the same rate. This situation is analogous to that encountered in the case of reversible chemical reactions (Section 4.8).

As the temperature of a liquid is raised, more molecules have enough energy to escape from the liquid state. The rate of vaporization increases. At equilibrium, the rates of vaporization and condensation are again equal, but the pressure exerted by the vapor is higher at the higher temperature. The vapor pressures of liquids increase with temperature.

Now let's see what's going on at the molecular level (Figure 7.8). As soon as molecules appear in the vapor state, some of them strike the liquid surface, are captured, and return to the liquid state. Condensation and vaporization both occur at the same time. We can represent these two opposing simultaneous processes by arrows pointing in opposite directions.

$$\text{Liquid} \underset{\text{condensation}}{\overset{\text{vaporization}}{\rightleftharpoons}} \text{Vapor}$$

If a liquid is placed in an open container, atmospheric pressure opposes the escape of molecules of the liquid. If the liquid is heated, its vapor pressure will increase until it is equal to atmospheric pressure. At that temperature, the liquid will begin to boil. Vaporization will take place not only at the surface but also in the body of the liquid, with vapor bubbles forming and rising to the surface. The **boiling point** of a liquid is the temperature at which its vapor pressure becomes equal to

Figure 7.8 ▶
Liquid-vapor equilibrium and vapor pressure illustrated. (a) Vaporization of a liquid begins. (b) Condensation begins as soon as the first vapor molecules appear, although in this illustration the rate of condensation is still less than the rate of vaporization. (c) The rates of vaporization and condensation have become equal. The maximum number of molecules that can be accommodated in the vapor state has been reached. The partial pressure exerted by these molecules—the vapor pressure of the liquid—remains constant.

Molecules in vapor state
Molecules undergoing vaporization
Molecules undergoing condensation

(a) (b) (c)

atmospheric pressure. Because the latter varies with altitude and weather conditions, boiling points of liquids do also (Figure 7.9). The cooking of foods requires that they be supplied with a certain amount of energy. When water boils at 100 °C, an egg can be placed in the water and soft-boiled in three minutes. At reduced atmospheric pressure, water boils at a lower temperature, and it contains less heat energy with which to cook the egg. It would take longer to prepare a soft-boiled egg on top of Mount Everest.

The boiling point is increased when external pressure is increased. Autoclaves and pressure cookers are based on this principle. Higher temperatures can be achieved at the higher pressures attained in these closed vessels. (Heat added to a liquid at atmospheric pressure will merely convert liquid to vapor. No increase in temperature occurs until all the liquid has vaporized.) The higher temperatures attained in a pressure cooker speed the chemical reactions involved in the cooking of a tough piece of meat, and those attained in an autoclave speed the killing of bacteria (even resistant bacterial spores). Table 7.1 gives the temperatures attainable with pure water at various pressures.

The boiling point of a liquid is a useful physical property, often used as an aid in identifying compounds. Because boiling point varies with pressure, it is necessary

Figure 7.9 ▶
The boiling point of water decreases with altitude because the atmospheric pressure decreases with altitude.

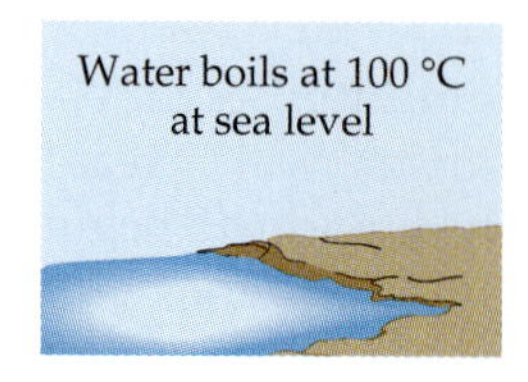
Water boils at 100 °C
at sea level

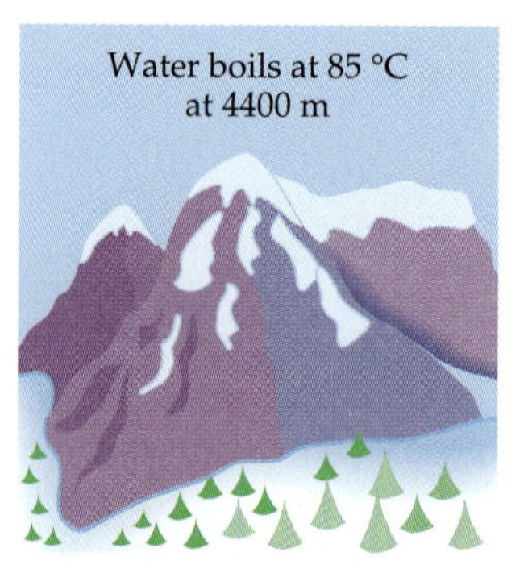
Water boils at 85 °C
at 4400 m

Pike's Peak, Colorado

Water boils at 71 °C
at 8800 m

Mount Everest, Tibet

to define the **normal boiling point,** that temperature at which a liquid boils under standard pressure (1 atm or 760 mmHg). Alternatively, one can specify the pressure at which the boiling point was determined. For example, the Handbook of Chemistry and Physics lists the boiling point of antipyrine (a pain reliever and fever reducer) as 319^{741}. This means that the compound boils at 319 °C under a pressure of 741 mmHg. Table 7.2 gives the normal boiling points of some familiar liquids.

Liquids can be purified by a process called **distillation.** Imagine a mixture of water and some nonvolatile material—that is, some material that will not vaporize readily. If the mixture is heated until it boils, the water will vaporize, but the nonvolatile material will not. The water vapor can then be condensed back to the liquid state and collected in a separate container. In such a distillation, the water is separated from the other component of the mixture and thereby purified.

Even if a mixture contains two or more volatile components, purification by distillation is possible. Consider a mixture of two components, one of which is somewhat more volatile than the other (for example, water and alcohol). At the boiling point of such a mixture, both components contribute some molecules to the vapor. In the vapor, there are more molecules of the more volatile component (alcohol in this example), because it is more easily vaporized, than of the less volatile component. When the vapor is condensed into another container, the resulting liquid will be richer in the more volatile component than the original mixture was. Thus a purer sample of the more volatile component is produced. Figure 7.10 shows a typical distillation apparatus.

Heat is required for the conversion of a liquid to a vapor. A liquid evaporating at room temperature absorbs heat from its surroundings. When our skin is wet, this cooling effect of evaporation can make us feel cool even on a warm day because the water evaporating from our skin removes heat. Volatile liquids, such as ethyl chloride (CH_3CH_2Cl), which boils at 12.5 °C, can be used as local anesthetics. Rapid evaporation from the skin removes enough heat to freeze a small area, rendering it insensitive to pain. Alcohol rubs also act to cool the skin by their evaporation.

The amount of heat required to vaporize a given amount of liquid can be measured. The quantity of heat required to vaporize 1 mol of a liquid at a constant pres-

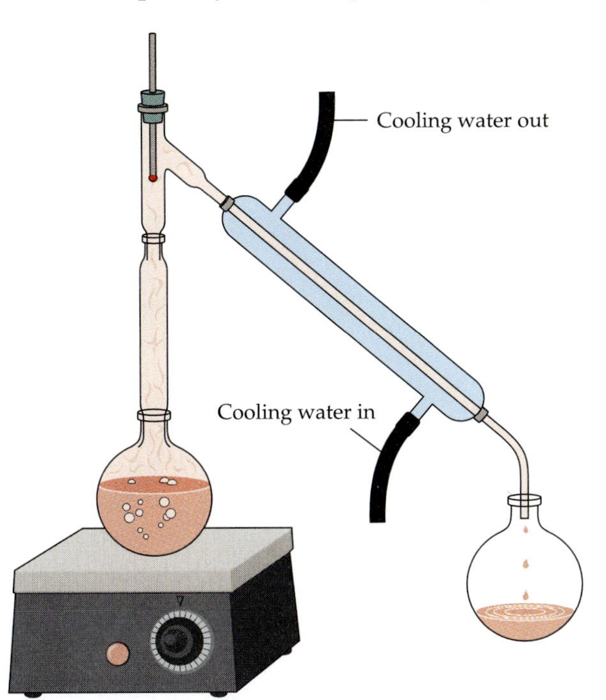

Figure 7.10
A distillation apparatus. A mixture is heated in the flask at the left. The vapors formed travel up the vertical column, and are then condensed in the cooled tube angled downward toward the right. The condensed liquid is then collected in the flask at the right.

Table 7.1 Boiling Points of Pure Water at Various Pressures

Pressure (mmHg)	Temperature (°C)
707	98
760	100
816	102
875	104
938	106
1004	108
1075	110
1283	115
1535	121

Table 7.2 Boiling Points of Various Compounds at 1 Atm

Compound	Boiling Point (°C)
Diethyl ether (an anesthetic)	34.6
Acetone (a solvent)	56.2
Methanol (wood alcohol)	64.5
Ethanol (grain alcohol)	78.3
Water	100.0
Mercury	356.6

sure is called the **molar heat of vaporization.** The heat of vaporization is characteristic of a given liquid. It depends to a large extent on the types of intermolecular forces in the liquid. Water, with molecules strongly associated through hydrogen bonding, has a heat of vaporization of 9.7 kcal/mol. Methane, with molecules held together by weak dispersion forces only, has a heat of vaporization of only 0.232 kcal/mol. Heats of vaporization of several liquids are given in Table 7.3.

Given the molar heat of vaporization, we can calculate the heat of vaporization in calories per gram or kilojoules per gram.

Example 7.4

The molar heat of vaporization of ammonia is 5200 cal/mol. What is the heat of vaporization in (**a**) calories per gram and (**b**) in kilojoules per gram?

Solution

a. The molar mass of ammonia (NH_3) is 17.0 g/mol. Therefore, the heat of vaporization is

$$\frac{5200\ \text{cal}/\cancel{\text{mol}}}{17.0\ \text{g}/\cancel{\text{mol}}} = 306\ \text{cal/g}$$

b. This requires the conversion factors, 1 cal = 4.184 J and 1000 J = 1 kJ

$$306\ \cancel{\text{cal}}/\text{g} \times \frac{4.184\ \cancel{\text{J}}}{1\ \cancel{\text{cal}}} \times \frac{1\ \text{kJ}}{1000\ \cancel{\text{J}}} = 1.28\ \text{kJ/g}$$

Practice Exercise

The molar heat of vaporization of chloroform ($CHCl_3$) is 7050 cal/mol. What is the heat of vaporization in calories per gram?

Example 7.5

How many kilojoules of heat are required to vaporize 425 g of water at its boiling point?

Solution

The heat of vaporization of water is 2260 J/g (Table 7.3) or 2.26 kJ/g.

$$425\ \cancel{\text{g}} \times 2.26\ \text{kJ}/\cancel{\text{g}} = 961\ \text{kJ}$$

Practice Exercise

How much heat is required to vaporize 425 g of chloroform at its boiling point? (Use your answer from the preceding practice exercise.)

Table 7.3 Heats of Vaporization (at the Normal Boiling Point) of Several Liquids

Compound	Molar Heat of Vaporization (cal/mol)	Heat of Vaporization (J/g)
Diethyl ether ($C_2H_5OC_2H_5$)	6,200	351
Benzene (C_6H_6)	7,300	393
Methanol (CH_3OH)	8,400	1,090
Water	9,700	2,260
Mercury	14,200	297

When a vapor condenses to a liquid, it gives up the same amount of heat energy as was absorbed in converting the liquid to a vapor. A refrigerator operates by alternately vaporizing and condensing a fluid. The heat required to vaporize the fluid is drawn from the refrigerated compartment. The heat is released to the outside atmosphere when the fluid is condensed back to the liquid state.

✓ Review Questions

7.13 What is the distinction between (**a**) vaporization and boiling? (**b**) the boiling point of a liquid and the normal boiling point of a liquid?

7.14 How is the heat of vaporization of a liquid related to intermolecular forces in the liquid?

7.15 How does temperature affect the vapor pressure of a liquid? Explain.

7.16 How does a pressure cooker work?

7.8 The Solid State

Solids resemble liquids in that the particles (atoms, molecules, or ions) in both are close together, making them virtually incompressible. But these two physical states differ significantly in the motion of their particles. In the liquid state, particles are in constant (if somewhat restricted) motion. In solids, there is little motion other than vibration about a fixed point. Consequently, diffusion in solids is generally extremely slow. An increase in temperature will increase the vigor of the vibrations in a solid. If the vibrations become violent enough, the solid will melt (Section 7.9).

Many familiar solids are crystalline. In a scientific sense, a crystal is a piece of a solid substance that has plane surfaces, sharp edges, and a regular geometric shape. The atoms, ions, or molecules that make up the crystal are assembled in a regular, repeating manner extending in three dimensions throughout the crystal. We can figure out the entire structure of a crystal from just a tiny portion of it, called a **unit cell.** The entire crystal can be generated by stacking unit cells much as we stack toy building blocks. Solids that lack this ordered arrangement, of which glass is an example, are called *amorphous solids.*

Repeating patterns are evident in strings of beads (one dimension: length) and in floor tiles (two dimensions: length and width). To describe crystals, we need to work with three-dimensional patterns. The framework on which we outline the pattern is called a **crystal lattice.** With 14 different lattices we could describe all crystalline solids, but we will limit our discussion to three lattices of the cubic type.

The simplest unit cell, the *simple cubic,* with structural particles (atoms, ions, or molecules) only at its corners, is quite rare. Most crystal structures are better described by a unit cell with more structural particles. The **body-centered cubic (bcc)** structure has an additional structural particle at the center of the cube. The **face-centered cubic (fcc)** structure has an additional particle at the center of each face. These three unit cells are shown in Figure 7.11. The common metals Fe, K, Na, and W have a bcc crystal structure, and Al, Cu, Pb, and Ag have an fcc structure.

Crystalline solids can also be classified by the types of intermolecular forces holding the particles together. The four main classes are ionic, molecular, covalent network, and metallic.

Ionic solids have ions at definite points in the lattice. We discussed a typical ionic solid, sodium chloride (NaCl), in Section 3.3. In the lattice, each chloride ion is surrounded by six sodium ions, and each sodium ion is surrounded by six chloride ions. Interionic forces are very strong; ionic solids consequently have high melting points and low vapor pressures, and are quite hard.

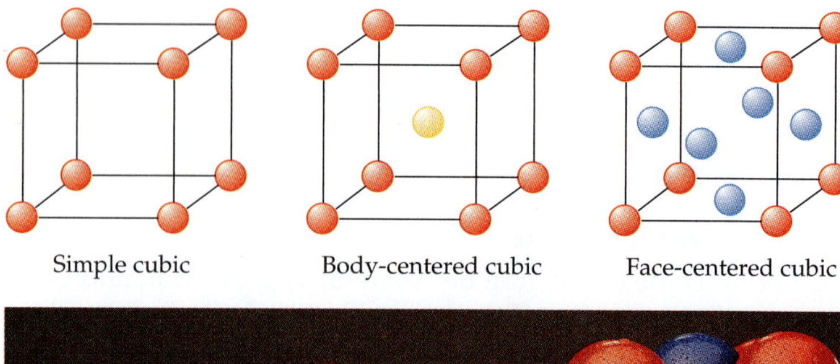

Figure 7.11 ► Three types of cubic crystal structures. Only the centers of spheres (atoms) are shown in the top row. The space-filling models in the bottom row show that certain of the spheres are in direct contact.

Molecular crystals have discrete covalent molecules at the lattice points. These are held together by rather weak dispersion forces, as in crystalline iodine, by dispersion forces plus dipolar forces, as in iodine chloride (ICl), or by hydrogen bonds, as in ice. Molecular solids often have lower melting points than ionic solids. Ice is an exceptional molecular solid; the H_2O molecules are strongly associated by hydrogen bonding.

Covalent network crystals have atoms at the lattice points. These are joined into extensive networks by covalent bonds; each crystal is in essence one large molecule. Covalent network solids are generally extremely hard and nonvolatile and melt (with decomposition) at very high temperatures. Diamond is a familiar example. Carbon atoms occupy the lattice points. Each is covalently bonded to four other carbon atoms. Silicon carbide (SiC), also called Carborundum, is another familiar compound with an extensive network of covalent bonds.

We can consider the lattice sites in metallic solids to be occupied by positive ions, formed when metal atoms, such as silver (Ag), lose their valence electrons. The electrons thus released are distributed throughout the lattice, almost like a fluid. These electrons, which can move freely about the lattice, make metals good conductors of heat and electricity. Some metals, such as sodium and potassium, are fairly soft and have low melting points. Others, such as magnesium and calcium, are hard and have high melting points. The extra valence electrons in calcium and magnesium atoms seem to lead to stronger forces between atoms. As a class, metals are malleable; that is, they can be shaped under the influence of pressure or heat. They can be rolled into bars, pressed into sheets, or extruded into wire.

Table 7.4 lists some characteristics of crystalline solids, and Figure 7.12 illustrates some examples.

✓ Review Questions

7.17 What is meant by the term crystal lattice?

7.18 Define or illustrate each of the following terms.

a. ionic crystal **b.** metallic solid
c. covalent network solid **d.** molecular solid

Table 7.4 Some Characteristics of Crystalline Solids

Particles in Crystal	Principal Attractive Force Between Particles	Melting Point	Electrical Conductivity of Liquid	Characteristics of the Crystal	Examples
		Ionic Crystals			
Positive and negative ions	Electrostatic attraction between ions (very strong)	High	High	Hard, brittle, most dissolve in polar solvents	NaCl, CaF_2, K_2S, MgO
		Covalent Network Crystals			
Atoms	Covalent bonds (very strong)	Generally do not melt	—	Very hard, insoluble	Diamond (C), SiC, AlN
		Metallic Crystals			
Positive ions plus mobile electrons	Metallic bonds (strong)	Most are high	Very high	Most are hard, malleable, ductile, good electrical and thermal conductors, insoluble unless a reaction occurs	Cu, Ca, Al, Pb, Zn, Fe, Na, Ag
		Molecular Crystals			
		Hydrogen-bonded			
Molecules with H on N, O, or F	Hydrogen bonds (intermediate)	Intermediate	Very low	Fragile, soluble in other H-bonding liquids	H_2O, HF, NH_3, CH_3OH
		Polar			
Polar molecules (no H bonds)	Electrostatic attraction between dipoles (rather weak)	Low	Very low	Fragile, soluble in other polar and many nonpolar solvents	HCl, H_2S, $CHCl_3$, ICl
		Nonpolar			
Atoms or nonpolar molecules	Dispersion forces only (weak)	Very low	Extremely low	Soft, soluble in nonpolar or slightly polar solvents	Ar, H_2, I_2, CH_4, CO_2, CCl_4

7.9 From Solid to Liquid: Melting (Fusion)

Solids can be changed to liquids; that is, they can be melted. The solid is heated, and the heat energy is absorbed by the particles of the solid. The energy causes the particles to vibrate in place with more and more vigor until, finally, the forces holding the particles in a particular arrangement are overcome. The solid has become a liquid. The temperature at which this happens is called the **melting point** of that solid. A high melting point is one indication that the forces holding a solid together are very strong.

The amount of heat required to convert 1 mol of a solid to a liquid at the melting point is called the **molar heat of fusion.** Generally the heat of fusion is much

▲ **Figure 7.12**
Examples of three types of crystalline substances: (a) ionic, (b) covalent, and (c) metallic.

Most of the heat energy that water absorbs is used to break hydrogen bonds, not to increase the kinetic energy of the H_2O molecules.

less than the heat of vaporization (Figure 7.13). The heat of fusion is the amount of energy that will disrupt the crystal lattice but still leave the particles in contact with one another and under the influence of their mutual attraction. A much larger amount of energy is required to vaporize the liquid because, in vaporization, the attraction between particles must be almost completely overcome. Table 7.5 gives some representative heats of fusion.

Example 7.6

The heat of fusion of water is 80 cal/g. How many calories of heat are required to melt 454 g of ice?

Solution

$$454\ \text{g} \times 80\ \text{cal/g} = 36{,}000\ \text{cal}$$

Example 7.7

The molar heat of fusion of naphthalene ($C_{10}H_8$) is 4610 cal/mol. What is the heat of fusion in kilojoules per gram?

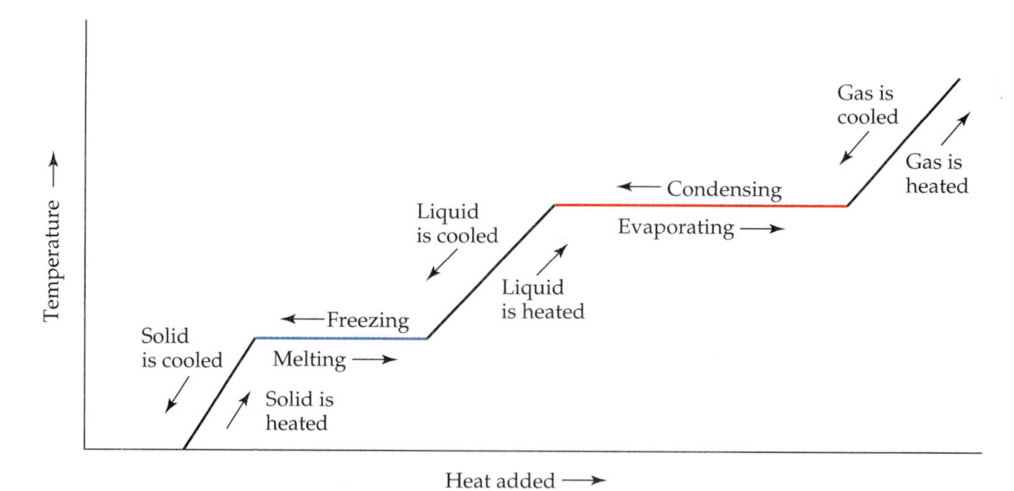

Figure 7.13 ►
This graph, followed from right to left, is called a cooling curve. (Read in the opposite direction, it is a heating curve.) Note that although *energy* (heat of fusion or heat of vaporization) is absorbed or released, the *temperature* does not change as a substance undergoes a change of state. The heat of fusion of a substance (blue line) is usually only a fraction of its heat of vaporization (red line).

Sublimation

Although few solids are as volatile as some familiar liquids, solids do vaporize, even if only to a limited extent. The passage of molecules directly from the solid to the vapor state (Figure 7.14) is called **sublimation.** The reverse process, the condensation of a vapor to a solid, is generally called deposition. A dynamic equilibrium is reached when the rates of sublimation and deposition become equal.

People living in cold climates are especially familiar with the phenomenon of sublimation. Snow disappears from the ground and ice from the windshield of an automobile even though the temperature stays below 0 °C. This occurs through sublimation, not melting; there is no liquid water at any point in the process. The vapor pressure of ice at 0 °C is 4.58 mmHg.

▲ **Figure 7.14**
Sublimation of iodine. Iodine has a high vapor pressure at atmospheric pressure and a temperature of 70 °C—well below its melting point of 114 °C. The purple iodine vapor $I_2(g)$ and the crystals of $I_2(s)$ on the cooler walls of the flask indicate that iodine *sublimes;* that is, it goes directly from the solid state to the vapor state. No liquid is present.

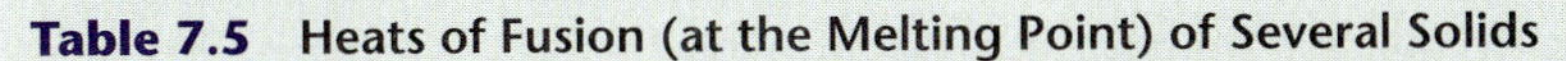

Table 7.5 Heats of Fusion (at the Melting Point) of Several Solids

Compound	Melting Point (°C)	Heat of Fusion (kJ/mol)	Heat of Fusion (cal/g)
Ammonia (NH_3)	−78	6.77	95
Water (H_2O)	0.0	6.01	80
Benzene (C_6H_6)	5.5	9.87	33
Copper (Cu)	1083	13.0	49
Sodium chloride (NaCl)	804	29.3	120
Tungsten (W)	3410	33.7	43

Solution

The molar mass of naphthalene ($C_{10}H_8$) is 128 g/mol. We can calculate the heat of fusion in calories per gram and use the equivalencies 1 kcal = 1000 cal and 1 kcal = 4.184 kJ to get the answer kilojoules per gram.

$$\frac{4610\ \text{cal/mol}}{128\ \text{g/mol}} \times \frac{1\ \text{kcal}}{1000\ \text{cal}} \times \frac{4.184\ \text{kJ}}{1\ \text{kcal}} = 0.151\ \text{kJ/g}$$

Practice Exercises

A. The heat of fusion of sodium chloride is 120 cal/g. How much heat, in kilocalories, is required to melt 454 g of sodium chloride?

B. The molar heat of fusion of calcium nitrate is 5120 cal/mol. How much heat, in kilojoules, is required to melt 227 g of calcium nitrate?

Review Question

7.19 Define or illustrate each of the following terms.

a. melting point **b.** molar heat of fusion

7.10 Water: A Most Unusual Liquid

Now that we have laid something of a theoretical foundation, let's look more closely at a very special liquid, water. Next to air, water is the most familiar substance on Earth. (It is the only liquid commonly found on the surface of our planet.) Even so, it is a most unusual compound. At room temperature it is the only liquid compound with a molar mass as low as 18 g/mol. Unlike most substances, the solid form of water (ice) is less dense than the liquid. This has immense consequences for life on this planet. Ice forms on the surfaces of lakes when the temperature drops below freezing, and this ice insulates the lower layers of water, enabling fish and other aquatic organisms to survive the winters of the temperate zones. If ice were denser than liquid water, it would sink to the bottom as it formed. The new surface water would freeze and, in its turn, sink to the bottom. This process would continue until eventually the lake was frozen from top to bottom. Even the deeper lakes of the northern latitudes would freeze solid in winter. Life in the northern lakes and rivers would be quite different from what it is now, if indeed there were life in those waters at all.

The relative densities of ice and liquid water have dangerous consequences for living cells. Because ice has a lower density than liquid water, 1 g of ice occupies a larger volume than 1 g of water. As ice crystals form in living cells, the expansion ruptures and kills the cells. The slower the cooling, the larger the crystals of ice and the more damage to the cell. The frozen food industry takes into account this property of water. Food is "flash frozen," that is, frozen rapidly to keep the ice crystals small and thus do minimum damage to the cellular structure of the food.

Nearly all other solids are more dense than their liquids. Consider, for example, a lead ball and molten lead or an iron ball and molten iron.

Liquid water has a higher density than most other familiar liquids. As a consequence, liquids that are less dense than water and insoluble in water float on its surface. A familiar problem in recent years has been the gigantic oil spills that occur when a tanker ruptures or when an offshore well gets out of control. The oil, floating on the surface of the water, covers the feathers of waterfowl and the coats of sea animals, such as the otter and the seal. The oil is often washed onto beaches, where it does considerable ecological and aesthetic damage (Figure 7.15). If oil were denser than water, it would sink to the bottom. The problem would be of a different nature, although not necessarily less acute.

▲ About 2% of Earth's water is in the form of ice.

Another unusual property of water is its high specific heat (Table 7.6). It takes 1 cal of heat to raise the temperature of 1 g of water 1 °C. That's almost ten times as much energy as is required to raise the temperature of the same amount of iron 1 °C. The specific heats of a number of common materials are given in Table 7.6.

The reason cooking utensils are made of iron, copper, aluminum, or glass is that these materials have low specific heats. Thus they heat up quickly. The reason the handle of a frying pan is usually made of wood or plastic is that these materials have high specific heats. When they are exposed to heat, their temperatures increase more slowly.

Figure 7.15 ▶
Oil coats the water of Prince William Sound, Alaska, where the *Exxon Valdez* ran aground on Bligh Reef in 1989, spilling 42 million liters of oil. One liter of oil can create a slick 2.5 hectares (6.2 acres) in size.

Table 7.6 Specific Heats of Some Common Substances

Substance	Specific Heat [cal/(g·°C)]	Specific Heat [J/(g·K)]
Water (liquid)	1.00	4.182
Water (solid)	0.50	2.1
Water (gas)	0.47	2.0
Ethanol	0.588	2.46
Wood	0.42	1.8
Aluminum	0.216	0.902
Glass	0.12	0.50
Iron	0.107	0.449
Copper	0.092	0.385
Silver	0.056	0.235

The high specific heat of water means not only that much energy is required to raise the temperature of water but also that much heat is given off by water for even a small drop in temperature. The vast amounts of water on the surface of the Earth thus act as a giant thermostat to moderate daily temperature variations. We need only consider the extreme temperature changes on the surface of the waterless moon to appreciate this important property of water. The temperature of the moon varies from just above the boiling point of water (100 °C) to about −175 °C, a range of 275 °C. In contrast, temperatures on Earth rarely fall below −50 °C (−58 °F) or rise above 50 °C (122 °F), a range of only 100 °C.

Example 7.8

How much energy is required to change 10.0 g of ice at −10 °C to steam at 100 °C?

Solution

Let's work this problem in parts. First calculate the energy required to raise the temperature of 10.0 g of ice from −10 °C to 0 °C, a change of 10 °C. The specific heat of ice is 0.50 cal/(g·°C). Recall from Chapter 1 that

$$\text{Heat absorbed or released} = \text{mass} \times \text{specific heat} \times \Delta T$$

$$= 10.0\ \text{g} \times 10\ {}^\circ\text{C} \times \frac{0.50\ \text{cal}}{1\ \text{g}\cdot{}^\circ\text{C}} = 50\ \text{cal}$$

Next, using the heat of fusion of water (80 cal/g), calculate the energy required to melt the ice at 0 °C.

$$10.0\ \text{g} \times 80\ \text{cal/g} = 800\ \text{cal}$$

Next, calculate the heat required to change the temperature of the water from 0 °C to 100 °C. The specific heat of liquid water is 1 cal/(g·°C), and ΔT is 100 °C.

$$\text{Heat absorbed} = 10.0\ \text{g} \times 100\ {}^\circ\text{C} \times \frac{1\ \text{cal}}{1\ \text{g}\cdot{}^\circ\text{C}} = 1000\ \text{cal}$$

Now calculate the amount of heat required to change the water (at 100 °C) to steam (at 100 °C). The heat of vaporization for water is 540 cal/g.

$$10.0\ \text{g} \times 540\ \text{cal/g} = 5400\ \text{cal}$$

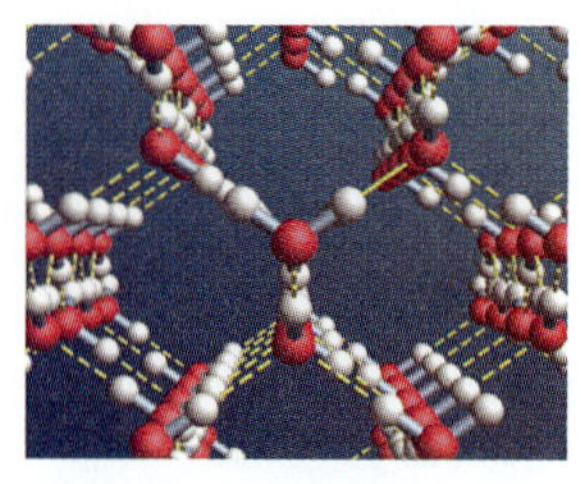

(a)

(b)

▲ **Figure 7.16**
Hydrogen bonds in ice. (a) Oxygen atoms are arranged in layers of distorted hexagonal rings. Hydrogen atoms lie between pairs of O atoms, closer to one (covalent bond) than to the other (hydrogen bond). (b) At the macroscopic level, this structural pattern is revealed in the hexagonal shapes of snowflakes.

Finally, total all the calculated values.

To raise the temperature of the ice from −10 °C to 0 °C	50 cal
To change the ice to liquid water	800 cal
To raise the temperature of the water from 0 °C to 100 °C	1000 cal
To change the water to steam	5400 cal
Total	7250 cal

Note that almost 75% of the total energy is used in vaporizing the water.

Practice Exercises

A. How much heat is required to increase the temperature of 80.0 g of water from 18.5 °C to 33.3 °C?

B. How much energy is required to change 100 g of ice at −10 °C to liquid water at 15 °C?

Water is also unusual in that it has an exceptionally high heat of vaporization; that is, a large amount of heat is required to evaporate a small amount of water. This is of enormous importance to animals. Large amounts of body heat, produced as a by-product of metabolic processes, can be dissipated by the evaporation of small amounts of water (perspiration) from the skin. The heat of vaporization of this water is obtained from the body, and the body is cooled. Conversely, when steam condenses, considerable heat is released. For this reason, steam causes more serious burns when it contacts the skin than does water at 100 °C. We previously mentioned that water's high specific heat modifies the climate. Water's high heat of vaporization also contributes to the climate-modifying effect of lakes and oceans. A large portion of the heat that would otherwise warm up the land is used instead to vaporize water from the surface of a lake or the sea. Thus, in summer it is cooler near a large body of water than in interior land areas.

All of these fascinating properties of water result from the unique structure of the water molecule (Section 3.13). Recall that the water molecule is polar. In the liquid state, water molecules are strongly associated by hydrogen bonding (see again Figure 7.4). These strong attractive forces account for the high heat of vaporization of water. They must be overcome by input of a large amount of energy if vaporization is to take place.

In the liquid state, water molecules are quite close together but randomly arranged. When water freezes, its molecules take on a more ordered arrangement with large hexagonal holes (Figure 7.16). This three-dimensional structure extends out for billions and billions of molecules. The holes account for the fact that ice is less dense than liquid water. The hexagonal arrangement allows water to assume forms of exquisite beauty as snowflakes.

Our bodies are about 65% water. Chemical reactions in living cells take place in water solutions. The importance of solutions is such that we devote the next chapter to the subject.

Review Questions

7.20 List three distinctive properties of water.

7.21 Why does ice float on liquid water?

Summary

Intermolecular forces are those acting among the particles making up any gas, liquid, or solid. These forces are usually negligible in gases but of prime importance in liquids and solids. The strongest force between particles is the *interionic force,* which holds together positive and negative ions in a solid. Polar molecules have intermolecular forces called **dipole forces.**

A **hydrogen bond** is a special type of dipole interaction found in molecules with a hydrogen bonded to a fluorine, oxygen, or nitrogen atom. This bond is the attraction between the partially positive H in one molecule and the partially negative F, O, or N in a neighboring molecule.

Dispersion forces result from momentary dipoles in neighboring molecules. These transient forces exist in all substances, but they are the only intermolecular forces holding nonpolar liquids and solids together.

Viscosity is a liquid's resistance to flow, which depends on the strength of the intermolecular forces. The molecules in the surface layer of a liquid experience only lateral and downward intermolecular forces, an imbalance of forces resulting in **surface tension.**

Vaporization is the change of state in which a liquid becomes a gas. The change of state in the opposite direction, gas to liquid, is **condensation.** The amount of heat needed to vaporize one mole of any liquid is the **molar heat of vaporization** for that liquid.

For a liquid in a closed container and at a given temperature, there is an equilibrium between vaporization and condensation at the liquid surface. The **boiling point** of a liquid is the temperature at which a liquid's vapor pressure equals atmospheric pressure. The **normal boiling point** is the temperature at which a liquid boils at 1 atm pressure.

Solids in which the particles are arranged in a regular array called a lattice are *crystalline solids.* The particles of a crystalline solid may be ions (an *ionic solid*), molecules (a *molecular solid*), covalently bonded atoms (a *covalent-network solid*) or metal atoms (a *metallic solid*). Solids in which the particles are arranged randomly are *amorphous solids.*

Fusion (melting) is the change of state in which a solid becomes a liquid. The change of state in the opposite direction, liquid to solid, is *solidification* (freezing). The temperature at which fusion occurs is the **melting point** of a solid, and the amount of heat needed to melt one mole of a solid is that solid's **molar heat of fusion.**

Key Terms

body-centered cubic (bcc) (7.8)
boiling point (7.7)
condensation (7.7)
crystal lattice (7.8)
dipole forces (7.3)
dispersion force (7.5)
distillation (7.7)
face-centered cubic (fcc) (7.8)
hydrogen bond (7.4)
intermolecular forces (page 175)
melting point (7.9)
molar heat of fusion (7.9)
molar heat of vaporization (7.7)
normal boiling point (7.7)
sublimation (7.9)
surface tension (7.6)
unit cell (7.8)
vaporization (7.7)
viscosity (7.6)

Problems

Some General Considerations

1. How do liquids and solids differ from gases in their compressibility, spacing of molecules, and intermolecular forces?
2. List four types of interactions between particles in the liquid and solid states. Give an example of each type.
3. Why does it take longer to boil an egg at high altitude than at sea level? Does it take longer to fry an egg at high altitude? Explain.
4. Why does steam at 100 °C cause more severe burns than liquid water at the same temperature?
5. In which process is energy absorbed by the material undergoing the change of state?
   a. melting or freezing
   b. condensation or vaporization
6. Label each arrow with the term that correctly identifies the process described.

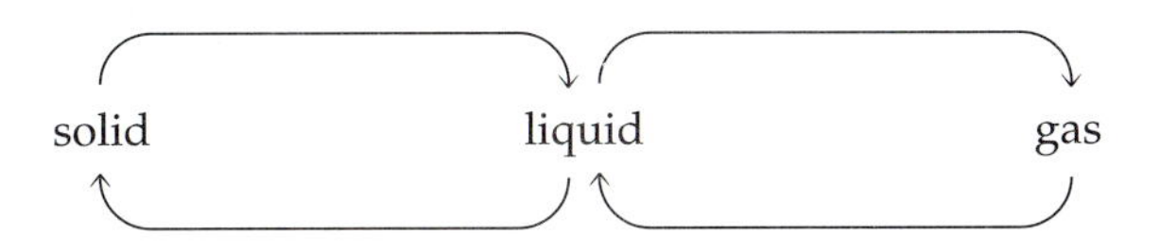

Intermolecular Forces

7. Which of the following would you expect to have the lower boiling point: carbon disulfide (CS_2) or carbon tetrachloride (CCl_4)? Why?
8. Which of the following would you expect to have the lower boiling point: phosphine (PH_3) or arsine (AsH_3)? Why?
9. Which of the following would you expect to have the higher boiling point: propane (C_3H_8) or ethanol (C_2H_5OH)? Why?
10. Which of the following would you expect to have the higher boiling point: water (H_2O) or carbon monoxide (CO)? Why?
11. Arrange the following substances in the expected order of increasing boiling point: H_2S, H_2Se, H_2Te. Give the reasons for your ranking.
12. Arrange the following substances in the expected order of increasing boiling point: H_2O, HCl, CH_4. Give the reasons for your ranking.

Heat of Vaporization

13. The heat of vaporization of bromine (Br_2) is 45 cal/g. What is the molar heat of vaporization of bromine in kilocalories per mole?

14. The heat of vaporization of ammonia (NH_3) is 327 cal/g. What is the molar heat of vaporization of ammonia in kilojoules per mole?

15. The molar heat of vaporization of acetic acid ($C_2H_4O_2$) is 5.81 kcal/mol. How many kilocalories of heat are required to vaporize 1.00 g of acetic acid?

16. The molar heat of vaporization of acetone (C_3H_6O) is 7.23 kcal/mol. How many kilojoules of heat are required to vaporize 5.80 g of acetone?

17. How many kilocalories of heat are required to raise the temperature of 25.0 g of H_2O from 18.0 to 60.0 °C? The specific heat of H_2O(l) is 1.00 cal/(g·°C).

18. How many kilojoules of heat are released when 42.4 g of CH_3OH(l) at 55.0 °C is cooled to 15.5 °C? The specific heat of CH_3OH(l) is 2.5 J/(g·°C).

Heat of Fusion

19. The molar heat of fusion of acetone (C_3H_6O) is 2.58 kcal/mol. How many kilocalories of heat are required to melt 7.75 g of acetone?

20. The heat of fusion of silver is 25 cal/g. Calculate the molar heat of fusion of Ag.

21. How many kilocalories of heat are required to melt a 355-g block of ice? (Refer to Table 7.5.)

22. How many kilojoules of heat are released when 275 mL of water freezes at 0 °C? (Refer to Table 7.5.)

Additional Problems

23. What is the value of the vapor pressure of a liquid at the normal boiling point of that liquid?

24. What types of interactions exist between molecules of each of the following?
a. H_2 **b.** NO **c.** HF

25. What is the difference in meaning of the terms polar molecule and polarizability of a molecule?

26. Of the substances NI_3, BF_3, and CH_3CH_3, only one is a gas at STP. Which do you think it is, and why?

27. Describe the principal features of the crystal lattices known as simple cubic, bcc, and fcc.

28. The following data are given for CCl_4: normal melting point, −23 °C; normal boiling point, 77 °C; density of liquid, 1.59 g/mL; heat of fusion, 0.78 kcal/mol. How much heat must be absorbed to convert 10.0 g of solid CCl_4 to liquid at −23 °C?

29. In which of the following compounds would hydrogen bonding be an important intermolecular force?

```
a. H—S                    H
      \                   |
       H            b. H—C—N—H
                          |  |
                          H  H

      H                   H
      |                   |
c. H—C—F            d. H—C—O
      |                   |  \
      H                   H   H

      H  H                H     H
      |  |                |     |
e. H—C—C—H          f. H—C—O—C—H
      |  |                |     |
      H  H                H     H
```

30. There is a rule of thumb that says that for many liquids the molar heat of vaporization is approximately 21 times the normal boiling point in kelvins. Use this rule to calculate the molar heat of vaporization for benzene, which has a boiling point of 353 K. How does your calculated value compare with the experimental value given in Table 7.3?

31. The specific heat of silver is 0.06 cal/(g·°C). That of gold is 0.03 cal/(g·°C). Which metal will be hotter (that is, will be at a higher temperature) if both absorb the same amount of heat?

32. The molar heats of vaporization of ethanol (C_2H_6O) and ethyl acetate ($C_4H_8O_2$) are 9.39 kcal/mol and 7.77 kcal/mol, respectively. Which liquid has the stronger intermolecular forces?

33. How much energy will be expended in changing 100 g of ice at −5 °C to steam at 100 °C?

34. How much heat is required to convert 10.0 g of ice at −12.0 °C to steam at 130 °C?

35. To obtain water each day, a bird in winter eats 5 g of snow at 0 °C. How many kcal (food Calories) of energy does it take to melt this snow and warm the liquid to the bird's body temperature of 40 °C?

36. To obtain water on a winter hike, a woman decides to eat snow. How many extra kilocalories (food Calories) of food would she have to take in each day to raise the 1500 g of snow that she needs from −10 °C to 0 °C, melt it, then raise the liquid water from 0 °C to her body temperature of 37 °C?

Solutions

Fish, plants, and other marine organisms reside within a water solution that contains all the necessary ingredients to support life.

Learning Objectives/Study Questions

1. What is a solution? What are the different types of solutions?
2. What is the difference between a soluble substance and a miscible substance?
3. What is the difference between a saturated solution and a supersaturated solution?
4. What factors determine the solubility of ionic compounds? Of covalent compounds?
5. How do temperature and pressure affect the solubility of gases in water?
6. What are different ways of expressing the concentration of a solution? When is each used?
7. What is a colligative property? How do colligative properties explain the use of antifreeze in radiators or of salt to melt snow and ice?
8. What are osmosis and osmotic pressure? What is their relevance to human biology and to medicine?
9. What are colloids? How do the properties of a colloid compare to those of a solution or suspension? Give some examples of colloids.
10. What is dialysis? What is its relevance to medicine?

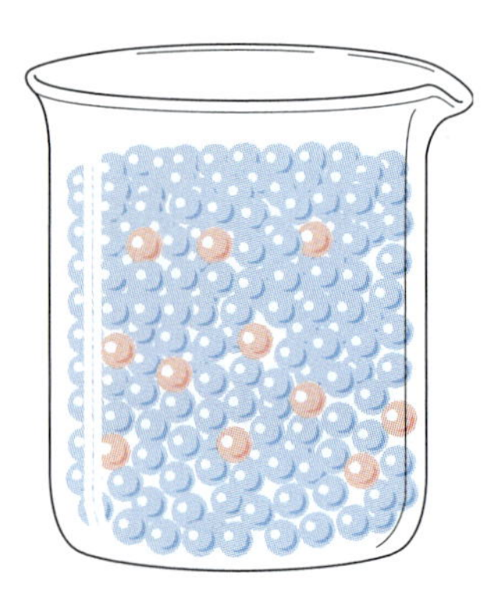

▲ **Figure 8.1**
In a solution, the solute molecules (pink) are randomly distributed among the solvent molecules (blue).

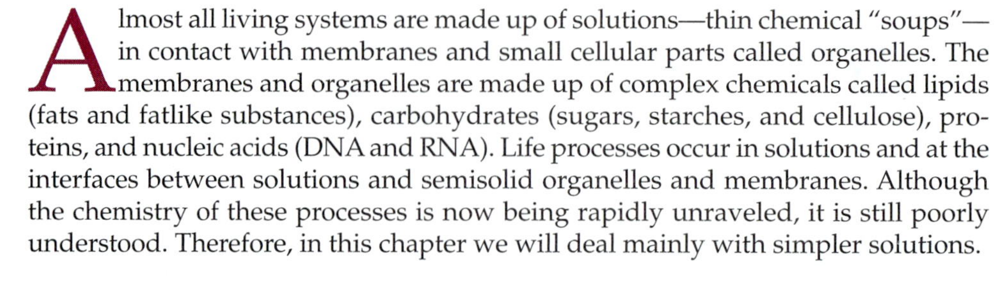

Almost all living systems are made up of solutions—thin chemical "soups"—in contact with membranes and small cellular parts called organelles. The membranes and organelles are made up of complex chemicals called lipids (fats and fatlike substances), carbohydrates (sugars, starches, and cellulose), proteins, and nucleic acids (DNA and RNA). Life processes occur in solutions and at the interfaces between solutions and semisolid organelles and membranes. Although the chemistry of these processes is now being rapidly unraveled, it is still poorly understood. Therefore, in this chapter we will deal mainly with simpler solutions.

8.1 Some Types of Solutions

Put a teaspoonful of sugar in a cup of water and stir until the sugar has all dissolved. Taste the sweetened water from one side of the cup and then from the other. Use a straw to taste it from the center and the bottom. If you mixed it thoroughly, the water has the same degree of sweetness throughout; it is *homogeneous.* You could add more sugar to make the solution sweeter or less sugar to make it less sweet. You could boil away the water and recover the solid sugar because the water and sugar have not reacted chemically.

A **solution** is a homogeneous mixture of two or more substances. A solution of sugar in water does not consist of tiny grains of solid sugar dispersed among droplets of liquid water. Rather, individual sugar molecules are randomly distributed among water molecules in much the same way that pink marbles can be distributed among blue ones. Because a solution is *homogeneous,* the composition and physical and chemical properties are identical in all parts of it.

The components of a solution are one or more **solutes,** the substance(s) being dissolved, and the **solvent,** the substance doing the dissolving (Figure 8.1). The solute is usually the component present in lesser quantities, and the solvent is usually present in the greatest quantity. There are many solvents: Hexane dissolves grease. Ethanol dissolves many drugs. Isopentyl acetate, a component of banana oil, is a solvent for the glue used in making model airplanes. Water is no doubt the most familiar solvent, dissolving as it does many common substances such as sugar, salt, and ethanol. A solvent need not be a liquid. Air is a solution of O_2, argon, water vapor, and other gases in $N_2(g)$. Steel is a solution of carbon in iron—a solid in a solid. The most common types of solutions are listed in Table 8.1. We focus our discussion in this chapter on **aqueous solutions,** those in which water is the solvent.

▲ Carbonated water is a solution of $CO_2(g)$ in water. The carbonated water is bottled under pressure; when the bottle is opened, bubbles of $CO_2(g)$ escape.

✓ Review Question

8.1 Define or explain—and, where possible, illustrate—the following terms.
a. solution **b.** solvent **c.** solute **d.** aqueous solution

Table 8.1 Types of Solutions

Solute	Solvent	Solution	Example
Gas	Gas	Gas	Air (O_2 in N_2)
Gas	Liquid	Liquid	Club soda (CO_2 in H_2O)
Liquid	Liquid	Liquid	Wine (alcohol in H_2O)
Solid	Liquid	Liquid	Saline solution (NaCl in H_2O)
Solid	Solid	Solid	14-karat gold (Ag in Au)

8.2 Qualitative Aspects of Solubility

We say that sugar is *soluble* in water. Just what does this mean? Can we dissolve a teaspoonful of sugar in a cup of water? Can we dissolve 10 teaspoonfuls, or 100 teaspoonfuls? We know from everyday experience that there is a limit to the quantity of sugar we can dissolve in a given volume of water. Nevertheless, we still find it convenient to say that sugar is soluble in water because an appreciable quantity dissolves.

Some substances, such as water and alcohol, can be mixed in all proportions. We say that such substances are completely **miscible.** As with sugar, however, most "soluble" substances are limited in the quantity that will dissolve in a given solvent. For others, which we call *insoluble,* that limit is near zero. Put an iron nail in a beaker of water. There is no apparent change. We say that iron is insoluble in water. Even insolubility is relative, however. A method sensitive enough would show that some iron had dissolved. The quantity might well be regarded as insignificant, and that is the sense in which the term *insoluble* is used. Such terms as "soluble" and "insoluble" are useful, but they are imprecise and must be used with care.

Two other imprecise but sometimes useful terms are "dilute" and "concentrated." A **dilute solution** is one that contains a little bit of solute in lots of solvent. A **concentrated solution** is one in which lots of solute is dissolved in a relatively small quantity of solvent. If we dissolve a few milliliters of ethylene glycol, or antifreeze (Chapter 14), in several liters of water, the dilute solution is quite "thin"—little changed in appearance from that of pure water. However, if we dissolve 1 L of ethylene glycol in 1 L of water, the concentrated solution is rather syrupy—similar to pure ethylene glycol in consistency.

The terms "dilute" and "concentrated" are sometimes used in a quantitative way for solutions of acids and bases. We specify their meanings in this context in Section 10.1.

Review Question

8.2 Define or explain—and, where possible, illustrate—the following terms.

a. soluble **b.** insoluble **c.** miscible
d. dilute solution **e.** concentrated solution

8.3 Solubility of Ionic Compounds

In Chapter 7, we saw that the unique structure of water results in relatively strong forces between water molecules. Further, we examined the different types of forces that exist between identical molecules in pure liquids. Now let's look at the types of forces that exist between the solute and solvent molecules in solutions. The solubility of a given solute depends on the relative attraction between particles in the pure substances and in the solution.

Water is a good solvent for compounds of the Group 1A elements: Almost all their compounds are soluble in water. Examples are sodium chloride (NaCl), sodium sulfate (Na_2SO_4), potassium phosphate (K_3PO_4), and lithium bromide (LiBr). Further, nearly all nitrate salts are soluble, as are compounds that contain the ammonium ion. Silver nitrate ($AgNO_3$), mercury(II) nitrate [$Hg(NO_3)_2$], aluminum nitrate [$Al(NO_3)_3$], ammonium chloride (NH_4Cl), ammonium sulfate [$(NH_4)_2SO_4$], and ammonium phosphate [$(NH_4)_3PO_4$] are examples. Why do these compounds dissolve in water? In essence, three things must happen. The attractive forces holding the ions of the solute together must be overcome. Similarly, the attractive forces holding at least some of the water molecules together must be overcome. Finally,

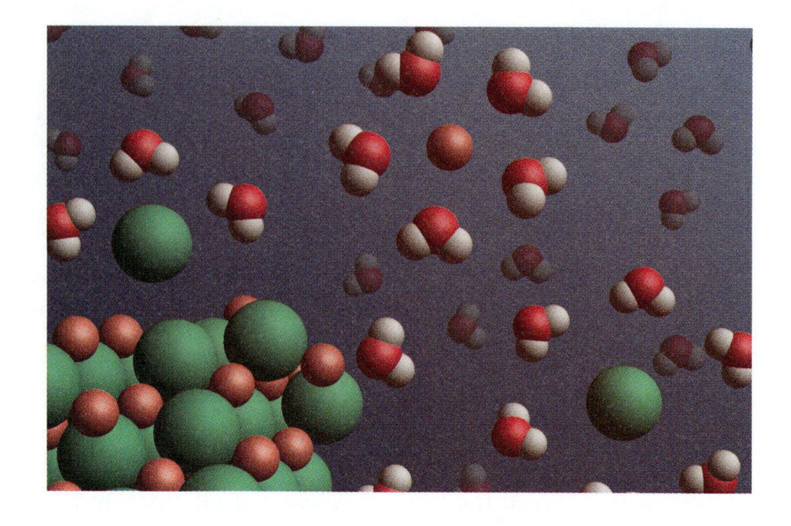

Figure 8.2 ▶
Ion-dipole forces in the dissolving of an ionic crystal. Water dipoles attract ions in the crystal lattice, causing them to enter the solution. The ion-dipole forces persist within the solution as well; the ions are *hydrated.*

the solute and solvent molecules must interact; that is, they must attract one another. *Hydration* is the process in which water molecules surround the solute ions.

The polarity of water molecules enables them to attract (and be attracted by) ions. They align with the positive ends of their dipoles pointing toward negative ions (anions) and the negative ends pointing toward the positive ions (cations). Still, the attraction between a dipole and an ion is not as strong as that between two ions. To compensate for their weaker attractive power, several water molecules surround each ion, and in this way the many *ion-dipole* interactions overcome the *ion-ion* interactions (Figure 8.2).

In some solids, the forces holding the ions together are so strong that they cannot be overcome by the hydration of the ions. Many solids in which both ions are doubly or triply charged are essentially insoluble in water. Examples are calcium carbonate (Ca^{2+} and CO_3^{2-}), aluminum phosphate (Al^{3+} and PO_4^{3-}), and barium sulfate (Ba^{2+} and SO_4^{2-}). The large electrostatic forces between the ions hold the particles together despite the attraction of solvent molecules. Table 8.2 summarizes the solubilities of some common ionic compounds.

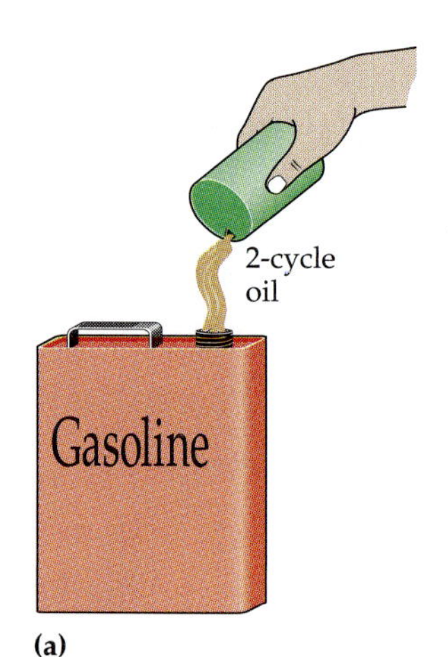

(a)

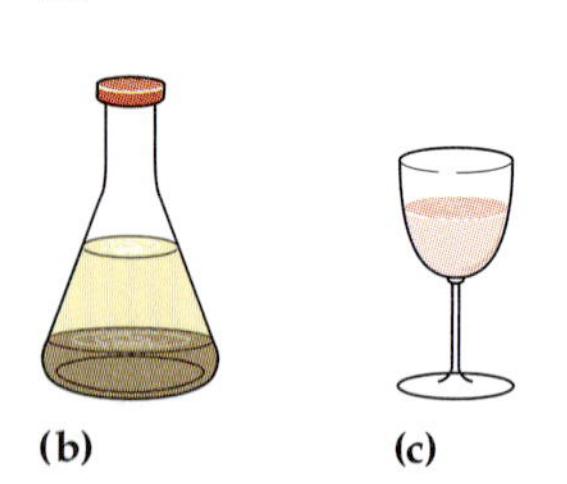

(b) (c)

▲ **Figure 8.3**
"Like dissolves like." (a) Lawn mowers with two-cycle engines are fueled and lubricated with a solution of nonpolar lubricating oil in nonpolar gasoline. (b) In Italian dressing, polar vinegar and nonpolar olive oil are mixed. The two liquids do not form a solution; they separate on standing. (c) Wine is a solution of polar ethyl alcohol in polar water.

✓ **Review Question**

8.3 Explain the interactions that occur when NaCl is dissolved in water.

8.4 Solubility of Covalent Compounds

An old but helpful chemical rule states that *like dissolves like.* This means that nonpolar (or slightly polar) solutes dissolve best in nonpolar solvents, and polar solutes dissolve best in polar solvents (Figure 8.3). The rule works well for nonpolar substances. Fats, oils, and greases (nonpolar or only slightly polar) dissolve well in nonpolar solvents such as toluene, C_7H_8. The forces that hold nonpolar molecules together are generally weak. Thus a small quantity of energy suffices to pull apart molecules of pure solute and to disrupt the attractive forces between molecules of pure solvent. This energy can be balanced by the energy released through the interaction of solute and solvent molecules, although this too is slight.

Water solubility of covalent compounds depends mainly on the ability of water to form *hydrogen bonds* to the solute molecules. Therefore, molecules containing a high proportion of nitrogen or oxygen atoms will dissolve in water because these are the usual elements that can form hydrogen bonds (Section 7.4). One example is methyl alcohol (CH_3OH), which is completely miscible with water. Methylamine

Table 8.2 General Rules for the Water Solubilities of Common Ionic Compounds

Compounds that are *soluble:*

Nitrates and acetates.

Alkali metal (Group 1A) salts and ammonium salts.

Compounds that are mostly *soluble:*

Chlorides, bromides, and iodides, *except* for those of Pb^{2+}, Ag^{+}, and Hg_2^{2+}.

Sulfates, *except* for those of Sr^{2+}, Ba^{2+}, Pb^{2+}, and Hg_2^{2+}. ($CaSO_4$ is slightly soluble.)

Compounds that are mostly *insoluble:*

Carbonates, hydroxides, phosphates, and sulfides, *except* for ammonium compounds and those of the Group 1A metals. (The hydroxides and sulfides of Ca^{2+}, Sr^{2+}, and Ba^{2+} are slightly to moderately soluble.)

(CH_3NH_2) is also quite soluble in water. Figure 8.4 gives the structures of these molecules (and others mentioned in this section) and shows how they interact with water by forming hydrogen bonds.

There need not be a hydrogen atom on the oxygen (or nitrogen) atom of the solute molecules. Acetaldehyde (CH_3CHO) is completely miscible with water, even though none of the hydrogen atoms in the molecule is attached to the oxygen (see Figure 8.4). The hydrogen bonds depicted in the drawing incorporate hydrogen atoms covalently bonded to the oxygen of the water molecules. We will have many occasions to discuss the importance of hydrogen bonding in later chapters of this text.

Some fairly complex molecules, such as those of the sugars, are quite soluble in water. Glucose ($C_6H_{12}O_6$) contains six carbon atoms, but its six oxygen atoms permit it to hydrogen bond to many water molecules, thereby making it water-soluble.

Hydrogen bonding, then, is the important factor in water solubility. Polarity alone is not enough. Methyl chloride (CH_3Cl) and methyl alcohol (CH_3OH) have about the same polarity, yet methyl chloride is essentially insoluble in water while methyl alcohol is completely miscible with water. Methyl chloride does not engage in hydrogen bonding, and methyl alcohol does. A few polar substances, such as hydrogen chloride (HCl), dissolve in water because they react chemically to form ions; these are discussed in subsequent chapters.

Alcohols:

CH_3OH	$CH_3(CH_2)_2CH_2OH$	$CH_3(CH_2)_{10}CH_2OH$
Methyl alcohol (1 C, 1 O; soluble in water)	Butyl alcohol (4 C, 1 O; slightly soluble in water)	Lauryl alcohol (12 C, 1 O; essentially insoluble in water)

Amines:

CH_3NH_2	$CH_3(CH_2)_2CH_2NH_2$	$CH_3(CH_2)_{10}CH_2NH_2$
Methylamine (1 C, 1 N; soluble in water)	Butylamine (4 C, 1 N; slightly soluble in water)	Laurylamine (12 C, 1 N; essentially insoluble in water)

Miscellaneous compounds:

CH_3CHO	$CH_3(CH_2)_2CH_3NH_2$	$C_{12}H_{22}O_{11}$
Acctaldehyde (1 C, 1 O; soluble in water)	Butane (4 C, 0 O; insoluble in water)	Sucrose (12 C, 11 O; soluble in water)

Figure 8.4
Nitrogen and oxygen atoms in covalent molecules can accept hydrogen bonds from water molecules.

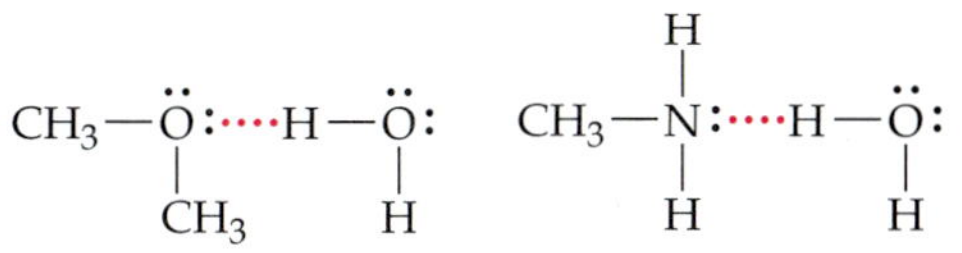

Covalent molecules with O—H or N—H groups can donate hydrogen bonds to water molecules.

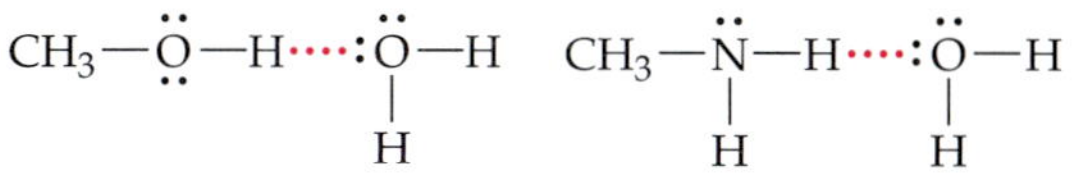

Molecules with four or fewer C atoms per N or O atom are usually soluble in water.

Terminology of Aqueous Systems

The importance of water as a solvent is reflected in the number of terms that have been coined to describe systems involving water. For example, the general term for the interaction of solvent with solute is *solvation,* but there is a special term for the interaction of water with a solute: *hydration.*

Certain compounds, such as calcium sulfate, tend to hold on to some water molecules even when they crystallize from solution. These compounds with their bound water molecules are called **hydrates.** The formulas for hydrates are written in a way that indicates the number of attached water molecules. Plaster of Paris is $(CaSO_4)_2 \cdot H_2O$ (one water molecule for every two calcium sulfate units). If more water is available, $CaSO_4 \cdot 2H_2O$ is formed (two water molecules for every $CaSO_4$ unit). When a plaster cast is formed to immobilize a broken bone, the first hydrate is converted to the second. The powdery plaster of Paris changes to the rigid, protective material of the cast.

If a hydrate is heated strongly enough, the bound water can be driven off to produce the **anhydrous** compound—that is, the compound without water. Some hydrates lose their bound water simply on standing in dry air. Such compounds are said to be **efflorescent.** Other compounds form hydrates by picking up water from the atmosphere. These are described as **hygroscopic.** And finally, some compounds are so good at pulling water molecules from the air that they eventually dissolve in the water thus accumulated. These compounds are said to be **deliquescent.**

▶ Anhydrous copper(II) sulfate is white, but copper(II) sulfate pentahydrate is a brilliant blue.

Review Questions

8.4 Benzene, C_6H_6, is a nonpolar solvent. Motor oil is a nonpolar mixture. Would you expect (**a**) NaCl to dissolve in benzene? (**b**) motor oil to dissolve in water? (**c**) motor oil to dissolve in benzene? Explain.

8.5 Ethyl alcohol, CH_3CH_2OH, is soluble in water; ethyl chloride, CH_3CH_2Cl, is insoluble in water. Explain.

8.6 Explain the difference between: (**a**) hygroscopic and deliquescent; (**b**) hydrate and anhydrous; (**c**) hygroscopic and efflorescent.

8.5 Dynamic Equilibria

For most substances, there is a limit to how much solute can be dissolved in a given volume of solvent. This limit, called the solubility of particular solute, varies with the nature of the solute and the solvent. Solubilities are often expressed in terms of grams of solute per 100 g of solvent. Since solubility varies with temperature, it is necessary to indicate the temperature at which the solubility is measured. For example, at 20 °C, 100 g of water will dissolve up to 109 g of sodium hydroxide (NaOH). At 50 °C, 145 g of NaOH will dissolve in 100 g of water. In a shorthand method, the solubility of sodium hydroxide is expressed as 109^{20} and 145^{50} (the 100 g of water is understood).

The solubility of sodium chloride, or common table salt (NaCl), is 36 g per 100 g of water at 20 °C. Suppose we place 40 g of NaCl in 100 g of water. What happens? Initially, many of the sodium (Na^+) ions and chloride (Cl^-) ions leave the surface of the crystals and wander about at random through the solvent. Some of the ions in their wanderings return to a crystal surface where they can be trapped, becoming once more a part of the crystal lattice. As more and more salt dissolves, an increasing number of "wanderers" return and are trapped again in the solid state. Eventually (when 36 g of NaCl has dissolved), the number of ions leaving the surface of the undissolved crystals just equals the number returning. A condition of **dynamic equilibrium** is established. The *net* quantity of sodium chloride in solution remains the same despite the fact that there is still a lot of activity as ions come and go from the surface of the crystals. The net quantity of undissolved crystals also remains constant (in this example, 4 g), although individual crystals may change in shape and size as ions leave one part of the crystal to enter solution while others are deposited at another part of the lattice. Some small crystals may disappear as others grow larger, yet the net quantity of undissolved salt does not change. The rate of dissolution just equals the rate of regrowth.

A solution that contains all the solute that it can at equilibrium and at a given temperature is said to be a **saturated solution.** One that contains less than this quantity is an **unsaturated solution.** A solution with 24 g of NaCl in 100 g of water at 20 °C is unsaturated because it could dissolve 12 g more at that temperature.

Equilibrium is established at a given temperature. If the temperature changes, solute will dissolve or separate until equilibrium is established at the new temperature. Consider once more a sodium hydroxide solution. If we add 145 g of NaOH to 100 g of water at 20 °C, 109 g of the NaOH will dissolve, leaving 36 g as undissolved solute. If the solution is then warmed, more solute will dissolve. Finally, at 50 °C all 145 g of NaOH is in solution.

Most solid compounds are increasingly soluble as the temperature is raised (Figure 8.5). This should not be surprising. As the temperature goes up, the kinetic energy of all the particles is increased. More ions are knocked loose from the lattice and go into solution. Further, it is more difficult for the crystal to recapture the ions that return to its surface, because they are moving at higher speeds. There are

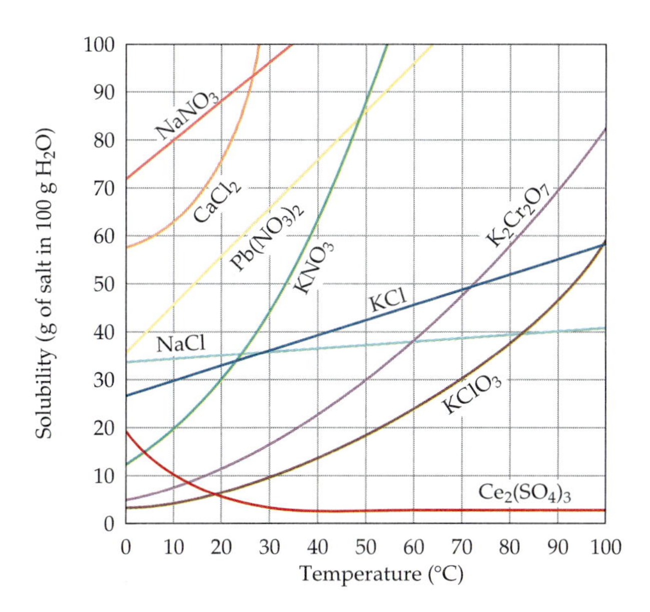

Figure 8.5
The solubilities of several salts in water as a function of temperature.

▲ **Figure 8.6**
Seeding a supersaturated solution. Addition of a seed crystal induces rapid crystallization of excess solute from a supersaturated solution.

a few exceptions to this general rule of increased solubility at higher temperatures. Note that the solubility of NaCl changes little over the indicated range of temperatures, and the solubility of $Ce_2(SO_4)_3$ decreases.

If a saturated solution (with excess solid solute present) is cooled, more solute precipitates until the equilibrium is once again established at the lower temperature. For example, consider a saturated solution of lead nitrate at 90 °C. Each 100 g of water has 120 g of $Pb(NO_3)_2$ dissolved. When the solution is cooled to 20 °C, the solution at equilibrium can contain only 54 g of $Pb(NO_3)_2$. The excess, 66 g, will separate as a **precipitate,** an insoluble or nearly insoluble solid that separates from a solution. This increases the quantity of undissolved solute.

Now consider what would happen if one started to cool a saturated solution of lead nitrate with *no* excess solute present. Would lead nitrate precipitate? It might. Then again, it might not. There is no equilibrium—no crystals to capture the wandering ions. One might well be able to cool the solution to 20 °C without precipitation. Such a solution, containing solute in excess of what it could contain if it were at equilibrium, is said to be a **supersaturated solution.** This system is not stable because it is not at equilibrium. Solute may precipitate when the solution is stirred or if the inside of the container is scratched with a glass rod. Addition of a "seed" crystal of solute will nearly always result in the sudden precipitation of all the excess solute. Equilibrium is established rather rapidly when there is a crystal to which the ions can attach themselves (Figure 8.6).

Review Questions

8.7 Explain the differences between an unsaturated solution, a saturated solution, and a supersaturated solution.

8.8 Precipitation is induced in a supersaturated solution by the addition of a seed crystal. When no more solid crystallizes, is the solution saturated, unsaturated, or supersaturated? Explain.

8.6 Solutions of Gases in Water

Solutions of gases in water are more common than you might think. First, there is the familiar case of carbonated beverages, solutions of $CO_2(g)$ in water, often with added flavors and sweeteners. Other examples include blood, which contains dissolved $O_2(g)$ and $CO_2(g)$; formalin, an aqueous solution of formaldehyde gas

Some Supersaturated Solutions in Nature

Supersaturated solutions are not just laboratory curiosities; they occur naturally. Jelly is one example; the principal solute in many jellies is sucrose (cane sugar). If jelly is left to stand, the sucrose crystallizes. We say, not very scientifically, that the jelly has "turned to sugar." Supersaturated sugar solutions are fairly common in foods. Honey is another example. Sugar often crystallizes from honey that has been standing for a long time.

Some wines have high concentrations of potassium hydrogen tartrate, $KHC_4H_4O_6$. When chilled, the solution may become supersaturated. After standing for some time, crystals may form and settle out. Modern wineries solve this problem—and render the wine less acidic—by a process known as cold stabilization. The wine is chilled to near 0 °C, a temperature below that commonly found in refrigerators. Tiny seed crystals of $KHC_4H_4O_6$ are added to the supersaturated wine. When crystallization is complete, the excess crystals are filtered off. At one time, winemaking was the principal source of potassium hydrogen tartrate, the "cream of tartar" used in baking.

▲ Jelly is a supersaturated solution of sugars. Upon standing for a long time, the sugars sometimes crystallize.

(HCHO) used as a biological preservative; and a variety of household cleaners that are aqueous solutions of $NH_3(g)$. In addition, *all* natural waters contain dissolved $O_2(g)$ and $N_2(g)$ and traces of other gases.

Unlike most solid solutes, gases generally become *less* soluble in water as the temperature increases. The gas molecules acquire more kinetic energy as the temperature increases and are "driven off" from solution.

The water solubility of oxygen at atmospheric pressure and 20 °C is only about 0.0043 g/100 g H_2O, but even this limited quantity is essential to aquatic life. Fish depend on dissolved air for $O_2(g)$. Moreover, the fact that the solubility *decreases* with temperature (Figure 8.7) explains why many fish (trout, for example) cannot survive in warm water—they don't get enough oxygen. At 30 °C, the quantity of dissolved $O_2(g)$ in water is only about one-half of what is found at 0 °C.

The world's major ocean fisheries are located in *cold* regions such as the Bering Sea and the Grand Banks of Newfoundland.

At constant temperature the solubility of a gas in water is directly proportional to the pressure of the gas in equilibrium with the aqueous solution (Section 6.9). The higher the pressure, the more gas will dissolve in a given volume of water. Figure 8.8 shows how the solubility of oxygen in water at 25 °C varies with pressure. A moderate pressure of $CO_2(g)$ above the beverage in a soft drink bottle keeps a lot of the gas dissolved in the water. When the bottle is opened, this pressure is released and dissolved $CO_2(g)$ escapes, causing the familiar fizzing.

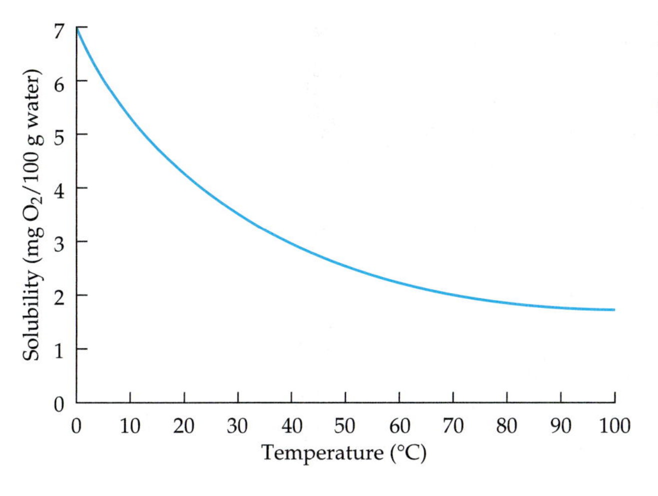

◀ **Figure 8.7**
The solubility of oxygen gas at 1 atm pressure as a function of temperature.

Figure 8.8 ▶
The solubility of oxygen gas at 25 °C as a function of pressure.

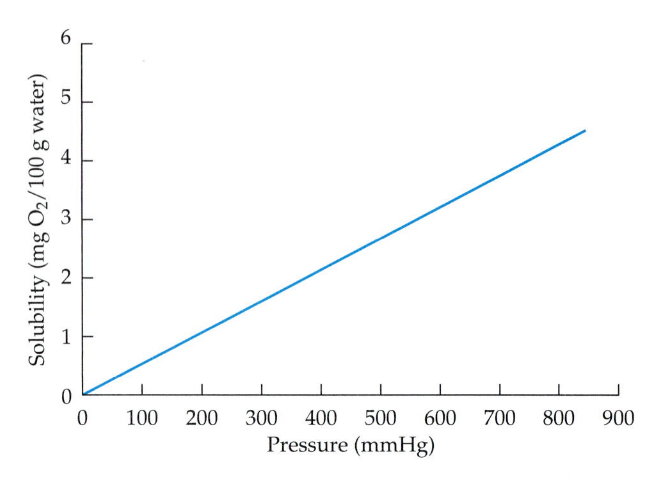

✓ Review Questions

8.9 When the cap is removed from a bottle of soda pop, carbon dioxide gas escapes. Explain why this occurs.

8.10 Why are the world's major ocean fisheries located in cold areas?

8.7 Solution Concentrations

Most reactions of interest to us, including those in our bodies, take place in solution. A good cook may well get by with concentrations expressed as "a pinch of salt in a pint of water," but scientific work generally requires more precise measurement of quantities. We have already discussed one quantitative unit: grams of solute per 100 g of solvent.

Molarity

The concentration unit that chemists use most is *molarity.* Substances enter into chemical reactions according to *molar* ratios. For reactions involving solutions, the amount of solute is usually measured in moles. The quantity of solution is usually measured in liters or milliliters. The **molarity (M)** is the amount of solute, in moles, per liter of solution.

$$\text{Molarity (M)} = \frac{\text{moles of solute}}{\text{liters of solution}}$$

Example 8.1

Calculate the molarity of a solution made by dissolving 3.50 mol of NaCl in enough water to produce 2.00 L of solution.

Solution

$$\text{Molarity (M)} = \frac{\text{moles of solute}}{\text{liters of solution}} = \frac{3.50\text{ mol NaCl}}{2.00\text{ L solution}} = 1.75\text{ M NaCl}$$

We read "1.75 M NaCl" as "1.75 molar NaCl."

Practice Exercises

A. Calculate the molarity of a solution that has 0.0500 mol of NH_3 in 5.75 L of solution.

B. Calculate the molarity of a solution made by dissolving 0.750 mol of H_3PO_4 in enough water to produce 775 mL of solution.

We cannot determine moles of a substance directly; we usually work with a given mass and divide by the molar mass of the substance, as illustrated in Example 8.2.

Example 8.2

What is the molarity of a solution in which 333 g of potassium hydrogen carbonate ($KHCO_3$) is dissolved in enough water to make 10.0 L of solution?

Solution

First, prepare the setup that would convert from mass of $KHCO_3$ to moles of $KHCO_3$.

$$333 \cancel{\text{g KHCO}_3} \times \frac{1 \text{ mol KHCO}_3}{100.1 \cancel{\text{g KHCO}_3}} = 3.33 \text{ mol KHCO}_3$$

Now use this value as the numerator in the defining equation for molarity. The solution volume, 10.0 L, is the denominator.

$$\text{Molarity} = \frac{3.33 \text{ mol KHCO}_3}{10.0 \text{ L solution}} = 0.333 \text{ M KHCO}_3$$

Practice Exercise

Calculate the molarity of each of the following solutions.

a. 18.0 mol of H_2SO_4 in 2.00 L of solution

b. 3.00 mol of KI in 2.39 L of solution

c. 0.206 mol of HF in 752 mL of solution

d. 0.522 g of HCl in 0.592 L of solution

e. 4.98 g of $C_6H_{12}O_6$ in 224 mL of solution

f. 10.5 g of C_2H_5OH in 24.7 mL of solution

Frequently we need to know the *mass* of solute required to prepare a given volume of solution of a given molarity. In such calculations we can use molarity as a conversion factor between moles of solute and liters of solution. Thus, in Example 8.3, "6.67 M NaOH" means 6.67 mol of NaOH per liter of solution, expressed as the conversion factor

$$\frac{6.67 \text{ mol NaOH}}{1 \text{ L solution}}$$

Example 8.3

How many grams of NaOH are required to prepare 0.500 L of 6.67 M NaOH?

Solution

First we calculate the moles of NaOH.

$$0.500 \cancel{\text{L solution}} \times \frac{6.67 \text{ mol NaOH}}{1 \cancel{\text{L solution}}} = 3.34 \text{ mol NaOH}$$

Then we use the molar mass to calculate the grams of NaOH.

$$3.34 \cancel{\text{mol NaOH}} \times \frac{40.01 \text{ g NaOH}}{1 \cancel{\text{mol NaOH}}} = 133 \text{ g NaOH}$$

Practice Exercise

How many grams of potassium hydroxide are required to prepare each of the following solutions?

a. 2.00 L of 6.00 M KOH

b. 100.0 mL of 1.00 M KOH

c. 10.0 mL of 0.100 M KOH

d. 33.0 mL of 2.50 M KOH

Quite often, solutions of known molarity are available. How would you calculate the *volume* you would need in order to get a certain number of moles of solute? We can again rearrange the definition of molarity to obtain

$$\text{Liters of solution} = \frac{\text{moles of solute}}{\text{molarity}}$$

Example 8.4

How many liters of 12.0 M HCl solution would one have to take to get 0.425 mol of HCl?

Solution

$$\text{Liters of HCl solution} = \frac{\text{moles of solute}}{\text{molarity}} = \frac{0.425 \text{ mol HCl}}{12.0 \text{ M HCl}} =$$

$$\frac{0.425 \cancel{\text{mol HCl}}}{12.0 \cancel{\text{mol HCl}}/\text{L}} = 0.0354 \text{ L}$$

We would need 0.0354 L (35.4 mL) of the solution to have 0.425 mol.

Practice Exercises

A. How many liters of 15.0 M aqueous ammonia (NH_3) solution do you need to get 0.445 mol of NH_3?
B. How many milliliters of 15.0 M HNO_3 solution do you need to get 0.245 mol of HNO_3?

Remember that molarity is moles per liter of *solution,* not per liter of solvent. To make a liter of a 1 M solution, we weigh out 1 mol of solute and place it in a volumetric flask, which is a standard piece of laboratory glassware designed to contain a precisely specified volume of liquid (Figure 8.9). Enough water is added to dissolve the solute, and then more water is added to bring the volume up to the mark that indicates 1 L of *solution.* (Simply adding 1 mol of solute to 1 L of water would, in most cases, give more than 1 L of solution.)

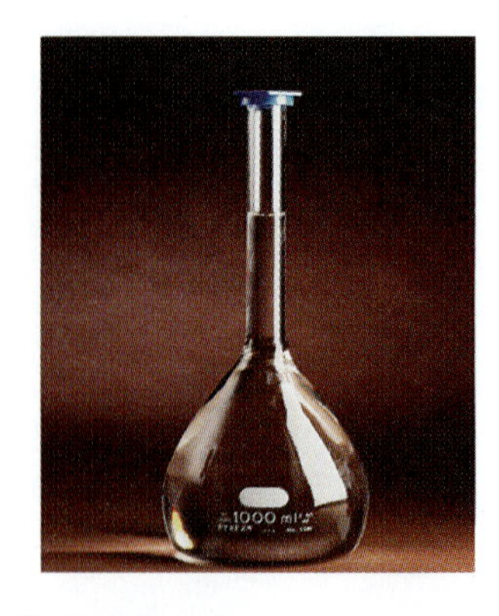

▲ **Figure 8.9**
The 1.000 M NaOH solution in this flask was made by dissolving 1.000 mol of NaOH (40.01 g of NaOH) in water, and then carefully diluting to a final volume of 1.000 L.

Percent Concentrations

For many practical applications, we often express solution compositions in percentage composition. Then, if we require a precise quantity of solution, we simply measure out a mass or volume. If both the solute and solvent are liquids, **percent by volume** is often used, because liquid volumes are so easily measured.

$$\% \text{ by volume} = \frac{\text{volume of solute}}{\text{volume of solution}} \times 100\%$$

Example 8.5

What is the percent by volume of a solution made by dissolving 235 mL of ethanol in enough water to make exactly 500 mL of solution?

Solution

$$\% \text{ by volume} = \frac{235 \text{ mL ethanol}}{500 \text{ mL solution}} \times 100\% = 47.0\%$$

Practice Exercise

What is the percent by volume of a solution made by dissolving 11.7 mL of ethanol in enough water to make 25.0 mL of solution?

Ethanol used for medicinal purposes is generally of a grade referred to as USP (an abbreviation of United States Pharmacopoeia, the official publication of standards for pharmaceutical products). USP ethanol is 95% CH_3CH_2OH by volume.

Example 8.6

Describe how to make 775 mL of a 40.0% by volume solution of acetic acid.

Solution

Let's begin by rearranging the equation for percent by volume to solve for volume of solute.

$$\text{Volume of solute} = \frac{\%\text{ by volume} \times \text{volume of solution}}{100\%}$$

Substituting, we get

$$= \frac{40.0\% \times 775\text{ mL acetic acid}}{100\%} = 310\text{ mL acetic acid}$$

Take 310 mL of acetic acid and add enough water to make 775 mL of solution.

Practice Exercise

Describe how to make 67.5 mL of a 33.0% by volume solution of acetic acid.

Many commercial solutions are labeled with the concentration in **percent by mass.** For example, sulfuric acid is sold as a solution that is 35.7% H_2SO_4 for use in storage batteries, 77.7% H_2SO_4 for the manufacture of phosphate fertilizers, and 93.2% H_2SO_4 for pickling steel. Each of these figures is a percent by mass: 35.7 g of H_2SO_4 per 100 g of sulfuric acid solution, and so on.

$$\%\text{ by mass} = \frac{\text{mass of solute}}{\text{mass of solution}} \times 100\%$$

Example 8.7

What is the percent by mass of a solution of 25.0 g of NaCl dissolved in 475 g (475 mL) of water?

Solution

$$\%\text{ by mass} = \frac{\text{mass of NaCl}}{\text{mass of solution}} \times 100\%$$

The mass of the solution is 25.0 g + 475 g = 500g.

$$\%\text{ by mass} = \frac{25.0\text{ g NaCl}}{500\text{ g solution}} \times 100\% = 5.00\%\text{ NaCl}$$

Practice Exercise

What is the percent by mass of a solution of 9.40 g of H_2O_2 dissolved in 335 g (335 mL) of water?

Example 8.8

Describe how to prepare 750 g of an aqueous solution that is 2.50% NaOH by mass.

Solution

Let's begin by rearranging the equation for percent by mass to solve for mass of solute.

$$\text{Mass of solute} = \frac{\%\text{ by mass} \times \text{mass of solution}}{100\%}$$

Substituting, we get

$$\text{Mass of NaOH} = \frac{2.50\cancel{\%} \times 750\text{ g}}{100\cancel{\%}} = 18.8\text{ g NaOH}$$

The required mass of water is 750 g solution minus 18.8 g NaOH.

$$\text{Mass } H_2O = 750\text{ g solution} - 18.8\text{ g NaOH} = 731\text{ g } H_2O$$

To make 750 g of solution, weigh out 18.8 g of NaOH and add it to 731 g of water.

Practice Exercise

Describe how to prepare 275 g of an aqueous solution that is 5.50% glucose by mass.

The unit **mass/volume percent** is also used in medicine. For example, a solution of sodium chloride used in intravenous injections has the composition 0.89% (mass/vol) NaCl; that is, it contains 0.89 g of NaCl per 100 mL of solution. A volume of 100 mL—one-tenth of a liter—is also a *deciliter.* If the mass of solute is expressed in *milligrams,* the mass/volume concentration unit is *milligrams per deciliter* (mg/dL). A blood cholesterol reading of 187, for example, means 187 mg cholesterol/dL blood. The use of milligrams per deciliter avoids the sometimes cumbersome use of decimal numbers. For example, in mass/volume percent the cholesterol reading just cited is 0.187%.

Consider these comparisons. One cent in $10,000 is one ppm, and one cent in $10,000,000 is one ppb. Five dollars is about one ppt of the current U.S. national debt. A single individual in the city of San Diego represents about one ppm, and a single person in the People's Republic of China, about one ppb.

Mass/volume percent and mass/mass percent are nearly the same for *dilute aqueous* solutions, because their densities are approximately 1.00 g/mL. That is, 100 mL of such a solution weighs 100 g.

For extremely dilute solutions, concentrations are often expressed in parts per million (ppm), parts per billion (ppb), or even parts per trillion (ppt). For example, in fluoridated drinking water, fluoride ion is maintained at about 1 ppm. A typical level of the contaminant chloroform in municipal drinking water taken from the lower Mississippi River is 8 ppb. For solutions, these are mass/volume units: 1 ppm is 1 mg/L, 8 ppb is 8 mg/L, and so on.

Note that for percent concentrations, the mass of the solute needed doesn't depend on what the solute is. A 10% by mass solution of NaOH contains 10 g of NaOH per 100 g of total solution. Similarly, 10% HCl and 10% $(NH_4)_2SO_4$ and 10% $C_{110}H_{190}N_3O_2Br$ each contain 10 g of the specified solute per 100 g of solution. For *molar* solutions, however, the mass of solute in a solution of specified molarity is different for different solutes. A liter of a 0.10 M solution requires 4.0 g (0.10 mol) of NaOH, 3.7 g (0.10 mol) of HCl, 13.2 g (0.10 mol) of $(NH_4)_2SO_4$, and 166 g (0.10 mol) of $C_{110}H_{190}N_3O_2Br$.

✓ Review Question

8.11 Explain how to calculate each of the following concentrations: (**a**) percent by mass; (**b**) percent by volume; (**c**) molarity; (**d**) mass/volume percent.

Setting Environmental Standards

Current concern over clean air, the purity of drinking water, and soil contamination centers on trace quantities of potentially dangerous compounds. For example, benzene has been shown to produce leukemia symptoms in laboratory animals and humans. The U.S. Supreme Court has dealt with the question of whether the concentrations of benzene in the air breathed by workers should be limited to 10 ppm or 1 ppm. The Court decided that industries could not be required to lower the concentration from 10 to 1 ppm unless the higher concentration was *proven* to be dangerous.

As technology becomes more sophisticated, our ability to detect minute quantities of materials increases. This increase in the sensitivity of analytical techniques raises important questions for which there are no set answers. When a substance is first detected in the ppb or ppt range, is it a new contaminant in the environment, or has it been there all along at levels that were previously undetectable? What is the relationship between the level at which a substance can be detected and the level at which it is injurious to the health of individuals or the environment?

8.8 Colligative Properties of Solutions

Solutions have higher boiling points and lower freezing points than the corresponding pure solvents. The antifreeze in automobile cooling systems is there precisely because of these effects. If water alone were used as the engine coolant, it would boil away in the heat of summer and freeze in the depths of northern winters. Addition of antifreeze to the water raises the boiling point of the coolant and also prevents the coolant from freezing solid when the temperature drops below 0 °C. Salt is thrown on icy sidewalks and streets because the salt lowers the freezing point of the ice. The ice melts because the outdoor temperature is no longer low enough to maintain it as a solid.

The extent to which freezing points and boiling points are affected by solutes is related to the number of solute particles present in solution. The higher the concentration of solute particles, the more pronounced the effect. **Colligative properties** of solutions are those properties, like boiling point elevation and freezing point depression, that depend on the number of solute particles present in solution. For living systems, perhaps the most important colligative property is osmotic pressure. Osmosis is a phenomenon we shall discuss in detail in the next section.

Before we discuss osmosis, however, we must first consider a rather subtle aspect of solute concentration. See if you can answer the following questions. How many solute particles are there in 1 L of a 1 M glucose ($C_6H_{12}O_6$) solution? How many in 1 L of a 1 M sodium chloride (NaCl) solution? How many in 1 L of a 1 M

◀ Salt (sodium chloride) lowers the melting point of ice, accounting for its use on streets and sidewalks in winter in northern climates.

calcium chloride ($CaCl_2$) solution? The answer "should" be 6×10^{23}, right? All of the solutions contain 1 mol of their respective solute compounds. However, the question did not ask for the number of *formula units;* it asked for the number of *solute particles.* Glucose is a covalent compound; its atoms are all firmly tied together in molecules. In the glucose solution, each solute particle is a glucose molecule, and there *are* 6×10^{23} of these. But in the sodium chloride solution, each formula unit of NaCl consists of a separate sodium ion (Na^+) and chloride ion (Cl^-) in solution. When sodium chloride dissolves in water, individual ions are carried off into solution by solvent molecules. So 6×10^{23} formula units of NaCl produce 12×10^{23} particles in solution—6×10^{23} sodium ions plus 6×10^{23} chloride ions. Each calcium chloride unit produces three particles in solution—one calcium ion plus two chloride ions. Thus, the effect of a 1 M NaCl solution on colligative properties is twice that of a 1 M glucose solution. A calcium chloride solution has about three times the effect of a glucose solution of the same molarity.

One mole of *particles* of any solute lowers the freezing point of 1 kg of water by 1.86 °C. The *nature* of the solute is immaterial. The freezing point of 1 kg of water containing 1 mol (46 g) of C_2H_5OH freezes at −1.86 °C as does 1 kg of water containing 1 mol (60 g) of $C_2H_4(OH)_2$, and so on. However, 1 mol NaCl depresses the freezing point almost twice 1.86 °C and 1 mol $CaCl_2$ almost three times 1.86 °C.

When colligative properties (specifically osmotic pressure, Section 8.9) are discussed, concentration is often reported in terms of *osmolarity* (osmol/L). An **osmole (osmol)** is a mole of solute particles. A 1 M NaCl solution contains 2 osmol of solute per liter of solution; a 1 M $CaCl_2$ solution contains 3 osmol/L. The osmolarity of a 1 M glucose solution is 1 osmol/L. The concentration of body fluids is typically reported in milliosmols per liter (mosmol/L).

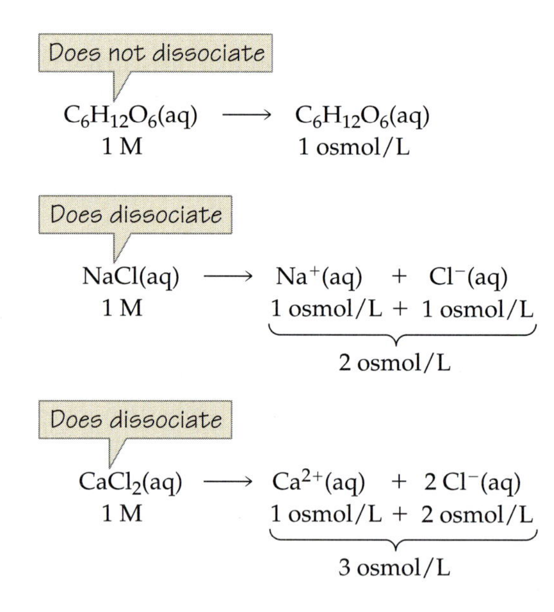

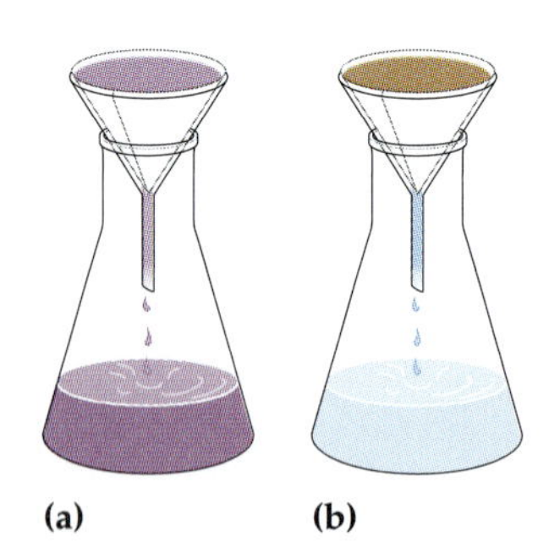

▲ (a) The aqueous solution is colored by its solute, potassium permanganate [$KMnO_4(aq)$]. After passing through the filter paper, the solution is still colored; the $K^+(aq)$ and $MnO_4^-(aq)$ ions pass through the pores of the paper. (b) A suspension of sand in water can be separated by filtration. The suspended sand particles are retained by the filter paper; a colorless filtrate passes through the paper.

✓ Review Question

8.12 Define or explain—and, where possible, illustrate—the following terms.
a. colligative property **b.** osmole **c.** osmolarity

8.9 Solutions and Cell Membranes: Osmosis

Everyday experience tells us that we can separate coffee grounds from brewed coffee by passing the mixture through filter paper. However, the filter paper does not remove the caffeine from the brewed coffee. Paper is *permeable* to water and other solvents and to solutes in solution. We also know that some materials are *impermeable* to liquids and solutions as well as to solids. Water and solutions do not pass through the metal walls of cans nor through the glass walls of beakers. Are there, perhaps, materials with intermediate properties? Materials that will pass solvent molecules but not solute molecules? Materials that are permeable to some solutes but not others? There are indeed. **Semipermeable membranes** are films of a material containing a network of submicroscopic holes or pores through which small solvent molecules can pass but which severely restrict the flow of solute particles. These may be natural materials of animal or vegetable origin, such as pig's bladder and parchment, or synthetic materials such as cellophane. Cell membranes, the lining of the digestive tract, and the walls of blood vessels are all semipermeable; they allow certain substances to go through while holding others back.

Consider two compartments separated by a semipermeable membrane. Place pure water in one compartment and a sugar solution in the other. The volume of liquid in the compartment containing sugar will increase, and the volume in the pure water compartment will decrease.

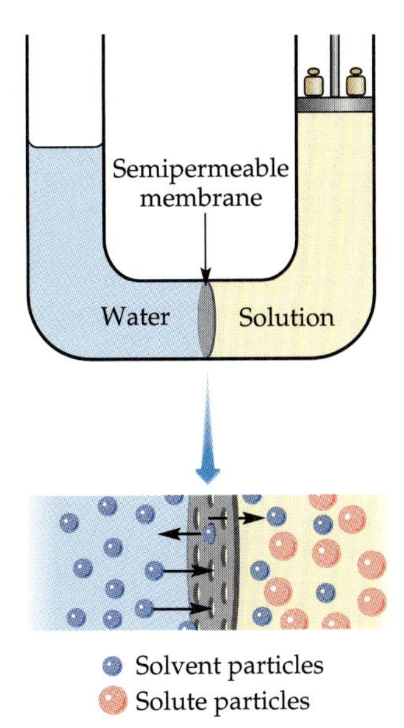

▲ **Figure 8.10**
The sieve model of osmosis holds that the semipermeable membrane has pores large enough to permit the passage of water and other small molecules but too small to allow the passage of larger molecules.

Look at Figure 8.10. On both sides of the membrane all molecules are moving about at random, occasionally bumping against the membrane. If a water molecule happens to hit one of the pores, it passes through the membrane into the other compartment. When the much larger sugar molecule strikes a pore, it bounces back instead of through. The more sugar molecules there are in solution (that is, the more concentrated the solution), the smaller the chance that a water molecule from the sugar solution side will strike a pore and move into the water solution side. Thus, in our example, there will be a *net* flow of water from the left compartment into the right compartment. This net diffusion of water through a semipermeable membrane is called **osmosis.** During osmosis, there is always a *net* flow of solvent from the more dilute solution (or pure solvent) into the more concentrated solution (Figure 8.11). For example, the net diffusion of water is *from* a 5% sugar solution into a 10% solution. As the liquid level in the right compartment builds up and that in the left compartment drops, the increased quantity of fluid in the right compartment exerts pressure that makes it more difficult for additional water molecules to enter that compartment. (See Section 6.3 on how a barometer works.) Eventually the buildup of pressure is sufficient to prevent further *net* flow of water into that compartment. The rates at which water molecules move back and forth are equal.

Instead of allowing the liquid level to build up and stop the net flow of water, the same thing can be accomplished by applying an external pressure to the compartment containing the more concentrated solution. The pressure needed to just prevent the net flow of solvent from the dilute solution to the concentrated one is called the **osmotic pressure.** The magnitude of the osmotic pressure depends only on the concentration of solute particles—that is, on the osmolarity of the solution. The more particles (the higher the osmolarity), the greater the osmotic pressure.

For a 20% sucrose solution, osmotic flow could lift a column of solution to a height of 150 m. The solution has an osmotic pressure of about 15 atm.

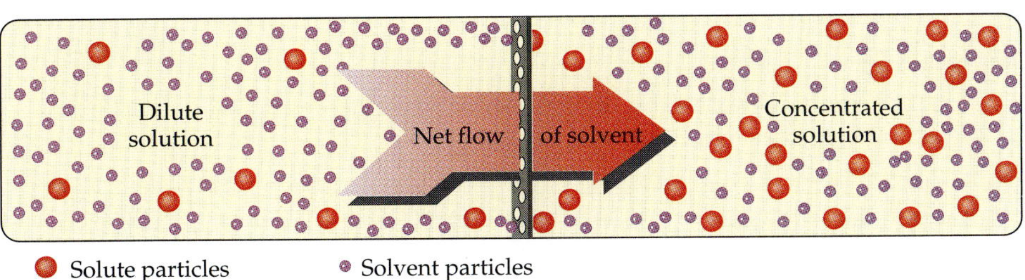

▲ **Figure 8.11**
Net flow through a semipermeable membrane occurs spontaneously in only one direction, from the compartment containing the more dilute solution (or pure solvent) into the region that has the more concentrated solution. Remember that ordinarily the terms *dilute* and *concentrated* refer to concentration of the *solute.* The net flow of *solvent* is from the sector in which the solvent is more concentrated to the region where it is less concentrated.

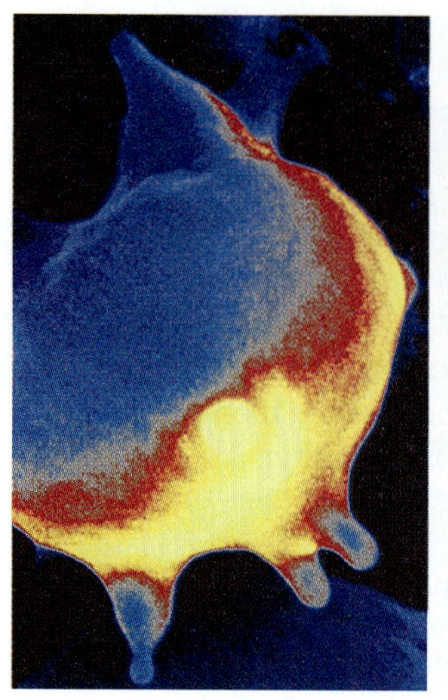
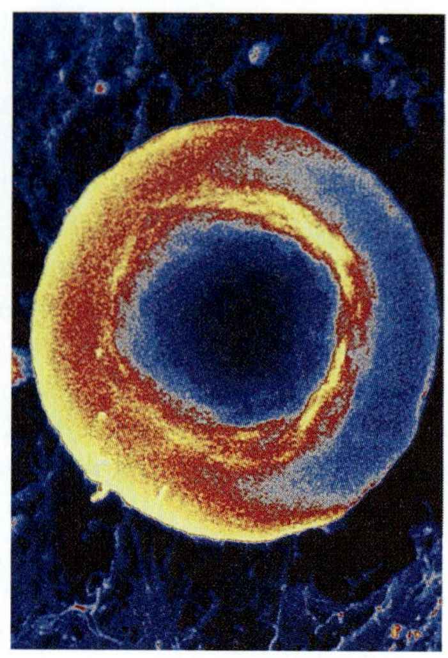
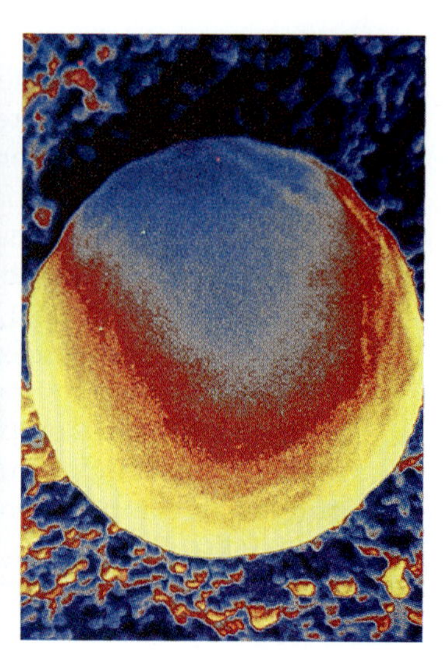

Figure 8.12 ▶
Normal human red blood cells are shown in the center photograph. A cell placed in a hypertonic solution (left) is wrinkled and shriveled. Placed in a hypotonic solution (right), the cell is swollen. The cells are shown in false color at a magnification of about 2000X.

You can think of osmotic pressure or osmolarity as a measure of the tendency of a solution to draw solvent into itself.

Examples of osmosis are found in living organisms everywhere. Cells are much like semipermeable bags filled with solutions of ions, small and large molecules, and still larger cell components. Cell function and survival depend on maintenance of the same osmotic pressure inside the cell and outside in the extracellular fluid. If we place cells, such as red blood cells, in a solution that is 0.89% NaCl (mass/vol), there is no net flow of water through the cell membranes and the cells are stable. The fluids inside the cells have the same osmotic pressure as the sodium chloride solution. A solution having the same osmotic pressure as body fluids is an **isotonic solution.** If the concentration of NaCl in a saline solution is greater than 0.89%, a net flow of water *out* of the cells causes them to shrink (Figure 8.12). The saline solution is a **hypertonic solution;** it has a higher osmotic pressure than red blood cells. If the concentration of NaCl is less than 0.89%, water flows *into* the cells, and they may burst. The solution is a **hypotonic solution;** it has a lower osmotic pressure than red blood cells.

One application of osmosis, called *reverse osmosis,* is based on *reversing* the normal net flow of water molecules through a semipermeable membrane. That is, by applying to a solution a pressure *exceeding* the osmotic pressure, water can be driven from a solution into pure water. In this way, pure water can be extracted from brackish water, seawater, or industrial wastewater. The success of reverse osmosis, widely used in ships at sea and in water-poor nations of the Middle East, requires the use of membranes that can withstand high pressures.

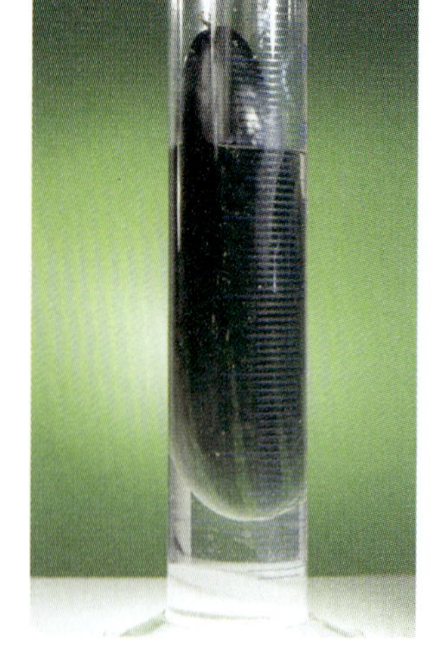

▲ A cucumber is made into a pickle by placing it in a concentrated salt solution. Water flows from the cucumber into the salt solution.

Example 8.9

What are the osmolarities of the two isotonic solutions (glucose and sodium chloride) mentioned in the medical applications essay (page 215)?

Solution

The concentration of glucose is 0.31 M. Glucose is a molecular substance; it does not ionize in aqueous solution. The osmolarity of a 0.31 M glucose solution is therefore 0.31 osmol/L (or 310 mosmol/L). NaCl is ionic; each formula unit provides two particles (1 Na^+ and 1 Cl^-) in aqueous solution. The osmolarity of a 0.16 M NaCl solution is therefore 0.32 osmol/L (or 320 mosmol/L).

Practice Exercise

What is the osmolarity of each of the following solutions?

a. 0.50 M $CaCl_2(aq)$ **b.** 0.15 M $KNO_3(aq)$ **c.** 1.32 M $C_6H_{12}O_6(aq)$

Medical Applications of Osmosis

The rupture of a cell by a *hypotonic* solution is called *plasmolysis.* If the cell is a red blood cell, the more specific term is *hemolysis.* The shrinkage of a cell in a *hypertonic* solution, called *crenation,* can lead to the death of a cell.

In replacing body fluids intravenously, the replacement fluid should be isotonic. Otherwise hemolysis or crenation may result and the patient's well-being may be jeopardized. As we have noted, an 0.89% NaCl (mass/vol) solution, called physiological saline, is isotonic with the fluid in red blood cells. The "D5W" referred to by television's doctors and paramedics is a 5.5% solution of glucose (also called dextrose, D) in water (W). It also is isotonic with the fluid in red blood cells. The 0.89% NaCl (mass/vol) is about 0.16 M, and the 5.5% glucose solution is approximately 0.31 M.

There are limits to intravenous feeding because a patient can handle at most about 3 L of water in a day. If an isotonic solution of 5.5% glucose is used, 3.0 L of this solution supplies only about 160 g of glucose, yielding an energy value of about 640 kcal (640 food Calories) per day. This is woefully inadequate. Even a resting patient requires about 1400 kcal per day. A person suffering from serious burns, for example, may require as many as 10,000 kcal per day. With carefully formulated solutions containing other vital nutrients as well as glucose, the feeding of a patient can be increased to about 1200 kcal per day. This still falls short of the requirements of many seriously ill people.

A way to get around this problem is to use solutions that are about six times as concentrated as isotonic solutions. But instead of being administered through a vein in an arm or a leg, this solution is infused directly through a tube into the superior vena cava, a large blood vessel leading to the heart (Figure 8.13). The large volume of blood flowing through this vein quickly dilutes the solution to levels that do not damage the blood. With this technique, patients have been given 5000 kcal per day and have even gained weight.

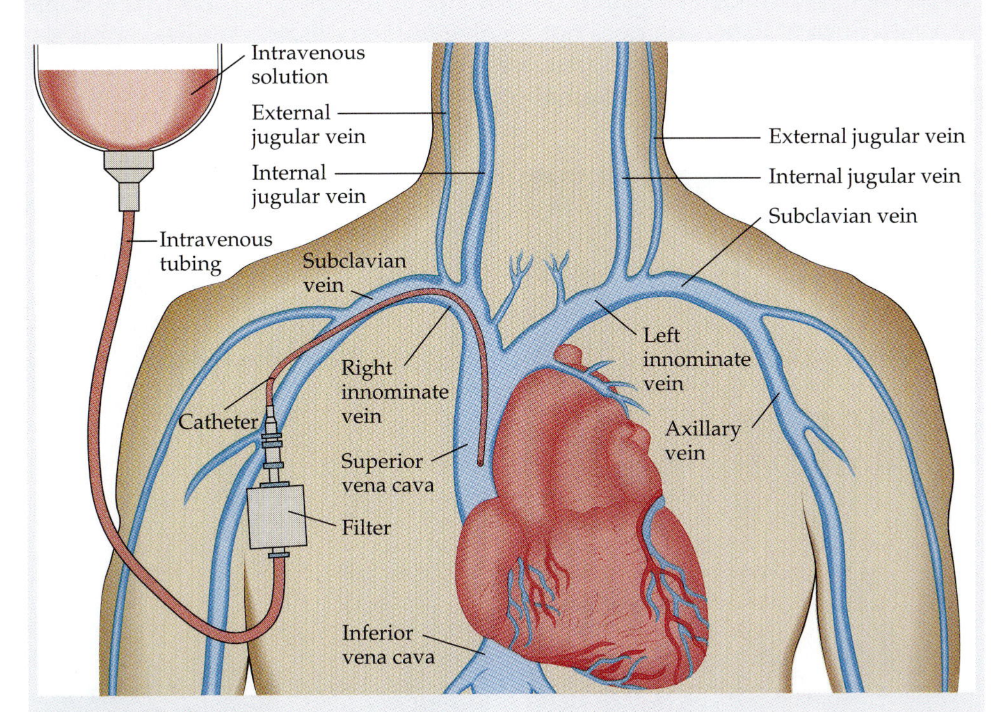

Figure 8.13 ▲
Concentrated nutrient solution is sometimes infused directly into the superior vena cava. The solution is quickly diluted to near-isotonic concentration by the large volume of blood that flows through the vein.

✓ Review Questions

8.13 (**a**) What is a semipermeable membrane? (**b**) What happens when a semipermeable membrane is placed between two solutions of glucose—one that has a concentration of 1 M and the other has a concentration of 0.001 M? (**c**) What is osmotic pressure?

8.14 Explain the differences between an isotonic solution, a hypotonic solution, and a hypertonic solution.

8.10 Colloids

The particles in a solution—atoms, ions, or molecules—are of submicroscopic size. Once the solute and solvent are thoroughly mixed, the *solute does not settle out;* molecular motion keeps the particles randomly distributed. The mixture is *homogeneous.* For example, the sugar in a bottle of apple juice does not settle to the bottom, and the last drop is just as sweet as, but no sweeter than, the first.

If we try to dissolve sand (silica, SiO_2) in water, the two substances may momentarily appear to be mixed, but the sand rapidly settles to the bottom of the container. The temporary dispersion of sand in water is called a *suspension.* We can separate the sand and water by pouring the suspension into a funnel fitted with filter paper; water passes through the paper and sand remains behind. The mixture of water and sand is obviously *heterogeneous;* part of it is clearly sand with one set of properties and part of it is water with another set of properties.

Is there no halfway point between true solutions, with particles of atomic, ionic, or molecular size, and suspensions, with gross chunks of insoluble matter? Yes, there is something else: *colloids.*

Even though silica (SiO_2) is insoluble in water, it's possible to prepare a dispersion of finely divided silica in water with up to 40% SiO_2 by mass that is stable for years. Such a dispersion does not involve ordinary grains of silica, nor is the suspended silica of ionic or molecular size. The dispersion is called a colloidal mixture. Figure 8.14 gives another example of a colloidal mixture.

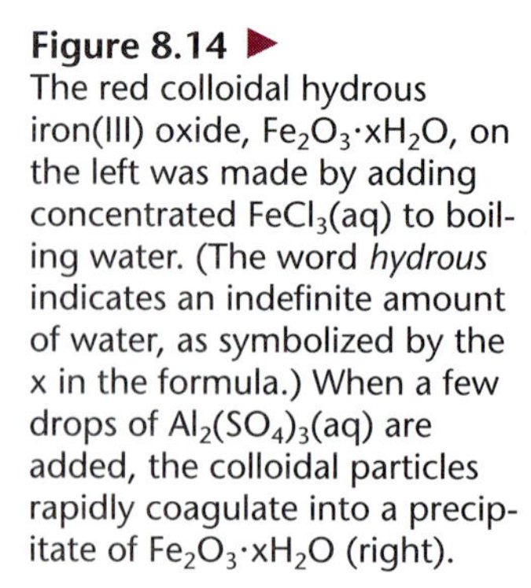

Figure 8.14 ▶ The red colloidal hydrous iron(III) oxide, $Fe_2O_3 \cdot xH_2O$, on the left was made by adding concentrated $FeCl_3(aq)$ to boiling water. (The word *hydrous* indicates an indefinite amount of water, as symbolized by the x in the formula.) When a few drops of $Al_2(SO_4)_3(aq)$ are added, the colloidal particles rapidly coagulate into a precipitate of $Fe_2O_3 \cdot xH_2O$ (right).

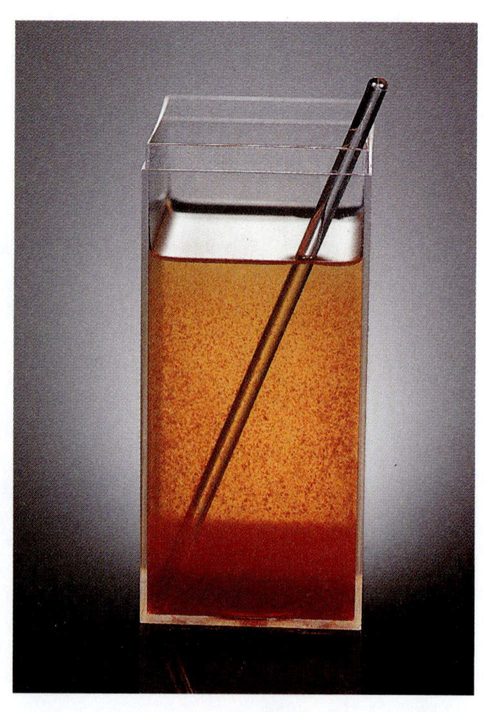

A material is called a colloid, not because of the kind of matter in a dispersion, but because of the extent of its subdivision. True solutions have solute and solvent particles less than about 1 nm in diameter. Ordinary suspensions have particle dimensions of about 1000 nm or more. A **colloid** has dispersed matter with one or more dimensions (length, width, or thickness) in the range from about 1 nm to 1000 nm.

Recall that 1 nm = 1×10^{-9} m. Simple optical microscopes are not able to resolve particles smaller than about 1000 nm. Colloidal particles generally cannot be seen under a microscope.

There are eight different kinds of colloids, based on the physical state of the particles themselves (the dispersed phase) and of the "solvent" (the dispersion medium). These are listed, together with examples, in Table 8.3. Much of living matter consists of colloidal particles, mainly in the form of sols and emulsions. Substances with high molar masses, such as starches and proteins, generally form colloidal dispersions rather than true solutions in water.

As you might expect, the properties of colloidal dispersions are different from those of true solutions and suspensions (Table 8.4). Colloids often appear milky or cloudy. Even those that appear clear reveal their colloidal nature by scattering a beam of light passed through them. This phenomenon, first studied by John Tyndall in 1869, is known as the **Tyndall effect** (Figure 8.15). The spectacular sunsets often seen in the desert are caused, at least in part, by the preferential scattering of blue light as sunlight passes through dust-laden air (an aerosol). The transmitted light is deficient in the color blue; it appears a reddish color.

In most colloidal dispersions the particles are charged, due to the adsorption of ions on the surface of the particle. A given colloid will preferentially adsorb only one kind of ion (either positive or negative) on its surfaces; thus all the particles of a given colloidal dispersion bear like charges. Because like charges repel, the particles tend to stay away from one another. They cannot come together and form particles large enough to settle out. These colloids can be made to coalesce and separate out by addition of ions of opposite charge, particularly those doubly or triply charged. Aluminum salts, which contain Al^{3+} ions, work well for breaking up colloids in which the particles are negatively charged. (See again Figure 8.14.)

Table 8.3 Types of Colloidal Dispersions

Type	Particle Phase	Medium Phase[b]	Example
Foam	Gas	Liquid	Whipped cream
Solid foam	Gas	Solid	Floating soap
Aerosol	Liquid	Gas	Fog, hair spray
Liquid emulsion	Liquid	Liquid	Milk, mayonnaise
Solid emulsion	Liquid	Solid	Butter
Smoke	Solid	Gas	Fine dust or soot in air
Sol[a]	Solid	Liquid	Starch solutions, jellies
Solid sol	Solid	Solid	Pearl

[a]Sols that set up in semisolid, jellylike form are called gels.

[b]By their very nature, gas mixtures always qualify as solutions. The size of particles in gas mixtures and their homogeneity fulfill the requirements of a true solution.

Table 8.4 Properties of Solutions, Colloids, and Suspensions

Property	Solution	Colloid	Suspension
Particle size	0.1–1.0 nm	1–1000 nm	>1000 nm
Settles on standing?	No	No	Yes
Filter with paper?	No	No	Yes
Separate by dialysis?	No	Yes	Yes
Homogeneous?	Yes	Borderline	No

(a)

(b)

▲ **Figure 8.15**
The Tyndall effect. (a) The flashlight beam is invisible as it passes through a true solution (left). However, it is visible as it passes through colloidal iron(III) oxide (right). (b) Light passing through fog, a colloidal dispersion of water in air, also illustrates the Tyndall effect.

Some colloids are stabilized by addition of a material that provides a protective coating. Oil is ordinarily insoluble in water, but it can be emulsified by soap. The soap molecules form a negatively charged layer about the surface of each tiny oil droplet. These negative charges keep the oil particles from coming together and separating out (see Section 20.2). In a similar manner, bile salts emulsify fats in the digestive tract, keeping them dispersed as tiny particles that are more efficiently digested. Milk is an emulsion in which fat droplets are stabilized by a coating of casein, a protein. Casein, soap, and bile salts are examples of **emulsifying agents,** substances that stabilize emulsions.

Review Questions

8.15 Define or explain—and, where possible, illustrate—the following terms.
- **a.** suspension **b.** colloid **c.** Tyndall effect
- **d.** dialysis **e.** emulsifying agent **f.** crenation
- **g.** plasmolysis **h.** hemolysis

8.16 Contrast a true solution and a colloid in terms of (**a**) the size of the solute particles; (**b**) the nature of the distribution of solute and solvent particles; (**c**) the color and clarity of the solution; (**d**) the Tyndall effect.

8.11 Dialysis

To live, an organism must take in food and get rid of toxic wastes. The nutrients necessary to life must enter cells, and wastes must leave the cells. Generally, then, cell membranes must permit the passage not only of water molecules but also of other small molecules and ions. At the same time, it is important that large molecules and colloidal particles not be lost from the cell. Membranes that pass small molecules and ions while holding back large molecules and colloidal particles are called *dialyzing membranes.* The process is called **dialysis.** It differs from osmosis in that osmotic membranes pass only solvent molecules. In dialysis, molecules and ions always diffuse from areas of higher concentration to areas of lower concentration.

Dialyzing membranes are used in laboratories to purify colloidal dispersions by removing smaller molecules and ions. The mixture to be purified is placed in a bag made of a dialyzing membrane. The bag is placed in a container of pure water (Figure 8.16). The ions and small molecules pass out through the membrane,

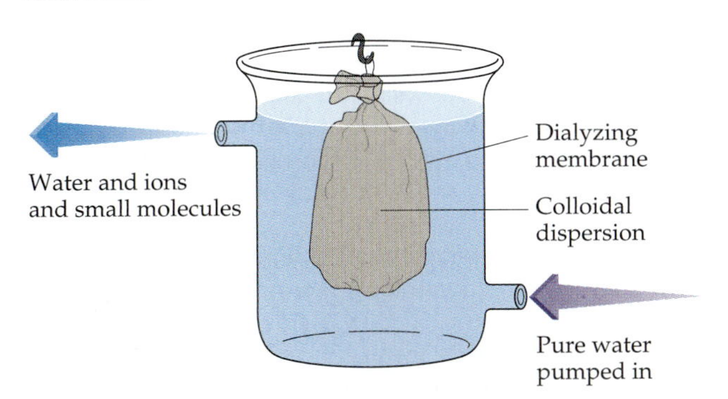

Figure 8.16
A simple apparatus for dialysis.

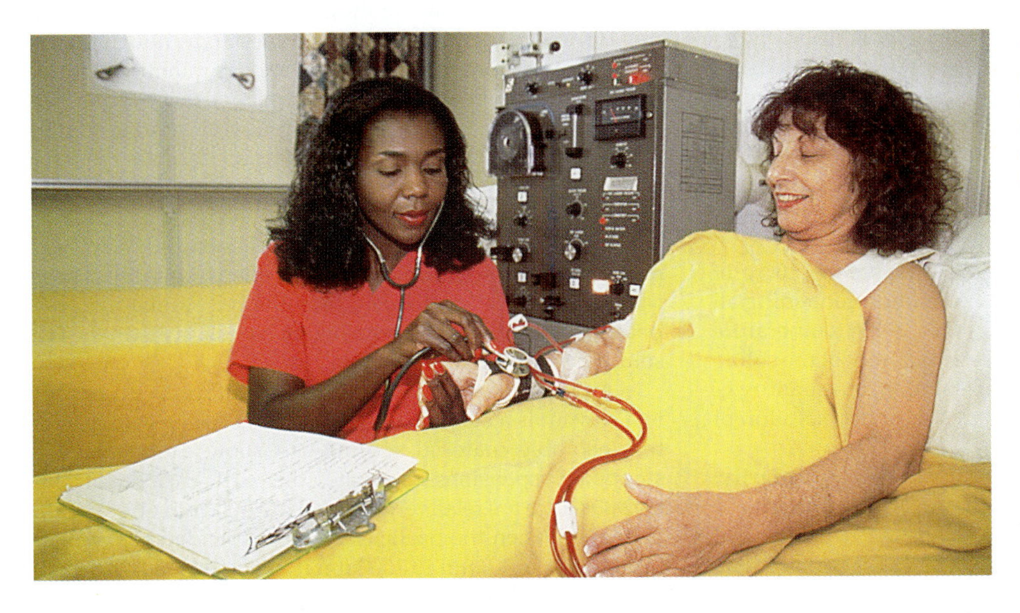

In renal dialysis, elaborate machinery substitutes for human kidneys that are no longer capable of removing toxic waste products from the blood.

leaving the colloidal particles behind. Pure water is continuously pumped past the bag, and the unwanted small particles are carried away. The dialyzing membrane may be an animal bladder, or it may be an artificial film of cellophane or collodion, a semisynthetic plastic made by treatment of cellulose with nitric acid, alcohol, and camphor.

The kidneys are a complex dialyzing system responsible for the removal of certain potentially toxic waste products from the blood. By first gaining an understanding of the function of living kidneys, which we discuss in Section 28.7, scientists have been able to construct artificial ones. Artificial kidneys are more elaborate in structure than the simple apparatus shown in Figure 8.16, but their principle of operation is the same.

Many substances—medications, poisons, anesthetics—act by changing the permeability of membranes. Methyl mercury, a powerful poison that acts on the nervous system, is thought to act by making the membranes of nerve cells leakier than normal. A person going into shock has leaky capillaries that allow proteins and other colloidal particles as well as fluids to escape into the spaces between cells. If untreated, the cells die from lack of oxygen and nutrients. As research increases our understanding of cellular membranes, we will undoubtedly gain a better understanding of drug action and of poisoning, health, and disease.

Review Question

8.17 Compare and contrast osmosis and dialysis with respect to the kinds of particles that pass through the semipermeable membrane.

Summary

A **solution** is a homogeneous mixture of two or more substances. In a solution, the **solvent** is usually present in a larger quantity, and one or more **solutes** in lesser quantities. An **aqueous solution** is one in which the solvent is water.

A substance is **soluble** in a solvent if some appreciable quantity of the substance dissolves in the solvent. Any substance that has unlimited solubility in a solvent is **miscible** with that solvent. Any substance that does not dissolve significantly in a solvent is **insoluble** in that solvent. In general, like dissolves like. Polar solutes are soluble in polar solvents and insoluble in nonpolar solvents. Nonpolar solutes are soluble in nonpolar solvents and insoluble in polar solvents.

Compounds of Group 1A elements, nitrate compounds, and ammonium compounds are soluble in water. The ionic bonds in these compounds are weak enough to be overcome by the attractive force exerted by the polar water molecules. Once the bonds are broken, the ions are *hydrated.* Many compounds containing two doubly- or triply-charged ions are insoluble in water.

Solubility is measured in grams of solute per 100 g of solvent at a specified temperature. A solution containing all the solute it can hold at equilibrium at a given temperature is **saturated.** One that contains a lesser quantity of the solute is **unsaturated.** In a **supersaturated** solution, the quantity of solute is greater than the quantity in a saturated solution of the same solute.

Solution concentration can be expressed as **molarity:** moles of solute per liter of solution.

A related concentration unit is *osmolarity:* osmoles of solute per liter of solution, where an **osmole** is 1 mol of solute particles. Concentrations can also be expressed as percentages. When both solute and solvent are liquids, the concentration is given as **percent by volume.** When the solute is a solid, the concentration is given as **percent by mass.**

A **colligative property** of a solution is one governed by the number of solute particles present in the solution—in other words, governed by the osmolarity of the solution.

Any barrier that allows solute and solvent particles to pass is *permeable.* Any barrier that does not allow either to pass is *impermeable.* A barrier that allows some solute particles to pass but not others, or that lets solvent particles pass but not solute particles, is *semipermeable.*

In **osmosis,** water passes through a semipermeable membrane. The direction of flow is from the side having the higher water concentration (lower solute concentration) to the side having the lower water concentration (higher solute concentration). As water passes through a semipermeable membrane, the pressure in the compartment it enters builds up. At some point this pressure becomes high enough to prevent any more water from entering, and the quantity of water passing into the compartment is the same as that passing out. The pressure at which this equilibrium occurs is the **osmotic pressure** of the solution. Osmotic pressure is directly proportional to osmolarity.

The solutions surrounding living cells are characterized according to concentration. Solutions that have the same osmolarity as the solution inside the cell are **isotonic.** Those that have a lower osmolarity than the cell solution are **hypotonic,** and those that have a higher osmolarity than the cell solution are **hypertonic.**

An insoluble substance mixed as uniformly as possible in a solvent is a **suspension.** A suspension is unstable, with the insoluble material settling out soon after mixing is stopped. Suspensions form when the particles of the insoluble substance are relatively large, about 1000 nm or greater in diameter. When the particles being suspended are smaller (ranging from 1.0 nm to 1000 nm in diameter), mixing causes a *colloidal dispersion,* or **colloid** to form. The particles in a colloid do not settle out when mixing is stopped. A colloidal dispersion can be stabilized by coating the dispersed particles with an **emulsifying agent.**

Key Terms

anhydrous (8.4)
aqueous solution (8.1)
colligative property (8.8)
colloid (8.10)
concentrated solution (8.2)
deliquescent (8.4)
dialysis (8.11)
dilute solution (8.2)
dynamic equilibrium (8.5)
efflorescent (8.4)
emulsifying agent (8.10)
hydrate (8.4)
hygroscopic (8.4)
hypertonic solution (8.9)
hypotonic solution (8.9)
isotonic solution (8.9)
mass/volume percent (8.7)
miscible (8.2)
molarity (M) (8.7)
osmole (osmol) (8.8)
osmosis (8.9)
osmotic pressure (8.9)
percent by mass (8.7)
percent by volume (8.7)
precipitate (8.5)
saturated solution (8.5)
semipermeable membrane (8.9)
solute (8.1)
solution (8.1)
solvent (8.1)
supersaturated solution (8.5)
Tyndall effect (8.10)
unsaturated solution (8.5)

Problems

Solubility of Ionic Compounds

1. Without referring to Table 8.2, indicate which of the following compounds you would expect to be soluble in water. Explain each answer. You may use the periodic table.
   a. NaBr b. $Ca(NO_3)_2$
   c. $BaCO_3$ d. $FePO_4$
2. Without referring to Table 8.2, indicate which of the following compounds you would expect to be soluble in water. Explain each answer. You may use the periodic table.
   a. $(NH_4)_2CO_3$ b. RbCl
   c. $PbSO_4$ d. LiOH

Dynamic Equilibria

3. In a dynamic equilibrium involving a saturated solution, describe the two processes for which the rates are equal.
4. Use the kinetic-molecular theory to explain why most solid solutes become more soluble in water with increasing temperature but gases generally become less soluble.
5. Use data from Figure 8.5 to determine (**a**) whether a solution containing 48 g of KNO_3 per 100 g of water at 40 °C is saturated, unsaturated, or supersaturated, and (**b**) the approximate temperature to which a mixture of 60 g of KNO_3 and 100 g of water must be heated so that the KNO_3 is completely dissolved.
6. Use data from Figure 8.5 to determine (**a**) the mass percent of $KClO_3$ in a saturated aqueous solution at 50 °C, and (**b**) the molarity of saturated $Ce_2(SO_4)_3(aq)$ at 50 °C.

Solubility of Covalent Compounds

7. Which of the following is likely to be soluble in water? Explain.
   a. CH_3COOH b. HCHO
   c. CH_3CH_2SH d. $CH_3CH_2CH_3$
8. Which of the following is likely to be soluble in water? Explain.
   a. CH_3CH_2Cl b. CH_3CH_3
   c. HCOOH d. $C_5H_{10}O_5$

Molarity

9. Calculate the molarity of each of the following solutions.
   a. 6.00 mol of HCl in 2.50 L of solution
   b. 0.00700 mol of Li_2CO_3 in 10.0 mL of solution
10. Calculate the molarity of each of the following solutions.
   a. 2.50 mol of H_2SO_4 in 5.00 L of solution
   b. 0.200 mol of C_2H_5OH in 18.4 mL of solution
11. Calculate the molarity of each of the following solutions.
   a. 8.90 g of H_2SO_4 in 100.0 mL of solution
   b. 439 g of $C_6H_{12}O_6$ in 1.25 L of solution
12. Calculate the molarity of each of the following solutions.
   a. 44.3 g of KOH in 125 mL of solution
   b. 2.46 g of $H_2C_2O_4$ in 750.0 mL of solution
13. How many grams of solute are needed to prepare each of the following solutions?
   a. 2.00 L of 1.00 M NaOH
   b. 10.0 mL of 4.25 M $C_6H_{12}O_6$
14. How many grams of solute are needed to prepare each of the following solutions?
   a. 250 mL of 2.50 M $K_2Cr_2O_7$
   b. 20.0 mL of 0.0100 M $KMnO_4$
15. What volume of 6.00 M NaOH is required to contain 1.25 mol of NaOH?
16. What volume of 2.50 M NaOH is required to contain 1.05 mol of NaOH?
17. What volume of 0.0250 M $KMnO_4$ must one take to get 8.10 g of $KMnO_4$?
18. What volume of 4.25 M $C_6H_{12}O_6$ must one take to get 205 g of $C_6H_{12}O_6$?

Percent Concentration

19. What is the volume percent concentration of each of the following solutions?
   a. 35.0 mL of water in 725 mL of an ethanol-water solution
   b. 78.9 mL of acetone in 1.55 L of an acetone-water solution
20. What is the volume percent concentration of each of the following solutions?
   a. 58.0 mL of water in 625 mL of an ethanol-water solution
   b. 79.1 mL of methanol in 755 mL of a methanol-water solution
21. What is the mass percent concentration of each of the following solutions?
   a. 4.12 g of NaOH in 100.0 g of water
   b. 5.00 mL of ethanol ($d = 0.789$ g/mL) in 50.0 g of water
22. What is the mass percent concentration of each of the following solutions?
   a. 175 mg of NaCl/g solution
   b. 275 mL of methanol ($d = 0.791$ g/mL) in 1 kg of water
23. Describe how you would prepare 775 g of an aqueous solution that is 10.0% NaCl *by mass.*
24. Describe how you would prepare 125 g of an aqueous solution that is 5.50% KOH *by mass.*
25. Describe how you would prepare exactly 2.00 L of an aqueous solution that is 2.00% acetic acid *by volume.*
26. Describe how you would prepare exactly 500 mL of an aqueous solution that is 30.0% isopropyl alcohol *by volume.*
27. Describe how you would prepare 250.0 mL of an aqueous solution of $MgSO_4$ that is 1.5% (mass/vol).
28. Describe how you would prepare 2.00 L of an aqueous solution of $AlCl_3$ that is 2.15% (mass/vol).
29. On average, glucose makes up about 0.10% of human blood, by mass. What is the approximate concentration in mg/dL?
30. A vinegar sample has a density of 1.01 g/mL and contains 5.88% acetic acid by mass. What mass of acetic acid is contained in a 1-L bottle of the vinegar?

Osmolarity and Colligative Properties

31. How many solute particles does each formula unit of each of the following compounds give in aqueous solution? How many osmol are there in one mole of each compound?
   a. KCl b. CH_3OH c. $(NH_4)_2SO_4$
32. How many solute particles does each formula unit of each

of the following compounds give in aqueous solution? How many osmol are there in one mole of each compound?
a. $CaBr_2$ **b.** NaOH **c.** $Al(NO_3)_3$

33. For each pair of solutions at 25 °C, indicate which member has the higher osmotic pressure.
a. 0.1 M $NaHCO_3$ or 0.05 M $NaHCO_3$
b. 1 M NaCl or 1 M glucose

34. For each pair of solutions at 25 °C, indicate which member has the higher osmotic pressure.
a. 1 M NaCl or 1 M $CaCl_2$ **b.** 1 M NaCl or 3 M glucose

35. How many moles of each of the following are needed to provide 1.0 osmol?
a. KCl **b.** CH_3OH **c.** $(NH_4)_2SO_4$

36. How many moles of each of the following are needed to provide 1.0 osmol?
a. $CaBr_2$ **b.** NaOH **c.** $Al(NO_3)_3$

37. For each pair of solutions at 25 °C, indicate which member has the higher osmotic pressure.
a. 1.0 osmol/L NaCl or 1.0 osmol/L $CaCl_2$
b. 1.0 osmol/L $CaCl_2$ or 2.0 osmol/L glucose ($C_6H_{12}O_6$)

38. For each pair of solutions at 25 °C, indicate which member has the higher osmotic pressure.
a. 1.0 osmol/L $NaHCO_3$ or 0.50 osmol/L $NaHCO_3$
b. 2.0 osmol/L CH_3OH or 1.0 osmol/L $CaCl_2$

Colloids

39. Compare and contrast a colloidal dispersion and a suspension.

40. Is it possible to have a colloidal dispersion of one gas in another? Explain.

Additional Problems

41. Arrange the following in order of increasing concentration: 1% by mass, 1 mg/dL, 1 ppb, 1 ppm, 1 ppt.

42. Is there a way in which solute can be crystallized from an unsaturated solution without changing the solution temperature? Explain.

43. Are there any exceptions to the general rule that a supersaturated solution can be made to deposit excess solute by cooling? Explain.

44. The solubility of $O_2(g)$ in water is 4.43 mg O_2/100 g H_2O at 20 °C when the gas pressure is maintained at 1 atm. What is the molarity of the saturated solution?

45. Consider 1.0 mol of each of the following: $Al_2(SO_4)_3$, CH_3COOH, CH_3OH, $MgBr_2$, and NaCl. Which would lower the freezing point of 1.0 L of water the most? The least? Explain.

46. If two containers of a gas at different pressures are connected, the net diffusion of gas is from the *higher* to the lower pressure. The net diffusion in osmosis is from the solution of *lower* to that of higher concentration. How can you explain this apparent difference?

47. What is meant by the term *reverse osmosis?* Give an example of its use.

48. The text describes an isotonic solution of NaCl as being about 0.16 M, whereas an isotonic solution of glucose is about 0.31 M. Explain why the concentrations of these isotonic solutions are not the same.

49. What would be the effect on red blood cells if they were placed in (**a**) 5.5% NaCl(aq) (**b**) 0.92% glucose(aq)?

50. Use the kinetic-molecular theory to explain why NaCl dissolves in water.

51. Pickles are made by soaking cucumbers in a salt solution. Which has the higher osmotic pressure, the salt solution or the liquid in the cucumber? Explain.

52. Two aqueous solutions of glucose ($C_6H_{12}O_6$) are separated by a semipermeable membrane. Solution A has 3.00% glucose (mass/volume) and solution B is 0.10 M glucose. In which direction will a net flow of water occur, from A to B or from B to A?

53. Aluminum sulfate is commonly used to coagulate or precipitate colloidal suspensions of clay particles in municipal water treatment plants. Why do you suppose this electrolyte is more effective than sodium chloride?

54. An aqueous solution is prepared by dissolving 11.3 mL of CH_3OH ($d = 0.793$ g/mL) in enough water to produce 73.5 mL of solution. What is the percent of CH_3OH, expressed as (**a**) volume percent and (**b**) mass-volume percent?

55. You have a stock solution of 6.0 M HCl. How many moles of HCl are there in the following quantities of solution?
a. 1.0 L **b.** 100 mL **c.** 1.0 mL

56. On the average, glucose ($C_6H_{12}O_6$) makes up about 0.10% by weight of human blood. How much glucose is there in 1.0 kg of blood?

57. A cyanide solution, made by adding 1.0 lb of sodium cyanide (NaCN) to 1.0 ton (2000 lb) of water, is used to leach gold from its ore. What is the concentration of the cyanide solution in each of the following units of concentration?
a. percent by mass **b.** ppm
c. g/kg **d.** mol/L

58. Analysis determined that the concentration of the pesticide Alar ($C_6H_{12}N_2O_3$) in a batch of apple juice was 3.0 ppm. What is the molarity of the Alar in the juice? (Assume a density of 1.00 g/mL).

59. Calcium chloride is often used to melt sidewalk ice because it is thought to be less harmful to trees than sodium chloride. If someone has been using 2.0 pounds of NaCl per application for his sidewalk, how much $CaCl_2$ will he need?

60. Complete the following table.

Concentration of Solute (g/L)	Molarity (mol/L)	Molar Mass of Solute (g/mol)
98	1.0	—
32	0.50	—
0.74	0.010	—
—	0.10	26
—	0.025	80
120	—	40
17	—	68

Acids and Bases I

Most foods are either neutral or slightly acidic. We use bases—called antacids—to counter acid indigestion.

Learning Objectives/Study Questions

1. What is Arrhenius's definition of an acid? A base?
2. What is the Brønsted-Lowry definition of an acid? A base?
3. What do the terms "strong" and "weak" mean when applied to an acid or a base?
4. How are common acids and bases named?
5. How is an acidic anhydride formed? A basic anhydride?
6. What are the reactants and products in a neutralization reaction?
7. What is an ionic equation? A net ionic equation?
8. What bases react with acids to form carbon dioxide gas?
9. How does an antacid work?
10. Why are strong acids and bases so harmful?

The chemistry in our bodies is complex. It includes digesting food and shedding tears. We also practice a complicated chemistry in everyday life, from taking antacids to baking bread. Central to much of this chemistry are two special kinds of compounds called *acids* and *bases.* Our bodies produce and consume acids and bases. We eat them and drink them. We use vinegar (acetic acid) in cooking and in salad dressings, and our beverages are made tart by citric acid or phosphoric acid. We treat excess stomach acid (hydrochloric acid) with bases such as aluminum hydroxide and magnesium hydroxide. We clean with ammonia and lye (sodium hydroxide), two familiar bases. Some of these and other acids and bases used around the home are shown in Figure 9.1.

Special senses recognize four tastes, which are related to acid-base chemistry. Acids taste *sour,* and bases taste *bitter.* Salts—ionic compounds formed in acid-base reactions—taste *salty. Sweet* tastes are related to acid-base chemistry in a more subtle way. To taste sweet, a compound must have a hydrogen-bond donor group (acidic), a hydrogen-bond acceptor group (basic), and a nonpolar group, all arranged in the proper geometry to fit the sweet taste receptor (see Selected Topic D).

Acids, bases, and other substances in the laboratory are often toxic; they should *never* be tasted.

Acid rain is an environmental problem around the world. Bitter, undrinkable alkaline (basic) water is often all that is available in desert areas. Much of our understanding of air and water pollution depends on our knowledge of acids and bases.

In this chapter, we consider some qualitative properties of acids and bases—what they are, how they are named, how they are produced, and how they react chemically. In Chapter 10, we consider some of the more quantitative aspects of acid-base chemistry.

9.1 Acids and Bases: Definitions and Properties

Historically, acids and bases have been classified according to some distinctive properties. *Acids* are substances that exhibit the following properties when dissolved in water.

(a)

(b)

Figure 9.1 ▶
(a) Some common acids: toilet bowl cleaner, vinegar, aspirin, beer, fruits, and fruit and tomato juices. (b) Some common bases: drain cleaners, oven cleaner, quinine water, and so on.

◀ Litmus is a dye obtained from lichens. Litmus paper is often used to determine whether a material is acidic or basic. The soil sample on the right is acidic and turns blue litmus red. The soil sample on the left is basic and turns red litmus blue.

- Acids taste sour.
- Acids produce a prickling or stinging sensation on the skin.
- Acids turn the color of the indicator dye litmus from blue to red.
- Acids react with many metals, such as magnesium, zinc, and iron, to produce ionic compounds and hydrogen gas.
- Acids react with bases, thereby losing their acidic properties.

Many foods taste sour because they are acidic. Vinegar contains about 5% acetic acid. Citrus fruits and many fruit-flavored drinks contain citric acid. Lactic acid is formed in sour milk and is responsible for the tart taste of yogurt. Phosphoric acid is used to impart tartness to beer and some soft drinks.

Bases exhibit the following properties when dissolved in water.

- Bases taste bitter.
- Bases feel slippery or soapy on the skin.
- Bases turn the color of the indicator dye litmus from red to blue.
- Bases react with acids, thereby losing their basic properties.

The Israelites, in their journey from Egypt to Canaan, came upon the bitter waters at Marah (Exodus 15:23). Like the waters in other arid areas, these were no doubt alkaline or basic.

The Arrhenius Theory

In 1887 the Swedish chemist Svante Arrhenius proposed that an **acid** is a molecular substance that *ionizes* (breaks up into ions) in aqueous solution into H^+ ions and anions, and a **base** is a substance that produces OH^- in aqueous solution. A base can either contain OH^-, as does an ionic hydroxide, or it can *ionize* to produce OH^-. The essential reaction between an acid and a base, *neutralization,* is the combination of H^+ and OH^- to form water. The cation of the base and the anion of the acid form an ionic compound, a **salt.**

Many medicines taste bitter because they contain *alkaloid* (baselike) compounds. Quinine, caffeine, and the antihistamines are familiar examples.

How can you tell when a compound is an acid or a base? Experimentally you can dissolve some of the compound in water and check for some of the properties listed above. Because tasting can be hazardous (some acids and bases are poisonous, and nearly all are corrosive unless they have been highly diluted), the simplest test is to use litmus paper. No matter whether we start with the red form or the blue form, moist litmus paper is blue in a base and red in an acid.

The Arrhenius theory has its limitations. We now know, for example, that a simple free proton does not exist in water solution because H^+ has such a high positive charge density that it immediately seeks out a negative charge. It finds a lone pair of electrons on the O atom of an H_2O molecule and attaches itself to form a *hydronium ion,* H_3O^+.

In acid-base chemistry, a "proton" is an ionized hydrogen atom (H^+).

$$H_2O + H^+ \longrightarrow [H_3O]^+$$

Water — Hydrogen ion (proton) — Hydronium ion

The H^+ ion is probably associated with several H_2O molecules—for example, four H_2O molecules in the ion $H(H_2O)_4{}^+$ or $H_9O_4{}^+$. The hydronium ion is quite reactive; it can readily transfer a proton to another molecule or ion. For most purposes, however, we will use H^+ and ignore its hydration. We will talk about protons when we mean their sources (hydronium ions). This simplification is permissible as long as we understand what we are doing. The properties of acids are therefore those of the H^+ ion. It is the H^+ ion that turns litmus red, tastes sour, and reacts with active metals and bases.

The Arrhenius theory is also limited in that it applies only to reactions in aqueous solution, and it does an inadequate job of explaining the source of the OH^- in the ionization of bases such as ammonia, NH_3. Like many scientific theories, it has been supplanted by a better one based on newer data.

The Brønsted-Lowry Theory

The shortcomings of the Arrhenius theory were largely overcome by a theory proposed independently, in 1923, by J. N. Brønsted in Denmark and T. M. Lowry in Great Britain. In their theory, an acid is a **proton donor** and a base is a **proton acceptor.**

The Brønsted-Lowry theory describes the ionization of hydrogen chloride in this way:

$$\underset{\text{Acid}}{HCl(g)} + \underset{\text{Base}}{H_2O} \longrightarrow H_3O^+(aq) + Cl^-(aq)$$

The main feature of the Brønsted-Lowry notation, also suggested by Figure 9.2, is that the overall reaction consists of a proton transfer from an acid to a base. In this case, H_2O acts as a base by accepting a proton from HCl, an acid.

Conjugate Pairs

By the Brønsted-Lowry theory, the products of an acid-base reaction are also an acid and a base. The overall reaction thus consists of *two* combinations of acids and bases called *conjugate pairs.* For example, we can represent the ionization of acetic acid as

$$\underset{\text{Acid (1)}}{CH_3COOH} + \underset{\text{Base (2)}}{H_2O} \rightleftharpoons \underset{\text{Acid (2)}}{H_3O^+} + \underset{\text{Base (1)}}{CH_3COO^-}$$

- An acid-base conjugate pair differs in structure only by a proton (H^+): The conjugate acid of a species is that species *plus* a proton; the conjugate base of a species is that species *minus* a proton.

CH_3COOH acts as an acid, noted as acid(1) above, by donating a proton to H_2O. In the reverse reaction, CH_3COO^-, noted as base(1) above, acts as a base by accepting a proton from H_3O^+. CH_3COO^- is the **conjugate base** [base(1)] of the acid CH_3COOH. Similarly, H_3O^+ is the **conjugate acid** of H_2O.

The Brønsted-Lowry theory describes the behavior of ammonia as a base in this way:

$$\underset{\text{Base (1)}}{NH_3} + \underset{\text{Acid (2)}}{H_2O} \rightleftharpoons \underset{\text{Acid (1)}}{NH_4{}^+} + \underset{\text{Base (2)}}{OH^-}$$

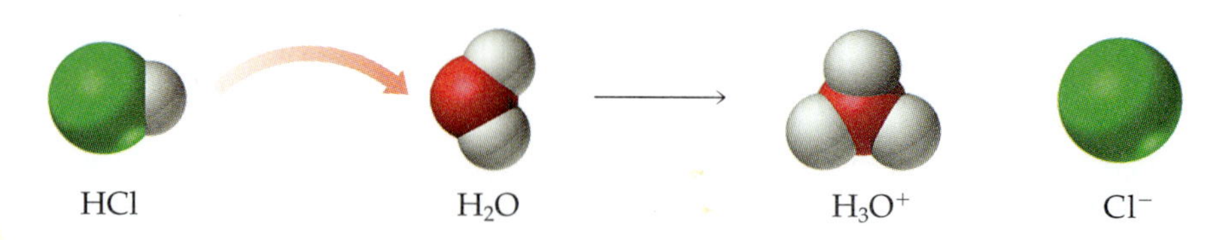

Figure 9.2 ▶ Ionization of HCl as a Brønsted-Lowry acid. The red arrow represents proton transfer from the HCl molecule to the H_2O molecule. Water molecules accept protons; water is a Brønsted-Lowry base in this reaction.

NH_3 acts as a base, noted as base(1), by accepting a proton from H_2O. In the reverse reaction, NH_4^+, noted as acid(1), loses a proton to OH^-; NH_4^+ is the *conjugate acid* of NH_3. Similarly, OH^- is the *conjugate base*—base(2)—of the acid H_2O—acid(2). OH^- accepts a proton from NH_4^+ and H_2O donates a proton to NH_3. Note that in this reaction H_2O is a acid, whereas it is a base in the ionization of CH_3COOH. A substance that can act either as an acid or a base is *amphiprotic*.

Review Questions

9.1 List five general properties of acidic aqueous solutions and four general properties of basic aqueous solutions. Which, if any, of these properties remain after an acid and a base neutralize each other?

9.2 What ion is responsible for the properties of acidic aqueous solutions? What ion is responsible for the properties of basic aqueous solutions?

9.3 Suggest some ways in which you might determine whether a particular water solution is acidic or basic in the Arrhenius sense.

9.4 Write the formula for the (**a**) conjugate acid of OH^-; (**b**) conjugate base of NH_4^+; (**c**) conjugate acid of H_2O; (**d**) conjugate base of HCOOH.

9.2 Strong and Weak Acids

The term *strong* refers only to the degree of ionization; a strong acid is completely ionized. The amount of solute in the solution is specified by its concentration. Both strong and weak acids can be either *concentrated* (relatively large amount of acid solute in a given volume of solution) or *dilute* (relatively little acid solute in a given volume of solution).

Acids that are completely ionized in water solution are called **strong acids.** To represent the ionization of a strong acid, such as hydrochloric acid, we can write the equation

$$HCl(g) \xrightarrow{H_2O} H^+(aq) + Cl^-(aq)$$

(We write H_2O over the arrow to indicate that the gaseous HCl is dissolved in water.) The reaction is complete; essentially no HCl molecules remain. There are only $H^+(aq)$ and $Cl^-(aq)$ ions in the hydrochloric acid solution. In 0.0010 M HCl(aq), for example, the molar concentrations (indicated by square brackets, [ ]) are

$$[H^+] = 0.0010\ M \qquad [Cl^-] = 0.0010\ M \qquad [HCl] = 0$$

Determining concentrations in an aqueous solution of a weak electrolyte, such as $[H^+]$, $[F^-]$, and [HF] in HF(aq), requires a kind of calculation that we consider in Chapter 10.

Most acids, because they are only partially ionized in aqueous solution, are called **weak acids.** Unlike that of a strong acid, the ionization of a weak acid in aqueous solution is a *reversible* reaction, represented in an equation with a double arrow. Hydrogen fluoride, a gas, ionizes in water as a weak acid.

most acids are weak

$$HF(aq) \overset{H_2O}{\rightleftharpoons} H^+(aq) + F^-(aq)$$

The aqueous solution is called hydrofluoric acid. When this reversible reaction reaches equilibrium, most of the HF molecules remain nonionized. In 1 M HF(aq), fewer than 1% of the molecules ionize. Because the predominant solute species are HF molecules, an aqueous solution of hydrofluoric acid is best represented as HF(aq), *not* as $H^+(aq) + F^-(aq)$.

Hydrochloric acid and hydrofluoric acid each have one ionizable H atom per molecule. They are **monoprotic acids,** as are nitric acid (HNO_3) and hydrocyanic acid (HCN). Sulfuric acid (H_2SO_4) and carbonic acid (H_2CO_3) each have two ionizable H atoms; they are **diprotic acids.** Phosphoric acid (H_3PO_4) has three ionizable H atoms; it is a **triprotic acid.** Acids that have more than one ionizable H atom per molecule, such as H_2CO_3, H_2SO_4, and H_3PO_4, are known as **polyprotic acids.** You should not assume that all the hydrogen atoms in a molecule are acidic, meaning that they are released as H^+ in water solutions. For example, none of the hydrogens of methane

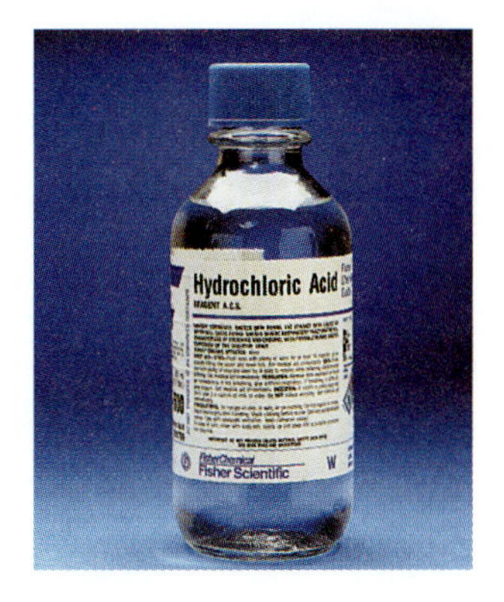

▲ Hydrogen chloride is a gas, usually represented as HCl(g). Hydrochloric acid, represented as HCl(aq), is an aqueous solution of hydrogen ions (H^+) and chloride ions (Cl^-).

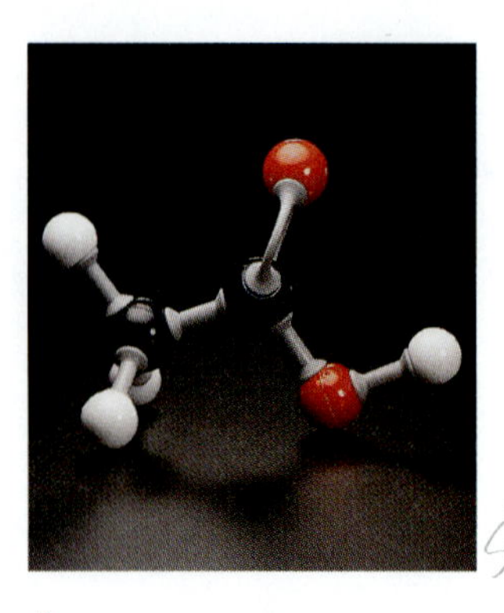

▲ A ball-and-stick model of an acetic acid molecule. Only the H atom (white) bonded to the O atom (red, lower right) is ionizable. The three H atoms attached to the C atom (black, left) are not ionizable.

(CH_4) is given up in aqueous solution. Only one of the hydrogen atoms in acetic acid ($C_2H_4O_2$) is acidic. For this reason, the formula for acetic acid is frequently written CH_3COOH or $HC_2H_3O_2$ to emphasize that only one proton is released.

So, you may wonder, how can you tell from its formula if a substance is an acid and, if so, whether it's a strong or weak acid? Chemists generally identify *ionizable* hydrogen atoms by writing formulas in one of two ways.

1. *We write a molecular formula with ionizable H atoms first.* HNO_3, H_2SO_4, and H_3PO_4 are acids with one, two, and three ionizable H atoms, respectively. Methane, CH_4, has four H atoms, but they are *not* ionizable; CH_4 is *not* an acid. From its name alone we know that acetic acid is an acid. When its formula is written as $HC_2H_3O_2$, we see that it has four H atoms, but only one that is ionizable—the H atom that is shown first.
2. *In organic chemistry, we often use formulas that show the ionizable hydrogen atoms last.* In Chapter 16, we consider a family of acids called *carboxylic acids.* A more informative way to write the formula of acetic acid is CH_3COOH. This formula shows that the three nonionizable H atoms are bonded to a C atom. The fourth H atom is bonded to an O atom, and it is this one that is ionizable. Acetic acid is the most familiar of the family of carboxylic acids and probably the most familiar of all the weak acids. Other water-soluble carboxylic acids include formic acid ($HCOOH$), propionic acid (CH_3CH_2COOH), and butyric acid ($CH_3CH_2CH_2COOH$). In each of these, only the H atom on the O atom is ionizable.

There are only a few *common* strong acids, listed in Table 9.1. Unless you are given information to the contrary, assume that any other acid is a weak acid. Some of the hazards associated with strong acids are discussed in Section 9.8. If dilute enough, solutions of some strong acids can be harmless. Hydrochloric acid, for example, is produced in dilute solutions in our stomachs, where it aids in the digestion of certain foodstuffs.

Review Questions

9.5 Which of the following would you expect to be Brønsted-Lowry acids, and which would you expect to be Brønsted-Lowry bases? Explain.

a. CH_3CH_2COOH **b.** H_2S **c.** CN^-
d. $CH_3NH_3^+$ **e.** $HCOOH$ **f.** $CH_3CH_2CH_2NH_2$.

9.6 Give examples of a monoprotic, a diprotic, and a triprotic acid.

9.7 What are the characteristic features of a polyprotic acid? Is CH_4 a polyprotic acid? Explain.

Table 9.1 Common Strong Acids

Name	Formula
Hydrochloric acid	$HCl(aq)$
Hydrobromic acid	$HBr(aq)$
Hydriodic acid	$HI(aq)$
Nitric acid	$HNO_3(aq)$
Sulfuric acid	$H_2SO_4(aq)$[a]
Perchloric acid	$HClO_4(aq)$

[a]Only the first ionization of H_2SO_4 is complete; HSO_4^- is only partly ionized and is classified as a weak acid.

9.3 Names of Some Common Acids

In Chapter 3 you learned the names of some common anions. Now we can use the following scheme to relate the names of acids to the anions produced when the acids ionize.

	Acid name	Example	Anion name	Example
Increasing	*hydro__ic acid*	hydrochloric acid (HCl)	*__ide*	chloride ion (Cl^-)
number of	*hypo__ous acid*	hypochlorous acid (HClO)	*hypo__ite*	hypochlorite ion (ClO^-)
oxygen	*__ous acid*	chlorous acid ($HClO_2$)	*__ite*	chlorite ion (ClO_2^-)
atoms ↓	*__ic acid*	chloric acid ($HClO_3$)	*__ate*	chlorate ion (ClO_3^-)
	per__ic acid	perchloric acid ($HClO_4$)	*per__ate*	perchlorate ion (ClO_4^-)

Note that in the molecular formula for the acid the number of H atoms is the same as the number of H^+ ions required to balance the electric charge of the anion. That is, one H^+ is needed to balance the -1 charge on the Cl^- ion, so we add one H atom in HCl. Similarly, we need two H atoms in H_2SO_4, and three in H_3PO_4. The scheme for naming acids is further illustrated by the examples in Table 9.2. Note that most of the names have the same stem in the acid as in the anion. A few have extra letters—for example, the *ur* in sulfuric and sulfurous and the *or* in phosphoric. These you should learn as special cases if you haven't done so already.

Example 9.1

The formula for phosphite ion is PO_3^{3-}. What is the formula for phosphorous acid?

Solution

It takes three H^+ ions to balance the electric charge on the PO_3^{3-} ion. We therefore need three H atoms in the molecular acid. The formula for phosphorous acid is H_3PO_3.

Practice Exercise

The formula for arsenate ion is AsO_4^{3-}. What is the formula for arsenic acid?

Table 9.2 Names of Some Common Acids and Their Salts

		Sodium Salt	
Formula of Acid	**Name of Acid**	**Formula**	**Name**
HNO_2	Nitr*ous* acid	$NaNO_2$	Sodium nitr*ite*
HNO_3	Nitr*ic* acid	$NaNO_3$	Sodium nitr*ate*
$HC_2H_3O_2$[a]	Acet*ic* acid	$NaC_2H_3O_2$	Sodium acet*ate*
H_2S	*Hydro*sulfur*ic* acid	Na_2S	Sodium sulf*ide*
H_2SO_3[b]	Sulfur*ous* acid	Na_2SO_3	Sodium sulf*ite*
H_2SO_4[b]	Sulfur*ic* acid	Na_2SO_4	Sodium sulf*ate*
H_3PO_4[b]	Phosphor*ic* acid	Na_3PO_4	Sodium phosph*ate*

[a]Acetic acid is also represented as CH_3COOH and sodium acetate as CH_3COONa.

[b]Table 3.4 also lists anions of these acids in which only one of the H atoms is replaced by Na (one or two in the case of H_3PO_4).

Example 9.2

The formula for selenic acid is H_2SeO_4. What is the formula for selenious acid? For sodium selenate? For sodium selenite?

Solution

An acid with a name ending in *-ous* has one less O atom than an acid with a name ending in *-ic*. The formula for selenious acid is therefore H_2SeO_3. Sodium selenate is a salt of selenic acid; its formula is Na_2SeO_4. Sodium selenite is a salt of selenious acid; its formula is Na_2SeO_3.

Practice Exercises

A. The formula for iodic acid is HIO_3. What is the formula for hypoiodous acid? For sodium iodate? For sodium iodite? For sodium hypoiodite?

B. The formula for potassium permanganate is $KMnO_4$. What is the formula for permanganic acid?

9.4 Some Common Bases

The Arrhenius definition of a base is a substance which produces hydroxide ions, OH^-, in aqueous solution (Section 9.1). Many bases are ionic compounds that contain Group 1A or Group 2A cations along with hydroxide ions. Each has enough OH^- ions to balance the charge of the accompanying cation, and they are named like other ionic compounds:

$NaOH$ = sodium hydroxide (commonly called *lye*)

KOH = potassium hydroxide

$Ca(OH)_2$ = calcium hydroxide (commonly called *slaked lime*)

Soluble ionic hydroxides such as NaOH are obviously bases. Moreover, because they are completely ionized, they are **strong bases.**

$$NaOH(s) \xrightarrow{H_2O} Na^+(aq) + OH^-(aq)$$

(In this case, H_2O over the arrow indicates that solid NaOH *dissolves* in water rather than reacting with it.)

Other compounds produce OH^- ions by *reacting* with water, not just by dissolving in it. They are also bases. The equation below shows that gaseous ammonia dissolves in water and reacts with water to produce an equilibrium mixture of ions and molecules.

$$NH_3(g) + H_2O \rightleftharpoons NH_4^+(aq) + OH^-(aq)$$

As in the case of acetic acid, most of the ammonia molecules in the aqueous solution remain nonionized. Ammonia is therefore a **weak base.**[1] Most molecular substances that act as bases are *weak* bases.

Many familiar household cleaning agents contain ammonia, an ingredient easily identified by its characteristic odor. Plain ammonia solutions are about 3.5% NH_3 (mass/volume). Sudsy ammonia solutions contain detergents and often have less NH_3.

How can an ammonia molecule act as a base—a proton acceptor? Recall that the N atom of ammonia has a lone pair of electrons. The lone pair can be used to attach a proton. When a proton leaves a water molecule, it leaves behind the electron pair that joined it to the O atom. The water molecule becomes a negatively charged hydroxide ion (see also Figure 9.3).

[1] In Arrhenius's time, chemists generally believed that a substance must *contain* OH groups to be a base; $NH_3(aq)$ was thought to be NH_4OH (ammonium hydroxide). Even though there is no compelling evidence for the existence of NH_4OH, this formula is still often seen as a representation of $NH_3(aq)$.

$$\mathrm{H{:}\overset{\displaystyle H}{\underset{\displaystyle H}{\ddot{N}}}{:} + H{:}\underset{\displaystyle H}{\ddot{O}}{:} \rightleftharpoons \left[H{:}\overset{\displaystyle H}{\underset{\displaystyle H}{\ddot{N}}}{:}H\right]^{+} + \left[{:}\ddot{O}{:}H\right]^{-}}$$

Base   Acid   Ammonium ion   Hydroxide ion

The concept of a base also includes not only hydroxide ion and molecules such as ammonia but also ions such as carbonate (CO_3^{2-}) and bicarbonate (HCO_3^-), (see Section 9.7). For example, carbonate ions react with water to yield a basic solution.

$$CO_3^{2-}(aq) + H_2O \rightleftharpoons HCO_3^-(aq) + OH^-(aq)$$

The family of organic compounds called **amines** (Chapter 17) have molecules in which one or more of the H atoms of NH_3 are replaced by a hydrocarbon group. Some examples are *methylamine,* CH_3NH_2; *dimethylamine,* $(CH_3)_2NH$; *trimethylamine,* $(CH_3)_3N$; and *ethylamine,* $CH_3CH_2NH_2$. The amines with one to four carbon atoms are water soluble and, like ammonia, are *weak* bases.

$$CH_3NH_2(aq) + H_2O \rightleftharpoons CH_3NH_3^+(aq) + OH^-(aq)$$

How can we recognize a base and how do we know whether the base is weak or strong? An ionic compound containing OH^- ions is obviously a base; if it is water soluble, it will form a strongly basic solution. NaOH and KOH are strong bases. On the other hand, methanol (CH_3OH) is *not* a base. The OH group is not present as OH^-; it is covalently bonded to the C atom. Similarly, acetic acid (CH_3COOH) is *not* a base. It does not contain OH^-, nor does it produce $OH^-(aq)$ in water; rather, it produces $H^+(aq)$ and is therefore an *acid.* To identify a weak base, you usually need a chemical equation for the ionization reaction. However, you can identify many by using these facts: There are only a few *common* strong bases, listed in Table 9.3. The most common weak bases are ammonia and the amines. Most bases are molecular substances and do not contain hydroxide ion. They produce it by reacting with water.

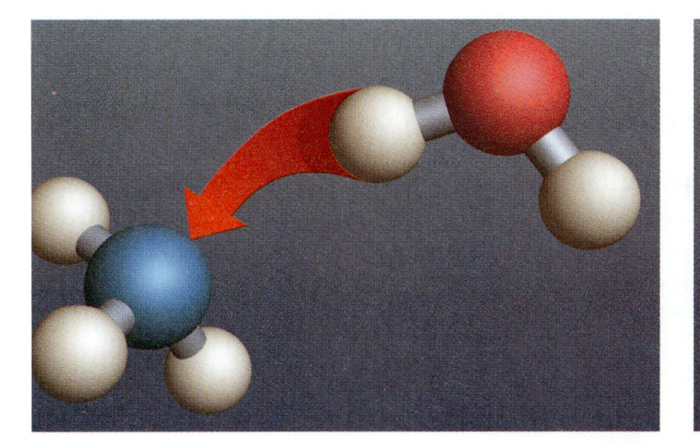

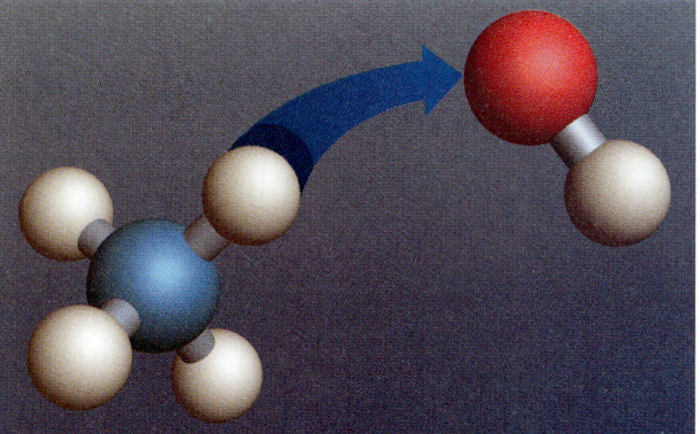

$$NH_3 + H_2O \rightleftharpoons NH_4^+ + OH^-$$

▲ **Figure 9.3**
Ionization of NH_3 as a Brønsted-Lowry base. The red arrow represents proton transfer in the forward reaction, and the reverse reaction is indicated by the blue arrow. Only relatively few ammonia molecules react, however; most remain unchanged in solution as NH_3 molecules. There are therefore relatively few hydroxide ions and ammonium ions; ammonia is a weak base.

Table 9.3 Strong Bases

Name	Formula
Alkali metal hydroxides	
Lithium hydroxide	$LiOH(aq)$
Sodium hydroxide	$NaOH(aq)$
Potassium hydroxide	$KOH(aq)$
Rubidium hydroxide	$RbOH(aq)$
Cesium hydroxide	$CsOH(aq)$
Alkaline earth hydroxides	
Calcium hydroxide	$Ca(OH)_2(aq)$
Strontium hydroxide	$Sr(OH)_2(aq)$
Barium hydroxide	$Ba(OH)_2(aq)$

The idea of an acid as a proton *donor* and a base as a proton *acceptor* greatly expands our concept of acids and bases, but this broader concept still *includes* the more limited definitions. It's just that the broader concept is useful in a greater variety of situations.

Review Question

9.8 What is aqueous ammonia? Why is it sometimes called ammonium hydroxide?

9.5 Acidic and Basic Anhydrides

Many acids are made by the reaction of nonmetal oxides with water. For example, sulfur trioxide reacts with water to form sulfuric acid.

Nonmetal oxide — Acid

$$SO_3 + H_2O \longrightarrow H_2SO_4$$

Similarly, carbon dioxide reacts with water to form carbonic acid.

$$CO_2 + H_2O \longrightarrow H_2CO_3$$

Nonmetal oxides are called **acidic anhydrides.** *Anhydride* means "without water." These reactions explain why rainwater is acidic, particularly when the rain forms in air that is polluted with sulfur oxides (see page 237).

Many common bases can be made from metal oxides. For example, calcium oxide (lime) reacts with water to form calcium hydroxide (slaked lime).

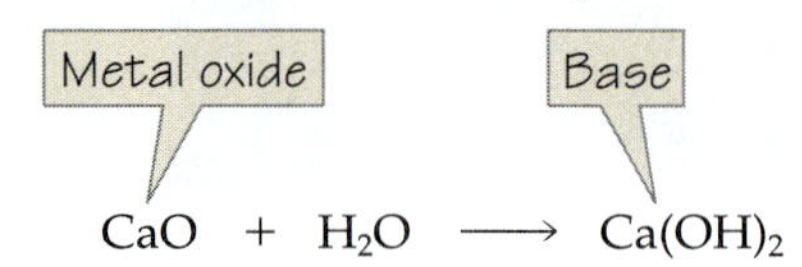

Another example is the reaction of lithium oxide with water to form lithium hydroxide.

$$Li_2O + H_2O \longrightarrow 2\ LiOH$$

In general, metal oxides react with water to form bases (Figure 9.4). These metal oxides are called **basic anhydrides.**

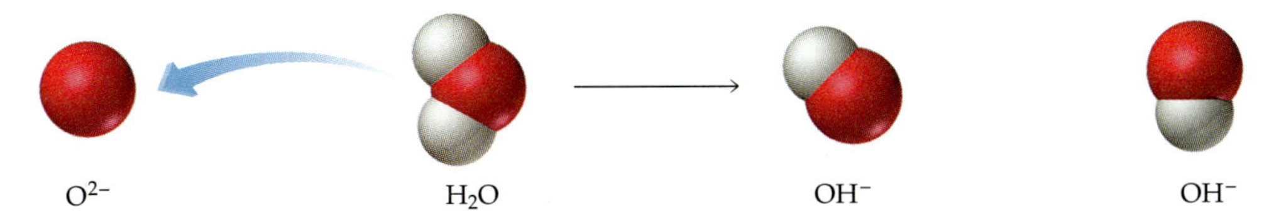

▲ **Figure 9.4**
Metal oxides are basic because the oxide ion reacts with water to form two hydroxide ions.

Example 9.3

Give the formula and name for (**a**) the acid formed when sulfur dioxide reacts with water and (**b**) the base formed by the addition of water to barium oxide (BaO).

Solution

a. First, write the formula for sulfur dioxide (SO_2). Then write an equation for its reaction with water.

$$SO_2 + H_2O \longrightarrow H_2SO_3$$

The acid is H_2SO_3 (sulfurous acid).

b. Barium oxide is BaO. The equation for the reaction of BaO with water is

$$BaO + H_2O \longrightarrow Ba(OH)_2$$

The base is $Ba(OH)_2$ (barium hydroxide).

Practice Exercises

A. Give the formula and name for the acid formed when dinitrogen pentoxide reacts with water. (Hint: Two molecules of acid are formed.)

B. Give the formula and name for the base formed by the addition of water to potassium oxide.

✓ Review Question

9.9 What is an acidic anhydride? A basic anhydride?

9.6 Neutralization Reactions

In the reaction of an acid and a base, called **neutralization,** the identifying characteristics of the acid and base cancel out or neutralize each other. The acid and base are converted to an aqueous solution of an ionic compound, called a salt. If we use conventional formulas for the acid and base, we can write what we might call a "complete formula" equation[2] for a neutralization reaction as follows:

$$HCl(aq) + NaOH(aq) \longrightarrow NaCl(aq) + H_2O$$

But this "complete formula" equation is not always the best way to show what happens in the neutralization. It is often better to write the equation in *ionic* form, indicating the actual ions and molecules present in solution.

[2]The "complete formula" equation is often called a "molecular" equation, but this term is misleading. Many of the formulas written in such equations—for example, NaCl(aq)—represent formula units, not actual molecules.

$$\underbrace{H^+(aq) + \cancel{Cl^-(aq)}}_{\text{Acid}} + \underbrace{\cancel{Na^+(aq)} + OH^-(aq)}_{\text{Base}} \longrightarrow \underbrace{\cancel{Na^+(aq)} + \cancel{Cl^-(aq)}}_{\text{Salt}} + \underbrace{H_2O}_{\text{Water}}$$

When we eliminate "spectator" ions—those that just "look on" and appear unchanged in the ionic equation—we find that the equation reduces to the even more informative **net ionic equation.**

$$H^+(aq) + OH^-(aq) \longrightarrow H_2O$$

The essence of a neutralization reaction, then, is that H^+ ions from an acid and OH^- ions from a base combine to form water. If the spectator ions form a soluble salt, they remain in solution. If the water is evaporated, the soluble salt is left as a solid. Example 9.4 provides additional examples of the different types of equations described here.

Example 9.4

Barium nitrate is used to produce a green color in fireworks. It can be made by the reaction of nitric acid with barium hydroxide. Write (**a**) "complete formula," (**b**) ionic, and (**c**) net ionic equations for this neutralization reaction.

Solution

a. Write chemical formulas for the substances involved in the reaction, and balance the equation.

$$HNO_3(aq) + Ba(OH)_2(aq) \longrightarrow Ba(NO_3)_2(aq) + H_2O \quad (\textit{not balanced})$$

$$2\,HNO_3(aq) + Ba(OH)_2(aq) \longrightarrow Ba(NO_3)_2(aq) + 2\,H_2O \quad (\textit{balanced})$$

b. Now, represent the strong acid, strong base, and soluble salt with the formulas of their ions and the nonelectrolyte water with its molecular formula.

$$2\,H^+(aq) + \cancel{2\,NO_3^-(aq)} + \cancel{Ba^{2+}(aq)} + 2\,OH^-(aq) \longrightarrow \cancel{Ba^{2+}(aq)} + \cancel{2\,NO_3^-(aq)} + 2\,H_2O$$

c. Cancel the spectator ions (Ba^{2+} and NO_3^-) in the above equation.

$$H^+(aq) + OH^-(aq) \longrightarrow H_2O$$

Practice Exercises

A. Calcium hydroxide is used to neutralize a waste stream of hydrochloric acid. Write (**a**) "complete formula," (**b**) ionic, and (**c**) net ionic equations for this neutralization reaction.

B. Ammonia [$NH_3(aq)$] is used to neutralize phosphoric acid to make ammonium phosphate. Write (**a**) "complete formula," (**b**) ionic, and (**c**) net ionic equations for this neutralization reaction.

9.7 Reactions of Acids with Carbonates and Bicarbonates

Add vinegar to baking soda, and you get a vigorous fizzing action. Some antacid preparations are designed to effervesce. What causes the fizz? Generally, it is the evolution of $CO_2(g)$ by the reaction of a carbonate or a bicarbonate salt with an acid. Carbonates and bicarbonates are salts of carbonic acid (H_2CO_3). This diprotic acid is a very weak one; it holds on to its protons quite tightly. (Notice the seeming contradiction: it's the weak acids that hold tightly to their protons and the strong acids that don't.) Conversely, carbonate ions and bicarbonate ions pick up protons readily to form carbonic acid. Further, carbonic acid is quite unstable. In a reaction unrelated to its acidity, carbonic acid decomposes to carbon dioxide and water.

$$H_2CO_3(aq) \longrightarrow H_2O + CO_2(g)$$

You can't purchase a bottle of carbonic acid; it is too unstable to be isolated and bottled. It can be formed in solution, but as soon as it is formed, it tends to decompose.

If sodium bicarbonate is dissolved in water, a solution of sodium ions and bicarbonate ions is formed.

$$NaHCO_3(s) \xrightarrow{H_2O} Na^+(aq) + HCO_3^-(aq)$$

If hydrochloric acid is added, the bicarbonate ions come into contact with the hydrogen ions from the acid and immediately acquire protons to form carbonic acid.

$$\cancel{Na^+(aq)} + HCO_3^-(aq) + H^+(aq) + \cancel{Cl^-(aq)} \longrightarrow H_2CO_3(aq) + \cancel{Na^+(aq)} + \cancel{Cl^-(aq)}$$

The carbonic acid is unstable and decomposes, forming carbon dioxide and water. The gaseous carbon dioxide bubbles out of solution (the fizz). If the remaining solution is evaporated, sodium chloride is left. Even if acid is poured on solid sodium bicarbonate, carbon dioxide is released. The complete formula equation is

$$NaHCO_3(s \text{ or } aq) + HCl(aq) \longrightarrow NaCl(aq) + CO_2(g) + H_2O$$

The net ionic equation is simpler.

$$HCO_3^-(aq) + H^+(aq) \longrightarrow H_2CO_3(aq) \longrightarrow H_2O + CO_2(g)$$

Similarly, if hydrochloric acid is added to sodium carbonate, bubbles of carbon dioxide and a solution of sodium chloride are formed.

$$Na_2CO_3(s \text{ or } aq) + 2\,HCl(aq) \rightarrow 2\,NaCl(aq) + CO_2(g) + H_2O$$

◀ The leavening action of baking soda. When acidified, here with citric acid from a lemon, baking soda ($NaHCO_3$) reacts to produce carbonic acid (H_2CO_3), which decomposes to carbon dioxide and water. The carbon dioxide gas produces a "lift" when dough is baked.

The net ionic equation shows that the doubly negative carbonate ion picks up two protons to form carbonic acid which then breaks down to form carbon dioxide and water.

$$CO_3^{2-}(aq) + 2\,H^+(aq) \rightarrow H_2CO_3(aq) \rightarrow H_2O + CO_2(g)$$

✓ Review Question

9.10 Write the equation for the decomposition of carbonic acid. What is the anhydride of carbonic acid?

9.8 Acids, Bases, and Human Health

Strong acids and bases act as corrosive poisons, causing damage on contact with living cells. Their action is nonspecific—all cells, regardless of type, are damaged. These poisons produce chemical burns. Once the offending agent is neutralized or removed, the injuries are similar to burns from heat, and they are often treated the same way. Ingestion of any strong acid causes corrosive damage to the digestive tract. As little as 10 milliliters of concentrated (98%) H_2SO_4, taken internally, can be fatal.

Sulfuric acid (H_2SO_4) is by far the leading chemical product of U.S. industry. About 40 billion kilograms are produced annually. Most of this acid is used by industry, much of it in the conversion of phosphate rock to soluble compounds for use as fertilizer. Only small quantities of sulfuric acid are used in or around the home, for example, in automobile batteries and in one type of drain cleaner.

Sulfuric acid is a powerful dehydrating agent. It takes up water from cellular fluid, rapidly killing the cell. The sulfuric acid molecules dehydrate by reacting with water in the cells to form hydronium ions and hydrogen sulfate ions.

$$H_2SO_4 + H_2O \rightarrow H_3O^+ + HSO_4^-$$

Hydration of these ions and other secondary reactions may also be involved in the dehydration process.

Aerosol mists of sulfuric acid in polluted air (page 237) break down cells in the alveoli of the lungs. The alveoli lose their resilience, which makes it difficult for them to remove carbon dioxide. Such lung damage may contribute to pulmonary emphysema, a condition characterized by increasing shortness of breath. Most emphysema is caused by cigarette smoking, but air pollution is also known to be a factor in the relatively few cases in nonsmokers.

Hydrochloric acid (also called muriatic acid) is used in homes to clean calcium carbonate deposits from toilet bowls. It is used in building construction to remove excess mortar from bricks. In industry, it is used to remove scale or rust from metals. Concentrated solutions (about 38% HCl) cause severe burns, but dilute solutions are considered safe enough for use around the home if handled carefully. The gastric juice in your stomach contains around 0.5% hydrochloric acid.

Lime (CaO) is the cheapest and most widely used commercial base. It is made by heating limestone ($CaCO_3$) to drive off CO_2.

$$CaCO_3(s) + \text{heat} \rightarrow CaO(s) + CO_2(g)$$

With an annual output of about 16 billion kilograms, lime is fifth in industrial chemical production in the United States. It is used to make mortar and cement and also to "sweeten" acidic soil.

▲ Treating acidic soil with lime makes it "sweeter" (less acidic).

Sodium hydroxide (commonly known as lye) is the strong base most often used in the home. It is used as an oven cleaner and in products intended to open clogged

Acid Rain

Natural rainwater is slightly acidic because water falling through the air dissolves carbon dioxide and forms carbonic acid. During thunderstorms the rainwater can be much more acidic, due to nitric acid formed by lightning. Polluted air leads to precipitation that is still more acidic.

Scottish chemist Robert Angus Smith (1817–1884) coined the term *acid rain* in 1856 after studying the rainfall in Manchester. He found that the air, heavily polluted from the burning of coal, produced rain that was abnormally acidic. The term *acid rain* has persisted, and today it usually refers to precipitation that is more acidic than that formed by dissolving carbon dioxide from the air.

Limestone and marble, important building stones, are mainly calcium carbonate ($CaCO_3$). Marble is also used in statues, monuments, and sculptures. The calcium carbonate in it is readily attacked by acids in the atmosphere or in rain. Acid precipitation results mainly from the burning of sulfur-containing fossil fuels in electric power plants. The sulfur combines with oxygen to form sulfur dioxide.

$$S + O_2 \rightarrow SO_2$$

Some of the sulfur dioxide reacts further with oxygen to form sulfur trioxide.

$$2\,SO_2 + O_2 \rightarrow 2\,SO_3$$

The sulfur trioxide then reacts with water to form sulfuric acid.

$$SO_3 + H_2O \rightarrow H_2SO_4$$

The sulfuric acid, in the form of an aerosol mist or diluted by rainwater, furnishes the hydrogen ions that dissolve the marble and limestone (Figure 9.5).

$$CaCO_3(s) + 2\,H^+(aq) \rightarrow Ca^{2+}(aq) + CO_2(g) + H_2O$$

The acid mists and acidic rainwater also attack and dissolve many metals. Damage to our buildings, automobiles, and other structures and machines from air pollution amounts to billions of dollars per year. Even then the story is not complete. These acid pollutants are also damaging to human health, as we see in Section 9.8.

▲ Burning of sulfur-containing coal in electric power plants results in the release of large quantities of acidic sulfur oxides into the atmosphere. The oxides ultimately are converted to sulfuric acid and fall as acid rain.

▲ **Figure 9.5**
Marble statues are slowly eroded by the action of acid rain on the marble (calcium carbonate).

drains. Sodium hydroxide destroys tissue rapidly, causing severe chemical burns. Several detergent additives, such as carbonates, silicates, and borates, form strongly basic solutions when dissolved in water. These, too, can cause corrosive damage to tissues, particularly those of the digestive tract and the eyes.

Both acids and bases, even in dilute solutions, disrupt the structure of protein molecules in living cells. Generally, the unfolded proteins are not able to carry out the functions of the original proteins. In living cells, proteins require an optimum acidity and fail to function properly if the acidity changes much in either direction (see Chapter 21).

Many acids and bases are quite useful, but when misused they can be damaging to human health. We shall learn more about the effects of acids and bases on human health in Chapter 26.

✓ Review Questions

9.11 What is the leading chemical product of U.S. industry?

9.12 What is the effect of strong acids on clothing?

9.13 What is the effect of strong acids and strong bases on skin? How do corrosive acids and bases destroy living tissue?

Antacids: A Basic Remedy

The stomach secretes hydrochloric acid as an aid in the digestion of food. The stomach lining is normally protected from the corrosive effects of the acid by a mucosal lining. Holes can develop in the lining, however, that allow the acid to attack the underlying tissue. These lesions, called *ulcers,* frequently bleed and are often quite painful. Over the years ulcers have been treated by restrictive diets and by preparations called **antacids,** basic compounds that neutralize the stomach acid. Today we know that most ulcers are caused by bacteria and the preferred treatment includes antibiotics. Nevertheless, antacids are still widely used to treat heartburn, a condition in which stomach acid refluxes into the esophagus, which has no protective lining. The antacid neutralizes the acid and relieves the burning sensation.

Many brands of antacids are available in the United States. Often aggressively advertised, their sales are about a billion dollars annually. Despite the many brand names, antacids contain only a few different ingredients, typically calcium carbonate, aluminum hydroxide, magnesium hydroxide, or sodium bicarbonate (Table 9.4). (Drugs such as Prilosec, Zantac, Tagamet, and Pepcid AC are not antacids. Rather, these drugs inhibit the release of hydrochloric acid by the stomach.) Let's look at some popular antacids from the standpoint of acid-base chemistry.

Sodium bicarbonate ($NaHCO_3$), commonly called baking soda, is an old standby antacid. It is probably safe and effective for occasional use by most people. Overuse can make the blood too alkaline, a condition called *alkalosis* (see Section 10.8). Sodium bicarbonate is not recommended for those with hypertension (high blood pressure) because high concentrations of sodium ion tend to aggravate the condition. Alka-Seltzer contains sodium bicarbonate, citric acid, and aspirin. When placed in water, its bicarbonate ions react with hydrogen ions from the citric acid, producing the familiar fizz.

$$HCO_3^-(aq) + H^+(aq) \rightarrow CO_2(g) + H_2O$$

The aspirin in Alka-Seltzer can be harmful to people with ulcers and other stomach disorders. (You can make your own aspirin-free "Alka-Seltzer." Simply place half a teaspoon of baking soda in a glass of orange juice. [What is the acid and what is the base in this reaction?])

Calcium carbonate, $CaCO_3$, is another common antacid ingredient. It is safe in small amounts, but regular use can cause constipation. Also, calcium carbonate apparently can cause *increased* acid secretion after a few hours. Temporary relief may be achieved at the expense of a worse problem later. Tums and many other antacid tablets are simply flavored calcium carbonate. The neutralization reaction is

$$CaCO_3(s) + 2\,H^+(aq) \rightarrow Ca^{2+}(aq) + CO_2(g) + H_2O$$

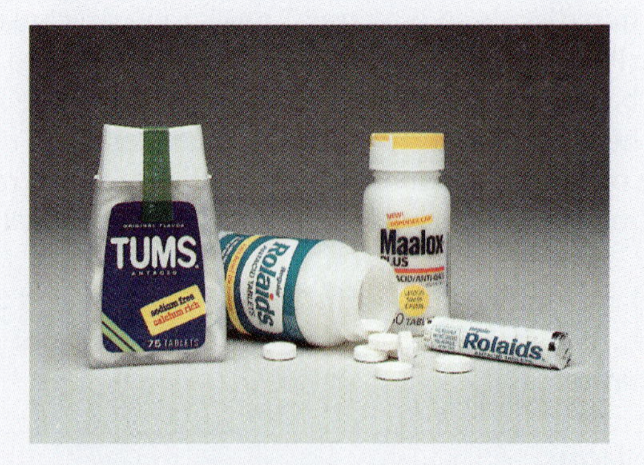

▲ A great variety of antacid formulations are available to the consumer. The main ingredients comprise a short list of basic compounds.

Table 9.4 Some Common Antacids

Commercial Product	Antacid Ingredient(s)
Alka-Seltzer	$NaHCO_3$, citric acid, aspirin
Amphojel	$Al(OH)_3$
Baking soda	$NaHCO_3$
DiGel	$CaCO_3$
Maalox	$Al(OH)_3$, $Mg(OH)_2$
Milk of magnesia	$Mg(OH)_2$
Mylanta	$Al(OH)_3$, $Mg(OH)_2$
Rolaids	$AlNa(OH)_2CO_3$
Rolaids Sodium-Free[a]	$CaCO_3$, $Mg(OH)_2$
Tums	$CaCO_3$

[a]Sodium-containing antacids are not recommended for people with hypertension (high blood pressure).

Aluminum hydroxide [$Al(OH)_3$], another popular antacid ingredient, is the only antacid in Amphojel but occurs in combination in many popular products. Like calcium carbonate, it can cause constipation in large doses. The neutralization reaction is

$$Al(OH)_3(s) + 3\,H^+(aq) \rightarrow Al^{3+}(aq) + 3\,H_2O$$

There is concern that antacids containing aluminum ions deplete the body of essential phosphate ions. The aluminum phosphate formed is insoluble and is excreted from the body.

$$Al^{3+}(aq) + PO_4^{\;3-}(aq) \rightarrow AlPO_4(s)$$

Magnesium compounds constitute the fourth category of antacids. These include magnesium carbonate ($MgCO_3$) and magnesium hydroxide [$Mg(OH)_2$]. Milk of magnesia is a suspension of magnesium hydroxide in water. It is sold under several brand names, including Phillips'. In small doses, magnesium compounds act as antacids. In large doses, they act as laxatives. Magnesium ions are poorly absorbed in the digestive tract. Rather, these small, dipositive ions draw water into the colon (large intestine), causing the laxative effect.

Claims of "fast action" for antacids are almost meaningless. All acid-base reactions are almost instantaneous. Some tablets may dissolve a little more slowly than others. You can speed their action by chewing them.

Some popular antacids, including Maalox and Mylanta, combine aluminum hydroxide, which tends to cause constipation, and a magnesium compound, which acts as a laxative. These tend to counteract one another.

Rolaids contains the complex substance aluminum sodium dihydroxy carbonate [$AlNa(OH)_2CO_3$]. Both the hydroxide ion and the carbonate ion consume acid. Sodium-free Rolaids contains calcium carbonate and magnesium hydroxide.

Antacids interact with other medications. Anyone taking any type of medication should consult a physician before taking an antacid. Generally, antacids are safe and effective for occasional use in small amounts. All antacids are basic compounds and generally may be chosen on the basis of price. Anyone with severe or repeated attacks of indigestion should consult a physician. Self-medication in such cases might be dangerous.

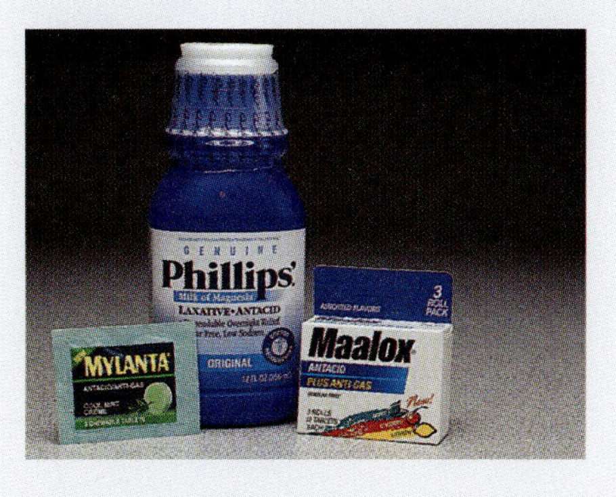

◀ A bottle of milk of magnesia. Water-insoluble hydroxides, such as $Mg(OH)_2$ (milk of magnesia) and $Al(OH)_3$, are used as antacids. One should *never* use a water-soluble ionic hydroxide, because in high concentrations $OH^-(aq)$ is strongly basic. It causes severe burning and scarring of tissue.

Review Questions

9.14 Name some of the active ingredients in antacid tablets.

9.15 Should a person who has hypertension be advised to use baking soda or milk of magnesia as an antacid? Why?

9.16 How does magnesium hydroxide, in large doses, act as a laxative? Would other magnesium compounds have similar effects?

Acids, Bases, and Eyes

The covering of the front of the eye is the *cornea,* a transparent tissue consisting of a matrix of protein molecules arranged in layers that do not interfere with the passage of light. Damage to the cornea can cause blindness even though the rest of the eye is perfectly normal.

As we shall see (Chapter 21), many substances *denature* proteins. The white of an egg, which consists largely of a watery composite of the protein albumin, provides a familiar example. Cooking denatures the albumin, converting the transparent egg white to an opaque solid.

The cornea of the eye is readily damaged by small quantities of acids and bases. Even in low concentrations that would not seriously damage the skin, acids and bases can denature the protein in the cornea, rendering it opaque and thus causing blindness. Now you know why your laboratory instructor "makes a fuss" about wearing goggles!

Summary

Arrhenius defined an **acid** as any substance that yields hydrogen ions (H^+) in aqueous solution. Because a hydrogen *atom* is one proton and one electron, a hydrogen *ion* is simply a proton. More generally, an **acid** is any substance that donates protons in a chemical reaction. In aqueous solution, a proton is always associated with a water molecule, forming a hydronium ion (H_3O^+). Every acid has a **conjugate base,** and every base has a **conjugate acid.** In an acid-base reaction, the forward reaction is between an acid and a base; the reverse reaction is between a conjugate base and a conjugate acid.

A **strong acid** is completely ionized when dissolved in water. In a **weak acid,** only some of the molecules ionize, forming relatively few H^+ ions.

Arrhenius defined a **base** as any substance that yields hydroxide ions (OH^-) in aqueous solution. A more general definition of **base** is any substance that accepts protons in a chemical reaction.

A **strong base** is one that dissociates completely when dissolved in water. A **weak base** is one that ionizes only slightly in water and so produces relatively few hydroxide ions.

Nonmetal oxides react with water to form acids and are **acidic anhydrides.** Metal oxides react with water to form bases and are called **basic anhydrides.**

A **salt** is a compound containing the anion of an acid and the cation of a base. A reaction between an acid and a base to form water and a salt is a **neutralization.** Carbonate (CO_3^{2-}) and bicarbonate (HCO_3^-) salts react with acids to produce the weak acid carbonic acid (H_2CO_3), which decomposes to gaseous carbon dioxide and water.

Key Terms

acid (9.1)
acidic anhydride (9.5)
amine (9.4)
antacid (9.8)
base (9.1)
basic anhydride (9.5)
conjugate acid (9.1)
conjugate base (9.1)
diprotic acid (9.2)
monoprotic acid (9.2)
net ionic equation (9.6)
neutralization (9.6)
polyprotic acid (9.2)
proton acceptor (9.1)
proton donor (9.1)
salt (9.1)
strong acid (9.2)
strong base (9.4)
triprotic acid (9.2)
weak acid (9.2)
weak base (9.4)

Problems

Brønsted-Lowry Acids and Bases

1. For each of the following identify the conjugate acid-base pairs by using acid(1), base(1), and so on.
   a. $HClO_2 + H_2O \rightleftharpoons H_3O^+ + ClO_2^-$
   b. $HSO_4^- + NH_3 \rightleftharpoons NH_4^+ + SO_4^{2-}$
   c. $HCO_3^- + OH^- \rightleftharpoons CO_3^{2-} + H_2O$
2. For each of the following identify conjugate acid-base pairs as acid(1), base(1), and so on.
   a. $HSO_4^- + F^- \rightleftharpoons HF + SO_4^{2-}$
   b. $NH_4^+ + Cl^- \rightleftharpoons NH_3 + HCl$
   c. $HCl + CH_3COO^- \rightleftharpoons CH_3COOH + Cl^-$

Strong and Weak Acids and Bases

3. In aqueous solution, which of the following substances are *strong* acids, which are *weak* acids, and which are neither?
   a. CH_3OH b. HBr c. HCOOH d. HNO_2
4. In aqueous solution, which of the following substances are *strong* bases, which are *weak* bases, and which are neither?
   a. CH_3NH_2 b. CH_3OH
   c. LiOH d. $Ba(OH)_2$
5. Identify each of the following substances as a strong acid, weak acid, strong base, weak base, or salt.
   a. Na_2SO_4 b. KOH
   c. $CaCl_2$ d. CH_3CH_2COOH
6. Identify each of the following substances as a strong acid, weak acid, strong base, weak base, or salt.
   a. HI b. NH_3
   c. NH_4I d. $Ca(OH)_2$
7. Based on the descriptions given, classify each of the following compounds as a strong acid, weak acid, weak base, or strong base.
   a. Thallium hydroxide (TlOH) is ionic in the solid state and is quite soluble in water.
   b. Hydrogen perchlorate ($HClO_4$) reacts completely with water to form hydronium ions and perchlorate ions.
8. Based on the descriptions given, classify each of the following compounds as a strong acid, weak acid, weak base, or strong base.
   a. Hydrogen sulfide (H_2S) gas reacts slightly with water to form relatively few hydrogen ions and hydrogen sulfide ions (HS^-).
   b. Methylamine (CH_3NH_2) gas reacts slightly with water to form relatively few hydroxide ions and methylammonium ions ($CH_3NH_3^+$).
9. According to the equation, is phenol, C_6H_5OH, an acid or a base? Should it be classified as strong or weak?

$$C_6H_5OH + H_2O \rightleftharpoons C_6H_5O^- + H_3O^+$$

10. According to the equation, is aniline, $C_6H_5NH_2$, an acid or a base? Should it be classified as strong or weak?

$$C_6H_5NH_2 + H_2O \rightleftharpoons C_6H_5NH_3^+ + OH^-$$

Names of Acids and Bases

11. Give the formulas for the following acids and bases.
   a. hydrochloric acid b. sulfuric acid
   c. carbonic acid d. lithium hydroxide
   e. magnesium hydroxide f. potassium hydroxide
12. Give the formulas for the following acids and bases.
   a. nitric acid b. sulfurous acid
   c. phosphoric acid d. hydrosulfuric acid
   e. calcium hydroxide
13. Name the following acids and bases.
   a. NaOH b. H_3PO_4 c. HNO_3
   d. H_2SO_3 e. $Ca(OH)_2$ f. H_2S
14. Name the following acids and bases.
   a. HCl b. H_2SO_4 c. LiOH
   d. H_2CO_3 e. $Mg(OH)_2$
15. What are the names of (**a**) Br^- and HBr(aq) and (**b**) of NO_2^- and HNO_2(aq)?
16. What are the names of (**a**) PO_3^{3-} and H_3PO_3(aq) and (**b**) of ClO^- and HClO?
17. Se^{2-} is selenide ion. What is the name of H_2Se(aq)?
18. $C_2O_4^{2-}$ is oxalate ion and $C_6H_5CO_2^-$ is benzoate ion. What are the names of (**a**) $H_2C_2O_4$(aq) and (**b**) $C_6H_5CO_2H$(aq)?

Ionization of Acids and Bases

19. Write equations showing the ionization of the following acids.
   a. HI(aq) b. CH_3CH_2COOH(aq)
   c. HNO_2(aq) d. $H_2PO_4^-$(aq)
20. Write equations showing the ionization of the following as Brønsted-Lowry weak acids.
   a. HOClO b. $CH_3CH_2CH_2COOH$
   c. HCN d. C_6H_5COOH
21. Write equations showing the ionization of the following acids and bases.
   a. HNO_3(aq) b. KOH(aq)
   c. HCOOH(aq) d. CH_3NH_2(aq)
22. Write equations showing the ionization of the following acids and bases.
   a. $HC_2O_4^-$(aq) b. $Ba(OH)_2$(aq)
   c. $HClO_2$(aq) d. $C_6H_5NH_2$(aq)
23. Use the definitions to identify the first compound in each equation as an acid or a base. (Hint: What is *produced* by the reaction?)
   a. $C_5H_5N + H_2O \longrightarrow C_5H_5NH^+ + OH^-$
   b. $CH_3COCOOH + H_2O \longrightarrow CH_3COCOO^- + H_3O^+$
   c. $C_6H_5O^- + H_2O \longrightarrow C_6H_5OH + OH^-$
24. Use the definitions to identify the first compound in each equation as an acid or a base.
   a. $C_6H_5SH + H_2O \longrightarrow C_6H_5S^- + H_3O^+$
   b. $CH_3NH_2 + H_2O \longrightarrow CH_3NH_3^+ + OH^-$
   c. $C_6H_5SO_2NH_2 + H_2O \longrightarrow C_6H_5SO_2NH^- + H_3O^+$

Acidic and Basic Anhydrides

25. Give the formula for the compound formed (**a**) when sulfur trioxide reacts with water and (**b**) when potassium oxide reacts with water. In each case, indicate whether the product is an acid or a base.
26. Give the formula for the compound formed (**a**) when magnesium oxide reacts with water and (**b**) when carbon dioxide reacts with water. In each case, indicate whether the product is an acid or a base.
27. What is the anhydride of H_2SeO_4?
28. What is the anhydride of $Sr(OH)_2$?

Neutralization Reactions

29. Write the "complete formula" equation and the net ionic equation for the reaction of sodium hydroxide with hydrochloric acid.
30. Write the "complete formula" equation and the net ionic equation for the reaction of lithium hydroxide with nitric acid.
31. Write the "complete formula" equation and the net ionic equation for the reaction of 1 mol of calcium hydroxide with 2 mol of hydrochloric acid.
32. Write the "complete formula" equation and the net ionic equation for (**a**) the reaction of 1 mol of sulfuric acid with 2 mol of potassium hydroxide and (**b**) for the reaction of 1 mol of phosphoric acid with 3 mol of sodium hydroxide.

Reactions of Acids with Carbonates and Bicarbonates

33. Write the net ionic equation for the reaction of a bicarbonate salt with an acid.
34. Write the net ionic equation for the reaction of a carbonate salt with an acid.
35. Write the "complete formula" equation and the net ionic equation for the reaction of solid sodium bicarbonate salt with acetic acid.
36. Write the "complete formula" equation and the net ionic equation that describes the action of sulfuric acid on marble.

Additional Problems

37. Can a substance be a Brønsted-Lowry acid if it does not contain H atoms? Are there any characteristic atoms that must be present in a Brønsted-Lowry base?
38. According to the Arrhenius theory, all acids have one element in common. What is that element? Are all compounds containing that element acids? Explain.
39. According to the Arrhenius theory, are all compounds containing OH groups bases? Explain.
40. Write equations showing how hydrogen phosphate ion, HPO_4^{2-}, can act either as a Brønsted-Lowry acid or as a Brønsted-Lowry base.
41. Slaked lime, $Ca(OH)_2(s)$, can be used to reduce excess acidity in natural waters such as lakes. Write a net ionic equation for the reaction that occurs.
42. With continued use, automatic coffeemakers often develop mineral deposits ($CaCO_3$). The manufacturer's instructions generally call for removing the deposit by treatment with vinegar. Write a net ionic equation for the reaction that occurs. (Hint: Recall that vinegar contains acetic acid, CH_3COOH.)
43. A paste of sodium hydrogen carbonate (sodium bicarbonate) and water can be used to relieve the pain of an ant bite. The irritant in the ant bite is formic acid (HCOOH). Write a net ionic equation for the reaction that occurs.
44. Decide which of the following aqueous solutions has the highest concentration of H^+ ion: 0.10 M HCl, 0.10 M H_2SO_4, 0.10 M CH_3COOH, or 0.10 M NH_3.
45. Acids with three or more H atoms can exist in two forms. The *ortho* form has two more H atoms and one more O atom than the *meta* form. For example, H_3PO_4 is sometimes called orthophosphoric acid; metaphosphoric acid is HPO_3. Orthosilicic acid is H_4SiO_4. What is the formula for metasilicic acid? For sodium metasilicate?
46. Boric acid is H_3BO_3. What is the formula for metaboric acid? For sodium metaborate? (See Problem 45.)

Acids and Bases II

◀ Laboratory reagent bottles are labeled as "CON" (for concentrated) or "DIL" (for dilute). Most acid and base solutions are made starting with these stock solutions.

Learning Objectives/Study Questions

1. What is the relationship between the moles of a solute in a concentrated solution and after it has been diluted?
2. What is an acid-base titration? How is it accomplished?
3. What is the ion product for water? How is it used?
4. What is the pH scale? How can we calculate the concentration of hydrogen ion from a pH value?
5. What is meant by the ionization constant for a weak acid or weak base?
6. How can we calculate the concentration of hydrogen ion from an ionization constant and a given concentration of a weak acid or weak base?
7. What is a buffer? How does a buffer work?
8. How does the body maintain the pH of blood and other fluids?
9. How can we use the Henderson-Hasselbalch equation to calculate the pH of a buffer?

Our bodies are largely solutions—quite special solutions, of course, but solutions nonetheless. The many solutes in our blood and other body fluids are maintained in delicate balances, particularly those solutes that are acids and bases. If the acidity of the blood changes very much, its capacity to carry oxygen is diminished. Because many bodily processes produce acids, the control of acidity is literally a matter of life and death.

Our bodies have developed a marvelously complex yet efficient mechanism for maintaining the proper acid-base balance. Before we can talk about this mechanism in a meaningful way, however, we need to develop a few concepts—concepts that are more quantitative in nature than those developed in Chapter 9. It is important to know how exact concentrations of acids and bases are determined and expressed. We must understand the concept of pH, particularly as it relates to the chemistry of the blood and other body fluids. Most important, we must see just how, through substances called buffers, the level of acidity is controlled.

Finally, after describing qualitatively the equilibriums established when a weak acid, a weak base, a salt of either, or some combination of these is dissolved in water, we will treat these equilibriums in a more quantitative fashion.

10.1 Concentrations of Acids and Bases

In diluting acid solutions, always add acid to water. If water is added to acid, the heat of reaction can cause the water to boil, spattering the acid and causing burns.

A common laboratory task is to make a more dilute solution from a more concentrated one. This process, called **dilution,** is a procedure to make solutions of a desired concentration. The principle of dilution is that *addition of solvent does not change the amount of solute in a solution but does change its concentration.* Examples of commercially available *concentrated* acid and base solutions include concentrated sulfuric acid (18 M), concentrated hydrochloric acid (12 M), concentrated ammonia (15 M), and concentrated nitric acid (16 M).

Recall the definition of molarity.

$$\text{Molarity (M)} = \frac{\text{moles of solute}}{\text{liters of solution } (V)}$$

We can solve this relationship for moles of solute.

▶ Visualizing the dilution of a solution. When pure water is added to a solution of methanol (CH_3OH) (left), a more dilute solution is produced (right). However, the diluted solution contains the same number of CH_3OH molecules as the original solution.

$$\text{Moles of solute} = \text{molarity (M)} \times \text{liters of solution } (V)$$
$$= \text{M} \times V$$

Because the number of moles of solute does not change upon dilution, we can write

$$(\text{Moles of solute})_{\text{conc}} = (\text{moles of solute})_{\text{dil}}$$

From the rearranged relationship we can substitute to obtain

$$\text{M}_{\text{conc}} \times V_{\text{conc}} = \text{M}_{\text{dil}} \times V_{\text{dil}}$$

Although the derivation of this expression assumes solution volumes in liters, any unit of volume can be used as long as it is the same on both sides of the equation.

The following examples illustrate the types of calculations involved in dilution problems.

Example 10.1

What volume of an 18.0 M H_2SO_4 stock solution would you use to prepare 2.00 L of 4.00 M H_2SO_4?

Solution

First, let's solve the dilution formula for the unknown quantity, the volume of the concentrated solution.

$$V_{\text{conc}} = \frac{\text{M}_{\text{dil}} \times V_{\text{dil}}}{\text{M}_{\text{conc}}}$$

Then we substitute the known quantities.

$$V_{\text{conc}} = \frac{4.00\ \cancel{\text{M}} \times 2.00\ \text{L}}{18.0\ \cancel{\text{M}}} = 0.444\ \text{L}$$

To make the solution, we measure out 0.444 L (444 mL) of 18.0 M H_2SO_4 and add it to enough water to make 2.00 L of solution.

Practice Exercise

A stock bottle of aqueous ammonia indicates that the solution is 15.0 M NH_3. What volume of this concentrated solution is needed to make 1.25 L of 2.50 M NH_3 solution?

Example 10.2

What is the molarity of a solution made by diluting 100.0 mL of a 15.7 M HNO_3 stock solution to 1.250 L?

Solution

First solve the dilution formula for the unknown quantity, the molarity of the dilute solution.

$$\text{M}_{\text{dil}} = \frac{\text{M}_{\text{conc}} \times V_{\text{conc}}}{V_{\text{dil}}}$$

$$\text{M}_{\text{dil}} = \frac{15.7\ \text{M} \times 100.0\ \cancel{\text{mL}}}{1.250\ \cancel{\text{L}}} \times \frac{1\ \cancel{\text{L}}}{1000\ \cancel{\text{mL}}} = 1.26\ \text{M}$$

The diluted solution will consist of 1.26 M HNO_3.

Practice Exercise

What is the molarity of a solution made by diluting 25.0 mL of a 23.5 M formic acid (HCOOH) solution to a volume of 775 mL?

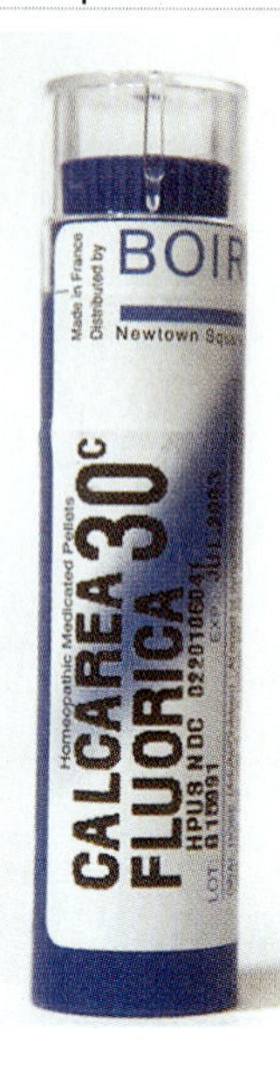

Homeopathic Dilutions

Homeopathic medicine is based on the idea of similarity between the disease the patient has and the symptoms induced in healthy individuals by the medicine. For example, a homeopathic practitioner might treat diarrhea with a small dose of a laxative. Most homeopathic remedies are prepared at extremely high dilutions. Each dilution is usually 1:100 and six successive such dilutions (denoted 6c) are usual. The concentration is thus diluted to 10^{-12} (one trillionth) of that of the original material. Even higher dilutions of 30c (10^{-60}) are common.

If one started with a 1 M solution of the material, the 6c dilution would be 10^{-12} M. The 30c dilution would be 10^{-60} M! A 10^{-23} M solution has only 6 molecules of material. Thus the 30c dilution would contain far less than one molecule of the material in a liter of the homeopathic remedy. Most scientists are skeptical of the validity of homeopathic medicine.

✓ Review Questions

10.1 What special meaning is conveyed when we refer to a *concentrated* acid or base?

10.2 Is the concentration of a solution changed by dilution? Is the number of moles of solute changed by dilution?

10.2 Acid-Base Titrations

The concept of molarity gives us a way to obtain a specific number of moles of a solute by measuring out a given volume of solution. The volume measurement is often made in a *buret,* a graduated, long glass tube calibrated to deliver precise volumes of solution through a stopcock valve. A buret is the chief instrument used in a **titration,** a procedure in which two reactants in solution react in the precise proportions shown by the chemical equation for the reaction.

Consider, for example, a typical acid-base titration. A measured volume of a solution of an acid of unknown concentration is transferred to a flask. Then, a solution of a base of *known* concentration is added carefully from a buret until the reaction of the acid with the base is just complete. The point at which the acid is just neutralized is called the **equivalence point** of the titration. At that point, the number of moles of OH^- added equals the number of moles of H^+ that were in the sample of acid. Figure 10.1 illustrates the titration of an acid solution of unknown concentration with a sodium hydroxide solution of known concentration.

An acid-base *indicator* is a substance that has the property of changing color when an acid or base is added to it.

The equivalence point of a titration reaction often is determined with an **indicator dye,** a substance that changes color as the reaction is completed. Litmus, mentioned in Section 9.1, is one such indicator. A better indicator for some acid-base titrations is phenolphthalein. In the titration pictured in Figure 10.1, phenolphthalein is colorless in the hydrochloric acid solution and remains so until the equivalence point is reached. With the addition of as little as one drop of NaOH(aq) beyond the equivalence point, however, the reaction mixture becomes basic and the phenolphthalein indicator turns pink. The titration is stopped when the pink color appears, the *end point* of the titration. (An indicator is chosen so that the end point and equivalence point coincide.)

The following examples illustrate calculations involving acid-base titrations.

(a)

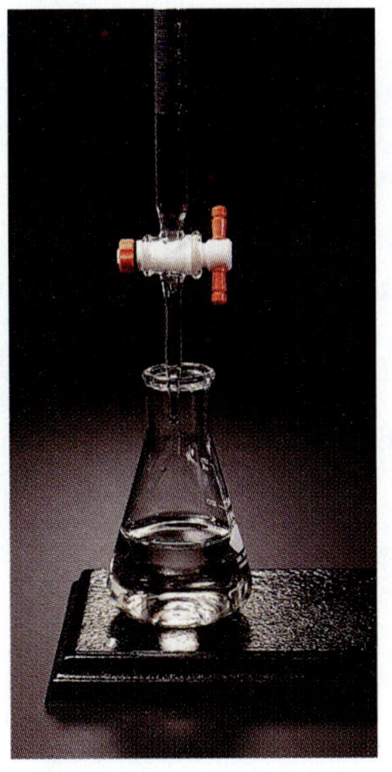
(b)

(c)

◀ **Figure 10.1**
The amount of acid or base in a solution is determined by *titration.* In the experiment shown here, (a) a precisely measured volume of a HCl(aq) solution of unknown concentration is added from a volumetric pipet to a quantity of water in a small flask. Phenolphthalein indicator solution (in the dropper bottle) is also added to the flask. (b) An NaOH(aq) solution of known concentration is slowly added from a previously filled buret. (c) As long as the acid is in excess, the solution remains colorless. When the acid is just neutralized, an additional drop of NaOH(aq) causes the solution to become slightly basic; the phenolphthalein indicator turns a faint pink color. This is the equivalence point of the titration

Example 10.3

Sodium hydroxide reacts with hydrochloric acid.

$$\text{NaOH(aq)} + \text{HCl(aq)} \rightarrow \text{NaCl(aq)} + \text{H}_2\text{O}$$

A flask contains 20.00 mL of HCl(aq) of unknown concentration. It is just neutralized by addition of 10.25 mL of 0.2010 M NaOH. What is the molarity of the acid?

Solution

First, convert the quantity of NaOH to moles, using the molarity (moles/liter) as a conversion factor.

$$10.25\ \cancel{\text{mL NaOH(aq)}} \times \frac{1\ \cancel{\text{L NaOH(aq)}}}{1000\ \cancel{\text{mL NaOH(aq)}}} \times \frac{0.2010\ \text{mol NaOH}}{1\ \cancel{\text{L NaOH (aq)}}} = 0.002060\ \text{mol NaOH}$$

Then, use the mole ratio from the chemical equation to relate amounts, in moles, of NaOH and HCl.

$$0.002060\ \cancel{\text{mol NaOH}} \times \frac{1\ \text{mol HCl}}{1\ \cancel{\text{mol NaOH}}} = 0.002060\ \text{mol HCl}$$

Finally, we divide this amount of HCl by the volume of HCl(aq) to get the molarity of the HCl(aq).

$$\frac{0.002060\ \text{mol HCl}}{20.00\ \cancel{\text{mL}}} \times \frac{1000\ \cancel{\text{mL}}}{1\text{L}} = 0.1030\ \text{M}$$

The HCl(aq) is 0.1030 M.

Practice Exercise

Calculate the molarity of an HNO_3 solution if 23.0 mL of it requires 18.7 mL of 0.108 M NaOH for neutralization.

$$\text{HNO}_3\text{(aq)} + \text{NaOH(aq)} \rightarrow \text{NaNO}_3\text{(aq)} + \text{H}_2\text{O}$$

Example 10.4

A flask contains 25.00 mL of $H_2SO_4(aq)$ of unknown concentration. Titration of the sample requires 25.20 mL of 0.1000 M NaOH(aq) for neutralization. What is the molarity of the sulfuric acid solution?

Solution

First we must write the equation for the reaction of sodium hydroxide with sulfuric acid.

$$2\,NaOH(aq) + H_2SO_4(aq) \rightarrow Na_2SO_4(aq) + 2\,H_2O$$

Next we convert the quantity of NaOH to moles, using the molarity (moles/liter) as a conversion factor.

$$0.02520\ \cancel{L\ NaOH(aq)} \times \frac{0.1000\ mol\ NaOH}{1\ \cancel{L\ NaOH(aq)}} = 0.002520\ mol\ NaOH$$

Then use the mole ratio from the chemical equation to relate moles of H_2SO_4 to moles of NaOH.

$$0.002520\ \cancel{mol\ NaOH} \times \frac{1\ mol\ H_2SO_4}{2\ \cancel{mol\ NaOH}} = 0.001260\ mol\ H_2SO_4$$

Finally use the definition of molarity to write

$$\frac{0.001260\ mol\ H_2SO_4}{25.00\ \cancel{mL}} \times \frac{1000\ \cancel{mL}}{1\ L} = 0.05040\ M\ H_2SO_4$$

Practice Exercises

A. What volume of 0.1000 M HCl is required to just neutralize 40.00 mL of 0.5020 M $Ca(OH)_2$ solution?

$$2\,HCl(aq) + Ca(OH)_2(aq) \rightarrow CaCl_2(aq) + 2\,H_2O$$

B. A flask contains 20.00 mL of KOH(aq) of unknown concentration. Titration of the solution requires 15.62 mL of 0.1104 M $H_2SO_4(aq)$. Calculate the molarity of the KOH(aq).

✓ Review Questions

10.3 Define or illustrate each of the following terms.

a. equivalence point **b.** indicator **c.** titration

10.4 Describe how you would determine the concentration of a solution of an acid by using a solution of a base of known concentration.

10.5 Would 1.00 mol of NaOH neutralize (**a**) 3.00 mol of H_3PO_4? (**b**) 0.500 mol of H_2SO_4?

10.6 Would 0.50 mol of $Ca(OH)_2$ neutralize all of the acid in (**a**) 0.50 mol of H_3PO_4? (**b**) 1.00 mol of HCl?

10.3 The pH Scale

When we think of water, we think of H_2O molecules. But even the purest water isn't all H_2O. About 1 molecule in 500 million transfers a proton to another water molecule, yielding a hydronium ion and a hydroxide ion.

$$H_2O + H_2O \rightleftharpoons H_3O^+ + OH^-$$

Acids, Bacteria, and Ulcers

For many years the conventional wisdom was that ulcers were caused by excessive hydrochloric acid in the stomach and that this excess acidity was caused by stress, spicy foods, alcohol, anti-inflammatory drugs such as aspirin, and cigarette smoking. In the 1980s, however, a young Australian physician, Barry J. Marshall, showed that most ulcers are caused by bacteria, *Helicobacter pylori,* that live in the mucus layer that lines the stomach. The bacteria, not the excess acid, erode the stomach lining. Other studies have confirmed Marshall's findings and have also linked these bacteria to stomach cancer. Ulcer treatment has therefore largely changed from diet, antacids, and drugs that block acid secretion, to antibiotics.

The stomach's mucus layer helps to protect its cells from its strong acid contents. The mucus also provides some protection for the bacteria, which ordinarily would be killed by the acid. But the bacteria have their own defense; they produce urease, an enzyme that catalyzes the breakdown of urea, a waste product of the cells' metabolism, to carbon dioxide and ammonia.

$$H_2O + \underset{\text{Urea}}{CO(NH_2)_2} \xrightarrow{\text{urease}} CO_2 + 2\,NH_3$$

The ammonia neutralizes the stomach acid in the vicinity of the bacteria. The practice of acid-base chemistry is not limited to humans.

In other words, water is in equilibrium with hydronium ion and hydroxide ion, or—ignoring the hydration of the proton—a water molecule ionizes into a hydrogen ion and a hydroxide ion.

$$H_2O \rightleftharpoons H^+ + OH^-$$

No matter which way we represent the process, the equilibrium point lies *far to the left.* The experimentally determined equilibrium concentrations in pure water at 25 °C are

$$[H^+] = [OH^-] = 1.0 \times 10^{-7}\,M$$

Keep in mind that the bracketed symbols are shorthand for the concentration of the enclosed ions in moles per liter. Thus $[H^+]$ is read as the molarity of hydrogen ion and $[OH^-]$ as the molarity of hydroxide ion. The product of these ion concentrations, called the **ion product of water (K_w),** has the following value at 25 °C.

$$K_w = [H^+][OH^-] = (1.0 \times 10^{-7})(1.0 \times 10^{-7}) = 1.0 \times 10^{-14}$$

This ion product relationship is of considerable importance. It doesn't just describe self-ionization in pure water, but it applies to *all aqueous solutions*—that is, to solutions of acids, bases, salts, or even molecular substances. As an illustration, consider Example 10.5.

Example 10.5

What are the $[H^+]$ and the $[OH^-]$ in a solution that is 0.00015 M HCl?

Solution

Because HCl is a strong acid, its ionization is complete.

$$HCl \rightarrow H^+ + Cl^-$$

And because the $[H^+]$ produced by the HCl (0.00015 M) is so much greater than that found in pure water (0.0000001 M), we can state with assurance that in this solution

$$[H^+] = 0.00015 \text{ M} = 1.5 \times 10^{-4} \text{ M}$$

So, solving the K_w expression for the unknown quantity, we can now calculate $[OH^-]$ in the solution.

$$[OH^-] = \frac{K_W}{[H^+]} = \frac{1.0 \times 10^{-14}}{1.5 \times 10^{-4}} = 6.7 \times 10^{-11} \text{ M}$$

Practice Exercises

A. What are the $[H^+]$ and the $[OH^-]$ in a solution that is 0.0022 M HNO_3?
B. What are the $[H^+]$ and the $[OH^-]$ in a solution that is 0.025 M NaOH?

The *H* in pH stands for "hydrogen" and the *p* refers to "power or potential."

Exponential notation (1.5×10^{-4}) is a convenient way to express a small quantity such as 0.00015, but there's an even more convenient way. In 1909 the Danish biochemist Søren P. L. Sørenson proposed that the number in the exponent be used to express acidity. His practice, which came to be known as the **pH scale,** is still used today. Thus pH is defined as the negative of the logarithm of $[H^+]$. The reaction of most people on first encountering this definition of pH is "You call this more convenient?" Well, it really is. To determine the pH of a solution that has $[H^+] = 1.0 \times 10^{-n}$, the value of n is the value of the pH. A solution that has a hydrogen ion concentration of 1×10^{-4} M has a pH of 4. It is easier to say, "The pH is 4" than to say, "The hydrogen ion concentration is 1×10^{-4} M." They mean the same thing.

Example 10.6

What is the pH of a solution that has $[H^+] = 1.0 \times 10^{-5}$ M?

Solution

Given that $[H^+] = 1.0 \times 10^{-5}$ M, we need only to plug that value into the pH equation and solve for pH.

$$\text{pH} = -\log[H^+] = -\log(1.0 \times 10^{-5}) = -(-5.00) = 5.00$$

Practice Exercise

What is the pH of a solution that has $[H^+] = 1.0 \times 10^{-9}$ M?

Steps for Calculating pH on a Calculator

Step 1: Enter the entire value for the $[H^+]$, using exponential mode and change-of-sign key for 10^{-n}.
Step 2: Press the LOG key.
Step 3: Press the +/− (change-of-sign) key.

In Example 10.6, where $[H^+] = 1.0 \times 10^{-5}$ M, following these steps gives a pH value of 5.00. *Step 1:* Enter 1.0, press EXP, enter 5, and press +/−. *Step 2:* Press the LOG key. *Step 3:* Press +/−. You should have a value of 5.00. (Calculator brands differ somewhat; be sure you practice enough on yours to be confident of your answers.)

Example 10.7

What is the pH of a solution that has $[H^+] = 4.5 \times 10^{-3}$ M?

Solution

$$pH = -\log[H^+] = -\log(4.5 \times 10^{-3}) = -(-2.35) = 2.35$$

Practice Exercise

What is the pH of a solution that has $[H^+] = 2.7 \times 10^{-9}$ M?

To determine the $[H^+]$ corresponding to a given pH, we do an inverse calculation.

Example 10.8

What is the $[H^+]$ in a solution with pH = 2.19?

Solution

$$-\log[H^+] = pH = 2.19$$

$$\log[H^+] = -2.19$$

$$[H^+] = \text{antilog}(-2.19) = 10^{-2.19} = 6.5 \times 10^{-3}\text{ M}$$

(On a calculator, enter 2.19 and press the +/− key, the INV (or 2nd) key, and then the LOG key. The "inverse logarithm" is commonly called the *antilog*.)

Practice Exercise

What are the $[H^+]$ and $[OH^-]$ in a solution that has a pH of 10.79?

Just as we use pH to indicate a hydrogen ion concentration, we can use **pOH** to express a hydroxide ion concentration.

$$pOH = -\log[OH^-]$$

A 2.5×10^{-3} M NaOH solution has $[OH^-] = 2.5 \times 10^{-3}$ M and

$$pOH = -\log(2.5 \times 10^{-3}) = -(-2.60) = 2.60$$

The relationship between pH and pOH at 25 °C is

$$pH + pOH = 14.00$$

Thus, the pH of 2.5×10^{-3} M NaOH is

$$pH = 14.00 - pOH = 14.00 - 2.60 = 11.40$$

In pure water at 25 °C, where $[H^+] = [OH^-] = 1.0 \times 10^{-7}$ M, pH and pOH are both 7.00. Pure water and all aqueous solutions with pH = 7.00 are *neutral.* If the pH is less than 7.00, the solution is *acidic;* if the pH is above 7.00, the solution is *basic* or *alkaline.* As a solution becomes more acidic, $[H^+]$ increases and pH decreases. As a solution becomes more basic, $[H^+]$ decreases and pH increases.

Figure 10.2 gives the pH values of a number of familiar materials. Note that pH is on a logarithmic scale, and that every unit change in pH represents a *tenfold* change in $[H^+]$. Thus, lemon juice (pH ≈ 2.3) is somewhat more than ten times as

The symbol ≈ means "approximately equal to."

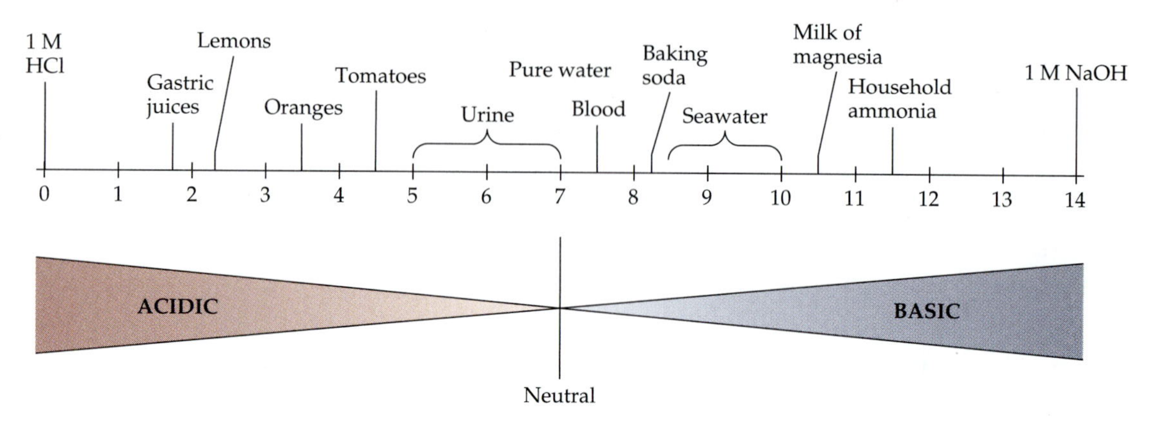

Figure 10.2 ▶
The pH scale. The pH values of common substances range from about 0 to 14.

acidic as orange juice (pH ≈ 3.5) and somewhat more than 100 times as acidic as tomato juice (pH ≈ 4.5).

Various dyes, some synthetic and others from plant and animal sources, have the property of changing color over a range of pH values. We noted in Chapter 9 that litmus, a vegetable dye, turns red in acidic solutions and blue in basic solutions. Actually, the change occurs over values of pH from 4.5 to 8.3. Certain combinations of indicator dyes will exhibit a whole range of colors as the pH changes from strongly basic to strongly acidic (Figure 10.3). By selecting the proper indicator, we can determine the pH of almost any clear, colorless aqueous solution.

More accurate measurements of pH can be made electrically with pH meters (Figure 10.4). Generally these instruments can measure pH to a precision of about 0.01 pH unit. Also, colors and turbidity generally do not interfere with electrical measurement of pH as they do with indicator color changes. Thus pH meters can be used on blood, urine, and other complex mixtures without prior treatment of the specimen.

▲ **Figure 10.3**
These pH papers change colors over a wide range of pH values.

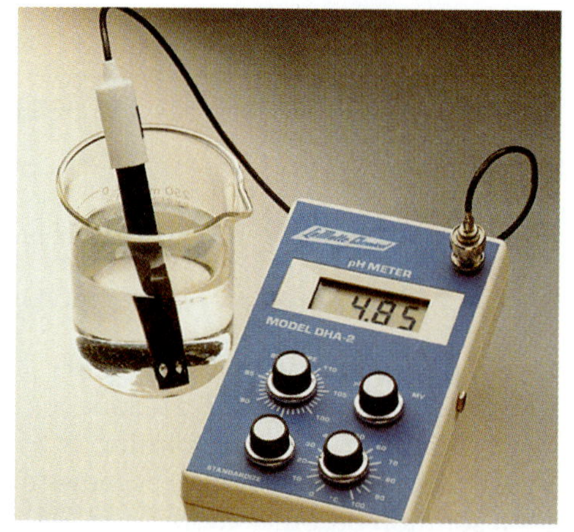

▲ **Figure 10.4**
A pH meter uses electrochemical reactions (Chapter 11) to measure the pH of a solution in comparison with a standard.

✓ Review Questions

10.7 Define or illustrate

a. K_w **b.** pH **c.** pOH **d.** pH meter

10.8 Indicate whether each pH value represents an acidic, basic, or neutral solution.

a. 11.2 **b.** 4.6 **c.** 7.0 **d.** 3.4

10.4 Equilibrium Calculations

We have seen (Section 9.2) that ionization of a weak acid or base is a reversible reaction, which typically reaches equilibrium when only a small percentage of molecules have ionized. We will now treat these equilibria in a more quantitative fashion. First, we need to look briefly at equilibria in general.

Equilibrium Constant Expressions

The proportions of reactants and products at equilibrium are determined by a simple relationship. Let's write a generalized reaction.

$$a\,\mathrm{A(g)} + b\,\mathrm{B(g)} \rightleftharpoons c\,\mathrm{C(g)} + d\,\mathrm{D(g)}$$

In this equation, A and B are reactants and C and D are products. The small letters are the coefficients for the substances. The relationship at a given temperature is given by the expression

$$K = \frac{[\mathrm{C}]^c \times [\mathrm{D}]^d}{[\mathrm{A}]^a \times [\mathrm{B}]^b}$$

The quantities in brackets stand for molar concentration at equilibrium. The quantity K is the **equilibrium constant,** and the entire expression is called the **equilibrium constant expression.** The coefficients in the generalized reaction become exponents in the equilibrium constant expression.

Let's consider a specific reaction.

$$\mathrm{H_2(g) + Cl_2(g) \rightleftharpoons 2\,HCl(g)}$$

The equilibrium constant expression for this reaction is

$$K = \frac{[\mathrm{HCl}]^2}{[\mathrm{H_2}] \times [\mathrm{Cl_2}]}$$

Similarly, for the reaction

$$\mathrm{N_2(g) + 3\,H_2(g) \rightleftharpoons 2\,NH_3(g)}$$

the equilibrium expression is

$$K = \frac{[\mathrm{NH_3}]^2}{[\mathrm{N_2}] \times [\mathrm{H_2}]^3}$$

Example 10.9

Write the equilibrium constant expression for each of the following reactions.

a. $\mathrm{H_2(g) + F_2(g) \rightleftharpoons 2\,HF(g)}$

b. $\mathrm{2\,NO(g) + O_2(g) \rightleftharpoons 2\,NO_2(g)}$

c. $\mathrm{CO(g) + H_2O(g) \rightleftharpoons CO_2(g) + H_2(g)}$

Solution

a. $K = \frac{[HF]^2}{[H_2] \times [F_2]}$ b. $K = \frac{[NO_2]^2}{[NO]^2 \times [O_2]}$

c. $K = \frac{[CO_2] \times [H_2]}{[CO] \times [H_2O]}$

Practice Exercise

Write the equilibrium constant expression for each of the following reactions.

a. $2\ HI(g) \rightleftharpoons H_2(g) + I_2(g)$ b. $3\ O_2(g) \rightleftharpoons 2\ O_3(g)$

c. $Xe(g) + 2\ F_2(g) \rightleftharpoons XeF_4(g)$

All of the preceding examples involve gaseous reactants and products. Next we apply a similar treatment to substances in aqueous solution. Special cases involving a solid reactant or product are considered in Section 11.7.

Ionization of Weak Acids

Now let's take another look at the ionization of acetic acid. The reaction is

$$CH_3COOH(aq) \rightleftharpoons H^+(aq) + CH_3COO^-(aq)$$

At equilibrium, the equilibrium constant expression[1] is

$$K_a = \frac{[H^+] \times [CH_3COO^-]}{[CH_3COOH]}$$

This constant is called the **acid ionization constant, K_a,** for the weak acid.

The strength of an acid is related to the degree of ionization. The greater the degree of ionization, the stronger the acid. (Strong acids, such as HCl, are essentially 100% ionized in dilute solution.) Also, the greater the degree of ionization, the larger the K_a. Table 10.1 lists the K_a values for several common acids. Don't forget: The larger the K_a (the less negative the exponent), the stronger the acid.

Ionization constants can be calculated from measurements of the hydrogen ion concentration, but a much more common type of problem involves calculating $[H^+]$ or pH from tabulated K_a values as shown in the following examples.

Example 10.10

Calculate the $[H^+]$ in a 0.10 M solution of acetic acid.

Solution

First, write the equation for the ionization.

$$CH_3COOH(aq) \rightleftharpoons H^+(aq) + CH_3COO^-(aq)$$

Then write the equilibrium constant expression.

$$K_a = \frac{[H^+] \times [CH_3COO^-]}{[CH_3COOH]} = 1.8 \times 10^{-5}$$

(The K_a of acetic acid is obtained from Table 10.1.)

---

[1]Recall that the reaction of acetic acid with water gives a hydronium ion and an acetate ion.

$$CH_3COOH + H_2O \rightleftharpoons H_3O^+ + CH_3COO^-$$

The equilibrium constant expression for this reaction would include $[H_2O]$. If you recognize that the hydronium ion is simply a hydrated proton, you will see that the two K expressions are the same.

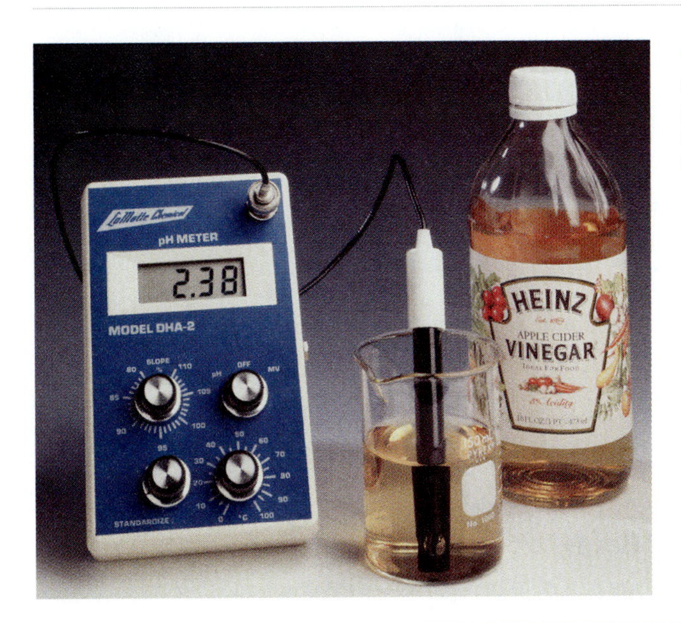

◀ We can show that acetic acid is a weak acid by measuring the pH of vinegar. Vinegar is approximately 1 M acetic acid. A strong acid of the same molarity would have a much lower pH, about pH 1.

For each CH_3COOH molecule that ionizes, one H^+ ion and one CH_3COO^- ion are formed. We can therefore let $x = [H^+] = [CH_3COO^-]$.

Because each H^+ formed represents one CH_3COOH ionized, the equilibrium concentration of CH_3COOH will be $0.10 - x$. Substitution yields

$$\frac{(x)(x)}{0.10 - x} = 1.8 \times 10^{-5}$$

The amount of CH_3COOH ionized is very small compared with the total amount of CH_3COOH that we started with, so let's assume that

$$0.10 - x \approx 0.10 = 1.0 \times 10^{-1}$$

We then have

$$\frac{x^2}{1.0 \times 10^{-1}} = 1.8 \times 10^{-5}$$

$$x^2 = (1.8 \times 10^{-5})(1.0 \times 10^{-1})$$

$$= 1.8 \times 10^{-6}$$

$$x = 1.3 \times 10^{-3}\,\text{M} = [H^+]$$

Table 10.1 Ionization Constants of Some Weak Acids in Water at 25 °C

Acid	Simplified Ionization Equilibrium	Ionization Constant, K_a
Inorganic Acids		
Nitrous acid	$HNO_2 \rightleftharpoons H^+ + NO_2^-$	7.2×10^{-4}
Hydrofluoric acid	$HF \rightleftharpoons H^+ + F^-$	6.6×10^{-4}
Hypochlorous acid	$HOCl \rightleftharpoons H^+ + OCl^-$	2.9×10^{-8}
Hypobromous acid	$HOBr \rightleftharpoons H^+ + OBr^-$	2.5×10^{-9}
Hydrocyanic acid	$HCN \rightleftharpoons H^+ + CN^-$	6.2×10^{-10}
Carboxylic Acids		
Chloroacetic acid	$CH_2ClCOOH \rightleftharpoons H^+ + CH_2ClCOO^-$	1.4×10^{-3}
Formic acid	$HCOOH \rightleftharpoons H^+ + HCOO^-$	1.8×10^{-4}
Benzoic acid	$C_6H_5COOH \rightleftharpoons H^+ + C_6H_5COO^-$	6.3×10^{-5}
Acetic acid	$CH_3COOH \rightleftharpoons H^+ + CH_3COO^-$	1.8×10^{-5}

Because we assumed that $0.10 - x \approx 0.10$, we should check our assumption. Is 0.10 changed significantly by subtracting 0.0013? To two significant figures, 0.10 is unchanged by subtracting 0.0013; no significant error was introduced.

Practice Exercises

A. Calculate the $[H^+]$ in a 0.033 M solution of hypobromous acid.

$$HOBr(aq) \rightleftharpoons H^+(aq) + OBr^-(aq)$$

B. Calculate the $[H^+]$ in a 0.0035 M solution of hypoiodous acid (HOI), $K_a = 2.3 \times 10^{-11}$.

Approximations such as those made in the preceding example are good only if the degree of ionization is small. Such an assumption should always be checked.

Equilibria Involving Weak Bases

Equilibrium constant expressions can also be written for weak bases. For ammonia, the reaction is

$$NH_3(aq) + H_2O \rightleftharpoons NH_4^+(aq) + OH^-(aq)$$

and the equilibrium constant expression is

$$K = \frac{[NH_4^+] \times [OH^-]}{[H_2O] \times [NH_3]}$$

Only a tiny fraction of the water in a dilute solution reacts, so we can assume that the concentration of water is constant. Ignoring water, we get a new equilibrium constant expression.

$$K_b = \frac{[NH_4^+] \times [OH^-]}{[NH_3]}$$

The new constant is called the **base ionization constant (K_b).** K_b values for several weak bases are given in Table 10.2.

Table 10.2 Ionization Constants of Some Weak Bases in Water at 25°C

Base	Ionization Equilibrium	Ionization Constant, K_b
Inorganic Bases		
Ammonia	$NH_3 + H_2O \rightleftharpoons NH_4^+ + OH^-$	1.8×10^{-5}
Hydrazine	$H_2NNH_2 + H_2O \rightleftharpoons H_2NNH_3^+ + OH^-$	8.5×10^{-7}
Hydroxylamine	$HONH_2 + H_2O \rightleftharpoons HONH_3^+ + OH^-$	9.1×10^{-9}
Amines		
Dimethylamine	$(CH_3)_2NH + H_2O \rightleftharpoons (CH_3)_2NH_2^+ + OH^-$	5.9×10^{-4}
Ethylamine	$CH_3CH_2NH_2 + H_2O \rightleftharpoons CH_3CH_2NH_3^+ + OH^-$	4.3×10^{-4}
Methylamine	$CH_3NH_2 + H_2O \rightleftharpoons CH_3NH_3^+ + OH^-$	4.2×10^{-4}
Pyridine	$C_5H_5N + H_2O \rightleftharpoons C_5H_5NH^+ + OH^-$	1.5×10^{-9}
Aniline	$C_6H_5NH_2 + H_2O \rightleftharpoons C_6H_5NH_3^+ + OH^-$	4.2×10^{-10}

Example 10.11

Calculate the hydroxide ion concentration in a 0.010 M solution of aniline.

Solution

We get the equation for the ionization from Table 10.2.

$$C_6H_5NH_2(aq) + H_2O \rightleftharpoons C_6H_5NH_3^+(aq) + OH^-(aq)$$

Then we can write the equilibrium constant expression.

$$K_b = \frac{[C_6H_5NH_3^+] \times [OH^-]}{[C_6H_5NH_2]} = 4.2 \times 10^{-10}$$

Let $x = [OH^-] = [C_6H_5NH_3^+]$. Assume $[C_6H_5NH_2] = 0.010 - x \approx 0.010 = 1.0 \times 10^{-2}$. Substituting, we have

$$\frac{(x)(x)}{1.0 \times 10^{-2}} = 4.2 \times 10^{-10}$$

$$x^2 = 4.2 \times 10^{-12}$$

$$x = 2.0 \times 10^{-6}\text{M} = [OH^-]$$

Checking our assumption, we see that x is small compared with 0.010.

Practice Exercise

Calculate the $[OH^-]$ in a 0.15 M solution of hydroxylamine.

10.5 Salts in Water: Acidic, Basic, or Neutral?

When an acid reacts with a base, the products are a salt and water. The process is called neutralization, but is the solution neutral? What if we simply take a salt and dissolve it in water? Would the solution be acidic, basic, or neutral? Although we might well expect a salt solution to be neutral, not all of them are. There are many neutral salts, including the familiar sodium chloride and others such as potassium sulfate and sodium nitrate. Some, such as ammonium nitrate and aluminum chloride, are acidic. Others, such as sodium acetate and potassium cyanide, are basic.

How do we tell if a salt is acidic, basic, or neutral? Simple. Just dissolve some of the salt in water and test the solution with an indicator (Figure 10.5) or with a pH meter. That is the experimental way, and who can argue with it? There is another way, though. We can *predict* whether a solution of a salt will be acidic, basic, or neutral by considering the relative strengths of the acid and base from which the salt was made. It is usually more convenient to apply a rule than to go into a laboratory and do an experiment. The rules are as follows:

1. The salt of a strong acid and a strong base forms a neutral solution.
2. The salt of a strong acid and a weak base forms an acidic solution.
3. The salt of a weak acid and a strong base forms a basic solution.
4. The salt of a weak acid and a weak base may form an acidic, basic, or (by chance) neutral solution.

To apply these rules, you must be able to recognize strong acids, strong bases, weak acids, and weak bases (recall Sections 9.2 and 9.4). The following examples will illustrate the process.

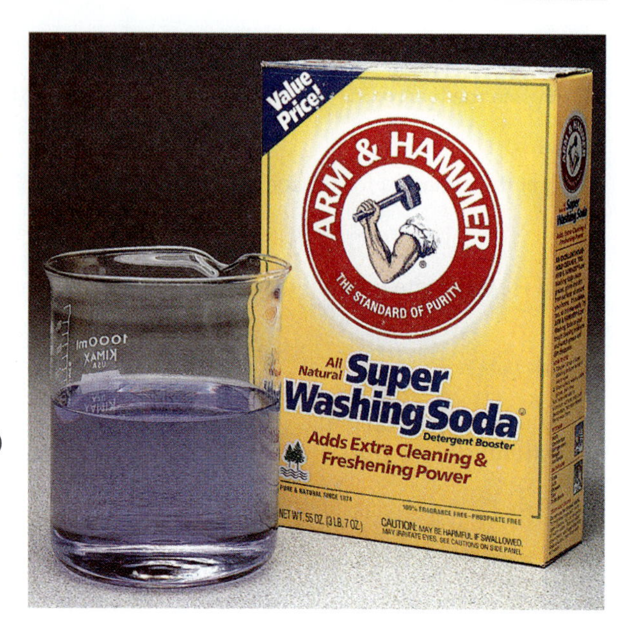

Figure 10.5 ▶
This sodium carbonate solution contains a few drops of thymolphthalein indicator. Thymolphthalein is blue when the pH of the solution is greater than 10.6 and colorless when the pH is less than 9.4. The blue color of the indicator shows that the pH of this solution is greater than 10.6. The solution is rather strongly basic as a result of the reaction of CO_3^{2-} as a base with water

$$CO_3^{2-}(aq) + H_2O \rightleftharpoons HCO_3^-(aq) + OH^-(aq)$$

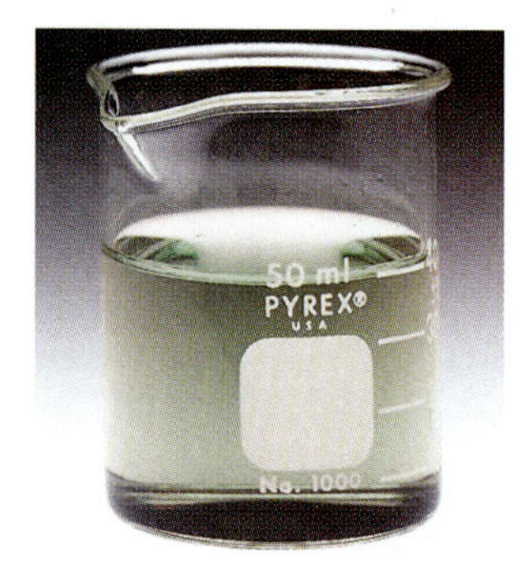

▲ The faint green color of the bromthymol blue indicator indicates that this NaCl(aq) is neutral. Bromthymol blue is yellow below pH 7, green at pH 7, and blue above pH 7.

▲ The yellow color of the bromthymol blue indicator indicates that this $NH_4Cl(aq)$ is acidic.

Example 10.12

For each of the following salts, predict whether it forms an acidic, basic, or neutral solution.

a. NaCl **b.** NH_4Cl **c.** CH_3COONa **d.** CH_3COONH_4

Solution

a. In any aqueous solution, there will always be H^+ (actually, H_3O^+) and OH^-. NaCl dissociates in water to yield $Na^+ + Cl^-$. Therefore, in an aqueous solution of NaCl, there will be the following four ions:

$$Na^+, Cl^-, H^+, OH^-$$

Combine the positive ion (cation) of the salt with OH^-, and the negative ion (anion) of the salt with H^+.

$$Na^+ + OH^- = NaOH$$

$$Cl^- + H^+ = HCl$$

Then determine the strength of the base and the acid thus formed. NaOH is a strong base, and HCl is a strong acid. Therefore, NaCl is a salt of a strong acid and a strong base, and (according to rule 1) it forms a neutral solution.

b. The four ions in solution are

$$NH_4^+, Cl^-, H^+, OH^-$$

Combining the appropriate ions, we have

$$NH_4^+ + OH^- = NH_4OH$$

$$H^+ + Cl^- = HCl$$

NH_4OH (really aqueous NH_3) is a weak base, and HCl is a strong acid. Consequently, NH_4Cl is a salt of a weak base and a strong acid, and (rule 2) it forms an acidic solution.

c. The four ions in solution are

$$Na^+, CH_3COO^-, H^+, OH^-$$

Combining the appropriate ions, we have

$$Na^+ + OH^- = NaOH$$

$$CH_3COO^- + H^+ = CH_3COOH$$

NaOH is a strong base, and CH_3COOH (acetic acid) is a weak acid. Therefore, CH_3COONa is a salt of a weak acid and a strong base, and (rule 3) it forms a basic solution.

d. The base, NH_4OH (aqueous NH_3), is a weak one. The acid, CH_3COOH, is also weak. There is no way to tell from the rules whether the solution is acidic, basic, or neutral (rule 4).

Practice Exercise

For each of the following salts, predict whether it forms an acidic, basic, or neutral solution.

a. KNO_3 **b.** Li_2CO_3 **c.** $(NH_4)_2CO_3$ **d.** $(NH_4)_2SO_4$

▲ The blue color of the bromthymol blue indicator indicates that this $CH_3COONa(aq)$ is basic.

Review Questions

10.9 Which of the following aqueous solutions would you expect to have a pH greater than 7, and which would you expect to have a pH lower than 7? (**a**) 0.25 M HCl; (**b**) 0.37 M NH_3; (**c**) 1.0 M CH_3COOH; (**d**) 0.050 M CH_3COONa

10.10 Which of the following 0.05 M aqueous solutions would you expect to have the *lowest* pH? (**a**) K_2SO_4; (**b**) K_2CO_3; (**c**) KNO_2; (**d**) KOH

10.11 Arrange the following 0.10 M aqueous solutions in order of *increasing* pH. (**a**) HCl; (**b**) KOH; (**c**) CH_3COOH; (**d**) CH_3COONa

10.6 Buffers: Control of pH

Who cares about the pH of a salt solution anyway? You do, that's who, for some of these salts play a vital role in the control of the pH of body fluids. If they should fail to function, so will you. Our bodies are acid factories. Our stomachs produce hydrochloric acid. Our muscles produce lactic acid. Starches and sugars produce pyruvic acid when metabolized. Carbon dioxide from respiration produces carbonic acid in the blood. Our bodies must eliminate or neutralize these acids, because excess acidity in the wrong place would kill us rather quickly.

A **buffer solution** is one in which the pH remains nearly constant even if acid or base is added. It contains a weak acid and one of its salts (or a weak base and one of its salts), usually in approximately equal concentrations. For example, 1 L of a solution that contains 1.00 mol of acetic acid (CH_3COOH) and 1.00 mol of sodium acetate (CH_3COONa) acts as a buffer at pH 4.74, an acidic value.[2] This buffer can absorb significant amounts of additional acid or base without appreciable change in pH (Figure 10.6).

How does a buffer work? It may seem strange that a solution can absorb acid or base without the pH changing appreciably. The explanation, however, is fairly simple. It follows Le Châtelier's principle (Section 4.11). The buffer solution has a

[2]This may seem a bit confusing because we have just stated the fact that a solution of sodium acetate is slightly basic (that is, has a pH greater than 7). A solution containing *only* the salt *is* basic. The buffer solution consists not only of a salt of acetic acid but also of some acetic acid itself. This acid makes the buffer solution acidic.

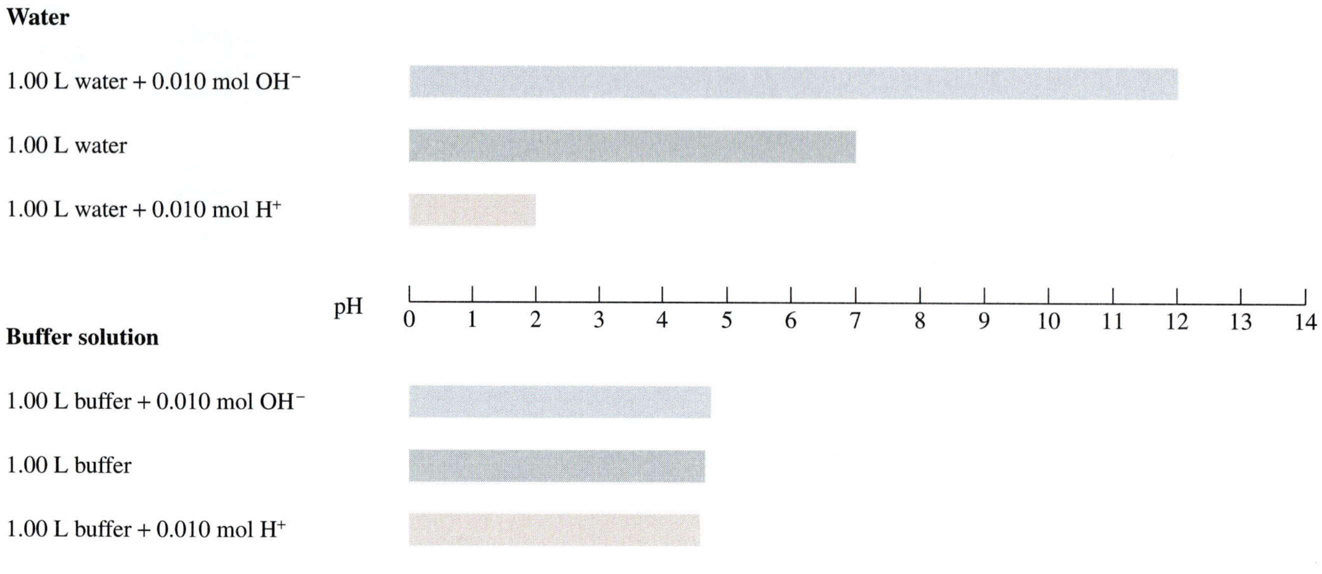

Water

1.00 L water + 0.010 mol OH^-

1.00 L water

1.00 L water + 0.010 mol H^+

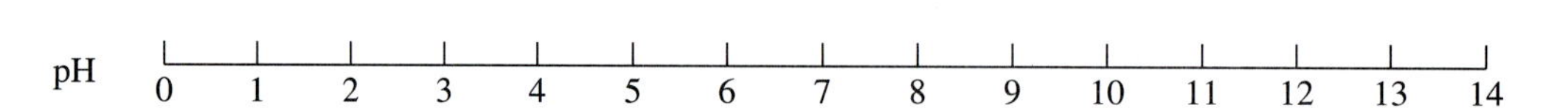

Buffer solution

1.00 L buffer + 0.010 mol OH^-

1.00 L buffer

1.00 L buffer + 0.010 mol H^+

▲ **Figure 10.6**
This represents buffer action. The addition of 0.010 mol of H^+ or of OH^- to 1.00 L produces a huge change in the pH of pure water and practically no change in the pH of a solution that is 1.00 M in both CH_3COOH and CH_3COONa. The acetic acid-sodium acetate solution is a buffer solution, and pure water has no buffering ability at all.

large reservoir of both acid molecules and the anions from the salt. If a strong acid is added, the hydrogen ions from the added acid will donate protons to the anions of the buffer to form the weak acid and water.

$$H^+ + CH_3COO^- \rightleftharpoons CH_3COOH$$

Although the reaction is reversible to a slight extent, most of the protons are removed from the solution as they are added, and the pH changes hardly at all.

When a strong base is added, the hydroxide ions will react with the hydrogen ions formed in the solution by the acetic acid of the buffer.

$$OH^- + H^+ \rightleftharpoons H_2O$$

The added hydroxide ions are tied up, and the hydrogen ions removed from the solution are immediately replaced by further ionization of the acetic acid in the buffer.

⟶ Removing H^+ ions shifts the equilibrium to the right.

$$CH_3COOH \rightleftharpoons CH_3COO^- + H^+$$

The concentration of hydrogen ions returns to approximately the original value, and the pH is only slightly changed.

There are many important buffer solutions. Most biochemical reactions, whether they occur in a laboratory or in our bodies, are carried out in buffered solutions (Figure 10.7). The buffers that control the pH of our blood will be discussed in the next section. Table 10.3 lists some buffers of interest and the pH ranges in which they operate.

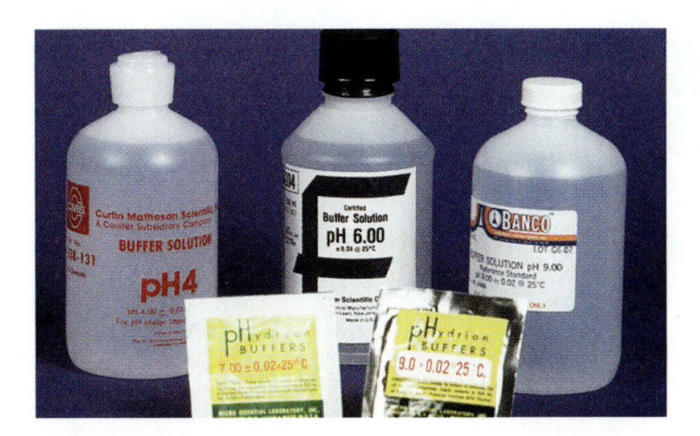

Figure 10.7
Buffers are available as prepackaged solutions and as powdered ingredients for preparing buffer solutions of specific pH values.

✓ Review Questions

10.12 Use acetic acid and acetate ion to explain how a buffer controls pH.

10.13 Use ammonia and ammonium chloride to explain how a buffer controls pH.

10.14 If acid is added to an unbuffered solution, will the pH increase or decrease?

Calculations Involving Buffers

We can calculate the hydrogen ion concentration in a buffer system containing both a weak acid and a salt of that acid.

If sodium acetate is added to a solution of acetic acid, ionization of the acid is considerably decreased because the concentration of acetate ion in the system is increased.

← Added acetate ions causes the equilibrium to shift to the left.

$$CH_3COOH(aq) \rightleftharpoons H^+(aq) + CH_3COO^-(aq)$$

The acetate ions from sodium acetate react with the H^+, forming more un-ionized acetic acid molecules. A greater proportion of the total acetic acid is in the molecular form; ionization is decreased. This is an example of the **common ion effect.** If an ion common to those in the equilibrium is added, the degree of ionization is decreased. A new equilibrium is established with more acetate ions but fewer H^+ ions. The value of K_a remains unchanged.

Example 10.13

What is the $[H^+]$ in a solution that is 0.10 M acetic acid and 0.10 M sodium acetate?

Table 10.3 Some Important Buffers

Buffer Components	Buffer System Names	pH[a]
$CH_3CHOHCOOH/CH_3CHOHCOO^-$	Lactic acid/lactate ion	3.86
CH_3COOH/CH_3COO^-	Acetic acid/acetate ion	4.74
$H_2PO_4^-/HPO_4^{2-}$	Dihydrogen phosphate ion/monohydrogen phosphate ion	7.20
H_2CO_3/HCO_3^-	(Carbon dioxide) carbonic acid/bicarbonate ion	6.46[b]
NH_4^+/NH_3	Ammonium ion/ammonia	9.25

[a]The values listed are for solutions that are 0.1 M in each compound at 25 °C.

[b]This value includes dissolved CO_2 molecules as undissociated H_2CO_3. The value for H_2CO_3 alone is about 3.8.

▶ The common ion effect. Both solutions contain bromophenol blue indicator. The yellow color indicates that the pH is less than 3.0 in 1.00 M CH_3COOH, and the blue-violet color indicates that the pH is greater than 4.6 in the solution that is 1.00 M in both CH_3COOH and CH_3COONa. The presence of CH_3COO^-, a common ion, raises the pH by about two units and reduces $[H^+]$ about 100-fold in the solution that has both acetic acid and sodium acetate as solutes.

bromophenol blue indicator:	pH < 3.0	pH > 4.6
	yellow	blue-violet

Solution

Use the K_a expression for acetic acid.

$$K_a = \frac{[H^+] \times [CH_3COO^-]}{[CH_3COOH]} = 1.8 \times 10^{-5}$$

Let $x = [H^+]$. Sodium acetate is completely ionized; thus the concentration of acetate ion equals the sum of $[CH_3COO^-]$ from the salt and $[CH_3COO^-]$ from the ionization of acetic acid.

$$[CH_3COO^-] = 0.10 + x \approx 0.10 = 1.0 \times 10^{-1}$$

The concentration of acetic acid at equilibrium is the original concentration minus the amount ionized.

$$[CH_3COOH] = 0.10 - x \approx 0.10 = 1.0 \times 10^{-1}$$

Substituting these values into the K_a expression gives

$$K_a = \frac{(x)(1.0 \times 10^{-1})}{1.0 \times 10^{-1}} = 1.8 \times 10^{-5}$$

$$x = 1.8 \times 10^{-5} M = [H^+]$$

Checking our assumption, was x small enough to be ignored when added to or subtracted from 0.10? Yes, 1.8×10^{-5} is negligibly small compared with 1.0×10^{-1}. (This buffer solution tends to keep the hydrogen ion concentration constant at a value of 1.8×10^{-5} mol/L.)

Practice Exercise

Calculate the $[H^+]$ in a solution that is 0.033 M in HOBr and 0.033 M in NaOBr.

The ionization of a weak base is also decreased by the addition of a common ion. Adding ammonium chloride to a solution of ammonia decreases the amount of hydroxide ion in the system.

$$NH_3(aq) + H_2O \rightleftharpoons NH_4^+(aq) + OH^-(aq)$$

Example 10.14

Calculate the $[OH^-]$ and $[H^+]$ of a solution that is 0.50 M NH_3 and 0.10 M NH_4^+.

Solution

Use the K_b expression for NH_3 and obtain the K_b value from Table 10.2.

$$K_b = \frac{[NH_4^+] \times [OH^-]}{[NH_3]} = 1.8 \times 10^{-5}$$

If we let $y = [OH^-]$, then $[NH_4^+] \approx 0.10 = 1.0 \times 10^{-1}$ and $[NH_3] \approx 0.50 = 5.0 \times 10^{-1}$. Substituting these values into the K_b expression gives

$$K_b = \frac{(1.0 \times 10^{-1})(y)}{5.0 \times 10^{-1}} = 1.8 \times 10^{-5}$$

$$y = 9.0 \times 10^{-5} = [OH^-]$$

Now use the ion product for water to calculate $[H^+]$.

$$K_W = [H^+][OH^-] = 1.0 \times 10^{-14}$$

$$[H^+] = \frac{K_W}{[OH^-]} = \frac{1.0 \times 10^{-14}}{9.0 \times 10^{-5}} = 1.1 \times 10^{-10}$$

Practice Exercise

Calculate the $[H^+]$ of a solution that is 0.25 M HCN and 0.17 M KCN.

Buffered solutions are widely used in analytical chemistry, medicine, and biochemistry and in many industrial applications such as leatherworking and dyeing. A rearrangement of the equilibrium constant expression, called the **Henderson-Hasselbalch equation,** is used to calculate the approximate pH of a buffer. For a weak acid HA, the equation is

$$pH = pK_a + \log\frac{[A^-]}{[HA]}$$

pK_a is a logarithmic term similar to pH.

$$pK_a = -\log K_a$$

Because $[A^-]$ is assumed to come only from the salt, and [HA] is the concentration of the weak acid (ignoring its slight ionization), the Henderson-Hasselbalch equation is sometimes written as

$$pH = pK_a + \log\frac{[\text{salt}]}{[\text{acid}]}$$

If large quantities of acid or base are added, the capacity of a buffer is exceeded. In general, the more concentrated its buffer components, the more added acid or base a solution can neutralize. As a rule, a buffer is most effective against both added H^+ and added OH^- if $[HA] = [A^-]$. The pH of such a buffer is called the *optimum pH* of the buffer and it has a value equal to pK_a. That is,

$$pH = pK_a + \log\frac{[A^-]}{[HA]} = pK_a + \log 1$$

$$pH = pK_a \text{ (at the optimum pH)}$$

The pH range over which a buffer solution is effective generally is about one pH unit on either side of pH = pK_a. For example, the pK_a for acetic acid is

$$pK_a = -\log K_a = -\log(1.8 \times 10^{-5}) = 4.74$$

The effective pH range for an acetic acid–acetate ion buffer is $3.74 < \text{pH} < 5.74$.

Example 10.15

What is the pH of a solution that is 0.20 M H_2S and 0.20 M HS^-? The K_a for H_2S is 1×10^{-7}. (Ignore the second ionization of H_2S.)

$$H_2S(aq) \rightleftharpoons H^+(aq) + HS^-(aq)$$

Solution

Because $[H_2S] = [HS^-]$, the Henderson-Hasselbalch equation reduces to

$$\text{pH} = pK_a$$

$$pK_a = -\log K_a = -\log(1 \times 10^{-7}) = 7$$

$$\text{pH} = pK_a = 7$$

Practice Exercise

What is the pH of a solution that is 0.50 M HF and 0.50 M F^-?

The concentration of HA and A^- are not always equal. Example 10.16 illustrates how the Henderson-Hasselbalch equation is used to calculate the pH values for such buffer solutions.

Example 10.16

What is the pH of a solution that is 0.10 M HCN and 0.50 M NaCN?

$$HCN(aq) \rightleftharpoons H^+(aq) + CN^-(aq)$$

Solution

The Henderson-Hasselbalch equation for HCN/NaCN is

$$\text{pH} = pK_a + \log\frac{[\text{salt}]}{[\text{acid}]}$$

$$\text{pH} = pK_a + \log\frac{[\text{NaCN}]}{[\text{HCN}]}$$

The K_a for HCN (Table 10.1) is 6.2×10^{-10}.

$$\text{pH} = -\log(6.2 \times 10^{-10}) + \log\frac{0.50}{0.10}$$

$$= -\log 6.2 \times 10^{-10} + \log 5.0$$

$$= -0.79 - (-10.00) + 0.70$$

$$= 9.91$$

Practice Exercise

What is the pH of a solution that is 0.40 M HNO_2 and 0.25 M NO_2^-?

Review Questions

10.15 Write the equation for the ionization of each of the following acids.

a. HBO_2 **b.** $HClO_2$ **c.** $HC_9H_7O_4$ **d.** H_2Se (first ionization only)

10.16 Write equations for the ionization of each of the following bases (that is, for the reaction of the base with water to form ions).

a. $C_4H_9NH_2$ **b.** $C_{11}H_{21}O_4N$ **c.** C_3H_5N

10.17 Explain what is meant by the *common ion effect.* How is this effect involved in the functioning of buffer solutions?

10.18 Which of the following will *suppress* the ionization of formic acid, HCOOH: (**a**) NaCl; (**b**) KOH; (**c**) $(HCOO)_2Ca$; or (**d**) Na_2CO_3? Explain.

10.7 Buffers in Blood

The pH of the blood of higher animals is held remarkably constant. In humans, blood plasma normally varies from 7.35 to 7.45 in pH. Should the pH rise above 7.8 or fall below 6.8, due to starvation or disease, the person may suffer irreversible damage to the brain or even die. Fortunately, human blood has at least three buffering systems, of which the bicarbonate/carbonic acid (HCO_3^-/H_2CO_3) system is the most important.

If acids enter the blood, hydrogen ions are taken up by the bicarbonate ions to form undissociated carbonic acid.

⟶ Added H^+ causes the equilibrium to shift to right.

$$HCO_3^- + H^+ \rightleftharpoons H_2CO_3$$

The carbonic acid is a weak acid, and its equilibrium with hydrogen ions and bicarbonate ions lies toward the undissociated form. As long as there is sufficient bicarbonate to take up the added acid, the pH will change little.

If bases enter the bloodstream, they will react with hydrogen ions to form water.

⟶ Added OH^- causes the equilibrium to shift to right.

$$OH^- + H^+ \rightleftharpoons H_2O$$

More carbonic acid molecules will ionize to replace the removed hydrogen ions.

⟶ Removing H^+ causes the equilibrium to shift to right.

$$H_2CO_3 \rightleftharpoons H^+ + HCO_3^-$$

Further, as the carbonic acid molecules are used up, more carbonic acid can be formed from the large reservoir of dissolved carbon dioxide in the blood. (The CO_2 in the blood is in equilibrium with CO_2 in the lungs.)

$$CO_2 + H_2O \rightleftharpoons H_2CO_3$$

Thus, bicarbonate/carbonic acid buffers the blood against either added base or added acid. However, the concentration of HCO_3^- in blood is ten times that of H_2CO_3, so the buffering system is most effective against acids.

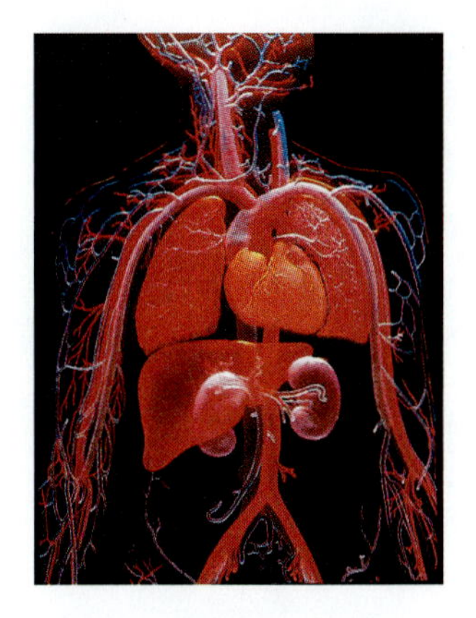

▲ Blood carried by the human circulatory system is maintained at a pH of about 7.4 in a highly buffered system.

Another blood buffer is the dihydrogen phosphate/monohydrogen phosphate ($H_2PO_4^-/HPO_4^{2-}$) system. Any acid reacts with monohydrogen phosphate ions to form dihydrogen phosphate ions.

$$HPO_4^{2-} + H^+ \rightleftharpoons H_2PO_4^-$$

The dihydrogen phosphate ion is a weak acid and exists in equilibrium with hydrogen ion and monohydrogen phosphate ion.

$$H_2PO_4^- \rightleftharpoons H^+ + HPO_4^{2-}$$

Any base that comes into the blood will react with hydrogen ions to form water. However, more dihydrogen phosphate will ionize to replace these hydrogen ions, leaving the pH essentially unchanged.

Proteins act as a third type of blood buffer. These complex molecules (Chapter 21) contain —COO^- groups, which, like acetate ions (CH_3COO^-), can act as proton acceptors. Proteins also contain —NH_3^+ groups, which, like ammonium ions (NH_4^+), can donate protons. If acid comes into the blood, hydrogen ions can be neutralized by the —COO^- groups.

$$\text{Protein—}COO^- + H^+ \rightleftharpoons \text{Protein—}COOH$$

If base is added, it can be neutralized by the —NH_3^+ groups.

$$\text{Protein—}NH_3^+ + OH^- \rightleftharpoons \text{Protein—}NH_2$$

These three buffers (and perhaps others) act to keep the pH of the blood constant. Buffers can be overridden by large amounts of acid or base; their capacity is not infinite. The blood buffers can be overwhelmed if the body's metabolism goes badly amiss.

Review Questions

10.19 Name three buffer systems operating in the blood.

10.20 What groups in proteins react with added acid and base?

10.8 Acidosis and Alkalosis

Have your muscles ever hurt after prolonged physical activity? If so, you have had your blood buffers somewhat overloaded. Muscle contraction produces lactic acid. This acid ionizes somewhat more strongly than carbonic acid and thus tends to lower the pH of the blood (it tends to release more hydrogen ions into the blood). Moderate amounts of lactic acid can be handled by the blood buffers. For bicarbonate, the reaction is

$$\underset{\text{Lactic acid}}{CH_3CHOHCOOH} + HCO_3^- \rightleftharpoons \underset{\text{Lactate ion}}{CH_3CHOHCOO^-} + H_2CO_3$$

Excessive amounts of lactic acid overload the buffers, however, and the pH is lowered. Nerve cells respond to the increased acidity by sending a message of pain to the brain.

If the pH of the blood falls below 7.35, the condition is called **acidosis.** If the pH of the blood rises above 7.45, **alkalosis** sets in. These pathological conditions can be caused by faulty respiration or by metabolic problems. In severe cases of starvation, the body gets its energy by the oxidation of stored fats. The products of fat metabolism are acidic, and prolonged starvation leads to acidosis (Section 26.4). Fad diets, such as those that severely limit the intake of carbohydrates, can also lead to acidosis.

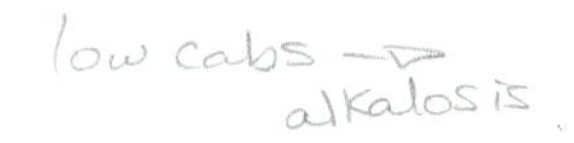

The body's excretory system tries to compensate for acidosis or alkalosis by selectively excreting certain compounds. Conversely, the conditions can be brought on by kidney failure or other excretory problems. We discuss these pathological problems further in later chapters.

Review Questions

10.21 Define or illustrate

a. acidosis **b.** alkalosis

10.22 If someone is suffering from alkalosis, is the blood pH too high or too low?

Summary

Acid and base concentrations are usually expressed in molarity (moles/liter). When a concentrated solution of acid or base is diluted, the concentration of the dilute solution can be calculated from the relationship

$$V_{conc} \times M_{conc} = V_{dil} \times M_{dil}$$

The concentration of an acid or base in a solution can be determined by **titration.** An **indicator,** a substance that is one color in acidic solutions and another color in basic solutions, is used to determine the end point of the titration. At the **equivalence point** of the titration, the number of moles of base is just equal to the number of moles of acid. The unknown molarity is calculated from the known molarity and volumes.

Some of the H_2O molecules in any sample of water dissociate into H^+ and OH^- ions. Pure water at 25 °C contains 1×10^{-7} mol of H^+ ions and 1×10^{-7} mol of OH^- ions. The product of these two concentrations is the **ion product of water,** K_w. The ion product is a constant at a given temperature, and so if you add H^+ to water, thereby raising the value of $[H^+]$, the value of $[OH^-]$ must decrease. This decrease occurs as some of the OH^- ions (formed when H_2O molecules dissociated) recombine with H^+ to reform H_2O.

The **pH scale** is a way of expressing H^+ concentrations. The pH of any solution is the exponent (ignoring the minus sign) of the H^+ concentration when that concentration is expressed in the form 1×10^{-x}. The **pOH** of a solution is the exponent of the OH^- concentration when that concentration is expressed in the form 1×10^{-x}. Because $[H^+][OH^-]$ is 1×10^{-14} (at 25 °C), pH + pOH = 14.

A neutral solution at 25 °C has a pH of 7, an acidic solution a pH below 7, and a basic solution a pH above 7.

When a salt is dissolved in water, the resulting solution may be neutral, acidic, or basic. It is neutral when the salt is made up of the anion of a strong acid and the cation of a strong base (NaCl), acidic when the salt is made up of the anion of a strong acid and the cation of a weak base (NH_4Cl), and basic when the salt is made up of the anion of a weak acid and the cation of a strong base (KNO_2). When the salt anion is from a weak acid and the cation from a weak base, the resulting solution may be neutral, acidic, or basic, depending on the relative strengths of the acid and base.

A **buffer solution,** made either from a weak acid and one of its salts or from a weak base and one of its salts, maintains a constant pH even when small amounts of acids or bases are added. Different acid/salt or base/salt combinations produce buffers that are effective in different pH ranges. A large amount of added acid or added base can overwhelm a buffer, allowing the pH to change.

The bicarbonate/carbonic acid buffer is the principal one maintaining the pH of blood. Several pathological conditions in humans can result in blood buffers being overwhelmed. A blood pH above 7.45 is the condition known as **alkalosis,** and a blood pH below 7.35 is **acidosis.**

We can use equilibrium constant expressions and tabulated ionization constants for weak acids (K_a) and weak bases (K_b) to calculate $[H^+]$ and pH values for their solutions. The Henderson-Hasselbalch equation can be used to calculate the pH of a buffer solution.

Key Terms

acid ionization constant (K_a) (10.4)
acidosis (10.8)
alkalosis (10.8)
base ionization constant K_b) (10.4)
buffer solution (10.6)
common ion effect (10.6)
dilution (10.1)
equilibrium constant (10.4)
equilibrium constant expression (10.4)
equivalence point (10.2)
Henderson-Hasselbalch equation (10.6)
indicator dye (10.2)
ion product of water (K_W) (10.3)
pH scale (10.3)
pOH (10.3)
titration (10.2)

Problems

Dilution of Solutions

1. What volume of (a) 12.0 M HCl is required to make 2.00 L of 1.00 M HCl? (b) 1.04 M Na_2CO_3 is required to make 0.500 L of 1.00 M Na_2CO_3?
2. What volume of (a) 8.89 M HBr is required to make 2.00 L of 1.00 M HBr? (b) 19.1 M NaOH is required to make 2.00 L of 6.00 M NaOH?
3. What volume of concentrated (18.0 M) sulfuric acid would be required to make each of the following?
   a. 1.25 L of 6.00 M solution
   b. 575 mL of 0.100 M solution
4. You have a stock solution of 12.0 M HCl. How would you prepare the following solutions?
   a. 125 mL of 1.25 M HCl solution
   b. 5.00 L of 6.00 M HCl solution
   c. 1.50 L of 1.00 M HCl solution

Acid-Base Titrations

5. Calculate the molarity of an HCl solution if 20.0 mL of it requires 33.2 mL of 0.150 M NaOH for neutralization.
6. Calculate the molarity of an HNO_3 solution if 30.0 mL of it requires 18.3 mL of 0.104 M KOH for neutralization.
7. Calculate the molarity of a $Ca(OH)_2$ solution if 18.5 mL of it requires 28.2 mL of 0.0302 M HCl for neutralization. The products are $CaCl_2(aq)$ and H_2O.
8. Calculate the molarity of a $H_2C_2O_4$ solution if 12.5 mL of it requires 25.7 mL of 0.0995 M NaOH for neutralization. The products are $Na_2C_2O_4(aq)$ and H_2O.
9. A 20-mL sample of gastric fluid is neutralized by 25 mL of 0.10 M NaOH. What is the molarity of HCl in the fluid? Assume that all the acidity of the gastric fluid is due to HCl.
10. When the stomach isn't being stimulated by food to make more, it produces 0.0023 mol of HCl and 30 to 60 mL of total juices per hour. What range of concentrations of HCl in the stomach does this represent?
11. How many milliliters of 0.100 M H_2SO_4 are required to react with 10.3 mL of 0.404 M $NaHCO_3$?

$$H_2SO_4(aq) + 2\,NaHCO_3(aq) \rightarrow Na_2SO_4(aq) + 2\,H_2O + 2\,CO_2(g)$$

12. How many milliliters of 0.110 M H_2SO_4 are required to react with 30.0 mL of 0.0887 M $Ba(OH)_2$?
13. How many milliliters of 0.0195 M HCl are required to titrate (a) 25.00 mL of 0.0365 M KOH(aq), (b) 10.00 mL of 0.0116 M $Ca(OH)_2(aq)$, and (c) 20.00 mL of 0.0225 M $NH_3(aq)$?
14. How many milliliters of 0.0108 M $Ba(OH)_2(aq)$ are required to titrate (a) 20.00 mL of 0.0265 M $H_2SO_4(aq)$, (b) 25.00 mL of 0.0213 M HCl(aq), and (c) 10.00 mL of 0.0868 M $CH_3COOH(aq)$?

pH and pOH

15. What is the pH of each of the following solutions?
    a. 1.0×10^{-2} M HCl   b. 1.0×10^{-4} M HNO_3
16. What is the pH of each of the following solutions?
    a. 0.00010 M HCl   b. 0.00010 M HNO_3
    c. 0.10 M HBr
17. What is the pOH of each of the following solutions?
    a. 1.0×10^{-2} M NaOH   b. 1.0×10^{-3} M KOH
18. What is the pOH of each of the following solutions?
    a. 0.0010 M NaOH   b. 0.010 M KOH
19. What is the pOH of each of the solutions in Problem 15?
20. What is the pH of each of the solutions in Problem 17?

Note: Problems 21–28 require the use of a table of logarithms or a scientific calculator with a base 10 logarithm function.

21. Calculate the pH values of solutions with the following molar hydrogen ion concentrations.
    a. 3.3×10^{-3} b. 5.7×10^{-5} c. 8.1×10^{-4}
22. Calculate the pH values of (a) a 3.6×10^{-2} M HCl solution and (b) an 8.8×10^{-4} M HNO_3 solution.
23. Calculate the pH of a blood solution that has a hydrogen ion concentration of 4.6×10^{-8} M.
24. Calculate the pH of a urine sample that has a hydrogen ion concentration of 2.3×10^{-6} M.
25. Calculate the pH of an ammonia solution that has a hydrogen ion concentration of 2.0×10^{-12} M.
26. Calculate the pH of a sample of gastric juice that has a hydrogen ion concentration of 0.12 M.
27. What is the hydrogen ion concentration of a urine sample that has a pH of 5.10?
28. What is the hydrogen ion concentration of a lemon juice sample that has a pH of 2.31?

Salt Solutions: Acidic, Basic, or Neutral?

29. Write an equation for the equilibrium established when CH_3COO^- is placed in water.
30. Write an equation for the equilibrium established when NH_4^+ is placed in water.
31. Classify the aqueous solution of each of these salts as acidic, basic, or neutral.
    a. KCl   b. NaCN   c. NH_4CN
32. Classify the aqueous solution of each of these salts as acidic, basic, or neutral.
    a. CH_3COOK b. $(NH_4)_2SO_4$ c. Na_2SO_4
33. A weak acid is titrated with a strong base. Is the solution at the equivalence point acidic, basic, or neutral? Explain.
34. A weak base is titrated with a strong acid. Is the solution at the equivalence point acidic, basic, or neutral? Explain.

Equilibria in Solutions of Weak Acids and Weak Bases

Note: You may refer to Table 10.1 for K_a values and to Table 10.2 for K_b values as necessary.

35. Write an equilibrium constant expression for each of the following reactions.
    a. $HOCl(aq) \rightleftharpoons H^+(aq) + OCl^-(aq)$

b. $HC_6H_7O_6(aq) \rightleftharpoons H^+(aq) + C_6H_7O_6^-(aq)$
c. $HCOOH(aq) \rightleftharpoons H^+(aq) + HCOO^-(aq)$

36. Write an equilibrium constant expression for each of the following reactions.
a. $C_5H_5N(aq) + H_2O \rightleftharpoons C_5H_5NH^+(aq) + OH^-(aq)$
b. $C_2H_5NH_2(aq) + H_2O \rightleftharpoons C_2H_5NH_3^+(aq) + OH^-(aq)$
c. $HPO_4^{2-}(aq) + H_2O \rightleftharpoons H_2PO_4^-(aq) + OH^-(aq)$

37. Calculate the $[H^+]$ and $[OH^-]$ in each of these solutions.
a. 0.010 M CH_3COOH b. 0.20 M C_6H_5COOH
c. 0.50 M HCN

38. Calculate the $[H^+]$ and $[OH^-]$ in each of these solutions.
a. 0.10 M HCOOH b. 0.15 M HF
c. 0.050 M HNO_2

39. Calculate the $[H^+]$ and $[OH^-]$ in each of these solutions.
a. 0.025 M NH_3 b. 0.10 M CH_3NH_2
c. 0.10 M $C_6H_5NH_2$

40. Calculate the $[H^+]$ and $[OH^-]$ in each of these solutions.
a. 0.010 M $(CH_3)_2NH$ b. 0.15 M H_2NNH_2
c. 0.030 M $HONH_2$

Buffer Solutions

41. What is $[H^+]$ in each of the following buffer solutions?
a. 0.25 M HCN and 0.25 M KCN
b. 0.20 M HF and 0.20 M NaF
c. 0.033 M C_6H_5COOH and 0.045 M $C_6H_5COO^-$

42. What is $[H^+]$ in each of the following buffer solutions?
a. 0.20 M HCN and 0.20 M KCN
b. 0.50 M HF and 0.20 M NaF
c. 0.40 M C_6H_5COOH and 0.20 M $C_6H_5COO^-$

43. Calculate $[OH^-]$ in a solution that is 0.40 M NH_3 and 0.040 M NH_4^+. What is $[H^+]$ in the solution?

44. Calculate $[OH^-]$ in a solution that is 0.040 M NH_3 and 0.020 M NH_4^+. What is $[H^+]$ in the solution?

Note: Use the Henderson-Hasselbalch equation to solve problems 45–54.

45. Calculate the pH of a solution that is 0.040 M HCN and 0.040 M CN^-.

46. Calculate the pH of a solution that is 0.20 M HCOOH and 0.20 M $HCOO^-$.

47. Calculate the pH of a solution that is 0.15 M benzoic acid and 0.15 M benzoate ion.

48. Calculate the pH of a solution that is 0.20 M HF and 0.50 M F^-.

49. Calculate the pH of a solution that is 0.15 M HCOOH and 0.60 M $HCOO^-$.

50. Calculate the pH of a solution that is 0.11 M benzoic acid and 0.96 M benzoate ion.

51. Calculate the pH of a solution that is 0.350 M CH_3CH_2COOH and 0.0786 M CH_3CH_2COOK.

$$CH_3CH_2COOH(aq) \rightleftharpoons H^+(aq) + CH_3CH_2COO^-(aq)$$

$$K_a = 1.3 \times 10^{-5}$$

52. Calculate the pH of a solution that is 0.132 M diethylamine, $(CH_3CH_2)_2NH$, and 0.145 M diethylammonium chloride, $(CH_3CH_2)_2NH_2Cl$.

$$(CH_3CH_2)_2NH(aq) + H_2O \rightleftharpoons (CH_3CH_2)_2NH_2^+(aq) + OH^-(aq)$$

$$K_b = 6.9 \times 10^{-4}$$

53. What is the pH of a buffer solution that is 0.10 M HOC_6H_5 and 0.10 M $NaOC_6H_5$? The K_a for HOC_6H_5 is 1.0×10^{-10}.

54. What is the pH of a buffer solution that is 0.050 M $C_5H_{11}COOH$ and 0.050 M $C_5H_{11}COO^-$? The K_a for $C_5H_{11}COOH$ is 1.0×10^{-8}.

Additional Problems

55. Vinegar is an aqueous solution of acetic acid, CH_3COOH. A 10.00-mL sample of a particular vinegar requires 31.45 mL of 0.2560 M KOH for its titration. What is the molarity of acetic acid in the vinegar? (Hint: Write a net ionic equation for the titration reaction.)

56. Most window cleaners are aqueous solutions of ammonia. A 10.00-mL sample of a particular window cleaner requires 39.95 mL of 1.008 M HCl for its titration. What is the molarity of ammonia in the window cleaner? (Hint: Write a net ionic equation for the titration reaction.)

57. What volume of $CO_2(g)$, measured at STP, is produced by the reaction of excess $H_2SO_4(aq)$ with 148 g of $Na_2CO_3(s)$?

$$Na_2CO_3(s) + H_2SO_4(aq) \rightarrow Na_2SO_4(aq) + H_2O + CO_2(g)$$

58. What volume of $SO_2(g)$, measured at 764 mmHg and 26 °C, is produced by the reaction of excess HCl(aq) with 212 g of $Na_2SO_3(s)$?

$$Na_2SO_3(s) + 2\,HCl(aq) \rightarrow 2\,NaCl(aq) + H_2O + SO_2(g)$$

59. Which of the following 0.010 M solutions has the highest $[H^+]$: $CH_3CH_2COOH(aq)$, HI(aq), $NH_3(aq)$, $H_2SO_4(aq)$, or $Ba(OH)_2$? Explain.

60. A railroad tank car carrying 1.5×10^3 kg of concentrated sulfuric acid derails and spills its load. The acid is 93.2% H_2SO_4 and has a density of 1.84 g/mL. What mass, in kilograms, of sodium carbonate (soda ash) is needed to neutralize the acid? (Hint: What is the neutralization reaction?)

61. Household ammonia, used as a window cleaner and for other cleaning purposes, is $NH_3(aq)$. A 31.08-mL portion of 0.9928 M HCl(aq) is required to neutralize the NH_3 present in a 5.00-mL sample of $NH_3(aq)$ in a titration. The reaction is

$$NH_3(aq) + HCl(aq) \rightarrow NH_4Cl(aq)$$

What is the molarity of NH_3 in the NH_3(aq) sample?

62. A buffer solution contains a Brønsted-Lowry acid as one component and a Brønsted-Lowry base as the other. (**a**) Can the two components be HCl and NaOH? Explain. (**b**) Acetic acid contains both CH_3COOH and CH_3COO^-. Can an acetic acid solution alone be considered a buffer? Explain.

63. Calculate the pH of 1.50 M formic acid, HCOOH.

64. Calculate the pH of a solution of pyridine that has 1.25 g in 125 mL of water solution. (Hint: First calculate the molarity of pyridine.)

65. What molarity of hydrazoic acid, HN_3, is required to produce an aqueous solution with a pH of 3.10?

$$HN_3(aq) \rightleftharpoons H^+(aq) + N_3^-(aq) \quad K_a = 1.9 \times 10^{-5}$$

Electrolytes

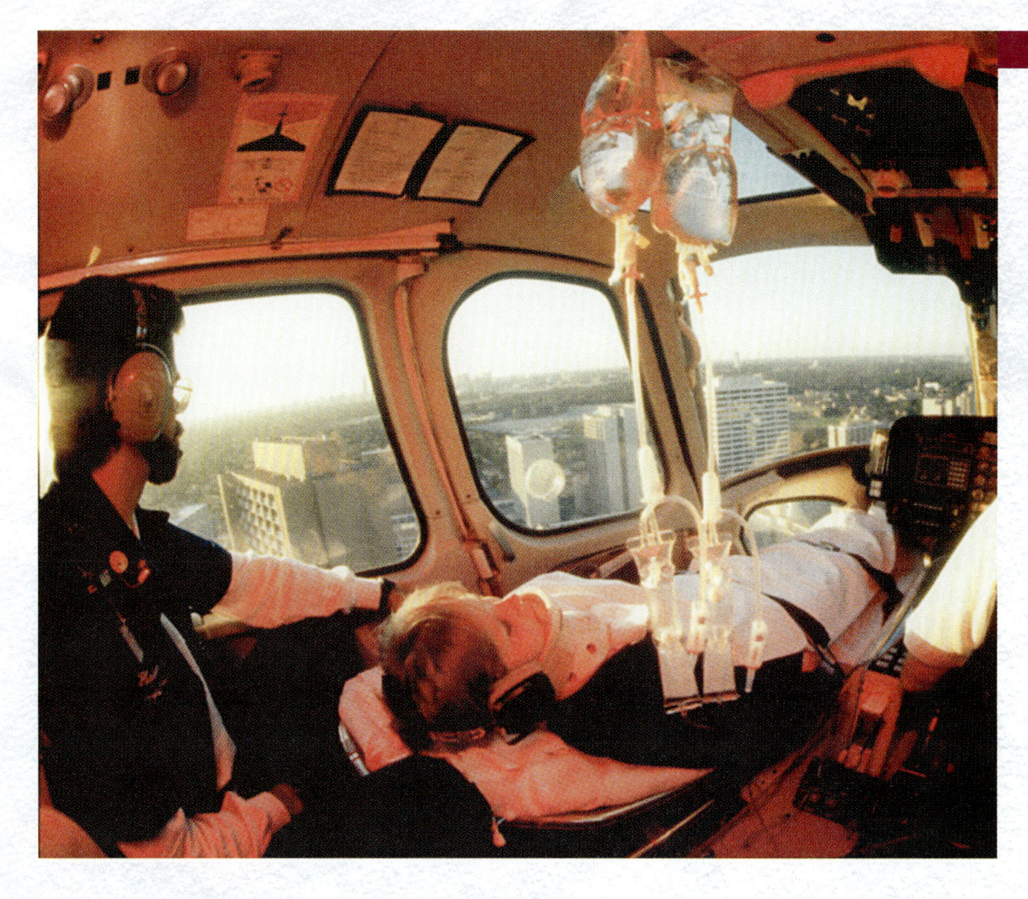

Electrolytes—substances (solutes) that provide ions in an aqueous solution—are important in batteries and in many chemical processes. Electrolytes are also critical in human health. Severely injured patients are usually given intravenous solutions containing electrolytes and buffers to stabilize them during transport to a medical center.

Learning Objectives/Study Questions

1. What are electrolytes? How are strong and weak electrolytes distinguished?
2. What is the difference between the terms *ionization* and *ion dissociation?*
3. What is electrolysis?
4. What is an electrochemical cell?
5. What is the activity series of metals? How is it used to predict chemical reactions?
6. What is the solubility product relationship? How is it used to predict whether or not precipitation will occur?
7. What are some of the minerals important to life?

We first introduced ions in Chapter 3. Since then we have discussed ionic bonds, ionic solids, and solutions of ions. The nineteenth-century scientists whose experiments laid the foundation for the ionic theory were working mainly for the joy of discovery and the recognition gained with success. We can only speculate as to whether they imagined the importance of ions in living systems.

The fluids in our bodies are like the salt water of the oceans. During fetal development, we have gill-like organs, hands that look like fins, and a shape much like that of a fish. Until we are born and breathe the air, we float in the watery darkness of the womb. Even after birth, fluids bathe our tissues and transport the materials that keep them alive. Messages are sent to and from the brain in the form of electric signals, often carried by ions through cellular and intercellular fluids. Certain ions are essential to the proper functioning of all living organisms. They must be present in proper concentrations, however. Too few or too many can be dangerous. We will discuss some of these ions and their properties in this chapter.

11.1 Early Electrochemistry

In 1800 an Italian physicist, Alessandro Volta, invented a battery that produced an electric current. Electrical phenomena had already been subjected to considerable study before this time. (In 1752 Benjamin Franklin determined that lightning was a form of electricity by flying a kite in a thunderstorm.) Volta's battery provided a convenient source of electricity, one that could be used by other scientists who wished to study the interaction of matter and electricity.

You may find it helpful to review oxidation-reduction reactions in Chapter 5.

Within six weeks of its invention two English chemists, William Nicholson and Anthony Carlisle, used a battery to decompose water into hydrogen gas and oxygen gas. They did this by passing an electric current through a water sample. Humphry Davy, another English chemist, used an electric current to liberate potassium metal from potassium hydroxide (KOH), sodium metal from sodium hydroxide (NaOH), and metallic magnesium, strontium, barium, and calcium from their respective compounds.

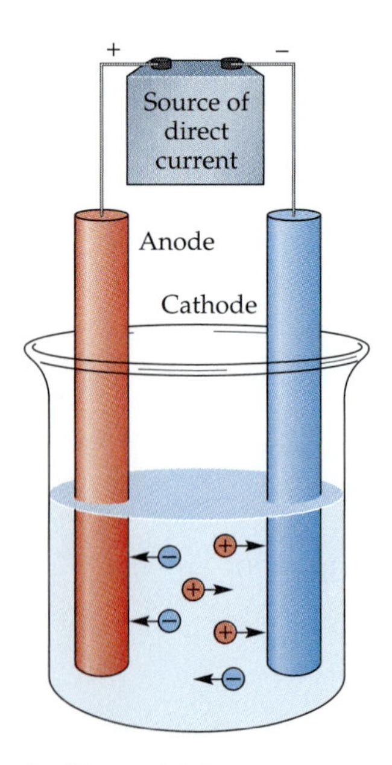

▲ **Figure 11.1**
When electric current is passed through an electrolyte, positive ions (cations) move to the cathode and negative ions (anions) move to the anode. Reduction occurs at the cathode, and oxidation takes place at the anode.

Davy's protégé, Michael Faraday, named the process of splitting compounds by means of electricity **electrolysis,** a word of Greek origin that literally means "releasing by electricity." Faraday thought up many of the terms used in electrochemistry today. We introduced some of these terms in Chapter 2, but we offer a brief review here. The carbon rods or metal strips that are connected to a source of electricity (the battery) and inserted into solutions under study are called **electrodes** (Figure 11.1). Electric current, as we now know, is a flow of electrons. In electrolysis, oxidation-reduction reactions occur at the electrodes. *Oxidation* occurs at the **anode,** where substances give up electrons to the positively charged electrode. *Reduction* occurs at the **cathode,** where substances pick up electrons from the negatively charged electrode.

In his studies, Faraday placed electrodes in certain solutions and melted ionic solids. When this was done, the electric "circuit" was completed. The battery provided the driving force (the electric potential or voltage), and electric current flowed from one electrode to the other through the solution and then back to the first electrode—that is, around the circuit. Faraday hypothesized that the electric current was carried through the solution (or melt) by atoms that had electric charges. He called these charged atoms *ions,* a term we have already used extensively. Recall that positively charged ions, which are attracted to the negatively charged electrode (the cathode), are called **cations,** and negatively charged ions, which are attracted to the positively charged anode, are called **anions.** (In electrolysis *anions* migrate to the *anode,* and *cations* migrate to the *cathode.*)

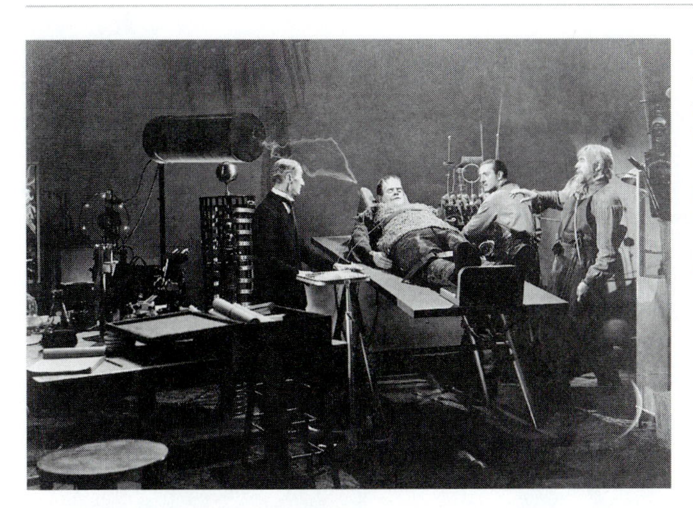

◀ The electrochemical research of Faraday and others was of great interest to scientists concerned with the fundamental structure of matter. A fictional experiment that related electricity to life captured the imagination of the general public. The experiment, portrayed in this photograph, was described by 21-year-old Mary Wollstonecraft Shelley in her 1818 novel, *Frankenstein.* The apparatus shown here was used by Dr. Frankenstein to generate the electricity that brought his monster to life.

✓ Review Question

11.1 Define or illustrate the following terms.

a. cathode **b.** anode **c.** cation
d. anion **e.** electrolysis **f.** electrode

11.2 Electrical Conductivity

The most familiar electric current flows in metallic wires. The valence electrons of the metal atoms flow through the wire, while the nucleus and inner-level electrons of the atom remain nearly fixed in place. Most nonmetals are nonconductors. The valence-level electrons of these elements are tightly bound or are shared with neighboring atoms in covalent compounds; they are not free to roam around. Solid ionic compounds do not conduct electricity. Although the ions have electric charges, they occupy fixed positions in a crystal lattice and are unable to move very much in an electric field. When the solid is melted or dissolved in water, the lattice is broken down and the ions are free to move about. A substance that conducts electricity when melted or in solution is called an **electrolyte.**

How do we tell when a substance is an electrolyte? One way is to dissolve some of the compound in water and test it with a conductivity apparatus. Figure 11.2, which compares the electrical conductivities of three solutions of equal molarity, illustrates three categories based on the following observations.

- *The light bulb lights up brightly.* The electrical conductivity is *high,* indicating the presence in the solution of a significant quantity of ions. The substance in solution—the solute—is called a **strong electrolyte.** (Even a dilute solution of a strong electrolyte gives high electrical conductivity.)
- *The bulb doesn't light.* There are few, if any, ions present. The solute is present in *molecular* form and is called a **nonelectrolyte.**
- *The bulb lights but glows only dimly.* The electrical conductivity is *low,* corresponding to a low concentration of ions. The solute is present only partly in ionic form and is called a **weak electrolyte.** (Even a concentrated solution of a weak electrolyte gives low electrical conductivity.)

If we were to test a large number of water-soluble substances by the method suggested in Figure 11.2, we would arrive at the following generalizations.

- Water-soluble ionic compounds and a few molecular compounds (such as the strong acids listed in Table 10.1) are *strong electrolytes.*
- Most molecular compounds are either *nonelectrolytes* or *weak electrolytes.*

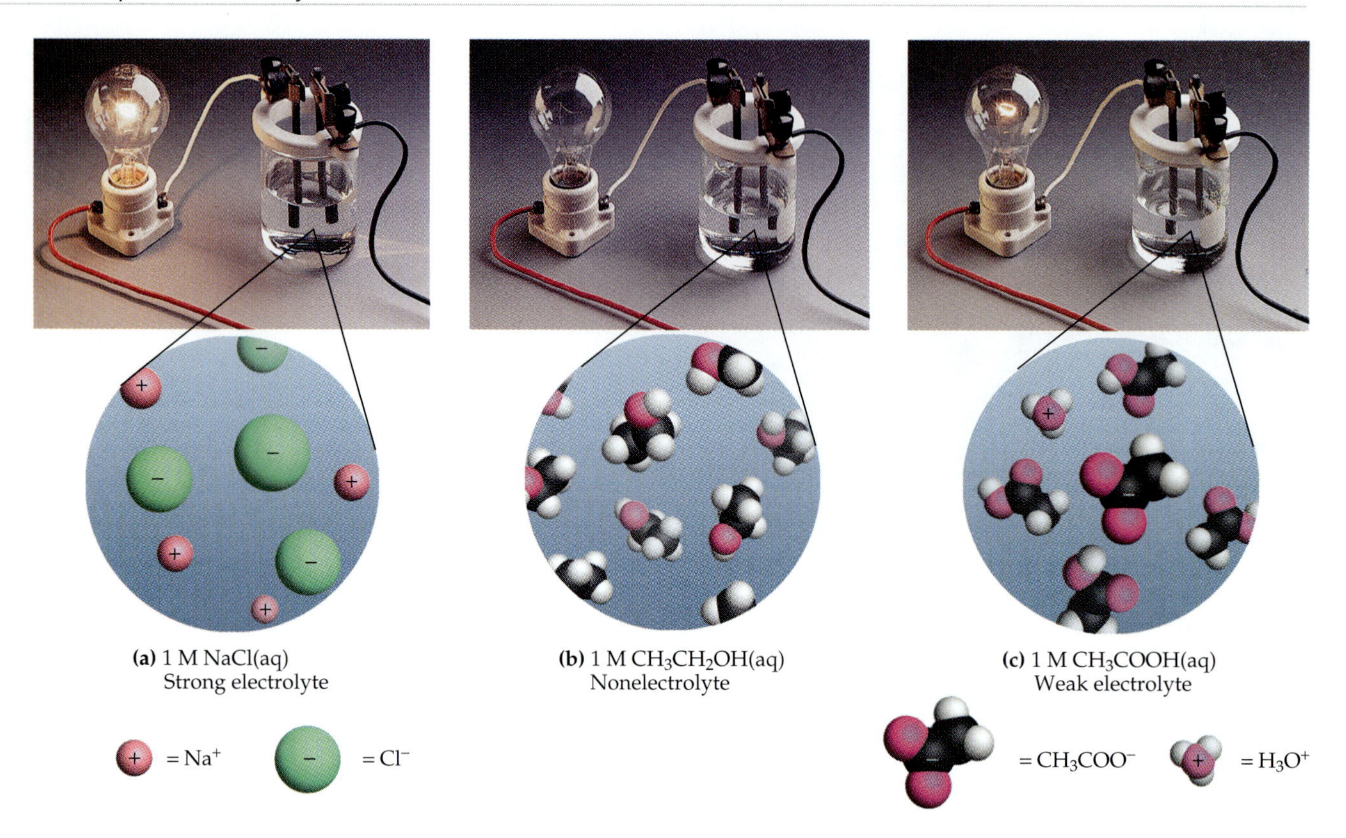

▲ **Figure 11.2**
Electrolytic properties of aqueous solutions. For electric current to flow, cations and anions must be present and free to flow between the graphite electrodes.

The weak acids and weak bases (Chapter 9) are also weak electrolytes. Most water-soluble organic compounds are nonelectrolytes (for example, alcohols and sugars), but the carboxylic acids (weak acids) and amines (weak bases) are weak electrolytes. Table 11.1 lists some familiar compounds and classifies them according to the behavior of their aqueous solutions toward an electric current.

Table 11.1 A Selection of Strong Electrolytes, Weak Electrolytes, and Nonelectrolytes

Compound Name	Formula	Kind of Compound	Electrical Conductivity
Hydrochloric acid	HCl	Strong acid	Strong
Sulfuric acid	H_2SO_4	Strong acid	Strong
Sodium hydroxide	$NaOH$	Strong base	Strong
Sodium chloride	$NaCl$	Salt	Strong
Calcium nitrate	$Ca(NO_3)_2$	Salt	Strong
Acetic acid	CH_3COOH	Weak acid	Weak
Ammonia	NH_3	Weak base	Weak
Methylamine	CH_3NH_2	Weak base	Weak
Methyl alcohol	CH_3OH	Molecular liquid	None
Ethyl alcohol	CH_3CH_2OH	Molecular liquid	None
Sugar (sucrose)	$C_{12}H_{22}O_{11}$	Molecular solid	None

Review Questions

11.2 Distinguish among a strong electrolyte, a weak electrolyte, and a nonelectrolyte.

11.3 Solid sodium chloride does not conduct electricity, but the molten salt does. Explain.

11.4 Formaldehyde (CH_2O) is a nonelectrolyte. What type of bonding exists in formaldehyde?

11.3 The Theory of Electrolytes: Ionization and Dissociation

In 1887 Svante Arrhenius proposed a general theory to explain the properties of electrolytes. We encountered a part of that theory in Chapter 9 in our study of acids and bases. Further, Arrhenius's theory has been modified somewhat through the years to account for new data. The modernized theory is summarized here.

1. An electrolyte, when dissolved in water, separates into ions. For salts and strong bases, which are already ionic in the solid state, we can summarize the process by equations such as

$$NaCl(s) \xrightarrow{H_2O} Na^+(aq) + Cl^-(aq)$$

$$Na_2SO_4(s) \xrightarrow{H_2O} 2\,Na^+(aq) + SO_4^{2-}(aq)$$

$$KOH(s) \xrightarrow{H_2O} K^+(aq) + OH^-(aq)$$

   These processes are called **dissociation,** the separation of existing ions. For acids and some bases, the process is actually one of *ion formation* or **ionization.**

$$HCl(g) + H_2O \longrightarrow H_3O^+(aq) + Cl^-(aq)$$

$$NH_3(g) + H_2O \longrightarrow NH_4^+(aq) + OH^-(aq)$$

2. When an ionic solid dissociates, or when a polar molecule ionizes, the algebraic sum of all positive charges and all negative charges is 0.

$$K_3PO_4(s) \longrightarrow 3\,K^+(aq) + PO_4^{3-}(aq)$$

$$3(+1) + (-3) = 0$$

   Solutions as a whole are therefore electrically neutral.
3. Each ion, regardless of size, charge, or shape, has the same effect on boiling-point elevation, freezing-point depression, and osmotic pressure (Chapter 8) as an undissociated molecule would have. This assumption holds quite well for dilute solutions but is not strictly true for concentrated ones, in which each ion is rather closely surrounded by others of opposite charge. The movement of the ions is thus hindered; they are not completely free. This decreases the expected activity somewhat. For example, the freezing-point depression of a solution of 1 mol of NaCl in 1 kg of water is 1.8 times (not quite the expected 2.0 times) that of a nonelectrolyte at the same concentration.
4. Weak electrolytes react with water to a limited extent and hence provide only a limited number of ions in solution.
5. Nonelectrolytes exist in molecular form in solution; they produce few, if any, ions.

Example 11.1

Classify the following as strong electrolytes, weak electrolytes, or nonelectrolytes in solution. Explain.

a. $MgCl_2$ **b.** $CH_3CH_2CH_2OH$ **c.** CH_3NH_2

Solution

a. $MgCl_2$, derived from the metal magnesium and the nonmetal chlorine, is an ionic compound, made up of Mg^{2+} ions and Cl^- ions. According to the solubility rules in Chapter 8, it is soluble in water and is therefore a strong electrolyte.

b. $CH_3CH_2CH_2OH$ has an oxygen atom that can form a hydrogen bond to water. It is soluble in water but it is a molecular compound. It does not ionize; it is a nonelectrolyte.

c. CH_3NH_2 is an amine, one of the weak bases we identified in Chapter 9. It is therefore also a weak electrolyte.

Practice Exercise

Classify the following as strong electrolytes, weak electrolytes, or nonelectrolytes in solution. Explain.

a. $NaNO_3$ **b.** CH_3CH_2COOH **c.** CH_3OCH_3

Review Questions

11.5 What is the difference in the way the terms ionization and ion dissociation are used?

11.6 State the main ideas of the modernized version of Arrhenius's theory of electrolytes.

11.7 Hydrogen chloride is a covalent molecule. Why does a solution of the gas in water conduct electricity?

11.4 Electrolysis: Chemical Change Caused by Electricity

If sodium chloride is heated until it melts, the molten salt conducts electricity, as one can see by testing the melt in the conductivity apparatus shown in Figure 11.2. In addition to causing the light bulb to glow, the electric current, as it passes through the melt, causes observable chemical changes. Yellow-green chlorine gas forms at the anode. At the cathode, silvery metallic sodium is formed and is rapidly vaporized by the hot melt. The molten, ionic sodium chloride is decomposed by electrical energy into elemental sodium and chlorine.

$$2\,NaCl + \text{energy} \longrightarrow 2\,Na + Cl_2$$

In crystalline form, sodium chloride does not conduct electricity. The ions occupy relatively fixed positions in the lattice and do not move very much, even under the influence of an electric potential. When sodium chloride melts, the ions are freed to move around. When a battery is connected to the melt through a pair of electrodes, the sodium ions are attracted to the electron-rich cathode, where they pick up electrons (Figure 11.3).

$$Na^+ + e^- \longrightarrow Na \quad (\text{Reduction: } \textit{at the cathode})$$

The chloride ions migrate to the electron-poor anode, where they give up electrons.

$$2\,Cl^- \longrightarrow Cl_2 + 2\,e^- \quad (\text{Oxidation: } \textit{at the anode})$$

The battery in the circuit keeps this exchange of electrons from leading to neutralized electrodes. Electrons picked up by the anode from chloride ions are immediately shunted, under the influence of the battery, to the cathode, which has been

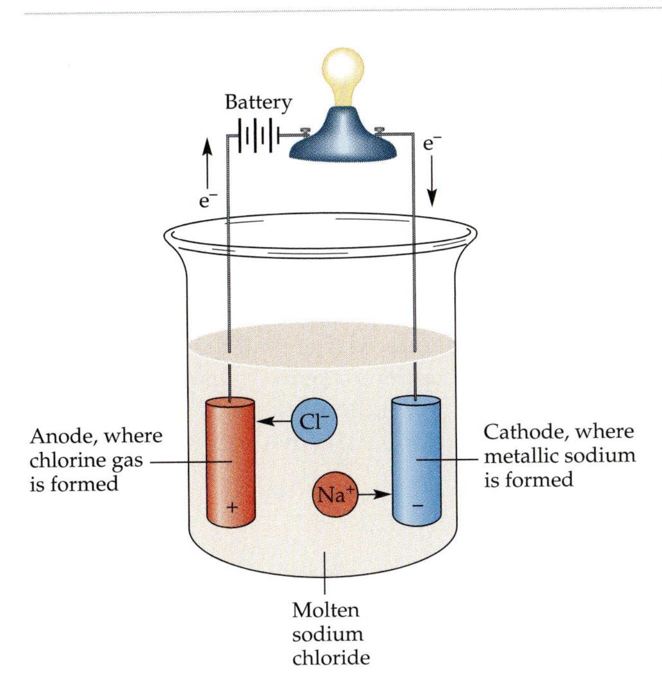

Figure 11.3
The electrolysis of molten sodium chloride. The electric current in the melt is carried by Na^+ and Cl^- ions.

losing electrons to sodium ions. Current flows as long as sodium ions and chloride ions are present in the melt. When all of the salt has been converted to elemental sodium and chlorine, the circuit is broken and current ceases.

Electrolysis, a process of using electricity to bring about chemical change, is useful for the preparation and purification (refining) of many metals. It is also used for coating one metal with another, an operation called *electroplating*. Usually the object to be electroplated, such as a spoon, is cast of a cheaper metal. It is then coated with a thin layer of a more attractive and more corrosion-resistant metal, such as gold or silver. The cost of the finished product is far less than that of a corresponding item made entirely of silver or gold. A cell for the electroplating of silver is shown in Figure 11.4. The silver bar is connected as the anode, and the spoon as the cathode. A solution of silver nitrate is used as the electrolyte. Under the influence of an external voltage, the silver ions (Ag^+) are attracted to the cathode (spoon), where they pick up electrons and are deposited as silver atoms.

Electrolytic processes are used in the production of Al, Li, K, Na, and Mg and in the refining of Cu.

$$Ag^+ + e^- \longrightarrow Ag$$

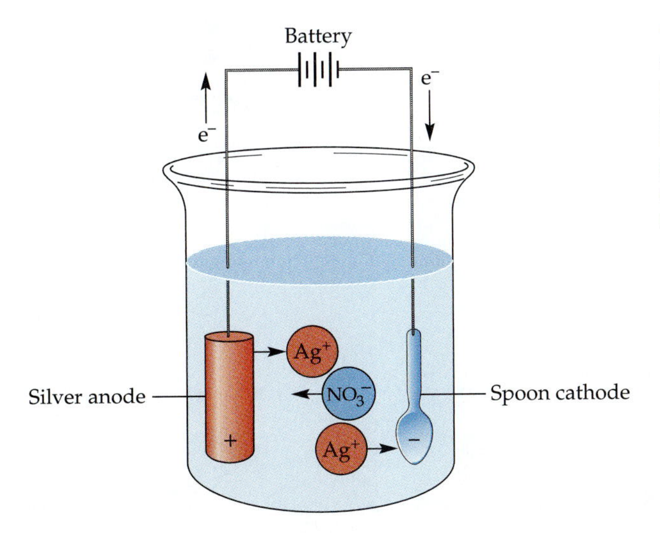

Figure 11.4
An electrochemical cell for the plating of silver. Silver atoms give up electrons at the anode, becoming Ag^+ ions. Silver ions gain electrons at the cathode and are plated out as silver metal.

At the anode, electrons are removed from the silver bar. Some of the silver atoms lose electrons to become silver ions.

$$Ag \longrightarrow Ag^{+} + e^{-}$$

The net process is one in which the silver from the bar is transferred to the spoon. The thickness of the deposit can be controlled by regulating the amount of current flow and the duration of the process.

Electrolysis finds some biological applications, including the removal of unwanted hair. In this process, a tiny wire needle is used to supply a mild electric current to the hair root. The chemical changes engendered there kill the living follicle. This is perhaps the only permanent method of hair removal. In the hands of a skilled technician, the method can be clean and safe. It is tedious, however, for each hair root must be treated individually. Similar procedures are sometimes used for the removal of warts or other growths.

Review Question

11.8 Is an object to be electroplated made the anode or the cathode in an electrolytic cell? Why?

11.5 Electrochemical Cells: Batteries

Electricity can cause chemical change. Conversely, chemical change can produce electricity. A device that uses a chemical reaction to produce electricity is called an **electrochemical cell.** Dry cells, storage batteries, and fuel cells all convert chemical energy to electrical energy.

When a strip of zinc metal is placed in a solution of copper(II) sulfate, the following reaction takes place.

$$Zn(s) + Cu^{2+}(aq) \longrightarrow Zn^{2+}(aq) + Cu(s)$$

The zinc atoms give up their valence electrons to the copper(II) ions. (The sulfate ions are not changed in the reaction and were omitted from the equation.) The zinc metal dissolves, going into solution as Zn^{2+} ions. The copper ions precipitate from the solution as copper metal (Figure 11.5). The Zn is oxidized; the Cu^{2+} ions are reduced. Electrons are transferred directly from zinc atoms to copper(II) ions. To em-

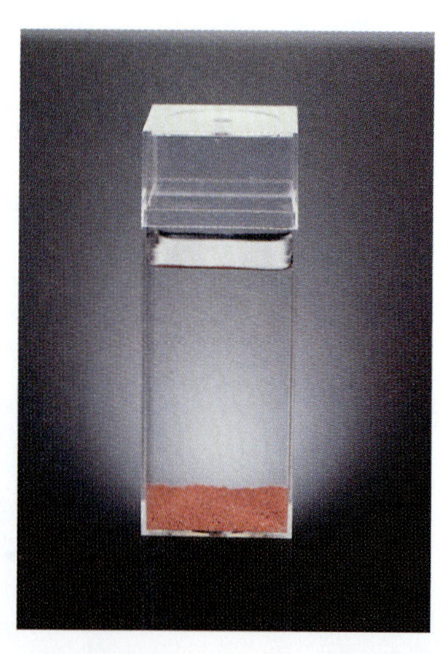

Figure 11.5 ▶
The photograph on the left shows a blue solution of Cu^{2+}(aq) ions and a sample of zinc metal. When the Zn is added to the Cu^{2+}(aq) solution, the more active Zn displaces the less active copper from solution. The products of the displacement reaction (right) are a reddish-brown precipitate of copper metal and a colorless solution of Zn^{2+}(aq) ions. The equation for the reaction is

$$Cu^{2+}(aq) + Zn(s) \longrightarrow Cu(s) + Zn^{2+}(aq)$$

phasize this transfer of electrons, the reaction can be split into two half-reactions (the oxidation portion and the reduction portion).

$$Zn \longrightarrow Zn^{2+} + 2\,e^- \ (\textit{oxidation half-reaction})$$

$$Cu^{2+} + 2\,e^- \longrightarrow Cu \ (\textit{reduction half-reaction})$$

If we place copper ions in one compartment and zinc metal in another, physically separating them (Figure 11.6), electrons have to pass through a wire to get from the zinc metal to the copper ions. This flow of electrons constitutes an *electric current*, and it can be used to run a motor or light a bulb.

In the cell pictured in Figure 11.6, there are two separate compartments. One contains zinc metal in a solution of zinc sulfate, and the other contains copper metal in a solution of copper(II) sulfate (which is blue). Zinc atoms give up electrons much more readily than copper atoms, so electrons flow away from the zinc and toward the copper. The zinc metal slowly dissolves as zinc atoms give up electrons to form zinc ions. The electrons flow through the wire to the copper, where copper ions pick up the electrons to become copper atoms. As time goes by, the zinc bar will slowly disappear, while the copper bar will get bigger, and the blue solution will gradually lose its color. Meanwhile something else also happens. Sulfate ions move from the blue copper sulfate solution to the zinc sulfate solution. Because more and more positively charged zinc ions are being added to the compartment at the left, while fewer and fewer copper ions remain in the compartment at the right, some of the negative sulfate ions must move from the right to the left in order to keep the two solutions electrically neutral. Notice that the porous partition in the cell allows the sulfate ions to move through it. (If the sulfate ions were unable to move through the barrier, the cell would not work.) Each time a zinc atom gives up two electrons, a copper ion picks up two electrons, and one sulfate ion moves from the right compartment to the left.

Why bother with the porous plate? Why not just put both solutions and both electrodes in a single compartment so the sulfate ions don't have to move through

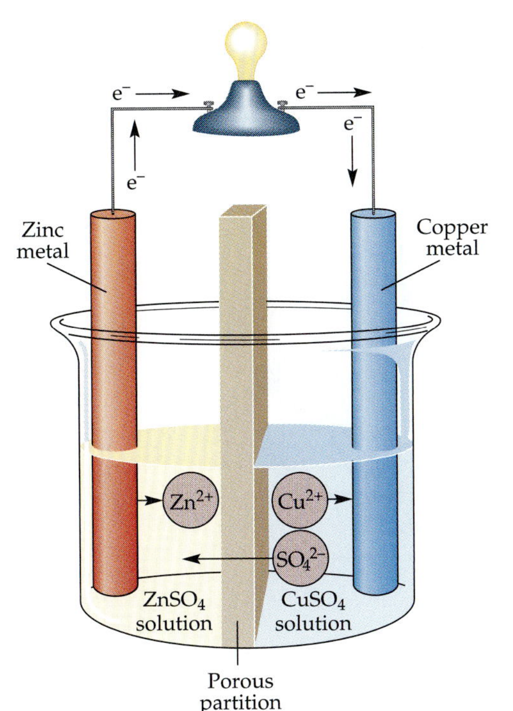

Figure 11.6
A simple electrochemical cell. The cell is described in the text.

pores? If we were to do that, the electrons wouldn't have to make the trip through the wire; they could just be passed from zinc atoms to copper ions, which would now be in contact with one another. And unless the electrons went through the wire, the light bulb wouldn't glow (nor would battery-operated portable radios, hand calculators, or cardiac pacemakers work).

If 1.0 M solutions of $ZnSO_4$ and $CuSO_4$ are used in the cell pictured in Figure 11.6, the system will produce about 1.1 volts at 25 °C. (The *volt* is a measure of electrical potential, or of the tendency of electrons to flow [see Section 11.6]).

The batteries used in most flashlights and portable electronic devices are *primary* batteries, so called because the cell reactions are irreversible and the batteries cannot be recharged. During use, reactants are converted to products, and when the reactants are used up, the battery is "dead." A typical example, the "dry" cell, is diagrammed in Figure 11.7(a). A zinc container is the anode and an inert carbon (graphite) rod is the cathode. The electrolyte is a moist paste of MnO_2, NH_4Cl, $ZnCl_2$, and carbon black. There is no free-flowing liquid in the cell, and it is for this reason that we call it a "dry" cell. Zinc metal is oxidized to Zn^{2+} at the anode. A simplified form of the overall reaction is

$$Zn(s) + 2\,MnO_2(s) + H_2O \longrightarrow Zn^{2+}(aq) + Mn_2O_3(s) + 2\,OH^-(aq)$$

Alkaline cells have KOH or NaOH as an electrolyte. They contain up to 50% more energy than ordinary dry cells.

In everyday life we call a device that stores chemical energy for later release of electricity a *battery.* A flashlight "battery" consists of a single cell with two electrodes in contact with one or more electrolytes. An automobile battery is a true battery in the sense that it consists of several simple electrochemical cells connected to one another.

The lead–acid storage battery used in automobiles is a *secondary* battery. The cell reaction can be reversed and the battery restored to near its original condition. The battery can be used through repeated cycles of discharging and recharging. Figure 11.8 shows a portion of a cell in the lead-acid battery. Several anodes and several cathodes are connected together in each cell to increase its current-delivering capacity. Each cell has a voltage of 2.05 V, and six cells are connected together in "series" fashion, + to −, to form a 12-volt battery.

The anodes in the lead-acid storage cell are of a lead alloy, and the cathodes are of a lead alloy impregnated with lead dioxide. The electrolyte is dilute sulfuric acid. The net reaction on discharge is

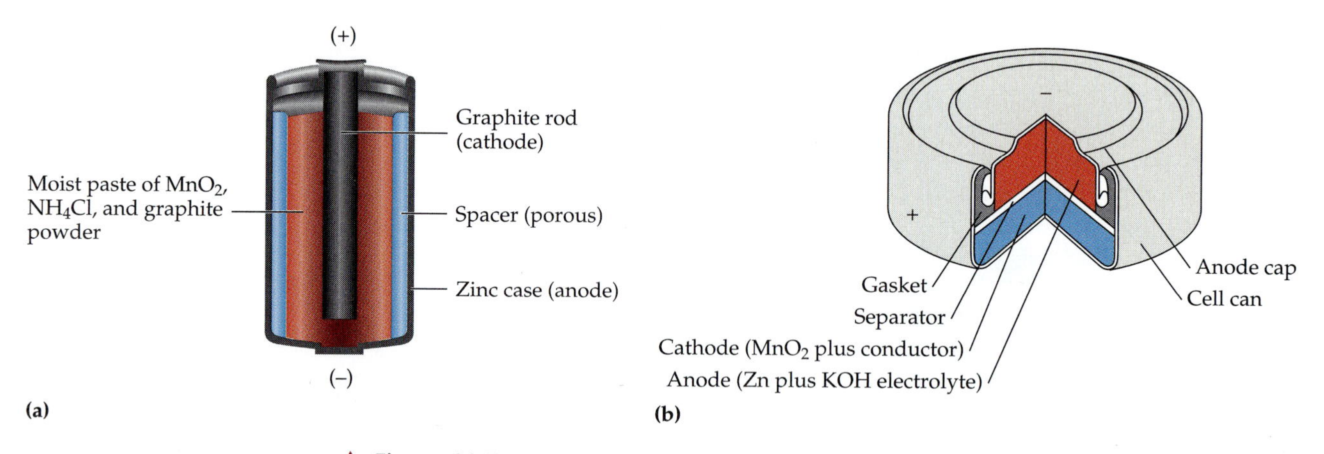

▲ **Figure 11.7**
(a) Cross section of a zinc–carbon dry cell. (b) Cutaway view of a miniature alkaline Zn-MnO_2 dry cell.

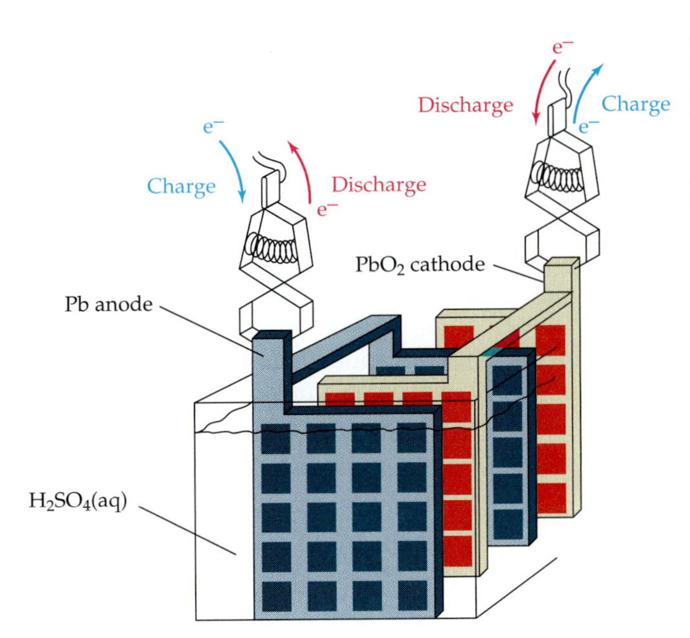

Figure 11.8
A lead–acid (storage) cell. The composition of the electrodes, the cell reaction, and the cell voltage are described in the text. Shown here are two anode plates and two cathode plates in "parallel" connections. This type of connection increases the surface area of the electrodes and the capacity of the cell to deliver current.

$$Pb(s) + PbO_2(s) + 4\,H^+(aq) + 2\,SO_4^{2-}(aq) \longrightarrow 2\,PbSO_4(s) + 2\,H_2O$$

As the cell reaction proceeds, $PbSO_4(s)$ precipitates and partially coats both electrodes; the water formed dilutes the $H_2SO_4(aq)$. The cell is *discharged.* By connecting the cell to an external electric energy source (of greater than 12 V), we can force electrons to flow in the opposite direction. The net cell reaction is reversed and the battery is recharged.

$$2\,PbSO_4(s) + 2\,H_2O \longrightarrow Pb(s) + PbO_2(s) + 4\,H^+(aq) + 2\,SO_4^{2-}(aq)$$

These lead storage batteries are durable, but they are heavy and contain corrosive sulfuric acid.

Other important commercial batteries include: mercury cells (zinc anode/HgO cathode), which are used in small electronic devices such as calculators and hearing aids; silver oxide cells (Zn anode/Ag_2O cathode), which are used in many watches and cameras; nickel–cadmium or "Ni–Cad" cells (Cd anode/NiO cathode), which are rechargeable and popular for portable radios and cordless appliances; and lithium cells (Li anode/MnO_2 cathode), which are very light in weight and used in devices such as pacemakers. All of these cells have a potassium hydroxide paste between the electrodes.

In a *fuel cell,* a gaseous fuel and oxygen gas flow over separate inert electrodes in contact with an electrolyte such as KOH(aq) at high temperatures. The cell reaction is

$$\text{Fuel} + \text{oxygen} \longrightarrow \text{oxidation (combustion) products}$$

The hydrogen-oxygen cell (Figure 11.9), which has been widely used in space vehicles, operates about 200 °C. The reactions are

Oxidation: $2\{H_2(g) + 2\,OH^-(aq) \longrightarrow 2\,H_2O + 2\,e^-\}$

Reduction: $O_2(g) + 2\,H_2O + 4\,e^- \longrightarrow 4\,OH^-(aq)$

Net: $2\,H_2(g) + O_2(g) \longrightarrow 2\,H_2O$

Automobiles using hydrogen–oxygen fuel cells produce no undesirable products (beyond humid air) and are classified as zero-emission vehicles. Future fuel cells, based on hydrocarbon fuels, hold great promise in the fight against air pollution.

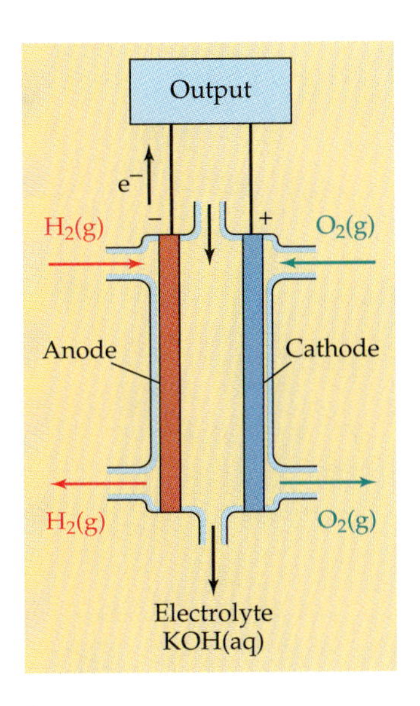

Figure 11.9
Cross section of a hydrogen–oxygen fuel cell. The electrodes are porous to allow easy access of the gaseous reactants to the electrolyte. Also, the electrode material catalyzes the electrode reactions.

Review Question

11.9 How does an electrolytic cell differ from an electrochemical cell?

11.6 The Activity Series

The zinc–copper electrochemical cell produces a voltage of about 1.1 V. This value does not depend on the size of the cell or on the size of the electrodes. The voltage measures the force with which electrons are moved around the circuit, and therefore it measures the tendency of this reaction to occur. Thus electrochemical cells give a quantitative measure of the relative tendencies of various redox reactions to occur.

If zinc metal is treated with hydrochloric acid, a reaction occurs to produce hydrogen gas and zinc chloride (see again Figure 5.5). The net reaction is

$$Zn(s) + 2\,HCl(aq) \longrightarrow ZnCl_2(aq) + H_2(g)$$

In fact, zinc will react with any acid to yield zinc ions and hydrogen gas. The reaction of zinc with sulfuric acid is

$$Zn(s) + H_2SO_4(aq) \longrightarrow ZnSO_4(aq) + H_2(g)$$

In these reactions with acids, zinc gives up electrons (is oxidized) and hydrogen ions gain electrons (are reduced).

$$Zn(s) \longrightarrow Zn^{2+}(aq) + 2\,e^- \; (\textit{oxidation half-reaction})$$

$$2\,H^+(aq) + 2\,e^- \longrightarrow H_2(g) \; (\textit{reduction half-reaction})$$

In contrast, if copper metal is treated with hydrochloric acid or sulfuric acid, no reaction occurs. Why? Because copper does not readily give up electrons to hydrogen ions. We say that zinc is *more active* than hydrogen and that copper is *less active* than hydrogen.

Zinc is one of a number of *active metals* that react with acids in this way. Not all these metals react at the same rate, however. Some, such as sodium and potassium, react violently with plain water—without any added acid. The reaction may produce enough heat to ignite the hydrogen gas, causing an explosion. With sodium, the by-product is a solution of sodium hydroxide.

$$2\,Na(s) + 2\,H_2O \longrightarrow 2\,NaOH(aq) + H_2(g)$$

A similar reaction occurs with potassium and the other Group 1A metals.

$$2\,K(s) + 2\,H_2O \longrightarrow 2\,KOH(aq) + H_2(g)$$

$$2\,Cs(s) + 2\,H_2O \longrightarrow 2\,CsOH(aq) + H_2(g)$$

Similar reactions involving members of the same chemical family, such as the alkali metals, are a part of the framework that helps make chemistry more understandable. We discuss the chemistry of some of the families of elements in Selected Topic A. However, a word of caution is in order. Even members of a family differ. They react at different rates. Sometimes even the *kind* of reaction is different, but that will not concern us here.

In contrast to the alkali metals, metals such as zinc and iron do not react appreciably when placed in water. However, if the temperature is raised and zinc is brought into contact with steam, then hydrogen gas is produced. Tin will not react with steam, but it will release hydrogen on contact with acids. Silver, mercury, and gold won't even react with acids to form hydrogen gas.

The **activity series of metals** (Table 11.2) has the most reactive metals at the top and the least reactive at the bottom. The position of a metal in the table reflects

Table 11.2 An Activity Series of Metals

Relative ease of oxidation (increases upward)	*Reduced Form* Metal atom	$\xrightarrow{\text{oxidation}}$	*Oxidized Form* Metal ion + *n* e⁻	
	Li	$\longrightarrow$	$Li^{+} + e^{-}$	React with cold water, steam, or acids, releasing hydrogen gas
	K	$\longrightarrow$	$K^{+} + e^{-}$	
	Ca	$\longrightarrow$	$Ca^{2+} + 2\,e^{-}$	
	Na	$\longrightarrow$	$Na^{+} + e^{-}$	
	Mg	$\longrightarrow$	$Mg^{2+} + 2\,e^{-}$	React with steam or acids, releasing hydrogen gas
	Al	$\longrightarrow$	$Al^{3+} + 3\,e^{-}$	
	Zn	$\longrightarrow$	$Zn^{2+} + 2\,e^{-}$	
	Cr	$\longrightarrow$	$Cr^{3+} + 3\,e^{-}$	
	Fe	$\longrightarrow$	$Fe^{2+} + 2\,e^{-}$	
	Cd	$\longrightarrow$	$Cd^{2+} + 2\,e^{-}$	React with acids, releasing hydrogen gas
	Ni	$\longrightarrow$	$Ni^{2+} + 2\,e^{-}$	
	Sn	$\longrightarrow$	$Sn^{2+} + 2\,e^{-}$	
	Pb	$\longrightarrow$	$Pb^{2+} + 2\,e^{-}$	
	H_2	$\longrightarrow$	$2\,H^{+} + 2\,e^{-}$	
	Cu	$\longrightarrow$	$Cu^{2+} + 2\,e^{-}$	Do not react with acids to release hydrogen gas
	Ag	$\longrightarrow$	$Ag^{+} + e^{-}$	
	Hg	$\longrightarrow$	$Hg^{2+} + 2\,e^{-}$	
	Au	$\longrightarrow$	$Au^{3+} + 3\,e^{-}$	

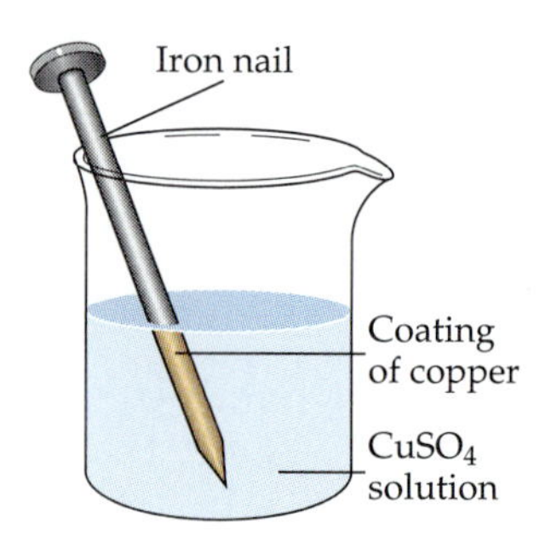

▲ **Figure 11.10**
An iron nail placed in a solution of $CuSO_4(aq)$ becomes coated with copper. The equation for the reaction is

$$Fe(s) + Cu^{2+}(aq) \longrightarrow Cu(s) + Fe^{2+}(aq)$$

its tendency to give up electrons to form ions. All of those listed above hydrogen in the series (the *active metals*) will react with acids to produce hydrogen gas. These metals will give their electrons to the hydrogen ions produced by the acids, thus becoming ions. Metals below hydrogen in the series will not give up electrons to hydrogen ions and hence will not liberate hydrogen gas from acids.

A lot of chemical information is summarized in Table 11.2. Remember that the arrangement of the table indicates how readily these metallic elements give up their electrons. When one of the metals is brought into contact with the ions of another metal, one of two things can happen. Either the metal can transfer electrons to the ions (a reaction occurs) or the ions will not accept the electrons (no reaction occurs). We can use the table to predict what will happen. A metal can transfer electrons to the ions of any metal that appears *lower* in the table. For example, an iron nail placed in a dilute solution of copper(II) sulfate ($CuSO_4$) gradually becomes coated with copper metal (Figure 11.10). Since iron atoms have a greater tendency than copper atoms to give up electrons, the iron atoms transfer electrons to the copper ions.

$$Fe(s) + Cu^{2+}(aq) \longrightarrow Fe^{2+}(aq) + Cu(s)$$

We can observe the copper metal plating out of solution as it is formed. Some of the iron dissolves as iron(II) ions, but this is not as obvious to the observer.

If we use another system, it is possible to see both processes as they occur. A strip of copper metal placed in a solution of silver nitrate ($AgNO_3$) becomes coated with crystals of metallic silver (Figure 11.11). The silver ions take electrons from the more reactive copper atoms.

$$2\ Ag^{+}(aq) + Cu(s) \longrightarrow 2\ Ag(s) + Cu^{2+}(aq)$$

▲ **Figure 11.11**
The displacement of $Ag^{+}(aq)$ by Cu(s). (a) A coil of copper wire is immersed in $AgNO_3(aq)$. (b) The shiny deposit that forms on the copper wire is metallic silver. The blue color of the solution indicates the presence of $Cu^{2+}(aq)$.

We can also see that the copper is dissolving because a solution of copper ions is blue, whereas a solution of silver ions is colorless. Thus, as the silver metal crystallizes out of solution, the copper ions form and slowly turn the solution blue.

Note that in one of the examples copper loses electrons and in the other it gains them. This is because there are elements both above and below copper in the activity series, and what happens depends only on the *relative tendencies of the elements involved to gain or lose electrons.*

Corrosion

An oxidation–reduction reaction of particular economic importance is the corrosion of metals. Perhaps 20% of all the iron and steel production in the United States each year goes to replace corroded items. Let's look first at the corrosion of iron.

In moist air, iron is oxidized, particularly at a nick or scratch.

$$\mathrm{Fe} \longrightarrow \mathrm{Fe^{2+}} + 2\,\mathrm{e^-}$$

In order for iron to be oxidized, oxygen must be reduced.

$$\mathrm{O_2} + 2\,\mathrm{H_2O} + 4\,\mathrm{e^-} \longrightarrow 4\,\mathrm{OH^-}$$

The net result, initially, is the formation of insoluble iron(II) hydroxide.

$$2\,\mathrm{Fe(s)} + \mathrm{O_2(g)} + 2\,\mathrm{H_2O} \longrightarrow 2\,\mathrm{Fe(OH)_2(s)}$$

This product is usually further oxidized to iron(III) hydroxide.

$$4\,\mathrm{Fe(OH)_2(s)} + \mathrm{O_2(g)} + 2\,\mathrm{H_2O} \longrightarrow 4\,\mathrm{Fe(OH)_3(s)}$$

The latter, sometimes written as $Fe_2O_3 \cdot 3H_2O$, is the familiar iron rust.

Oxidation and reduction often occur at separate points on the metal surface. Electrons are transferred through the iron metal. The circuit is completed by an electrolyte in aqueous solution. In the snowbelt, this solution is often the slush from road salt and melting snow. The metal is pitted in an anodic area, where iron is oxidized to Fe^{2+}. These ions migrate to the cathodic area, where they react with the hydroxide ions formed by reduction of oxygen.

$$\mathrm{Fe^{2+}(aq)} + 2\,\mathrm{OH^-(aq)} \longrightarrow \mathrm{Fe(OH)_2(s)}$$

As indicated above, this iron(II) hydroxide is then oxidized to $Fe(OH)_3$, or rust. This process is diagrammed in Figure 11.12. Notice that the anodic area is protected from oxygen by the water film, whereas the cathodic area is exposed to air.

Aluminum is more reactive than iron (see Table 11.2). Therefore we might expect it to corrode more rapidly, but billions of beer cans testify to the fact that it doesn't. How can this be? Freshly prepared aluminum quickly forms a thin, hard surface film of aluminum oxide. This film is impervious to air and protects the underlying metal from further oxidation. Corrosion can sometimes be a problem with aluminum, however. Certain

▶ Corrosion of iron is an electrochemical reaction of great economic importance, as is illustrated by this rusted steel support column of a bridge. Overall, corrosion costs about $100 billion dollars a year in the United States.

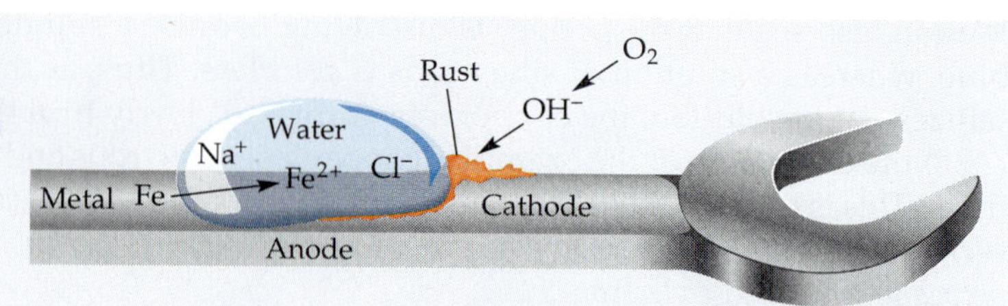

Figure 11.12 ▶ The corrosion of iron requires water, $O_2(g)$ from the air, and an electrolyte.

substances, such as salt, can interfere with the protective oxide coating on aluminum, allowing the metal to oxidize. Mag wheels on automobiles have cracked, and planes with aluminum landing gear have had their wheels sheared off because of this problem.

The tarnish on silver is due to the oxidation of the silver surface by hydrogen sulfide (H_2S) in the air. It produces a film of black silver sulfide (Ag_2S) on the metal surface. You can use a silver polish to remove the tarnish, but in doing so, you also lose part of the silver. An alternative method uses aluminum metal to reduce the silver ions back to silver metal.

$$3\,Ag^+(aq) + Al(s) \longrightarrow 3\,Ag(s) + Al^{3+}(aq)$$

This reaction also requires an electrolyte. Sodium bicarbonate ($NaHCO_3$) is usually used. The tarnished silver is placed in contact with aluminum foil and covered with a solution of sodium bicarbonate. A precious metal is conserved at the expense of a cheaper one.

Oxidation of Iron: Warm Hands and Fresh Foods

Generally we think of the rusting of iron as undesirable. However, this chemical property of iron—the tendency to rust—can be used to our advantage in some cases. Following are two examples.

A portable hand warmer has powdered iron metal in a sealed plastic pouch. The instructions say: "Open wrapper and expose warmer to air." How does it work? When the plastic wrapper is opened, a chemical reaction occurs. The reaction is quite complex, but in a simplified way can be represented as Fe metal oxidized by oxygen (from the air) to iron(III) oxide. The iron is powdered to provide maximum surface area for the reaction to occur. The reaction is exothermic, and the heat released by the reaction warms the hands.

Packaged foods spoil because oxygen from the air reacts with fats, turning them rancid. Some food processing plants vacuum-pack their products to minimize exposure to oxygen. The Japanese use packets of iron powder to absorb any free oxygen inside the bag, thus keeping the food fresh. As in the hand warmer, the iron is oxidized, principally to iron(III) oxide.

✓ Review Questions

11.10 What is meant by the activity series of metals? How is it used?

11.11 Why are water, oxygen, and an electrolyte all required for the corrosion of iron?

11.12 How does silver tarnish? How can the tarnish be removed without the loss of silver?

11.7 Precipitation: The Solubility Product Relationship

Recall our discussion of the solubiltiy of ionic compounds in Section 8.3.

So far we have focused our attention on strong electrolytes, compounds that supply high concentrations of ions in solution. There are also ionic compounds whose solubilities in water are quite low. For example, the solubility of barium sulfate

($BaSO_4$) is 0.0002 g per 100 g of water at 25 °C, so slight that barium sulfate is a very weak electrolyte in aqueous solution. The fact that a salt is a weak electrolyte or a nonelectrolyte does not necessarily mean that it is unimportant in the chemistry of living systems. On the contrary, for some physiological processes, the relative insolubility of certain salts is of critical importance. To understand these processes, it is necessary to consider a new way of describing solubility and a way to determine when a salt will start precipitating from solution because its solubility has been exceeded.

When solid barium sulfate is added to water, an equilibrium is established between the undissolved solute and the ions.

Recall that a precipitate is in dynamic equilibrium with dissolved solute in solution (Section 8.5).

$$BaSO_4(s) \rightleftharpoons Ba^{2+}(aq) + SO_4{}^{2-}(aq)$$

An extremely small amount goes into solution. Even though barium salts are quite toxic, large amounts of barium sulfate can be swallowed or given by enema because very little will dissolve in the solutions of the body. Because barium sulfate is opaque to X-rays, technicians can use it to outline the stomach or intestines for X-ray photographs. The undissolved barium salt scattered throughout the intestines blocks the X-rays. The X-ray film is unexposed in these areas, which therefore appear white in the developed negative. The barium sulfate is later voided from the body unchanged.

Even though we say that barium sulfate is insoluble, the small amount that does dissolve can be measured. The concentration of barium ions (Ba^{2+}) and sulfate ions ($SO_4{}^{2-}$) in a saturated solution of barium sulfate is 0.00001 M (or 1×10^{-5} M) each at 25 °C.

The product of the concentrations of the ions in a saturated solution, $[Ba^{2+}][SO_4{}^{2-}]$, is a constant.

$$K_{sp} = [Ba^{2+}][SO_4{}^{2-}]$$

At 25 °C, the product is

$$(1 \times 10^{-5})(1 \times 10^{-5}) = 1 \times 10^{-10}$$

This value is called the **solubility product constant (K_{sp})** for barium sulfate.

Example 11.2

Write a K_{sp} expression for equilibrium in a saturated aqueous solution of iron(III) phosphate, $FePO_4$.

Solution

First, we write an equation for the dissociation of $FePO_4$ in water.

$$FePO_4(s) \rightleftharpoons Fe^{3+}(aq) + PO_4{}^{3-}(aq)$$

The K_{sp} expression is simply

$$K_{sp} = [Fe^{3+}][PO_4{}^{3-}]$$

Practice Exercises

A. Write a K_{sp} expression for equilibrium in a saturated aqueous solution of copper sulfide, CuS.

B. Write a K_{sp} expression for equilibrium in a saturated aqueous solution of magnesium fluoride, MgF_2. (Hint: As in other equilibrium constant expressions, coefficients in the balanced equation appear as exponents in the K_{sp} expression.)

K_{sp} values for many salts are known; a few are given in Table 11.3. These values are useful in predicting whether precipitation will occur upon mixing various ions. For example, if a solution containing barium ions is mixed with one containing

▲ Addition of KI(aq) to $Pb(NO_3)_2(aq)$ yields a yellow precipitate of lead(II) iodide [$PbI_2(s)$].

Table 11.3 Selected Solubility Product Constants at 25 °C

Compound	Formula	K_{sp}
Barium carbonate	$BaCO_3$	5.1×10^{-9}
Barium sulfate	$BaSO_4$	1.1×10^{-10}
Calcium carbonate	$CaCO_3$	2.8×10^{-9}
Copper(II) sulfide	CuS	6×10^{-37}
Lead chromate	$PbCrO_4$	2.8×10^{-13}
Magnesium carbonate	$MgCO_3$	3.5×10^{-8}
Mercury(II) sulfide	HgS	2×10^{-53}
Silver acetate	$AgC_2H_3O_2$	2.0×10^{-3}
Silver bromide	$AgBr$	5.0×10^{-13}
Silver chloride	$AgCl$	1.8×10^{-10}

sulfate ions, precipitation will occur if the product $[Ba^{2+}][SO_4^{2-}]$ is greater than 1×10^{-10}. Conversely, barium sulfate in water will continue to dissociate until the product is just equal to 1×10^{-10}. At that value the solution will be saturated. If $[Ba^{2+}][SO_4^{2-}]$ is less than 1×10^{-10}, no barium sulfate precipitate will be formed. The solution will be unsaturated in barium sulfate.

As with the ion product of water, K_w (Section 10.3), the *product* of the ion concentrations is the important thing. If barium sulfate is simply placed in water, the concentration of barium ion equals the concentration of sulfate ion. It is, however, possible to prepare a solution in which the two concentrations are not equal (Example 11.3). The barium ion concentration could be higher than the sulfate ion concentration, or vice versa. Precipitation will always occur when the *product* of the two concentrations exceeds 1×10^{-10}, but precipitation will not necessarily coincide with concentrations of 1×10^{-5} M for each ion.

Example 11.3

If 0.001 mol of $BaCl_2(s)$ and 0.0001 mol of $Na_2SO_4(s)$ are added to 1 L of water, will $BaSO_4$ precipitate? (Assume no volume change.)

Solution

$BaCl_2$ and Na_2SO_4 are soluble salts; both completely dissociate.

$$\underset{0.001\ \text{mol}}{BaCl_2} \longrightarrow \underset{0.001\ \text{mol}}{Ba^{2+}} + \underset{0.002\ \text{mol}}{2\,Cl^-}$$

$$\underset{0.0001\ \text{mol}}{Na_2SO_4} \longrightarrow \underset{0.0002\ \text{mol}}{2\,Na^+} + \underset{0.0001\ \text{mol}}{SO_4^{2-}}$$

The 0.001 mol of $BaCl_2$ will yield 0.001 M Ba^{2+} (as well as 0.002 M Cl^-). The 0.0001 mol of Na_2SO_4 will yield 0.0001 M SO_4^{2-} (and 0.0002 M Na^+). The product of the barium ion and sulfate ion concentrations is

$$[Ba^{2+}][SO_4^{2-}] = (1 \times 10^{-3})(1 \times 10^{-4}) = 1 \times 10^{-7}$$

Remember that the smaller negative exponent corresponds to the larger number. For example, $10^{-1} = 0.1$ and $10^{-2} = 0.01$.

Because 1×10^{-7} is greater than the K_{sp} value of 1×10^{-10}, $BaSO_4$ will precipitate until the product of the $[Ba^{2+}]$ and $[SO_4^{2-}]$ is just equal to 1×10^{-10}. The sodium ions and chloride ions in solution will not precipitate as sodium chloride salt because sodium chloride is quite soluble in water.

Practice Exercises

A. Will a precipitate form if 0.100 mol of $AgNO_3(s)$ is added to 1.00 L of 1.0×10^{-6} M NaCl(aq)? (Assume no volume change.)

B. Will a precipitate form if 2.74×10^{-5} mol of $Na_2CrO_4(s)$ is added to 225 mL of 1.5×10^{-4} M $AgNO_3(aq)$? For $Ag_2CrO_4(s)$, $K_{sp} = 2.4 \times 10^{-12}$. (Assume no volume change.)

The solubility product principle can be used in the preparation of certain compounds. If silver nitrate solution is mixed with sodium chloride solution, slightly soluble silver chloride will precipitate. The silver chloride can be removed by filtration—that is, by pouring the mixture through a porous paper that will trap the solid silver chloride but permit passage of the solution. Further, if equimolar quantities of silver nitrate and sodium chloride are used, the filtrate (the solution going through the filter paper) can be evaporated, and the other ions can be obtained as sodium nitrate in crystalline form. The solubility product (K_{sp}) of silver chloride, $[Ag^+][Cl^-]$, is 1.8×10^{-10} at 25 °C. This means that a very small number of silver ions and chloride ions in solution will pass through the filter paper. But the vast majority of these ions will be trapped on the paper as the solid salt.

Precipitation is important in the formation of many minerals in nature. Geologists are not the only ones concerned with the formation of mineral deposits, however. Our teeth and bones are largely calcium phosphate salts. One such salt is $Ca_3(PO_4)_2$, sometimes called tricalcium phosphate. Teeth and bones are formed by the precipitation of calcium phosphate salts from solution. In order for this precipitation to occur, the concentrations of ions must exceed the solubility product in the immediate area of deposition. If we assume that the precipitation reaction is

$$3\,Ca^{2+}(aq) + 2\,PO_4^{3-}(aq) \rightleftharpoons Ca_3(PO_4)_2(s)$$

the solubility product is

$$K_{sp} = [Ca^{2+}]^3[PO_4^{3-}]^2$$

Each calcium phosphate unit provides three calcium ions (Ca^{2+}) and two phosphate ions (PO_4^{3-}).

At 37 °C (normal body temperature), the solubility product constant for calcium phosphate is about 4×10^{-27}. In the blood, the concentration of free calcium ions is about 0.0012 M (1.2×10^{-3} M), and the concentration of phosphate ions is about 1.6×10^{-8} M. Plugging these values into the solubility product relationship, we get

$$K_{sp} = (1.2 \times 10^{-3})^3 \times (1.6 \times 10^{-8})^2 = (1.7 \times 10^{-9}) \times (2.6 \times 10^{-16}) = 4.4 \times 10^{-25}$$

Because this value is *larger* than the solubility product (4×10^{-27}), we expect precipitation of calcium phosphate and the subsequent growth of bone and teeth.

In growing children, the foregoing description may closely represent the actual mechanism. In adults, however, teeth and bones are no longer growing. What keeps them from getting ever larger? Metabolic processes in the cells keep the pH at the area of growth a little below 7.4. The hydrogen ions that are present tie up phosphate ions as hydrogen phosphate ions.

$$PO_4^{3-}(aq) + H^+(aq) \rightleftharpoons HPO_4^{2-}(aq)$$

The phosphate ion concentration is reduced to a value that just satisfies the solubility product relationship at the constant value of 4.0×10^{-27}. Normal bone and tooth maintenance takes place with neither growth nor diminution.

Let's look at two pathological conditions of the mouth and teeth. If the salivary glands are removed or destroyed, the teeth are no longer bathed by saliva. At the usual pH values, saliva provides just the right concentration of calcium ions and phosphate ions to prevent dissolution of the teeth. With the salivary glands gone, the concentration of the ions rapidly falls below the solubility product constant. When this happens, more calcium phosphate dissolves, and the teeth, if not removed, erode away.

A similar situation occurs when children suffer from chronic acidosis. The blood pH may be as low as 7.1, and the pH in the immediate areas of bone formation is probably even lower. The additional hydrogen ions tie up phosphate ions, lowering their concentration. Bone growth is greatly hindered, and the child's skeleton is badly formed.

Tooth decay (caries) can also be related to solubility. *Mucin,* a glycoprotein (protein with attached carbohydrates), forms a film, called plaque, on teeth. If it is not removed by brushing and flossing, buildup of plaque continues. Food and bacteria trapped in the plaque metabolize carbohydrates, producing lactic acid. Saliva does not penetrate plaque and hence cannot buffer against the buildup of the acid. The pH at the surface of the tooth may go as low as 4.5, and the concentration of phosphate ions in solution is rapidly depleted as hydrogen phosphate ions are formed. The calcium phosphate of the tooth dissolves to replenish the phosphate ion in solution, leaving a cavity in the tooth.

Teeth are also eroded in people suffering from *bulimia,* a condition characterized by binge eating followed by vomiting. Hydrochloric acid vomited from the stomach acts in the mouth to tie up phosphate ions. The pH may drop as low as 1.5, and erosion can be much more rapid than in caries.

Knowledge of solubility product relationships can also be used to bring slightly soluble salts into solution. If a relatively concentrated solution of calcium chloride ($CaCl_2$) is added to moderately concentrated sodium oxalate ($Na_2C_2O_4$), a precipitate of calcium oxalate (CaC_2O_4) is formed. If hydrochloric acid is now added to the precipitate, the solid dissolves because the oxalate ions ($C_2O_4^{2-}$) are tied up by the hydrogen ions (from the HCl) to form soluble, slightly ionized oxalic acid.

$$C_2O_4^{2-}(aq) + 2\,H^+(aq) \rightleftharpoons H_2C_2O_4(aq)$$

This decreases the concentration of oxalate ions so that the solubility product constant for calcium oxalate is no longer exceeded. The precipitate dissolves.

Some kidney stones are primarily calcium oxalate. Unfortunately, strong hydrochloric acid solutions cannot be used to dissolve kidney stones while they are still in the kidneys. The acid would be much too corrosive to cells. There are chemicals, however, that have shown modest success in dissolving kidney stones. If the stones are calcium oxalate, removing foods containing oxalates, such as chocolate, spinach, and black tea, from the diet sometimes helps.

Review Questions

11.13 What is a solubility product constant? How is it used to predict whether or not a precipitate will form when various ions are placed in the same solution?

11.14 What causes tooth decay? How are each of the following related to erosion of the teeth?

a. bulimia **b.** removal or destruction of the salivary glands

11.15 What happens to bone formation in children who suffer from chronic acidosis? Why?

11.8 The Salts of Life: Minerals

A variety of inorganic compounds, called *minerals,* are necessary for the proper growth and repair of our tissues. It is estimated that these minerals represent about 4% of human body weight. Some of these, such as the chlorides (Cl^-), phosphates

(PO_4^{3-}), bicarbonates (HCO_3^-), and sulfates (SO_4^{2-}), occur in the blood and other body fluids. Others, such as iron (as Fe^{2+}) in hemoglobin and phosphorus in the nucleic acids (DNA and RNA), are constituents of complex organic compounds.

We will discuss electrolytes in the body in more detail in Chapter 28.

Our bodies are built of relatively few elements (Figure 11.13). Carbon, hydrogen, nitrogen, and oxygen are called the *structural elements;* they are the main elements from which proteins, fats, and carbohydrates are made. These four and sodium, magnesium, potassium, phosphorus, sulfur, chlorine, and calcium make up the bulk of living matter. Other essential elements are present in only trace amounts. For example, a 70-kg human has only about 4 g of iron (about 60 ppm by mass), yet proper levels of iron are essential to good health. Other trace elements play similar vital roles.

Minerals serve a variety of functions. Perhaps the most dramatic is that of iodine. A small amount of iodine is necessary for the proper functioning of the thyroid gland. A deficiency of iodine has serious effects, of which *goiter* is perhaps the best known. Iodine is available in seafood. To guard against iodine deficiency, a small amount (0.02% by mass) of potassium iodide (KI) is added to table salt (NaCl). Iodized salt has greatly reduced the incidence of goiter.

Iron(II) ions (Fe^{2+}) are necessary for the proper functioning of the oxygen-transporting compound hemoglobin. Lack of sufficient iron results in iron deficiency *anemia,* a shortage of oxygen transporting material and thus of oxygen supplied to the body tissues. Symptoms include weakness and fatigue. Foods especially rich in iron compounds include red meats and liver.

Our diets contain ample phosphate ions, but calcium ions become the limiting factor in the development of *osteoporosis,* the brittle bone disease. After about the age of 35, people tend to consume less calcium and/or it is absorbed from foods less efficiently. Nutritionists recommend that we eat foods (dairy products) that are rich in calcium or take calcium supplements.

Calcium and phosphorus are necessary for the proper development of bones and teeth. Growing children need about 1.5 g of each per day. These elements are available in plentiful quantities in milk. Adults also need these elements. For example, calcium ions are necessary for the coagulation of blood (to stop bleeding) and for maintaining the rhythm of the heartbeat. Phosphorus compounds are necessary for carbohydrate metabolism (Chapter 25) and many other essential functions, some of which we will encounter in subsequent chapters.

Sodium chloride in moderate amounts is essential to life. It is important in the exchange of fluids between cells and plasma, for example. The presence of salt increases water retention. A high volume of retained fluids can cause swelling (*edema*) and high blood pressure (*hypertension*). Over 120 million prescriptions are written each year in the United States for diuretics, drugs that induce urination in an attempt to reduce the volume of retained fluids. Another 30 million prescriptions are written for potassium (K^+) supplements to replace the potassium that is washed out in the excess urine. As many as 60 million people in the United States suffer from hypertension, and most physicians agree that our diets generally contain too much salt. The American Heart Association recommends that adults limit their salt intake to no more than 3 g (3000 mg) per day, a 50% reduction from our present levels of salt intake.

Iron, copper, zinc, cobalt, manganese, molybdenum, calcium, and magnesium are essential to the proper functioning of certain critical enzymes. The functions of some other minerals are quite complex. Some things are known about how they operate, but a great deal remains to be learned about the role of inorganic chemicals in our bodies.

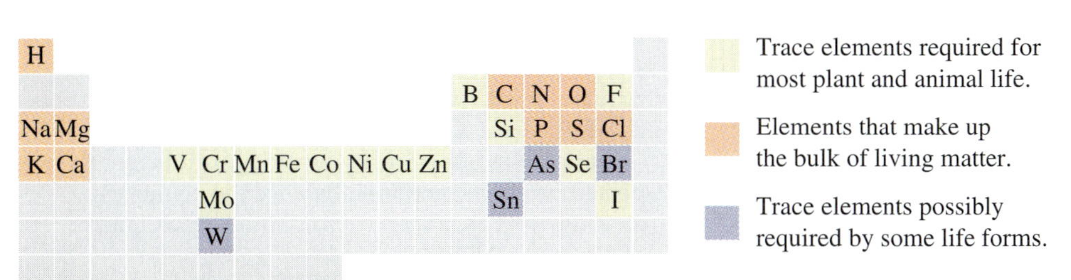

◀ **Figure 11.13**
The elements in living matter. Nine of the 11 most abundant essential elements are also the most abundant in seawater. The other two, nitrogen and phosphorus, are among the 20 most abundant elements in seawater. Humans have been described as "walking sacks of seawater."

Review Questions

11.16 Both goiter and anemia are deficiency diseases. How is each prevented?

11.17 Describe the roles of calcium and phosphorus in the proper development of bones and teeth.

11.18 What is hypertension? What is a diuretic and how is it used to treat hypertension?

Summary

Electrolysis is the process of chemical change caused by an electric current passed through a solution containing ions or a molten ionic substance.

An electrolytic cell consists of a source of electricity (usually a battery) and two solid **electrodes:** an **anode** where oxidation takes place and a **cathode** where reduction occurs.

Substances that conduct electricity when either dissolved in water or melted are **electrolytes.** Electrolytes can be weak (poor conductors in solution) or strong (good conductors in solution). In general, the more ions a substance produces when mixed with water, the stronger an electrolyte it is. **Nonelectrolytes** are substances that do not conduct electricity when dissolved in water or melted.

Dry cells, storage batteries, and fuel cells are devices that convert chemical energy to electrical energy. The devices resemble an electrolytic cell except that there is no external source of electricity and the species being reduced and oxidized are separated by a physical barrier. Each electrode is a metal sitting in a solution of some salt of that metal. Because of the barrier, the electrons produced in the oxidation are forced to travel through a metal wire to get to the species being reduced. This movement of electric charge through the wire is an electric current that can be used to do work.

Whether or not a metal is oxidized in an electrolysis depends on the relative tendency of the two metals involved to give up electrons. The ranking of this tendency is called the **activity series for metals.** Any member of the series gives up its electrons to ions of any metal listed below it.

An ionic solid that is sparingly soluble in water is characterized by its **solubility product constant** (K_{sp}). As long as the product of the ion concentrations in a solution of a sparingly soluble solid stays below K_{sp}, no precipitate is formed. As soon as the addition of ions causes the ion product to exceed K_{sp}, the solid precipitates.

Various inorganic compounds, called *minerals,* are necessary for growth and repair of tissues. These minerals make up about 4% of body weight.

Key Terms

activity series of metals (11.6)
anion (11.1)
anode (11.1)
cathode (11.1)
cation (11.1)
dissociation (11.3)
electrochemical cell (11.5)
electrode (11.1)
electrolysis (11.1)
electrolyte (11.2)
ionization (11.3)
nonelectrolyte (11.2)
solubility product constant (11.7)
strong electrolyte (11.2)
weak electrolyte (11.2)

Problems

Strong Electrolytes, Weak Electrolytes, and Nonelectrolytes

1. Classify the following as strong electrolytes, weak electrolytes, or nonelectrolytes in solution.
   a. KCl **b.** HNO_3 **c.** H_2CO_3
   d. Na_2SO_4 **e.** KOH
2. Classify the following as strong electrolytes, weak electrolytes, or nonelectrolytes in solution.
   a. CH_3OH **b.** $Ca(NO_3)_2$ **c.** CCl_4
   d. $CaCl_2$ **e.** NH_3
3. Lead chromate ($PbCrO_4$) is an ionic compound, yet a saturated aqueous solution of lead chromate does not conduct electricity. Explain.
4. Classify each of the following electrolytes as an acid, a base, or a salt. Write equations to show what happens when each is dissolved in water.
   a. $Ca(OH)_2$ **b.** HBr **c.** $Sr(NO_3)_2$
5. Why does a solution of 1 mol of NaCl in 1 kg of water freeze at a lower temperature than a solution of 1 mol of sugar in 1 kg of water? Why is the freezing-point depression for the salt solution not quite twice that for the sugar solution?
6. Which freezes at a lower temperature, 1.0 M NaCl(aq) or 1.0 M $CaCl_2$(aq)? Explain.

Activity Series

7. Complete the following by writing formulas for the expected products, and then balance the equations.
   a. $Ca(s) + HCl(aq) \longrightarrow$ **b.** $Ni(s) + HCl(aq) \longrightarrow$
   c. $Mg(s) + HNO_3(aq) \longrightarrow$
8. Complete the following by writing formulas for the expected products, and then balance the equations.
   a. $Zn(s) + H_2SO_4(aq) \longrightarrow$ **b.** $Al(s) + H_2SO_4(aq) \longrightarrow$
   c. $Pb(s) + HNO_3(aq) \longrightarrow$

9. Complete the following by writing formulas for the expected products, and then balance the equations.
   a. $Na(s) + H_2O \longrightarrow$ b. $Ba(s) + H_2O \longrightarrow$
10. Complete the following by writing formulas for the expected products, and then balance the equations.
   a. $Ca(s) + H_2O \longrightarrow$ b. $K(s) + H_2O \longrightarrow$
11. Complete the following equations for those reactions that can occur. You may refer to Table 11.2.
   a. $Mg(s) + Cu^{2+}(aq) \longrightarrow$ b. $Ag(s) + Pb^{2+}(aq) \longrightarrow$
   c. $Fe(s) + Zn^{2+}(aq) \longrightarrow$ d. $Al(s) + Ni^{2+}(aq) \longrightarrow$
12. Complete the following equations for those reactions that can occur. You may refer to Table 11.2
   a. $Cr(s) + Na^{+}(aq) \longrightarrow$ b. $Au(s) + Ag^{+}(aq) \longrightarrow$
   c. $Sn(s) + K^{+}(aq) \longrightarrow$ d. $Ca(s) + Al^{3+}(aq) \longrightarrow$

Solubility Product and Precipitation Criteria

Note: You may refer to Table 11.3 for the K_{sp} values.

13. Will a precipitate form if 0.001 mol of $MgCl_2$ and 0.001 mol of Na_2CO_3 are added to 1.0 L of water? The net ionic reaction is

$$Mg^{2+}(aq) + CO_3^{2-}(aq) \rightleftharpoons MgCO_3(s)$$

14. Will a precipitate form if 0.001 mol of $AgNO_3$ and 0.0001 mol of NaCl are added to 1.0 L of water? The net ionic reaction is

$$Ag^{+}(aq) + Cl^{-}(aq) \rightleftharpoons AgCl(s)$$

15. Will a precipitate occur if 1×10^{-6} mol of $Pb(NO_3)_2$ and 1×10^{-5} mol of Na_2CrO_4 are added to 1.0 L of water? The net ionic reaction is

$$Pb^{2+}(aq) + CrO_4^{2-}(aq) \rightleftharpoons PbCrO_4(s)$$

16. If a concentrated solution of sodium acetate is mixed with a solution of silver nitrate, a precipitate of silver acetate is formed. The precipitate dissolves readily when nitric acid is added. Write equations to explain what happens.

Additional Problems

17. Write an equation that shows why hydrogen iodide (HI), a gas, forms an aqueous solution that conducts electricity.
18. Why does aluminum corrode more slowly than iron, even though aluminum is more reactive than iron?
19. What new substances are formed when electricity is passed through the following?
   a. molten KBr   b. molten LiCl   c. molten Al_2O_3
20. Boiler scale ($CaCO_3$) is insoluble in water, yet it readily dissolves in hydrochloric acid. Write the equation that explains what happens.
21. Both $Ba(OH)_2$ and H_2SO_4 are strong electrolytes. When equimolar solutions of the two are mixed, the resulting solution is essentially nonconducting. Write the equation for the reaction and explain the observation.
22. Nineteen centuries ago, the Romans added calcium sulfate to wine. It clarifies the wine and also removes any dissolved lead. If 1.0 L of wine is saturated with calcium sulfate, the concentration of SO_4^{2-} is 0.014 M. What concentration of Pb^{2+} would remain in solution? The K_{sp} for lead sulfate is 1.1×10^{-8}, and the net ionic equation is

$$Pb^{2+}(aq) + SO_4^{2-}(aq) \rightleftharpoons PbSO_4(s)$$

23. Phosphates can be removed from water by precipitation, using iron(III) sulfate.

$$2\,PO_4^{3-}(aq) + Fe_2(SO_4)_3(aq) \longrightarrow 2\,FePO_4(s) + 3\,SO_4^{2-}(aq)$$

An estimated 7.0 metric tons per day of phosphates enters the East Anglian (United Kingdom) water system. How much iron(III) sulfate would be required to remove this phosphate?

24. Hard water is about 2×10^{-4} M in Ca^{2+}. Water is fluoridated with 1 g of F^- in 10^3 L of water. Will CaF_2 ($K_{sp} = 2 \times 10^{-10}$) precipitate upon fluoridation of hard water?

SELECTED TOPIC 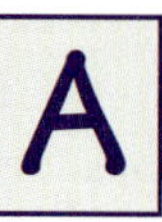

Inorganic Chemistry

In this selected topic, we take a look at some of the inorganic chemistry that is important to living systems. It almost seems a contradiction; in everyday life *inorganic* means nonliving and not derived from life, and *organic* means living, having the characteristics of life, or derived from life. So how could inorganic chemicals be important to living organisms?

In the old days chemists used the terms *organic* and *inorganic* in much the same way as everyone else. They believed that, while they could make many different inorganic chemicals, they could not hope to make organic compounds in the laboratory. They believed, with everyone else, that only living organisms, within their cells, could make organic compounds. Some mysterious *vital force*, they thought, was necessary for the synthesis of organic substances.

A series of experiments in the early 1800s caused chemists to discard the vital force theory. Perhaps the most important single step was made in 1828 by Friedrich Wöhler while he was a medical student at the University of Heidelberg. He attempted to prepare ammonium cyanate by heating a mixture of two inorganic substances, lead cyanate and ammonium hydroxide. To his surprise, instead of ammonium cyanate, he obtained crystals of urea, a well-known organic compound.

Urea is synthesized in the liver, transported to the kidneys, and excreted in the urine, from which it had been isolated in 1780. Wöhler correctly concluded that ammonium cyanate is formed initially, but heat causes it to rearrange to urea. Both urea and ammonium cyanate have the formula CN_2H_4O, but the atoms are arranged differently.

$$\underset{\text{Ammonium cyanate}}{NH_4^+CNO^-} \xrightarrow{\text{Heat}} \underset{\text{Urea}}{H_2N\overset{\overset{\displaystyle O}{\|}}{C}NH_2}$$

Although a few die-hard vitalists held out for several decades, the vital force theory was practically dead by the middle of the nineteenth century. **Organic chemistry** is now defined as the chemistry of the compounds of carbon. **Inorganic chemistry** is the chemistry of all the other elements. We devote several later chapters to organic chemistry. Here we examine some properties of the various groups of elements and relate these properties to electronic structure.

A.1 Using the Periodic Table to Write Electronic Configurations

In Chapter 2, we used an order-of-filling of orbitals (Figure 2.13) to write electronic configurations. We don't have to memorize such a chart, however, because we can deduce configurations directly from the periodic table. All we need to know is which subshells fill in different regions of the periodic table. To assist in this, let's refer to the four blocks of elements shown in Figure A.1.

- ***s*-block:** The *ns* subshell (the *s* subshell of the valence shell) fills by the aufbau (build-up) process. These are *main group* elements.
- ***p*-block:** The *np* subshell (the *p* subshell of the valence shell) fills. These are also *main group* elements.
- ***d*-block:** The $(n - 1)d$ subshell (the *d* subshell of the next-to-outermost shell) fills. These are *transition* elements found in the main body of the periodic table.
- ***f*-block:** The $(n - 2)f$ subshell (the *f* subshell of the second-from-outermost shell) fills. To keep the periodic table at a convenient width of 18 members, these elements are placed below the main body of the table. The 4*f* subshell fills in the *lanthanide* series, and the 5*f* subshell fills in the *actinide* series.[1]

The approach we use here is generally called *descriptive chemistry*. We start with the *s*-block elements, and then follow with sections that deal with the *p*-block and *d*-block, respectively. We will not deal extensively with the *f*-block elements.

As a brief review, recall (Chapter 2) that the more metallic the element is, the easier it is to remove the outermost electrons, and that metallic character *decreases* from left to right in the periodic table and *increases* from top to bottom with increasing atomic size. The alkali metals are all quite metallic, and the halogens are all distinctly nonmetallic. In the middle of the table we see intermediate behavior. Group 4A has carbon, a

[1]The lanthanides and actinides are sometimes called the *inner-transition* elements because they fall within series of *d*-block elements. The lanthanide series follows lanthanum ($Z = 57$) in the periodic table, and the actinide series follows actinium ($Z = 89$).

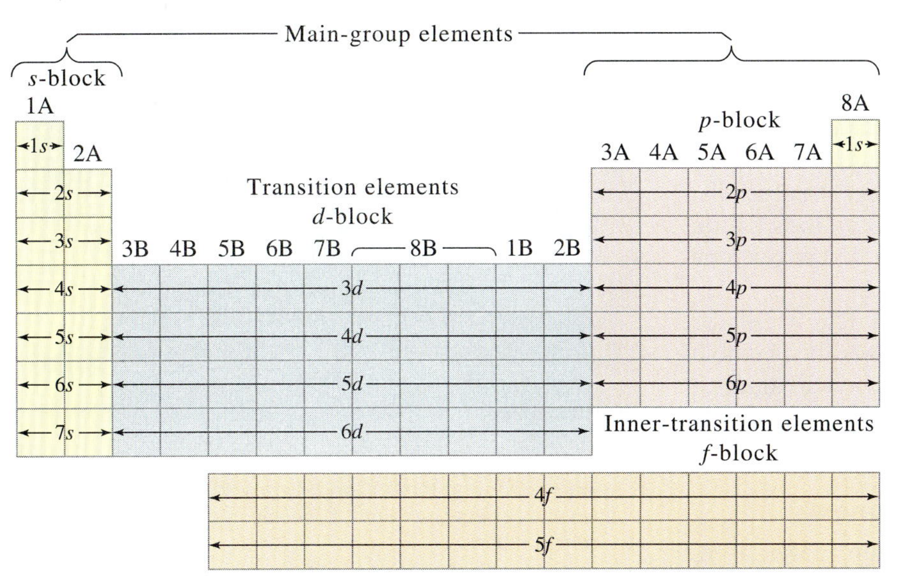

Figure A.1
The periodic table and the order of filling of subshells. Read through this periodic table, starting at the upper left, and you will discover the same order of filling of subshells as was shown in Figure 2.13. Note that helium ($Z = 2$) is an *s*-block element, but it is grouped with the *p*-block elements because we place it in Group 8A, with the other noble gas elements that it so strongly resembles.

nonmetal, at the top, and two metals, tin and lead, at the bottom. In between are the *metalloids*—metal-like elements—silicon and germanium. The following brief description of the blocks of elements will emphasize periodic trends.

A.2 The *s*-Block Elements

Four of the 14 elements that comprise the *s*-block are somewhat unusual cases. The first is hydrogen. A hydrogen atom has only one proton in its nucleus, and in the ground state its only electron is in the $1s^1$ configuration. We usually place hydrogen in Group 1A of the periodic table, but it actually is a nonmetal with little resemblance to the active metals of Group 1A. We discussed the chemistry of hydrogen separately in Chapter 5.

The second special case is helium. Because the helium atom has electrons only in its 1*s* orbital ($1s^2$), it is an *s*-block element. In its properties, however, helium closely resembles the other noble gases, which have the outer-shell electronic configuration ns^2np^6. They are *p*-block elements, and we will consider helium with the *p*-block elements (Section A.3) despite its electronic configuration.

The remaining two special cases, francium and radium, do have physical and chemical properties that resemble those of the other *s*-block members, but they are rare and highly radioactive. We will not consider them further here.

Group 1A: The Alkali Metals

All the **alkali metals** (Group 1A)—lithium, sodium, potassium, rubidium, cesium, and francium—have the valence electronic configuration ns^1. In the elemental form, the alkali metals are soft solids with low melting points. When freshly cut, these metals are bright and shiny, but they tarnish readily as they become oxidized by oxygen in the atmosphere. They are the most reactive of the metals: for example, they all react vigorously with water to evolve hydrogen gas. In chemical reactions, the atoms of the alkali metals tend to give up one electron each and form 1+ ions. Using symbols, we write the process for sodium as follows.

$$\text{Na(s)} \longrightarrow \text{Na}^+\text{(aq)} + \text{e}^-$$

Each alkali metal ion has the same electronic configuration as the noble gas that immediately precedes it in the periodic table. For example, the sodium ion, Na^+, has the same configuration as a neon atom.

Na $1s^22s^22p^61s^1$	Na^+ $1s^22s^22p^6$	Ne $1s^22s^22p^6$
Sodium atom	Sodium ion	Neon atom

All alkali metals form oxides of the general formula M_2O. All common compounds of the alkali metals are soluble in water.

Lithium salts are found in certain naturally occurring brines. Lithium carbonate (Li_2CO_3) is used in medicine to level out the dangerous "manic" highs that occur in manic-depressive psychoses. Some practitioners also recommend lithium carbonate for the depression stage of the cycle. It appears to act by affecting

Sodium, a soft metal, can be cut with a knife. The freshly cut surface is silvery, but this active metal is soon covered with a thick oxide coating.

the transport of chemical substances across cell membranes in the brain.

Sodium salts, including table salt (NaCl), are quite common. The balance of sodium ions and potassium ions is important in many functions of living tissues, including transmission of nerve impulses. In animals, potassium ions are the principal positive ions inside cells, and sodium ions are the principal positive ions in the extracellular fluid (see Chapter 28).

Potassium ion is also an essential nutrient for plants. It is generally abundant and is readily available to plants except in soil depleted by high-yield agriculture. The usual form of potassium in commercial fertilizers is potassium chloride (KCl).

Group 2A: The Alkaline Earth Elements

The six **alkaline earth metals**—beryllium, magnesium, calcium, strontium, barium, and radium—all have the valence electronic configuration ns^2. In the elemental form these metals are fairly soft and reactive. Except for beryllium, alkaline earth atoms show a tendency to give up two electrons, forming 2+ ions, as illustrated here for magnesium.

$$\mathrm{Mg(s)} \longrightarrow \mathrm{Mg^{2+}(aq)} + 2\,\mathrm{e^-}$$

The magnesium ion, Mg^{2+}, has the same electronic configuration as neon.

Mg $1s^22s^22p^61s^2$	Mg^{2+} $1s^22s^22p^6$	Ne $1s^22s^22p^6$
Magnesium atom	Magnesium ion	Neon atom

All alkaline earth metals form oxides of the general type MO. (Mendeleev based his periodic table to a large degree on the fact that elements within a group formed oxides and hydrides with the same general formula.)

Beryllium is something of an oddball member of the family. Unlike the others, it does not react with water. The metal itself is rather hard, rigid, and strong. Its lightness makes it valuable in structural alloys, but beryllium is poisonous in all its forms.

Magnesium ions are essential to both plants and animals. In plants, magnesium ions are incorporated in chlorophyll molecules; Mg^{2+} is therefore essential to photosynthesis. Both calcium and magnesium are essential for proper functioning of the nerves that control muscles.

Calcium ions are necessary for the proper development and maintenance of bones and teeth, clotting of blood, and maintenance of a regular heartbeat. For this reason growing children are usually encouraged to drink milk, a rich source of calcium. Adults also can obtain the calcium they need through dairy products or mineral supplements.

▲ Calcium, as Ca^{2+} ions, is necessary for the development of strong bones and teeth. Good dietary sources of calcium include milk, cheese, sardines, broccoli, and calcium carbonate antacid tablets.

Calcium carbonate (limestone) and other rocks containing calcium ions (Ca^{2+}), magnesium ions (Mg^{2+}), or iron ions (Fe^{2+} or Fe^{3+}) are widely distributed in nature. Water containing calcium, magnesium, or iron ions is known as *hard water*. The ions react with soaps to form curdy precipitates sometimes called bathtub ring (see Chapter 20).

A.3 The *p*-Block Elements

Groups 3A through 8A make up the *p*-block. The *p*-block includes the noble gases (except helium), all the nonmetals except hydrogen, all the metalloids, and even a few metals, such as aluminum, tin, and lead.

Three of the *p*-block elements—O, Si, and Al—are the most abundant in Earth's crust. Six of them—C, N, O, P, S, and Cl—are among the elements making up the bulk of living matter. Five others—B, F, Si, Se, and I—are required in trace amounts by most plant and animal life. Two of the *p*-block elements that can occur in the free state—C and S—have been known from prehistoric times, and three of them—Sn, Sb, and Pb—have been known for several thousand years. Most of the *p*-block elements, however, were not discovered until the eighteenth and nineteenth centuries.

In our discussion of the *p*-block elements, we will relate the properties of the elements to their positions in the periodic table. We will describe important compounds of the elements, comment on their uses, and discover some ways in which we encounter them in daily life. In this discussion we will use fundamental principles to explain chemical phenomena.

Group 3A: Boron and Aluminum

Group 3A consists of the metalloid boron and the metals aluminum, gallium, indium, and thallium. All have the electronic configuration ns^2np^1. Boric acid (H_3BO_3) is a familiar ingredient of mild antiseptic eye rinses. The other elements in Group 3A are typical metals and tend to form 3+ ions, as illustrated here for aluminum.

$$Al(s) \longrightarrow Al^{3+}(aq) + 3\,e^-$$

The aluminum ion, Al^{3+}, has the same electronic configuration as the noble gas neon.

Al $1s^22s^22p^63s^23p^1$	Al^{3+} $1s^22s^22p^6$	Ne $1s^22s^22p^6$
Aluminum atom	Aluminum ion	Neon atom

Aluminum is the most abundant metal in the Earth's crust, but it is tightly bound in compounds. Much energy, mainly electricity, is required to extract aluminum from its principal ore, bauxite [aluminum oxide (Al_2O_3)].

Aluminum is strong and light, with a density only one-third that of steel. Although it is considerably more active than iron, aluminum corrodes much more slowly. Freshly prepared aluminum metal reacts with oxygen to form a hard, transparent film of Al_2O_3 over its surface. The thin film protects the metal from further oxidation. Iron, on the other hand, forms an oxide coating that is porous and flaky and encourages further oxidation. Iron rusts away; aluminum does not.

Group 4A: Some Compounds of Carbon

Group 4A is made up of the nonmetal carbon, the metalloids silicon and germanium, and the metals tin and lead. All have the electronic configuration ns^2np^2. Of these, carbon is easily the most important. Carbon forms thousands—perhaps millions—of compounds with hydrogen. These compounds, called hydrocarbons, are discussed in detail in Chapter 13. Hydrocarbons and compounds derived from them are called organic compounds, and several later chapters are devoted to their study, a field called organic chemistry. A few simple compounds of carbon, however, are often considered to be inorganic. Among these are carbon monoxide (CO), carbon dioxide (CO_2), and such minerals as limestone and marble (calcium carbonate, $CaCO_3$).

▲ Aluminum is widely used in aircraft and boats.

Carbon exists in two main allotropic forms. **Allotropes** are modifications of an element that can exist in more than one form in the same physical state. Solid elemental carbon exists as graphite (the "lead" of pencils) and diamond (the precious jewel). Coal and related solid fuels are composed of varying amounts of elemental carbon, from about 6% in peat up to 88% or more in anthracite. When burned in sufficient oxygen, the carbon in coal combines with the oxygen to form carbon dioxide.

$$C(s) + O_2(g) \longrightarrow CO_2(g)$$

With less oxygen available, poisonous carbon monoxide (CO) forms.

$$2\,C(s) + O_2(g) \longrightarrow 2\,CO(g)$$

Carbon monoxide is an invisible, odorless, tasteless gas. The normal function of hemoglobin is to transport oxygen. Carbon monoxide binds tenaciously to hemoglobin—once on, it refuses to get off. The hemoglobin is thus prevented from binding and transporting oxygen (see Section 28.2).

Group 5A: Some Nitrogen Compounds

Group 5A includes the nonmetals nitrogen and phosphorus, the metalloids arsenic and antimony, and the metal bismuth. Their atoms have the electronic configuration ns^2np^3. We limit our discussion to the chemistry of nitrogen, an element essential to both plant and animal life in that it is a vital component of proteins, nucleic acids, and other biochemicals.

Although nitrogen makes up 78% of the atmosphere, the molecules of nitrogen gas (N_2) cannot be used directly by higher plants or by animals. They first

◄ Diamond and graphite are allotropic forms of carbon.

The Greenhouse Effect and Global Warming

Carbon dioxide is a product of both combustion and respiration. Generally it is regarded as innocuous. Certainly any immediate effect on us is slight. But what about long-term effects?

No matter how cleanly engines and factories operate, they produce CO_2 when they burn fossil fuels. The concentration of CO_2 in the atmosphere has increased 18% in this century, an increase largely attributed to the burning of fossil fuels. The concentration continues to increase, and at an accelerating rate, because of the increased burning of carbon fuels. The burning of forests adds CO_2 to the atmosphere, and the clearing of forests eliminates trees and other plants that would otherwise remove CO_2 by photosynthesis. The quantity of carbon dioxide in the atmosphere could double by the year 2030 if we continue our present practices.

The environmental impact of CO_2 and certain other gases is often called the **greenhouse effect.** These gases are transparent to visible light; they let the sun's rays through to warm the surface of Earth. When Earth tries to reradiate energy as infrared radiation into outer space, some of the energy is trapped by molecules of CO_2 and other greenhouse gases.

The greenhouse effect keeps Earth warm enough to be habitable, but human activities add 25 billion metric tons of carbon dioxide to the atmosphere each year, with 22 billion metric tons coming from the burning of fossil fuels. About 15 billion metric tons are removed by plants, the soil, and the oceans, leaving a net addition of 10 billion metric tons per year. The concentration of carbon dioxide is therefore increasing at a rate of 1 ppm per year. This leads many scientists to fear that it will enhance the greenhouse effect and lead to **global warming.**

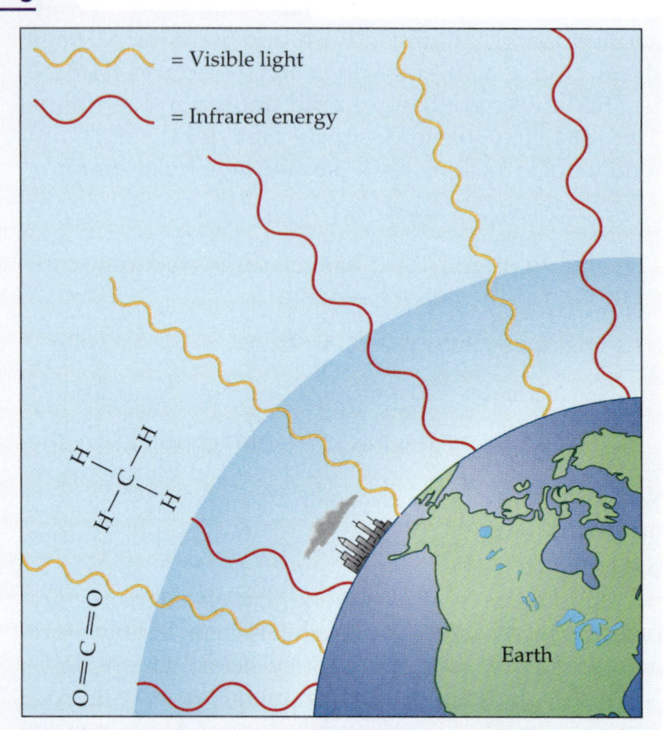

▲ The greenhouse effect. Sunlight passing through the atmosphere is absorbed, warming Earth's surface. The warm surface emits infrared radiation. Some of this radiation is absorbed by CO_2, H_2O, and other gases and is retained in the atmosphere as thermal energy.

Methane, chlorofluorocarbons (CFCs), and other trace gases also contribute to the greenhouse effect. Methane is 20 to 30 times and CFCs 20,000 times as effective as carbon dioxide at holding heat in Earth's atmosphere. Water vapor also acts as a greenhouse gas. When released into the atmosphere, however, water soon falls back to Earth as rain. It therefore affects the climate mostly at the local level.

Most atmospheric scientists predict that increases in greenhouse gases in the atmosphere will lead to global warming, but they often differ in their estimates of its magnitude and effect. The U.S. Environmental Protection Agency estimated that the planet would warm just 1 degree Celsius by the year 2050 and only 2 degrees by 2100, about half the previous estimates for warming in the twenty-first century. That was the good news. The bad news came in a report from a United Nations panel of scientists projecting growing deserts, dying forests, and flooded coastal areas as a result of global warming.

have to be "fixed"—that is, converted to a compound that can be used by plants. Certain types of bacteria convert atmospheric nitrogen to nitrates. Lightning also serves to "fix" nitrogen by causing it to combine with oxygen to form nitrogen monoxide (NO), commonly called nitric oxide, and nitrogen dioxide (NO_2).

First N_2 and O_2 react, forming NO.

$$N_2 + O_2 \xrightarrow{\text{lightning}} 2\,NO$$

NO reacts further with O_2, forming NO_2.

$$2\,NO + O_2 \longrightarrow 2\,NO_2$$

NO_2 reacts with water to form nitric acid (HNO_3).

$$3\,NO_2 + H_2O \longrightarrow 2\,HNO_3 + NO$$

The nitric acid falls in rainwater, adding to the supply of available nitrates in the oceans and the soil.

Scientists have learned how to fix nitrogen, and huge quantities are fixed industrially in the manufacture of nitrogen fertilizers. This has greatly increased our food supply, because the availability of fixed nitrogen is often the limiting factor in the production of food. In 1912 American farmers produced an average of 26 bushels of corn per acre. Today the yield per acre

▲ The high temperatures in a lightning bolt cause nitrogen and oxygen to combine. This nitrogen fixation during electrical storms contributes greatly to the nitrogen compounds available to plants.

is almost 100 bushels. This fourfold increase is due in large part to the increased use of nitrogen fertilizers. Not all the consequences of industrial fixation are favorable, however; excessive runoff of nitrogen fertilizer has led to serious water-pollution problems in some areas.

Group 6A: Compounds of Oxygen and Sulfur

Group 6A includes the nonmetals oxygen, sulfur, and selenium, the metalloid tellurium, and the radioactive metal polonium. Atoms of the Group 6A elements have the valence electronic configuration ns^2np^4. We limit our discussion to the chemistry of oxygen and sulfur.

Oxygen occurs in the atmosphere mainly as the diatomic molecule O_2. Oxygen is involved in oxidation processes (Chapter 5) such as combustion (rapid burning) and rusting and other forms of corrosion, and in

Photochemical Smog

Any high-temperature combustion of a fuel in air causes some nitrogen to combine with oxygen, forming nitrogen monoxide (nitric oxide).

$$N_2 + O_2 \longrightarrow 2\,NO$$

The greatest sources of atmospheric NO are in the exhaust fumes of high-compression, high-temperature internal combustion automobile engines and in the stack gases of fossil fuel–burning electrical power plants. NO is considered an air pollutant mainly because it participates in various reactions that yield other pollutants. It reacts with O_2 and with other pollutants in the air to form red-brown nitrogen dioxide (NO_2), an irritant to the eyes and respiratory system. However, NO_2 is objectionable mostly for its chemical reactions, including its decomposition.

$$NO_2 + \text{sunlight} \longrightarrow NO + O \quad \text{(Atomic oxygen)}$$

The oxygen atoms produced by this *photochemical* (light-induced) decomposition are highly reactive. They can react with many substances that are generally available in polluted air. For example, they may react with O_2 molecules to form ozone.

$$O + O_2 \longrightarrow O_3$$

Atomic oxygen (O) + Molecular oxygen (O_2) ⟶ Ozone (O_3)

Photochemical smog is characterized by a concentration of ozone (O_3) in air that is considerably higher than normal. It also has NO and NO_2, collectively called NO_x, unburned hydrocarbons, and several other components produced by the action of sunlight.

Ozone in the stratosphere protects us from harmful ultraviolet radiation, but ground-level ozone is the main cause of breathing difficulties that some people experience during smog episodes. Ozone also causes rubber to crack and deteriorate. Photochemical smog reduces visibility, and its components can cause heavy damage to crops. By forming nitric acid, NO and NO_2 contribute to the acidity of rainwater (Section 9.7), which accelerates the corrosion of metals and building materials. They also produce crop damage, although the specific effects of NO_x are difficult to separate from those of other pollutants.

▲ Photochemical smog is characterized by an amber haze, like that seen in this view across the city of Buenos Aires, Argentina.

respiration. Generally, oxygen reacts with metals by acquiring electrons and forming oxide ions (O^{2-}), which have the same electronic configuration as neon.

$O\ 1s^2 2s^2 2p^4$	$O^{2-}\ 1s^2 2s^2 2p^6$	$Ne\ 1s^2 2s^2 2p^6$
Oxygen atom	Oxide ion	Neon atom

Oxygen reacts rapidly with more active metals. Magnesium, for example, burns with a brilliant white flame when ignited in air, forming magnesium oxide (MgO).

$$2\,Mg + O_2 \longrightarrow 2\,MgO + \text{heat} + \text{light}$$

At room temperature, oxidation on the surface of freshly prepared magnesium metal forms a thin, transparent coating of magnesium oxide that is impervious to air and prevents further oxidation. Such oxide coatings are common on metals such as aluminum and titanium, making it possible for us to use otherwise quite reactive metals in utensils and machines.

Oxygen also reacts with many nonmetals. Sulfur, for example, burns in air to form sulfur dioxide, a choking, acrid gas.

$$S(s) + O_2(g) \longrightarrow SO_2(g)$$

When dissolved in water, sulfur dioxide reacts to form sulfurous acid.

$$SO_2(g) + H_2O \longrightarrow H_2SO_3(aq)$$

Small amounts of oxygen occur as *ozone* (O_3), an allotropic form of oxygen. Ozone is quite unstable. At room temperature, it breaks down slowly to ordinary oxygen.

$$2\,O_3(g) \longrightarrow 3\,O_2(g) + \text{heat}$$

Ozone is formed by electrical discharges through oxygen and by ultraviolet lamps. The pungent odor around electrical equipment is due to ozone. Ozone also is formed in photochemical smog. The ozone shield in the upper atmosphere protects us from harmful ultraviolet radiation.

Sulfur occurs in nature in both the combined and elemental forms. Free sulfur occurs as S_8, a ring of eight atoms. For simplicity, however, sulfur is often represented in equations only by the letter S, as if it were monatomic. Sulfur atoms can accept two electrons to form sulfide ions (S^{2-}), which have the same electronic configuration as argon. Sulfur gains electrons from the more active metals, such as calcium, to form sulfides, such as calcium sulfide (CaS).

Group 7A: The Halogens

Fluorine, chlorine, bromine, iodine, and astatine make up Group 7A, the **halogen** family. All halogens are nonmetals, and their atoms have the electronic configuration ns^2np^5. In the elemental form, all the halogens exist as diatomic molecules. The atoms can achieve a noble gas configuration by gaining an electron to form a negative ion. For example, fluorine atoms gain an electron to form F^-, an ion with the same electronic configuration as neon.

$$F_2 + 2\,e^- \longrightarrow 2\,F^-$$

This tendency, combined with that of many metals to give up electrons readily, is responsible for the ability of the halogens to form many salts, including sodium chloride, the familiar table salt. Indeed, the word halogen is derived from Greek words meaning "salt former." Our discussion is limited to the first four halogens, because astatine is rare and highly radioactive.

Fluoride salts, in moderate to high concentrations, are acute poisons. Small amounts of fluoride ion, however, are essential for our well-being. Concentrations of 0.7 to 1.0 part per million, by mass, of sodium fluoride have been added to the drinking water of many communities. Fluoridation results in a reduction in the incidence of dental caries (cavities) by strengthening tooth enamel.

Chlorine in the elemental form (Cl_2) is used to kill bacteria in water-treatment plants. Sodium hypochlorite (NaOCl) is an ingredient of common household bleaching solutions. Chlorine is present in our bodies as chloride ion (Cl^-), which is essential for preserving the acid-base balance and osmotic pressure and for the production of hydrochloric acid by our stomachs.

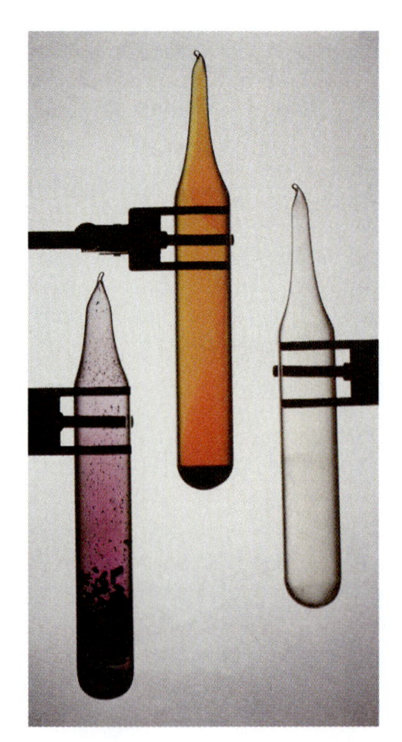

◀ At room temperature, chlorine (Cl_2, right) is a greenish-yellow gas; bromine (Br_2, middle) is a reddish-brown liquid that readily vaporizes; and iodine (I_2, left) is a dark solid that sublimes to form a violet vapor.

Industrial Smog

Polluted air associated with industrial activities is often called **industrial smog.** The presence of SO_2 and SO_3, collectively referred to as SO_x, is an important characteristic of this type of smog. Coal, especially soft coal from the eastern United States, has a relatively high sulfur content. When this coal is burned, sulfur compounds in it also burn, forming sulfur dioxide. Sulfur dioxide is readily absorbed in the respiratory system. It is a powerful irritant and is known to aggravate the symptoms of people who suffer from asthma, bronchitis, emphysema, and other lung diseases.

Some of the sulfur dioxide reacts further with oxygen in air to form sulfur trioxide (SO_3). Sulfur trioxide then reacts with water to form sulfuric acid (H_2SO_4). Fine droplets of this acid form an aerosol mist that is even more irritating to the respiratory tract than sulfur dioxide.

Usually industrial smog is also characterized by high levels of particulate matter, solid and liquid particles with dimensions of a few micrometers. The largest particles often are visible in air as dust and smoke. Particulate matter consists mainly of soot (unburned carbon) and the mineral matter that occurs in coal and does not burn. Much of this solid mineral matter is carried into the air by the tremendous draft created by the fire. When inhaled, this particulate matter contributes to respiratory problems in animals and humans.

The harmful effects of sulfur dioxide and particulate matter may be considerably magnified by their interaction. A certain level of sulfur dioxide, without the presence of particulate matter, might be reasonably safe. A certain level of particulate matter, without sulfur dioxide around, might be fairly harmless. But combine these same levels, and the effect might well be deadly. *Synergistic effects* such as this are quite common whenever certain chemicals are brought together. For example, some forms of asbestos are carcinogenic, and about 35 or 40 of the chemicals in cigarette smoke are carcinogens. Asbestos workers who smoke develop cancer at a much greater rate than do people who are exposed to one carcinogen but not the other.

◀ Industrial smog is often characterized by visible smoke and contains high levels of sulfur oxides. This copper smelter emitted 900 tons of SO_2 daily before it ceased operation in January of 1987.

When the pollutants in industrial smog come into contact with the alveoli of the lungs, the cells are broken down. The alveoli lose their resilience, and it becomes difficult for them to expel carbon dioxide. Such lung damage leads to—or at least contributes to—pulmonary emphysema, a condition characterized by an increasing shortness of breath.

The oxides of sulfur and the aerosol mists of sulfuric acid are damaging to plants. Leaves become bleached and splotchy when exposed to sulfur oxides. The yield and quality of farm crops can be severely affected. These compounds are also major ingredients in the production of acid rain.

Several compounds of bromine are of importance. Silver bromide (AgBr) is sensitive to light and is used in photographic film. Sodium bromide (NaBr) and potassium bromide (KBr) have been used medicinally as sedatives. Bromide ions depress the central nervous system. Unfortunately, prolonged intake can cause mental deterioration and other problems. Bromides have been largely replaced by other, presumably safer, sedatives.

Iodine is an essential nutrient. Compounds of iodine are necessary for the proper action of the thyroid gland. Iodine is also used as a topical antiseptic. Tincture of iodine is a solution of elemental iodine (I_2) in a mixture of alcohol and water. Iodine-releasing compounds, called iodophors, are often used as antiseptics in hospitals.

The halogens also react with hydrogen to form hydrogen halides. For example, chlorine reacts with hydrogen to form hydrogen chloride.

$$H_2(g) + Cl_2(g) \longrightarrow 2\,HCl(g)$$

In water solutions these compounds form acids. Hydrochloric acid is a familiar example. This acid is present in the stomach and is involved in the digestive process.

Group 8A: The Noble Gases

In the last decade of the nineteenth century, a group of elements was discovered that made up an entirely new family, one completely unexpected by Mendeleev and his contemporaries. Nonetheless, this new group, called the **noble gases,** fit neatly between the highly active nonmetals of Group 7A and the very reactive alkali metals (Group 1A). In the usual form of the periodic table, the noble gases are placed at the far right as Group 8A.

The six noble gases are helium, neon, argon, krypton, xenon, and radon. All are found to some extent in the atmosphere. Argon is abundant, making up nearly 1% of the atmosphere by volume. Xenon, on the other hand, makes up only 91 parts per billion of the atmosphere.

The noble gases rarely enter into chemical reactions. This lack of reactivity is a reflection of their electronic configurations. The helium atom has a filled first energy level ($1s^2$), and the other noble gases have the valence-shell electronic configuration ns^2np^6. They neither lose nor gain an electron readily. The helium atom, for example, has no affinity for an additional electron because this would have to be accommodated in the $2s$ orbital, which is at a much higher energy than the filled $1s$ orbital. Noble gases occur naturally only in elemental form and only as monatomic species. These elements were once called the "inert gases," but since 1962 a few compounds of krypton and xenon have been prepared. As yet, no compounds have been made of the lighter noble gases, helium, neon, and argon.

Helium is found in natural gas deposits, where it was formed from radioactive elements within the Earth. Helium is used to fill balloons and blimps. Its lifting power is more than 90% that of hydrogen, the lightest of all the gases, and it has the advantage of being nonflammable. Helium is also used to provide an inert atmosphere for the welding of metals that otherwise might be attacked by oxygen in the air. Liquid helium is used to achieve extremely low temperatures; it boils at only 4.2 K.

Neon is used in lighted signs for advertising. A tube with electrodes is shaped into letters or symbols and is filled with neon at low pressure. When an electric current is passed through the tube, the atoms emit their characteristic orange glow.

Argon, the most plentiful of the noble gases, is used to fill incandescent lightbulbs. Unlike nitrogen and oxygen, it does not react with the tungsten filament. Fluorescent lights are filled with a mixture of argon and mercury vapor.

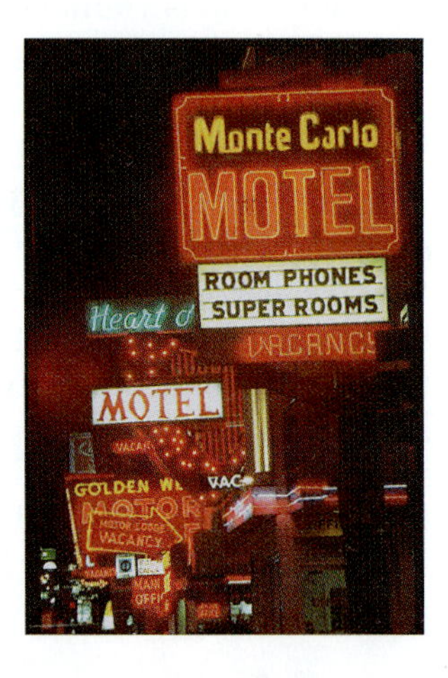

▶ The orange-red color of neon signs results from changes in electronic energy levels in neon atoms.

Krypton and xenon are too expensive to have many important commercial applications, although krypton has found some use in lightbulbs. Radon, although exceedingly rare in the atmosphere, can be collected from the radioactive decay of radium. Sealed in small vials, it is used for radiation therapy of certain malignancies. It seeps from the ground and accumulates in well-insulated buildings with poor ventilation; it is thought to be a minor cause of lung cancer. (Cigarette smoking is the main cause.)

A.4 The *d*-Block Elements

The *d*-block elements are the B groups of the periodic table. All are metals in the elemental form. They conduct electricity and have a characteristic metallic luster. Their metallic character ranges from highly metallic at the left end of the *d*-block to somewhat less metallic members at the right. Because they provide a transition from the highly metallic character of the *s*-block metals to the less metallic character of the *p*-block metals, the *d*-block (and *f*-block) metals are often called *transition metals*. A more precise definition is that a **transition metal** is an element in which *d*-orbital vacancies are found either in the metal atoms or in one or more of the metal ions. In this sense, Zn, Cd, and Hg are not transition metals because they do not have *d*-orbital vacancies in their atoms or ions.

Recall that the third period ends with argon ($Z = 18$), which has eight electrons in its valence shell. The fourth period begins with potassium ($Z = 19$), which has the valence electronic configuration $4s^1$. It continues with calcium ($Z = 20$), which has two valence electrons in its fourth shell ($4s^2$). Recall, however, that the $2n^2$ rule predicts a maximum of 18, not 8, electrons for the third energy level ($2n^2 = 2 \times 3^2 = 2 \times 9 = 18$). The fourth energy level has begun to fill before the third one is full.

However, with the next element, scandium, the third shell resumes filling.

$$\text{Sc}\ (Z = 21)\ [\text{Ar}]\ 3d^14s^2$$

The first transition series (from scandium to zinc) corresponds to a filling of the $3d$ subshell. With gallium ($Z = 31$), we return to an A group element. All inner shells are filled, and the next electron enters the outer shell.

Physical properties vary widely in the transition series. For example, mercury (Hg) is a liquid at room temperature (its melting point is −38 °C). Tungsten (W), however, melts at 3410 °C. Chemically, the transition elements also exhibit a variety of properties. Most form more than one kind of simple ion. Iron is a familiar example. It readily forms both iron(II) and iron(III) ions (Fe^{2+} and Fe^{3+}, respectively).

▲ Some solutions of transition metal salts are brightly colored. The colors are due to the following ions (left to right): Mn^{2+}, Fe^{2+}, Co^{2+}, Ni^{2+}, Cu^{2+}, and Zn^{2+}.

Many compounds of the transition metals are colored, some brilliantly so. For a given element, the color is different for different ions. Aqueous solutions of Fe^{2+} are often pale green; those of Fe^{3+} may be yellow. With polyatomic ions, color variation often is greater still. An aqueous solution containing manganese(II) ions (Mn^{2+}) is a faint pink. Solutions of MnO_4^{2-} are an intense green, and those of MnO_4^- are deep purple.

Transition metals known to be essential to life are iron, copper, zinc, cobalt, manganese, vanadium, chromium, nickel, tungsten, and molybdenum. Others are sometimes found in body tissues but have not been shown to be essential. Still others, most notably cadmium and mercury, are dangerous poisons.

Chromium plays a role in stabilizing glucose levels in the blood and in metabolizing carbohydrates and lipids. Cobalt is a component of vitamin B_{12}, a deficiency of which leads to pernicious anemia. Vitamin B_{12} is found only in animal products. One hazard of a strict vegetarian diet is vitamin B_{12} deficiency.

Iron is essential for the proper functioning of hemoglobin, the red protein molecule involved in oxygen transport (see Section 28.2). This iron must be in the form of the 2+ ion. If it is changed to the 3+ form, the resulting compound (methemoglobin) is incapable of carrying oxygen. The oxygen-deficiency disease that results is called *methemoglobinemia*. In infants, this condition is called the blue-baby syndrome.

We will not consider the *f*-block elements here.

Key Terms

alkali metal (A.2)
alkaline earth metal (A.2)
allotropes (A.3)
d-block (A.1)
f-block (A.1)
global warming (A.3)
greenhouse effect (A.3)
halogen (A.3)
industrial smog (A.3)
inorganic chemistry (page 294)
noble gas (A.3)
organic chemistry (page 294)
p-block (A.1)
photochemical smog (A.3)
s-block (A.1)
transition metal (A.4)

Review Questions

1. Define each of the following terms.
   a. inorganic chemistry **b.** noble gas
   c. halogen **d.** alkali metal
   e. alkaline earth metal **f.** transition element
2. What two alkali metal ions play major roles in maintaining fluid balance in the body?
3. What important molecule in plants incorporates magnesium?
4. Name two functions of calcium ions in the body.
5. What is hard water? How does it affect the action of soaps?
6. What alkali metal compound is used to treat manic-depressive psychoses?
7. Which alkaline earth metal is most different from the others? What are some of those differences?
8. What is the most abundant metal in the Earth's crust?
9. Why is aluminum replacing steel wherever possible in automobiles?
10. What condition(s) lead(s) to formation of carbon monoxide during combustion?
11. How does carbon monoxide exert its poisonous effect?
12. What is the greenhouse effect? How do greenhouse gases contribute to global warming?
13. What are allotropes? Name two sets of allotropes.
14 How do oxygen atoms, oxygen molecules, and ozone differ in structure and properties?
15. What are the health effects associated with ozone in the stratosphere and at ground level? Why are they not the same?
16. How does the burning of sulfur-containing coal lead to acid rain?
17. What are the halogens? Why are they so called?
18. What is the effect on tooth enamel of small amounts of fluoride?
19. Name one use for each of the following substances.
    a. chlorine **b.** iodine **c.** NaF
    d. AgBr **e.** NaOCl **f.** KCl
20. What is the outstanding chemical property of the noble gases? What structural feature accounts for this property?
21. What properties of helium makes it preferred to hydrogen for filling blimps, even though hydrogen has greater lifting power?

22. What properties of argon makes it useful in an electric lightbulb?
23. How does a neon sign work?
24. What is the origin of helium in natural gas wells?
25. Why are the noble gases no longer called the "inert gases"?
26. List three distinguishing characteristics of transition metals.
27. List four transition metals essential to life. Indicate their functions in the body.
28. What is synergism? Indicate one specific example of a synergistic effect concerning air pollution.
29. What was the vital force theory? How was it overthrown?
30. What is photochemical smog? What chemical compound starts the formation of photochemical smog by absorbing sunlight?

Problems

Electronic Configurations and the Periodic Table

31. What kind of subshell (*s*, *p*, *d*, or *f*) is being filled in each of the following regions of the periodic table?
    a. Groups 1A and 2A **b.** Groups 3A through 8A
32. What kind of subshell (*s*, *p*, *d*, or *f*) is being filled in each of the following regions of the periodic table?
    a. the transition elements
    b. the lanthanides and actinides
33. Referring only to the periodic table, indicate what similarity in electronic configuration fluorine and chlorine share. How do their electronic configurations differ? How do the electronic configurations of oxygen and fluorine differ?
34. Referring only to the periodic table, indicate what similarity in electronic configuration carbon and silicon share. How do their electronic configurations differ. How do the electronic configurations of carbon and nitrogen differ?
35. What similarities in properties are shared by the alkali metals? What structural feature do atoms of these elements share?
36. What similarities in properties are shared by the alkaline earth metals? What structural feature do atoms of these elements share?
37. What are the differences in electronic configurations that distinguish main-group and transition elements?
38. Without referring to any tables or listing in the text, mark an appropriate location for each of the following in the blank periodic table provided: **(a)** the fourth period noble gas; **(b)** the *d*-block element having one 3*d* electron; **(c)** a *p*-block element that is a metalloid; **(d)** a metal that forms the oxide M_2O_3.

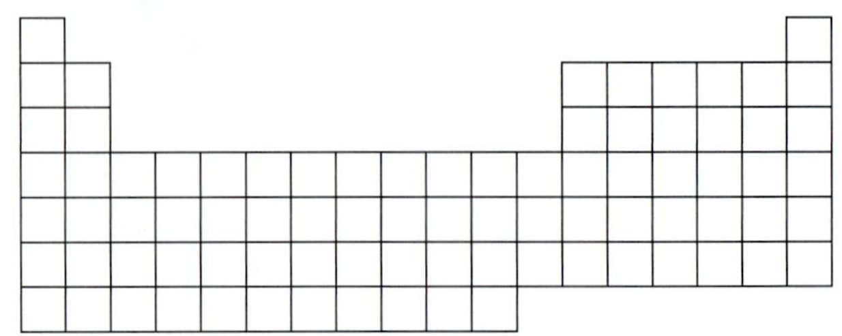

Chemical Equations

39. Complete and balance the following equations.
    a. $Li(s) + O_2(g) \longrightarrow$ **b.** $Mg(s) + O_2(g) \longrightarrow$
    c. $S(s) + O_2(g) \longrightarrow$
40. Complete and balance the following equations.
    a. $Ca(s) + O_2(g) \longrightarrow$ **b.** $SO_2(g) + H_2O \longrightarrow$
    c. $Ca(s) + S(s) \longrightarrow$

Additional Problems

41. Give the electronic configuration for the electrons beyond the xenon core of the hafnium (Hf) atom. Relate this electronic configuration to the position of hafnium in the periodic table.
42. Give the electronic configuration for the electrons beyond the xenon core of the mercury (Hg) atom. Relate this electronic configuration to the position of mercury in the periodic table.
43. Give Lewis symbols for gallium (Ga) and gallium ion.
44. Write a series of equations representing the natural fixation of atmospheric nitrogen in an electrical storm.
45. Write a series of equations showing how the burning of sulfur-containing coal leads to acid rain.
46. Write equations representing **(a)** the production of nitrogen monoxide (nitric oxide) in an automobile engine and **(b)** the action of sunlight on nitrogen dioxide.

The Atomic Nucleus

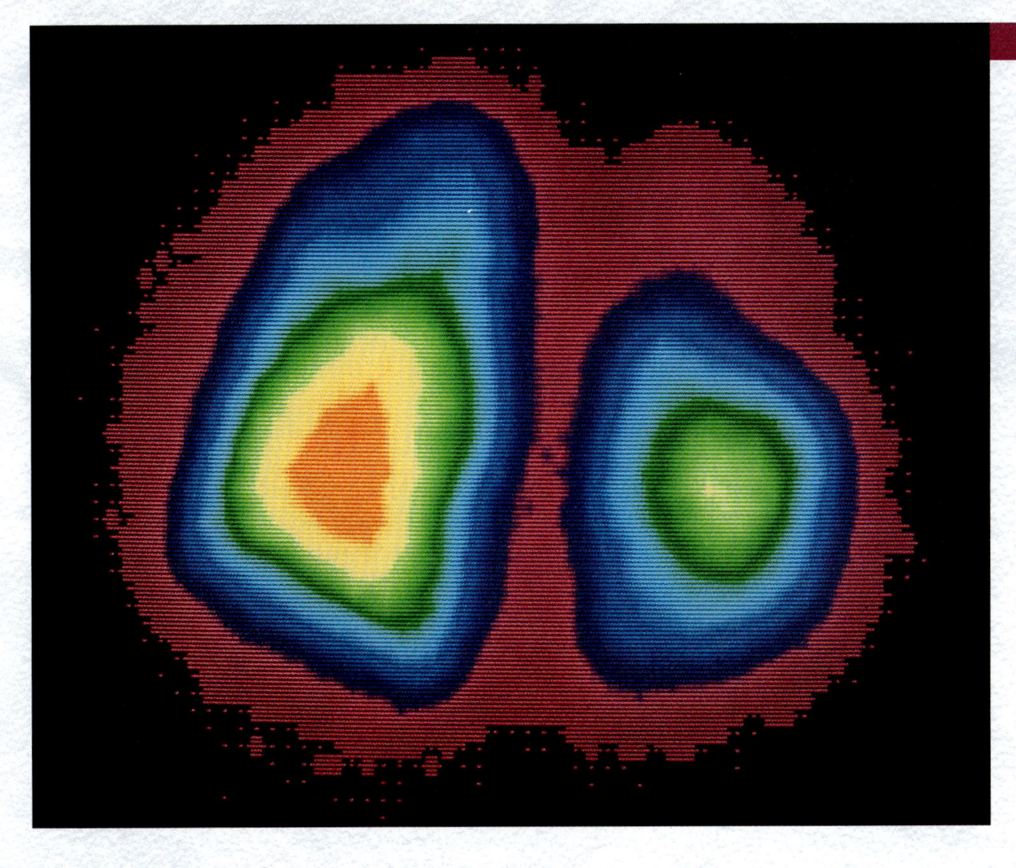

Knowledge of the properties of atomic nuclei makes it possible to obtain images that correspond to the distribution of various biochemical processes in the body. In this photograph, gamma rays from a radioactive isotope provide a false color image of the blood flow in a patient's lungs. Blood flow to the left lung has been reduced by a tumor.

Learning Objectives/Study Questions

1. What are the different types of radioactivity?
2. Which type of radioactivity is most hazardous when outside the body? Inside the body?
3. How is radioactivity measured?
4. What is ionizing radiation?
5. What is meant by the half-life of a radioisotope?
6. How are radioisotopes used to date materials? How is this used in archeology and geology?
7. What is transmutation?
8. What is nuclear fission? Nuclear fusion?
9. How are radioisotopes used in medicine?

Chemists often focus only on the valence electrons of atoms, for it is usually only these outer electrons that are involved in chemical reactions. Most of this text is devoted to the study of chemical reactions. For this chapter, though, let us turn our attention to that tiny speck of matter called the nucleus. The volume of an entire atom is about 10,000 times that of its nucleus, yet it is the nucleus that holds the power that became the symbol of much of the twentieth century.

Nuclear power confronts us with a great paradox. Throughout the cold war with the Soviet Union, there was great fear that nuclear power unleashed in wrath would destroy cities and perhaps civilizations. Controlled nuclear power, however, promised to provide the energy necessary to run our cities and maintain our civilization. Today we are afraid that terrorists will obtain a nuclear weapon and use it to threaten the destruction of a city, and even the peaceful uses of nuclear power are controversial. As citizens of the nuclear age, we have difficult decisions to make. Nuclear bombs can kill, but nuclear medicine saves lives. Diseases once regarded as incurable can be diagnosed and treated effectively with radioactive isotopes. Applications of nuclear chemistry to biology, industry, and agriculture have improved the human condition significantly. The use of radioisotopes in biological and agricultural research has led to increased crop production, which provides more food for a hungry world.

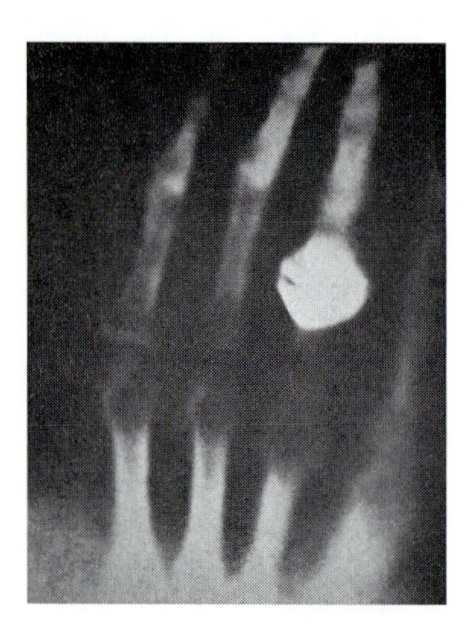

▲ The first-ever X-ray of a human being was made by Wilhelm Roentgen (1845–1923) shortly after his discovery of X-rays in 1895. It shows the hand of his wife, with the ring she was wearing.

12.1 Discovery of Radioactivity

For nearly all ordinary *chemical* reactions, the existence of isotopes (Chapter 2) can be ignored. Chemical reactions involve the valence electrons of atoms, and it matters little if there are differences in the numbers of neutrons buried deep in the atomic nucleus. *Nuclear reactions,* however, do involve the nucleus, and isotopes are of utmost importance, as we shall see.

The discovery of the first nuclear reactions was triggered by the discovery of X-rays, even though X-rays are not nuclear phenomena. In 1895 a German scientist, Wilhelm Roentgen, found that he could produce a form of radiation that passed right through solid materials. This mysterious radiation was like visible light in that it was a form of pure energy and resulted from electrons moving from one energy level to another. It was unlike visible light in that you could not see it. For want of a better name, Roentgen called the radiation **X-rays.** X-rays are high-energy radiation; they contain more energy than the most energetic visible light. This high energy gives X-rays great penetrating power.

X-rays are high-energy radiation.

The medical community immediately recognized the significance of the penetrating power of X-rays. X-ray pictures were used, for example, to locate bullets in wounds for expedient removal by surgery.

Other scientists also found Roentgen's discovery fascinating. One, French physicist Antoine Henri Becquerel (1852–1908), had been studying fluorescence, a phenomenon in which certain substances, after being exposed to strong sunlight, continue to glow even when taken into a dark room. Becquerel wondered if any substances give off X-rays when they fluoresce. To find out, he tested many substances, including uranium compounds. He found that uranium did emit invisible, penetrating rays, which had nothing to do with either fluorescence or X-rays. He had discovered a totally new phenomenon.

Recall (Chapter 2) that Marie Curie named this new phenomenon **radioactivity.** Working with her husband, Pierre, until his death in a traffic accident in 1906, she went on to discover several new radioactive elements, including radium and polonium.

Review Question

12.1 How are X-rays like visible light? How do they differ?

Marie Sklodowska Curie (1867–1934) was born in Poland. She went to Paris as a young woman to study for her doctor's degree in mathematics and physics. There she met and married French physicist Pierre Curie (1859–1906). Pierre was killed when run over by a horse-drawn carriage three years after he, Marie, and Becquerel shared the Nobel Prize in physics. Marie continued to work with radioactive substances and was awarded the Nobel Prize in chemistry in 1911. The Curies are honored on France's 500-franc note.

12.2 Types of Radioactivity

Scientists soon realized that uranium and other radioactive elements gave off three types of radiation. When passed through a strong magnetic or electric field, one portion was deflected in one direction, another portion was deflected in the opposite direction, and a third was not deflected at all (Figure 12.1). Ernest Rutherford (Chapter 2) assigned names to the three types of radioactivity. The portion deflected the least was found to be a stream of tiny particles identical to helium nuclei (^4_2He); these were named **alpha (α) particles.** The portion deflected in the opposite direction and to a greater extent was shown to be streams of electrons which were named **beta (β) particles.** Because beta particles have a negative charge, the symbol is sometimes written as β^-. Because its mass is only 1/1837 that of a proton, the mass number is taken to be 0 and the symbol $_{-1}^{0}e$ is used. The third, undeflected portion of radioactivity was shown to be electromagnetic radiation similar to, but having a different wavelength than visible light and X-rays, that is, pure radiant energy. Named **gamma (γ) rays,** this radiation has the highest energy and is the most penetrating form of radiation yet discovered. Gamma rays have no mass and no charge.

Alpha (α) particle: ^4_2He

Beta (β^-) particle: $_{-1}^{0}e$

The emission of an alpha or beta particle is often accompanied by the emission of a gamma (γ) ray.

Two other types of radioactivity, discovered later, are also of interest. The **positron (β^+)** is a particle equal in mass but opposite in charge to the electron. It is represented as $_{+1}^{0}e$. **Electron capture (E.C.)** is a process in which a nucleus absorbs an electron from an inner electron shell, usually the first or second. When an electron from a higher shell drops to the level vacated by the captured electron, an X-ray is released; E.C. is always accompanied by X-radiation. Once inside the nucleus, the captured electron combines with a proton to form a neutron. Positron-emitting isotopes and those which undergo electron capture are quite important in medical applications (Sections 12.9 and 12.10).

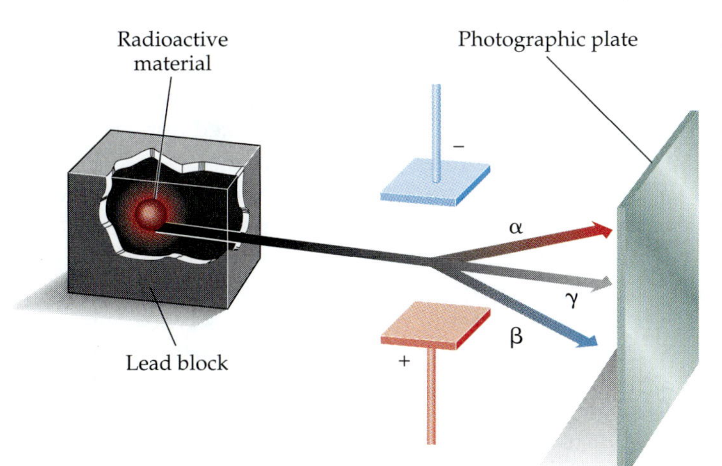

Figure 12.1
Three types of radiation emanate from radioactive material enclosed in a lead block. When the radiation is passed through an electric field, it splits into three beams. One beam is attracted to the negative plate (−); it is composed of positively charged alpha (α) particles. The second beam is attracted to the positive plate (+); it is a beam of beta (β) particles. The third beam, called gamma (γ) rays, is not deflected by an electric field.

We can describe the forms of radioactivity in detail, but what is radioactivity? How is it produced? The answer is that some nuclei are unstable as they occur in nature; they undergo **radioactive decay.** Radium atoms with a mass number of 226, for example, break down spontaneously, giving off alpha particles. Because alpha particles are identical to helium nuclei, this process can be summarized by the equation

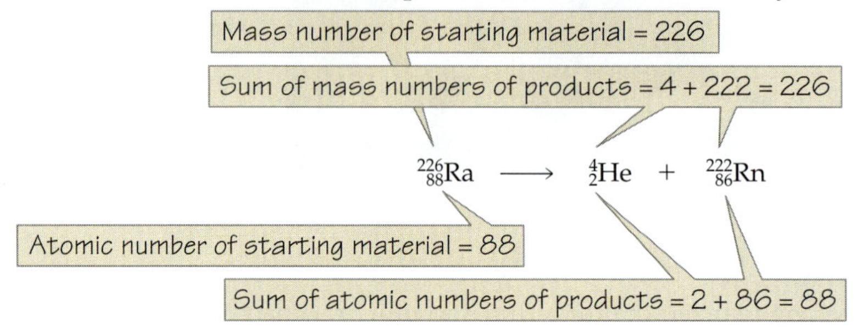

The atomic number (86) identifies the new element as radon (Rn). Note that the mass number of the starting material must equal the total of the mass numbers of the products. The same is true for atomic numbers. We use the symbol $^{4}_{2}He$ for the alpha particle (rather than α) because it allows us to check the balance of mass and atomic numbers more readily.

Tritium, one of the heavy isotopes of hydrogen (Section 2.3), also has unstable nuclei. Like all hydrogen nuclei, the tritium nucleus contains one proton. Unlike the most common isotope of hydrogen, however, the tritium nucleus contains two neutrons, and its mass is therefore 3 u and its symbol is $^{3}_{1}H$. Tritium decomposes by *beta decay.* Because a beta particle is identical to an electron, this process can be written as

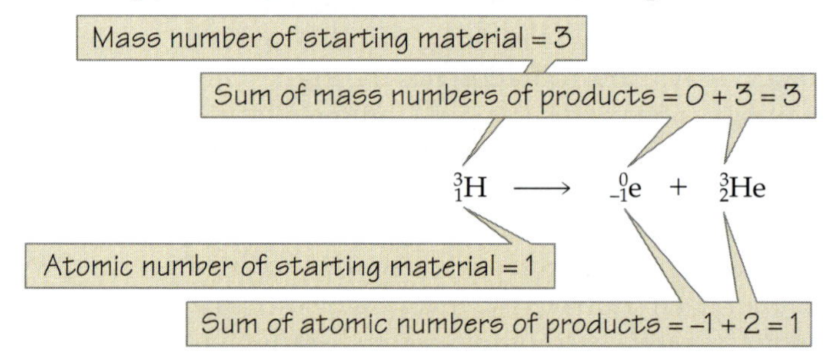

The atomic number (2) identifies the product isotope as helium.

How can the original nucleus, which contains only a proton and two neutrons, emit an electron? We can envision one of the neutrons in the original nucleus splitting into a proton and an electron.

$$^{1}_{0}n \longrightarrow {}^{1}_{1}p + {}^{0}_{-1}e$$

The new proton is retained by the nucleus, increasing the atomic number of the product by 1. The almost massless electron or beta particle is kicked out, leaving the product nucleus with essentially the same mass as the original.

In *gamma decay,* no particle is emitted; gamma rays are pure radiant energy. Gamma emission involves no change in atomic number or mass. The process is analogous to the emission of light from an atom. Visible light is emitted when an electron changes from a higher to a lower energy level. In gamma emission, a nucleus in a higher energy state drops to a lower energy state. Because this type of emission involves no particle, no equation is needed. Gamma decay is particularly useful when **radioisotopes** (radioactive isotopes) are used for diagnostic purposes in medicine. We discuss such uses in Section 12.9.

Fluorine-18 decays by positron emission.

$$^{18}_{9}F \longrightarrow {}^{0}_{+1}e + {}^{18}_{8}O$$

In this case, we envision a proton in the nucleus changing into a neutron and a positron.

$$^{1}_{1}p \longrightarrow {}^{1}_{0}n + {}^{0}_{+1}e$$

When the positron is emitted, the original radioactive nucleus has one less proton and one more neutron than it had before. Therefore, the mass number of the product nucleus is the same, but its atomic number is 1 less than that of the original nucleus. The emitted positron quickly encounters an electron (in any ordinary matter there are numerous electrons), both particles are annihilated, and two gamma rays are produced.

$$^{0}_{+1}e + {}^{0}_{-1}e \longrightarrow 2\gamma$$

Iodine-125, used in medicine to diagnose pancreatic function and intestinal fat absorption, decays by electron capture (E.C.).

$$^{125}_{53}I + {}^{0}_{-1}e \longrightarrow {}^{125}_{52}Te \quad \text{(followed by X-radiation)}$$

Note that electron capture has the same effect on the nucleus as positron emission. Obviously, though, the X-rays emitted as a result of E.C. differ from the gamma rays formed during positron annihilation.

The three types of particulate radioactive decay are pictured in Figure 12.2, and all five types of decay are summarized in Table 12.1.

Example 12.1

Write balanced nuclear equations for each of the following processes. In each case, indicate what new element is formed.

a. Plutonium-239 emits an alpha particle when it decays.
b. Protactinium-234 undergoes beta decay.
c. Carbon-10 emits a positron when it decays.

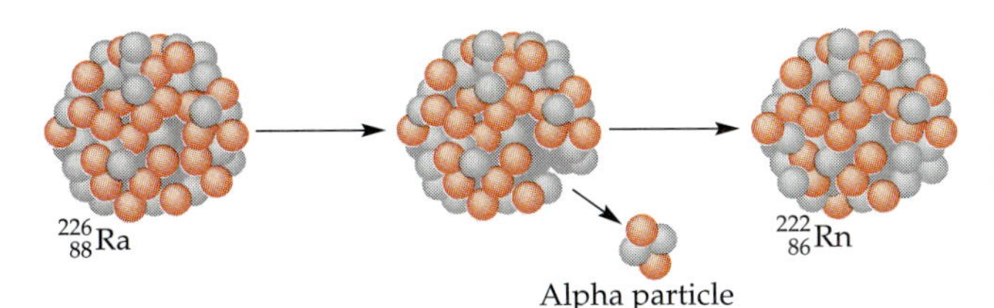

(a) Nuclear changes accompanying alpha decay.

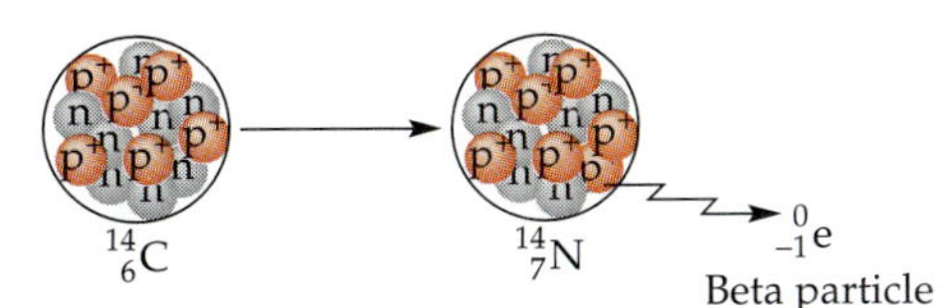

(b) Nuclear changes accompanying beta decay.

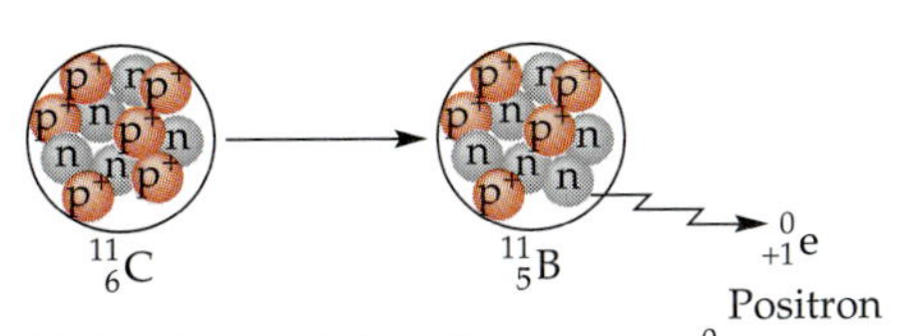

(c) Nuclear emission of a positron, $^{0}_{+1}e$.

Figure 12.2 Nuclear emission of (a) an alpha particle, (b) a beta particle, and (c) a positron.

Table 12.1 Major Types of Radioactive Decay

The Radiation					The Emitting (Absorbing) Nucleus	
Type	Symbol	Mass No.	Charge	Penetrating Power	Change in Mass Number	Change in Atomic Number
Alpha	α or ${}^{4}_{2}He$	4	2+	Slight	Decreases by 4	Decreases by 2
Beta	β, β^- or ${}^{0}_{-1}e$	0	1−	Intermediate	No change	Increases by 1
Gamma	γ	0	0	Great	No change	No change
Positron	β^+ or ${}^{0}_{+1}e$	0	1+	(Note 1)	No change	Decreases by 1
Electron capture	(E.C.)	0	—	(Note 2)	No change	Decreases by 1

(1) The penetrating power of a positron is quite limited because when a positron comes into contact with an electron, the two particles annihilate each other and are replaced by gamma radiation.

(2) Once inside the nucleus, the captured electron combines with a proton to form a neutron. Electron capture is always accompanied by X-radiation.

Solution

a. We start by writing the symbol for plutonium-239 and a partial equation showing that one of the products is an alpha particle (helium nucleus).

$$ {}^{239}_{94}Pu \longrightarrow {}^{4}_{2}He + ? $$

Mass and charge are conserved. The new element must have a mass of 239 − 4 = 235 and a charge of 94 − 2 = 92. The nuclear charge (atomic number) of 92 identifies the element as uranium (U).

$$ {}^{239}_{94}Pu \longrightarrow {}^{4}_{2}He + {}^{235}_{92}U $$

b. Write the symbol for protactinium-234 and a partial equation showing that one of the products is a beta particle (electron).

$$ {}^{234}_{91}Pr \longrightarrow {}^{0}_{-1}e + ? $$

The new element still has the mass number of 234. It must have a nuclear charge of 92 in order for the total charge to be the same on each side of the equation. The nuclear charge identifies the new atom as another isotope of uranium (U).

$$ {}^{234}_{91}Pr \longrightarrow {}^{0}_{-1}e + {}^{234}_{92}U $$

c. Write the symbol for carbon-10 and a partial equation showing that one of the products is a positron.

$$ {}^{10}_{6}C \longrightarrow {}^{0}_{+1}e + ? $$

To balance the equation, a particle with a mass number of 10 and an atomic number of 5 (boron) is required.

$$ {}^{10}_{6}C \longrightarrow {}^{0}_{+1}e + {}^{10}_{5}B $$

Practice Exercise

Write balanced nuclear equations for each of the following processes. In each case, indicate what new element is formed.

a. Fermium-250 undergoes alpha decay.
b. Selenium-85 undergoes beta decay.
c. Gold-188 decays by positron emission.
d. Iridium-192 undergoes electron capture.

Review Questions

12.2 What type of radiation is emitted when a nucleus undergoes (**a**) an increase of one unit in its atomic number? (**b**) a decrease of two units in its atomic number?

12.3. What two radioactive processes produce a decrease of one unit in the atomic number of the nucleus?

12.4 When a nucleus emits a gamma ray, what changes occur in the mass number and atomic number of the nucleus?

12.3 Penetrating Power of Radiation

Radioactive materials can be dangerous because the radiation emitted by decaying nuclei can damage living tissue. The ability of the radiation to inflict injury depends, in part, on its penetrating power.

All other things being equal, the more massive the particle, the less its penetrating power. Of alpha, beta, and gamma radiation, alpha particles are the least penetrating. Alpha particles are helium nuclei, each having a mass number of 4. Beta particles are more penetrating than alpha particles. The electrons that make up the stream of beta particles are assigned a mass number of 0. Beta particles are so much lighter than alpha particles that their mass is usually ignored. Gamma rays, high-energy radiation with truly no mass, are the most penetrating form of nuclear radiation.

Alpha radiation—least penetrating

Gamma rays—most penetrating

That the biggest particles make the least headway may seem contrary to common sense. Consider that penetrating power reflects the ability of the radiation to make its way through a sample of matter. It is as if you were trying to roll some rocks through a field of boulders. The alpha particle acts as if it were a boulder itself. Because of its size, it cannot get very far before it bumps into and is stopped by other boulders. The beta particle acts like a small stone. It can sneak between boulders and perhaps ricochet off one or another until it has made its way farther into the field (Figure 12.3). The gamma ray can be compared to a grain of sand that can get through the smallest openings: although the sand grain may brush against some of the boulders, it can, in general, make its way through most of the field without being stopped.

If all other things are equal, the penetrating power is determined by the mass of the particle. However, we must also consider other factors. For example, the faster a particle moves or the more energetic the radiation is, the more penetrating power it has.

In evaluating the danger of a radioactive substance, we also must consider where it is. If it is *outside* the body, alpha particles, with their low penetrating power, are least dangerous; they are stopped by the dead cells of the outer layer of skin. Beta particles also usually are stopped before reaching vital organs. Gamma rays readily pass through tissues; an external gamma source can be quite dangerous. When the radioactive source is *inside* the body, the situation is reversed. Alpha particles can do great damage. They don't penetrate very far, but all are trapped within the body, which must then absorb all the energy they release. The damage they do is

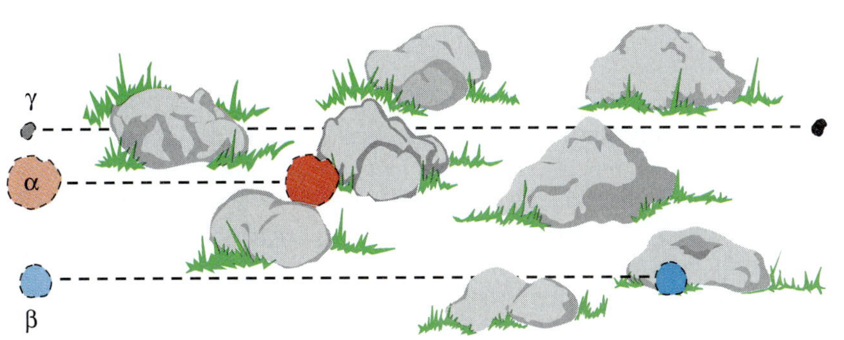

Figure 12.3
Shooting radioactive particles through matter is like rolling rocks through a field of boulders; the larger rocks are stopped more readily.

Table 12.2 Factors to Consider in Protection from Nuclear Radiation

1. Distance: The more distant the source, the greater the safety.
2. Sample size: The smaller the radiating sample, the greater the safety.
3. Type of radiation: The less penetrating the radiation, the greater the safety. Thus, for external sources safety increases in the order γ, β, α.
4. Half-life: The longer the half-life, the greater the safety. (Over an identical time period, the activity of a given quantity of an isotope with a long half-life is less than that of the same quantity of an isotope with a short half-life.)
5. Time: Generally, the shorter the time of exposure, the greater the safety.
6. Frequency: The fewer the exposures, the greater the safety.

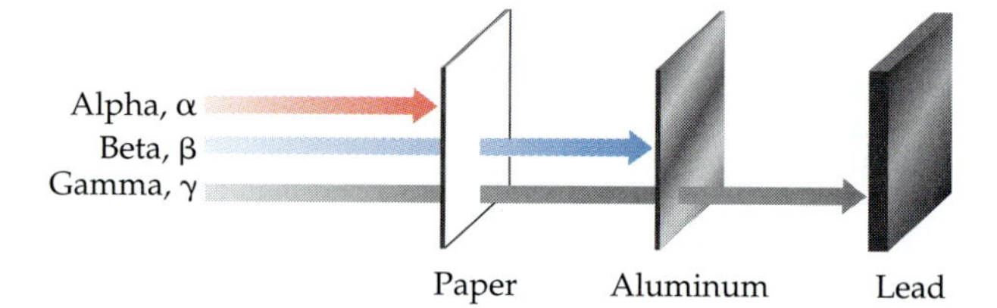

Figure 12.4 ▶ The relative penetrating powers of alpha, beta, and gamma radiation. Alpha particles are stopped by a sheet of paper and beta particles by a sheet of aluminum foil. It takes a block of lead several centimeters thick to stop gamma rays.

concentrated in a very small area. Beta particles travel farther, distributing their damage over a somewhat larger area. Tissue may recover from limited damage spread over a large area; it is less likely to survive concentrated damage.

People working with radioactive materials can protect themselves in several ways (Table 12.2). The simplest is to move away from the source, because intensity of radiation decreases with the square of the distance from the source. Workers can also be protected by shielding. A sheet of paper will stop most alpha particles. A block of wood or a thin sheet of aluminum will stop beta particles. But it takes several meters of concrete or several centimeters of lead to stop gamma rays (Figure 12.4).

Review Questions

12.5 List two ways in which workers can protect themselves from the radioactive materials with which they work.

12.6 A pair of gloves would be sufficient to shield the hands from which type of radiation: the heavy alpha particles or the massless gamma rays?

12.7 Heavy lead shielding is necessary as protection from which type of radiation: alpha, beta, or gamma?

12.4 Radiation Measurement

Several units are used to measure radioactivity. The *rate* at which nuclear disintegration occurs in a particular sample is measured in **curies (Ci),** named in honor of Marie Curie. A curie is 3.7×10^{10} disintegrations per second. A sample with an activity of hundreds or thousands of curies might be used as a source of externally applied radiation for the treatment of cancer. A sample with an activity of 10 millicuries (mCi) can be taken internally for diagnostic purposes by an adult, whereas a sample administered internally to a child might be measured in microcuries (μCi).

The *effect of radiation on matter* can be measured in several ways. The **roentgen (R)** is a measure of the ability of a source of X-rays or gamma rays to ionize an air sample. The **rad** (*r*adiation *a*bsorbed *d*ose) measures the amount of energy absorbed by any form of matter from any ionizing radiation. A dose of 1 rad is the amount of radiation that causes 1 kg of a substance to absorb 0.01 J of energy. For our purposes, the roentgen and the rad are about equivalent because 1 R generates about 1 rad of energy when absorbed in muscle tissue. A whole-body exposure of about 500 rads would kill most people. Alpha particles, beta particles, gamma rays, and X-rays are forms of **ionizing radiation**—that is, they cause the formation of ions from neutral particles. In the body, ionizing radiation most often interacts with water molecules. The reactive particles formed from water attack other molecules essential to proper cell function, thus damaging living tissue.

Primitive organisms such as yeast, bacteria, and viruses have much greater ability to withstand radiation than do mammals.

Some cells are more susceptible to radiation than others. Cells that are constantly and rapidly replaced are affected most. These include the intestinal mucosa, germ cells, embryonic cells, blood cells, and the organs responsible for producing blood cells, such as the bone marrow. Damage to reproductive cells show up as abnormalities in the descendants of affected persons.

The **rem** (*r*oentgen *e*quivalent, *m*an) is a measure of the relative biological damage produced by a particular dose of radiation. One rem measures the effect when 1 roentgen is absorbed. The International Commission on Radiological Protection recommends that adults whose occupations expose them to ionizing radiation limit their exposure to 5 rem in any one year.

Devices for measuring radiation range from the simple to the sophisticated. Individuals who work with radioactive materials wear *film badges* on their pockets or at their waist or as rings (Figure 12.5). The film in these badges reacts to radiation from radioactive isotopes or X-ray sources just as photographic film reacts to visible light. Radiation exposes the film in the badge, alerting the wearer to the potential danger. Sophisticated electronic devices, such as the *Geiger counter* (Figure 12.6), measure the ionizing effects of radiation and translate them into an observable signal (a meter reading or a clicking sound or a flashing light). Other detectors provide a permanent visual record of the intensity of the radiation. Such detectors play an important role in medical diagnosis (see Section 12.9).

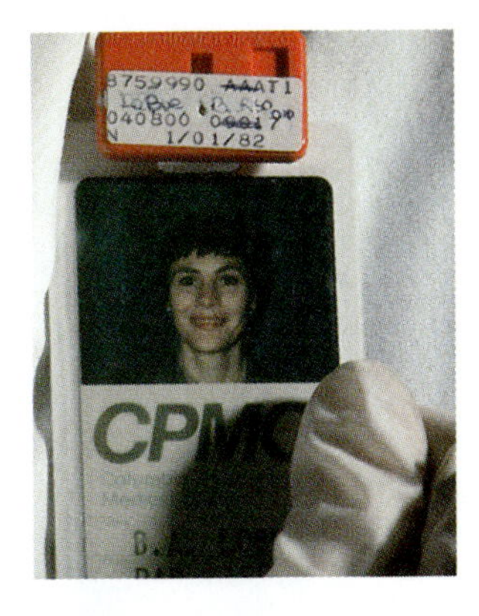

▲ **Figure 12.5**
People who work around radioactive materials wear film badges. Radiation clouds the photographic film, indicating the worker's degree of exposure over the time the film is worn.

Review Questions

12.8 Why is the rem a more satisfactory unit for measuring radiation dosage than the rad?

12.9 Describe briefly the meaning of each of the following terms.
a. ionizing radiation **b.** curie **c.** roentgen

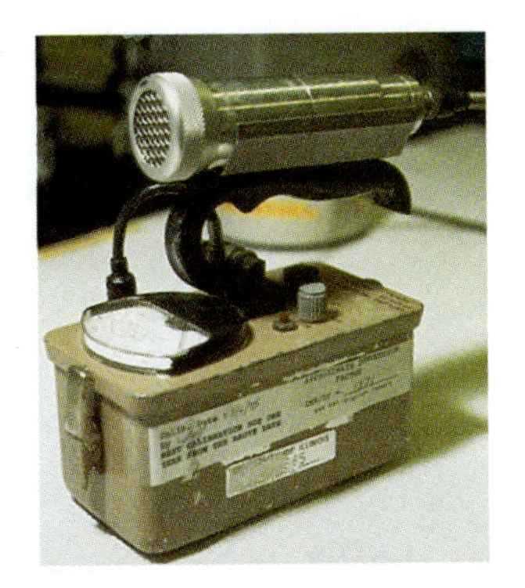

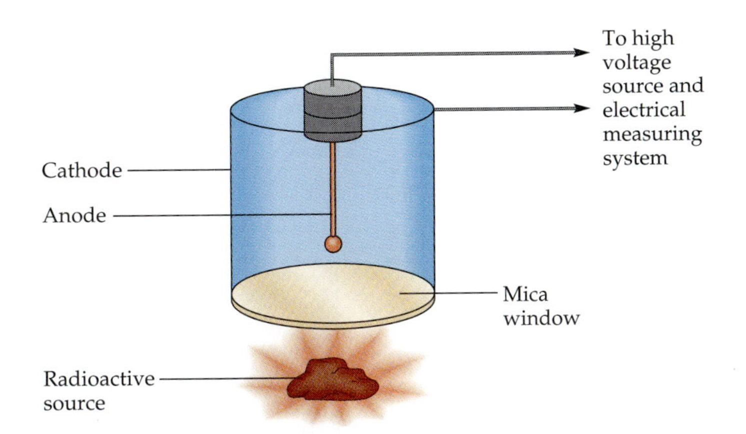

▲ **Figure 12.6**
A typical Geiger counter. The schematic at the right shows that radiation enters the tube through the mica window. Ions produced by the radiation cause an electrical discharge through the gas (usually argon) in the tube. Each pulse of electric current is counted as it passes through the circuit.

12.5 Half-Life

Radioactivity results when nuclei decay. We cannot predict when a particular nucleus will decay, but we can accurately predict the *rate of decay* of large numbers of radioactive atoms. (Life insurance companies cannot tell exactly when you will die, but their business depends on being able to predict how many of their clients will die over a particular period of time.)

Half-life ($t_{1/2}$) is the time in which one-half of the radioactive atoms present undergo decay. The half-life of an element can be very long (millions of years) or extremely short (tiny fractions of a second). The half-life of uranium-238 is 4.5 billion years; that of boron-9 is 8×10^{-19} s.

A radioactive isotope is characterized by a quantity called its **half-life ($t_{1/2}$)**, the period of time in which one-half of the original number of atoms undergo radioactive decay. Suppose, for example, that you had 16 billion atoms of the radioactive hydrogen-3 isotope (tritium), which has a half-life of 12.3 years. In 12.3 years, one-half, or 8 billion, of the original 16 billion atoms would undergo radioactive decay. In another 12.3 years, half of the remaining 8 billion atoms would decay. That is, after two half-lives, 4 billion atoms or one-fourth of the original number of atoms would remain unchanged. Two half-lives do not make a whole! The concept of half-life is shown graphically in Figure 12.7.

We can't say exactly when *all* the tritium atoms will have decayed. For many practical purposes, we can assume that nearly all the radioactivity is gone after about 10 half-lives. For the tritium sample considered here, 10 half-lives would be 123 years, at which time only about 0.1% of the original atoms would still be present.

Much of the concern over nuclear power centers on the long half-lives of some isotopes that can be released in a nuclear accident or that are simply left as by-products of the normal operation of a nuclear reactor. People fear that an accident, such as that at Chernobyl, Ukraine, could render large areas uninhabitable for hundreds of years. Even with no accidents, normal operation of a reactor produces nuclear wastes that must be safely stored for thousands of years.

We can calculate the fraction of the original isotope that remains after a given number of half-lives from the relationship

After 10 half-lives, the activity is $1/2^{10} = 1/1024$ of the original value; about one-thousandth of the original activity remains.

$$\text{Fraction remaining} = \frac{1}{2^n}$$

where n is the number of half-lives.

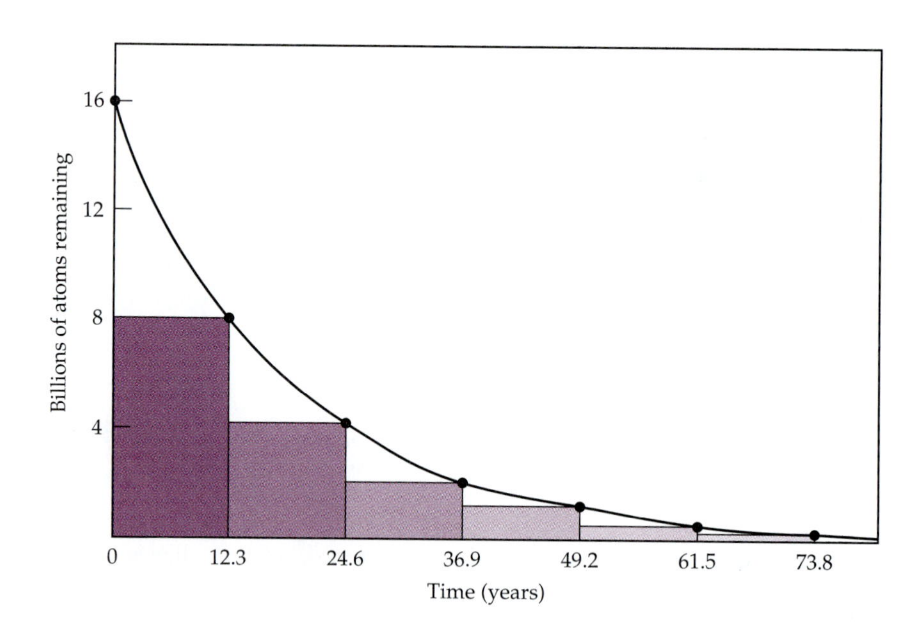

Figure 12.7 ▶ The radioactive decay of tritium (hydrogen-3). Each colored block represents one half-life.

Example 12.2

You obtain a new 484-mg sample of cobalt-60, which has a half-life of 5.25 years. How much cobalt-60 remains after 15.75 years (three half-lives)?

Solution

The fraction remaining after three half-lives ($n = 3$) is

$$\text{Fraction remaining} = \frac{1}{2^n} = \frac{1}{2^3} = \frac{1}{8}$$

The amount of cobalt-60 remaining is (1/8)(484 mg) = 60.5 mg.

Practice Exercises

A. You have 1.224 mg of freshly prepared gold-189, which has a half-life of 30 min. How much of the gold-189 sample remains after 2.5 h (five half-lives)?

B. You obtain a 20.0-mg sample of mercury-190, which has a half-life of 20 minutes. How much of the mercury-190 sample remains after 2.0 hours?

12.6 Radioisotopic Dating

The half-lives of certain isotopes can be used to estimate the ages of rocks and archaeological artifacts. Uranium-238 decays with a half-life of 4.5 billion years. The initial products of this decay are also radioactive, and breakdown continues until an isotope of lead (lead-206) is formed. By measuring the relative amounts of uranium-238 and lead-206, scientists can estimate the age of a rock. Some of the older rocks on the Earth have been found to be from 3.0 to 4.5 billion years old. Moon rocks and meteorites have been dated at a maximum age of about 4.5 billion years. Thus the age of the Earth is generally estimated to be about 4.5 billion years.

The dating of artifacts often involves radioactive carbon-14. This isotope is formed in the upper atmosphere by the bombardment of ordinary nitrogen atoms by neutrons from cosmic rays.

$$^{14}_{7}\text{N} + ^{1}_{0}\text{n} \longrightarrow ^{14}_{6}\text{C} + ^{1}_{1}\text{H}$$

This process leads to a nearly constant concentration of carbon-14 in atmospheric CO_2. Living plants and animals incorporate this isotope into their tissues. When an organism dies, however, no more carbon-14 is incorporated, and that already present decays—with a half-life of 5730 years—to nitrogen-14. By measuring the carbon-14 activity remaining in an artifact of plant or animal origin, we can determine the age of the artifact. For example, a sample that has half the carbon-14 activity of new plant material is 5730 years old; it has been dead for one half-life. Similarly, an artifact with 25% of the carbon-14 activity of new plant material is 11,460 years old; it died two half-lives ago.

Carbon-14 dating, as outlined here, assumes that the formation of the isotope was constant over the years. This is not quite the case. However, for the most recent 7000 years or so, carbon-14 dates have been correlated with those obtained from the annual growth rings of trees. Calibration curves have been constructed from which accurate dates can be determined. Generally, carbon-14 can be used to date objects up to about 50,000 years old with reasonable accuracy. Objects older than 50,000 years have too little of the isotope left for accurate measurement.

Charcoal from the fires of an ancient people, dated by determining the carbon-14 activity, is used to estimate the ages of other artifacts found at the same archaeological site. Carbon-14 dating also has been used to detect forgeries of supposedly ancient artifacts. For example, the Shroud of Turin, an old piece of linen cloth about

▲ **Figure 12.8**
The image on the Shroud of Turin is best seen in a photographic negative, as shown here. The shroud was thought by some to be the burial shroud of Jesus Christ.

Table 12.3 Several Isotopes Useful in Radioactive Dating

Isotope	Half-Life (years)	Useful Range	Dating Applications
Carbon-14	5730	500 to 50,000 years	Charcoal, organic material
Tritium (Hydrogen-3)	12.3	1 to 100 years	Aged wines
Potassium-40	1.3×10^9	10,000 years to the oldest Earth samples	Rocks, the Earth's crust, the moon's crust
Rhenium-187	4.3×10^{10}	4×10^7 years to oldest samples in the universe	Meteorites
Uranium-238	4.5×10^9	10^7 years to the oldest Earth samples	Rocks, the Earth's crust

4 meters long that bears a faint human likeness (Figure 12.8), has been alleged since about 1350 C.E. to be part of the burial shroud of Christ. However, carbon-14 dating studies by three different nuclear laboratories indicate that the flax used in making the cloth was not grown until sometime between 1260 and 1390 C.E. These studies show that the cloth could not have existed at the time of Christ. In contrast, the Dead Sea Scrolls were shown by carbon-14 dating to be authentic records from a civilization that existed about 2000 years ago.

Tritium, the radioactive isotope of hydrogen, with a half-life of 12.3 years, is useful for dating items up to about 100 years old. One application is the dating of brandies. These alcoholic beverages are quite expensive when aged for 10 to 50 years. Tritium dating can be used to check the truth of advertising claims about the most expensive brandies.

Many other isotopes are useful for estimating the ages of objects and materials. Several of the more important ones are listed in Table 12.3.

Example 12.3

A piece of fossilized wood has a carbon-14 activity that is one-sixteenth that of new wood. How old is the artifact? The half-life of carbon-14 is 5730 years.

Solution

The carbon-14 has gone through four half-lives ($n = 4$).

$$\text{Fraction remaining} = \frac{1}{2^n} = \frac{1}{2^4} = \frac{1}{16}$$

It is therefore about $4 \times 5730 = 22{,}920$ years old.

Practice Exercise

How old is a piece of charcoal that has a carbon-14 activity one-fourth that of new wood? The half-life of carbon-14 is 5730 years.

Artificial transmutation:

Artificial—not occurring in nature

Transmutation—the changing of one element into another

(The change of one element into another by natural radioactive decay may be considered natural transmutation.)

12.7 Artificial Transmutation and Induced Radioactivity

The forms of radioactivity we have discussed so far occur in nature. For example, the alpha decay of radioactive elements in the Earth's crust over billions of years has produced the helium we use to fill balloons today (Figure 12.9). By bombarding stable nuclei with high-energy alpha particles, neutrons, or other subatomic particles, scientists can bring about nuclear reactions not encountered in nature. These reactions are called **artificial transmutations** because in the process one element is changed into another.

Ernest Rutherford studied the bombardment of the nuclei of a variety of light elements with alpha particles. One such experiment, in which he bombarded nitrogen nuclei, resulted in the production of protons.

$$^{14}_{7}N + ^{4}_{2}He \longrightarrow ^{17}_{8}O + ^{1}_{1}H$$

(Recall that the hydrogen nucleus is a proton, hence the alternative symbol $^{1}_{1}H$ for the proton.) Note that the mass numbers and the atomic numbers are balanced in the equation.

Irène Curie (daughter of Marie and Pierre) and her husband, Frédéric Joliot, studied the bombardment of aluminum nuclei with alpha particles. The reaction yielded neutrons and an isotope of phosphorus.

$$^{27}_{13}Al + ^{4}_{2}He \longrightarrow ^{30}_{15}P + ^{1}_{0}n$$

Much to their surprise, the target continued to emit particles after the bombardment was halted. The phosphorus isotope was radioactive, emitting positrons ($^{0}_{+1}e$).

$$^{30}_{15}P \longrightarrow ^{0}_{+1}e + ^{30}_{14}Si$$

Nuclear medicine depends on the availability of a broad range of radioisotopes, and many of these are artificially produced. Later in this chapter we will look into some aspects of nuclear medicine.

▲ **Figure 12.9**
The balloons in this parade are filled with helium gas, a product of the alpha decay of radioactive elements.

▲ Frédéric (1900–1958) and Irène (1897–1956) Joliot Curie discovered artificially induced radioactivity in 1934. (Frédéric Joliot changed his name to Joliot Curie when he married Irène in order to perpetuate the Curie name.) They received the Nobel Prize in chemistry in 1935.

Example 12.4

Write a balanced equation for the nuclear reaction in which potassium-39 is bombarded with neutrons, producing chlorine-36.

Solution

We are given two reactants and a product. Writing nuclear symbols, we get an expression that is not balanced; we need an additional product.

$$^{39}_{19}K + ^{1}_{0}n \longrightarrow ^{36}_{17}Cl + ?$$

To balance the expression, we need a particle with a mass number of 4 and an atomic number of 2; that is, an alpha particle.

$$^{39}_{19}K + ^{1}_{0}n \longrightarrow ^{36}_{17}Cl + ^{4}_{2}He$$

Practice Exercise

Write a balanced equation for the nuclear reaction in which technetium-97 is produced by bombarding molybdenum-96 with a deuteron (hydrogen-2 nucleus).

✓ Review Question

12.10 What is an artificial transmutation? How is it accomplished?

12.8 Fission and Fusion

When a large unstable nucleus is bombarded with relatively slow-moving neutrons, the large nucleus breaks apart, leaving two medium-sized nuclei and releasing more neutrons. This kind of nuclear reaction is called **nuclear fission.** A typical fission reaction is

In nuclear fission, the nucleus breaks apart to form two nuclei of roughly equal size.

$$^{235}_{92}U + ^{1}_{0}n \longrightarrow ^{90}_{38}Sr + ^{143}_{54}Xe + 3\,^{1}_{0}n$$

Because bombardment with neutrons triggered the reaction in the first place, the product neutrons can cause more of the large nuclei to undergo fission. Neutrons produced by this second wave of reactions will trigger more reactions, and so on. Thus nuclear fission is a chain reaction (Figure 12.10). Each reaction in the chain

Figure 12.10 ▶
A uranium-235 nucleus is split when it is struck by a relatively slow-moving neutron. An unstable uranium-236 nucleus is formed, but it breaks into two fragments with the release of several neutrons. These neutrons can induce the fission of other uranium-235 nuclei. Fission of 1 g of uranium-235 yields 22,600 kW of energy.

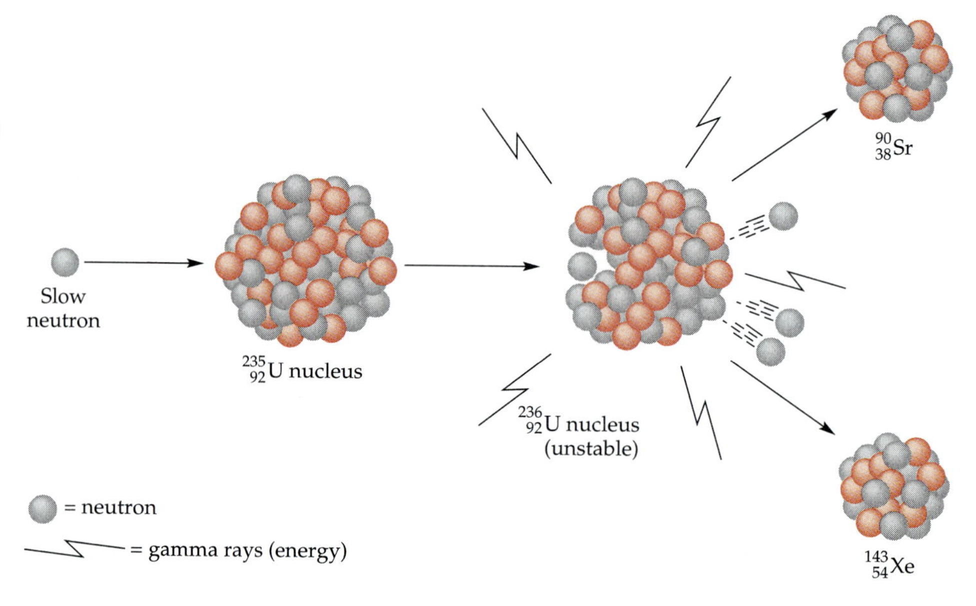

releases energy. If the fission is carried out in a controlled manner, the energy released can be used to generate electric power, as is done in a nuclear reactor. The fission can also be carried out in such a way that all the energy is released in one gigantic explosion, as in a nuclear bomb. (A nuclear reactor cannot explode like a nuclear bomb; special conditions are required for an effective bomb.) The products of fission, however, whether from a bomb or a reactor, are radioactive. When a bomb explodes, these products are thrown into the atmosphere and eventually reach the ground as radioactive "fallout." In a reactor, these products must be periodically removed and stored.

At present, more than 20% of the electric power generated in the United States is produced by nuclear power plants.

In nuclear fusion, two or more nuclei are combined to form one larger nucleus.

Energy is also released by the combination of two smaller nuclei into a larger one, a process called **nuclear fusion.** The fusion is accompanied by the release of vast amounts of energy. A typical fusion reaction is illustrated in Figure 12.11.

Our sun is powered by the *fusion* of atomic nuclei, and its fuel supply—mostly hydrogen-1—will last for billions of years. On Earth, scientists have unleashed the extraordinary energy of uncontrolled fusion reactions in hydrogen bombs. Controlled fusion for the production of electricity faces daunting challenges. It is difficult to force the positively charged nuclei close enough to fuse, because the positively charged nuclei repel one another strongly. Close approach requires enormously high temperatures of over 40,000,000 °C, at which gases are completely ionized into *plasma,* a mixture of nuclei and electrons. The plasma must be confined at a very high density long enough for the fusion to occur. No vessel can withstand such temperatures; plasma is usually confined in a magnetic field.

If perfected, nuclear fusion will offer distinct advantages over nuclear fission for power generation. The two most important advantages are (1) greater energy production per fusion event than per fission event and (2) the fact that the radioactive waste produced in nuclear fusion is much more limited and of relatively short half-life. The half-life of tritium, for example, is only 12.3 years. Long-term storage of nuclear waste is not required.

▲ The mushroom-shaped cloud that accompanies a nuclear explosion often serves as a symbol for the nuclear age.

✓ Review Questions

12.11 Briefly compare nuclear fission and nuclear fusion.

12.12 Which subatomic particles are responsible for carrying on the chain of reactions that are characteristic of nuclear fission?

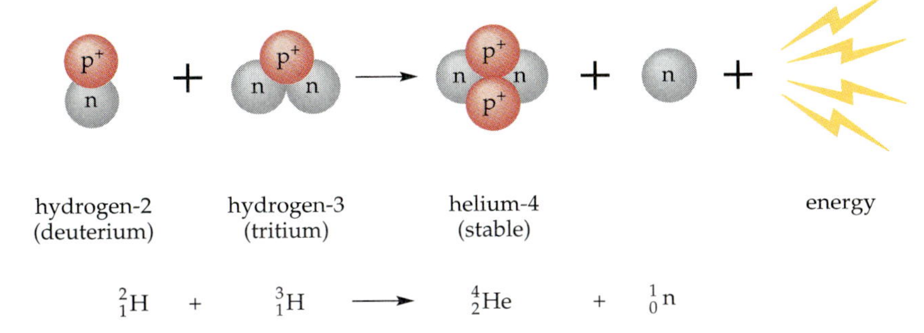

Figure 12.11
The fusion reaction that is most promising for development of a controlled nuclear fusion reactor is the deuterium-tritium (DT) reaction. A deuterium ($^{2}_{1}H$) nucleus fuses with a tritium ($^{3}_{1}H$) nucleus to form a helium nucleus plus a neutron, and a considerable amount of energy is released.

12.13 What dangers are there in using nuclear fission to generate power? What are the chief advantages of nuclear fusion over nuclear fission as a power source?

12.14 Why are such high temperatures required for nuclear fusion reactions?

12.9 Nuclear Medicine

Radioisotopes are used in nuclear medicine in two distinct ways: therapeutic and diagnostic. Radiation therapy is an attempt to treat or cure disease, such as cancer, with radiation. Some forms of cancer are particularly susceptible to radiation therapy. Radiation is carefully aimed at the cancerous tissue, and exposure of normal cells is minimized (Figure 12.12). If the cancer cells are killed by the destructive effects of the radiation, the malignancy is halted. But persons undergoing radiation therapy often get quite sick from the treatment, with nausea and vomiting the usual symptoms of radiation sickness. (Remember that the intestinal mucosa is particularly susceptible to radiation.) Thus the aim of radiation therapy is to destroy the cancerous cells before too much damage is done to healthy tissue. Radiation is most lethal to rapidly reproducing cells, and this is precisely the characteristic of cancer

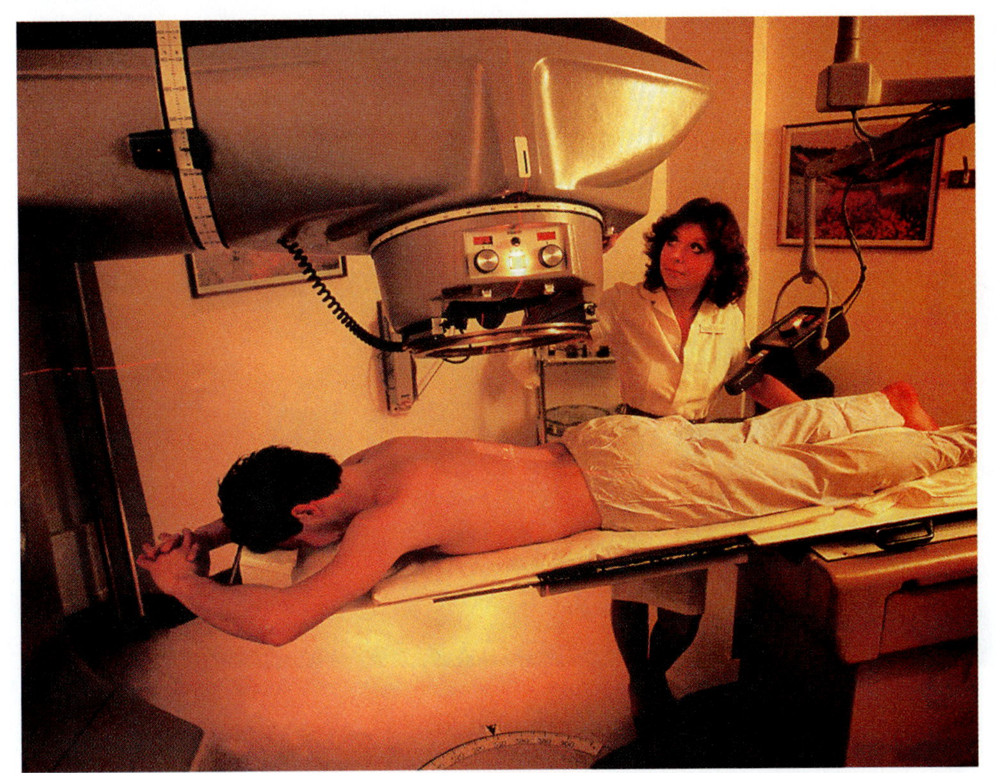

Figure 12.12
A cobalt-60 unit for radiation therapy.

cells that allows the therapy to be successfully applied. Some tumors, such as prostate cancer, are treated by implanting a radioactive substance directly into the tumor. This minimizes the exposure of other tissues to the radiation.

When used for diagnostic purposes, radioisotopes provide information about the type or extent of illness. Radioactive iodine-131 is used to determine the size, shape, and activity of the thyroid gland. It is also used therapeutically to treat thyroid cancer and to control a hyperactive thyroid. In either case, the patient drinks a solution of potassium iodide, KI, incorporating iodine-131. The body concentrates iodide in the thyroid. Large doses are used to treat thyroid cancer; the radiation from the isotope concentrates in the thyroid cancer cells even if the cancer has spread to other parts of the body. For diagnostic purposes, however, only a small dose is needed. A detector is set up to translate readings into a permanent visual record showing the differential uptake of the isotope. The "picture" that results is called a *photoscan*, and it can pinpoint the location of a tumor in that area of the body.

Rosalyn Yalow shared the 1977 Nobel Prize in medicine for her work with radioisotopes. The technique she developed, known as radioimmunoassay (RIA), is extremely sensitive and can be used to detect substances not detectable by other methods. RIA uses a radionuclide that selectively binds to a biochemical compound, such as a drug or hormone.

A radioisotope widely used in medical imaging is gadolinium-153, used to determine bone mineralization. Its popularity is an indication of the large number of people, mostly women, who suffer from *osteoporosis* (reduction in the quantity of bone) as they grow older. Gadolinium-153 gives off two characteristic radiations, a gamma ray and an X-ray. A scanning device compares these radiations after they pass through bone. Bone densities are then determined by differences in absorption of the rays.

Technetium-99^m is used in a variety of diagnostic tests (Figure 12.13). The *m* stands for *metastable*, which means that this isotope will give up some energy to become a more stable version of the same isotope (same atomic number, same atomic mass). The energy it gives up is the gamma ray needed to detect the isotope

$$^{99m}_{43}\text{Tc} \longrightarrow {}^{99}_{43}\text{Tc} + \gamma$$

Technetium-99^m is a pure gamma emitter; it produces no alpha or beta particles, which could cause unnecessary damage to the body. It also has a short half-life (about 6 hours), which means that the radioactivity does not linger in the body long after the scan has been completed. Because of its short half-life, technetium-99^m is produced on site by the decay of molybdenum-99.

$$^{99}_{42}\text{Mo} \longrightarrow {}^{99m}_{43}\text{Tc} + {}^{0}_{-1}\text{e} + \gamma$$

A container of the molybdenum isotope is obtained, and the decay product, technetium-99^m, is "milked" from the container as needed.

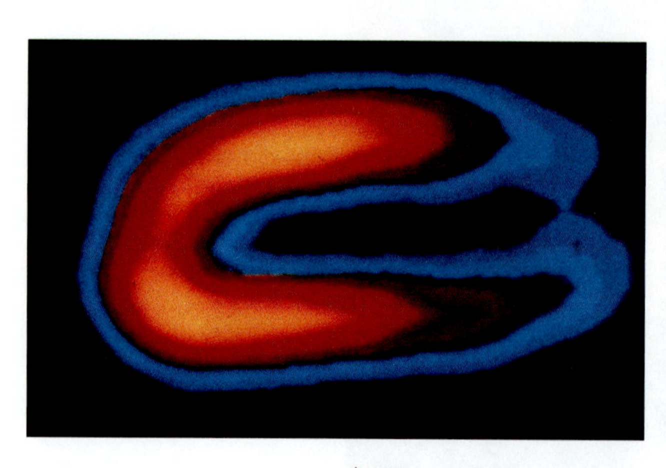
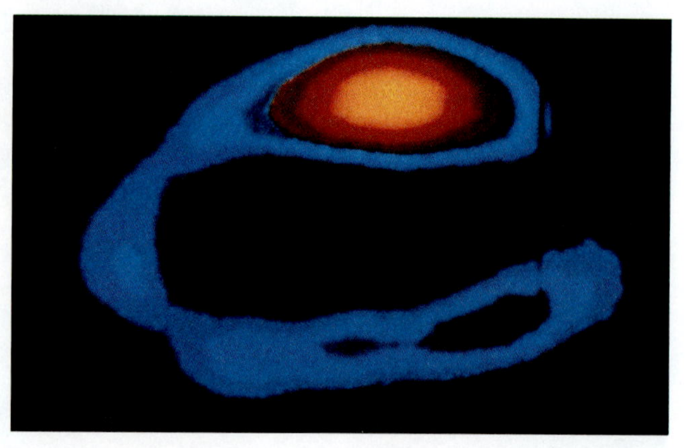

▲ **Figure 12.13**
Blood flow patterns in a healthy heart (left) and a damaged heart (right). The highlighted images from a technetium-99^m compound indicate regions receiving adequate blood flow.

Table 12.4 Some Radioisotopes and Their Application in Medicine

Isotope	Name	Radiation	Uses
^{14}C	Carbon-14	β^-	Radioimmunoassay
^{51}Cr	Chromium-51	E.C., γ	Determination of volume of red blood cells and total blood volume
^{57}Co	Cobalt-57	E.C., γ	Determination of uptake of vitamin B_{12}
^{60}Co	Cobalt-60	β^-, γ	Radiation treatment of cancer
^{153}Gd	Gadolinium-153	E.C., γ	Determination of bone density
^{131}I	Iodine-131	β^-, γ	Detection of thyroid malfunction; measurement of liver activity and fat metabolism; treatment of thyroid cancer
^{192}Ir	Iridium-192	E.C., γ	Radiation treatment for breast cancer
^{59}Fe	Iron-59	β^-, γ	Measurement of rate of formation and lifetime of red blood cells
^{32}P	Phosphorus-32	β^-	Detection of skin cancer or cancer of tissue exposed by surgery
^{238}Pu	Plutonium-238	α, γ	Power pacemakers in patients having irregular heartbeat
^{226}Ra	Radium-226	α, γ	Radiation therapy for cancer
^{24}Na	Sodium-24	β^-, γ	Detection of constrictions and obstructions in the circulatory system
^{85m}Sr	Strontium-85^m	γ	Bone scans
^{99m}Tc	Technetium-99^m	γ	Imaging of brain, thyroid, liver, kidney, lung, and cardiovascular system
^{3}H	Tritium	β^-	Determination of total body water

Table 12.4 is a list of some radioisotopes in common use in medicine. The list is necessarily incomplete, but even this abbreviated tabulation should give you an idea of the importance of radioisotopes in medicine. The claim that nuclear science has saved many more lives than nuclear bombs have destroyed is not an idle one.

Review Questions

12.15 Explain how radioisotopes can be used for therapeutic purposes.

12.16 Which radioisotope has been used extensively for treatment of overactive or cancerous thyroid glands?

12.17 Describe the use of a radioisotope as a diagnostic tool in medicine.

12.18 What are some of the characteristics that make technetium-99^m such a useful radioisotope for diagnostic purposes?

12.10 Medical Imaging

Medical imaging provides a means of looking at internal organs without resorting to surgery. The history of medical imaging dates back to the discovery of X-rays at the turn of the century. Penetrating X-rays were used almost immediately to visualize skeletal structure. Radiation from an external X-ray source passes through the body (except where it is absorbed by more dense structure, such as bone) and exposes film, thus providing a picture that distinguishes the more dense structures from the less dense tissue. Softer tissue can be visualized by introducing material that absorbs X-rays into the area to be studied. For example, compounds of barium have been used to view portions of the digestive tract (Figure 12.14).

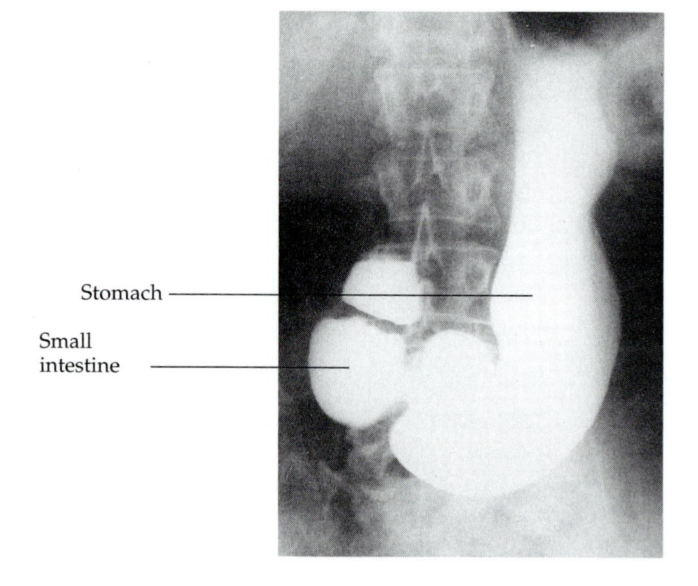

Figure 12.14 ▶
Barium sulfate ($BaSO_4$) is insoluble in water and opaque to X-rays. The salt can be swallowed by the patient, and an X-ray photographic outline of the stomach taken.

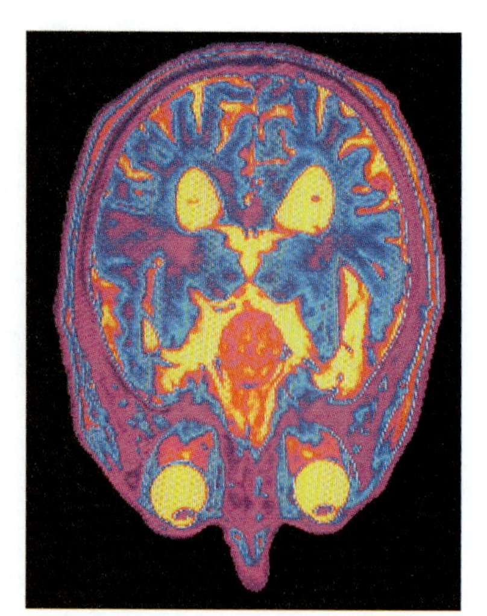

▲ **Figure 12.15**
This image was made using computer tomography (CT), a scanning technique that uses X-rays rather than radioisotopes. It shows a view, through the skull, of a pituitary tumor (red, lower center).

Positron emission tomography reveals metabolic changes that occur in the brain, which depends on glucose for most of its energy. Changes in how this sugar is metabolized by the brain may signal a disease such as cancer, epilepsy, Parkinson's disease, or schizophrenia.

Radioisotopes can also be used to visualize internal organs. In this technique, the source of the radiation is inside the body, and the radiation (usually gamma rays) is detected as it emerges from the body. Both X-ray technology and nuclear imaging have been coupled with computer technology to provide versatile and powerful imaging techniques. In computed tomography (referred to as CT or, sometimes, CAT scanning), many X-ray readings are obtained, processed by a computer, and then displayed. The resulting pictures present cross-sectional slices of a portion of the body. A series of these pictures gives a three-dimensional view of organs such as the brain (Figure 12.15). CT scans are widely used for detecting tumors and cancer.

Positron emission tomography (PET) can be used to measure dynamic processes occurring in the body, such as blood flow or the rate at which oxygen or glucose is being metabolized. PET scans are used to pinpoint the area of brain damage that triggers severe epileptic seizures and to detect hard-to-spot tumors and tiny blockages in blood vessels. Compounds incorporating positron-emitting isotopes, such as fluorine-18, are inhaled or injected prior to the scan. The emitted positron immediately encounters an electron in a substance in the cell. The positron and electron are mutually annihilated, producing two gamma rays.

$$^{18}_{9}\text{F} \longrightarrow {}^{18}_{8}\text{O} + {}^{0}_{+1}\text{e}$$

$${}^{0}_{+1}\text{e} + {}^{0}_{-1}\text{e} \longrightarrow 2\gamma$$

The gamma rays exit the body in exactly opposite directions and are recorded by detectors positioned on opposite sides of the patient. If the recorders are set so that two simultaneous gamma rays must be "seen," natural background gamma rays are ignored. A computer is then used to calculate the point within the body at which the annihilation of the positron and electron occurs, and an image of that area is produced (Figure 12.16).

✓ Review Questions

12.19 What form of radiation is detected in CT scans? In PET scans?

12.20 What is the advantage of using nonionizing radiation for medical imaging? Name two imaging techniques that do not use ionizing radiation.

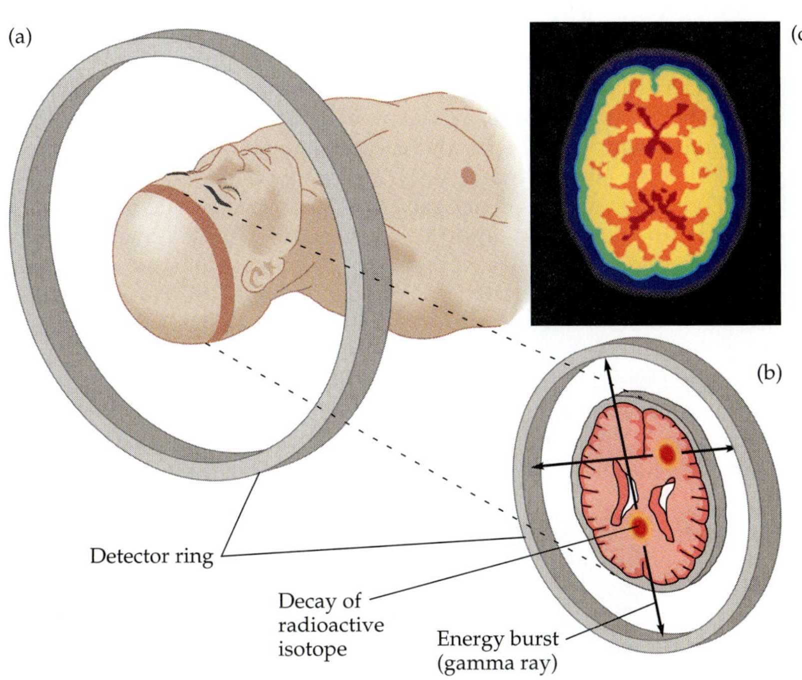

Figure 12.16
Positron emission tomography (PET) uses a positron-emitting radioisotope incorporated in a compound such as glucose, a simple sugar, which is given to a patient (a). The geometry of the X-rays emitted when the positron is annihilated by an electron provides an image of a "slice" of the head (b). The numbers of decays at each location are indicated by colors on a television screen (c).

Ultrasonography and Magnetic Resonance Imaging (MRI)

Both X-rays and nuclear radiation are ionizing radiations. Both cause some tissue damage, but modern techniques keep this damage to a minimum. Other imaging techniques use nonionizing radiation. We will look at two such techniques here.

Ultrasonography uses high-frequency sound waves. These waves are bounced off tissue, and their echo is recorded. A computer processes the data to produce an image of the tissue. (The technique is related to sonar detection of submarines.) Because it involves no ionizing radiation, ultrasonography is used extensively in obstetrics to follow fetal development (Figure 12.17).

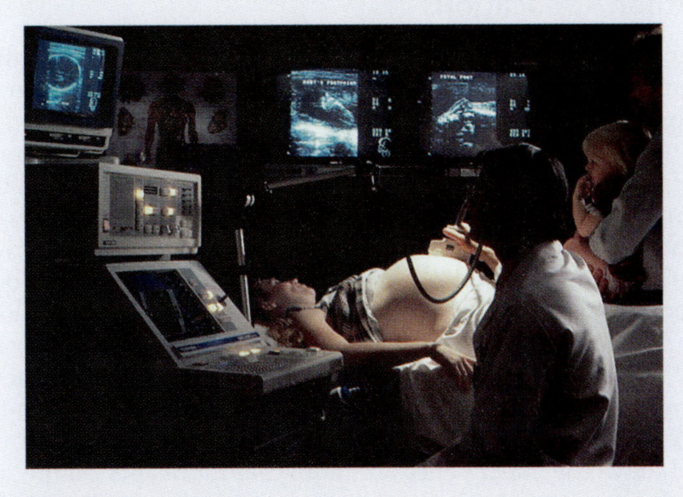

Figure 12.17
An ultrasonogram test on a woman who is eight months pregnant.

Another nonionizing method is *magnetic resonance imaging* (MRI), a technique based on the fact that some nuclei behave as if they were little magnets. Placed in a strong magnetic field (for example, between the poles of a much more powerful magnet), the nuclei line up in a certain manner. When supplied with the right amount of energy, the nuclei absorb the energy and flip over in the magnetic field. The energy required to flip the nuclei is provided by radio-frequency, nonionizing radiation. An instrument detects the absorption of the energy by the nuclei and produces a computer-enhanced image of the tissue in which the nuclei reside. The MRI technique provides not only images of organs but also information about the metabolic activity in particular tissues. MRI is particularly adept at detecting small tumors, blockages in blood vessels, damage to vertebral discs, and problems in joints.

Most of these computer-based technologies are quite expensive, ranging into the millions of dollars for a single installation. The great advantage of these techniques is that they provide information that could otherwise be obtained only by subjecting the patient to the risks of surgery or, in some instances, is not available by any other means.

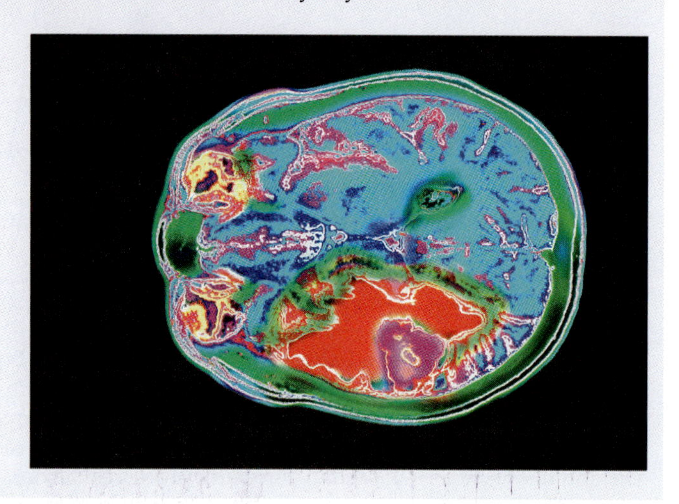

▶ This false-color MRI scan shows a tumor in one of the hemispheres of a human brain.

Table 12.5 Some Peacetime Uses of Nuclear Energy

Chemical analysis (by neutron activation)
Dating old artifacts
Flow rate indicators in pipes
Generation of electricity
Manufacture of semiconductors
Manufacture of radioisotopes
Medical diagnoses
Modifying plastics (cross-linking of polymer chains)
Power for desalination of seawater
Power for spacecraft, ships, and submarines
Preservation of foods
Radiation therapy (cancer)
Radioactive tracers in scientific research
Radioimmunoassays
Smoke detectors
Sterilization of insects
Sterilization of medical supplies
Synthesis of new elements
Thickness gauges (sheet metal, plastic films)
Wear testing (engines, tires)

12.11 Other Applications

In the last several decades nuclear research has emphasized peaceful rather than military applications. Aside from the uses already mentioned, many other applications have been found (see Table 12.5).

Elements that do not appear in nature can be prepared through transmutation. Examples of synthetic elements include elements 43, 61, and 85 and the transuranium elements. Chemical and biochemical reaction mechanisms (the ways in which atoms interact to make new products) can often be determined only with radioactive tracers. The calcium uptake in metabolism has been found to be 90% in the young but only 40% in older animals and humans. Polycythemia vera (formation of too many red blood cells) has been studied with the aid of iron-59. It was found that ten times as much iron as the body can use is assimilated by patients with this disease. Similar studies have shown that the uptake of trace elements by trees is quite pronounced in the winter. Thus, zinc moves up the trunk of a tree at the rate of about 2 ft/day. The effectiveness of industrial lubricants can be measured by monitoring the concentrations of metal residues in the oils. By tagging the metal with radioactive isotopes, concentrations as low as 10^{-19} g/L can be detected.

Radioisotopes are also used as sources for the irradiation of foodstuffs as a method of preservation (Figure 12.18). The radiation destroys microorganisms that

Figure 12.18 Gamma irradiation delays the decay of mushrooms. Those on the right were irradiated; those on the left were not.

cause food spoilage. Irradiated food shows little change in taste or appearance. Some people are concerned about possible harmful effects of chemical substances produced by the radiation, but there is no good evidence of harm to laboratory animals fed irradiated food, nor are there any known adverse effects in humans in countries where irradiation has been used for several years. There is no residual radioactivity in the food after the sterilization process.

Radioisotopes have been used extensively in basic scientific research. The mechanism of photosynthesis was worked out in large part by using carbon-14 as a tracer. For example, to determine how plants make the sugar glucose from carbon dioxide and water, the plant is exposed to radioactive carbon dioxide. The compounds formed from these starting materials and their order of formation are then followed by determining which new compounds become radioactive, and in what sequence. Using data from radioactive tracer experiments, scientists determine metabolic pathways in plants, animals, and humans.

Review Questions

12.21 What are radioactive tracers? How are they used?

12.22 What basic principle underlies radiation processing of foods?

12.12 The Nuclear Age Revisited

We live in an age in which fantastic forces have been unleashed. The threat of nuclear war was a constant specter from World War II until the collapse of the Soviet Union in 1991. Nuclear bombs have been used to destroy cities—and men, women, and children. Science and scientists have been greatly involved in it all. Would the world be a better place had we not discovered the secrets of the atomic nucleus? Consider this: More lives have been saved through nuclear medicine than have been destroyed by nuclear bombs. Also, nuclear power, even with its attendant problems, still may be one of our best hopes for a plentiful energy supply without the risk of global warming for decades to come. Imagine yourself to be John Dalton, who proposed the atomic theory (Chapter 2) nearly two centuries ago. Dalton was a gentle Quaker school teacher whose own formal education ended when he was 11 years old. Could he have anticipated the chain of scientific developments that would follow from his original speculations on the nature of matter?

Summary

Some atoms have an unstable nucleus; they emit radiation as they move toward a more stable nuclear configuration. This phenomenon is called **radioactivity,** and the isotopes that undergo the process are called **radioisotopes.** We consider five kinds of radioactivity in this chapter:

- An **alpha particle** (α) has a mass four times that of a hydrogen atom and a charge of +2; it is identical to the nucleus of a helium-4 atom.
- A **beta particle** (β) is an electron and so is nearly massless and carries a charge of −1.
- **Gamma rays** (γ) are radiation rather than particles and so have no mass or charge.
- A **positron** (β^+) has the same mass as the electron, but carries a charge of +1.
- In **electron capture (E.C.),** a nucleus absorbs an electron from an inner electron shell. An X-ray is emitted when an electron from a higher shell drops to the level vacated by the captured electron.

In alpha decay, a helium nucleus is emitted.

In beta decay, a neutron in the starting material breaks up into a proton plus an electron. This newly formed nuclear electron is emitted from the atom.

In gamma decay, no particles leave the nucleus; the starting material and the product are the same. The nucleus drops to a lower energy as a gamma ray (γ) is given off.

A positron is formed in the nucleus when one of the nuclear protons transmutes to a neutron plus a positron during the decay.

Gamma rays are the most penetrating of the three types of nuclear emissions. Alpha particles are the least penetrating.

The **curie (Ci)** is the unit used to measure the rate at which a radioisotope decays. The **roentgen (R)** measures the ability of a radioactive source to ionize air. The **rad** measures how much energy is absorbed by an object that is exposed to radiation. Because different types of radiation cause different amounts of damage to living tissue, exposure levels are measured not in rads but in a unit called the **rem.**

Alpha particles, beta particles, and gamma radiation are all **ionizing radiation,** which means they can knock orbital electrons out of neutral atoms and thereby create ions.

The **half-life** of a radioactive isotope is the length of time it takes for half the atoms in a sample of that isotope to decay.

Atoms of one element can be changed into those of another by bombarding the starting nuclei with alpha particles, neutrons, or other particles. This process is called **artificial transmutation.**

In **nuclear fission,** a large, unstable nucleus splits into two medium-sized nuclei. Fission is initiated by bombarding the large nuclei with neutrons from an external source. Additional neutrons are formed during the fission, and these can initiate further fissions. The ever-growing number of fission events is called a chain reaction. A controlled fission reaction can be used to produce electric power; when the reaction is allowed to proceed uncontrolled, the result is a nuclear explosion.

Nuclear fusion is a reaction in which two smaller nuclei come together (they *fuse*) to form a larger nucleus. An out-of-control fusion is a thermonuclear explosion; scientists have yet to develop a controlled fusion process.

Radioisotopes are used in medicine to treat some types of cancer and as a diagnostic tool. The most useful element for the latter is technetium-99^m, which emits only gamma radiation.

Key Terms

alpha (α) particle (12.2)
artificial transmutation (12.7)
beta (β) particle (12.2)
curie (Ci) (12.4)
electron capture (E.C.) (12.2)
gamma ray (γ) (12.2)
half-life (12.5)
ionizing radiation (12.4)
nuclear fission (12.8)
nuclear fusion (12.8)
positron (β^+) (12.2)
rad (12.4)
radioactive decay (12.2)
radioactivity (12.1)
radioisotope (12.2)
rem (12.4)
roentgen (R) (12.4)
X-ray (12.1)

Problems

Nuclear Symbols

1. Supply a name for each of the following.
 a. ${}^4_2\text{He}$ **b.** β^- **c.** ${}^1_0\text{n}$ **d.** ${}^2_1\text{H}$

2. Supply a symbol for each of the following.
 a. gamma ray **b.** tritium
 c. positron **d.** carbon-14

Nuclear Equations

3. Write a balanced nuclear equation for each of the following reactions.
 a. Lead-209 undergoes beta decay.
 b. Thorium-225 undergoes alpha decay.

4. Write a balanced nuclear equation for each of the following reactions.
 a. Americium-241 undergoes alpha decay.
 b. Copper-67 undergoes beta decay.

5. Sulfur-31 decays by positron emission. Write a balanced nuclear equation for this reaction.

6. A radioactive isotope decays by alpha emission to produce bismuth-211. What was the original element?

7. Write a balanced nuclear equation for the emission of a neutron by bromine-87.

8. Write a balanced nuclear equation for the emission of a proton by magnesium-21.

9. When magnesium-24 is bombarded with a neutron, a

proton is ejected. What new isotope is formed? [Hint: Write a balanced nuclear equation.]

10. When nitrogen-14 is bombarded with an alpha particle, a proton is ejected. What new isotope is formed?

11. Complete the following nuclear equations.

a. ${}^{10}_{5}B + {}^{1}_{0}n \longrightarrow ? + {}^{1}_{1}H$

b. ${}^{121}_{51}Sb + ? \longrightarrow {}^{121}_{52}Te + {}^{1}_{0}n$

c. ${}^{59}_{27}Co + {}^{1}_{0}n \longrightarrow {}^{56}_{25}Mn + ?$

12. Complete the following nuclear equations.

a. ${}^{154}_{62}Sm + {}^{1}_{0}n \longrightarrow ? + 2\ {}^{1}_{0}n$

b. $? + {}^{4}_{2}He \longrightarrow {}^{133}_{57}La + 4\ {}^{1}_{0}n$

c. ${}^{246}_{96}Cm + {}^{13}_{6}C \longrightarrow {}^{254}_{102}No + ?$

13. Write an equation to represent each of the following nuclear processes.

a. The reaction of two deuterons to produce helium-3.

b. The production of ${}^{243}_{97}Bk$ by the α-particle bombardment of ${}^{241}_{95}Am$.

c. The bombardment of ${}^{121}_{51}Sb$ by α particles to produce a neutron and ${}^{124}_{53}I$, followed by its radioactive decay by positron emission.

14. What is the new nucleus formed in each of the following processes?

a. Lead-196 goes through two successive E.C. processes.

b. Bismuth-215 decays through two successive β^- emissions.

c. Protactinium-231 decays through four successive α emissions.

Half-Life

15. The half-life of iodine-131 is 8.04 days. We follow the activity of a sample of iodine-131 while it falls to one-eighth of its initial value. How long will this take?

16. A 128-mg sample of technetium-99^m, half-life of 6.0 h, is used in a medical procedure. How much of the technetium-99^m sample remains after 24 h?

17. The half-life of molybdenum-99 is 67 h. How much time passes before a sample with an activity of 160 disintegrations per minute has decreased to an activity of 5.0 disintegrations per minute.

18. Krypton-81^m is used for lung ventilation studies. Its half-life is 13 s. How long will it take the activity of this isotope to reach one-fourth of its original value?

Additional Problems

19. What changes would you look for to establish whether a particular radioactive decay was by gamma emission or electron capture? Explain.

20. When a nucleus emits a neutron, what changes occur in the mass number and atomic number of the nucleus?

21. A sample of 16.0 mg of nickel-57, with a half-life of 36.0 hours, is produced in a nuclear reactor. How much of the nickel-57 sample remains after 7.5 days?

22. One proposal for reducing nuclear wastes is to use neutron bombardment to convert long-lived radioisotopes to ones with shorter half-lives. For example, technetium-99, a major by-product of nuclear weapons production, has a half-life of 210,000 years. Neutron bombardment converts it to technetium-100, which decays with a 16-s half-life to ruthenium-100. Write nuclear equations for these reactions.

23. In 1994 German scientists announced the synthesis of an isotope of element $Z = 110$. When bombarded with nickel-62 nuclei, a lead-208 nucleus yields a nucleus with mass number 269 plus one neutron. Only five atoms of the new element were formed, but it was identified by its decay pattern of two positrons and three successive α decays ending with nobelium-257. Write nuclear equations for the synthesis of element 110 and its subsequent decay reactions.

24. Ionizing radiation can be used to measure and control the thickness of paper, textiles, or rubber sheeting produced in a continuous process. The intensity of the beam of radiation is diminished as it passes through the material. The counting rate when the radiation strikes a Geiger Müller tube is directly related to the thickness. Which type of radiation—α, β, or γ—do you think is best suited to this application? Explain.

25. The activity of one radiation source is 500 Ci and that of another is 10 mCi. To be used properly, one source is taken into the body and the other remains outside the body during treatment. Which is likely to be the internal source, and which the external?

26. Adult patient A takes internally an iodine-131 sample with an activity of 150 mCi and adult patient B takes internally a dose of 15 *m*Ci. In which patient is the iodine-131 being used to treat a malignancy, and in which patient is the isotope being used for imaging the thyroid gland?

27. About 2 mCi of thallium-201 is given by intravenous administration for imaging the heart. It is estimated that the total body radiation dose in humans is about 0.07 rad per mCi of thallium-201. How does the radiation dose used in this procedure compare with the lethal dose for humans?

28. Radioactive radon gas in homes is thought to be a lung cancer risk. One set of reactions involves two successive alpha decays from radon-222. What products are formed?

29. Selenium-82 undergoes a rare reaction in which two beta particles are emitted. What is the product?

30. C. E. Bemis and colleagues at Oak Ridge National Laboratory confirmed the synthesis of element 104, the half-life of which was only 4.5 s. Only 3000 atoms of the element were created in the tests. How many atoms were left after 4.5 s? After 9.0 s?

31. Abandoned salt mines are often cited as good places to store nuclear waste. What are the pros and cons of such disposal?

32. Discuss the impact of nuclear science on the following.

a. war and peace **b.** medicine

c. our energy needs

33. Cesium is one of the components of nuclear fallout. Why is it a particularly dangerous threat to the environment?

34. Nuclear wastes typically need to be stored for 20 half-lives to be safe. This translates into hundreds of years of storage time. (For instance, cesium-137 and strontium-90 have half-lives of about 30 years.) Would the shooting of such wastes into outer space be a responsible solution?

CHAPTER 13 Hydrocarbons

Offshore drilling for natural gas and crude oil. Petroleum is the chief raw material for the production of most organic compounds: plastics, pesticides, drugs, detergents, and many other important commodities.

Learning Objectives/Study Questions

1. What is organic chemistry? In general, how do organic compounds differ from inorganic compounds?
2. Why are there so many more organic compounds than inorganic compounds?
3. What are hydrocarbons? What structural features identify alkanes? Alkenes? Alkynes? Aromatic hydrocarbons?
4. How are alkanes, alkenes, alkynes, and aromatic hydrocarbons named using IUPAC nomenclature?
5. What are the physical and chemical properties of alkanes, alkenes, alkynes, and aromatic hydrocarbons?
6. What is an alkyl group?
7. What are some products from reactions of alkanes? How are they formed?
8. What are the major reactions of alkenes? What are the products of those reactions?
9. What are polymers? How are they formed and utilized?
10. What is petroleum and how is it utilized?

Scientists of the eighteenth and nineteenth centuries studied compounds obtained from plants and animals and labeled them *organic* because they were isolated from organized (living) systems. Compounds isolated from rocks and ores, from the atmosphere and oceans, were labeled *inorganic* because they were obtained from nonliving systems. The early chemists believed that only living organisms could synthesize organic compounds. Although by the middle of the nineteenth century a number of "organic" compounds had been prepared using ordinary laboratory techniques, the labels *organic* and *inorganic* remained.

Today **organic chemistry** is defined simply as the chemistry of the compounds of carbon.[1] It may seem strange that we divide chemistry into two branches, one that considers only compounds of one element and one that covers the 100 plus remaining elements. However, this division seems more reasonable when one considers that of 20 million or so compounds that have been characterized, the overwhelming majority contain carbon.

The Chemical Abstract Service of the American Chemical Society has recorded about 20 million known chemical compounds. New chemical substances are added at the rate of approximately one new listing every nine seconds! About 95% of all the compounds are organic substances.

Carbon differs from the other elements in several ways. Carbon atoms have a unique ability to form stable, covalent bonds with each other and with atoms of other elements in infinite variations. The molecules thus produced may contain from only one to over a million carbon atoms. So complex is the chemistry of carbon that we shall approach its study by dividing its millions of compounds into families. We will study one family at a time, beginning with the simpler members of each family. Eventually we shall consider those molecules that are organic in the original sense. These complex, carbon-containing molecules determine the forms and functions of living systems and are the subject of the discipline we call biochemistry.

Science has come a long way since Wöhler's synthesis of urea in 1828 (page 294). Before that year, scientists believed they could not synthesize even the simplest organic molecule. Now, through genetic engineering (Chapter 23), scientists can modify organisms, make multiple copies (by cloning), and use them to synthesize

◀ The word *organic* has different meanings. Organic fertilizer is organic in the original sense—it is derived from living organisms. Organic foods generally are foods grown without synthetic pesticides or fertilizers. Organic chemistry is the chemistry of compounds of carbon.

[1]This definition is not always followed strictly; several carbon-containing compounds, such as the following, often are considered inorganic:

Carbon monoxide (CO)	Carbonates (e.g., Na_2CO_3)	Thiocyanates (e.g., NaSCN)
Carbon dioxide (CO_2)	Bicarbonates (e.g., $NaHCO_3$)	Cyanates (e.g., KOCN)
Carbon disulfide (CS_2)	Cyanides (e.g., KCN)	Carbides (e.g., CaC_2)

Table 13.1 Contrasting Properties of Organic and Inorganic Compounds

Organic	Benzene (C_6H_6)	Inorganic	Sodium chloride (NaCl)
Low melting points	5.5 °C	High melting points	801 °C
Low boiling points	80 °C	High boiling points	1413 °C
Low solubility in water; high solubility in nonpolar solvents	Insoluble in water; soluble in gasoline	High solubility in water; low solubility in nonpolar solvents	Soluble in water; insoluble in gasoline
Flammable	Highly flammable	Nonflammable	Nonflammable
Solutions do not conduct electricity	Nonconductive	Solutions conduct electricity	Conductive in aqueous solution
Exhibit covalent bonding	Covalent bonds	Exhibit ionic bonding	Ionic bonds

medicines and other materials. In addition, thousands of carbon compounds each day silently carry out vital chemical reactions within our bodies.

Organic compounds, like inorganic ones, obey all the natural laws. Often there is no clear distinction in chemical or physical properties between organic and inorganic molecules. Nevertheless, it is useful to compare typical members of each class, as in Table 13.1. (Keep in mind, however, that there are exceptions to every category in this table.) To further illustrate typical organic–inorganic differences, the table also lists properties of the inorganic compound sodium chloride (NaCl, common table salt) and the organic compound benzene (C_6H_6, a solvent once widely used to strip furniture for refinishing; see Section 13.14).

Alkanes: Saturated Hydrocarbons

We begin our study of organic chemistry with the **hydrocarbons,** compounds containing only two elements, carbon and hydrogen. There are several different kinds of hydrocarbons: They are distinguished by the types of bonding between carbon atoms and by the properties that result from that bonding. Let's look first at a family that has only carbon-to-carbon single bonds. Later in the chapter we will examine hydrocarbons with double bonds, triple bonds, and with a special kind of bonding called aromaticity.

13.1 Alkanes: Structures and Names

We introduced the hydrocarbon methane (CH_4) in Chapter 3. Methane is merely the first member of a series of related compounds with an increasing number of carbon atoms. The next member of the series is ethane (C_2H_6). These compounds are called **alkanes** or **saturated hydrocarbons.** *Saturated,* in this case, means that each carbon atom is bonded to four other atoms, the highest possible number, with no double or triple bonds in the molecules. We generally represent the structures of methane and ethane as follows.

```
     H             H   H
     |             |   |
  H—C—H        H—C—C—H
     |             |   |
     H             H   H

  Methane        Ethane
```

Alkanes, often called *saturated hydrocarbons,* are hydrocarbons in which there are only single bonds.

These flat representations do not accurately portray bond angles or molecular geometry. Models, shown in Figure 13.1, attempt to portray the three-dimensional molecular shapes.

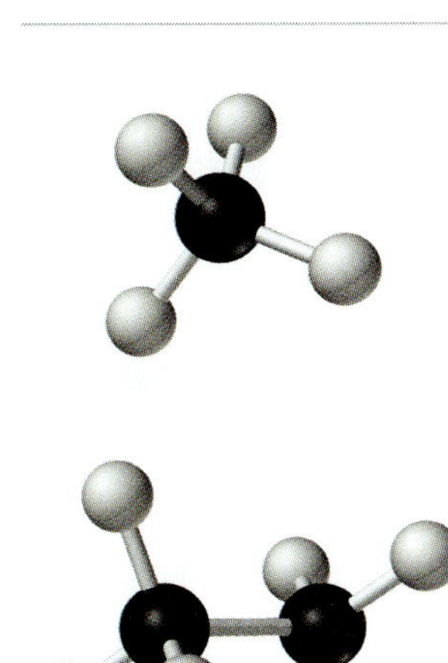

◀ **Figure 13.1**
Ball-and-stick and space-filling models of methane and ethane.

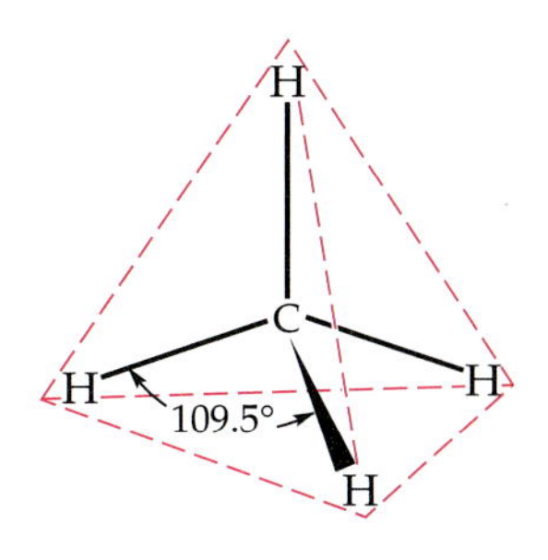

▲ Recall from Section 3.12 that the VSEPR theory correctly predicts a tetrahedral shape for the methane molecule.

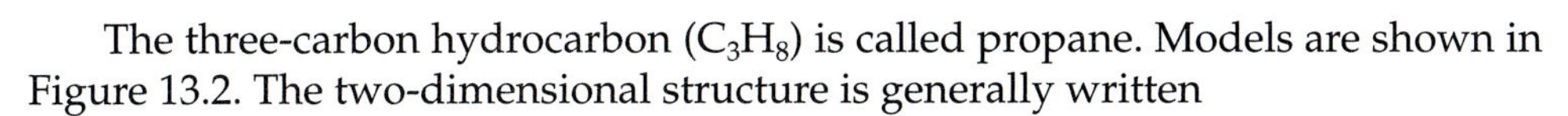

The three-carbon hydrocarbon (C_3H_8) is called propane. Models are shown in Figure 13.2. The two-dimensional structure is generally written

```
    H  H  H
    |  |  |
H—C—C—C—H
    |  |  |
    H  H  H
```

Propane

Homology

Methane, ethane, and propane are the beginning of a series of compounds in which any two adjacent members differ by one carbon atom and two hydrogen atoms—that is, by a CH_2 unit (Figure 13.3). Compounds related in this way make up a **homologous series.** The members of such a series, called *homologs*, have properties that vary in a regular and predictable manner. The principle of *homology* gives organization to organic chemistry in much the same way that the periodic table gives organization to inorganic chemistry. Instead of a bewildering array of individual carbon compounds, we can study a few members of a homologous series, and from them deduce some of the properties of other compounds in the series.

Isomers

Continuing the homologous series of alkanes, we add another carbon atom and two hydrogen atoms to C_3H_8 to get C_4H_{10}. To write a structure corresponding to this formula, we merely string four carbon atoms in a row,

—C—C—C—C—

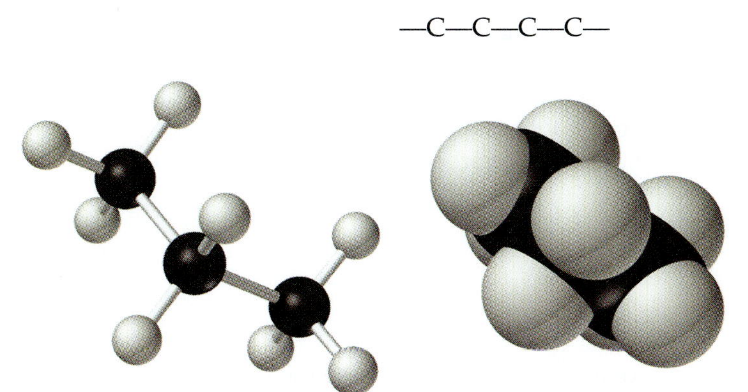

◀ **Figure 13.2**
Ball-and-stick and space-filling models of propane.

```
    H            H  H
    |            |  |
H—C—H      H—C—C—H
    |            |  |
    H            H  H

 Methane       Ethane
  (CH4)        (C2H6)

       H  H  H
       |  |  |
   H—C—C—C—H
       |  |  |
       H  H  H

       Propane
       (C3H8)
```

▲ **Figure 13.3**
Members of a homologous series. Each succeeding formula incorporates one carbon atom and two hydrogen atoms more than the previous formula.

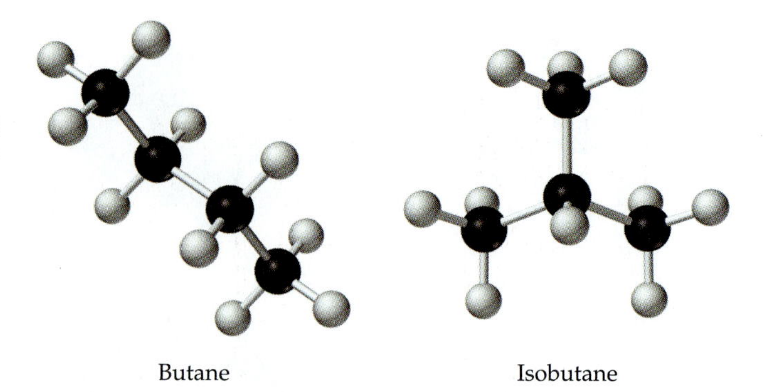

Figure 13.4 ▶ Ball-and-stick models of butane and isobutane. These two compounds are isomers; both have the molecular formula C_4H_{10}.

and then we add enough hydrogen atoms to give each carbon atom four bonds:

```
    H  H  H  H
    |  |  |  |
 H—C—C—C—C—H
    |  |  |  |
    H  H  H  H
```

The compound butane has this structure, but there is another way to put four carbon atoms and ten hydrogen atoms together. Place three of the carbon atoms in a row, and then branch the fourth one off the middle carbon:

```
 —C—C—C—
     |
     C
```

Now we add enough hydrogen atoms to give each carbon four bonds.[2]

```
    H  H  H
    |  |  |
 H—C—C—C—H
    |  |  |
    H  |  H
       |
    H—C—H
       |
       H
```

Just as our structural theory predicts, there is a hydrocarbon that corresponds to this structure. There are two compounds, then, that have the molecular formula C_4H_{10}. One boils at −1 °C; the other at 12 °C. Different compounds having the same molecular formula are called **isomers.** To give the two butanes unique names, we call the one with the continuous carbon chain *butane* and the one with the branched chain *isobutane* (Figure 13.4).

A continuous chain of carbon atoms is also referred to as a *straight* chain.

The three smallest homologs of the alkane series do not exist in isomeric forms because there is only one way to arrange the atoms in each formula so that each carbon atom has four bonds. It is important to realize that bending a chain does *not* mean an isomer is formed. With butane, for example, there are two (and only two) ways to arrange the carbon and hydrogen atoms. Butane molecules have a continuous four-carbon chain. This chain may be bent as in Figure 13.4, but it is still continuous. The formula of isobutane shows a continuous chain of three carbon atoms only, with the fourth attached as a branch off the middle carbon of the continuous chain.

```
       C     C
       |     |
 C—C—C  =  C—C—C  =  C—C—C—C  =  C—C—C  =  C—C—C
                                 |              |
                                 C              C
```

[2]We drew the bond to the fourth carbon longer than the others to avoid congestion. In the actual molecule, all the bonds are the same length.

One more step up the homologous alkane series gets us to C_5H_{12}. There are three compounds with this molecular formula. Collectively, these compounds are called pentanes. Compound I is pentane because it has all five carbon atoms in a continuous chain. We can call compound II isopentane because, like isobutane, it has a single carbon atom branched off the second carbon of the continuous chain. But what shall we call compound III? Let's name it the way the chemists did when it was discovered in 1870. Because the other two pentanes were characterized first, this one was called neopentane (from the Greek *neos*, new).

Alkanes can be represented by the general formula C_nH_{2n+2}, in which n is the number of carbon atoms.

```
                                                              H
                                                              |
                                                            H-C-H
                                                              |
  H H H H H              H H H H                          H   |   H
  | | | | |              | | | |                          |   |   |
H-C-C-C-C-C-H          H-C-C-C-C-H                      H-C---C---C-H
  | | | | |              | | | |                          |   |   |
  H H H H H              H | H H                          H   |   H
                           |                                  |
                         H-C-H                              H-C-H
                           |                                  |
                           H                                  H
```

I. Pentane   II. Isopentane   III. Neopentane

Condensed Structural Formulas

The formulas we have used so far, showing all the carbon and hydrogen atoms and how they are attached to one another, are called **structural formulas.** They convey much more information than simple molecular formulas. For example, the molecular formula C_4H_{10} expresses only the number of atoms in the molecule. It doesn't distinguish between butane and isobutane. The structural formulas

```
  H H H H                   H H H
  | | | |                   | | |
H-C-C-C-C-H       and     H-C-C-C-H
  | | | |                   | | |
  H H H H                   H | H
                              |
                            H-C-H
                              |
                              H
```

identify the specific isomers by showing the order of attachment of the various atoms.

Just as one can use a set of boards to build more than one kind of structure (for example, a fence or a bridge), one can construct more than one kind of molecule (structural formula) from a given set of atoms (molecular formula). Because these molecules differ in the way the atoms are joined together (that is, in the order of attachment of atoms or groups to one another), this type of isomerism is called structural isomerism.

Unfortunately, structural formulas are difficult to type and take up a lot of space. Chemists often use **condensed structural formulas** to alleviate these problems. The condensed formulas show the hydrogen atoms right next to the carbon atoms to which they are attached. For example, the two butanes become

```
CH3—CH2—CH2—CH3      and      CH3—CH—CH3
                                  |
                                  CH3
```

Sometimes condensed formulas are further simplified by omitting even more of the bond lines:

```
CH3CH2CH2CH3   and   CH3CHCH3   or   (CH3)3CH
                        |
                        CH3
```

The parentheses in condensed structural formulas indicate that the enclosed grouping of atoms attaches to the adjacent carbon.

Alkyl Groups

In the next section we consider a systematic way of naming hydrocarbons. First, however, we need to consider a way to name certain groups of carbon and hydrogen atoms that frequently occur together.

Table 13.2 Common Alkyl Groups

Parent Hydrocarbon		Alkyl Group		Condensed Formula
Methane	H H—C—H H	Methyl	H H—C— H	CH_3—
Ethane	H H H—C—C—H H H	Ethyl	H H H—C—C— H H	CH_3CH_2— or C_2H_5—
Propane	H H H H—C—C—C—H H H H	Propyl	H H H H—C—C—C— H H H	$CH_3CH_2CH_2$— or C_3H_7—
		Isopropyl	H | H H—C—C—C—H H H H	| CH_3CHCH_3 or $(CH_3)_2CH$—

The group of atoms that results when one hydrogen atom is removed from an alkane is called an **alkyl group.** Thus the general formula for an alkyl group is C_nH_{2n+1}, and it is named by replacing the *-ane* suffix of the parent hydrocarbon with *-yl.* For example, the CH_3— group, derived from methane (CH_4) by subtracting one hydrogen atom, is the *methyl group.* There are many alkyl groups, but the ones we use most frequently are listed in Table 13.2. Notice that two alkyl groups can be derived from propane. These are called the propyl group and isopropyl group, respectively. Remember that there is only one alkane named propane, but a chain of three carbon atoms can be attached to a longer chain in two ways. The attachment can be at an end carbon of the three-carbon chain (a propyl group), or it can be at the middle carbon (an isopropyl group).

Alkyl groups are not independent molecules, they are merely parts of molecules that we consider as a unit in order to name compounds systematically.

Review Questions

13.1 Briefly identify the important distinctions between:

a. an alkane and alkyl group

b. a straight-chain and a branched-chain alkane

13.2 What compounds contain fewer carbon atoms than propane and are homologs of propane?

13.2 IUPAC Nomenclature

Naming hydrocarbons seems complex, but there is some system to the process. To bring order to the chaotic naming of newly discovered compounds, an international meeting of chemists was held at Geneva, Switzerland, in 1892. The meeting resulted in a simple, unequivocal system for naming organic compounds. The system is still used today and is updated periodically by the group, now known as the International Union of Pure and Applied Chemistry (IUPAC). The set of rules, used worldwide, is known as the **IUPAC System of Nomenclature.** (Names that we used earlier, such as isobutane, isopentane, and neopentane, are called *common names.*)

The stem of a hydrocarbon name indicates the number of carbon atoms in the molecule (Table 13.3). For five or more carbon atoms, the stems are derived from Greek or Latin names for the numbers.

The first 10 straight-chain (unbranched) alkanes are shown in Table 13.4. Notice that the number of isomers increases rapidly with increasing carbon number. There are five hexanes, nine heptanes, and 18 octanes. Even more striking are the 366,319 possible isomers of the alkane whose molecular formula is $C_{20}H_{42}$ and the 62,491,178,805,831 possible isomers of the alkane whose molecular formula is $C_{40}H_{82}$. Not all of these have been isolated or characterized.

The IUPAC rules for naming the alkanes are as follows.

1. Saturated hydrocarbons are named according to the longest continuous chain of carbon atoms in the molecule (rather than the total number of carbon atoms). This longest chain is the parent compound. The suffix *-ane* indicates that the molecule is a saturated hydrocarbon.
2. The name of the parent hydrocarbon is modified by noting what alkyl groups are attached to the chain.
3. The chain is numbered, and the position of each substituent alkyl group is indicated by the number of the carbon atom to which it is attached. The chain is numbered in such a way that the substituents occur on the carbon atoms with the lowest numbers. Note that hyphens are used to separate numbers from names of substituents; numbers are separated from each other by commas.
4. Names of the substituent groups are placed in alphabetical order before the name of the parent compound. If the same alkyl group appears more than once, the numbers of all the carbon atoms to which it is attached are expressed. If the same group appears more than once on the same carbon, the number of that carbon is repeated as many times as the group appears. The number of identical groups is indicated by the Greek prefixes *di-*, *tri-*, *tetra-*, and so on. These prefixes are NOT considered in determining the alphabetical order of the substituents. For example, ethyl is listed before dimethyl; the di- is simply ignored). The last alkyl group named is prefixed to the name of the parent alkane to form one word.

Table 13.5 contains some examples of the IUPAC system for naming organic compounds. The best way to learn how to name alkanes is by working out examples, not just by memorizing rules. It's easier than it sounds. Try the following.

Table 13.3 Stems That Indicate the Number of Carbon Atoms in Organic Molecules

Stem	Number
Meth-	1
Eth-	2
Prop-	3
But-	4
Pent-	5
Hex-	6
Hept-	7
Oct-	8
Non-	9
Dec-	10

A *substituent* is any atom or group that substitutes for a hydrogen atom in an organic molecule.

Table 13.4 Straight-Chain Alkanes

Name	Molecular Formula	Condensed Structural Formula	Number of Possible Isomers
Methane	CH_4	CH_4	1
Ethane	C_2H_6	CH_3CH_3	1
Propane	C_3H_8	$CH_3CH_2CH_3$	1
Butane	C_4H_{10}	$CH_3(CH_2)_2CH_3$	2
Pentane	C_5H_{12}	$CH_3(CH_2)_3CH_3$	3
Hexane	C_6H_{14}	$CH_3(CH_2)_4CH_3$	5
Heptane	C_7H_{16}	$CH_3(CH_2)_5CH_3$	9
Octane	C_8H_{18}	$CH_3(CH_2)_6CH_3$	18
Nonane	C_9H_{20}	$CH_3(CH_2)_7CH_3$	35
Decane	$C_{10}H_{22}$	$CH_3(CH_2)_8CH_3$	75

Table 13.5 Examples of IUPAC Nomenclature

Condensed Structural Formula	Rewritten and Numbered	IUPAC Name
$CH_3CH_2CH(CH_3)CH_2CH(CH_3)_2$	CH_3 (on C-4), CH_3 (on C-2) $\overset{6}{C}H_3-\overset{5}{C}H_2-\overset{4}{C}H-\overset{3}{C}H_2-\overset{2}{C}H-\overset{1}{C}H_3$	2,4-Dimethylhexane NOT 3,5-Dimethylhexane (rule 3)
$(C_2H_5)_2CHCH(CH_3)CH_2CH_2CH_3$	CH_3 (on C-4) $\overset{1}{C}H_3-\overset{2}{C}H_2-\overset{3}{C}H-\overset{4}{C}H-\overset{5}{C}H_2-\overset{6}{C}H_2-\overset{7}{C}H_3$ CH_2-CH_3 (on C-3)	3-Ethyl-4-methylheptane NOT 4-Methyl-5-ethylheptane (rule 3)
$(CH_3)_2CHCH(C_3H_7)_2$	CH_3 (on CH) $CH_3-CH-\overset{4}{C}H-\overset{3}{C}H_2-\overset{2}{C}H_2-\overset{1}{C}H_3$ $\overset{5}{C}H_2-\overset{6}{C}H_2-\overset{7}{C}H_3$ (on C-4)	4-Isopropylheptane NOT 2-Methyl-3-propylhexane (rule 1)
$(CH_3)_2CHCH_2CH_2C(CH_3)_3$	CH_3 (on C-2) $\overset{6}{C}H_3-\overset{5}{C}H-\overset{4}{C}H_2-\overset{3}{C}H_2-\overset{2}{C}-\overset{1}{C}H_3$ CH_3 (on C-5), CH_3 (on C-2)	2,2,5-Trimethylhexane NOT 2,5-Trimethylhexane (rule 4)
$(C_2H_5)_2CHC(CH_3)(C_2H_5)_2$	CH_3 (on C-3) $\overset{6}{C}H_3-\overset{5}{C}H_2-\overset{4}{C}H-\overset{3}{C}-\overset{2}{C}H_2-\overset{1}{C}H_3$ CH_2-CH_3 (on C-4), CH_2-CH_3 (on C-3)	3,4-Diethyl-3-methylhexane NOT 3,4-Ethyl-3-methylhexane (rule 4)

Example 13.1

Name each of the following compounds.

a. $CH_3-CH_2-CH-CH-CH_3$ with CH_3 on each CH

b. $CH_3-CH-CH_2-CH-CH_3$ with CH_2-CH_3 on the first CH and CH_3 on the second CH

c. $CH_3CH_2CH_2CH_2-C-CH_2CH_2CH_3$ with $H-C-CH_3$ bearing CH_3 above C, and CH_3 below C

$$\begin{array}{c} CH_3 \\ | \\ H-C-CH_3 \\ | \\ CH_3CH_2CH_2CH_2-C-CH_2CH_2CH_3 \\ | \\ CH_3 \end{array}$$

Solution

a. The longest continuous chain (LCC) has five carbon atoms, and so the parent compound is pentane (rule **1**). There are methyl groups (rule **2**) attached to the second and third carbon atoms of the pentane chain (not the third and fourth; rule **3**). The name is therefore 2,3-dimethylpentane.

b. The LCC has six carbon atoms; the parent compound is hexane (rule **1**). Note that the LCC is not necessarily the chain drawn straight across the page; here the longest chain is bent. There are methyl groups (rule **2**) attached to the second and fourth carbon atoms. The correct name is 2,4-dimethylhexane, NOT 2-ethyl-4-methylpentane.

```
      4    3    2    1
CH3—CH—CH2—CH—CH3
   5|        |
    CH2      CH3
   6|
    CH3
```

Note: Structural formulas are *usually* presented with the longest chain written horizontally. This example emphasizes the point of searching out the longest chain.

c. The longest continuous chain (LCC) has eight carbon atoms, and so the parent compound is octane (rule **1**). There are a methyl and an isopropyl group, (rule **2**) both attached to the fourth carbon atom. The correct name is 4-isopropyl-4-methyloctane.

Practice Exercise

Name the following compounds:

a.
```
CH3CH2CHCH2CH2CH3
       |
       CH3
```

b.
```
   CH3   CH3
   |     |
CH3CHCH2CHCH3
```

c.
```
CH3CH2CHCH2CH2CH3
       |
       CH2CH3
```

d.
```
CH3CH2CH2CHCH2CH2Cl
           |
         H—C—CH3
           |
           CH3
```

Example 13.2

Draw the structural formula for each of the following compounds.

a. 2,3-dimethylbutane

b. 4-isopropyl-2-methylheptane

Solution

In drawing structural formulas, always start with the parent chain.

a. The parent chain is butane, indicating four carbon atoms in the LCC.

```
—C—C—C—C—
```

Then add the groups at their proper positions. You can number the parent chain from either direction as long as you are consistent; just don't change directions in the middle of the problem. The name indicates two methyl groups, one on the second carbon atom and one on the third.

```
 1  2  3  4
—C—C—C—C—
   |  |
   CH3 CH3
```

Finally, fill in all the hydrogen atoms, keeping in mind that each carbon atom must have four bonds.

```
CH3—CH—CH—CH3
    |   |
    CH3 CH3
```

b. The parent chain is heptane in this case, indicating seven carbon atoms in the LCC.

```
—C—C—C—C—C—C—C—
```

Adding the groups (carbon skeletons only) at their proper positions gives

```
     C   C—C—C
     |     |
 C—C—C—C—C—C—C
 1   2   3   4   5   6   7
```

Filling in all the hydrogen atoms gives the following condensed structural formula.

```
                 CH3   CH3
                    \ /
      CH3            CH
       |             |
   CH3CH—CH2—CH—CH2CH2CH3
```

Practice Exercise

Draw the structural formulas for the following compounds:

a. 4-propylheptane
b. 3-ethyl-2-methylpentane
c. 3-isopropyl-3-methyloctane

Review Questions

13.3 How many carbon atoms are there in (**a**) ethane, (**b**) heptane, (**c**) butane, and (**d**) nonane?

13.4 What is a substituent? How is the location of a substituent indicated in IUPAC nomenclature?

13.5 Briefly identify the important distinctions between:
a. a common name and an IUPAC name
b. a propyl group and an isopropyl group

13.3 Properties of Alkanes

Alkanes serve as the basis of comparison for the properties of many other families of organic compounds. Because alkane molecules are nonpolar, they have relatively simple physical properties. Because they have no special reactive sites (called functional groups; Section 14.1), they undergo relatively few chemical reactions.

Physical Properties

Alkane molecules are nonpolar. Therefore they are insoluble in water but soluble in nonpolar and slightly polar solvents. Alkanes are commonly used as solvents for organic substances of low polarity, such as fats, oils, and waxes.

Nearly all alkanes are less dense than water; their densities are less than 1.0 g/mL. Alkanes are colorless and tasteless, and many of them are odorless. The odor associated with natural gas is not from methane or any of the other alkanes, but rather from a sulfur-containing odorant purposely added to the gas to allow the detection of leaks. Table 13.6 lists some properties of the first ten straight-chain alkanes.[3] The physical states of substances at room temperature (about 20 °C or 68 °F) can be determined from their melting points and boiling points. If the melting point of a substance is above 20 °C, the substance exists as a solid at room temperature. For example, octane melts at −57 °C and boils at 125 °C. It is obviously a liquid at 20 °C. Propane boils at −42 °C; it is a gas at room temperature.

Similarly, the density of a substance indicates whether the substance is lighter or heavier than an equal volume of water. We know that oil and grease do not mix

▲ This oil slick produced by the *Exxon Valdez* oil spill in Alaska in 1989 provides a reminder that hydrocarbons and water don't mix.

[3]Organic chemistry textbooks generally include many tables of physical properties. The data in these tables are not meant to be memorized but are given to provide a numerical description of some physical characteristics of organic molecules and to serve as standards of purity.

Table 13.6 Physical Properties of Some Alkanes

Name	Molecular Formula	Melting Point (°C)	Boiling Point (°C)	Density (g/mL at 20 °C)	Physical State (at 20 °C)
Methane	CH_4	−182	−164	—	Gas
Ethane	C_2H_6	−183	−89	—	Gas
Propane	C_3H_8	−190	−42	—	Gas
Butane	C_4H_{10}	−138	−1	—	Gas
Pentane	C_5H_{12}	−130	36	0.626	Liquid
Hexane	C_6H_{14}	−95	69	0.659	Liquid
Heptane	C_7H_{16}	−91	98	0.684	Liquid
Octane	C_8H_{18}	−57	125	0.703	Liquid
Nonane	C_9H_{20}	−51	151	0.718	Liquid
Decane	$C_{10}H_{22}$	−30	174	0.730	Liquid

with water but rather float on the surface. They do not mix with water because they are *insoluble* in water; they float on top of the water because they are *less dense* than water. We also use densities to convert a given mass of a liquid into the corresponding volume of that liquid, or vice versa. (Recall that density equals the mass of a substance divided by its volume.)

Table 13.6 indicates that the first four members of the alkane series are gases. Natural gas is composed chiefly of methane, which has a density of about 0.65 g/L. The density of air is about 1.29 g/L. Because natural gas is less dense than air, it rises. When a natural-gas leak is detected and shut off in a room, the gas can be removed by opening an upper window. On the other hand, the three constituents of bottled gas are much heavier than air. Propane has a density of 1.6 g/L, and the butane isomers have densities of about 2.0 g/L. If bottled gas escapes into a room, it collects near the floor. This presents a much more serious fire hazard than a natural-gas leak because it is more difficult to rid the room of the heavier gas.

The data in Table 13.6 show that the boiling points of the straight-chain alkanes increase with increasing molar mass. This general rule holds true for the straight-chain homologs of all the families of organic compounds. Larger molecules are able to wrap around one another and interact, and thus more energy is required to separate them. In general, a straight-chain isomer has a higher boiling point than its branched-chain isomers. The straight-chain compounds can be likened to strands of spaghetti. The molecules can be packed closely, resulting in relatively strong intermolecular dispersion forces of attraction (Section 7.5). Dispersion forces depend on the total area of contact available between two molecules. The greater this area of contact, the greater the attractive force, and therefore the greater amount of heat needed to separate the molecules and reach the boiling point.

Branched-chain hydrocarbons are more compact than their straight-chain isomers. Consequently, there is less surface area to interact. The dispersion forces between molecules are weaker, and the molecules can more easily escape from the liquid. Table 13.7 lists the melting points and boiling points of the isomers of butane and pentane. Notice that there is no trend that enables us to predict melting points. In general, more symmetrical isomers tend to have higher melting points than less symmetrical isomers (because symmetrical molecules can pack closer together in the crystal lattice). Contrast 2,2-dimethylpropane with the less symmetrical pentanes in Table 13.7. Contrast, also, the very symmetrical octane isomer $(CH_3)_3CC(CH_3)_3$ with the straight-chain octane: the former is a solid and melts at 101 °C; the latter is a liquid whose melting point is −57 °C.

Table 13.7 Physical Properties of the Isomers of Butane and Pentane

IUPAC Name	Condensed Structural Formula	Melting Point (°C)	Boiling Point (°C)
Butane	$CH_3(CH_2)_2CH_3$	−138	−1
2-Methylpropane	$CH_3CH(CH_3)_2$	−159	−12
Pentane	$CH_3(CH_2)_3CH_3$	−130	36
2-Methylbutane	$CH_3CH_2CH(CH_3)_2$	−160	30
2,2-Dimethylpropane	$(CH_3)_4C$	−17	9

Physical properties of the alkanes are not only important in and of themselves, but also because of their contributions to the properties of other families of organic and biological compounds. For example, large portions of the structures of lipids consist of nonpolar alkyl groups (Figure 13.5). Lipids include the dietary fats and fatlike compounds called phospholipids and sphingolipids (Chapter 20) that serve as structural components of living tissues. These compounds have both polar and nonpolar groups, enabling them to bridge the gap between water-soluble and water-insoluble phases. This characteristic is essential for the selective permeability of cell membranes.

Physiological Properties

Because they can dissolve lipids in cell membranes, alkanes can have a profound physiological effect. Their physiological properties vary in a regular way as we proceed through the homologous series. Methane appears to be physiologically inert. We probably could breathe a mixture of 80% methane and 20% oxygen without ill effect. The mixture would be highly flammable, however, and we couldn't have fires or allow a spark of any kind in such an atmosphere.

In high concentrations, the other gaseous alkanes and the vapors of volatile liquid alkanes act as anesthetics. Inhaling ("sniffing") gasoline or solvents for their

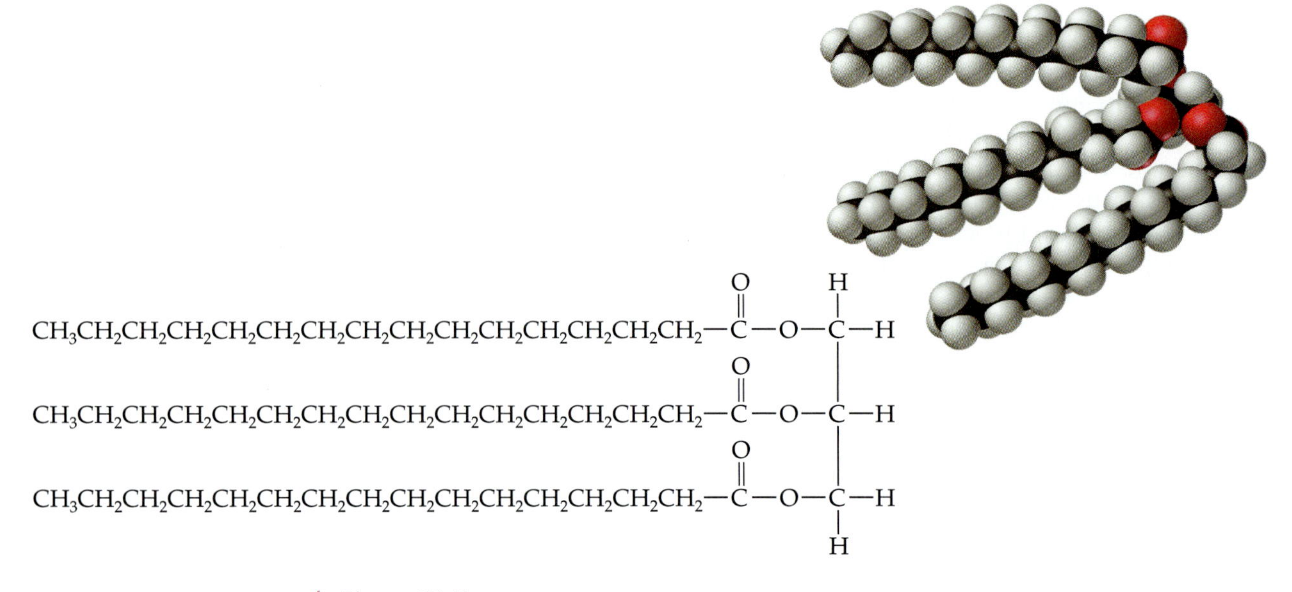

▲ **Figure 13.5**
Tripalmitin, a typical fat molecule, has long hydrocarbon chains typical of most lipids.

intoxicating effect is a major health problem. It can lead to liver, kidney, or brain damage or even immediate death by asphyxiation from excluding oxygen.

The lighter liquid alkanes (C_5 to about C_{12}) have varied effects on the body, depending on the part exposed. On the skin, alkanes dissolve body oils, and repeated contact can cause dermatitis (inflammation of the skin). Swallowed, alkanes do little harm while in the stomach. In the lungs, however, they cause "chemical pneumonia" by dissolving fatlike molecules from cell membranes in the alveoli. The cells become less flexible, and the alveoli are no longer able to expel fluids. The buildup of fluids is similar to that which occurs in bacterial or viral pneumonia. People who swallow gasoline, petroleum distillates, or other liquid alkane mixtures should not be made to vomit, as this would increase the chance of their getting the alkanes into the lungs.

Heavier liquid alkanes (those above about C_{17}), when applied to the skin, act as emollients (skin softeners). Such alkane mixtures as mineral oil can be used to replace natural skin oils washed away by frequent bathing or swimming. Petroleum jelly (Vaseline is one brand) is a semisolid mixture of hydrocarbons that can be applied as an emollient or simply as a protective film. Water and aqueous solutions such as urine will not dissolve such a film, which explains why petroleum jelly protects a baby's tender skin from diaper rash.

Review Questions

13.6 Briefly identify the important distinctions between natural gas and bottled gas.

13.7 What is the danger in swallowing a liquid alkane?

13.8 Distinguish between lighter and heavier liquid alkanes in terms of their effects on the skin.

13.4 Chemical Properties: Reactions of Alkanes

The alkanes generally do not react with most laboratory acids, bases, oxidizing agents, or reducing agents. In fact, the alkanes undergo so few reactions that they are sometimes called *paraffins* (from the Latin *parum affinis*, little affinity).

The alkanes do undergo a few important reactions, most notably combustion and halogenation. When mixed with oxygen at room temperature, alkanes give no apparent reaction. However, when a flame or spark supplies sufficient energy (the activation energy) to get things started, an exothermic (heat-producing) reaction, called **combustion,** proceeds vigorously. For methane, the reaction is

$$CH_4 + 2\,O_2 \longrightarrow CO_2 + 2\,H_2O + \text{heat}$$

If the reactants are adequately mixed and there is sufficient oxygen, the only products are carbon dioxide, water, and the all-important heat (for cooking foods, heating homes, and drying clothes). Conditions are rarely ideal, however, and other products are frequently formed. When the oxygen supply is limited, carbon monoxide is a by-product:

$$2\,CH_4 + 3\,O_2 \longrightarrow 2\,CO + 4\,H_2O$$

This reaction is responsible for dozens of deaths each year from unventilated or improperly adjusted gas heaters. (Similar reactions with similar results occur with kerosene heaters.)

Review Question

13.9 Why are alkanes sometimes call paraffins?

13.5 Halogenated Hydrocarbons

Alkanes react with chlorine and bromine in the presence of ultraviolet light or at a high temperature to yield organochlorine and organobromine compounds. Fluorine combines explosively with most hydrocarbons, whereas iodine is relatively unreactive. In general, **halogenated hydrocarbons** have one or more hydrogen atoms replaced by halogen atoms:

$$
\begin{array}{ccc}
 & H & \\
 & | & \\
H- & C & -Cl \\
 & | & \\
 & H &
\end{array}
\qquad
\begin{array}{ccccccc}
 & H & & Br & & Br & \\
 & | & & | & & | & \\
H- & C & - & C & - & C & -H \\
 & | & & | & & | & \\
 & H & & H & & H &
\end{array}
\qquad
\begin{array}{ccccccc}
 & Br & & H & & Cl & \\
 & | & & | & & | & \\
H- & C & - & C & - & C & -H \\
 & | & & | & & | & \\
 & H & & CH_3 & & H &
\end{array}
$$

Replacement of only one hydrogen atom gives an **alkyl halide.** These compounds are often called by *common names* (which differ from official IUPAC names) that consist of two parts. The first is the name of the alkyl group; the second is the stem of the name of the halogen, with the ending *-ide.*

Example 13.3

Give common names for each of the following alkyl halides.

a. CH_3CH_2Br **b.** $(CH_3)_2CHCl$

Solution

a. The alkyl group (CH_3CH_2–) is an ethyl group and the halogen is bromine. The common name is therefore ethyl bromide.

b. The alkyl group has three carbon atoms, with a chlorine atom attached to the middle one. It is an isopropyl group, and the common name is isopropyl chloride.

Practice Exercise

Give common names for each of the following alkyl halides.

a. CH_3I **b.** $CH_3CH_2CH_2F$

In the IUPAC system, we follow the rules for naming alkanes (Section 13.2), noting that halogen substituents are indicated by the prefixes fluoro-, chloro-, bromo-, and iodo-. The prefix is used with the name of the parent alkane, with numbers to indicate the position of the halogen, if necessary.

Example 13.4

Give the IUPAC name for each of the following compounds.

a.
$$
\begin{array}{ccccc}
^{1} & ^{2} & ^{3} & ^{4} & ^{5} \\
CH_3 & CH & CH_2 & CH_2 & CH_3 \\
 & | & & & \\
 & Cl & & &
\end{array}
$$

b.
$$
\begin{array}{cccccc}
^{1} & ^{2} & ^{3} & ^{4} & ^{5} & ^{6} \\
CH_3 & CH & CH_2 & CH & CH_2 & CH_3 \\
 & | & & | & & \\
 & CH_3 & & Br & &
\end{array}
$$

Solution

a. The parent alkane has five carbon atoms in the LCC; it is pentane (rule **1**). A chloro group is attached (rule **2**) to the second carbon atom of the chain. The IUPAC name is 2-chloropentane (rule **3**).

b. The parent alkane is hexane (rule **1**). Methyl and bromo groups are attached (rule **2**) to the second and fourth carbon atoms. Listing the substituents in alphabetical order (rule **4**) gives the name 4-bromo-2-methylhexane.

Practice Exercise

Give the IUPAC name for each of the following compounds.

a.
$$
\begin{array}{ccccccc}
CH_3 & - & CH & - & CH & - & CH_3 \\
 & & | & & | & & \\
 & & CH_3 & & Cl & &
\end{array}
$$

b.
$$
\begin{array}{ccccccc}
CH_3 & - & CH & - & CH & - & CH_2CH_2Br \\
 & & | & & | & & \\
 & & CH_3 & & Cl & &
\end{array}
$$

Halogenated Hydrocarbons and Ozone Depletion

Alkanes substituted with both fluorine and chlorine atoms have been used as the dispersing gases in aerosol cans, as foaming agents for plastics, and as refrigerants. Two of the best known of these **chlorofluorocarbons (CFCs)** are listed in Table 13.8.

At room temperature the chlorofluorocarbons are either gases or low-boiling-point liquids. They are essentially insoluble in water and inert toward most other substances. These properties make them ideal for many uses, but their inertness allows these compounds to persist in the environment.

Chlorofluorocarbons contribute to the greenhouse effect in the lower atmosphere, and they diffuse into the stratosphere, where they are broken down by ultraviolet radiation. Chlorine atoms (free radicals) formed in this process break down the ozone (O_3) molecules that protect the Earth from harmful ultraviolet radiation.

$$O_3 + \text{ultraviolet light} \longrightarrow O_2 + O$$

$$CF_2Cl_2 + \text{ultraviolet light} \longrightarrow CF_2Cl\cdot + Cl\cdot$$

$$Cl\cdot + O_3 \longrightarrow ClO\cdot + O_2$$

$$\cdot ClO + O \longrightarrow Cl\cdot + O_2$$

Note that the last step yields another chlorine atom that can break down another molecule of ozone. The third and fourth steps are repeated many times; thus the decomposition of one molecule of chlorofluorocarbon can result in the destruction of many molecules of ozone. (Other compounds, such as nitrogen oxides from agricultural activities and from aircraft that fly in the stratosphere also contribute to depleting the ozone shield.)

Research into this mechanism by Mario Molina, Sherwood Rowland, and Paul Crutzen led them to predict an ozone "hole" and laid the groundwork for its discovery in 1985 over the South Pole. They shared the 1995 Nobel Prize in chemistry for sounding the alarm about the depletion of the Earth's protective ozone layer.

Destruction of ozone is especially worrisome over Antarctica, where the thinning has caused a "hole" in the ozone layer. A similar but less dramatic thinning has been detected in the Arctic, and satellite data show that the ozone layer overall has been thinning at a rate of 0.5% per year since 1978. Every 1% drop in ozone means about 1.5% more ultraviolet radiation reaching the surface of the earth, and the U.S. National Research Council predicts a 2–5% increase in skin cancer for each 1% depletion of the ozone layer.

Worldwide action to reduce use of CFCs and related compounds began in 1987. The concentration of CFCs in the stratosphere has now peaked and will decline over the next century. The CFCs and other chlorine- or bromine-containing ozone-destroying compounds are being replaced with more benign substances. Fluorocarbons (HFCs), such as CH_2FCF_3, which have no Cl or Br to form radicals, are one alternative. Another is hydrochlorofluorocarbons (HCFCs), such as $CHCl_2CF_3$. HCFC molecules break down more readily in the troposphere, and fewer ozone-destroying molecules reach the stratosphere.

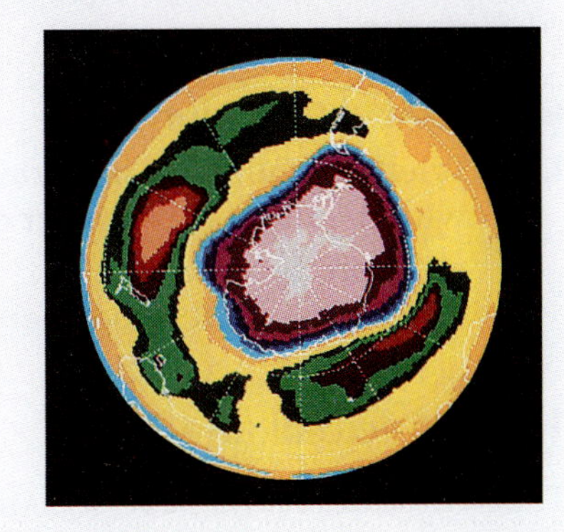

▶ The ozone hole over Antarctica. Ozone in the upper atmosphere shields Earth's surface from ultraviolet radiation from the sun, which can cause skin cancer in humans and is also harmful to other animals and to some plants. Ozone "holes" in the upper atmosphere (the gray, pink, and purple areas at the center) are large areas of substantial ozone depletion. They occur mainly over Antarctica from late August through early October and fill in about mid-November. Ozone depletion has also been noted over the Arctic regions.

Table 13.8 Some Halogenated Hydrocarbons

Formula	Common name	IUPAC name	Some important uses
		Derived from methane	
CH_3Cl	Methyl chloride	Chloromethane	Refrigerant; manufacture of silicones, methyl cellulose, and synthetic rubber
CH_2Cl_2	Methylene chloride	Dichloromethane	Laboratory and industrial solvent
$CHCl_3$	Chloroform	Trichloromethane	Industrial solvent
CCl_4	Carbon tetrachloride	Tetrachloromethane	(see text)
$CBrF_3$	Halon-1301	Bromotrifluoromethane	Fire extinguisher systems
CCl_3F	CFC-11	Trichlorofluoromethane	Foaming plastics
CCl_2F_2	CFC-12	Dichlorodifluoromethane	Refrigerant
		Derived from ethane	
CH_3CH_2Cl	Ethyl chloride	Chloroethane	Local anesthetic
$ClCH_2CH_2Cl$	Ethylene dichloride	1,2-dichloroethane	Solvent for rubber
CCl_3CH_3	Methylchloroform	1,1,1-trichloroethane	Solvent for cleaning computer chips and molds for shaping plastics

A wide variety of interesting and often useful compounds have one or more halogen atoms per molecule. For example, methane can react with chlorine, replacing one, two, three, or all four hydrogen atoms by chlorine. Several halogenated products derived from methane and ethane are listed in Table 13.8, along with some of their uses.

Many of the chlorinated hydrocarbons are suspected carcinogens (cancer-causing substances), and exposure to them can cause severe liver damage. Some were once widely used in consumer products. For example, carbon tetrachloride was once used as a dry-cleaning solvent and in fire extinguishers. It is no longer recommended for either use. Even when breathed in small amounts, the vapor can cause serious illness if the exposure is prolonged. Use of a carbon tetrachloride fire extinguisher in conjunction with water to put out a fire can be deadly. Carbon tetrachloride reacts with water at high temperatures to form deadly phosgene ($COCl_2$) gas.

▲ Ethyl chloride (chloroethane) is used as a spray-on local anesthetic to treat athletic injuries.

Ethyl chloride is used as an external local anesthetic. When sprayed on the skin, it begins to evaporate, and this cools the area enough to make it insensitive to pain. It can also be used as an emergency general anesthetic.

Methyl bromide (CH_3Br) has long been used as a component of weed killers and insecticide preparations, particularly to sterilize soil. Other bromine-containing compounds are widely used in fire extinguishers and as fire retardants on clothing and other materials. Because of their toxicity and other effects on the environment, scientists are engaged in designing safer substitutes for many of these halogenated compounds.

✓ Review Question

13.10 Briefly identify the important distinctions among the following, especially with reference to the effect of each on the ozone layer.

a. chlorofluorocarbon **b.** fluorocarbon
c. hydrochlorofluorocarbon

13.6 Cycloalkanes

The hydrocarbons we have encountered so far have been open-ended chains. Carbon atoms can also be joined in a *ring* or *cycle*. The simplest cyclic hydrocarbon has the formula C_3H_6. Each of the three carbon atoms has two hydrogen atoms attached

(Figure 13.6). The compound is called cyclopropane. In addition to having an interesting structure, cyclopropane has some intriguing properties. It is a potent, quick-acting anesthetic with few undesirable side effects. It is no longer used in surgery, however, because it forms explosive mixtures with air at nearly all concentrations.

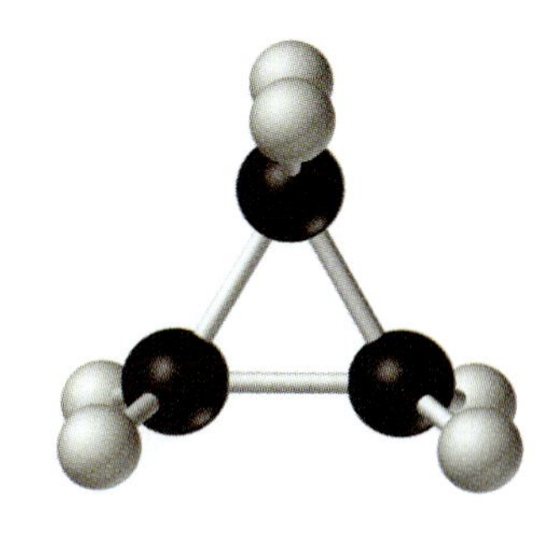

▲ **Figure 13.6**
Ball-and-stick model of cyclopropane.

Names of cycloalkanes are formed by addition of the prefix *cyclo-* to the name of the open-chain compound having the same number of carbon atoms as there are in the ring. Thus the name for the cyclic compound C_4H_8 is cyclobutane. Names and structures of several cycloalkanes are given in Figure 13.7. The carbon atoms in cyclic compounds form regular geometric figures, and in the abbreviated structures shown in the figure, each corner of the geometric figure represents a carbon plus as many hydrogen atoms as needed to give the carbon four bonds. For example, the triangle stands for C_3H_6,

H_2
C
H_2C-CH_2

and the hexagon for C_6H_{12}.

H_2
C
H_2C CH_2
H_2C CH_2
CH_2

Some cyclic compounds have attached substituent groups. If there is only one substituent on the ring, the substituent does not have to be numbered because all positions on the ring are equivalent. When there is more than one substituent, numbers are required. The ring carbon atoms are numbered so that the carbon atoms bearing the substituents have the lowest numbers.

Differences in geometric configuration lead to a major distinction between the cyclic and noncyclic alkanes. Groups can rotate freely about the carbon–carbon bonds of open–chain alkanes, but in cyclic compounds the ring structure holds the carbon atoms rigidly in place. Free rotation cannot occur without disruption of the ring structure. For cyclic compounds, the position of any substituent relative to the ring becomes extremely important. (The substituent is situated either above or below the ring.) A new form of isomerism is possible, which we discuss in Section 18.4. For now, let it suffice to say that these dimethylcyclopropanes are not the same, but are isomers.

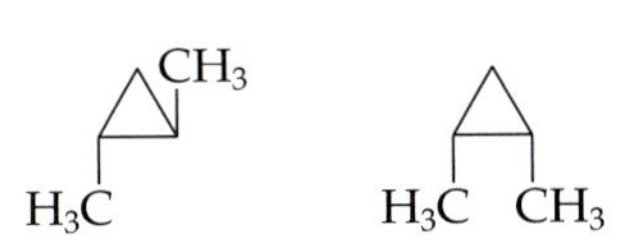

Example 13.5

Draw structures for each of the following compounds.

a. cyclooctane
b. ethylcyclohexane
c. 1,1,2-trimethylcyclobutane

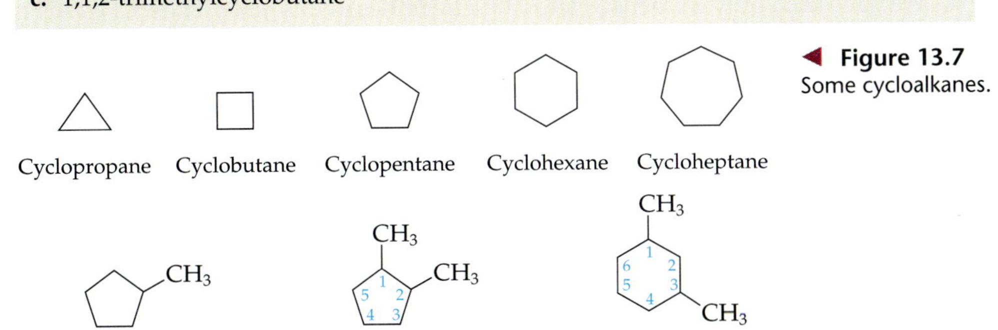

◀ **Figure 13.7**
Some cycloalkanes.

Solution

a. 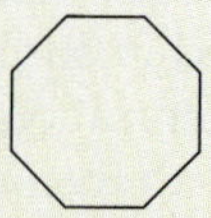 (cyclooctane ring) b. 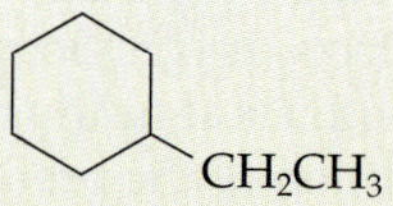 (cyclohexane ring with CH_2CH_3) c. 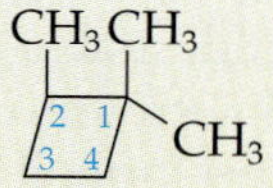 CH_3 CH_3 on a four-membered ring (positions 2 1 / 3 4) with CH_3

Practice Exercise

Draw structures for each of the following compounds.

a. cyclopentane
b. 1-ethyl-2-methylcyclopentane
c. 1-ethyl-1,2,5,5-tetramethylcycloheptane

The physical, chemical, and physiological properties of cyclic hydrocarbons are generally quite similar to those of the corresponding open-chain compounds. Cycloalkanes (with the exception of cyclopropane, which has a highly strained ring) act very much like noncyclic alkanes. Like all other hydrocarbons, cyclic hydrocarbons burn. Those with five- and six-membered rings occur in petroleum from certain areas; California crude, for instance, is particularly rich in these compounds. Cyclic structures containing five or six atoms, such as cyclopentane and cyclohexane, are particularly stable. We will see in Chapter 19 that some carbohydrates (sugars) form five- or six-membered rings in solution.

Alkenes and Alkynes: Unsaturated Hydrocarbons

Not all hydrocarbons are as resistant to reaction as are the alkanes. In fact, some are quite reactive. Hydrocarbons with double bonds are called *alkenes,* and those with triple bonds are called *alkynes.* Collectively, alkenes and alkynes are called **unsaturated hydrocarbons,** because they have fewer hydrogen atoms than does an alkane with the same number of carbon atoms, as is indicated in the following general formulas.

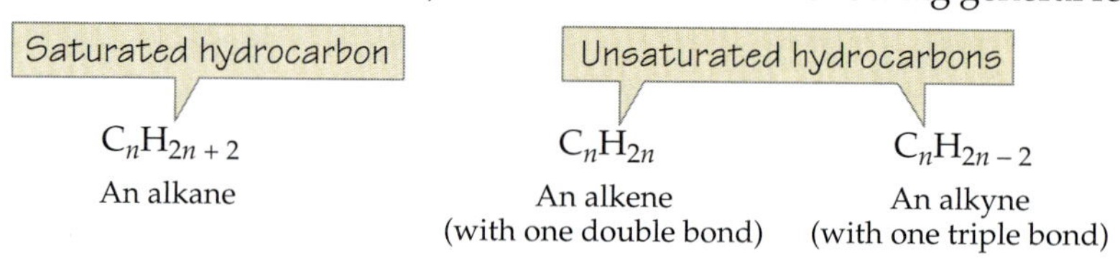

13.7 Alkenes: Structures and Names

An **alkene** (note the *-ene* ending) is characterized by the presence of a carbon–carbon double bond. Names, structures, and physical properties of a few representative alkenes are given in Table 13.9.

We have used only condensed structural formulas in this table. Thus, $CH_2{=}CH_2$ stands for

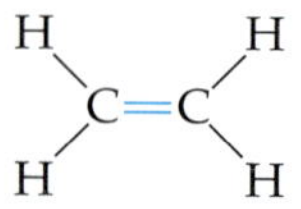

$$\begin{matrix} H & & H \\ & C{=}C & \\ H & & H \end{matrix}$$

The double bond is shared by the two carbon atoms and does not involve the hydrogen atoms, although the condensed formula does not make this point obvious. Note that the molecular formula for the two-carbon alkene is C_2H_4 whereas that for ethane, the two-carbon alkane, is C_2H_6.

The first two alkenes in Table 13.9, ethene and propene, are most often called by their common names, ethylene and propylene, respectively. Ethylene ($CH_2{=}CH_2$)

Table 13.9 Physical Properties of Some Selected Alkenes

IUPAC Name	Molecular Formula	Condensed Structure	Melting Point (°C)	Boiling Point (°C)
Ethene	C_2H_4	$CH_2{=}CH_2$	−169	−104
Propene	C_3H_6	$CH_3CH{=}CH_2$	−185	−47
1-Butene	C_4H_8	$CH_3CH_2CH{=}CH_2$	−185	−6
1-Pentene	C_5H_{10}	$CH_3CH_2CH_2CH{=}CH_2$	−138	30
1-Hexene	C_6H_{12}	$CH_3(CH_2)_3CH{=}CH_2$	−140	63
1-Heptene	C_7H_{14}	$CH_3(CH_2)_4CH{=}CH_2$	−119	94
1-Octene	C_8H_{16}	$CH_3(CH_2)_5CH{=}CH_2$	−102	121

is a major commercial chemical. The U.S. chemical industry produces about 25 billion kg of ethylene annually, making it the most important of all synthetic organic chemicals. More than half of this ethylene goes into the manufacture of polyethylene, one of the most familiar plastics (Section 13.10). Another one-sixth is converted to ethylene glycol (Chapter 14), the principal component of most brands of antifreeze for automobile radiators.

Propylene ($CH_3CH{=}CH_2$) is also an important industrial chemical. It is converted to plastics, isopropyl alcohol (Chapter 14), and a variety of other products.

Although there is only one alkene with the formula C_2H_4 (ethene) and only one with the formula C_3H_6 (propene), there are several alkenes with the formula C_4H_8. Common names are hardly helpful in naming the many isomers of the higher alkenes. For the most part the IUPAC system is used for these compounds. Some of the IUPAC rules for alkenes are as follows.

1. The longest chain of atoms *containing the double bond* is the parent compound. The name has the same stem as the alkane having the same number of carbon atoms, but ends in *-ene*. Thus the compound $CH_3CH{=}CH_2$ is named *propene*.
2. When it is necessary to indicate the position of the double bond, the first carbon of the two that are doubly bonded is given the lowest possible number. The compound $CH_3CH{=}CHCH_2CH_3$, for example, has the double bond between the second and third carbon atoms. Its name is 2-pentene.
3. Substituent groups are named as with alkanes. Their position is indicated by a number. Thus,

$$\begin{array}{l} CH_3\underset{\displaystyle CH_3}{\underset{|}{C}}HCH_2CH{=}CHCH_3 \end{array}$$

is 5-methyl-2-hexene. Note that the numbering of the parent chain is always done in such a way as to give the double bond the lowest number, even if that requires a substituent to have a higher number. The double bond has priority in numbering.

Table 13.10 shows the IUPAC names of some representative alkenes and cycloalkenes. After studying the table, try the following examples.

Example 13.6

Name each of the following compounds.

a. $CH_3CH{=}CHCH_2\underset{\displaystyle CH_3}{\underset{|}{C}}H{-}\underset{\displaystyle CH_3}{\underset{|}{C}}HCH_3$

b. $CH_2{=}\underset{\displaystyle CH_2CH_3}{\underset{|}{C}}{-}CH_2CH_3$

Table 13.10 IUPAC Names of Representative Alkenes and Cycloalkenes

Condensed Structural Formula	Rewritten and Numbered	IUPAC Name
$CH_3(CH_2)_2CH{=}CH_2$	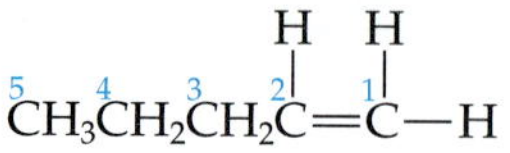 $\overset{5}{C}H_3\overset{4}{C}H_2\overset{3}{C}H_2\overset{2}{C}(H){=}\overset{1}{C}(H){-}H$	1-Pentene
$CH_3CH{=}CHCH_2CH_3$	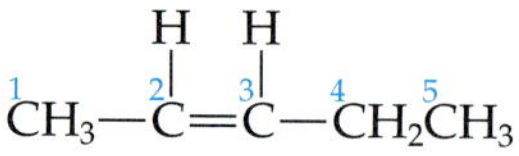 $\overset{1}{C}H_3{-}\overset{2}{C}(H){=}\overset{3}{C}(H){-}\overset{4}{C}H_2\overset{5}{C}H_3$	2-Pentene
$(CH_3)_2CHC(CH_3){=}CHCH_3$	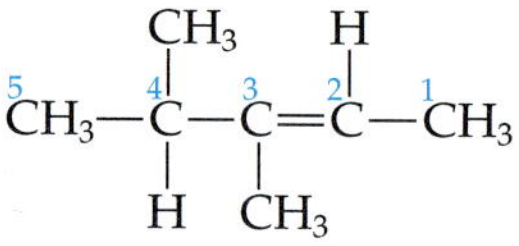 $\overset{5}{C}H_3{-}\overset{4}{C}(CH_3)(H){-}\overset{3}{C}(CH_3){=}\overset{2}{C}(H){-}\overset{1}{C}H_3$	3,4-Dimethyl-2-pentene
$CH_3CH_2CH(CH_3)C(C_3H_7){=}CH_2$	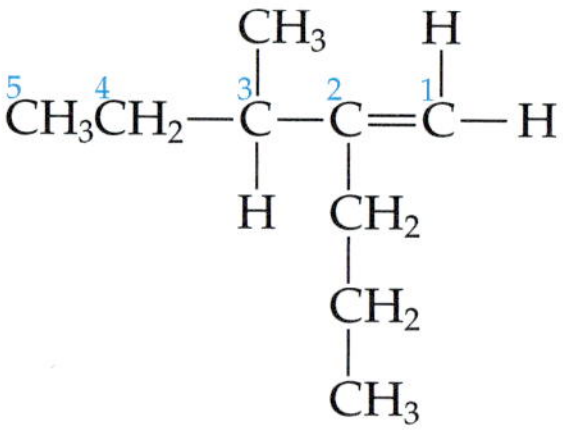 $\overset{5}{C}H_3\overset{4}{C}H_2{-}\overset{3}{C}(CH_3)(H){-}\overset{2}{C}(CH_2CH_2CH_3){=}\overset{1}{C}(H){-}H$	3-Methyl-2-propyl-1-pentene
 (cyclopentene ring)	 (cyclopentene ring)	Cyclopentene
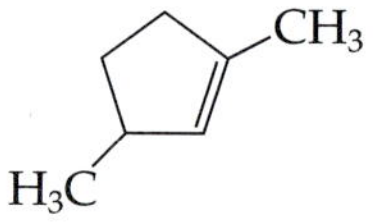 (cyclopentene ring with CH_3 and H_3C)	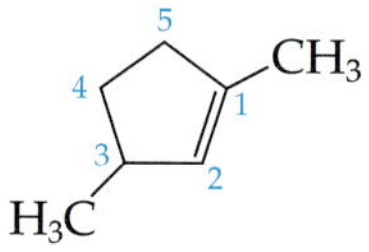 (ring numbered 1–5, CH_3 on C1, H_3C on C3)	1,3-Dimethylcyclopentene
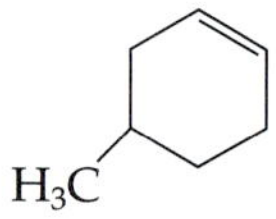 (cyclohexene ring with H_3C)	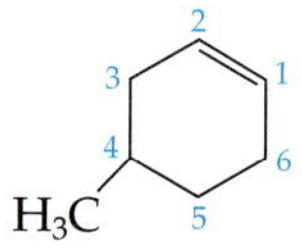 (ring numbered 1–6, H_3C on C4)	4-Methylcyclohexene
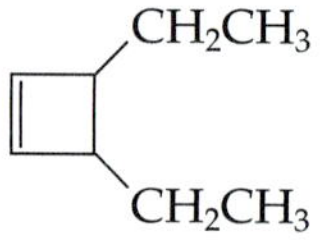 (cyclobutene ring with two CH_2CH_3)	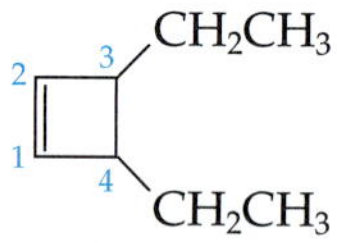 (ring numbered 1–4, CH_2CH_3 on C3 and C4)	3,4-Diethylcyclobutene

Solution

a. The longest chain containing the double bond has seven carbon atoms; the compound is a *heptene* (rule **1**). To give the first carbon of the double bond the lowest number (rule **2**), we number from the left:

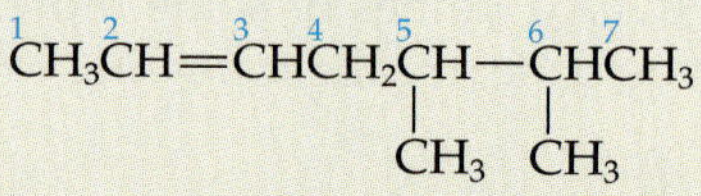

$$\overset{1}{C}H_3\overset{2}{C}H{=}\overset{3}{C}H\overset{4}{C}H_2\underset{CH_3}{\overset{5}{C}H}{-}\underset{CH_3}{\overset{6}{C}H}\overset{7}{C}H_3$$

The compound is a 2-heptene. There are methyl groups on the fifth and sixth carbon atoms (rule **3**). The compound is 5,6-dimethyl-2-heptene.

b. The longest chain *containing the double bond* has four carbon atoms; the parent compound is a *butene* (rule **1**). (The longest chain overall has five carbon atoms, but it does not contain the double bond; the parent name is not *pentene*.) To give the first carbon of the double bond the lowest number (rule **2**), we number from the left; the compound is a 1-butene. There is an ethyl group on the second carbon atom (rule **3**). The compound is 2-ethyl-1-butene.

Practice Exercise

Name the following compounds.

a. $CH_3C{=}CHCHCH_2CH_3$ (with CH_3 on the second carbon and CH_2CH_3 on the fourth carbon)

b. CH_3

c. $(CH_3)_2C{=}CHC(CH_3)_3$

Example 13.7

Draw the structural formula for 3,4-dimethyl-2-pentene.

Solution

First write the parent chain of five carbon atoms:

$$C{-}C{-}C{-}C{-}C$$

Then add the double bond between the second and third carbon atoms:

$$\overset{1}{C}{-}\overset{2}{C}{=}\overset{3}{C}{-}\overset{4}{C}{-}\overset{5}{C}$$

Now add the groups at their proper positions and add enough hydrogen atoms to give each carbon atom a total of four bonds.

$$CH_3{-}CH{=}\underset{}{\overset{CH_3}{\overset{|}{C}}}{-}\overset{CH_3}{\overset{|}{CH}}{-}CH_3$$

Practice Exercise

Draw structural formulas for the following compounds.

a. 3-ethyl-2-methyl-1-hexene **b.** 3-isopropylcyclopentene

The double bond of alkene molecules, like the ring structure of cycloalkanes, imposes geometric restrictions. Recall that, according to VSEPR theory (Chapter 3), each doubly bonded carbon atom lies in the center of an equilateral triangle (to minimize the repulsive forces among the three regions of electron density). The carbon atoms of a double bond and the two atoms bonded to each carbon all lie in a single plane (Figure 13.8). Free rotation about doubly bonded carbon atoms is *not* possible without rupturing the bond. Therefore, the relative positions of substituent groups located above or below the double bond become significant. The nomenclature employed in situations such as this, as well as the consequences of restricted rotation, are discussed in Chapter 18.

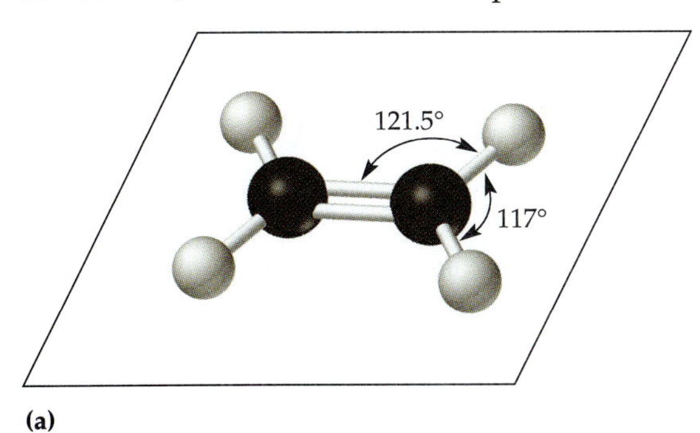

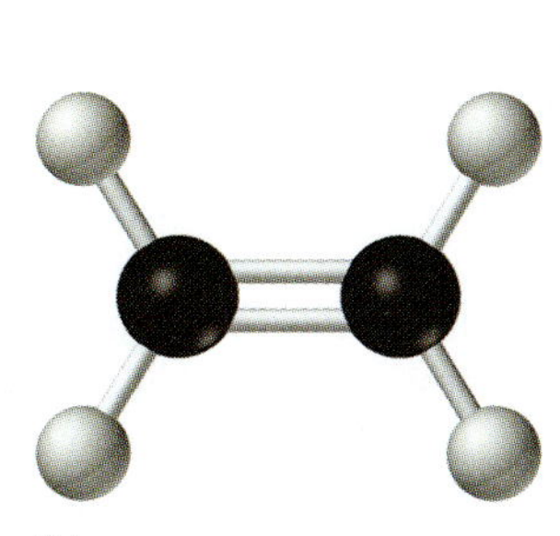

Figure 13.8 ▶
(a) Planar configuration of the carbon-to-carbon double bond. (b) Ball-and-spring model of ethylene (ethene).

▲ The bright red color of tomatoes is due to lycopene, a polyene.

Review Question

13.11 Briefly identify the important distinctions between:

- **a.** a saturated hydrocarbon and an unsaturated hydrocarbon
- **b.** an alkene and an alkane

13.8 Properties of Alkenes

The physical properties of alkenes are quite similar to those of the alkanes. Table 13.9 shows that the boiling points of the straight-chain alkenes increase with increasing molar mass, just as they did with the alkanes. For molecules with the same number of carbon atoms and the same general shape, the boiling points usually differ only slightly, just as we would expect for substances whose molar mass differs by only 2 u (two H atoms).

Like all other hydrocarbons, the alkenes are insoluble in water but soluble in organic solvents.

The physiological properties of the alkenes are similar to those of the alkanes. Ethylene has been used as an inhalation anesthetic. Like the gaseous alkanes, ethylene can cause unconsciousness and even death by asphyxiation. Few people ever encounter large amounts of liquid and solid (or mixtures of liquid and solid) alkenes. They would probably act on or in our bodies much as the alkanes do.

Alkenes occur widely in nature. Ripening fruits and vegetables give off ethylene, which triggers further ripening. Fruit processors artificially introduce ethylene to hasten the normal ripening process; 1 kg of tomatoes can be ripened by exposure to as little as 0.1 mg of ethylene for 24 hours. Unfortunately, the tomatoes don't taste much like those that ripen on the vine.

Other alkenes that occur in nature include 1-octene, a constituent of lemon oil, and octadecene ($C_{18}H_{36}$), found in fish liver. Dienes (which have two double bonds) and polyenes (three or more double bonds) are also common. Butadiene ($CH_2{=}CH{-}CH{=}CH_2$) is found in coffee. Lycopene and the carotenes are isomeric polyenes ($C_{40}H_{56}$) that give the attractive red, orange, and yellow colors to watermelons, tomatoes, carrots, and other vegetables and fruits. Vitamin A, essential to good vision, is derived from a carotene (Special Topic F). The world would be a much less colorful place without alkenes.

Review Question

13.12 What physiological effect is shared by ethylene and cyclopropane?

13.9 Chemical Properties: Reactions of Alkenes

Like the alkanes—and all other hydrocarbons—the alkenes burn. We can write an equation for the combustion of ethene that is similar to the one for alkane combustion:

$$C_2H_4 + 3\,O_2 \longrightarrow 2\,CO_2 + 2\,H_2O + \text{heat}$$

While this is a hazard when working with alkenes, they are not commercially important as fuels. Rather, they are valued mainly for **addition reactions.** In these reactions, one of the bonds in the double bond is broken, and each of the involved carbon atoms then bonds to another atom or group. The two carbon atoms that were originally joined by the double bond are still attached by a single bond. Perhaps the simplest addition reaction is that which occurs with hydrogen in the presence of a catalyst such as nickel, platinum, or palladium.

Alkenes typically undergo addition reactions in which all the atoms of the reactants are incorporated into a single product.

$$\mathrm{H_2C{=}CH_2} + \mathrm{H{-}H} \xrightarrow{\mathrm{Ni}} \mathrm{H{-}\underset{H}{\overset{H}{C}}{-}\underset{H}{\overset{H}{C}}{-}H}$$

Ethene — Hydrogen — Ethane

The product is an alkane having the same carbon skeleton as the alkene. This addition of hydrogen to an unsaturated molecule is called **hydrogenation.** (Use of hydrogenation to convert unsaturated vegetable oils to saturated fats is discussed in Chapter 20.)

Alkenes readily add halogen molecules. Indeed, the reaction with bromine is often used to test for alkenes (Figure 13.9). Solutions of bromine are brownish red. When we add a bromine solution to an alkene, the color of the solution disappears because the alkene reacts with the bromine:

$$\mathrm{H_2C{=}CH_2} + \mathrm{Br{-}Br} \longrightarrow \mathrm{H{-}\underset{Br}{\overset{H}{C}}{-}\underset{Br}{\overset{H}{C}}{-}H}$$

Ethene — Bromine (brownish red) — 1,2-Dibromoethane (colorless)

Another important addition reaction is that between alkenes and water. This reaction, called **hydration,** requires the presence of a mineral acid, such as sulfuric acid, as a catalyst:

$$\mathrm{H_2C{=}CH_2} + \mathrm{H{-}OH} \xrightarrow{\mathrm{H_2SO_4}} \mathrm{H{-}\underset{H}{\overset{H}{C}}{-}\underset{OH}{\overset{H}{C}}{-}H}$$

Ethene — Water — Ethyl alcohol

Vast quantities of ethyl alcohol, for use as an industrial solvent, are made from ethylene. This alcohol is structurally identical to that used in alcoholic beverages. However, federal law requires that all drinking alcohol be produced by the natural process called fermentation. (See Section 14.4 for a more extensive discussion of the hydration reaction.)

Example 13.8

Write equations for the reaction between $CH_3CH{=}CHCH_3$ and each of the following.
a. H_2 (Ni catalyst) **b.** Br_2 **c.** H_2O (H_2SO_4 catalyst).

Solution

In each reaction, the reagent adds across the double bond:

a. $$\mathrm{CH_3CH{=}CHCH_3 + H_2 \xrightarrow{Ni} CH_3\underset{H}{CH}{-}\underset{H}{C}HCH_3 \ \text{or}\ CH_3CH_2CH_2CH_3}$$

b. $$\mathrm{CH_3CH{=}CHCH_3 + Br_2 \longrightarrow CH_3\underset{Br}{CH}{-}\underset{Br}{C}HCH_3}$$

c. $$\mathrm{CH_3CH{=}CHCH_3 + H_2O \xrightarrow{H_2SO_4} CH_3\underset{H}{CH}{-}\underset{OH}{C}HCH_3 \ \text{or}\ CH_3CH_2\underset{OH}{C}HCH_3}$$

Practice Exercise

Write equations for the reaction of (cyclopentene) with each of the following.
a. H_2 (Ni catalyst) **b.** Cl_2 **c.** H_2O (H_2SO_4 catalyst)

(a)

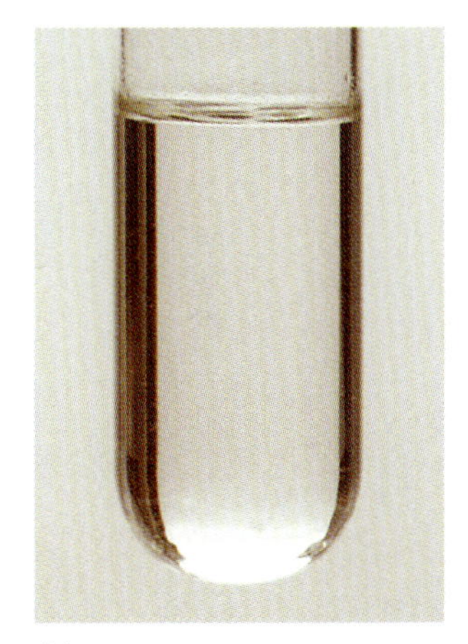

(b)

▲ **Figure 13.9**
(a) When bromine is added to a saturated hydrocarbon such as cyclohexane, the brownish-red color remains because the bromine dissolves in the cyclohexane but does not react with it. (b) When bromine is added to an unsaturated hydrocarbon such as cyclohexene, the brownish-red color disappears because the bromine adds to the double bond of the cyclohexene. This reaction serves as a convenient test for the presence of a carbon-to-carbon double bond in a molecule.

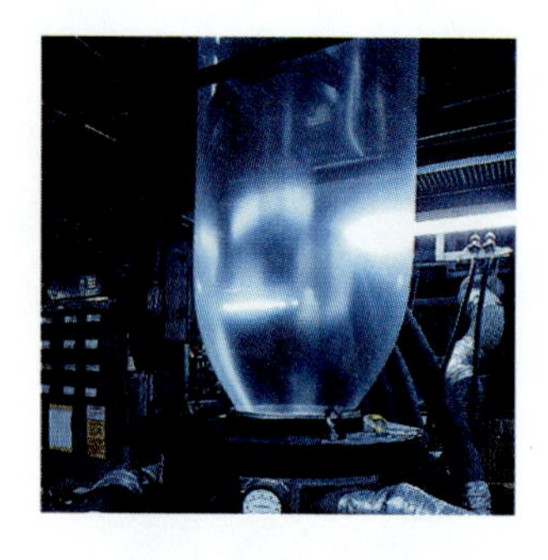

▲ A giant bubble of tough, transparent plastic film emerges from a die of an extruding machine. The film is used in packaging, consumer products, and food services.

13.10 Polymerization

The most important commercial reaction of alkenes is *polymerization*. The giant molecules assembled from much smaller ones are called **polymers** (from the Greek *poly,* many, and *meros,* parts). Polymer formation involves the hooking together of many smaller molecules, building blocks called **monomers** (from the Greek *monos,* one, and *meros,* parts). A polymer is as different from its monomer as a long strand of spaghetti is from a tiny speck of flour. For example, polyethylene, the familiar waxy material used to make plastic bags, is made from the monomer ethylene, which is a gas.

There are two general types of polymerization reactions: addition polymerization and condensation polymerization (Chapter 16). In **addition polymerization,** the monomers add to one another in such a way that the polymer contains all the atoms of the starting monomers. Under high pressure and temperature and in the presence of a catalyst, ethylene molecules are made to join together in long chains (Figure 13.10). The polymerization can be represented by the reaction of a few monomer units:

$$\cdots + \mathrm{H_2C{=}CH_2} + \mathrm{H_2C{=}CH_2} + \mathrm{H_2C{=}CH_2} + \cdots$$

$$\longrightarrow \cdots \begin{array}{cccccccccccc} & \mathrm{H} & & \mathrm{H} & & \mathrm{H} & & \mathrm{H} & & \mathrm{H} & & \mathrm{H} \\ & | & & | & & | & & | & & | & & | \\ \cdots & \mathrm{C} & - & \mathrm{C} & - & \mathrm{C} & - & \mathrm{C} & - & \mathrm{C} & - & \mathrm{C} \cdots \\ & | & & | & & | & & | & & | & & | \\ & \mathrm{H} & & \mathrm{H} & & \mathrm{H} & & \mathrm{H} & & \mathrm{H} & & \mathrm{H} \end{array}$$

The dotted lines in the formula of the product are like etceteras: they indicate that the structure extends for many units in each direction. Notice that all the atoms—two carbon and four hydrogen atoms—of each monomer molecule are incorporated into the polymer structure.

Many natural materials such as proteins (Chapter 21), natural rubber, cellulose and starch (Chapter 19), and complex silicate minerals are polymers. Artificial materials such as fibers, films, plastics, semisolid resins, and synthetic rubbers are also polymers. More than half of the compounds produced by the chemical industry are synthetic polymers (Table 13.11). Many are mundane—plastic bags, food wrap,

Figure 13.10 ▶ The formation of polyethylene. In the synthesis of polyethylene, many monomer units join together to form huge polymer molecules. In this computer-generated representation, the yellow dots indicate a new bond forming as an ethylene molecule (upper left) is added to the growing polymer chain.

Table 13.11 Some Addition Polymers

Monomer	Polymer	Polymer Name	Some Uses
$H_2C{=}CH_2$ Ethylene	$\left[-CH_2-CH_2-\right]_n$	Polyethylene	Plastic bags, bottles, toys, electrical insulation
$H_2C{=}CH-CH_3$ Propylene	$\left[-CH_2-CH(CH_3)-\right]_n$	Polypropylene	Indoor-outdoor carpeting, bottles, luggage
$H_2C{=}CH-C_6H_5$ Styrene	$\left[-CH_2-CH(C_6H_5)-\right]_n$	Polystyrene	Simulated wood furniture, styrofoam insulation, cups, toys, packing materials
$H_2C{=}CH-Cl$ Vinyl chloride	$\left[-CH_2-CH(Cl)-\right]_n$	Poly(vinyl chloride), PVC	Plastic wrap, simulated leather (Naugahyde), bags for intravenous drugs, garden hoses, rainwear, floor covering
$H_2C{=}CCl_2$ 1,1-Dichloroethene (Vinylidene chloride)	$\left[-CH_2-CCl_2-\right]_n$	Poly(vinylidene chloride), Saran	Food wrap, seatcovers
$F_2C{=}CF_2$ Tetrafluoroethylene	$\left[-CF_2-CF_2-\right]_n$	Polytetrafluoroethylene, Teflon	Nonstick coating for cooking utensils, electrical insulation, lubricant, bearings
$H_2C{=}CH-CN$ Cyanoethylene (Acrylonitrile)	$\left[-CH_2-CH(CN)-\right]_n$	Polyacrylonitrile, Orlon, Acrilan, Creslan, Dynel	Yarns, wigs, paints
$H_2C{=}CH-O-C({=}O)-CH_3$ Vinyl acetate	$\left[-CH_2-CH(O-C({=}O)-CH_3)-\right]_n$	Poly(vinyl acetate), PVA	Adhesives, textile coatings, chewing gum resin, paints
$H_2C{=}C(CH_3)-C({=}O)-O-CH_3$ Methyl methacrylate	$\left[-CH_2-C(CH_3)(C({=}O)-O-CH_3)-\right]_n$	Poly(methyl methacrylate), Lucite, Plexiglas	Glass substitute, bowling balls

The Many Uses of Polyethylene

Polyethylene was invented shortly before the start of World War II. It proved to be tough and flexible and an excellent electrical insulator. It could withstand both high and low temperatures. Before long, it was used for insulating cables in radar, a top-secret invention that helped British pilots spot enemy aircraft earlier than they could by sight alone. Without polyethylene the British could not have had effective radar, and without radar the Battle of Britain might have been lost. The invention of this simple plastic may have changed the course of history.

Today there are two principal kinds of polyethylene produced by the use of different catalysts and different reaction conditions. *High-density polyethylenes (HDPE)* have largely linear molecules that pack closely together. These linear molecules can assume a fairly ordered, crystalline structure; this structure gives high-density polyethylenes greater rigidity and higher tensile strength than other ethylene plastics. Linear polyethylenes are used for such things as threaded bottle caps, toys, bottles, and milk jugs.

Low-density polyethylenes (LDPE), on the other hand, have many side chains branching off the main chain. The branches prevent the molecules from packing closely together and assuming a crystalline structure. Low-density polyethylenes are waxy, bendable plastics that have lower melting points than high-density polyethylenes. Objects made of HDPE hold their shape in boiling water, while those made of LDPE become severely deformed at this temperature. Low-density polyethylenes are used to make plastic bags, plastic film, squeeze bottles, electric-wire insulation, and many common household products.

▶ These polyethylene bottles were heated in the same oven for the same length of time. The one that melted has branched polyethylene molecules; the other has unbranched molecules.

toys, and tableware—but there are also polymers that conduct electricity, amazing new adhesives, and synthetic materials stronger than steel but much lighter in weight. In many applications, substitution of plastics for natural materials saves energy. Their use in place of metals in automobiles and airplanes saves weight and therefore saves fuel. Plastics present some environmental problems, but they have become such an important part of our daily lives that we would find it difficult to live without them.

Review Questions

13.13 List several uses of polyethylene.

13.14 What is addition polymerization?

13.15 What structural feature usually characterizes molecules used as monomers in addition polymerization?

13.11 Alkynes

▲ **Figure 13.11** Ball-and-spring model of acetylene.

In an alkene, carbon atoms in a double bond share two pairs of electrons. In an **alkyne,** carbon atoms share three pairs of electrons, forming a triple bond. The simplest alkyne is known by its common name, acetylene (C_2H_2; see Figure 13.11). Its structure is

$$H{-}C{\equiv}C{-}H$$

About 10% of all acetylene produced is used in oxyacetylene torches for cutting and welding metals. The flame from such a torch can be very hot. Most acetylene, however, is converted to chemical intermediates that are used to make vinyl and acrylic plastics, fibers, resins, and a variety of other chemical products.

The common name, acetylene, sounds very much like ethylene or propylene. Remember, however, that acetylene is an alkyne, whereas ethylene and propylene are alkenes.

The alkynes are similar to the alkenes in both physical and chemical properties. For example, they undergo many of the typical addition reactions of alkenes. Like ethene, acetylene has been used as an anesthetic for surgery. At higher concentrations it causes narcosis and asphyxia. The IUPAC nomenclature for alkynes parallels that of the alkenes, except that the family ending is *-yne* rather than *-ene*. The IUPAC name for acetylene is ethyne.

Review Question

13.16 Briefly identify the important distinctions between an alkene and an alkyne. How are they similar?

13.12 Benzene

In 1825 Michael Faraday isolated a hydrocarbon from a sample of illuminating gas made from whale oil. The compound was later given the name benzene because it could be obtained from benzoic acid (Chapter 16).

Benzene is a liquid that smells like gasoline, boils at 80 °C, and freezes at 5.5 °C.

The molecular formula of benzene was found to be C_6H_6, for which there are many possible structural formulas, including 1,3,5-cyclohexatriene. This formula would be a slightly lopsided hexagon having three double bonds in a six-membered ring (Figure 13.12a). The high degree of unsaturation would imply a very high reactivity. Chemists soon discovered, however, that benzene is unreactive and behaves more like an alkane than an alkene. It does not react readily with bromine, which, as mentioned in Section 13.9, is a test for unsaturation. In fact, all the evidence from laboratory work pointed to a molecule in which all six carbon atoms were equivalent.

In 1865 the German chemist Friedrich August Kekulé proposed a structure that could account for many of the known chemical properties of benzene. His theory was that the benzene molecule consists of a cyclic, hexagonal, planar structure of six carbon atoms with alternate single and double bonds. Each carbon atom is bonded to only one hydrogen atom. He accounted for the equivalence of all six carbon atoms by suggesting that the double bonds are not static but rather are mobile and oscillate from one position to another:

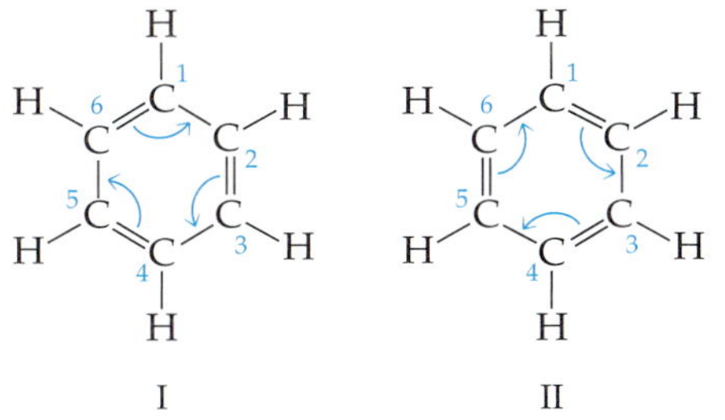

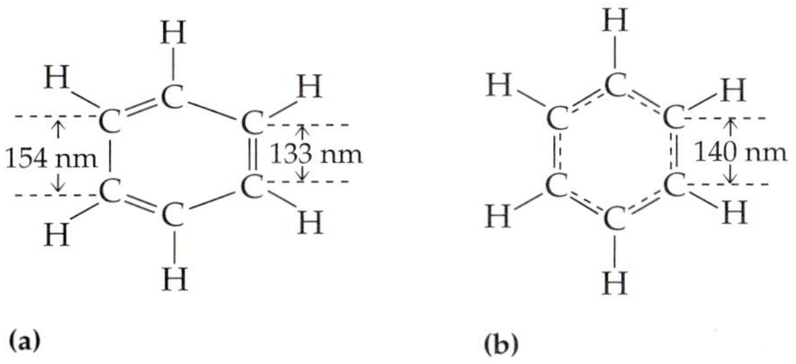

I   II

(a) 154 nm, 133 nm   (b) 140 nm

Figure 13.12
(a) The hypothetical 1,3,5-cyclohexatriene molecule would be a lopsided hexagon because double bonds are shorter than single bonds. (b) The actual benzene molecule is a regular hexagon having all sides 140 nm long.

Kekulé's Dream

Many years after proposing the cyclic structure for benzene, Kekulé told the story that one evening he fell asleep while sitting in front of a fire. He dreamed about chains of atoms having the forms of twisting snakes. Suddenly one of the snakes caught hold of its own tail, forming a whirling ring. Kekulé awoke, freshly inspired, and spent the remainder of the night working on his now-famous hypothesis. He is said to have written, "Let us learn to dream, gentlemen, and then perhaps we shall learn the truth."

Structures I and II differ from each other only in the positions of the double bonds, and although they satisfy the requirements for equivalent hydrogen atoms, they do not explain why benzene does not behave like an unsaturated hydrocarbon. Furthermore, X-ray diffraction measurements indicate that all the carbon–carbon bonds in benzene are the same length, 140 nm (Figure 13.12b). This value falls between the length of a carbon–carbon double bond (133 nm) and that of a carbon–carbon single bond (154 nm).

To accommodate all these findings, chemists have postulated that benzene exhibits resonance. **Resonance** is a word used to describe the phenomenon in which no single classical Lewis structure adequately accounts for the experimentally observed properties of a molecule (such as bond energies and bond distances). According to the resonance theory, the benzene molecule is a **resonance hybrid** of structures I and II. That is, neither of the two structures actually exists, but taken together they represent the true structure of the molecule. We depict this situation with a double-headed arrow ($\leftrightarrow$) that connects the contributing structures (Figure 13.13a). This resonance arrow should be clearly distinguished from the pair of half-headed arrows ($\rightleftharpoons$) that indicate an equilibrium condition.

Although the term *resonance* is much used in organic chemistry, it is not unique to organic compounds. Many inorganic compounds (e.g., O_3, SO_2, and NO_3^-) also exhibit resonance.

Some chemists combine the two contributing forms into the single structure depicted in Figure 13.13b. The inner circle indicates that the valence electrons are shared equally by all six carbon atoms (that is, the electrons are *delocalized*, or spread out, over all the carbon atoms). This method is a valid shortcut device for writing benzene, but it does not adequately account for all the electrons in the molecule. The representation of benzene that contains alternating single and double bonds (*resonance* or *Kekulé forms*) is the best model for keeping track of electrons. The circle within a hexagon better describes the molecule by indicating the equal sharing of the electrons and the identical bond lengths within the molecule. In this text, we shall use the hexagon with inscribed circle. Remember that when either representation is used, it is understood that each corner of the hexagon is occupied by one carbon atom. Attached to each carbon atom is one hydrogen atom, and all the atoms lie in the same plane. Any other atom, or group of atoms, substituted for a hydrogen atom must be shown to be bonded to a particular corner of the hexagon.

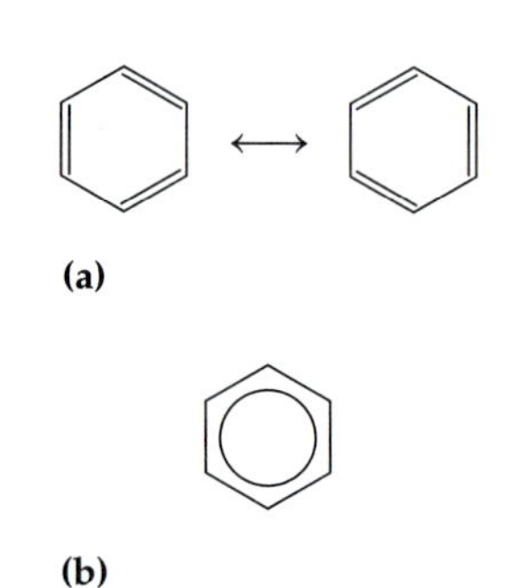

▲ Figure 13.13 Two methods of representing the structure of benzene. (a) Kekulé structures; (b) hexagon with inscribed circle.

Taken literally, the term "resonance" is a misnomer, indicating oscillation from one structure to another. *There is no oscillation.* A mule, the offspring of a male donkey and a mare, can be considered a hybrid[4] of a donkey and a horse. However, it is not a horse at one instant and a donkey at another. It is always a mule, a distinctive animal that has some of the characteristics of a horse and some of a donkey. Likewise, any molecule that exhibits resonance has *one* real structure; it is never any of the separate Lewis structures used to describe it. The difficulty arises because we cannot adequately portray the molecule, not because of the molecule itself.

[4]Perhaps a better analogy is as follows: A rhinoceros can be thought of as a resonance hybrid of a unicorn and a dragon. Neither the unicorn nor the dragon actually exist, but they have characteristic features that are incorporated into the rhinoceros. This analogy relates to the idea that the structures we draw are not true representations of any known molecule.

13.13 Structure and Nomenclature of Aromatic Compounds

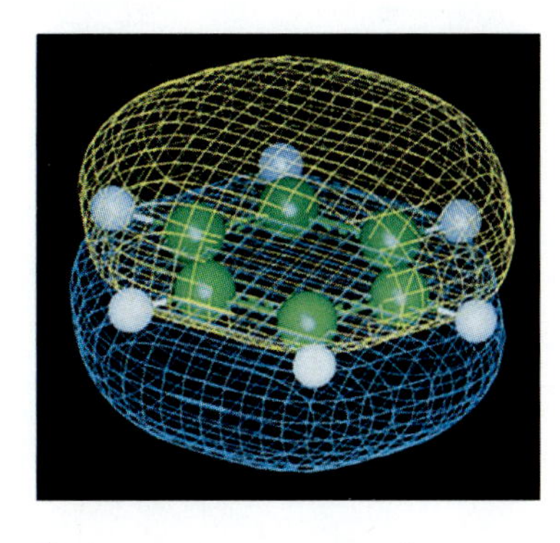

▲ A computer-generated structure of benzene. The yellow and blue structures represent the orbitals occupied by the six delocalized electrons.

Benzene and similar compounds are referred to as **aromatic hydrocarbons.** This is because quite a few of the first benzene-like substances to be discovered had strong aromas. Even though many benzene derivatives have turned out to be odorless, the name stuck. Today the term "aromatic" in organic chemistry is applied to any compound that contains a benzene ring or has certain properties similar to those of benzene. (All the nonaromatic hydrocarbons that we have considered—the alkanes, cycloalkanes, alkenes, and alkynes—are referred to collectively as **aliphatic compounds** to distinguish them from aromatic compounds. "Aliphatic" originally meant that the source of the compound was a fat. Today, however, it simply means "not aromatic.")

Aromatic hydrocarbons have both common names and systematic names. Some aromatic compounds are referred to exclusively as derivatives of benzene, whereas others are more frequently denoted by their common names. Note that in the following structures, it is immaterial whether the substituent is written at the top, side, or bottom of the ring: a hexagon is symmetrical, and therefore all positions are equivalent.

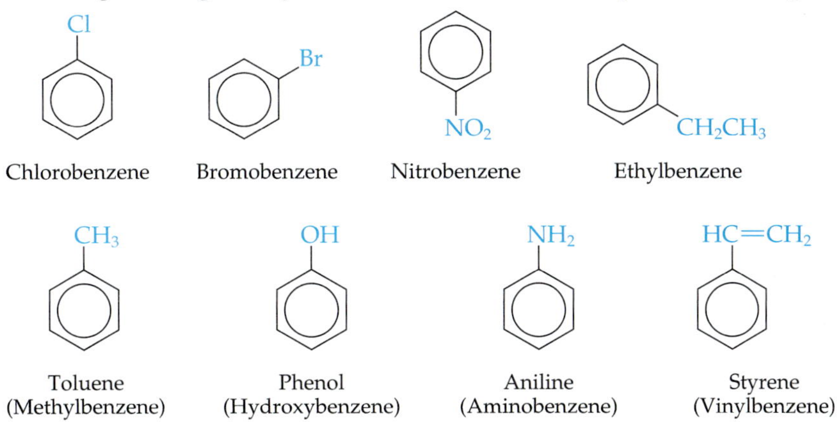

Chlorobenzene　Bromobenzene　Nitrobenzene　Ethylbenzene

Toluene (Methylbenzene)　Phenol (Hydroxybenzene)　Aniline (Aminobenzene)　Styrene (Vinylbenzene)

A complication arises when there is more than one substituent because now all the positions on the hexagon are no longer equivalent and relative positions must be designated. In the case of a disubstituted benzene, one nomenclature system uses the prefixes *ortho* (*o*-) for 1,2-disubstitution, *meta* (*m*-) for 1,3-disubstitution, and *para* (*p*-) for 1,4-disubstitution.

Cl, NO_2　Br, Br　F, I

o-Chloronitrobenzene　*m*-Dibromobenzene　*p*-Fluoroiodobenzene

Alternatively, the ring is numbered and the substituent names are listed in alphabetical order. The first substituent is given the lowest number. When a common name is used, the carbon atom that bears the group responsible for the name is carbon 1 (C-1):

CH_3 (1), CH_3 (3)　Cl (1), CH_2CH_3 (3)　CH_3 (1), Br (2)　HO (1), NO_2 (4)

m-Xylene (1,3-Dimethylbenzene)

m-Chloroethylbenzene (1-Chloro-3-ethylbenzene)

o-Bromotoluene (2-Bromotoluene) (1-Bromo-2-methylbenzene)

p-Nitrophenol (4-Nitrophenol) (1-Hydroxy-4-nitrobenzene)

If there are more than two substituents on a benzene ring, the numbering is determined by the requirement that the numbers give the smallest possible sum (substituents are listed in alphabetical order).

Occasionally an aromatic group is a substituent that is bonded to an aliphatic compound or to another aromatic ring. The group of atoms remaining when a hydrogen atom is removed from an aromatic compound is called an **aryl group.** The most common aryl group is the one derived from benzene (C_6H_5—). It is called *phenyl,* a name derived from *pheno,* an old name for benzene.[5]

$CH_3—CH—CH_2—CH_2—CH_2—CH_2—CH_3$

Phenyl group 2-Phenylheptane

Some common aromatic hydrocarbons consist of fused benzene rings.

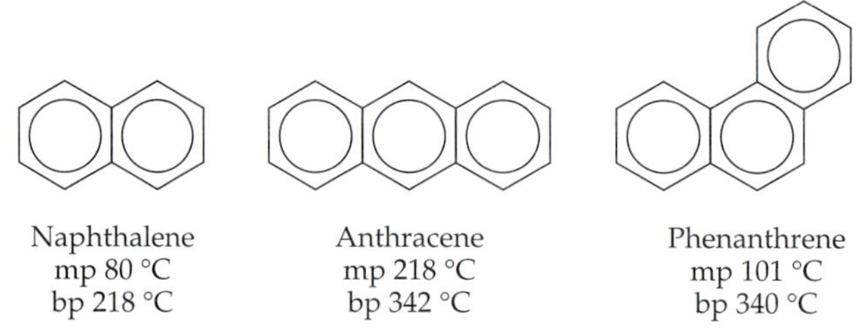

Naphthalene mp 80 °C bp 218 °C

Anthracene mp 218 °C bp 342 °C

Phenanthrene mp 101 °C bp 340 °C

These three substances are colorless, crystalline solids obtained from coal tar. Naphthalene has a pungent odor and is commonly used in mothballs. Anthracene is an important starting material in the manufacture of certain dyes. A large group of naturally occurring substances, the steroids (Chapter 20), contain the phenanthrene structure.

Polycyclic aromatic compounds do not exist in unprocessed coal itself but are formed by the intense heating needed to distill coal tar. For many years, it has been known that workers in coal-tar refineries are susceptible to a type of skin cancer known as tar cancer. Investigation has shown that a number of polycyclic aromatic hydrocarbons (PAHs) are carcinogens. One of the most active carcinogenic compounds, benzpyrene, occurs in coal tar and has been isolated from cigarette smoke, automobile exhaust gases, and charcoal-broiled steaks. It is estimated that more than 1000 tons of benzpyrene are emitted into the air over the United States each year. Only a few milligrams of benzpyrene are required to induce cancer in experimental animals.

The carcinogens seem not to be the PAHs themselves, but one or more of their metabolites. (As we shall learn in Chapter 24, metabolites are the products of chemical transformations in living cells.) Figure 13.14 indicates the conversion of benzpyrene via a multistep oxidation sequence to yield a highly carcinogenic diolepoxide metabolite.

Aromatic hydrocarbons containing only one benzene ring are generally liquids, and PAHs are generally solids. All are insoluble in water but soluble in organic

[5]This terminology is confusing because it would seem that the group derived from benzene should be called benzyl. The problem is compounded by the fact that another group is called benzyl. Replacement of one of the methyl hydrogen atoms of toluene gives the *benzyl* group, $C_6H_5CH_2$—.

—CH_2— —CH_2Br

Benzyl group Benzyl bromide

Epoxide = 3-membered cyclic ether (Chapter 14)

O_2 enzymes

HO OH

Benzpyrene

A diolepoxide

Diol = two alcohol groups (Chapter 14)

◀ Figure 13.14
Benzpyrene is metabolized in the body to produce an active carcinogen. The conversion of benzpyrene to the carcinogenic diolepoxide is catalyzed by enzymes in the liver known as cytochrome P450. These enzymes are involved in the metabolism of many drugs, converting them to products that are more water soluble and can be readily excreted through the urine.

solvents. Aromatic hydrocarbons are readily combustible. Unlike aliphatic hydrocarbons, which burn with a relatively clean flame, aromatic compounds burn with a sooty flame.

Review Questions

13.17 Briefly identify the important distinctions between:

a. an aromatic compound and an aliphatic compound
b. an alkyl group and an aryl group

13.18 Identify each substitution pattern as meta, ortho, or para.

a. 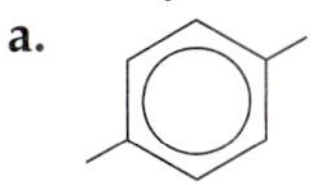 **b.** 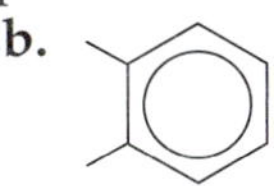 **c.** 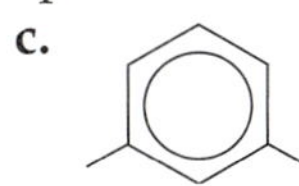

13.19 Indicate whether each compound is aromatic or aliphatic.

a.  **b.** 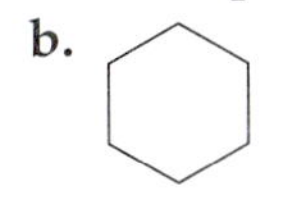 **c.** 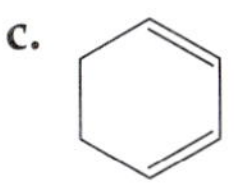 **d.** 

13.20 Describe a physiological effect of some polycyclic aromatic hydrocarbons.

13.14 Uses of Benzene and Benzene Derivatives

Most of the benzene used commercially comes from petroleum. Benzene is employed industrially as a starting material for the production of such products as detergents, drugs, dyes, insecticides, and plastics. Benzene was once widely used as an organic solvent, but it is now known to have both short- and long-term toxic effects. Inhalation of large concentrations of benzene can cause nausea and even death due to respiratory or heart failure. Repeated exposure leads to a progressive disease in which the ability of the bone marrow to make new blood cells is eventually destroyed. This results in a condition called *aplastic anemia,* in which there is a decrease in the numbers of both the red and white blood cells.

Unleaded gasolines contain about 2% benzene. For most people, exposure to benzene comes from inhaling gasoline fumes, or being around a cigarette smoker. For smokers, cigarettes overwhelm all other sources of benzene.

Because of these hazards, many chemical laboratories have replaced benzene with toluene as a general solvent.[6] Toluene is used in the production of dyes, drugs, and explosives and as a solvent in lacquers. It is commonly used as a preservative for urine specimens and is added to fuels to improve their octane rating. The explosive trinitrotoluene (TNT), is preferred to nitroglycerin (Chapter 14), because it is not sensitive to shock on jarring and must be exploded by a detonator.

[6]The maximum allowed concentration for an 8-hour exposure to toluene is 100 ppm, whereas the allowable exposure to benzene for the same period of time is 1 ppm. There are no toxic symptoms attributable to toluene until concentrations reach 200 ppm. Most commercial toluene, however, contains benzene as an impurity and toluene itself has been shown to cause birth defects, so pregnant women should avoid breathing toluene vapors. Everyone should avoid prolonged inhalation of toluene.

Nitrobenzene is used extensively in the manufacture of aniline, the parent compound of many dyes and drugs. Phenol (Chapter 14) containing a small amount of water is a liquid and in this form is referred to as carbolic acid. It is a good antiseptic and germicide, but its use is limited because of its toxicity.

Xylene is the common name for dimethylbenzene. Hence, there are three xylene isomers: *o*-, *m*-, and *p*-dimethylbenzene.

The xylenes are good solvents for grease and oil and are used for cleaning microscope slides and optical lenses and for removing wax from skis. The xylenes are also used to raise the octane rating of unleaded gasolines.

Substances containing the benzene ring are common in both animals and plants, although they are more abundant in the latter. Plants can synthesize the benzene ring from carbon dioxide, water, and inorganic materials. Animals cannot, but they are dependent on aromatic compounds for their survival and must therefore obtain these compounds from their food. Included among the compounds necessary for animal metabolism are several compounds that contain the benzene ring: the amino acids phenylalanine, tyrosine, and tryptophan, and vitamins such as vitamin K, riboflavin, and folic acid (Table 13.12). In addition, many drugs contain the benzene ring (Selected Topics B and C).

Review Question

13.21 What are some of the hazards associated with the use of benzene?

Table 13.12 Some Biologically Important Compounds That Contain a Benzene Ring

Name	Reference
Drugs	Selected Topic B
Aspirin	Figure B.2
Acetaminophen	Figure B.2
Ibuprofen	Figure B.2
Drugs	Selected Topic C
Amphetamine	Figure C.5
Lidocaine	Figure C.6
Diazepam (Valium)	Figure C.8
Amino Acids:	Chapter 21
Phenylalanine	Table 21.1
Tyrosine	Table 21.1
Tryptophan	Table 21.1
Vitamins:	Selected Topic F
Vitamin K	Section F.5
Riboflavin	Figure F.3
Folic acid	Figure F.7

Petroleum and Natural Gas

Our civilization runs to a large extent on hydrocarbons. **Petroleum** is a complex mixture of hydrocarbons (mainly alkanes) produced by the decomposition of animal and vegetable matter that has been entrapped for ages in the earth's crust. Most of the *petrochemicals* so vital to our modern economy are derived from these alkanes. A smaller group of chemicals, mainly aromatic compounds, is derived from coal.

Petroleum from the ground is of limited use until it is separated into fractions by boiling in a distillation column (Figure 13.15). The lighter molecules, those containing one to four carbon atoms each, come off the top of the column. The next fraction contains mostly molecules having from five to 12 carbon atoms

Crude oil and natural gas are the liquid and gaseous components of petroleum. **Natural gas** is about 80% methane and 10% ethane, and the remaining 10% is a mixture of the higher alkanes. Methane is a product of the bacterial decay of plant and marine organisms that have become buried beneath the Earth's surface. It is also produced by the microbial decomposition of organic matter in sewage treatment plants. Natural gas is the cleanest of the fossil fuels because it contains the least quantity of sulfur compounds and thus produces little sulfur dioxide when burned.

Propane and the butanes are familiar fuels. Although they are gases at ordinary temperatures and atmospheric pressure, they are liquefied under pressure and sold as liquefied petroleum gas (LPG). Liquefied butane can be seen in disposable butane cigarette lighters.

Gasoline, generally the petroleum fraction most in demand, is a mixture of hydrocarbons, typically alkanes ranging from C_5H_{12} to $C_{12}H_{26}$. It includes many isomeric forms, particularly for the higher members of the group, small amounts of other hydrocarbons, and some sulfur- and nitrogen-containing compounds.

The gasoline fraction of petroleum as it comes from the distillation column is called *straight-run gasoline.* It doesn't burn very well in modern, high-compression automobile engines. Chemists have learned how to modify it in a variety of ways to make it burn more smoothly, that is, to boost its **octane rating.** When this rating was established, the best performing fuel in a laboratory test engine was a compound called isooctane (one of 18 isomeric octanes tested). Isooctane was assigned a value of 100 octane. The unbranched-chain compound heptane caused a very bad knock and so was given an octane rating of 0. A gasoline rated 90 octane was one that performed the same as a mixture that was 90% isooctane and 10% heptane.

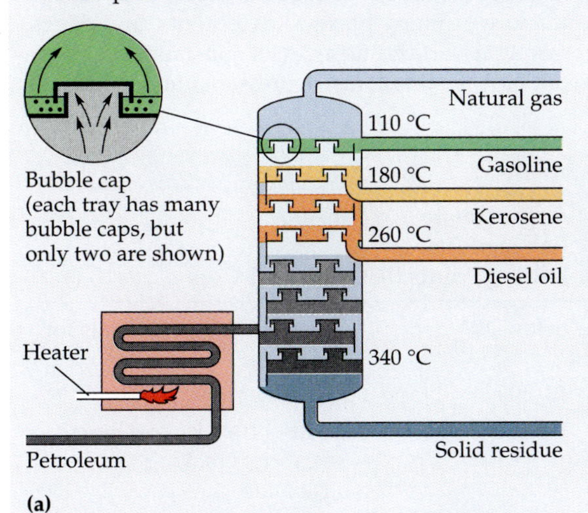

(a)

(b)

▲ **Figure 13.15**
(a) A schematic of an oil refinery distillation column. Petroleum is vaporized by heating with very hot steam at the bottom of a tall column. The components with higher boiling points condense quite low in the column, and those with lower boiling points move farther up. Fractions with different compositions are removed from the column at different heights.
(b) Petroleum distillation columns.

$$CH_3-\underset{\displaystyle CH_3}{\overset{\displaystyle CH_3}{\overset{|}{\underset{|}{C}}}}-CH_2-\overset{\displaystyle CH_3}{\overset{|}{C}H}-CH_3 \qquad CH_3CH_2CH_2CH_2CH_2CH_2CH_3$$

Isooctane Heptane

Branched-chain alkanes burn more evenly than straight-chain alkanes. Conversion of unbranched structures to highly branched ones, by heating the gasoline in the presence of catalysts such as sulfuric acid (H_2SO_4) and aluminum chloride ($AlCl_3$), improves the octane rating. Various additives have also been used to improve the antiknock quality of gasoline.

The use of tetraethyllead [$(CH_3CH_2)_4Pb$], formerly the most widely used additive, has been banned in most industrial nations because of its toxicity. To get high octane ratings in unleaded fuels, petroleum refineries use *catalytic reforming* to convert low-octane alkanes to high-octane aromatic compounds. For example, hexane (octane rating 25) is converted to benzene (106 octane).

$$\underset{(C_6H_{14})}{CH_3CH_2CH_2CH_2CH_2CH_3} \xrightarrow[\text{heat}]{\text{catalyst}} \underset{(C_6H_6)}{\text{benzene}} + 4\,H_2$$

Now, however, refineries are trying to reduce the aromatics in gasoline, and especially benzene, because of health concerns. Newer octane boosters include various alcohols and ethers, collectively called *oxygenates* because they all contain oxygen. Methyl *tert*-butyl ether, $(CH_3)_3COCH_3$, is perhaps the most important. Methanol, ethanol, and *tert*-butyl alcohol also are used. None of these is nearly as effective as tetraethyllead in boosting the octane rating. They must therefore be used in fairly large quantities. These oxygenates, however, not only improve the octane rating of gasoline, but they also decrease the concentration of carbon monoxide in the auto exhaust gas.

To obtain sufficient gasoline for modern transportation needs, petroleum fractions (Table 13.13) with higher boiling points are converted to gasoline by *cracking*, a process of heating in the absence of air. Cracking not only converts some molecules to ones in the gasoline range, but it also results in useful by-products such as unsaturated hydrocarbons, starting materials for the manufacture of many plastics, detergents, and drugs. Starting with petroleum, the chemist can create a dazzling array of substances having a wide variety of properties: painkillers, antibiotics, stimulants, depressants, and detergents to name just a few.

Table 13.13 Typical Petroleum Fractions

Fraction	Typical Range of Hydrocarbons	Approximate Range of Boiling Points (°C)	Typical Uses
Natural gas	CH_4 to C_4H_{10}	Below 40	Fuel, starting materials for plastics
Gasoline	C_5H_{12} to $C_{12}H_{26}$	40–200	Fuel, solvents
Kerosene	$C_{12}H_{26}$ to $C_{16}H_{34}$	175–275	Diesel fuel, jet fuel, home heating; cracking to gasoline
Heating oil	$C_{15}H_{32}$ to $C_{18}H_{38}$	250–400	Industrial heating; cracking to gasoline
Lubricating oil	$C_{17}H_{36}$ and up	Above 300	Lubricants
Residue	$C_{20}H_{42}$ and up	Above 350 (some decomposition)	Paraffin, asphalt, road tar, roofing tar

Summary

Organic chemistry is the chemistry of carbon compounds. Carbon atoms can form stable covalent bonds both with other carbon atoms and with atoms of other elements, and this property allows carbon to form the millions of carbon compounds we call *organic compounds.*

Carbon has four valence electrons and can form up to four covalent bonds. A carbon atom with four single bonds is a *saturated carbon.* (Joined to the maximum number of other atoms, the carbon is *saturated* with atoms.)

Hydrocarbons contain only hydrogen and carbon. Hydrocarbons in which each carbon is bonded to four other atoms are called **alkanes** or **saturated hydrocarbons.** They have the general formula C_nH_{2n+2}. Any given alkane differs from the next by one CH_2 unit. Any family of compounds in which adjacent members differ from each other by a definite factor is called a **homologous series,** and the individual members are *homologs.*

Two or more compounds having the same molecular formula but different structural formulas are **isomers** of each other. There are no isomeric forms for the three smallest alkanes, but beginning with butane (C_4H_{10}), all other alkanes have isomeric forms.

An **alkyl group** is a unit formed by removing one hydrogen atom from an alkane.

Alkanes are generally unreactive toward laboratory acids, bases, oxidizing agents, and reducing agents. They do burn (undergo combustion) and react with halogens by substitution of one or more halogen atoms for hydrogen to form **halogenated hydrocarbons.**

Cycloalkanes are hydrocarbons whose molecules are closed rings rather than straight or branched chains.

Any hydrocarbon containing either a double bond or a triple bond is an **unsaturated hydrocarbon. Alkenes** have a carbon-to-carbon double bond. The general formula for alkenes with one double bond is C_nH_{2n}. Alkenes can be straight chain, branched chain, or cyclic. **Alkynes** have a carbon-to-carbon triple bond.

More reactive than alkanes, alkenes undergo addition reactions across the double bond.

- Addition of hydrogen (**hydrogenation**): $H_2C{=}CH_2 + H_2 \longrightarrow H_3C{-}CH_3$
- Addition of halogen (**halogenation**): $H_2C{=}CH_2 + X_2 \longrightarrow X{-}CH_2{-}CH_2{-}X$, where X = F, Cl, Br, or I
- Addition of water (**hydration**): $H_2C{=}CH_2 + HOH \longrightarrow H{-}CH_2{-}CH_2{-}OH$

Alkenes also undergo **polymerization,** molecules joining together to form long-chain molecules.

$$\cdots H_2C{=}CH_2 + H_2C{=}CH_2 + H_2C{=}CH_2 + \cdots \longrightarrow \cdots CH_2CH_2{-}CH_2CH_2{-}CH_2CH_2\cdots$$

The reactant units are **monomers,** and the product is a **polymer.**

The cyclic unsaturated hydrocarbon *benzene,* C_6H_6, is an extremely stable ring, undergoing none of the reactions expected of alkenes. Because of this stability it is believed that the ring is not one of alternating single and double bonds but rather a **resonance hybrid** in which the "extra" electrons are *delocalized* over all six carbon atoms. Compounds containing one or more benzene rings (or other similarly stable resonance-hybrid units) are called *aromatic compounds.* (Any hydrocarbon not containing an aromatic unit is an **aliphatic compound.**)

One or more of the hydrogen atoms on a benzene ring can be replaced by other atoms. When two hydrogen atoms are replaced, the product name is based on the relative position of the replacement atoms (or atom groups). A 1,2-disubstituted benzene is designated an *ortho (o-)* isomer; 1,3-, a *meta (m-)* isomer; and 1,4-, a *para (p-)* isomer.

Key Terms

addition polymerization (13.10)
addition reaction (13.9)
aliphatic compound (13.13)
alkane (13.1)
alkene (13.7)
alkyl group (13.1)
alkyl halide (13.5)
alkyne (13.11)
aromatic hydrocarbon (13.13)
aryl group (13.13)
chlorofluorocarbon (CFC) (13.5)
combustion (13.4)
condensed structural formula (13.1)
halogenated hydrocarbon (13.5)
homologous series (13.1)
hydration (13.9)
hydrocarbon (13.1)
hydrogenation (13.9)
isomer (13.1)
IUPAC System of Nomenclature (13.2)
monomer (13.10)
natural gas (13.14)
octane rating (13.14)
organic chemistry (page 329)
petroleum (13.14)
polymer (13.10)
resonance (13.12)
resonance hybrid (13.12)
saturated hydrocarbon (13.1)
structural formula (13.1)
unsaturated hydrocarbon (13.7)

Problems

Organic Versus Inorganic

1. Classify each compound as organic or inorganic.
a. C_6H_{10} **b.** $CoCl_2$ **c.** $C_{12}H_{22}O_{11}$

2. Classify each compound as organic or inorganic.
a. CH_3NH_2 **b.** $NaNH_2$ **c.** $Cu(NH_3)_6Cl_2$

3. Which member of each pair has a higher melting point?
a. CH_3OH and NaOH **b.** CH_3Cl and KCl

4. Which member of each pair has a higher melting point?
a. $C_{20}H_{42}$ and $C_{40}H_{82}$ **b.** CH_4 and LiH

Alkanes: Structures and Names

5. Draw the structural formula for each of the following compounds.
   a. heptane   b. 3-methylpentane
   c. 2,2,5-trimethylhexane   d. 4-ethyl-3-methyloctane
6. Draw the structural formula for each of the following compounds.
   a. 2-methylpentane
   b. 4-ethyl-2-methylhexane
   c. 2,2,3,3-tetramethylbutane
   d. 4-ethyl-3-isopropyloctane
7. Name the following compounds by the IUPAC system.
   a. 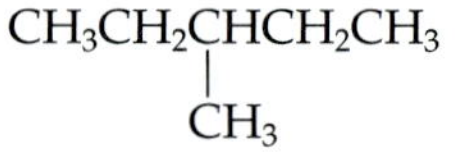 $CH_3CH_2CHCH_2CH_3$ (with CH_3 on the central CH)   b. 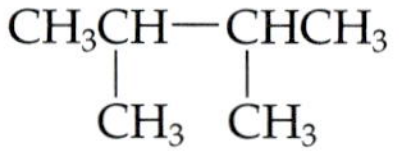 $CH_3CH{-}CHCH_3$ (with CH_3 CH_3 on each CH)
8. Name the following compounds by the IUPAC system.
   a. 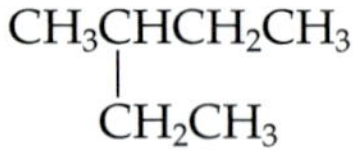 $CH_3CHCH_2CH_3$ (with CH_2CH_3 on CH)   b. 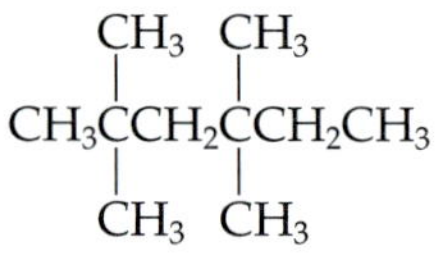 $CH_3CCH_2CCH_2CH_3$ (with CH_3 CH_3 above and CH_3 CH_3 below each C)
9. Draw the structural formulas for the following alkyl groups.
   a. ethyl   b. isopropyl
10. Draw the structural formulas for the following alkyl groups.
    a. methyl   b. propyl
11. Draw the structural formulas for both four-carbon alkanes (C_4H_{10}). Identify butane and isobutane, and give the IUPAC name for the latter.
12. Draw the structural formulas for the five isomeric hexanes (C_6H_{14}). Name each by the IUPAC system.

Alkanes: Physical Properties

13. Which member of each pair has the higher boiling point?
    a. pentane or butane
    b. $(CH_3)_2CHCH(CH_3)_2$ or $CH_3(CH_2)_4CH_3$
    c. cyclopentane or cyclohexane
    d. $CH_3(CH_2)_5CH_3$ or $CH_3(CH_2)_7CH_3$
14. Which member of each pair has the higher melting point?
    a. pentane or butane
    b. neopentane or pentane
    c. cyclopentane or cyclohexane
    d. $CH_3(CH_2)_5CH_3$ or $CH_3(CH_2)_7CH_3$

Cyclic Hydrocarbons

15. Name the following compounds by the IUPAC system.
    a. 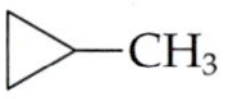 (cyclopropane)$-CH_3$ b. 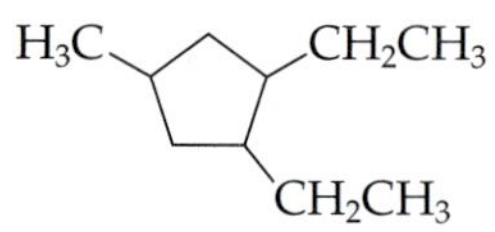 H_3C, CH_2CH_3, CH_2CH_3 (on cyclopentane ring)
    c. 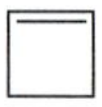 (cyclobutene)   d. 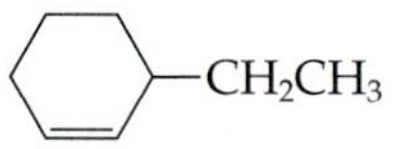 (cyclohexene)$-CH_2CH_3$
16. Name the following compounds by the IUPAC system.
    a. 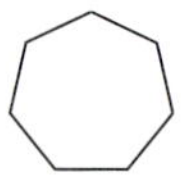 (cycloheptane)   b.  (cyclohexene)
    c. 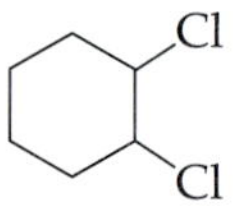 Cl, Cl (on cyclohexane ring)   d. 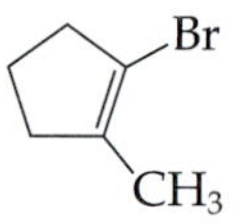 Br, CH_3 (on cyclopentene ring)   e. 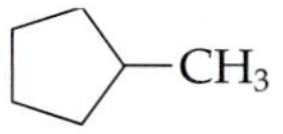 (cyclopentane)$-CH_3$
17. Draw the structural formula for each of the following compounds.
    a. ethylcyclobutane   b. 3-chlorocyclopentene
18. Draw the structural formula for each of the following compounds.
    a. 1,3-diethylcyclopentane
    b. 4-methylcyclohexene

Halogenated Hydrocarbons

19. Draw the structural formula for each of the following compounds.
    a. methyl chloride   b. chloroform
20. Draw the structural formula for each of the following compounds.
    a. ethyl bromide   b. carbon tetrachloride
21. Draw the structural formulas for the two isomers that have the molecular formula C_3H_7Br. Give the common name and the IUPAC name of each.
22. Draw the structural formulas for the four isomers that have the molecular formula C_4H_9Br. Give the IUPAC name of each.

Saturated Versus Unsaturated

23. Classify each of the following as saturated or unsaturated.
    a. 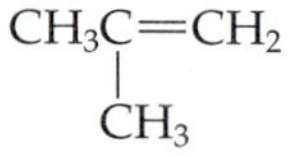 $CH_3C{=}CH_2$ (with CH_3 on C)
    b. $CH_3C{\equiv}CCH_3$ c. $(CH_3)_2CHCl$
24. Classify each of the following as saturated or unsaturated.
    a. 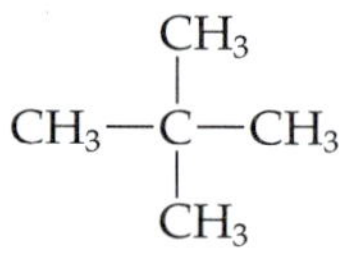 $CH_3{-}C{-}CH_3$ (with CH_3 above and CH_3 below C)   b. 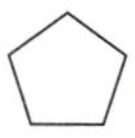 (cyclopentane)
    c. $(CH_3)_2C{=}CHCH_3$

Alkenes and Alkynes

25. Draw the structural formula for each of the following compounds.
    a. acetylene   b. cyclohexene
    c. 3-isopropyl-1-hexyne   d. 2,3-dimethyl-2-butene
26. Draw the structural formula for each of the following compounds.
    a. 1,2-dimethylcyclobutene   b. 3-ethyl-2-pentene
    c. cyclooctyne   d. 4-methyl-2-hexene
27. Draw the structural formula for each of the following compounds.
    a. 2-methyl-2-pentene   b. 5-methyl-1-hexene

28. Draw the structural formula for each of the following compounds.
a. 2-ethyl-1-butene **b.** 2,4,6,6-tetramethyl-2-heptene

29. Name the following compounds by the IUPAC system.

a. $CH_2{=}CCH_2CH_2CH_3$
$\quad\quad |$
$\quad\;\; CH_3$

b. $CH_3CH_2CH{=}CCH_3$
$\quad\quad\quad\quad\quad\quad |$
$\quad\quad\quad\quad\quad CH_3$

c. $CH_3C{=}CHCH_2CHCH_3$
$\quad\quad |\quad\quad\quad\quad\; |$
$\quad\; CH_3\quad\quad\;\; CH_3$

30. Name the following compounds by the IUPAC system.

a. $CH_3C{=}CHCH_3$
$\quad\quad |$
$\quad\; CH_3$

b. $CH_3CHClCH_2CH{=}CHCH_2CHCl_2$

c.
$\quad\; CH_3$
$\quad\quad |$
$CH_3C{-}CH{=}C{-}CH_2CH_3$
$\quad\quad |\quad\quad\quad |$
$\quad\; CH_3\quad\;\; CH_3$

Isomers

31. Indicate whether the structures in each set represent the same compound or isomers.

a. CH_3CH_3 and CH_3
$\quad\quad\quad\quad\quad\quad\quad |$
$\quad\quad\quad\quad\quad\quad CH_3$

b.
$\quad\; CH_3$
$\quad\quad |$
CH_3CH_2 and $CH_3CH_2CH_3$

c.
$\quad\quad\quad\; CH_3\; CH_3\quad\quad\quad CH_3\quad\quad\; CH_3$
$\quad\quad\quad\quad |\quad\quad |\quad\quad\quad\quad\; |\quad\quad\quad\quad |$
$CH_3CH_2CH{-}CH_2$ and $CH_2CH_2CHCH_3$

32. Indicate whether the structures in each set represent the same compound or isomers.

a.
$\quad\quad\quad\; CH_3\quad\quad\quad\quad\quad\quad\quad CH_3$
$\quad\quad\quad\quad |\quad\quad\quad\quad\quad\quad\quad\quad |$
$CH_3CH_2CHCH_2CH_3$ and $CH_3CHCH_2CH_2CH_3$

b.
$\quad\; CH_3\quad\quad\quad\quad\quad\quad\quad\quad\quad\quad CH_3$
$\quad\quad |\quad\quad\quad\quad\quad\quad\quad\quad\quad\quad\quad |$
$CH_3CHCH_2CH_3$ and CH_3CH_2CH
$\quad\quad\quad\quad\quad\quad\quad\quad\quad\quad\quad\quad\quad\quad\quad\; |$
$\quad\quad\quad\quad\quad\quad\quad\quad\quad\quad\quad\quad\quad\quad CH_3$

33. Indicate whether the structures in each set represent the same compound or isomers.

a. $CH_3CH{=}CHCH_3$ and $CH_3CH_2CH{=}CH_2$

b.
$\quad\; CH_3\; CH_3$
$\quad\quad |\quad\quad |$
$CH_3C{=}CCH_3$ and $CH_3C{=}CCH_3$
$\quad\quad\quad\quad\quad\quad\quad\quad\quad\quad\quad\; |\quad\quad |$
$\quad\quad\quad\quad\quad\quad\quad\quad\quad\quad\; CH_3\; CH_3$

c.
$\quad\quad\quad\quad\quad\quad\quad\quad\quad\quad CH_3$
$\quad\quad\quad\quad\quad\quad\quad\quad\quad\quad\quad |$
$CH_3CH_2CH_2CH{=}CCH_3$ and
$\quad\quad CH_3$
$\quad\quad\quad |$
$CH_3CHCH{=}CHCH_2CH_3$

34. Indicate whether the structures in each set represent the same compound or isomers.

a.
$\quad\quad\quad CH_3\quad\quad\quad\quad\quad\quad\quad\quad\quad\quad\quad CH_2CH_3$
$\quad\quad\quad\quad |\quad\quad\quad\quad\quad\quad\quad\quad\quad\quad\quad\quad\; |$
$CH_2{=}CCH_2CH_3$ and $CH_2{=}CCH_3$

b.
$CH_3\quad\quad\quad CH_2CH_3$
$\quad\quad \diagdown\quad \diagup$
$\quad\quad\; C{=}C$
$\quad\quad \diagup\quad \diagdown$
$CH_3CH_2\quad\quad CH_2CH_3$

and

$CH_3CH_2\quad\quad CH_2CH_3$
$\quad\quad \diagdown\quad \diagup$
$\quad\quad\; C{=}C$
$\quad\quad \diagup\quad \diagdown$
$CH_3\quad\quad\quad CH_2CH_3$

Chemical Reactions of Alkenes

35. Complete the following equations.

a. $(CH_3)_2C{=}CH_2 + Br_2 \longrightarrow$

b. $CH_2{=}C(CH_3)CH_2CH_3 + H_2 \xrightarrow{Ni}$

c. [cyclobutene ring bearing $-CH_2CH_3$ on each alkene carbon] $\xrightarrow[H_2SO_4]{H_2O}$

$-CH_2CH_3$

CH_2CH_3

36. Complete the following equations.

a. $CH_2{=}CHCH{=}CH_2 + 2\,H_2 \xrightarrow{Ni}$

b. $(CH_3)_2C{=}C(CH_3)_2 \xrightarrow[H_2SO_4]{H_2O}$

c. [cyclobutene ring]$-CH_3 + Cl_2 \longrightarrow$

37. Give the reagents required for the following transformations.

a. $H_2C{=}CHCH_3 \longrightarrow CH_3CH_2CH_3$

b. [cyclopentene] $\longrightarrow$ [cyclopentanol]

OH

38. Give the reagents required for the following transformations.

a.
$\quad\quad\quad\quad\quad\quad\quad\quad\quad\quad\quad\quad\quad\quad\quad\; OH$
$\quad\quad\quad\quad\quad\quad\quad\quad\quad\quad\quad\quad\quad\quad\quad\quad |$
$CH_3CH{=}CHCH_3 \longrightarrow CH_3CHCH_2CH_3$

b.
$\quad\quad\quad\quad\quad\quad\quad\quad\quad\quad\quad\quad\; Cl\quad Cl$
$\quad\quad\quad\quad\quad\quad\quad\quad\quad\quad\quad\quad\;\; |\quad\;\; |$
$CH_2{=}CHCH_3 \longrightarrow CH_2CHCH_3$

39. List the starting materials required to complete the following transformations.

a. $? \xrightarrow[Ni]{H_2}$ [cyclohexane]

b. 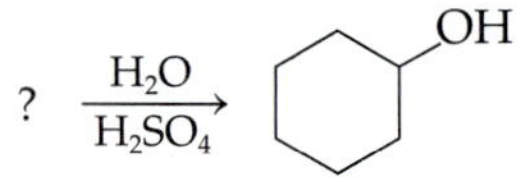 ? $\xrightarrow[H_2SO_4]{H_2O}$ (cyclohexanol, OH)

40. List the starting materials required to complete the following transformations.

a. 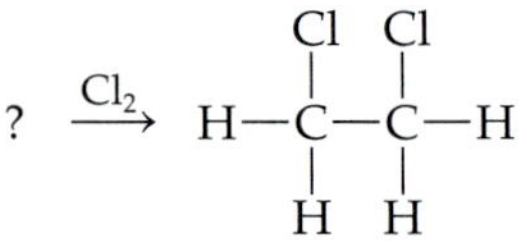 ? $\xrightarrow{Cl_2}$ H—C(Cl)(H)—C(Cl)(H)—H

b. 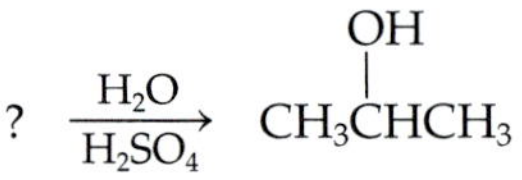 ? $\xrightarrow[H_2SO_4]{H_2O}$ $CH_3CH(OH)CH_3$

Aromatic Compounds

41. Draw the structural formula for each of the following compounds.
   a. toluene b. *m*-diethylbenzene
   c. 2,4-dinitrotoluene
42. Draw the structural formula for each of the following compounds.
   a. *p*-dichlorobenzene b. naphthalene
   c. 1,2,4-trimethylbenzene
43. Name the following compounds by the IUPAC system.

a. 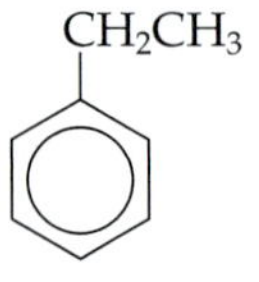 CH_2CH_3 b. 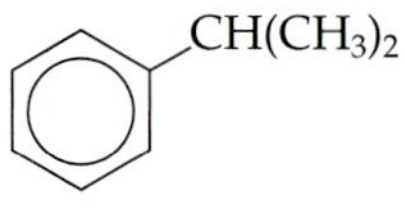 $CH(CH_3)_2$

c. 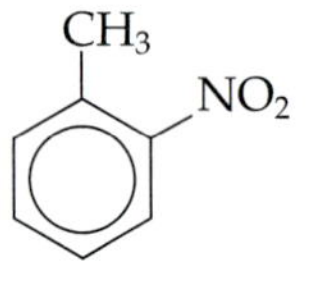 CH_3, NO_2 d. 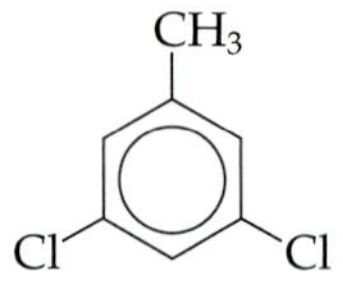 CH_3, Cl, Cl

44. Name the following compounds by the IUPAC system.

a. 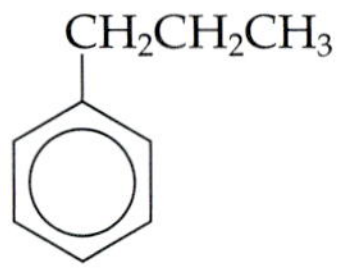 $CH_2CH_2CH_3$ b. 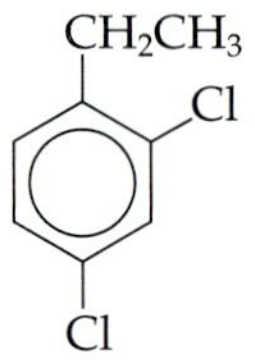 CH_2CH_3, Cl, Cl

c. 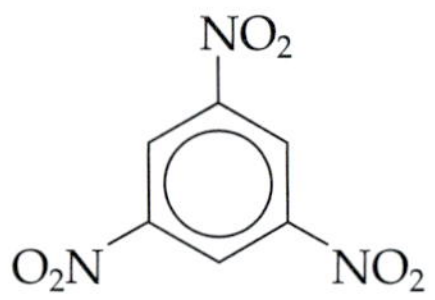 NO_2, O_2N, NO_2 d. 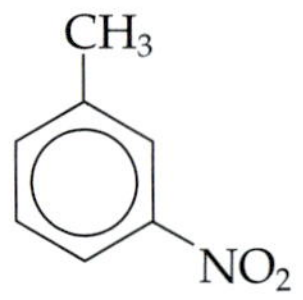 CH_3, NO_2

Additional Problems

45. What kinds of compounds make up the bulk of petroleum?
46. What is straight-run gasoline? List three ways to increase the octane rating of gasoline.
47. What is the main component of natural gas? What other compounds are present?
48. What chemical change occurs during catalytic reforming?
49. List five products made from petroleum.
50. How is natural gas formed in nature? What are some advantages of natural gas as a fuel?
51. You find an unlabeled jar containing a solid that melts at 48 °C. It ignites readily and burns cleanly. The substance is insoluble in water and floats on the surface of the water. Is the substance likely to be organic or inorganic?
52. Write the molecular formula for each of the following.

a. 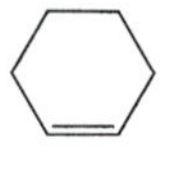 b. 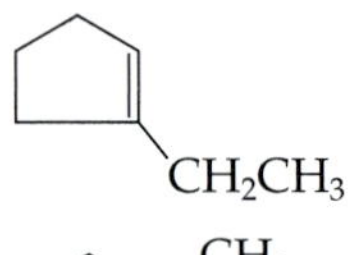 CH_2CH_3

c. 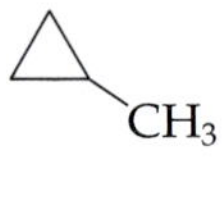 CH_3 d. 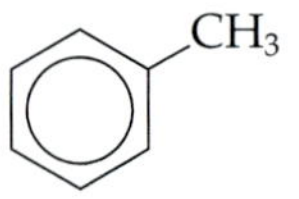 CH_3

53. Three isomeric pentenes, X, Y, and Z, can be hydrogenated to 2-methylbutane. Addition of chlorine to Y gives 1,2-dichloro-3-methylbutane and addition of chlorine to Z gives 1,2-dichloro-2-methylbutane. Write the structural formulas for the three isomers.
54. Pentane and 1-pentene are both colorless, low-boiling liquids. Give a simple test that distinguishes the two compounds. Indicate what you would observe.
55. What is wrong with each of the following names? Draw the structural formula and give the correct name for each compound.
   a. 2-dimethylpropane
   b. 2,3,3-trimethylbutane
   c. 2,4-diethylpentane
   d. 3,4-dimethyl-5-propylhexane
56. What is wrong with each of the following names? Draw the structural formula and give the correct name for each compound.
   a. 2-methyl-4-heptene b. 2-ethyl-3-hexene
   c. 2,2-dimethyl-3-pentene d. 4-bromocyclobutene
57. Write equations for the complete combustion of each of the following.
   a. natural gas (methane)
   b. a typical petroleum hydrocarbon (such as octane)
58. The complete combustion of benzene forms carbon dioxide and water:

$$C_6H_6 + O_2 \longrightarrow CO_2 + H_2O$$

Balance the equation. What mass of carbon dioxide is formed by the complete combustion of 39.0 g of benzene?
59. The density of a gasoline sample is 0.690 g/mL. On the basis of the complete combustion of octane, calculate the amount of carbon dioxide and water formed per gallon (3.78 L) of the gasoline when used in an automobile.

Alcohols, Phenols, and Ethers

Ethanol, the most familiar alcohol and the one found in alcoholic beverages, is made by fermentation of starches and sugars from various sources. Grapes are the principal source of the ethanol found in wine.

Learning Objectives/Study Questions

1. What is the general structure for an alcohol? A phenol? An ether?
2. What are functional groups? Why are they useful in the study of organic chemistry?
3. What structural feature is used to classify alcohols as primary, secondary, or tertiary?
4. How are alcohols named by the common and IUPAC systems?
5. Why are the boiling points of alcohols higher than those of ethers and alkanes of similar molar masses?
6. Why are alcohols and ethers of four carbons or less soluble in water while comparable alkanes are not?
7. How are alcohols prepared from alkenes? What is Markovnikov's rule?
8. How do various alcohols affect the human body?
9. What are the major reactions of alcohols?
10. What product is formed by the oxidation of a primary alcohol? A secondary alcohol? A tertiary alcohol?
11. Describe the structure and uses of some common polyhydric alcohols.
12. Describe the structure and uses of some phenols.
13. How does the structural difference between alcohols and ethers affect their physical characteristics and reactivity?
14. How are simple ethers named? Describe the structure and uses of some ethers.

The three families of organic compounds discussed in this chapter occur widely in nature. Humans have been quick to adapt some members of these families to their own use. The earliest written histories record that primitive peoples used the compound we know as alcohol. According to Genesis, Noah planted a vineyard after the Flood, drank wine from its grapes, and became drunk.

Human ingenuity may have reached some sort of peak in finding sources of *aqua vitae*, the water of life. Alcohol has been obtained from the fermentation of fruits, grains, potatoes, rice, and even cacti. It was prescribed as medicine in the twelfth century but has been most frequently used without such justification. What we know as alcohol is actually only one member of a family of organic compounds known by that name. The family also includes such familiar substances as cholesterol and the carbohydrates.

The pungent, antiseptic odor of phenol, the simplest member of another family of compounds we will introduce in this chapter, is likely to be familiar to anyone who has ever been in a hospital. And the name of the third family considered in this chapter, the ethers, has become almost synonymous with anesthesia.

14.1 General Formulas and Functional Groups

We discuss three families in one chapter for several reasons. Two of the families, alcohols and phenols, feature a hydroxyl (OH) group. Members of the third family, the ethers, are often made from alcohols and phenols. Further, the members of all three families can be considered organic derivatives of water.

General Formulas

The water molecule is bent, with a central oxygen atom attached to two hydrogen atoms. Replacing one hydrogen atom with an alkyl group, for which we use the symbol R, gives the general formula for the alcohol family.

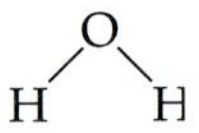

H–O–H (Water)   R–O–H (An alcohol)

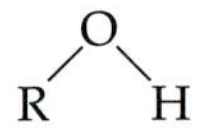

As shown in Figure 14.1(a), the alkyl group (R) may be methyl, ethyl, isopropyl, or even an aliphatic group too complicated to have a simple name. In any case, as long as the carbon attached to the **hydroxyl group (—OH)** is aliphatic, the compound is an **alcohol** and the general formula is ROH.

If the hydroxyl group is attached directly to an aromatic ring, properties are different enough that the compounds are considered to be a different family, called *phenols*, with the general formula ArOH (Figure 14.1(b)). Recall that Ar is an aryl group (Section 13.13).

The phenol family is discussed in Section 14.8.

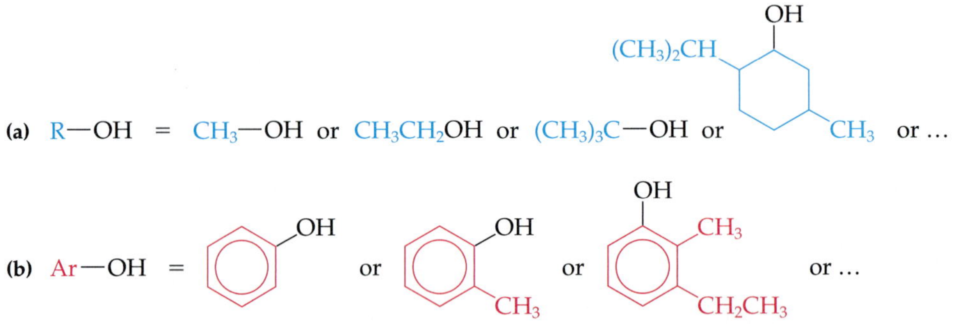

Figure 14.1 ▶ In the notation of organic chemistry, R represents any of a great variety of alkyl groups (a). Similarly, Ar represents any aryl (aromatic) group (b).

(a) R—OH = CH_3—OH or CH_3CH_2OH or $(CH_3)_3C$—OH or [cyclohexane ring bearing OH, $(CH_3)_2CH$ and CH_3] or ...

(b) Ar—OH = [benzene ring with OH] or [benzene ring with OH and CH_3] or [benzene ring with OH, CH_3 and CH_2CH_3] or ...

Ar—OH

A phenol

Replacement of both hydrogen atoms of water by alkyl or aryl groups gives rise to an *ether*.

The ether family is discussed in Section 14.9.

R—O—R'   R—O—Ar   Ar—O—Ar''

Ethers

Functional Groups

Phenols and alcohols are separate, if closely related, families. Nonetheless, both families have characteristic properties—both chemical and physical—largely determined by the hydroxyl group. Most of the typical reactions of alcohols and many of those of phenols take place at the hydroxyl group. Such a group of atoms, which confers characteristic properties on a family of organic compounds, is called a **functional group.**

The hydroxyl group of alcohols and phenols is one functional group. Most ethers are characterized by an alkoxy group (—OR). Because the —OR is bonded to another carbon group, the functional group of an ether is simply an oxygen atom bonded to two carbon atoms. We have already encountered other groups often considered to be functional groups, such as the carbon–carbon double bond (C═C) in alkenes and the carbon–carbon triple bond (C≡C) in alkynes. In both families, the multiple bond confers a particular chemical reactivity on the members: Both alkenes and alkynes undergo addition reactions.

The alkanes are characterized by their *lack* of a distinct functional group. Other functional groups will serve as unifying concepts for the next three chapters. Some of the more important functional groups are listed in Table 14.1. For ready reference, this table is also reproduced on the inside back cover.

Review Questions

14.1 What is a functional group? Give the structure of and name the functional group in (**a**) alkenes, (**b**) alcohols, and (**c**) ethers.

14.2 How are the structures of alcohols and phenols alike? How are they different?

14.2 Classification and Nomenclature of Alcohols

Some of the properties of alcohols depend on the arrangement of the carbon atoms in the molecule. Alcohols can be grouped into three classes based on their different structural arrangements. The classes are determined by the other atoms or groups attached to the carbon atom to which the hydroxyl group is attached.

1. A *primary (1°) carbon atom* is one attached directly either to no other carbons or to one other carbon:

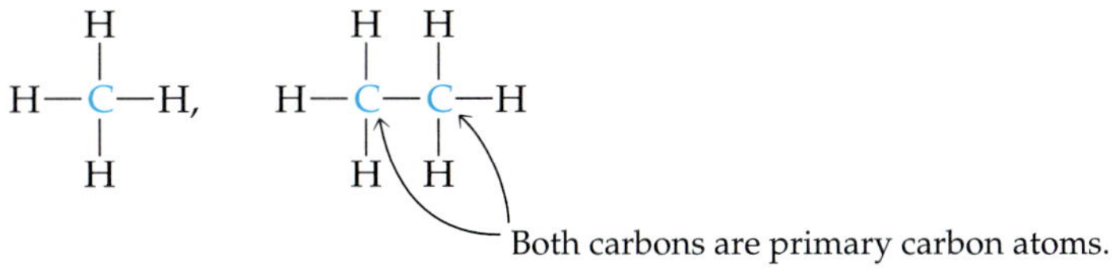

```
    H            H  H
    |            |  |
 H—C—H,      H—C—C—H
    |            |  |
    H            H  H
```

Both carbons are primary carbon atoms.

2. A *secondary (2°) carbon atom* is one attached to two other carbon atoms:

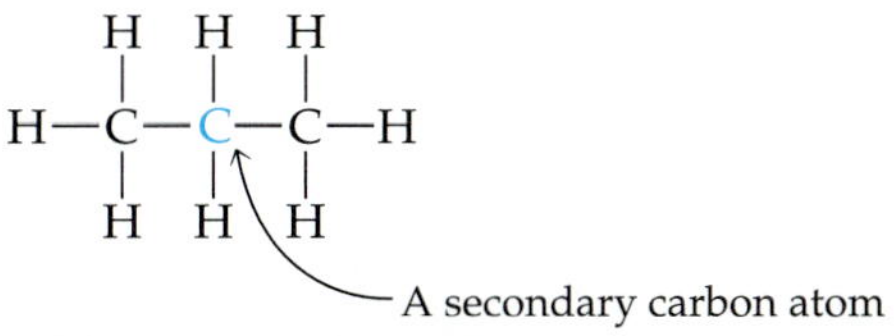

```
     H  H  H
     |  |  |
  H—C—C—C—H
     |  |  |
     H  H  H
```

A secondary carbon atom

Table 14.1 Selected Organic Functional Groups

Name of Family	Functional Group	General Formula(s) of Family
Alkane	None	R—H
Alkene	$>C=C<$	$R_2C=CR_2$
Alkyne	—C≡C—	R—C≡C—R
Alcohol	$-\overset{\vert}{\underset{\vert}{C}}-O-H$	R—O—H
Phenol	C_6H_5—O—H (benzene ring—O—H)	Ar —O—H
Ether	$-\overset{\vert}{\underset{\vert}{C}}-O-\overset{\vert}{\underset{\vert}{C}}-$	R—O—R
Aldehyde	$-\overset{O}{\overset{\Vert}{C}}-H$	$R-\overset{O}{\overset{\Vert}{C}}-H$
Ketone	$-\overset{O}{\overset{\Vert}{C}}-$	$R-\overset{O}{\overset{\Vert}{C}}-R$
Amine	$-\overset{\vert}{\underset{\vert}{C}}-\underset{\vert}{N}-$	$R-\overset{H}{\overset{\vert}{N}}-H$ $R-\overset{H}{\overset{\vert}{N}}-R$ $R-\overset{R}{\overset{\vert}{N}}-R$
Carboxylic acid	$-\overset{O}{\overset{\Vert}{C}}-O-H$	$R-\overset{O}{\overset{\Vert}{C}}-O-H$
Ester	$-\overset{O}{\overset{\Vert}{C}}-O-\overset{\vert}{\underset{\vert}{C}}-$	$R-\overset{O}{\overset{\Vert}{C}}-O-R$
Amide	$-\overset{O}{\overset{\Vert}{C}}-\underset{\vert}{N}-$	$R-\overset{O}{\overset{\Vert}{C}}-\underset{H}{\underset{\vert}{N}}-H$ $R-\overset{O}{\overset{\Vert}{C}}-\underset{H}{\underset{\vert}{N}}-R$ $R-\overset{O}{\overset{\Vert}{C}}-\underset{R}{\underset{\vert}{N}}-R$

3. A *tertiary (3°) carbon atom* is one attached to three other carbon atoms:

$$H-\overset{H}{\overset{\vert}{\underset{H}{\underset{\vert}{C}}}}-\overset{H-\overset{H}{\overset{\vert}{C}}-H}{\overset{\vert}{\underset{H}{\underset{\vert}{C}}}}-\overset{H}{\overset{\vert}{\underset{H}{\underset{\vert}{C}}}}-H$$

A tertiary carbon atom

We can therefore classify alcohols as follows:

- A **primary (1°) alcohol, RCH_2OH,** is one in which the hydroxyl group replaces one of the hydrogen atoms on a primary carbon.

Table 14.2 Classification and Nomenclature of Some Alcohols

Structural Formula	Class of Alcohol	Common Name	IUPAC Name
CH_3OH	Primary	Methyl alcohol	Methanol
CH_3CH_2OH	Primary	Ethyl alcohol	Ethanol
$CH_3CH_2CH_2OH$	Primary	Propyl alcohol	1-Propanol
$CH_3CH(OH)CH_3$	Secondary	Isopropyl alcohol	2-Propanol
$CH_3CH_2CH_2CH_2OH$	Primary	Butyl alcohol	1-Butanol
$CH_3CH_2CH(OH)CH_3$	Secondary	*sec*-Butyl alcohol	2-Butanol
$CH_3CH(CH_3)CH_2OH$	Primary	Isobutyl alcohol	2-Methyl-1-propanol
$CH_3—C(OH)(CH_3)—CH_3$	Tertiary	*tert*-Butyl alcohol	2-Methyl-2-propanol
(cyclohexane ring)—OH	Secondary	Cyclohexyl alcohol	Cyclohexanol
(benzene ring)—CH_2OH	Primary	Benzyl alcohol	Phenylmethanol

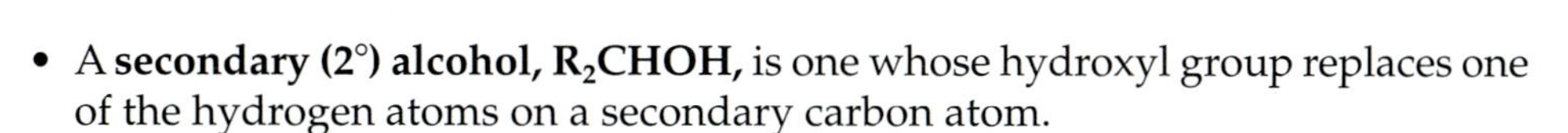

- A **secondary (2°) alcohol, R_2CHOH,** is one whose hydroxyl group replaces one of the hydrogen atoms on a secondary carbon atom.
- A **tertiary (3°) alcohol, R_3COH,** is one whose hydroxyl group replaces the hydrogen on a tertiary carbon.

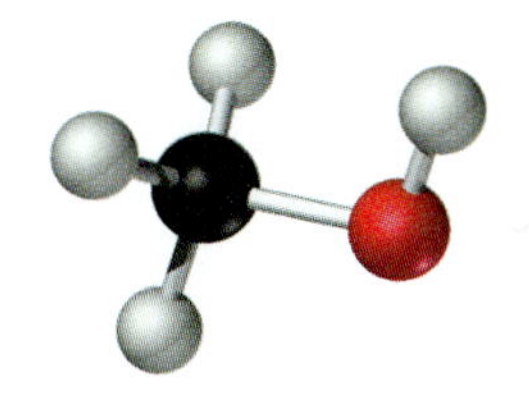

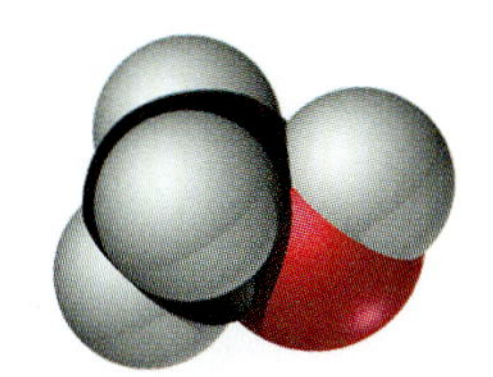

Methanol (CH_3OH)

▲ Ball-and-stick and space-filling models of methanol.

Table 14.2 presents the nomenclature and classification of some of the simpler alcohols.

As shown in the table, the common names of the lower members of the alcohol family are formed in a manner similar to that used for alkyl halides. The name of the alkyl group is followed by the word *alcohol* to indicate the presence of the hydroxyl group. The designations primary, secondary, and tertiary are not used in the IUPAC system of naming alcohols. Following are the IUPAC rules for naming alcohols.

1. The longest continuous chain (LCC) of carbons containing the —OH group is taken as the parent compound. The alcohol is named as a derivative of the alkane with the same number of carbon atoms. The chain is numbered from the end closer to the hydroxyl group.
2. The number that indicates the position of the hydroxyl group is prefixed to the name of the parent hydrocarbon, and the *-e* ending of the parent alkane is replaced by the suffix *-ol.* (When the hydroxyl group is on carbon 1 of a ring, the "1" is not needed in the name.) Substituents are named and numbered as usual.

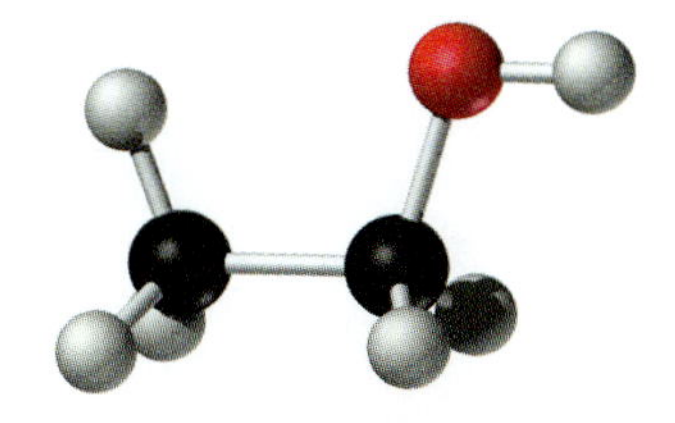

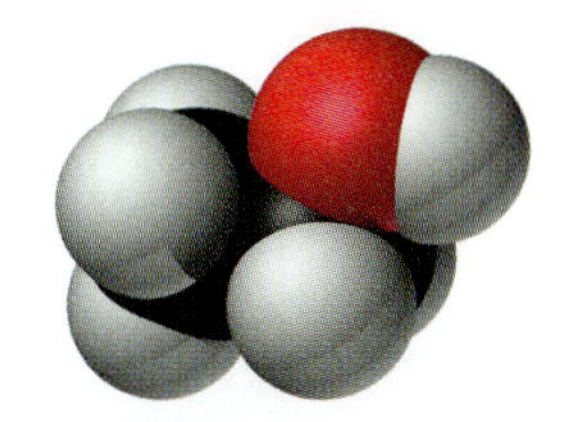

Ethanol (CH_3CH_2OH)

▲ Ball-and-stick and space-filling models of ethanol.

3. If more than one hydroxyl group appears in the same molecule (polyhydroxy alcohols), the suffixes *-diol*, *-triol*, etc., are used. In these cases, the *-e* ending of the parent alkane is retained.

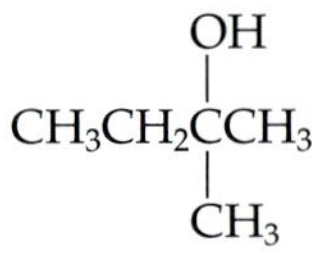

$CH_3CH_2C(OH)(CH_3)CH_3$
2-Methyl-2-butanol

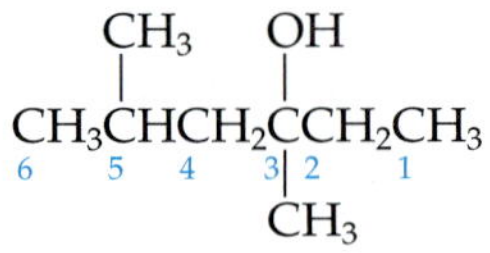

$CH_3CH(CH_3)CH_2C(OH)(CH_3)CH_2CH_3$ (numbered 6 5 4 3 2 1)
3,5-Dimethyl-3-hexanol

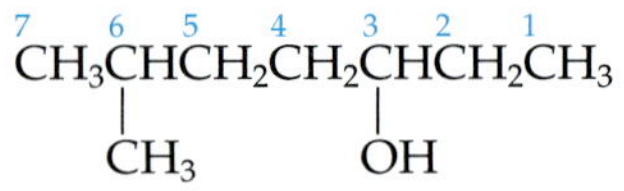

$CH_3CH(CH_3)CH_2CH_2CH(OH)CH_2CH_3$ (numbered 7 6 5 4 3 2 1)
6-Methyl-3-heptanol

Rules 1 and 2

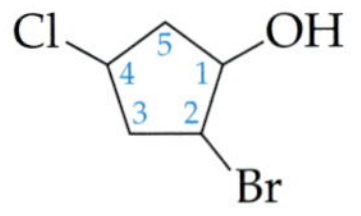

Cyclopentane ring (numbered 1–5) with Cl at C-4, OH at C-1, Br at C-2
2-Bromo-4-chlorocyclopentanol

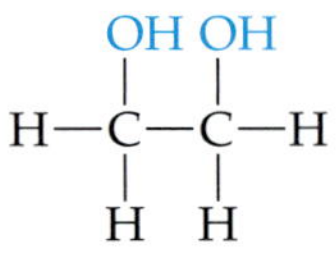

Rule 2

$HOCH_2CH_2OH$ (H—C(OH)(H)—C(OH)(H)—H)
1,2-Ethanediol
(Ethylene glycol)

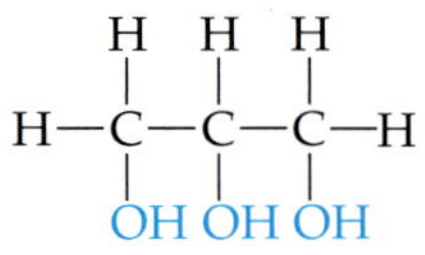

$HOCH_2CH(OH)CH_2OH$ (H—C(H)(OH)—C(H)(OH)—C(H)(OH)—H)
1,2,3-Propanetriol
(Glycerol)

Rule 3

Example 14.1

Give the IUPAC name for each of the following compounds.

a. 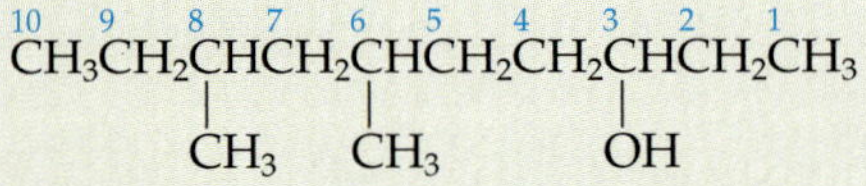 $CH_3CH_2CH(CH_3)CH_2CH(CH_3)CH_2CH_2CH(OH)CH_2CH_3$ (numbered 10 9 8 7 6 5 4 3 2 1)

b. $HOCH_2CH_2CH_2CH_2CH_2OH$

Solution

a. Ten carbon atoms in the LCC makes the compound a derivative of decane (rule **1**), and the OH on the third carbon makes it a 3-decanol (rule **2**). The carbons are numbered from the end closer to the –OH group. That fixes the two methyl groups at the sixth and eighth position. The name is 6,8-dimethyl-3-decanol (not 3,5-dimethyl-8-decanol).
b. Five carbon atoms in the LCC makes the compound a derivative of pentane. Two OH groups on the first and fifth carbon atoms gives the name 1,5-pentanediol (rule **3**).

Practice Exercise

Give the IUPAC name for each of the following compounds.

a. $CH_3CHOHCH_2OH$

b. 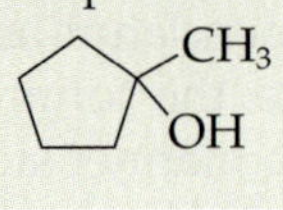 Cyclopentane ring with CH_3 and OH on the same carbon

14.3 Physical Properties of Alcohols

As we noted earlier, alcohols can be considered derivatives of water. This relationship becomes particularly apparent, especially for the lower homologs, when we discuss the physical and chemical properties of the alcohols. Replacement of a single hydrogen atom with a hydroxyl group greatly changes solubility and physical state, enabling the molecules to associate through hydrogen bonding (Chapter 7). Alcohols can form hydrogen bonds with water molecules as well, and so the lower homologs of the series are water soluble. Remember that the alkanes methane, ethane, and propane are gases and insoluble in water. In contrast, the alcohols methanol, ethanol, 1-propanol, and 2-propanol are liquids and are completely soluble in water.

Table 14.3 lists the molar masses and the boiling points of some common compounds. The table shows that substances with similar molar masses do not always have similar boiling points. The relatively high boiling points of alcohols result from strong intermolecular attractions. Recall that boiling point is a rough measure of the amount of energy necessary to separate a liquid molecule from its nearest neighbors. If the nearest neighbors are associated with that molecule through hydrogen bonds, a considerable amount of energy must be supplied to break those bonds. Only then can the molecule escape from the liquid into the gaseous state.

Figure 14.2 illustrates hydrogen bonding in water and in an alcohol. This diagram reveals why water boils at a higher temperature than methyl alcohol, even though water is a lighter molecule. The oxygen atom and both hydrogen atoms of the water molecule can participate in hydrogen bonding with up to four adjacent water molecules. Because the alkyl group of an alcohol molecule does not participate in hydrogen bonding, the molecule is associated with at most three other alcohol molecules. More energy is required to disrupt the greater number of intermolecular bonds in water than the fewer in methanol, and thus greater energy is needed to vaporize the water. (The energy required to break a hydrogen bond is about 5 kcal/mol. Although this is distinctly less than the energy required to break any of the intramolecular bonds in water or alcohol, it is still an appreciable amount of energy. Its significance is evident in the differences in boiling points.)

Polarity and hydrogen bonding are significant factors in the water solubility of alcohols. A common expression among chemists is "like dissolves like," implying that polar solvents dissolve polar solutes, and nonpolar solvents dissolve nonpolar solutes. Care must be taken, however, not to apply this generalization to all cases. All alcohol molecules are polar, yet not all are water soluble. Only the lower homologs of the alcohol family have an appreciable solubility in water. As the length of the carbon chain increases, water solubility decreases. On the other hand, all alcohols are soluble in most common nonpolar solvents such as toluene or hexane.

The differences in water solubility can be explained in the following manner. The hydroxyl group confers polarity and water solubility upon the alcohol molecule. The alkyl group confers nonpolarity and water insolubility. Whenever the hydroxyl group represents a substantial portion of a molecule, the molecule is water soluble. As the size of the alkyl group increases, the alcohols become more like alkanes and less like water; they become less soluble in water. Decyl alcohol ($CH_3CH_2CH_2CH_2CH_2CH_2CH_2CH_2CH_2CH_2OH$) is insoluble in water. The hydroxyl group's ability to form hydrogen bonds is almost totally overshadowed by the lack of attraction between water molecules and the long alkane chain. (Figure 14.3).

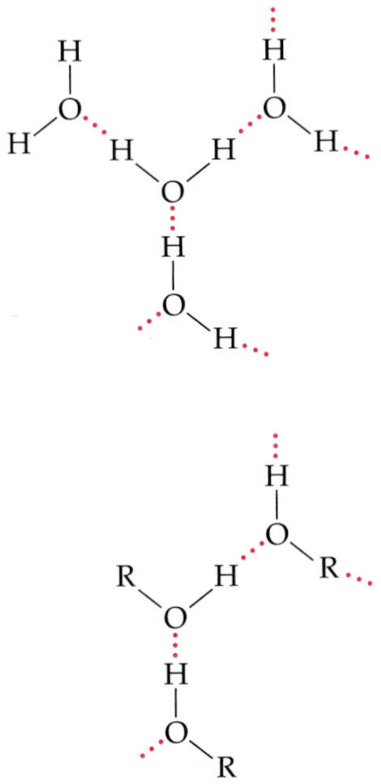

Table 14.3 Comparison of Boiling Points and Molar Masses

Formula	Name	Molar Mass	Boiling Point (°C)
CH_4	Methane	16	−164
HOH	Water	18	100
C_2H_6	Ethane	30	−89
CH_3OH	Methanol	32	65
C_3H_8	Propane	44	−42
CH_3CH_2OH	Ethanol	46	78
C_4H_{10}	Butane	58	−1
$CH_3CH_2CH_2OH$	1-Propanol	60	97

▲ **Figure 14.2** Intermolecular hydrogen bonding (top) in water and (bottom) in an alcohol.

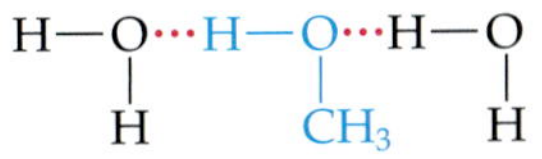

▲ Hydrogen bonding in a methanol-water mixture.

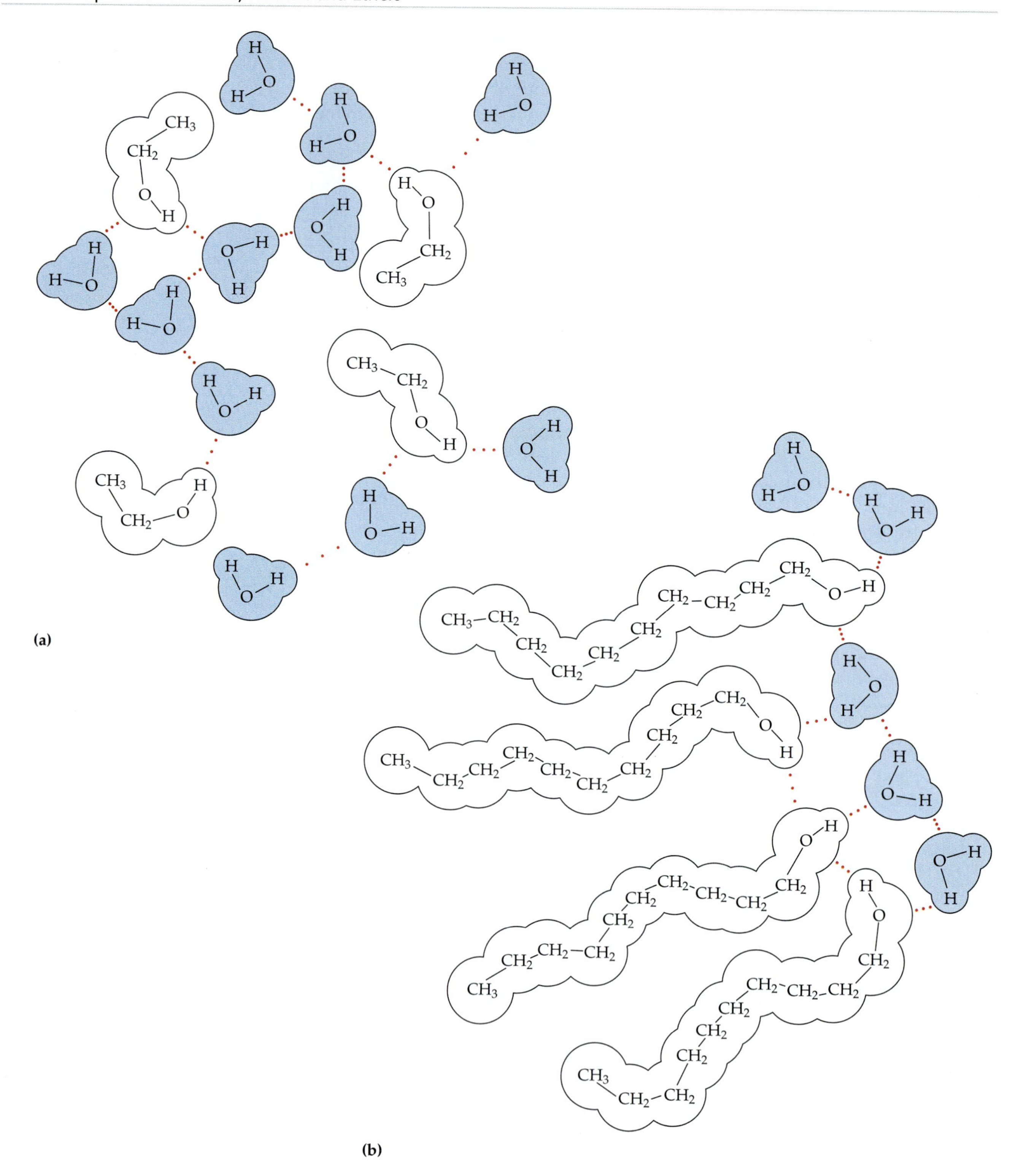

▲ **Figure 14.3**
(a) Hydrogen bonding between ethanol molecules and water molecules accounts for the solubility of ethanol in water. (b) Water molecules interact with 1-decanol molecules only near the hydroxyl end. The water molecules are unable to surround the 1-decanol molecules, and therefore 1-decanol is insoluble in water.

Table 14.4 Solubilities of the Butyl Alcohols in Water

Name IUPAC (Common)	Formula	Solubility (g/100 g H_2O)
1-Butanol (Butyl alcohol)	$CH_3CH_2CH_2CH_2OH$	8
2-Methyl-1-propanol (Isobutyl alcohol)	$(CH_3)_2CHCH_2OH$	11
2-Butanol (*sec*-Butyl alcohol)	$CH_3CH_2CH(OH)CH_3$	12.5
2-Methyl-2-propanol (*tert*-Butyl alcohol)	$(CH_3)_3COH$	Completely soluble

Table 14.4 lists the solubilities of the butyl alcohols in water. The four isomers have the same molar mass and the same functional group. The large differences in the water solubilities of these alcohols must therefore be attributed to some other factor. Actually, they result from the different geometric shapes of the molecules. The compact *tert*-butyl alcohol molecules experience weaker intermolecular attractions and therefore are more easily surrounded by water molecules. Hence *tert*-butyl alcohol has a lower boiling point (83 °C) than any of its isomers (all of which boil above 100 °C) and a higher solubility in water.

In summary, solubility depends on molecular shape as well as the molar mass and the balance of polar and nonpolar groups within a molecule. The more polar a molecule and the more compact its shape, the greater its water solubility. Molecules that can effectively form hydrogen bonds to water dissolve in water. Each functional group, such as the hydroxyl, that can form hydrogen bonds to water can carry into solution an alkyl group of up to four or five carbon atoms. Thus we frequently find that the borderline of water solubility in a family of organic compounds occurs at four or five carbon atoms.

Review Questions

14.3 Why is methanol more soluble in water than 1-hexanol?

14.4 Why does *tert*-butyl alcohol have a lower boiling point than 1-hexanol? Than 1-butanol?

14.4 Preparation of Alcohols

Many simple alcohols are made by the hydration of alkenes (Section 13.9). Ethanol is made by the hydration of ethylene in the presence of sulfuric acid.

$$H_2C{=}CH_2 + H{-}OH \xrightarrow{H^+} H{-}\overset{H}{\underset{H}{C}}{-}\overset{H}{\underset{OH}{C}}{-}H$$

In a similar manner, isopropyl alcohol is produced by the addition of water to propene (propylene).

$$H{-}\overset{H}{\underset{H}{C}}{-}\overset{H}{C}{=}CH_2 + H{-}OH \xrightarrow{H^+} H{-}\overset{H}{\underset{H}{C}}{-}\overset{H}{\underset{OH}{C}}{-}\overset{H}{\underset{H}{C}}{-}H$$

With 2-methylpropene, the product is 2-methyl-2-propanol:

$$H{-}\overset{H}{C}{=}\overset{CH_3}{C}{-}CH_3 + H{-}OH \xrightarrow{H^+} H{-}\overset{H}{\underset{H}{C}}{-}\overset{CH_3}{\underset{OH}{C}}{-}CH_3$$

Note that in the last two reactions the hydrogen atom of water goes on the carbon atom (of the two involved in the double bond) that has the most hydrogen atoms already bonded to it. The hydroxyl group goes on the carbon with fewer hydrogens. Thus addition of water to propylene always gives isopropyl alcohol, never propyl alcohol:

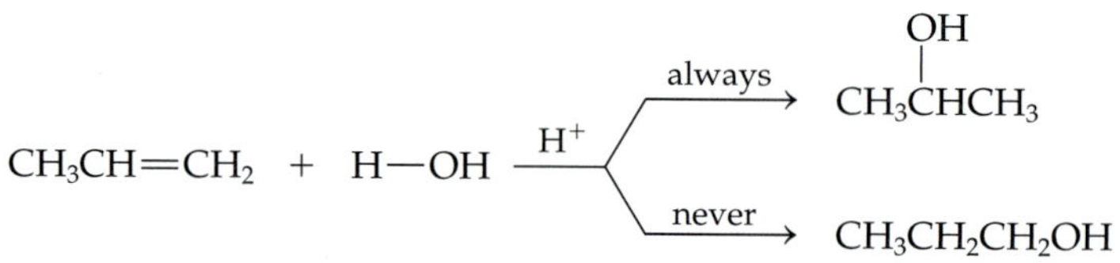

$$CH_3CH{=}CH_2 + H{-}OH \xrightarrow{H^+} \begin{cases} \xrightarrow{\text{always}} CH_3\underset{}{\overset{OH}{\overset{|}{C}}}HCH_3 \\ \xrightarrow{\text{never}} CH_3CH_2CH_2OH \end{cases}$$

The above rule, in a more general form, was first formulated in 1870 by Vladimir V. Markovnikov, a Russian chemist. It is widely known as **Markovnikov's rule.** Sometimes the rule is stated (somewhat facetiously) as "the rich get richer" or "them that has, gets."

Example 14.2

Which alcohols are formed by the hydration of (**a**) 2-methyl-1-pentene and (**b**) 1-methylcyclopentene?

Solution

First write out the structural formulas and count the number of hydrogen atoms directly bonded to each double–bond carbon. (Recall that in a cyclic hydrocarbon, each corner of the geometric figure represents a carbon atom *and* enough hydrogen atoms to give the carbon atom a total of four bonds.)

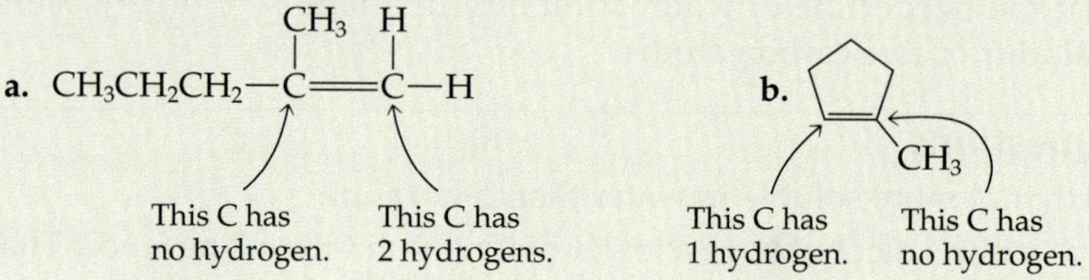

a. $CH_3CH_2CH_2{-}\overset{CH_3}{\overset{|}{C}}{=}\overset{H}{\overset{|}{C}}{-}H$

This C has no hydrogen. This C has 2 hydrogens.

b. (1-methylcyclopentene, ring with CH_3)

This C has 1 hydrogen. This C has no hydrogen.

According to Markovnikov's rule, the hydrogen goes to the carbon that has the most hydrogen atoms. The hydroxyl group goes to the other carbon:

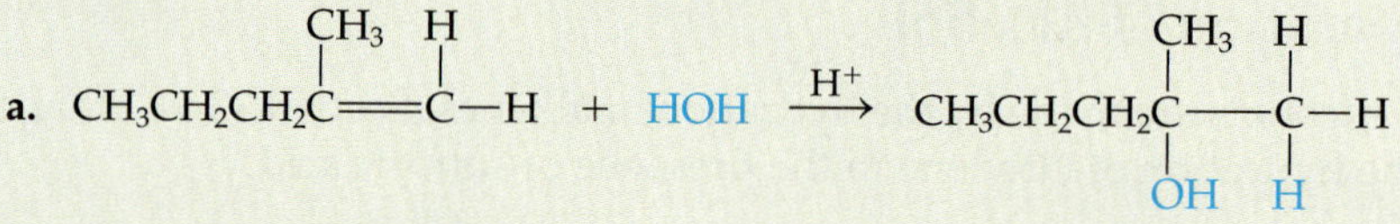

a. $$CH_3CH_2CH_2\overset{CH_3}{\overset{|}{C}}{=}\overset{H}{\overset{|}{C}}{-}H + HOH \xrightarrow{H^+} CH_3CH_2CH_2\underset{OH}{\underset{|}{\overset{CH_3}{\overset{|}{C}}}}{-}\underset{H}{\underset{|}{\overset{H}{\overset{|}{C}}}}{-}H$$

2-Methyl-1-pentene → 2-Methyl-2-pentanol

b. 1-Methylcyclopentene (ring with CH_3) $+ HOH \xrightarrow{H^+}$ 1-Methylcyclopentanol (ring with OH and CH_3)

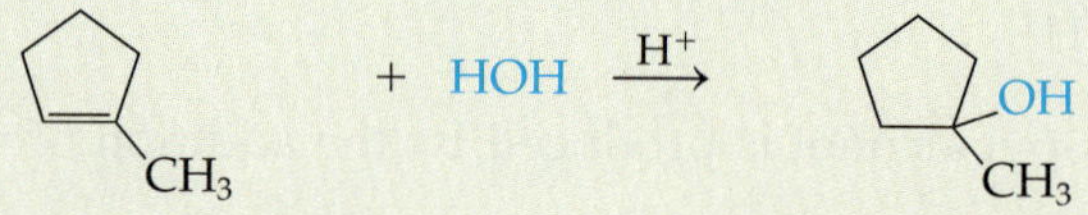

Practice Exercise

Write the equation for the hydration of each of the following alkenes and give the IUPAC name of the alcohol formed in each case.

a. 2-methyl-2-pentene **b.** 1-ethylcyclobutene

Alkene hydration is an important industrial source of alcohols. It is also quite common in biochemistry, and many hydroxy compounds in living systems are formed in this manner. The following reaction, for example, occurs in the Krebs cycle (Chapter 24).

$$
\underset{\text{Fumarate}}{^{-}OOC-\overset{\displaystyle H}{\overset{|}{C}}=\underset{\displaystyle H}{\underset{|}{C}}-COO^{-}} + HOH \xrightleftharpoons{\text{enzyme}} \underset{\text{Malate}}{^{-}OOC-\underset{\displaystyle H}{\underset{|}{\overset{\displaystyle H}{\overset{|}{C}}}}-\underset{\displaystyle OH}{\underset{|}{\overset{\displaystyle OH}{\overset{|}{C}}}}-COO^{-}}
$$

Methanol

Prior to 1923 methanol was prepared by the destructive distillation of wood. Sometimes it is still called *wood alcohol.* When wood is heated to 450 °C in the absence of air (Figure 14.4), it decomposes to charcoal and a volatile fraction that includes methanol, acetic acid, and acetone. The methanol, which comprises 2% to 3% of this fraction, can be separated from the other components by fractional distillation. A ton of wood produces about 35 lb of methanol. Today methanol is prepared more economically by combining hydrogen and carbon monoxide at high temperature and pressure in the presence of a zinc oxide–chromium oxide catalyst:

About 1.7 billion gallons of methanol are produced each year in the U.S. by catalytic reduction of carbon monoxide with hydrogen gas.

$$2\,H_2 + CO \xrightarrow[\text{ZnO, Cr}_2\text{O}_3]{\text{200 atm, 350 °C}} CH_3OH$$

Ethanol: Fermentation

The production of alcoholic spirits is one of the oldest known chemical reactions. Even in biblical times, ethanol was prepared by the fermentation of sugars or starch from various sources (potatoes, corn, wheat, rice, and so forth). Fermentation is catalyzed by enzymes found in yeast and proceeds by an elaborate multistep mechanism (Chapter 25). We can represent the overall process as follows.

$$\underset{\text{Starch}}{(C_6H_{10}O_5)_x} \xrightarrow{\text{enzymes}} \underset{\text{Glucose}}{C_6H_{12}O_6} \xrightarrow{\text{enzymes}} \underset{\text{Ethanol}}{2\,CH_3CH_2OH} + 2\,CO_2$$

On an industrial scale, either molasses from sugar cane or starches from various grains are fermented by yeast to ethanol. Most people are referring to ethanol when they say "alcohol," meaning an alcoholic beverage. The greatest use of ethanol is as a beverage. Wines contain about 12% ethanol by volume, champagnes 14 to 20%, beers and ciders about 4%, and whiskey, gin, and brandy 40 to 50%. The

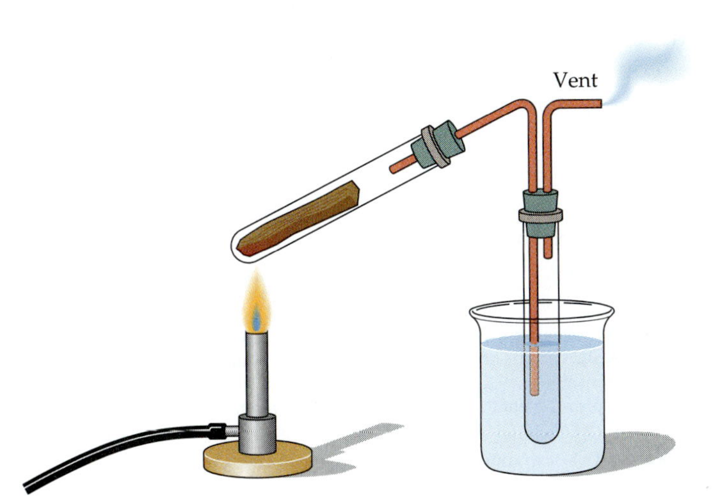

Figure 14.4 An apparatus for the destructive distillation of wood. The wood is heated in an enclosed tube, and the methanol is condensed in the second tube by the cold water in the beaker. Gases formed in the process are burned as they exit the vent tube.

▶ Ethanol can be made by fermentation of nearly any type of sugary or starchy material: *(left to right)* wine from rice, vodka from potatoes, aperitif from artichoke, raki from raisins, and wine from grapes.

alcoholic content of a beverage is indicated by a measure known as *proof spirit.*[1] The proof value is twice the alcoholic content by volume; whiskey that is 50% alcohol is 100 proof. Fermented alcohol can be concentrated to as much as 95% (190 proof) by distillation. Such grain alcohol is frequently used as a solvent for drugs meant for internal consumption.

✓ Review Questions

14.5 State Markovnikov's rule. Why is ethyl alcohol the only primary alcohol that can be prepared by the hydration of an alkene?

14.6 What is meant by the proof of an alcoholic beverage? What is the percent by volume of a beverage that is 86 proof?

14.7 Give two methods for preparation of methanol. Which method gave methanol its nickname?

14.5 Physiological Properties of Alcohols

People often want to know whether or not something is a *poison.* The question is difficult to answer. Toxicity depends on the nature of the substance, the amount, and the route by which it is taken into the body.

The simple alcohols are all poisonous to some degree. In an attempt to quantify the degree of toxicity, scientists use the term $\mathbf{LD_{50}}$ to indicate the *l*ethal *d*ose of a chemical to 50% of a population of test animals. Like humans, individual animals respond differently to various poisons. Some are killed by amounts much smaller than the LD_{50}; others survive considerably larger amounts. The LD_{50} term, then, is only approximate for animals. Extrapolation to human toxicities can introduce even larger errors. The dosage usually is expressed as the amount of tested substance per kilogram of body weight of the test animal. The smaller the LD_{50} value, the smaller the quantity of the substance required to kill the animal and therefore the more toxic the substance. Table 14.5 lists LD_{50} values for alcohols administered orally to rats.

Methanol

Note that Table 14.5 lists no LD_{50} for methanol. Methanol is not particularly toxic to horses, rats, and other animals that are deficient in the enzymes that oxidize alcohols to aldehydes. It is, however, dangerous[2] to humans and other primates

[1]The term has its origin in a seventeenth-century English method for testing whiskey. Dealers were perhaps too often tempted to increase profits by adding water to the booze. The whiskey to be tested was poured on gunpowder and the mixture was ignited. Because an ethanol–water solution will ignite when the alcohol concentration is about 50%, this solution scored 100 in the test, "proof" of the spirit content in the whiskey. The powder would not burn if the whiskey contained too much water.

[2]Toxicity studies in other animals cannot always be extrapolated to humans. In many cases, however, trends in toxicities can be judged from animal studies.

Table 14.5 Lethal Oral Doses (in Rats) for Some Alcohols

Alcohol	Structure	Boiling Point (°C)	LD_{50} (g/kg body weight)	Uses
Methanol	CH_3OH	64	—	Solvent, fuel additive
Ethanol	CH_3CH_2OH	78	7.06	Solvent, fuel additive, beverages
1-Propanol	$CH_3CH_2CH_2OH$	97	1.87	Solvent
2-Propanol (Isopropyl alcohol)	$CH_3CHOHCH_3$	82	5.8	Solvent, body rubs
1-Butanol	$CH_3CH_2CH_2CH_2OH$	118	4.36	Solvent
1-Hexanol	$CH_3(CH_2)_4CH_2OH$	156	4.59	—
1,2-Ethanediol (Ethylene glycol)	$HOCH_2CH_2OH$	198	8.54	Antifreeze
1,2,3-propanetriol (Glycerol)	$HOCH_2CHOHCH_2OH$	290 (decomposes)	>25	Moisturizer

because they have liver enzymes that oxidize primary alcohols to aldehydes (Section 14.6). Ethanol, for example, is oxidized to acetaldehyde:

$$\underset{\text{Ethanol}}{CH_3CH_2OH} \xrightarrow{\text{liver enzymes}} \underset{\text{Acetaldehyde}}{CH_3-\overset{\overset{\displaystyle O}{\|}}{C}-H}$$

The acetaldehyde is in turn oxidized to acetic acid, a normal constituent of cells. The acetic acid can then be oxidized to carbon dioxide and water.

Similarly, methanol is oxidized to formaldehyde:

$$\underset{\text{Methanol}}{CH_3OH} \xrightarrow{\text{liver enzymes}} \underset{\text{Formaldehyde}}{H-\overset{\overset{\displaystyle O}{\|}}{C}-H}$$

Formaldehyde reacts rapidly with the components of cells. It coagulates proteins in much the same way that an egg is coagulated by cooking. This property of formaldehyde accounts for much of the toxicity of methanol. The LD_{50} for formaldehyde administered orally to rats is 0.070 g per kilogram of body weight. For acetaldehyde under the same conditions, LD_{50} is 1.9 g per kilogram of body weight. Thus, formaldehyde is about 27 times as toxic to rats as acetaldehyde.

Although its short-term toxicity in humans is not terribly high, methanol can cause permanent blindness or death, even in small concentrations. Each year many accidents are attributed to this alcohol, which is frequently mistaken for its less harmful relative ethanol. Methanol should never be applied to the body, nor should its vapors be inhaled, because it is readily absorbed through the skin and respiratory tract.

Ingestion of as little as 15 mL of methanol can cause blindness. As little as 30 mL (1 fluid ounce) can be fatal, but the usual deadly dose is 100 to 250 mL.

The antidote for methanol poisoning has long been ethanol, administered intravenously. The ethanol preferentially loads up the liver enzymes in humans and other primates. If the enzymes are tied up oxidizing ethanol to acetaldehyde, they cannot catalyze the oxidation of the methanol to formaldehyde. This allows the unoxidized methanol to be gradually excreted from the body.

Despite its toxicity, methanol is a valuable industrial solvent. Its largest use is as the starting material for the commercial synthesis of formaldehyde. It is also used in windshield washer fluids and as a solvent for paint, gum, and shellac.

Some racing cars are fueled by methanol.

Ethanol

Ethyl alcohol is potentially toxic to humans. Rapid ingestion of 1 pint (about 500 mL) of pure ethanol would kill most people, and acute ethanol poisoning kills several hundred people each year—often those engaged in some sort of drinking contest. Ethanol freely crosses into the brain, where it depresses the respiratory control center, resulting in failure of the respiratory muscles in the lungs and hence suffocation. Ethanol is believed to act on the nerve cell membranes, causing a diminution in speech, thought, cognition, and judgment. Excessive ingestion over a long period of time leads to deterioration of the liver (cirrhosis) and loss of memory and may lead to strong physiological addiction. Addiction to ethanol (alcoholism) is the most serious drug problem in the United States. It has been estimated that there are about 40 times as many alcoholics (about 10 million) as there are heroin addicts in the United States. If ethanol is diluted (as in alcoholic beverages) and consumed in small quantities, it is relatively safe. The body possesses enzymes that have the capacity to metabolize it to carbon dioxide and water (Chapter 25).

Ethanol not intended for beverage purposes is commercially prepared by the hydration of ethylene, a by-product of the petroleum industry. The ethanol so produced is 95% ethanol and 5% water. The water that remains in this mixture cannot

Alcohol in the Blood

Contrary to popular belief, ethanol is a central nervous system depressant, not a stimulant. The illusory stimulation comes from depression of brain areas responsible for judgment. The resulting lack of inhibitions and restraints may cause one to feel "stimulated." People under the influence of alcohol suffer from diminished judgment and control of their actions and hence may endanger themselves and others, especially if they are driving. In some states, a blood alcohol concentration (BAC) of 0.080% (80 mg of alcohol in 100 mL of blood) is legal evidence of intoxication. Other states have a BAC of 0.10% as a legal limit. A BAC of 0.5–1% leads to coma and death (Table 14.6).

Ethanol in excess has many other harmful effects. Contrary to popular belief, ethanol does not keep a person warm in cold weather. Rather, it causes blood vessels to dilate and the resulting increased flow of blood through the capillaries beneath the skin imparts a feeling of warmth and a reddish hue to the skin, while actually accelerating the loss of body heat through the skin. About half of all the people treated in emergency rooms for frostbite are legally drunk. Heavy drinking alters brain cell function, causes nerve damage, and shortens the life span by contributing to diseases of the liver, cardiovascular system, and virtually every other organ of the body.

Table 14.6 Approximate Relationship Between Drinks Consumed, Blood Alcohol Concentration (BAC), and Effect for a 70-kg (154-lb) Moderate[a] Drinker

Number of Drinks[b]	BAC (% by volume)	Effect
2	0.05	Mild sedation; tranquillity
4	0.10	Lack of coordination
6	0.15	Obvious intoxication
10	0.30	Unconsciousness
20	0.50	Possible death

[a]An inexperienced drinker would be affected more strongly, or more quickly, than one who is ordinarily a moderate drinker. Conversely, an experienced *heavy* drinker would be affected less.

[b]Rapidly consumed 30-mL (1-oz) "shots" of 90 proof whiskey, 360-mL (12-oz) bottles of beer, or 150-mL (5-oz) glasses of wine.

be removed by ordinary distillation because 95% ethanol is a constant-boiling mixture (an **azeotrope**). This 95% ethanol is used in chemical laboratories as a solvent. Special procedures are required to produce 100% ethanol. One method is to dry the ethanol over calcium oxide for several hours and then distill the remaining ethanol, known as **absolute** (100%) **alcohol.**

An azeotropic mixture is a constant-boiling mixture of two components that are present in a fixed ratio. A mixture of 95% ethanol and 5% water is an azeotropic mixture that boils at 78 °C.

Ethanol is used as a solvent for perfumes, medicinal formulations (tinctures), lacquers, varnishes, and shellacs. It is widely used as an antiseptic in mouthwashes and aerosol disinfectants because it denatures enzymes in bacteria (Chapter 21). Ethanol is also employed in the synthesis of other organic compounds. When used for such industrial purposes, it is not subject to a federal tax (more than $20/gallon in most states). To ensure the legitimate use of tax-free ethanol, the government requires that it be treated with additives that make it unfit to drink. Such ethanol is known as **denatured alcohol.** Common denaturants are methanol and 2-propanol. These compounds are toxic but do not interfere with the solvent properties of the ethanol.

Ethanol is blended into some gasolines. Mixtures containing 80 to 90% unleaded gasoline and 10 to 20% alcohol generally work well. Ethanol can be obtained easily from grain and can be used to augment our fuel supplies. Brazil, which has an abundance of sugar cane (and no oil), has produced cars designed to burn only ethanol. In addition to extending the fuel, ethanol increases the octane rating of the gasoline with which it is blended.

Isopropyl Alcohol

A 70% solution of isopropyl alcohol is commonly called rubbing alcohol. It has a high vapor pressure, and its rapid evaporation from the skin produces a cooling effect. Rubbing alcohol, like ethanol, denatures bacterial enzymes and is used as an antiseptic to cleanse the skin before taking a blood sample or giving an injection. Isopropyl alcohol is toxic when ingested but, compared to methanol, is less readily absorbed through the skin. Though more toxic than ethanol, it causes fewer fatalities, because it often induces vomiting, in which case it doesn't stay down long enough to kill you. Much of the isopropyl alcohol produced industrially is used for the manufacture of acetone (Chapter 15) and the introduction of the isopropyl group into organic molecules.

✓ Review Questions

14.8 Define the following terms.

a. absolute alcohol **b.** azeotrope **c.** LD_{50}

14.9 What is denatured alcohol? Why is some alcohol denatured?

14.10 Why is methanol so much more toxic to humans than ethanol?

14.11 What chemical compound is used in the treatment of acute methanol poisoning? How does it work?

14.6 Chemical Properties of Alcohols

Chemical reactions in alcohols occur mainly at the functional group[3]. Some reactions, however, involve hydrogen atoms attached to the hydroxyl-bearing carbon or to an adjacent carbon. We discuss three major kinds of reactions of the alcohols. Dehydration and oxidation are considered here. Esterification is covered in Section 16.8. An outline of the reactions is provided in Figure 14.5.

[3]Some reactions of alcohols are simple substitution reactions. For example, in the reaction of 1-butanol with hydriodic acid, the OH group is replaced by an iodine atom.

$$CH_3CH_2CH_2CH_2OH + HI \longrightarrow CH_3CH_2CH_2CH_2I + HOH$$

Figure 14.5 ▶
Reactions of alcohols

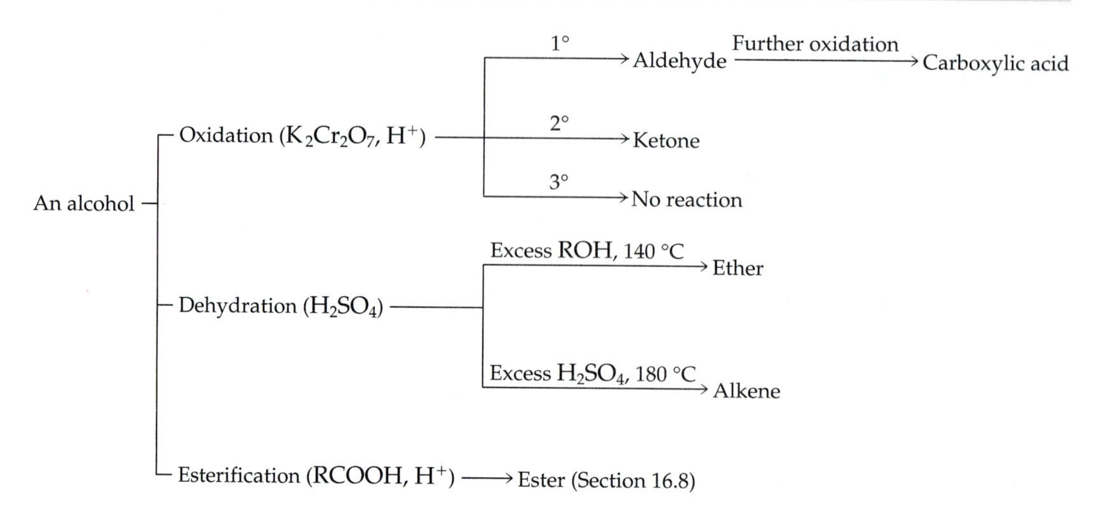

When a dehydration can yield more than one alkene, the actual product is a mixture of the possible isomers. In such cases the predominant isomer is the alkene with more alkyl groups attached to the double-bond carbons. For example, the dehydration of 2-butanol can give either 1-butene or 2-butene. The latter is the major product.

$$CH_3CH_2CH(OH)CH_3 \xrightarrow{\text{conc. } H_2SO_4} CH_3CH{=}CHCH_3 + HOH$$

Dehydration

Dehydration (removal of water) is usually accomplished by adding concentrated sulfuric acid to the alcohol and heating the resulting mixture. The hydroxyl group is removed from the alcohol carbon, and a hydrogen atom is removed from an adjacent carbon, giving an alkene:

$$\underset{\text{Ethanol}}{H{-}CH_2{-}CH_2{-}OH} \xrightarrow[\text{excess acid}]{\text{concd } H_2SO_4,\ 180\ °C} \underset{\text{Ethylene}}{H_2C{=}CH_2} + HOH$$

Under the proper conditions it is possible to perform a dehydration involving two molecules of alcohol. The hydroxyl group of one alcohol and only the hydrogen of the hydroxyl group of the second alcohol molecule are removed. The two organic groups remaining combine to form an ether molecule (Section 14.9):

$$\underset{\text{Two molecules of ethanol}}{CH_3CH_2OH + HOCH_2CH_3} \xrightarrow[\text{excess ethanol}]{\text{conc. } H_2SO_4} \underset{\text{Diethyl ether}}{CH_3CH_2{-}O{-}CH_2CH_3} + H_2O$$

Thus, depending on conditions, one can prepare either alkenes or ethers by dehydration of alcohols. At 180 °C and with an excess of H_2SO_4, dehydration of ethanol gives ethylene as the main product. At 140 °C and with an excess of ethanol, the main product of dehydration is diethyl ether.

Dehydration (and its reverse, hydration) reactions occur continuously in cellular metabolism. In these biochemical dehydrations, enzymes serve as catalysts instead of acids, and the reaction temperature is 37 °C instead of the elevated temperatures required in the laboratory. The following reaction occurs in glycolysis (Chapter 25).

$$\underset{\text{2-Phosphoglycerate}}{HOCH_2{-}CH(OPO_3^{2-}){-}COO^-} \xrightleftharpoons{\text{enzyme}} \underset{\text{Phosphoenolpyruvate}}{H_2C{=}C(OPO_3^{2-})COO^-} + HOH$$

Although the compounds involved are more complex than the ethanol and ethylene in our previous example, the reaction is not. We can ignore the other functional groups in the molecule because they remain unchanged in the product. Note that all that happens is that a hydrogen atom and a hydroxyl group are eliminated from the starting material, and the product contains a double bond. The point is that if you know the chemistry of a particular functional group, you know the chemistry of hundreds or even thousands of different compounds. Alcohols have a potential for undergoing dehydration: Big ones, little ones, and ones that incorporate other functional groups all dehydrate if conditions are right.

Dehydration to form simple ethers in biological systems is perhaps less common than the reaction to form an alkene. However, many important reactions, such as the formation of glycosides from sugars (Chapter 19), are at least technically dehydrations leading to etherlike compounds.

Oxidation

Primary and secondary alcohols are readily oxidized. We saw earlier how methanol and ethanol are oxidized by liver enzymes to form aldehydes. These reactions can also be carried out in the laboratory with chemical oxidizing agents. For example, in acid solution, potassium dichromate oxidizes ethyl alcohol to acetaldehyde.

The Breathalyzer test to detect drunk drivers is based on the color change associated with the reduction of dichromate ion. If a suspect's breath causes the color to change to green chromium(III) compounds, his/her blood contains more than the legal level of alcohol. Further tests are then carried out at the police station to confirm the suspicion.

$$8\,H^+ + \underset{\text{Dichromate ion}}{Cr_2O_7^{2-}} + 3\,CH_3CH_2OH \longrightarrow \underset{\text{Chromium(III) ion}}{2\,Cr^{3+}} + 3\,CH_3CHO + 7\,H_2O$$

Similarly, propyl alcohol is oxidized to propionaldehyde. The balanced equation is similar to the one for ethanol. Organic chemists often simplify such equations, showing only the change involving the organic molecules. Thus the oxidation of propanyl alcohol would be simplified to

$$\underset{\text{Propyl alcohol}}{CH_3CH_2CH_2OH} \xrightarrow[H^+]{K_2Cr_2O_7} \underset{\text{Propionaldehyde}}{CH_3CH_2\overset{\overset{\displaystyle O}{\|}}{C}-H}$$

The required inorganic reagents are written either above or below the arrow. The inorganic by-products are ignored (they're still formed, but we just ignore them in this form of the equation). In this way we focus attention on the organic starting material and product, rather than on balancing the complicated equations.

The abbreviated form of this particular equation indicates that a primary alcohol is oxidized to an aldehyde. We shall see (Chapter 15) that aldehydes are even more easily oxidized than alcohols and yield carboxylic acids. To isolate the aldehyde initially formed in the oxidation of the alcohol, it is removed from contact with the oxidizing agent by distilling it from the reaction mixture as it forms.

Secondary alcohols are oxidized to *ketones* (Chapter 15). Oxidation of isopropyl alcohol by dichromate ion gives acetone:

$$\underset{\substack{\text{Isopropyl alcohol}\\ \text{(a secondary alcohol)}}}{CH_3-\overset{\overset{\displaystyle OH}{|}}{CH}-CH_3} \xrightarrow[H^+]{K_2Cr_2O_7} \underset{\substack{\text{Acetone}\\ \text{(a ketone)}}}{CH_3-\overset{\overset{\displaystyle O}{\|}}{C}-CH_3}$$

Unlike aldehydes, ketones are relatively resistant to further oxidation and no special precautions are required to isolate the product of this reaction.

As we saw in the preceding section, alcohol oxidation is important in living organisms. Indeed, enzyme-controlled oxidation reactions provide the energy that cells need in order to do useful work. One step in the Krebs cycle (Chapter 24) involves the oxidation of the secondary alcohol group in isocitric acid to a ketone group.

$$\underset{\text{Isocitric acid}}{\begin{array}{l} \;\;\;\;\;\;CH_2-COOH \\ \;\;\;\;\;\;| \\ \;\;\;\;\;\;CH-COOH \\ \;\;\;\;\;\;| \\ HO-CH-COOH \end{array}} \xrightarrow{\text{enzyme}} \underset{\text{Oxalosuccinic acid}}{\begin{array}{l} \;\;\;\;\;CH_2-COOH \\ \;\;\;\;\;| \\ \;\;\;\;\;CH-COOH \\ \;\;\;\;\;| \\ O=C-COOH \end{array}}$$

Note that the overall reaction is identical to the conversion of isopropyl alcohol to acetone. The complexity of structure that distinguishes isocitric acid in no way interferes with the characteristic reaction of its secondary alcohol group.

Tertiary alcohols are resistant to oxidation, because the carbon atom bonded to a hydroxyl group is not also bonded to a hydrogen. The oxidation reactions we have described involve the formation of a carbon–oxygen double bond. Thus the hydroxyl carbon in the alcohol must be able to release one of its attached atoms to form the double bond with oxygen. The carbon–hydrogen bond is easily broken under oxidative conditions, but the carbon–carbon bond is not. Therefore tertiary alcohols are not easily oxidized.

Example 14.3

Write equations for the reactions of the following alcohols with $K_2Cr_2O_7$ and H_2SO_4.

a. 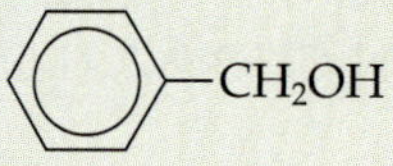 C_6H_5–CH_2OH b. 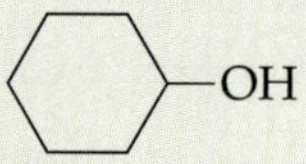 cyclohexyl–OH c. 1-methylcyclohexanol (ring C bearing OH and CH_3)

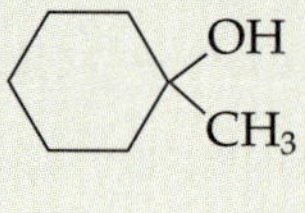

Solution

The first step is to recognize the class of each alcohol as primary, secondary, or tertiary. Therefore:

a. This alcohol has the OH group on a carbon atom that is attached to only one other carbon atom (one of the carbon atoms of the benzene ring); it is a primary alcohol. Oxidation forms first an aldehyde, and further oxidation forms a carboxylic acid.

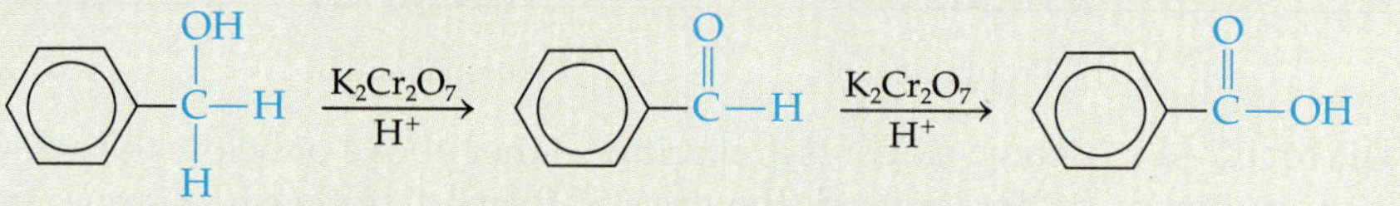

$$C_6H_5\text{–}CH(OH)\text{–}H \xrightarrow[H^+]{K_2Cr_2O_7} C_6H_5\text{–}C(=O)\text{–}H \xrightarrow[H^+]{K_2Cr_2O_7} C_6H_5\text{–}C(=O)\text{–}OH$$

b. This alcohol has the OH group on a carbon atom that is attached to two other carbon atoms in the cyclohexane ring. It is a secondary alcohol, and oxidation gives a ketone.

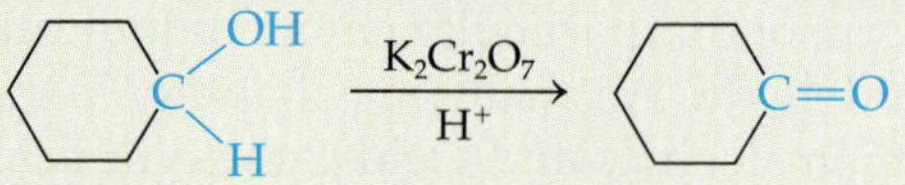

$$\text{cyclohexyl } C(OH)H \xrightarrow[H^+]{K_2Cr_2O_7} \text{cyclohexyl } C{=}O$$

c. This alcohol has the OH group on a carbon atom that is attached to three other carbon atoms (two in the cyclohexane ring and one in the methyl group. It is a tertiary alcohol, and oxidation with $K_2Cr_2O_7$ gives no reaction.

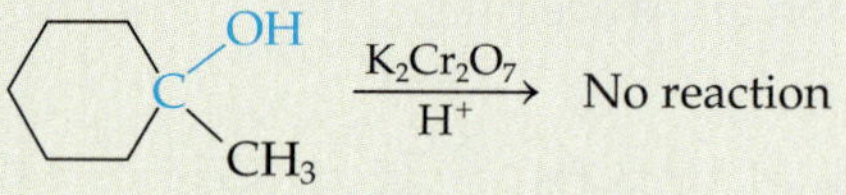

$$\text{cyclohexyl } C(OH)CH_3 \xrightarrow[H^+]{K_2Cr_2O_7} \text{No reaction}$$

Practice Exercise

Write equations for the reactions of the following alcohols with $K_2Cr_2O_7$ and H_2SO_4.

a. $CH_3CH_2CH_2CH_2CH_2OH$ b. $CH_3CH_2CH_2CH(OH)CH_3$ c. $CH_3CH_2C(OH)(CH_3)CH_3$

Review Questions

14.12 Name three major types of chemical reactions of alcohols.

14.13 Why do tertiary alcohols not undergo oxidation? Can a tertiary alcohol undergo dehydration?

14.14 In the preparation of diethyl ether from ethanol, why is it so critical to maintain the reaction temperature between 130 and 150 °C?

14.7 Multifunctional Alcohols: Glycols and Glycerol

The simple alcohols we have considered so far contain only *one* hydroxyl group each and are called *monohydric alcohols*. Several important alcohols contain *more than one* hydroxyl group per molecule and are called **polyhydric alcohols.** Those with two hydroxyl groups are said to be **dihydric alcohols,** and those with three hydroxyl groups are called **trihydric alcohols.**

Dihydric alcohols are also known as **glycols.** The most important of these is 1,2-ethanediol, usually called ethylene glycol.

Two carbon atoms: "ethylene"

$$HOCH_2CH_2OH$$

Two hydroxyl groups: "glycol"

Ethylene glycol is the main ingredient in most antifreeze mixtures for automobile radiators. Ethylene glycol is a sweet, colorless, somewhat viscous liquid. The two hydroxyl groups lead to extensive intermolecular hydrogen bonding. This results in a high boiling point (198 °C), and thus ethylene glycol does not boil away when used as antifreeze. It is also completely miscible with water. A solution of 60% ethylene glycol in water freezes at −49 °C (−56 °F) and thus protects the automobile radiator down to that temperature. Ethylene glycol is also used in the manufacture of polyester fiber and magnetic film used in tapes for recorders and computers.

Ethylene glycol is quite toxic. As with methanol, its toxicity is due to a metabolite. Liver enzymes oxidize the ethylene glycol to oxalate ion.

$$\underset{\text{Ethylene glycol}}{HOCH_2CH_2OH} \xrightarrow{\text{liver enzymes}} \underset{\text{Oxalate ion}}{{}^-O{-}\overset{\overset{O}{\|}}{C}{-}\overset{\overset{O}{\|}}{C}{-}O^-}$$

In the kidneys, the oxalate ion combines with calcium ion, precipitating as calcium oxalate (CaC_2O_4). These crystals cause renal damage and can lead to kidney failure and death. As with methanol poisoning, the usual treatment for ethylene glycol poisoning is ethanol, administered to load up and thus block the liver enzymes from catalyzing the conversion of ethylene glycol to oxalate ion.

Another common dihydric alcohol is 1,2-propanediol, usually called propylene glycol. The physical properties of this compound are quite similar to those of ethylene glycol. Its physiological properties, however, are quite different. Propylene glycol is essentially nontoxic, and it can be used as a solvent for drugs. It is also used as a moisturizing agent for foods. Like other alcohols, propylene glycol is oxidized by liver enzymes.

$$\underset{\text{Propylene glycol}}{CH_3{-}\overset{\overset{HO}{|}}{C}HCH_2OH} \xrightarrow{\text{liver enzymes}} \underset{\text{Pyruvate ion}}{CH_3{-}\overset{\overset{O}{\|}}{C}{-}\overset{\overset{O}{\|}}{C}{-}O^-}$$

In this case, however, the product is pyruvate ion, a normal intermediate in carbohydrate metabolism (Chapter 25).

1,2,3-Propanetriol, or glycerol (also called glycerin), is the most important trihydric alcohol. It is a sweet, syrupy liquid. Essentially nontoxic, it is a product of the hydrolysis of fats and oils. Glycerol has many uses, some of which are listed in Table 14.7.

Table 14.7 Some Uses of Glycerol

1. Ingredient in hand lotions and cosmetics
2. Additive in inks, tobacco products, and plastic clays to prevent dehydration (glycerol is hygroscopic)
3. Constituent of glycerol suppositories
4. Sweetening agent and solvent for medicines
5. Lubricant
6. Starting material in the production of plastics, surface coatings, and synthetic fibers
7. Source of nitroglycerin

The equation for the preparation of nitroglycerin shows that three molecules of nitric acid are required for every molecule of glycerin.

$$\begin{array}{c} H \\ | \\ H-C-OH \\ | \\ H-C-OH \\ | \\ H-C-OH \\ | \\ H \end{array} + 3\,HONO_2 \xrightarrow[10\text{–}20\ ^\circ C]{H_2SO_4} \begin{array}{c} H \\ | \\ H-C-ONO_2 \\ | \\ H-C-ONO_2 \\ | \\ H-C-ONO_2 \\ | \\ H \end{array} + 3\,H_2O$$

Glycerol (Glycerin) — Glycerol trinitrate (Nitroglycerin)

The glycerin must be very pure to ensure stability of the product.

Nitroglycerin is a pale yellow, oily liquid that detonates upon slight impact. The explosive power arises from the extremely rapid conversion of a small volume of liquid into a large volume of hot, expanding gases.

$$4\,C_3H_5(ONO_2)_3(l) \longrightarrow 6\,N_2(g) + 12\,CO_2(g) + 10\,H_2O(g) + O_2(g)$$

The reaction produces temperatures above 3000 °C and pressures above 2000 atm, leading to an enormous explosive wave that accounts for the damaging effect of the detonation. The shock wave can travel at speeds up to 9000 m/s (about 20,000 mi/h), causing the explosion to occur at a rate far faster than that of other chemical reactions.

Review Questions

14.15 What is a polyhydric alcohol? What is a glycol?

14.16 Why is ethylene glycol so much more toxic to humans than propylene glycol? What chemical compound is used in the treatment of acute ethylene glycol poisoning? How does it work?

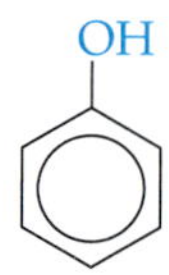

Phenol

14.8 Phenols

As we noted in Section 14.1, compounds in which a hydroxyl group is attached directly to an aromatic ring are called **phenols.** The parent compound, C_6H_5OH, is itself called phenol. It is a white crystalline compound that has a distinctive ("hospital smell") odor. Other compounds may be named as derivatives of phenol, but most of those of interest to us are best known by special nonsystematic names.

One of the major characteristics that distinguish phenols from alcohols is that phenols are slightly acidic ($K_a \approx 10^{-10}$) in water, whereas alcohols are neutral. Phenols can be neutralized by strong bases, but they are too weakly acidic to react with weak bases such as aqueous sodium bicarbonate. The latter reaction serves to distinguish the phenols from the carboxylic acids (Chapter 16), which do react with $NaHCO_3$.

$$CH_3CH_2OH + NaOH(aq) \longrightarrow \text{No reaction}$$

$$C_6H_5\text{–}OH + NaOH(aq) \longrightarrow C_6H_5\text{–}O^-Na^+(aq) + HOH$$

$$C_6H_5\text{–}OH + NaHCO_3(aq) \longrightarrow \text{No reaction}$$

The phenols generally are either low-melting-point solids or oily liquids. Most are only sparingly soluble in water. They are widely used as antiseptics (substances that kill microorganisms on living tissue) and as disinfectants (substances intended to kill microorganisms on inanimate objects such as furniture or floors).

Nitroglycerin

Nitroglycerin was first prepared in 1846 by the Italian chemist Ascanio Sobrero, who was lucky that he lived to tell of his discovery. Sobrero mixed nitric acid and glycerin, and the ensuing explosion nearly killed him. Fifteen years later the Swedish chemist and inventor Alfred Nobel discovered a method to prepare and transport the compound safely. He found that a type of diatomaceous earth, a claylike material, was capable of absorbing the nitroglycerin, thus rendering it insensitive to shock. The stabilized mixture was referred to as *dynamite*. Unless exploded by means of a percussion cap or a detonator containing lead azide [$Pb(N_3)_2$], it was quite stable.

With dynamite, the construction of canals, dams, highways, mines, and railroads became much easier. Its use as a weapon in warfare greatly disturbed the Nobel family, however, and Alfred Nobel's will established a trust fund to provide an annual award for an outstanding contribution toward peace. Other funds were set aside to offer annual awards in the fields of chemistry, physics, literature, and medicine or physiology.

It is surprising that a compound so sensitive to shock is also used as a drug to relieve angina pectoris—sharp chest pains caused by an insufficient supply of oxygen to the heart muscle. Nitroglycerin is administered in tablet form (mixed with nonactive ingredients), as an alcoholic solution (spirit of glyceryl trinitrate), or in the form of a patch from which the drug is absorbed through the skin. Nitroglycerin functions as a vasodilator. It relaxes cardiac muscle and smooth muscle in the smaller blood vessels, thus increasing the supply of blood (and hence oxygen) to the heart.

◄ Alfred Nobel (1833–1896), inventor of dynamite and founder of the famed Nobel Prizes.

The first widely used antiseptic was phenol, then called carbolic acid. Joseph Lister used it for antiseptic surgery in 1867. Phenol is quite toxic, however, and it can cause severe burns when applied to the skin. In the bloodstream it is a systemic poison—that is, one that is carried to and affects all parts of the body. Its severe side effects led to searches for safer antiseptics, a number of which have been found.

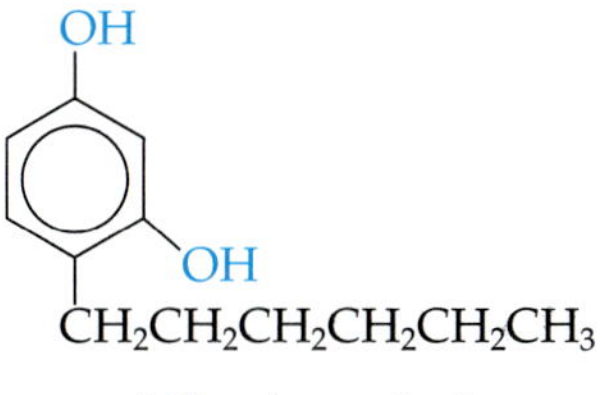

4-Hexylresorcinol

One of the most active phenolic antiseptics is 4-hexylresorcinol. It is much more powerful than phenol as a germicide and has fewer undesirable side effects. Indeed, it is safe enough to be used as the active ingredient in some mouthwashes and in throat lozenges.

The methyl derivatives of phenols are called *cresols*, compounds that were once widely used as ingredients in the wood preservative creosote.

OH, CH_3 — *o*-Cresol

OH, CH_3 — *m*-Cresol

HO, CH_3 — *p*-Cresol

All three dihydroxybenzenes, each of which has a nonsystematic name, have commercial significance. Two of them are important components of biochemical molecules. Hydroquinone occurs in a coenzyme (Chapter 22), and catechol forms part of the structure of certain neurotransmitters termed catecholamines (Selected Topic C).

Catechol   Resorcinol   Hydroquinone

The most important commercial reaction of phenols is the condensation with formaldehyde to yield phenolic polymers (Bakelite). Bakelite was used initially as an electrical insulator and later to form plastic parts for the automotive and radio industries.

Review Question

14.17 Distinguish between an antiseptic and a disinfectant.

14.9 Ethers

As indicated in Section 14.1, **ethers** may be considered to be derivatives of water in which both hydrogen atoms have been replaced by alkyl or aryl groups. They may also be considered derivatives of an alcohol in which the hydroxyl hydrogen has been replaced by an alkyl or aryl group:

$$\underset{\text{Water}}{H-O-H} \xrightarrow[\text{hydrogens}]{\text{replace both}} \underset{\text{Ether}}{R-O-R'} \xleftarrow[\text{hydrogen}]{\text{replace hydroxyl}} \underset{\text{Alcohol}}{R-O-H}$$

The general formula for the ethers is R—O—R′. When both R groups are the same, the compound is a *symmetrical* ether. When R and R′ are different, the ether is *unsymmetrical.*

Symmetrical Ether		*Unsymmetrical Ether*	
R—O—R	$H_3C—O—CH_3$	R—O—R′	$H_3C—O—CH_2CH_3$

Simple ethers are simply named. Just name the groups attached to oxygen and then add the generic name *ether.* For symmetrical ethers, the group name should be preceded by the prefix *di-*, as in dimethyl ether ($CH_3—O—CH_3$) and diethyl ether ($CH_3CH_2—O—CH_2CH_3$).

Ether molecules have no hydrogen atom on oxygen (that is, no OH group). Therefore the molecules in a pure liquid ether are incapable of intermolecular hydrogen bonding. Given their molar mass, then, the ethers have quite low boiling points. Indeed, ethers have boiling points about the same as those of alkanes of comparable molar mass and much lower than those of the corresponding alcohols (Table 14.8).

Ether molecules do have an oxygen atom, however, and so can hydrogen-bond with water molecules. Consequently, the ethers have about the same water solubilities as their isomeric alcohols. (For example, dimethyl ether and ethanol are completely soluble in water, whereas diethyl ether and 1-butanol are each soluble to the extent of 8 g/100 mL of water.)

Chemically, the ethers are quite inert. Like the alkanes, they do not react with the usual oxidizing agents, reducing agents, or bases. Their inertness makes ethers excellent solvents for organic materials. Often called simply "ether," diethyl ether

Table 14.8 Comparison of Boiling Points of Alkanes, Alcohols, and Ethers

Formula	Name	Molar Mass	Boiling Point (°C)
$CH_3CH_2CH_3$	Propane	44	−42
CH_3OCH_3	Dimethyl ether	46	−25
CH_3CH_2OH	Ethyl alcohol	46	78
$CH_3CH_2CH_2CH_2CH_3$	Pentane	72	36
$CH_3CH_2OCH_2CH_3$	Diethyl ether	74	35
$CH_3CH_2CH_2CH_2OH$	Butyl alcohol	74	117

is often used in the extraction of organic compounds from plant and animal materials or from mixtures of organic and inorganic substances. The volatile ether is then easily removed by evaporation, and the desired organic components are left behind. Ether is extremely hygroscopic. A freshly opened can of ether quickly picks up about 1 to 2% water from the moisture in the air. Special techniques are required in handling ether when reaction conditions call for anhydrous ether.

Use of diethyl ether in the laboratory produces unusual hazards. The compound is quite volatile and flammable. The vapors, which are heavier than air, form an explosive mixture with air. They can travel long distances along a tabletop or the floor to reach a flame or spark and set off an explosion. Hence open flames are not permitted in a laboratory in which ether is being used. Ether fires cannot be extinguished with water because the ether is less dense than water and floats on top of it. The use of carbon dioxide fire extinguishers is recommended.

Diethyl ether should not be stored in an ordinary refrigerator. Even at low temperatures it has sufficient vapor pressure to form an explosive mixture with air. A spark can ignite the vapors. Special explosion-proof refrigerators, with sealed electrical equipment to prevent contact between spark and flammable vapors, are required for safe storage of volatile flammable liquids.

Still another hazard with ethers is that, upon standing, they react with oxygen from the air to form peroxides.

$$\underset{\text{Diethyl ether}}{CH_3CH_2{-}O{-}CH_2CH_3} + O_2 \longrightarrow \underset{\text{A peroxide}}{CH_3\underset{\displaystyle O{-}O{-}H}{\underset{|}{C}}H{-}O{-}CH_2CH_3}$$

Peroxides are less volatile than ether and are concentrated in the residue left behind during a distillation or evaporation. The concentrated peroxides are highly explosive and sensitive to both shock and heat. People usually avoid these problems by buying only the amounts of ether sufficient for immediate use and by keeping containers tightly closed and away from strong light, which catalyzes peroxide formation. Ether suspected of containing peroxides should never be used.

The potency of an anesthetic is related to its solubility in olive oil: the more soluble in olive oil, the more potent as an anesthetic. This unusual observation has led many scientists to believe that anesthetics act by dissolving in the fatty membranes (Chapter 20) surrounding nerve cells. The resultant changes in the fluidity and shape of the membranes apparently decrease the ability of sodium ions to pass into the nerve cells, thereby blocking the firing of nerve impulses.

✓ Review Question

14.18 What precautions must be taken when using diethyl ether as a solvent in a laboratory experiment?

Anesthesia

A **general anesthetic** acts on the brain to produce unconsciousness as well as insensitivity to pain. A *local anesthetic* (Selected Topic C) renders one part of the body insensitive to pain yet leaves the patient conscious. The ideal general anesthetic should quickly make the patient unconscious but allow a quick return to consciousness, have few side effects, and be safe to handle.

Diethyl ether (CH_3CH_2—O—CH_2CH_3) was the first general anesthetic. It was introduced into surgical practice in 1846 by a Boston dentist, William Morton. Inhalation of ether vapor produces unconsciousness by depressing the activity of the central nervous system. Ether is relatively nontoxic because there is a fairly wide gap between the effective level of anesthesia and the lethal dose. The disadvantages are its high flammability and postanesthetic nausea and vomiting.

Dinitrogen monoxide (N_2O), also called nitrous oxide and laughing gas, was discovered by Joseph Priestley in 1772. Priestley noted its narcotic effect, and it soon came to be used widely at laughing gas parties among the nobility. Nitrous oxide was tried by Morton without success before he tried ether. Mixed with oxygen, nitrous oxide finds some use in modern anesthesia. It is quick-acting but not very potent. Concentrations of 50% or greater must be used to be effective. When nitrous oxide is mixed with ordinary air instead of oxygen, not enough oxygen gets into the patient's blood, and permanent brain damage can result.

Chloroform ($CHCl_3$) was introduced as a general anesthetic in 1847 and was used widely for years. Its popularity soared in 1853 after Queen Victoria gave birth to her eighth child while anesthetized by chloroform. Nonflammable, chloroform produces effective anesthesia, but it has serious drawbacks. For one, it has a narrow safety margin; the effective dose is close to the lethal dose. It also causes liver damage, and it must be protected from oxygen during storage to prevent the formation of deadly phosgene gas.

Modern anesthetics include fluorine-containing compounds such as halothane, enflurane, and methoxyflurane (Figure 14.6). These compounds are nonflammable and relatively safe for the patient. Their safety—particularly that of halothane—for operating-room personnel, however, has been questioned. For example, female operating-room workers suffer a higher rate of miscarriages than the general population.

Curare is the arrow poison used by South American Indian tribes. Large doses of curare kill by causing a complete relaxation of all muscles. Death occurs because of respiratory failure.

Modern surgical practice has moved away from the use of a single anesthetic. Generally, a patient is given an intravenous anesthetic such as thiopental (Selected Topic C) to produce unconsciousness. The gaseous anesthetic then is administered to provide insensitivity to pain and to keep the patient unconscious. A relaxant, such as curare, also may be employed. Curare and related compounds produce profound relaxation; thus, only light anesthesia is required. This practice avoids the hazards of deep anesthesia.

Figure 14.6 ▶ Three modern general anesthetics.

H—C(Br)(Cl)—C(F)(F)—F

Halothane

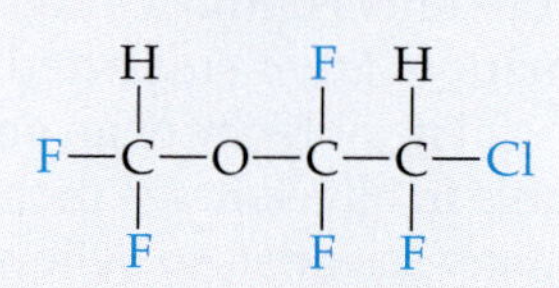

Enflurane

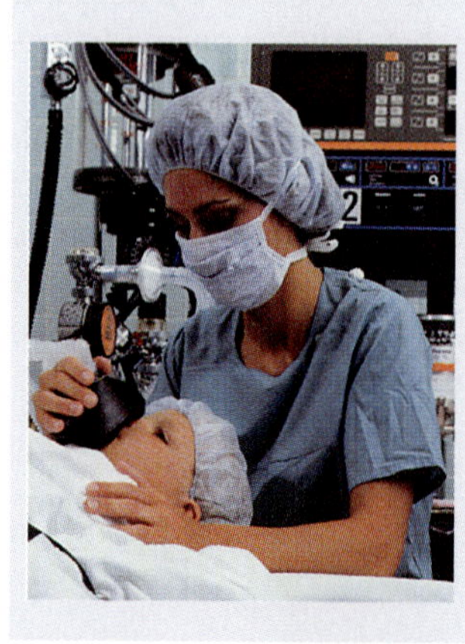

◀ Inhalant anesthetic agents are used to render patients unconscious and insensitive to pain before surgery.

H—C(Cl)(Cl)—C(F)(F)—O—C(H)(H)—H

Methoxyflurane

Summary

A **functional group** is any atom or atom group that confers characteristic properties to a family of compounds. The **hydroxyl group,** OH, is one example, and carbon–carbon double and triple bonds are two others.

A primary carbon atom is one attached to no other or one other carbon, a secondary carbon atom is one attached to two other carbons, and a tertiary carbon atom is one attached to three other carbons.

Alcohols are aliphatic compounds in which one or more hydrogen atoms have been replaced by a hydroxyl group. A **primary alcohol** has the hydroxyl group on a primary carbon; a **secondary alcohol**, on a secondary carbon atom; and a **tertiary alcohol,** on a tertiary carbon atom.

Alcohols can be synthesized by hydration of alkenes. **Markovnikov's rule** states that the H of HOH goes to the alkene carbon bonded to the higher number of hydrogen atoms and the OH to the other alkene carbon.

When a molecule of water is removed from an alcohol in a dehydration step, the result is either an alkene or an ether, depending on reaction conditions.

Primary alcohols are oxidized to aldehydes or carboxylic acids, and secondary alcohols are oxidized to ketones. Tertiary alcohols are not easily oxidized.

Alcohols containing one hydroxyl group are **monohydric alcohols,** those containing two are **dihydric alcohols,** and those containing three are **trihydric alcohols.** Dihydric alcohols are usually called **glycols.**

Phenols (ArOH) are compounds having the hydroxyl group attached to an aromatic ring.

Ethers (ROR, ROAr, ArOAr) are compounds in which an oxygen atom is joined to two organic groups.

Key Terms

absolute alcohol (14.5)
alcohol (ROH) (14.1)
azeotrope (14.5)
denatured alcohol (14.5)
dihydric alcohol (14.7)
ether (ROR′) (14.9)
functional group (14.1)
general anesthetic (14.9)
glycol (14.7)
hydroxyl group (14.1)
LD_{50} (14.5)
Markovnikov's rule (14.4)
phenol (ArOH) (14.8)
polyhydric alcohol (14.7)
primary (1°) alcohol (14.2)
secondary (2°) alcohol (14.2)
tertiary (3°) alcohol (14.2)
trihydric alcohol (14.7)

Problems

Alcohols: Names and Structural Formulas

1. Name each of the following.
   a. $CH_3CH_2CH_2CH_2CH_2CH_2OH$
   b. $CH_3CH_2CH_2CH_2CHOHCH_3$
2. Name each of the following.
   a. $CH_3CH_2CHOHC(CH_3)_3$ b. $(CH_3)_2CHCH_2OH$
3. Name each of the following.
   a. $CH_3CHOHCH_2CHCl_2$ b. $(CH_3)_2COHCBr_2CH_3$
4. Name each of the following.
   a. 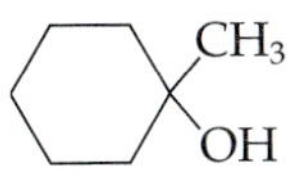 CH_3 / OH (on cyclohexane ring) b. $CH_3CH_2CH_2COH(CH_2CH_3)_2$
5. Give structural formulas for each of the following.
   a. 3-hexanol b. 3,3-dimethyl-2-butanol
6. Give structural formulas for each of the following.
   a. cyclopentanol b. 4-methyl-2-hexanol
7. Give structural formulas for each of the following.
   a. 4,5-dimethyl-3-heptanol
   b. 2-ethyl-1-phenyl-1-butanol
8. Give structural formulas for each of the following.
   a. 2-bromo-2-chlorocyclobutanol
   b. 3-phenylcyclopentanol

Physical Properties

(Answer Problems 9–12 without consulting tables.)

9. Arrange in order of increasing boiling point: ethanol, 1-propanol, methanol.
10. Arrange in order of increasing boiling point: butane, ethylene glycol, 1-propanol.
11. Arrange in order of increasing solubility in water: methanol, 1-butanol, 1-octanol.
12. Arrange in order of increasing solubility in water: pentane, propylene glycol, diethyl ether.

Preparation of Alcohols

13. Give the structure of the alkene from which each of the following alcohols is made by reaction with water in acidic solution:
    a. 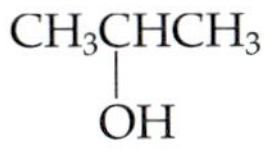 CH_3CHCH_3 with OH on C-2 b.  CH_3 / OH (on cyclopentane ring) c. $CH_3C(CH_3)_2—OH$ (CH₃ above and below C) 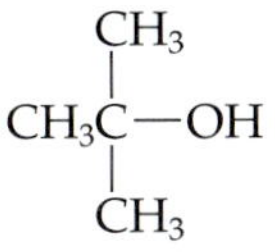
14. Give the structure of the alkene from which each of the following alcohols is made by reaction with water in acidic solution:

a. CH_3CH_2OH b. 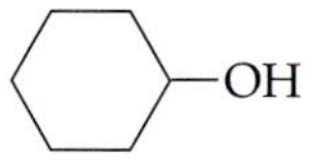 (cyclohexane ring)—OH

c. $CH_3CHCH_2CH_3$ (OH on C-2)

Chemical Reactions of Alcohols

15. Classify each of the following conversions as oxidation, dehydration, or hydration (only the organic starting material and product are shown):

a. $CH_3OH \longrightarrow H{-}\overset{H}{\underset{}{C}}{=}O$

b. $CH_3\overset{OH}{C}HCH_3 \longrightarrow CH_3CH{=}CH_2$

c. $CH_2{=}CHCH_2CH_3 \longrightarrow CH_3CHOHCH_2CH_3$

16. Classify each of the following conversions as oxidation, dehydration, or hydration (only the organic starting material and product are shown):

a. $CH_3\overset{OH}{C}HCH_3 \longrightarrow CH_3\overset{O}{\overset{\|}{C}}CH_3$

b. $HOOCCH{=}CHCOOH \longrightarrow HOOCCH_2\overset{OH}{C}HCOOH$

c. $2\,CH_3OH \longrightarrow CH_3OCH_3$

17. Each of the four isomeric butyl alcohols is treated with potassium dichromate in acid. Draw the product (if any) expected from each reaction.

18. Write an equation for the dehydration of 2-propanol to yield (**a**) an alkene and (**b**) an ether.

19. Draw the structural formula of the ether formed by the *intra*molecular dehydration of:

$$HOCH_2CH_2CH_2CH_2CH_2OH$$

20. Draw the structure of the alkene formed by the dehydration of cyclohexanol.

21. Give the structural formula of the product of each of the following reactions.

a. $CH_2{=}CHCH_2CH_3 \xrightarrow[H^+]{H_2O}$

b. 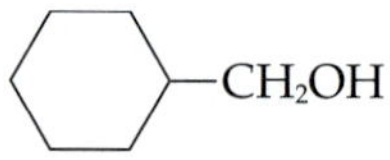 (cyclohexane ring)—$CH_2OH \xrightarrow[H^+]{K_2Cr_2O_7}$

22. Give the structural formula of the product of each of the following reactions.

a. $CH_3CHOHCH_3 \xrightarrow[H^+]{KMnO_4}$

b. 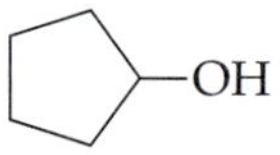 (cyclopentane ring)—OH $\xrightarrow[180\ °C]{\text{concd } H_2SO_4}$

23. What reagents are necessary to carry out the following conversions?

a. $CH_3CH{=}CH_2 \xrightarrow{?} CH_3\overset{OH}{C}HCH_3$

b. $CH_3\overset{}{C}HCH_3$ (OH on C-2) $\xrightarrow{?} CH_3\underset{\overset{\|}{O}}{C}CH_3$

c. $2\,CH_3CH_2OH \xrightarrow{?} CH_3CH_2OCH_2CH_2$

24. What reagents are necessary to carry out the following conversions?

a. $CH_2{=}\underset{CH_3}{C}{-}CH_3 \xrightarrow{?} CH_3\overset{OH}{\underset{CH_3}{C}}{-}CH_3$

b. $CH_3CH_2OH \xrightarrow{?} CH_2{=}CH_2$

c. $CH_3CH_2CH_2OH \xrightarrow{?} CH_3CH_2\overset{O}{\overset{\|}{C}}{-}H$

Polyhydric Alcohols

25. Give structural formulas for each of the following.
a. 1,4-pentanediol b. propylene glycol

26. Give structural formulas for each of the following.
a. 1,3-hexanediol b. glycerol

Phenols

27. Name each of the following.

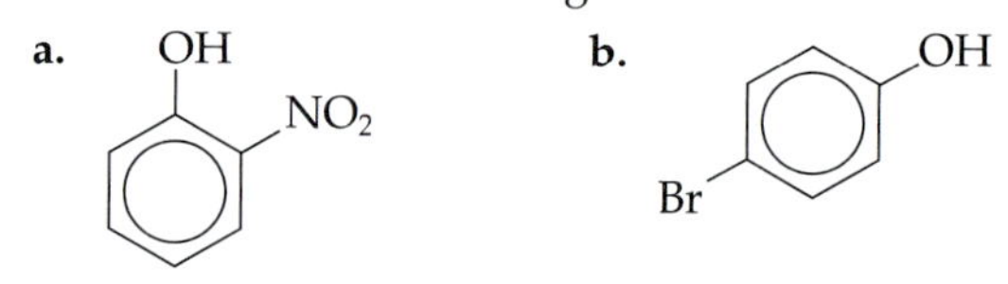

28. Name each of the following.

a. 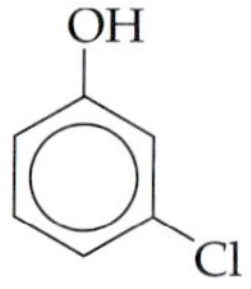 (benzene ring with OH and Cl in meta positions) b. 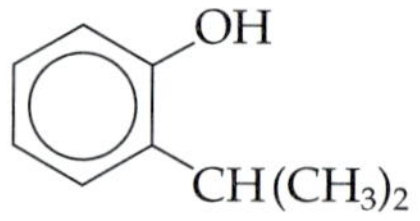 (benzene ring with OH and $CH(CH_3)_2$ in ortho positions)

29. Give structural formulas for each of the following.
a. *m*-iodophenol b. *p*-methylphenol (*p*-cresol)

30. Give structural formulas for each of the following.
a. 2,4,6-trinitrophenol (picric acid)
b. 3,5-diethylphenol

31. Write an equation for the reaction (if any) of phenol with aqueous (**a**) NaOH and (**b**) $NaHCO_3$.

32. Write an equation for the ionization of phenol in water.

Ethers: Names and Structural Formulas

33. Name each of the following.
 a. $CH_3CH_2CH_2OCH_2CH_2CH_3$
 b. 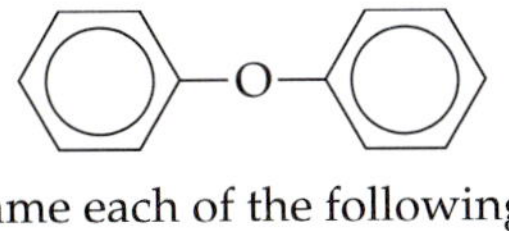 $C_6H_5{-}O{-}C_6H_5$ (two benzene rings joined by O)

34. Name each of the following.
 a. $CH_3CH_2OCH(CH_3)_2$
 b. 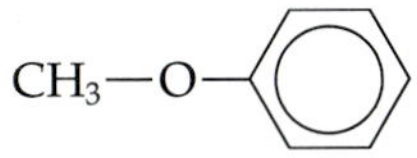 $CH_3{-}O{-}C_6H_5$ (benzene ring)

35. Give structural formulas for each of the following.
 a. ethyl methyl ether b. phenyl benzyl ether

36. Give structural formulas for each of the following.
 a. diisopropyl ether b. cyclopropyl propyl ether

Additional Problems

37. Ethyl alcohol, like rubbing alcohol (isopropyl alcohol), is often used for sponge baths. What property of alcohols makes them useful for this purpose?

38. What is a general anesthetic? Name some compounds used as general anesthetics.

39. Tetrahydrocannabinol (THC) is the principal active ingredient in marijuana. What functional groups are present in the THC molecule?

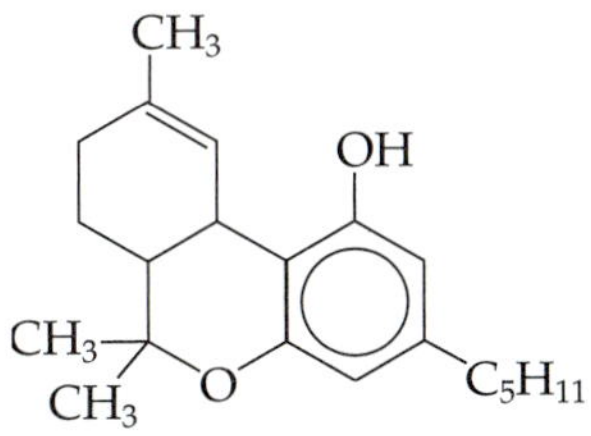

(Structure: CH_3, OH, CH_3, CH_3, O, C_5H_{11})

Tetrahydrocannabinol (THC)

40. Without consulting tables, arrange in order of increasing boiling point: 1-butanol, diethyl ether, and propylene glycol.

41. In addition to ethanol, the fermentation of grain produces other organic compounds collectively called fusel oils (FO). The four principal FO components are 1-propanol, isobutyl alcohol, 3-methyl-1-butanol, and 2-methyl-1-butanol. Draw a structural formula for each. (FO is quite toxic and accounts in part for hangovers.)

42. Give a name and one use for each of the following:
 a. CH_3OH b. CH_3CH_2OH
 c. $CH_3CHOHCH_3$ d. CH_2OHCH_2OH
 e. $CH_2OHCHOHCH_2OH$ f. 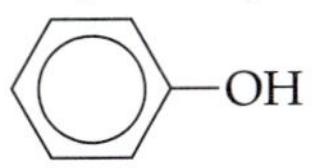 $C_6H_5{-}OH$ (benzene ring)

43. Give structural formulas for the eight isomeric pentyl alcohols ($C_5H_{12}O$). (Hint: Three are derived from pentane, four from isopentane, and one from neopentane.) Name the alcohols by the IUPAC system.

44. Classify the alcohols in Problem 43 as primary, secondary, or tertiary.

45. Draw and name the isomeric ethers that have the formula $C_5H_{12}O$.

46. Menthol is an ingredient in mentholated cough drops and nasal sprays. It produces a cooling, refreshing sensation when rubbed on the skin and so is used in shaving lotions and cosmetics. Thymol, the aromatic equivalent of menthol, is the flavoring constituent of thyme. (**a**) What is the IUPAC name of menthol? (**b**) Give two names for thymol.

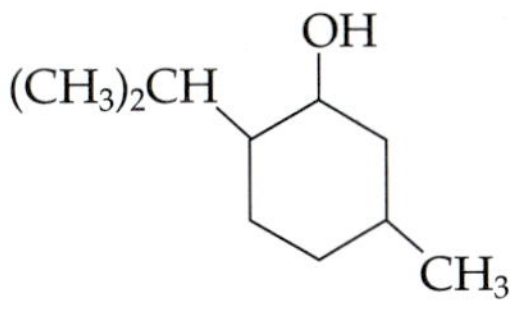

(Structures: OH, $(CH_3)_2CH$, CH_3 on cyclohexane ring — Menthol; OH, $(CH_3)_2CH$, CH_3 on benzene ring — Thymol)

Menthol 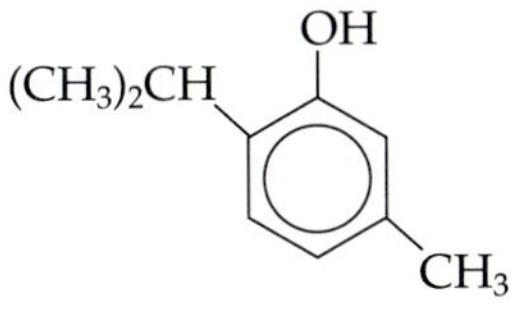 Thymol

47. Benzyl alcohol and *p*-cresol are isomers. Write the formula for each. Compare their solubilities in (**a**) water and (**b**) an aqueous solution of NaOH.

48. Write the equation for the production of ethanol by the addition of water to ethylene. How much ethanol can be made from 14.0 kg of ethylene?

49. The label on a bottle of light wine indicates that 100 mL of the wine furnishes 70 Cal (food calories), 0.2 g of protein, 5.77 g of carbohydrates, and 0.0 g of fat. Assuming that carbohydrates and proteins furnish 4 Cal/g each, that alcohol furnishes 7 Cal/g, and that no other caloric nutrients are present, (**a**) how many Calories are provided by the alcohol in a 100-mL serving of the wine? What percentage of the total Calories is provided by alcohol? (**b**) How many grams of alcohol are there in each 100-mL serving? (**c**) The density of alcohol is 0.789 g/mL. How many milliliters of alcohol are there in each 100-mL serving? What is the percent alcohol by volume?

50. Methanol is not particularly toxic to rats. If methanol were newly discovered and tested for toxicity in laboratory animals, what would you conclude about its safety for human consumption?

CHAPTER 15

Aldehydes and Ketones

The delightful aroma of butter is largely due to a compound with two ketone functional groups. The compound, sometimes called diacetyl, has the IUPAC name 2,3-butanedione. Many other spices and foods have flavors due to aldehydes and ketones.

Learning Objectives/Study Questions

1. What is the general structure for an aldehyde? A ketone?
2. How are the common names of aldehydes and ketones determined? How are aldehydes and ketones named using IUPAC nomenclature?
3. Why are the boiling points of aldehydes and ketones higher than those of ethers and alkanes of similar molar masses, but lower than those of comparable alcohols?
4. How do the solubilities of aldehydes and ketones of four carbons or less compare to the solubilities of comparable alkanes and alcohols in water?
5. How are aldehydes and ketones prepared?
6. What typical reactions take place with aldehydes and ketones?
7. What are some common aldehydes and ketones and their uses?

What do certain hormones, vanilla flavor, a biological tissue preservative, and fresh cucumbers have in common? The answer: molecules containing a carbonyl functional group, a group characteristic of aldehydes and ketones. As the preceding list indicates, these two families of compounds, which we consider in this chapter, are found in a most diverse company of products. Both the tempting aromas associated with cinnamon, vanilla, and fresh baked goods and the sickening smell of some rancid foods are associated with the carbonyl functional group.

With the aldehydes and ketones, we can study the carbonyl group in its simplest surroundings. In Chapter 16, we consider more complicated functional groups that incorporate the carbonyl group. And in later chapters, we will find this ubiquitous grouping of atoms in carbohydrates, fats, proteins, nucleic acids, hormones, vitamins, and the host of organic compounds critical to living systems. But first things first. Let's begin by focusing on the carbonyl group in aldehydes and ketones.

15.1 The Carbonyl Group: A Carbon–Oxygen Double Bond

The **carbonyl group** has a carbon atom joined to oxygen by a double bond.

$$-\overset{\displaystyle O}{\overset{\|}{C}}-$$

The carbon-to-oxygen double bond, like the carbon-to-carbon double bond in an alkene, tends to undergo addition reactions. Unlike the alkene double bond, however, the carbonyl bond involves an oxygen atom and thus is highly polar. That polarity confers special properties on aldehydes and ketones.

What is the difference between a ketone and an aldehyde? It appears to be rather trivial at first sight. In **ketones,** two carbon groups are attached to the carbonyl carbon. The following general formulas all represent ketones:

$$R-\overset{\displaystyle O}{\overset{\|}{C}}-R \qquad Ar-\overset{\displaystyle O}{\overset{\|}{C}}-R \qquad Ar-\overset{\displaystyle O}{\overset{\|}{C}}-Ar$$

In **aldehydes,** at least one of the attached groups must be a hydrogen atom. The following compounds are all aldehydes:

$$H-\overset{\displaystyle O}{\overset{\|}{C}}-H \qquad R-\overset{\displaystyle O}{\overset{\|}{C}}-H \qquad Ar-\overset{\displaystyle O}{\overset{\|}{C}}-H$$

The structures of aldehydes and ketones, with the carbon-to-oxygen double bond, are difficult to type. For this reason, the functional groups are often condensed on one line.

—CHO	—CO—
An aldehyde	A ketone

We use —CHO to identify an aldehyde rather than —COH which might be confused with an alcohol.

In these structures, the carbon-to-oxygen double bond is understood; it is still there, but not written out.

Because they contain the same functional group, aldehydes and ketones share many common properties. They are different from one another in other respects, however—different enough to warrant their classification into two families.

Review Question

15.1 How are aldehydes and ketones alike in chemical structure? How do they differ?

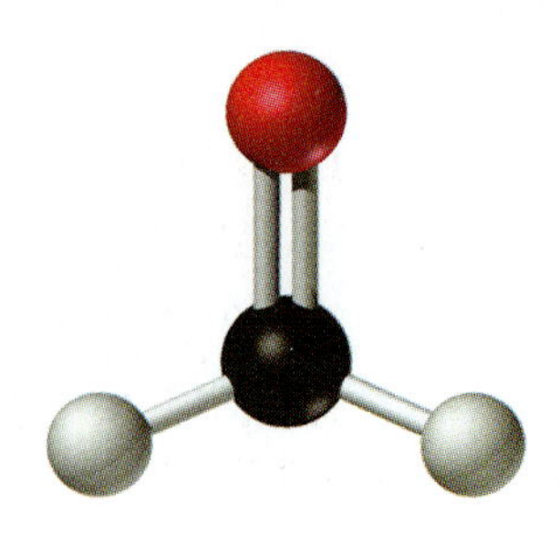

▲ Ball-and-stick model of formaldehyde (methanal).

15.2 Names of Aldehydes

The name aldehyde is derived from the two words *al*cohol *dehyd*rogenation because aldehydes can be obtained by the removal of hydrogen from an alcohol. Both common and IUPAC names are frequently used for aldehydes, with common names predominating for the lower homologs. The common names are taken from the names of the acids (Chapter 16) into which the aldehydes can be converted by oxidation (represented by [O]):

$$\underset{\text{Formaldehyde}}{\mathrm{H{-}\overset{\overset{\displaystyle O}{\|}}{C}{-}H}} \xrightarrow{[O]} \underset{\text{Formic acid}}{\mathrm{H{-}\overset{\overset{\displaystyle O}{\|}}{C}{-}OH}}$$

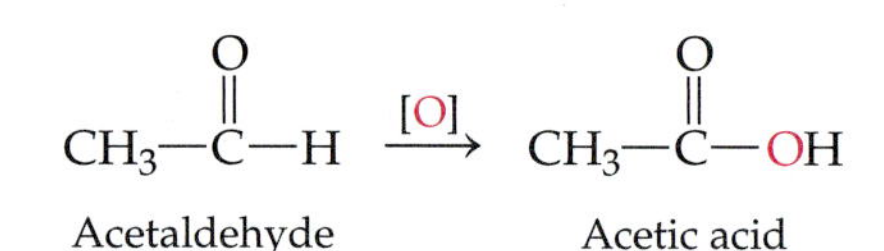

▲ Ball-and-stick model of acetaldehyde (ethanal).

The IUPAC names of aldehydes are derived from those of the corresponding alkanes. Select the longest continuous chain (LCC) of carbon atoms that contains the functional group. Take the name of the alkane having that number of carbon atoms, drop the *-e*, and add the ending *-al*.[1] The aldehyde functional group takes precedence over all the groups discussed so far. Thus, the carbonyl carbon of an aldehyde is always considered to be carbon 1 (C-1), and it is unnecessary to designate this group by number. Examples of aldehyde nomenclature are provided in Table 15.1.

$$\overset{5}{\mathrm{C}}{-}\overset{4}{\mathrm{C}}{-}\overset{3}{\mathrm{C}}{-}\overset{2}{\mathrm{C}}{-}\overset{1}{\mathrm{CHO}}$$

Example 15.1

Give the IUPAC name for $CH_3CH_2CH_2CH(CH_3)CHO$.

Solution

There are five carbon atoms in the LCC and a methyl group on the second carbon:

$$\overset{5}{\mathrm{CH_3}}\overset{4}{\mathrm{CH_2}}\overset{3}{\mathrm{CH_2}}\underset{\underset{\displaystyle CH_3}{|}}{\overset{2}{\mathrm{C}}}\mathrm{H}\overset{1}{\mathrm{C}}\mathrm{HO}$$

The name is 2-methylpentanal.

Practice Exercises

A. Give the IUPAC name for $CH_3C(CH_3)_2CH_2CHO$.
B. Give the structure and IUPAC name for the compound that has the common name *m*-bromobenzaldehyde.

Example 15.2

Write the structural formula for 7-chlorooctanal.

[1]Because the IUPAC ending for alcohols is *-ol*, there is occasionally confusion unless care is exercised in writing and pronouncing the IUPAC names of these two families. The one-carbon alcohol is methanol, with the ending pronounced like the *ol* in old. The one-carbon aldehyde is methanal, with the ending pronounced like the male name *Al*.

Table 15.1 Nomenclature of Aldehydes

Molecular Formula	Condensed Structural Formula	Common Name	IUPAC Name
CH_2O	$H-C(=O)-H$	Formaldehyde	Methanal
C_2H_4O	$CH_3C(=O)-H$	Acetaldehyde	Ethanal
C_3H_6O	$CH_3CH_2C(=O)-H$	Propionaldehyde	Propanal
C_4H_8O	$CH_3CH_2CH_2C(=O)-H$	Butyraldehyde	Butanal
C_4H_8O	$CH_3CH(CH_3)C(=O)-H$	Isobutyraldehyde	2-Methylpropanal
$C_3H_6O_3$	$CH_2(OH)-CH(OH)C(=O)-H$	Glyceraldehyde	2,3-Dihydroxypropanal
$C_5H_{10}O$	$CH_3CH_2CH_2CH_2C(=O)-H$	Valeraldehyde	Pentanal
$C_5H_{10}O$	$CH_3CH(CH_3)CH_2C(=O)-H$	Isovaleraldehyde	3-Methylbutanal (NOT 2-Methylbutanal)
C_7H_6O	$C_6H_5-C(=O)-H$	Benzaldehyde	Benzaldehyde (Phenylmethanal)
C_8H_8O	$C_6H_5-CH_2-C(=O)-H$	Phenylacetaldehyde	Phenylethanal
$C_8H_8O_3$	$HO-C_6H_3(OCH_3)-C(=O)-H$	Vanillin (odor of vanilla)	4-Hydroxy-3-methoxybenzaldehyde
C_9H_8O	$C_6H_5-CH=CH-C(=O)-H$	Cinnamaldehyde (odor of cinnamon)	3-Phenyl-2-propenal

Solution

From the "octan-" we know that the LCC has eight carbon atoms. There is a chlorine atom on the seventh carbon atom, numbering from the carbonyl group and counting the carbonyl carbon as C-1:

$$\overset{8}{C}H_3\overset{7}{C}H(Cl)\overset{6}{C}H_2\overset{5}{C}H_2\overset{4}{C}H_2\overset{3}{C}H_2\overset{2}{C}H_2\overset{1}{C}HO$$

Practice Exercises

A. Write the structural formula for 5-bromo-3-iodoheptanal.

B. Give the IUPAC name for glyceraldehyde, $HOCH_2CHOHCHO$. [Hint: As a substituent, the OH group is named *hydroxy*.]

15.3 Naming the Common Ketones

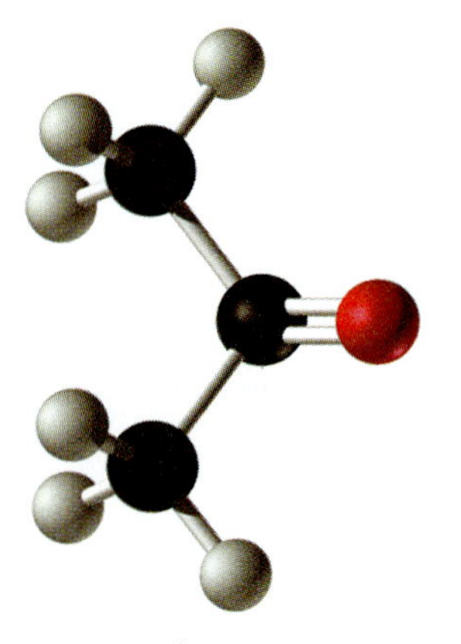

▲ Ball-and-stick model of acetone (propanone).

Because the carbonyl group in a ketone must be attached to two carbon groups, the simplest ketone has three carbon atoms. It is widely known as *acetone.* (It was first prepared from acetic acid.) The name is unique and does not correspond to the first in a series of similar common names.

Table 15.2 illustrates the nomenclature for ketones. Generally, the common names consist of the names of the groups attached to the carbonyl group, followed by the word *ketone.* (Note the similarity to the naming of ethers.) Another name for acetone, then, is *dimethyl ketone.* With four carbon atoms, we have ethyl methyl ketone. We can use common names for ketones if we can name the groups attached to the carbonyl carbon.

In the IUPAC system, the LCC containing the carbonyl carbon is the parent chain. The *-e* ending of the corresponding alkane name is dropped and replaced with *-one.* The IUPAC name for acetone thus is propanone and that for ethyl methyl ketone is butanone. In higher ketones a number indicates the position of the carbonyl carbon, and the chain is numbered so that the carbonyl carbon has the lowest possible number. In cyclic ketones it is understood that the carbonyl carbon is C-1.

Example 15.3

Write the structural formula for 4-methyl-3-hexanone.

Solution

The "hexan-" tells us that the LCC has six carbon atoms. The "3" means that the carbonyl carbon is C-3 in this chain, and the "4" tells us that there is a methyl group at C-4:

$$\overset{1}{C}H_3\overset{2}{C}H_2\overset{3}{C}(=O)-\overset{4}{C}H(CH_3)\overset{5}{C}H_2\overset{6}{C}H_2\overset{7}{C}H_3$$

Practice Exercise

Write the structural formula for 1,5-dibromo-4-ethyl-2-heptanone.

Example 15.4

a. Give the IUPAC name for the following compound.

(Structure: cyclopentanone ring, O= on one carbon, CH_3 on the ring carbon two positions away.)

Table 15.2 Nomenclature of Ketones

Molecular Formula	Condensed Structural Formula	Common Name	IUPAC Name
C_3H_6O	$CH_3C(=O)CH_3$	Acetone (Dimethyl ketone)	Propanone
C_4H_8O	$CH_3C(=O)CH_2CH_3$	Ethyl methyl ketone	Butanone
C_4H_6O	$CH_2{=}CHC(=O)CH_3$	Methyl vinyl ketone	3-Buten-2-one (NOT 1-Buten-3-one)
$C_5H_{10}O$	$CH_3CH_2C(=O)CH_2CH_3$	Diethyl ketone	3-Pentanone
$C_5H_{10}O$	$CH_3CH_2CH_2C(=O)CH_3$	Methyl propyl ketone	2-Pentanone
$C_5H_{10}O$	$CH_3CH(CH_3)C(=O)CH_3$	Isopropyl methyl ketone	3-Methyl-2-butanone (NOT 2-Methyl-3-butanone)
$C_6H_{10}O$	(cyclohexane ring)=O	Cyclohexanone	Cyclohexanone
C_8H_8O	(benzene ring)—C(=O)—CH_3	Acetophenone (Methyl phenyl ketone)	Phenylethanone
$C_{13}H_{10}O$	(benzene ring)—C(=O)—(benzene ring)	Benzophenone (Diphenyl ketone)	Diphenylmethanone

b. Give the common name and the IUPAC name for the following compound.

$$CH_3CH(CH_3)-C(=O)-CH(CH_3)CH_3$$

Solution

a. There are five carbon atoms in the ring, and the carbonyl carbon is always C-1. This means that the methyl group is on C-3 and the name is 3-methylcyclopentanone.

O=(cyclopentane ring, numbered 1–5)—CH_3 on C-3

b. Common name: Both the alkyl groups are isopropyl groups; the compound is diisopropyl ketone. IUPAC name: There are five carbon atoms on the LCC. The carbonyl carbon is C-3, and there are methyl groups on C-2 and C-4. The IUPAC name is 2,4-dimethyl-3-pentanone.

$$\overset{1}{C}H_3\overset{2}{C}H(CH_3)-\overset{3}{C}(=O)-\overset{4}{C}H(CH_3)\overset{5}{C}H_3$$

Practice Exercises

A. Give the common name and the IUPAC name for the following compound.

$$CH_3CH_2CH_2{-}\underset{\underset{O}{\|}}{C}{-}\underset{\underset{CH_3}{|}}{C}HCH_3$$

B. Give the the IUPAC name for the following compound. Consult Table 14.2 and propose a common name.

$$CH_3CH_2\underset{\underset{O}{\|}}{C}{-}CH_2\underset{\underset{CH_3}{|}}{C}HCH_3$$

15.4 Physical Properties of Aldehydes and Ketones

The carbon-to-oxygen double bond of the carbonyl group is quite polar. The electronegative oxygen has a much greater attraction for the bonding electron pairs than does the carbon. Thus, the electron density is greater at the oxygen end of the bond and less at the carbon end. The carbon is left with a partial positive charge and the oxygen with a partial negative charge:

$$\overset{\delta+}{C}{=}\overset{\delta-}{O}$$

This carbon–oxygen double bond is more polar than a carbon–oxygen single bond. As we noted in Chapter 7, charge separation in a molecule leads to dipole interactions. Indeed, double-bond polarity is great enough to affect the boiling points of aldehydes and ketones, whereas the polar single bonds in ethers have little effect on boiling points (Table 15.3). The dipolar forces in aldehydes and ketones however, are not comparable to the hydrogen bonding between molecules of an alcohol.

Formaldehyde is a gas at room temperature. Acetaldehyde boils at 20 °C; in an open vessel it boils away in a warm room. Most other common aldehydes are liquids at room temperature. Although the lower members of the homologous series have pungent odors, many higher aldehydes have pleasant odors and are used in perfumes and artificial flavorings.

The oxygen atom of the carbonyl group can accept a hydrogen bond from a water molecule.

$$C{=}O\cdots H{-}\underset{\underset{H}{|}}{O}$$

The hydrogen atoms of water molecules can form hydrogen bonds with the carbonyl oxygen; thus the solubility of aldehydes is about the same as that of alcohols and ethers. Formaldehyde and acetaldehyde are soluble in water; as the carbon chain increases, water solubility decreases. The borderline of solubility occurs at about four carbon atoms per oxygen atom. All aldehydes are soluble in organic solvents and, in general, are less dense than water.

The physical properties of the ketones are almost identical to those of the corresponding aldehydes. Acetone has a pleasant odor, and it is the only ketone that is completely soluble in water. Most of the higher homologs are colorless liquids,

Table 15.3 Boiling Points of Compounds Having Similar Molar Masses but Different Types of Intermolecular Forces

Compound	Family	Molar Mass	Type of Intermolecular Forces	Boiling Point (°C)
$CH_3CH_2CH_2CH_3$	Alkane	58	Dispersion only	−1
$CH_3OCH_2CH_3$	Ether	60	Weak dipole	6
$CH_3CH_2\overset{\overset{O}{\|}}{C}H$	Aldehyde	58	Strong dipole	49
$CH_3CH_2CH_2OH$	Alcohol	60	Hydrogen bonding	97

Table 15.4 Physical Properties of Selected Aldehydes and Ketones

Compound	Formula	Boiling Point (°C)	Solubility in Water (g/100 g H_2O)
Formaldehyde	HCHO	−21	Miscible
Acetaldehyde	CH_3CHO	20	Miscible
Propionaldehyde	CH_3CH_2CHO	49	16
Butyraldehyde	$CH_3CH_2CH_2CHO$	76	7
Valeraldehyde	$CH_3CH_2CH_2CH_2CHO$	103	Slightly soluble
Benzaldehyde	C_6H_5CHO	178	0.3
Acetone	CH_3COCH_3	56	Miscible
Ethyl methyl ketone	$CH_3COCH_2CH_3$	80	26
Methyl propyl ketone	$CH_3COCH_2CH_2CH_3$	102	6.3
Diethyl ketone	$CH_3CH_2COCH_2CH_3$	101	5

are slightly soluble in water, and, unlike the aldehydes, have rather bland odors. Table 15.4 lists some physical constants for several aldehydes and ketones.

Review Question

15.2 Why is the boiling point of butanal (76 °C) lower than the boiling point of 1-butanol (117 °C), but higher than the boiling point of diethyl ether (35 °C)?

15.5 Preparation of Aldehydes and Ketones

As we noted in Section 14.6, primary and secondary alcohols are oxidized to aldehydes and ketones, respectively. However, unless special care is taken or special reagents[2] chosen, the product aldehyde is further oxidized to a carboxylic acid. We shall see in Chapters 24 and 25 that the enzyme-catalyzed oxidation of alcohols to aldehydes and ketones is of great significance in biological systems.

$$\underset{\text{A primary alcohol}}{R{-}CH_2OH} \xrightarrow{[O]} \underset{\text{An aldehyde}}{R{-}\overset{\overset{\displaystyle O}{\|}}{C}{-}H}$$

$$\underset{\text{1-Butanol}}{3\,CH_3CH_2CH_2CH_2OH} + \underset{\text{(orange)}}{CrO_3} + 3\,HCl \xrightarrow[CH_2Cl_2]{\text{pyridine}} \underset{\text{Butanal}}{3\,CH_3CH_2CH_2\overset{\overset{\displaystyle O}{\|}}{C}{-}H} + \underset{\text{(green)}}{CrCl_3} + 3\,H_2O$$

$$\underset{\text{Benzyl alcohol}}{C_6H_5{-}\overset{\displaystyle OH}{\underset{\displaystyle H}{C}}{-}H} \xrightarrow[\text{pyridine}/CH_2Cl_2]{CrO_3/H^+} \underset{\text{Benzaldehyde}}{C_6H_5{-}\overset{\overset{\displaystyle O}{\|}}{C}{-}H}$$

Like the aldehydes, ketones are obtained by alcohol oxidation. However, the alcohol must be a secondary alcohol, and no special reagents are needed because the ketone is not susceptible to further oxidation. Although many oxidants are used, the most commonly employed are chromium(VI) compounds and sulfuric acid:

[2]Aldehydes can be prepared from primary alcohols using chromic oxide as an oxidizing agent, HCl, and the organic solvents pyridine and methylene chloride. In organic solvents, chromium(VI) compounds are mild oxidizing agents that can oxidize primary alcohols to aldehydes without oxidizing the aldehydes to acids.

$$\text{R—}\overset{\text{OH}}{\overset{|}{\text{CH}}}\text{—R}' \xrightarrow{[\text{O}]} \text{R—}\overset{\text{O}}{\overset{\|}{\text{C}}}\text{—R}'$$

A secondary alcohol — A ketone

$$\text{CH}_3\overset{\text{OH}}{\overset{|}{\text{CH}}}\text{CH}_3 \xrightarrow[\text{H}_2\text{SO}_4]{\text{K}_2\text{Cr}_2\text{O}_7} \text{CH}_3\overset{\text{O}}{\overset{\|}{\text{C}}}\text{CH}_3$$

Isopropyl alcohol — Acetone

$$\text{Cyclohexanol (C}_6\text{H}_{11}\text{—OH)} \xrightarrow[\text{H}_2\text{SO}_4]{\text{CrO}_3} \text{Cyclohexanone (C}_6\text{H}_{10}\text{=O)}$$

Cyclohexanol — Cyclohexanone

As we shall see in Chapters 24 and 25, the electrons released when alcohols are oxidized to carbonyl compounds are converted to a form of energy that can be used by living cells. These biochemical oxidation reactions are carried out at body temperature (37 °C) and are catalyzed by enzymes. Notice that in the following reaction the enzyme selectively catalyzes the oxidation of the secondary alcohol group to a ketone, but it does not oxidize the primary alcohol group to an aldehyde. We discuss enzyme specificity in Chapter 22.

$$^{2-}\text{O}_3\text{PO—CH}_2\text{—CH(OH)—CH}_2\text{—OH} \xrightleftharpoons{\text{dehydrogenase}} {}^{2-}\text{O}_3\text{PO—CH}_2\text{—}\overset{\text{O}}{\overset{\|}{\text{C}}}\text{—CH}_2\text{—OH}$$

Glycerol 3-phosphate — Dihydroxyacetone phosphate

Review Question

15.3 Account for the fact that the yield from oxidation of primary alcohols to aldehydes is usually lower than that of secondary alcohol oxidation to ketones.

15.6 Chemical Properties of Aldehydes and Ketones

Aldehydes and ketones are much alike in many of their reactions, just as we would expect from the presence of a carbonyl functional group in both. They differ greatly, however, in one most important type of reaction: oxidation.

Oxidation

As we have seen, aldehydes are readily oxidized to carboxylic acids, whereas ketones resist oxidation.

$$\text{RCHO} \xrightarrow{[\text{O}]} \text{RCOOH}$$

An aldehyde — A carboxylic acid

$$\text{RCOR}' \xrightarrow{[\text{O}]} \text{No reaction}$$

A ketone

The aldehydes are, in fact, among the most easily oxidized of organic compounds, and this fact helps chemists identify them. A sufficiently mild oxidizing agent can distinguish aldehydes not only from ketones but also from alcohols. Three such reagents have found widespread use. Each is an alkaline solution of an oxidizing metal ion kept in solution by a complexing agent.

Name	Oxidizing agent	Complexing agent	Positive test for aldehyde
Tollens's reagent	Ag^+	Ammonia (NH_3)	Silver mirror
Benedict's reagent	Cu^{2+}	Citrate ion ($C_6H_5O_7{}^{3-}$)	Brick-red precipitate
Fehling's reagent	Cu^{2+}	Tartrate ion ($C_4H_4O_6{}^{2-}$)	Brick-red precipitate

The silver ion in Tollens's reagent is kept in solution by complexing it with ammonia molecules.

$$H_3N—Ag^+—NH_3$$

Similarly, because Cu^{2+} forms an insoluble hydroxide in basic solution, citrate or tartrate ions are added to keep it in solution.

When Tollens's reagent oxidizes an aldehyde, the silver ion is reduced to free silver.

An aldehyde → RCHO; Silver mirror → Ag(s)

$$RCHO(aq) + 2\ Ag(NH_3)_2{}^+(aq) + 3\ OH^-(aq) \longrightarrow RCOO^-(aq) + 2\ Ag(s) + 4\ NH_3(aq) + 2\ H_2O$$

The silver, when deposited on a clean glass surface, produces a beautiful mirror. Indeed, the Tollens reaction is often used to silver mirrors. The reducing agent of choice for this is often the sugar glucose (which contains an aldehyde functional group) rather than a simple aldehyde. Ordinary ketones do not react with Tollens's reagent.

The copper(II) ion is the oxidizing agent and therefore must be the substance that is reduced, in this case to copper(I) oxide.

Both Benedict's and Fehling's reagents are blue because of the copper(II) ion complexes. A positive test for the aldehyde group is evidenced by a color change to brick red, indicating the presence of the copper(I) oxide.

An aldehyde → RCHO; A brick-red precipitate → $Cu_2O(s)$

$$RCHO(aq) + 2\ Cu^{2+}(aq\ complex) + 5\ OH^-(aq) \longrightarrow RCOO^-(aq) + Cu_2O(s) + 3\ H_2O$$

Although ketones resist oxidation by ordinary laboratory oxidizing agents, they do undergo combustion, as do aldehydes. Acetone is a common organic solvent. Neither it nor any other volatile, flammable organic solvent should be used around open flames, heating elements, or other possible sources of ignition.

$$\underset{\text{Acetone}}{CH_3COCH_3(l)} + \underset{\text{Oxygen (air)}}{4\ O_2(g)} \longrightarrow 3\ CO_2(g) + 3\ H_2O$$

◀ Benedict's test for aldehydes uses a solution of copper(II) ions (blue). Adding a small amount of glucose, a sugar with an aldehyde functional group, produces an olive-green color (center). Adding more glucose produces the brick-red precipitate of copper(I) oxide (right).

Reduction

A variety of reducing agents can convert aldehydes and ketones to the corresponding primary and secondary alcohols, respectively.

$$\underset{\text{An aldehyde or ketone}}{-\overset{\overset{\displaystyle O}{\|}}{C}-} \underset{\text{oxidation}}{\overset{\text{reduction}}{\rightleftharpoons}} \underset{\text{A 1° or 2° alcohol}}{-\overset{\overset{\displaystyle O-H}{|}}{\underset{|}{C}}-H}$$

Carbonyl compounds can also be reduced to alcohols by hydrogen gas in the presence of a metal catalyst (catalytic hydrogenation). However, this method suffers from the disadvantages that many of the catalysts (Pt, Pd, Ru) are expensive and other functional groups are also reduced.

$$\underset{\text{Acrolein (2-Propenal)}}{H_2C=CH-CHO} + 2\,H_2 \xrightarrow{Pt} \underset{\text{1-Propanol}}{H-CH_2-CH_2-CH_2-OH}$$

$$C_6H_9-\overset{\overset{\displaystyle O}{\|}}{C}-CH_3 \xrightarrow[Ni]{H_2} C_6H_{11}-\overset{\overset{\displaystyle OH}{|}}{\underset{\underset{\displaystyle H}{|}}{C}}-CH_3$$

For that reason sodium borohydride, $NaBH_4$, is often employed as the reducing agent. The $NaBH_4$ transfers a hydrogen as a hydride ion ($H:^-$) to the carbonyl carbon atom. Acid is then added, donating a proton (H^+) to the carbonyl oxygen atom. In this addition reaction, the $H:^-$ ion adds to the partially positive carbonyl carbon atom. In a subsequent step, H^+ ion from the acid adds to the negatively charged oxygen atom. The net result is the addition of hydrogen across the carbon-to-oxygen double bond, but the two hydrogen atoms come from different sources.

$$\underset{\text{An aldehyde}}{-\overset{\overset{\displaystyle O}{\|}}{C}-H} \xrightarrow{NaBH_4} -\overset{\overset{\displaystyle O^-}{|}}{\underset{\underset{\displaystyle H}{|}}{C}}-H \xrightarrow{H_3O^+} \underset{\text{A 1° alcohol}}{-\overset{\overset{\displaystyle O-H}{|}}{\underset{\underset{\displaystyle H}{|}}{C}}-H}$$

Two extremely important biochemical carbonyl reduction reactions, the reduction of acetaldehyde to ethanol and the reduction of pyruvate to lactate, are discussed in Chapter 25. In each case, the enzyme that catalyzes the reaction contains the coenzyme NADH, which is the reducing agent. (See Figure F.4 for the structure of the coenzyme.) NADH plays a role much like that of hydride ion in the sodium borohydride reduction.

$$\underset{\text{Acetaldehyde}}{CH_3-\overset{\overset{\displaystyle O}{\|}}{C}-H} + NADH + H^+ \underset{\text{dehydrogenase}}{\overset{\text{alcohol}}{\rightleftharpoons}} \underset{\text{Ethyl alcohol}}{CH_3-\overset{\overset{\displaystyle OH}{|}}{\underset{\underset{\displaystyle H}{|}}{C}}-H} + NAD^+$$

$$\underset{\text{Pyruvic acid}}{CH_3-\overset{\overset{\displaystyle O}{\|}}{C}-\overset{\overset{\displaystyle O}{\|}}{C}-OH} + NADH + H^+ \underset{\text{dehydrogenase}}{\overset{\text{lactic acid}}{\rightleftharpoons}} \underset{\text{Lactic acid}}{CH_3-\overset{\overset{\displaystyle OH}{|}}{\underset{\underset{\displaystyle H}{|}}{C}}-COOH} + NAD^+$$

Hydration of Carbonyl Compounds

Formaldehyde dissolves readily in water, reacting with the water to form a *hydrate*.

$$\mathrm{H{-}\overset{\overset{\displaystyle O}{\|}}{C}{-}H} + \mathrm{H{-}OH} \rightleftharpoons \mathrm{H{-}\overset{\overset{\displaystyle OH}{|}}{\underset{\underset{\displaystyle H}{|}}{C}}{-}OH}$$

Formaldehyde Water Formaldehyde hydrate

This is an addition reaction, analogous to the hydration of the carbon–carbon double bond of an alkene. The net result is that a hydrogen atom from water is added to the carbonyl oxygen and a hydroxyl group from water becomes attached to the carbonyl carbon. The reaction is reversible; the hydrate readily breaks down to re-form formaldehyde and water. At equilibrium at 20 °C, the hydrate predominates. There is only 1 molecule of free formaldehyde to about 10,000 of the hydrate.

Acetaldehyde is also hydrated in aqueous solution, but to a lesser extent than formaldehyde.

$$CH_3CHO + H_2O \rightleftharpoons CH_3CH(OH)_2$$

Acetaldehyde Acetaldehyde hydrate

At equilibrium about 58% of the molecules are in the hydrated form and 42% in the form of the free aldehyde. Generally higher aldehydes and ketones are even less hydrated in water. At equilibrium they exist primarily in the form of the free aldehyde (or ketone).

Addition of Alcohols to Carbonyl Groups

Alcohols add to the carbonyl group of aldehydes and ketones in much the same way as water does. The addition of 1 mol of an alcohol to 1 mol of an aldehyde yields a **hemiacetal**. (The corresponding reaction with a ketone forms a **hemiketal**.) In the presence of an acid catalyst, equilibrium is rapidly established, generally favoring the carbonyl compounds (the reactants). As with the hydrates, simple hemiacetals and hemiketals are generally not stable enough to be isolated.

Chloral Hydrate

In most cases it is impossible to isolate the hydrates from solution. Attempts to do so result in loss of water and regeneration of the carbonyl compound. The hydrate of trichloroacetaldehyde, commonly called chloral, is an exception.

$$CCl_3CHO + H_2O \longrightarrow CCl_3CH(OH)_2$$

Chloraldehyde Chloral hydrate

Chloral hydrate is a solid, soluble in water, and one of the very few stable organic compounds that possess two hydroxyl groups on the same carbon atom. It is a powerful sedative and soporific (sleep-inducing drug). Chloral hydrate has been widely used to quiet colicky babies, but it is quite toxic and is no longer recommended for that use. It is perhaps even better known in fictional mystery stories. Slipped into someone's drink, the mixture is called a "Mickey Finn" or "knockout drops." Such combinations of alcohol and chloral hydrate—two depressant drugs—are exceedingly dangerous. A little too much, and the unfortunate victim may be put to sleep permanently.

$$\mathrm{R{-}\overset{O}{\overset{\|}{C}}{-}H} + \mathrm{R'OH} \rightleftharpoons \mathrm{R{-}\underset{H}{\underset{|}{\overset{OH}{\overset{|}{C}}}}{-}OR'}$$

An aldehyde An alcohol A hemiacetal

$$\mathrm{R{-}\overset{O}{\overset{\|}{C}}{-}R'} + \mathrm{R'OH} \rightleftharpoons \mathrm{R{-}\underset{R'}{\underset{|}{\overset{OH}{\overset{|}{C}}}}{-}OR'}$$

A ketone An alcohol A hemiketal

When the alcohol and carbonyl groups occur within the same molecule, however, the equilibrium favors the formation of cyclic hemiacetals and hemiketals that have five- or six-membered rings. These cyclic compounds result from *intramolecular* interaction between the —OH and C=O groups. [The cyclization reactions are quite important in our discussion of the structures of simple sugars called monosaccharides (Chapter 19)].

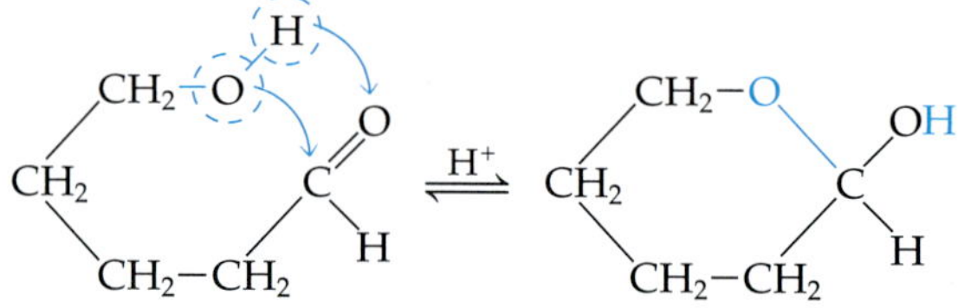

CH₂—O—H ... C(=O)H ⇌(H⁺) cyclic

$$\text{5-Hydroxypentanal: } \mathrm{HOCH_2CH_2CH_2CH_2CHO} \underset{}{\overset{H^+}{\rightleftharpoons}} \text{Cyclic hemiacetal (ring: } \mathrm{CH_2{-}O{-}CH(OH){-}CH_2{-}CH_2{-}CH_2}\text{)}$$

5-Hydroxypentanal Cyclic hemiacetal

In the presence of an anhydrous acid, such as HCl (g), hemiacetals react further with alcohols to form an **acetal**. Unlike hemiacetals and hydrates, acetals are stable in neutral or basic solutions and can be isolated.

$$\mathrm{R{-}\overset{O}{\overset{\|}{C}}{-}H} + \mathrm{R'OH} \rightleftharpoons \mathrm{R{-}\underset{H}{\underset{|}{\overset{OH}{\overset{|}{C}}}}{-}OR'} \xrightarrow[\text{dry HCl}]{\mathrm{R'OH}} \mathrm{R{-}\underset{H}{\underset{|}{\overset{OR'}{\overset{|}{C}}}}{-}OR'} + H_2O$$

An aldehyde An alcohol A hemiacetal (unstable) An acetal (stable)

First an alcohol molecule adds to the double bond of the aldehyde to form the hemiacetal. Then the hydroxyl group of the hemiacetal reacts with the hydroxyl hydrogen of a second alcohol molecule to eliminate a water molecule and form a second ether linkage. Note that the acetal has two ether linkages on the same carbon atom. A **ketal** is formed by a similar reaction when the starting carbonyl compound is a ketone rather than an aldehyde.

Acetals and ketals are resistant to oxidation. For this reason, acetal or ketal formation is often used to protect the functional group of aldehydes or ketones while other chemical operations are performed on the molecules. An aldehyde is converted to an acetal, an oxidation reaction is then carried out on another part of the molecule, and finally the aldehyde is regenerated from the acetal. The carbonyl group is easily regenerated by aqueous acid.

$$\mathrm{RCH(OR')_2} + \mathrm{H_2O} \xrightarrow{H^+} \mathrm{RCHO} + 2\,\mathrm{R'OH}$$

An acetal An aldehyde

Note that the regeneration of the aldehyde is the reverse reaction for the formation of the acetal, and that an aqueous solution favors the formation of the aldehyde, whereas anhydrous conditions favor formation of the acetal.

Chloramphenicol: A Cyclic Ketal

In an interesting application, the antibiotic chloramphenicol is treated with acetone to form a protective cyclic ketal that masks the bitter taste of the drug. The chloramphenicol carries both of the alcohol hydroxyl groups required to form this ketal.

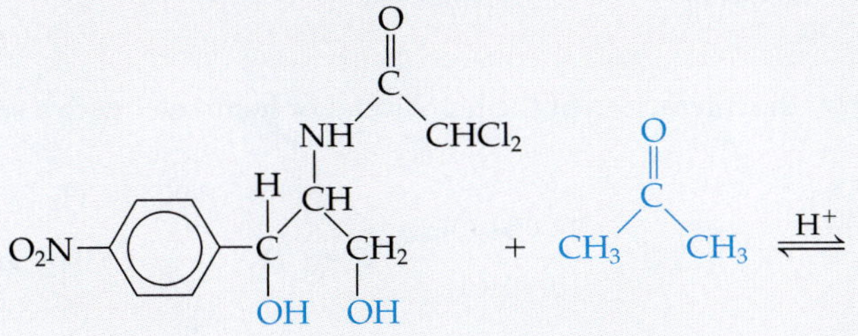

$$O_2N\text{-}C_6H_4\text{-}CH(OH)\text{-}CH(NHCOCHCl_2)\text{-}CH_2OH + CH_3COCH_3 \overset{H^+}{\rightleftharpoons}$$

Chloramphenicol (bitter) Acetone

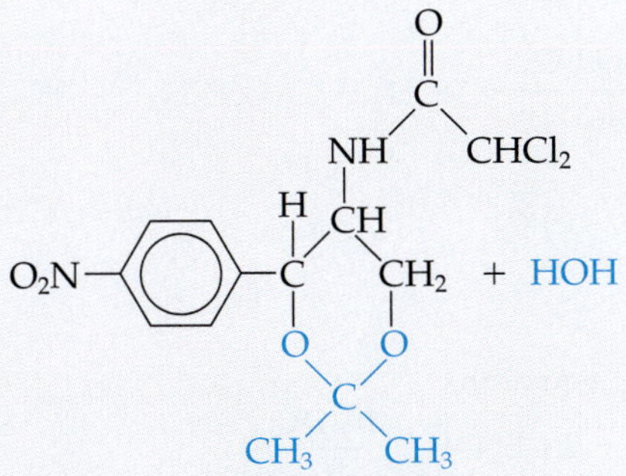

$$O_2N\text{-}C_6H_4\text{-}CH\text{-}CH(NHCOCHCl_2)\text{-}CH_2\text{ (ring: O}\text{-}C(CH_3)_2\text{-}O) + HOH$$

A cyclic ketal (not bitter)

Acids in the digestive tract convert the cyclic ketal back to chloramphenicol, a powerful but hazardous antibiotic. It is used only when other, less dangerous drugs are ineffective. In about 1 person in 20,000 to 40,000 (depending on dosage), chloramphenicol causes fatal aplastic anemia.

Example 15.5

Complete the following equations.

a. $CH_3CH_2CHO + 2\,CH_3CH_2OH \xrightarrow{\text{dry HCl}}$

b. $CH_3COCH_3 + 2\,CH_3OH \xrightarrow{\text{dry HCl}}$

Solution

First, recognize the type of reaction. The formulas are condensed structures of propanal (**a**) and acetone (**b**). The two moles of an alcohol and the dry HCl indicate acetal formation (**a**) and ketal formation (**b**). Recall that acetal and ketal formation each involves two steps. First the addition of 1 mol of alcohol to 1 mol of the carbonyl compound.

a.

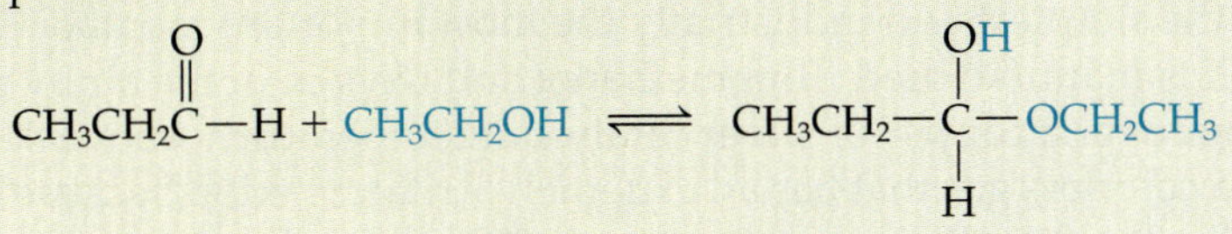

$$CH_3CH_2\overset{O}{\overset{\|}{C}}\text{—}H + CH_3CH_2OH \rightleftharpoons CH_3CH_2\text{—}\underset{H}{\overset{OH}{C}}\text{—}OCH_2CH_3$$

Propanal (Propionaldehyde) Ethanol A hemiacetal

b.

$$CH_3\overset{\overset{\displaystyle O}{\|}}{C}CH_3 + CH_3OH \rightleftharpoons CH_3-\underset{\underset{\displaystyle CH_3}{|}}{\overset{\overset{\displaystyle OH}{|}}{C}}-OCH_3$$

Propanone (Acetone) — Methanol — A hemiketal

This is followed by the interaction of the hemiacetal or hemiketal with a second mole of the alcohol.

a.

$$CH_3CH_2-\underset{\underset{\displaystyle H}{|}}{\overset{\overset{\displaystyle OH}{|}}{C}}-OCH_2CH_3 \xrightarrow[\text{dry HCl}]{CH_3CH_2OH} CH_3CH_2-\underset{\underset{\displaystyle H}{|}}{\overset{\overset{\displaystyle OCH_2CH_3}{|}}{C}}-OCH_2CH_3 + H_2O$$

A hemiacetal — An acetal

b.

$$CH_3-\underset{\underset{\displaystyle CH_3}{|}}{\overset{\overset{\displaystyle OH}{|}}{C}}-OCH_3 \xrightarrow[\text{dry HCl}]{CH_3OH} CH_3-\underset{\underset{\displaystyle CH_3}{|}}{\overset{\overset{\displaystyle OCH_3}{|}}{C}}-OCH_3 + H_2O$$

A hemiketal — A ketal

Practice Exercises

A. Complete the following equations.

a. $CH_3CH(CH_3)CHO + 2\,CH_3OH \xrightarrow{\text{dry HCl}}$

b. (cyclopentanone) $=O + 2\,CH_3CH_2OH \xrightarrow{\text{dry HCl}}$

B. Complete the following equation.

$CH_3CH_2CHO + HOCH_2CH_2OH \xrightarrow{\text{dry HCl}}$

Review Questions

15.4 Name three types of reactions shared by aldehydes and ketones. For each type, indicate the product from reaction of an aldehyde and from reaction of a ketone.

15.5 How can one distinguish between an aldehyde and a ketone? Describe one of these tests in detail.

15.6 An aldehyde can be converted to an alcohol or a carboxylic acid. What reagents are used for each reaction?

15.7 The formation of an acetal from an aldehyde and alcohol requires an acid catalyst, as does the formation of an aldehyde from an acetal. What conditions determine whether the acetal or the aldehyde is the predominant product?

▲ Large quantities of formaldehyde are used to make phenol-formaldehyde resins for gluing the wood sheets in plywood and for use as adhesives in other building materials.

15.7 Some Common Carbonyl Compounds

Formaldehyde is the simplest and industrially the most important member of the aldehyde family. It is manufactured from methanol and oxygen in the air by passing methanol vapor over a copper or silver catalyst at temperatures above 300 °C. Formaldehyde is a colorless gas that has an extremely irritating odor. Because of its reactivity, it is difficult to handle in the gaseous state. For many uses it is therefore dissolved in water and sold as a 37 to 40% aqueous solution (such a solution is called *formalin*).

Benzaldehyde Cinnamaldehyde Vanillin

cis-3-Hexenal *trans*-2-*cis*-6-Nonadienal

Figure 15.1
Some interesting aldehydes; benzaldehyde is an oil found in almonds; cinnamaldehyde is oil of cinnamon; vanillin gives vanilla its flavor; *cis*-3-hexenal provides an herbal odor; and *trans*-2-*cis*-6-nonadienal gives a cucumber odor.

The largest use of formaldehyde is as a reagent for the preparation of many other organic compounds, especially polymers. It is used in the manufacture of hard polymers such as Bakelite, Formica, and Melmac, and for the manufacture of adhesives.

Formaldehyde denatures proteins (Chapter 21), rendering them insoluble in water and resistant to bacterial decay. For this reason it is used in embalming solutions and in the preservation of biological specimens. Formalin is also used as a general antiseptic in hospitals to sterilize gloves and surgical instruments. However, its use as an antiseptic, preservative, and embalming fluid has declined in recent years because formaldehyde is suspected of being carcinogenic.

Formaldehyde is present in wood smoke. Because it kills bacteria, it is one of the compounds responsible for the preservative effect in smoked foods.

Acetaldehyde is an extremely volatile, colorless liquid. It is prepared by the catalytic (Ag) oxidation of ethyl alcohol or the catalytic ($PdCl_2$) oxidation of ethylene. It is a starting material for the preparation of many other organic compounds, such as acetic acid, ethyl acetate, and chloral. Acetaldehyde is formed as a metabolite in the fermentation of sugars and in the detoxification of alcohol in the liver (Chapter 25).

Aldehydes are the active components of many other familiar substances (Figure 15.1). Even the odor of green leaves is due in part to a carbonyl compound, *cis*-3-hexenal that, along with other carbonyl compounds (with related acetals, ketals, and alcohols), imparts a "green" herbal odor to shampoos and other cosmetics.

Acetone is the simplest and most important ketone. It is produced in large quantities by the catalytic (Ag) oxidation of isopropyl alcohol and is an important by-product of phenol production. Because it is miscible with water as well as with most organic solvents, acetone finds its chief use as an industrial solvent (for example, for paints and lacquers). It is the chief ingredient in some brands of nail polish remover. Acetone is also an important intermediate in the preparation of chloroform, iodoform, dyes, methacrylate polymers, and many other complex organic compounds.

Acetone is formed in the human body as a by-product of lipid metabolism. Normally it does not accumulate to an appreciable extent because it is oxidized to carbon dioxide and water. The normal concentration of acetone in the human body is less than 1 mg/100 mL of blood. In certain disease states, such as uncontrolled diabetes mellitus, the acetone concentration rises to higher levels. It is then excreted in the urine, where it is easily detected. In severe cases, its odor can be noted on the breath (Chapter 26).

▲ Cinnamaldehyde is the main flavor component of cinnamon.

Like aldehydes, ketones are the active components of many familiar substances (Figure 15.2). Also, as we shall see in Selected Topic E, several steroid hormones have the carbonyl functional group as an integral part of their structures. These

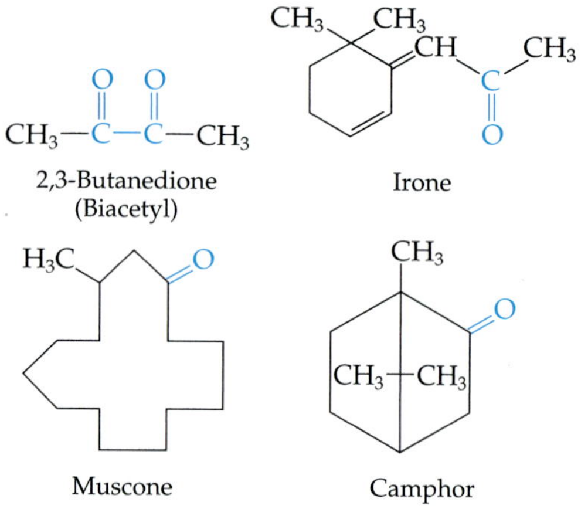

Figure 15.2 ▶
Some interesting ketones. 2,3-Butanedione is a butter flavoring (page 394), and irone is responsible for the odor of violets. Muscone is musk oil, an ingredient in perfumes, and camphor is used in some insect repellents.

$CH_3-\overset{O}{\overset{\|}{C}}-\overset{O}{\overset{\|}{C}}-CH_3$

2,3-Butanedione (Biacetyl)

Irone

Muscone

Camphor

include progesterone, a hormone secreted by the ovaries that stimulates the growth of cells in the uterus wall, preparing it for attachment of a fertilized egg, and testosterone, the main male sex hormone. These (and other) sex hormones affect our development and our lives in most fundamental ways.

Review Question

15.8 Name two aldehydes that serve as active principles in flavors or aromas.

Summary

The **carbonyl group,** a carbon-to-oxygen double bond, is the defining feature of **aldehydes** and **ketones.** In aldehydes at least one bond on the carbonyl group is a carbon–hydrogen bond, and in ketones both available bonds on the carbonyl carbon are carbon–carbon bonds.

Aldehydes are synthesized by oxidation of primary alcohols. The aldehyde can be further oxidized to a carboxylic acid. Ketones are prepared by oxidation of secondary alcohols. Both aldehydes and ketones undergo combustion and both can be reduced to alcohols. Aldehydes are oxidized to carboxylic acids by even mild oxidizing agents. Ketones are not oxidized by these reagents.

Aldehydes add water across the carbonyl double bond to form molecules known as *hydrates,* compounds with two OH groups on the same carbon atom. Aldehydes add an alcohol across the carbonyl group, forming a **hemiacetal.** When this alcohol addition takes place with a ketone, the product is a **hemiketal**. Hemiacetals and hemiketals have an OH group and an OR group on the same carbon atom.

In the presence of excess alcohol and dry HCl, the hemiacetal and hemiketal react further to form an **acetal** or a **ketal**. Both acetals and ketals have two OR groups on the same carbon atom. Alcohol addition can also take place intramolecularly in molecules that contain both a hydroxyl group and a carbonyl group. In this case the product is a *cyclic hemiacetal* or *cyclic hemiketal.*

Key Terms

acetal (15.6)
aldehyde (15.1)
carbonyl group (15.1)
hemiacetal (15.6)
hemiketal (15.6)
ketal (15.6)
ketone (15.1)

Problems

Names and Structural Formulas

1. Name each of the following:

a. 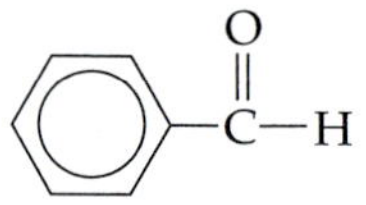 $C_6H_5-\overset{O}{\overset{\|}{C}}-H$

b. $CH_2OHCH_2\overset{O}{\overset{\|}{C}}-H$

c. $(CH_3)_3CCH_2CH_2\overset{O}{\overset{\|}{C}}-H$

d. 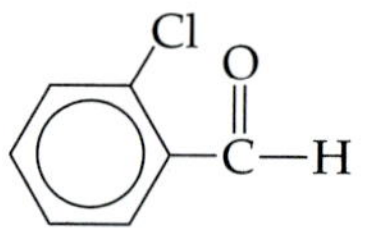 (2-Cl-C_6H_4)$-\overset{O}{\overset{\|}{C}}-H$

2. Name each of the following:
   a. $CH_3CH_2CH_2\overset{\overset{O}{\|}}{C}-H$
   b. $CH_3CHClCCl_2\overset{\overset{O}{\|}}{C}-H$
   c. $(CH_3CH_2)_2CH\overset{\overset{O}{\|}}{C}-H$
   d. 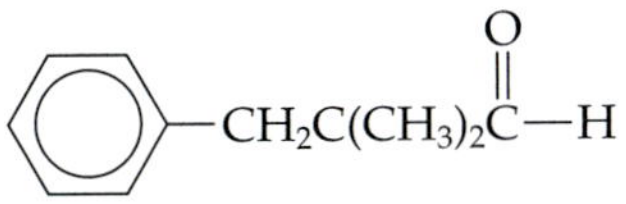 $C_6H_5-CH_2C(CH_3)_2\overset{\overset{O}{\|}}{C}-H$
3. Name each of the following:
   a. $CH_3CH_2\overset{\overset{O}{\|}}{C}CH_2CH(CH_3)_2$
   b. 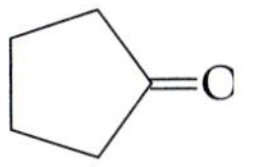 cyclopentane ring $=O$
   c. $CH_3\overset{\overset{O}{\|}}{C}CH_2CH_2CH_3$
   d. $(CH_3)_3C\overset{\overset{O}{\|}}{C}CHBrCH_3$
4. Name each of the following:
   a. $(CH_3)_2CH\overset{\overset{O}{\|}}{C}CHCl_2$
   b. $CH_3CH_2CH(CH_3)\overset{\overset{O}{\|}}{C}CH_3$
   c. 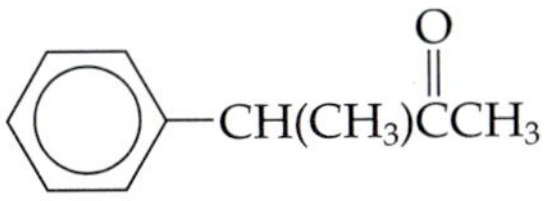 $C_6H_5-CH(CH_3)\overset{\overset{O}{\|}}{C}CH_3$
   d. 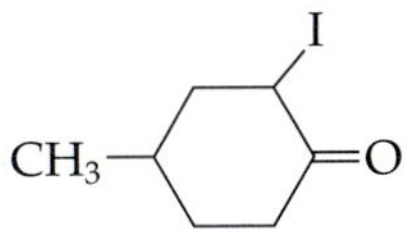 cyclohexane ring with CH_3-, I, and $=O$
5. Give structural formulas for each of the following:
   a. butyraldehyde b. 3-methylheptanal
   c. *p*-nitrobenzaldehyde
6. Give structural formulas for each of the following:
   a. 5-ethyloctanal b. 2-chloropropanal
   c. 2,5-dimethylhexanal
7. Give structural formulas for each of the following:
   a. 2-hexanone b. 3-bromo-2-heptanone
   c. 4-methylcyclohexanone
8. Give structural formulas for each of the following:
   a. 1-phenyl-2-butanone
   b. 2-iodo-2-methyl-4-octanone
   c. 2-hydroxy-3-pentanone

Physical Properties

9. Which compound has the higher boiling point: acetone or 2-propanol? Explain.
10. Which compound has the higher boiling point: butanal or 1-butanol? Explain.
11. Which compound has the higher boiling point: dimethyl ether or acetaldehyde? Explain.
12. Which compound has the higher boiling point: acetone or isobutane? Explain.

Preparation of Aldehydes and Ketones

13. Give the structures of the alcohols that could be oxidized to:
    a. 4-methylcyclohexanone
    b. 2,2-dimethylpropanal
    c. 3-bromopentanal
14. Give the structures of the alcohols that could be oxidized to:
    a. 2-pentanone b. phenylethanal
    c. *o*-methylbenzaldehyde

Chemical Reactions

15. Write the equations for the reactions of acetaldehyde with each of the following:
    a. 1 mol of CH_3OH
    b. 2 mol of CH_3OH, with dry HCl present
    c. 1 mol of $HOCH_2CH_2OH$, with dry HCl present
16. Write the equations for the reactions of acetaldehyde with each of the following:
    a. Cu^{2+} b. $K_2Cr_2O_7$
    c. hydrogen gas with a nickel catalyst
17. Write the equations for the reactions, if any, of acetone with the reagents in Problem 15.
18. Write the equations for the reactions, if any, of acetone with the reagents in Problem 16.
19. Indicate whether Tollens's reagent could be used to distinguish between the compounds in each of the following sets. Explain.
    a. 1-pentanol and pentanal
    b. 2-pentanol and 2-pentanone
    c. pentanal and 2-pentanone
    d. pentanal and pentane
    e. 2-pentanone and pentane
20. Assume that a *stronger* oxidizing agent, such as $K_2Cr_2O_7$, could be used as a test for distinguishing among compounds. For each set in Problem 19, indicate whether this reagent would distinguish between the two compounds. Explain.
21. What reagent would you use to distinguish between 2-pentanone and 2-pentanol? What would you observe when the reagent was added?
22. What reagent would you use to distinguish between 2-pentanone and pentanal? What would you observe when the reagent was added?
23. List the reagents necessary to carry out the following conversions.
    a. $CH_3CH_2\overset{\overset{O}{\|}}{C}-H \xrightarrow{?} CH_3CH_2\overset{\overset{O}{\|}}{C}-O^- + Ag(s)$
    b. $CH_3CH_2CH_2CH_2OH \xrightarrow{?} CH_3CH_2CH_2\overset{\overset{O}{\|}}{C}-H$
    c. $CH_3-\overset{\overset{O}{\|}}{C}-CH_3 \xrightarrow{?} CH_3-\underset{\underset{CH_3}{|}}{\overset{\overset{OCH_3}{|}}{C}}-OCH_3$
24. List the reagents necessary to carry out the following conversions.
    a. $(CH_3)_2CH\overset{\overset{OH}{|}}{C}HCH_3 \xrightarrow{?} (CH_3)_2CH\overset{\overset{O}{\|}}{C}CH_3$

b. 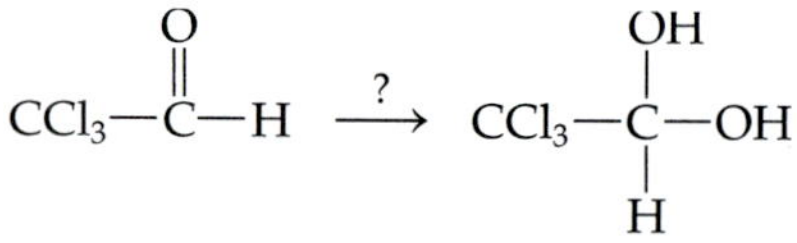 $CCl_3-\overset{\overset{O}{\|}}{C}-H \xrightarrow{?} CCl_3-\underset{\underset{H}{|}}{\overset{\overset{OH}{|}}{C}}-OH$

c. (cyclopentanone) $\xrightarrow{?}$ (cyclopentanol)

e. $C_6H_5-CH_2CH_2\underset{\underset{OH}{|}}{C}HOH$

f. $C_6H_5-\underset{\underset{OCH_2CH_3}{|}}{C}HOCH_2CH_3$

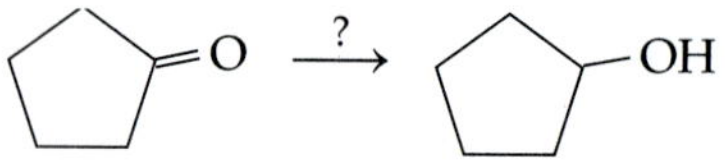

Hydrates, Hemiacetals, and Acetals

25. Which of the following compounds are hemiacetals?

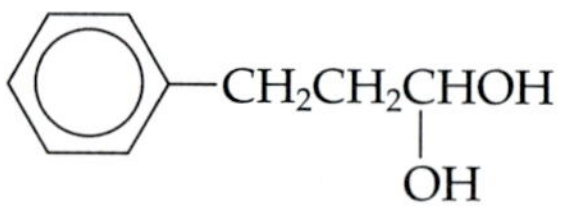

a. $CH_3CH_2\underset{\underset{OH}{|}}{C}HOCH_3$ b. $CH_3CH_2\underset{\underset{OCH_3}{|}}{C}HOCH_3$

c. $CH_3CH_2\underset{\underset{OH}{|}}{C}HOH$ d. 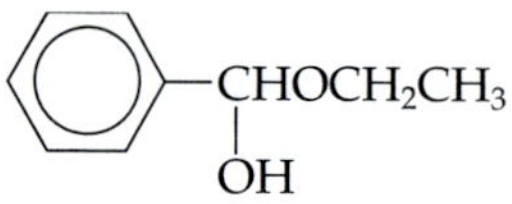 $C_6H_5-\underset{\underset{OH}{|}}{C}HOCH_2CH_3$

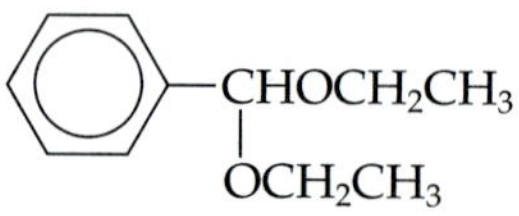

26. Which of the substances in Problem 25 are acetals?

27. Which of the substances in Problem 25 are hydrates?

28. Draw the hemiacetal formed from the intramolecular reaction of

$$HOCH_2CH_2CH_2CH_2\overset{\overset{O}{\|}}{C}-H$$

Additional Problems

29. Name the three functional groups on the vanillin molecule (Figure 15.1).

30. Name the three functional groups on the testosterone molecule.

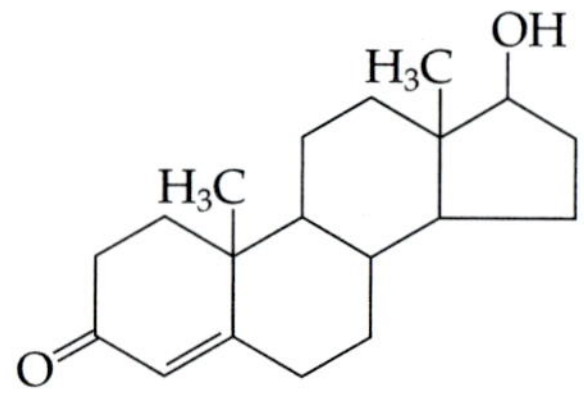

Testosterone

31. Draw structures, and give common and IUPAC names, for the four isomeric aldehydes having the formula $C_5H_{10}O$.

32. Draw structures, and give common and IUPAC names, for the three isomeric ketones having the formula $C_5H_{10}O$.

33. As we shall see in Chapter 19, 2,3-dihydroxypropanal and 1,3-dihydroxyacetone are important carbohydrates. Draw their structural formulas.

34. Glutaraldehyde (pentanedial) is a germicide that is replacing formaldehyde as a sterilizing agent. It is less irritating to the eyes, nose, and skin. Draw the structural formula of glutaraldehyde.

35. Chloral (CCl_3CHO), which forms a stable hydrate, also forms a stable hemiacetal. Give the structure of the hemiacetal formed by the reaction of chloral with methanol.

36. What is the effective chemical reagent in each of the following?

a. Benedict's solution b. Fehling's solution

c. Tollens's reagent

37. Which of the compounds in Figure 15.1 would give a positive Benedict's test?

Carboxylic Acids and Derivatives

The fragrance of many consumer products such as soaps, colognes, potpourri, and perfumes is quite dependent on esters, one of the families of compounds discussed in this chapter.

Learning Objectives/Study Questions

1. What is the general structure for a carboxylic acid? An ester? An amide? What is the functional group for each of these classes of organic molecules?
2. How are the common names of carboxylic acids, esters, and amides determined? How are these compounds named using IUPAC nomenclature?
3. Why are the boiling points of carboxylic acids higher than those of alcohols of similar molar mass? How do the boiling points of esters and amides compare with alcohols of similar molar mass?
4. How do the solubilities of carboxylic acids, esters, and amides of five carbons or less compare to the solubilities of comparable alkanes and alcohols in water?
5. How are carboxylic acids, esters, and amides prepared?
6. What typical reactions take place with carboxylic acids, esters, and amides?
7. What are some common carboxylic acids and esters and their uses?
8. What are phosphate esters? Why are they important?

Organic acids were known long before the inorganic acids were isolated. Primitive tribes were familiar with organic acids, such as the acetic acid they obtained when their fermentation reactions went awry and produced vinegar instead of alcohol. Naturalists of the seventeenth century knew that the sting of a red ant's bite was due to an organic acid that the pest injected into the wound. And it has long been recognized that the crisp, tart flavor of citrus fruits is produced by an organic compound appropriately called citric acid. The acetic acid of vinegar, the formic acid of red ants, and the citric acid of fruits all belong to the same family of compounds, the carboxylic acids.

Several derivatives of carboxylic acids are also important. The amides, of which proteins (Chapter 21) are perhaps the most spectacular example, and the esters, which include fats (Chapter 20), are two classes of acid derivatives we shall consider most carefully. Two synthetic fibers are also classed within these two families of derivatives. Nylon, like silk and wool, is a polyamide. Dacron is a polyester.

In this chapter, we look at simple carboxylic acids, esters, and amides. The more complex worlds of lipids and proteins we shall save for later chapters.

16.1 Carboxylic Acids and Their Derivatives: The Functional Groups

We devoted Chapter 15 to the carbonyl group, and there we noted that this functional group determines the chemistry of the aldehydes and ketones. The carbonyl group is also incorporated in carboxylic acids and several families of compounds derived from them. However, in these compounds, the carbonyl group is only one part of the functional group that characterizes these families.

The functional group of the **carboxylic acids** is the **carboxyl group.** This group can be considered a combination of the *carb*onyl group ($>C{=}O$) and the hydrox*yl* group (—OH), but it has characteristic properties of its own. As with aldehydes and ketones (Chapter 15), the carbon-to-oxygen double bond can be shown explicitly or the carboxyl group can be written in condensed form on one line.

$$\mathrm{R{-}\overset{\overset{\large O}{\|}}{C}{-}OH} \quad \text{or} \quad \mathrm{RCOOH} \qquad\qquad \mathrm{{-}\overset{\overset{\large O}{\|}}{C}{-}OH} \quad \text{or} \quad \mathrm{{-}COOH}$$

A carboxylic acid   A carboxyl group

The **amide** functional group has a nitrogen atom, instead of oxygen, attached to the carbonyl group. The properties of the amide functional group differ from those of the simple carbonyl group and from those of simple nitrogen-containing compounds, called amines (Chapter 17).

$$\mathrm{R{-}\overset{\overset{\large O}{\|}}{C}{-}\underset{|}{N}{-}} \quad \text{or} \quad \mathrm{RCO\underset{|}{N}{-}} \qquad\qquad \mathrm{{-}\overset{\overset{\large O}{\|}}{C}{-}\underset{|}{N}{-}} \quad \text{or} \quad \mathrm{{-}CO\underset{|}{N}{-}}$$

An amide   An amide group

The functional group of the **esters** looks a little like that of an ether and a little like that of a carboxylic acid. As you should now suspect, compounds in this group react neither like carboxylic acids nor like ethers, but rather make up a distinctive family.

$$\mathrm{R{-}\overset{\overset{\large O}{\|}}{C}{-}OR'} \quad \text{or} \quad \mathrm{RCOOR'} \qquad\qquad \mathrm{{-}\overset{\overset{\large O}{\|}}{C}{-}OR} \quad \text{or} \quad \mathrm{{-}COOR}$$

An ester   An ester group

Table 16.1 Carboxylic Acid Derivatives

Family	Functional Group	Example	Common Name	IUPAC Name
Carboxylic acid	$-C(=O)-OH$	$CH_3C(=O)-OH$	Acetic acid	Ethanoic acid
Amide	$-C(=O)-N<$	$CH_3C(=O)-NH_2$	Acetamide	Ethanamide
Ester	$-C(=O)-O-C\equiv$	$CH_3C(=O)-OCH_3$	Methyl acetate	Methyl ethanoate

We consider the amides and esters to be *derivatives* of carboxylic acids because the carboxyl's OH is replaced with another group. These functional groups are listed in Table 16.1. The table also offers an example (with common and IUPAC names) for each type of compound. We shall consider nomenclature in more detail as we take up each family separately.

Review Question

16.1 Draw the functional group in each of the following. How are they alike? How do they differ?

a. aldehydes **b.** ketones **c.** carboxylic acids
d. esters **e.** ethers **f.** amides

Carboxylic Acids

Carboxylic acids occur widely in nature, usually combined with alcohols or other groups in fats, waxes, and other familiar substances. They are components of foods, medicines, and many household products (Figure 16.1). An understanding of the chemistry of carboxylic acids is essential to the biochemistry that comes later in the text.

Figure 16.1
Carboxylic acids occur in many common household items. Vinegar contains acetic acid; aspirin is acetylsalicylic acid; vitamin C is ascorbic acid; lemons contain citric acid, and both spinach and Zud cleanser contain oxalic acid.

$H-\overset{\overset{O}{\|}}{C}-OH$

Formic acid

$CH_3-\overset{\overset{O}{\|}}{C}-OH$

Acetic acid

$CH_3CH_2-\overset{\overset{O}{\|}}{C}-OH$

Propionic acid

$CH_3CH_2CH_2-\overset{\overset{O}{\|}}{C}-OH$

Butyric acid

$C_6H_5-\overset{\overset{O}{\|}}{C}-OH$

Benzoic acid

16.2 Some Common Carboxylic Acids: Structures and Names

Most of the carboxylic acids we consider are derived from natural sources. Many of them are best known by common names that are based upon Latin and Greek names related to the source of the acid.

The simplest carboxylic acid is formic acid (HCOOH). It was first obtained by the distillation of ants (Latin *formica,* ant). The bite of an ant smarts because the ant injects formic acid as it bites. The stings of wasps and bees also contain formic acid (as well as other poisonous materials).

The next higher homolog is acetic acid (CH_3COOH). It can be made by the aerobic fermentation of a mixture of cider and honey. This produces a solution (vinegar) that contains 4 to 10% acetic acid, plus a number of other compounds that give vinegar its flavor. Acetic acid is probably the most familiar *weak* acid used in educational and industrial chemistry laboratories.

The third homolog, propionic acid (CH_3CH_2COOH), is seldom encountered in everyday life. The fourth member is more familiar, at least by its odor. If you've ever smelled rancid butter, you probably wish you hadn't—but you know what butyric ($CH_3CH_2CH_2COOH$) acid smells like. It is one of the most foul-smelling substances imaginable. Butyric acid can be isolated from butterfat or synthesized in the laboratory. It is one of the ingredients of body odor. Extremely small amounts of this and other chemicals enable bloodhounds to track fugitives.

The acid with the carboxyl group attached directly to a benzene ring is called benzoic acid (C_6H_5COOH). In general, carboxylic acids are represented by the formula RCOOH, where R can be either an alkyl or an aryl group.

Table 16.2 lists several members of the carboxylic acid family and the derivations of their common names. When common names are used, substituted acids are named by designating the position of the substituent group by the Greek letters α, β, γ, δ, and so forth, rather than numbers. These letters refer to the position of the carbon atom in relation to the carboxyl carbon.

$$\overset{\delta}{C}-\overset{\gamma}{C}-\overset{\beta}{C}-\overset{\alpha}{C}-COOH$$

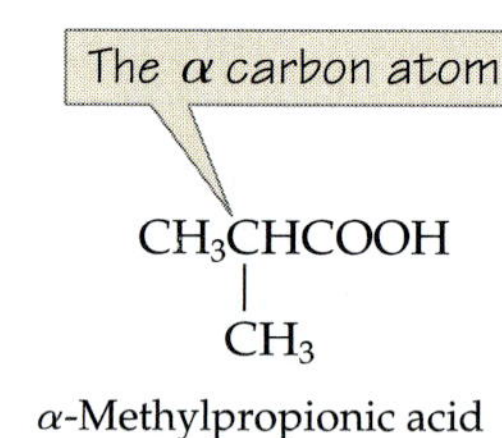

α-Methylpropionic acid

The β carbon atom

$CH_3\underset{\underset{OH}{|}}{C}HCH_2COOH$

β-Hydroxybutyric acid

In the IUPAC system, the parent hydrocarbon is taken to be the one that corresponds to the longest continuous chain (LCC) containing the carboxyl group. The -*e* ending of the parent alkane is replaced by the suffix -*oic,* and the word *acid.* As with aldehydes, the carboxyl carbon is designated C-1, and numbers are used to indicate any substituted carbons in the parent chain. Remember: Greek letters with common names, numbers with IUPAC names.

▲ Acetic acid is a familiar laboratory weak acid. It is also the active ingredient in vinegar.

Example 16.1

Give the common and IUPAC names for the following compound.

$$CH_3CH_2\underset{\underset{Br}{|}}{C}HCOOH$$

Table 16.2 Some Common Aliphatic Carboxylic Acids

Condensed Formula	IUPAC Name	Common Name	Derivation of Common Name
$HCOOH$	Methanoic acid	Formic acid	Latin *formica*, ant
CH_3COOH	Ethanoic acid	Acetic acid	Latin *acetum*, vinegar
CH_3CH_2COOH	Propanoic acid	Propionic acid	Greek *protos*, first, and *pion*, fat
$CH_3CH_2CH_2COOH$	Butanoic acid	Butyric acid	Latin *butyrum*, butter
$CH_3(CH_2)_3COOH$	Pentanoic acid	Valeric acid	Latin *valere*, powerful
$CH_3(CH_2)_4COOH$	Hexanoic acid	Caproic acid	Latin *caper*, goat
$CH_3(CH_2)_6COOH$	Octanoic acid	Caprylic acid	Latin *caper*, goat
$CH_3(CH_2)_8COOH$	Decanoic acid	Capric acid	Latin *caper*, goat
$CH_3(CH_2)_{10}COOH$	Dodecanoic acid	Lauric acid	Laurel tree
$CH_3(CH_2)_{12}COOH$	Tetradecanoic acid	Myristic acid	*Myristica fragrans* (nutmeg)
$CH_3(CH_2)_{14}COOH$	Hexadecanoic acid	Palmitic acid	Palm tree
$CH_3(CH_2)_{16}COOH$	Octadecanoic acid	Stearic acid	Greek *stear*, tallow

Notice in Table 16.2 that the higher carboxylic acids listed have an even number of carbons. As we shall see in Section 26.5, these acids, called fatty acids, are synthesized in nature by adding two carbon atoms at a time.

Solution

The LCC contains four carbon atoms; the compound is therefore named as a substituted butyric (or butanoic) acid.

$$\overset{4}{\underset{\gamma}{}}\ \overset{3}{\underset{\beta}{}}\ \overset{2}{\underset{\alpha}{}}\ \overset{1}{} \qquad CH_3CH_2\underset{\substack{|\\ Br}}{C}HCOOH$$

The bromine atom is at the alpha carbon in the common system, or at C-2 in the IUPAC system. The compound is α-bromobutyric acid or 2-bromobutanoic acid.

Practice Exercises

A. Give the IUPAC name for $ClCH_2CH_2CH_2CH_2CH_2COOH$

B. Give the IUPAC name for $(CH_3)_2CHCH_2CH_2COOH$.

Example 16.2

Draw the structural formula for α,β-dichloropropionic acid.

Solution

Propionic acid has three carbon atoms:

$$C{-}C{-}COOH$$

Two chlorine atoms must be attached to the parent chain, one at the alpha carbon and one at the beta carbon. Enough hydrogen atoms are attached to give each carbon atom four bonds.

$$\underset{\substack{|\\ Cl}}{\overset{\beta}{C}}H_2\underset{\substack{|\\ Cl}}{\overset{\alpha}{C}}HCOOH$$

Practice Exercises

A. Draw a structural formula for 4-bromo-5-methylhexanoic acid.

B. Draw a structural formula for γ-aminobutyric acid. (The amino group is $-NH_2$.)

The aliphatic dicarboxylic acids and their derivatives are quite important in biological systems. These acids are almost always referred to by their common names; a mnemonic for remembering these names is given in Table 16.3.

Table 16.3 Aliphatic Dicarboxylic Acids

Formula	IUPAC Name	Common Name	Mnemonic
HOOCCOOH	Ethanedioic acid	Oxalic acid	Oh
$HOOCCH_2COOH$	1,3-Propanedioic acid	Malonic acid	My
$HOOC(CH_2)_2COOH$	1,4-Butanedioic acid	Succinic acid	Such
$HOOC(CH_2)_3COOH$	1,5-Pentanedioic acid	Glutaric acid	Good
$HOOC(CH_2)_4COOH$	1,6-Hexanedioic acid	Adipic acid	Apple
$HOOC(CH_2)_5COOH$	1,7-Heptanedioic acid	Pimelic acid	Pie

Review Question

16.2 Give the common and IUPAC names for the straight-chain carboxylic acids containing the given number of carbon atoms.

a. 1 **b.** 2 **c.** 3 **d.** 4

16.3 Preparation of Carboxylic Acids

Few carboxylic acids occur free in nature. Many aliphatic acids, particularly those containing even numbers of carbon atoms, are found combined with glycerol in fats (Chapter 20). Some of the acids are available from the hydrolysis of fats and thus are called free fatty acids.

Carboxylic acids are the final oxidation products of aldehydes and/or primary alcohols. The acid produced contains the same number of carbon atoms as did the precursor aldehyde or alcohol.

General Equations

$$\underset{\text{A primary alcohol}}{RCH_2OH} \xrightarrow{[O]} \underset{\text{A carboxylic acid}}{RCOOH}$$

$$\underset{\text{An aldehyde}}{RCHO} \xrightarrow{[O]} \underset{\text{A carboxylic acid}}{RCOOH}$$

Specific Equation

$$\underset{\text{Ethanol}}{CH_3CH_2OH} \xrightarrow[H_2SO_4]{K_2Cr_2O_7} \underset{\text{Acetic acid}}{CH_3COOH}$$

Our bodies also oxidize alcohols to acids; the liver has enzymes that convert ethanol to acetic acid.

$$CH_3CH_2OH \xrightarrow{\text{Alcohol dehydrogenase}} CH_3CHO \xrightarrow{\text{Acetaldehyde dehydrogenase}} CH_3COOH$$

Acetic acid can be further oxidized to provide energy (Chapter 24) or converted to fat (Chapter 26).

Review Question

16.3 Valeric acid (pentanoic acid) can be prepared in an oxidation reaction from (**a**) what alcohol and (**b**) what aldehyde?

16.4 Physical Properties of Carboxylic Acids

The first nine members of the carboxylic acid series are colorless liquids with disagreeable odors. The odor of vinegar is that of acetic acid; the odor of rancid butter is primarily that of butyric acid. Caproic acid is present in the hair and the secretions of goats. The acids from C_5 to C_{10} all have "goaty" odors (odor of Limburger cheese). These acids are also produced by the action of skin bacteria on human sebum (skin oils), hence the odor of poorly ventilated locker rooms ("essence of old gym sneakers"). The acids above C_{10} are waxlike solids, and their odor diminishes with increasing molar mass and resultant decreasing volatility.

Even in the vapor phase, some of the hydrogen bonds between acid molecules are not broken. The structure of the carboxyl group permits two molecules to hydrogen-bond very strongly to one another:

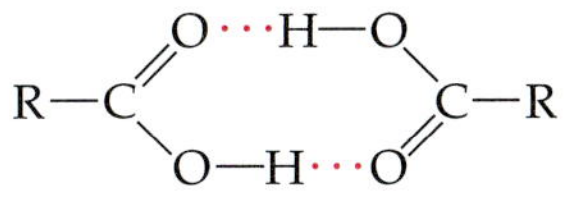

```
      O···H—O
     //      \
R—C           C—R
     \      //
      O—H···O
```

In many situations the interaction is so strong that the **dimer** (two-molecule unit) acts as a single particle. Because of this, when carboxylic acids are the solute, osmotic pressures, freezing-point depressions, and other colligative properties are frequently less than one would expect if the solute behaved as two separate molecules.

The carboxyl group readily hydrogen-bonds to water molecules. The acids having one to four carbon atoms are colorless liquids that are completely miscible with water. Solubility decreases with increasing number of carbon atoms: hexanoic acid [$CH_3(CH_2)_4COOH$] is soluble only to the extent of 1.0 g per 100 g of water; palmitic acid [$CH_3(CH_2)_{14}COOH$] is essentially insoluble. The aromatic carboxylic acids are odorless solids that are sparingly soluble in water. The carboxylic acids generally are soluble in such organic solvents as alcohol, toluene, methylene chloride, and diethyl ether.

Table 16.4 lists physical constants for the first ten straight-chain members of the carboxylic acid family. Notice that the melting points show no regular increase with increasing molar mass. Pure acetic acid freezes at 16.6 °C. Because this is only slightly below normal room temperature (about 20 °C), acetic acid solidifies when cooled only slightly. In the poorly heated laboratories of a century ago in northern North America and Europe, acetic acid often froze on the reagent shelf. For that reason, pure acetic acid (sometimes called concentrated acetic acid) came to be known as *glacial acetic acid,* a name that survives to this day.

Table 16.4 Physical Constants of Carboxylic Acids

Formula	Name of Acid	Melting Point (°C)	Boiling Point (°C)	Solubility (g/100 g H_2O)	K_a (25 °C)
$HCOOH$	Formic	8	100	Miscible	1.8×10^{-4}
CH_3COOH	Acetic	17	118	Miscible	1.8×10^{-5}
CH_3CH_2COOH	Propionic	−22	141	Miscible	
$CH_3(CH_2)_2COOH$	Butyric	−5	163	Miscible	
$CH_3(CH_2)_3COOH$	Valeric	−35	187	5	
$CH_3(CH_2)_4COOH$	Caproic	−3	205	1	
$CH_3(CH_2)_5COOH$	Enanthic	−8	224	0.24	1.5×10^{-5}
$CH_3(CH_2)_6COOH$	Caprylic	16	238	0.07	
$CH_3(CH_2)_7COOH$	Pelargonic	14	254	0.03	
$CH_3(CH_2)_8COOH$	Capric	31	268	0.02	

Review Questions

16.4 Would you expect butyric acid (butanoic acid) to be more or less soluble than 1-butanol in water? Explain.

16.5 Describe the hydrogen bonding in carboxylic acids, both acid-acid and acid-water. How does this influence their physical properties?

16.5 Chemical Properties of Carboxylic Acids: Neutralization

When we first defined an acid (Chapter 9), we noted that one of the properties was a sour taste, a characteristic we note in foods and beverages. Oil-and-vinegar salad dressings are sour because the vinegar contains acetic acid. Grapefruits and lemons are sour because they contain citric acid. Sour milk and yogurt contain lactic acid. The acids we eat are, for the most part, organic acids in both the original sense—they come from plant or animal matter—and the modern sense that they are compounds of carbon. Now many carboxylic acids are synthesized in the laboratory from petroleum products.

Water-soluble carboxylic acids form moderately acidic solutions. They ionize in water, and their aqueous solutions exhibit the typical properties of acids.

$$RCOOH + H_2O \rightleftharpoons RCOO^- + H_3O^+$$

The anion, $RCOO^-$, formed when a carboxylic acid dissociates is called the carboxylate anion.

For example, such a solution will change litmus from blue to red. Whether water-soluble or not, carboxylic acids react with aqueous solutions of sodium hydroxide, sodium carbonate, and sodium bicarbonate to form salts

$$RCOOH + NaOH(aq) \rightarrow RCOO^-Na^+(aq) + H_2O$$

$$2\,RCOOH + Na_2CO_3(aq) \rightarrow 2\,RCOO^-Na^+(aq) + H_2O + CO_2(g)$$

$$RCOOH + NaHCO_3(aq) \rightarrow RCOO^-Na^+(aq) + H_2O + CO_2(g)$$

In these reactions, the carboxylic acids act just the way inorganic acids act: they neutralize basic compounds. With solutions of carbonate and bicarbonate ions, they also form carbon dioxide gas.

The carboxylic acids are weak acids and therefore tend to ionize only slightly in aqueous solution. Below are some organic and inorganic acids ordered according to their relative acidities:

Strongest acid				Weakest acid
H_2SO_4, HNO_3, HCl >	RCOOH >	H_2CO_3 >	ArOH >	H_2O
Mineral acids	Carboxylic acids	Carbonic acid	Phenols	Water

The differences in relative acidities among organic compounds enables us to use solubility behavior to identify carboxylic acids. Carboxylic acids that are insoluble in water dissolve in aqueous hydroxide, carbonate, or bicarbonate solutions because they react to form ionic salts that are water soluble. Solution in aqueous sodium bicarbonate, with the formation of carbon dioxide bubbles, is characteristic of carboxylic acids.

Example 16.3

Write equations for the reactions of decanoic acid with **(a)** NaOH and **(b)** $NaHCO_3$.

Solution

a. Decanoic acid has ten carbon atoms. It reacts with NaOH to form a salt and water.

$$CH_3(CH_2)_8COOH + NaOH(aq) \rightarrow CH_3(CH_2)_8COO^-Na^+(aq) + H_2O$$

b. With $NaHCO_3$, the products are a salt, water, and carbon dioxide.

$$CH_3(CH_2)_8COOH + NaHCO_3(aq) \longrightarrow CH_3(CH_2)_8COO^-Na^+(aq) + H_2O + CO_2$$

Practice Exercises

A. Write equations for the reactions of benzoic acid with (**a**) NaOH and (**b**) $NaHCO_3$.

B. Write equations for the reactions of pentanoic acid with (**a**) KOH, (**b**) Na_2CO_3, and (**c**) $Ca(OH)_2$.

Carboxylic acid salts are named in the same manner as inorganic salts: The name of the cation is followed by the name of the organic anion. The name of the anion is obtained by dropping the *-ic* ending of the acid name and replacing it with the suffix *-ate*. This rule applies whether we are using common names or IUPAC names:

$CH_3COO^-Li^+$	$CH_3CH_2CH_2COO^-K^+$	C_6H_5—COO^-Na^+
Lithium acetate Lithium ethanoate	Potassium butyrate Potassium butanoate	Sodium benzoate

The sodium or potassium salts of long-chain carboxylic acids are called soaps. We discuss the chemistry of soaps in some detail in Chapter 20.

$$CH_3CH_2CH_2CH_2CH_2CH_2CH_2CH_2CH_2CH_2CH_2CH_2CH_2CH_2CH_2COO^-Na^+$$

Sodium palmitate (a soap)

Other organic salts have commercial significance as preservatives. They prevent spoilage by inhibiting the growth of bacteria and fungi. Calcium and sodium propionate are added to processed cheese and bakery goods; sodium benzoate is used as a preservative in cider, jellies, pickles, and syrups; and sodium and potassium sorbate are added to fruit juices, sauerkraut, soft drinks, and wine. (Look for these salts on ingredient labels the next time you are shopping.)

$(CH_3CH_2COO^-)_2Ca^{2+}$	$CH_3CH{=}CHCH{=}CHCOO^-K^+$
Calcium propionate	Potassium sorbate

Review Question

16.6 How does neutralization of a carboxylic acid differ from that of an inorganic acid? How are they similar?

Esters

Esters are interesting derivatives of the carboxylic acids. They are widely distributed in nature. Unlike the carboxylic acids from which they are derived, the esters generally have pleasant odors and are often responsible for the characteristic fragrances of fruits and flowers. Once a flower or fruit has been chemically analyzed, flavor chemists can attempt to duplicate the natural odor or taste. They are seldom completely successful, but they often get close enough for practical purposes. Esters are used in the manufacture of perfumes and as flavoring agents in the confectionery and soft drink industries. (A mixture of nine esters is used to produce an artificial raspberry flavor.)

▲ The distinctive aroma and flavor of oranges are due in part to octyl acetate, an ester formed from 1-octanol (octyl alcohol) and acetic acid.

Fats and vegetable oils (Chapter 20) are esters of long-chain fatty acids and glycerol. Esters of phosphoric acid (Section 16.10) are of the utmost importance to life.

The general formula for an ester is RCOOR′, where R may be a hydrogen atom, an alkyl group, or an aryl group and R′ may be alkyl or aryl but *not* hydrogen.

16.6 An Ester by Any Other Name . . .

Although esters are covalent compounds and salts are ionic, esters are named in a manner similar to that used for naming salts. The group name of the alkyl or aryl portion is given first and is followed by the name of the acid portion. In both common and IUPAC nomenclature, the *-ic* ending of the parent acid is replaced by the suffix *-ate* (Table 16.5).

Example 16.4

Give the common and IUPAC names for each of the following.

a. $CH_3CH_2C(=O)-OCH_2CH_2CH_2CH_3$

b. $C_6H_5-C(=O)-OCH(CH_3)_2$

c. $C_6H_5-C(=O)-O-C_6H_5$

Solution

a. The four-carbon alkyl group attached directly to oxygen is a butyl group.

Four-carbon group, attached to oxygen by an end carbon = butyl

$CH_3CH_2C(=O)-OCH_2CH_2CH_2CH_3$

Three-carbon group, derived from the acid= = propionate

The part of the molecule derived from the acid has three carbon atoms. It is called propionate (common) or propanoate (IUPAC). The ester is therefore butyl propionate (common) or butyl propanoate (IUPAC).

b. The three-carbon alkyl group attached directly to oxygen is attached by the middle carbon atom; it is an *isopropyl* group. The part derived from the acid (that is, the benzene ring and the carbonyl group) is benzoate. The ester is therefore isopropyl benzoate by either the common or IUPAC system.

Table 16.5 Nomenclature of Esters

Formula	Common Name	IUPAC Name
$HCOOCH_3$	Methyl formate	Methyl methanoate
CH_3COOCH_3	Methyl acetate	Methyl ethanoate
$CH_3COOCH_2CH_3$	Ethyl acetate	Ethyl ethanoate
$CH_3CH_2COOCH_2CH_3$	Ethyl propionate	Ethyl propanoate
$CH_3CH_2CH_2COOCH_2CH_2CH_3$	Propyl butyrate	Propyl butanoate
$CH_3CH_2CH_2COOCH(CH_3)_2$	Isopropyl butyrate	Isopropyl butanoate
$C_6H_5-COOCH_2CH_3$	Ethyl benzoate	Ethyl benzoate
$CH_3COO-C_6H_5$	Phenyl acetate	Phenyl ethanoate

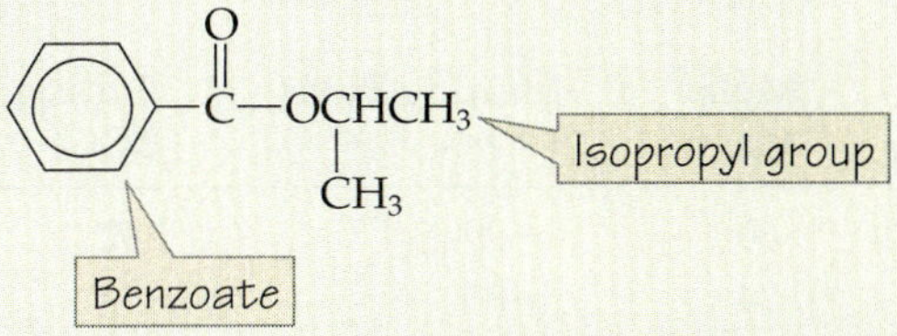

$$C_6H_5-\overset{\overset{\large O}{\|}}{C}-O\underset{\underset{\large CH_3}{|}}{C}HCH_3$$

Isopropyl group

Benzoate

c. The group attached to oxygen is a phenyl group. The acid portion corresponds to the benzoate group (from benzoic acid). Therefore, the compound is phenyl benzoate by either the common or IUPAC system.

Practice Exercise

A. Give the common and IUPAC names for each of the following.

a. $CH_3CH_2CH_2\overset{\overset{\large O}{\|}}{C}-OCH_2CH_3$ b. $CH_3CH_2\overset{\overset{\large O}{\|}}{C}OCH_2CH_2CH_2CH_3$

B. Give the IUPAC names for each of the following.

a. $CH_3CH_2CH_2CH_2CH_2COOCH_2CH_2CH_3$

b. $CH_3CH_2CH_2CH_2CH_2CH_2CH_2COOCH_3$

Example 16.5

Draw the structure for ethyl pentanoate.

Solution

It is easier to start with the portion from the acid. Draw the pentanoate (five-carbon) group first; keeping in mind that it includes the carbonyl group.

$$CH_3CH_2CH_2CH_2\overset{\overset{\large O}{\|}}{C}-O$$

Then simply attach the ethyl group to the bond that ordinarily holds the hydrogen atom in the carboxyl group.

$$CH_3CH_2CH_2CH_2\overset{\overset{\large O}{\|}}{C}-OCH_2CH_3$$

Practice Exercise

Draw the structure for (**a**) phenyl butanoate and (**b**) isopropyl hexanoate.

✓ **Review Question**

16.7 Of the families of compounds discussed in this chapter, which is known for its characteristically unpleasant odors? Which for its characteristically pleasant aromas?

16.7 Physical Properties of Esters

Ester molecules are polar but incapable of forming intermolecular hydrogen bonds with one another. Esters thus have considerably lower boiling points than the isomeric carboxylic acids. As one might expect, the boiling points of esters are about intermediate between those of ketones and ethers of comparable molar mass. Because ester molecules can form hydrogen bonds with water molecules, esters of low molar mass are somewhat water soluble. Borderline solubility occurs in those molecules that have three to five carbon atoms. Table 16.6 lists the physical properties of some common esters.

Table 16.6 Physical Properties of Esters

Formula	Name	Molar Mass	Melting Point (°C)	Boiling Point (°C)	Aroma
$HCOOCH_3$	Methyl formate	60	−99	32	
$HCOOCH_2CH_3$	Ethyl formate	74	−80	54	Rum
CH_3COOCH_3	Methyl acetate	74	−98	57	
$CH_3COOCH_2CH_3$	Ethyl acetate	88	−84	77	
$CH_3CH_2COOCH_3$	Methyl propionate	88	−88	80	
$CH_3CH_2COOCH_2CH_3$	Ethyl propionate	102	−74	99	
$CH_3CH_2CH_2COOCH_3$	Methyl butyrate	102	−85	102	Apple
$CH_3CH_2CH_2COOCH_2CH_3$	Ethyl butyrate	116	−101	121	Pineapple
$CH_3COO(CH_2)_4CH_3$	Pentyl acetate	130	−71	148	Pear
$CH_3COOCH_2CH_2CH(CH_3)_2$	Isopentyl acetate	130	−79	142	Banana
$CH_3COOCH_2C_6H_5$	Benzyl acetate	150	−51	215	Jasmine
$CH_3CH_2CH_2COO(CH_2)_4CH_3$	Pentyl butyrate	158	−73	185	Apricot
$CH_3COO(CH_2)_7CH_3$	Octyl acetate	172	−39	210	Orange

The most important use of esters is as industrial solvents. Ethyl acetate is commonly used to extract organic solutes from aqueous solutions. It is used, for example, to remove caffeine from coffee. Ethyl acetate is also used as a nail polish remover and is a major constituent of some paint removers. Cellulose nitrate is dissolved in ethyl acetate and butyl acetate to form lacquers. The solvent evaporates as the lacquer "dries," leaving a thin protective film on the lacquered surface. Esters having high boiling points are used as softeners (plasticizers) for brittle plastics.

Review Question

16.8 List two commercial uses of esters.

16.8 Preparation of Esters: Esterification

Some esters are prepared by direct esterification of the carboxylic acid. This conversion is accomplished by heating a carboxylic acid with an alcohol in the presence of a mineral acid catalyst.

$$R{-}\overset{\overset{\displaystyle O}{\|}}{C}{-}OH + R'OH \overset{H^+}{\rightleftharpoons} R{-}\overset{\overset{\displaystyle O}{\|}}{C}{-}OR' + H_2O$$

An alcohol molecule condenses with an acid molecule, splitting out water to form an ester. The reaction is reversible, and it soon comes to equilibrium. The composition of the equilibrium mixture is governed by the value of the equilibrium constant. Consideration of the law of chemical equilibrium and Le Châtelier's principle (Chapter 4) allows us to set reaction conditions that produce a maximum yield of the desired ester.

If the reaction involves an inexpensive alcohol, such as methanol, excess alcohol can be used to drive the reaction toward completion. In the preparation of methyl benzoate, for example, 10 mol of CH_3OH may be used for each mole of benzoic acid.

$$C_6H_5{-}COOH + \underset{(excess)}{CH_3OH} \overset{H^+}{\rightleftharpoons} C_6H_5{-}COOCH_3 + H_2O$$

Adding methanol forces the reaction to the right and can lead to a 75% yield of methyl benzoate.

Similarly, if the acid is cheap, it can be employed in excess. In the preparation of butyl acetate, acetic acid is used in a molar ratio of 2:1 (or greater).

$$\underset{\text{(excess)}}{CH_3COOH} + CH_3CH_2CH_2CH_2OH \overset{H^+}{\rightleftharpoons} CH_3COOCH_2CH_2CH_2CH_3 + H_2O$$

A third method of driving esterification toward completion involves removal of the product water as it is formed. This is easily accomplished if the acid, the alcohol, and the ester all boil at temperatures well above 100 °C:

$$\underset{\substack{\text{Butyric acid}\\ \text{(bp 164 °C)}}}{CH_3CH_2CH_2COOH} + \underset{\substack{\text{Butyl alcohol}\\ \text{(bp 118 °C)}}}{CH_3CH_2CH_2CH_2OH} \rightleftharpoons \underset{\substack{\text{Butyl butyrate}\\ \text{(bp 165 °C)}}}{CH_3CH_2CH_2COOCH_2CH_2CH_2CH_3} + \underset{\substack{\text{Water}\\ \text{(bp 100 °C)}}}{H_2O}$$

The water is distilled from the reaction mixture as it is formed, forcing the reaction to the right and increasing the yield of butyl butyrate.

In general, the reaction between an acid and an alcohol is extremely slow, and a catalyst must be employed to speed it up. Sulfuric acid is commonly used because it is both an acid and a good dehydrating agent. Thus it serves both to increase the rate of reaction (by acting as a catalyst), and to shift the equilibrium to the right (by reacting with water to form hydronium ions).

A commercially important esterification is **condensation polymerization,** in which a reaction occurs between a dicarboxylic acid and a dihydric alcohol (diol), with the elimination of water. Such a reaction yields an ester that contains a free carboxyl group at one end and a free alcohol group at the other end. Further condensation reactions then occur, producing **polyester** polymers. The most important polyester, polyethylene terephthalate (PET), is made from terephthalic acid and ethylene glycol monomers.

$$\underset{\text{Ethylene glycol}}{n\,HOCH_2CH_2OH} + \underset{\text{Terephthalic acid}}{n\,HOOC{-}C_6H_4{-}COOH} \longrightarrow \underset{\text{Polyethylene terephthalate}}{\left[OCH_2CH_2OCO{-}C_6H_4{-}CO \right]_n} + 2n\,H_2O$$

The polyester molecules make excellent fibers (Figure 16.2) that are spun into thread or yarn and marketed under such trade names as Dacron or Fortrel. Dacron polyester is used in permanent-press garments, carpets, tires, and many other products. A mesh of Dacron, which is biologically inert, is used in surgery to repair or replace diseased sections of blood vessels. PET is used to make bottles for soda pop and other beverages. It is also formed into films called Mylar. When magnetically coated, Mylar tape is used in audio and video cassettes.

▲ **Figure 16.2**
A liquid polymer is forced through the tiny holes of a device called a spinneret. The filaments solidify as they cool.

Review Question

16.9 What is esterification of a carboxylic acid? How does it differ from neutralization?

16.9 Chemical Properties of Esters: Hydrolysis

Esters are neutral compounds, unlike the acids from which they are formed. Esters typically undergo chemical reactions in which the alkoxy (—OR′) group is replaced by another group. One such reaction is **hydrolysis,** or splitting with water. Hydrolysis of esters is catalyzed by either acid or base. Acidic hydrolysis is simply the reverse of esterification. The ester is refluxed with a large excess of water containing a strong acid catalyst. However, the equilibrium is unfavorable for ester hydrolysis, and so the reaction never goes to completion.

General Equation:

$$R-\overset{\overset{\displaystyle O}{\|}}{C}-OR' + H_2O \overset{H^+}{\rightleftharpoons} R-\overset{\overset{\displaystyle O}{\|}}{C}-OH + R'OH$$

Specific Equation:

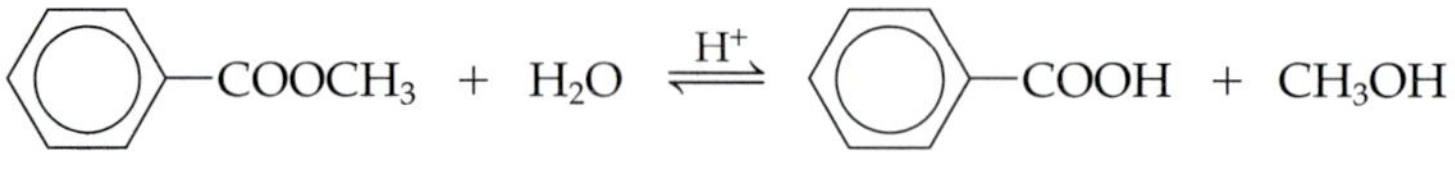

$$C_6H_5-COOCH_3 + H_2O \overset{H^+}{\rightleftharpoons} C_6H_5-COOH + CH_3OH$$

Methyl benzoate — Benzoic acid — Methanol

Example 16.6

Write an equation for the acidic hydrolysis of ethyl acetate and name the products.

Solution

Remember that in acidic hydrolysis, water splits the ester bond. The H of water joins to the oxygen in the —OR part of the original ester, and the OH joins to the carbonyl carbon:

$$CH_3\overset{\overset{\displaystyle O}{\|}}{C}-OCH_2CH_3 + H_2O \overset{H^+}{\rightleftharpoons} CH_3\overset{\overset{\displaystyle O}{\|}}{C}-OH + CH_3CH_2OH$$

The H of water goes on the O of the OR group.

Water splits the ester here.

The OH of water adds to the carbonyl carbon.

The products are acetic acid and ethanol.

Practice Exercise

Write an equation for the acidic hydrolysis of isopropyl propanoate.

When a base (such as sodium hydroxide or potassium hydroxide) is used to hydrolyze an ester, the reaction goes to completion because the carboxylic acid is removed from the equilibrium by its conversion to a salt. Carboxylate salts do not react with alcohols, so the reaction is essentially irreversible. Accordingly, ester hydrolysis is usually carried out in basic solution. Because soaps are prepared by the alkaline hydrolysis of fats and oils (Chapter 20), alkaline hydrolysis of all esters is called **saponification** (Latin *sapon,* soap; *facere,* to make). Note that in the saponification reaction, the base is a reactant and not simply a catalyst:

General Equation:

$$R-\overset{\overset{\displaystyle O}{\|}}{C}-OR' + NaOH(aq) \longrightarrow R-\overset{\overset{\displaystyle O}{\|}}{C}-ONa(aq) + R'OH$$

Specific Equation:

The alcohol portion of the ester ends up as the free alcohol.

$$CH_3\overset{\overset{\displaystyle O}{\|}}{C}-OCH_2CH_3 + NaOH(aq) \longrightarrow CH_3\overset{\overset{\displaystyle O}{\|}}{C}-O^-Na^+(aq) + CH_3CH_2OH$$

The acid portion of the ester ends up as the salt of the acid.

Example 16.7

Write an equation for the hydrolysis of methyl benzoate in a potassium hydroxide solution.

Solution

In basic hydrolysis, the molecule of base splits the ester linkage. The acid portion of the ester ends up as the *salt* of the acid. The alcohol portion of the ester ends up as the free alcohol.

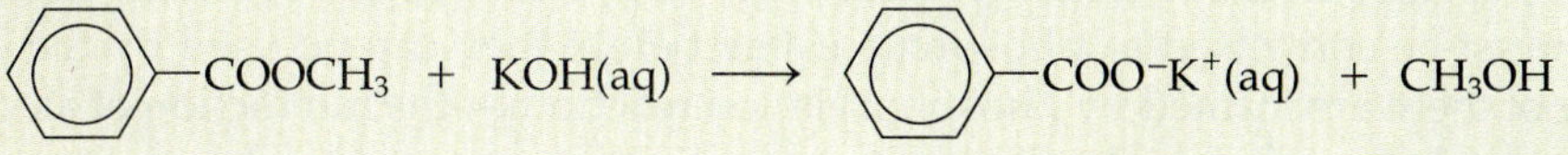

$$C_6H_5\text{—}COOCH_3 + KOH(aq) \longrightarrow C_6H_5\text{—}COO^-K^+(aq) + CH_3OH$$

Methyl benzoate — Potassium benzoate — Methanol

Practice Exercise

Write an equation for the hydrolysis of phenyl methanoate in a sodium hydroxide solution.

✓ Review Questions

16.10 How do acidic and basic hydrolysis of an ester differ in terms of (**a**) products obtained and (**b**) the extent of reaction?

16.11 What is saponification?

16.10 Esters of Phosphoric Acid

The esters we have discussed so far are derived from alcohols and carboxylic acids, but esters can also be derived from alcohols and inorganic acids, such as HNO_3, H_2SO_4, and H_3PO_4. We mentioned one such ester, nitroglycerin (glycerol trinitrate), formed from glycerol and nitric acid, in Section 14.7.

Perhaps the most important esters of inorganic acids in biochemistry are those of phosphoric acid and two of its anhydrides, pyrophosphoric acid[1] and triphosphoric acid. Phosphate or pyrophosphate esters are present in every plant and animal cell. They are biochemical intermediates in the transformation of food into usable energy (in the form of ATP). Phosphate esters are also important structural constituents of phospholipids (Chapter 20), nucleic acids (Chapter 23), and coenzymes (Selected Topic F).

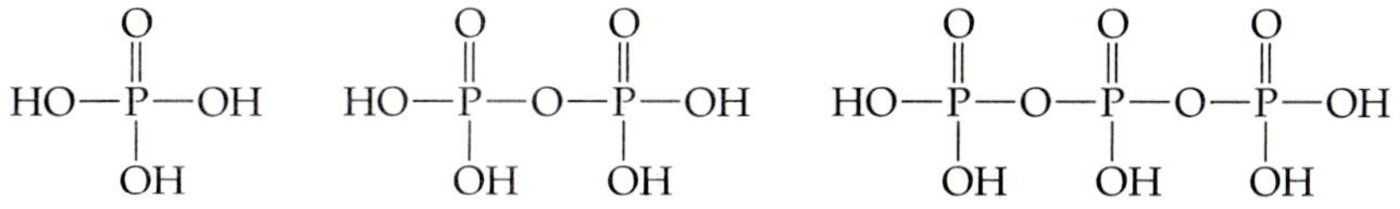

$$HO\text{—}P(=O)(OH)\text{—}OH \qquad HO\text{—}P(=O)(OH)\text{—}O\text{—}P(=O)(OH)\text{—}OH \qquad HO\text{—}P(=O)(OH)\text{—}O\text{—}P(=O)(OH)\text{—}O\text{—}P(=O)(OH)\text{—}OH$$

Phosphoric acid — Pyrophosphoric acid — Triphosphoric acid

Phosphoric acid is triprotic and can form monoalkyl, dialkyl, and trialkyl esters as one or more hydrogen atoms are replaced by alkyl groups.

[1]Pyrophosphoric acid is an anhydride derived from two molecules of phosphoric acid by removal of one molecule of water.

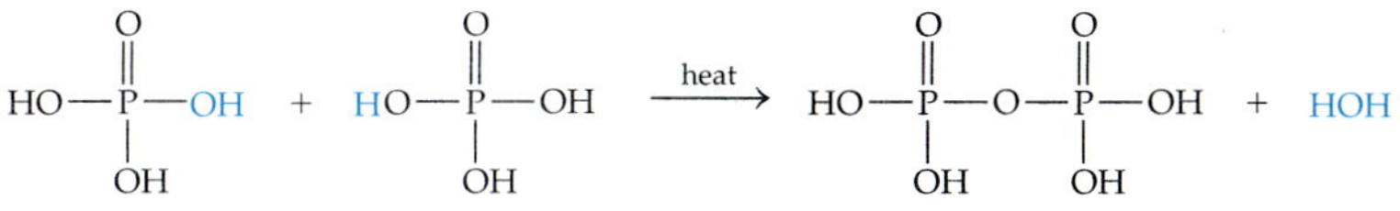

$$HO\text{—}P(=O)(OH)\text{—}OH + HO\text{—}P(=O)(OH)\text{—}OH \xrightarrow{\text{heat}} HO\text{—}P(=O)(OH)\text{—}O\text{—}P(=O)(OH)\text{—}OH + HOH$$

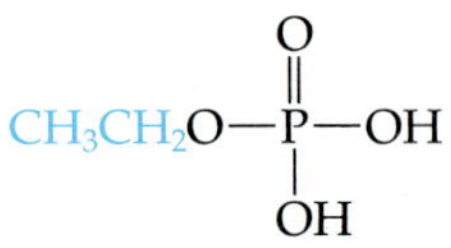

$CH_3CH_2O-P(=O)(OH)-OH$ — Ethyl dihydrogen phosphate

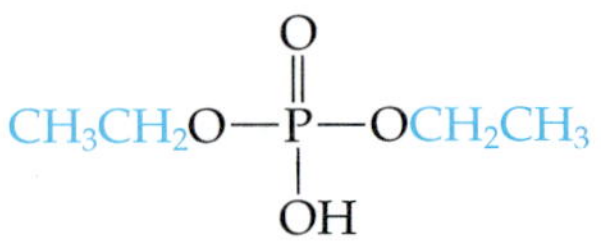

$CH_3CH_2O-P(=O)(OH)-OCH_2CH_3$ — Diethyl hydrogen phosphate

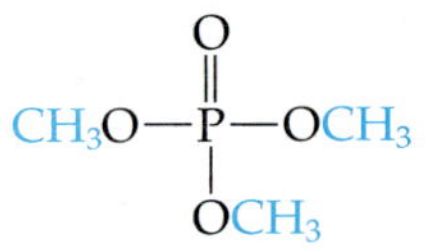

$CH_3O-P(=O)(OCH_3)-OCH_3$ — Trimethyl phosphate

A wide variety of naturally occurring phosphate esters are compounds of central importance in metabolism. The majority of the substances obtained from our food must be converted to phosphate esters before they can be used by the cells. Many of them are formed by phosphorylation reactions—the interaction of alcohols with anhydrides of phosphoric acid:

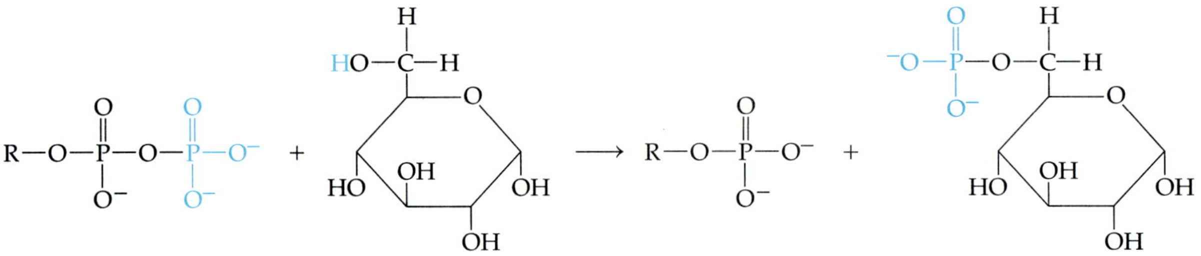

$$R-O-P(=O)(O^-)-O-P(=O)(O^-)-O^- + \text{Glucose} \longrightarrow R-O-P(=O)(O^-)-O^- + \text{Glucose 6-phosphate}$$

A polyphosphate Glucose Glucose 6-phosphate

Figure 16.3 presents only a few of the phosphate esters found in living cells. At physiological pH values (≈7), the phosphate groups are ionized.

Review Question

16.12 What compounds can be used to form phosphate esters?

Amides

The amide functional group has a nitrogen atom attached to a carbonyl carbon atom. If the two remaining bonds to nitrogen are attached to hydrogen atoms, the compound is called a *simple amide.* If one or both of the two remaining bonds to nitrogen are attached to alkyl or aryl groups, the compound is called a *substituted amide.* The carbonyl carbon-to-nitrogen bond is called the **amide linkage.** This bond is quite stable and is found in the repeating units of protein molecules (Chapter 21), in nylon, and in many other industrial polymers.

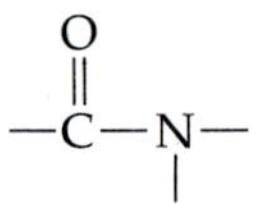

$-C(=O)-N<$ — The amide group

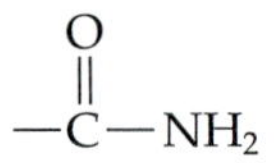

$-C(=O)-NH_2$ — A simple amide

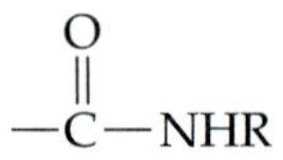

$-C(=O)-NHR$ — A substituted amide

16.11 Amides: Structures and Names

Amides are named as derivatives of carboxylic acids. The *-ic* ending of the common name or the *-oic* ending of the IUPAC name is replaced with the suffix -*amide.*

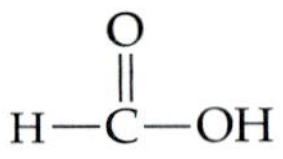

$H-C(=O)-OH$ — Formic acid, Methanoic acid

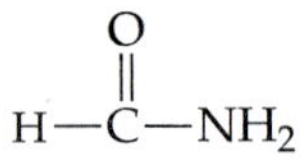

$H-C(=O)-NH_2$ — Formamide (Methanamide)

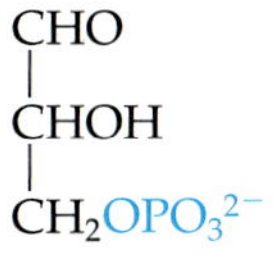

$$\begin{array}{l} CHO \\ | \\ CHOH \\ | \\ CH_2OPO_3^{2-} \end{array}$$

Glyceraldehyde-3-phosphate
(Section 25.1)

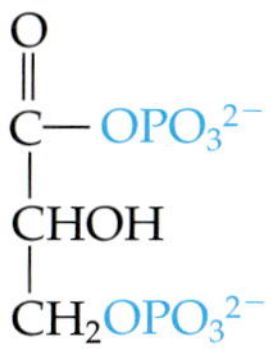

$$\begin{array}{l} \;O \\ \;\| \\ C-OPO_3^{2-} \\ | \\ CHOH \\ | \\ CH_2OPO_3^{2-} \end{array}$$

1,3-Bisphosphoglycerate
(Section 25.1)

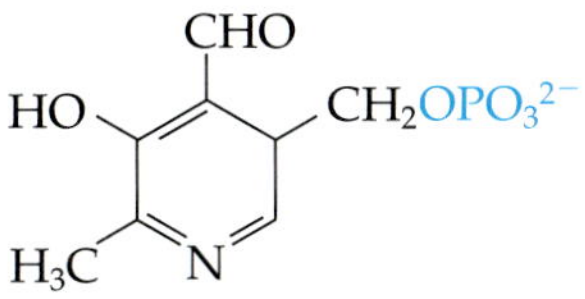

Pyroxidal phosphate
(Section F.6)

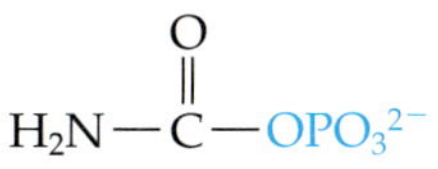

$$H_2N-\overset{\overset{\displaystyle O}{\|}}{C}-OPO_3^{2-}$$

Carbamoyl phosphate
(Section 27.3)

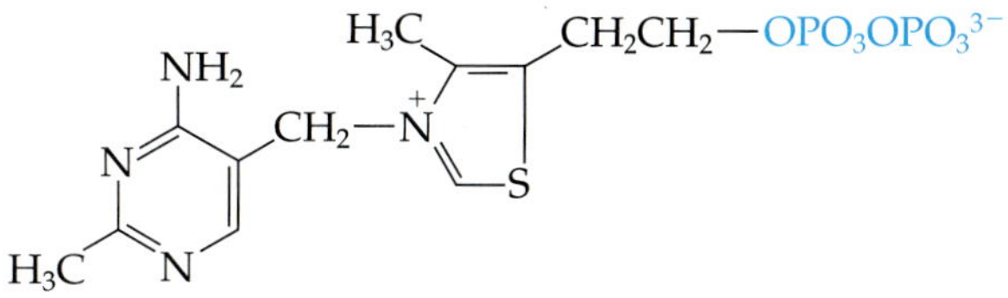

Thiamine pyrophosphate
(Section F.6)

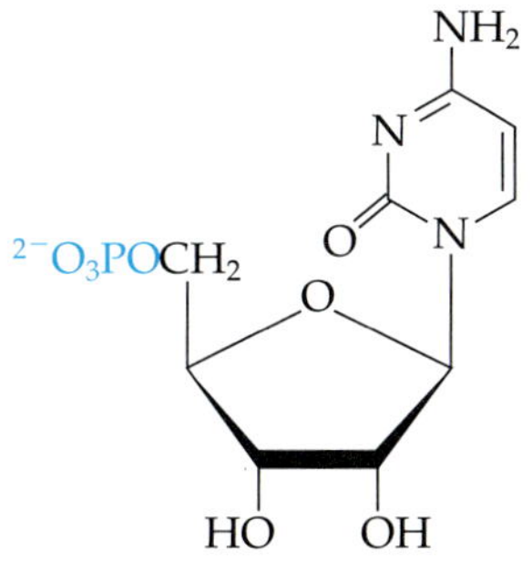

Cytidine monophosphate
(Section 23.1)

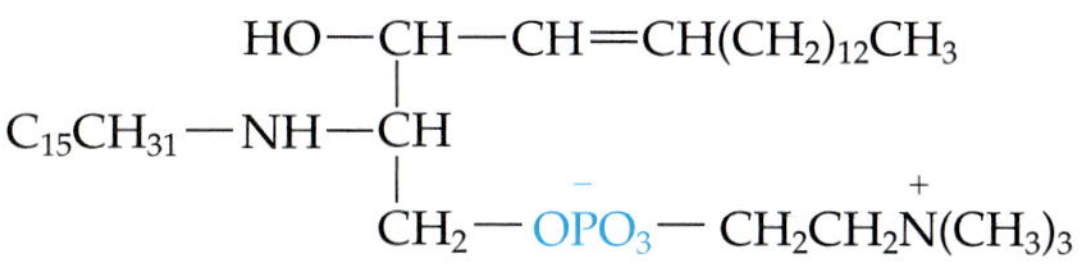

$$\begin{array}{l} \qquad\qquad\qquad\quad HO-CH-CH=CH(CH_2)_{12}CH_3 \\ \qquad\qquad\qquad\qquad\quad | \\ C_{15}CH_{31}-NH-CH \\ \qquad\qquad\qquad\qquad\quad | \\ \qquad\qquad\qquad\quad\; CH_2-OPO_3^{-}-CH_2CH_2\overset{+}{N}(CH_3)_3 \end{array}$$

Sphingomyelin
(Section 20.4)

▲ **Figure 16.3**
Some phosphate compounds of biological importance.

Example 16.8

Name each of the following compounds.

a. $CH_3-\overset{\overset{\displaystyle O}{\|}}{C}-NH_2$ **b.** 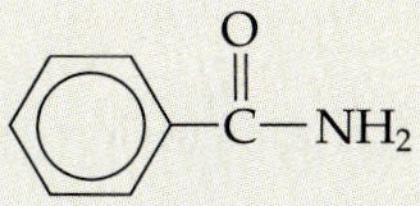 $C_6H_5-\overset{\overset{\displaystyle O}{\|}}{C}-NH_2$

Solution

a. This amide is derived from acetic acid; the OH of acetic acid is replaced by an NH_2 group. Dropping the *-ic* from *acetic* and attaching the ending *-amide* gives the name *acetamide* (or ethanamide in the IUPAC system).

b. This amide is derived from benzoic acid. Dropping the *-oic* and adding *-amide* gives benzamide.

Practice Exercises

A. Name each of the following compounds.

a. $CH_3CH_2CH_2-\overset{\overset{\displaystyle O}{\|}}{C}-NH_2$ **b.** $CH_3CH_2CH_2CH_2CH_2CH_2-\overset{\overset{\displaystyle O}{\|}}{C}-NH_2$

B. Draw a structural formula for pentanamide.

In substituted amides, alkyl groups attached to the nitrogen atom are named as substituents. Instead of using a Greek letter or a number to specify location, chemists indicate the group's attachment to nitrogen by a capital letter *N*. If the substituent on nitrogen is phenyl, the compound is named as an anilide. The *-ic* or *-oic* ending of the acid name is replaced with *-anilide* instead of *-amide:*

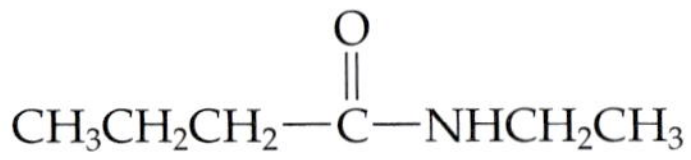

$$CH_3CH_2CH_2-\overset{\overset{\displaystyle O}{\|}}{C}-NHCH_2CH_3 \qquad H-\overset{\overset{\displaystyle O}{\|}}{C}-N(CH_3)_2 \qquad CH_3-\overset{\overset{\displaystyle O}{\|}}{C}-NH-C_6H_5$$

N-Ethylbutyramide  — *N,N*-Dimethylformamide 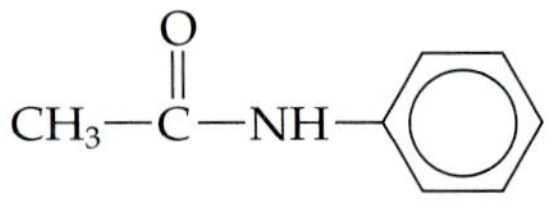 — Acetanilide

Example 16.9

Name the following compound.

$$CH_3-\overset{\overset{\displaystyle O}{\|}}{C}-NH\underset{\underset{\displaystyle CH_3}{|}}{C}HCH_3$$

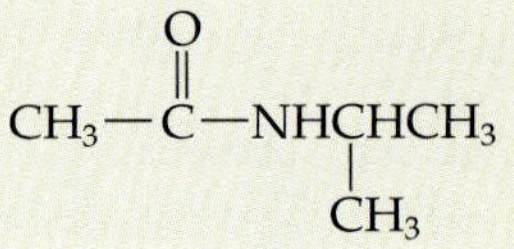

Solution

The acid portion of the molecule is derived from acetic acid. This compound is therefore named as a substituted acetamide.

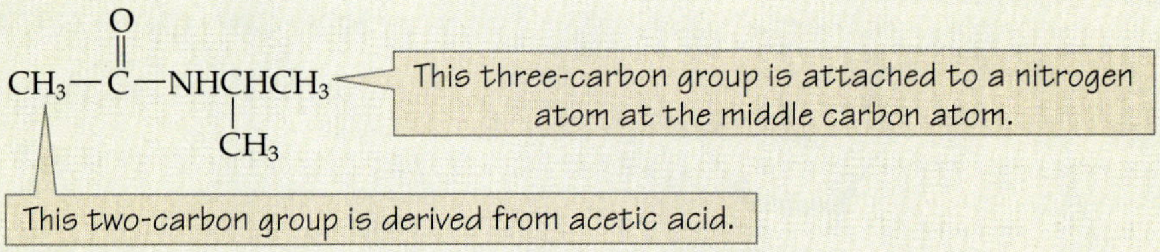

$$CH_3-\overset{\overset{\displaystyle O}{\|}}{C}-NH\underset{\underset{\displaystyle CH_3}{|}}{C}HCH_3$$

This three-carbon group is attached to a nitrogen atom at the middle carbon atom.

This two-carbon group is derived from acetic acid.

The substituent attached directly to the nitrogen atom is an isopropyl group. The name of the compound is *N*-isopropylacetamide. (The IUPAC name is *N*-isopropylethanamide.)

Practice Exercise

A. Name the following compound.

$$CH_3-\overset{\overset{\displaystyle O}{\|}}{C}-\underset{\underset{\displaystyle CH_3}{|}}{N}-CH_3$$

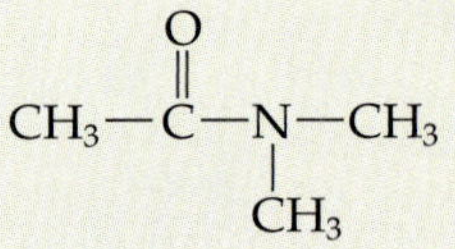

B. Draw a structural formula for *N,N*-dimethylpropanamide.

16.12 Physical Properties of Amides

With the exception of formamide, which is a liquid, all unsubstituted amides are solids (Table 16.7). Most amides are colorless and odorless. The lower members of the series are soluble in water, with borderline solubility occurring in those that have five or six carbon atoms. Like the esters, solutions of amides usually are neutral, neither acidic nor basic.

The amides have high boiling points and melting points. These characteristics and their water solubility result from the polar nature of the amide group and the formation of hydrogen bonds (Figure 16.4). (Similar hydrogen bonding plays a critical role in determining the structure and properties of proteins, DNA, RNA, and

Table 16.7 Physical Constants of Some Unsubstituted Amides

Formula	Name	Melting Point (°C)	Boiling Point (°C)	
$HCONH_2$	Formamide	2	193	Soluble in water
CH_3CONH_2	Acetamide	82	222	Soluble in water
$CH_3CH_2CONH_2$	Propionamide	81	213	Soluble in water
$CH_3CH_2CH_2CONH_2$	Butyramide	115	216	Soluble in water
$C_6H_5CONH_2$	Benzamide	132	290	Insoluble in water

other giant molecules so important to life processes.) Electrostatic forces and hydrogen bonding contribute to the strong intermolecular attractions found in the amides. Note, however, that disubstituted amides have no hydrogen atoms bonded to nitrogen and thus are incapable of intermolecular hydrogen bonding. *N,N*-Dimethylacetamide has a melting point of −20 °C, which is about 100 °C lower than the melting point of acetamide.

Review Question

16.13 Would you expect the boiling point of *N,N*-dimethylacetamide to be higher or lower than the boiling point of acetamide? Explain.

16.13 Synthesis of Amides

The addition of ammonia to a carboxylic acid results in the formation of an amide, but the reaction is very slow at room temperature. The ammonium salt of the acid is formed first; then water can be split out if the reaction temperature is maintained above 100 °C. The second step is reversible; the equilibrium favors salt formation. Continuous removal of the water shifts the equilibrium to the right:

$$\underset{\text{Acetic acid}}{CH_3COOH} + NH_3 \longrightarrow \underset{\text{Ammonium acetate}}{CH_3COO^-NH_4^+} \overset{\Delta}{\rightleftharpoons} \underset{\text{Acetamide}}{CH_3CONH_2} + H_2O$$

Recall (Section 16.8) that the condensation polymerization of a diol and a diacid forms a polyester. A similar reaction between a diacid and a diamine yields a **polyamide.** The two difunctional monomers often employed are adipic acid and 1,6-hexanediamine. The monomers condense by splitting out water to form a new product, which is still difunctional and thus can react further to yield a polyamide polymer.

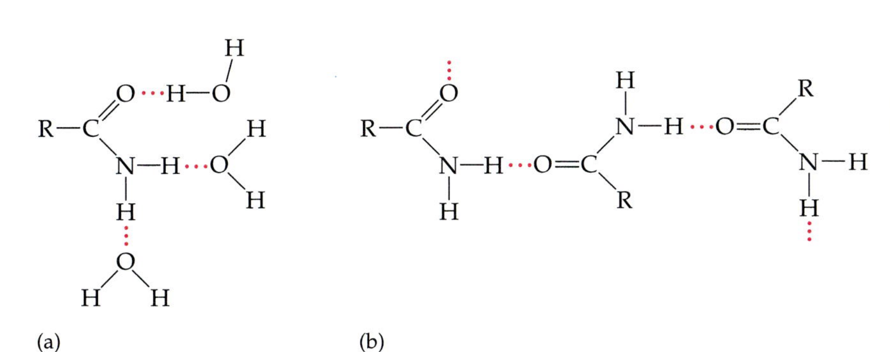

◄ **Figure 16.4**
(a) Hydrogen bonding of amides with water molecules. (b) Intermolecular hydrogen bonding in amides.

$$n\,H{-}\overset{\displaystyle H}{\overset{|}{N}}CH_2CH_2CH_2CH_2CH_2CH_2\overset{\displaystyle H}{\overset{|}{N}}{-}H \;+\; n\,HO{-}\overset{\displaystyle O}{\overset{\|}{C}}CH_2CH_2CH_2CH_2\overset{\displaystyle O}{\overset{\|}{C}}{-}OH \xrightarrow[10\text{ atm}]{270\ ^\circ C}$$

1,6-Hexanediamine — Adipic acid

$$-\underset{\displaystyle H}{\underset{|}{N}}\left[-\overset{\displaystyle O}{\overset{\|}{C}}-(CH_2)_4-\overset{\displaystyle O}{\overset{\|}{C}}-\underset{\displaystyle H}{\underset{|}{N}}-(CH_2)_6-\underset{\displaystyle H}{\underset{|}{N}}-\right]_n\overset{\displaystyle O}{\overset{\|}{C}}- \;+\; 2n\,HOH$$

Nylon 66
(a polyamide)

▲ In 1934 Wallace Carothers and coworkers at E. I. du Pont de Nemours and Company produced nylon, the first synthetic fiber. Today students carry out a variation of this polymerization reaction as a laboratory experiment.

Several different synthetic polyamides are known as *nylons.* Nylon 66 is the most common; the 66 designation refers to the number of carbon atoms in each monomer. Other nylons have varying numbers of carbon atoms in either the dicarboxylic acid or the diamine or both. Nylon 66 is stable in dilute acids or bases, has a high melting point (260 °C), and is extremely strong. Nylons are among the most widely used synthetic fibers; for example, they are used in ropes, sails, carpets, clothing, tires, brushes, and parachutes. They also can be molded into blocks for use in electrical equipment, gears, bearings, and valves.

Review Question

16.14 What reagents could be used to prepare nylon 6,10?

16.14 Chemical Properties of Amides: Hydrolysis

Generally, the amides resist hydrolysis in plain water, even upon prolonged heating. In the presence of added acid or base, however, hydrolysis proceeds at a moderate rate. Acidic hydrolysis of a simple amide gives a carboxylic acid and an ammonium salt, as shown in the following example.

$$CH_3CH_2\overset{\displaystyle O}{\overset{\|}{C}}-NH_2 + HCl(aq) + H_2O \longrightarrow CH_3CH_2\overset{\displaystyle O}{\overset{\|}{C}}-OH + NH_4^+Cl^-(aq)$$

Propionamide (Propanamide) — Propionic acid (Propanoic acid) — Ammonium chloride

Basic hydrolysis gives a salt of the carboxylic acid and ammonia, as shown in the following example.

$$CH_3CH_2\overset{\displaystyle O}{\overset{\|}{C}}-NH_2 + NaOH(aq) \longrightarrow CH_3CH_2\overset{\displaystyle O}{\overset{\|}{C}}-O^-Na^+ + NH_3$$

Propionamide (Propanamide) — Sodium propionate (Sodium propanoate) — Ammonia

It may be easier to see why the products of the two reactions differ if we consider the hydrolysis products that would form if the reaction could be carried out in the absence of added acid or base.

$$CH_3CH_2\overset{\displaystyle O}{\overset{\|}{C}}-NH_2 + H_2O \rightleftharpoons CH_3CH_2\overset{\displaystyle O}{\overset{\|}{C}}-OH + NH_3$$

Propionamide — Propionic acid — Ammonia

Products: an acid and a base: Not possible

The products of this reaction are an acid (the carboxylic acid) and a base (ammonia). If we carry out the hydrolysis in the presence of hydrochloric acid, some of the hydrochloric acid reacts with the ammonia to form ammonium chloride. If, instead, we use sodium hydroxide to speed the reaction, some of this base will react with the carboxylic acid to form the sodium salt of the carboxylic acid.

Note that the several hydrolysis reactions discussed in this chapter are closely related and should be considered as variations on a theme rather than as separate reactions. Note also that chemical principles are not confined within chapters. If acids react with bases to form salts in Chapter 9, they also react that way in this chapter. If a reaction under consideration produces an acidic product, that product always exhibits any and all properties of an acid.

Example 16.10

Write an equation for the hydrolysis of (**a**) butyramide in the presence of HCl and (**b**) benzamide in the presence of NaOH.

Solution

a. The acid hydrolysis of an amide gives the organic acid and an ammonium salt. Butyramide thus yields butyric acid and ammonium chloride.

$$CH_3CH_2CH_2\overset{\overset{O}{\|}}{C}{-}NH_2 + HCl(aq) + H_2O \longrightarrow CH_3CH_2CH_2\overset{\overset{O}{\|}}{C}{-}OH + NH_4^+Cl^-(aq)$$

b. The basic hydrolysis of an amide gives the salt of the carboxylic acid and ammonia. Benzamide thus yields sodium benzoate and ammonia.

$$C_6H_5{-}\overset{\overset{O}{\|}}{C}{-}NH_2 + NaOH(aq) \longrightarrow C_6H_5{-}\overset{\overset{O}{\|}}{C}{-}O^-Na^+(aq) + NH_3$$

Practice Exercise

Write an equation for the hydrolysis of (**a**) benzamide in the presence of HCl and (**b**) acetamide in KOH(aq).

The hydrolysis of amides is of more than theoretical interest. Digestion of proteins (Chapter 27) involves the hydrolysis of amide bonds. Clothes made of nylon have been known to disintegrate in air polluted with sulfuric acid mist. The acid catalyzes the hydrolysis of the amide bonds that hold the long chains of the nylon molecules together. Perhaps it is worth considering what the same polluted air does to the proteins of our lungs.

Review Question

16.15 How do acidic and basic hydrolysis of an amide differ in terms of products obtained?

Summary

Carboxylic acids (RCOOH) contain the functional group —COOH, called the **carboxyl group.** The fourth bond on the carboxyl carbon may be with a hydrogen atom (as in formic acid, HCOOH), an alkyl group (as in acetic acid, CH_3COOH), or an aryl group (as in benzoic acid, C_6H_5COOH).

A carboxylic acid is formed by oxidation of the aldehyde containing the same number of carbon atoms. Because aldehydes are formed from primary alcohols, these alcohols are also a starting material for carboxylic acids.

Carboxylic acids have strong, often disagreeable odors. They are highly polar molecules and readily form intermolecular hydrogen bonds. For this reason they have comparatively high boiling points. The hydrogen bonding in carboxylic acids is so strong that in many cases the acid exists not as a collection of individual molecules but rather as a collection of two-molecule units called **dimers.** Such dimers cause the boiling point of carboxylic acids to be unusually high.

Carboxylic acids are weak acids. They react with bases to form salts and with carbonate and bicarbonate salts to form carbon dioxide gas.

The compound formed when the OH of a carboxylic acid is replaced either by an OR or an OAr group is called an **ester (RCOOR′).** Esters are pleasant-smelling compounds that are responsible for the fragrances of flowers and fruits. They have lower-than-expected boiling points because, even though ester molecules are somewhat polar, they cannot form intermolecular hydrogen bonds. They can hydrogen-bond with water, however, and consequently the low molar mass esters are water soluble.

Esters can be synthesized by *direct esterification,* in which a carboxylic acid and an alcohol are combined under acidic conditions.

Esters are neutral compounds, and one of their most common reactions is **hydrolysis,** a reaction with water. When ester hydrolysis is carried out under acidic conditions, the reaction is essentially the reverse of esterification. When carried out under basic conditions, the process is called **saponification.**

Inorganic acids also react with alcohols to form esters. Some of the most important esters in biochemistry are those formed from phosphoric acid.

Organic compounds containing a carbonyl group bonded to a nitrogen atom are **amides,** and the carbon-nitrogen bond is an **amide linkage.**

Most amides are colorless and odorless, and the lighter ones are soluble in water. Because they are polar molecules, amides have comparatively high boiling and melting points. Amides are synthesized from carboxylic acids and ammonia. Amides are neutral compounds. They resist hydrolysis in water, but both acids and bases catalyze the reaction.

Key Terms

amide (16.1)
amide linkage (page 428)
carboxyl group (16.1)
carboxylic acid (16.1)
condensation polymerization (16.8)
dimer (16.4)
ester (16.1)
hydrolysis (16.9)
polyamide (16.13)
polyester (16.8)
saponification (16.9)

Problems

Carboxylic Acids: Names and Structural Formulas

1. Draw structural formulas for each of the following.
   a. heptanoic acid
   b. 3-methylbutanoic acid
   c. 2,3-dibromobenzoic acid
   d. *m*-isopropylbenzoic acid
2. Draw structural formulas for each of the following.
   a. *o*-nitrobenzoic acid **b.** *p*-chlorobenzoic acid
   c. 3-chloropentanoic acid **d.** 3-phenylpropanoic acid
3. Draw structural formulas for each of the following.
   a. oxalic acid **b.** β-hydroxybutyric acid
4. Draw structural formulas for each of the following.
   a. α-chloropropionic acid **b.** phenylacetic acid
5. Name each of the following.
   a. $(CH_3)_2CHCH_2COOH$
   b. $(CH_3)_3CCH(CH_3)CH_2COOH$
   c. $CH_2OHCH_2CH_2COOH$
   d. $(CH_3)_2CHCH_2CH(CH_3)COOH$
6. Name each of the following.
   a. $CH_3(CH_2)_8COOH$
   b. $(CH_3)_2CHCCl_2CH_2CH_2COOH$
   c. $CH_3CHOHCH(CH_2CH_3)CHICOOH$
   d. Br–C_6H_4–COOH (para-substituted benzene ring)

Salts: Names and Structural Formulas

7. Draw structural formulas for each of the following.
   a. potassium acetate **b.** calcium propanoate
8. Name each of the following.
   a. C_6H_5–C(=O)–O^- Li^+
   b. $CH_3CH_2CH_2C(=O){-}O^-NH_4^+$

Esters: Names and Structural Formulas

9. Draw structural formulas for each of the following.
   a. methyl acetate **b.** phenyl acetate
10. Draw structural formulas for each of the following.
    a. ethyl pentanoate
    b. ethyl 3-methylhexanoate
11. Draw structural formulas for each of the following.
    a. ethyl benzoate **b.** phenyl benzoate
12. Draw structural formulas for each of the following.
    a. ethyl butyrate **b.** isopropyl propionate
13. Name each of the following.
    a. C_6H_5–C(=O)–O–CH_3
    b. $CH_3{-}O{-}C(=O){-}H$
    c. $CH_3CH_2C(=O){-}O{-}CH_2CH_3$
14. Name each of the following.

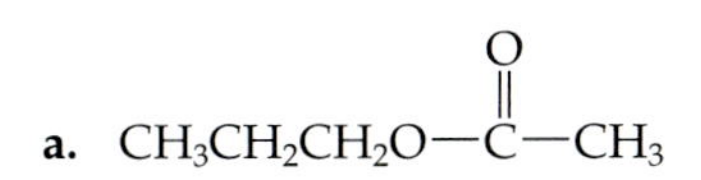

a. $CH_3CH_2CH_2O-\overset{O}{\overset{\|}{C}}-CH_3$

b. $CH_3CH_2\overset{O}{\overset{\|}{C}}-OCH_2CH_2CH_3$

c. $CH_3CH_2CH_2\overset{O}{\overset{\|}{C}}-O-C_6H_5$

Amides: Names and Structural Formulas

15. Draw structural formulas for each of the following.
  a. butanamide b. hexanamide
  c. *N*-methylacetamide

16. Draw structural formulas for each of the following.
  a. formamide b. propionamide
  c. *N,N*-dimethylbenzamide

17. Name each of the following.

a. $C_6H_5-\overset{O}{\overset{\|}{C}}-NH_2$

b. $CH_3CH_2CH(CH_3)\overset{O}{\overset{\|}{C}}-NH_2$ c. $CH_3\overset{O}{\overset{\|}{C}}-NH_2$

18. Name each of the following.

a. $CH_3CH_2\overset{O}{\overset{\|}{C}}-\underset{CH_3}{\underset{|}{N}}-CH_3$

b. $Cl-C_6H_4-\overset{O}{\overset{\|}{C}}-NH_2$

c. $CH_3CH_2CH_2\overset{O}{\overset{\|}{C}}-NH-C_6H_5$

Physical Properties

19. Which compound has the higher boiling point? Explain.

$CH_3CH_2CH_2OCH_2CH_3$ (I) $CH_3CH_2CH_2COOH$ (II)

20. Which compound has the higher boiling point? Explain.

$CH_3CH_2CH_2CH_2CH_2OH$ (I) $CH_3CH_2CH_2COOH$ (II)

21. Which compound has the higher boiling point? Explain.

$CH_3CH_2CH_2\overset{O}{\overset{\|}{C}}-NH_2$ (I) $CH_3\overset{O}{\overset{\|}{C}}-O-CH_2CH_3$ (II)

22. Which compound has the higher boiling point? Explain.

$CH_3CH_2CH_2\overset{O}{\overset{\|}{C}}-OH$ (I) $CH_3CH_2\overset{O}{\overset{\|}{C}}-O-CH_3$ (II)

23. Which compound is more soluble in water? Explain.

$CH_3\overset{O}{\overset{\|}{C}}-OH$ (I) $CH_3CH_2CH_2CH_3$ (II)

24. Which compound is more soluble in water? Explain.

$CH_3CH{=}CHCH_3$ (I) $CH_3\overset{O}{\overset{\|}{C}}-NH_2$ (II)

25. Which compound is more soluble in water? Explain.

$CH_3\overset{O}{\overset{\|}{C}}OCH_3$ (I) $CH_3CH_2CH_2CH_2\overset{O}{\overset{\|}{C}}OCH_2CH_3$ (II)

26. Which compound is more soluble in water? Explain.

$C_6H_5-\overset{O}{\overset{\|}{C}}-OH$ (I) $C_6H_5-\overset{O}{\overset{\|}{C}}-O^-\,Na^+$ (II)

Chemical Reactions

27. Write equations for the reactions of butyric acid with (**a**) aqueous NaOH and (**b**) aqueous $NaHCO_3$.

28. Write equations for the reactions of benzoic acid with (**a**) aqueous NaOH and (**b**) aqueous $NaHCO_3$.

29. Write an equation for the acid-catalyzed hydrolysis of ethyl acetate.

30. Write an equation for the base-catalyzed hydrolysis of ethyl acetate.

31. Write an equation for the acid-catalyzed hydrolysis of benzamide.

32. Write an equation for the base-catalyzed hydrolysis of benzamide.

33. Complete the following equations.

a. $CH_3CH_2\overset{O}{\overset{\|}{C}}-OH \xrightarrow{NaOH}$

b. $C_6H_4(COOH)_2$ (ortho) $\xrightarrow{\text{excess } NaHCO_3}$

34. Complete the following equations.

a. $HOOC-COOH \xrightarrow{\text{excess } NaOH}$

b. $C_6H_5-COOH \xrightarrow{KOH}$

35. Complete the following equations.

a. $C_6H_5\text{—}\overset{O}{\overset{\|}{C}}\text{—}OCH_2CH_2CH_3 + NaOH \longrightarrow$

b. $C_6H_{11}\text{—}OH + CH_3\overset{O}{\overset{\|}{C}}OH \xrightleftharpoons{H^+}$

36. Complete the following equations.

a. $C_6H_5\text{—}CH_2OH + CH_3\overset{O}{\overset{\|}{C}}OH \xrightleftharpoons{H^+}$

b. $CH_3\overset{O}{\overset{\|}{C}}\text{—}OCH(CH_3)_2 + KOH(aq) \longrightarrow$

37. Complete the following equations.

a. $CH_3\overset{O}{\overset{\|}{C}}\text{—}OH + CH_3CH_2CH_2OH \xrightleftharpoons{H^+}$

b. $HO\text{—}\overset{O}{\overset{\|}{C}}CH_2\overset{O}{\overset{\|}{C}}\text{—}OH + 2\,CH_3OH \xrightleftharpoons{H^+}$

38. Complete the following equations.

a. $CH_3CH_2CH_2O\text{—}\overset{O}{\overset{\|}{C}}\text{—}C_6H_5 + H_2O \xrightleftharpoons{H^+}$

b. $(CH_3)_2CH\text{—}\overset{O}{\overset{\|}{C}}\text{—}O\text{—}CH_2CH_3 + H_2O \xrightleftharpoons{H^+}$

39. Complete the following equations.

a. $CH_3\overset{O}{\overset{\|}{C}}\text{—}NH_2 + HCl + H_2O \longrightarrow$

b. $C_6H_5\text{—}\overset{O}{\overset{\|}{C}}\text{—}N(CH_3)\text{—}CH_3 + NaOH(aq) \longrightarrow$

40. Complete the following equations.

a. $CH_3CH_2\overset{O}{\overset{\|}{C}}\text{—}NH_2 + KOH(aq) \longrightarrow$

b. $C_6H_5\text{—}\overset{O}{\overset{\|}{C}}\text{—}NHCH_3 + HCl + H_2O \longrightarrow$

41. List the reagents necessary to carry out the following conversions.

a. $CH_3CH_2CH(CH_3)CH_2OH \xrightarrow{?} CH_3CH_2CH(CH_3)\overset{O}{\overset{\|}{C}}\text{—}OH$

b. $CH_3\overset{O}{\overset{\|}{C}}\text{—}H \xrightarrow{?} CH_3\overset{O}{\overset{\|}{C}}OH$

c. $C_6H_5\text{—}\overset{O}{\overset{\|}{C}}\text{—}OH \xrightarrow{?} C_6H_5\text{—}\overset{O}{\overset{\|}{C}}\text{—}O^-\,Na^+$

42. List the reagents necessary to carry out the following conversions.

a. $C_6H_5\text{—}\overset{O}{\overset{\|}{C}}\text{—}OH \xrightarrow{?} C_6H_5\text{—}\overset{O}{\overset{\|}{C}}\text{—}O^-\,NH_4^+$

b. $C_6H_5\text{—}\overset{O}{\overset{\|}{C}}\text{—}OH \xrightarrow{?} C_6H_5\text{—}\overset{O}{\overset{\|}{C}}\text{—}OCH_3$

43. List the reagents necessary to carry out the following conversions.

a. $CH_3CH_2CH_2CH_2CH_2OH \xrightarrow{?} CH_3CH_2CH_2CH_2CH_2O\overset{O}{\overset{\|}{C}}CH_3$

b. $CH_3CH_2CH_2\overset{O}{\overset{\|}{C}}OCH_3 \xrightarrow{?} CH_3CH_2CH_2\overset{O}{\overset{\|}{C}}O^-\,Li^+ + CH_3OH$

44. List the reagents necessary to carry out the following conversions.

a. $C_6H_5\text{—}\overset{O}{\overset{\|}{C}}\text{—}NH_2 \xrightarrow{?} C_6H_5\text{—}\overset{O}{\overset{\|}{C}}\text{—}OH + NH_4Br$

b. $CH_3CH_2\overset{O}{\overset{\|}{C}}\text{—}NH_2 \xrightarrow{?} CH_3CH_2\overset{O}{\overset{\|}{C}}O^-K^+ + NH_3$

Phosphorus Compounds

45. Draw the structural formulas for each of the following.
   a. diethyl hydrogen phosphate
   b. methyl dihydrogen phosphate
   c. triphosphoric acid

46. Name each of the following.

a. $HO\text{—}\overset{O}{\overset{\|}{P}}(OH)\text{—}O\text{—}\overset{O}{\overset{\|}{P}}(OH)\text{—}OH$

b. $CH_3CH_2O\text{—}\overset{O}{\overset{\|}{P}}(OH)\text{—}OH$

Additional Problems

47. All the following compounds are isomers. Circle and name the functional groups in each.

a. $CH_3CH_2CH_2\overset{\overset{\displaystyle O}{\|}}{C}OH$ **b.** $CH_3CH_2\overset{\overset{\displaystyle O}{\|}}{C}CH_2OH$

c. $CH_3\overset{\overset{\displaystyle O}{\|}}{C}CH_2CH_2OH$ **d.** $H\overset{\overset{\displaystyle O}{\|}}{C}CH_2CH_2CH_2OH$

e. $CH_3OCH_2CH_2\overset{\overset{\displaystyle O}{\|}}{C}H$ **f.** $CH_3CH_2OCH_2\overset{\overset{\displaystyle O}{\|}}{C}H$

g. $CH_3CH_2CH_2O\overset{\overset{\displaystyle O}{\|}}{C}H$ **h.** $CH_3OCH_2\overset{\overset{\displaystyle O}{\|}}{C}CH_3$

i. $CH_3CH_2O\overset{\overset{\displaystyle O}{\|}}{C}CH_3$ **j.** $CH_3CH_2\overset{\overset{\displaystyle O}{\|}}{C}OCH_3$

k. $CH_3\overset{\overset{\displaystyle O}{\|}}{C}-\overset{\overset{\displaystyle OH}{|}}{C}HCH_3$ **l.** $CH_3\underset{\underset{\displaystyle OH}{|}}{C}HCH_2\overset{\overset{\displaystyle O}{\|}}{C}H$

48. Arrange in order of increasing acidity, with the least-acidic compound first: benzoic acid, benzyl alcohol ($C_6H_5CH_2OH$), phenol, toluene.

49. Without consulting tables, arrange the following compounds in order of increasing boiling point: butyl alcohol, methyl acetate, pentane, propionic acid.

50. Name and draw structural formulas for all the isomeric amides that have the molecular formula C_4H_9NO.

51. Benzoic acid is insoluble in water. If the reactions described in Problem 28 were carried out in test tubes, what would you observe?

52. Offer explanations for the following facts.

a. Even though it has a higher molar mass, ethyl acetate has a lower boiling point (57 °C) than either methyl alcohol (65 °C) or formic acid (100 °C).

b. Sodium benzoate is soluble in water, whereas benzoic acid is insoluble.

c. The alkaline hydrolysis of esters is irreversible, whereas the acidic hydrolysis of esters is reversible.

d. Both acidic hydrolysis and alkaline hydrolysis of amides are irreversible.

53. From which alcohol might each acid be prepared via oxidation with acidic dichromate?

a. CH_3CH_2COOH **b.** $HOOCCOOH$

c. $HCOOH$ **d.** $(CH_3)_2CHCH_2COOH$

54. A lactone is a cyclic ester. What product is formed in each reaction?

a. 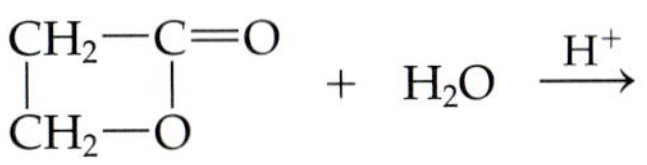
$$\begin{array}{l} CH_2-C{=}O \\ |\qquad\ \ | \\ CH_2-O \end{array} + H_2O \xrightarrow{H^+}$$

b. 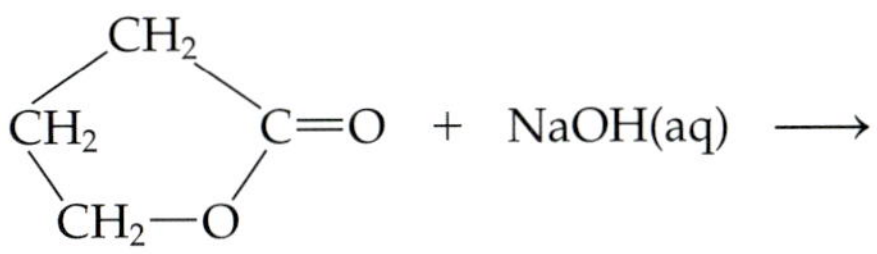
$$\begin{array}{c} \quad CH_2 \\ \diagup \qquad \diagdown \\ CH_2 \qquad\quad C{=}O \\ \diagdown \qquad \diagup \\ CH_2-O \end{array} + NaOH(aq) \longrightarrow$$

55. A lactam is a cyclic amide. What product is formed in each reaction?

a. 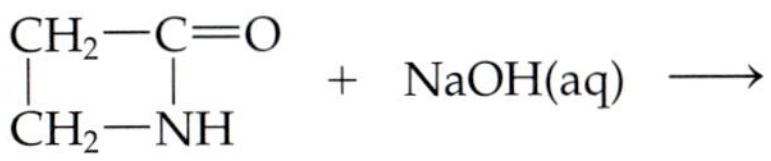
$$\begin{array}{l} CH_2-C{=}O \\ |\qquad\ \ | \\ CH_2-NH \end{array} + NaOH(aq) \longrightarrow$$

b. 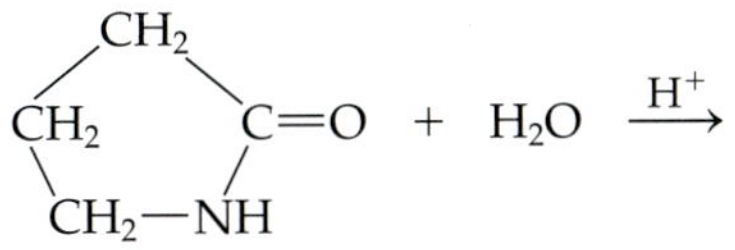
$$\begin{array}{c} \quad CH_2 \\ \diagup \qquad \diagdown \\ CH_2 \qquad\quad C{=}O \\ \diagdown \qquad \diagup \\ CH_2-NH \end{array} + H_2O \xrightarrow{H^+}$$

56. An ester with the molecular formula $C_6H_{12}O_2$ was hydrolyzed in aqueous acid to yield an acid (Y) and an alcohol (Z). Oxidation of the alcohol with potassium dichromate gave the identical acid (Y). What is the structural formula of the ester?

57. The neutralization of 125 mL of a 0.400 M NaOH solution requires 5.10 g of a monocarboxylic acid. Write all possible structural formulas for the acid.

58. If 3.00 g of acetic acid reacts with excess methanol, how many grams of methyl acetate are formed?

59. How many milliliters of a 0.100 M barium hydroxide solution are required to neutralize 0.500 g of dichloroacetic acid?

SELECTED TOPIC B

Drugs: Some Carboxylic Acids, Esters, and Amides

According to the broadest definition, a **drug** is any substance that affects an individual in such a way as to bring about physiological, emotional, or behavioral change. In order to have the desired effect, the active form of the drug must arrive at the appropriate receptors on the targeted tissues. Drug activity depends on a variety of factors, including the way the drug is metabolized, its steric fit to the drug receptor site (see Section 18.5), and the nature of the groups on the unbound end of the molecule. Many of the chemical parameters we have studied, including acidity and basicity, electronegativity, size and shape, and bond angles, play an important role.

People have long sought relief from pain and discomfort, and **analgesics,** drugs that relieve pain, are an important class of drugs. Many analgesics are carboxylic acids or their derivatives, esters or amides. Classes of pain-relieving drugs include **nonsteroidal anti-inflammatory drugs (NSAIDs)** such as aspirin, opioid agents such as morphine, both local and general anesthetics, dissociative anesthetics that inhibit receptors in the brain, and central nervous system depressants, including ethanol. In this selected topic, we discuss NSAIDs and the opiates. We also look at the amide LSD and at marijuana.

B.1 Aspirin and Other Salicylates

The most widely used analgesic is aspirin, often considered the prototype of NSAIDs. As the name suggests, most NSAIDs have anti-inflammatory as well as analgesic and **antipyretic** (fever reducing) effects. The development of aspirin is interesting and instructive. Soon after acetic anhydride[1] became available, in the nineteenth century, chemists began to use it to introduce an acetyl group (CH_3CO) into a variety of physiologically active compounds. The process, called acetylation, introduces structural modifications that often change the properties of drugs. Two such cases, aspirin and heroin, are described here. They serve as examples of many similar processes of drug development.

The first successful synthetic pain relievers were derivatives of salicylic acid (Figure B.1). Salicylic acid was first isolated from willow bark in 1860, although an English clergyman named Edward Stone had reported to the Royal Society as early as 1763 that an extract of willow bark was useful in reducing fever. The use of willow bark was also common among several American Indian tribes. Salicylic acid, which is both a carboxylic acid and a phenol, is itself a good analgesic and antipyretic, but it is sour and irritating when taken orally. Chemists sought to modify the structure of the mole-

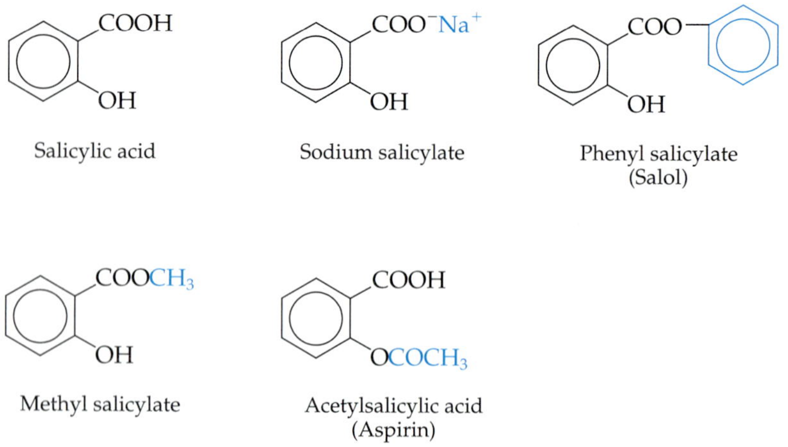

COOH, OH — Salicylic acid

COO^-Na^+, OH — Sodium salicylate

COO—C_6H_5, OH — Phenyl salicylate (Salol)

$COOCH_3$, OH — Methyl salicylate

COOH, $OCOCH_3$ — Acetylsalicylic acid (Aspirin)

◀ **Figure B.1**
Salicylic acid and some of its derivatives.

[1]Acetic anhydride can be considered as being formed when one molecule of water is removed from two molecules of acetic acid.

$$CH_3\overset{O}{\overset{\|}{C}}—OH + HO—\overset{O}{\overset{\|}{C}}CH_3 \longrightarrow CH_3\overset{O}{\overset{\|}{C}}—O—\overset{O}{\overset{\|}{C}}CH_3 + HOH$$

cule to remove this undesirable property while retaining (or even improving) the desirable properties.

The first modification was simple neutralization of the acid. Sodium salicylate was first used as a pain reliever in 1875. It was less unpleasant to swallow, but it proved to be highly irritating to the lining of the stomach. Phenyl salicylate (salol) was introduced in 1886. It passed unchanged through the stomach. In the small intestine, it was hydrolyzed to the desired salicylic acid, but phenol, which is rather toxic, also was formed.

Methyl salicylate, an ester, is an oil found in many plants and has a fragrance associated with wintergreen. It is the active ingredient in rub-on analgesics such as Ben-Gay. Methyl salicylate causes a mild burning sensation when applied to the skin, thus serving as a counterirritant for sore muscles. It also readily penetrates the skin, after which it is hydrolyzed to salicylic acid, which helps relieve the soreness.

Acetylsalicylic acid, to which the German Bayer Company assigned the trade name Aspirin, was first introduced in 1899, and soon became the most used drug in the world. Billions of aspirin tablets (under many trade names) are now consumed worldwide each year.

Aspirin relieves minor aches and pains, suppresses inflammation, reduces fever, and acts as an **anticoagulant** (inhibiting blood clotting). How does it exert these diverse effects? The most widely accepted explanation of the actions of the NSAIDs is that they inhibit an enzyme (cyclooxygenase) necessary for the synthesis of prostaglandins (see Selected Topic E). Aspirin does this by acetylating an amino acid, serine; other NSAIDs have other mechanisms of inhibition.

Among their many functions, prostaglandins are involved in inflammation, increased blood pressure, and smooth muscle contraction. Elevated concentrations of prostaglandins appear to activate pain receptors in the tissues, making the tissues more sensitive to any pain stimulus. **Pyrogens,** fever-inducing substances produced and released by leukocytes and other circulating cells, usually use prostaglandins as secondary mediators.[2] Prostaglandins also play a role in blood-clot formation by promoting platelet aggregation.

Thus, inhibition of prostaglandin synthesis can counter these effects. Aspirin and other NSAIDs do not cure whatever is causing the pain; they merely kill the messenger. Aspirin often is the initial drug of choice for treatment of arthritis, a disease characterized by the inflammation of joints and connective tissues. Small daily doses seem to lower the risk of coronary heart attack and stroke, presumably by inhibiting platelet aggregation in the arteries. For people with heart problems, many doctors now recommend a daily aspirin tablet, with dosage ranging from 81 mg (a baby aspirin, one-quarter of an adult tablet) to a full adult tablet. Of the 80 million aspirin tablets consumed each day in the United States, most of them are taken to reduce the chance of heart attack and stroke.

Aspirin is not effective for severe pain—for example, pain from a migraine headache. Like all other drugs, aspirin is somewhat toxic. It is the drug most often involved in the accidental poisoning of children. The toxicities of aspirin and other drugs are listed in Table B.1. Prolonged use, as for arthritic pain, can lead to gastrointestinal disorders, probably by damaging the protective mucosal lining of the stomach and increasing gastric acid secretion. Its anticoagulant properties mean that aspirin should not be used by people facing surgery, childbirth, or other hazards involving the possible loss of blood (nonuse should start a week before the hazard).

Still another hazard associated with aspirin use is allergic reaction. In some people, an allergy to aspirin can cause skin rashes, asthmatic attacks, and even loss of consciousness. Some doctors claim that the allergic reaction may be delayed three to five hours, so the victim may not associate the reaction with aspirin. Susceptible individuals must be careful to avoid aspirin by itself or in any combination with other drugs.

Use of aspirin to treat children feverish with flu or chicken pox is associated with **Reye's syndrome.** This syndrome is characterized by vomiting, lethargy, confusion, and irritability. Fatty degeneration of the liver and other organs can lead to death unless treatment is

Table B.1 Acute Toxicities of Chemicals Presently or Formerly Used in Over-the-Counter Drugs

Chemical Compound	LD_{50}[a]
Acetaminophen	338[b]
Acetanilide[c]	800
Aspirin	1500
Caffeine	355 (246)
Ibuprofen	(1050)
Methyl salicylate	887
Ketoprofen	101
Naproxen	534
Phenacetin[c]	1650
Piroxicam	360[b]

[a]LD_{50} values are for oral administration of the drug in rats in milligrams per kilogram of body weight (unless otherwise noted). Values in parentheses are for male rats.
[b]Orally in mice.
[c]No longer used.

[2]Aspirin doesn't affect some pyrogens that do not work through prostaglandins.

begun promptly. Just how aspirin use enhances the onset of Reye's syndrome is not known, but the correlation is quite strong. Aspirin products bear a warning not to use aspirin to treat children with fevers.

Aspirin is a single chemical compound. Like other compounds, its properties are invariant. An ordinary aspirin *tablet* contains 325 mg of acetylsalicylic acid, held together with inert binders such as starch. Aspirin has been tested extensively. The conclusions of impartial studies are invariably the same: the only significant difference between brands is price.

"Buffered" aspirin contains antacids but is not truly buffered. A *buffer* reacts with either acid or base and keeps the pH of a solution essentially constant. The antacids neutralize acid only; they do not buffer. Some people experience mild stomach irritation when they take aspirin on an empty stomach. Eating a little food first or drinking a full glass of water with aspirin is just as effective as taking "buffered" tablets.

"Extra-strength" and "maximum-strength" aspirin tablets usually have 500 mg of aspirin rather than the usual 325 mg. They have no other active ingredients. Simple arithmetic tells us that three plain aspirin tablets are equal to two extra-strength tablets in dosage, and they are usually lower in price.

B.2 Aspirin Substitutes and Combination Pain Relievers

A large variety of NSAIDs have been developed (Figure B.2). Most of them have an acidic center, an aromatic ring, and an alkyl group or second ring. For aspirin, activity appears to depend on the salicylate anion. The various drugs differ in the strengths of their analgesic, antipyretic, and anti-inflammatory effects, and in their toxicities.

People allergic to aspirin or susceptible to bleeding generally take acetaminophen, an amide. This compound gives pain relief and fever reduction comparable to the action of aspirin. Unlike aspirin, however, it is not effective against inflammation, so it is of limited use to people with arthritis. Neither does it induce bleeding, so it is often used to relieve the pain that follows minor surgery. (Its action on prostaglandin synthesis is thought to occur more in the central nervous system than in the periphery, and this may explain some of its different actions.) The fact that acetaminophen does not promote bleeding probably accounts for the fact that it is used in hospitals more frequently than aspirin. Acetaminophen usually costs more than aspirin, especially in the form of highly advertised brands such as Tylenol. Overuse of acetaminophen is linked to liver and kidney damage, especially in those who drink a lot of alcohol.

Derivatives of propanoic acid, the 2-arylpropanoic acids ("profens"), are an important class of NSAIDs. These include ibuprofen, naproxen, and ketoprofen (see again Figure B.2). They may be a bit more effective than aspirin in treating inflammation, and they relieve mild pain and reduce fevers. They are usually much more expensive than aspirin.

Many products on the market are combinations of aspirin, acetaminophen, and other drugs. Are these combinations really more effective than aspirin or acetaminophen alone? The question has no simple answer. Let's look first at some history of some of the more familiar "combination pain relievers."

For many years the most familiar combination was aspirin, phenacetin, and caffeine (APC). Phenacetin has about the same effectiveness as aspirin in reducing fever and relieving minor aches and pains. It has been

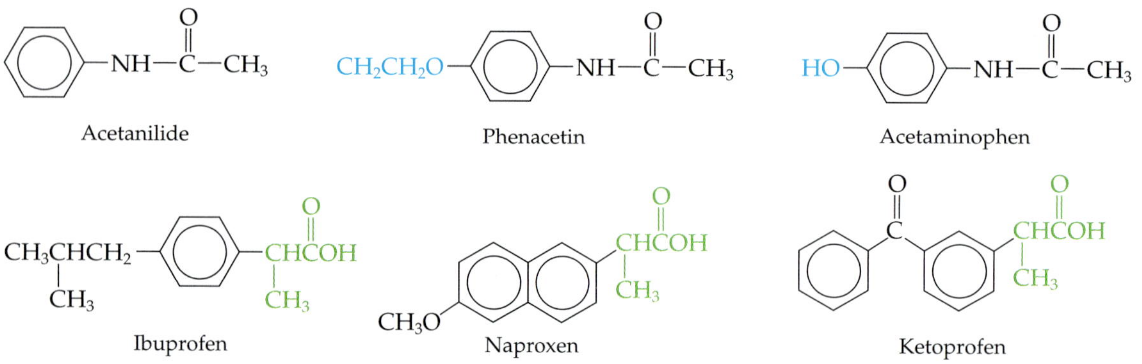

Acetanilide

Phenacetin

Acetaminophen

Ibuprofen

Naproxen

Ketoprofen

▲ **Figure B.2**
Some aspirin substitutes. Phenacetin and acetaminophen (Tylenol) are derivatives of acetanilide. Ibuprofen (Advil, Nuprin), naproxen (Aleve), and ketoprofen (Orudis KT) have long been available by prescription but are now available over-the-counter; all three are derivatives of propionic acid. Phenacetin and acetanilide are no longer used.

implicated in damage to the kidneys, in blood abnormalities, and as a likely carcinogen, however. The U.S. FDA banned further use of phenacetin in 1983.

Anacin and Excedrin were once APC formulations. Anacin now contains only aspirin and caffeine, and Excedrin adds acetaminophen to this combination. Caffeine is a mild stimulant found in coffee, tea, and cola syrup. There is no reliable evidence that caffeine significantly enhances the effect of aspirin. In fact, evidence indicates that for fever reduction, caffeine *counteracts* the action of aspirin. Combinations containing caffeine are therefore *less effective* than plain aspirin for this use.

O CH3
CH3 N N
O N N
CH3

Caffeine

Many other combination products are available. Brands and formulations change frequently. Extensive studies show repeatedly that, for most people, plain aspirin is the cheapest, safest, and most effective product. However, people vary in their response, and a particular person might find acetaminophen or ibuprofen more effective than aspirin for relieving pain.

(a)

(b)

▲ The opium poppy flower (a) and seed pod (b).

B.3 Opium Alkaloids

Many plants produce **alkaloids,** nitrogen-containing compounds that usually have physiological activity. Nicotine from tobacco and caffeine from coffee are familiar alkaloids. The opium poppy, *Papaver somniferum*, produces a number of alkaloids, including morphine and codeine. These and related compounds are called opiates (Figure B.3). The opiates are the most effective pharmaceutical treatment for severe pain. They are often called **narcotic** analgesics because they induce *narcosis*, a state of profound stupor.

Morphine, first isolated in 1805, is the most important narcotic analgesic in medicine, used by prescription for relief of severe pain. It is still obtained from opium, and it remains the standard against which new analgesics are measured.

Morphine acts by fitting specific receptor sites on neurons in the brain and spinal cord (Figure B.4). The existence of these opiate receptors was first demonstrated in 1973 by Solomon Snyder and Candace Pert at Johns Hopkins University School of Medicine.

Why should the human nervous system have receptors for a plant-derived drug such as morphine? When several investigators searched for morphine-like substances (opioids) produced by the human body, they found not one but several such substances, called **endorphins** ("endogenous morphines"). Each was a short peptide chain composed of amino acid units (Chapter 21). Those with five amino acid units are called **enkephalins.** There are two enkephalins, and they differ only in the amino acid at the end of the chain. *Leu*-enkephalin has the amino acid sequence Tyr-Gly-Gly-Phe-Leu, and *Met*-enkephalin is Tyr-Gly-Gly-Phe-Met.[3] Other endorphins have chains of about 30 amino acids.

Endorphins are released as a response to pain deep in the body. It appears that analgesia from acupuncture or biofeedback stems from release of the brain opioids. Endorphin release also has been used to explain other phenomena once thought to be largely psychological. A soldier wounded in battle may feel no pain until the skirmish is over, because the body has secreted its own painkiller. The production of these compounds during strenuous athletic activity also may explain the "high" reported by distance runners.

The enkephalins have been synthesized in the laboratory and shown to be potent pain relievers. Their use in medicine is quite limited, however, because, after being injected, they are rapidly broken down by the enzymes that hydrolyze proteins. Enkephalin analogs more resistant to hydrolysis are a subject of considerable research.

[3]The three-letter symbols are abbreviations for particular amino acids; see Table 21.1.

Figure B.3
Morphine and some related compounds.

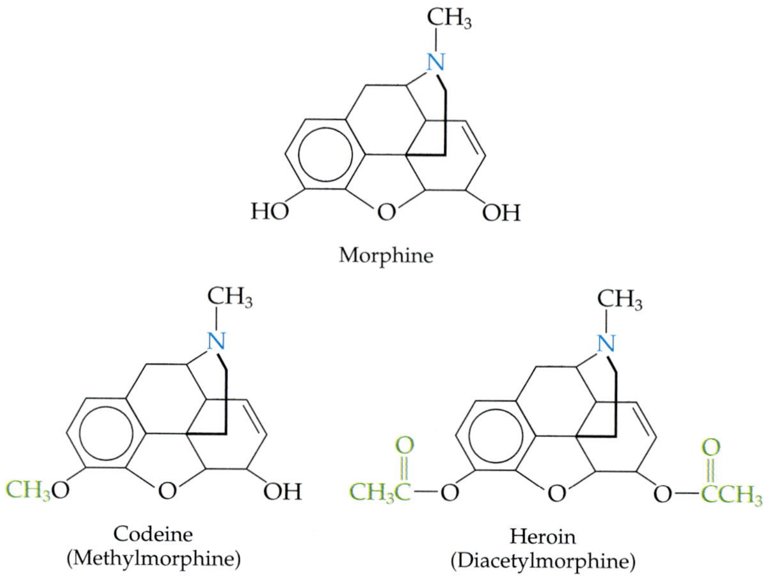

Morphine

Codeine
(Methylmorphine)

Heroin
(Diacetylmorphine)

Unfortunately, both the natural enkephalins and the synthetic analogs seem to be addictive.

Thus morphine is still the most used analgesic for relief of severe pain. It also induces lethargy, drowsiness, confusion, euphoria, chronic constipation, and depression of the respiratory system. It can be addictive if administered in amounts greater than the prescribed doses or for a period longer than the prescribed time. Its use has been under the control of the federal government since the Harrison Act of 1914.

When pure, morphine is an insoluble, odorless, white, crystalline solid having a bitter taste. It is the principal alkaloid (about 10% by weight) of raw opium, which is the dried juice from the unripened seed pod of the poppy plant. Raw opium was used in many of the patent medicines of a century ago. Ayer's Cherry Pectoral, Jayne's Expectorant, Pierce's Golden Medical Discovery, and Mrs. Winslow's Soothing Syrup were but a few.

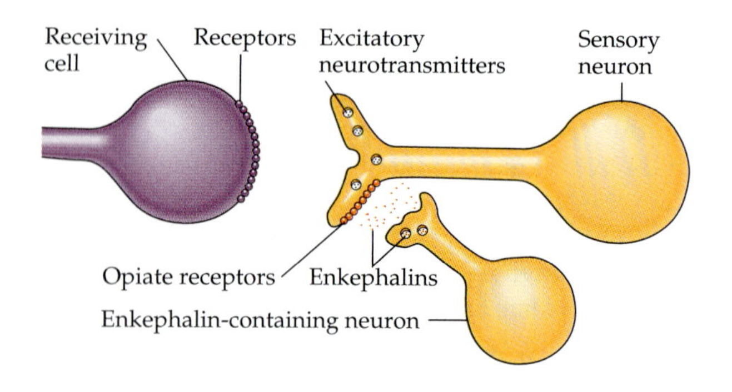

Figure B.4
A proposed mechanism for opiate activity. Enkephalins bind to the opiate receptors and block the release of neurotransmitters that would convey the pain message to the brain.

B.4 Semisynthetic Opiates

Slight changes in the molecular structure of morphine produce altered physiological properties (see again Figure B.3). Replacement of the phenolic OH group by a methoxy (CH_3O) group produces codeine. Codeine occurs in opium to an extent of about 0.5%, but a much larger quantity is synthesized by methylating the more abundant morphine molecules.

Codeine is similar to morphine in its physiological action, but is less potent, has less tendency to induce sleep, and is thought to be less addictive. In amounts of less than 2.2 mg/mL, codeine is exempt from stringent narcotics regulations and is used in a few "controlled substance" cough syrups.

Reaction of morphine with acetic anhydride produces diacetylmorphine; both OH groups (alcohol and phenol) are converted to esters. This morphine derivative was first prepared by chemists at the Bayer Company of Germany in 1874 and assigned the trade name heroin. It received little attention until 1890, when it was proposed as an antidote for morphine addiction. Shortly thereafter Bayer widely advertised heroin as a sedative for coughs, often in the same ads as aspirin. It soon was found, however, that heroin induced addiction more quickly than morphine and that heroin addiction was harder to cure.

The physiological action of heroin is similar to that of morphine, except that heroin seems to produce a stronger feeling of euphoria for a longer period of time. Heroin is not legal in the United States, even by prescription. It has, however, been advocated for use in pain relief in terminal cancer patients and has been so used in Britain.

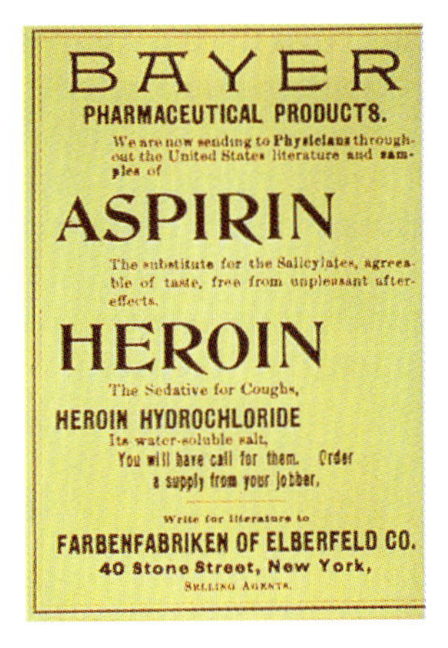

▶ Heroin was regarded as a safe medicine in 1900; it was widely used as a sedative for coughs.

Deaths from heroin usually are attributed to overdose, most frequently when the user is not aware of the amount of heroin in a given amount of powder. As an illustration, the office of the chief medical examiner for New York City analyzed 132 samples of drugs, supposedly heroin, that had been confiscated on the streets. Twelve contained no heroin at all. The remaining 120 varied from 1 to 77% heroin. A user expecting a dose of 1 unit might actually get 77 times as much—a catastrophic overdose.

B.5 Synthetic Narcotics: Analgesia and Addiction

Much research has gone into developing a nonaddictive analgesic as effective as morphine. Perhaps the best known of the synthetic narcotics is meperidine (Figure B.5), known by the trade name Demerol. Meperidine is somewhat less effective than morphine, but may cause less nausea and constipation. Repeated use, unfortunately, does lead to addiction.

Another synthetic narcotic is methadone, which is effective for oral use in cancer patients. This drug has been widely used to treat heroin addiction. Like heroin, methadone is highly addictive. However, when taken orally, it does not induce the sleepy stupor characteristic of heroin intoxication, and a person on methadone usually is able to hold a productive job. In some places methadone is available free in clinics. If an addict who has been taking methadone reverts to heroin, the methadone in the body effectively blocks heroin's euphoric rush and so reduces the temptation to use heroin.

Methadone maintenance is not a perfect answer. Injected methadone produces an effect similar to that of heroin, and it has been diverted for illegal use in this manner. And an addict on methadone is still an addict.

Chemists have synthesized thousands of morphine analogs. Only a few have shown significant analgesic activity. Most are addictive. Morphine acts by binding to receptors in the nervous system. A molecule that mimics the action of a drug is called an **agonist.** An **antagonist** blocks the action of a drug. Morphine antagonists are molecules that block the action of morphine, most likely by blocking the receptors. Some molecules have both agonist and antagonist effects. These substances show promise as analgesics. An example is pentazocine (Talwin). It is less addictive than morphine and yet effective for relief of moderate pain. There is some hope that an effective, nonaddictive analgesic will be developed, but to date the two effects seem inseparable.

Pure antagonists such as naloxone are of value in treating opiate addicts. Overdosed addicts can be brought back from death's door by an injection of naloxone. Long-acting antagonists can block the action of heroin for as much as a month, thus aiding an addict in overcoming his or her addiction.

B.6 LSD: A Hallucinogenic Drug

A most interesting amide is the *N,N*-diethylamide of lysergic acid, better known as LSD (from the German *lysergsaure diethylamid*). The physiological properties of LSD were discovered quite accidentally in 1943 when Albert Hofmann, a chemist at the Sandoz Laboratories in Switzerland, unintentionally ingested some. He later took 250 *mg*, which he considered a small dose, to verify that LSD had caused the symptoms he had experienced. Hofmann had a very rough time for the next few hours, exhibiting such symptoms as visual disturbance and schizophrenic behavior.

Meperidine (Demerol)

Methadone

Pentazocine (Talwin)

Naloxone

▶ **Figure B.5**
Some synthetic narcotics. Meperidine and methadone are opiate agonists. Naloxone is an opiate antagonist. Pentazocine has both agonist and antagonist properties.

Lysergic acid diethylamide
(LSD)

Lysergic acid is obtained from ergot, a fungus that grows on rye.[4] It is easily converted to the diethylamide. Note that a part of the LSD structure resembles that of serotonin (Selected Topic C). LSD seems to act as a serotonin agonist.

LSD is a potent drug, as indicated by the small amount required for a person to experience its fantastic effects. The usual dose is probably about 10 to 100 μg. No wonder Hofmann had a bad time with 250 *mg*! To give you an idea of how small 10 μg is, let's compare that amount of LSD with the amount of aspirin in one tablet—one aspirin tablet contains 325,000 μg of aspirin.

LSD is classed with heroin and is not used medically, although it was tried as an aid to psychotherapy in the 1950s and 1960s. Is LSD a dangerous drug? Certainly it is in the sense that it sometimes induces dangerous hallucinations such as the belief that one can fly from a tall building. It has been reported to induce long-term psychoses and other mental problems. Reports in the 1960s indicated that LSD damages chromosomes, especially those of the leukocytes (white blood cells). The reports received wide publicity, but additional studies produced mixed results.

B.7 Marijuana: Some Chemistry of Cannabis

The active ingredient of marijuana is not a carboxylic acid, ester, nor amide. However, it has three other functional groups that we have discussed—phenol, ether, and alkene—and we consider it here because of its significance. (In the United States, marijuana is second only to alcohol as an intoxicant.) Even though many books have been written about marijuana, all we know for certain about the drug fills only a few pages. Let's look at some of its chemistry.

The plant *Cannabis sativa* has long been useful. The stems yield tough fibers for making rope, and the plant has been used as a drug in tribal religious rituals. The term **marijuana** refers to a preparation made from the leaves, flowers, seeds, and small stems of the plant. These are generally dried and smoked. Marijuana contains a variety of chemical substances, but the principal active ingredient is tetrahydrocannabinol (THC). Actually, there are several active cannabinoids in marijuana; only one is shown here.

Tetrahydrocannabinol
(THC)

Marijuana plants vary considerably in THC content. Potency depends on the genetic variety of plant, not to any significant extent on the climate or the soil where it is grown. The potency of cultivated plants has increased substantially since the 1970s. THC content of between 4 and 8% is now common.

The physiological effects of marijuana are difficult to measure, partly because of variable THC content. A marijuana variety of standard potency is grown and supplied for controlled clinical studies. With this standard product, some effects can be measured in reproducible experiments. Smoking *Cannabis* increases the pulse rate, distorts the sense of time, and impairs some complex motor functions. Psychic effects vary greatly and include a euphoric sense of lightness, hallucinations, anxiety, and an impression of brilliance, although studies have shown no mind-expanding effects. Marijuana seems to heighten one's enjoyment of food, with users relishing beans as much as they normally would enjoy steak.

The long-term effects of marijuana use are more difficult to evaluate. Some people claim that smoking

◀ The marijuana plant.

[4]Several useful drugs are obtained from the ergot fungus. Ergotamine shrinks blood vessels in the brain; it is used to treat migraine headaches. Ergonovine causes small blood vessels to contract. It is used to induce uterine contractions and thus reduce bleeding after childbirth. Both compounds, like LSD, are amides of lysergic acid.

marijuana leads to the use of harder drugs, but others disagree. More heroin addicts start drug use with alcohol than with marijuana.

There is some evidence—both direct and indirect—that marijuana causes brain damage. Studies in rats indicate that marijuana causes brain lesions. Heavy users often are lazy, passive, and mentally sluggish. It is difficult, however, to prove that these are effects of marijuana use. There are millions of users, and even if the observed brain damage is due to marijuana, it is less extensive than that caused by alcohol.[5]

Some people claim that excessive use of marijuana leads to psychoses. The drug has long been known to induce short-term psychotic episodes in those already predisposed and in others who take excessive amounts. Long-term psychoses, however, occur at the same rate among regular marijuana users as among the general population.

Heavy marijuana use has been reported to cause gynecomastia (enlarged breasts) in some men, but THC also has been shown to cause an initial rise in testosterone levels. There is a slight structural similarity between THC and the female hormones (Selected Topic E). Some studies indicate that THC binds weakly to estrogen receptors.

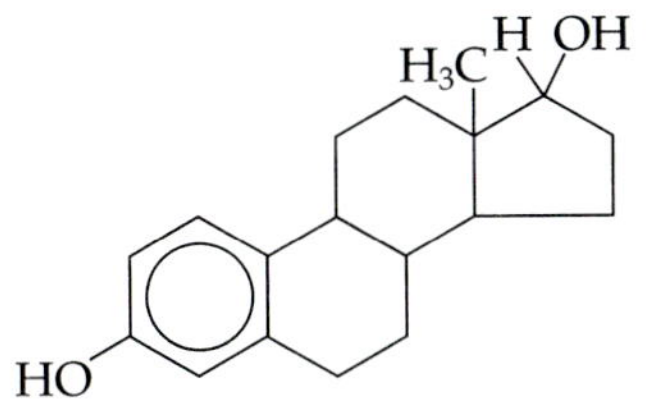

H_3C H OH

HO

Estradiol, a female hormone

Chemists have isolated the active components in marijuana and synthesized them. They can monitor the THC content of marijuana as well as the THC levels in the bloodstream and can identify the breakdown products. They have not yet, however, determined its long-term effect on body chemistry.

Unlike alcohol, THC persists in the bloodstream for several days because it is soluble in fats. This persistence indicates that some of a given dose may still be active when another dose is taken. This might account for the fact that an experienced pot smoker can get high on a dose that doesn't affect a novice.

Marijuana has some legitimate medical uses and has been approved for such use in several states. It reduces pressure in the eyes of people who have glaucoma. If not treated, the buildup of pressure eventually causes blindness. Also, marijuana relieves the nausea that afflicts cancer patients undergoing radiation treatment and chemotherapy and that of heavily medicated AIDS patients.

[5]The gene for a THC receptor has been cloned (see Chapter 23). The receptors are found in movement-control centers, thus explaining the loss of coordination in those intoxicated with the drug. The memory and cognition areas of the brain also are rich in THC receptors. This expains why marijuana users often do poorly on tests. There are few receptors in the brain stem, where breathing and heartbeat are controlled. This correlates with the fact that it is difficult to get a lethal dose of marijuana.

B.8 Thalidomide: Revival of a Discarded Drug

The importance of thorough drug evaluation—and the quandary of balancing risks and benefits—is shown in the thalidomide story. This amide-like drug was introduced for use as a tranquilizer in the 1950s.

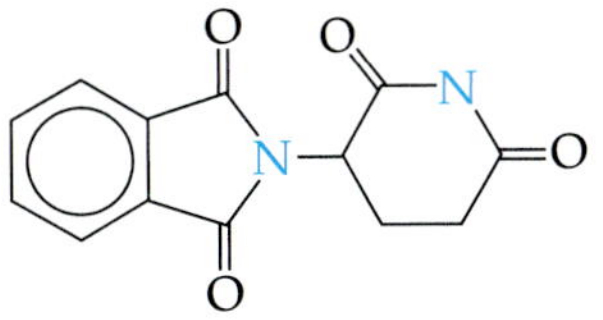

O O N N O O

Thalidomide

It was considered so safe, based on laboratory studies, that it often was prescribed for pregnant women. In Germany it was available without a prescription. It took several years for the human population to provide evidence that laboratory animals had not. The drug had a disastrous effect on developing human embryos. Women who had taken the drug during the first 12 weeks of pregnancy had babies that suffered from phocomelia, a condition characterized by shortened or absent arms and legs, and other physical defects. The drug was used widely in Germany and Great Britain, and these two countries bore the brunt of the tragedy. The United States escaped relatively unscathed because an official of the FDA had blocked its approval because she believed that there was evidence to doubt the drug's safety. In 1962, the drug was banned worldwide. In recent years, however, tests with this drug have shown that it has unique anti-inflammatory and immunomodulatory (modifying the immune response) properties. After much debate, it was approved in September 1998 for use in a debilitating skin condition of leprosy. Thalidomide is subject to very strict prescribing restrictions, aimed at preventing its use by pregnant women. Studies currently in progress are evaluating thalidomide's effects in Kaposi's sarcoma (a complication of AIDS), graft-versus-host reactions in organ transplants and autoimmune diseases, ulcers in AIDS, and tuberculosis.

Key Terms

agonist (B.5)
alkaloid (B.3)
analgesic (page 438)
antagonist (B.5)
anticoagulant (B.1)
antipyretic (B.1)
drug (page 438)
endorphin (B.3)
enkephalin (B.3)
marijuana (B.7)
narcotic (B.3)
nonsteroidal anti-inflammatory drugs (NSAIDs) (page 438)
pyrogen (B.1)
Reye's syndrome (B.1)

Review Questions

1. Define, explain, or give an example of each of the following terms.
   a. drug **b.** analgesic **c.** antipyretic
   d. pyrogen **e.** Reye's syndrome
2. Give the structure and chemical name for aspirin. In what ways may one brand of aspirin differ from another? In what ways must brands of aspirin be the same?
3. What is an alkaloid? Name several alkaloids.
4. What is a narcotic? Name several narcotics.
5. How does morphine act to relieve pain? What is a morphine agonist? A morphine antagonist?
6. How does heroin differ in structure from morphine? How does it differ in physiological properties?
7. How does methadone work as a treatment for heroin addiction?
8. What is an endorphin? An enkephalin?
9. What are the effects of a hallucinogenic drug?
10. How are endorphins related to (**a**) the anesthetic effect of acupuncture and (**b**) the absence of pain in a badly wounded soldier?
11. What are some of the problems in the clinical evaluation of (**a**) LSD and (**b**) marijuana?
12. How might marijuana have a feminizing effect on males?
13. Why is tetrahydrocannabinol retained in the body for several days?
14. A "maximum-strength" aspirin has 1000 mg of aspirin per two-tablet dose. How many "regular" 325-mg aspirin tablets would you take to get about the same amount of aspirin?

Problems

15. To what family of organic compounds does ibuprofen belong?
16. What alkyl group is attached to the *para* position of the benzene ring of the ibuprofen molecule? Can you see how the generic name ibuprofen was derived?
17. List the reagents necessary to carry out the following conversions.

a.

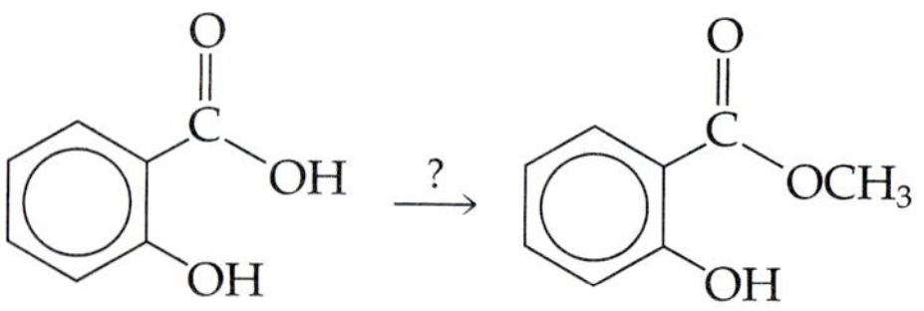

b.

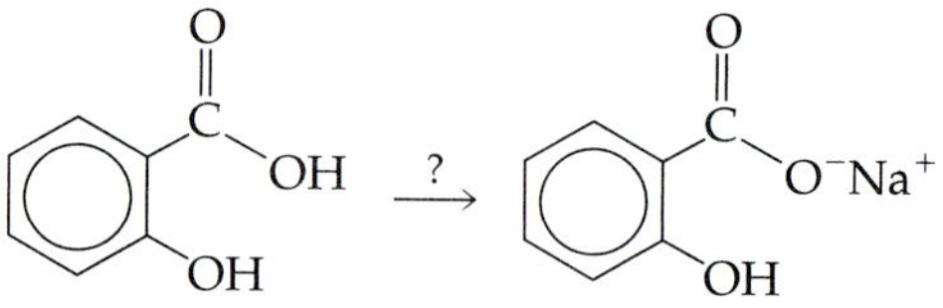

18. List the reagents necessary to carry out the following conversions.

a.

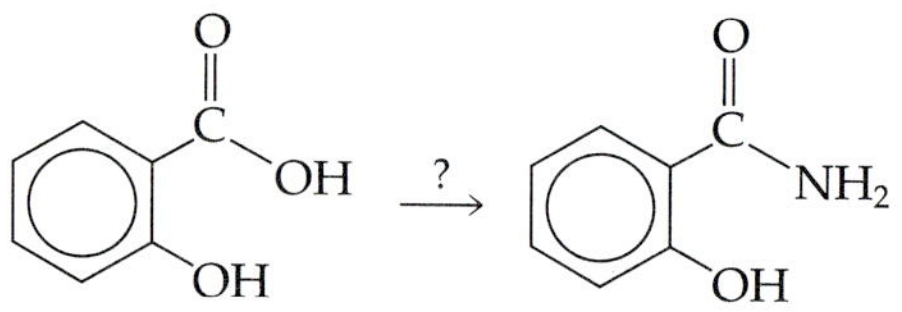

b.

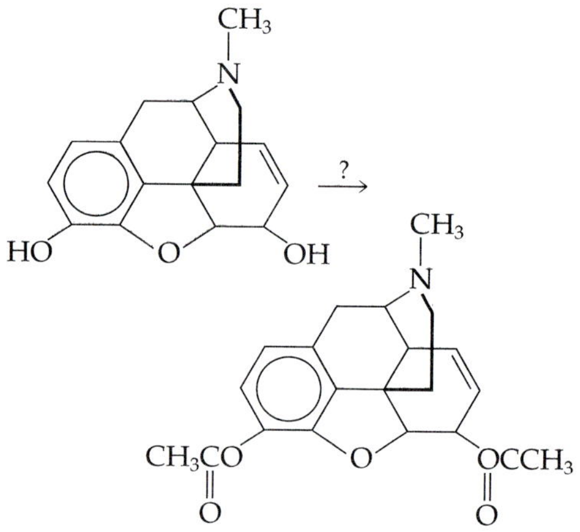

Additional Problems/Projects

19. In what areas of the brain are THC receptors found? How does the location of these receptors explain some of the effects of THC as a drug?
20. Compare effective doses of LSD and aspirin in terms of moles.
21. Examine the labels of at least five "combination" pain relievers (e.g., Excedrin, Empirin, Anacin). Make a list of the ingredients in each. Look up the properties (medical use, dosage, side effects, toxicity) in a reference work such as *The Merck Index*.
22. Do a cost analysis on at least five brands of plain aspirin, calculating the cost per 650-mg dose.

Amines and Derivatives

From herbal medicine to modern pharmaceuticals, amines play a prominent role in treatment of disease.

Learning Objectives/Study Questions

1. What is the general structure for an amine? What is the functional group for this class of organic molecules?
2. What structural feature is used to classify amines as primary, secondary, or tertiary?
3. How are the names of amines determined?
4. Why are the boiling points of primary and secondary amines higher than those of alkanes or ethers of similar molar mass, but lower than those of alcohols? How do the boiling points of tertiary amines compare with alcohols, alkanes, and ethers of similar molar mass?
5. How do the solubilities of amines of five carbons or less compare to the solubilities of comparable alkanes and alcohols in water?
6. What typical reactions take place with amines?
7. What are heterocyclic amines?

If carbon compounds are the basis of life, nitrogen compounds are the bases of life (pun intended). Recall from Chapter 9 that ammonia is a nitrogen-containing weak base. An **amine** is an alkyl (or aryl) derivative of ammonia, and like ammonia, amines are weak bases. The functional group of an amine is simply the nitrogen atom with its lone pair of electrons. Amines occur in living (and especially in once-living but now-decaying) organisms.

Nitrogen is an essential constituent of many physiologically active compounds. All enzymes—indeed, all proteins—contain nitrogen. Many vitamins and hormones contain nitrogen, as do most drugs. Nitrogenous bases are part of the complex structure of the compounds that carry our genetic heritage, the nucleic acids DNA and RNA (bases in acids!). In this chapter, we discuss the amines generally, and in Selected Topic C, we discuss a number of related nitrogen-containing compounds that exhibit interesting physiological effects. The discussion of proteins and nucleic acids we shall save for later chapters.

In subsequent chapters we shall also consider the implications of the following facts. Plants can take inorganic nitrogen, usually in the form of nitrate or ammonium salts, and combine it with carbon compounds from photosynthesis to make all the organic nitrogen compounds they require. Animals are not quite so clever. They require in their diet some preformed organic nitrogen compounds that are essential to their health but that they themselves cannot synthesize.

17.1 Structure and Classification of Amines

In Chapter 14 we noted that alcohols and ethers can be considered derivatives of water. In a similar manner, amines are derived from ammonia. Amines are classified according to the number of carbon atoms bonded directly to the *nitrogen* atom. A **primary (1°) amine** has one alkyl (or aryl) group on the nitrogen atom, a **secondary (2°) amine** has two, and a **tertiary (3°) amine** has three (Figure 17.1).

This use of the terms *primary, secondary,* and *tertiary* is somewhat different from that used to classify alcohols (Section 14.2). For example, consider structures I and II in Figure 17.2. Compound I is a primary amine because only one alkyl group is attached to the nitrogen. Compound II looks quite similar to compound I, but it is a *secondary* alcohol. When determining whether an alcohol is primary, secondary, or tertiary, we count the number of carbon atoms bonded to the *carbon that bears the OH* group, not to the oxygen itself. Note another difference in structures III and IV. Compound III is a secondary amine, but compound IV is an ether (*not* an alcohol, secondary or otherwise). When there is only one carbon group attached to oxygen, the compound is an alcohol; when there are two, the compound is an ether. In contrast, whether there are one, two, or three alkyl or aryl groups attached to nitrogen, the compounds are all classified as amines.

Let's look at the structure of ammonia and the amines a little more closely. In ammonia, three hydrogen atoms are bonded to nitrogen. The nitrogen atom also has a lone pair of electrons.

▶ Figure 17.1 Amines are derived from ammonia in a manner similar to that in which alcohols and ethers are derived from water.

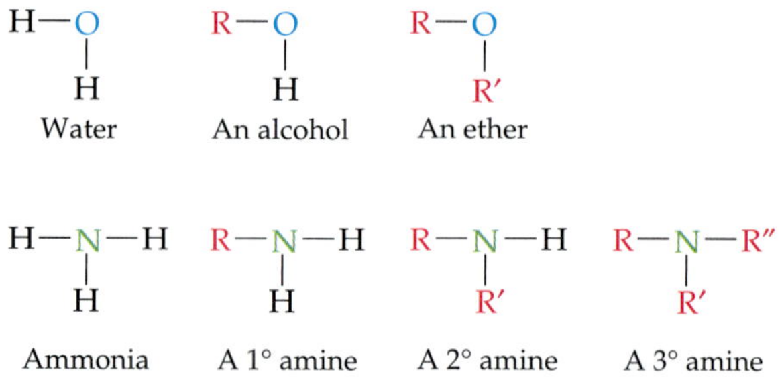

H—O—H	R—O—H	R—O—R′	
Water	An alcohol	An ether	
H—N(—H)—H	R—N(—H)—H	R—N(—R′)—H	R—N(—R′)—R″
Ammonia	A 1° amine	A 2° amine	A 3° amine

$$\mathrm{H{-}\ddot{N}{-}H}$$
$$\mathrm{\ \ \ \ |}$$
$$\mathrm{\ \ \ \ H}$$

Ammonia

Ammonia can undergo reactions in which it shares that lone pair. It can accept a proton and form the ammonium ion.

$$\mathrm{H{-}\ddot{N}(H){-}H} + \mathrm{H^+} \longrightarrow \left[\mathrm{H{-}N(H)(H){-}H}\right]^+$$

This, of course, is how ammonia reacts as a base. An amine also has a lone pair of electrons, and it too acts as a base. We will consider this reaction in more detail in Section 17.4, but for now let's concentrate on the fact that in the ammonium ion, the nitrogen atom is bonded to *four* hydrogen atoms. Just as amines are derived from ammonia by replacement of one or several of the hydrogens with alkyl or aryl groups, so too can substituted ammonium ions be derived from the simple ammonium ion. Any or all of the hydrogens on the ammonium ion can be replaced by alkyl (or aryl) groups:

$$\left[\mathrm{R{-}N(H)(H){-}H}\right]^+ \quad \left[\mathrm{R{-}N(R)(H){-}H}\right]^+ \quad \left[\mathrm{R{-}N(R)(H){-}R}\right]^+ \quad \left[\mathrm{R{-}N(R)(R){-}R}\right]^+$$

$\mathrm{CH_3{-}CH(NH_2){-}CH_3}$

I

Primary amine

$\mathrm{CH_3{-}CH(OH){-}CH_3}$

II

Secondary alcohol

$\mathrm{CH_3{-}N(H){-}CH_3}$

III

Secondary amine

$\mathrm{CH_3{-}O{-}CH_3}$

IV

Ether

▲ **Figure 17.2**
Organic nitrogen-containing compounds are classified in a different manner from organic oxygen-containing compounds.

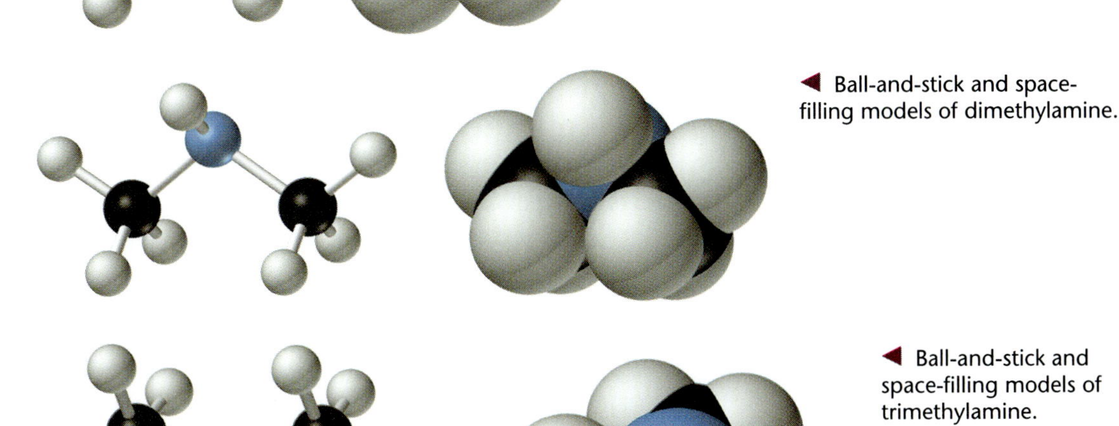

◀ Ball-and-stick and space-filling models of methylamine.

◀ Ball-and-stick and space-filling models of dimethylamine.

◀ Ball-and-stick and space-filling models of trimethylamine.

The ion in which all four hydrogens are replaced by alkyl groups is a *quaternary* ammonium ion. Compounds that incorporate this type of ion are called **quaternary (4°) ammonium salts.**

✓ Review Questions

17.1 Why is isopropyl alcohol classified as a secondary alcohol, while isopropylamine is classified as a primary amine?

17.2 Define primary, secondary, and tertiary amine. Compare this terminology with that used for alcohols.

17.2 Naming Amines

The common names for simple aliphatic amines merely specify the alkyl groups attached to nitrogen and add the suffix *-amine*. The **amino group (NH_2)** is named as a substituent in amines that incorporate other functional groups and amines in which the alkyl groups cannot be simply named.

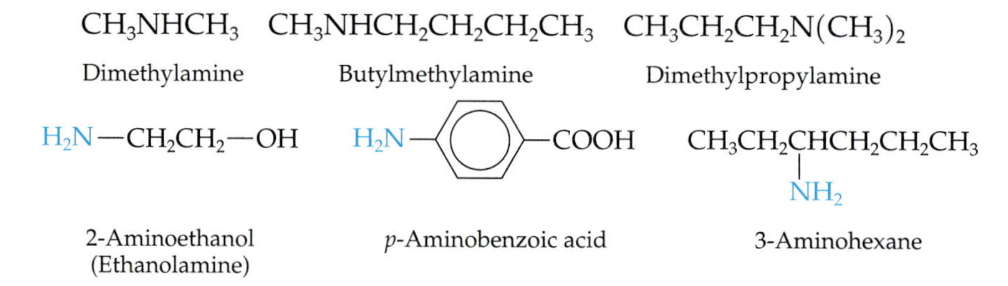

Example 17.1

Name and classify each of the following.

a. CH_3CHNH_2 with CH_3 on the CH carbon ($CH_3\underset{\displaystyle CH_3}{\underset{|}{C}}HNH_2$)

b. $CH_3CH_2NHCH_2CH_3$

c. $CH_3NHCH_2CH_2CH_3$

d. $CH_3CH_2\underset{\displaystyle CH_3}{\underset{|}{N}}CH_3$

Solution

a. There is only one alkyl group attached to nitrogen; the amine is primary.

$$CH_3\underset{\displaystyle CH_3}{\underset{|}{C}}H—NH_2$$

The three-carbon group is isopropyl; thus the name is isopropylamine.

b. There are two ethyl groups attached to the nitrogen.

$$CH_3CH_2—NH—CH_2CH_3$$

The compound is diethylamine, a secondary amine.

c. There are a methyl group and a propyl group on the nitrogen.

$$CH_3—NH—CH_2CH_2CH_3$$

The compound is methylpropylamine, a secondary amine.

d. There are two methyl groups and one ethyl group on the nitrogen. The compound is ethyldimethylamine, a tertiary amine.

Practice Exercises

A. Name and classify each of the following.

a. 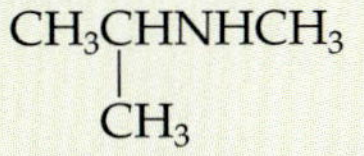 $CH_3CHNHCH_3$ with CH_3 on the CH carbon

b. 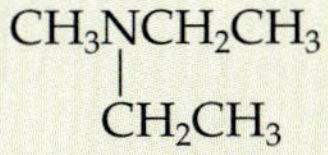 $CH_3NCH_2CH_3$ with CH_2CH_3 on the N

B. Name and classify each of the following. Refer to Table 14.4, if necessary.

a. $CH_3CH_2CH_2CH_2NH_2$

b. 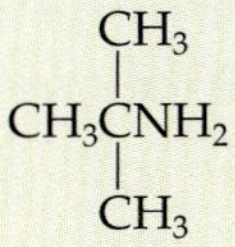 CH_3CNH_2 with CH_3 above and CH_3 below the C

Example 17.2

Give the structural formula for and classify (**a**) isopropyldimethylamine and (**b**) dipropylamine.

Solution

a. The name indicates that there are an isopropyl group and two methyl groups attached to nitrogen; the amine is tertiary.

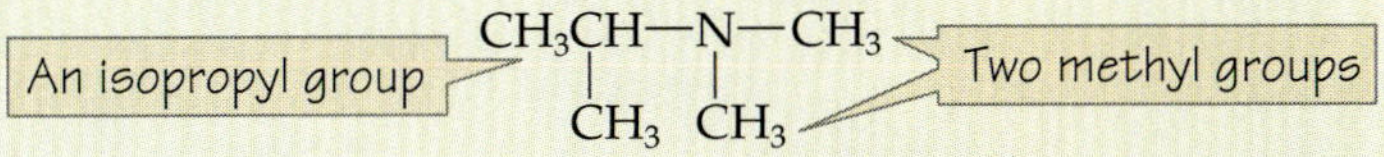

$CH_3CH—N—CH_3$, with CH_3 on the CH and CH_3 on the N

An isopropyl group

Two methyl groups

b. The name indicates that there are two propyl groups attached to nitrogen; the amine is secondary. (The third bond on nitrogen goes to a hydrogen atom.)

$$CH_3CH_2CH_2—NH—CH_2CH_2CH_3$$

Practice Exercise

Give the structural formula for (**a**) ethylisopropylamine and (**b**) propyldiethylamine.

The primary amine in which the nitrogen is attached directly to a benzene ring has the special name *aniline.* Aryl amines are named as derivatives of aniline. Compounds in which the nitrogen is attached to both a benzene ring and an alkyl group are also named as derivatives of aniline. The alkyl groups are named first, and their attachment at the nitrogen atom is indicated by the capital letter *N*.

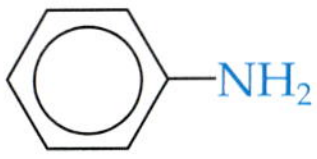

NH_2 · O_2N NH_2 · NH_2 Cl · H CH_3 N · CH_3 CH_3 N

Aniline  · *p*-Nitroaniline 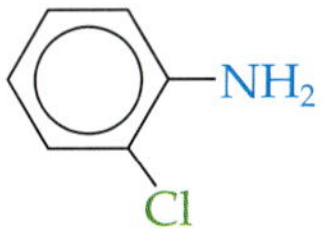 · *o*-Chloroaniline 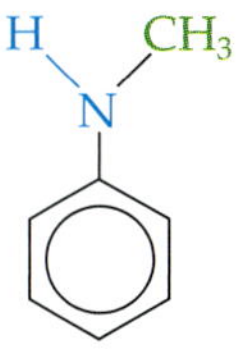 · *N*-Methylaniline 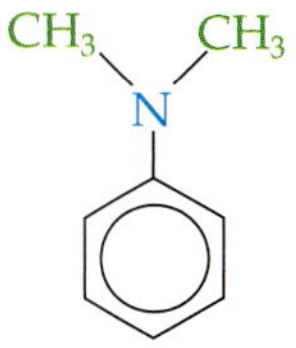 · *N,N*-Dimethylaniline

Example 17.3

Name the following compound.

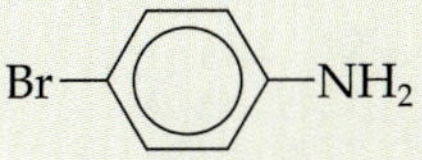

Br—(benzene ring)—NH_2

Solution

The benzene ring with an NH_2 group is aniline. The compound is named as a derivative of aniline: 4-bromoaniline or *p*-bromoaniline.

Practice Exercise

Name the following compound.

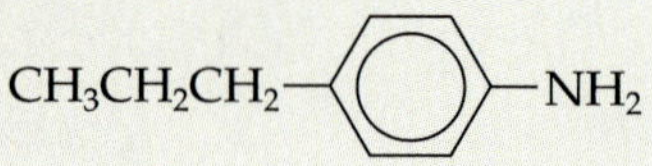

$$CH_3CH_2CH_2-C_6H_4-NH_2$$

Example 17.4

Draw the structural formulas for (**a**) *m*-ethylaniline and (**b**) *N*-ethylaniline.

Solution

Both compounds are derivatives of aniline.

a. This compound is a primary amine having an ethyl group located *meta* to the *amino* ($-NH_2$) group.

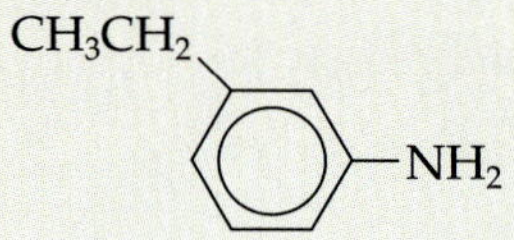

$$CH_3CH_2-C_6H_4-NH_2$$

b. This compound is a secondary amine in which the ethyl group is attached at the nitrogen atom.

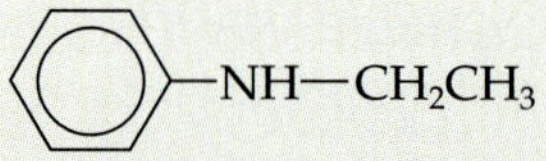

$$C_6H_5-NH-CH_2CH_3$$

Practice Exercise

Draw the structural formulas for diphenylamine (*N*-phenylaniline) and triphenylamine (*N,N*-diphenylaniline).

Example 17.5

Draw the structural formula for 2-amino-3-methylpentane.

Solution

Always start with the parent compound and draw the five-carbon pentane chain. Then attach a methyl group at the third carbon atom and an amino group at the second.

$$CH_3CH(NH_2)-CH(CH_3)CH_2CH_3$$

Practice Exercise

Draw the structural formula for 2-amino-3-ethyl-1-phenylheptane.

Ammonium ions in which one or more hydrogens are replaced with alkyl groups are named in a manner analogous to that used for simple amines. The alkyl groups are named as substituents, and the parent species is regarded as the ammonium ion.

Example 17.6

Name the following ions.

a. $CH_3NH_3^+$ **b.** $(CH_3)_2NH_2^+$ **c.** $(CH_3)_3NH^+$ **d.** $(CH_3)_4N^+$

Solution

The ions have (**a**) one, (**b**) two, (**c**) three, and (**d**) four methyl groups attached to a nitrogen atom. Their names are therefore (**a**) methylammonium, (**b**) dimethylammonium, (**c**) trimethylammonium, and (**d**) tetramethylammonium ions.

Practice Exercise

Name the following ions.

a. $CH_3CH_2NH_3^+$ **b.** $(CH_3CH_2)_3NH^+$
c. $(CH_3CH_2CH_2)_2NH_2^+$ **d.** $(CH_3CH_2CH_2CH_2)_4N^+$

Ions in which one hydrogen of the ammonium ion is replaced by a benzene ring are named as anilinium ions instead of ammonium ions. Such ions are prepared from the corresponding aniline, and the name reflects this fact.

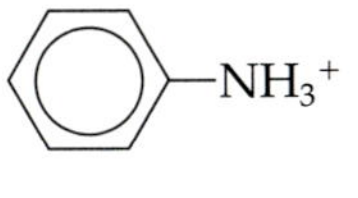

$C_6H_5-NH_3^+$ Anilinium ion

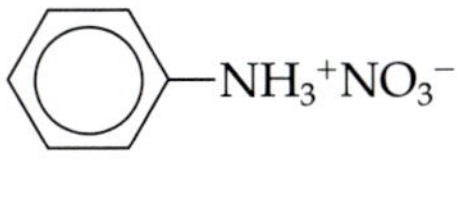

$C_6H_5-NH_3^+NO_3^-$ Anilinium nitrate

17.3 Physical Properties of Amines

Primary and secondary amines have hydrogens bonded to nitrogen and are therefore capable of intermolecular hydrogen bonding (Figure 17.3a). These forces are not as strong as those between alcohol molecules (which have hydrogens bonded to oxygen, a more electronegative element than nitrogen). Amines boil at higher temperatures than alkanes but at lower temperatures than alcohols of comparable molar mass. For example, compare the boiling point of CH_3NH_2 (−6 °C) with those of CH_3CH_3 (−89 °C) and CH_3OH (65 °C). Tertiary amines have no hydrogen bonded to nitrogen and therefore cannot form intermolecular hydrogen bonds. They have boiling points comparable to those of the ethers (Table 17.1).

Amines are generally more soluble than alcohols with the same number of carbon atoms because the nitrogen atom in the amines is a better hydrogen bond acceptor (more basic) than the oxygen atom of alcohols.

All three classes of amines can hydrogen-bond to water (Figure 17.3b). Amines of low molar mass are quite soluble in water, the borderline of water solubility coming at five or six carbon atoms.

Amines have interesting (!) odors. The simple ones smell very much like ammonia. Higher aliphatic amines smell like decaying fish. Or perhaps we should put

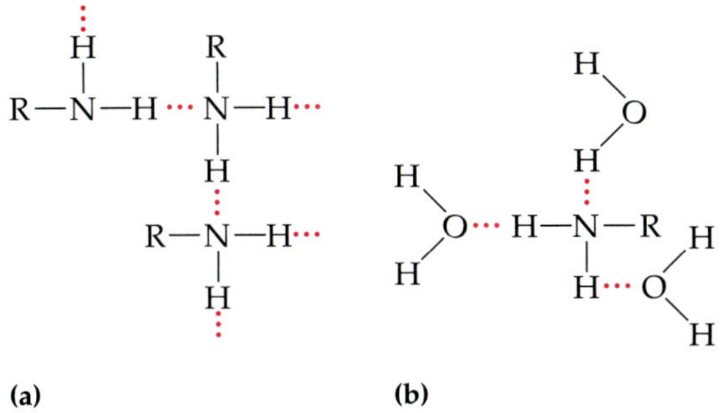

(a) (b)

Figure 17.3 (a) Intermolecular hydrogen bonding in amines. (b) Hydrogen bonding between amine and water molecules.

Table 17.1 Physical Properties of Some Amines and Comparable Oxygen-Containing Compounds

Name	Formula	Class	Molar mass	Boiling Point (°C)	Solubility at 25 °C (g/100 g H_2O)
Butylamine	$CH_3CH_2CH_2CH_2NH_2$	1°	73	78	Miscible
Diethylamine	$(CH_3CH_2)_2NH$	2°	73	55	Miscible
Butyl alcohol	$CH_3CH_2CH_2CH_2OH$	—	74	118	8
Dipropylamine	$(CH_3CH_2CH_2)_2NH$	2°	101	111	4
Triethylamine	$(CH_3CH_2)_3N$	3°	101	90	14
Dipropyl ether	$(CH_3CH_2CH_2)_2O$	—	102	91	0.25

$H_2NCH_2CH_2CH_2CH_2NH_2$

1,4-Diaminobutane
(Putrescine)

$H_2NCH_2CH_2CH_2CH_2CH_2NH_2$

1,5-Diaminopentane
(Cadaverine)

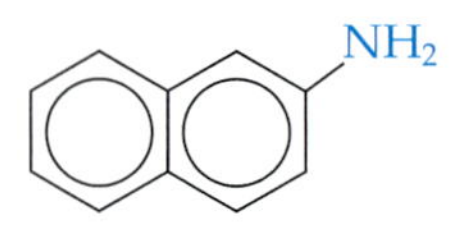

β-Naphthylamine

▲ **Figure 17.4**
Some amines of interest. Putrescine and cadaverine have odors indicated by their names. β-Naphthylamine is a carcinogen.

it the other way around: Decaying fish give off odorous amines. The stench of rotting fish is due in part to putrescine and cadaverine (Figure 17.4), two compounds that are diamines. They arise from the decarboxylation of ornithine and lysine, respectively, amino acids found in animal cells (Chapter 21).

Aromatic amines generally are quite toxic. They are readily absorbed through the skin, and workers must exercise caution when handling these compounds. Several aromatic amines, including β-naphthylamine (Figure 17.4), are potent carcinogens.

✓ Review Questions

17.3 Contrast the physical properties of amines with those of alcohols and alkanes.

17.4 Describe the odors of amines.

17.4 Amines as Bases

Like ammonia, the amines are weak bases. The lone electron pair on nitrogen can accept a proton from water to form substituted ammonium ions and hydroxide ions.

General Equation:

$$R-\ddot{N}(R)-R + H_2O \rightleftharpoons \left[R-N(H)(R)-R\right]^+ + OH^-$$

Specific Equation:

$$\underset{\text{Methylamine}}{CH_3NH_2(aq)} + H_2O \rightleftharpoons \underset{\text{Methylammonium ion}}{CH_3NH_3^+(aq)} + OH^-(aq)$$

Like that for ammonia, this equilibrium strongly favors the nonionized form.

Simple aliphatic amines are somewhat more basic than ammonia, although still much less basic than compounds such as sodium hydroxide. Aromatic amines, such as aniline, are much weaker bases than ammonia.

Nearly all amines, including those that are not very soluble in water, will react with strong acids to form water-soluble salts.

$$\underset{\text{Octylamine (insoluble)}}{CH_3(CH_2)_6CH_2NH_2(l)} + HNO_3(aq) \longrightarrow \underset{\text{Octylammonium nitrate (soluble)}}{CH_3(CH_2)_6CH_2NH_3^+NO_3^-(aq)}$$

Amine salts are named like other salts: The name of the cation is followed by that of the anion. Remember that the ions formed from aliphatic amines are named as substituted ammonium ions.

Example 17.7

Name the following salt.

$$[CH_3NH_2CH_2CH_3]^+CH_3COO^-$$

Solution

The cation has two groups, methyl and ethyl, attached to nitrogen.

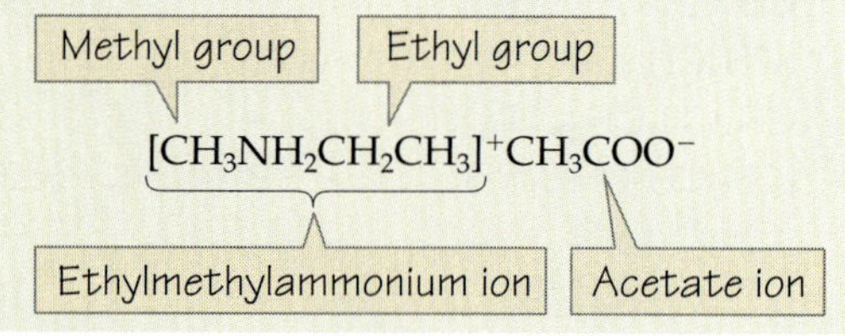

The cation is the ethylmethylammonium ion, and the anion is the acetate ion. The salt is therefore ethylmethylammonium acetate.

Practice Exercise

Name the following salt.

$$(CH_3CH_2CH_2)_4N^+I^-$$

Salts of aniline are named as anilinium compounds. An older system, still in use for naming drugs, calls the salt of aniline and hydrochloric acid "aniline hydrochloride." By this older system, the formula of the compound is frequently drawn to correspond to the name. Keep in mind that these compounds are really ionic—they are salts—even though the name and formula seem to indicate a loose association of molecules. The properties of the compounds (solubility, for example) are those characteristic of salts.

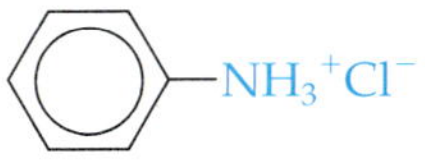

$$C_6H_5{-}NH_3^+Cl^-$$

Anilinium chloride

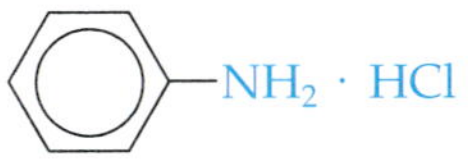

$$C_6H_5{-}NH_2 \cdot HCl$$

"Aniline hydrochloride"

To facilitate injection in aqueous solution and/or for ease of transport through the blood, pharmaceutical companies often convert an insoluble amine into the more soluble amine salt. For instance, procaine is soluble only to the extent of 0.5 g in 100 g of water. The hydrochloride is soluble to the remarkable degree of 100 g in 100 g of water. Procaine hydrochloride, perhaps better known by the trade name Novocain, is widely used as a local anesthetic.

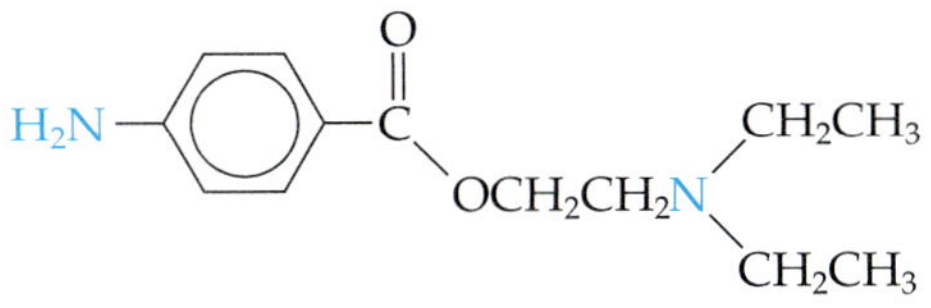

$$H_2N{-}C_6H_4{-}C(=O){-}OCH_2CH_2N(CH_2CH_3)_2$$

Procaine

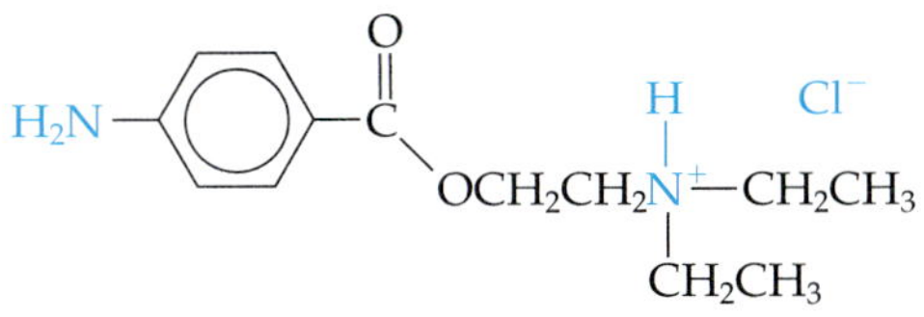

$$H_2N{-}C_6H_4{-}C(=O){-}OCH_2CH_2\overset{H}{\overset{|}{N^+}}(CH_2CH_3){-}CH_2CH_3 \quad Cl^-$$

Procaine hydrochloride (Novocain)

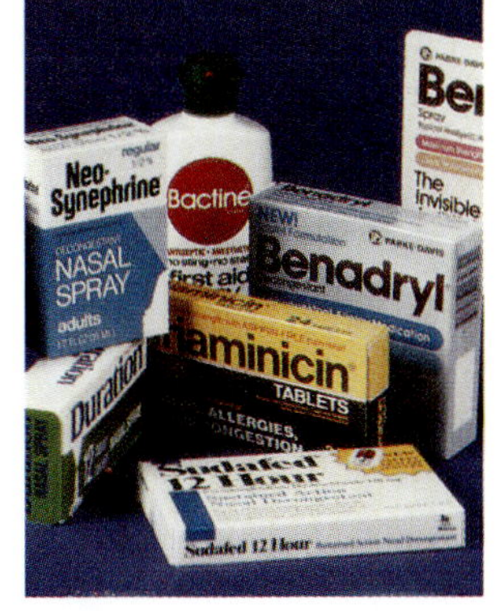

▲ Amine hydrochloride salts are the main active ingredients in a variety of over-the-counter drugs.

We also use the chemistry of amines when we put lemon juice on fish. The unpleasant fishy odor is due to amines. The citric acid in the juice converts the amines to nonvolatile salts, thus reducing the odor.

✓ Review Questions

17.5 Explain the basicity of amines.

17.6 What chemical reaction occurs when lemon juice is added to fish? Why is this desirable?

Basic Buffers

Recall that a weak acid and its salt (for example, acetic acid and sodium acetate) can be used in the preparation of buffer solutions (Chapter 10). Similarly, amines and their salts can also be used to make buffers. Whereas the acid/acid salt combination yields a solution buffered at acidic pH values, the amine/amine salt buffer stabilizes the pH in a basic range. An important example is tris(hydroxymethyl)aminomethane, often called simply tris. This compound and its salt, tris hydrochloride, buffer in the range of pH 7–9. Amine buffers find wide use in the cosmetics and textile industries, in cleaning compounds, and in biochemical research. Tris is also used in the treatment of metabolic acidosis (Chapter 26).

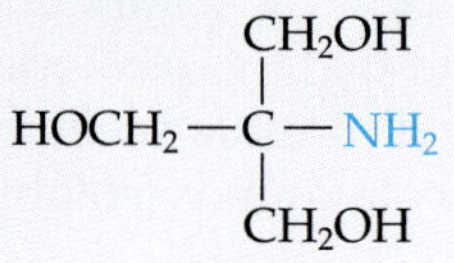

$$HOCH_2-\underset{CH_2OH}{\overset{CH_2OH}{\underset{|}{\overset{|}{C}}}}-NH_2$$

"Tris"

$$HOCH_2-\underset{CH_2OH}{\overset{CH_2OH}{\underset{|}{\overset{|}{C}}}}-NH_3^{+}Cl^{-}$$

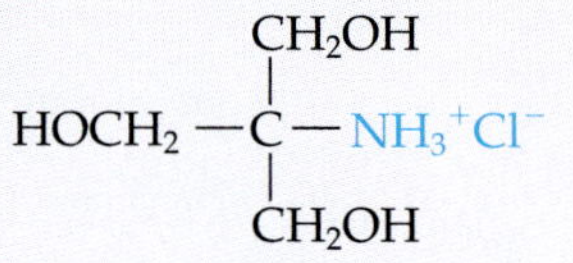

Tris hydrochloride

17.5 Other Chemical Properties of Amines

In the preceding section, we saw that water-soluble amines give basic solutions and that amines generally react with mineral acids to form salts. In Chapter 16 we saw that carboxylic acids react with ammonia when heated to produce unsubstituted amides. In a similar reaction, carboxylic acids react with amines to produce substituted amides.

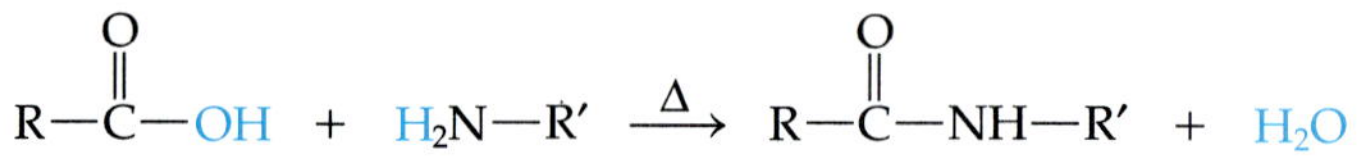

$$R-\overset{O}{\overset{\|}{C}}-OH + H_2N-R' \xrightarrow{\Delta} R-\overset{O}{\overset{\|}{C}}-NH-R' + H_2O$$

In chemistry laboratories, this reaction is carried out using special reactive derivatives of the carboxylic acid instead of the acid itself. We shall see in Chapter 21 that, in living organisms, enzymes catalyze the formation of the amide bond, also called the peptide bond, between the amino group of one amino acid and the carboxyl group of another. The resulting substituted polyamides are proteins and polypeptides.

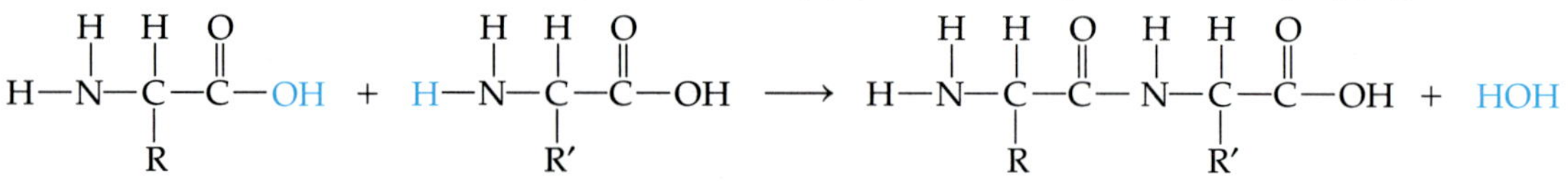

$$H-\overset{H}{\overset{|}{N}}-\underset{R}{\underset{|}{\overset{H}{\overset{|}{C}}}}-\overset{O}{\overset{\|}{C}}-OH + H-\overset{H}{\overset{|}{N}}-\underset{R'}{\underset{|}{\overset{H}{\overset{|}{C}}}}-\overset{O}{\overset{\|}{C}}-OH \longrightarrow H-\overset{H}{\overset{|}{N}}-\underset{R}{\underset{|}{\overset{H}{\overset{|}{C}}}}-\overset{O}{\overset{\|}{C}}-\overset{H}{\overset{|}{N}}-\underset{R'}{\underset{|}{\overset{H}{\overset{|}{C}}}}-\overset{O}{\overset{\|}{C}}-OH + HOH$$

Two additional reactions of amines are important to those studying biological processes. First, amines react with nitrous acid. The nature of the products depends on the class of the amine. Primary amines give a quantitative yield of nitrogen gas.

$$R-NH_2\ (1°) + HNO_2 \rightarrow N_2(g) + \text{other products}$$

Because nitrous acid is unstable, it is usually made *in situ* (right in the reaction vessel) by addition of hydrochloric acid to sodium nitrite.

$$NaNO_2(aq) + HCl(aq) \rightarrow HNO_2(aq) + NaCl(aq)$$

When a primary amine is added to this solution, bubbles of nitrogen gas can be seen escaping. These bubbles indicate that the amine is primary, because secondary and tertiary amines do not release nitrogen gas. Therefore this reaction serves as a qualitative test for primary amines. It can be used for the quantitative determination of primary amino groups by carefully measuring the amount of the original amine and the volume of the nitrogen produced (corrected to standard temperature and pressure). One molecule of nitrogen (N_2) is liberated for each free amino group. This procedure is referred to as the Van Slyke method and is a classical method for quantitating amino acids and proteins.

Classical laboratory bench procedures have been superceded to a large extent by spectroscopic methods.

Secondary amines react with nitrous acid to form oily *N*-nitroso compounds.

$$\underset{\displaystyle R}{\underset{|}{R{-}N{-}H}}\ (2^\circ) + HO{-}N{=}O \longrightarrow \underset{\displaystyle R}{\underset{|}{R{-}N{-}N{=}O}} + HOH$$

The appearance of an oil when an amine is added to a solution of nitrous acid indicates that the amine is probably secondary. However, the test is no longer recommended because most *N*-nitroso compounds are potent carcinogens.

Tertiary amines also react with nitrous acid. Generally the only product is a salt. Sometimes, however, the tertiary amine is cleaved to a secondary one, which then can form a nitroso derivative.

Nitrites in the Diet

Because our diet contains sodium nitrite and our stomachs contain hydrochloric acid, nitrosoamines are formed from secondary amines in the breakdown products of the food we eat. Sodium nitrite is used as a food additive in cured meats such as frankfurters, ham, and cold cuts. The NO_2^- serves two functions. It inhibits growth of bacteria, especially *Clostridium botulinum,* which produces the potentially fatal food poisoning known as *botulism.* It also preserves the red color of the meat and thereby its appetizing appearance. However, this benefit is not without its risk. It has been postulated that nitrites found in prepared meats are responsible for the high rates of stomach cancer in countries with a high consumption of these products.

In the 1980s the U.S. Department of Agriculture required meat firms to reduce the sodium nitrite added to preserve bacon from 200 to 120 ppm. The amount of nitrite added to ham and frankfurters was also reduced, and other preservatives are being used. The meat industry points out, however, that there is no evidence to link the ingestion of nitrites and the incidence of stomach and intestinal cancer. They say that the amount of nitrite that occurs naturally in foods far surpasses the amount that is used as an additive. Nitrosoamines have also been found in alcoholic beverages, cosmetics, and pesticides.

◀ Sodium nitrite is added to luncheon meats and frankfurters to prevent spoilage and to provide an attractive red color to the meat.

To detect the presence of amines, their reaction with ninhydrin is often used. Some amines react with ninhydrin to form a purple to blue anion.

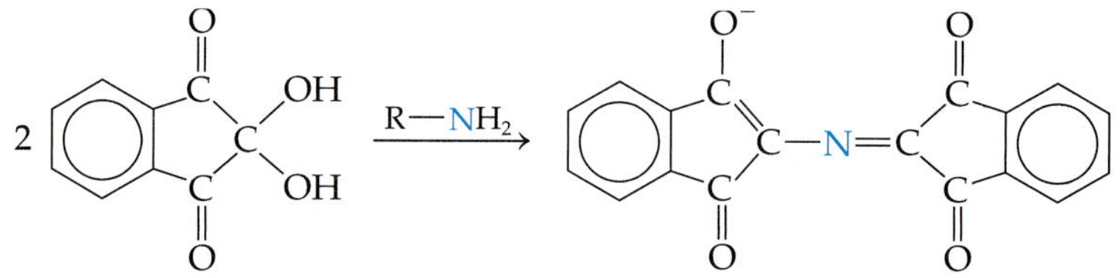

The ninhydrin test is especially suited to the detection of amino acids. Mixtures of amino acids, such as those that result from the hydrolysis of proteins, can be separated by a process called paper chromatography. At the conclusion of such a separation, the individual amino acids are scattered at different locations on the paper chromatogram. The amino acids cannot be seen, however, until the chromatogram is sprayed with ninhydrin and the colored ions become visible. Then the position of the constituents of the original mixture can be compared with those of known amino acids, and an identification can be made on this basis.

Review Questions

17.7 Describe the formation of the amide bond and explain its importance in living systems.

17.8 Describe two laboratory tests to detect amines. Which will distinguish between a primary and a secondary amine?

17.9 Why are nitrite food additives considered by some people to be health hazards?

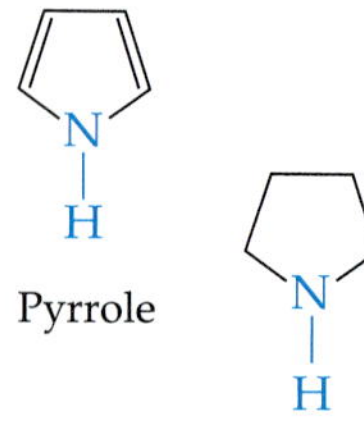

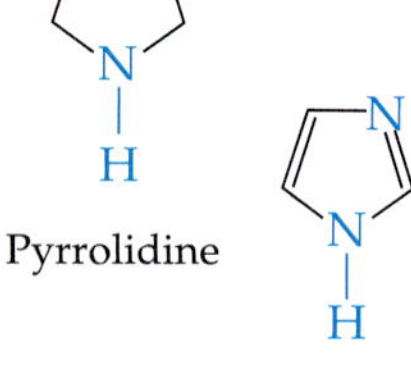

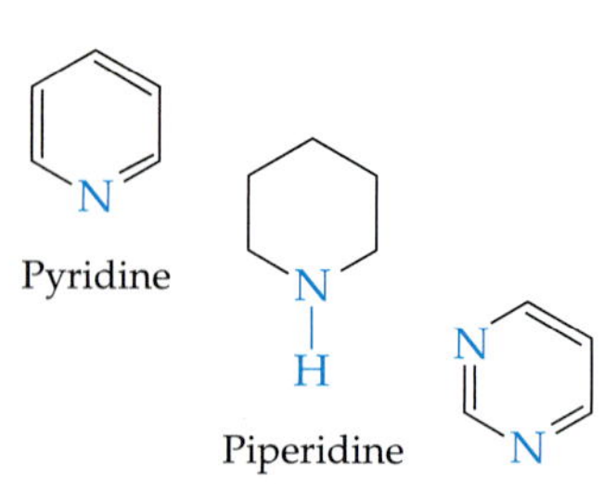

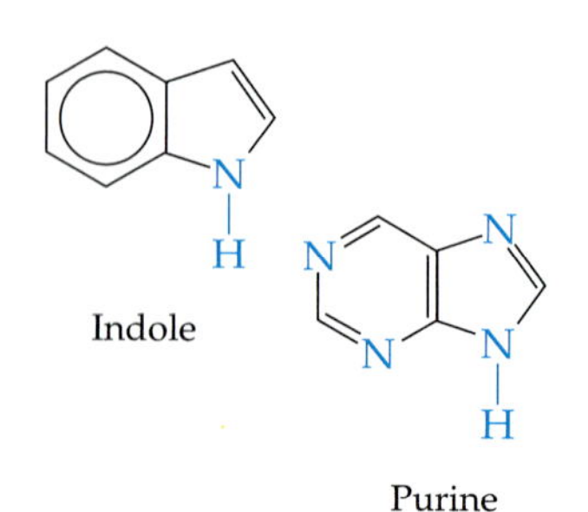

▲ **Figure 17.5**
Some heterocyclic amines.

17.6 Heterocyclic Amines

Recall that a variety of cyclic hydrocarbons were introduced in Chapter 13. All the atoms in the rings of these compounds are carbon atoms. There are other cyclic compounds in which nitrogen, oxygen, sulfur, or some other atom is incorporated in the ring. These are called **heterocyclic compounds** (Greek *heteros,* other). A few containing nitrogen are shown in Figure 17.5.

The compounds pyrrole and pyrrolidine each have four carbon atoms and one nitrogen atom in a ring. Pyrrole is an aromatic compound and has properties similar to those of benzene. Pyrrolidine, which contains four more hydrogen atoms than does pyrrole, behaves like an aliphatic amine. Imidazole also has a five-membered ring, but it contains two nitrogen atoms and only three carbon atoms. Like pyrrole, imidazole has aromatic properties.

Pyridine and piperidine each have five carbon atoms and one nitrogen atom. Pyridine is aromatic; piperidine is aliphatic. Another six-membered aromatic heterocyclic compound is pyrimidine, which has two nitrogen atoms and four carbon atoms.

Other heterocyclic compounds have two rings that share two carbon atoms, and thus a common "side" (as does naphthalene, Chapter 13). Indole has a benzene ring fused with a pyrrole ring. Purine has a pyrimidine ring sharing two carbons with an imidazole structure. Bases related to purine and pyrimidine make up a part of the structure of the nucleic acids, compounds that compose the genetic material of cells and that direct protein synthesis. Nucleic acids are discussed in Chapter 23, in which we encounter one of the truly outstanding examples of the critical importance of the shapes of molecules and of molecular structure in general.

Review Question

17.10 What is a heterocyclic compound?

Alkaloids

Many heterocyclic amines occur naturally in plants. Like most other amines, these compounds are basic. They are called **alkaloids** (Selected Topic B), a name that means "like alkalis." Knowledge of many of these, at least in their crude forms, dates back to antiquity. Opium, which contains about 10% morphine as the principal alkaloid, has been used for thousands of years, although morphine was not isolated until 1805.

▲ *Conium maculatum* (poison hemlock)

When the Greek philosopher Socrates was accused of corrupting the youth of Athens in 399 B.C.E., he was given the choice of exile or death. He chose the latter and implemented his decision by drinking a cup of hemlock. His hemlock was probably prepared from the fully grown but unripened fruit of *Conium maculatum*, or poison hemlock. The fruit would probably have been carefully dried and then brewed into a tea. Hemlock contains several alkaloids, but the principal one is coniine. Coniine causes nausea, weakness, paralysis, and ultimately—as in the case of Socrates—death.

Hemlock tea, anyone?

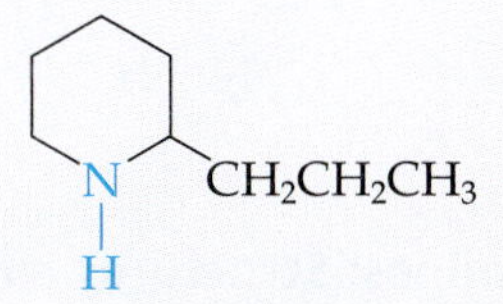

Coniine

▲ Jacques Louis David's painting *The Death of Socrates* (1787) shows Socrates about to drink the cup of hemlock to carry out the death sentence decreed by the rulers of Athens.

Summary

Amines are nitrogen-containing organic molecules derived from an ammonia molecule. A **primary amine (RNH_2)** has one organic group bonded to the nitrogen, a **secondary amine (R_2NH)** has two organic groups bonded to the nitrogen, and a **tertiary amine (R_3N)** has three. The nitrogen atom can bond a fourth organic group through its unshared electron pair, forming a *quaternary ammonium ion* (R_4N^+).

Amines are basic compounds. Aliphatic amines are more basic than ammonia and aromatic amines are less basic than ammonia. Amines react with strong acids to produce ammonium salts and with carboxylic acids, upon heating, to yield amides.

Primary, secondary, and tertiary amines can be distinguished from each other by how they react with nitrous acid. Primary amines yield gaseous nitrogen (the *Van Slyke method*). Secondary ones yield *N*-nitroso products that cause an oily film on the solution surface, and tertiary ones yield salts.

A cyclic compound in which the ring contains one or more noncarbon atoms is called a **heterocyclic compound.** There are many heterocyclic amines, including the physiologically important purine and pyrimidine. **Alkaloids** are heterocyclic amines found in many plants. Morphine in the opium plant and coniine in hemlock are examples.

Key Terms

alkaloid (17.6)
amine (page 448)
amino group (NH_2) (17.2)
heterocyclic compound (17.6)
primary (1°) amine (17.1)
quaternary (4°) ammonium salt (17.1)
secondary (2°) amine (17.1)
tertiary (3°) amine (17.1)

Problems

Classification of Compounds

1. Classify as an amine, an amide, both, or neither:

a. 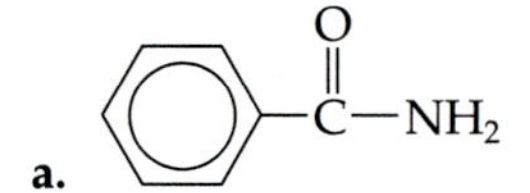

b. 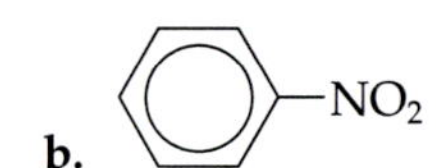

c. $H_2NCH_2CH_2CH_2\overset{O}{\overset{\|}{C}}NH_2$

e. N—H

f. —NH_2

2. Classify the following as an amine, an amide, both, or neither:

a. $(CH_3)_4N^+I^-$ b. $CH_3CH_2NH\overset{O}{\overset{\|}{C}}CH_2CH_3$

c. $CH_3CH_2NHCH_2CH_3$

3. Identify the following compounds as amines, alcohols, phenols, or ethers. Classify any amines and alcohols as primary (1°), secondary (2°), or tertiary (3°).

a. $CH_3CH_2CH_2OH$ b. $CH_3CH_2CH_2NH_2$

c. $CH_3\overset{OH}{\overset{|}{C}}HCH_3$ d. $CH_3\overset{NH_2}{\overset{|}{C}}HCH_3$

e. O f. —OH

4. Identify the following compounds as amines, alcohols, phenols, or ethers. Classify any amines and alcohols as primary, secondary, or tertiary.

a. $CH_3CH_2NHCH_2CH_3$ b. $CH_3CH_2OCH_2CH_3$

c. $CH_3—\underset{CH_3}{\underset{|}{N}}—CH_3$ d. $CH_3—\underset{CH_3}{\underset{|}{\overset{OH}{\overset{|}{C}}}}—CH_3$

Amines: Structures and Names

5. Draw structural formulas for each of the following.
   a. dimethylamine b. diethylmethylamine
   c. 2-amino-1-cyclobutanol d. 2-aminoethanol
6. Draw structural formulas for each of the following.
   a. 3-aminopentane b. 1,6-diaminohexane
   c. cyclohexylamine d. ethylphenylamine
7. Draw structural formulas for each of the following.
   a. aniline b. *m*-bromoaniline
   c. pyrimidine d. *N*-ethylaniline
8. Draw structural formulas for each of the following.
   a. pyridine b. purine
   c. *N*,*N*-dimethylaniline d. 3,5-dichloroaniline
9. Name each of the following compounds.
   a. $CH_3CH_2CH_2NH_2$ b. $(CH_3)_2CHNHCH_3$
   c. $(CH_3CH_2)_3N$ d. $CH_3CH(NH_2)CH_2CH_2CH_3$
10. Name each of the following compounds.

a. O_2N—⬡—NH_2

b. ⬡—$NHCH_2CH_3$

Names and Formulas of Amine Salts

11. Draw structural formulas for each of the following.
    a. anilinium bromide
    b. tetramethylammonium chloride
12. Draw structural formulas for each of the following.
    a. ethylmethylammonium chloride
    b. anilinium nitrate

13. Name the following compounds:
   a. $[CH_3CH_2NH_2CH_2CH_3]^+Br^-$
   b. $(CH_3CH_2)_4N^+I^-$
14. Name the following compounds:
   a. 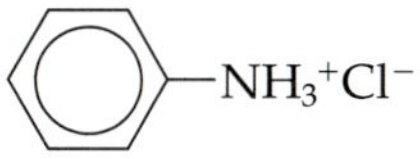 $C_6H_5{-}NH_3^+Cl^-$
   b. $(CH_3)_4N^+NO_2^-$

Physical Properties

15. Which compound has the higher boiling point, butylamine or pentane? Explain.
16. Which compound has the higher boiling point, butylamine or butyl alcohol? Explain.
17. Which compound has the higher boiling point, trimethylamine or propylamine? Explain.
18. Which compound has the higher boiling point, CH_3NH_2 or $CH_3CH_2CH_2CH_2CH_2NH_2$? Explain.
19. Which compound is more soluble in water, $CH_3CH_2CH_3$ or $CH_3CH_2NH_2$? Explain.
20. Which compound is more soluble in water, $CH_3CH_2CH_2NH_2$ or $CH_3CH_2CH_2CH_2CH_2CH_2CH_2NH_2$? Explain.
21. Which compound is more soluble in water? Explain.

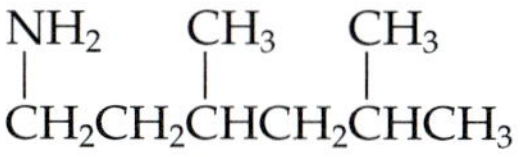

$CH_2(NH_2)CH_2CH(CH_3)CH_2CH(CH_3)CH_3$ or $CH_2(NH_2)CH_2CH(NH_2)CH_2CH(NH_2)CH_3$

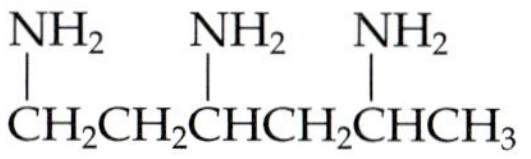

22. Which compound is more soluble in water? Explain.

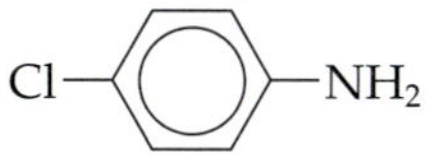

$Cl{-}C_6H_4{-}NH_2$ or $C_6H_5{-}NH_3^+Cl^-$

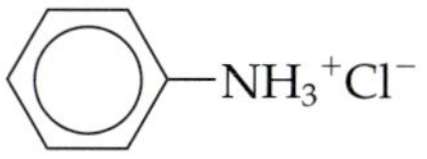

Chemical Reactions

23. Draw the structural formula of the salt formed in each of the following reactions.
   a. $CH_3NH_2 + HBr \longrightarrow$
   b. 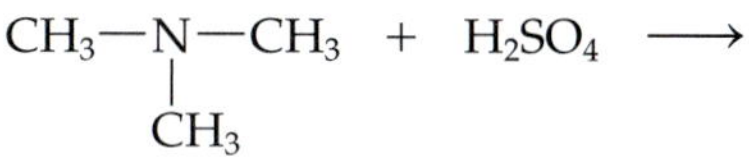 $(CH_3)_3N + H_2SO_4 \longrightarrow$
24. Draw the structural formula of the salt formed in each of the following reactions.
   a. 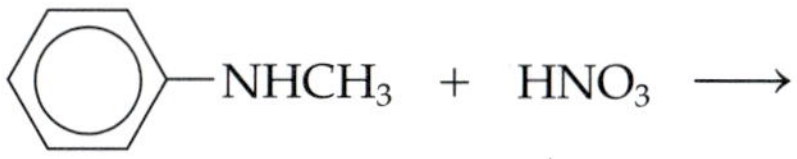 $C_6H_5{-}NHCH_3 + HNO_3 \longrightarrow$
   b. 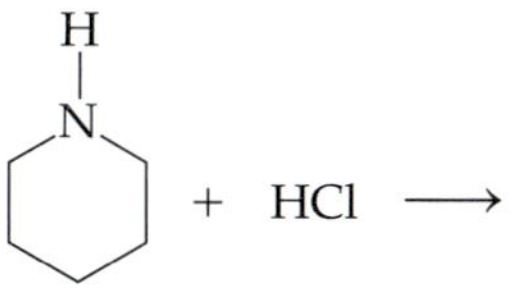 piperidine (six-membered ring with N–H) $+ HCl \longrightarrow$
25. Draw the amide, if any, derived from (**a**) hexanoic acid and butylamine and (**b**) benzoic acid and aniline.
26. Draw the amide, if any, derived from each of the following.
   a. 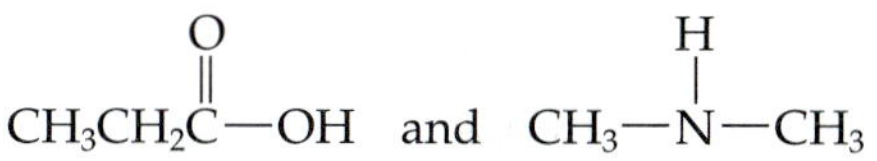 $CH_3CH_2C(=O){-}OH$ and $CH_3{-}NH{-}CH_3$
   b. 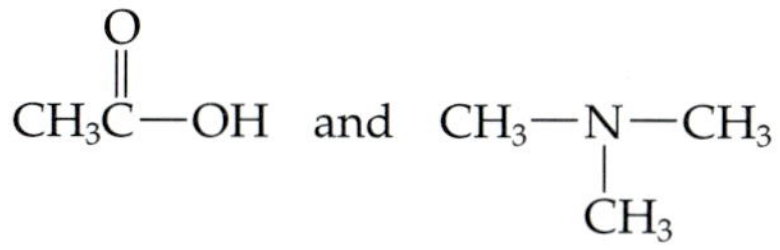 $CH_3C(=O){-}OH$ and $CH_3{-}N(CH_3){-}CH_3$
27. Draw the carboxylic acid and amine from which each of the following amides was formed.
   a. 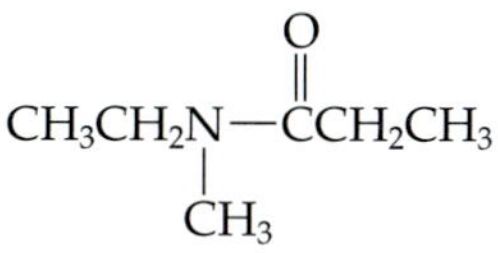 $CH_3CH_2N(CH_3){-}C(=O)CH_2CH_3$
   b. 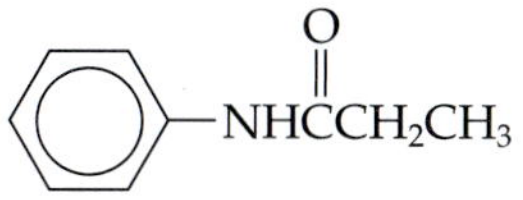 $C_6H_5{-}NHC(=O)CH_2CH_3$
28. Draw the carboxylic acid and amine from which each of the following amides was formed.
   a. 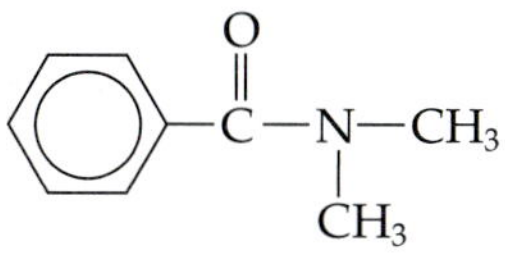 $C_6H_5{-}C(=O){-}N(CH_3){-}CH_3$
   b. 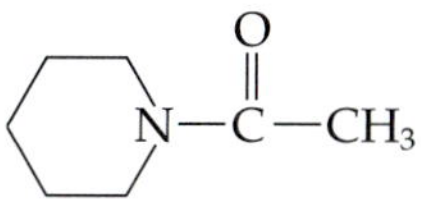 piperidine ring $N{-}C(=O){-}CH_3$
29. Draw the structural formula for the principal organic product formed in the following reaction.

$$C_6H_5{-}C(=O){-}O^-NH_4^+ \xrightarrow{\text{heat}}$$

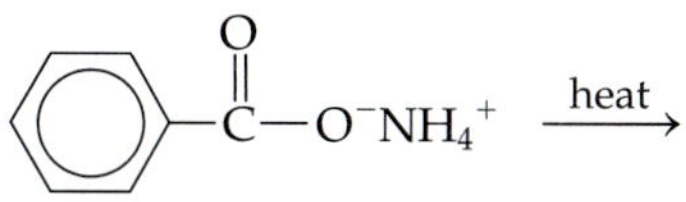

30. Draw the structural formula for the principal organic product formed in the reaction

$$CH_3NHCH_2CH_3 + HNO_2 \rightarrow$$

31. List the reagents needed to carry out the following conversions.
   a. 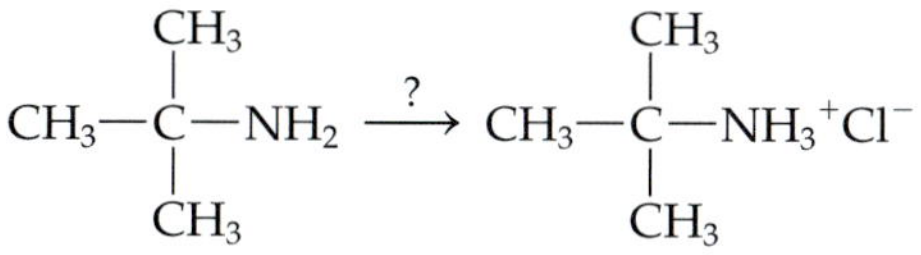 $CH_3{-}C(CH_3)_2{-}NH_2 \xrightarrow{?} CH_3{-}C(CH_3)_2{-}NH_3^+Cl^-$
   b. 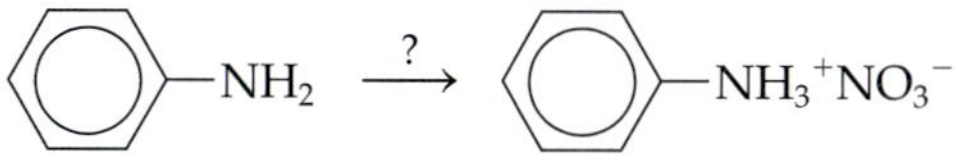 $C_6H_5{-}NH_2 \xrightarrow{?} C_6H_5{-}NH_3^+NO_3^-$
32. List the reagents needed to carry out the following conversions.
   a. 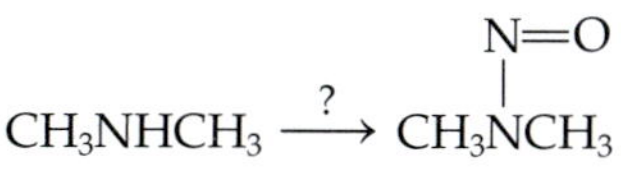 $CH_3NHCH_3 \xrightarrow{?} CH_3N(N{=}O)CH_3$
   b. 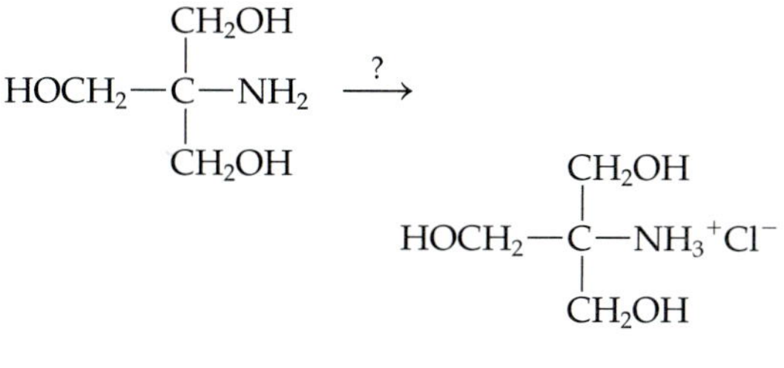 $HOCH_2{-}C(CH_2OH)_2{-}NH_2 \xrightarrow{?} HOCH_2{-}C(CH_2OH)_2{-}NH_3^+Cl^-$

Additional Problems

33. Tell whether each of the following compounds forms an acidic, basic, or neutral solution in water.

a. $CH_3CH_2NH_2$ **b.** CH_3CH_2OH

c. 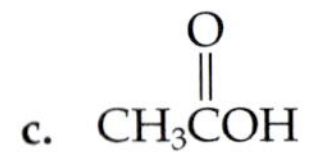 $CH_3\overset{O}{\overset{\|}{C}}OH$ **d.** $CH_3\overset{O}{\overset{\|}{C}}NH_2$

34. Amine X is insoluble in water but dissolves readily in aqueous hydrochloric acid. Explain.

35. Draw structural formulas for the eight isomeric amines that have the molecular formula $C_4H_{11}N$. Give each a common name, and classify it as primary, secondary, or tertiary.

36. Draw structural formulas for the five isomeric amines that have the molecular formula C_7H_9N and contain a benzene ring. Name each compound, and classify it as primary, secondary, or tertiary.

37. A carboxylic acid group and an amino group combined to form the amide functional group in the following compound. Draw the two starting materials for the reaction.

$$? + ? \longrightarrow H_2N{-}CH_2\overset{O}{\overset{\|}{C}}{-}NHCH_2\overset{O}{\overset{\|}{C}}{-}OH$$

38. Write equations for the reaction of anthranilic acid (*o*-aminobenzoic acid) with each of the following.

a. NaOH **b.** HCl

c. H_2SO_4 **d.** CH_3OH, H^+

39. How many milliliters of 0.150 M hydrochloric acid are required to neutralize 0.250 g of diethylamine?

40. Contrast the physical properties of amines with those of amides.

SELECTED TOPIC 

Brain Amines and Related Drugs

Many of the more than fifty naturally occurring molecules that act on the vertebrate nervous system are amines. Drugs affecting the nervous system typically act by influencing the activity of these molecules. It is not surprising, therefore, that many neuroactive drugs are amines. Some are rather simple amines. Others, including some we discuss elsewhere, are alkaloids (Selected Topic B). In this selected topic, we look at some amines and related compounds that affect our mental state, render us insensitive to pain, put us to sleep, and calm our anxieties. First, however, let's take a look at nerve cells and how they work.

C.1 Some Chemistry of the Nervous System

The nervous system is made up of billions of interconnected **neurons** (nerve cells).[1] The brain operates with a power output of about 25 W and has capacity for about 10 trillion bits of information. There are over 50 major types of neurons, which vary a great deal in shape and size. The most common type, the multipolar neuron, is shown in Figure C.1. The essential parts are the cell body, the axon, and the dendrites. Neurons conduct impulses in both the central nervous system (CNS: brain and spinal cord) and the peripheral nervous system (PNS: cranial and spinal nerves, which leave the CNS to innervate organs). The PNS includes both sensory input and the motor output, which in turn consists of neurons supplying skeletal muscles and the autonomic (involuntary) neurons that carry messages between the CNS and all organs that act involuntarily (such as the heart, digestive organs, lungs, and glands).

Although the axons on a given neuron may be up to 100 cm long, there is no continuous pathway from the brain to an organ. Transmission of information within a neuron is electrical in nature, but nearly all communication from one neuron to another is chemical. When a stimulus initiates an electric impulse (called an action potential) in a neuron, the signal travels (propagates) along the neuron until it reaches the end of the axon. The impulse stimulates secretion of **neurotransmitters,** specific chemicals that carry the signal to the next neuron across tiny, fluid-filled gaps called **synapses** (see again Figure C.1).[2] Each neurotransmitter has a specific function and fits one or more specific receptor sites on the receptor cell (or muscle or gland). Neurotransmitters are inactivated after acting at the receptors by being taken back up into the neurons that released them or by being degraded. They are often prepared for more action right in the nerve terminal.

The many neurotransmitters are often classified structurally as amines, amino acids, and peptides (such as the endorphins we mentioned in Selected Topic B) (Table C.1), but recently identified transmitters include the purines and small molecules such as NO (page 468). They are also classified by activity as excitatory or inhibitory, but many transmitters can have either effect, depending on the receptor.

Drugs can affect the action of neurotransmitters in a variety of ways. They may prevent or enhance synthesis or release of the transmitter, often by acting on an enzyme. Many drugs (and some poisons) act by mimicking the action of the neurotransmitter. Others act by blocking the receptor and preventing the neurotransmitter from acting on it. The activity of a transmitter is also changed by blocking its reuptake or accelerating its breakdown. It was long thought that each neuron released only one neurotransmitter; it is now known that more than one type of transmitter can coexist in some neurons.

C.2 Brain Amines

We all have our ups and downs in life. These moods probably result from multiple causes, but a variety of chemical compounds formed in the brain are most likely involved. The major amines acting in the central nervous system are the three catecholamines, serotonin, and histamine. Collectively, these and a few others are called biogenic amines or monoamines, and the neurons are called monoaminergic.

Catecholamines

The catecholamines (Figure C.2) are an important group of neurotransmitters synthesized in the nerve

[1] The density of neurons in the human visual cortex is 50,000/mm^3. The volume of the adult human brain is about 1500 cm^3.

[2] The motor cortex of the macaque monkey has 9.6×10^8 neurons per cubic millimeter, and EACH neuron has 60,000 synapses.

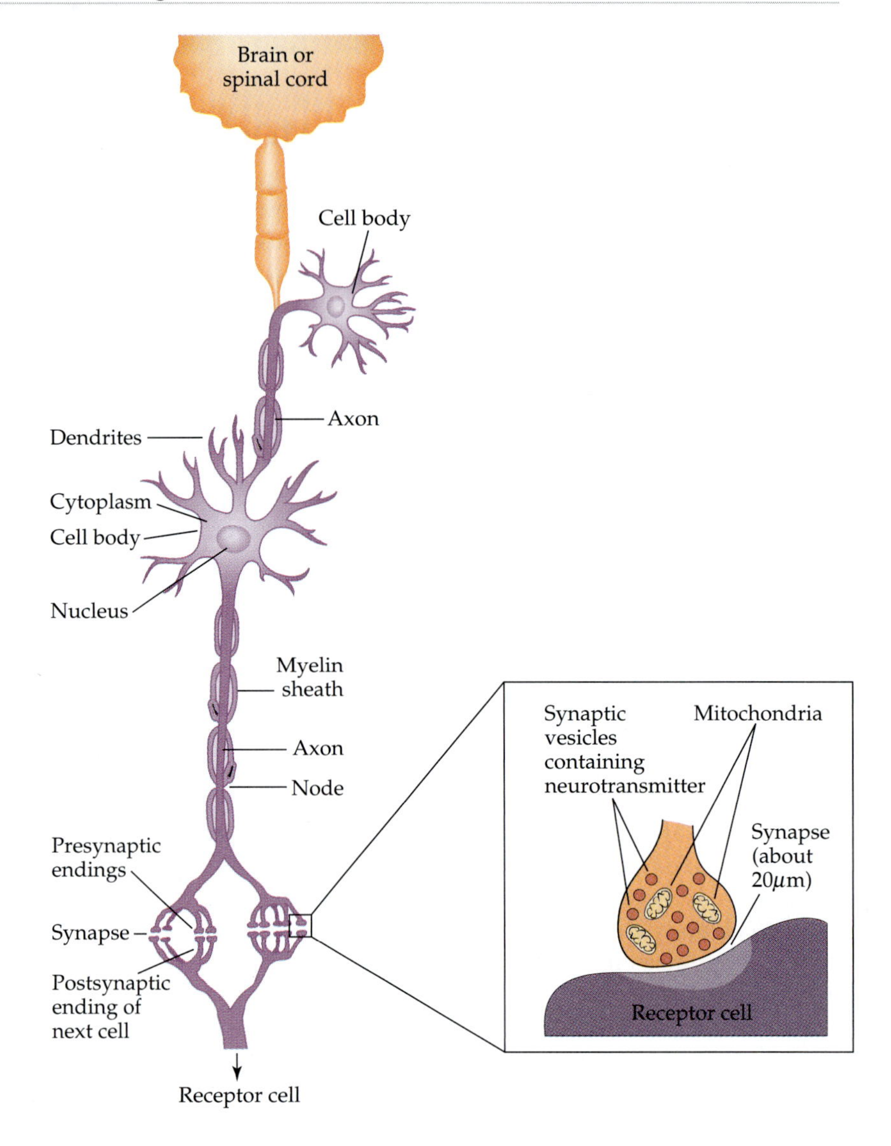

Figure C.1
Diagram of a human nerve cell, showing the pathway by which messages are transmitted from one neuron to another neuron or a receptor cell in a gland or an organ. When an electric signal reaches the presynaptic nerve ending, neurotransmitter molecules are released from the vesicles. They migrate across the synapse to the receptor cell where they fit specific receptor sites. (Multipolar neurons, like the one shown here, having multiple short branches plus an axon, are the most common type of neuron and are found throughout the nervous system.)

Table C.1 Amine and Amino Acid Neurotransmitters

Transmitter	Location	Effect at Synapse
Amino Acids[a]		
Gamma-aminobutyric acid (GABA)	Spinal cord	Inhibitory
Glutamate and Aspartate	Everywhere in CNS	Excitatory
Glycine	Spinal cord and brainstem	Inhibitory
Monoamines		
Dopamine	Cell bodies at all levels of CNS	Inhibitory or excitatory[b]
Epinephrine (Adrenaline)	Chiefly in periphery	Excitatory or inhibitory[b]
Norepinephrine (Noradrenaline)	From cells in brain stem to rest of CNS	Excitatory or inhibitory[b]
Histamine	(See text)	
5-Hydroxytryptamine (serotonin)	From cells in midbrain to all CNS	Usually inhibitory; can be excitatory
Quaternary ammonium salt		
Acetylcholine	Everywhere in CNS	Excitatory or inhibitory[b]

[a]Amino acids are discussed in Chapter 21.
[b]Depending on location and receptor type

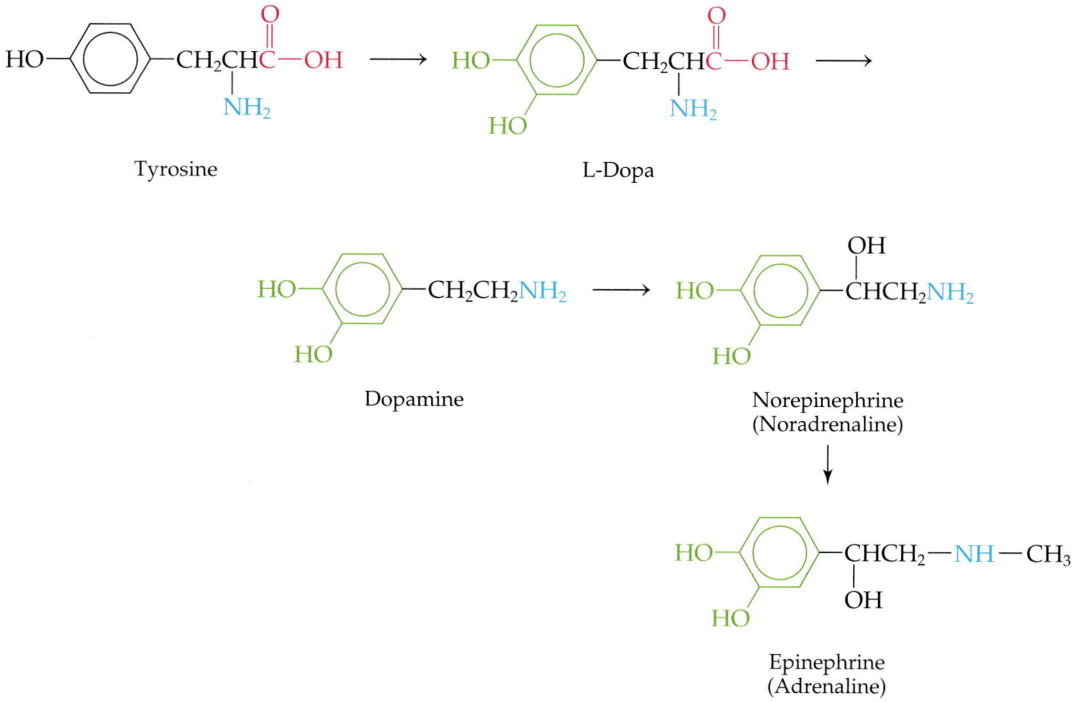

Tyrosine → L-Dopa → Dopamine → Norepinephrine (Noradrenaline) → Epinephrine (Adrenaline)

Figure C.2 The biosynthesis of norepinephrine and epinephrine from tyrosine.

terminals from the amino acid tyrosine.[3] Their name stems from the catechol ring (green), a benzene with two adjacent (ortho) hydroxyl groups; they are also called adrenergic transmitters. They include dopamine, norepinephrine (NE; noradrenaline), and epinephrine (adrenaline). Epinephrine is a minor transmitter in the CNS, but norepinephrine plays a role in many pathways. Neurons containing catecholamines innervate large areas of the cerebral cortex and are prominent in the long ascending/descending pathways from the brain through the spinal cord. The catecholamines and serotonin (see below) are widely distributed within the brain and appear to have a broad role in controlling the "tone" of the cortex, which may translate into mood. An excess of NE causes a person to be elated—perhaps even hyperactive. In large excess, NE induces a manic state. Dopamine plays a role in behavioral and emotional stability and in voluntary movement coordination, especially in the control of detailed motion (such as grasping small objects).

At least nine types of receptors are activated by NE; these are classed as alpha or beta types. Dopamine has at least three types. There are profound differences in how norepinephrine and various drugs act at the varied receptors. NE agonists (drugs that enhance or mimic its action) are stimulants. NE antagonists (drugs that block the action of NE) slow down various processes. Dopamine agonists increase motor activity.

All of the catecholamines lie in one synthetic path (see again Figure C.2). Each step is catalyzed by one or more enzymes. The intermediate L-dopa also has physiological activity and has been used successfully in the symptomatic treatment of Parkinson's disease. The action of catecholamines at the synapse is terminated by reuptake into the nerve terminal, where they are repackaged or degraded. An important enzyme for the degradation of monoamines is monoamine oxidase (MAO), which catalyzes their deamination.

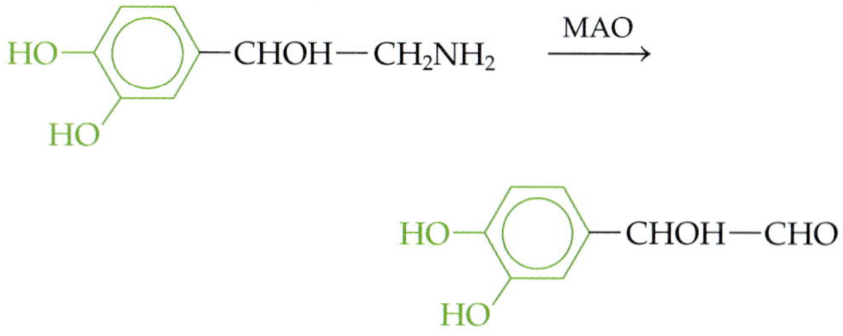

$$\text{(HO)}_2\text{C}_6\text{H}_3\text{—CHOH—CH}_2\text{NH}_2 \xrightarrow{\text{MAO}} \text{(HO)}_2\text{C}_6\text{H}_3\text{—CHOH—CHO}$$

Epinephrine and, to a smaller extent, norepinephrine, are also important as hormones outside the nervous system. Synthesized in the adrenal medulla, they regulate many aspects of metabolism, including carbohydrate metabolism. Many organs, generally those under involuntary (autonomic) control, respond to both circulating (hormonal) and neural catecholamines. Responses are usually stimulatory and include increased heart rate and blood pressure, shift of blood to skeletal muscle, increased blood glucose, and dilation

[3]Contrary to the claims of some who advocate amino acids as nutritional supplements, tyrosine is usually available to the first enzyme of the pathway, tyrosine hydroxylase, in such large amounts that changes in plasma levels of this amino acid usually do not affect catecholamine levels.

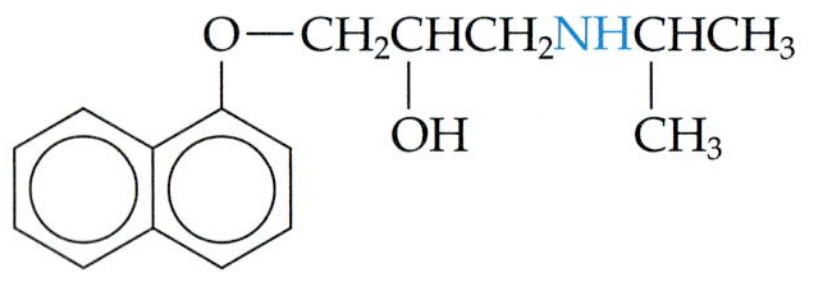

Propranolol
(Inderal)

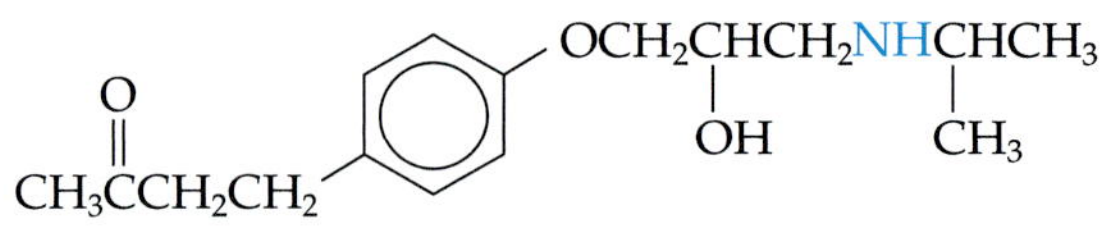

Metoprolol
(Lopressor)

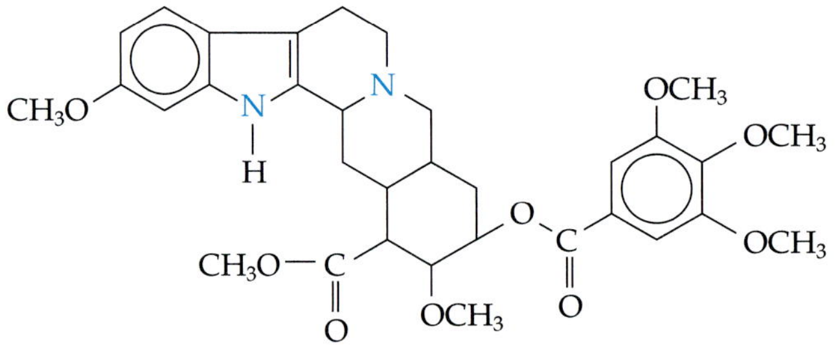

Reserpine

Figure C.3
Three drugs based on epinephrine and norepinephrine.

of lung bronchioles. The hormonal and nervous systems are coordinated and can act as a unit. When a person is under stress or frightened, this coordinated discharge prepares the body for fight or flight. Because culturally imposed inhibitions prevent fighting or fleeing in most modern situations, the adrenaline-induced supercharge is not used. This sort of frustration has been implicated in some forms of mental illness.

The biogenic amine theory of depression suggests that depression is caused by a deficiency in monoamines. It is likely that an imbalance among neurotransmitters is responsible for both bipolar (manic-depression) and monopolar (major depression) forms. Probably norepinephrine, dopamine, serotonin, and acetylcholine are all involved. One theory for the development of exogenous depression (commonly arising after stressful events) is that large quantities of biogenic amines released in response to stress lead to overworked synapses, which eventually become unable to release sufficient transmitter.

Because epinephrine and norepinephrine affect both the central nervous system and several vital organs, such as the cardiovascular system, drugs based on either of them can have a wide variety of side effects. Three examples are given in Figure C.3. Reserpine depletes nerve endings throughout the body of norepinephrine, serotonin, and dopamine by interfering with the ability of catecholamine storage vesicles to take up and store the transmitters. Reserpine was one of the first drugs for hypertension, and is still used. However, it can cause sedation, depression, and Parkinson-like symptoms. Other NE antagonists called beta blockers reduce the stimulant action of epinephrine and NE on various kinds of cells by occupying the receptors. Propranolol (Inderal) is used to treat cardiac arrhythmias, angina, and hypertension by lessening slightly the force of the heart beat. Unfortunately, it also causes lethargy and depression. Metoprolol (Lopressor) acts selectively on cells in the heart. It can be used by hypertensive patients who have asthma because it does not act on receptors in the bronchi.

Serotonin

Another monoamine neurotransmitter is 5-hydroxytryptamine, often called serotonin. It is produced in the nerve terminal (and in the gastrointestinal tract and blood platelets) from the amino acid tryptophan (Figure C.4).[4] Serotonin-containing neurons are found in the midbrain, hypothalamus, and brain stem. Serotonergic axons extend from the brain stem to the cerebral cortex. When released, serotonin tends to diffuse over several cells. The effect, excitatory or inhibitory, depends on the receptor type. Serotonin is removed from the synapse rapidly by reuptake into the terminal, where it is repackaged into vesicles. Like the catecholamines,

[4] The synthesis of 5-hydroxytryptamine (serotonin) is affected by levels of its precursor, the amino acid tryptophan, in the plasma. This has led to attempts to treat depression by dietary means, and may account for the effect of milk in helping persons to sleep.

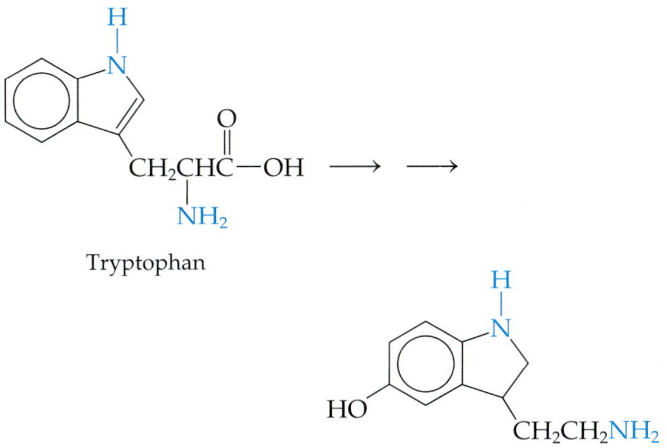

$$\text{Tryptophan: indole-}CH_2CH(NH_2)C(=O){-}OH \longrightarrow \longrightarrow \text{Serotonin: HO-indole-}CH_2CH_2NH_2$$

Tryptophan

Serotonin

▲ **Figure C.4**
The biosynthesis of serotonin from tryptophan.

it can be deaminated in a reaction catalyzed by monoamine oxidase.

Serotonin is involved in mood, appetite, sleep, sensory perception, and the regulation of body temperature. Its exact role in mental illness is not clear, but considerable evidence suggests that a depletion of serotonin contributes to depression. Electroconvulsive therapy for depression increases the number and activity of serotonin receptors, and drugs that block reuptake of serotonin are effective agents against depression. A metabolite of serotonin, 5-hydroxyindoleacetic acid (5-HIAA) is found in unusually *low* levels in the spinal fluid of violent suicide victims.[5] Serotonin agonists are used to treat depression, anxiety, and obsessive-compulsive disorder. Serotonin antagonists are used to treat migraine headaches and to relieve the nausea caused by cancer chemotherapy.

Histamine

Histamine is formed by decarboxylation of the amino acid L-histidine.

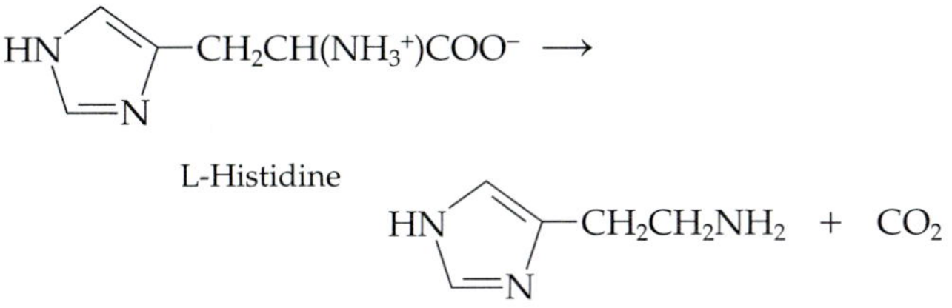

$$\text{imidazole-}CH_2CH(NH_3^+)COO^- \longrightarrow \text{imidazole-}CH_2CH_2NH_2 + CO_2$$

L-Histidine

Histamine

Unlike the other biogenic amines, it lacks a hydroxyl group. It is widely distributed in the central nervous system but apparently has less importance there than elsewhere in the body. Outside the nervous system, it mediates allergic reaction, inflammation, and acid production in the stomach. Many antihistamines formulated to act outside the nervous system can cause drowsiness.

Amino Acids

About 15 amino acids have been proposed as neurotransmitters! Thus far, firm evidence exists for a role for glutamate and aspartate as excitatory transmitters and for gamma-aminobutyric acid (GABA) and glycine as inhibitory ones. Note that the excitatory amino acids are dicarboxylic, whereas the inhibitory ones have a single carboxylate group.

$$^-OOCCH_2CH_2CH(NH_3^+)COO^-$$

Glutamate

$$^-OOCCH_2CH(NH_3^+)COO^-$$

Aspartate

$$CH_2(NH_3^+)COO^-$$

Glycine

There are at least three different glutamate receptors in the brain. One of these, the N-methyl-D-aspartate (NMDA) receptor, may play a role in short-term memory, in the brain's reaction to decreased oxygen (as in a stroke), and in integration of emotions and cognition. Glutamate is thought to play a role in modulation of dopamine action.

Another probable dopamine modulator, γ-aminobutyric acid (GABA) is found throughout the CNS, probably in 25–45% of all CNS nerve terminals. It is synthesized from glutamate by decarboxylation of the α-carboxylate group.

$$^-OOCCH_2CH_2CH(NH_3^+)COO^- \longrightarrow$$

Glutamate

$$^-OOCCH_2\overset{\alpha}{C}H_2\overset{\beta}{C}H_2\overset{\gamma}{C}H_2NH_3^+ + CO_2$$

GABA

Dysfunction of GABA systems is thought to play a role in anxiety. Loss of GABA synapses occurs in Huntington's disease.

The amino acid glycine functions as an inhibitory transmitter at several sites in the brain stem and spinal cord, but it also plays an excitatory role as coagonist with glutamate at NMDA receptors.

Acetylcholine

Acetylcholine, an important transmitter with widespread effects, is a quaternary ammonium compound.

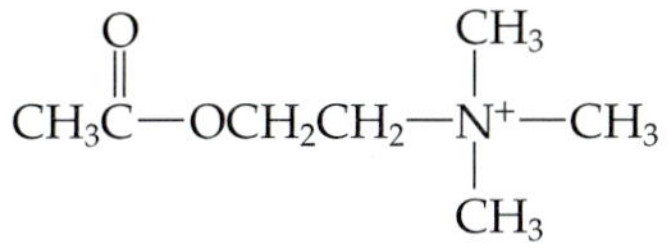

$$CH_3\overset{O}{\overset{\|}{C}}{-}OCH_2CH_2{-}N^+(CH_3)_3$$

[5] Levels of 5-HIAA also are low in murderers and other violent offenders. Levels of this serotonin metabolite are higher than normal in persons with obsessive-compulsive disorders, sociopaths, and people who have guilt complexes.

Toxic Gases and the Learning Process

Gases such as carbon monoxide (CO), nitrogen monoxide (NO, commonly called nitric oxide), and hydrogen sulfide (H_2S) have long been known for their toxic effects. Imagine the surprise when scientists discovered that brain cells make NO and use it for communication.

A living cell constantly senses its environment. It changes in response to information flowing into and out of the cell. Most chemical messengers are complex substances such as norepinephrine and serotonin. But NO and possibly CO and H_2S are also chemical messengers. NO is formed in cells from the amino acid arginine (Chapter 21) in an enzyme-catalyzed reaction. NO is produced in the cerebellum, hippocampus, and other areas of the brain. As a small molecule, it can diffuse widely. It is synthesized and inactivated rapidly, often increasing in response to inflammation. Although it is involved in the formation of long-term memories, it may also contribute to cell death after a stroke. NO kills invading microorganisms, probably by deactivating iron-containing enzymes in much the same way that CO destroys the oxygen-carrying capacity of hemoglobin. NO also helps to regulate blood pressure by direct action on blood vessels.

Carbon monoxide is formed in cells by the enzyme-catalyzed oxidation of heme. Like NO, it seems to be involved in the long-term potentiation of learning. H_2S has been found in the brain cells of rats, but there is as yet no proof that the cells make it. There is evidence that H_2S stimulates certain receptors that strengthen the connections between brain cells, an indication of long-term learning. Learning can be a gas!

Nearly all fibers leaving the nervous system are cholinergic (that is, they use acetylcholine as the transmitter). Many early studies of neurotransmitters looked at the action of acetylcholine at the neuromuscular junction, where the receptors on the muscle lie exactly opposite the release sites.

In the brain, the cell bodies of cholinergic neurons are in the midbrain. Many of these innervate a region of the brain called the hippocampus, which appears to be involved in formation of short-term memory. Acetylcholine is diminished in the brains of patients with Alzheimer's disease, and much research has focused on prolonging the effects of this transmitter in such patients. Drugs approved for symptomatic treatment reversibly inhibit acetylcholinesterase, the enzyme that breaks down the transmitter to acetyl-CoA and choline in the synapse. Organophosphate cholinesterase inhibitors (nerve gases), on the other hand, are potent poisons whose binding to the enzyme is essentially irreversible.

Neurotransmitters and Mental Illness

Nearly one out of every ten people in the United States suffers from mental illness. Over half the patients in hospitals are there because of mental problems. Brain function is characterized by finely controlled integration among the neurotransmitter systems, and dysfunction in any one of them can lead to profound alteration in mental or emotional function. When the biochemistry of the brain is more fully understood, mental illness may be cured (or at least alleviated) by administration of drugs. In subsequent sections, we see just how far we have already come in learning to control our moods with drugs. As is true for so many things, the potential for good that such compounds represent is matched by a potential for abuse.

Drugs affecting mental processes, often called psychoactive drugs, fall into three major categories: antipsychotic agents (neuroleptics), antidepressants, and antianxiety drugs (anxiolytics). Other classifications include sedatives and stimulants. They are also described in terms of their enhancement or antagonism of specific neurotransmitters.

C.3 Antianxiety Agents and Sedatives: GABA Modulators

When someone refers to "tranquilizers," he or she usually means the antianxiety agents, or anxiolytics. These are prescribed for persons with symptoms ranging from severe chronic anxiety to acute stress-induced tension. They are also prescribed for treatment of sleep problems.

Benzodiazepines

The most widely used antianxiety agents are the benzodiazepines, compounds that feature a seven-member heterocyclic ring (Figure C.5). Of these, perhaps the best known are diazepam (Valium), lorazepam (Ativan), and alprazolam (Xanax). These drugs are among the most widely prescribed medications in the United States.[6] They are also one of several classes of anticonvulsants and are used for treatment of epilepsy and prevention of seizures during some medical/surgical procedures.

[6]Certain benzodiazepines, such as flurazepam (Dalmane) and triazolam (Halcion), are used to treat insomnia. They do help a person fall asleep, but the kind of sleep achieved is not restful. These pills may be useful to help someone get through tough times, but they do not cure insomnia or correct the conditions that cause it.

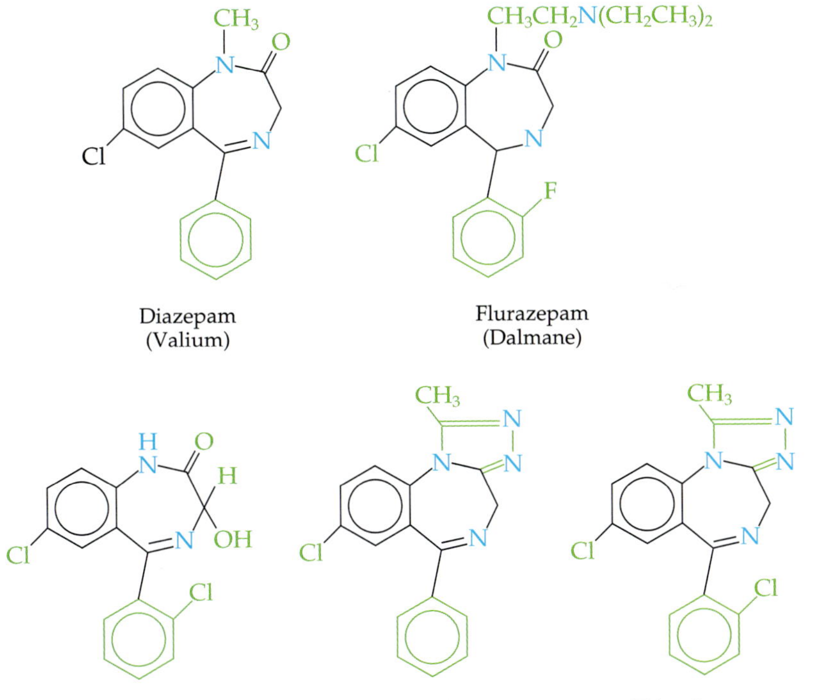

Diazepam
(Valium)

Flurazepam
(Dalmane)

Lorazepam
(Ativan)

Alprazolam
(Xanax)

Triazolam
(Halcion)

Figure C.5
Some benzodiazepine drugs.

The benzodiazepines act specifically at GABAergic synapses, enhancing the inhibitory action of the amino acid transmitter. The drug action is complex, occurring through a change in conformation of the GABA receptor. The seven-member ring appears essential for activity. At high doses the drugs affect nearly all other CNS transmitters.

After 20 years of use, benzodiazepines were found to be addictive. People trying to get off the drugs after prolonged use go into painful withdrawal. Tolerance also develops to their anticonvulsant activity.

Buspirone

A newer antianxiety agent, buspirone (Buspar), also alters GABA (and monoamine) transmission, but in a different, more selective manner.

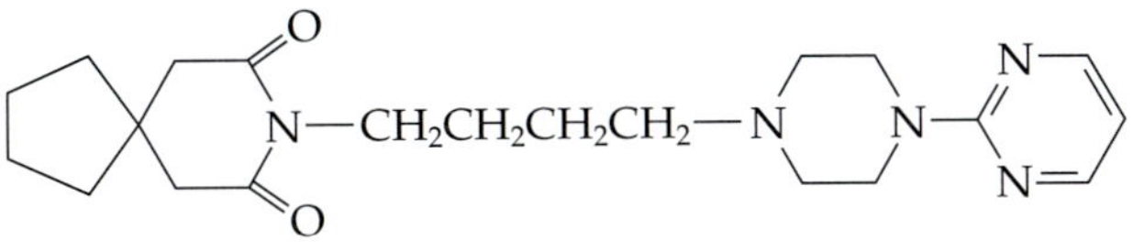

Buspirone (Buspar)

Buspirone inhibits serotonin activity but enhances dopamine and NE activity. Unlike the benzodiazepines, it does not have anticonvulsant, sedative, or addictive properties. Study of the mode of action of this drug may further an understanding of anxiety itself.

Barbiturates

Barbiturates also modulate activity at the GABA receptor, but at a different specific site. Like the benzodiazepines, barbiturates enhance GABAergic transmission by causing a change in the conformation of the receptor. However, they not only increase GABA activity but also can produce GABA-like effects in the absence of the transmitter. They also decrease oxidative metabolism in the brain.

Barbiturates are cyclic amides with a far simpler structure than that of the benzodiazepines (Figure C.6). As a family of compounds, the barbiturates display a wide variety of properties. They can be employed to produce mild sedation, deep sleep, and even death. Although sometimes classed with the antianxiety agents, today they are seldom used for anxiety because only a small increase in dose causes CNS depression. Modern tranquilizers are much safer.

The barbiturates were once used in small doses as sedatives. The dose for sedation was generally a few milligrams. In larger doses (about 100 mg), barbiturates induce sleep. They were once the sleeping pills so widely used—and abused—by middle-class, often middle-aged, people. Barbiturates were once the drugs of choice for suicides—news reports listed the cause of death as "an overdose of sleeping pills." There is also potential for accidental overdose.

Several thousand barbiturates have been synthesized through the years, but only a few have found

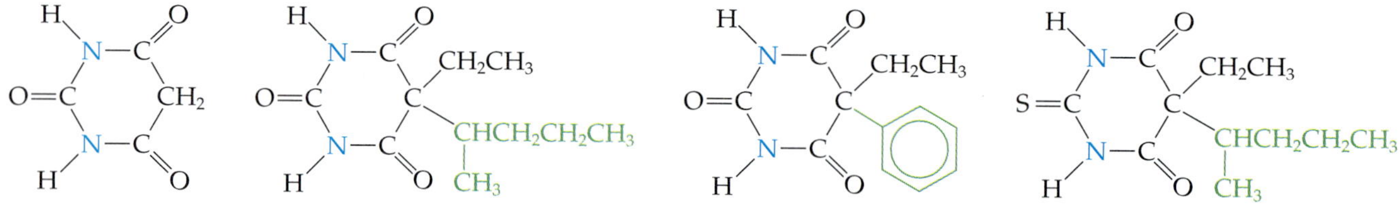

Barbituric acid

Pentobarbital (Nembutal)

Phenobarbital (Luminal)

Thiopental (Pentothal)

▲ **Figure C.6**
Barbituric acid and some barbiturate drugs.

widespread use in medicine.[7] Pentobarbital (Nembutal) is employed as a short-acting hypnotic drug. Phenobarbital (Luminal) is a long-acting drug. It, too, is a hypnotic and can be used as a sedative. It is employed widely as an anticonvulsant for epileptics and brain-damaged people. Thiopental (Pentothal)[8] is used widely in anesthesia.

The barbiturates are especially dangerous when ingested along with ethyl alcohol. This combination produces an effect much more drastic than just the sum of the effects of two depressants. The effect of a barbiturate is enhanced by as much as 200-fold when taken with an alcoholic beverage. This effect of one drug enhancing the action of another is called **synergism.**

The barbiturates are strongly addictive. Habitual use leads to tolerance, which means that ever-larger doses are required to get the same degree of intoxication. Barbiturates are legally available by prescription only, but they have been a part of the illegal drug scene also. They are known as "downers" because of their depressant, sleep-inducing effects.

The side effects of barbiturates are similar to those of alcohol. Abuse leads to hangovers, drowsiness, dizziness, and headaches. Withdrawal symptoms are often severe, accompanied by convulsions and delirium.

Other Antianxiety Drugs

Several over-the-counter drugs, often called "nighttime pain relievers," claim to be able to help us with our anxieties. Such products usually contain a little aspirin or acetaminophen plus an antihistamine. The latter has a side effect of making one drowsy.

[7]Barbituric acid was first synthesized in 1864 by Adolph von Baeyer, a young student of Friedrich August Kekulé (Chapter 13). The term barbiturates, according to Willstätter, came about because, at the time of the discovery, von Baeyer was infatuated with a girl named Barbara. The word comes from *Barbara* and *urea.* Curiously, in the United States, the names of the barbiturates end in *-al* even though they are ketones rather than aldehydes (Figure C.6). The British spelling uses the suffix *-one.*

[8]Thiopental has been investigated as a possible "truth drug." It does seem to help psychiatric patients recall traumatic experiences. It also helps uncommunicative individuals talk more freely. It does not, however, prevent one from withholding the truth or even from lying. No true truth drug exists.

The hectic pace of life in the modern world has driven people to seek rest and relaxation in chemicals. Ethyl alcohol is undoubtedly the most widely used tranquilizer. The drink before dinner—to "unwind" from the tensions of the day—is a part of the American way of life. Many people, however, seek their relief in other chemical forms.

C.4 Antidepressants: Monoamine Enhancers

Depression is a widespread mood disorder for which a well-founded treatment rationale exists. The biogenic amine theory of depression (see Section C.2) leads to the hypothesis that if monoamine depletion causes depression, restoration of normal monoamine levels should restore normal function. The concentration of a monoamine in the synapse may be increased by inhibiting its breakdown enzyme, monoamine oxidase, or the reuptake mechanism. Both approaches have been used successfully to treat depression. The major classes of antidepressant drugs are the monoamine oxidase (MAO) inhibitors, tricyclic antidepressants, selective serotonin reuptake inhibitors, selective NE reuptake inhibitors, and some newer "atypical" antidepressants.

The MAO inhibitors prevent breakdown of the enzyme that degrades the catecholamines and serotonin. This results in mood elevation and blood pressure lowering. The older inhibitors, such as phenelzine (Nardil), are irreversible inhibitors with a long duration of action and many side effects, including insomnia and agitation.

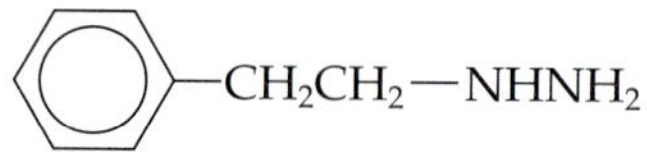

Phenelzine (Nardil)

The tricyclic (three-ring) antidepressants (Figure C.7), such as desipramine (Norpramin) and imipramine (Tofranil), block reuptake of both NE and serotonin into the nerve terminals, thus prolonging their action. However, the clinical effects are not com-

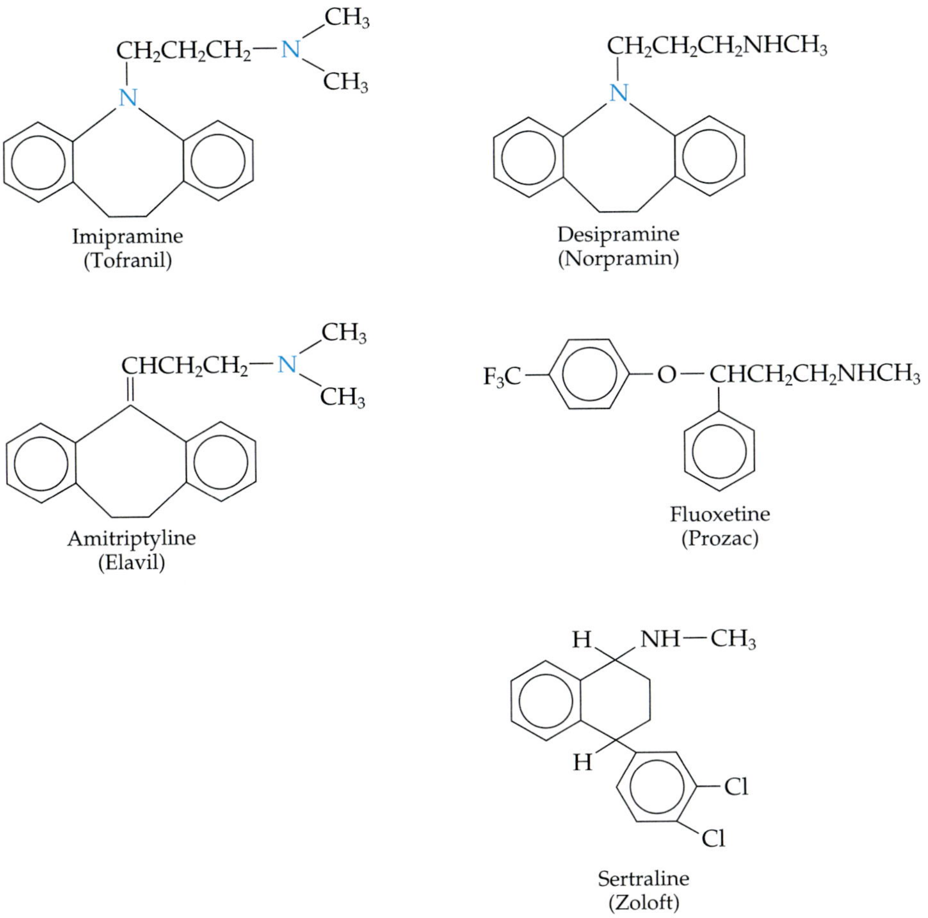

$CH_2CH_2CH_2-N(CH_3)_2$

Imipramine
(Tofranil)

$CH_2CH_2CH_2NHCH_3$

Desipramine
(Norpramin)

$CHCH_2CH_2-N(CH_3)_2$

Amitriptyline
(Elavil)

$F_3C-C_6H_4-O-CHCH_2CH_2NHCH_3$

Fluoxetine
(Prozac)

H, $NH-CH_3$, H, Cl, Cl

Sertraline
(Zoloft)

Figure C.7 Some antidepressant drugs. Imipramine, desipramine, and amitriptyline are tricyclic antidepressants. Fluoxetine and sertraline are selective serotonin reuptake inhibitors (SSRIs).

pletely consistent with this reuptake mechanism. Although the tricyclic antidepressants have been around since the 1950s, they have been only somewhat successful. The range in which the dose is both safe and effective is quite narrow. Undesirable side effects include nausea, headache, dizziness, loss of appetite, or grogginess, as well as more serious problems such as jaundice or high blood pressure. Nevertheless, they have been widely used for treatment of depression, panic and obsessive-compulsive disorders, and chronic pain.

(It is interesting to note that slight changes in structure can result in profound changes in properties. Replacing the sulfur atom of the antipsychotic promazine with a CH_2CH_2 group produces the antidepressant imipramine. Another common tricyclic is amitriptyline (Elavil), in which the ring nitrogen atom is replaced by a carbon atom.)

The selective serotonin reuptake inhibitors (SSRIs), including fluoxetine (Prozac) and sertraline (Zoloft), have become very popular for treatment of depression. Doctors also prescribe them to help people cope with gambling problems, obesity, fear of public speaking, or premenstrual syndrome (PMS). They are safer than the older antidepressants and more easily tolerated. However, they do not work for everyone, and they can cause nausea or insomnia. Research continues to seek antidepressants that restore normal brain biochemistry without adverse effects.

Lithium, given as the carbonate salt, is the drug of choice for bipolar (manic-depressive) illness. Its mechanism of action is unknown.

C.5 Antipsychotic Agents: Dopamine Antagonists

Abnormal behavior caused by drugs such as the amphetamines (Section C.6) is inhibited by antipsychotic (neuroleptic) agents. This knowledge led to the hypothesis that schizophrenia results from abnormal release of dopamine in circuits responsible for emotional and behavioral integration. Nearly all drugs successful in treating psychosis are dopamine antagonists, binding to and blocking specific receptors. Current research into the disease views the action of dopamine within the context of its interactions with other neurotransmitter systems, particularly the amino acids glutamate and GABA. Many antipsychotics also affect the serotonin system.

There are approximately 100 antipsychotic agents, which are grouped into seven classes, of which we will consider representatives of two. The largest group is the phenothiazines, which include the classic drug chlorpromazine (Figure C.8). The potent agent haloperidol is in the butyrophenone class. When the various drugs are ranked by antipsychotic effectiveness, there is a general, but not absolute trend for the most potent to be the most specific dopamine antagonists.

It is interesting to look at the development of the antipsychotics. This process has involved exploration of herbal remedies and observation of side effects in drugs used for other purposes.

For centuries the people of India used the snakeroot plant, *Rauwolfia serpentina,* to treat a variety of ailments including fever, snakebite and other poisonings, and—most important—maniacal forms of mental illness. Western scientists became interested in the plant near the middle of the twentieth century—after disdaining such remedies as quackery for many generations.

In 1952 rauwolfia was introduced into American medical practice to treat hypertension (high blood pressure) by Robert Wilkins of Massachusetts General Hospital. In the same year, Emil Schlittler of Switzerland isolated an active alkaloid, which he named reserpine (see again Figure C.3). Reserpine was found not only to reduce blood pressure (as we noted in Section C.2) but also to bring about sedation. The latter finding attracted the interest of psychiatrists, who found reserpine so effective that by 1953 it had replaced electroshock therapy for 90% of psychotic patients.

Also in 1952 chlorpromazine (Thorazine) was tried as a tranquilizer on psychotic patients in the United States. The drug had been tested in France as an antihistamine. Medical workers there noted that it calmed mentally ill people who were being treated for allergies. Chlorpromazine was found to be quite effective in controlling the symptoms of schizophrenia. It truly revolutionized mental illness therapy.

The antipsychotic drugs (so-called "major tranquilizers") have been one of the real triumphs of chemical research. They greatly reduced the number of patients confined to mental hospitals by controlling schizophrenic symptoms to the extent that 95% of all patients no longer need hospitalization.

C.6 Stimulant Drugs: Dopamine Enhancers

Among the more widely known stimulant drugs are a variety of synthetic amines related to phenylethylamine (Figure C.9). Note the similarity between these molecules and epinephrine and norepinephrine; all are derived from the basic phenylethylamine structure. The amphetamines promote the release of dopamine and norepinephrine from the axon terminal. They also block their reuptake and metabolism by MAO enzymes. Thus they enhance the alerting and mood-elevating effects of dopamine and norepinephrine. The effect on dopaminergic neurons predominates.

Amphetamine was once extensively used for weight reduction. It has also been employed for treating mild depression and narcolepsy, a rare form of sleeping sickness. Mainly by blocking the reuptake of norepinephrine by peripheral nerves, amphetamine induces excitability, restlessness, tremors, insomnia, dilated pupils, increased pulse rate and blood pressure, hallucinations, and psychoses. Chronic use, particularly at high doses, can lead to tolerance and physical dependence. It is no longer recommended for weight reduction. Studies showed that any weight loss was usually only temporary. The greatest problem, however, was the diversion of vast quantities of this inexpensive drug into the illegal drug market.

▶ Figure C.8
Some antipsychotic drugs. Promazine and chlorpromazine are phenothiazines. The phenothiazine (two benzene rings connected through a nitrogen atom and a sulfur atom) part of the structures is shown in red. Haloperidol is a butyrophenone. The butyrophenone (a four-carbon ketone group attached to a benzene ring) part of the structure is shown in blue.

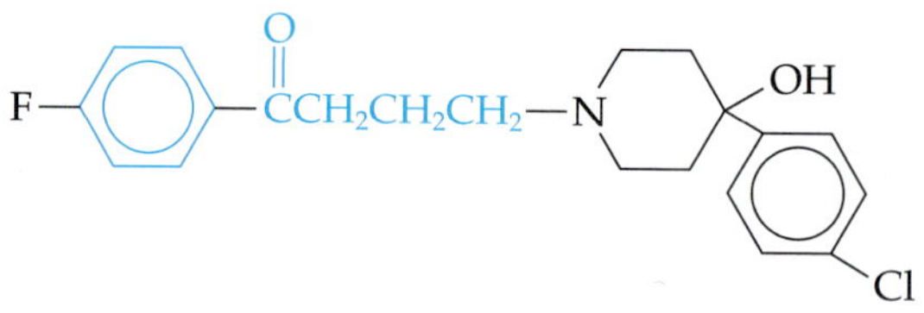

Haloperidol (Haldol)

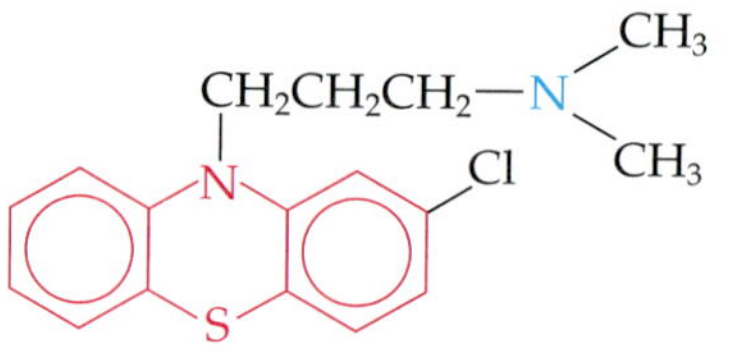

Chlorpromazine (Thorazine)

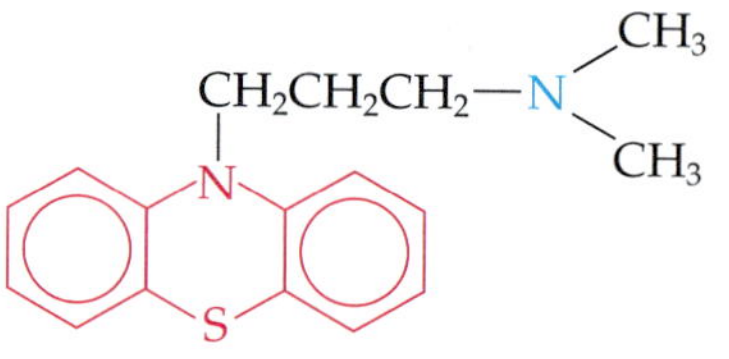

Promazine

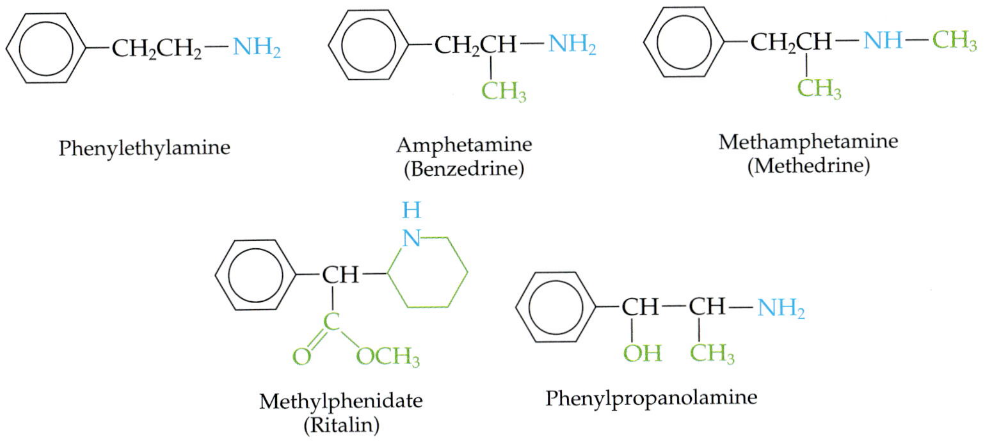

Phenylethylamine

Amphetamine (Benzedrine)

Methamphetamine (Methedrine)

Methylphenidate (Ritalin)

Phenylpropanolamine

Figure C.9 Phenylethylamine and related compounds.

Table C.2 Toxicities of Various Drugs[a]

Drug	LD_{50} (mg/kg body weight)	Method of Administration	Experimental Animal
Local anesthetics			
Lidocaine	292	Oral	Mice
Procaine	45	Intravenous	Mice
Cocaine	17.5	Intravenous	Rats
Barbiturates			
Barbital	600	Oral	Mice
Pentobarbital	118	Oral	Rats
Phenobarbital	162	Oral	Rats
Amobarbital	212	Subcutaneous	Mice
Thiopental	149	Intraperitoneal	Mice
Narcotics			
Morphine	500	Subcutaneous	Mice
Heroin	21.8	Intravenous	Mice
Meperidine	170	Oral	Rats
Stimulants			
Caffeine	355	Oral	Rats
Nicotine	230	Oral	Mice
Nicotine	0.3	Intravenous	Mice
Amphetamine	180	Subcutaneous	Rats
Methamphetamine	70	Intraperitoneal	Mice
Mescaline	370	Intraperitoneal	Rats

[a]Comparisons of toxicities in different animals—and extrapolation to humans—are at best crude approximations. The method of administration can have a profound effect on the observed toxicity. *Source:* Susan Budavari (Ed.), *The Merck Index*, 12th ed. Rahway, NJ: Merck and Co., 1996.

Amphetamine[9] and methamphetamine have been widely abused. Methamphetamine has a more pronounced psychological effect than amphetamine. Generally, the "speed" that abusers inject into their veins is methamphetamine.[10] Such injections, at least initially, are said to give the abuser a euphoric rush. Shooting methamphetamine is quite dangerous, however, because the drug is relatively toxic (see Table C.2).

Another amphetamine derivative, phenylpropanolamine, is widely used as an over-the-counter appetite suppressant. Like its relatives, this compound is a stimulant. Studies show that it is at best marginally effective as a diet aid, and it poses a threat to people with hypertension. Nevertheless, sales of phenylpropanolamine are 1 billion tablets, or $150 million, each year.

[9]Amphetamine is not a single compound but a mixture of two stereoisomers (Chapter 18) marketed under the trade name Benzedrine.

[10]Like other amine drugs (Chapter 17), the amphetamines normally are used in the form of hydrochloride salts. A freebase form of methamphetamine is used like crack cocaine (Section C.6) for smoking. This form is called "ice" because it is a clear crystalline solid that resembles the solid form of water.

One controversial use of amphetamines is their employment in the treatment of attention deficit disorder (ADD) in children. The drug of choice is often methylphenidate (Ritalin). The drug acts as a stimulant for adults but seems to have the opposite effect on children. It calms them and helps them filter out extraneous stimuli, enabling them to focus their attention on the task at hand. This use has been criticized as "leading to drug abuse" and as "solving the teacher's problem, not the child's."

Cocaine

Although not a phenylethylamine derivative, cocaine also acts as a stimulant by blocking reuptake of dopamine. High levels of dopamine are therefore available to stimulate the pleasure centers of the brain. The enhancement of dopamine action is thought to be responsible for cocaine's "high" and its addictive properties. After the binge, dopamine is depleted in less than an hour. This leaves the user in a pleasureless state and (often) craving more cocaine.

Cocaine was first used as a local anesthetic and is still used topically in some urogenital procedures. It acts directly on sensory nerves to block conduction (Section C.7). The drug is obtained from the leaves of a shrub that grows almost exclusively on the eastern slopes of the Andes Mountains. Many of the Indians living in and around the area of cultivation chew coca leaves—mixed with lime and ashes—for their stimulant effect.

Cocaine is used as the salt cocaine hydrochloride and in the form of broken lumps of the free base, called *crack cocaine.* Because it is soluble in water, cocaine hydrochloride is readily absorbed through the watery mucous membranes of the nose when snorted. Crack is more volatile than the hydrochloride. It vaporizes at the temperature of a burning cigarette. When smoked, cocaine reaches the brain in 15 seconds.

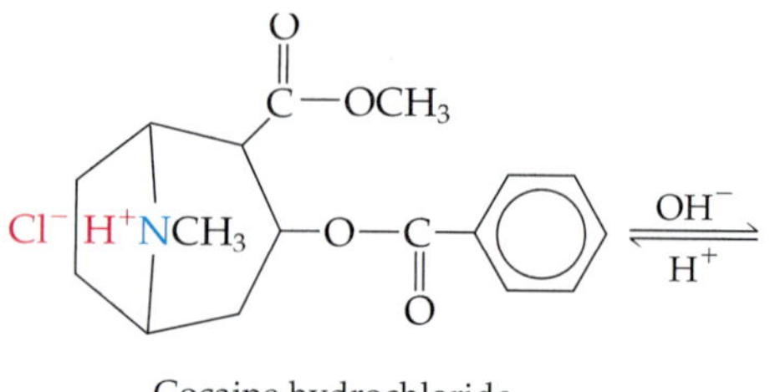

Cocaine hydrochloride

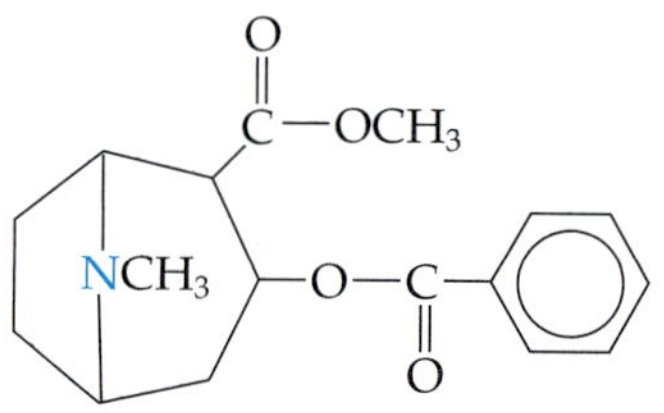

Cocaine (crack)

Use of cocaine increases stamina and reduces fatigue, but the effect is short-lived. Stimulation is followed by depression. Once quite expensive and limited to use mainly by the wealthy, cocaine is now available in cheap and potent forms. Its direct constrictive effect on blood vessels, coupled with a local anesthetic effect, makes it more dangerous than amphetamines. Hundreds, including several well-known athletes, have died from cocaine overdose.

Caffeine

The methylated xanthines, caffeine, theophylline, and theobromine, are stimulants found in coffee, tea and some soft drinks.[11] Their mechanism of action is not well understood but is thought to be related to blockage of receptors for adenosine, a purine (see Section 23.1) recently recognized as a neurotransmitter. The effective dose of caffeine is about 200 mg, corresponding to about two cups of strong coffee or tea. Caffeine is also available in tablet form as a stay-awake drug, such as No-Doz and Vivarin.

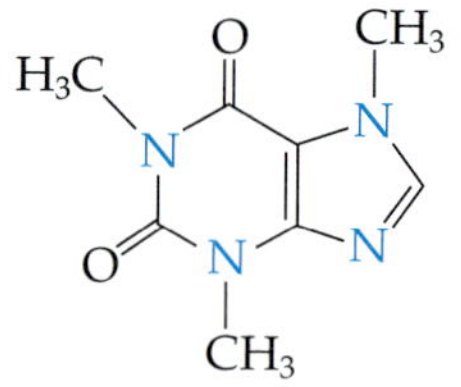

Caffeine

Is caffeine addictive? The "morning grouch" syndrome and headaches upon withdrawal indicate that it is mildly so.

Nicotine

Nicotine acts as a stimulant by yet another mechanism—probably as an acetylcholine agonist. (Some cholinergic receptors are selectively sensitive to nicotine and are called nicotinic receptors: most of these, however, are at the neuromuscular junction.) This drug is taken by smoking or chewing tobacco. Its stimulant effect seems transient, as this initial response is followed by depression. Nicotine is highly toxic to animals (Table C.2). It is especially deadly when injected; the lethal dose for a human is estimated to be about 50 mg. Nicotine has been used in agriculture as a contact insecticide.

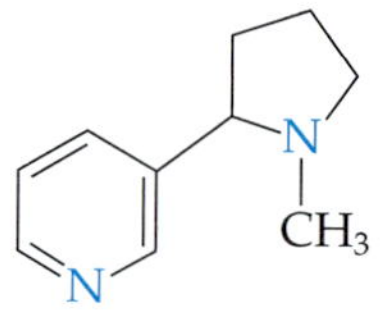

Nicotine

[11] Each year a million kilograms of caffeine are added to food in the United States. Most of it goes into soft drinks.

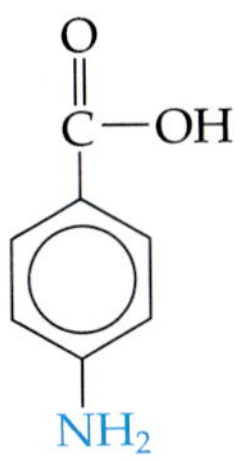

para-Aminobenzoic acid
(a)

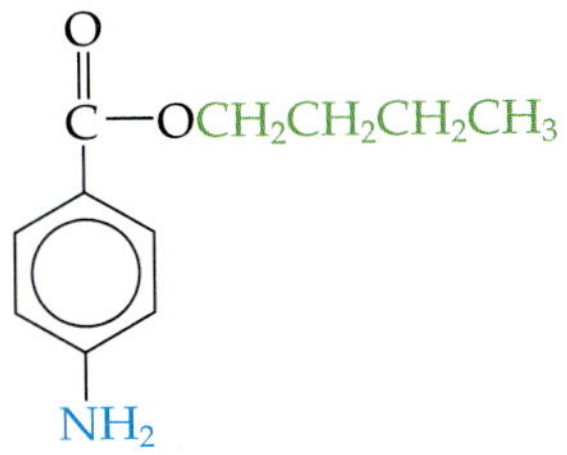

Butyl *para*-aminobenzoate
(Butesin)
(b)

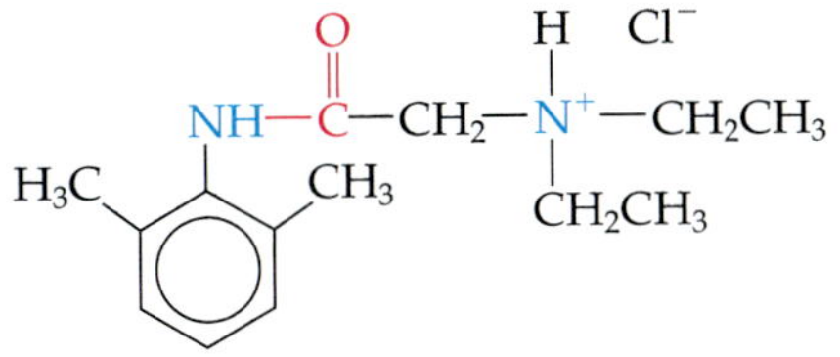

Lidocaine (Xylocaine)
(c)

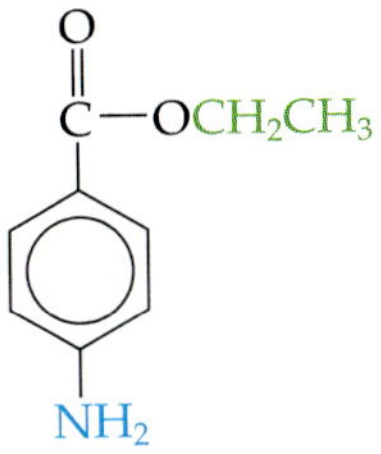

Ethyl *para*-aminobenzoate
(Benzocaine)
(b)

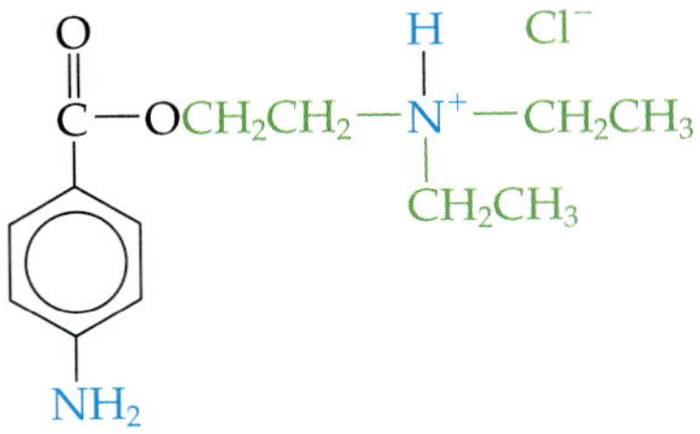

Procaine
(Novocain)
(b)

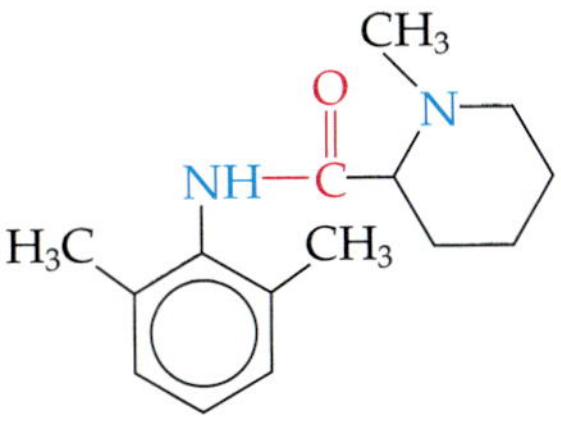

Mepivacaine
(c)

▲ **Figure C.10**
Some local anesthetics. Three (b) are derived from *p*-aminobenzoic acid (a). The other two (c) contain amide functional groups. All usually are used in the form of the hydrochloride salt, which is more soluble in water than the free base.

Is nicotine addictive? Casual observation of a person trying to quit smoking seems to indicate that it is,[12] and the various tobacco trials and hearings in recent years seem to have established the fact beyond doubt.

C.7 Anesthesia Revisited

An **anesthetic** is any substance that causes either unconsciousness or insensitivity to pain. We discussed inhalant anesthetics in Chapter 14 and mentioned thiopental as an intravenous anesthetic in Section C.3. Here we look briefly at two other types.

Local Anesthetics

For dental work and minor surgery it is usually desirable to deaden the pain in one part of the body only. **Local anesthetics** are drugs applied to nerve tissue that block transmission of pain messages to the brain. They act on all kinds of nerve cells and on all parts of the nervous system.

The first local anesthetic to be used successfully was cocaine. Its structure was determined by Richard Willstätter in 1898. Even before Willstätter's work, there were attempts to develop synthetic compounds having similar properties.

Certain esters of *p*-aminobenzoic acid (PABA) act as local anesthetics (Figure C.10). The ethyl and butyl esters are applied as ointments to relieve the pain of burns and open wounds.

More powerful in their anesthetic action are a series of PABA derivatives that have a nitrogen atom substituted in the alkyl group of the ester. Perhaps the best known of these is procaine (Novocain), first synthesized by Alfred Einhorn in 1905. Procaine can be injected as a local anesthetic, or it can be injected into the spinal column to deaden the entire lower portion of the body.

The local anesthetic of choice nowadays is often lidocaine or mepivacaine. Each compound is highly effective and yet has a fairly low toxicity (Table C.2). Note that lidocaine and mepivacaine are not derivatives of *p*-aminobenzoic acid, but they do share some structural features with the compounds that are.

Dissociative Anesthetics

Ketamine, like thiopental an intravenous anesthetic, is called a **dissociative anesthetic**—it induces hallucinations similar to those reported by people who have had near-death experiences. They seem to remember observing their rescuers from a vantage point above it all, or moving through a dark tunnel toward a bright light. Unlike thiopental, ketamine seems to affect associative pathways before it hits the brain stem, perhaps by blocking the effects of the excitatory amino acid glutamate or possibly by binding to the NMDA channel.

Closely related to ketamine (Figure C.11) is phencyclidine (PCP; "Angel Dust"), which also binds to the

[12]Clonidine, a drug used to treat high blood pressure, reduces nicotine withdrawal symptoms.

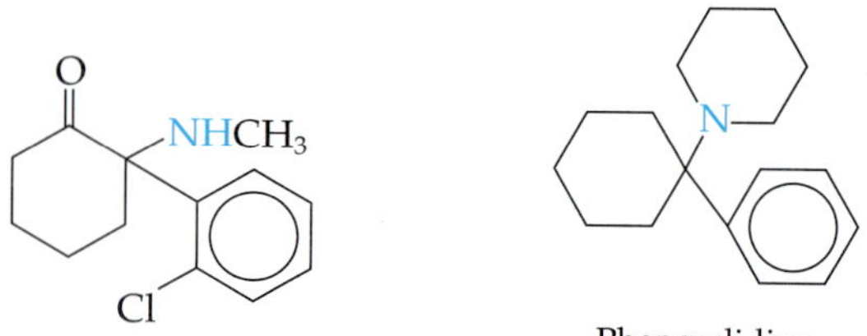

Ketamine

Phencyclidine (PCP)

▲ **Figure C.11**
Two dissociative anesthetics.

NMDA channel. PCP is soluble in fat and has no appreciable water solubility. It is stored in fatty tissue and released when the fat is metabolized; this accounts for the "flashbacks" commonly experienced by users. It was tested and found to be too dangerous for human use, but it has been used as an animal tranquilizer and marketed for this purpose under the trade name Sernylan.

At times PCP is an important part of the illegal drug scene. It is cheap and easily prepared. Many users experience bad "trips" with PCP. About 1 in 1000 develops a severe form of schizophrenia. Laboratory tests show that PCP depresses the immune system. Despite these well-known problems, every few years a new crop of young people appears on the scene to be victimized by this hog tranquilizer.

From a pharmacological viewpoint, PCP is of interest because it acts in various ways. It can both stimulate and depress the central nervous system; it is a hallucinogen and an analgesic. Few drugs seem to induce so wide a range of effects. The immediate danger is that PCP is quite toxic, particularly when mixed with alcohol (it triggers violent, psychotic behavior). PCP is considered more dangerous than heroin because medically nothing is known about its effects, nor are there drugs available to counteract it. It is easy to overdose on PCP to the point of convulsions.

Key Terms

anesthetic (C.7)
barbiturate (C.3)
dissociative anesthetic (C.7)
local anesthetic (C.7)
neuron (C.1)
neurotransmitter (C.1)
synapse (C.1)
synergism (C.3)

Review Questions

1. Define or identify each of the following terms.
   a. neuron **b.** synapse **c.** neurotransmitter
2. Which three naturally occurring amines are presently considered to play major roles in the biochemistry of mental health? What are their proposed roles?
3. Which amino acids serve as precursors for the amines of Problem 2? How might these amines relate our mental state to our diet?
4. How do amphetamines exert a stimulant effect?
5. How does cocaine exert a stimulant effect?
6. What is a local anesthetic? How does a local anesthetic work?
7. What is a general anesthetic? How does a general anesthetic work?
8. Name two dissociative anesthetics. How do they work?
9. What do we mean when we say that (**a**) amphetamines are "uppers" and (**b**) barbiturates are "downers"?
10. What is synergism? Give an example.
11. Drugs such as reserpine deplete nerve endings of norepinephrine and dopamine. How might these drugs be useful for treating manic patients?
12. Electroconvulsive therapy (shock treatment) induces the release of norepinephrine. What sort of mental problems are treated with this therapy?

Problems

Structural Formulas and Functional Groups

13. What is the basic structure common to all barbiturate molecules? How is this structure modified to change the properties of individual barbiturate drugs?
14. Examine the structure of the reserpine molecule (Figure C.3) and identify the following.
    a. five ether functional groups
    b. two amine functional groups
    c. two ester functional groups
15. Acebutolol (Sectral) is used as a drug for the treatment of heart disease (angina and arrhythmias) and hypertension. There are five functional groups in the compound. Name the five families of organic compounds to which acebutolol could be assigned.

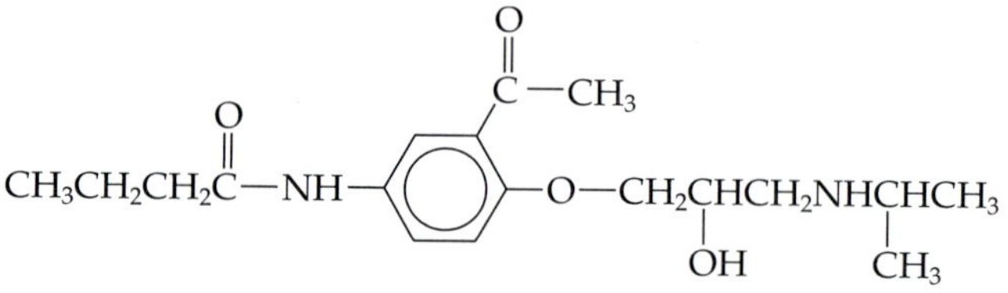

$CH_3CH_2CH_2C(=O)-NH-C_6H_3(C(=O)-CH_3)-O-CH_2CH(OH)CH_2NHCH(CH_3)CH_3$

16. Labetalol (Labelol) is used as a drug for the treatment of angina and hypertension. Circle the four functional groups in the molecule, and name the families of organic compounds that incorporate these functional groups.

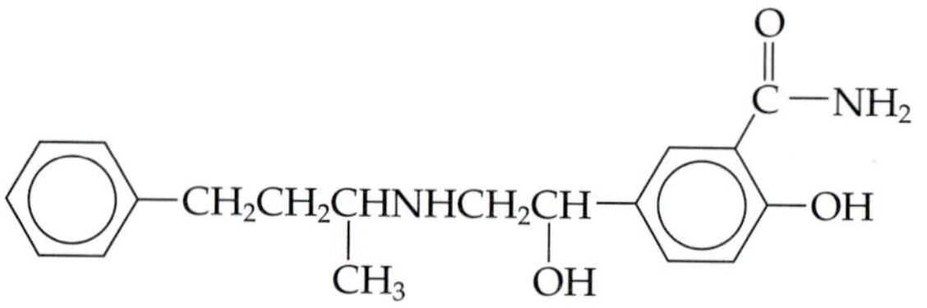

$C_6H_5-CH_2CH_2CH(CH_3)NHCH_2CH(OH)-C_6H_3(C(=O)-NH_2)-OH$

Toxicities

17. When administered intravenously to rats, procaine and cocaine have LD_{50} values of 50 mg/kg and 17.5 mg/kg, respectively. Which drug is more toxic?
18. If the minimum lethal dose (MLD) of amphetamine is 5 mg per kilogram of body weight, what would be the MLD for a 70-kg person? Can toxicity studies on animals always be extrapolated to humans?

Additional Problems

19. Cocaine is usually used in the form of the salt cocaine hydrochloride and sniffed up the nose, where it is readily absorbed through the watery mucous membranes. Some prefer to take their cocaine by smoking it (mixed with tobacco, for example). Before smoking, the cocaine hydrochloride must be converted back to the free base (that is, to the molecular form). Explain the choice of dosage form for each route of administration.
20. Haloperidol (Figure C.8) is one of the most widely prescribed antipsychotic drugs. What five functional groups are present in the molecule?

CHAPTER 18

Stereoisomerism

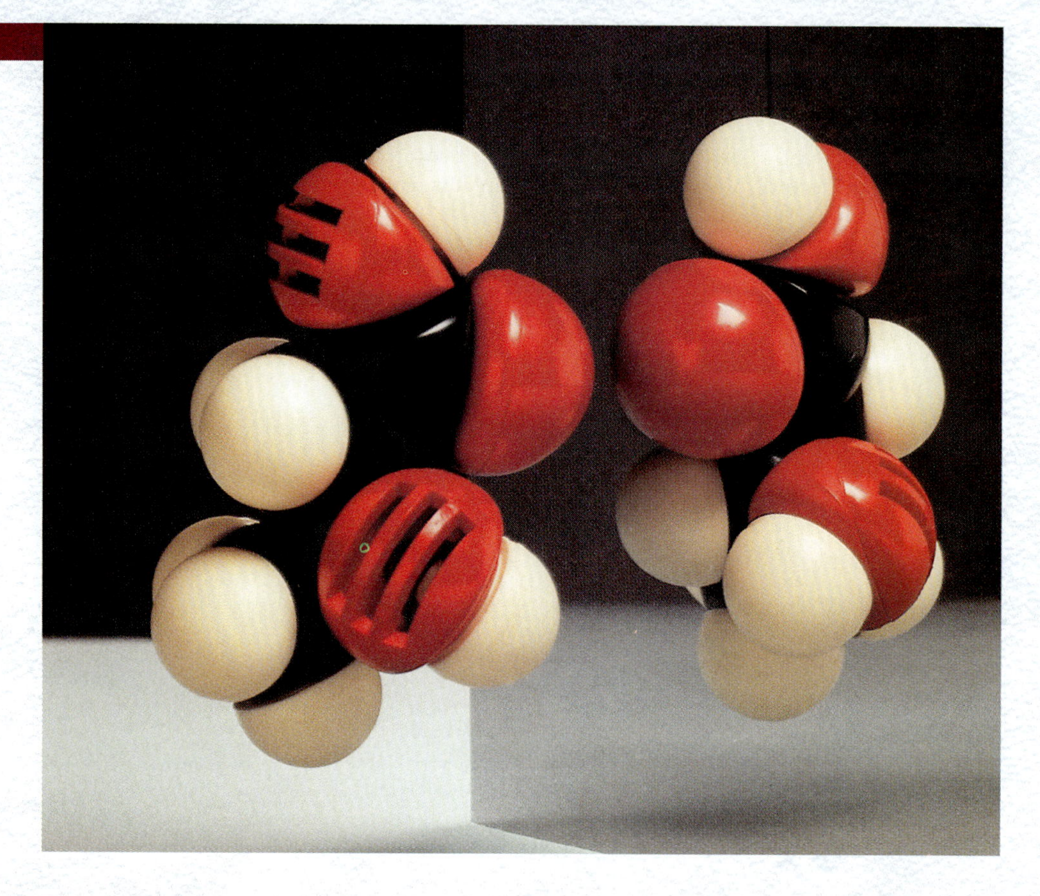

A space-filling model and its mirror image. Three-dimensional shapes of molecules such as these are vital to our understanding of life processes.

Learning Objectives/Study Questions

1. What are the differences between structural isomers and stereoisomers? Between enantiomers and diastereomers?
2. What is polarized light and how is it used to characterize optically active compounds?
3. What is a chiral center?
4. How can you determine the number of possible stereoisomers for a compound?
5. Why is an understanding of stereoisomers important in the study of biochemistry?

Isomerism is the phenomenon whereby two or more *different* compounds are represented by *identical* molecular formulas. There are two main categories of isomers: structural isomers and stereoisomers.

As the term implies, **structural isomers** are compounds that have different structural formulas. Two types of structural isomers have been mentioned—positional isomers and functional-group isomers. *Positional isomers* result from the presence of an atom or a group of atoms at different positions on the carbon chain. Examples of such isomers were discussed in the sections on alkanes, alkyl halides, and alcohols (Table 18.1). Two molecules that have the same molecular formula but contain different functional groups are *functional-group isomers.* We provide examples of functional-group isomers of alcohols and ethers, aldehydes and ketones, and acids and esters in Table 18.2.

Some chemists refer to structural isomers as constitutional isomers.

Table 18.1 Examples of Positional Isomers

Alkanes	
$CH_3CHCH_2CH_2CH_2CH_2CH_3$ $\quad |$ $\quad CH_3$	$CH_3CH_2CHCH_2CH_2CH_2CH_3$ $\quad |$ $\quad CH_3$
2-Methylheptane	3-Methylheptane
Alkyl Halides	
$CH_3CHCH_2CH_2CH_3$ $\quad |$ $\quad Cl$	$CH_3CH_2CHCH_2CH_3$ $\quad |$ $\quad Cl$
2-Chloropentane	3-Chloropentane
Alcohols	
$CH_3CH_2CH_2CH_2OH$	$CH_3CH_2CHCH_3$ $\quad |$ $\quad OH$
1-Butanol	2-Butanol

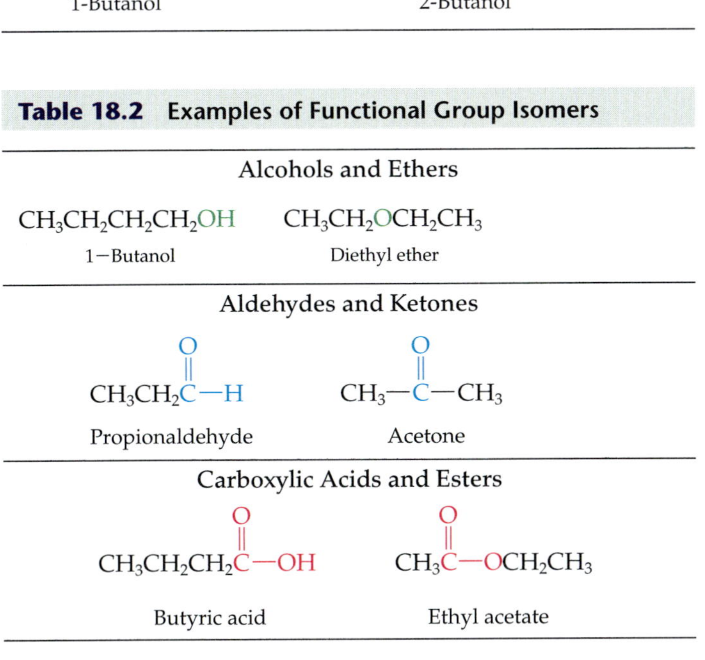

Table 18.2 Examples of Functional Group Isomers

Alcohols and Ethers	
$CH_3CH_2CH_2CH_2OH$	$CH_3CH_2OCH_2CH_3$
1–Butanol	Diethyl ether
Aldehydes and Ketones	
$CH_3CH_2\overset{\overset{O}{\parallel}}{C}—H$	$CH_3—\overset{\overset{O}{\parallel}}{C}—CH_3$
Propionaldehyde	Acetone
Carboxylic Acids and Esters	
$CH_3CH_2CH_2\overset{\overset{O}{\parallel}}{C}—OH$	$CH_3\overset{\overset{O}{\parallel}}{C}—OCH_2CH_3$
Butyric acid	Ethyl acetate

The second major category of isomerism is *stereoisomerism*, or space isomerism. Unlike structural isomers, stereoisomers have the identical order of atoms and identical functional groups. They differ only with respect to the spatial arrangement of atoms or groups of atoms within the molecule. Therefore, **stereoisomers** are isomers with a single structural formula that differ in the arrangement of atoms in three-dimensional space. Much of what we know about stereoisomers comes from their effect on plane-polarized light. Let's take a look at the nature of this light before we examine the molecules that act upon it.

18.1 Polarized Light and Optical Activity

Let's begin by establishing a convention for drawing waves. A wave is something that goes up and down or increases and decreases or varies regularly in some such way. Here's a wave coming toward you while moving to the right and left across the page.

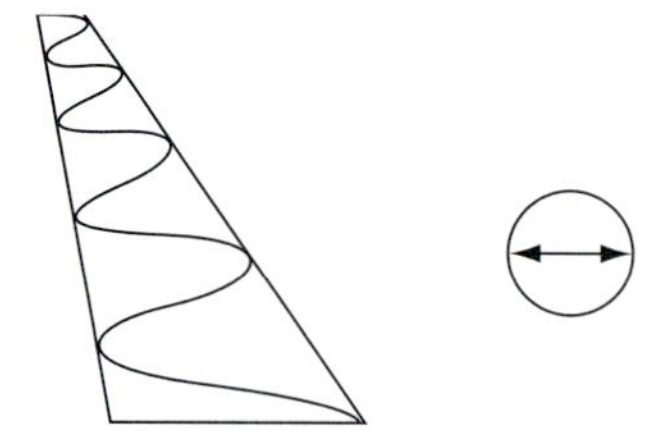

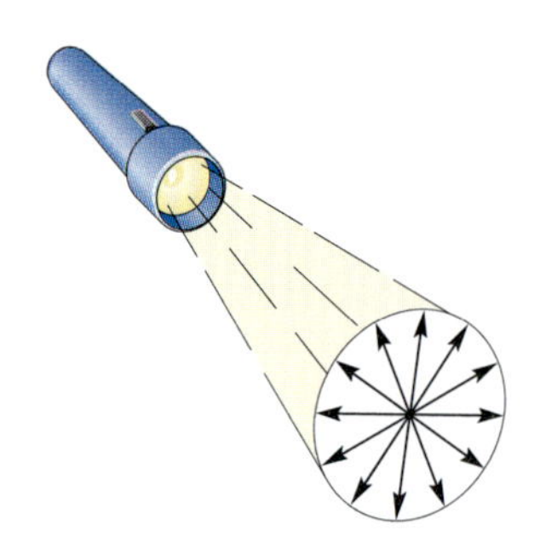

▲ **Figure 18.1**
The beam of light in this illustration is not polarized.

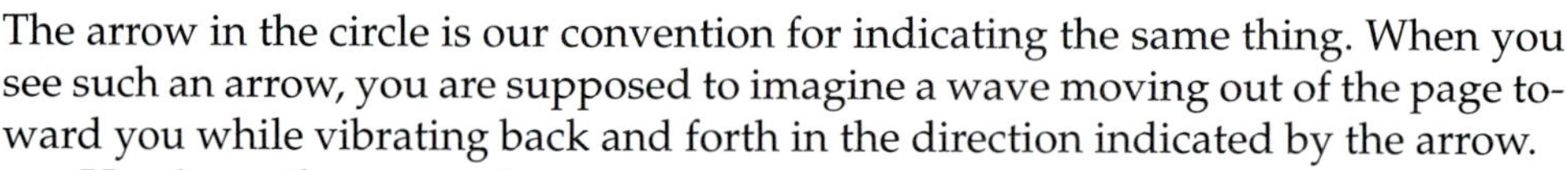

The arrow in the circle is our convention for indicating the same thing. When you see such an arrow, you are supposed to imagine a wave moving out of the page toward you while vibrating back and forth in the direction indicated by the arrow.

Here's another example.

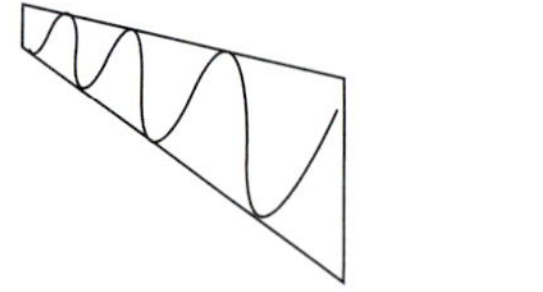

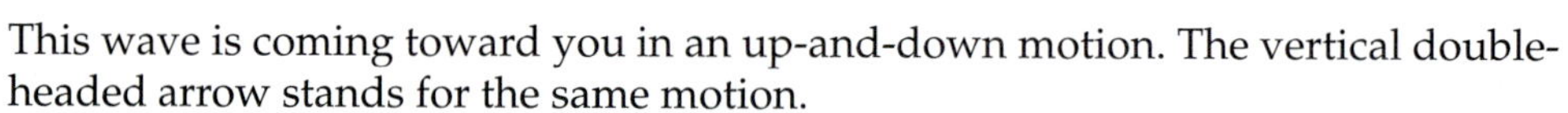

This wave is coming toward you in an up-and-down motion. The vertical double-headed arrow stands for the same motion.

Ordinary light may be described as a wave; it has characteristics associated with wavelike motion. A beam of ordinary light can be pictured as a bundle of waves, some of which move up and down, some sideways, and others at all conceivable angles (Figure 18.1). While such light is quite ordinary, we can also describe it as *nonpolarized* light, in contrast to polarized light, the subject of this section.

The waves of **polarized light** vibrate in a single plane. Both of the beams of light shown in Figure 18.2 are polarized. The two polarized beams differ only in the angle of the plane of polarization (represented in our drawings by the orientation of the double-headed arrows).

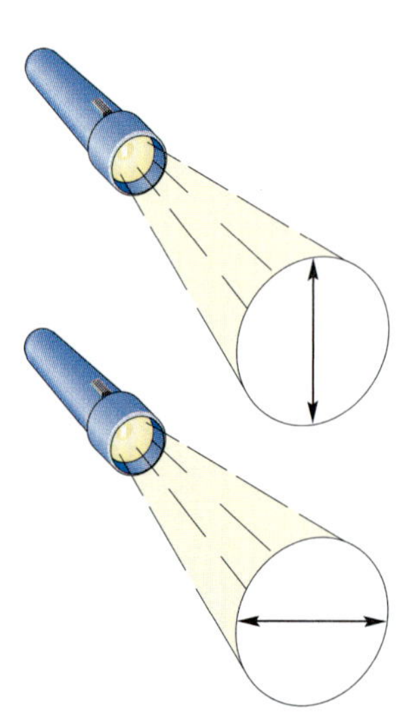

▲ **Figure 18.2**
Both of the beams of light in this illustration are polarized. The planes of polarization differ.

Sunlight, in general, is not polarized, nor is the light from an ordinary light bulb, nor the beam of light from an ordinary flashlight. One way to polarize light is to pass ordinary light through Polaroid sheets, such as those used for the lenses of some sunglasses. These lenses are made by carefully orienting organic compounds in plastic to produce a material that permits only light vibrating in a single plane to pass through (Figure 18.3). To the eye, polarized light doesn't "look" any different from nonpolarized light. We can detect polarized light, however, by using a second sheet of polarizing material (Figure 18.4).

Certain substances act on polarized light by rotating the plane of vibration. Such substances are said to be **optically active.** The extent of optical activity is measured

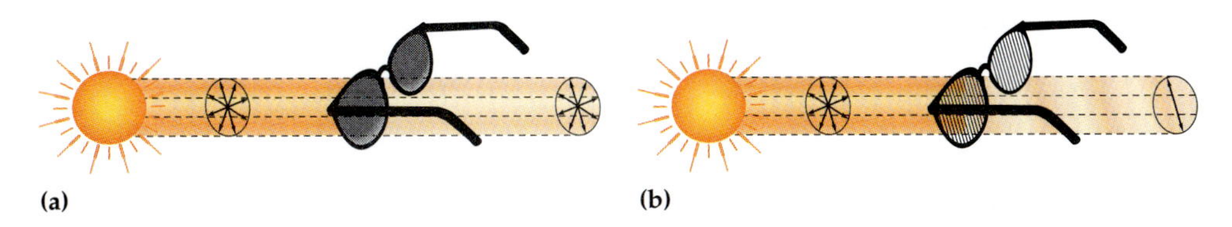

(a) (b)

▲ **Figure 18.3**
Ordinary sunglasses (a) dim light by preventing some of it from passing through the lenses. They do not discriminate among light waves vibrating at different angles; rather they filter all the waves to some extent. Light reaching the eyes through these glasses is nonpolarized. Sunglasses with Polaroid lenses (b) selectively pass light waves vibrating in a single plane. Light waves vibrating in other planes do not pass through. Light reaching the eyes through Polaroid sunglasses is plane-polarized.

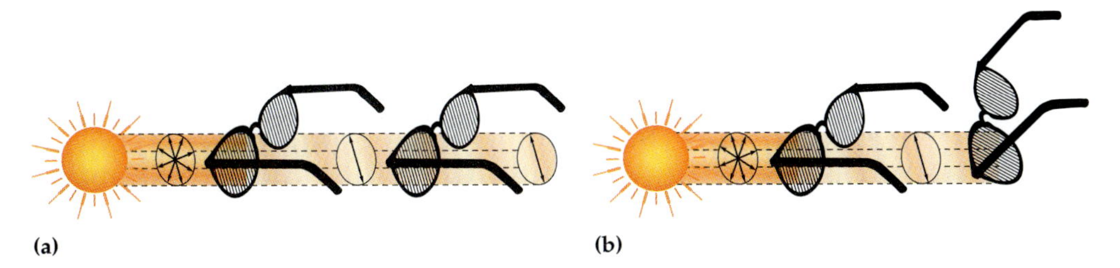

(a) (b)

▲ **Figure 18.4**
(a) The light that passes through the first Polaroid lens is polarized. The second pair of glasses is oriented like the first; its Polaroid lens therefore passes the polarized light. (b) The second pair of glasses is oriented 90° to the first. The plane of polarization of the light that made it through the first Polaroid lens is oriented incorrectly for the second lens. No light gets through the second lens.

in an instrument called a **polarimeter** (Figure 18.5). A polarimeter has two polarizing lenses, a *polarizer* and an *analyzer.* With the sample tube empty or containing distilled water, maximum light reaches the observer's eye when the polarizer and the analyzer are aligned so that both pass light vibrating in the same plane. When an optically active substance[1] is placed in the sample tube, that substance rotates the plane of polarization of the light passing through it. The polarized light emerging from the sample tube is vibrating in a different direction than when it entered the tube. To see the maximum amount of light when the sample is in place, the observer must rotate the analyzing lens to accommodate this change in the plane of polarization. The angle of rotation, indicated by a pointer on the analyzing lens, corresponds to the change in the plane of polarization caused by the sample.

The size of the rotation angle depends not only on the structure of the optically active material but also on the length of the sample tube, the concentration of the solution, and even the color of light used and its temperature. However, scientists have agreed to certain standard conditions for reporting the angle of rotation. When a rotation is calculated and reported with these conditions taken into account, the value is referred to as the **specific rotation** $[\alpha]$ and is a physical constant as characteristic of the material as are its melting point, boiling point, density, and solubility. For example, the specific rotation of an aqueous solution of sucrose is $+66.5°$.

Some optically active substances rotate the plane of polarized light to the right (clockwise from the observer's point of view). These compounds are said to be **dextrorotatory** (Latin *dexter,* right); substances that rotate light to the left (counterclockwise) are **levorotatory** (Latin *laevus,* left) (Figure 18.6). To denote the direction of rotation, a positive sign (+) is given to dextrorotatory substances and a negative sign (−) to levorotatory substances. Sucrose is said to be dextrorotatory because it

[1]The sample may be a pure gas, a pure liquid, a pure crystalline solid, or a solute dissolved in an appropriate solvent.

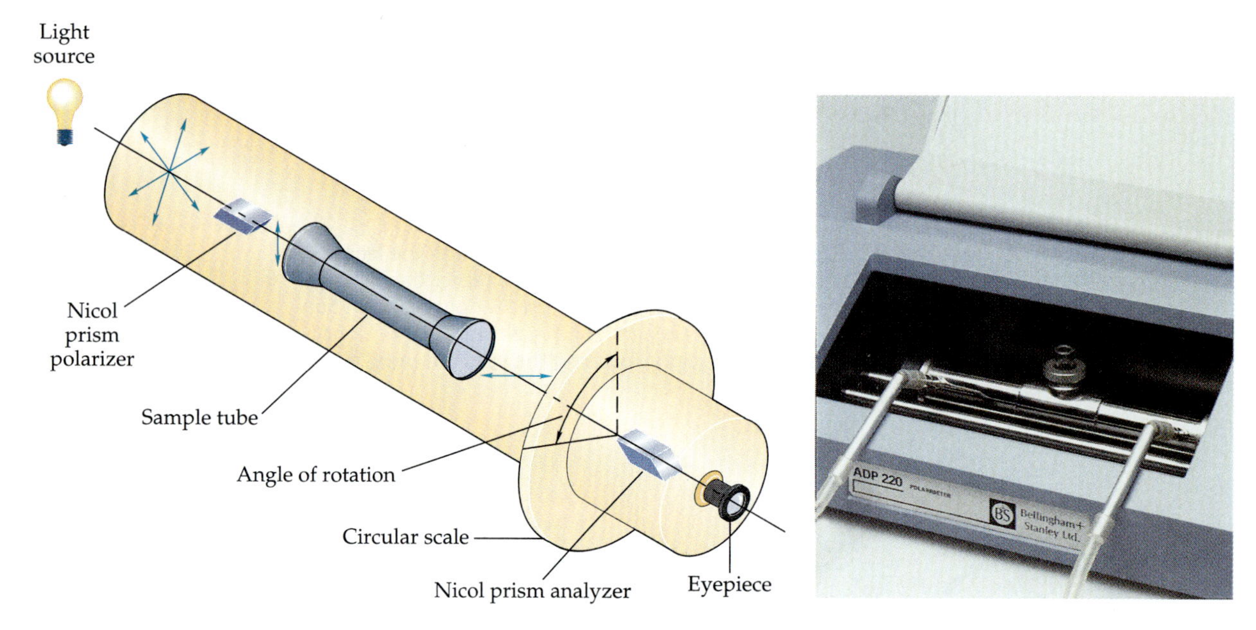

▲ **Figure 18.5**
A polarimeter.

► **Figure 18.6**
Direction of rotation of analyzer.

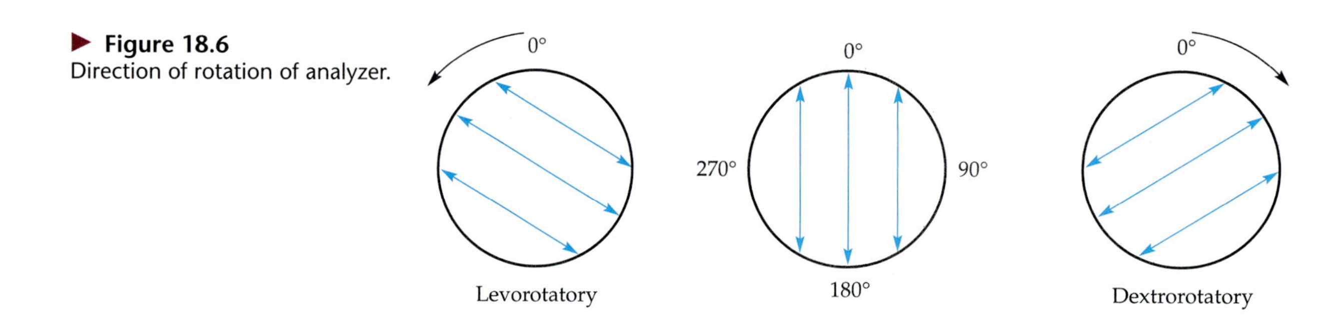

rotates plane-polarized light 66.5° in the clockwise direction, and it is designated as (+)-sucrose.

Review Questions

18.1 What is polarized light?

18.2 What is meant by (**a**) an optically active substance? (**b**) a dextrorotatory substance? (**c**) a levorotatory substance?

18.3 How does a polarimeter work? What is meant by the specific rotation of a substance?

18.2 Chiral Centers

So far we have described optically active compounds and defined certain terms for use in dealing with them, but we have not yet answered certain fundamental questions:

1. Why are some compounds optically active?
2. What are the spatial arrangements of the atoms in these compounds?
3. How do these compounds differ from one another and from compounds that are not optically active?

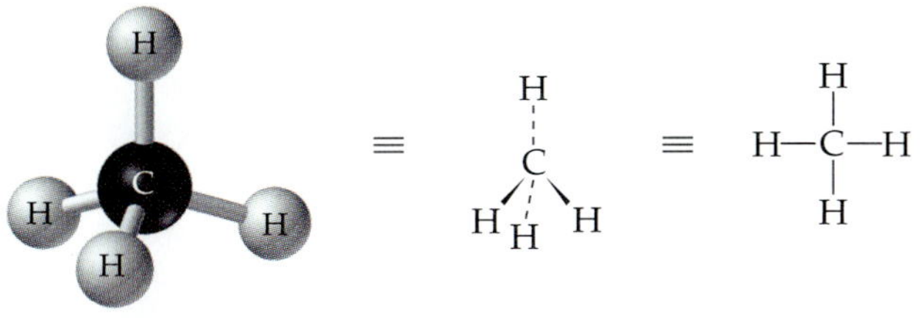

(a)

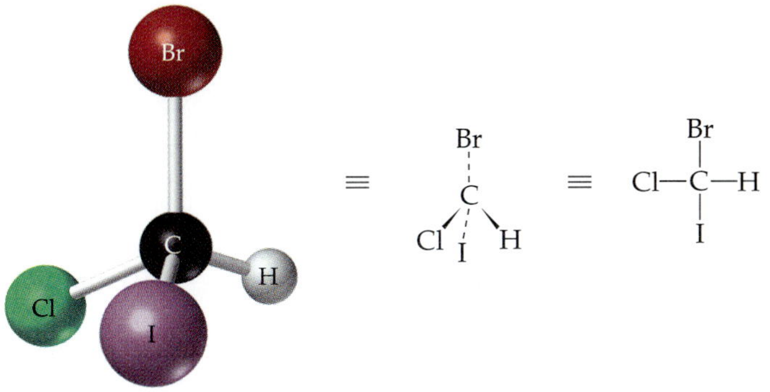

(b)

◀ **Figure 18.7**
Tetrahedral configurations of (a) methane and (b) a trisubstituted derivative of methane.

Many chemists were involved in the discovery and explanation of optically active isomers, including Louis Pasteur (1822–1895), Johannes Wislicenus (1835–1902), Jacobus van't Hoff (1852–1911), and Joséph-Achille Le Bel (1847–1930). Based on the experimental work of Pasteur and Wislicenus, van't Hoff and Le Bel postulated the *tetrahedral carbon atom,* the key to the explanation of optical activity in organic compounds.

Recall that the methane molecule is tetrahedral. If any or all hydrogen atoms are replaced by other atoms or groups of atoms, the tetrahedral arrangement about the central carbon atom is retained (Figure 18.7). If additional carbon atoms are added and if they all are joined by single bonds, the configuration about each of the carbon atoms is tetrahedral.

Consider the two generalized molecules shown in Figure 18.8. Compound I contains two identical substituents and two different substituents bonded to the central carbon atom, and compound II contains four dissimilar substituents. Compound I is a symmetrical molecule; that is, a plane of symmetry passes through b, d, and the central carbon atom. Compound II is a nonsymmetrical molecule; you can't draw a plane of symmetry anywhere through the molecule.

To understand better why compound I is symmetrical and compound II is nonsymmetrical, we can place both compounds in front of a mirror and attempt to impose the mirror images upon the original molecules (Figure 18.9). A symmetrical molecule is *superimposable* on (is identical to) its mirror image, whereas a nonsymmetrical molecule cannot be superimposed upon its mirror image. In doing the superimposing, you can twist and turn the bonds but none of them can be broken.[2]

[2]These molecules, like any others, can be turned upside down or spun about or tipped forward or backward, just as you can stand on your feet or on your head or lie on your back, and so forth. If you had a mole on your right arm, however, no matter what position you assumed, the mole would still be on your right arm. Your mirror image would always have a left-arm mole. No amount of spinning or turning would cause the mole to change from your right to your left arm. The molecules pictured in Figure 18.9 are distinguished in the same way. If, for these molecules, we call atom b the *head* and the side to which atom e is attached the *front,* then one compound always has atom d on its right side and the other always has d on its left side.

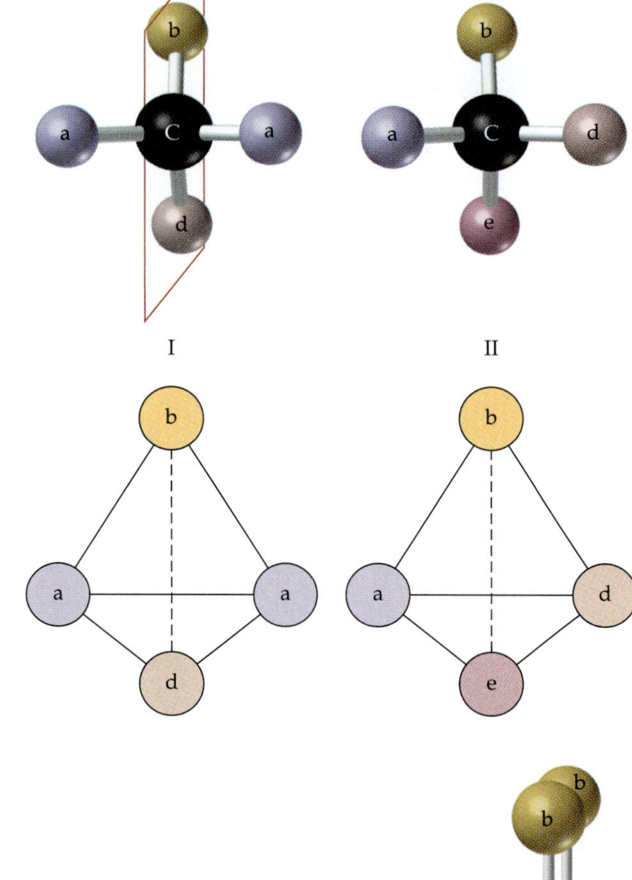

▶ **Figure 18.8**
Compound I has two similar and two dissimilar substituents. Compound II has four dissimilar substituents.

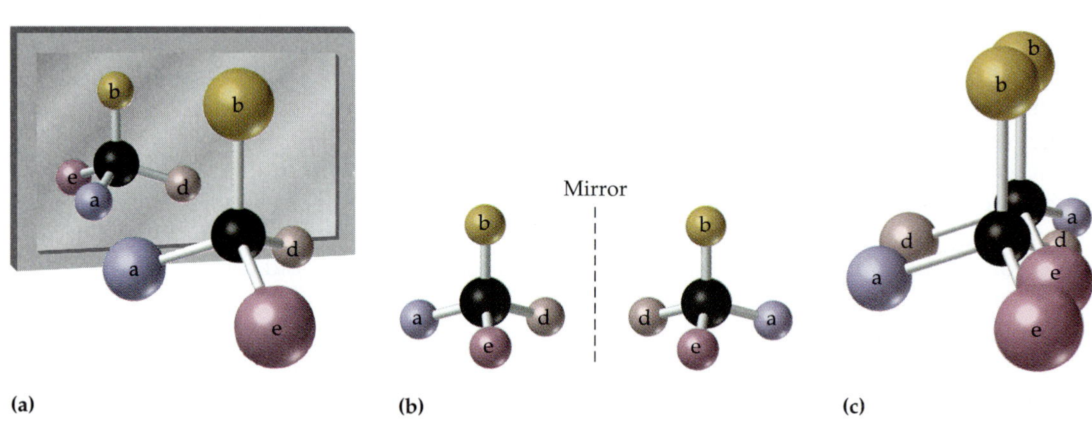

▶ **Figure 18.9**
(a) and (b) Mirror images of a nonsymmetrical molecule. (c) A nonsymmetrical molecule cannot be superimposed on its mirror image.

A useful analogy can be drawn between *nonsuperimposable* (nonmatching) mirror-image compounds and your right and left hands (Figure 18.10a). Regardless of how you twist and turn them, you cannot superimpose your right hand upon your left, or vice versa. This is because, although your right and left hands are nearly perfect mirror images of each other, they are *not* identical. This difference becomes immediately apparent if you try to place your right hand in a left-hand glove. The general property of "handedness" is called **chirality** (Greek *cheir,* hand). An object that is not superimposable upon its mirror image is said to be *chiral.* An object that is superimposable upon (that is, identical to) its mirror image is **achiral** (Figure 18.10b).

The word *chiral* (pronounced KYE-ral) is now widely used; it has largely displaced the earlier terms dissymmetric and asymmetric.

A **chiral center** is an atom in a molecule or ion that has *four different groups* attached. A carbon atom with four different groups attached is a *chiral carbon atom.* A chiral molecule is *not* identical with (that is, *not* superimposable on) its mirror image. A molecule containing only one chiral center is always chiral. Some molecules that contain more than one chiral center may be achiral, as we shall see in Section 18.3.

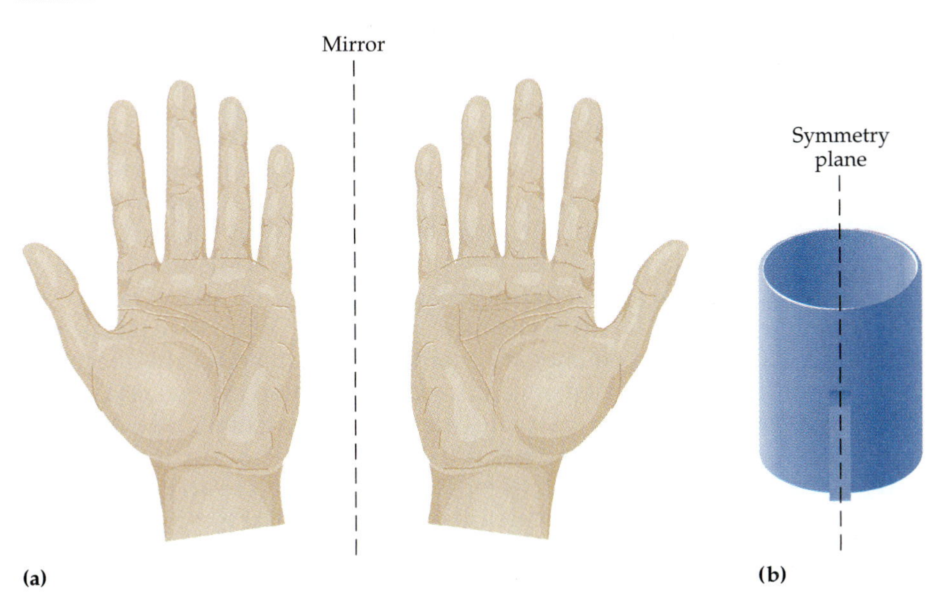

Figure 18.10
(a) The left and right hands are nonsuperimposable mirror images; they are chiral. (b) This coffee mug is achiral; a plane of symmetry passes through it. The cup is superimposable on its mirror image.

Molecules that are nonsuperimposable (nonidentical) mirror images of each other are called **enantiomers** (Greek *enantios*, opposite). Enantiomers have identical physical properties except one: they rotate plane-polarized light the same number of degrees—*but in opposite directions.* Enantiomers also have identical chemical properties except when they react with other chiral substances.

Let's look at some real molecules that have chiral carbons. We can represent lactic acid as follows.

$$CH_3\underset{\displaystyle |\atop \displaystyle OH}{C}HCOOH$$

Examination of the structure reveals that the second carbon atom is a chiral carbon atom because it has four different groups attached: a hydrogen atom, a carboxyl group, a hydroxyl group, and a methyl group. There should be *two* lactic acids. Are there? Yes! (Why, it's enough to give you faith in chemical theory!) One lactic acid is found in sour milk. It is levorotatory and can be designated as (−)-lactic acid. Another lactic acid is found in muscle tissue, particularly after exercise. It is dextrorotatory and is called (+)-lactic acid. How are the two related? They are enantiomers—one is the mirror image of the other. The following perspective drawings are an attempt to represent the enantiomers in three dimensions.

H; H_3C ⟋ ◯ ⟍ COOH; OH	⋮	H; HOOC ⟋ ◯ ⟍ CH_3; OH
	Mirror	

Enantiomers can also be drawn using "flat" formulas:

$$H_3C-\overset{\displaystyle H}{\underset{\displaystyle OH}{C}}-COOH \qquad HOOC-\overset{\displaystyle H}{\underset{\displaystyle HO}{C}}-CH_3$$

These flat drawings require more of *you.* You have to visualize that the horizontal bonds project out of the plane of the page toward you and the vertical bonds project behind the page away from you. Because they are easier to draw, we shall use these "flat" projection formulas (often called **Fischer projections**) to represent stereoisomers.

In what ways do the actual lactic acid enantiomers differ from each other? In many respects, they seem more alike than different. All of their physical properties are identical save one, the direction in which they rotate the plane of polarized light. One has a specific rotation of +2.6°; the other, −2.6°. Only the sign is different. Simple chemical properties are also the same. Both form acidic solutions. Both neutralize bases. Both form esters. Indeed, when reacting with *achiral* molecules, the two enantiomers exhibit identical chemical properties. Common reagents such as water, hydroxide ion, and ethanol are achiral; they contain no chiral centers. However, when the lactic acid isomers react with other chiral molecules, they behave differently. They react at different rates and to different extents, and the products formed have different properties. (Wislicenus worked out much of the chemistry of the lactic acids in 1873. His work underlay that of van't Hoff and Le Bel, who independently proposed the tetrahedral nature of the carbon atom the next year.)

In living cells, reactions are controlled by enzymes, and enzymes are chiral. Enantiomers generally behave quite differently in living cells. Enantiomers may have different tastes and smells. One may be an effective drug and the other worthless. One may be essential to health and the other toxic. Quite literally, the difference may be a matter of life or death.

When lactic acid is made from pyruvic acid in the laboratory, it shows *no* optical activity.

$$CH_3-\overset{\overset{\displaystyle O}{\|}}{C}-COOH \xrightarrow[Ni]{H_2} CH_3-\underset{\underset{\displaystyle OH}{|}}{\overset{\overset{\displaystyle H}{|}}{C}}-COOH \qquad HOOC-\underset{\underset{\displaystyle HO}{|}}{\overset{\overset{\displaystyle H}{|}}{C}}-CH_3$$

Pyruvic acid

Racemic lactic acid
[50%(+)-lactic acid and 50%(−)-lactic acid]

"*Mirror, mirror on the wall, who is the enantiomerest of them all?*"

▲ Enantiomers are nonsuperimposable mirror images.

How can this be? The lactic acid has a chiral center, so why isn't it optically active? The answer: In syntheses of this sort, the (+) and (−) forms are formed in exactly equal amounts. Such a mixture of enantiomers is called a **racemic mixture.** It shows no optical activity because it contains equal amounts of molecules with equal but opposite rotatory power. Everything cancels out. Racemic lactic acid is designated (±). A racemic mixture may exhibit physical properties different from those of the pure enantiomers (Table 18.3).

In summary, for compounds that contain only one chiral carbon atom, there are always two isomers that are nonsuperimposable mirror images (enantiomers), a dextrorotatory form and a levorotatory form. A convenient method of drawing the enantiomer of an optically active compound is to maintain the positions of two of the substituents (usually the larger ones) about the chiral center and invert the positions of the other two.

Table 18.3 Properties of Lactic Acids

Form	Melting Point (°C)	Specific Rotation	pK_a
(+)	53	+2.6°	3.8
(−)	53	−2.6°	3.8
(±)	16.8	0	3.8

Example 18.1

2-Methyl-1-butanol exists in two optically active forms. The specific rotation of one is +5.756°, and for the other it is −5.756°. Draw structural formulas for these enantiomers.

Solution

Write the structural formula for one enantiomer and identify the chiral center. (To allow for valid comparisons, the convention is to draw the carbon chain vertically.) Then generate the other enantiomer by interchanging the positions of two of the groups about the chiral center while maintaining the positions of the other two groups.

$$\begin{array}{ccc} & CH_3 & \\ & | & \\ & CH_2 & \\ & | & \\ H- & C & -CH_3 \\ & | & \\ & CH_2OH & \end{array} \qquad \begin{array}{ccc} & CH_3 & \\ & | & \\ & CH_2 & \\ & | & \\ CH_3- & C & -H \\ & | & \\ & CH_2OH & \end{array}$$

It is not possible to tell which isomer is dextrorotatory and which is levorotatory merely by inspection of the structural formulas. The distinction can be made only by measuring the optical rotation of each compound in a polarimeter.

Practice Exercise

Draw structural formulas for the enantiomers of 2-butanol.

✓ Review Questions

18.4 What is meant by a (**a**) chiral center? (**b**) chiral carbon atom?

18.5 Does a molecule and its mirror image always make up a pair of enantiomers? Explain.

18.6 In what ways do enantiomers resemble each other? How do they differ from each other?

18.7 What is a racemic mixture?

18.8 Compare (+)-lactic acid and (−)-lactic acid with respect to
- **a.** boiling point
- **b.** melting point
- **c.** specific rotation
- **d.** solubility in H_2O
- **e.** reaction with ethanol
- **f.** reaction with (+)-*sec*-butylamine

18.9 (−)-Menthol melts at 43 °C, boils at 212 °C, and has a density of 0.890 g/cm^3 and a specific rotation of −50°. List the corresponding properties of (+)-menthol.

18.10 Would (+)-nicotine and (−)-nicotine react the same way with (**a**) HCl? (**b**) (+)-lactic acid? (**c**) (−)-tartaric acid?

18.3 Multiple Chiral Centers

A molecule may have more than one chiral center. Indeed, simple carbohydrate molecules generally have several each. Giant molecules, such as starch, cellulose, and the proteins, may have several hundred or even several thousand chiral centers. Let us look first, however, at molecules with just two.

Consider 2,3-pentanediol.

$$\begin{array}{ccccc} CH_3- & CH & - & CH & -CH_2CH_3 \\ & | & & | & \\ & OH & & OH & \end{array}$$

There are *four ways* in which the groups can be arranged about the two chiral centers (Figure 18.11). Note that structures I and II are enantiomers; they are nonsuperimposable mirror images of one another. Compounds III and IV are another pair of enantiomers. What is the relationship, though, between structures II and III? They are stereoisomers because they differ only in their spatial arrangement; they are

▶ Figure 18.11
The four stereoisomeric 2,3-pentanediols. (a) Perspective drawings. (b) Flat projection formulas. Note that in Fischer projections of molecules with more than one chiral center, the carbon chain is arranged vertically.

(a)

I	II	III	IV
CH_3	CH_3	CH_3	CH_3
H, OH	HO, H	H, OH	HO, H
H, OH	HO, H	HO, H	H, OH
C_2H_5	C_2H_5	C_2H_5	C_2H_5

I and II: Enantiomers; III and IV: Enantiomers

(b)

I	II	III	IV
CH_3	CH_3	CH_3	CH_3
H—C—OH	HO—C—H	H—C—OH	HO—C—H
H—C—OH	HO—C—H	HO—C—H	H—C—OH
C_2H_5	C_2H_5	C_2H_5	C_2H_5

I and II: Enantiomers; III and IV: Enantiomers

not enantiomers, however, because they are not mirror images. Stereoisomers that are not enantiomers are called **diastereomers.** Note that II and IV are also diastereomers, as are I and III and I and IV. Diastereomers generally have *different* physical properties (such as boiling point and solubility, as well as specific rotation). Unlike enantiomers, they can be separated by distillation or fractional crystallization.

The first chemist to postulate the existence of multiple stereoisomeric forms was van't Hoff (first Nobel prize in chemistry, 1901). He formulated a statement (**van't Hoff's rule**) that makes it possible to predict the total number of possible stereoisomers for a molecule containing more than one chiral center. *The maximum number of different configurations is* 2^n, where n is the number of chiral carbon atoms. The rule is best illustrated with an example.

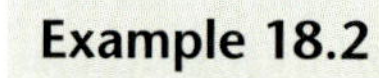

▲ Louis Pasteur (1822–1895), a French chemist who invented the process, now called pasteurization, of heating milk to kill bacteria and retard spoilage. Pasteur's work also led to the germ theory of disease and to immunization procedures. He also discovered the stereoisomerism associated with enantiomers.

Example 18.2

Draw all the possible stereoisomers of 2-methyl-1,3-butanediol.

Solution

First draw the structural formula and note the number of chiral carbon atoms. Using the formula 2^n, calculate the number of possible stereoisomers.

$$CH_3-\underset{\large OH}{\underset{|}{CH}}-\underset{\large CH_3}{\underset{|}{CH}}-CH_2OH$$

There are two chiral carbon atoms (red) and therefore $2^n = 2^2 = 4$ possible stereoisomers.

Because two configurations are possible for each chiral carbon atom, there are two sets of enantiomers:

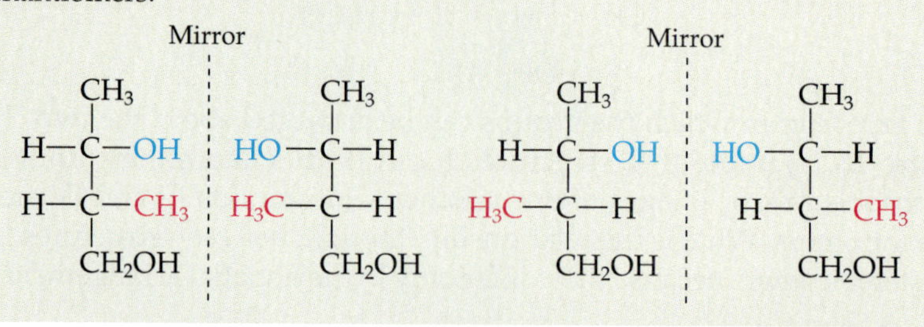

Practice Exercise

Draw all the possible stereoisomers of 2,3-dibromobutanal.

As we mentioned in Chapter 8, one of the problems in winemaking is that tartaric acid salts precipitate from the wine upon standing. While studying winemaking, Pasteur noticed that these salts form two types of crystals, one of which was the mirror image of the other. After laboriously separating the crystals, he found that they exhibited optical activity even when dissolved in water. He concluded that optical activity must therefore be a property of the individual molecules, and not just a property of the crystals.

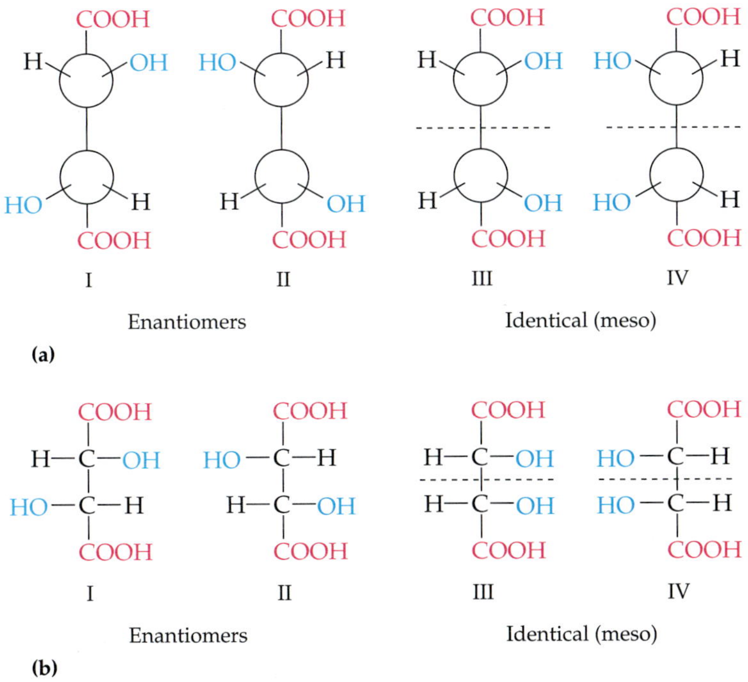

Let us consider one more set of compounds before we move on to the next section. These compounds, the tartaric acids, were involved in the earliest studies relating structure and optical activity. The investigator was Louis Pasteur, and the compounds he studied in 1848 included a new type of stereoisomer.

Tartaric acid, like 2,3-pentanediol, contains two chiral carbon atoms.

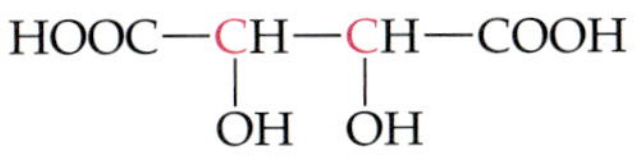

$$HOOC-\underset{OH}{\underset{|}{CH}}-\underset{OH}{\underset{|}{CH}}-COOH$$

There is, however, one notable difference between tartaric acid and 2,3-pentanediol. In tartaric acid, each chiral carbon is attached to the *same* four groups: COOH, OH, H, and CH(OH)COOH. In 2,3-pentanediol, one chiral carbon is attached to a methyl group, whereas the other is attached to an ethyl group. This difference is significant, as we shall see.

Writing out the perspective formulas of tartaric acid (Figure 18.12), we see a pair of enantiomers (I and II). The other apparent pair (III and IV), however, are not enantiomers; they are not even isomers. They are, in fact, molecules of the same compound. The structures can be superimposed by rotating one of them 180° in the plane of the page (Figure 18.13). It is important to stress once more that enantiomers are not simply mirror images of one another; instead, they are *nonsuperimposable* mirror images. Every molecule has a mirror image. An achiral molecule is superimposable upon (identical with) its mirror image; a chiral molecule is not. Structures III and IV in Figure 18.12 are mirror images, but they are superimposable and therefore identical. The corresponding structures for 2,3-pentanediol in Figure 18.11 are not superimposable because the CH_3 group and C_2H_5 group are different.

The *single* compound represented by structures III and IV in Figure 18.12 is termed a **meso compound.** It is a diastereomer of compound I and of compound II.

COOH, H, OH, HO, H, COOH — I; COOH, HO, H, H, OH, COOH — II; COOH, H, OH, H, OH, COOH — III; COOH, HO, H, HO, H, COOH — IV

Enantiomers (I, II) — Identical (meso) (III, IV)

(a)

I	II	III	IV
COOH	COOH	COOH	COOH
H—C—OH	HO—C—H	H—C—OH	HO—C—H
HO—C—H	H—C—OH	H—C—OH	HO—C—H
COOH	COOH	COOH	COOH

Enantiomers (I, II) — Identical (meso) (III, IV)

(b)

◀ **Figure 18.12**
The three stereoisomeric tartaric acids. (a) Perspective drawings. (b) Flat projection formulas. Note that the meso form has an internal plane of symmetry.

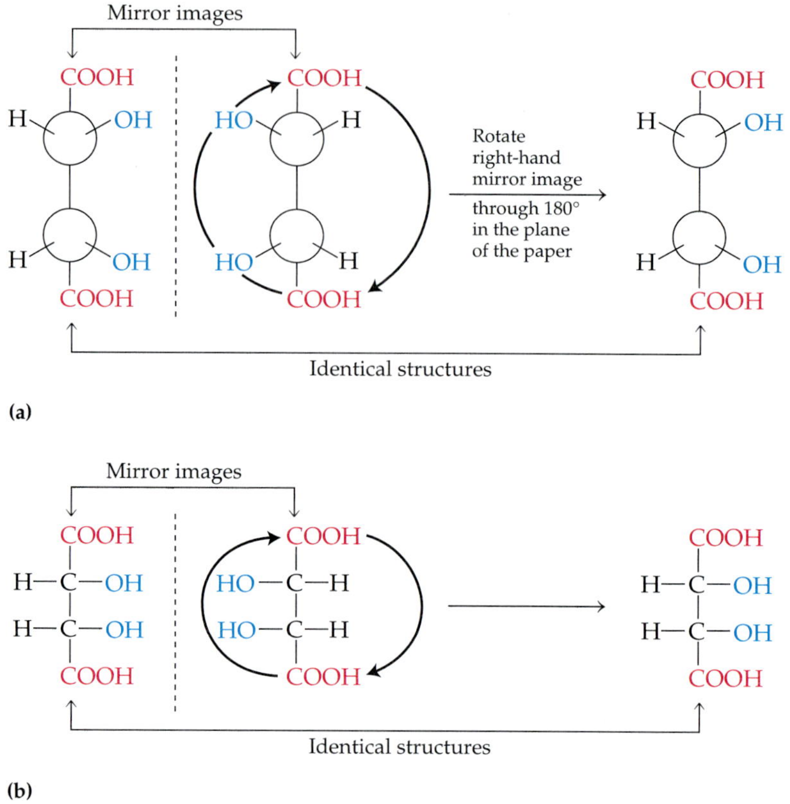

Figure 18.13 *meso*-Tartaric acid is superimposable on its mirror image. It exists *not* as a pair of enantiomers but only as a single compound. (a) Perspective drawings. (b) Flat projection formulas.

All meso compounds are optically inactive.[3] They contain at least two chiral centers but have an internal symmetry plane; in Figure 18.12 we have indicated this mirror plane in structures III and IV by a dashed line. The meso molecule as a whole is *not* chiral. Table 18.4 lists all the forms of tartaric acid.

Let us summarize here the various conditions that result in a lack of optical activity.

- Achiral compounds, for example ethanol (CH_3CH_2OH), have no chiral center and therefore are not optically active.
- A meso compound contains chiral centers, but it is not chiral because it also has an internal mirror plane. It, too, is optically inactive.
- A racemic mixture contains chiral molecules but is not optically active because there are equimolar amounts of dextrorotatory molecules and levorotatory ones; the effects are canceled out.

---

[3] A meso compound is superimposable on its mirror image even though it contains chiral carbon atoms. For this reason, it is incorrect to say that all molecules that contain chiral carbon atoms are chiral and thus optically active. Another example of a meso compound is ribitol, whose structural formula is

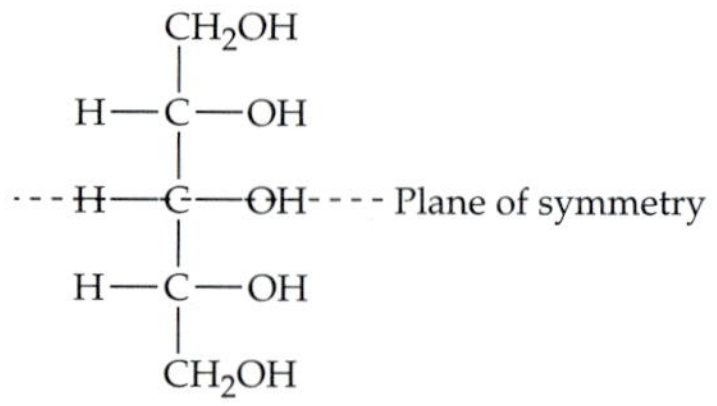

CH_2OH
H—C—OH
- - - H—C—OH - - - Plane of symmetry
H—C—OH
CH_2OH

Structures that have meso forms have fewer stereoisomers than predicted by van't Hoff's rule. Tartaric acid, for example, has two chiral carbon atoms but exists in only three stereoisomeric forms, including one that is meso.

Table 18.4 Properties of the Tartaric Acids

Form	Melting Point (°C)	Specific Rotation	pK_a	Solubility (g/100 g H_2O)
(+)	168–170	+12.0°	2.93	133
(−)	168–170	−12.0°	2.93	133
(±)	206	0	2.96	21
Meso	140	0	3.11	125

Review Questions

18.11 How does a meso compound differ from a racemic mixture?

18.12 What are diastereomers? Can a meso compound be a diastereomer? An enantiomer?

18.4 Geometric Isomerism (Cis–Trans Isomerism)

We mentioned in Chapter 13 that cyclic molecules and those containing a carbon–carbon double bond have certain restrictions placed upon them. These restrictions lead to another kind of stereoisomerism.

In Chapter 13, we identified three isomers of butene, C_4H_8:

$$\underset{\substack{\text{1-Butene}\\ \text{I}}}{CH_3CH_2CH{=}CH_2} \qquad \underset{\substack{\text{2-Butene}\\ \text{II}}}{CH_3CH{=}CHCH_3} \qquad \underset{\substack{\text{2-Methylpropene}\\ \text{III}}}{CH_3{-}\overset{\overset{\displaystyle CH_3}{|}}{C}{=}CH_2}$$

Experimental evidence has shown, however, that there are not three but rather four different butene molecules, all having distinctly different physical properties. The fourth isomer also has structure II. Although these two isomeric 2-butenes are structural isomers of I and III, they are not structural isomers of one another. Our knowledge of stereoisomerism leads us to the conjecture that these two molecules might differ in the spatial configuration of their atoms. Because they have different physical properties, however, they cannot be enantiomers. Therefore a different type of configurational explanation must be sought to account for this phenomenon. The explanation is based upon the geometric arrangement of the carbon–carbon double bond.

Recall that the two carbon atoms of a C=C double bond and the four atoms attached to them are all in the same plane and that rotation around the double bond is prevented. This is in contrast to the free rotation of carbon atoms linked to one another by single bonds. Ball-and-stick models indicate two ways to arrange the atoms of 2-butene that are in keeping with its structural formula (Figure 18.14). These three-dimensional models are more simply represented as

CH_3		CH_3	CH_3		H
	C=C			C=C	
H		H	H		CH_3
	IIa			IIb	
	cis-2-Butene mp −139 °C bp 4 °C			*trans*-2-Butene mp −106 °C bp 1 °C	

In structure IIa, the two methyl groups lie on the same side of the molecule, and this compound is the **cis isomer** (Latin *cis*, on this side). The methyl groups of structure IIb are on opposite sides of the molecule; it is the **trans isomer** (Latin *trans*, across). Because of the restriction on free rotation about the double bond, the structures are

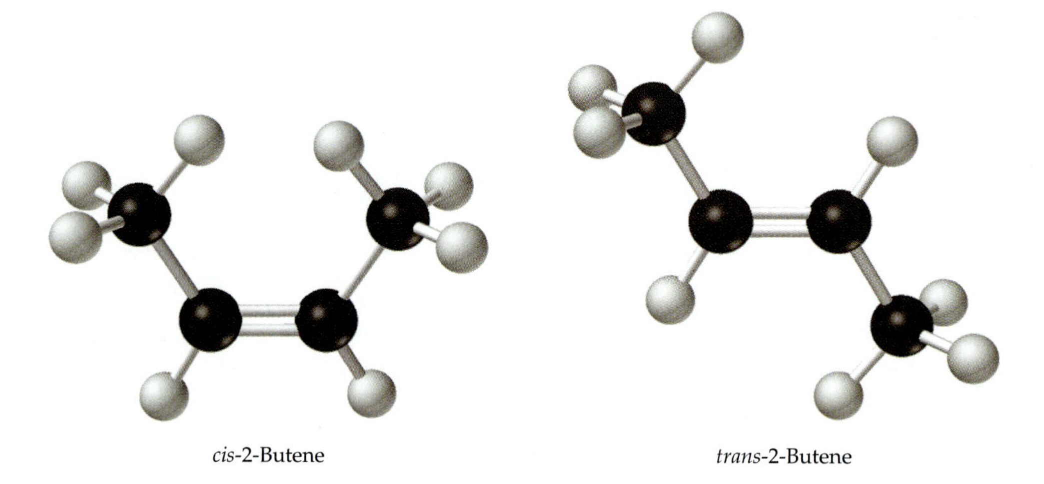

▶ Figure 18.14
cis-2-Butene and *trans*-2-butene.

clearly nonsuperimposable and hence not identical. *cis*-2-Butene and *trans*-2-butene are geometric isomers of each other.

Geometric isomers are compounds that have different configurations because of the presence of a rigid structure in the molecule. They are diastereomers—stereoisomers that are not enantiomers. For alkenes, there are *only two* geometric isomers (cis and trans) that correspond to each double bond.

We can draw two *seemingly* different propenes:

$$\underset{\text{IV}}{\begin{array}{ccc} H & & CH_3 \\ & C{=}C & \\ H & & H \end{array}} \qquad \underset{\text{V}}{\begin{array}{ccc} H & & H \\ & C{=}C & \\ H & & CH_3 \end{array}}$$

However, these two structures are not really different from each other. If you could pick either molecule up from the page and flip it over top to bottom, you would see that the two formulas are identical:

$$\underset{\text{IV}}{\begin{array}{ccc} H & & CH_3 \\ & C{=}C & \\ H & & H \end{array}} \qquad \underset{\text{V (flipped over)}}{\begin{array}{ccc} H & & CH_3 \\ & C{=}C & \\ H & & H \end{array}}$$

The same thing *cannot* be done with cis and trans isomers. If we start with drawings of the two 2-butenes (VI and VII),

$$\underset{\substack{\text{VI} \\ \textit{cis}\text{-2-Butene}}}{\begin{array}{ccc} CH_3 & & CH_3 \\ & C{=}C & \\ H & & H \end{array}} \qquad \underset{\substack{\text{VII} \\ \textit{trans}\text{-2-Butene}}}{\begin{array}{ccc} CH_3 & & H \\ & C{=}C & \\ H & & CH_3 \end{array}}$$

and then flip the trans isomer top to bottom, the resulting structure is still clearly different from the cis isomer:

$$\underset{\text{VI}}{\begin{array}{ccc} CH_3 & & CH_3 \\ & C{=}C & \\ H & & H \end{array}} \qquad \underset{\text{VII (flipped over)}}{\begin{array}{ccc} H & & CH_3 \\ & C{=}C & \\ CH_3 & & H \end{array}}$$

As propene proves, the mere presence of a double bond is not the only criterion (nor is it a necessary one) for geometric isomerism. There are two requirements for such isomerism:

1. Rotation must be restricted in the molecule.
2. There must be two nonidentical groups on *each* doubly bonded carbon.

In structures VI and VII, the doubly bonded carbon on the left has a hydrogen group and a methyl group (two different groups), and the doubly bonded carbon on the right has a hydrogen group and a methyl group (two different groups). Thus 2-butene exists as cis and trans isomers. Propene (structures IV and V) has a doubly bonded carbon with two hydrogen atoms (two identical groups) attached. The second requirement for geometric isomerism is not fulfilled, therefore, and this compound does *not* exist as cis and trans isomers. One of the doubly bonded carbons in propene does have two different groups attached, but the rules require that *both* carbons have two different groups.

In general, when two identical (or nearly identical) substituents are on the same side of the double bond, the compound is the cis isomer. The trans isomer is the one in which similar groups are on opposite sides of the double bond:

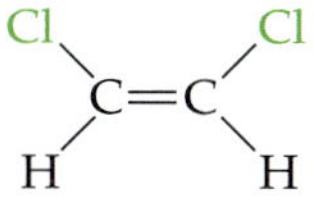

$(Cl)(H)C{=}C(Cl)(H)$

cis-1,2-Dichloroethene
mp −80 °C
bp 60 °C

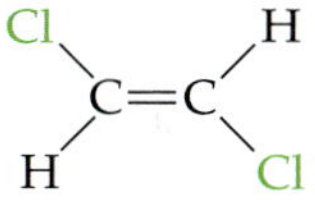

$(Cl)(H)C{=}C(H)(Cl)$

trans-1,2-Dichloroethene
mp −50 °C
bp 48 °C

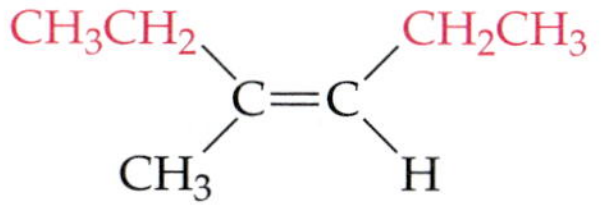

$(CH_3CH_2)(CH_3)C{=}C(CH_2CH_3)(H)$

cis-3-Methyl-3-hexene

$(CH_3)(CH_3CH_2)C{=}C(CH_2CH_3)(H)$

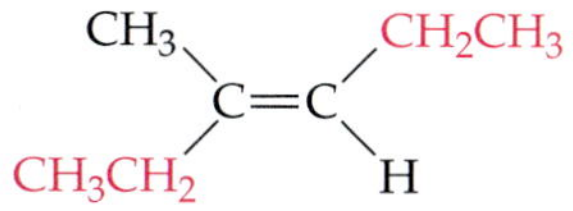

trans-3-Methyl-3-hexene

Example 18.3

Draw all alkenes with the formula C_5H_{10} and indicate which ones exist as cis and trans isomers. Give the IUPAC name for each isomer.

Solution

First we draw the various possible carbon skeletons, incorporating a double bond (no bond angles are implied):

$CH_2{=}CHCH_2CH_2CH_3$

1-Pentene
VIII

$CH_3CH{=}CHCH_2CH_3$

1-Pentene
IX

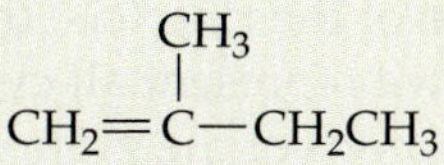

$CH_2{=}C(CH_3){-}CH_2CH_3$

2-Methyl-1-butene
X

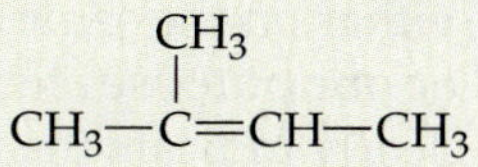

$CH_3{-}C(CH_3){=}CH{-}CH_3$

2-Methyl-2-butene
XI

$CH_2{=}CH{-}CH(CH_3){-}CH_3$

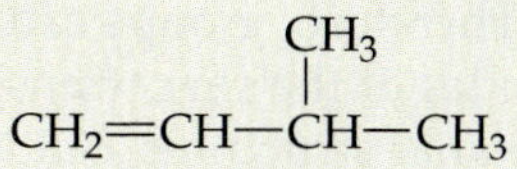

3-Methyl-1-butene
XII

All five structures have a double bond and thus meet rule 1. However, only IX meets rule 2 and exists as cis and trans isomers:

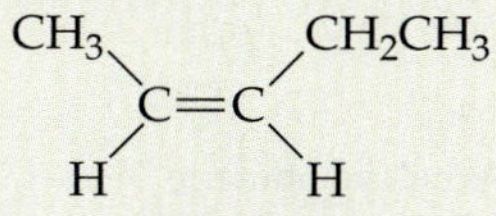

$(CH_3)(H)C{=}C(CH_2CH_3)(H)$

cis-2-Pentene

$(CH_3)(H)C{=}C(H)(CH_2CH_3)$

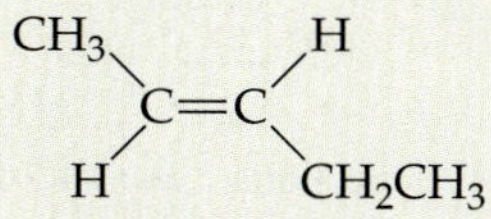

trans-2-Pentene

Structures VIII, X, and XII each have two hydrogen atoms on one of their doubly bonded carbon atoms, and structure XI has two methyl groups on one of its doubly bonded carbons. All fail rule 2.

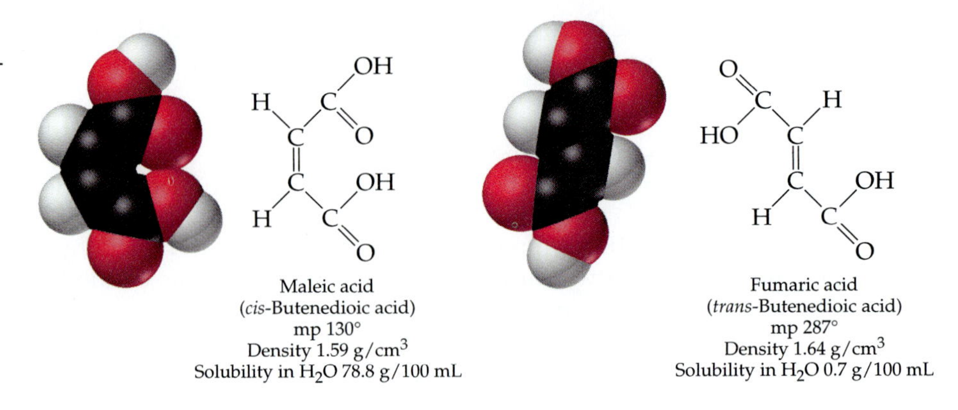

▶ Figure 18.15
Space-filling models, structural formulas, and properties of maleic acid and fumaric acid.

Practice Exercise

Draw and name all alkenes having the formula $C_3H_4Br_2$ and indicate which ones exist as cis and trans isomers.

Maleic and fumaric acids are classic examples of geometric isomers that have widely different chemical and physical properties (Figure 18.15). Because of the proximity of its carboxyl groups, maleic acid readily loses water to form an anhydride upon gentle heating:

$$\text{Maleic acid} \xrightarrow{\Delta} \text{Maleic anhydride} + H_2O$$

Fumaric acid is incapable of anhydride formation under the same reaction conditions. If it is heated to high temperatures (≈300 °C), it rearranges to form maleic acid, which then loses water to form the anhydride. We shall see in Chapter 25 that fumaric acid is the isomer produced and utilized by enzymes in living cells.

Recall from Chapter 13 that the bonding in cycloalkanes also imposes geometric constraints on the groups bonded to the ring carbon atoms. Common to all ring structures is the inability of groups to rotate about any of the ring carbon–carbon bonds. Therefore, groups can be either on the same side of the ring (cis) or on opposite sides of the ring (trans). For our purposes here, we represent all cycloalkanes as planar structures, but we clearly indicate the positions of the groups, either above or below the plane of the ring.

trans-1,2-Dibromocyclopropane   *cis*-1,2-Dibromocyclopropane   *trans*-1,2-Dimethylcyclobutane   *cis*-1,2-Dimethylcyclobutane

trans-1-Chloro-3-iodocyclopentane   *cis*-1-Chloro-3-iodocyclopentane   *trans*-4-Ethylcyclohexanol   *cis*-4-Ethylcyclohexanol

✓ Review Question

18.13 What are geometric isomers? Are geometric isomers enantiomers or diastereomers?

18.5 Biochemical Significance

Molecular configurations are of the utmost importance in biochemistry. We have already noted the example of the two enantiomers of lactic acid (Section 18.2). The dextrorotatory form is isolated from muscle tissue, and the levorotatory isomer is found in yeast and some bacteria. Laboratory synthesis of lactic acid from either acetaldehyde or pyruvic acid produces a racemic mixture of (±)-lactic acid; it is impossible to synthesize chemically only one of the two chiral forms using achiral substances.

The obvious question, then, is how do muscle cells synthesize only (+)-lactic acid, and yeast only the (−)-isomer? The explanation arises many times during the study of biochemistry. Enzymatic control is the answer.

$$\underset{\text{(−)-Lactic acid}}{H{-}\overset{\overset{COOH}{|}}{\underset{\underset{CH_3}{|}}{C}}{-}OH} \xleftarrow[\text{in yeast}]{\text{lactic acid dehydrogenase}} \underset{\text{Pyruvic acid}}{\overset{\overset{COOH}{|}}{\underset{\underset{CH_3}{|}}{C}}{=}O} \xrightarrow[\text{in muscle}]{\text{lactic acid dehydrogenase}} \underset{\text{(+)-Lactic acid}}{HO{-}\overset{\overset{COOH}{|}}{\underset{\underset{CH_3}{|}}{C}}{-}H}$$

Enzymes are biological catalysts that are chiral organic compounds. Almost every organic compound that occurs in living organisms is one enantiomer of a pair. Foods and medicines must have the proper molecular configurations if they are to be beneficial to the organism. For example, the popular flavoring agent Accent is levorotatory monosodium glutamate.[4] The dextrorotatory form of this salt would not enhance the flavor of meat because our taste buds could not recognize it (Figure 18.16). Similarly, the natural form of epinephrine is levorotatory. It has a

MSG was once used heavily in Chinese foods, and many people thought MSG was responsible for a set of symptoms (including headaches and a feeling of weakness) known as Chinese-restaurant syndrome. Experts who reviewed 230 studies concluded that MSG was not the cause of the syndrome, and this "not guilty" finding has been confirmed by further studies.

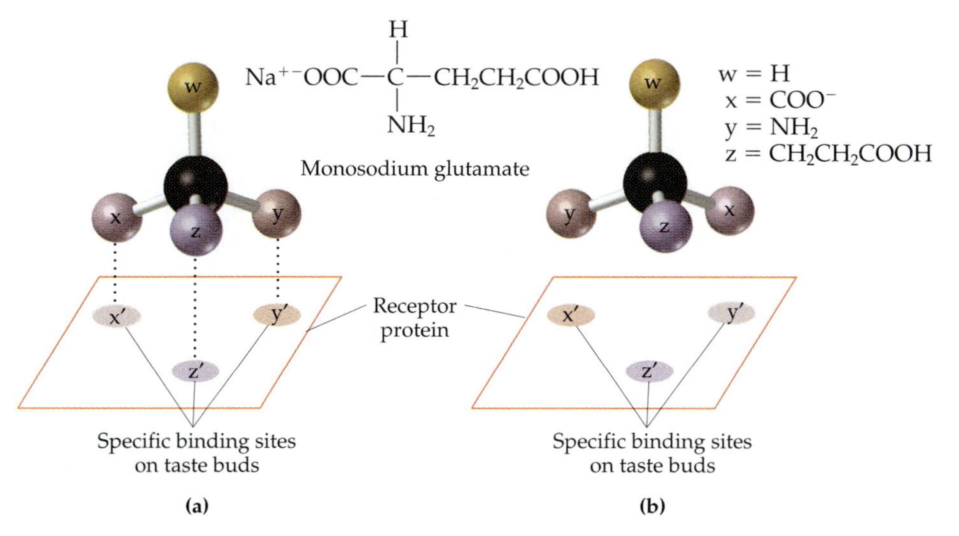

Figure 18.16
(a) The levorotatory stereoisomer of monosodium glutamate fits precisely at bonding sites on a receptor protein of our taste buds. (b) The dextrorotatory isomer does not fit and therefore cannot bind to the receptor sites.

[4]L-Monosodium glutamate (MSG) does not itself impart any taste, but it enhances the flavor of foods to which it is added. Although glutamates are found naturally in proteins, there is evidence that huge excesses can be harmful. MSG can numb portions of the brains of laboratory animals. It also may cause birth defects when administered in large amounts.

Biologically active drugs are often only one of a dozen or so stereoisomers. For example, the anticancer drug paclitaxel (Taxol), first obtained from the Pacific yew tree, has 11 chiral centers, each of which must have the proper handedness. Using van't Hoff's rule, we calculate that the active form is one of $2^{11} = 2048$ possible stereoisomers. Laboratory synthesis of paclitaxel is exceedingly complex, but it has been accomplished, first by R. A. Holton and colleagues in 1994.

physiological activity about 15 to 20 times greater than that of its dextrorotatory enantiomer.

In Section C.3 we mentioned that the stimulant drug amphetamine is not a pure compound. Rather, it is a mixture of two enantiomers:

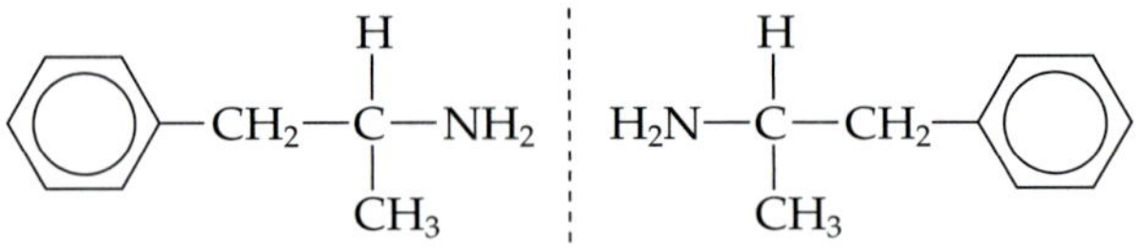

$$C_6H_5{-}CH_2{-}\underset{CH_3}{\overset{H}{C}}{-}NH_2 \qquad H_2N{-}\underset{CH_3}{\overset{H}{C}}{-}CH_2{-}C_6H_5$$

(+)-Amphetamine (Dextroamphetamine) (−)-Amphetamine

The dextrorotatory form is a stronger stimulant than its levorotatory isomer. Dexedrine is the trade name for the pure dextro isomer. Benzedrine is the trade name for a mixture of the two isomers in equal amounts. Dexedrine is two to four times as active as Benzedrine.

The subtle differences in structural configurations of organic molecules are of primary importance to life. We shall deal with the vitally important stereoselectivity of enzymes only after examining the compositions of the three major classes of biochemical compounds—carbohydrates, lipids, and proteins. It is necessary to observe strictly the proper configurational formulas of these compounds. If enzymes can recognize such subtle differences of shape and structure, so must we.

Pheromones

Because of the dangers of pesticides, scientists have been searching for alternative methods to control insects. One method involves the use of **pheromones,** chemicals used for communication between members of the same insect species. Insects emit pheromones for a variety of purposes, such as sending an alarm, social regulation, attracting a mate, trail marking, and territorial marking.

For insect control, sex-attractant pheromones are the most important. The females of many insect species depend upon an attractant to lure males for mating. These chemicals are remarkably powerful; a few drops can attract males within a range of two miles. Traps baited with the sex attractant, plus an insecticide, can be used to lure the males of that species to their deaths. One such compound is trimedlure, which has been found to be strongly attractive to the male Mediterranean fruit fly. Trimedlure has eight possible stereoisomers, which vary considerably in their ability to attract male flies. The fly is most strongly drawn to the isomer in which the methyl and ester groups are trans to each other.

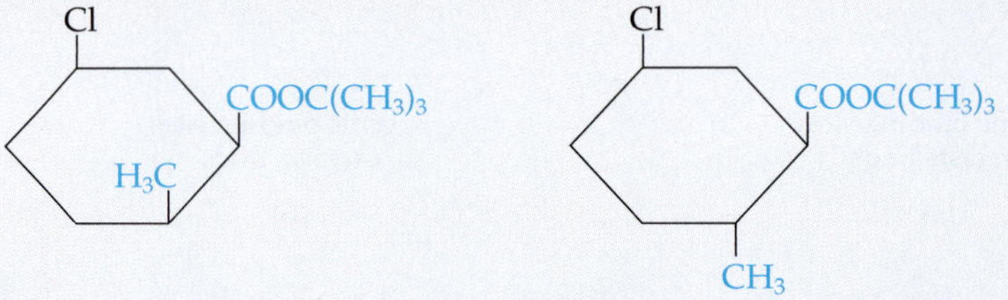

Cl, $COOC(CH_3)_3$, H_3C — Methyl and ester groups are cis.

Cl, $COOC(CH_3)_3$, CH_3 — Methyl and ester groups are trans.

Sex attractants of more than 30 insects have been identified. In most cases, just one of the possible stereoisomers is physiologically active (Figure 18.17).

In a few insect species, including the boll weevil, the male emits the pheromone.

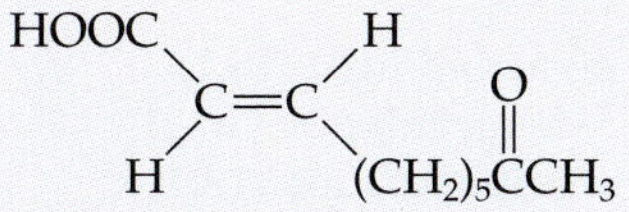

9-Oxo-*trans*-2-decenoic acid
(a)

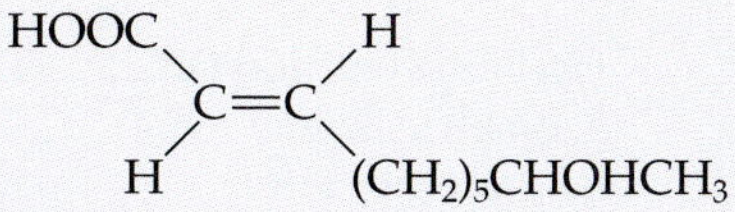

9-Hydroxy-*trans*-2-decenoic acid
(b)

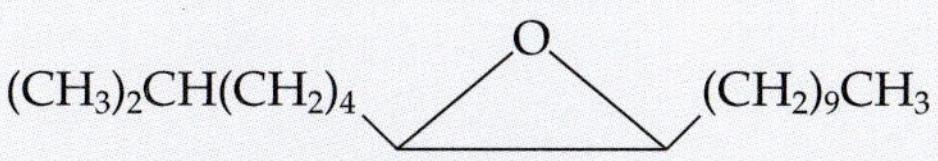

cis-7,8-Epoxy-2-methyloctadecane
(Disparlure)
(c)

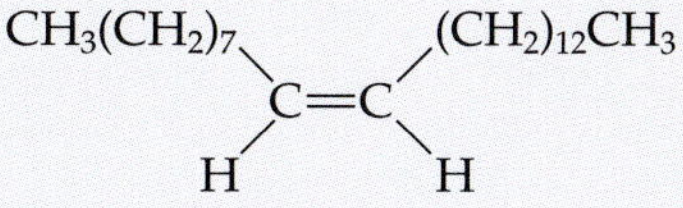

cis-9-Tricosene
(Muscalure)
(d)

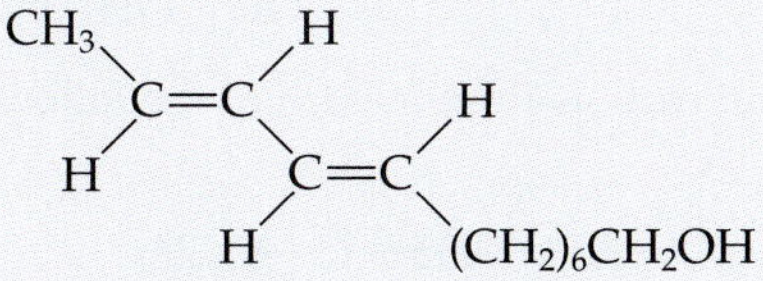

trans-8-*trans*-10-Dodecadien-1-ol
(e)

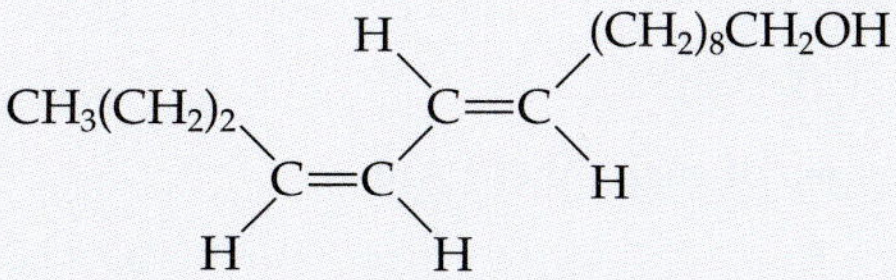

trans-10-*cis*-12-Hexadecadien-1-ol
(Bombykol)
(f)

▲ **Figure 18.17**
Sex attractants of some female insects. (a) and (b) Queen honeybee, (c) gypsy moth, (d) common housefly, (e) coddling moth, (f) silkworm moth.

Summary

Structural isomers share an identical molecular formula but differ from one another either in the position of a certain substituent or in how the same atoms are combined into different functional groups. In **stereoisomers,** both the order in which atoms are joined and the functional groups are identical; all that differs is the three-dimensional spatial arrangement of the atoms.

Ordinary (nonpolarized) light has waves vibrating in all directions. **Polarized light** has most waves filtered out so that all the remaining ones vibrate in the same plane. Substances that rotate the plane of vibration of polarized light are **optically active.** Those that rotate the plane to the right are **dextrorotatory** (+) compounds, those that rotate it leftward are **levorotatory** (−) compounds.

A **chiral center** is one bonded to four different groups, and one chiral center in a molecule makes the whole molecule chiral. A chiral molecule and its mirror image are *nonsuperimposable,* and two molecules that are nonsuperimposable mirror images of each other are a special type of *stereoisomer* called **enantiomers.**

Physical properties of the members of an enantiomeric pair are identical save one: One member rotates polarized light in one direction, and the other rotates polarized light in the opposite direction. Chemical properties of the members of an enantiomeric pair are the same when they react with **achiral** substances, but differ when they react with other *chiral* compounds.

A **racemic mixture** has equimolar amounts of an enantiomeric pair; this mixture has no optical activity because the rotation caused by one enantiomer is exactly canceled by the rotation caused by the other.

Diastereomers are stereoisomers that are not enantiomers. The maximum number of possible stereoisomers in a chiral molecule is given by **van't Hoff's rule** as 2^n, where n is the number of chiral carbons. Meso compounds are *not* optically active because of their internal symmetry plane. The rotation caused by one part of the molecule is canceled by the rotation caused by the comparable part of the molecule on the other side of the symmetry plane.

Geometric isomers are stereoisomers in which molecules differ from each other only in their configuration around a rigid part of the molecule, such as a carbon–carbon double bond or a ring structure. The molecule having two identical atoms or groups on the same side of the molecule is the **cis isomer;** the one having the two groups on opposite sides of the molecule is the **trans isomer.** Geometric isomers are diastereomers but not enantiomers.

Key Terms

achiral (18.2)
chiral center (18.2)
chirality (18.2)
cis isomer (18.4)
dextrorotatory (18.1)
diastereomers (18.3)
enantiomers (18.2)
Fischer projection (18.2)
geometric isomer (18.4)
levorotatory (18.1)
meso compound (18.3)
optically active (18.1)
pheromone (18.5)
polarimeter (18.1)
polarized light (18.1)
racemic mixture (18.2)
specific rotation (18.1)
stereoisomer (page 480)
structural isomer (page 479)
trans isomer (18.4)
van't Hoff's rule (18.3)

Problems

Chirality

1. Are these structures mirror images of each other? Are they superimposable?

$H-C(CH_3)(CH_3)-OH$ $\quad$ $HO-C(CH_3)(CH_3)-H$

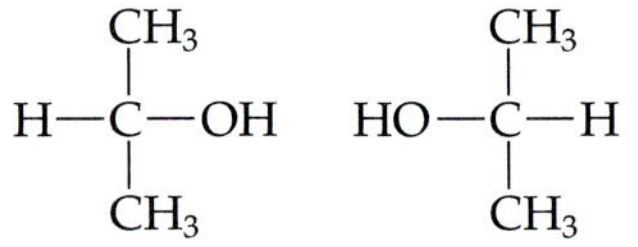

2. Are these structures mirror images of each other? Are they superimposable?

$CH_3CH_2-C(H)(OH)-CH_3$ $\quad$ $CH_3-C(H)(OH)-CH_2CH_3$

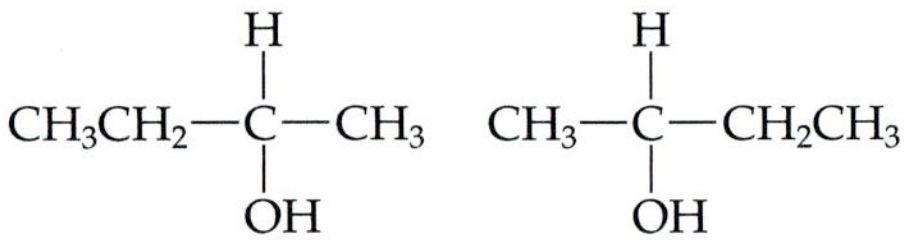

3. Use an asterisk (*) to indicate which of the carbon atoms shown in color are chiral centers.

a. $C_6H_5-CH_2CH(CH_3)-NH_2$

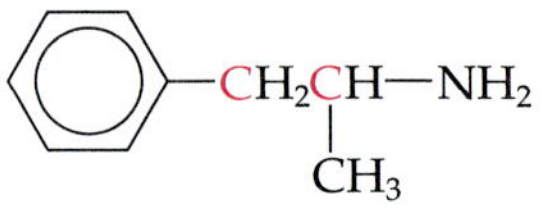

b. $H_2NC(=O)OCH_2C(CH_2CH_2CH_3)(CH_3)CH_2OC(=O)NH_2$

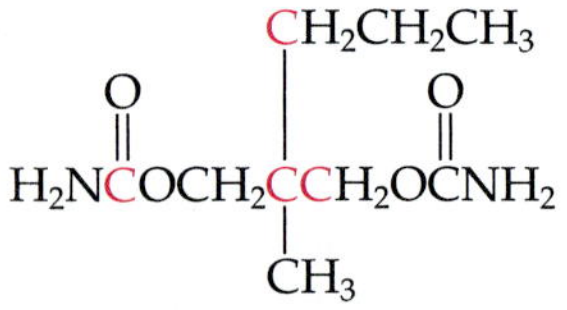

4. Circle each chiral carbon atom in each of the following.

a. $HO(HO)C_6H_3-CH(OH)CH_2-NH-CH_3$

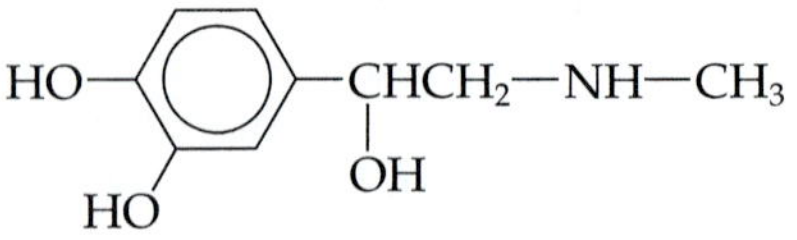

Epinephrine

b. $HO-C(=O)-CH_2-CH_2-CH(NH_2)-C(=O)O^-Na^+$

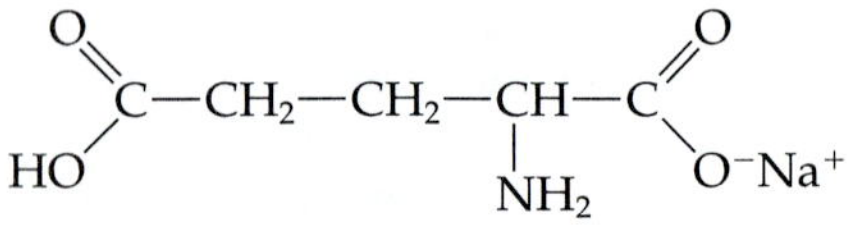

MSG

5. Circle each chiral carbon atom in each of the following.

a. $CH_3CH(OH)CH_2OH$ $\quad$ **b.** $CH_3CH(NH_2)COOH$

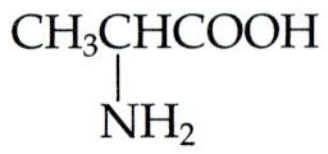

c. $C_6H_5CH_2CH(NH_2)CH_3$ $\quad$ **d.** $CH_3CH(Br)CH_2CH_3$

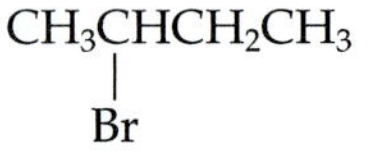

e. $CH_3CH(OH)CHO$ $\quad$ **f.** $CH_3CH(OH)-CH(OH)CH_3$

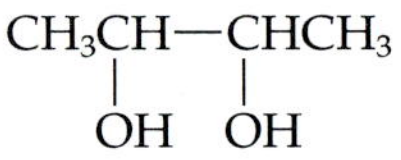

6. Circle each chiral carbon atom in each of the following.

a. $H_2N-CH(CH_2COOH)-C(=O)-NH-CH(CH_2C_6H_5)-C(=O)-OCH_3$

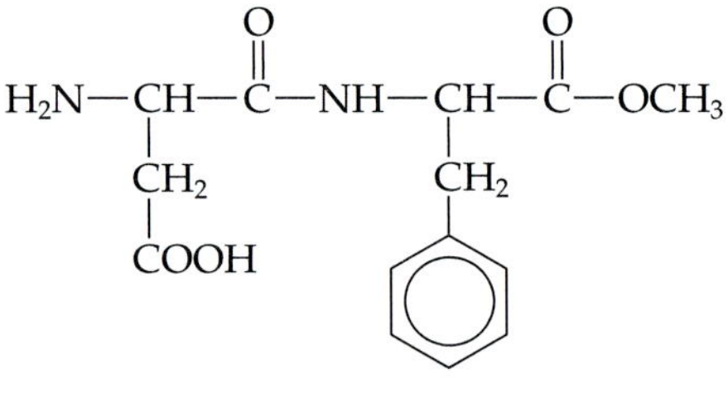

Aspartame

b. $CH_3-N^+(H)(CH_3)-CH(CH_3)CH_2-C(C_6H_5)_2-C(=O)CH_2CH_3$ $\quad Cl^-$

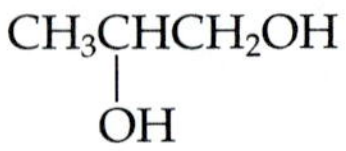

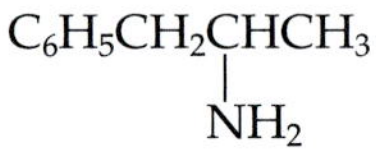

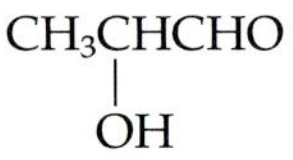

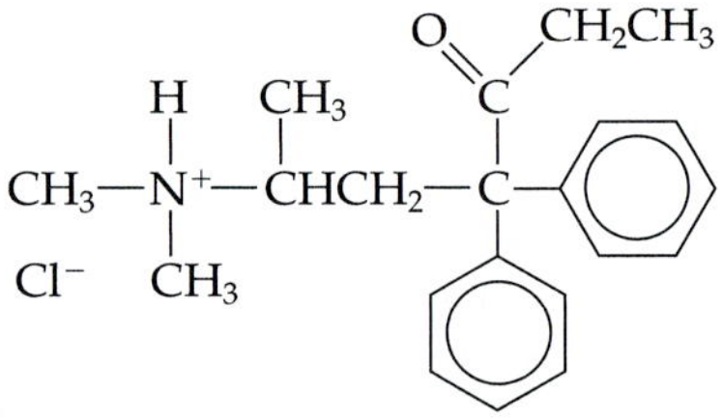

Methadone

7. Indicate whether each of the following compounds is chiral.

a. $CH_3CH(NH_2)CH_2CH_2CH_3$

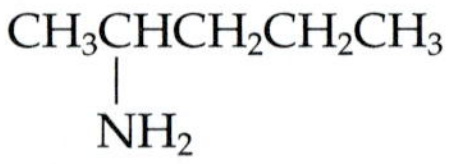

b. $H_2N-C_6H_4-C(=O)OCH_2CH(CH_3)_2$

c. Five-membered ring: $C(=O)-O-C(H)(-C(H)(OH)-CH_2OH)-C(OH)=C(HO)-$ (ring atoms: C=O, O, C–H, C–OH, HO–C; side chain C(H)(OH)–CH₂OH)

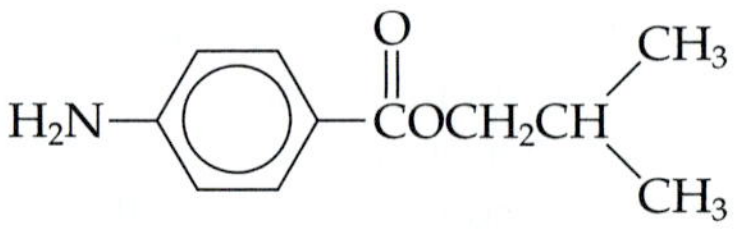

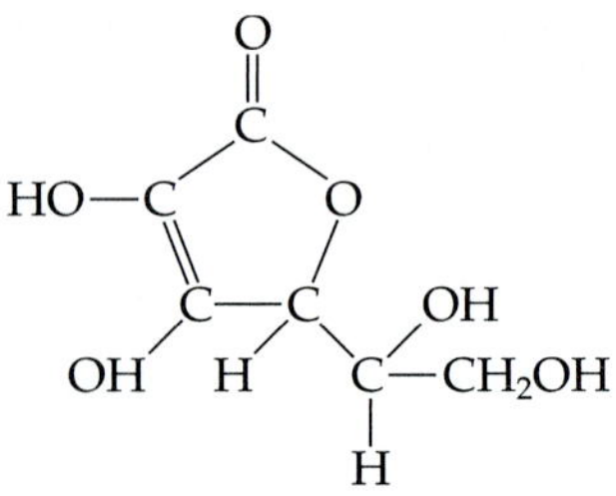

$$\begin{array}{l} \qquad\qquad\qquad\quad O \\ \qquad\qquad\qquad\quad \| \\ \textbf{d. } CH_3CHCH_2OCHCH_3 \\ \qquad\quad | \qquad\quad | \\ \qquad CH_3 \quad\;\; CH_3 \end{array}$$

8. Indicate whether each of the following molecules as drawn is chiral.

$$\textbf{a. } \begin{array}{c} CH_3 \\ | \\ H{-}C{-}OH \\ | \\ CH_3 \end{array} \qquad \textbf{b. } \begin{array}{c} \quad H \;\; H \\ \quad | \quad | \\ CH_3{-}C{-}C{-}CH_3 \\ \quad | \quad | \\ \quad Br \; Br \end{array}$$

$$\textbf{c. } \begin{array}{c} \quad H \;\; Br \\ \quad | \quad | \\ CH_3{-}C{-}C{-}CH_3 \\ \quad | \quad | \\ \quad Br \;\; H \end{array} \qquad \textbf{d. } \begin{array}{c} \quad H \;\; H \\ \quad | \quad | \\ CH_3{-}C{-}C{-}CH_3 \\ \quad | \quad | \\ \quad H \;\; Cl \end{array}$$

9. Draw the enantiomers of (**a**) 2-butanol and (**b**) 2-methylpropanoic acid.
10. Draw the enantiomers of (**a**) 2,3-dihydroxypropanal and (**b**) 2-bromobutanoic acid.
11. Draw Fischer projections (flat) for 2,3-dichlorobutanal. Label pairs of enantiomers.
12. Draw Fischer projections (flat) for the stereoisomers of bromochlorofluoromethane.

Multiple Chiral Centers

13. Which of the following can exist in the meso form?

$$\textbf{a. } \begin{array}{l} CH_2OH \\ | \\ CHOH \\ | \\ CHOH \\ | \\ CH_2OH \end{array} \qquad \textbf{b. } \begin{array}{l} CHO \\ | \\ CHOH \\ | \\ CHOH \\ | \\ CH_2OH \end{array} \qquad \textbf{c. } \begin{array}{l} COOH \\ | \\ CHOH \\ | \\ CHOH \\ | \\ CH_2OH \end{array} \qquad \textbf{d. } \begin{array}{l} CH_3 \\ | \\ CHCl \\ | \\ CHCl \\ | \\ CH_3 \end{array}$$

14. Which of the following can exist in the meso form?

$$\textbf{a. } \begin{array}{l} CH_3CH{-}CHCH_3 \\ \qquad | \qquad | \\ \quad\; OH \quad OH \end{array} \qquad \textbf{b. } \begin{array}{l} CH_3CH{-}CHCH_2CH_3 \\ \qquad | \qquad | \\ \quad\;\; Br \quad Br \end{array}$$

$$\textbf{c. } \begin{array}{l} HOOCCH{-}CHCOOH \\ \qquad\quad | \qquad | \\ \qquad\; OH \quad OH \end{array}$$

Van't Hoff's Rule

15. How many stereoisomers are there for each of the following?

$$\textbf{a. } \begin{array}{l} CH_3CHCH_2CHCH_2CH_3 \\ \qquad | \qquad\quad | \\ \quad\;\; OH \qquad OH \end{array}$$

$$\textbf{b. } \begin{array}{l} CH_3CH{-}CH{-}CHCH_2CH_3 \\ \qquad | \qquad | \qquad | \\ \quad\;\; Br \quad\; Br \quad\; Br \end{array}$$

16. How many stereoisomers are there for the following?

$$\begin{array}{l} CH_2{-}CH{-}CH{-}CH{-}C{-}CH_2 \\ \; | \qquad | \qquad | \qquad | \quad \| \quad | \\ OH \quad OH \quad OH \quad OH \;\; O \;\; OH \end{array}$$

Geometric (Cis–Trans) Isomers

17. Write the formulas of the geometric isomers for the following compounds. Label them cis and trans. If there are no geometric isomers, write None.
    a. 2-bromo-2-pentene **b.** 3-hexene
    c. 4-methyl-2-pentyne **d.** 1,1-dibromo-1-butene
    e. 2-butenoic acid ($CH_3CH{=}CHCOOH$)
    f. 4-methyl-2-pentene
18. Write the formulas of the geometric isomers for the following compounds. Label them cis and trans. If there are no geometric isomers, write None.
    a. 2,3-dimethyl-2-pentene
    b. 1,1-dibromo-2-ethylcyclopropane
    c. 1,2-dibromocyclohexene
    d. 2-chlorocyclohexanol
    e. 1-bromo-3-chlorocyclobutane
    f. 1,2,3-trimethylcyclopropane

Additional Problems

19. Write formulas for pairs of isomers that represent the following.
    a. positional isomers **b.** functional group isomers
    c. enantiomers **d.** diastereomers
    e. noncyclic geometric isomers
    f. cyclic geometric isomers
20. What is a pheromone? How are pheromones used in insect control?
21. 1-Butene reacts with HCl to form 2-chlorobutane. Does the reactant have a chiral center? Does the product have a chiral center? Would 2-chlorobutane formed in this manner show optical activity? Why or why not?
22. Draw Fischer projections (flat) for 2,3-dibromo-1-butanol. Label pairs of enantiomers.
23. Draw Fischer projection formulas for (+), (−), and meso forms of 2,3-butanediol. Which is meso? Can you tell which is (+) and which is (−)?
24. In Chapter 14, Problem 43, you are asked to draw structures for the eight isomeric pentyl alcohols. Which of these could exist in enantiomeric forms? (That is, which molecules have chiral centers?) Draw Fischer projection formulas for each pair of enantiomers.
25. One isomer of 3-phenyl-2-butanol is shown on page 500. Draw the Fischer projection formulas for the remaining three stereoisomers. Predict boiling point and specific rotation for any of the other three isomers that you can. Include a statement explaining why you were not able to predict some of the properties.

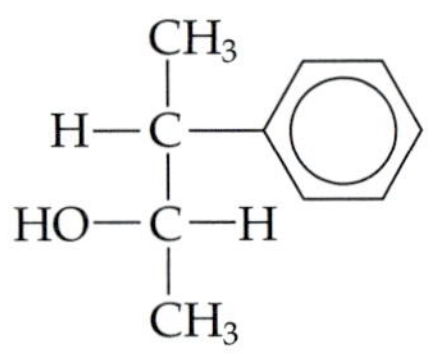

CH_3
H—C—(phenyl)
HO—C—H
CH_3

bp (25 mmHg) 118°C
[α] +30.9

26. 1,2-Dimethylcyclobutane exists as cis and trans isomers. One isomer is chiral, the other is achiral. Draw mirror images for both sets of geometric isomers and identify which isomers are identical and which are enantiomers. Are *cis*- and *trans*-1,2-dimethylcyclobutane diastereomers?
27. For 1,2-dibromocyclopropane, there are three stereoisomers. Draw their structures.
28. Draw the other geometric isomers for all the pheromones shown in Figure 18.17.
29. Urushiol is an unsaturated phenolic compound that is the active agent in poison ivy and poison sumac. Draw all its geometric isomers.

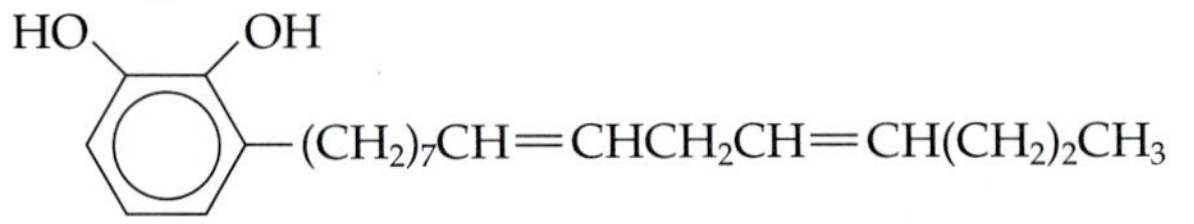

HO OH

$-(CH_2)_7CH{=}CHCH_2CH{=}CH(CH_2)_2CH_3$

30. If the four bonds of carbon were directed toward the corners of a square, how many isomers of CH_2BrCl would exist? Draw them.
31. Draw and name all the geometric isomers (both noncyclic and cyclic) corresponding to the molecular formula C_5H_{10}.
32. Only one aldehyde isomer of pentanal is optically active. What is its formula? Draw the two enantiomers of this compound.
33. Give the formula for the smallest noncyclic alkane that could be optically active.
34. An alcohol has the molecular formula $C_4H_{10}O$.
    a. Write all the possible structural formulas for the alcohol.
    b. If the alcohol can be separated into two optically active forms, which of the formulas in part (a) is correct?
35. β-Hydroxybutyric acid occurs in the urine of diabetics in amounts up to 30 g/day. Draw the structure of both enantiomers. (Hint: Place the carbon atoms in a vertical column with the carboxyl group at the top.)
36. Lysergic acid diethylamide (LSD) contains two chiral carbon atoms (Selected Topic B). Identify which carbon atoms are chiral and then draw structural formulas of the four stereoisomers of LSD. [Only one of these four isomers, (+)-lysergic acid diethylamide, is physiologically active.]

Project

37. Examine the labels of some common household products (detergents, foods, drugs, sprays, cosmetics). Write structural formulas for the chemical compounds contained in these products.

SELECTED TOPIC 

Chemistry of the Senses

We perceive the world around us through our five senses: sight, sound, smell, taste, and touch. In this selected topic we'll look at some of the chemistry involved in three of these senses: sight, smell, and taste.

D.1 Vision: Cis-Trans Isomerism

The retina of the human eye contains approximately 125 million rods and 6 million cones. **Rods** are very light-sensitive photoreceptor cells and function in dim light, but they do not discriminate among different colors of light. **Cones** are the photoreceptor cells that provide us with color vision and function in bright light. Both rods and cones contain an outer and an inner segment. The outer segment contains a stack of flattened membrane sacs known as *discs*. These discs contain organic compounds known as **visual pigments** called rhodopsins that are made up of a protein, opsin, and an aldehyde, 11-*cis*-retinal (Figure D.1). The rods have one type of opsin. Humans have three types of cone pigments, each absorbing light of a different wavelength (color); in each pigment, the retinal structure is identical, but the opsins differ.

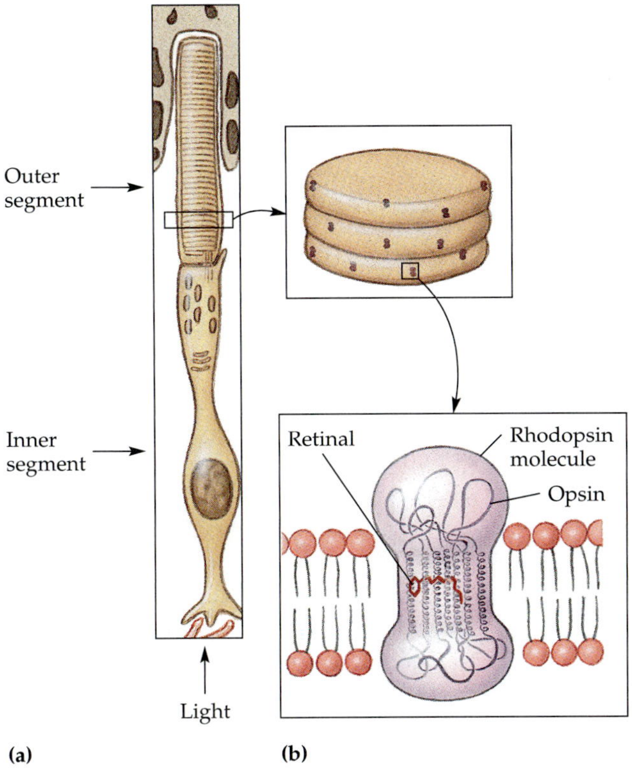

Outer segment

Inner segment

Light

(a)

Retinal

Rhodopsin molecule

Opsin

(b)

▲ **Figure D.1**
(a) The structure of a rod. (b) The structure of a rhodopsin molecule.

In the dark, the outer segments of the photoreceptor cells contain open ion channels that allow sodium ions to flow into these cells. The ion channels are kept open in the presence of a phosphate ester (Section 16.10), cyclic guanosine monophosphate (cGMP). When light strikes the retina, a complex series of reactions is initiated by the isomerization of 11-*cis*-retinal, a derivative of vitamin A (Section F.2), to the trans configuration:

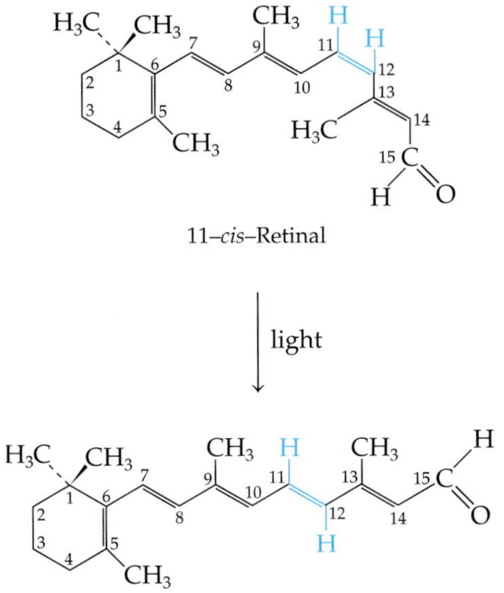

11–*cis*–Retinal

light

All–trans retinal

This reaction is termed a **photochemical isomerization** because the energy of light causes the geometric change.

The formation of all-*trans*-retinal leads to its dissociation from the protein opsin. Opsin then activates a second protein known as transducin (Figure D.2), which in turn activates a third protein known as phosphodiesterase. Phosphodiesterase is an enzyme (or biological catalyst; Chapter 22) that catalyzes the hydrolysis of cGMP to GMP:

$$\text{Cyclic GMP} \xrightarrow[\text{cGMP phosphodiesterase}]{H_2O \quad H^+} \text{5'-GMP}$$

Cyclic GMP

5′-GMP

Once the cGMP is hydrolyzed, it can no longer keep the sodium ion channels open. This reduces the concentration of Na^+ in the photoreceptor cells and leads to a decrease in the release of a specific neurotransmitter. A nerve impulse is triggered and transmitted via the optic nerve to the brain. This is interpreted by the brain as an image—in other words, we see.

After the photoreceptor cell has absorbed light, there must be a way to convert the all-*trans*-retinal back to 11-*cis*-retinal. This is also an enzyme-catalyzed reaction. Once reformed, the 11-*cis*-retinal can again bind to opsin to form rhodopsin.

Some retinal is lost during the regeneration of rhodopsin. It must be replaced by vitamin A stored in the eye or obtained from the bloodstream.

D.2 Smell and Taste

The senses of smell (*olfaction*) and taste (*gustation*) are known as the chemical senses because in both, receptor cells are activated or stimulated by chemicals (rather than by light, as in vision, or sound waves, as in hearing). They are also linked in the enjoyment and recognition of food as illustrated in Figure D.3.

Odors are detected by olfactory receptors located on the surface of the olfactory organs in the nasal cavity on either side of the septum (Figure D.4). To be

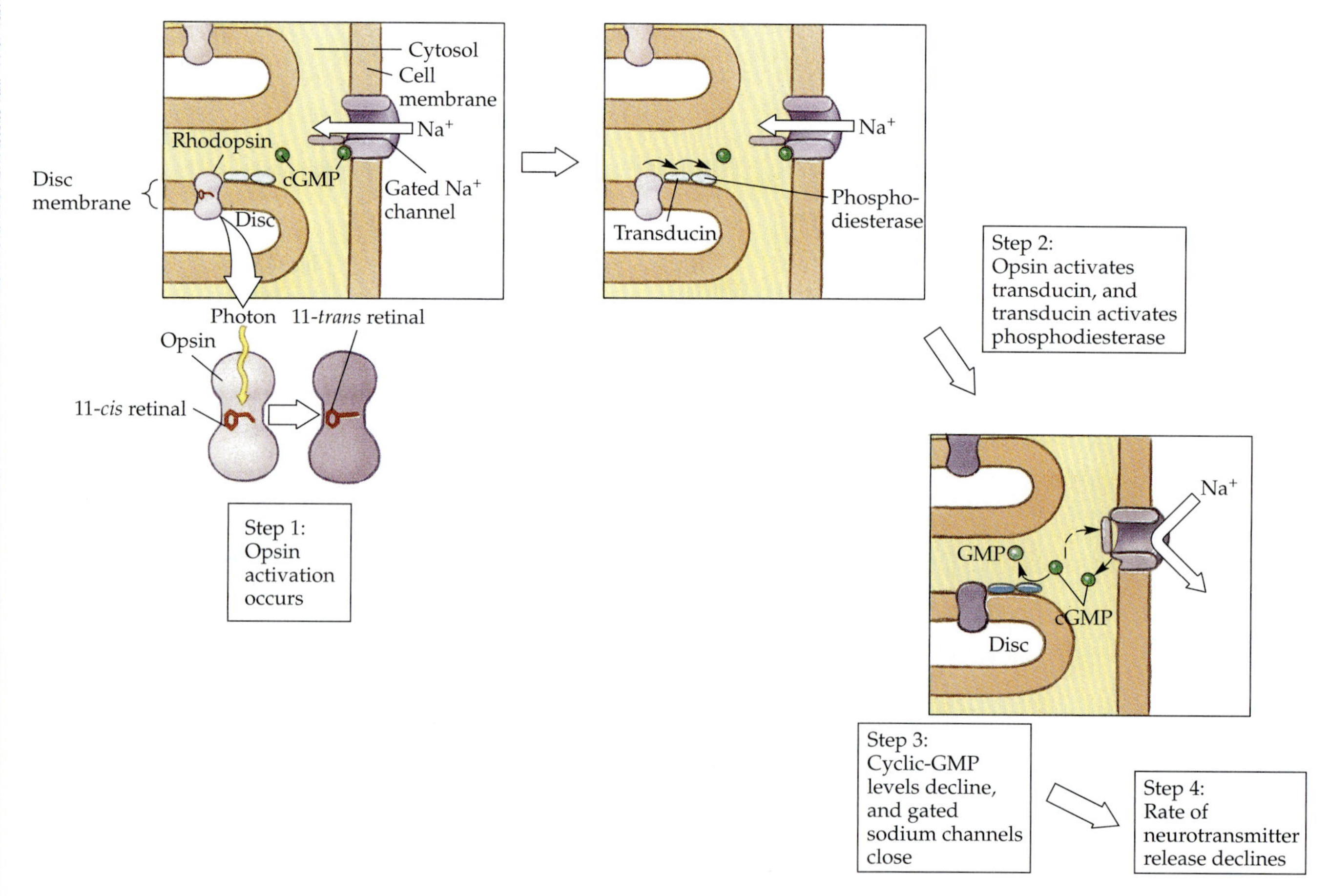

▲ **Figure D.2**
Photoreception.

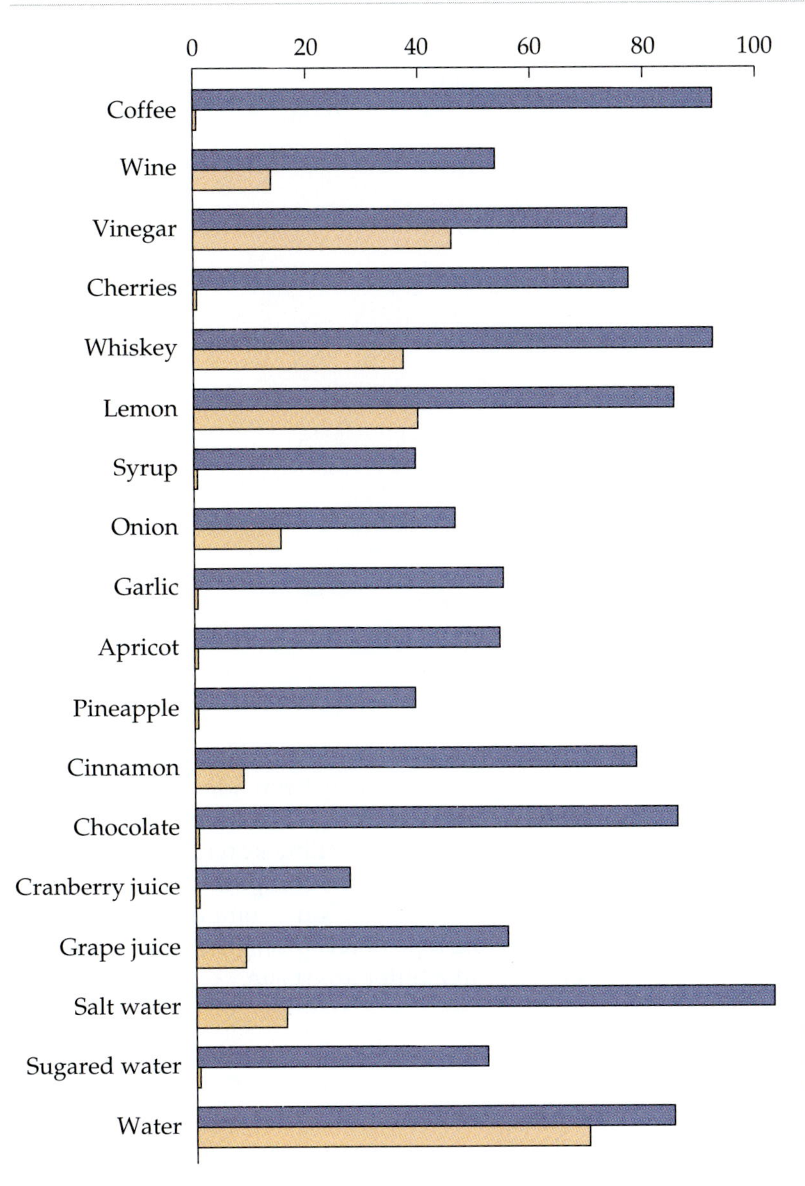

◀ **Figure D.3**
Demonstration of the degree of recognizability of foods when tasted with and without the involvement of smell. The yellow line represents how readily a substance was recognized when the olfactory receptors were blocked.

detected, a substance or odorant must be volatile to some extent; that is, some molecules must be in the vapor or gas phase (Section 7.7). Proteins located on the surface of the olfactory receptors bind to **odorants,** chemicals capable of stimulating these receptors. This binding initiates a series of steps that results in the opening of Na^+ channels, triggering a nerve impulse that is sent to the brain.

Even though scientists have proposed the existence of the olfactory receptors, it has only been in the past few years that an olfactory receptor has been isolated and shown to bind to a specific odorant. Continued research is concerned with demonstrating the exact interaction between the odorant and olfactory receptor, as well as isolating other olfactory receptors and demonstrating what types of molecules they bind.

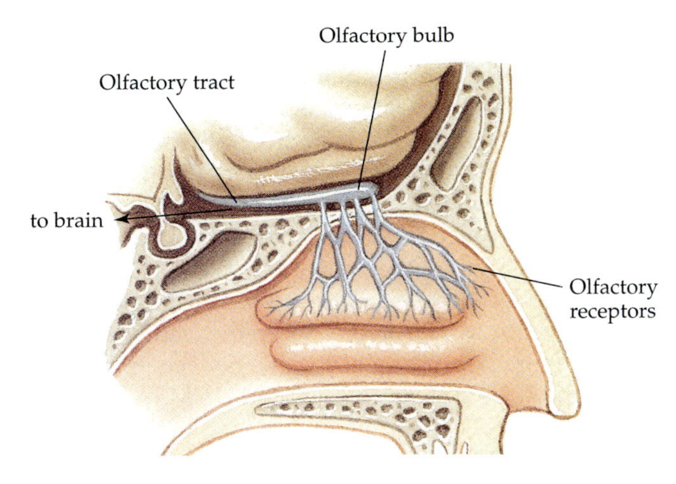

▲ **Figure D.4**
The structure of the olfactory organ on the right side of the nose.

Freshly baked bread smells quite different from rotting fish, due to the different volatile compounds produced by each. But is there a link between the shape or size of a molecule and its perceived odor? Efforts have been made to define odors on the basis of a small set of primary scents, analogous to the primary colors. Early researchers defined seven primary smells and hypothesized that there would be seven types of olfactory receptors with distinctive shapes and affinities for specific molecules. For example, it was thought that molecules with a camphor-like smell had in common a round shape and a diameter of 0.6 nm. However, the binding of odorants to olfactory receptors is a complex process, and it is now estimated that there are 30–50 "primary smells." The sensitivity of the olfactory system is remarkable: Humans can detect odorants within the range of 10^{-4} to 10^{-13} M.

The structure of an odorant is important. For example, the organic compound carvone exists as a pair of stereoisomers (Chapter 18). (−)-Carvone is a primary odorant in spearmint, while (+)-carvone is a major odorant in caraway seeds. However, researchers have not yet been able to identify a link between a particular structural feature on a molecule and the odor detected by the nose.

CH_3, O, H, C, H_3C, CH_2 — (−)-Carvone

CH_3, O, H, C, H_3C, CH_2 — (+)-Carvone

There are fewer primary taste sensations than primary odors. Five have been identified: sweet, sour, bitter, salty, and umami. Umami is the taste characteristic of meat broths and is produced by the binding of amino acids, especially glutamate (particularly the flavor enhancer monosodium glutamate, MSG), small peptides, and nucleotides to specific taste receptors. The taste receptors are found on the taste buds distributed over the surface of the tongue as well as portions of the pharynx and larynx. Taste receptors for the various primary tastes are located on different regions of the tongue, as shown in Figure D.5.

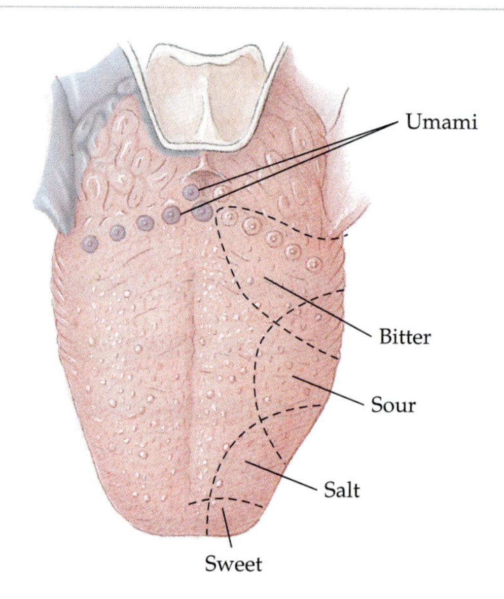

▲ **Figure D.5**
Distribution of taste receptors for the five primary tastes.

As with olfactory receptors, it has been difficult to isolate specific taste receptors from the taste buds and demonstrate the binding of specific molecules to these receptors. A taste receptor that binds to MSG, leading to the umami taste, was discovered in 1996. Other possible taste receptors have been identified, but researchers have not determined what compounds bind to these receptors. Work continues on identifying structural features that cause one compound to taste sweet, while another tastes bitter. Some compounds may bind to more than one type of receptor, giving, for example, a sweet taste but a bitter aftertaste.

Key Terms

cones (D.1)
odorants (D.2)
photochemical isomerization (D.1)
rods (D.1)
visual pigments (D.1)

Review Questions

1. What chemical transformation occurs in the retinal molecule when light is absorbed?
2. What is the relationship of vitamin A to the visual process?
3. Why are smell and taste known as the "chemical senses"?
4. What is an odorant?
5. What is the difference in structure between (+)-carvone and (−)-carvone?
6. What are the five primary taste sensations?

Carbohydrates

These foods are excellent sources of carbohydrates. The pastas and legumes contain primarily complex carbohydrates (starches), while fruits contain sugars.

Learning Objectives/Study Questions

1. What are carbohydrates? What is the difference between mono-, di-, and polysaccharides?
2. What are the structures of the most commonly occurring monosaccharides? Be able to classify them as aldoses or ketoses and as trioses, pentoses, or hexoses.
3. What is the difference between a D and an L sugar?
4. What is mutarotation? How does it occur?
5. What are the structures of sucrose, lactose, and maltose, the most common disaccharides? What monosaccharides make up each of these disaccharides?
6. Compare and contrast starch, glycogen, and cellulose.

On any cereal box or other processed food you will see a Nutrition Facts box telling you how many calories are in each serving of that food, the size of a serving, and so forth. You will also find listed the amounts of fat, carbohydrate, and protein in each serving. These three kinds of compounds, along with the nucleic acids, make up the four major types of macromolecules found in living organisms. In this chapter, we begin the study of **biochemistry**—the chemistry of the molecules and reactions found in living organisms—by looking at carbohydrates. In subsequent chapters, we will study the other three: lipids (which includes fats), proteins, and nucleic acids. Carbohydrates are the body's primary source of energy and must be included in any well-balanced diet.

Green plants are capable of synthesizing the sugar glucose from carbon dioxide and water, utilizing solar energy in the process known as photosynthesis.

$$6\,CO_2 + 6\,H_2O + \underset{\text{(From solar energy)}}{686\text{ kcal}} \longrightarrow \underset{\text{Glucose}}{C_6H_{12}O_6} + 6\,O_2$$

This glucose can be used by the plant for energy, or it can be converted to a more complex carbohydrate (starch) to serve as a source of energy for later use, perhaps as nourishment for the plant's seeds. Some of the glucose is converted to cellulose, the structural material of plants. We can gather and eat the parts of plants that store energy—seeds, roots, tubers, fruits—and use some of that energy for ourselves. In addition to providing energy for both plants and animals, carbohydrates provide the building blocks for the synthesis of proteins, lipids, and nucleic acids.

In contrast, most animal tissue contains a comparatively small amount of carbohydrate (less than 1% in humans). Animals cannot synthesize carbohydrates from carbon dioxide and water, and are therefore dependent upon the plant kingdom as a source of these vital compounds. We use carbohydrates not only for our food (about 60 to 65% by mass of the average diet) but also for our clothing (cotton, linen, rayon), shelter (wood), fuel (wood), and paper (wood).

Carbohydrates are compounds containing carbon, hydrogen, and oxygen. They include the starches and fiber (complex carbohydrates), the sweet-tasting compounds called sugars, and structural materials such as cellulose. The term *carbohydrate* has its origin in a misinterpretation of the molecular formulas of many of these substances. For example, the formula for glucose is $C_6H_{12}O_6$, but we could also represent this molecule as a "carbon hydrate"[$C_6 \cdot 6H_2O$]. Carbohydrates are not hydrates of carbon, however. They are alcohols—they all contain the hydroxyl (—OH) functional group. Most also contain either a real or latent carbonyl (C=O) group. (By a *latent* carbonyl group, we mean a functional group such as a hemiacetal or an acetal [Section 15.6] that can be converted to a carbonyl group.)

A formal definition of carbohydrates is difficult. Chemically, carbohydrates are polyhydroxy aldehydes or ketones or compounds that can be hydrolyzed to form such compounds.

✓ Review Question

19.1. What is meant by a latent carbonyl group?

Monosaccharides

Simple carbohydrates, those that cannot be further hydrolyzed, are called **monosaccharides.** The naturally occurring monosaccharides contain three to seven carbon atoms per molecule. Two or more monosaccharides can link together to form chains that contain from two to several hundred monosaccharide units. Prefixes are used to indicate the number of such units in the chains. *Disaccharide* molecules have two monosaccharide units, *trisaccharide* molecules have three, and so on. Chains with many monosaccharide units joined together are called *polysaccharides.* All these higher saccharides can be hydrolyzed back to their constituent monosaccharides.

Because carbohydrate molecules contain several —OH groups, the molecules can form an extensive network of intermolecular hydrogen bonds. Because of this extensive bonding, monosaccharides and disaccharides are crystalline solids at room temperature. They have relatively high melting points and often char before melting. Carbohydrate molecules also can form hydrogen bonds to water molecules, and the mono- and disaccharides are readily soluble in water. For example, 100 g of glucose dissolves in 100 mL of water at 25 °C.

19.1 General Terminology and Stereochemistry

The general names for the monosaccharides are obtained in a manner analogous to the naming of organic compounds by the IUPAC system. The number of carbon atoms is denoted by the appropriate stem, and *-ose* is the generic ending for any sugar. For example, the terms triose, tetrose, pentose, and hexose signify three-, four-, five-, and six-carbon monosaccharides, respectively. In addition, those monosaccharides that contain an aldehyde group are called **aldoses;** those containing a ketone group are **ketoses.** By combining these terms, both the type of carbonyl group and the number of carbon atoms in the molecule are easily expressed. Thus monosaccharides are generally referred to as aldotetroses, aldopentoses, ketopentoses, ketoheptoses, and so forth. Glucose and fructose are specific examples of an aldohexose and a ketohexose, respectively.

Glucose
(an aldohexose)

Fructose
(a ketohexose)

The simplest sugars are the trioses. Two trioses derived by oxidation of the triol glycerol are important intermediates in the chemical reactions of the body (see Section 25.1). Dihydroxyacetone is a ketotriose, and glyceraldehyde is an aldotriose (Figure 19.1a). Dihydroxyacetone does not contain a chiral center, but glyceraldehyde does and thus exists in two optically active forms. Except for the direction in which they rotate plane-polarized light, these two enantiomers have identical physical properties. One enantiomer has a specific rotation of +8.7°, while the other has a specific rotation of −8.7°.

The German chemist Emil Fischer (1852–1919) initiated the convention of projecting the formulas onto a two-dimensional plane so that the aldehyde group is written at the top, with the hydrogen and hydroxyl written to the right or left. (Formulas of chiral molecules represented in this manner are referred to as Fischer projections, Fischer models, or Fischer configurations.) Arbitrarily, Fischer then decided that the formula of glyceraldehyde in which the hydroxyl group is positioned to the right of the chiral carbon atom represents the dextrorotatory isomer. He assigned the letter D as its prefix. The levorotatory isomer, in which the —OH group is positioned to the left of the chiral carbon atom, was accordingly assigned the letter L as its prefix.[1]

The two forms of glyceraldehyde are especially important because the more complex sugars can be considered to be derived from them. They therefore serve as a reference point for designating and drawing all other monosaccharides. Sugars whose Fischer projections terminate in the same configuration as D-glyceraldehyde are designated as **D sugars;** those derived from L-glyceraldehyde are designated as **L sugars.** Humans cannot obtain energy from L sugars because of the stereospecificity of the enzymes which are utilized (see Section 22.4).

The letters D and L often mislead the beginning student. It must be emphasized that these prefixes serve only to signify the *absolute configuration* of a molecule. A D sugar is one that has the same configuration about the *penultimate* carbon atom as D-glyceraldehyde has (H on the left, OH on the right). The letters do not in any way refer to the optical rotation of the molecule. The direction in which a molecule rotates

The penultimate (next to last) carbon has been chosen, by convention, to be the reference carbon atom. It is the chiral carbon atom farthest from the aldehyde or ketone group.

[1]Fischer's arbitrary assignment proved to be correct. In 1951, chemists, with the aid of X-ray crystallography, determined the absolute configurations of the glyceraldehyde enantiomers and found that the D isomer is indeed dextrorotatory.

Figure 19.1
(a) Structures of the trioses. D- and L-glyceraldehyde are mirror images of each other and represent a pair of enantiomers. (b) The D and L designations do not refer to the direction in which a molecule rotates plane-polarized light. This is indicated with the symbols (+) and (−). D-Ribose and D-2-deoxyribose are D sugars but rotate light in a counterclockwise direction.

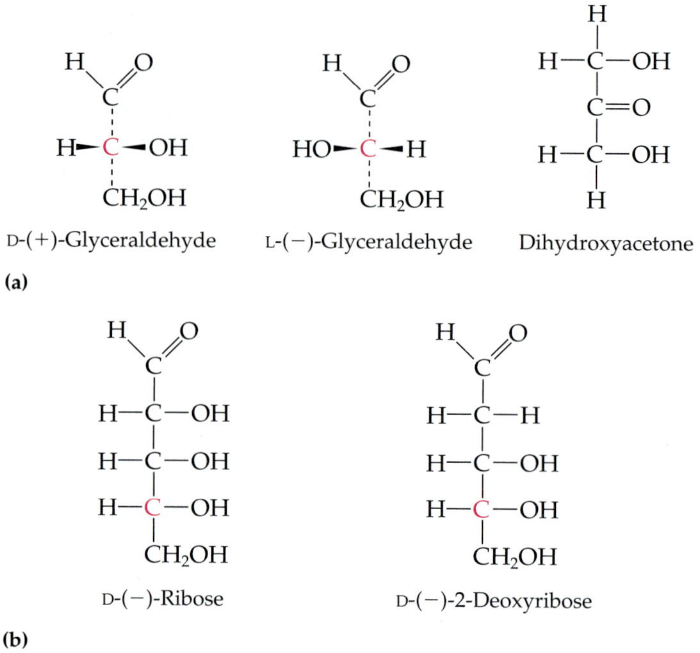

D-(+)-Glyceraldehyde L-(−)-Glyceraldehyde Dihydroxyacetone

(a)

D-(−)-Ribose D-(−)-2-Deoxyribose

(b)

plane-polarized light is a specific property of each optically active molecule and is specified with the symbols (+) and (−) (see Section 18.1). D-Glyceraldehyde just happens to be dextrorotatory. A compound derived from D-glyceraldehyde may have *either* (+) or (−) optical rotation, depending on any additional chiral carbon atoms.

For example, consider the two aldopentoses D-ribose, the sugar unit found in ribonucleic acids (RNA), and D-2-deoxyribose, the sugar unit found in deoxyribonucleic acids (DNA). As the 2-*deoxy* implies, this sugar does not have an oxygen on the second carbon atom (see Figure 19.1b). Note that both of these D sugars are levorotatory.

Review Questions

19.2 Define each of the following terms: (**a**) monosaccharide; (**b**) polysaccharide; (**c**) disaccharide.

19.3 Draw a (**a**) ketopentose; (**b**) aldopentose; (**c**) ketoheptose.

19.4 What do the prefixes D and L mean? How are the D and L designations determined for a given sugar?

19.2 Hexoses

Aldohexoses contain four chiral carbon atoms, and thus there are eight enantiomeric pairs, or 16 isomers. Fortunately, we will only be concerned with the three isomers most commonly found in nature: D-(+)-glucose, D-(+)-mannose, and D-(+)-galactose. Ketohexoses have three chiral carbon atoms, giving four enantiomeric pairs, or eight isomers, but we will consider only one of these, D-(−)-fructose. All of the sugars we discuss in the remainder of this chapter belong to the D family. If no family designation is given, you can assume that the compound is a D sugar.

Glucose

D-Glucose is the most abundant sugar found in nature. It is commonly found in fruits, especially in ripe grapes, and for this reason it is often referred to as *grape sugar.* It is also known as *dextrose,* a name that derives from the fact that it is dextrorotatory.

Most of the carbohydrates we eat are eventually converted to glucose in a series of metabolic pathways that produce energy for our cells. Glucose is the circulating carbohydrate of animals; hence the name *blood sugar.* Normal blood sugar values range from 70 to 105 mg/dL glucose, and normal urine may contain anywhere from a trace to 20 mg/dL glucose. Glucose can be given intravenously (as a 5% mass-volume solution) to patients who are unable to take food orally.

Commercially, glucose is made by the hydrolysis of starch. In the United States, cornstarch is used in the process; thus, glucose is also known as *corn sugar.* In Europe, the starch is obtained from potatoes. Glucose is only 74% as sweet as table sugar (sucrose), but it has the same caloric value per gram.

The structure of D-glucose is given in Figure 19.2. This formula follows our convention of writing the aldehyde group at the top and the primary alcohol group at the bottom. Glucose is a D sugar because the hydroxyl group at the fifth carbon (the chiral center farthest from the carbonyl group) is on the right. In fact, all the hydroxyl groups except the one at the third carbon are to the right.

Mannose

D-Mannose (Figure 19.2) is a component of the polysaccharide mannan, found in some berries and in "vegetable ivory," the endosperm (white fleshy part) of palm nuts. Note that the configuration of mannose differs from that of glucose only at the second carbon atom.

Galactose

D-Galactose is formed by the hydrolysis of lactose, a disaccharide composed of a glucose unit and a galactose unit. Galactose does not occur in nature in the uncombined state. The galactose needed by the human body for the synthesis of lactose (in the mammary glands) is obtained by the metabolic conversion of D-glucose to D-galactose. Galactose is also an important constituent of the glycolipids (see Section 20.4) that occur in the brain and in the myelin sheath of nerve cells. For this reason

Figure 19.2 Structures of four important hexoses pictured with a food source in which each is commonly found. The first three sugars are aldoses; fructose is a ketose.

▶ These buttons are made of vegetable ivory.

it is also known as *brain sugar.* Figure 19.2 shows the structure of galactose. Notice that the configuration differs from that of glucose only at the fourth carbon atom.

Fructose

D-Fructose, whose structure is shown in Figure 19.2, is the only naturally occurring ketohexose. Note that from the third through the sixth carbon atoms, its structure is the same as that of glucose. It occurs, along with glucose and sucrose, in honey (which is 40% fructose) and sweet fruits. Fructose (Latin *fructus,* fruit) is also referred to as *levulose* because it has a specific rotation that is strongly levorotatory (−92.4°). It is the sweetest sugar (1.7 times sweeter than sucrose). Many nonsugars, however, are several hundred or several thousand times as sweet (see Table 19.1 and the boxed essay "Artificial Sweeteners"). Fructose is the only sugar found in the semen of bulls and men. It is formed in the prostate gland as the major energy source for spermatozoa.

High-fructose corn syrup is made by using enzymes to convert much of the glucose in the syrup to fructose.

Artificial Sweeteners

Although sweetness is commonly associated with most mono- and disaccharides, it is not a specific property of carbohydrates. Many sugars are sweet to varying degrees, but several organic compounds have been synthesized that are far superior as sweetening agents—they are referred to as high-intensity sweeteners. These synthetic compounds are useful for those persons (e.g., diabetics) who must control their carbohydrate intake. They are noncaloric or used in such small quantities that they do not add significantly to the caloric value of a food item.

The first artificial sweetener, saccharin, was discovered by accident in 1879. Saccharin passes through the body unchanged, and thus is noncaloric. It is 500 to 700 times sweeter than sucrose. After its discovery saccharin was used until it was banned in the early 1900s. However, during the sugar-short years of World War I, the ban was lifted, and it was used once again. One drawback to the use of saccharin was its bitter metallic aftertaste. This was initially overcome by combining it with cyclamate, another artificial sweetener, discovered in 1937. This combination has no aftertaste.

In the 1960s and 1970s several clinical tests with laboratory animals implicated both cyclamate and saccharin as carcinogenic. The results from the cyclamate tests were available first, and cyclamate was banned in the United States in 1969. Then in 1977 a major study was released in Canada indicating that saccharin increased the incidence of bladder cancer in rats. The FDA proposed a ban on saccharin, which raised immediate public opposition because saccharin was the only artificial sweetener still available for use by diabetics and calorie-conscious individuals who wanted their diet sodas. The public opposition led Congress to pass the Saccharin Study and Labeling Act in 1977, which permitted the use of saccharin as long as any product containing it was labeled with a consumer warning regarding its possible elevation of the risk of bladder cancer. (It is interesting that the reverse is true in Cana-

Table 19.1 Relative Sweetness of Some Compounds (Sucrose = 100)

Compound	Relative Sweetness
Lactose	16
Maltose	33
Glucose	74
Sucrose	100
Fructose	173
Aspartame	16,000
Acesulfame K	20,000
Saccharin	50,000
Sucralose	60,000

da: cyclamate is not banned, while saccharin is!) The FDA is currently reviewing the ban on cyclamate as 75 additional studies have failed to show any carcinogenic effect of cyclamate.

In 1965 a third artificial sweetener was discovered—aspartame. Aspartame is a white crystalline compound that is about 160 times sweeter than sucrose and has no aftertaste. Approved for use in 1981, aspartame is used as a sweetener for a wide variety of foods because it can blend well with other food flavors. In the body aspartame is hydrolyzed to the amino acids aspartic acid and phenylalanine, and methanol. There has been repeated controversy regarding the safety of aspartame, partly based on the fact that the body metabolizes the released methanol to formaldehyde. It should be noted, though, that a glass of tomato juice has six times as much methanol as a similar amount of a diet soda containing aspartame. The only documented risk with aspartame use is for individuals with the genetic disease phenylketonuria (PKU). These individuals lack the enzyme needed to metabolize the phenylalanine released when aspartame is broken down by the body. Thus, all products containing aspartame must carry a warning label. Because aspartame is also hydrolyzed when heated, it is not heat-stable and is not used for baked goods.

Acesulfame K was discovered just two years after aspartame (1967) and approved for use in the United States in 1988. It is 200 times sweeter than sugar and, unlike aspartame, is heat-stable. It has no lingering aftertaste.

The newest artificial sweetener to gain approval by the FDA (April 1998) for use in the United States is sucralose. It is a white crystalline solid that is approximately 600 times sweeter than sucrose. It is synthesized from sucrose and has three chlorine atoms substituted for three hydroxyl groups. It is noncaloric because it passes through the body unchanged. It is heat-stable, so it can be used in baking.

All of the extensive clinical studies completed to date have indicated that in moderate amounts these artificial sweeteners approved for use in the United States are safe to consume by healthy individuals.

▲ In 1997 the average American consumed more than 54 gallons of soft drinks for a total of more than 14 billion gallons! Nearly 25% of the soft drinks sold were diet soft drinks. Aspartame is the most commonly used artificial sweetener in diet soft drinks.

Saccharin: $N^{-}Na^{+}$, SO_2

Cyclamate: $SO_3^{-}Na^{+}$

Aspartame: $H_3N^{+}-CH(CH_2COO^{-})-C(=O)-NH-CH(CH_2C_6H_5)-C(=O)-OCH_3$

Acesulfame K: $N^{-}K^{+}$, H_3C

Sucralose: CH_2OH, Cl, CH_2Cl, CH_2Cl, OH

▲ Structures of artificial sweeteners.

Review Questions

19.5 Identify these sugars by their proper names: (**a**) blood sugar; (**b**) levulose; (**c**) dextrose; (**d**) brain sugar.

19.6 Identify each of the following sugars as an aldose or a ketose: (**a**) glyceraldehyde; (**b**) ribose; (**c**) galactose; (**d**) fructose.

19.3 Cyclic Structures of Monosaccharides

So far we have represented the monosaccharides as polyhydroxy aldehydes and ketones. These representations are useful in studying structural relationships and reactions (Section 19.4) of these simple sugars. However, in Section 15.6 we mentioned that aldehydes and ketones react with alcohols to form hemiacetals and hemiketals:

$$R{-}\overset{\overset{\large O}{\|}}{C}{-}H \;+\; R'OH \;\rightleftharpoons\; R{-}\underset{\underset{\large OR'}{|}}{\overset{\overset{\large OH}{|}}{C}}{-}H$$

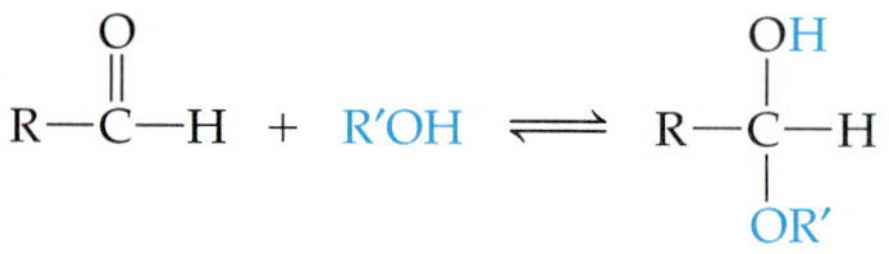

Aldehyde Alcohol Hemiacetal

You should not be surprised to find that hydroxyl groups and carbonyl groups conveniently located on the same molecule react with one another. Consequently, monosaccharides larger than tetroses exist mainly as cyclic hemiacetals or hemiketals (Figure 19.3).

You might wonder why the aldehyde reacts with the hydroxyl group on carbon 5 rather than the hydroxyl group on carbon 2 next to it. You'll recall from Section 13.6 that cyclic alkanes containing five or six carbons in the ring are the most stable. The same is true for cyclic hemiacetals and hemiketals—rings with five or six atoms in the ring are the most stable. As shown in Figure 19.4a, when a monosaccharide such as glucose forms a cyclic structure, the carbonyl oxygen may be pushed either up or down, giving rise to two cyclic structures. The structure on the left, with the hydroxyl on the first carbon projected downward, represents what is called the *alpha (α) form.* The structure on the right, with the hydroxyl on the first carbon pointed upward, is the *beta (β) form.* These two isomers are referred to as **anomers;** they differ in structure around the **anomeric carbon**—that is, the carbon that was the carbonyl carbon in the straight-chain form.

Crystalline glucose may exist in either the alpha or the beta form. The two forms have different properties. The alpha form melts at 146 °C and has a specific rotation of +112°, while the beta form melts at 150 °C and has a specific rotation of +18.7°. In solution, an equilibrium mixture is formed (Figure 19.4a). You can start out with

Figure 19.3
Cyclization of D-glucose. The Fischer projection (a) is rearranged into a three-dimensional representation (b). Reaction of the hydroxyl group on carbon 5 with the aldehyde group on carbon 1 gives the cyclic hemiacetal (c).

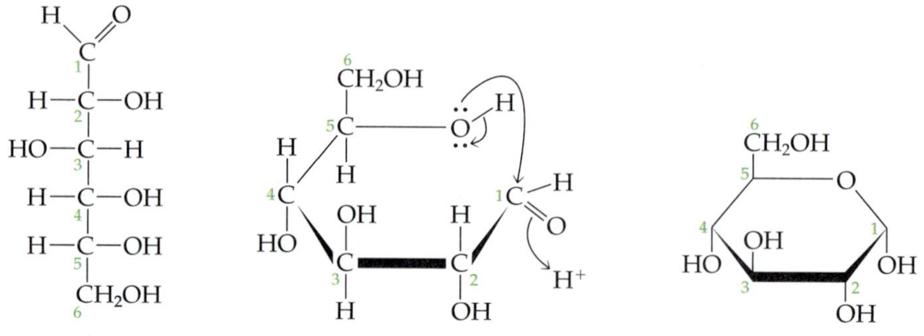

(a) Fischer projection

(b) Three-dimensional representation

(c) Cyclic hemiacetal

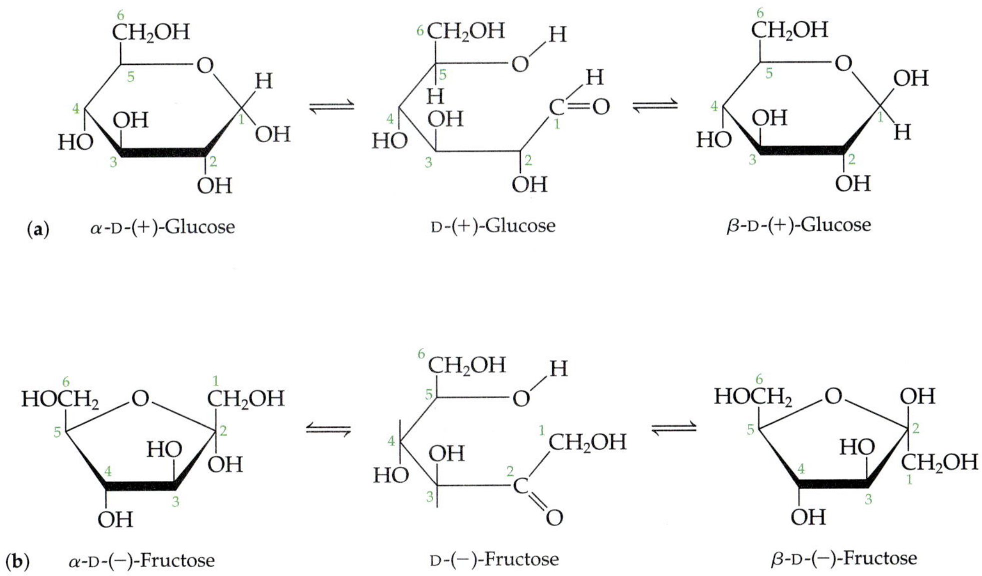

(a) α-D-(+)-Glucose ⇌ D-(+)-Glucose ⇌ β-D-(+)-Glucose

(b) α-D-(−)-Fructose ⇌ D-(−)-Fructose ⇌ β-D-(−)-Fructose

Figure 19.4 In aqueous solution, monosaccharides exist as an equilibrium mixture of three forms. This interconversion is known as mutarotation. Shown is the mutarotation of (a) D-glucose and (b) D-fructose. Note that the open-chain form of D-glucose has the aldehyde group that takes part in reactions typical of aldehydes (Section 15.6). Such reactions will shift the equilibrium to produce more free aldehyde from the cyclic forms.

either pure crystalline anomer, but as soon as it is dissolved in water, the hemiacetal group opens to form the free carbonyl and then closes to form either the alpha or beta anomer. This process is continuously repeated. This interconversion is referred to as **mutarotation** (Latin *mutare,* to change). At equilibrium, the mixture is about 36% alpha, 64% beta, and less than 0.02% of the open-chain aldehyde form. The observed rotation of this solution is +52.7°. Even though there is only a small amount of the open-chain aldehyde form, this is enough so that solutions of these compounds can give characteristic reactions of aldehydes. As the small amount of free aldehyde is used up in a reaction, the cyclic forms open up to yield more free aldehyde. Thus, *all* the molecules may eventually react, even though very little free aldehyde is present at any given time.

The cyclic forms of sugars in Figures 19.3 and 19.4, and elsewhere in this book, use a convention first suggested by the English chemist W. N. Haworth (1883–1950). The molecules are drawn as planar hexagons with darkened edges toward the viewer. Ring carbon atoms and the hydrogen atoms directly attached to them are not shown. The position of the hydroxyl groups, either above or below the plane of the ring, is sufficient to define the correct configuration of the molecule. Any group of atoms written to the right in the Fischer projection appears below the plane of the ring, and any group written to the left appears above the plane in the Haworth projections.

Intramolecular hemiacetal formation is not unique to glucose. It occurs in galactose, mannose, and the naturally occurring aldopentoses and aldoheptoses. Fructose and other ketoses of five carbons or more form intramolecular hemiketals. Galactose and mannose form six-membered cyclic structures analogous to those drawn for glucose, while the most common cyclic structure for fructose is a five-member ring (Figure 19.4b). The difference between the alpha and beta forms of the sugars may seem trivial, but keep in mind that such differences are often crucial in biochemical reactions. We shall encounter some examples of this principle later in this chapter (Section 19.6) and when we discuss enzyme specificity (Chapter 22).

Review Questions

19.7 Define each of the following terms: (**a**) mutarotation; (**b**) anomer; (**c**) anomeric carbon.

19.8 How can it be shown that a solution of α-D-glucose exhibits mutarotation?

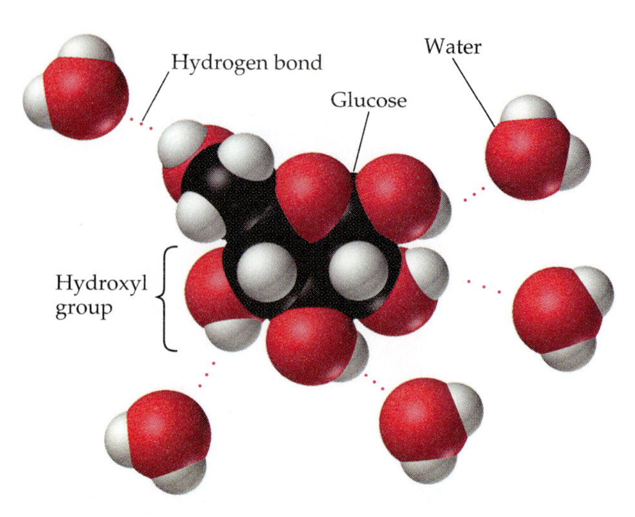

▶ **Figure 19.5**
Extensive hydrogen bonding between sugar molecules and water molecules leads to the considerable water solubility of most sugars.

19.4 Properties of Monosaccharides

Glucose, mannose, galactose, and fructose are crystalline solids at room temperature. With five hydroxyl groups per molecule, these sugars are quite soluble in water (Figure 19.5). Chemically, these monosaccharides undergo the reactions to be expected from their functional groups. The hydroxyl groups react to form esters and ethers. These reactions, though, are more important commercially for the polysaccharide cellulose (Section 19. 6) than for the simpler sugars.

One important reaction is the oxidation of the aldehyde group, which is one of the most easily oxidized organic functional groups. This can be accomplished by any mild oxidizing agent. Tollens's and Benedict's reagents (Section 15.6) are frequently used. The Tollens's test is based on the reduction of silver ions, and the Benedict's test involves the reduction of a copper complex. Any carbohydrate capable of this reduction without first undergoing hydrolysis is said to be a **reducing sugar.**[2]

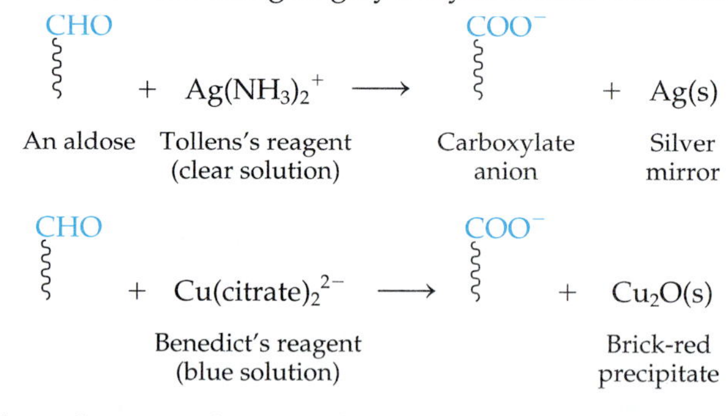

These reactions have been used as simple and rapid diagnostic tests for the presence of glucose in blood or urine. For example, Clinitest tablets, which are used in clinical laboratories to test for sugar in the urine, contain Cu(II) ions and are based on Benedict's test. A green color indicates very little sugar, whereas a brick-red color indicates sugar in excess of 2 g/100 mL of urine.

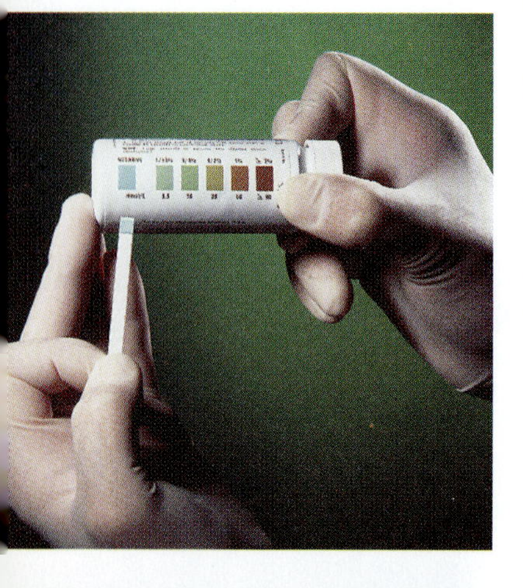

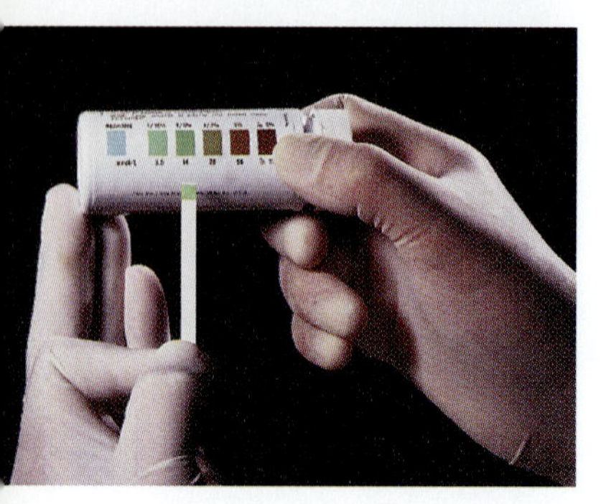

▲ The results of a urinalysis test for glucose in urine. The sample at the top shows the result for normal urine; the sample on the bottom is from a urine sample with a high glucose content.

Review Questions

19.9 Are all monosaccharides soluble in water? Explain.

19.10 What is a reducing sugar?

[2]Ketoses also give a positive test. In alkaline solution, an equilibrium exists between the ketoses and the aldoses in a reaction known as tautomerism. Because the oxidizing reagents commonly used to detect reducing sugars are prepared in a basic solution, all monosaccharides act as reducing sugars. It is possible to distinguish between aldoses and ketoses in acidic solutions.

Disaccharides and Polysaccharides

The carbohydrates we have studied so far are all monosaccharides. The monosaccharide units can join together in chains. Two monosaccharide molecules can join together by eliminating a water molecule to form a *disaccharide.* Three monosaccharide molecules can join to form a *trisaccharide,* and so on. Many monosaccharide molecules can join to form a *polysaccharide.*

19.5 Disaccharides

In Section 15.6 we noted that an acetal is formed when an alcohol reacts with a hemiacetal in the presence of an acid, while a ketal is formed when an alcohol reacts with a hemiketal. **Disaccharides** ($C_{12}H_{22}O_{11}$) are composed of two monosaccharide units joined by an acetal linkage formed when the hemiacetal group of one monosaccharide reacts with the hydroxyl group of a second monosaccharide. When the two monosaccharides combine, the carbon–oxygen–carbon linkage is called a **glycosidic linkage.**

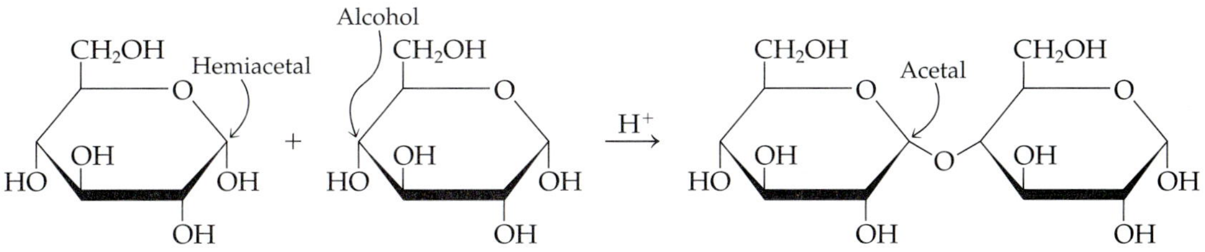

Alcohol
Hemiacetal
CH_2OH ... CH_2OH $\xrightarrow{H^+}$ Acetal

α-D-Glucose + α-D-Glucose → α-Maltose

The disaccharides differ from one another in their monosaccharide constituents and in the type of glycosidic linkage connecting them. There are three common disaccharides—maltose, lactose, and sucrose. All three are white crystalline solids at room temperature. Sucrose is quite soluble in water (200 g in 100 mL), and lactose is moderately soluble (20 g in 100 mL). We'll consider each of these sugars in more detail.

Maltose

Maltose occurs to a limited extent in sprouting grain. Its major source, however, is the partial hydrolysis of starch. In the manufacture of beer, maltose is liberated by the action of malt (germinating barley) on starch, and for this reason it is often referred to as *malt sugar.* Maltose is about 30% as sweet as sucrose. The human body is unable to utilize maltose or any other disaccharide directly because the molecules are too large to pass through cell membranes. Therefore, the disaccharide must first be broken down by hydrolysis into its two constituent monosaccharide units. In the body, this hydrolysis reaction is catalyzed by enzymes. The same hydrolysis reaction can be carried out in the laboratory with dilute acid as a catalyst, but the reaction rate is much slower and requires high temperatures. Hydrolysis of maltose, either by the action of the enzyme *maltase* or by means of an acid catalyst, produces two molecules of D-glucose.

$$\text{Maltose} \xrightarrow{H^+ \text{ or maltase}} 2\ \text{D-Glucose}$$

Maltose is a reducing sugar, and it exhibits mutarotation (Figure 19.6). The formula of maltose must therefore incorporate two glucose molecules in such a way that one free hemiacetal hydroxyl group exists. The glucose units in maltose are

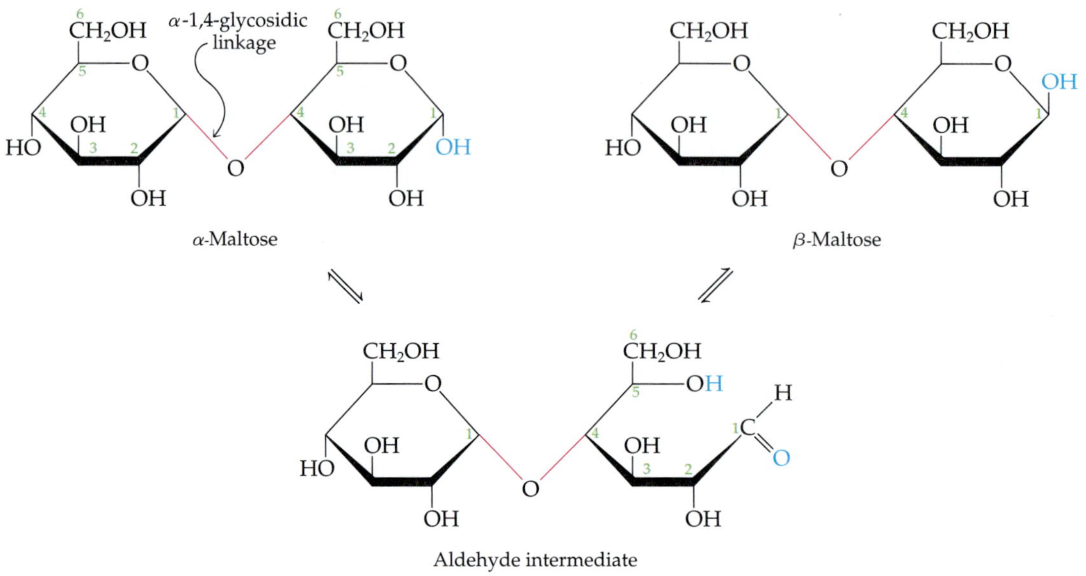

α-1,4-glycosidic linkage

α-Maltose

β-Maltose

Aldehyde intermediate

▲ **Figure 19.6**
Equilibrium mixture of maltose isomers.

Another way to express the glycosidic linkage is α (1 → 4) glycosidic linkage, which shows that the bond is from carbon 1 of the first monosaccharide to carbon 4 of the second.

joined in a head-to-tail fashion through an alpha linkage from C-1 of one glucose molecule to C-4 of the second glucose molecule (that is, an α-1,4-glycosidic linkage).

Lactose

Lactose is known as *milk sugar* because it occurs in the milk of humans, cows, and other mammals. Human milk contains about 7.5% lactose, whereas cow's milk, which is not as sweet, contains about 4.5% lactose. One of the few carbohydrates associated exclusively with the animal kingdom, lactose is synthesized only by mammary tissue. (Most other carbohydrates are plant products.) Lactose is produced commercially from whey, a by-product in the manufacture of cheese. Lactose is one of the lowest ranking sugars in terms of sweetness (about one-sixth as sweet as sucrose). It is a reducing sugar and exhibits mutarotation. The alpha form of the sugar is commercially important as an infant food and in the production of penicillin. Drug dealers use lactose to "cut" their heroin and thus increase their profits.

Lactose is composed of one molecule of D-galactose and one molecule of D-glucose joined by a β-1,4-glycosidic bond. The two monosaccharides are obtained from lactose by acid hydrolysis or by catalytic action of the enzyme *lactase:*

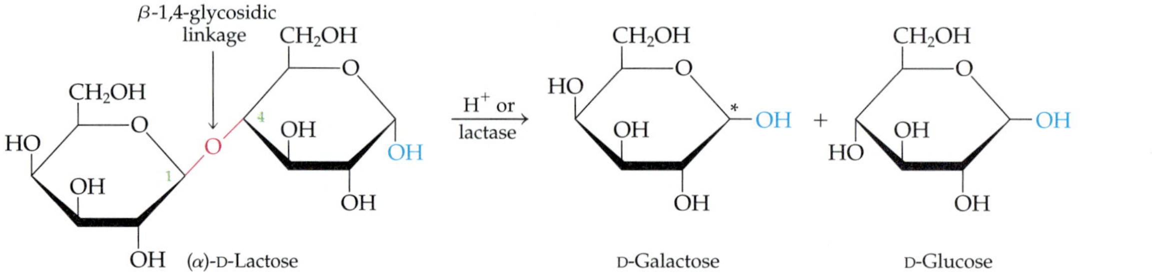

β-1,4-glycosidic linkage

(α)-D-Lactose $\xrightarrow{H^+ \text{ or lactase}}$ D-Galactose + D-Glucose

(*We use this convention for writing the hydroxyl group on the hemiacetal carbon when we do not wish to specify either the α or the β isomer.)

Many adults, and some children, have a deficiency of lactase. These individuals are said to be **lactose intolerant** because they cannot digest the lactose found in milk.

Lactose Intolerance and Galactosemia

People who suffer from lactose intolerance are unable to digest the lactose found in milk. Lactose makes up about 40% of an infant's diet during the first year of life. Infants and small children have one form of lactase in their small intestines and therefore can easily digest lactose. However, adults have a less active form of the enzyme, and about 70% of the world's adult population have some lactase deficiency, which means lactose cannot be hydrolyzed in the small intestine to galactose and glucose. For some people the inability to synthesize sufficient enzyme increases with age. Up to 20% of the U.S. population suffers some degree of lactose intolerance.

The unhydrolyzed lactose passes into the colon, and its presence tends to draw water from the interstitial fluid into the intestinal lumen by osmosis. At the same time, intestinal bacteria may act on the lactose to produce organic acids and gases. The intake of water, plus the bacterial decay products, leads to the abdominal distention, cramps, and diarrhea that are symptoms of the condition.

Symptoms disappear if milk or other products containing lactose are limited or excluded from the diet. Many food stores now carry special brands of milk that have been pretreated with lactase to hydrolyze the lactose. When milk is cooked or fermented, the lactose is at least partially hydrolyzed. For this reason people with lactose intolerance may still be able to enjoy cheese, yogurt, or cooked foods containing milk with little or no problem. Lactose intolerance is most commonly treated by lactase preparations (e.g., Lactaid®), which are available in liquid and tablet form at drugstores and grocery stores. These are taken orally with dairy foods to assist in their digestion.

In the genetic disease galactosemia, either of the two enzymes needed to convert galactose to glucose is lacking. The blood galactose level is markedly elevated, and galactose is found in the urine. An infant with galactosemia experiences a lack of appetite, weight loss, diarrhea, and jaundice. The disease may result in impaired liver function, cataracts, mental retardation, and even death. If recognized in early infancy, the effects of galactosemia can be eliminated by removing milk and all other sources of galactose from the diet. As the child grows older, he or she normally develops an alternate pathway for metabolizing galactose, and thus the need to restrict milk is not permanent. The incidence of galactosemia in the United States is 1 in every 65,000 newborn babies.

A more serious problem is the genetic disease **galactosemia,** which results from the absence of an enzyme that converts galactose to glucose. These two medical problems are discussed in more detail in the boxed essay "Lactose Intolerance and Galactosemia."

Maple syrup is the concentrated sap of the sugar maple tree. It is a solution of sugars—about 65% sucrose with small amounts of glucose and fructose.

Sucrose

Sucrose is known as beet sugar, cane sugar, table sugar, or simply sugar. It is probably the largest-selling pure organic compound in the world. As its names imply, sucrose is obtained from sugar canes and sugar beets (whose juices are 14–20% sucrose) by evaporation of the water and recrystallization. The dark brown liquid that remains after crystallization of the sugar is sold as molasses.

The sucrose molecule is unique among common disaccharides in having an α-1, β-2-glycosidic (head-to-head) linkage. Because this acetal linkage involves the hydroxyl group on the carbon in position 1 of α-D-glucose and the hydroxyl group on carbon 2 of β-D-fructose, it ties up the anomeric carbons of both glucose and fructose.

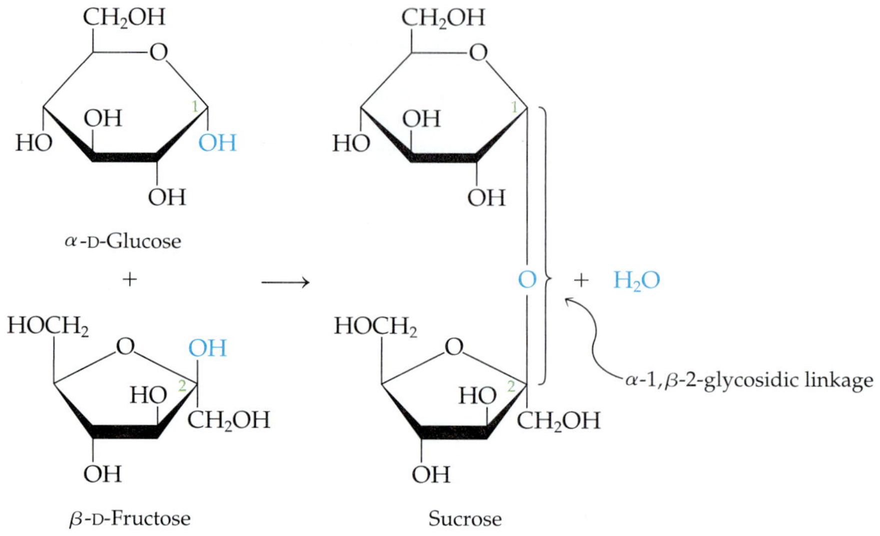

α-D-Glucose + β-D-Fructose ⟶ Sucrose + H_2O

α-1,β-2-glycosidic linkage

This bonding bestows certain properties upon sucrose that are quite different from those of the other disaccharides we have just discussed. As long as the sucrose molecule remains intact, it cannot "uncyclize" to form the open-chain structure. Thus sucrose is incapable of mutarotation and exists in only one form both in the solid state and in solution. Because the α-1, β-2-glycosidic linkage involves the reducing hydroxyl ends of both monosaccharides, sucrose does not undergo reactions that are typical of aldehydes and ketones. Therefore, sucrose is a nonreducing sugar.

The hydrolysis of sucrose in dilute acid or the presence of the enzyme *sucrase* gives an equimolar mixture of glucose and fructose. This 1:1 mixture is referred to as **invert sugar.**[3]

This hydrolysis reaction has several practical applications. Because sucrose can exist in only one molecular configuration, it readily crystallizes from a solution. Invert sugar has a much greater tendency to remain in solution. In the manufacture of jelly and candy and in the canning of fruit, crystallization of the sugar is undesirable. Therefore, conditions leading to the hydrolysis of sucrose are employed in these processes (see "Chocolate-Covered Cherries"). Because fructose is sweeter than sucrose, the hydrolysis adds to the sweetening effect. Bees carry out this reaction when they make honey.

The average American consumes more than 100 pounds of sucrose every year. About two-thirds of this amount is ingested in soft drinks, presweetened cereals, and other highly processed foods. The widespread use of sucrose has generated much adverse publicity. Various health magazines have reported that excess sugar causes cancer, heart disease, migraine headaches, hyperactivity in children, obesity, and tooth decay. Only the latter two claims have been substantiated. As we shall see in Chapter 26, carbohydrates are converted to fat when the caloric intake exceeds the body's requirements. Sucrose does cause tooth decay by serving as part of the plaque that sticks to a tooth. The bacteria contained within the plaque use sucrose both as an adhesive and as a food source. We shall see in Chapter 25 that bacteria can decompose sugar to lactic acid, which corrodes the teeth and leads to gum destruction. Thus the amount of plaque formed can be decreased by reducing your intake of sucrose. (The best way to remove plaque already formed is by daily flossing.)

[3]Sucrose is dextrorotatory (specific rotation +66.5°). During hydrolysis, the observed rotation decreases and eventually becomes negative. This is because fructose has a high negative specific rotation (−92.4° at equilibrium) that more than balances the positive rotation of glucose (+52.7°). Because the sign of rotation changes during the reaction, the process is known as *inversion*, another name for sucrase is *invertase*, and the products are called *invert sugar.*

Chocolate-Covered Cherries

If you are a devotee of chocolate-covered cherries, you may wonder how the manufacturer surrounds the cherry with liquid without making a hole in the chocolate covering. The secret is a chemical reaction that takes place after the candy is made. The process uses an enzyme, sucrase (also known as invertase), to convert sucrose into invert sugar (a 1:1 mixture of glucose and fructose).

Before being dipped in chocolate, the cherries are coated with a sugary paste containing sucrase. Once the paste hardens, the cherries are dipped in chocolate and then stored for one to two weeks. During this storage period the sucrase catalyzes the hydrolysis of sucrose (which is crystallized) into glucose and fructose, which have a greater tendency to remain in the solution formed from the cherry's juice, rather than crystallize. This reaction also increases the sweetness of the candy because fructose is sweeter than sucrose.

The preparation of the chocolate for the coating is also important. It must be of an optimum particle size to please the palate. If the particles are too small, the chocolate will feel slimy when eaten; if too large, it will feel gritty.

▲ The liquid interior of a chocolate-covered cherry is a result of the hydrolysis of sucrose.

Some food faddists claim that raw sugar is much better for you than refined sugar. Raw sugar does contain a few trace minerals, but hardly enough to make it a lot more desirable than refined sugar. People in the United States probably consume too much sugar—whether raw or refined.

✓ Review Questions

19.11 Identify these sugars by their proper names: (**a**) milk sugar; (**b**) table sugar; (**c**) malt sugar.

19.12 For the abbreviated trisaccharide drawn below, indicate whether each glycosidic linkage is alpha or beta.

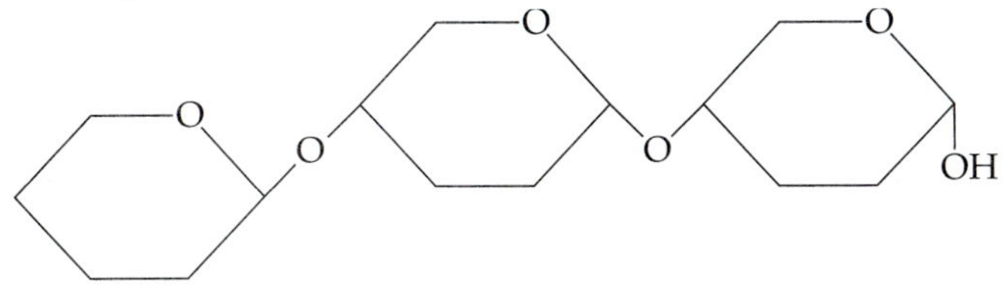

19.6 Polysaccharides

The polysaccharides are the most abundant carbohydrates in nature and serve a variety of functions, such as energy storage or structural components of plant cells. **Polysaccharides** are high-molar mass (25,00 to 15,000,000 daltons) polymers of monosaccharides joined together by glycosidic linkages. The three most abundant polysaccharides are starch, glycogen, and cellulose. These three are also referred to as *homopolymers* because each yields only one type of monosaccharide upon complete hydrolysis. *Heteropolymers* may contain sugar acids, amino sugars, or noncarbohydrate substances in addition to monosaccharides. Heteropolymers are common in nature (gums, pectins, and other substances) but we will not discuss them further. The polysaccharides are nonreducing carbohydrates, are not sweet-tasting (probably because of their limited solubility), and do not undergo mutarotation.

▲ **Figure 19.7**
Starch forms water-insoluble granules, such as those that make up the bulk of the cells in a potato.

▲ Note the deep blue-violet color that forms as a solution containing iodine is added to a solution of cornstarch.

Starch

Starch is the most important source of carbohydrate in the human diet and accounts for more than 50% of our carbohydrate intake. Starch occurs in plants in the form of granules (Figure 19.7). Starch granules are particularly abundant in seeds (especially the cereal grains) and tubers, where they serve as a storage form of carbohydrate. The breakdown of starch to glucose nourishes the plant during periods of reduced photosynthetic activity. We often think of potatoes as a "starchy" food, and yet other plants contain a much greater percentage of starch (potatoes 15%, wheat 55%, corn 65%, and rice 75%).

Starch is a mixture of two polymers, **amylose** and **amylopectin,** which can be separated from each other by physical and/or chemical methods. Natural starches consist of about 10–30% amylose and 70–90% amylopectin. Amylose is a linear polysaccharide composed entirely of D-glucose units joined by α-1,4-glycosidic linkages, as in maltose (Figure 19.8a). Thus amylose might be thought of as either polymaltose or polyglucose. There may be 60 to 300 glucose units per chain. Experimental evidence indicates that amylose is not a straight chain of glucose units. Rather, it is coiled like a spring with six glucose monomers per turn (Figure 19.8b). When coiled in this fashion, amylose has just enough room in its core to accommodate an iodine molecule. The characteristic blue-violet color that starch gives when treated with iodine is due to the formation of the amylose-iodine complex. The test is sensitive enough to detect even minute amounts of starch in solution.

Amylopectin is a branched-chain polysaccharide composed of glucose units linked primarily by α-1,4-glycosidic bonds but with occasional α-1,6-glycosidic bonds, which are responsible for the branching. It has been estimated that there may be 300 to 6000 glucose units in amylopectin and that branching occurs about

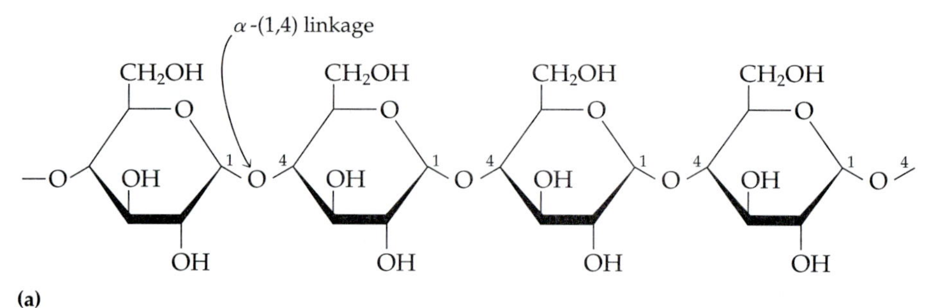

(a)

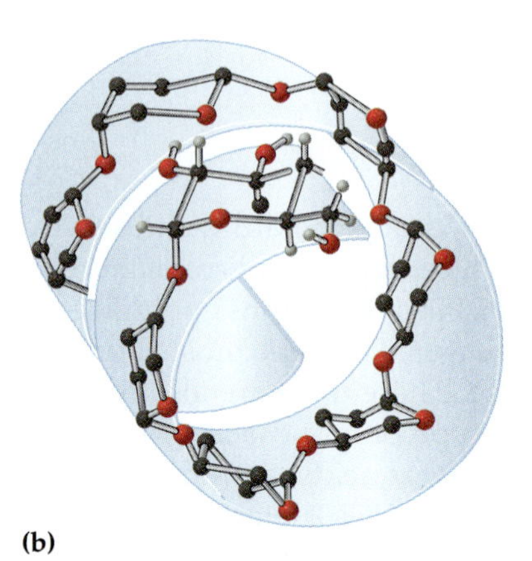

(b)

▲ **Figure 19.8**
(a) Amylose is a linear chain of α-D-glucose chains joined by α-(1,4)-glycosidic bonds. (b) Because of hydrogen bonding, amylose forms a spiral structure that contains six glucose units per turn.

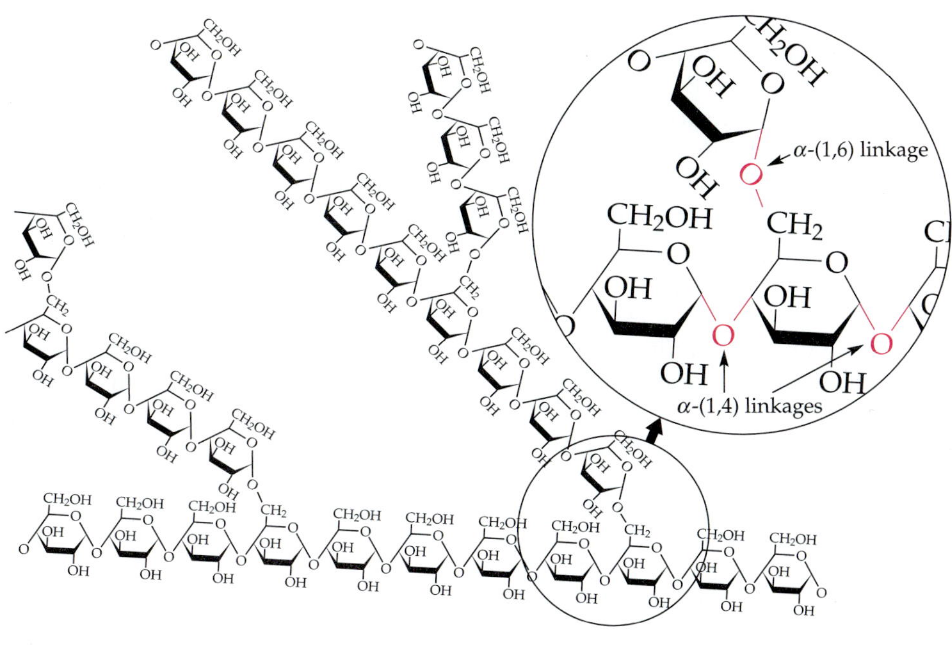

(a)

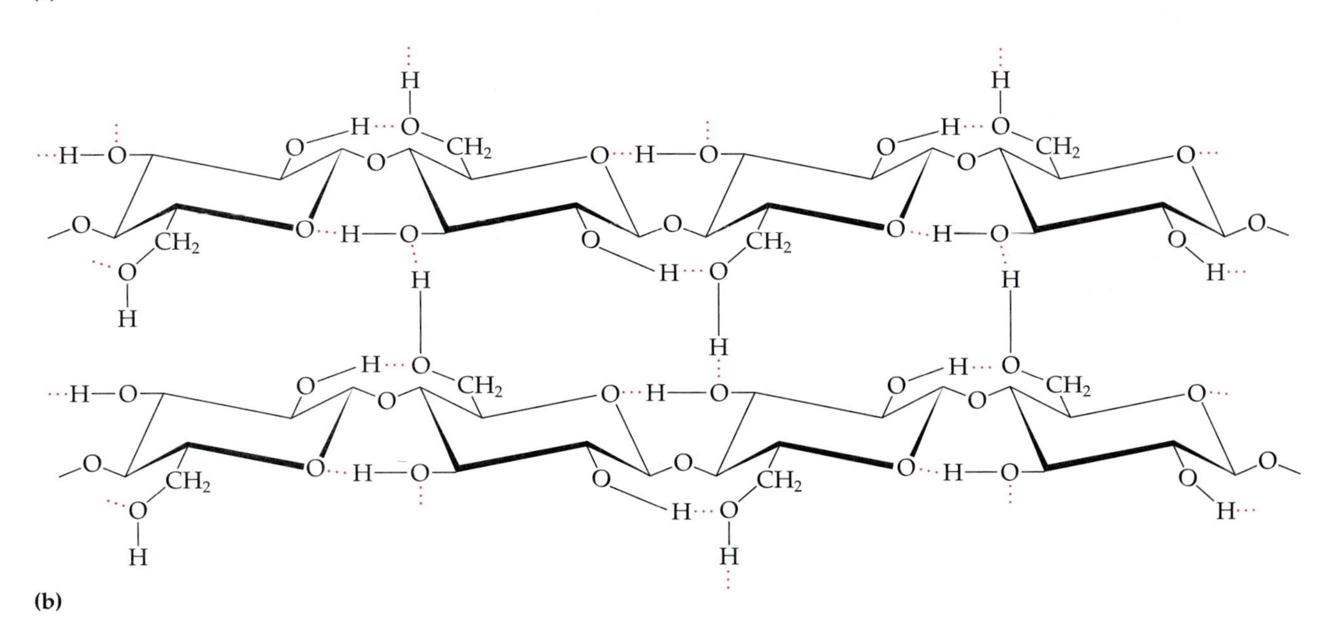

(b)

▲ **Figure 19.9**
(a) Representation of the branching in amylopectin and glycogen. (b) The conformation of cellulose. Note the extensive hydrogen bonding.

once every 25 to 30 units (Figure 19.9a). The helical structure of amylopectin is disrupted by the branching of the chain, so instead of the deep blue-violet color amylose gives with iodine, amylopectin produces a less intense reddish brown.

Commercial starch is a white powder. The complete hydrolysis of starch (amylose and amylopectin) yields, in successive stages: dextrins ⟶ maltose ⟶ glucose. Heating in the presence of dilute acid will hydrolyze starch. In the human body several enzymes, known collectively as amylases, degrade starch sequentially into usable glucose units.

Dextrins are glucose polysaccharides of intermediate size. The shine and stiffness imparted to clothing by starch are due to the presence of dextrins formed when the clothing is ironed. Because of their characteristic stickiness upon wetting, dextrins are used as adhesives on stamps, envelopes, and labels, as binders to hold pills and tablets together, and as pastes. Because dextrins are more easily digested than starch, they are used extensively in the commercial preparation of infant foods. A dried mixture of dextrins, maltose, and milk is used in the preparation of malted milk.

Glycogen

About 70% of the total glycogen in the body is stored in muscle cells. Although the percent of glycogen (by mass) is higher in the liver, the much greater mass of skeletal muscle stores a greater total amount of glycogen.

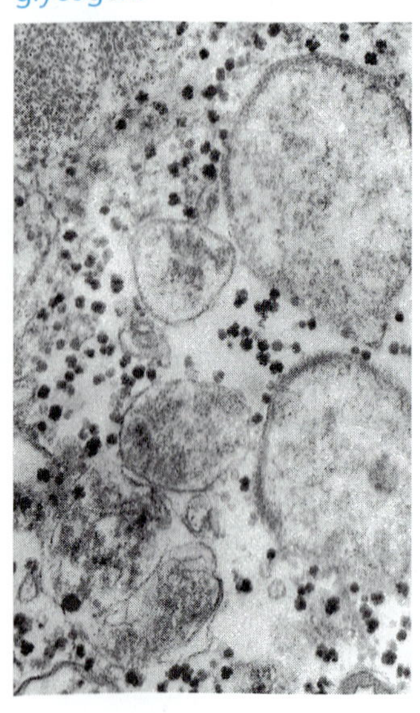

▲ **Figure 19.10**
Electron micrograph of glycogen granules in a rat muscle cell.

Glycogen, often called animal starch, is the reserve carbohydrate of animals. Practically all mammalian cells contain some stored carbohydrate in the form of glycogen. However, it is especially abundant in the liver (4–8% of tissue by mass) and in skeletal muscle cells (0.5–1.0%). Like starch in plants, glycogen molecules appear as granules in liver and muscle cells (Figure 19.10). When fasting, animals draw upon these glycogen reserves on the first day to obtain the glucose needed to maintain metabolic balance.

Glycogen is structurally quite similar to amylopectin, but glycogen is more highly branched (8–12 glucose units between branches, versus about 25 in amylopectin) and has shorter branches than those in amylopectin. When treated with iodine, glycogen gives a reddish-brown color. Glycogen can be broken down into its D-glucose subunits by acid hydrolysis or by the same enzymes that catalyze the breakdown of starch. In animals, the enzyme phosphorylase catalyzes the breakdown of glycogen to phosphate esters of glucose (see Section 25.5).

Cellulose

Cellulose is a fibrous carbohydrate found in all plants; it is the structural component of plant cell walls. Because the Earth is covered with vegetation, cellulose is the most abundant of all carbohydrates, accounting for over 50% of all the carbon found in the vegetable kingdom. Cotton fibrils and filter paper are almost entirely cellulose (about 95%), wood is about 50% cellulose, and the dry weight of leaves is about 10–20% cellulose. From an industrial and economic standpoint, cellulose is the most important carbohydrate. The largest use of cellulose is in the manufacture of paper and paper products. Although the use of synthetic fibers is increasing, rayon (made from cellulose) and cotton still account for over 70% of textile production.

Like amylose, cellulose is a linear polymer of glucose. It differs, however, in that the glucose units (about 2000 to 3000) are joined by β-1,4-glycosidic linkages, giving a more

▶ **Figure 19.11**
Electron micrograph of the cell wall of an alga. The wall consists of successive layers of cellulose fibers in parallel arrangement.

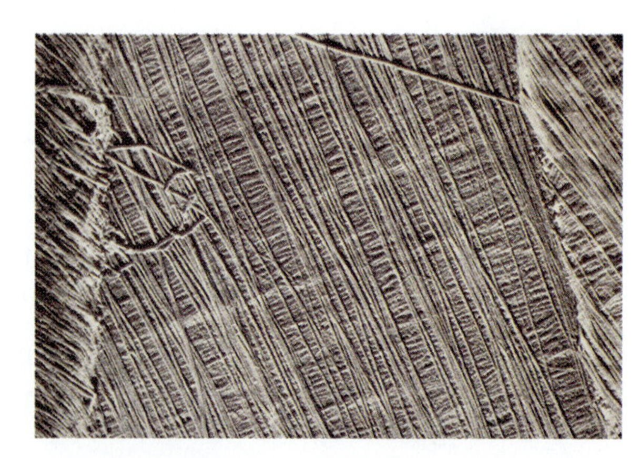

Dietary Fiber

The importance of dietary fiber in the human diet has received a great deal of attention, with several dietary studies citing its benefits in reducing the risk of colon cancer and heart disease, as well as in better control of blood sugar levels in diabetic patients. Dietary fiber is obtained from plant sources and can be divided into two categories, based on its behavior in water. *Soluble fiber* is suspended in water and includes pectins and gums, which are polysaccharides. Good sources of soluble fiber are oats, psyllium seed husks, and fruits. *Insoluble fiber* is not suspended in water and is primarily composed of celluose. Nearly all fiber-containing foods have a higher amount of insoluble fiber. As we will see, each type of fiber is important for different reasons.

The link between a lowered incidence of colon cancer and an increase of fiber in the diet was first noted in studies showing that people in developed countries are much more likely to get colon cancer than people in underdeveloped nations. People in developed countries eat diets rich in highly processed, low-fiber foods, while those in more "primitive" areas eat high-fiber diets, which lead to frequent and robust bowel movements. Low-fiber diets therefore result in less frequent bowel action, with high retention times for feces in the colon. Bacteria act upon the material in the colon. (Indeed, bacteria, both living and dead, make up about one-third of the dry weight of feces.) With a high-fiber diet, the materials seldom remain in the colon for more than one day. With a low-fiber diet, the retention time can be as long as three days, allowing for prolonged bacterial activity that produces a high level of mutagenic chemicals. Chemicals that are mutagenic often are also carcinogenic.

A clinical study published in 1999 in the *New England Journal of Medicine* failed to show a link between the incidence of colon cancer and the intake of dietary fiber. This contradicts earlier clinical studies and epidemiological studies. It is possible that other substances, such as antioxidants (Section F.8), are present in higher amounts in foods high in fiber and these lower the risk of colon cancer.

The link between dietary fiber and a lowered incidence of heart disease is related to the effect of dietary fiber on cholesterol levels. An increase of soluble fiber in a low-fat diet will reduce high cholesterol levels by 5% or more.

Other clinical studies have shown that a high-fiber diet is useful in the treatment of diabetes. It has long been known that diabetics ought to limit ingestion of rapidly digested sugars such as glucose and sucrose and eat more carbohydrates that are slowly digested (e.g., starches high in amylopectin). The presence of fiber in the diet reduces the rate of absorption of glucose, and therefore the peak blood sugar concentration is lowered.

What is the recommended amount of fiber in a healthful diet? The American Dietetics Association (ADA) recommends that adults include 20 to 35 g of dietary fiber each day or 10 to 13 g of dietary fiber for every 1,000 calories consumed. The actual average intake of dietary fiber in the United States is about half this amount—14 to 15 g/day. Recommended levels for children are still being determined. Based on limited clinical data, the recommendation for dietary fiber intake for children two years of age and older is to increase their dietary fiber intake to an amount equal to or greater than their age plus five. Thus, the minimum amount of dietary fiber for a 14-year-old would be 19 g/day. By the time the individual is 20, he or she should include sufficient dietary fiber in the diet to meet the recommended levels for adults.

▲ These foods are excellent sources of dietary fiber.

extended structure than amylose (Figure 19.8b). The linear nature of the cellulose chains allows a great deal of hydrogen bonding between hydroxyl groups on adjacent chains. As a result, the chains are closely packed into fibers (Figure 19.11), and there is little interaction with water or with any other solvent. Cotton and wood, for example, are completely insoluble in water and have considerable mechanical strength. Because there is no helical structure, cellulose does not bind to iodine to give a colored product.

Cellulose yields D-glucose upon complete acid hydrolysis, and yet humans (and all other vertebrates) cannot utilize cellulose as a source of glucose. We can eat potatoes, but we can't eat grass. Our digestive juices lack enzymes that can hydrolyze the β-glycosidic linkages found in cellulose. However, there are microorganisms that can digest cellulose because they make the enzyme cellulase, which catalyzes the hydrolysis of cellulose. The presence of these microorganisms in the digestive tracts of herbivorous animals (cows, horses, sheep, and so forth) allows these higher animals to degrade the cellulose from plant material into glucose for energy. Termites also contain cellulase-secreting microorganisms and thus can subsist on a wood diet. This once again demonstrates the extreme stereospecificity of biochemical processes.

Although it has no nutritive value, cellulose makes up the greater part of dietary fiber (see boxed essay). The fibrous portions of plants (stems, peels, and seeds) are rich in fiber. The bran of cereal grains, celery, beans, apples, raspberries, and figs are good sources of dietary fiber. Because so much of the carbon in the biosphere exists as cellulose, considerable research has gone into converting it to glucose or some other form of food for humans. There are enzymes that will degrade cellulose, but to do so efficiently requires some pretreatment of the fibers.

Review Questions

19.13 What purposes do starch and cellulose serve in plants?

19.14 What purpose does glycogen serve in animals?

Summary

Carbohydrates, compounds containing carbon, hydrogen, and oxygen, include sugars, starch, glycogen, and cellulose. All carbohydrates contain alcohol functional groups, and either an aldehyde or a ketone group (often in the form of a hemiacetal, acetal, hemiketal, or ketal). The simplest carbohydrates are **monosaccharides.** Those that can be hydrolyzed to two monosaccharide units are **disaccharides,** and those that can be hydrolyzed to many monosaccharide units are **polysaccharides.** Most sugars are either monosaccharides or disaccharides. Cellulose, glycogen, and starch are polysaccharides.

A sugar is designated as being a D sugar or an L sugar according to how, in a Fischer projection of the molecule, the hydrogen and hydroxyl group are attached to the *penultimate* carbon, which is the carbon immediately before the terminal alcohol carbon. If the structure at this carbon is the same as that of D-glyceraldehyde (—OH to the right), the sugar is a D sugar; if the configuration at the penultimate carbon is that of L-glyceraldehyde (—OH to the left), the sugar is an L sugar. The D/L notation in no way relates to whether the sugar is dextrorotatory or levorotatory. An L sugar or a D sugar can rotate the plane of polarized light either to the right or to the left. The direction of optical activity is indicated by a plus or a minus sign following the D/L part of the name.

Monosaccharides of five carbons or more readily form cyclic hemiacetal structures with the carbonyl carbon reacting with a hydroxyl group on a carbon three or four carbons away:

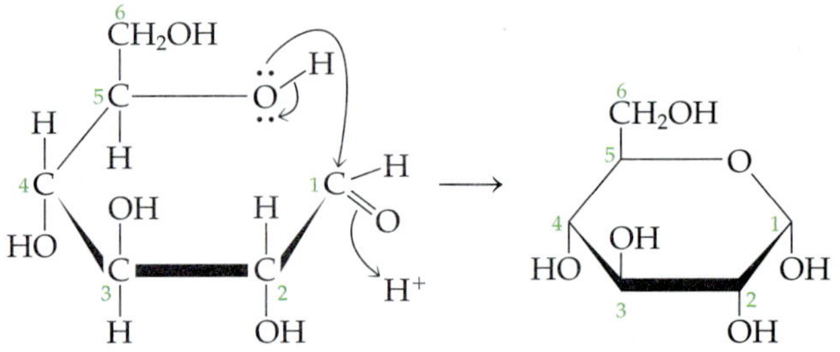

Glucose in solution exists as an equilibrium mixture of three forms: α-, β-, and open-chain. In Haworth projections, the *alpha* form is drawn with the hydroxyl group on the "former" carbonyl carbon (**anomeric carbon**) pointing downward; the *beta* form, with the hydroxyl group pointing upward; these two compounds are stereoisomers and are given the more specific term of **anomers:**

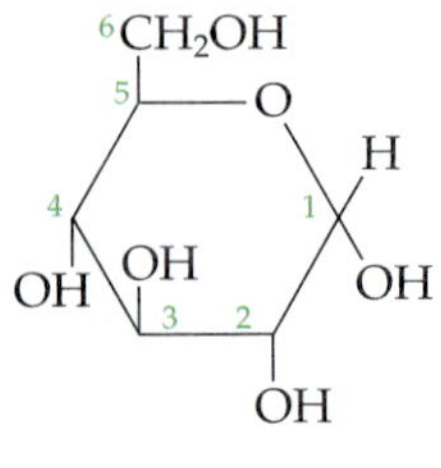

α-Glucose

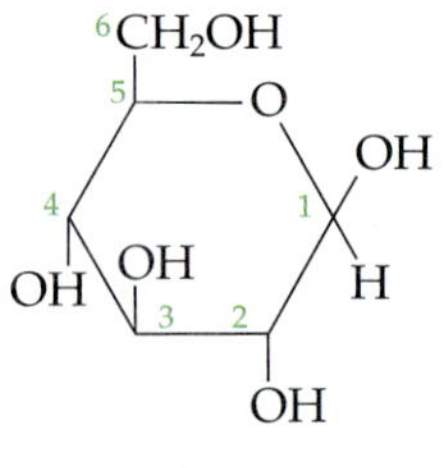

β-Glucose

Any solid reducing sugar can be all alpha or all beta. Once the sample is dissolved in water, however, the hemiacetal ring opens up into the open-chain form and then closes to form either the α- or the β-anomer. These interconversions occur back and forth until an equilibrium mixture is achieved, and the process is called **mutarotation.**

Monosaccharides form esters and ethers at the hydroxyl groups, and the carbonyl group is easily oxidized by Tollens's or Benedict's reagents (as well as others). Any mono- or disaccharide containing a latent carbonyl group is a **reducing sugar.**

The disaccharide *maltose* contains two glucose units joined in an α-1,4-glycosidic linkage. The disaccharide *lactose* comprises a galactose unit and a glucose unit joined via a β-1,4-glycosidic linkage. Both maltose and lactose contain a hemiacetal, so they are reducing sugars; they also undergo mutarotation.

The disaccharide *sucrose* (table sugar) consists of a glucose unit and a fructose unit joined by an acetal linkage. The linkage is designated an α-1, β-2-glycosidic linkage because it involves the carbon-1 hydroxyl group of glucose and the carbon-2 hydroxyl group of fructose. Sucrose is not a reducing sugar because there is no latent carbonyl group, and it cannot undergo mutarotation because of the restrictions imposed by this linkage.

Starch, the principal carbohydrate of plants, is made up of the polysaccharides **amylose** (10–30%) and **amylopectin** (70–90%). Ingested by humans, starch is hydrolyzed to glucose and used as the body's energy source. *Glycogen* is the polysaccharide animals use to store excess ingested carbohydrates. Similar in structure to amylopectin, glycogen is hydrolyzed to glucose whenever the animal needs energy for some metabolic process. The polysaccharide *cellulose* is the structural component of plant cells. It is a linear polymer of glucose units joined by β-1,4-glycosidic linkages. It is indigestible in the human body but digestible by many herbivores and microorganisms.

Key Terms

aldose (19.1)
amylopectin (19.6)
amylose (19.6)
anomers (19.3)
anomeric carbon (19.3)
biochemistry (page 506)
carbohydrate (page 506)
D sugar (19.1)
disaccharide (19.5)
galactosemia (19.5)
glycosidic linkage (19.5)
invert sugar (19.5)
ketose (19.1)
L sugar (19.1)
lactose intolerant (19.5)
monosaccharide (page 506)
mutarotation (19.3)
polysaccharide (19.6)
reducing sugar (19.4)

Problems

Monosaccharides: Terminology, Stereochemistry, and Cyclic Structures

1. Identify each of the following sugars as an aldose or ketose and then as a triose, tetrose, pentose, or hexose.
   a. D-glucose **b.** D-deoxyribose
   c. L-fructose **d.** D-mannose
2. Identify each of the following sugars as an aldose or ketose and then as a triose, tetrose, pentose, or hexose.
   a. L-glyceraldehyde **b.** dihydroxyacetone
   c. L-galactose **d.** D-fructose
3. Specify whether each of the following is a D sugar or an L sugar:

   a.
   H—C=O
   H—C—OH
   HO—C—H
   H—C—OH
   CH_2OH

   b.
   CH_2OH
   C=O
   H—C—OH
   HO—C—H
   CH_2OH

4. Specify whether each of the following is a D sugar or an L sugar.

   a.
   H—C=O
   HO—C—H
   HO—C—H
   CH_2OH

   b.
   H—C=O
   HO—C—H
   HO—C—H
   H—C—OH
   CH_2OH

5. Why are (+)-glucose and (−)-fructose both classified as D sugars?
6. How does ribose differ from deoxyribose?
7. From memory, draw formulas for the open-chain forms of D-glucose and D-mannose.
8. From memory, draw formulas for the open-chain forms of D-galactose and D-fructose.
9. Draw the cyclic structure for α-D-glucose. Identify the anomeric carbon.
10. Draw the cyclic structure for β-D-fructose. Identify the anomeric carbon.
11. Knowing that mannose differs from glucose only in the configuration at the second carbon, draw the cyclic structure for β-D-mannose.

12. Knowing that galactose differs from glucose only in the configuration at the fourth carbon, draw the cyclic structure for β-D-galactose.

Properties and Reactions of Monosaccharides

13. Which of the following gives a positive Benedict's test?
 a. L-galactose b. levulose c. D-mannose
14. Which of the following gives a positive Benedict's test?
 a. D-ribose b. grape sugar c. D-fructose
15. D-glucose can be oxidized at C-1 to form D-gluconic acid, at C-6 to yield D-glucuronic acid, and at both C-1 and C-6 to yield D-glucaric acid. Draw structures of these three oxidation products.

Disaccharides

16. For each of these abbreviated sugar formulas, indicate whether the glycosidic linkage is alpha or beta.
 a. CH_2OH CH_2OH O O OH
 b. CH_2OH O CH_2OH O OH O
17. For each of these abbreviated sugar formulas, indicate whether the glycosidic linkage is alpha or beta.
 a. CH_2OH CH_2OH O CH_2OH O O OH
 b. CH_2OH O CH_2OH O CH_2OH O
18. a. What is the orientation of the hydroxyl group at the hemiacetal carbon of structures **a** and **b** in Problem 16?
 b. Which structure (if any) shown in Problem 16 is *not* a reducing sugar?
19. a. What is the orientation of the hydroxyl group at the hemiacetal carbon of structures **a** and **b** in Problem 17?
 b. Which structure (if any) shown in Problem 17 is *not* a reducing sugar?
20. Melibiose is a disaccharide that occurs in some plant juices. Its structure is:

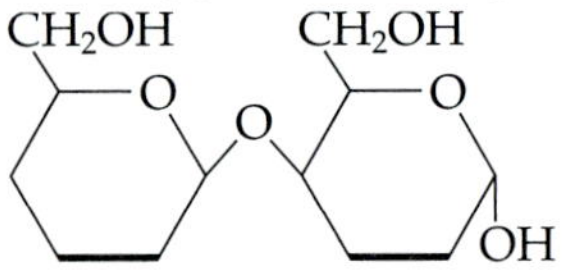

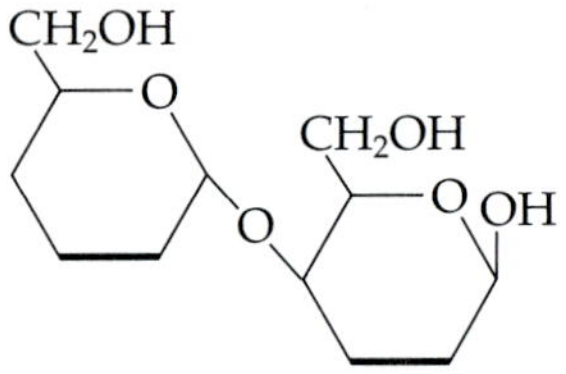

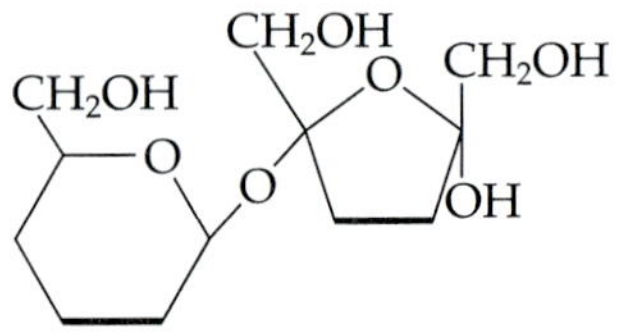

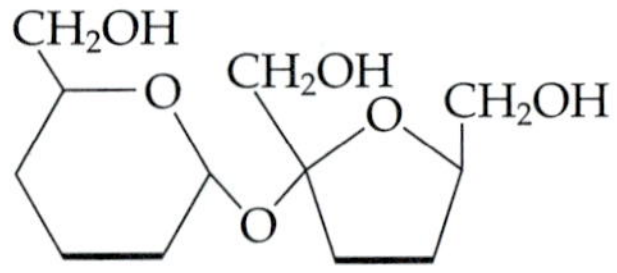

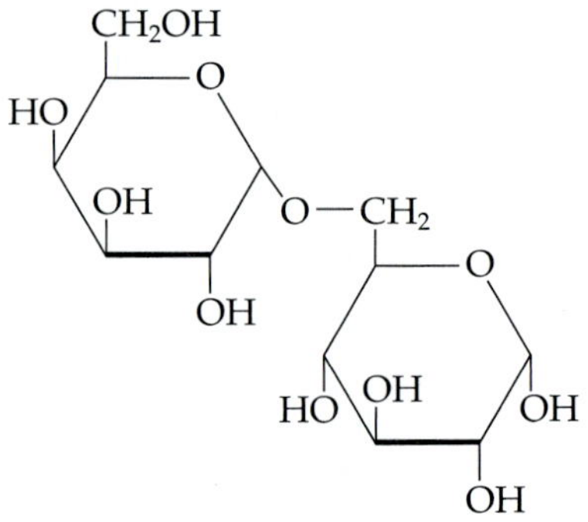

 CH_2OH O HO OH O—CH_2 O OH HO OH OH OH

 a. What monosaccharide units are incorporated in melibiose?
 b. What type of linkage (α or β) joins the two rings of melibiose?
 c. Why is melibiose a reducing sugar?
 d. Circle the hemiacetal carbon and indicate whether the hydroxyl group is α or β.
21. Gentiobiose is a disaccharide composed of two glucose units joined by a β-1,6-glycosidic linkage. Draw the structure of gentiobiose.
22. What monosaccharide(s) is (are) obtained from the hydrolysis of each of the following?
 a. lactose b. sucrose c. maltose
23. List the reagents necessary for the following conversion:

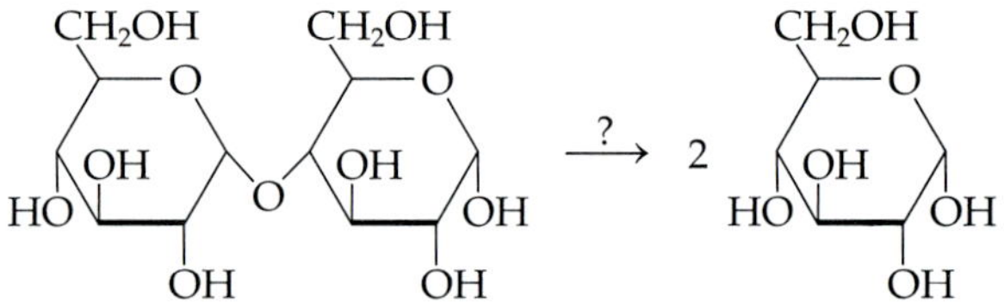

 CH_2OH O OH HO OH O CH_2OH O OH OH OH $\xrightarrow{?}$ 2 CH_2OH O OH HO OH OH

Polysaccharides

24. What monosaccharide(s) is (are) obtained from the hydrolysis of each of the following?
 a. starch b. cellulose c. glycogen
25. How do amylose and amylopectin differ from each other? How are they similar?
26. How do amylose and cellulose differ from each other? How are they similar?
27. How do amylopectin and glycogen differ from each other? How are they similar?

Additional Problems

28. Which of the following gives a positive Benedict's test?
 a. maltose b. glycogen c. D-glucose
 d. sucrose e. D-glyceraldehyde f. cellulose
29. What structural characteristics are necessary if a disaccharide is to be a reducing sugar? Draw the structure of a hypothetical nonreducing disaccharide composed of two aldohexoses.
30. Raffinose is a trisaccharide (found in beans and sugar beets) containing D-galactose, D-glucose, and D-fructose. The enzyme α-galactase catalyzes the hydrolysis of raffinose to

galactose and sucrose. Draw the structure of raffinose. (The linkage from galactose to the glucose unit is α-1,6.)

31. List the reagents necessary for the following conversions:

a. H–C(=O)–C(H)(OH)... $\xrightarrow{?}$ ...

(Structures: $HC(=O)$, $HO-C-H$, $H-C-OH$, CH_2OH → $^-O-C(=O)$, $HO-C-H$, $H-C-OH$, CH_2OH)

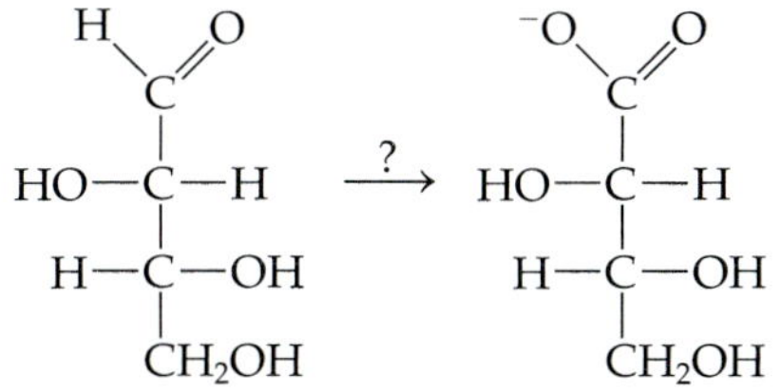

b. (Pyranose ring with CH_2OH, O, OH, HO, OH, OH) $\xrightarrow{?}$ (Pyranose ring with CH_2OH, O, OH, HO, OCH_3, OH)

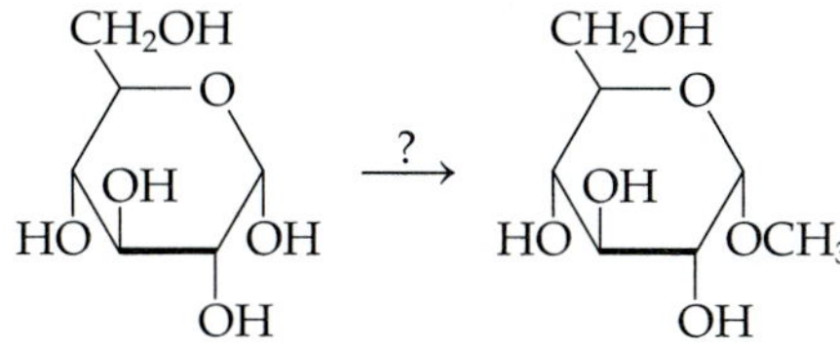

32. In the schematic below, Glc represents glucose. What substance is indicated? Justify your answer.

· · · Glc-Glc

· · · Glc-Glc-Glc-Glc

· · · Glc-Glc-Glc-Glc-Glc-Glc-Glc-Glc-Glc

· · · Glc-Glc-Glc-Glc-Glc-Glc-Glc-Glc-Glc-Glc-Glc-Glc-Glc

· · · Glc-Glc-Glc-Glc-Glc-Glc-Glc-Glc

33. Xylulose, found in the urine of humans suffering from a genetic condition called *pentosuria,* has the structure shown below. Classify xylulose as fully as possible.

CH_2OH
$C{=}O$
$H-C-OH$
$HO-C-H$
CH_2OH

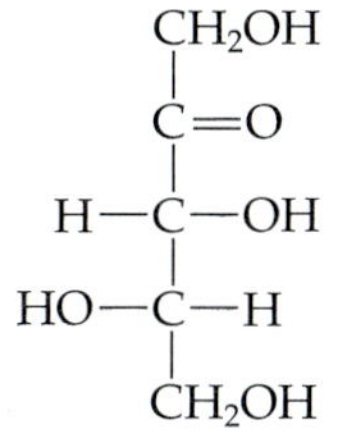

34. Erythrulose, which can be prepared from D-fructose, has the structure shown below. Classify erythrulose as fully as possible.

CH_2OH
$C{=}O$
$HO-C-H$
CH_2OH

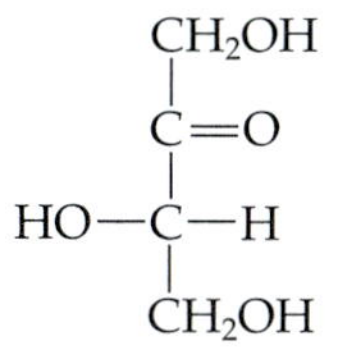

35. What monosaccharide units make up the disaccharide lactulose (below)?

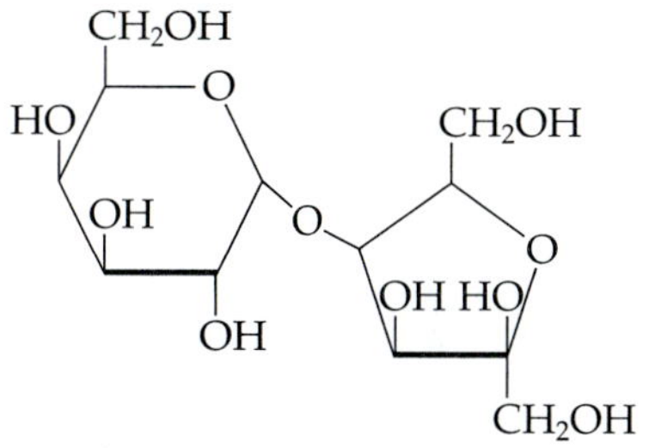

(Structure: pyranose ring with CH_2OH, O, HO, OH, OH linked via O to furanose ring with CH_2OH, O, OH, HO, CH_2OH)

36. Which artificial sweetener(s): **(a)** are currently approved for use in the United States; **(b)** has a bitter, metallic aftertaste; **(c)** was most recently approved for use in the U.S.; **(d)** contains potassium.

37. If 3 mmol (3×10^{-3} mol) of saccharin, sucralose, aspartame, and acesulfame K were each dissolved in 500 mL of water, which solution would have the sweetest taste? Which solution would have the least sweet taste?

38. Identify two functional groups found in saccharin, aspartame, acesulfame K, and sucralose (structures are found in the boxed essay "Artificial Sweeteners").

39. Why does a deficiency of lactase lead to cramps and diarrhea?

40. How does galactosemia differ from lactose intolerance in terms of the cause of the disease and its symptoms and severity?

41. The Acme Chemical Company couldn't read your order for sucrase and sent lactase instead. Because lactase can hydrolyze a disaccharide you decide to use it in place of sucrase to make chocolate-covered cherries. What will be the result? Explain your answer.

42. a. Many clinical studies have indicated that the incidence of colon cancer can be decreased with an increase of what type of dietary fiber in the diet?

b. Blood cholesterol can be lowered by an increase of what type of dietary fiber in the diet? What foods are good sources of this type of fiber?

43. Why is a high-fiber diet beneficial for a person with diabetes?

44. Look at the Nutritional Facts box on several cereals and canned foods at the grocery or deli. **(a)** Which cereals provide the most fiber per serving? **(b)** What canned goods are good sources of fiber? **(c)** Does the Nutritional Facts box distinguish between soluble and insoluble fiber?

CHAPTER 20

Lipids

Canola oil, a polyunsaturated oil, is obtained from the canola plant. Canola is a name given to a genetically modified rapeseed plant that produces a nutritionally superior seed and oil. The harvested seeds are small and round and may be black, brown, or yellow in color. Upon crushing, the seeds yield 40–44% oil. The seed resembles turnip, mustard, cabbage, and broccoli seeds, to which canola is closely related.

Learning Objectives/Study Questions

1. How are lipids defined? How are they classified?
2. What is a fatty acid? What is the difference between a saturated, monounsaturated, and polyunsaturated fatty acid?
3. Why are fats and oils referred to as triglycerides (or triacylglycerols)? What determines if a triglyceride is a fat or oil?
4. What does the iodine number tell you about a triglyceride?
5. Why is it important for a soap to have both a hydrophilic and a hydrophobic end?
6. What are the functions of phospholipids, glycolipids, and sphingolipids? What distinguishing characteristics are used to place lipids into one of these categories?
7. What are the major components of cell membranes and how are they arranged?
8. What are the functions of steroids?
9. What is the link between cholesterol, lipoproteins, and cardiovascular disease?

Fats and oils, components of many foods, belong to a class of biomolecules known as lipids. Gram for gram, fats and oils pack about twice the caloric content of carbohydrates. Although this may be bad news for the dieter, it says something about the efficiency of nature's designs. The body has a limited capacity for storing carbohydrates. It can tuck away a bit of glycogen in the liver or in muscle tissue, but carbohydrates, primarily in the form of glucose, are meant to serve the body's *immediate* energy needs. If we intend to store energy reserves, then the more energy we can pack into a given space, the better off we are. The oxidation of fats supplies about 9 kcal/g, whereas the oxidation of carbohydrates supplies only 4 kcal/g. The body, an efficient organism, is geared to store fats, and its capacity for doing so is astounding. Most of us store enough energy in the form of fats to last a month or so, but there is a recorded instance of a man weighing 486 kg (1070 lb). If all that energy were stored as carbohydrate, he would have weighed 1000 kg (2200 lb) or more!

Do not confuse the term *oil,* used here to refer to a particular group of lipids, with the hydrocarbon petroleum oils.

Fats and other lipids have other functions besides energy storage. Lipids are important components of brain and nervous tissue, and serve as protective padding and insulation for vital organs. Without lipids in our diets, we'd be deficient in the fat-soluble vitamins A, D, E, and K. Most important, lipids are a major component of the membranes of each of the 10 trillion cells in our bodies.

Lipids are not classified by functional group, as were carbohydrates, but by a physical property—solubility. Lipids are generally insoluble in water, but are soluble in organic solvents such as hexane, acetone, and dichloromethane. Compounds isolated from body tissues are classified as **lipids** if they are more soluble in organic solvents than in water. Thus it is not surprising that there are many different kinds of lipids. Included in this category are esters of glycerol and the fatty acids (or phosphoric acid), compounds that incorporate sugar units or a complicated amino alcohol called sphingosine, and steroids such as cholesterol. Because of this broad variation in structure, we can't present a general formula for lipids. We shall, instead, consider one subclass at a time and try to point out similarities and differences in structure as we go along. Figure 20.1 indicates one scheme for classifying the lipids.

▲ A sampling of foods rich in lipids.

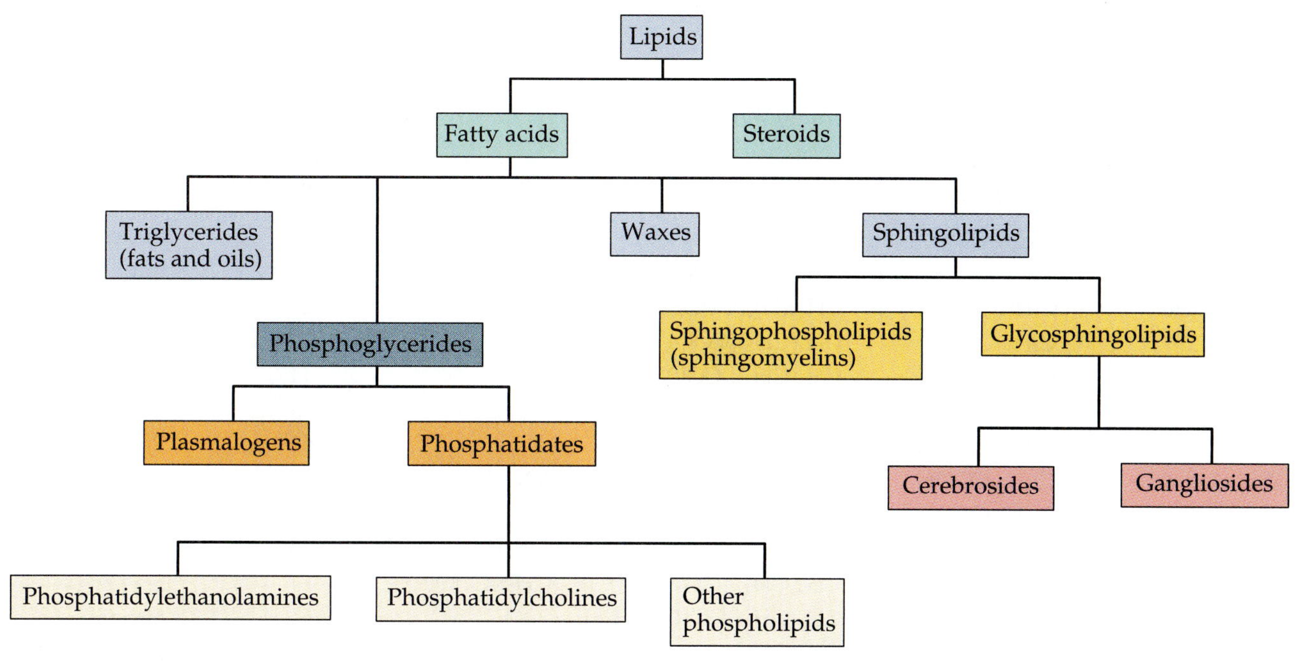

▲ **Figure 20.1**
A scheme that organizes the major types of lipids on the basis of structural relationships.

Review Question

20.1 Define the term lipid.

20.1 Fatty Acids

Fatty acids are carboxylic acids found as structural components of fats and oils. More than 70 have been identified in nature. Nearly all contain an even number of carbon atoms, and they are generally unbranched. They are rarely found free in living organisms, but usually exist as components of fats, oils, and other lipids.

Fatty acids can be classified by the presence and number of carbon–carbon double bonds. **Saturated fatty acids** contain no carbon–carbon double bonds, **monounsaturated fatty acids** contain one carbon–carbon double bond, and **polyunsaturated fatty acids** are those that have two or more double bonds. The double bonds in the unsaturated fatty acids generally exist in the *cis* configuration. Table 20.1 lists some common fatty acids and one important source for each.

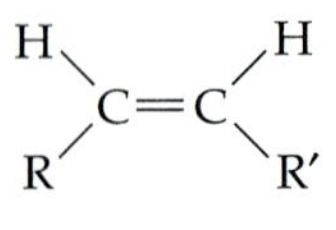

cis configuration

H, R′, C=C, R, H

trans configuration

The tetrahedral bond angles of carbon require that the chain of saturated fatty acid molecules assume a zigzag configuration (Figure 20.2a), but the molecule viewed as a whole is relatively straight (Figure 20.2b). Such molecules pack closely together into a crystal lattice (Figure 20.2c), a capability that gives fatty acids and the fats derived from them relatively high melting points. In an unsaturated fatty acid, each *cis* carbon–carbon double bond results in a severe bend in the molecule (Figure 20.2d). These molecules don't stack neatly, and so the attractions between molecules are smaller. Consequently, the unsaturated fatty acids (and unsaturated fats) have lower melting points. Most are liquids at room temperature.

Review Questions

20.2 Use Table 20.1 to find an example of a: **(a)** saturated fatty acid; **(b)** a polyunsaturated fatty acid; **(c)** a monounsaturated fatty acid.

Table 20.1 Some Common Fatty Acids Found in Natural Fats

Name	Abbreviated Formula[a]	Condensed Structure	Melting Point (°C)	Source
Butyric acid	C_3H_7COOH	$CH_3CH_2CH_2COOH$	−8	Butter
Caproic acid	$C_5H_{11}COOH$	$CH_3(CH_2)_4COOH$	−3	Butter
Caprylic acid	$C_7H_{15}COOH$	$CH_3CH_2)_6COOH$	17	Coconut oil
Capric acid	$C_9H_{19}COOH$	$CH_3(CH_2)_8COOH$	31	Coconut oil
Lauric acid	$C_{11}H_{23}COOH$	$CH_3(CH_2)_{10}COOH$	44	Palm kernel oil
Myristic acid	$C_{13}H_{27}COOH$	$CH_3(CH_2)_{12}COOH$	58	Oil of nutmeg
Palmitic acid	$C_{15}H_{31}COOH$	$CH_3(CH_2)_{14}COOH$	63	Palm oil
Palmitoleic acid	$C_{15}H_{29}COOH$	$CH_3(CH_2)_5CH{=}CH(CH_2)_7COOH$	0.5	Cod liver oil
Stearic acid	$C_{17}H_{35}COOH$	$CH_3(CH_2)_{16}COOH$	70	Beef tallow
Oleic acid	$C_{17}H_{33}COOH$	$CH_3(CH_2)_7CH{=}CH(CH_2)_7COOH$	16	Olive oil
Linoleic acid	$C_{17}H_{31}COOH$	$CH_3(CH_2)_3(CH_2CH{=}CH)_2(CH_2)_7COOH$	−5	Soybean oil
Linolenic acid	$C_{17}H_{29}COOH$	$CH_3(CH_2CH{=}CH)_3(CH_2)_7COOH$	−11	Fish oil
Arachidonic acid	$C_{19}H_{31}COOH$	$CH_3(CH_2)_4(CH{=}CHCH_2)_4CH_2CH_2COOH$	−50	Liver

[a]Saturated fatty acids have the general formula $C_nH_{2n+1}COOH$; unsaurated fatty acids are of the form $C_nH_{2n-1}COOH$, $C_nH_{2n-3}COOH$, $C_nH_{2n-5}COOH$, and so on.

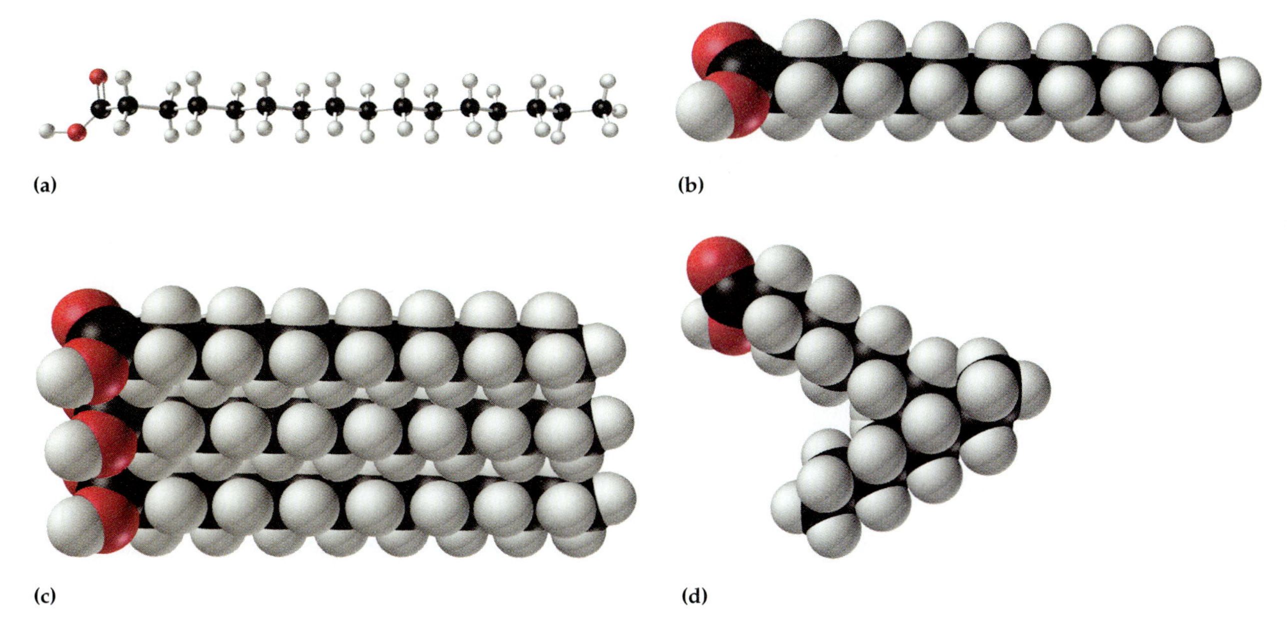

▲ **Figure 20.2**
(a) A ball-and-stick model of a palmitic acid molecule. (b) A space-filling model of a palmitic acid molecule. (c) These saturated acid molecules stack nicely in a crystal lattice. (d) Palmitoleic acid, with its *cis* double bond, will not fit neatly into a crystalline arrangement.

20.3 Draw the structures of each of the following fatty acids (be sure any C=C double bonds are drawn with the correct geometry): **(a)** myristic acid; **(b)** oleic acid; **(c)** linoleic acid.

20.2 Fats and Oils

Fats and oils are the most abundant lipids found in nature. They provide energy for living organisms, but also serve important functions as heat insulators and padding.

Structure and Properties of Triglycerides

Fats and oils are called **triglycerides** because they are *esters* composed of *three fatty acids* joined to *glycerol*, a *trihydroxy alcohol:*

The systematic name for these esters is *triacylglycerols*, but they are commonly called by the old name, triglycerides.

$$
\begin{array}{ccccccc}
\text{Acid} & + & \text{Alcohol} & \longrightarrow & \text{Ester} & + & \text{Water} \\
\text{RCOOH} & & H_2C{-}OH & & H_2C{-}O{-}\overset{\displaystyle O}{\overset{\|}{C}}{-}R & & H_2O \\
\text{R'COOH} & + & HC{-}OH & \xrightarrow{\text{catalyst}} & HC{-}O{-}\overset{\displaystyle O}{\overset{\|}{C}}{-}R' & + & H_2O \\
\text{R''COOH} & & H_2C{-}OH & & H_2C{-}O{-}\overset{\displaystyle O}{\overset{\|}{C}}{-}R'' & & H_2O \\
\text{Three fatty acids} & & \text{Glycerol} & & \text{A triglyceride} & &
\end{array}
$$

If all three hydroxyl groups of the glycerol molecule are esterified with the same fatty acid, the resulting ester is called a *simple triglyceride.* Although some simple triglycerides have been synthesized in the laboratory, they rarely occur in nature. The triglycerides obtained from naturally occurring fats and oils contain two or three different fatty acid components and are thus termed *mixed triglycerides.*

$$\begin{array}{l} H-\overset{\displaystyle H}{\underset{|}{\overset{|}{C}}}-O-\overset{\displaystyle O}{\overset{\|}{C}}-CH_2(CH_2)_{15}CH_3 \\ H-\underset{|}{\overset{|}{C}}-O-\overset{\displaystyle O}{\overset{\|}{C}}-CH_2(CH_2)_{15}CH_3 \\ H-\underset{\displaystyle H}{\underset{|}{\overset{|}{C}}}-O-\overset{\displaystyle O}{\overset{\|}{C}}-CH_2(CH_2)_{15}CH_3 \end{array}$$

Glyceryl stearate
(Tristearin)
(a simple triglyceride)

$$\begin{array}{l} H-\overset{\displaystyle H}{\underset{|}{\overset{|}{C}}}-O-\overset{\displaystyle O}{\overset{\|}{C}}-CH_2(CH_2)_9CH_3 \\ H-\underset{|}{\overset{|}{C}}-O-\overset{\displaystyle O}{\overset{\|}{C}}-CH_2(CH_2)_{13}CH_3 \\ H-\underset{\displaystyle H}{\underset{|}{\overset{|}{C}}}-O-\overset{\displaystyle O}{\overset{\|}{C}}-(CH_2)_7CH{=}CH(CH_2)_7CH_3 \end{array}$$

Glyceryl lauropalmitooleate
(a mixed triglyceride)

A triglyceride is called a **fat** if it is a solid at 25 °C and an **oil** if it is a liquid at the same temperature. (These differences in melting points reflect differences in the degree of unsaturation and number of carbons in the constituent fatty acids.) Furthermore, triglycerides obtained from animal sources are usually solids, while those of plant origin are generally oils. Therefore we commonly speak of animal fats and vegetable oils. Coconut and palm oils, which are highly saturated, and fish oils, which are relatively unsaturated, are notable exceptions to the general rule.

No single formula can be written to represent the naturally occurring fats and oils because they are highly complex mixtures of molecules in which many different fatty acids are represented. Table 20.2 shows the fatty acid compositions of some common fats and oils. Notice that there is a fairly wide range of values: The composition of

Table 20.2 Fatty Acid Components of Some Common Fats and Oils

	Component Fatty Acids (%)[a]						
	Lauric (C_{12})	Myristic (C_{14})	Palmitic (C_{16})	Stearic (C_{18})	Oleic (C_{18})	Linoleic (C_{18})	Linolenic (C_{18})
Fats							
Butter	1–4	8–13	25–32	8–13	22–29	2–4	
Tallow		2–3	24–32	20–25	37–43	2–3	
Lard		1–2	25–30	12–16	40–50	3–8	
Edible Oils							
Coconut oil[b]	44–50	13–18	7–10	1–4	5–8	1–3	
Palm oil[c]		1–6	32–47	1–6	40–52	2–11	
Olive oil	0–1	0–2	7–20	2–3	53–86	4–22	
Peanut oil		0–1	6–11	3–6	40–65	17–38	
Cottonseed oil		0–3	17–23	1–3	23–44	34–55	
Corn oil		1–2	8–12	2–5	29–49	34–56	
Soybean oil		0–1	6–10	2–5	20–30	50–60	2–10
Safflower oil			6–7	2–3	12–14	75–80	0–2
Nonedible Oil							
Linseed oil		0–1	5–9	4–7	9–29	8–29	45–67

[a]Totals less than 100% indicate the presence of lower or higher acids in small amounts.

[b]Coconut oil is highly saturated. It contains an unusually high percentage (53 to 70%) of the low-melting C_8, C_{10}, and C_{12} saturated fatty acids. Coconut oil is a liquid in the warmer, tropical climates, but at room temperature in the temperate zone it is a solid.

[c]Palm oil is highly saturated because of the large percentage of palmitic acid.

any given fat or oil varies depending on the plant or animal species as well as dietetic and climatic factors. To cite just two examples, lard from corn-fed hogs is more highly saturated than lard from peanut-fed hogs, and linseed oil from cold climates is more unsaturated than linseed oil from warm climates. Palmitic acid is the most abundant of the saturated fatty acids, and oleic acid is the most abundant unsaturated fatty acid. Unsaturated fatty acids predominate over saturated ones in most plants and animals.

Tallow is a fat obtained from cattle and sheep. Lard is a fat obtained from pigs.

Fats and oils are often classified according to the degree of unsaturation of the fatty acids they incorporate. Saturated fats contain a high proportion of saturated fatty acids, while unsaturated fats contain a high proportion of unsaturated fatty acids. Saturated fats have been implicated, along with cholesterol (Section 20.6), in one type of *arteriosclerosis* (hardening of the arteries). There is a strong correlation between diets rich in saturated fats and incidence of the disease. This correlation has led to a concern over the relative amounts of saturated and unsaturated fats in our diets. Most nutritionists are advising us to use olive oil and canola oil as our major source of dietary lipids. These oils contain a high percentage of monounsaturated fatty acids, which have been shown to lower LDL cholesterol (see Section 20.7).

The degree of unsaturation of a fat or an oil is usually measured in terms of the **iodine number.** Recall that chlorine and bromine add readily to carbon–carbon double bonds (Section 13.9). Iodine also adds, but less readily:

$$\mathrm{(H)C{=}C(H)} + I_2 \longrightarrow -\overset{\overset{\Large H}{|}}{\underset{\underset{\Large I}{|}}{C}}-\overset{\overset{\Large H}{|}}{\underset{\underset{\Large I}{|}}{C}}-$$

Butter is prepared from milk by churning its cream, a process that causes the fat globules to coalesce; the liquid that remains is called buttermilk.

The iodine number of a fat or oil is the number of grams of iodine that react with 100 g of fat or oil. The more double bonds in a lipid, the more iodine is required for the addition reaction; thus a high iodine number means a high degree of unsaturation. Representative iodine numbers are given in Table 20.3. Notice the generally lower values for the animal fats (butter, tallow, lard) compared with those for the vegetable oils.

Table 20.3 Typical Iodine Numbers for Some Fats and Oils[a]

Fat or Oil	Iodine Number
Coconut oil	8–10
Butter	25–40
Beef tallow	30–45
Palm oil	37–54
Lard	45–70
Olive oil	75–95
Peanut oil	85–100
Cottonseed oil	100–117
Corn oil	115–130
Fish oils	120–180
Soybean oil	125–140
Safflower oil	130–140
Sunflower oil	130–145
Linseed oil	170–205

[a]Most oils are from plant sources. Three fats and one oil (in color) come from animals.

Contrary to popular belief, *pure* fats and oils are colorless, odorless, and tasteless. The characteristic colors, odors, and flavors associated with these lipids are imparted by foreign substances that have been absorbed by the lipids and are soluble in them. For example, the yellow color of butter is due to the presence of the pigment carotene; the taste of butter is a result of two compounds, butanedione ($CH_3COCOCH_3$) and 3-hydroxy-2-butanone [$CH_3COCH(OH)CH_3$], produced by bacteria in the ripening of the cream. Fats and oils are lighter than water, having densities of about 0.8 g/cm^3. They are poor conductors of heat and electricity and therefore serve as excellent insulators for the body, slowing the loss of heat through the skin.

Reactions of Fats and Oils

Fats and oils undergo a variety of chemical reactions; the most important is hydrolysis. Triglycerides are esters. They can be hydrolyzed in either an acidic or basic solution. Acid hydrolysis is of little importance, however, because it is difficult to dissolve fats in acidic solutions. Basic hydrolysis is of considerable importance in the making of soap and is discussed in Section 20.3. When we eat fats, they are hydrolyzed by enzymes (lipases) in our bodies to glycerol and fatty acids. This process is discussed in Chapter 24.

The double bonds in fats and oils can undergo hydrogenation or oxidation. Hydrogenation of vegetable oils to produce semisolid fats is an important process

▲ Hydrogenation of an oil. A solid vegetable fat like Crisco® is made by bubbling hydrogen through a vegetable oil in the presence of a catalyst such as nickel.

in the food industry. The chemistry of this conversion process is essentially identical to the catalytic hydrogenation reaction described for alkenes in Section 13.9:

$$\overset{H}{\diagdown}\mathrm{C{=}C}\overset{H}{\diagup} + H_2 \xrightarrow{\mathrm{Ni}} -\overset{\displaystyle H}{\underset{\displaystyle H}{\overset{|}{\underset{|}{C}}}}-\overset{\displaystyle H}{\underset{\displaystyle H}{\overset{|}{\underset{|}{C}}}}-$$

If reaction conditions are properly controlled, it is possible to prepare a fat with a desirable physical consistency (soft and pliable). In this manner, inexpensive and abundant vegetable oils (cottonseed, corn, soybean) are converted into oleomargarine and cooking fats. The consumer would get much greater unsaturation by using the oils directly, but most people would rather spread margarine than pour oil on their toast. Table 20.4 lists iodine numbers for butter and various kinds of margarines.

In the preparation of margarine, the partially hydrogenated oils are mixed with water, salt, and nonfat dry milk. Flavoring agents, coloring agents, and vitamins A and D are added to approximate the taste of butter. (Preservatives and antioxidants are also added.) The peanut oil in most commercial peanut butter has been partially hydrogenated to prevent the oil from separating out.

Many individuals have switched to margarine or vegetable shortening because of concern about the role of saturated animal fats in raising blood cholesterol and clogging arteries. However, during the hydrogenation of vegetable oils, an isomerization reaction produces some *trans* fatty acids. Recent studies have shown that these *trans* fatty acids, like saturated fatty acids, raise cholesterol levels and increase the incidence of coronary heart disease. Figure 20.3a shows that these *trans* fatty acids do not have the bend in their structure seen with the *cis* fatty acids (see Figure 20.2c) and thus pack well together in the same way that the saturated fatty acids do (Figure 20.3b). Consumers are now being advised to use polyunsaturated oils and soft or liquid margarine and to reduce their total fat consumption to less than 30% of total calorie intake each day.

On standing at room temperature in contact with moist air, fats and oils soon turn rancid. This rancidity, characterized by a disagreeable odor, results from hydrolysis and oxidation. Hydrolysis of the ester bonds releases volatile fatty acids. Butter, for example, yields foul-smelling butyric, caprylic, and capric acids. Microorganisms present in the air furnish the lipases that catalyze the process. Hy-

Table 20.4 Iodine Numbers and Comparative Unsaturation Ratings of Butter and Various Margarines

Food Product	Iodine Number	Comparative Unsaturation[a]
Butter	27	100
Margarines		
Hard type A	68	252
Hard type B	72	267
Hard type C	77	285
Soft type D	84	311
Soft type E	88	326
Liquid type F	90	333
Liquid type G	93	344

[a]Calculated by dividing the iodine number of the substance by the iodine number of butter and multiplying the result by 100.

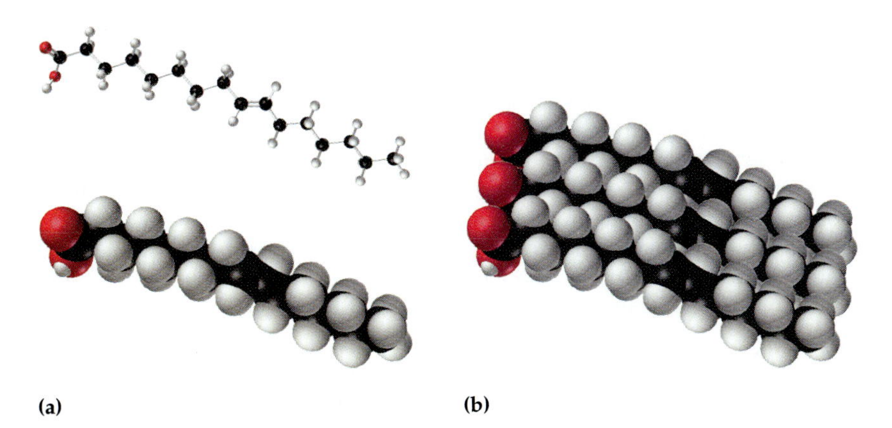

Figure 20.3
(a) A ball-and-stick model (*top*) and a space-filling model (*bottom*) of a *trans*-palmitoleic acid molecule. (b) These *trans*-fatty acid molecules stack nicely in a crystal lattice.

drolytic rancidity can easily be prevented by storing a fat or oil covered in a refrigerator. The stale, sweaty odor of unwashed skin results when fats and oils excreted by the body are first hydrolyzed and then oxidized by oxygen in the air to carboxylic acids of 5 to 10 carbons in length.

Oxidation of the unsaturated fatty acid components also produces a variety of volatile, odorous compounds. The structural unit

$$\sim CH{=}CH{-}CH_2{-}CH{=}CH\sim$$

in linoleic and linolenic acids is readily oxidized. One particularly offensive product, formed by the cleavage of both double bonds, is a compound called malonaldehyde.

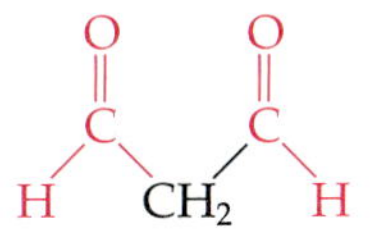

Malonaldehyde

Rancidity is a major concern of the food industry, and food chemists are always seeking new and better substances to act as **antioxidants.** Such compounds are added in very small amounts (0.001 to 0.01%) to suppress rancidity. They have a greater affinity for oxygen than lipids present in the food and thus function by preferentially depleting the supply of absorbed oxygen. We shall see in Selected Topic F that two vitamins (C and E) have antioxidant properties.

Reducing Fat Intake

We often hear of the need for a low fat diet to reduce the risk of cancer, heart disease, and other problems associated with obesity. But fats provide texture, flavor, and a creamy "mouthfeel" to foods. Artificial or high-intensity sweeteners have been around for over a century (see boxed essay in Chapter 19), but the same is not true for fat replacers. The first fat replacers, introduced in the 1960s, used carbohydrates as the primary ingredient. Carrageenan, a seaweed derivative approved for use in food in 1961, was initially used as an emulsifier, stabilizer, and thickener in food. In the early 1990s it was used as a fat replacer. Protein-based fat substitutes came along in the early 1990s. Unlike many of the initial carbohydrate-based products that were first used for other purposes, these were specifically designed to replace fat in foods. Microparticulated proteins, such as Simplesse®, are made from whey protein or milk and egg protein. The carbohydrate- and protein-based fat replacers can give the "mouthfeel," bulk, and texture of fats, but cannot be used for frying.

In January 1996 the first true fat-based fat replacer was approved—olestra (marketed as Olean® by Procter & Gamble). Olestra is composed of a sucrose core with six to eight fatty acids attached. Because olestra is a much larger molecule than a triglyceride, it is too large to be hydrolyzed by lipases and digested or absorbed by the body or metabolized by the microorganisms in the intestinal tract. Thus it adds no fat or calories to foods. Olestra is not broken down or degraded when it is exposed to high temperatures; thus it can be used for frying foods.

Olestra does have some drawbacks to its use. Clinical studies have shown that it may cause intestinal cramps and loose stools in some individuals. It also reduces the absorption of fat-soluble nutrients, such as vitamins A, D, E, and K and carotenoids, from foods eaten at the same time. Because of these concerns, the Food and Drug Administration (FDA) requires that foods containing olestra be fortified with vitamins A, D, E, and K and that the following statement appear on the package: "This product contains olestra. Olestra may cause abdominal cramping and loose stools. Olestra inhibits the absorption of some vitamins and other nutrients. Vitamins A, D, E, and K have been added."

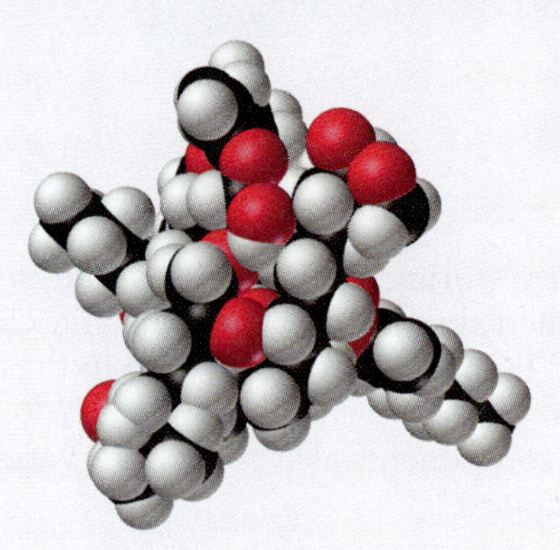

Olestra

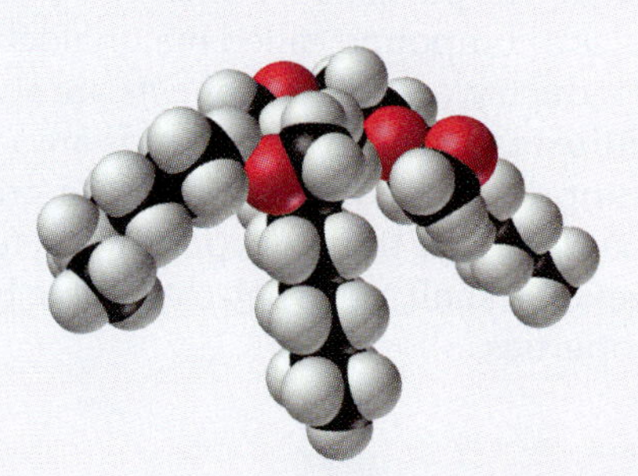

Triglyceride

◄ Molecular models of olestra and a trizglyceride. Olestra's core is sucrose with six to eight fatty acids attached.

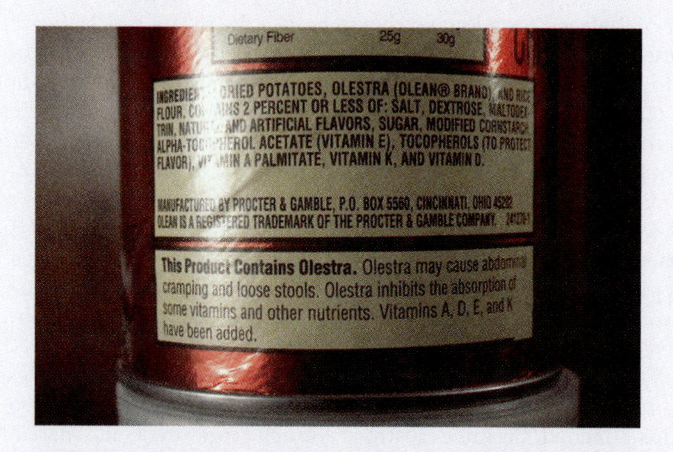

▲ The nutrition label on a food product containing olestra, including the statement required by the FDA and the listing of olestra (as the trade name Olean®).

Review Questions

20.4 Define each of the following terms: **(a)** triglyceride; **(b)** iodine number; **(c)** antioxidant.

20.5 What functions does fat serve in the body?

20.6 Write an equation for the complete hydrogenation of triolein (glyceryl trioleate).

20.3 Soaps

Animal fats are available in large quantities as a by-product of the meatpacking industry. Fatty acids and many other long-chain organic compounds are derived from these fats. The most important derivatives are soaps. The reaction that converts animal fats to soaps is called **saponification,** which we discussed originally as a reaction of esters (Section 16.9). Soapmaking is one of the oldest organic syntheses known, second only to the production of ethyl alcohol (fermentation). Although the Phoenicians (600 B.C.E.) and the Romans made soap from animal fat and wood ash, the widespread production of soap did not begin until the 1700s.

The old method of soap production consisted of treating molten tallow with a slight excess of alkali in large open vats. The mixture was heated, and steam was bubbled through it. After saponification was completed, the soap was precipitated by the addition of sodium chloride, then filtered and washed several times with water. It was then dissolved in water and reprecipitated by the addition of more sodium chloride. The glycerol was also recovered from the aqueous wash solutions.

Today most soaps are prepared by a continuous hydrolysis of triglycerides (frequently tallow and/or coconut oil) by water under high pressures and temperatures [700 lb/in.^{2} (~50 atm or 5000 kPa) and 200 °C]. Sodium carbonate is used to convert the fatty acids to their sodium salts (soap molecules):

$$\begin{array}{l} CH_2OOC(CH_2)_nCH_3 \\ | \\ CHOOC(CH_2)_nCH_3 \\ | \\ CH_2OOC(CH_2)_nCH_3 \end{array} \xrightarrow[\text{heat, pressure}]{H_2O} \text{Glycerol} + \underset{\text{Fatty acids}}{3\ CH_3(CH_2)_nCOOH} \xrightarrow{Na_2CO_3} \underset{\text{Sodium salts of fatty acids (soap)}}{3\ CH_3(CH_2)_nCOO^-Na^+}$$

The crude soap is used as industrial soap without further processing. Pumice or sand may be added to produce scouring soap. Other ingredients such as dyes or perfumes are added to produce colored or fragrant soaps. A floating soap is produced by blowing air through molten soap. Such a soap is not necessarily purer than other soaps; it merely contains more air. Ordinary soap is a mixture of the sodium salts of various fatty acids. Potassium soaps (soft soap) are more expensive but produce a finer lather and are more soluble. They are used in liquid soaps, shampoos, and shaving creams.

Dirt and grime usually adhere to skin, clothing, and other surfaces because they are combined with body oils, cooking fats, lubricating greases, and a variety of similar substances that act like glues. Because oils are not miscible with water, washing with water alone does little good. Soap helps in the process because soap molecules have a dual nature. One end is ionic and dissolves in water; the other end is like a hydrocarbon and dissolves in oils. Often, the ionic end is referred to as **hydrophilic** (water-soluble) and the nonpolar end as **hydrophobic** (repelled by water). We can illustrate the cleansing action of soap schematically (Figure 20.4). The hydrocarbon "tails" dissolve in the oil; the ionic "heads" remain in the aqueous phase. In this manner, the oil is broken into tiny droplets called *micelles* (Section 20.5) and dispersed throughout the solution. The droplets don't coalesce because of the repulsion of the charged groups (the carboxylate anions) on their surfaces. With the

The suffix *phil* means to love, and the suffix *phob* means to fear or hate. Hydrophobic substances are sometimes said to be *lipophilic.*

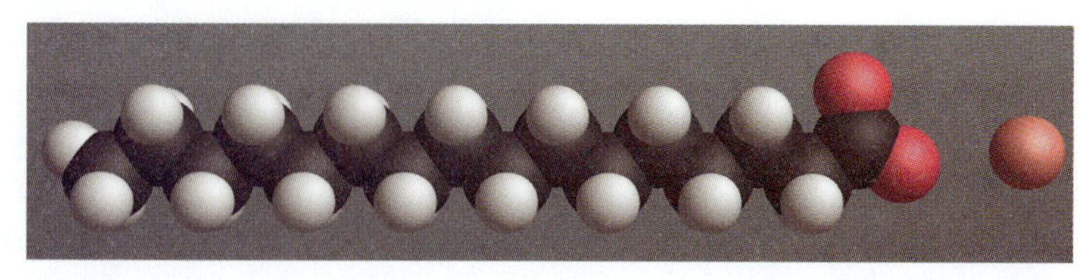

(a)

(b)

▶ **Figure 20.4**
Cleaning action of soap visualized. (a) Sodium palmitate, $CH_3(CH_2)_{14}COO^-Na^+$, a typical soap. (b) A microscopic view of soap action. In an oil droplet suspended in water, the hydrocarbon "tails" of the soap molecules are immersed in the oil and the ionic ends extend into the water. The attractive forces between these ionic "heads" and the water molecules cause the oil droplet to be suspended in the water.

oil no longer "gluing" it to the surface (skin, cloth, dish), the dirt can be easily removed when the soap solution is rinsed away.

Synthetic detergents have largely replaced soaps for cleaning clothes and for many other purposes because soaps have two rather serious shortcomings. One is that in acidic solutions, soaps are converted to free fatty acids:

$$\underset{\text{A soap}}{CH_3(CH_2)_{16}COO^-Na^+} + H^+ \longrightarrow \underset{\text{A fatty acid}}{CH_3(CH_2)_{16}COOH} + Na^+$$

The fatty acids, unlike soap, don't have an ionic end. Lacking the necessary dual nature, they can't emulsify the oil and dirt; that is, they do not exhibit any cleaning action. What is more, these fatty acids are insoluble in water and separate out as a greasy scum. To counteract this lack of cleaning action in acidic solution, various alkaline substances are added to laundry soap formulations to keep the pH high. These basic compounds include carbonates and silicates.

The second and more serious disadvantage of soap is that it doesn't work well in hard water. Hard water is water that contains certain metal ions, particularly magnesium, calcium, and iron ions. The soap anions react with these metal ions to form greasy, insoluble curds:

$$\underset{\text{Soap (soluble)}}{2\,CH_3(CH_2)_{16}COO^-Na^+} + Ca^{2+} \longrightarrow \underset{\text{Bathtub ring (insoluble)}}{[CH_3(CH_2)_{16}COO^-]_2Ca^{2+}} + 2\,Na^+$$

Waxes

Waxes are esters formed from long-chain fatty acids and long-chain monohydroxy alcohols. (Household paraffin wax, which is a mixture of high-molar-mass hydrocarbons, has waxlike properties but is not a wax according to our definition.) The general formula for a wax, then, is the same as that of a simple ester. For a wax, however, R and R′ are limited to alkyl groups containing large numbers of carbon atoms.

$$R-C(=O)-O-R'$$

A wax
(a simple ester)

$$CH_3(CH_2)_{24}C(=O)-O-(CH_2)_{29}CH_3$$

Myricyl cerotate
(found in carnauba wax)

Most natural waxes are mixtures of such esters. Many also contain free alcohols, hydrocarbons, and esters of diprotic acids, hydroxy acids, and diols. All have similar properties; they feel "waxy," are insoluble in water, and melt at temperatures above body temperature (37 °C) and below the boiling point of water (100 °C).

Waxes are not as easily hydrolyzed as triglycerides and therefore are useful as protective coatings. Plant waxes on the surfaces of leaves, stems, flowers, and fruits protect the plant from dehydration and from invasion by harmful microorganisms. Carnauba wax, largely myricyl cerotate ($C_{25}H_{51}COOC_{30}H_{61}$), is obtained from the leaves of certain Brazilian palm trees and is used extensively in floor waxes, automobile waxes, and furniture polish.

Animal waxes also serve as protective coatings. They are found on the surfaces of feathers, skin, and hair and help to keep these surfaces pliable and water-repellent. Earwax, for example, protects the delicate lining of the inner ear. The waxy coating on the feathers of waterbirds (ducks, gulls) helps them to stay afloat. If this wax is dissolved as a result of the bird swimming in an oil slick, the feathers become wet and heavy; the bird cannot maintain its buoyancy and will drown.

Beeswax is the material from which bees construct honeycombs. Upon saponification, beeswax yields alcohols and fatty acid salts. It is used in such household products as candles and shoe polish. Commercial operations that harvest beeswax can always leave behind enough honey for the bees to live on. Obtaining spermaceti, however, is a little harder on the creature that produces it. Spermaceti (mp 42–50 °C) crystallizes when oil from the head of the sperm whale is cooled, and so whales must be killed before the product can be obtained. The principal constituent of spermaceti is cetyl palmitate. Esters of lauric, myristic, and palmitic acids are also present, as are esters of higher alcohols. Whales and other marine species store these waxes as metabolic fuels. Once widely used in ointments, cosmetics, soaps, and candles, spermaceti is now in short supply because the sperm whale has been hunted almost to extinction.

$$CH_3(CH_2)_{14}C(=O)-O-(CH_2)_{15}CH_3$$

Cetyl palmitate

Lanolin is a wax from sheep's wool. It is a mixture of esters and polyesters of 33 alcohols and 36 fatty acids. Some of the alcohols are steroids (similar to cholesterol). Lanolin is used as a base for ointments and cosmetic lotions.

Many natural waxes have been replaced by synthetic materials, mainly polymers. By careful control of the molar masses of these polymers, the properties of natural waxes can be closely duplicated. For example, Carbowax, a polymer of ethylene glycol ($HOCH_2CH_2OH$), is available in a wide range of average molecular weights. Synthetic waxes are used in cosmetics and ointments and in certain industrial processes.

▲ The sperm whale was widely hunted prior to 1981 for the spermaceti obtained from its head.

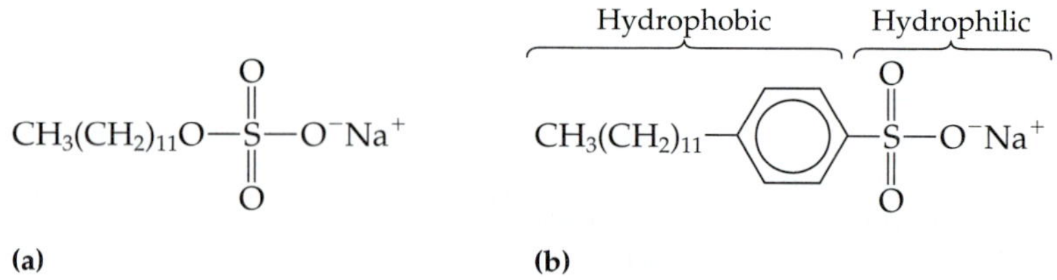

▶ **Figure 20.5**
(a) Sodium lauryl sulfate is a detergent used in toothpastes and shampoos. (b) Sodium dodecyl benzenesulfonate is a detergent used in laundry detergents.

(a) $CH_3(CH_2)_{11}O{-}S(=O)_2{-}O^-Na^+$

(b) Hydrophobic: $CH_3(CH_2)_{11}{-}C_6H_4{-}$ Hydrophilic: $S(=O)_2{-}O^-Na^+$

These deposits make up the familiar "bathtub ring." They leave freshly washed hair sticky and are responsible for the "telltale gray" of clothes laundered with soap.

The term **detergent** is a general one meaning any cleansing agent. Even though soaps fall under this broad definition, the popular use of the word generally refers to *synthetic detergents.* Synthetic detergents have the desirable property of not forming precipitates with the ions of hard water. There are close to a thousand synthetic detergents commercially available in the United States, and worldwide production exceeds 25 million tons. Although most synthetic detergents are now made from petroleum, early ones were made from fats. Some, such as sodium lauryl sulfate (sodium dodecyl sulfate) shown in Figure 20.5a, still are. Sodium lauryl sulfate is widely used in specialty products such as toothpastes and shampoos. Sodium dodecyl benzenesulfonate (Figure 20.5b) is found in many laundry detergents.

✓ Review Questions

20.7 Distinguish between terms in the following pairs: **(a)** hard water and soft water; **(b)** hard soap and soft soap; **(c)** soaps and synthetic detergents.

20.8 Describe how soaps clean.

20.4 Membrane Lipids

The most polar lipids are found in the membranes surrounding both individual cells and the organelles within the cell. These polar lipids contain both hydrophilic and hydrophobic groups, a characteristic important in the structure of membranes, as we shall see in Section 20.5. **Phospholipids** are phosphorus-containing polar lipids. The most abundant lipids in membranes are phospholipids known as phosphoglycerides (or glycerophospholipids).

Glycolipids are sugar-containing lipids that are confined to the outer surface of the cell membrane. Glycolipids provide cells with distinguishing surface markers that may serve in cellular recognition and cell-to-cell communication. **Sphingolipids,** which are also classified as phospholipids or glycolipids, contain the unsaturated amino alcohol sphingosine rather than glycerol. Diagrammatic structures of representative membrane lipids are given in Figure 20.6.

Phosphoglycerides

In phosphoglycerides a phosphoric acid unit is bonded to the third carbon of glycerol. This phosphoric acid unit is also esterified with another alcohol molecule, usually an amino alcohol (Figure 20.7a). Two main types of phosphoglycerides are plasmalogens and phosphatidates. *Plasmalogens* contain a hydrocarbon chain (rather than a fatty acid) linked to the first carbon of the glycerol backbone through an ether linkage and a fatty acid linked to the second carbon of the glycerol backbone (Figure 20.7c). Plasmalogens are important constituents of the membranes of nerves and muscles. The **phosphatidates** are esters of glycerol in which there are two fatty acid groups and a phosphoric acid unit esterified with another alcohol molecule.

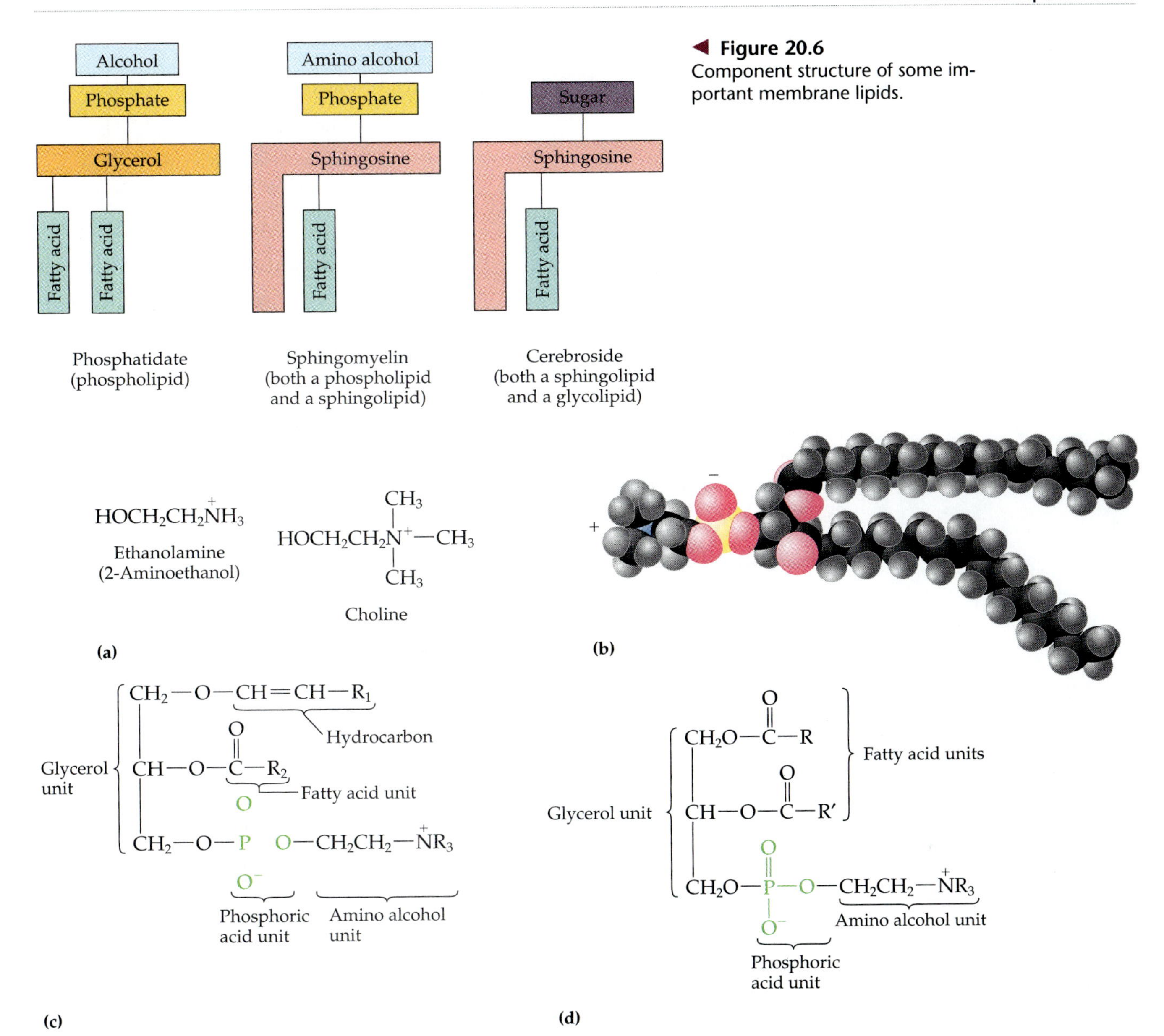

◄ **Figure 20.6**
Component structure of some important membrane lipids.

▲ **Figure 20.7**
(a) Amino alcohols commonly found in phosphoglycerides. (b) Space-filling model of a phosphoglyceride. (c) Structural formula of a plasmalogen; R_1 and R_2 represent straight-chain alkyl groups that usually contain ten or more carbons. (d) Structural formula of a phosphatidate.

Notice that the phosphatidate molecule is identical to a triglyceride up to the phosphoric acid unit (Figure 20.7d).

Two common phosphatidates are shown in Figure 20.8. When the phosphatidate contains the *ethanolamine* structural group, the compounds are called phosphatidylethanolamines or **cephalins.** The cephalins are found in brain tissue and nerves. They are also involved in blood clotting. When choline is the amino alcohol unit, the compounds are called phosphatidylcholines or **lecithins.** Lecithins occur in all living organisms. They, too, are important constituents of nerve and brain tissue. Egg yolks are especially rich in lecithins. Commercial-grade lecithins isolated from soybeans are widely used in foods as emulsifying agents. An *emulsifying agent* is used to stabilize an **emulsion**—a dispersion of two liquids that do not

▲ The mixture at the bottom contains an egg yolk that acts as an emulsifying agent to stabilize the emulsion of oil in water.

► **Figure 20.8**
Two common phosphatides.

$$\begin{array}{l} CH_2O\text{—Fatty acid 1} \\ | \\ CHO\text{—Fatty acid 2} \\ | \\ CH_2O\text{—Phosphate—}OCH_2CH_2\overset{+}{N}H_3 \end{array}$$

Phosphatidylethanolamine (a cephalin)

$$\begin{array}{l} CH_2O\text{—Fatty acid 1} \\ | \\ CHO\text{—Fatty acid 2} \\ | \\ CH_2O\text{—Phosphate—}OCH_2CH_2\text{—}\overset{+}{N}(CH_3)_3 \end{array}$$

Phosphatidylcholine (a lecithin)

normally mix, such as oil and water. Many foods are emulsions. Milk is an emulsion of butterfat in water. The stabilizing agent in milk is a protein called casein. Mayonnaise is an emulsion of salad oil in water, stabilized by lecithins present in egg yolk.

Sphingolipids

Sphingomyelins are the "simplest" sphingolipids, each containing a fatty acid, phosphoric acid, sphingosine, and choline (Figure 20.9). Because they contain phosphoric acid, they are also classified as phospholipids. Sphingomyelins are important constituents of the myelin sheath that surrounds the axon of a nerve cell. Multiple sclerosis is one of several diseases related to a fault in the myelin sheath.

Most animal cells contain sphingolipids called **cerebrosides** (Figure 20.10). Cerebrosides are composed of sphingosine, a fatty acid, and galactose or glucose. The structure resembles that of a sphingomyelin, except that a sugar unit is found in place of the choline phosphate group. Cerebrosides are important constituents of the membranes of nerve and brain cells. Gaucher disease, a hereditary affliction, results from the inability to remove glucose from glucocerebrosides. Large amounts of these cerebrosides accumulate, causing enlargement of the liver and the spleen (Section 23.7).

Related sphingolipids, the **gangliosides,** have an even more complex structure, usually containing a branched chain of three to eight monosaccharides and/or

$$\begin{array}{l} CH_3(CH_2)_{12}CH{=}CHCH\text{—}OH \\ \qquad\qquad\qquad\quad | \\ \qquad\qquad\qquad\quad CH\text{—}NH_2 \\ \qquad\qquad\qquad\quad | \\ \qquad\qquad\qquad\quad CH_2OH \end{array}$$

Sphingosine

(a)

Sphingosine unit: $CH_3(CH_2)_{12}CH{=}CHCH\text{—}OH$

$CH\text{—}NH\text{—}C(=O)(CH_2)_9CH{=}CH(CH_2)_9CH_3$ (Fatty acid unit: $C(=O)(CH_2)_9CH{=}CH(CH_2)_9CH_3$)

$CH_2O\text{—}P(=O)(O^-)\text{—}OCH_2CH_2\text{—}\overset{+}{N}(CH_3)_3$ (Phosphoric acid unit: $P(=O)(O^-)$; Choline unit: $OCH_2CH_2\overset{+}{N}(CH_3)_3$)

(b)

▲ **Figure 20.9**
(a) Sphingosine is an amino alcohol found in all sphingolipids. (b) A sphingomyelin.

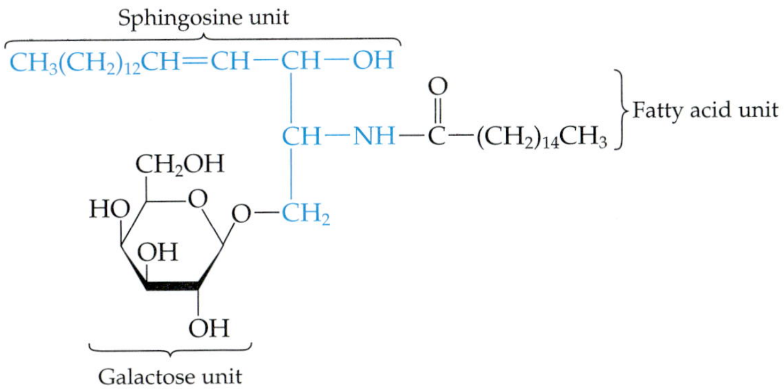

Sphingosine unit

$CH_3(CH_2)_{12}CH{=}CH{-}CH{-}OH$

$CH{-}NH{-}C({=}O){-}(CH_2)_{14}CH_3$ Fatty acid unit

CH_2OH

HO O O—CH_2

OH

OH

Galactose unit

Figure 20.10 Cerebrosides are sphingolipids that contain a sugar unit.

substituted sugars. There is considerable variation in their sugar components, and about 130 varieties of gangliosides have been identified. It is the sequence of sugars that most often determines cell-to-cell recognition and communication (e.g., blood group antigens). They are most prevalent in the outer membranes of nerve cells, although they also occur in smaller quantities in the outer membranes of most other cells. Both cerebrosides and gangliosides contain sugar groups and therefore are also classified as *glycolipids*.

The A, B, AB, and O blood types differ in the structure of the oligosaccharides bound to the surface of the red blood cells (Chapter 28).

Review Questions

20.9 Draw the complete structure of a phosphatidylethanolamine with palmitic acid at C-1 and oleic acid at C-2.

20.10 What general structural feature do phosphatidates share with soaps and detergents?

20.5 Cell Membranes

The components of a living cell are enclosed within a membrane. Both plant cells (Figure 20.11) and animal cells (Figure 20.12) have cell nuclei that contain nucleic acids (Chapter 23). Everything between the cell membrane and the nuclear membrane—including the fluids and a variety of subcellular components such as the mitochondria and ribosomes—is called the **cytoplasm.** We will encounter several components of the cytoplasm in following chapters. Note that plant cells have rigid *cell walls* that surround and protect the cell, but animal cells have only the cell membrane. Membranes of all cells have a similar structure, but membrane function varies tremendously from one organism to another and from one cell to another within a single organism. This diversity arises mainly from the different proteins and phospholipids in the membrane.

When polar lipids, such as phospholipids, are placed in water, they disperse and form clusters of molecules in any one of three arrangements: *micelles*, *monolayers*, and *bilayers*. **Micelles** are aggregations of molecules that contain both polar and nonpolar groups. The hydrocarbon "tails" of these lipids, being hydrophobic, are directed inward away from the water; the hydrophilic "heads" are directed outward into the water. Each micelle may contain thousands of lipid molecules (Figure 20.13a). Polar lipids also form *monolayers*, which are layers one molecule thick on the surface of water (Figure 20.13b). The polar heads stick into the water, and the nonpolar tails stick up into the air. **Bilayers** are layers that have hydrophobic tails sandwiched between hydrophilic heads sticking outward into the water (Figure 20.13c). *It is such layers that make up every cell membrane.*

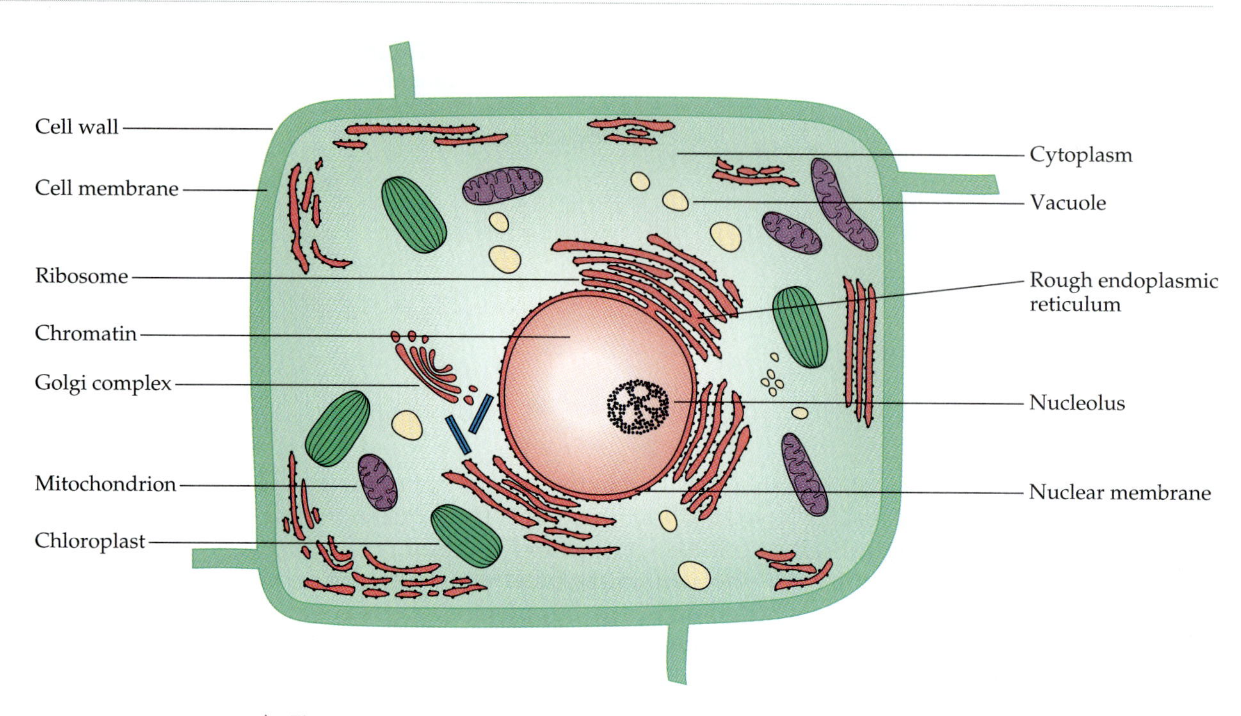

▲ **Figure 20.11**
An idealized plant cell. Not all the structures shown here occur in every type of plant cell.

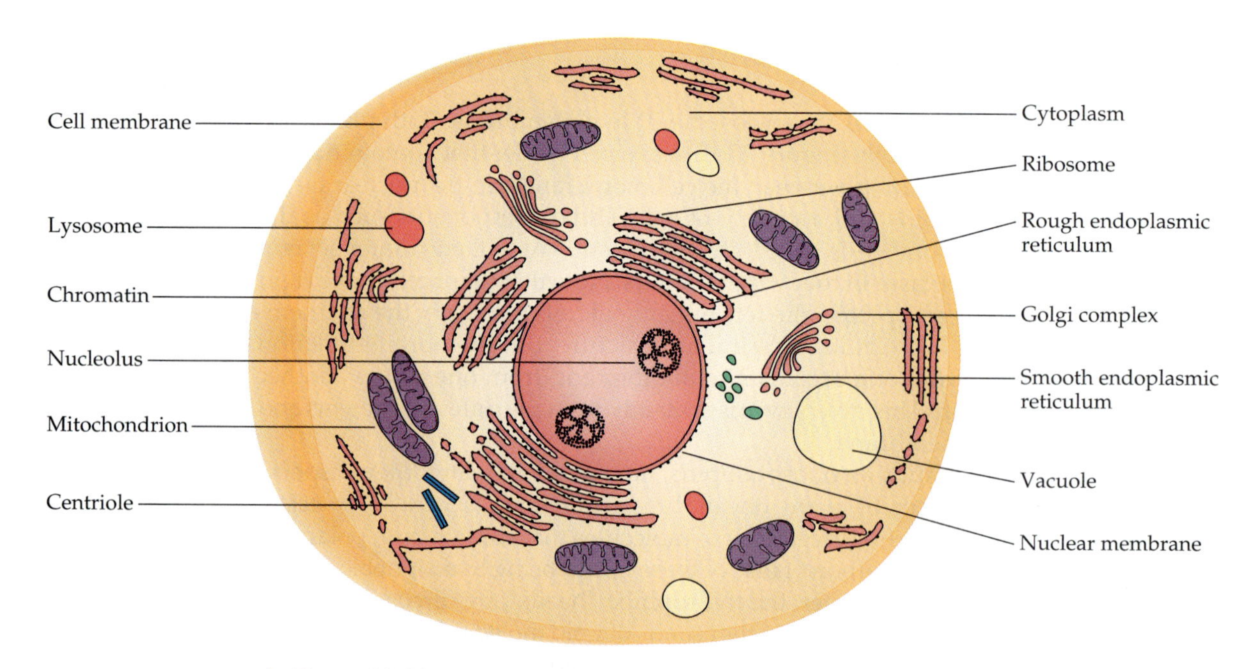

▲ **Figure 20.12**
An idealized animal cell. The entire range of structures shown here seldom occurs in a single animal cell. Each kind of tissue has cells specific to the function of that tissue. Muscle cells are different from neurons and neurons differ from red blood cells, and so on.

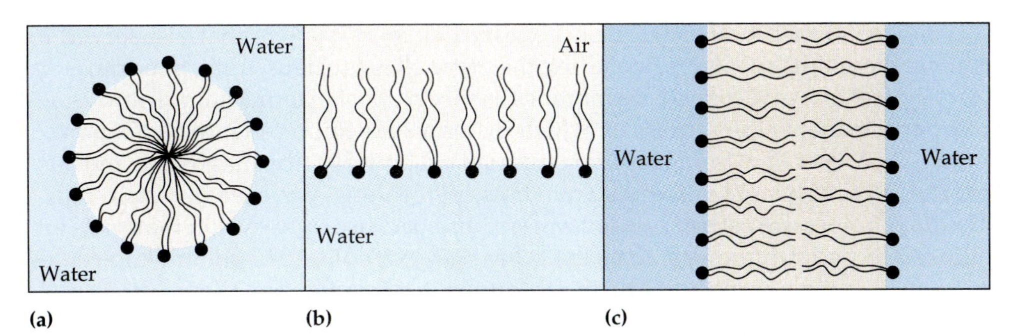

▶ **Figure 20.13**
(a) Micelle, (b) monolayer, and (c) bilayer formed when polar lipids are added to water.

The three major classes of lipid molecules in the membrane bilayer of animal cells are phospholipids, glycolipids (about 5%), and cholesterol (see Section 20.6). Roughly equal numbers of phospholipid and cholesterol molecules are present in the membranes of these cells. As shown in Figure 20.14, the lipid bilayer consists of two rows of phospholipid molecules arranged tail to tail. The hydrophobic tails (the fatty acid portions) interact by means of dispersion forces. This interaction is relatively weak because of the presence of unsaturated fatty acids. As a result, the lipid portion is not rigid but is quite fluid and allows movement within the membrane. The hydrophobic tails are isolated from the aqueous environment that exists within and outside the cells. The polar portions of the phospholipids project from the inner and outer surfaces of the membrane and interact with water molecules. Cholesterol is also a key moderator of membrane fluidity. It is inserted between the fatty acid chains, disrupting the packing of the hydrophobic chains and decreasing the strength of the attractive forces.

Cell membranes become quite rigid at low temperatures as their constituent phospholipids tend to solidify. Warm-blooded animals have no problem with this; they maintain membrane fluidity simply by maintaining body temperature. Membrane rigidity does present a problem, however, to cold-blooded animals that stay active in cold weather. To adapt to the cold, some of these cold-blooded animals are able to make their cell membranes more fluid by increasing the degree of unsaturation of membrane phospholipids, thus lowering the solidification temperature.

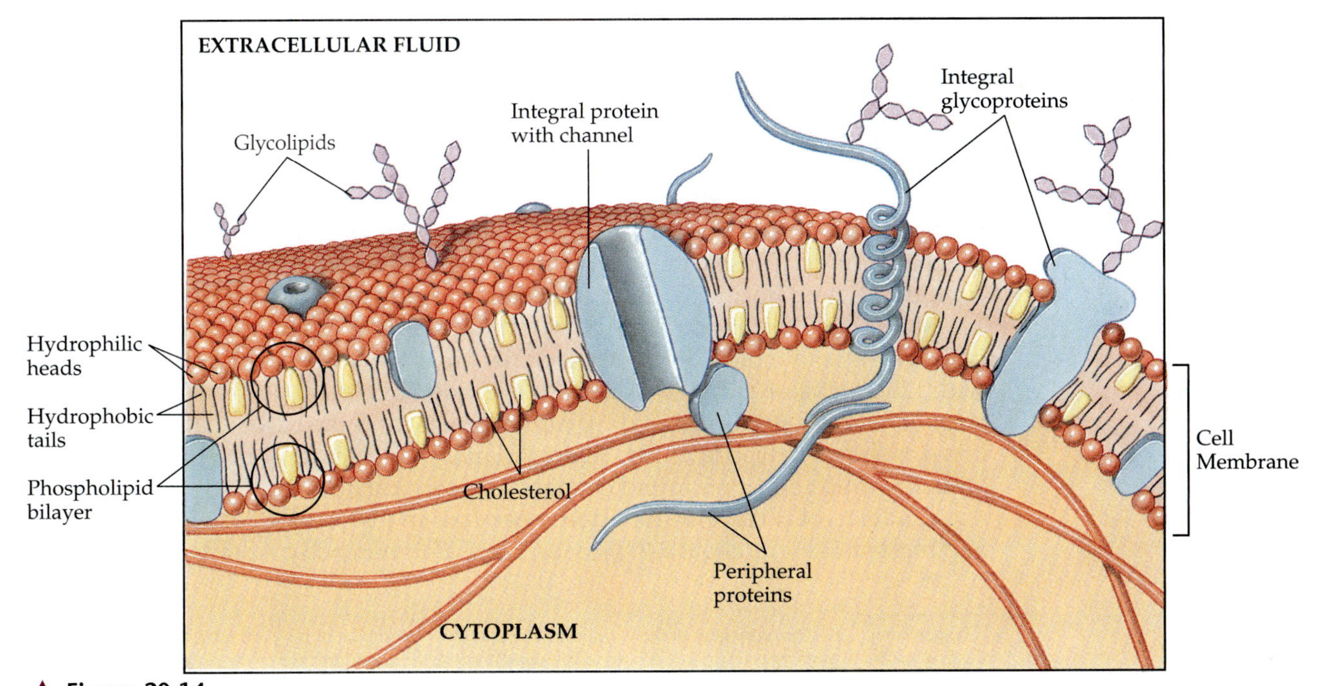

▲ **Figure 20.14**
Schematic diagram of a cell membrane. The membrane is a phospholipid bilayer with embedded cholesterol and protein molecules. Short polysaccharide chains are attached to the outer surface.

Biological membranes were once viewed as inert barriers that just served to contain the cellular contents, including the organelles (nucleus, mitochondria, etc.). Now, however, we recognize membranes as receptors of external stimuli and as active participants in the process by which materials are imported into and exported from cells. Most cell membranes are composed of about 50% (by mass) lipid and 50% protein, although large variations from these percentages can exist in certain cells.[1] The membranes are referred to as *semipermeable* because only certain materials are allowed to pass from one side to the other. For example, lipid-soluble molecules can cross the membrane by diffusing through the phospholipid bilayer; large molecules that are not lipid-soluble cannot diffuse through the membrane.

If membranes were composed only of lipids, they would act as barriers to the passage of ions or polar molecules. Instead, the passage of polar species across the membrane is facilitated by proteins that move about in the lipid bilayer. There are two classes of proteins in the cell membrane. **Integral proteins** span the hydrophobic interior of the bilayer, whereas **peripheral proteins** are more loosely associated with the membrane surface (Figure 20.14). Peripheral proteins appear to be attached to integral proteins and/or the polar head groups of phospholipids by hydrogen bonds and electrostatic forces.

Small ions and water-soluble molecules enter and leave the cell by way of channels through the integral proteins. There are special proteins that facilitate the passage of certain molecules (e.g., hormones and neurotransmitters). A specific interaction occurs between the carrier protein and the molecule being transported.

Review Questions

20.11 Define each of the following terms: **(a)** monolayer; **(b)** micelle; **(c)** bilayer; **(d)** semipermeable

20.12 Distinguish between an integral protein and a peripheral protein. What is a key function of integral proteins?

20.6 Steroids: Cholesterol and Bile Salts

Steroids with hormonal activity are discussed in detail in Selected Topic E.

All the lipids discussed so far are *saponifiable*. They react with aqueous alkali to yield simpler components such as glycerol, fatty acids, amino alcohols, or sugars. Any lipids extracted from cellular material, however, contain a small but important fraction that does not react with alkali. The most important nonsaponifiable lipids are the **steroids.** These compounds include the bile salts, cholesterol and related compounds, and hormones such as cortisone and the sex hormones.

More than 40 steroids have been found in nature. They occur in plants, animals, yeasts, and molds, but not in bacteria. They may exist either free or combined with fatty acids or carbohydrates. All steroids have a structure with four fused rings. The rings are designated by capital letters, and the carbon atoms are numbered as shown in Figure 20.15a. Slight variations in structure or in the nature of substituent groups effect profound changes in biological activity.

Cholesterol (Figure 20.15b) does not occur in plants, but it is the best known and most abundant (about 240 g) steroid in the human body. About one-half of the

[1]The percentage of each component of the membrane is related to the function of the particular tissue. For example, the myelin sheath of nerve cells can contain up to 80% lipid because the major function of the membrane is insulation and protection. The inner mitochondrial membrane, on the other hand, is unique in having a large protein component (75 to 80%). The proteins (enzymes) within the membrane play an integral role in the energy-conversion function of the mitochondria (see Section 24.5). It is important to emphasize that these are percents by *mass.* Lipid molecules are much smaller than proteins; hence there are always more lipid than protein molecules. In an average membrane there are 50 lipid molecules for one protein molecule.

(a) Steroid skeleton

(b) Cholesterol

(c) Cholic acid (a bile acid)

▲ **Figure 20.15**
(a) Four-fused ring steroid skeleton showing letter designations for each ring and the numbering of the carbons. (b) Cholesterol. (c) Cholic acid, which is synthesized from cholesterol.

body's cholesterol is interspersed among the phospholipid molecules in cell membranes (recall Figure 20.14). Much of the rest is converted to cholic acid (Figure 20.15c), which is used in the formation of bile salts. Cholesterol is also an important precursor in the biosynthesis of the sex hormones, adrenal hormones, and vitamin D. Excess cholesterol not utilized by the body is released from the liver and transported by the blood to the gallbladder. Normally it stays in solution and is secreted into the intestine (in the bile) to be eliminated. Sometimes the cholesterol precipitates in the gallbladder, producing gallstones. Its name is derived from this source (Greek *chole*, bile; *stereos*, solid).

Bile is a yellowish-green liquid (pH 7.8–8.6) that is produced in the liver. Bile serves as a route for the excretion of drugs, of end products from hemoglobin breakdown (see Section 28.2), and of heavy metal ions. Its composition is given in Table 20.5. The most important constituents of bile are bile salts, which are sodium salts of amide-like combinations of bile acids (Figure 20.16a) and glycine or the rare amino acid taurine (Figure 20.16b). They are synthesized from cholesterol in the liver, stored in the gallbladder, and then secreted in bile into the small intestine (Figure 20.17). In the gallbladder the composition of bile gradually changes as water is absorbed and the other components become more concentrated.

Because they contain both hydrophobic and hydrophilic groups, bile salts are highly effective detergents and emulsifying agents—they break down large fat globules into smaller ones and keep these smaller globules suspended in the aqueous digestive environment (see Section 24.2). Enzymes can then hydrolyze the fat molecules more effectively. Thus the major function of bile salts is to aid in the digestion of dietary lipids. They also aid in the absorption of fatty acids, cholesterol, and

▲ An assortment of human gallstones, which are nearly pure cholesterol. (The dime gives an indication of relative sizes.)

If the gallbladder becomes infected, inflamed, or perforated, it is often surgically removed. This does not seriously affect digestion because bile is still produced by the liver, but it is more dilute and its secretion into the duodenum is not as closely tied to the arrival of food.

Table 20.5 Composition of Bile

Component	Percent
Water	97
Bile salts	0.7
Inorganic salts	0.7
Bile pigments	0.2
Fatty acids	0.15
Lecithin	0.1
Fat	0.1
Cholesterol	0.06

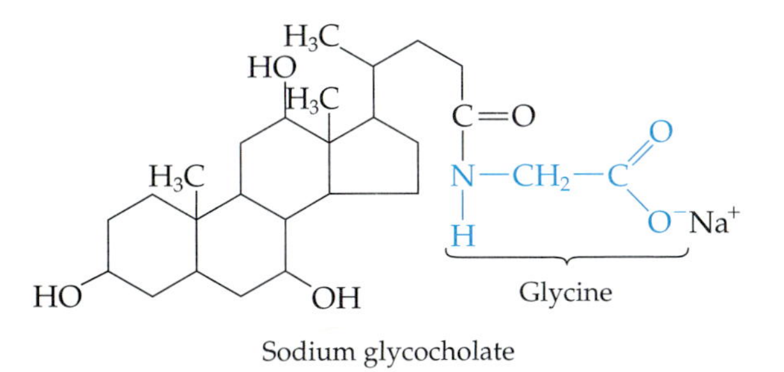

(b)

◀ **Figure 20.16**
(a) Sodium glycocholate is a bile salt synthesized from cholic acid and glycine. (b) Taurine replaces glycine in some bile salts.

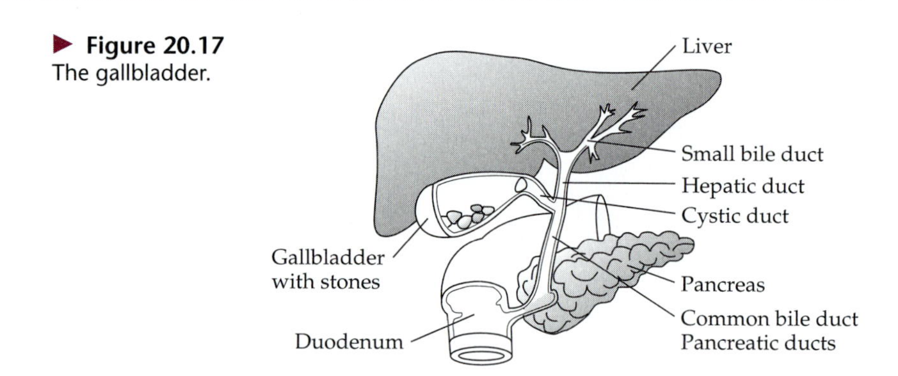

▶ **Figure 20.17**
The gallbladder.

the fat-soluble vitamins by forming complexes (micelles) that can diffuse into cells lining the intestines.

Review Questions

20.13 Define each of the following terms: **(a)** steroid; **(b)** bile; **(c)** bile salts.

20.14 Distinguish between a saponifiable lipid and a nonsaponifiable lipid.

20.7 Cholesterol and Cardiovascular Disease

We get half of our total fats, three-fourths of all our saturated fats, and all of our cholesterol from animal products such as meat, milk, cheese, and eggs. Advertising claims that a vegetable oil contains no cholesterol are simply silly; *no* vegetable product contains cholesterol.

In the past 30 years few subjects in the nutrition field have attracted as much public attention as cholesterol. Today, everything from margarine and vegetable oils to egg substitutes and meat analogs is advertised on the basis that it contains little or no cholesterol. Cholesterol is believed to be a primary factor in the development of atherosclerosis, coronary artery disease, and stroke—major health problems in the United States today. The American Heart Association reported that in 1995 cardiovascular diseases (which include stroke and heart attack) claimed 960,592 lives in the United States. This was 41.5% of all deaths reported that year. The second leading cause of death that year was cancer, with 538,455 deaths reported.

Scientists agree that elevated cholesterol levels in the blood, as well as high blood pressure and cigarette smoking, are associated in humans with an increased risk of heart attack. A long-term investigation by the National Institutes of Health (NIH) showed that among men aged 30 to 49, the incidence of coronary artery disease was five times greater for those whose cholesterol levels were above 260 mg/100 mL of serum than for those with cholesterol levels of 200 mg/100 mL or less.

The cholesterol content of blood varies considerably with age, diet, and sex. Young adults average about 170 mg of cholesterol per 100 mL of blood, whereas males at age 55 may have 250 mg/100 mL or higher (because the rate of cholesterol metabolism decreases with age). Females tend to have lower blood cholesterol levels than males.

To understand the link between cardiovascular disease and cholesterol levels it is important to understand how cholesterol (and other lipids) is transported in the body. Because lipids such as cholesterol are not soluble in water, they cannot be transported in the blood (an aqueous medium) unless they are complexed with water-soluble proteins as **lipoproteins** (Figure 20.18). Lipoproteins generally are classified according to their density and composition (see Table 20.6). There are four broad categories: *chylomicrons, very low-density lipoproteins* (VLDL), *low-density lipoproteins* (LDL), and *high-density lipoproteins* (HDL). The density of lipoproteins is determined by the relative contents of protein and lipid. Because lipids are less dense than proteins, lipoproteins containing a greater amount of lipid are less dense than those containing a greater proportion of protein. The chylomicrons contain the largest proportion of lipids, while the HDLs contain the least.

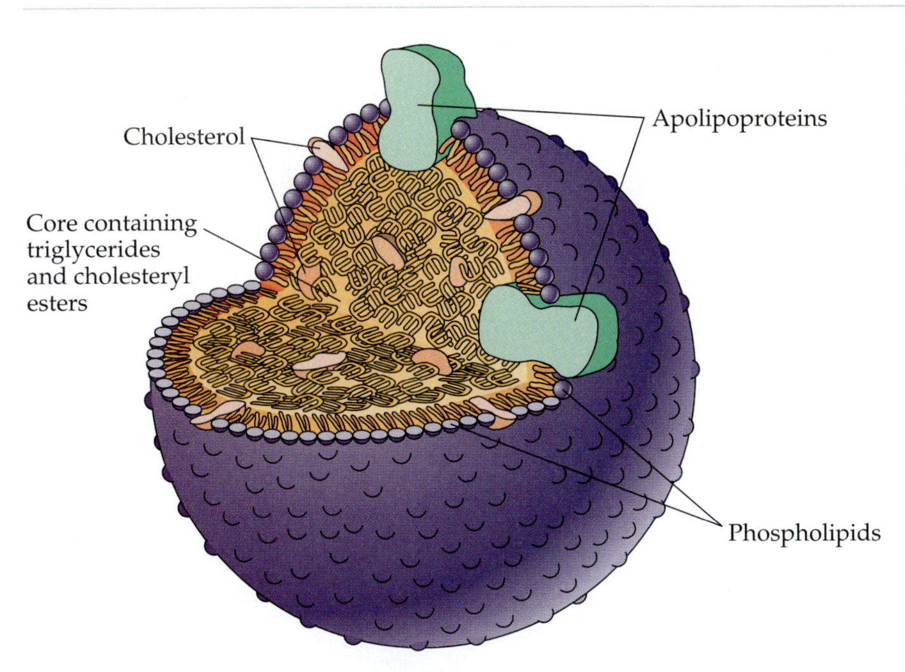

Figure 20.18
Structure of a lipoprotein. Lipoproteins are roughly spherical and contain a core of triglycerides and cholesteryl esters. The surface of the lipoprotein is a layer of phospholipids embedded with cholesterol and proteins known as apolipoproteins.

LDLs are the major cholesterol-carrying lipoproteins, whereas chylomicrons and VLDLs are the major carriers of triglycerides in the blood.

Research on cholesterol and its role in heart disease indicates that atherosclerosis and coronary artery disease are associated with elevated levels of serum LDLs, rather than with serum lipoproteins in general. One of the most fascinating discoveries in this field is that cholesterol bound to HDLs reduces a person's risk of developing coronary artery disease. On the other hand, cholesterol bound to LDLs increases that risk. Thus the serum LDL:HDL ratio is a better predictor of coronary artery disease risk than the level of serum cholesterol. Persons who, because of hereditary or dietary factors, have high LDL:HDL ratios in their blood have a higher incidence of coronary artery disease.

The LDLs transport cholesterol directly from the liver to cells that need it. LDL receptor proteins on the cell membranes are vital to the uptake of cholesterol from the blood (see Figure 20.19). These receptors are absent or deficient in an inherited condition called *familial hypercholesterolemia.* Studies of individuals with this disease have provided much of the information about the role of LDLs and coronary artery disease. When the LDL receptors are absent or deficient, cholesterol levels in the blood rise dramatically, and the excess cholesterol that can't enter cells is deposited in certain tissues, particularly in skin, tendons, and arteries. In the arteries, the LDLs are broken down enzymatically to cholesterol, cholesterol esters, and protein. The cholesterol and cholesterol esters are then oxidized in the artery wall, becoming major parts of an atherosclerotic plaque (Figure 20.20). Individuals who are homozygous for familial hypercholesterolemia (about 1 in 1 million) have

Table 20.6 Compositions of Lipoproteins Isolated from Normal Subjects

Lipoprotein Class	Density Range (g/mL)	Composition (wt %) Protein	Triglyceride	Cholesterol Free	Cholesterol Ester	Phospholipid
Chylomicrons	<0.94	1–2	85–95	1–3	2–4	3–6
VLDL	0.94–1.006	6–10	50–65	4–8	16–22	15–20
LDL	1.006–1.063	18–22	4–8	6–8	45–50	18–24
HDL	1.063–1.21	45–55	2–7	3–5	15–20	26–32

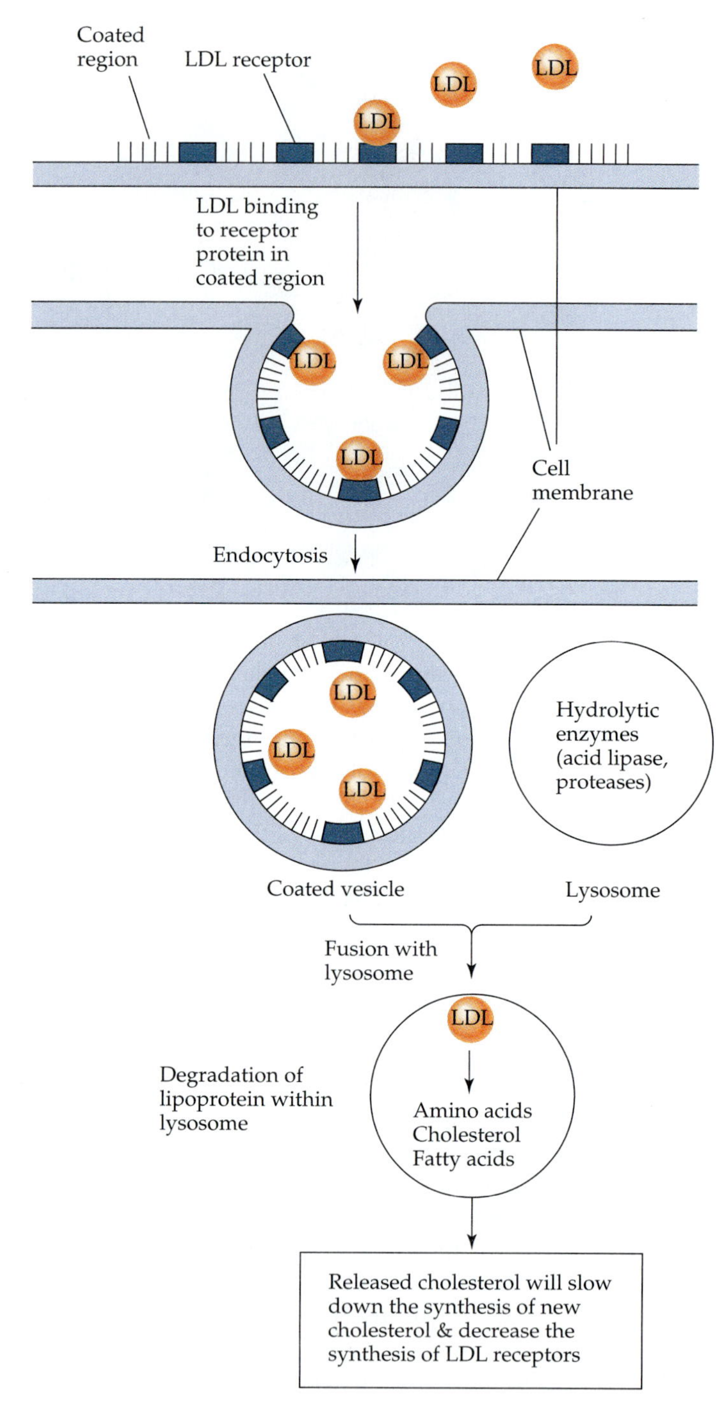

Figure 20.19
Uptake of LDLs into cells by specific LDL receptors (proteins). These receptors are located on the cell membrane in areas that are coated with a specific protein. Once LDL is bound to the LDL receptor, this complex is brought into the cell through a process known as endocytosis in which a coated vesicle is formed. This coated vesicle fuses with a lysosome that contains many hydrolytic enzymes that degrade the lipoprotein, releasing cholesterol as well as amino acids and fatty acids.

essentially no functional LDL receptors, and they die of coronary artery disease before age 20. About 1 individual in 500 is heterozygous for this condition; with about half the number of LDL receptors, such persons often develop atherosclerosis by age 50, though they can enjoy a normal life span.

How do HDLs reduce the risk of developing coronary artery disease? No one knows for sure, but one role of HDLs appears to be the transport of excess cholesterol from various tissues to the liver, where it can be metabolized. Therefore HDLs aid in removing cholesterol from blood and the smooth muscle cells of the arterial wall.

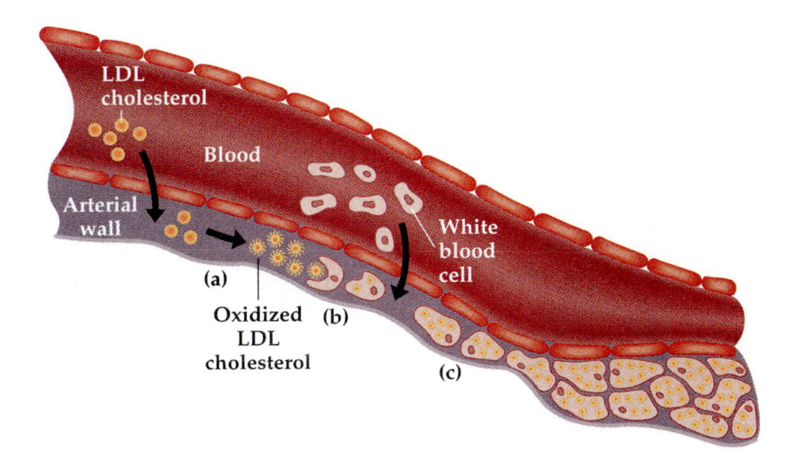

Figure 20.20
The constriction of arteries. (a) LDL cholesterol accumulates and is oxidized. (b) The oxidation products, which can damage arterial tissue, attract white blood cells. (c) The white blood cells becomes engorged with the oxidation products and accumulate within the arterial wall, eventually narrowing the artery.

Assuming that HDL helps to protect the body against coronary artery disease, what can be done to increase serum levels of this lipoprotein? One way is by sustained exercise (e.g., aerobic training). Another way is to lose weight. Finally, it has been reported that HDL levels in the blood can be increased by drinking alcohol *in moderation* (less than two glasses/day).

This does not suggest that traditional college age students should begin drinking! Their risk of coronary artery disease is very low and alcohol provides no benefit.

Dietary modification can help lower total cholesterol and improve the LDL:HDL ratio. The average American consumes about 600 mg of cholesterol from animal products each day. The human body synthesizes about 1 g of cholesterol each day, mostly in the liver; all 27 carbon atoms in the cholesterol molecule are derived from acetyl-CoA molecules (see Section 24.3). The plasma cholesterol level controls the synthesis of cholesterol by the liver. When the cholesterol level in the blood exceeds 150 mg/100mL, the rate of cholesterol biosynthesis is halved. Hence, if cholesterol is present in the diet, a feedback mechanism suppresses its synthesis in the liver. However, this is not a 1:1 ratio. The reduction in biosynthesis does not equal the amount of cholesterol ingested. Fasting also inhibits cholesterol synthesis because of the limited availability of acetyl-CoA. Conversely, diets high in carbohydrate or fat tend to accelerate cholesterol synthesis because they increase the amount of acetyl-CoA in the liver.

One large egg contains approx. 215 mg of cholesterol. Most health authorities recommend a maximum of 300 mg/day of cholesterol in the diet.

Dietary substitution of unsaturated for saturated fat can help lower both serum cholesterol and risk of heart disease. The lipids of fish and poultry contain relatively more unsaturated fatty acids than the lipids of beef, lamb, and pork. There is also statistical evidence that fish oils can prevent heart disease. Researchers at the University of Leiden in the Netherlands have found that Greenlanders, who eat a lot of fish, have a low risk of heart disease despite a diet high in total fat and cholesterol. The probable effective agents are the polyunsaturated fatty acids such as eicosapentaenoic acid and docosahexaenoic acid (so-called omega-3 fatty acids):

$$CH_3CH_2(CH{=}CHCH_2)_5CH_2CH_2COOH$$

ω-3-Eicosapentaenoic acid

$$CH_3CH_2(CH{=}CHCH_2)_6CH_2CH_2COOH$$

ω-3-Docosahexaenoic acid

Eicosapentaenoic acid and docosahexaenoic acid are referred to as omega-3 fatty acids because the endmost double bond is three carbons from the methyl end of the chain. (Omega, ω, is the last letter in the Greek alphabet, thus the omega position is the one farthest from the carboxyl group.)

Other studies have shown that diets with added fish oil lead to lower cholesterol and triglyceride levels in the blood.

Review Questions

20.15 What physical property is used in classifying lipoprotein complexes?

20.16 High levels of which lipoprotein complex are an indication that an individual has a greater risk of heart disease?

Atherosclerosis

Atherosclerosis is the most common form of arteriosclerosis, a thickening and toughening of arterial walls. Atherosclerosis consists of the formation of lipid deposits in the middle layer of large and middle-sized arteries. This plaque formation may begin with damage to the lining of the blood vessels followed by accumulation of wound-healing tissue or with accumulation of lipid-laden cells. In the latter mechanism, LDL cholesterol lodges in arterial walls and is oxidized by free radicals. White blood cells migrate into the arterial cells and attempt to cleanse the cells by consuming the oxidation products. The enlarged white blood cells accumulate within the arterial lining, causing plaques that narrow the arteries (Figure 20. 21). That tissue buildup and the resulting constriction reduce blood flow, diminish oxygen supply, and lead to high blood pressure. (Note that high blood pressure is also aggravated by other factors such as lack of exercise, obesity, heredity, stress, and smoking.) Occasionally, the plaque ruptures through the lining into the interior of the artery. This rupture stimulates the blood platelets to initiate blood clots. These clots further obstruct the vessels, and may completely block the artery. Arterial clots are responsible for the most serious consequence of atherosclerosis: heart attack.

Heart attacks occur when one of the coronary arteries (arteries that supply the heart muscle itself) is blocked, either by increase in plaque size or by clot movement. If a blood clot suddenly breaks loose, it may be carried to a narrower part of the artery, obstructing blood flow. Deprived of nutrients and oxygen, the heart muscle once served by the blocked artery rapidly and painfully dies. If the area is small, the heart may be able to continue functioning and the patient will recover. Death of large areas of heart muscle is almost instantly fatal. Although heart attacks are the major cause of death from atherosclerosis, cholesterol deposits and clots form in arteries throughout the body. A clot or a plaque deposit that obstructs an artery supplying the brain can cause a stroke, with the same results as if the artery had burst.

A heart attack usually does not happen suddenly. The body has an early warning system. The American Heart Association has compiled a list of early warning signals:

1. One of the first signs is pressure or pain in the middle of the chest. That's where the heart is, not on the left as many believe .
2. The pain can get worse and spread to the shoulders, neck or arms.
3. A sensation of pressure, fullness, or squeezing may occur in the abdomen and is often mistaken for indigestion. This is observed more often in women than in men.
4. Pain may occur in any one or a combination of these areas at the same time. It could go away and return later. The pain is often accompanied by sweating, nausea, vomiting, and/or shortness of breath.

At the first sign of these symptoms, go to the nearest hospital emergency room at once! Get there as fast as possible. Have someone take you or dial 911 (where available) and tell the operator that there is a heart attack victim at your location. In the hospital, insist that it is an emergency—that you have chest pain (or other symptoms) and may be having a heart attack.

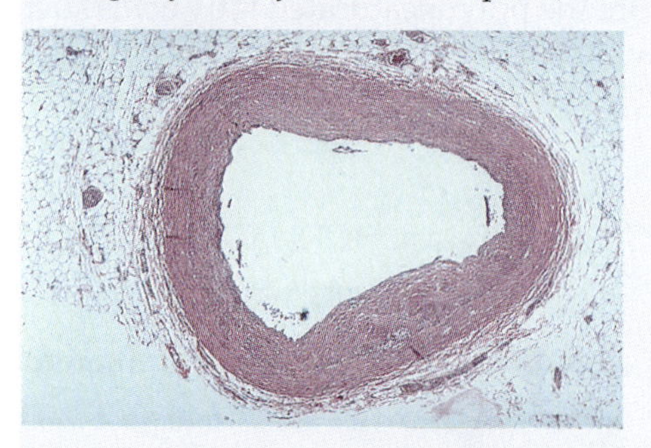

(a)

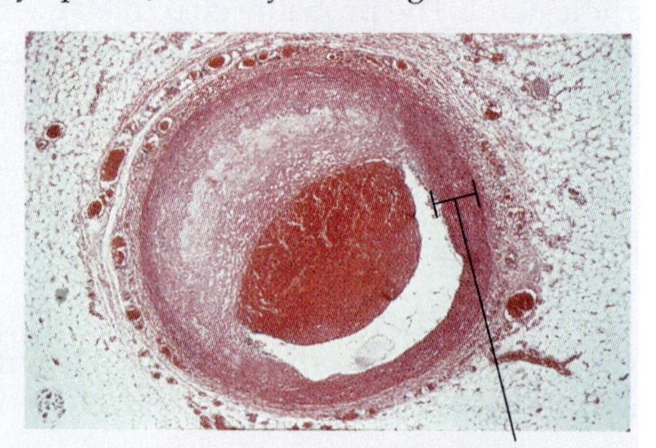

(b)

▲ Figure 20.21
(a) A section of a normal artery. (b) A section of a coronary artery narrowed by plaque formation.

Summary

Lipids, found in the body tissue of all organisms, are compounds that are more soluble in organic solvents than in water. **Fatty acids** are carboxylic acids that generally contain an even number of 4–20 carbons in a straight chain. **Saturated fatty acids** do not contain any carbon–carbon double bonds. **Monounsaturated fatty acids** contain a single carbon–carbon double bond, while **polyunsaturated fatty acids** contain more than one. The lipids known as **fats** and **oils** are classified as triacylglycerols or, more commonly, as **triglycerides**—esters composed of three fatty acids joined to the trihydroxy alcohol glycerol.

Fats are triglycerides that are solid at room temperature, and oils are triglycerides that are liquid at room temperature. Fats are found mainly in animals, and oils, mainly in plants. *Saturated triglycerides* are those containing mainly saturated fatty acid chains (few C=C bonds); *unsaturated triglycerides* contain mainly unsaturated fatty acid chains. The **iodine number** of a triglyceride is a measure of how unsaturated the molecule is: the higher the iodine number the higher the number of C=C bonds in the molecule.

Saponification is the process whereby a triglyceride is hydrolyzed in a basic solution to form glycerol and three carboxylate anions or soap molecules. The carboxylate end of the soap molecule is **hydrophilic** (water-loving) and is soluble in water; the hydrocarbon part of the molecule is **hydrophobic** (water-hating) and thus more soluble in organic solvents than in water. Soap cleans when its hydrophobic end dissolves any lipid attached to dirt particles. Pulled by the hydrophilic end, plus the hydrophobic end, the attached lipid goes into the wash water. The "degreased" dirt particle then leaves the surface to which it is clinging and disperses in the wash water.

Phospholipids are lipids that contain phosphorus. In **phosphatidates,** the phosphorus is joined to an amino alcohol unit. The phosphatidates known as **cephalins** contain an ethanolamine group, while the **lecithins** contain choline as the amino alcohol group. **Sphingolipids** are lipids for which the precursor is the amino alcohol sphingosine, rather than glycerol. A **glycolipid** has a sugar substituted at one of the —OH groups of either glycerol or sphingosine.

Every living cell is enclosed by a *cell membrane* made up of two layers of polar lipids arranged with their hydrophobic ends facing each other in a lipid **bilayer**. In animal cells, the bilayer contains mainly phospholipids, glycolipids, and the steroid cholesterol. Imbedded in the bilayer are **integral proteins,** and loosely associated with the bilayer along the interior surface of the cell are **peripheral proteins.** Any integral protein having either or both of its hydrophilic ends poking out of the bilayer and thus exposed to the liquid inside the cell or the interstitial fluid (or to both) is called a *transmembrane protein.* The purpose of integral and peripheral proteins is to transport ionic and polar substances back and forth across the cell membrane as the cell either needs these substances for metabolism or needs to get rid of them as waste products.

Most lipids can be saponified, but some cannot be. **Steroids** are one type of nonsaponifiable lipid. The steroid **cholesterol** is found in animal cells but never in plant cells. It is a main component of all cell membranes and a precursor for hormones, vitamin D, and bile salts. **Bile** is a yellowish-green liquid, secreted by the gallbladder into the small intestine, that is needed for the proper digestion of lipids. The most important constituents of bile are the bile salts.

Because lipids are not soluble in the blood, they are transported as protein complexes known as **lipoproteins**. Lipoproteins are classified according to their density and composition into four categories: *chylomicrons, very low-density lipoproteins* (VLDL), *low-density lipoproteins* (LDL), and *high-density lipoproteins* (HDL). Research indicates that cholesterol bound to HDLs reduces a person's risk of developing coronary disease, while cholesterol bound to LDLs increases the risk. HDL levels can be increased by sustained aerobic exercise and for overweight individuals, losing weight.

Key Terms

antioxidant (20.2)
bilayer (20.5)
bile (20.6)
cephalin (20.4)
cerebroside (20.4)
cholesterol (20.6)
cytoplasm (20.5)
detergent (20.3)
emulsion (20.4)
fat (20.2)
fatty acid (20.1)
ganglioside (20.4)
glycolipid (20.4)
hydrophilic (20.3)
hydrophobic (20.3)
integral protein (20.5)
iodine number (20.2)
lecithin (20.4)
lipid (page 529)
lipoprotein (20.7)
micelle (20.5)
monounsaturated fatty acid (20.1)
oil (20.2)
peripheral protein (20.5)
phosphatidates (20.4)
phospholipid (20.4)
polyunsaturated fatty acid (20.1)
saponification (20.3)
saturated fatty acid (20.1)
sphingolipid (20.4)
sphingomyelin (20.4)
steroid (20.6)
triglyceride (20.2)

Problems

Fatty Acids

1. Classify each fatty acid as saturated or unsaturated and indicate the number of carbon atoms in each.
   a. caproic acid **b.** oleic acid **c.** stearic acid

2. Classify each fatty acid as saturated or unsaturated and indicate the number of carbon atoms in each.
   a. palmitic acid **b.** linolenic acid
   c. myristic acid

3. Write structural formulas for the following:
   a. lauric acid b. palmitoleic acid
   c. potassium palmitate
4. Without referring to Table 20.1, arrange the following fatty acids in order of increasing melting point. Justify your order.
   linoleic acid linolenic acid oleic acid stearic acid
5. Why do unsaturated fatty acids have lower melting points than saturated fatty acids?

Fats, Oils, and Soaps

6. Write structural formulas for the following:
   a. triolein
   b. a triglyceride likely to be found in cottonseed oil
7. Write structural formulas for the following:
   a. tristearin
   b. a triglyceride likely to be found in beef tallow
8. What triglyceride is formed by the complete hydrogenation of triolein? Of trilinolein?
9. Write out the reaction (using structural formulas) for the hydrolysis of tripalmitin in a basic solution.
10. Which would you expect to have a higher iodine number—tristearin or triolein? Explain.
11. Which would you expect to have a higher iodine number—corn oil or beef tallow? Explain.
12. a. What compound with a disagreeable odor is formed when butter becomes rancid?
    b. What leads to its formation?
    c. How can rancidity be prevented?
13. Write the equation (using structural formulas) for the saponification of glyceryl trilaurate.
14. a. What structural features are necessary for a compound to be a good detergent?
    b. What advantages do synthetic detergents have over soap?

Membrane Lipids and Cell Membranes

15. Which of the lipids in Figure 20.22 would be classified as phospholipids, glycolipids, or sphingolipids? (Some lipids may be given more than one classification.)
16. Which of the lipids in Figure 20.23 would be classified as phospholipids, glycolipids, or sphingolipids? (Some lipids may be given more than one classification.)
17. Draw the structure of the cerebroside that has palmitic acid as its fatty acid and glucose as its sugar.
18. Determine whether or not each of the following can diffuse through the lipid bilayer of a cell membrane. Justify your answer.
    a. glucose b. NaCl
    c. $CH_3CH_2CH_2OCH_2CH_2CH_3$ d. CH_3CH_2OH

Steroids

19. Which of the following compounds are classified as steroids?
    a. tristearin b. cholesterol c. lecithin
20. Which of the following compounds are classified as steroids?
    a. Vitamin D b. cephalin c. cholic acid
21. Draw the basic steroid skeleton and identify each ring with the appropriate letter designation.
22. Why are bile salts important for digesting lipids? What is their function?

Cholesterol and Cardiovascular Disease

23. Distinguish between VLDLs, LDLs, and HDLs in terms of the structure (composition) of each and the function of each.
24. How did studies with individuals having familial hypercholesterolemia help researchers determine the link between LDL levels and coronary artery disease?

(a) (b) (c)

▲ Figure 20.22

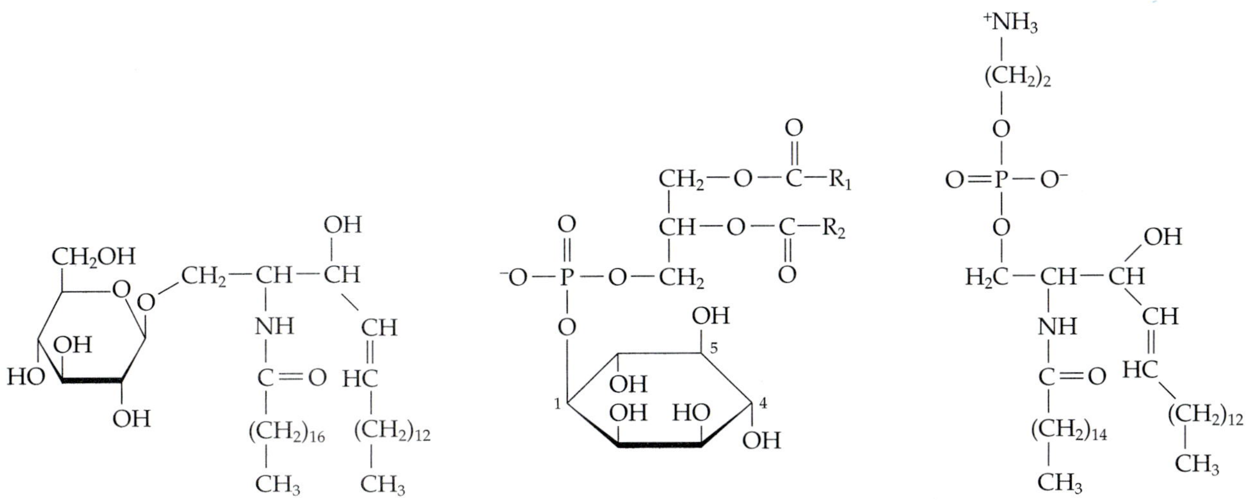

(a) (b) (c)

▲ Figure 20.23

Additional Problems

25. The melting point of elaidic acid is 52 °C.
    - a. What trend is observed when comparing the melting points of elaidic acid, oleic acid, and stearic acid? Explain why this trend is observed.
    - b. Would you expect the melting point of *trans*-hexadecenoic acid to be lower or higher than that of elaidic acid? Explain.

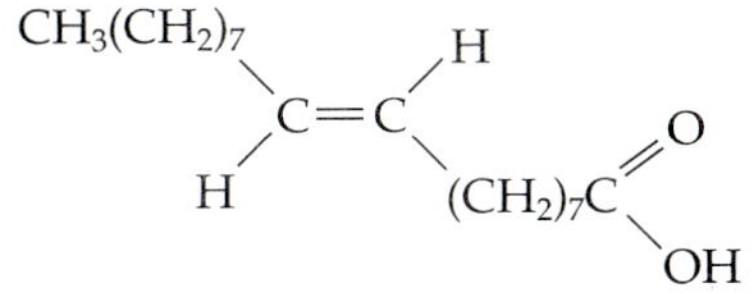

$CH_3(CH_2)_7CH{=}CH(CH_2)_7COOH$

Elaidic acid

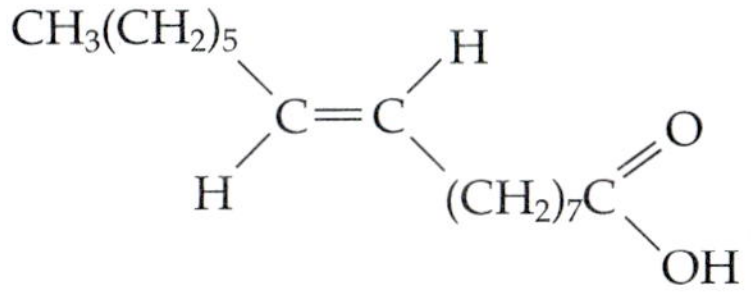

$CH_3(CH_2)_5CH{=}CH(CH_2)_7COOH$

trans-Hexadecenoic acid

26. Eicosapentaenoic acid (EPA) [$C_{20}H_{30}O_2$] is found in the brain and retina. If the first double bond occurs at C-5 (the carboxyl carbon is C-1), what is its probable structure? Would you expect its melting point to be higher or lower than that of arachidonic acid? Explain.
27. Examine the labels on three brands of margarine and two brands of shortening and list the oils used in the various brands.
28. How many mixed triglycerides are possible by combining stearic acid and oleic acid with glycerol?
29. Construct a typical lecithin molecule starting with one molecule each of glycerol, palmitic acid, oleic acid, phosphoric acid, and choline. Circle all the ester bonds.
30. In cerebrosides, what type of bond joins the fatty acid to sphingosine? What type of bond joins the sugar unit to sphingosine?
31. Phosphatidylinositol is a phosphatide found in cell membranes. Draw a structural formula for this phosphatide, given the structure of the alcohol inositol. (Linkage of the alcohol to the phosphate group occurs at the colored hydroxyl group.)

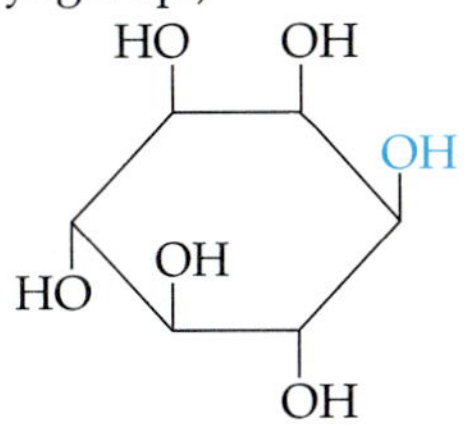

Inositol

32. Serine is an amino acid that has the structure:

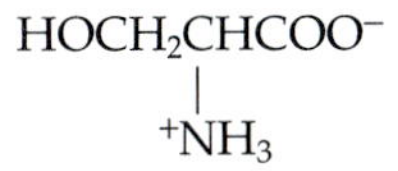

$HOCH_2CH(NH_3^+)COO^-$

Give the structure for phosphatidylserine.

33. *E. coli* bacteria were cultured at different temperatures, and the fatty acid composition of their cell membranes was determined. The data shown on the next page was obtained from cells cultured at 10 °C and at 40 °C. What is the importance of the observed variation in the percentages of each type of fatty acid?

Fatty Acid Composition of *E. coli* Cells Cultured at Two Different Temperatures

	Percentage of total fatty acids	
Fatty Acid	**10 °C**	**40 °C**
Palmitic Acid	18	48
Palmitoleic Acid	26	9
Oleic Acid	38	12

34. How does the structure of cholic acid differ from that of cholesterol? Which compound would you expect to be more polar? Justify your answer.

35. What role do LDL receptors play in controlling the levels of blood cholesterol?

36. Discuss the roles of cholesterol, saturated fats, and fish oils in atherosclerosis.

37. What benefits are there to the use of olestra? What are the drawbacks?

38. Can waxes be converted to soaps? Explain.

39. A principal wax in spermaceti is an ester of palmitic acid and cetyl alcohol (1-hexadecanol). Draw the structure of the ester.

40. How is an arterial plaque formed? How can the formation of an arterial plaque lead to a heart attack?

41. What are the warning signs of a heart attack?

SELECTED TOPIC 

Hormones

Humans and all other multicellular organisms must have a way for cells to communicate with each other—intercellular communication. Informational signals must be sent from one cell to adjacent cells or to cells or tissues at a greater distance. Table E.1 outlines the ways that cells and tissues communicate with each other. Selected Topic C discussed neurotransmitters, needed for synaptic communication. In this special topic we consider the molecules needed for paracrine and endocrine communication. A distinguishing characteristic of these compounds is their production of dramatic effects at very low concentrations. In **paracrine communication** the chemical messengers, known as **paracrine factors,** move from one cell to an-

Table E.1 Mechanisms of Intercellular Communication

Mechanism	Transmission	Chemical Mediators	Distribution of Effects
Direct communication	Through gap junctions from cytoplasm to cytoplasm	Ions, small solutes, lipid-soluble materials	Limited to adjacent cells that are directly interconnected by interlocking membrane proteins
Paracrine communication	Through extracellular fluid	Paracrine factors	Primarily limited to local area, where concentrations are relatively high; target cells must have appropriate receptors
Endocrine communication	Through the circulatory system	Hormones	Target cells are primarily in other tissues and organs and must have appropriate receptors
Synaptic communication	Across synaptic clefts	Neurotransmitters	Limited to very specific area; target cells must have appropriate receptors

other within a single tissue, such as the liver. In **endocrine communication** chemical messengers, known as **hormones,** are released in one tissue and transported through the circulatory system to one or more other tissues. Some compounds may act both as paracrine factors and as hormones.

E.1 The Endocrine System

Hormones are synthesized in the endocrine glands (Figure E.1) and enter the circulatory system as chemical messengers. Hormones released in one part of the body can lead to profound physiological changes in other parts of the body (heart, liver, muscle, kidney, etc.). They initiate, speed up, or slow down a variety of reactions. In this way they control growth, metabolism, reproduction, and many other functions of body and mind. Hormones can affect: (1) the permeability of cell membranes, (2) the rate of enzymatic reactions, and (3) the rate of synthesis of certain proteins.

If we consider hormones as messengers, the pituitary gland must be viewed as the central dispatcher or control. Many of the pituitary hormones control the production of additional hormones by other endocrine glands. Shutting down the central control ultimately shuts down much of the endocrine system. The pituitary itself responds to hormone signals from the hypothalamus. The hypothalamus secretes hormones, called releasing factors (or regulatory hormones), that trigger the production of the pituitary hormones. The hypothalamus is triggered by nerve impulses. The sequence of events is outlined in Figure E.2.

Let us take just a moment to consider the extraordinary complexity of this system. For example, in response to neural stimulation, the hypothalamus produces corticotropin-releasing factor (CRF) [also known as corticotropin regulatory hormone (CRH)]. The CRF reaches the pituitary and causes that gland to produce adrenocorticotropic hormone (ACTH). ACTH then stimulates the release of glucocorticoids (cortisol and related compounds) by the adrenal gland. Among other effects, glucocorticoids slow the migration of particular cells of the immune system to an injury site, giving rise to an anti-inflammatory effect. As the levels of glucocorticoids build up, a feedback mechanism causes the hypothalamus to slow down its production of CRF. This, in turn, slows the production of ACTH by the pituitary gland.

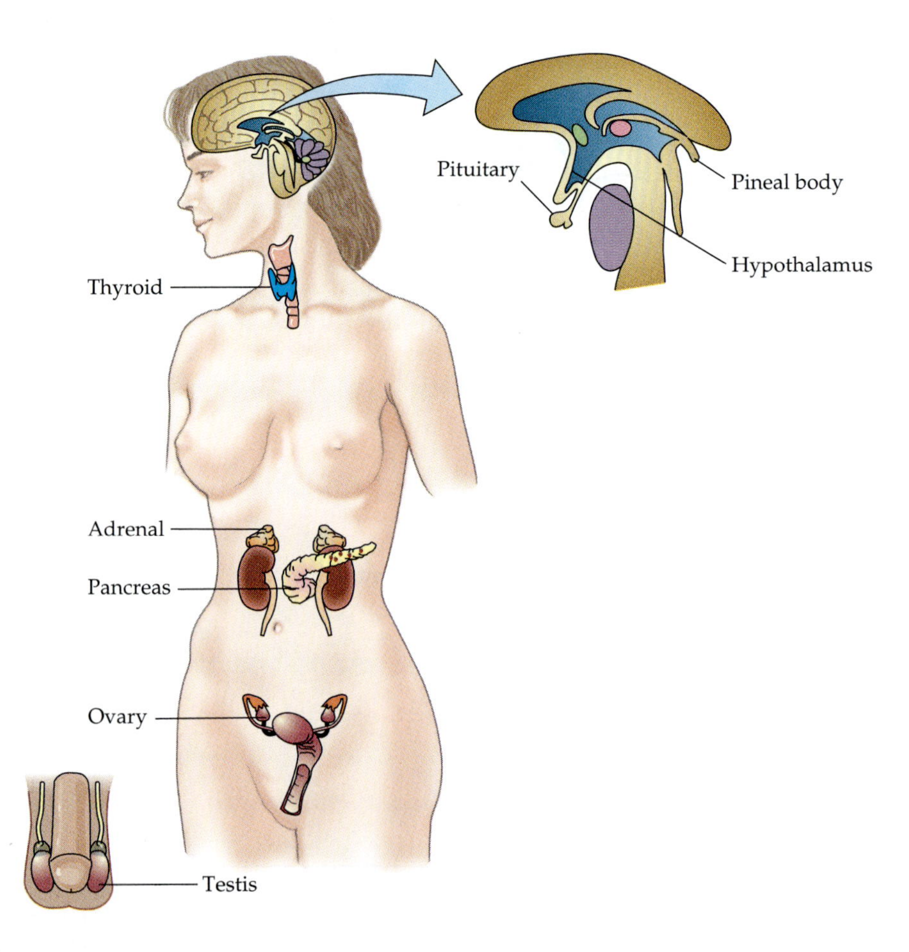

Figure E.1
Location of major endocrine glands in the human body.

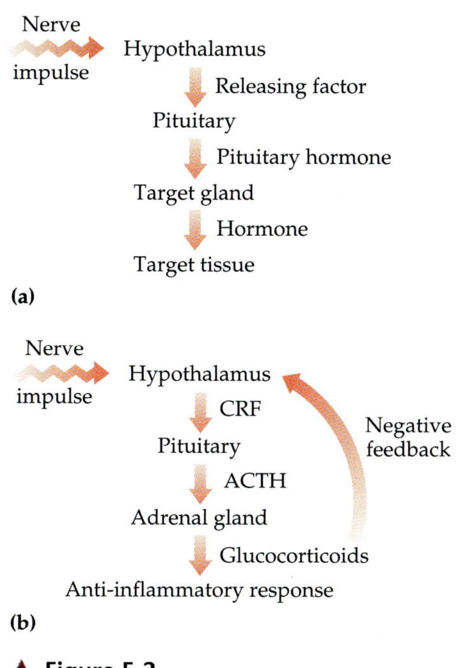

▲ **Figure E.2**
(a) General example and (b) specific example of the sequence of events in the release of hormones.

In a healthy individual all of this happens routinely, in perfect balance and with no conscious direction. And that's just one of the myriad interrelated biochemical processes that life requires. That it all works, and works so well, is nothing short of miraculous.

Many important human hormones and their physiological effects are listed in Table E.2. Hormones can be classified on the basis of chemical structure: (1) amino acid derivatives; (2) peptide hormones; and (3) steroid hormones. We shall discuss several peptide hormones, including vasopressin and oxytocin, in Chapter 21. The peptide hormones insulin and glucagon, which are produced in the pancreas, are considered in Chapter 25. We will be focusing on the steroid hormones in this selected topic.

The steroid hormones (e.g., adrenocortical hormones, sex hormones) are lipids. They are soluble in the lipid components of the cell membrane and can easily diffuse into cells. Inside the cell they combine with specific receptor molecules in the cytoplasm. They may influence enzymatic reactions directly, or the steroid-receptor complex can enter the nucleus. In the nucleus, steroids bind to specific sites on DNA, where they increase the rate of synthesis of mRNA (see Section 23.5), thus increasing the rate of synthesis of cellular enzymes. Because the primary effect of steroid hormones is regulating protein synthesis, their effects are generally much slower than those of other hormones (i.e., hours rather than minutes).

E.2 Adrenocortical Hormones

The outer part, or cortex, of the adrenal[1] gland (Figure E.3) uses cholesterol to produce a mixture of many steroids essential to life. These steroids constitute a family of hormones of which aldosterone and cortisol are the major representatives.

Aldosterone is called a *mineralocorticoid.* This name alludes to its function in regulating the exchange of sodium, potassium, and hydrogen ions. Although aldosterone acts on most cells in the body, it is particularly effective in enhancing the rate of reabsorption of sodium ions in the kidney tubule and in increasing the secretion of potassium ions and/or hydrogen ions by the tubule. Because the concentration of sodium ions is the major factor in water retention in the tissues, aldosterone also promotes water retention and reduces urine output. It thus supplements the action of vasopressin (Section 21.5).

Aldosterone

Cortisol and its keto derivative, cortisone, are called *glucocorticoids.* These hormones regulate a number of key metabolic reactions (e.g., they increase glucose production and mobilize fatty acids and amino acids). They also inhibit the inflammatory response of tissue to injury or stress (e.g., chemical irritants, exposure to radiation or to extreme temperature). In an inflammatory response the blood vessels in the surrounding area become dilated to bring extra blood to the affected area. (This accounts for the redness associated with inflammation.) Dilated blood vessels are more permeable, and fluid tends to leave the blood and enter the damaged tissue, causing swelling. Prostaglandins (Section E.4) contribute to the inflammatory response by: (1) promoting vasodilation of the blood vessels, (2) increasing the permeability of the capillaries, and (3) stimulating the pain receptors. It is thought that the glucocorticoids exert their anti-inflammatory effects by inhibiting an enzyme (phospholipase) necessary for the synthesis of arachidonic acid, a prostaglandin precursor.

[1]The term *adrenal* comes from the gland's location in the body, *ad*jacent to the *renal* (kidney).

Table E.2 Some Human Hormones and Their Physiological Effects

Name	Gland and Tissue	Chemical Nature[a]	Effect
Various releasing and inhibitory factors	Hypothalamus	Peptide	Trigger or inhibit release of pituitary hormones
Human growth hormone (HGH)	Pituitary, anterior lobe	Peptide	Controls the general body; controls bone growth
Thyroid-stimulating hormone (TSH)	Pituitary, anterior lobe	Peptide	Stimulates growth of the thyroid gland and production of thyroxine
Adrenocorticotrophic hormone (ACTH)	Pituitary, anterior lobe	Peptide	Stimulates growth of the adrenal cortex and production of corticol hormones
Follicle-stimulating hormone (FSH)	Pituitary, anterior lobe	Peptide	Stimulates growth of follicles in ovaries of females, sperm cells in testes of males
Luteinizing hormone (LH)	Pituitary, anterior lobe	Peptide	Controls production and release of estrogens and progesterone from ovaries, testosterone from testes
Prolactin	Pituitary, anterior lobe	Peptide	Maintains the production of estrogens and progesterone, stimulates the formation of milk
Vasopressin	Pituitary, posterior lobe	Peptide	Stimulates contractions of smooth muscle; regulates water uptake by the kidneys
Oxytocin	Pituitary, posterior lobe	Peptide	Stimulates contraction of the smooth muscle of the uterus; stimulates secretion of milk
Parathyroid	Parathyroid	Peptide	Controls the metabolism of phosphorus and calcium
Thyroxine	Thyroid	Amino acid derivative	Increases rate of cellular metabolism
Insulin	Pancreas, beta cells	Peptide	Increases cell usage of glucose; increases glycogen storage
Glucagon	Pancreas, alpha cells	Peptide	Stimulates conversion of liver glycogen to glucose
Cortisol	Adrenal gland, cortex	Steroid	Stimulates conversion of proteins to carbohydrates
Aldosterone	Adrenal gland, cortex	Steroid	Regulates salt metabolism; stimulates kidneys to retain Na^+ and excrete K^+
Epinephrine (adrenaline)	Adrenal gland, medulla	Amino acid derivative	Stimulates a variety of mechanisms to prepare the body for emergency action, including the conversion of glycogen to glucose
Norepinephrine (noradrenaline)	Adrenal gland, medulla	Amino acid derivative	Stimulates sympathetic nervous system; constricts blood vessels, stimulates other glands
Estradiol	Ovary, follicle	Steroid	Stimulates female sex characteristics; regulates changes during menstrual cycle
Progesterone	Ovary, corpus luteum	Steroid	Regulates menstrual cycle; maintains pregnancy
Testosterone	Testes	Steroid	Stimulates and maintains male sex characteristics

[a]Peptide hormones are discussed in Chapter 21 and Chapter 25.

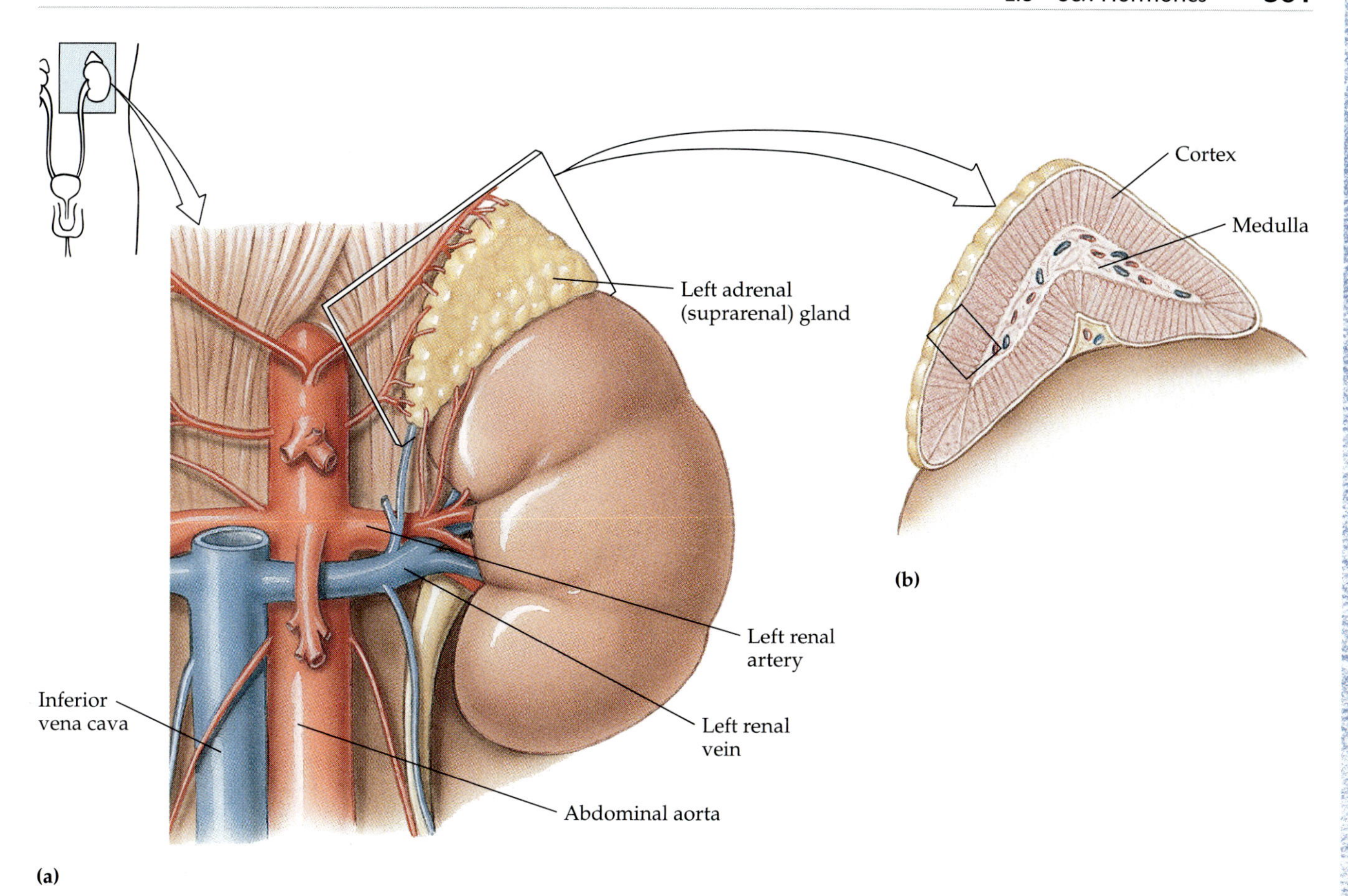

▲ **Figure E.3**
(a) Superficial view of the left kidney and adrenal gland. (b) An adrenal gland in section.

Cortisol (Hydrocortisone)

Cortisone

Prednisolone

Glucocorticoids are used as drugs for immunosuppression after transplant operations and in the treatment of severe skin allergies and autoimmune diseases, such as rheumatoid arthritis. The hormones or their analogs are injected, taken orally, or applied directly to the site of inflammation. Prolonged use of cortisone can have serious side effects, including high blood pressure, wasting of muscles, and resorption of bone. Several synthetic analogs, including prednisolone, methylprednisolone, triamcinolone, and dexamethasone, have been developed in an effort to minimize side effects through variations in dose or duration of effect.

E.3 Sex Hormones

The sex hormones are a class of steroids secreted by the gonads (ovaries or testes), the placenta, and the adrenal glands. The primary male sex hormones (called **androgens**), *testosterone* and *androstenedione*, are produced in the testes (and in lesser amounts in the adrenal cortex and the ovaries). They control the primary sexual characteristics of males, that is, the development of the male genital organs and the continued production of sperm. Androgens are also responsible for the development of secondary male characteristics, such as facial hair, deep voice, and muscle strength. Men generally have larger muscles than women because men have more testosterone.

Two sex hormones are of particular importance in females. *Progesterone* prepares the uterus for pregnancy and prevents the further release of eggs from the ovaries during pregnancy. The **estrogens** are mainly responsible for the development of female secondary sexual characteristics, such as breast development and increased deposition of adipose tissue in the breasts, buttocks, and thighs.

Progesterone (biosynthesized from cholesterol) is the precursor for the synthesis of the glucocorticoids, mineralocorticoids, androgens, and estrogens (Figure E.4). Both males and females produce androgens and estrogens, differing in the amounts of secreted hormones, not in the presence or absence of one or the other. Notice that the male and female hormones exhibit only very slight structural differences. Their physiological effects, however, differ enormously.

Sex hormones—both natural and synthetic—are sometimes used therapeutically. For example, a woman who has had her ovaries removed may be given female hormones to compensate for those no longer produced by the ovaries. Some of the earliest chemical compounds employed in cancer chemotherapy were sex

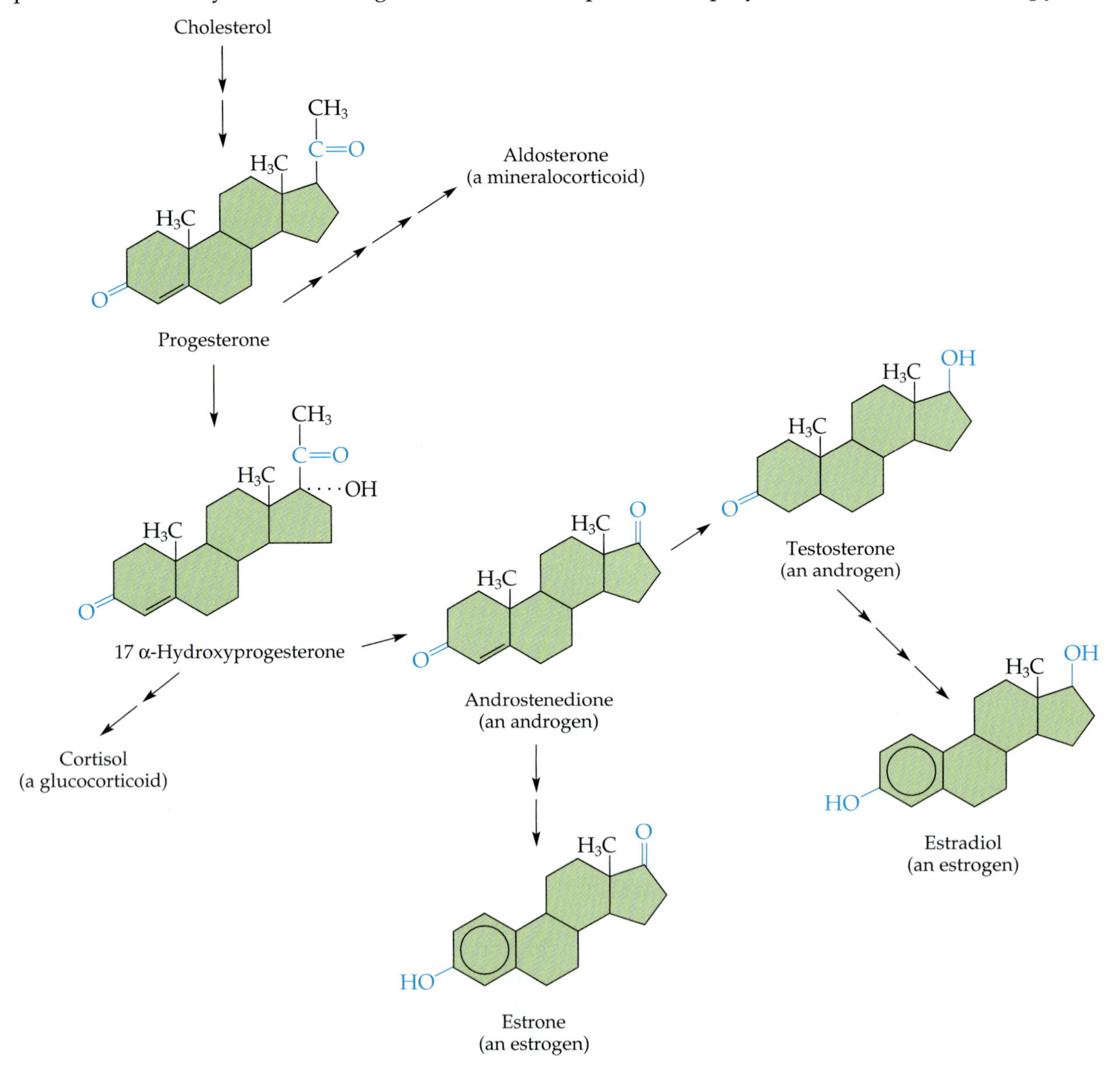

▲ **Figure E.4**
Biosynthesis of the major steroid hormones from cholesterol. The structures of the important sex hormones are included.

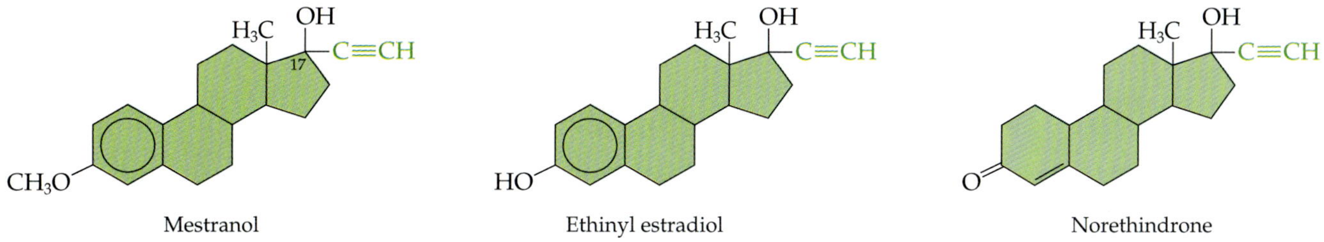

▲ **Figure E.5**
Synthetic derivatives of the female sex hormones. (a) Analogs of the estrogens. (b) Analog of progesterone.

hormones. Testosterone was used to treat carcinoma of the breast in females, and estrogens were given to males to treat carcinoma of the prostate. Sex hormones are also important in sex-change operations. Before surgery, hormones are administered to promote the development of the proper secondary sexual characteristics.

Anabolic Steroids

There has been a great deal of controversy generated by the widespread use of **anabolic steroids** by athletes. The drugs in question are synthetic androgens that stimulate protein synthesis (especially in skeletal muscle cells) without affecting the sex glands. Testosterone injected intramuscularly has been shown to increase muscle mass and strength in normal men. However, testosterone is not very active if taken orally because it is metabolized in the liver. The incorporation of a methyl group at C-17 prevents this metabolism. Introduction of a second double bond in ring A produces a compound, methandienone (Dianabol), that has anabolic activity (stimulation of protein synthesis) but little of the virilizing effects of testosterone.

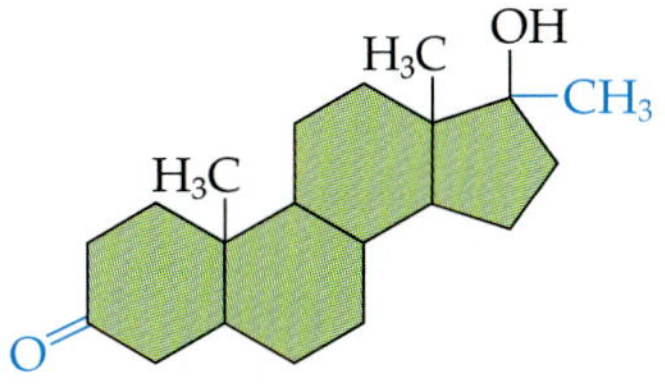

Methandienone
(Dianabol)

Anabolic steroids do increase strength—at least for a while—but have many side effects. In males, side effects include testicular atrophy and loss of function, impotence, acne, liver damage, edema (swelling), elevated cholesterol levels, and growth of breasts. Liver cancer is now showing up at an alarming rate in athletes who began using steroids in the 1960s. Anabolic steroids act as male hormones (androgens). They make women more masculine. They help women build larger muscles, but they also induce balding, development of extra body hair, deepening of the voice, and menstrual irregularities.

Chemists can detect the presence of synthetic steroids in the body by monitoring the urine and testing for degradation products. Anabolic steroids are marketed for use in the treatment of hypogonadism or delayed puberty in males, endometriosis in women, anorexia, and the wasting associated with AIDS. (Wasting means there is an involuntary loss of more than 10% of normal body weight.)

Another steroid, androstenedione (Andro), made the headlines in the summer of 1998 after a reporter spotted a bottle of Andro in the locker of the St. Louis Cardinal's home-run slugger Mark McGwire. Androstenedione is not banned in major league baseball, nor the NHL and NBA. However, it is banned by the International Olympic Committee, the NFL, and the NCAA. Androstenedione is converted to testosterone in the body. What is not known is how much of the androstenedione taken orally is absorbed by the body and converted to testosterone. No clinical studies have been conducted to determine whether androstenedione improves athletic performance or if it increases the risk of heart attack, cancer, or liver dysfunction with long-term use. Many medical experts believe that more research must be done before its use can be recommended.

Conception and Contraceptives

When taken regularly, synthetic derivatives of the female sex hormones prevent ovulation. The oral contraceptives are usually mixtures of analogs of progesterone and the estrogens (Figure E.5).[2] For example, Ortho-Novum® contains 1 mg or less of norethindrone and 0.035 mg of ethinyl estradiol, while Norinyl® 1 + 50 tablets contain 1 mg of norethindrone and 0.05 mg of mestranol.

Norethindrone and related compounds are called **progestins** because they mimic the action of progesterone. The progestin acts by establishing a state of false

[2]The prevention of ovulation is also effected by administration of progesterone and estradiol, but these hormones must be injected into the body for maximum results. It is the C≡C group in the synthetic analogs that confers upon them the ability to be taken orally and to function in the same manner as the steroids produced by the body.

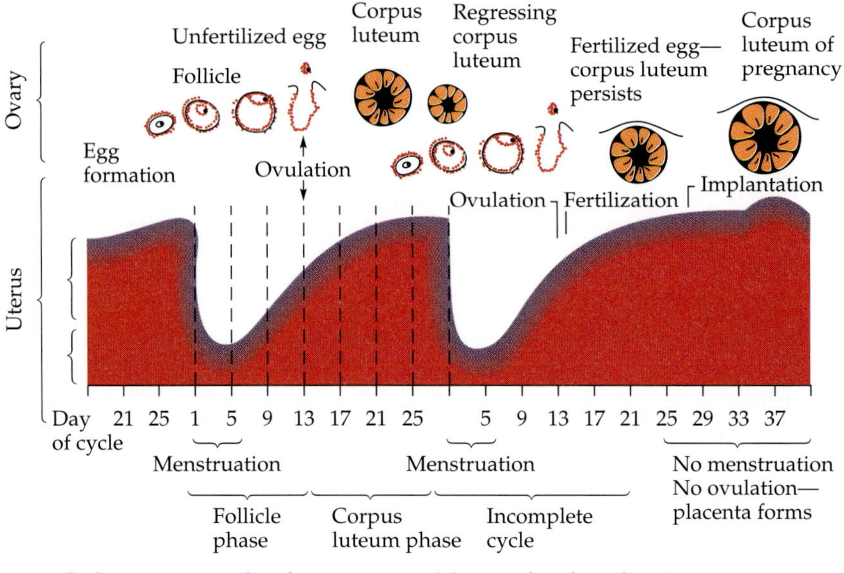

Figure E.6 Changes in the ovary and the uterus during the menstrual cycle. Both pregnancy and the pseudopregnancy caused by birth control pills prevent ovulation.

pregnancy (Figure E.6). A synthetic estrogen is added to regulate the menstrual cycle. A woman does not ovulate when she is pregnant (or in the state of false pregnancy established by the progestin). Because the woman does not ovulate, she cannot conceive.

Oral contraceptives have been used in the United States since 1960 by millions of women. They appear to be safe in most cases, but some women experience hypertension, acne, or abnormal bleeding. The pills increase the risk of blood clotting in some women, but so does pregnancy. Blood clots can clog arteries and cause death by stroke or heart attack. The death rate associated with birth control pills is about 1–2 in 100,000, approximately 1/10 of that associated with childbirth (10–15 per 100,000). The FDA advises women over 40, and any women who smoke, to use some other method of contraception. For these women the risk of using birth control pills is greater than the risk of childbirth.

Progesterone is essential for the maintenance of pregnancy. If its action is blocked, pregnancy cannot be established or maintained. Rousel-Uclaf, a French subsidiary of Hoescht, has developed a drug that blocks the action of progesterone. Mifepristone (RU-486), the "morning after" pill, is available in France and China, where it has replaced a substantial number of surgical abortions. A woman who wishes to abort a pregnancy takes three mifepristone tablets, followed in a few days by an injection or oral dose of a prostaglandin (Section E.4). The lining of the uterine wall and the implanted fertilized egg are sloughed off and pregnancy is terminated.

Mifepristone

RU-486 has other medical uses, too. It can increase uterine contractions when labor has stalled during childbirth. It appears to trigger lactation in mothers and to increase milk production. It seems to slow the growth of certain types of cancer. It is also being studied as a treatment for Cushing's syndrome, which results from excessive production of cortisone. Because of ethical and political issues related to abortions, RU-486 has not yet been approved for use in the United States.

In 1990 the FDA approved the use of an under-the-skin implant for birth control. Levonorgesterel (Norplant®), the progestin used in some birth control pills,

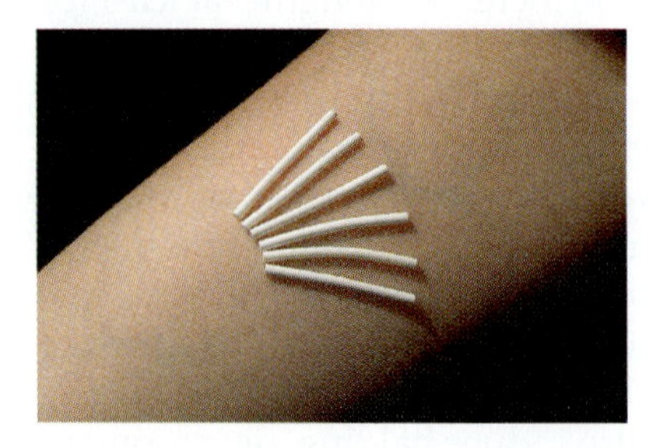

Norplant capsules are placed under the skin of a woman's upper arm and prevent pregnancy for five years.

is incorporated into six plastic capsules, each about 2.5 cm long and the diameter of a matchstick. These are implanted under the skin of a woman's upper arm. Release of the drug over time prevents pregnancy for five years with a failure rate of only 0.2%, essentially the same rate as birth control pills. It has about the same side effects as the progestin-only minipill. In 1992 medroxyprogesterone acetate (Depo-Provera®), a derivative of progesterone, was approved for use as a contraceptive. It is injected into the buttocks or arm muscle every three months. It prevents conception by inhibiting ovulation and causing changes in the uterine lining that make it less likely for pregnancy to occur.

Why do females have to bear the responsibility for contraception? Why not a pill for males? Recent studies have shown that many men—including a majority of the younger ones—are willing to share the risks and responsibility of contraception. Nevertheless, there are biological reasons for females to bear the burden: women get pregnant when contraception fails, and in females, contraception has only to interfere with one monthly event—ovulation. On the other hand, males produce sperm continuously.

Despite these arguments, a good deal of research has gone into male contraceptives, but results so far have not been very successful. Estrogens would work, but they would bring about the development of female characteristics in men, including a complete loss of interest in sexual relations with women. Gossypol, a pigment found in cottonseed that suppresses sperm production, has been used in China as a contraceptive for males. Current research is aimed at reducing its toxic side effects, which, at high doses, include accumulation of fluid in the lungs, shortness of breath, and paralysis.

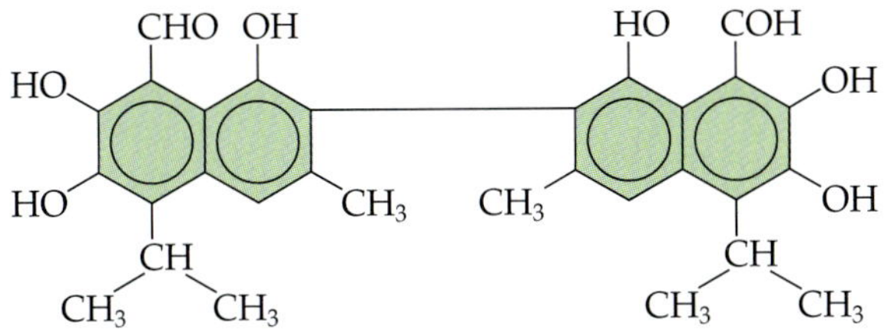

Gossypol

A drug called danazol (testosterone enanthate) has also been tested and found safe and effective. Administered along with testosterone, it suppresses sperm production, and the effect is reversed when the drug is withdrawn. Like the "Pill" for women, danazol often causes weight gain in men. The main problem, however, is that the drug is too expensive for widespread use. Perhaps the best approach for a male contraceptive is the development of a drug that would block sperm growth and/or transport.

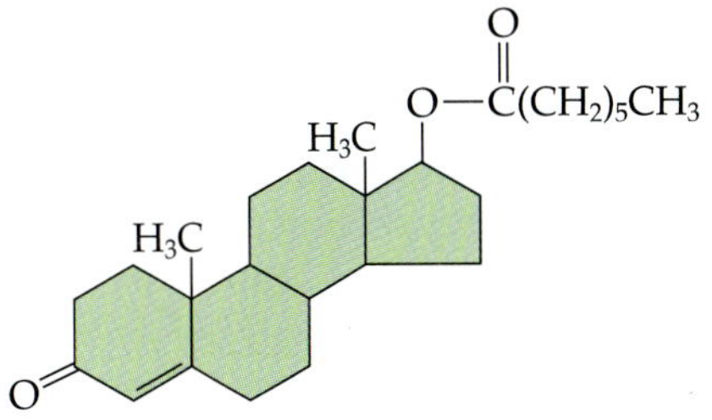

Testosterone enanthate
(Danazol)

The availability of vasectomies—simple surgical procedures that block the emission of sperm—has lessened the demand for a male contraceptive. A major drawback to vasectomy, however, is that it is often irreversible. A safe, cheap, and effective male contraceptive is still a goal of biochemical research.

The control of human reproduction is an issue subject to great moral, political, and legal controversy. Much more research is needed to increase our understanding of the biochemistry and physiology of reproduction. Armed with new knowledge, chemists might be able to design drugs acceptable to people of divergent views.

Hormone Replacement Therapy

In most women menstrual cycles continue into the late forties or early fifties, at which time they become increasingly irregular and finally cease altogether. *Menopause* is said to have occurred when a woman has twelve months without a period. Follicles no longer mature, ovulation does not occur, and the plasma estrogen concentration sharply decreases.

Lowering of estrogen levels leads to modification of female secondary sexual characteristics and loss of the feedback control by estrogens over the secretion of follicle-stimulating hormone (FSH) and luteinizing hormone (LH). The uncontrolled release of these hormones by the pituitary gland seems to be responsible for some of the unpleasant sensations that many menopausal women experience (e.g., hot flashes, irritability, anxiety, and fatigue). These discomforts can be overcome by the administration of daily doses of estrogens or estrogen substitutes. Many clinical studies have shown that long-term use of these drugs also reduces the risk of brittle bones (osteoporosis) and heart disease, and possibly also Alzheimer's disease and colon cancer.

The use of estrogen alone or estrogen and progesterone (or its derivatives) together by menopausal women is referred to as **hormone replacement therapy** (HRT). Many women and their physicians have been wary of estrogen because of reports linking it to an elevated risk of breast and uterine cancers. If progesterone is prescribed along with estrogen, there is a decreased risk of cancer of the uterine lining. Heart disease is the leading cause of death in women, killing more women each year than all cancers combined. Osteoporosis disables many postmenopausal women. Thus HRT is beneficial to a large number of women. However, the protracted effects are not yet known. Several large scale studies have been done and others are still being carried out to determine whether the benefits of long-term hormone replacement therapy outweigh the risks. Pharmaceutical companies are also actively developing such alternative therapies as selective estrogen receptor modulators, which appear to have estrogen-like effects in bone and the cardiovascular system, but not in the breast and uterus.

A study published by the *New England Journal of Medicine* on June 19, 1997, looked at the relationship between the length of time women used hormone replacement therapy and the balance of risks and benefits. This study found that the benefits outweighed the risks during the first ten years of hormone replacement. However, after ten years the benefits diminished drastically, particularly for women who were at low risk for cardiovascular diseases. There were still overall benefits, but not to the same extent. The diminished benefits were due to an increased risk of breast cancer mortality. Further studies are still needed to determine if the risk of breast cancer is increased for women taking progesterone along with estrogen as compared to women using estrogen alone. In addition, more studies are needed to determine whether changes in lifestyle such as not smoking, lowering cholesterol, and increasing physical activity can provide the same benefits as hormone replacement therapy without the side effects. In considering hormone replacement therapy, women should consult with their doctors to be fully informed of the risks versus the benefits.

E.4 Prostaglandins

Paracrine factors transfer information from cell to cell within a single tissue, exerting their primary effects on

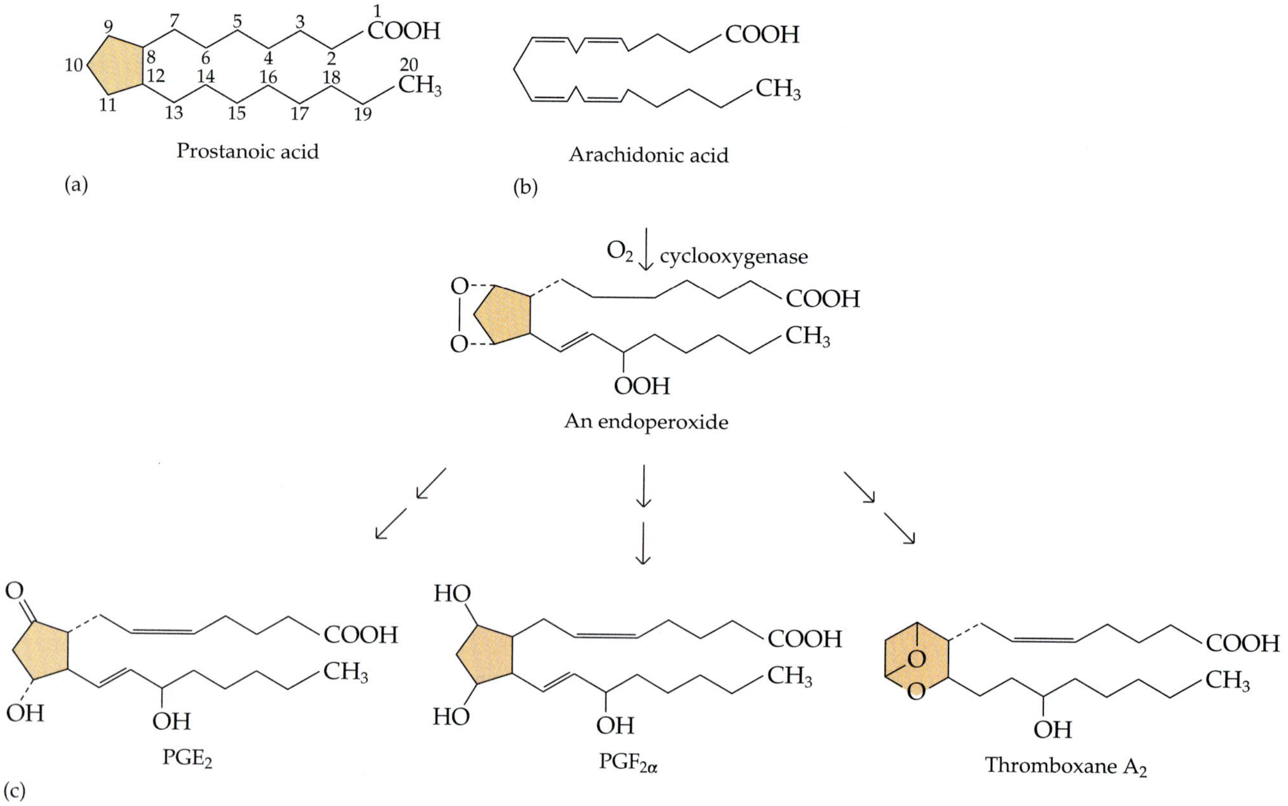

▲ **Figure E.7**
The arachidonic acid cascade. (a) Prostanoic acid is the parent compound of the prostaglandins. (b) Arachidonic acid is released from membrane phospholipids and is the precursor for the synthesis of (c) the prostaglandins and the related thromboxanes.

tissues in which they are synthesized. They can also enter the circulation and have secondary effects in other tissues and organs, thus acting as hormones. **Prostaglandins** are paracrine factors that were originally isolated from semen found in the prostate gland. In the mature male the prostate gland secretes about 0.1 mg/day of prostaglandins. However, prostaglandins are synthesized by most mammalian tissues and affect almost all organs in the body.

Prostaglandins are a family of unsaturated fatty acids, each containing 20 carbon atoms and having the same basic skeleton as prostanoic acid (Figure E.7a). The major classes are PGA, PGB, PGE, and PGF followed by a subscript that denotes the number of double bonds outside the five-carbon ring. Prostaglandins are not stored as such in cells. Rather, they are synthesized on demand from arachidonic acid (Figure E.7b), a 20-carbon polyunsaturated fatty acid that is released from phospholipids in cell membranes (by the action of the enzyme phospholipase). Arachidonic acid is converted to an endoperoxide by an enzyme complex called prostaglandin cyclooxygenase. The endoperoxide intermediate can then be transformed to the prostaglandins or to a related groups of compounds, the thromboxanes. This sequence of reactions is sometimes referred to as the *arachidonic acid cascade* (Figure E.7).

The prostaglandins are among the most potent biological substances known. Slight structural changes are responsible for quite distinct biological effects; however, all prostaglandins exhibit some ability to induce smooth muscle contraction, to lower blood pressure, and to contribute to the inflammatory response. We mentioned earlier that certain steroid drugs (cortisol and its synthetic analogs) exert their antiinflammatory action by inhibiting the release of arachidonic acid from membrane phospholipids (i.e., they inhibit *phospholipase*). On the other hand, aspirin and the other nonsteroidal antiinflammatory agents (e.g., indomethacin—Indocin; ibuprofen—Motrin, Advil, Nuprin) obstruct the synthesis of prostaglandins by inhibiting cyclooxygenase.

Their wide range of physiological activity has led to the synthesis of hundreds of prostaglandins and their analogs. Derivatives of PGE_2 are now in use in the United States to induce labor. Other prostaglandins have been employed clinically to lower or increase blood pressure, to inhibit stomach secretions, to relieve nasal congestion, to provide relief from asthma, and to prevent the formation of the blood clots associated with heart attacks and strokes. Thromboxane A_2 (made in blood platelets) induces blood clotting by stimulating blood platelet aggregation. Recall (Section B.1) that one of the side effects of aspirin is a prolonged bleeding time. This is due to the inhibition of platelet aggregation by blocking biosynthesis of thromboxane A_2.

A major clinical use of prostaglandins and their analogs is in induction of abortion. Their mechanism is uncertain, but it is different from that of the steroids. The prostaglandins cause regression of the corpus luteum, uterine contractions, and abortion of the embryo. Because they induce abortion, prostaglandins would have to be taken only once a month or only if a menstrual period were missed. Practically every major pharmaceutical company now has active prostaglandin research programs to develop new syntheses and to discover new natural sources of prostaglandins.

Key Terms

anabolic steroid (E.3)
androgens (E.3)
endocrine communication (page 558)
estrogens (E.3)
hormone replacement therapy (E.3)
hormones (page 558)
paracrine communication (page 557)
paracrine factors (page 557)
progestins (E.3)
prostaglandins (E.4)

Review Questions

1. How do paracrine factors differ from hormones?
2. What is the sequence of events that results in the release of hormones from an endocrine gland?
3. What gland produces releasing factors?
4. What gland is the target of releasing factors?
5. Give an example of a hormone that would be classified as a(n):
   a. steroid **b.** amino acid derivative **c.** peptide
6. What are the differences in biological function between the mineralocorticoids and the glucocorticoids?
7. Define and give an example of:
   a. hormone **b.** androgen
   c. estrogen **d.** progestin
8. What is the general structural classification of the compounds incorporated in birth control pills?
9. How do birth control pills work?
10. What structural feature gives a synthetic steroid sex hormone its oral effectiveness?
11. What fatty acid is the precursor of the prostaglandins?
12. List some therapeutic uses of prostaglandins.

CHAPTER 21 Proteins

Hemoglobin, one of the most studied proteins, is found in red blood cells. It transports oxygen from the lungs to cells in the body where it is needed, picking up carbon dioxide and transporting it back to the lungs, where it is expelled by breathing out.

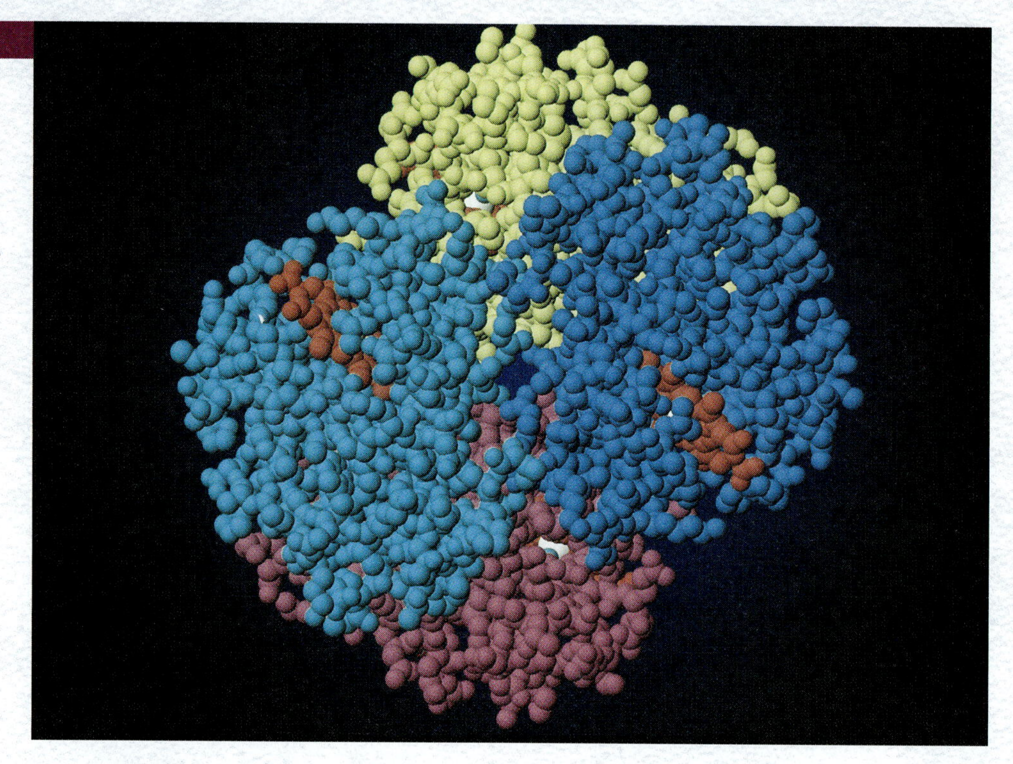

Learning Objectives/Study Questions

1. What are proteins? Why are they important?
2. What is an amino acid? How are amino acids classified, based on the characteristics of their side chains?
3. What is the difference between D- and L-amino acids?
4. How does an amino acid act like an acid? Like a base? Define the isoelectric pH of an amino acid.
5. How can electrophoresis be used to separate amino acids or proteins?
6. How is a peptide or protein formed from amino acids? Why is the sequence of amino acids in a protein important?
7. Describe the four levels of protein structure.
8. What are the two main types of secondary structure found in proteins?
9. What types of attractive interactions hold proteins in their most stable three-dimensional structure?
10. Do all proteins have quaternary structure? Explain.
11. What is meant by protein denaturation? How can proteins be denatured?

Proteins are an important class of biomolecules, serving a variety of roles in living organisms. Muscle tissue is largely protein, as are skin and hair. Proteins are present in the blood, in the brain, and even in tooth enamel. Each type of cell makes its own specific kinds of proteins, as well as proteins common to all or most cells. Whereas carbohydrates and lipids are used primarily as energy sources, the primary function of proteins is body building and maintenance. Lipids and carbohydrates are stored by the body as energy reserves, but proteins are not stored to any appreciable extent. In times of starvation the body must degrade muscle tissue, which contains large amounts of protein, to provide the amino acids needed to make vital enzymes and other proteins.

In an increasingly crowded world, the availability of protein is a growing concern. The rich nations have it—in the form of beefsteak, fish, or fowl. The poor nations need it, but can't afford meat or other high-quality proteins. A nation's use of sulfuric acid has long been considered an indication of its industrial development. Perhaps its consumption of protein is a better indication of the quality of life of its people, for without protein no nation can have the healthy, vigorous people vital to progress.

Proteins may be defined as compounds of high molar mass consisting largely or entirely of chains of amino acids. Their molar masses may range from several thousand to several million daltons. In addition to carbon, hydrogen, and oxygen, all proteins contain nitrogen and sulfur, and many also contain phosphorus and traces of other elements. The composition of most proteins is remarkably constant at about 51% carbon, 7% hydrogen, 23% oxygen, 16% nitrogen, 1–3% sulfur, and less than 1% phosphorus. We shall look at the properties and reactions of amino acids and then look at peptides and proteins, which are composed of amino acid units covalently linked together.

The **dalton** is a unit of mass used by biochemists and biologists. It is equivalent to the atomic mass unit: 1/12 the mass of an atom of carbon-12, 1.66×10^{-24} g. A 30,000 dalton protein has a molar mass of 30,000 u.

Amino Acids

The proteins in all living species, from bacteria to humans, are constructed from the same basic set of 20 amino acids. As the name indicates, an **amino acid** contains an amino group (Section 17.2) attached to a carboxylic acid (Section 16.1). The amino acids in proteins are α-amino acids, which means the amino group is attached to the α-carbon of the carboxylic acid. The amino acids are known exclusively by their common names, because the IUPAC names are too cumbersome. Asparagine was the first amino acid to be isolated (1806) and was given its name because it was obtained from protein found in asparagus juice. Glycine, the major amino acid found in gelatin, received its name because of its sweet taste (Greek *glykys*, sweet).

$$H_3\overset{+}{N}-\underset{\displaystyle R}{\underset{|}{CH}}-C\begin{matrix}\nearrow O \\ \searrow O^-\end{matrix}$$

An α-amino acid

21.1 General Properties of Amino Acids

In addition to the amino group, attached to the α-carbon of most amino acids is a side chain, called the R group. Each amino acid has unique characteristics as a result of the size, shape, solubility, and ionization properties of its R group. As we shall see in Section 21.8, the side chains of amino acids exert a profound effect on the conformation and the biological activity of proteins. Amino acids can be classified in several ways. We can group them into four classes according to the nature of the functional group on the side chains at neutral pH: (1) nonpolar, (2) polar but neutral, (3) negatively charged, and (4) positively charged. The structures and names of the 20 amino acids, their one- and three-letter abbreviations, and certain of their distinctive features are given in Table 21.1.

Table 21.1 Naturally Occurring Amino Acids

Name	Abbrev.	Structural Formula (at pH 6)	Molar Mass	Distinctive Features
1. Amino Acids with a Nonpolar R-Group				
Glycine	Gly (G)	$H-C(H)(NH_3^+)-C(=O)O^-$	75	The only amino acid lacking a chiral carbon.
Alanine	Ala (A)	$H_3C-C(H)(NH_3^+)-C(=O)O^-$	89	
Valine*	Val (V)	$(H_3C)_2CH-C(H)(NH_3^+)-C(=O)O^-$	117	Most animals cannot synthesize branched-chain amino acids. They are therefore essential in the diet.
Leucine*	Leu (L)	$(H_3C)_2CH-CH_2-C(H)(NH_3^+)-C(=O)O^-$	131	
Isoleucine*	Ile (I)	$H_3C-CH_2-CH(CH_3)-C(H)(NH_3^+)-C(=O)O^-$	131	
Phenylalanine*	Phe (F)	$C_6H_5-CH_2-C(H)(NH_3^+)-C(=O)O^-$	165	
Tryptophan	Trp (W)	(indol-3-yl)$-CH_2-C(H)(NH_3^+)-C(=O)O^-$ (ring NH)	204	A heterocyclic amino acid (a derivative of indole).
Methionine*	Met (M)	$H_3C-S-CH_2-CH_2-C(H)(NH_3^+)-C(=O)O^-$	149	Contains a sulfur atom in the nonpolar side chain and is important as a donor of methyl groups.
Proline	Pro (P)	(pyrrolidine ring, HN^+ with H)$-C(=O)O^-$	115	Contains a secondary amino group rather than a primary amino group and so is referred to as an *α-imino acid.*
2. Amino Acids with a Polar but Neutral R-Group				
Serine	Ser (S)	$HO-CH_2-C(H)(NH_3^+)-C(=O)O^-$	105	Occurs at the active site of many enzymes. The hydroxyl group may take part in the usual alcoholic reactions such as ester formation.

Table 21.1 Continued

Name	Abbrev.	Structural Formula (at pH 6)	Molar Mass	Distinctive Features
Threonine*	Thr (T)	$(HO)(H_3C)CH{-}C(H)(NH_3^+){-}C(=O)O^-$	119	Named for its similarity to the sugar threose (contains two chiral carbons).
Cysteine	Cys (C)	$HS{-}CH_2{-}C(H)(NH_3^+){-}C(=O)O^-$	121	Often occurs in proteins in its oxidized form, cystine.
Tyrosine	Tyr (Y)	$HO{-}C_6H_4{-}CH_2{-}C(H)(NH_3^+){-}C(=O)O^-$	181	The *p*-hydroxy derivative of phenylalanine.
Asparagine	Asn (N)	$(O{=})(H_2N)C{-}CH_2{-}C(H)(NH_3^+){-}C(=O)O^-$	132	The amide of aspartic acid.
Glutamine	Gln (Q)	$(O{=})(H_2N)C{-}CH_2{-}CH_2{-}C(H)(NH_3^+){-}C(=O)O^-$	146	The amide of glutamic acid.
3. Amino Acids with a Negatively Charged R-Group				
Aspartic acid	Asp (D)	$(O{=})(^-O)C{-}CH_2{-}C(H)(NH_3^+){-}C(=O)O^-$	132	At physiological pH the carboxyl groups are ionized, and these amino acids are also known as aspartate and glutamate.
Glutamic acid	Glu (E)	$(O{=})(^-O)C{-}CH_2{-}CH_2{-}C(H)(NH_3^+){-}C(=O)O^-$	146	
4. Amino Acids with a Positively Charged R-Group				
Lysine*	Lys (K)	$H_3\overset{+}{N}{-}CH_2CH_2CH_2CH_2{-}C(H)(NH_3^+){-}C(=O)O^-$	147	
Arginine*	Arg (R)	$(H_2N)(H_2\overset{+}{N}{=})C{-}NH{-}CH_2CH_2CH_2{-}C(H)(NH_3^+){-}C(=O)O^-$	175	Almost as strong a base as NaOH.
Histidine*	His (H)	imidazole ring ($H\overset{+}{N}$, NH)$\,{-}CH_2{-}C(H)(NH_3^+){-}C(=O)O^-$	155	The only amino acid whose R group has a pK_a (6.0) near physiological pH. Thus the imidazole ring can be charged (+) or uncharged in the physiological pH range.

*An essential amino acid; see Section 27.1.

Figure 21.1
Some biologically important nonprotein amino acids.

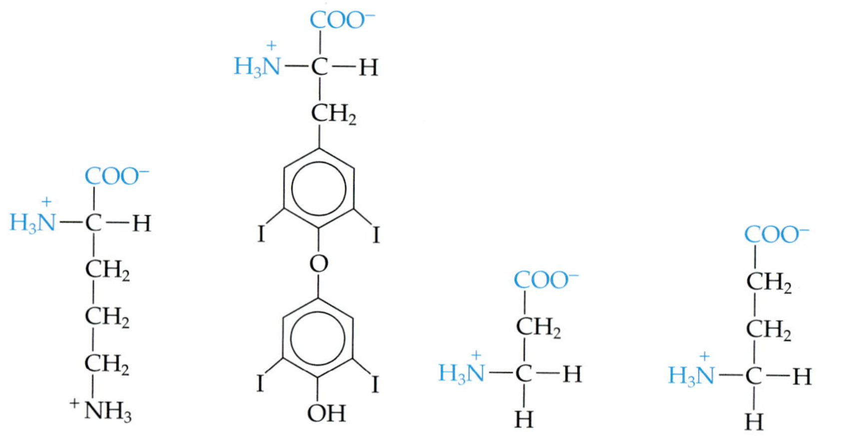

$$\begin{array}{c} COO^- \\ | \\ H_3\overset{+}{N}-C-H \\ | \\ CH_2 \\ | \\ CH_2 \\ | \\ CH_2 \\ | \\ {}^+NH_3 \end{array}$$

Ornithine (see Section 27.3)

Thyroxine (thyroid hormone)

$$\begin{array}{c} COO^- \\ | \\ CH_2 \\ | \\ H_3\overset{+}{N}-C-H \\ | \\ H \end{array}$$

β-Alanine (component of coenzyme A)

$$\begin{array}{c} COO^- \\ | \\ CH_2 \\ | \\ CH_2 \\ | \\ H_3\overset{+}{N}-C-H \\ | \\ H \end{array}$$

γ-Aminobutyrate (GABA, inhibitory neurotransmitter— see Section 27.5)

The amino acids are colorless, nonvolatile, crystalline solids, melting with decomposition at temperatures above 200 °C. Glycine, alanine, proline, threonine, lysine, and arginine are quite soluble in water, while the other amino acids are sparingly soluble. All amino acids are insoluble in nonpolar organic solvents. Their properties diverge widely from those of their unsubstituted carboxylic acid analogs. Organic acids of comparable molar mass are liquids or low-melting solids that are soluble in organic solvents but have limited solubility in water. In fact, the properties of the amino acids are more similar to those of inorganic salts than to those of amines or organic acids. The saltlike character of the amino acids indicates that the structure of amino acids in the solid state and in neutral solution is best represented by a dipolar ion (also called **zwitterion**) structure.

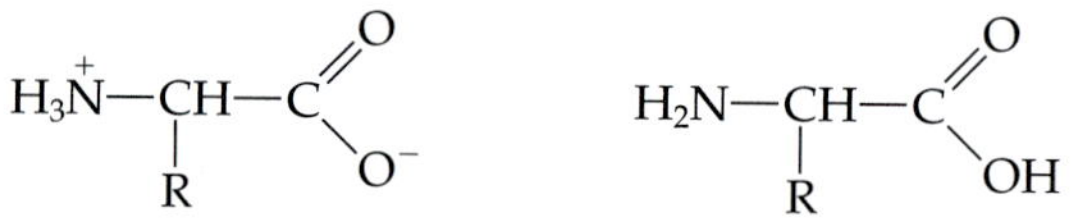

Zwitterion (ionized; electrically neutral)

Uncharged (not an accurate representation of amino acid structure)

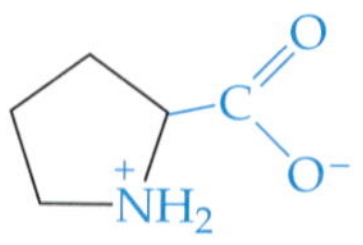

Proline

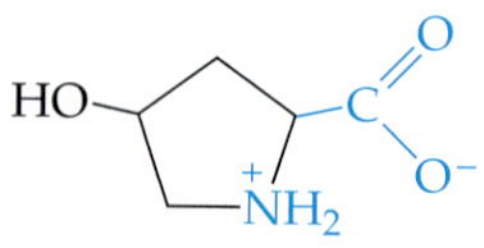

Hydroxyproline

Several other amino acids (e.g., hydroxyproline), which occur to some extent in certain proteins, are all derivatives of these 20 amino acids and are modified *after* incorporation into the protein chain. In addition, there are more than 150 other amino acids of physiological importance that are not derived from proteins. Most have been isolated from plants, but some are found in animals. These amino acids perform important biological functions (e.g., as intermediates in metabolic pathways), either as single molecules or combined in molecules of relatively small size (Figure 21.1).

Configuration

Notice in Table 21.1 that glycine is the only amino acid whose α carbon atom is *not* a chiral center. Therefore, with the exception of glycine, the amino acids are optically active and may exist in either the D- or the L-enantiomeric form. Once again, the reference compound for the assignment of configuration is glyceraldehyde (Section 19.1). (The amino group of the amino acid takes the place of the hydroxyl group of glyceraldehyde.)

$$\underset{\text{L-(−)-Glyceraldehyde}}{\begin{array}{c} H\diagdown\ \diagup\!\!\!\!\diagup O \\ C \\ | \\ HO-C-H \\ | \\ CH_2OH \end{array}} \qquad \underset{\text{L-Amino acid}}{\begin{array}{c} {}^{-}O\diagdown\ \diagup\!\!\!\!\diagup O \\ C \\ | \\ H_3\overset{+}{N}-C-H \\ | \\ R \end{array}} \qquad \underset{\text{D-Amino acid}}{\begin{array}{c} {}^{-}O\diagdown\ \diagup\!\!\!\!\diagup O \\ C \\ | \\ H-C-\overset{+}{N}H_3 \\ | \\ R \end{array}}$$

It is interesting to note that the naturally occurring sugars belong to the D series, whereas nearly all known plant and animal proteins are composed entirely of L-amino acids. However, certain bacteria contain D-amino acids in their cell walls. *Streptococcus faecalis* requires D-alanine, and *Staphylococcus aureus* needs D-glutamic acid. Several antibiotics (e.g., actinomycin D and the gramicidins) contain varying amounts of D-leucine, D-phenylalanine, and D-valine. Recall that the letters D and L refer only to a specific configuration (Section 19.1), not to the direction of optical rotation. For example, in a neutral solution L-alanine is dextrorotatory, whereas L-serine is levorotatory. The optical rotation of any amino acid is very much dependent upon the pH of the solution.

✓ Review Questions

21.1 What is the general structure for an α-amino acid?

21.2 Identify the amino acid that fits each description: **(a)** first isolated from asparagus; **(b)** most abundant amino acid in gelatin; **(c)** does not have a chiral center; **(d)** almost as strong a base as NaOH.

21.2 Reactions of Amino Acids

Amino acids can act either as acids or as bases. Indeed, amino acids, both free and incorporated into proteins, act as buffers in living organisms. In the presence of added acid, the carboxylate group of the zwitterion captures a proton to form a positive ion. If base is added, proton removal from the amino group of the zwitterion forms a negative ion. In both instances, the amino acid acts to maintain the pH of the system—that is, to tie up added acid (H^+) or base (OH^-).

Addition of an acid:

$$H_3\overset{+}{N}-\underset{\underset{R}{|}}{CH}-COO^- + H^+ \longrightarrow H_3\overset{+}{N}-\underset{\underset{R}{|}}{CH}-COOH$$

Addition of a base:

$$H_3\overset{+}{N}-\underset{\underset{R}{|}}{CH}-COO^- + OH^- \longrightarrow H_2N-\underset{\underset{R}{|}}{CH}-COO^- + H_2O$$

At some intermediate pH value, an amino acid exists almost entirely as the zwitterion. The particular pH at which an amino acid exists in solution as a zwitterion is called the **isoelectric pH**. At the isoelectric pH, the positive and negative charges on an amino acid balance, and the molecule as a whole is electrically neutral. Each amino acid has a characteristic isoelectric pH. The neutral amino acids (with un-ionizable side chains) have isoelectric pH values ranging from 5.0 to 6.5. Basic amino acids (those with side chains that are positively charged at neutral pH) have relatively high isoelectric pH values. Acidic amino acids (those with side chains that are negatively charged at neutral pH) have quite low isoelectric pH values (Table 21.2).

By adjusting the pH of a solution, one can vary the net charge on an amino acid, thereby causing amino acids to migrate at different rates in an electric field. This process of separating mixtures of amino acids is called **electrophoresis**. In a

Table 21.2 Isoelectric pH Values of Some Representative Amino Acids

Amino Acid	Type	Isoelectric pH
Alanine	Neutral, nonpolar	6.0
Valine	Neutral, nonpolar	6.0
Serine	Neutral, polar	5.7
Threonine	Neutral, polar	6.5
Aspartic acid	Acidic	3.0
Glutamic acid	Acidic	3.1
Histidine	Basic	7.6
Lysine	Basic	9.8
Arginine	Basic	10.8

typical experiment, a paper strip saturated with a buffer solution at a specified pH is suspended between two reservoirs of the buffer (Figure 21.2). A sample of the solution of amino acids is applied to the center of the paper, and an electric potential is applied between the two buffer solutions. Any amino acids having an isoelectric pH equal to the pH of the buffer have no net charge and do not migrate. Any amino acids with a net negative charge move toward the positive electrode, and any with a net positive charge migrate toward the negative electrode (Figure 21.3). Amino acids of different sizes move at different rates, even if both have the same electric charge. After a period of time, the various amino acids are separated into individual spots on the paper.

Simple chemical tests can be used to identify amino acids because these compounds undergo reactions characteristic of carboxylic acids and amines. The ninhydrin test (Section 17.5) gives a purple color when amino acids are present and serves as a qualitative test for amino acids. This and similar reactions are quite important in the detection and separation of amino acids. The most important reaction

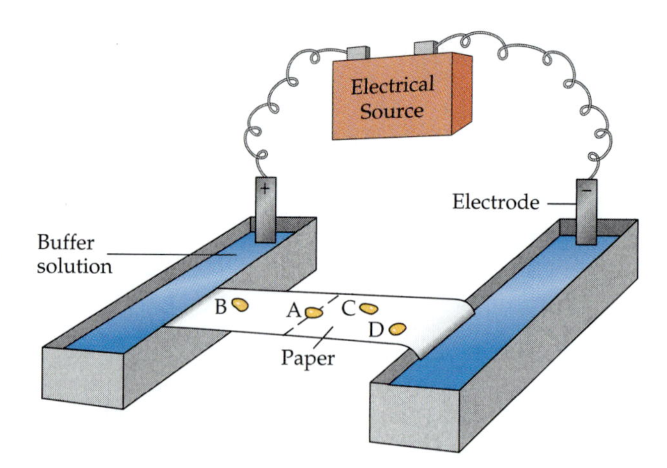

Figure 21.2 An electrophoresis apparatus. Amino acid A is in the zwitterion form and has not migrated; B exists as an anion and has therefore moved toward the positive electrode; C and D are in the cationic form and have accordingly moved toward the negative electrode.

$$R-\overset{\displaystyle H}{\underset{\displaystyle {}^{+}NH_3}{C}}-C\!\!\begin{array}{l}{//}O\\ \backslash OH\end{array} \quad \underset{OH^-}{\overset{H^+}{\rightleftharpoons}} \quad R-\overset{\displaystyle H}{\underset{\displaystyle {}^{+}NH_3}{C}}-C\!\!\begin{array}{l}{//}O\\ \backslash O^-\end{array} \quad \underset{H^+}{\overset{OH^-}{\rightleftharpoons}} \quad R-\overset{\displaystyle H}{\underset{\displaystyle NH_2}{C}}-C\!\!\begin{array}{l}{//}O\\ \backslash O^-\end{array}$$

Cationic form at pH values below isoelectric pH
Net charge = +1

Zwitterion form at isoelectric pH
Net charge = 0

Anionic form at pH values above isoelectric pH
Net charge = −1

Figure 21.3 Acid-base behavior of neutral amino acids.

of all, however, is the polymerization reaction that forms peptides and proteins (Sections 21.3 and 23.6).

Review Questions

21.3 Define the following terms: **(a)** zwitterion; **(b)** isoelectric pH; **(c)** electrophoresis.

21.4 **(a)** Write a structural formula for the anion formed when glycine reacts with a base. **(b)** Write a structural formula for the cation formed when glycine reacts with an acid.

Peptides

Two or more amino acids can be joined together to form a **peptide.** When fifty or more amino acids are linked together, we call this larger molecule a protein. By convention, we begin writing the structure of peptides and proteins with the amino acid whose amino group is free (the *N-terminal end*). The other end has a free carboxyl group and is referred to as the *C-terminal end.*

21.3 The Peptide Bond

In Section 17.5 we discussed the reaction between amines and carboxylic acids to form substituted amides. If we heat the salt of an amine and a carboxylic acid, an amide is formed:

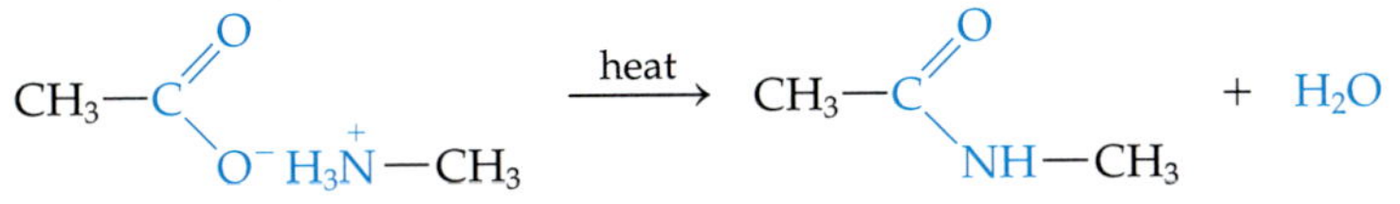

$$CH_3-C(=O)-O^- \; H_3\overset{+}{N}-CH_3 \xrightarrow{\text{heat}} CH_3-C(=O)-NH-CH_3 + H_2O$$

Methylammonium acetate — *N*-Methylacetamide

Similarly, the amino group on one amino acid molecule can react with the carboxyl group on another. A molecule of water is split out and an amide linkage is formed:

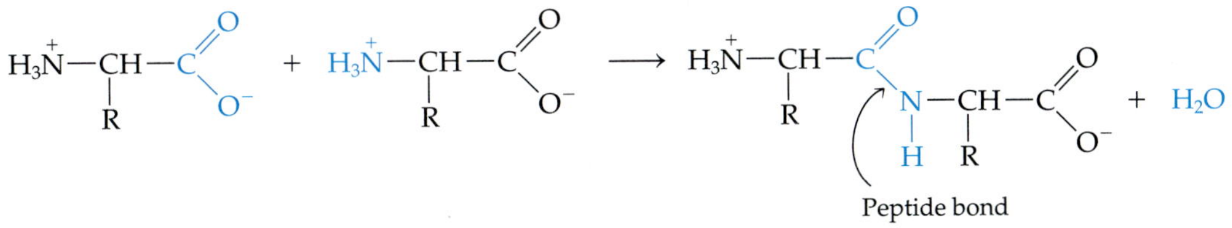

$$H_3\overset{+}{N}-\underset{R}{CH}-C(=O)-O^- + H_3\overset{+}{N}-\underset{R}{CH}-C(=O)-O^- \longrightarrow H_3\overset{+}{N}-\underset{R}{CH}-C(=O)-\underset{H}{N}-\underset{R}{CH}-C(=O)-O^- + H_2O$$

Peptide bond

The amide linkage is called a **peptide bond** when it joins two amino acid units. Note that the product molecule still has a reactive amino group on the left and a carboxyl group on the right. Each of these can react further to join more amino acid units. This process can continue until thousands of units have joined to form large proteins.

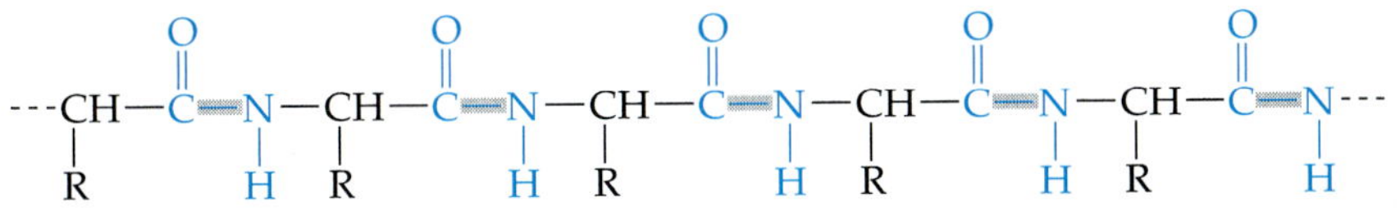

$$\cdots\underset{R}{CH}-\overset{O}{\overset{\|}{C}}-\underset{H}{N}-\underset{R}{CH}-\overset{O}{\overset{\|}{C}}-\underset{H}{N}-\underset{R}{CH}-\overset{O}{\overset{\|}{C}}-\underset{H}{N}-\underset{R}{CH}-\overset{O}{\overset{\|}{C}}-\underset{H}{N}-\underset{R}{CH}-\overset{O}{\overset{\|}{C}}-\underset{H}{N}\cdots$$

When only two amino acids are joined, the product is called a dipeptide, and when three amino acids are combined, the substance is a tripeptide. Each amino acid in the peptide, with the exception of the C-terminal amino acid, is named as an acyl group in which the suffix *-ine* is replaced by *-yl*.

Prefixes (di, tri, tetra, etc.) are used to indicate the number of amino acids that are joined together. The general term **peptide** is often used to designate a combination of an unspecified number of amino acids.

$H_3\overset{+}{N}-CH_2-COO^- + H_3\overset{+}{N}-CH(CH_2C_6H_5)-COO^- \longrightarrow H_3\overset{+}{N}-CH_2-C(=O)-NH-CH(CH_2C_6H_5)-COO^-$

Glycine Phenylalanine Glycylphenylalanine (a dipeptide)

$H_3\overset{+}{N}-CH(CH_2OH)-C(=O)-NH-CH(CH_3)-C(=O)-NH-CH(CH_2SH)-COO^-$

Serylalanylcysteine (a tripeptide)

Combinations with more than 10 amino acid units are often simply called **polypeptides**. Chains of about 50 amino acids or more are called proteins or polypeptides. A protein may be composed of one or more polypeptide chains (Section 21. 9).

Review Questions

21.5 Distinguish between the N-terminal amino acid and the C-terminal amino acid.

21.6 Amino acid units in a protein are connected by peptide bonds. What is another name for the functional group linking the amino acids?

21.7 Draw structures for: **(a)** valylthreonine; **(b)** tyrosylglutamine.

21.4 The Sequence of Amino Acids

For peptides and proteins to be physiologically active, it is not enough that they incorporate certain amounts of specific amino acids. The order, or *sequence*, in which the amino acids are connected is also of critical importance. Glycylalanine is different from alanylglycine.

$H_3\overset{+}{N}-CH_2-C(=O)-NH-CH(CH_3)-COO^-$ $H_3\overset{+}{N}-CH(CH_3)-C(=O)-NH-CH_2-COO^-$

Glycylalanine (Gly-Ala) Alanylglycine (Ala-Gly)

Although the difference seems minor, the two substances behave differently in the body.

As the length of a peptide chain increases, the number of possible sequential variations becomes very large. This potential for many different arrangements is exactly what one needs in a material that makes up such diverse materials as hair, skin, eyeballs, toenails, and a thousand different enzymes (the topic of Chapter 22). To appreciate their enormous complexity, consider that proteins contain fifty or more amino acids. It is estimated that the human body contains over 100,000 different protein molecules, all of which are characterized by different sequential arrangements of the 20 fundamental building blocks.

Just as we can make millions of different words with our 26-letter English alphabet, we can make millions of different proteins with the 20 different amino acids. Just as one can write gibberish with the English alphabet, one can make nonfunctional proteins by putting together the *wrong sequence* of amino acids. Sometimes a seemingly minor change can have a disastrous effect. Some people have hemoglobin with a single incorrect amino acid (valine replacing glutamic acid) unit out of about 300. That "minor" error is responsible for sickle cell anemia, an inherited condition that ordinarily proves fatal (Section 28.2). We will further discuss the effects of errors in amino acid sequences in Chapter 23.

Review Question

21.8 Define or describe the following terms: **(a)** peptide bond; **(b)** tripeptide; **(c)** polypeptide.

21.5 Peptide Hormones

Several naturally occurring peptides possess significant biological activity. In Section C.5, we said that the brain produces a variety of peptides, several of which act like morphine to relieve pain. Rather, we should say that morphine acts like these peptides because the brain was making peptides long before people discovered the painkilling effect of the juice of the opium poppy.

Many hormones are peptides, known as peptide hormones (Section E.1). The hormones *oxytocin* and *vasopressin* are cyclic nonapeptides produced by the pituitary gland. Notice that seven of the nine amino acids are identical in both peptides, yet their physiological effects are markedly different. Oxytocin stimulates lactation and causes the contraction of smooth muscles in the uterine wall. It is often administered at childbirth to induce labor. Oxytocin was the first naturally occurring peptide to be synthesized in the laboratory (1953). The synthetic peptide was found to have the same physiological properties as the corresponding natural one.

Ile-Tyr-Cys
S
S
Gln-Asn-Cys-Pro-Leu-Gly

Oxytocin

Phe-Tyr-Cys
S
S
Gln-Asn-Cys-Pro-Arg-Gly

Vasopressin

Vasopressin is called *antidiuretic hormone* (ADH) because it acts on the kidneys to reduce the amount of water excreted. (**Diuretics** are substances that increase the volume of urine, and therefore any substance that reduces the volume of urine is an antidiuretic.)[1] Thus the major function of vasopressin is to increase water reabsorption in the kidney. After drinking alcohol, many people excrete more urine than can be accounted for by the volume of water in their drinks. It is thought that alcohol inhibits the secretion of vasopressin, so that the kidney reabsorbs less water. A deficiency of vasopressin, or an inability of the kidney to respond to vasopressin, results in diabetes insipidus, in which too much urine is excreted ($>$10 L/day). This disease (which is treated by administering vasopressin) should not be confused with diabetes mellitus (see Chapter 25). In addition, vasopressin stimulates the contractions of muscles in the walls of blood vessels and thus increases the blood pressure. It has been used to overcome low blood pressure caused by shock following surgery. Table 21.3 offers a comparison of the effects of vasopressin and oxytocin.

Table 21.3 A Comparison of the Effect of Oxytocin and Vasopressin

Structure or Function Affected	Vasopressin	Oxytocin
Water diuresis	Inhibits	Has no effect on
Blood pressure	Elevates	Slightly lowers
Coronary arteries	Constricts	Slightly dilates
Intestinal contractions	Stimulates	Has questionable effect on
Uterine contractions	Stimulates	Stimulates
Ejection of milk	Slightly stimulates	Stimulates

[1]Diuretics are often given to people with high blood pressure (see Section 28.7) to cause the loss of water and sodium ions, both of which contribute to the elevated blood pressure.

Remember the great similarity in the structures of these compounds as you look at the table.

Vasopressin is not the only peptide hormone that can affect blood pressure. Bradykinin, a nonapeptide produced in the blood by the cleavage of a larger protein precursor has the amino acid sequence:

Arg–Pro–Pro–Gly–Phe–Ser–Pro–Phe–Arg

Bradykinin is a potent biochemical that lowers blood pressure, stimulates smooth muscle tissue, increases capillary permeability, and causes pain. The reverse peptide,

Arg–Phe–Pro–Ser–Phe–Gly–Pro–Pro–Arg

has been synthesized. It shows none of the activity of bradykinin.

The octapeptide angiotensin II is produced in the kidneys from the decapeptide angiotensin I.

Asp–Arg–Val–Tyr–Ile–His–Pro–Phe [angiotensin II]

This substance is the most powerful vasoconstrictor known; thus its effects oppose those of vasopressin and bradykinin. Angiotensin II acts to maintain blood pressure. Clinical research indicates that some forms of hypertension probably involve overproduction of angiotensin II. Drugs, such as captopril, that act to block the formation of angiotensin II from angiotensin I are important in the control of hypertension.

Review Questions

21.9 What is the physiological function of oxytocin?

21.10 How does bradykinin affect blood pressure?

Proteins

Each of the many naturally occurring proteins has its own characteristic composition, amino acid sequence, and three-dimensional shape. In the past 40 years scientists have determined the amino acid sequences and three-dimensional conformation of many proteins. This information can help the understanding of how each protein performs its specific function in the body.

21.6 Classification of Proteins

Because of their great complexity, protein molecules cannot possibly be classified systematically in the same way as the carbohydrates and lipids are categorized—that is, on the basis of structural similarities. *One way to classify proteins is based on structure and solubility*. Some proteins, such as those that make up hair, skin, muscles, and connective tissue, are fiberlike. These **fibrous proteins** are insoluble in water. They usually serve structural, connective, and protective functions. Examples of fibrous proteins are keratins, collagens, myosins, and elastins. Hair and the outer layer of skin are composed of keratin. Connective tissues contain collagen. Myosins are muscle proteins and are involved in contraction and extension of muscles. Elastins are found in the elastic tissue of artery walls and in ligaments.

Globular proteins, the other major class, are soluble in aqueous media. The protein chains of globular proteins are folded so that the molecule as a whole is roughly spherical. The mixtures of globular proteins and water are actually colloidal dispersions rather than true solutions. Familiar examples of globular proteins include egg albumin from egg whites and serum albumin in blood, which plays a major role in transporting fatty acids, as well as in maintaining a proper balance of osmotic pressures in the body (Section 28.6). Hemoglobin and myoglobin, which are important for binding oxygen, are also globular proteins.

The structure of proteins is generally discussed at four organizational levels. The **primary structure** of a protein refers to the number and sequence of the amino acids in its polypeptide chain(s), beginning with the free amino group. The primary structure is "held" together by the peptide bonds that form when the amino acids are linked. The primary structure of the enzyme lysozyme, composed of 129 amino acids, is shown in Figure 21.4.

Protein molecules aren't just arranged at random as tangled threads. The chains are held together in unique conformations. The term **secondary structure** refers to the fixed arrangement of the polypeptide backbone. The term **tertiary structure** refers to the unique three-dimensional shape that results from the precise folding and bending of the protein backbone. The tertiary structure of a protein is intimately involved with the proper biochemical functioning of that protein, as we shall see in the next chapter. Figure 21.5 shows a ribbon model and a space-filling model for the three-dimensional structure of lysozyme.

We can relate these three levels of organization to a more familiar object. Think of the coiled cord on a telephone receiver. The cord starts out as a long, straight wire (Figure 21.6). We'll call that the primary structure. The wire is coiled into a helical arrangement—its secondary structure. When the receiver is hung up, the coiled cord folds into a particular pattern. That would be its tertiary structure.

Some proteins contain more than one polypeptide chain (as subunits), and this multichain arrangement causes the molecule to have another level of structure. The **quaternary structure** of a protein describes the way in which the subunits are packed together in the protein molecule. Hemoglobin is the most familiar example of a protein having quaternary structure. The four polypeptide chains are arranged in a specific pattern (see Figure 21.16). We consider hemoglobin in much greater detail in Chapter 28. A schematic representation of the four levels of protein structure is shown in Figure 21.7.

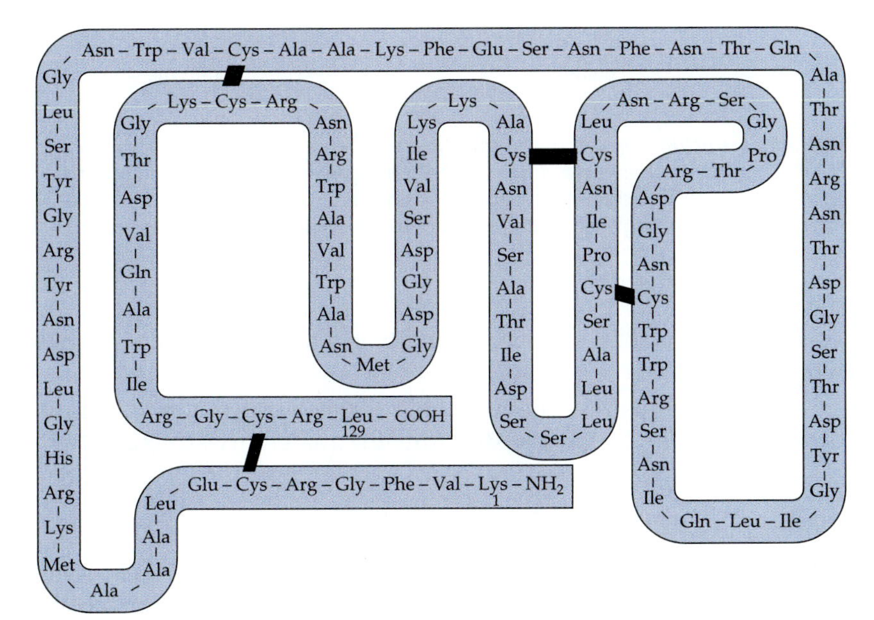

Figure 21.4 The amino acid sequence of lysozyme, a protein enzyme found in humans, plants, and the whites of eggs. It destroys invading bacteria by catalyzing the cleavage of polysaccharide chains that form part of the bacterial cell wall. Without a rigid cell wall, a sudden influx of water bursts the bacterial cell. The thick black lines represent disulfide linkages (see page 586).

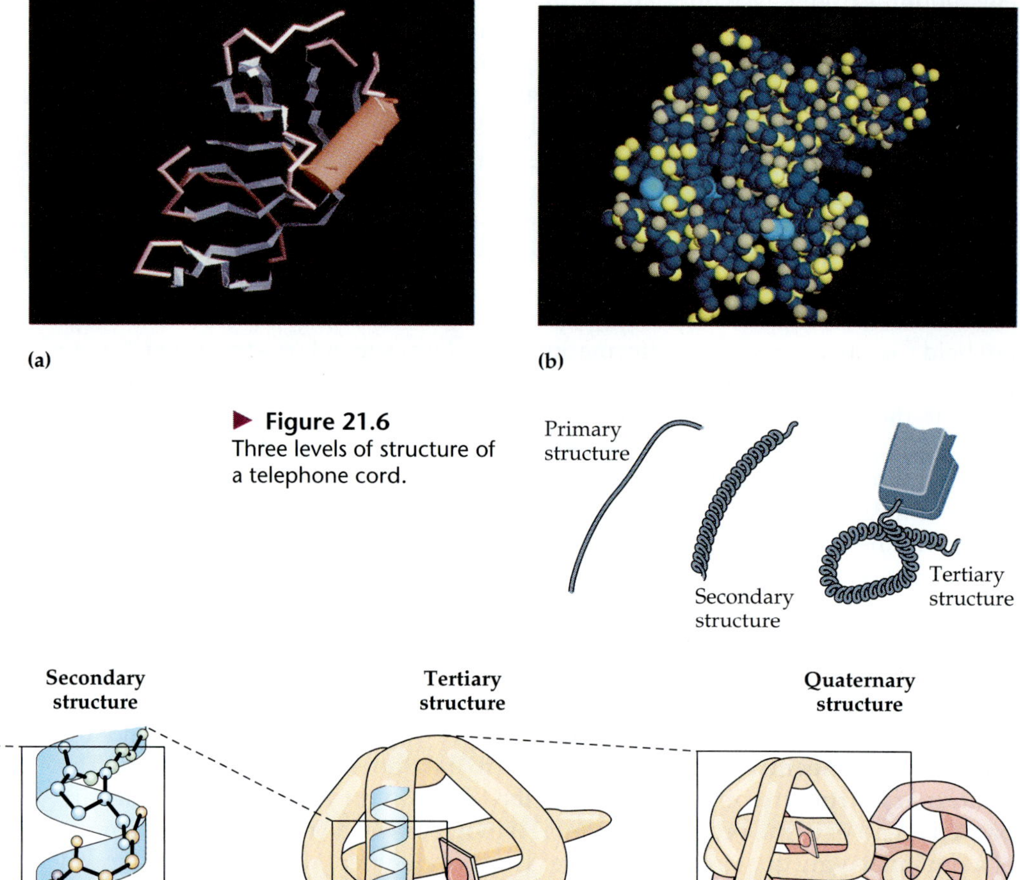

▶ **Figure 21.5**
(a) A ribbon model of lysozyme. (b) A space-filling model of the three-dimensional structure of lysozyme.

▶ **Figure 21.6**
Three levels of structure of a telephone cord.

▲ **Figure 21.7**
Levels of structure in proteins. The *primary structure* consists of the specific amino acid sequence. The resulting peptide chain can form an α-helix, one type of *secondary structure.* This helical segment is incorporated into the *tertiary structure* of the folded polypeptide chain. The single polypeptide chain is one of the subunits that make up the *quaternary structure* of a protein such as hemoglobin that consists of four polypeptide chains.

✓ Review Questions

21.11 Which class of proteins shows greater solubility in aqueous solution—fibrous or globular?

21.12 Define, describe, or illustrate the following terms: **(a)** primary structure; **(b)** secondary structure; **(c)** tertiary structure; **(d)** quaternary structure.

21.7 Secondary Structure of Proteins

Two major considerations are involved in the secondary structure of proteins. The first involves the manner in which the protein chain is folded and bent; the second involves the nature of the attractive forces that stabilize this structure.

Based upon X-ray studies, Linus Pauling and Robert Corey postulated that some proteins have a spiral shape (that is, they are shaped like a helix). The helix is stabilized by hydrogen-bond formation between the carbonyl oxygen of one amino acid and the amide hydrogen four amino acids up the chain (located on the next turn of the helix). This *intrachain* hydrogen-bonded structure is designated as a right-handed α-helix. X-ray data indicate that the helix makes one turn for every 3.6 amino acids, and that the side chains of these amino acids project outward from the coiled backbone (Figure 21.8). The α-keratins, found in hair and wool, are exclusively α-helical in conformation.

Not all proteins assume a helical conformation. Some proteins, such as gamma globulin, chymotrypsin, and cytochrome c, have little or no helical structure. Other proteins, such as hemoglobin and myoglobin, are helical in certain regions of the polypeptide chain; the remaining portions assume random conformations. The polypeptide chains of structural proteins such as silk fibroin and certain enzymes such as carboxypeptidase A and lysozyme are aligned side by side in a sheetlike arrangement. In these proteins, segments of the polypeptide chains lie next to one another and run either parallel or antiparallel, with *interchain* hydrogen bonding connecting the adjacent strands (Figure 21.9). This structural arrangement is designated as the β-pleated sheet conformation, and it occurs when two extended polypeptide chains (or two separate regions on the same chain) are aligned side by side.

The physical characteristics of wool and silk are a result of their structural conformations. Wool is very flexible and extensible. It can stretch to twice its normal length without breaking, and the fiber will return to its original state upon release of tension. (Think of how the coiled cord on a telephone can be stretched.) The stretching process involves breaking hydrogen bonds along turns of the α-helix (covalent bonds remain intact). The disulfide bonds (page 586) between helices, together with re-formed hydrogen bonds, provide the forces that operate to restore the helix when tension is released.

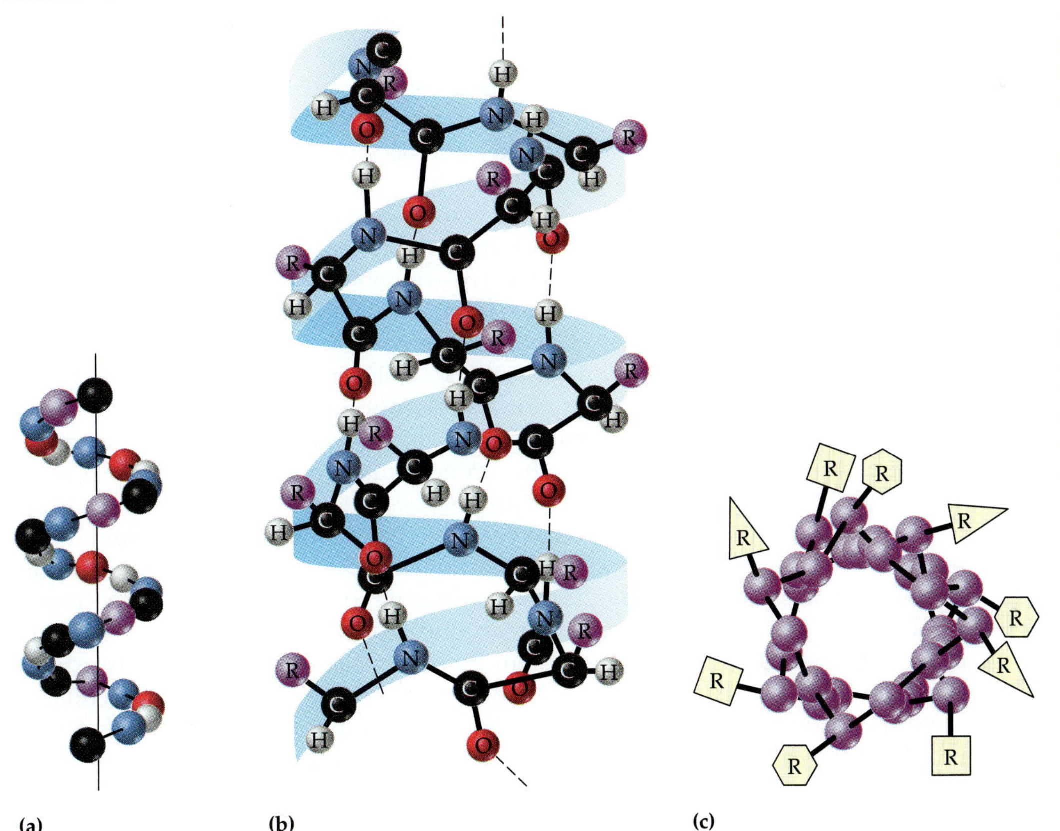

Figure 21.8 Representations of an α-helix. (a) The skeletal representation shows only the atoms in the polypeptide backbone. (b) The ball-and-stick model shows the intrachain hydrogen bonding between the carbonyl oxygens and amide hydrogens. Each turn of the helix contains 3.6 amino acids. (c) Top view of an α-helix. Note that the side chains of each amino acid point out from the helix.

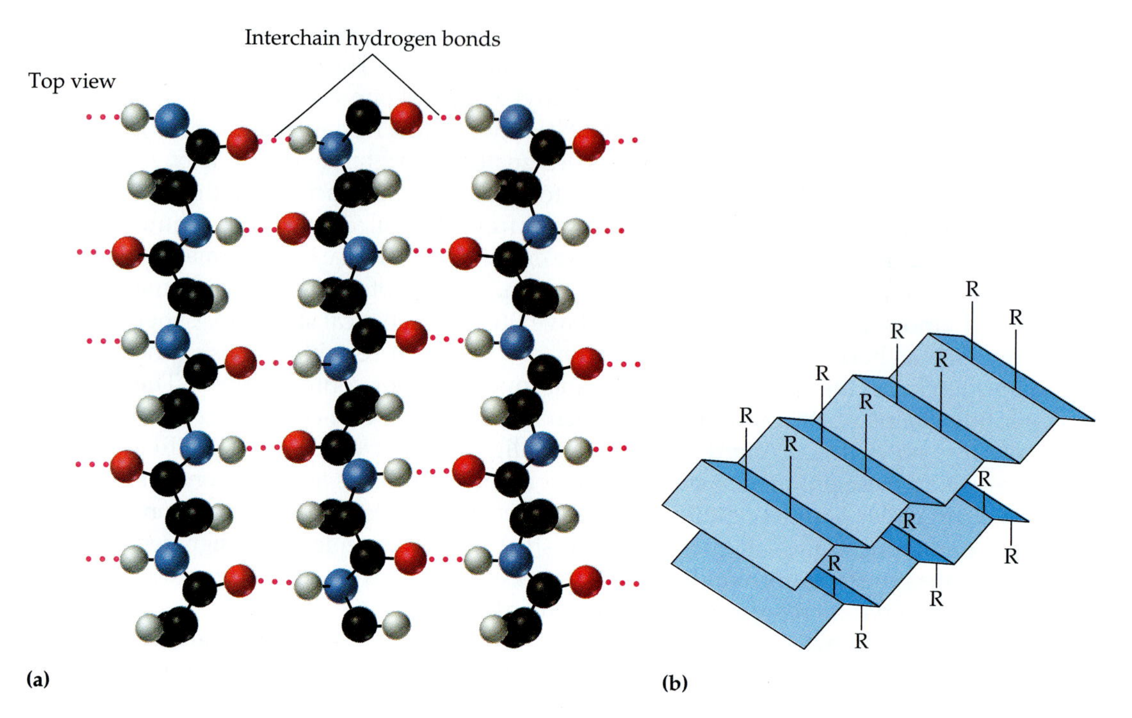

▲ **Figure 21.9**
Pleated sheet conformation of protein chains. (a) Ball-and-stick model. (b) Schematic drawing emphasizing the pleats. The peptide bonds lie in the plane of the pleated sheet. The side chains extend above or below the sheet and alternate along the chain. The protein chains are held together by interchain hydrogen bonds.

Silk, on the other hand, is already stretched out in the pleated sheet conformation. Silk fibers have the hydrogen-bonded layers arranged one over the other. The properties of silk—strength, flexibility, and resistance to stretching—are a consequence of its structure. Breaking the fibers involves rupturing thousands of hydrogen bonds or breaking covalent bonds, and because the chains are already fully extended, the fibers cannot be stretched easily.

The principal protein of connective tissues is *collagen* (Figure 21.10). The most abundant protein in higher vertebrates, it comprises about 25% of all the protein in the human body. Most of the organic portions of skin, bones, tendons, and teeth are collagen. Like other fibrous proteins, collagen is not readily digestible. Treatment with boiling water converts it to *gelatin*, which is not only water-soluble but also digestible. Nutritionally speaking, the gelatin derived from collagen is a poor quality protein because it lacks many of the essential amino acids (see Section 27.1). The cooking of meat converts part of the tough connective tissue to gelatin, making the meat more tender.

Approximately one-third of all the amino acids in collagen are glycine, while another third are proline and hydroxyproline. Because these three make up such a high percentage of the amino acids in collagen, the peptide chains cannot adopt a regular α-helical or β-sheet structure. Instead, each peptide chain is wound about its own axis in a unique left-handed helix, referred to as the collagen helix (Figure 21.11). Three of these chains then wrap around one another like three strands of a rope and are cross-linked by interchain hydrogen bonding. This structure is referred to as tropocollagen. In tendons, collagen fibers are formed by tropocollagen molecules arranged in parallel bundles that have nearly the tensile strength of steel wire but little or no capacity to stretch. In bones and teeth, the collagen cables form the matrix upon which the network of calcium salts is built.

Collagen chains are also somewhat cross-linked by covalent bonds. As an animal grows older, the extent of cross-linking increases and the meat gets tougher.

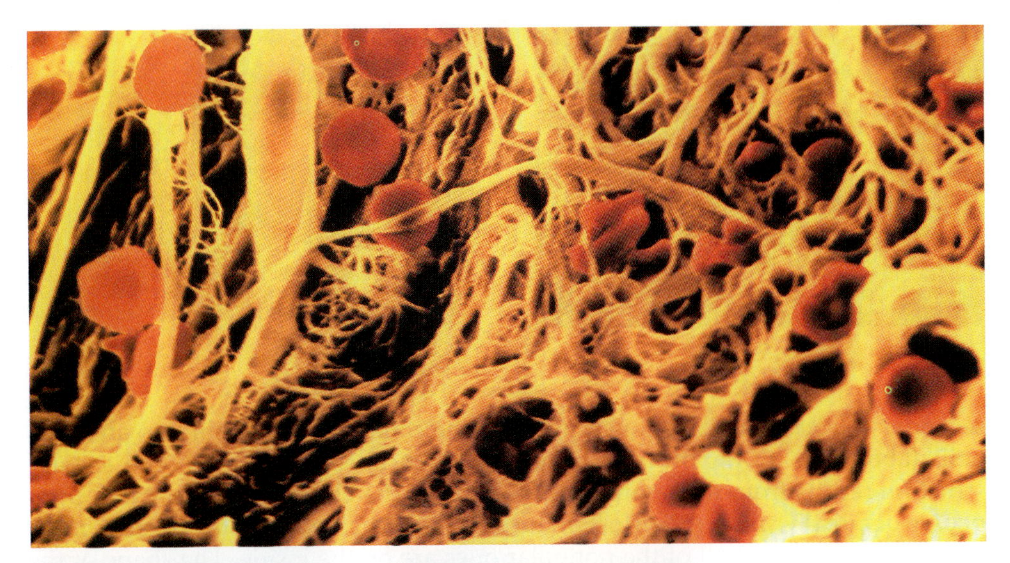

Figure 21.10
Electron micrograph of human connective tissue, showing collagen fibers (yellow strands) and red blood cells.

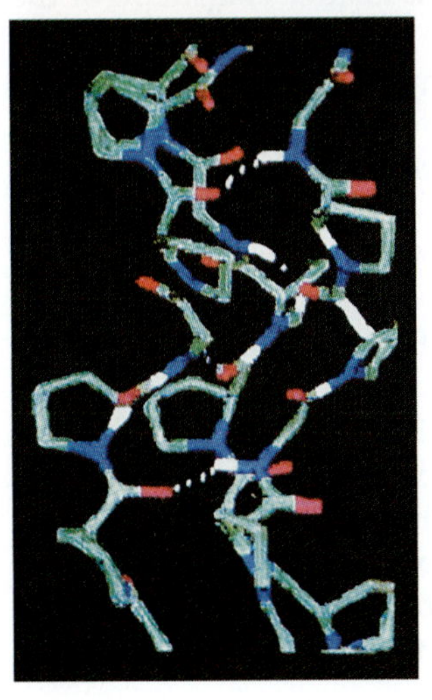

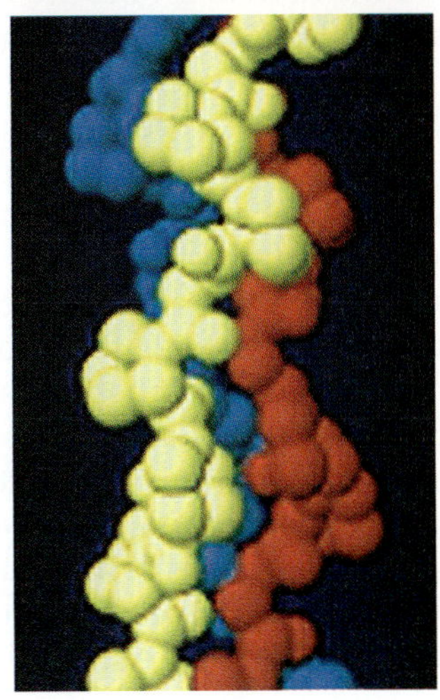

(a) (b)

Figure 21.11
(a) A stick model of the collagen triple helix that consists of the repeated sequence Gly-Pro-Pro in each of the polypeptide chains. Hydrogen bonds between the backbone atoms of the three chains are indicated. (b) A space-filling model of the collagen triple helix.

Collagen is of considerable commercial importance. The process of tanning increases the degree of cross-linking, converting skin to leather. The soluble gelatin derived from collagen is used in food, film emulsions, and glue, among other things.

Review Questions

21.13 What is the predominant attractive force that stabilizes the formation of secondary structure in proteins?

21.14 What name is given to the secondary structure of: **(a)** silk; **(b)** keratin or wool protein; **(c)** collagen?

21.8 Tertiary Structure of Proteins

Myoglobin, a globular protein, can bind molecular oxygen and store the oxygen in muscle cells until it is needed. Myoglobin is particularly abundant in marine mammals such as whales, seals, and porpoises, enabling them to remain under water for prolonged periods. In humans, myoglobin is found mainly in skeletal and cardiac muscle. It is a single polypeptide chain consisting of 153 amino acids.

During the 1950s John C. Kendrew was able, through the use of X-ray diffraction studies, to elucidate the secondary and tertiary structures of myoglobin. The

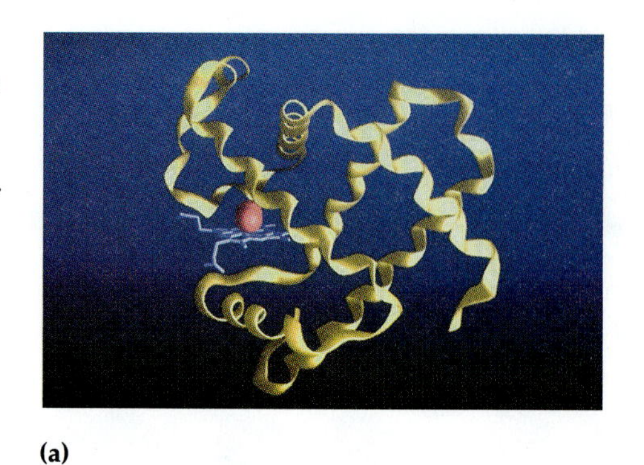

(a)

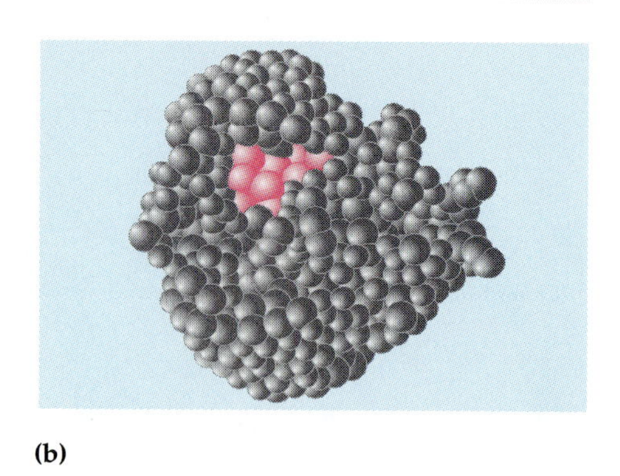

(b)

▶ **Figure 21.12**
Myoglobin structure. The heme group with bound oxygen is shown in pink. (a) Ribbon model of the polypeptide chain. The coils represent sections of α-helical structure. (b) Space-filling model.

$H_2C{=}CH$ CH_3
H_3C $CH{=}CH_2$
N N
Fe
N N
H_3C CH_3
CH_2 CH_2
CH_2COOH CH_2COOH

▲ **Figure 21.13**
Heme is a complex organometallic compound present in both myoglobin and hemoglobin.

secondary structure of myoglobin involves the coiling of this chain into an alpha helix (about 70% of the protein strand has a helical conformation). The tertiary structure results from the nonuniform folding of the chain to form a stable compact structure (Figure 21.12). Most of the polar side chains are on the outside of the molecule, and almost all of the nonpolar ones are on the inside. The shape of the final molecule includes a hole that nicely accommodates a **heme** unit (an organometallic complex that is the oxygen-binding component of myoglobin; see Figure 21.13). The tertiary structure also brings two amino acid side chains into position to anchor the heme unit to the protein portion of the myoglobin chain.

The linkages responsible for the tertiary structure of a protein such as myoglobin are a function of the nature of the amino acid side chains within the molecule. Globular proteins are extremely compact, almost spherical in shape. As was observed with myoglobin, such proteins have most of their nonpolar side chains directed toward the interior of the molecule (the hydrophobic or nonaqueous region) and the majority of their polar side chains projected outward from the surface of the molecule toward the aqueous environment. The resulting picture is very similar to that of a micelle, which was discussed in connection with the properties of cell membranes (Section 20.5). The major types of linkages or attractive interactions that are important for the formation and stability of the tertiary structure of proteins are shown in Figure 21.14. Table 21.4 indicates the relative strengths of these interactions.

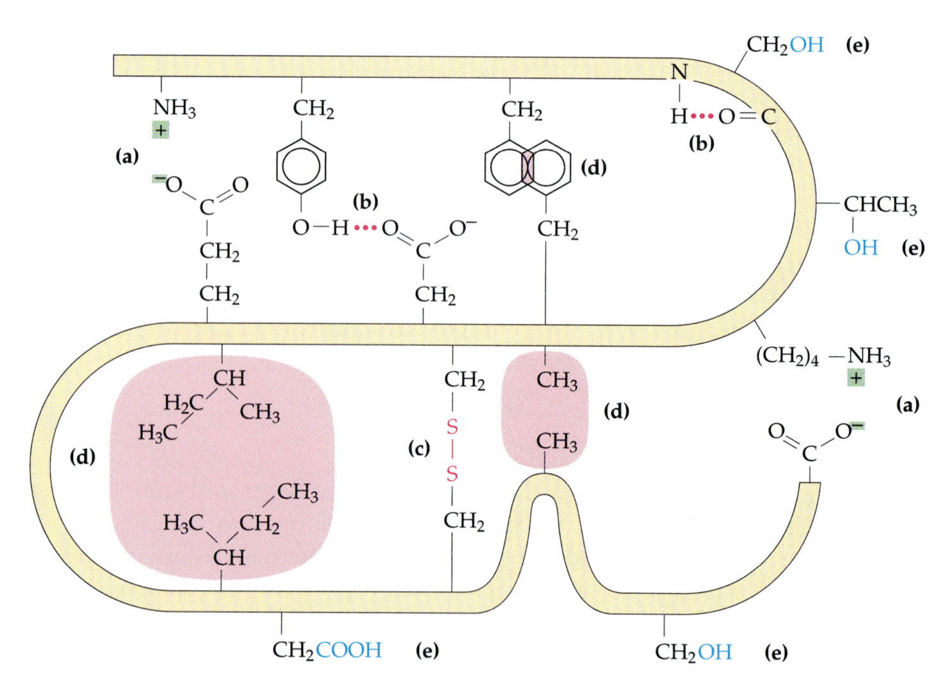

▶ **Figure 21.14**
Bonds that stabilize the tertiary structure of proteins: (a) ionic bonds, (b) hydrogen bonds, (c) disulfide linkages, (d) dispersion forces, and (e) polar groups that interact with water.

Table 21.4 Noncovalent Bonds and Interactions in Polypeptides

Example	Type of Bond	Approximate Stabilization Energy (kcal/mol)
$>C{=}O\cdots H{-}N<$	Hydrogen bond between peptides	2–5
$-\overset{\mid}{\underset{\mid}{C}}-O(H)\cdots H-O-C<$	Hydrogen bond between neutral groups	2–5
$-C(O)(O)^{-}\cdots H-O-$	Hydrogen bond between neutral and charged groups	2–5
$>C{=}O\cdots HO-C_6H_4-$	Hydrogen bond between peptide and R group	2–5
$-\overset{+}{N}H_3\;\;{}^{-}O_2C-$	Ionic bond between charged groups (strongly dependent on distance)	<10
$-CH_3\;\;CH_3-$	Dispersion forces	0.3
(stacked aromatic rings)	Dispersion forces—stacking of aromatic rings	1.5
$(H_3C)(CH_3)C(H)-CH_2-\;\;\;(H_3C)(CH_3)C(H)-CH_2-$	Dispersion forces	1.5
$H_2C-\overset{+}{N}H_3\;\;H_2\overset{+}{N}{=}C(NH_2)-NH-$	Repulsive interactions between similarly charged groups (strongly dependent on distance)	<–5

Ionic Bonding

Ionic bonds, sometimes called salt linkages, result from electrostatic interactions between positively and negatively charged groups on the side chains of the basic and acidic amino acids. For example, the mutual attraction between an aspartic acid carboxylate ion and a lysine ammonium ion helps to maintain a particular folded area of the protein.

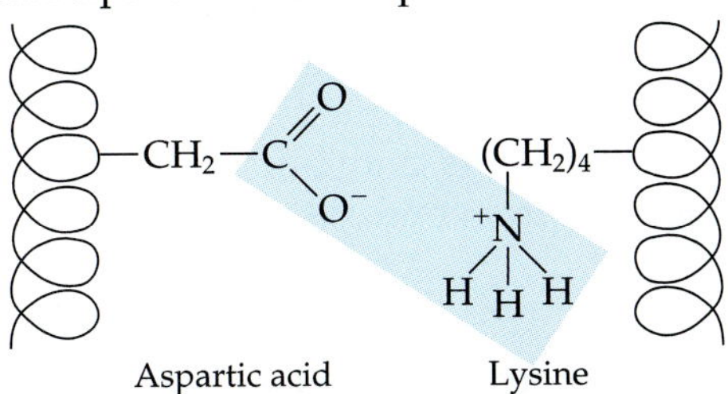

Hydrogen Bonding

Hydrogen bonds are formed principally between the side chains of the polar amino acids or between a carboxyl oxygen and a hydrogen donor group. The hydrogen-bonding capabilities of the terminal amino group of lysine and the terminal carboxyl groups of aspartic acid and glutamic acid are pH-dependent. These groups can serve as both hydrogen-bond acceptors and hydrogen-bond donors only over a certain range of pH. Hydrogen bonds (as well as ionic attractions) are extremely important in both the intra- and intermolecular interactions of proteins.

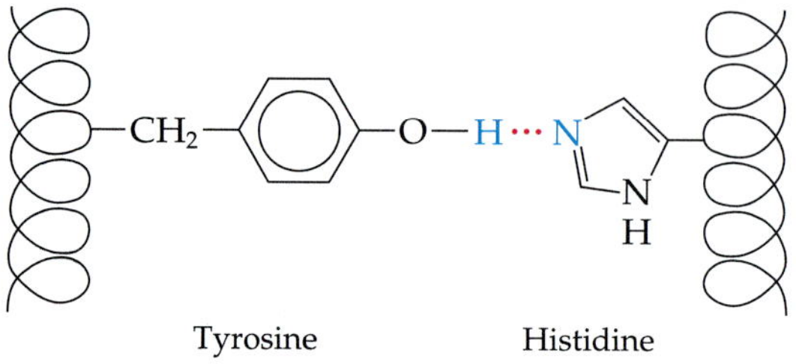

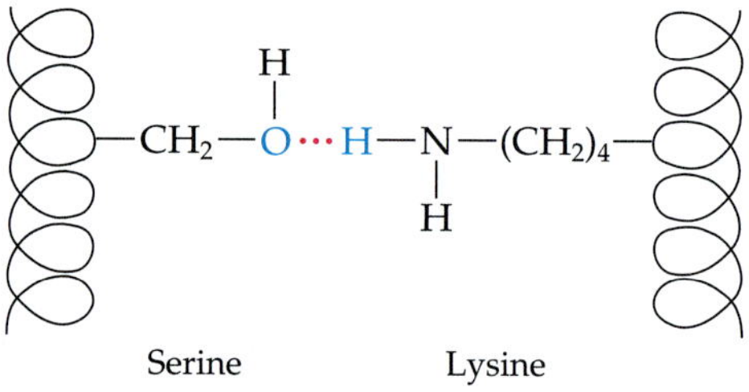

$—CH_2—C_6H_4—O—H \cdots N$ (imidazole ring, N—H)

Tyrosine Histidine

$—CH_2—O(H) \cdots H—N(H)—(CH_2)_4—$

Serine Lysine

A significant feature of hydrogen bonds is that they are highly directional. The strongest hydrogen bond results when the hydrogen donor and the acceptor atom are colinear. If the acceptor atom is at an angle to the covalently bonded hydrogen atom, the hydrogen bond is much weaker.

Disulfide Linkages

Two cysteine residues may come in proximity as the protein molecule folds. A disulfide linkage results from the subsequent oxidation of the highly reactive sulfhydryl (—SH) groups to form cystine.

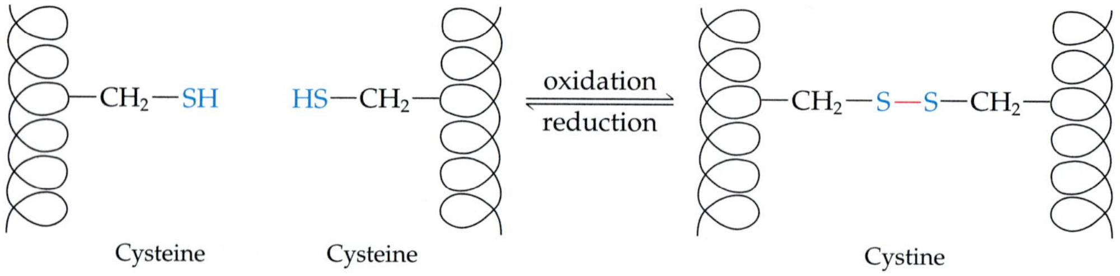

$$—CH_2—SH \quad HS—CH_2— \underset{\text{reduction}}{\overset{\text{oxidation}}{\rightleftharpoons}} —CH_2—S—S—CH_2—$$

Cysteine Cysteine Cystine

Intrachain **disulfide linkages** are found in many proteins, such as lysozyme (Figure 21. 4), and act to stabilize the tertiary structure. Interchain disulfide bonds are also important forces that link two separate polypeptide chains, thus stabilizing quaternary structure. Insulin is an example of a protein with interchain disulfide bonds (Figure 21.15). Disulfide bonds, both intrachain and interchain, are found in proteins which are transported from one cell to another.

Dispersion Forces

Dispersion forces arise when a normally nonpolar atom becomes momentarily polar due to an uneven distribution of electron charge density, leading to an instantaneous

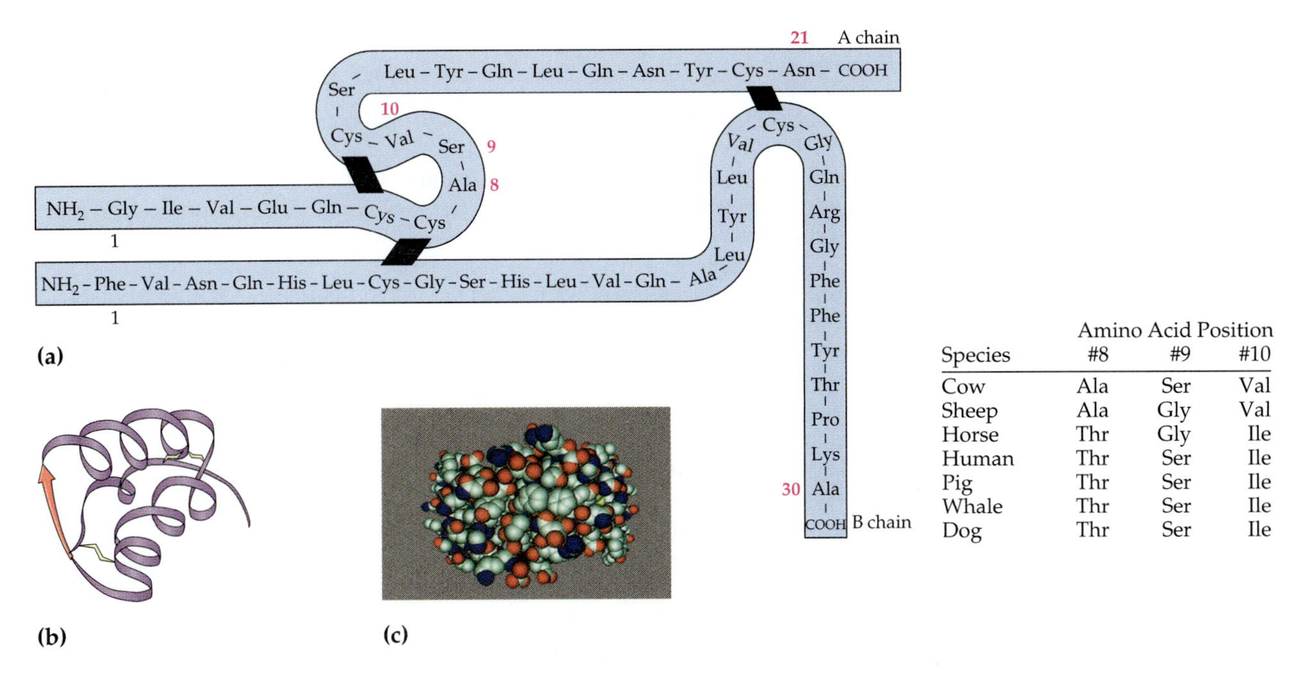

Species	Amino Acid Position #8	#9	#10
Cow	Ala	Ser	Val
Sheep	Ala	Gly	Val
Horse	Thr	Gly	Ile
Human	Thr	Ser	Ile
Pig	Thr	Ser	Ile
Whale	Thr	Ser	Ile
Dog	Thr	Ser	Ile

▲ **Figure 21.15**
(a) The amino acid sequence of bovine insulin, a protein hormone produced in the pancreas. Insulin from other mammalian species has the same structure except for amino acid positions 8, 9, and 10 of the A chain, which differ as shown in the table. In addition, human insulin differs from all the others at position 30 of the B chain, where threonine replaces alanine. (b) A ribbon model and (c) a space-filling model of the three-dimensional structure of insulin.

dipole that induces a shift of electron density in a neighboring nonpolar atom (Section 7.5). Dispersion forces are weak but can be important when other types of interactions are either missing or minimized. This is the case for fibroin, the major protein in silk, in which a high proportion of amino acids in the protein have nonpolar side chains.

The term *hydrophobic interactions* is often misused. It does not refer to the attractions between nonpolar groups; these are dispersion forces. Hydrophobic interactions arise because water molecules tend to hydrogen-bond with other water molecules (or groups in proteins capable of hydrogen bonding). Because nonpolar groups cannot form hydrogen bonds, the protein folds in such a way that these groups are buried in the interior part of the protein structure, minimizing their contact with water.

✓ Review Questions

21.15 Name the four types of interactions that maintain the tertiary structure of proteins.

21.16 Two cysteine units joined through a disulfide linkage are sometimes considered a different amino acid, called cystine. Draw the structure of cystine.

21.9 Quaternary Structure of Proteins

The quaternary structure of a protein is formed and stabilized by the same kinds of forces (ionic bonds, hydrogen bonds, interchain disulfide bonds, dispersion forces, and hydrophobic interactions) that are involved in forming and maintaining the tertiary structure. One of the best studied proteins having a quaternary structure is hemoglobin, a protein that binds and transports oxygen through the bloodstream. The structure of the hemoglobin molecule was determined by Max Perutz in a research

Prions

In 1997 the Nobel Prize in physiology or medicine was given to Stanley B. Prusiner, a professor at the University of California, San Francisco, for his work in identifying the cause of several fatal nervous system disorders that include scrapie and mad cow disease in animals and Creutzfeldt-Jakob disease in humans. Prusiner has hypothesized that these diseases are transmitted by abnormal proteins known as prion proteins, rather than by the usual infective agents such as viruses or bacteria.

Initial work by other researchers, performed in the late 1960s through the early 1980s, showed that the replication of the infective agent that causes scrapie did not involve nucleic acids (such as RNA or DNA). In addition, electron micrographs indicated that a particular protein might be important. Then, in 1982, Prusiner and his coworkers isolated and characterized the infectious agent from scrapie-infected brain tissue and characterized it as a protein, which they called *prion protein*.

Subsequent work showed that this protein was also present in noninfected animals. However, the prion protein can adopt one of two three-dimensional structures: normal and infectious (see below). The normal form is soluble in water and is predominantly α-helical in structure, while the infectious form is insoluble in water and has less α-helical structure and a more extended β-pleated sheet region. Prusiner has proposed that the infectious prion protein infects a cell and converts the normal prion proteins present to the abnormal or infectious structure. There is still controversy regarding the prion hypothesis, and some researchers believe that the prion protein must have a virus or viral-type material associated with it. Prions and the diseases they are proposed to cause continue to be an active area of research.

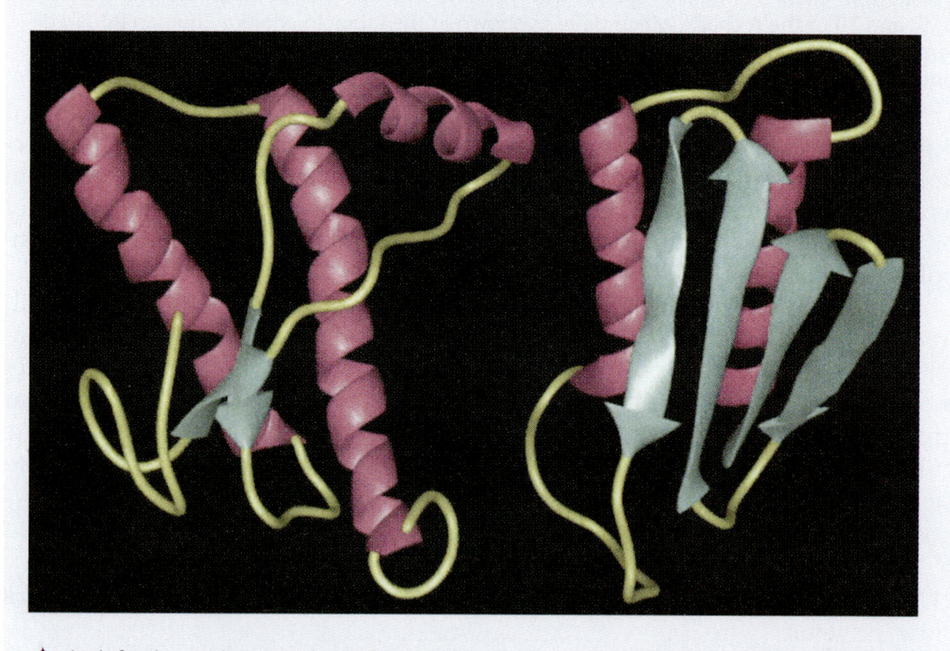

▲ An infectious prion protein (right) is formed when a normal prion protein (left) refolds. In the refolding process, some of the α-helical regions (purple coils) unfold, forming an extended β-sheet region (flat green arrows).

Perutz had to overcome many obstacles, not all of them the technical problems of working with such a large protein. He was born in Austria, but carried out his scientific work in Britain. During World War II he was placed in an internment camp in Britain for half a year as an enemy alien, and then he was asked to help build an aircraft carrier made of ice!

project that took 19 years, beginning in the late 1930s. John Kendrew worked with Max Perutz, and the two men shared the Nobel Prize in chemistry in 1962 for their work in determining the three-dimensional structures of hemoglobin and myoglobin.

Hemoglobin consists of four polypeptide chains—two identical alpha chains (141 amino acids each) and two identical beta chains (146 amino acids each), as shown in Figure 21.16. Each chain is very similar in structure to the single polypeptide chain of myoglobin. Because each chain contains a heme group, one hemoglobin molecule

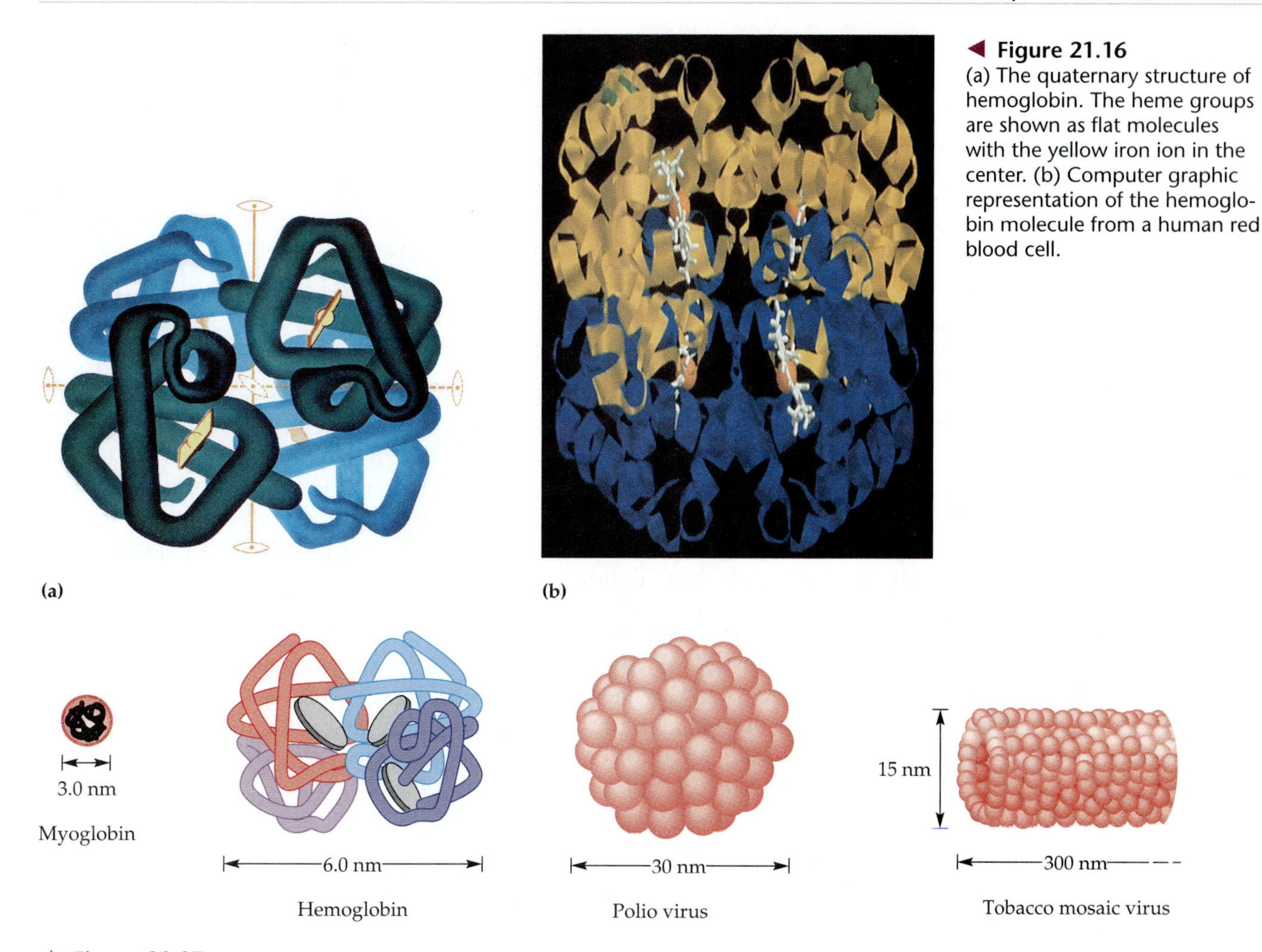

Figure 21.16
(a) The quaternary structure of hemoglobin. The heme groups are shown as flat molecules with the yellow iron ion in the center. (b) Computer graphic representation of the hemoglobin molecule from a human red blood cell.

Figure 21.17
Schematic diagrams of the sizes and structures of several proteins.

can bind four molecules of oxygen. The four hemoglobin subunits are held together by non-covalent surface interactions between the polar side chains, as well as by dispersion forces and hydrophobic interactions.

The protein coats of several viruses are composed almost entirely of polypeptide subunits arranged in a highly ordered conformation. The polio virus contains 130 polypeptide chains. The tobacco mosaic virus contains a grand total of about 345,000 amino acids arranged in 2130 individual polypeptide chains that are assembled around a central core of nucleic acids (Figure 21.17). The average molar mass of each polypeptide chain is about 18,000 daltons.

Review Questions

21.17 Distinguish between tertiary and quaternary structure.

21.18 **(a)** How do the structures of myoglobin and hemoglobin differ? **(b)** What are the functions of these two proteins?

21.10 Electrochemical Properties of Proteins

When amino acids combine to form the polypeptide chain(s) of protein molecules, the majority of their amino and carboxyl groups are tied up in the peptide bonds. However, the side chains of aspartic acid, glutamic acid, histidine, cysteine, lysine,

Table 21.5 Isoelectric pH Values of Various Proteins

Protein	Isoelectric pH
Pepsin	<1.1
Silk fibroin	2.2
Pepsinogen	3.7
Casein	4.6
Egg albumin	4.7
Serum albumin	4.8
Urease	5.0
Insulin	5.3
Fibrinogen	5.5
Catalase	5.6
Hemoglobin	6.8
Myoglobin	7.0
Ribonuclease	9.5
Cytochrome c	10.6
Lysozyme	11.0

and arginine all retain their acidic and basic R groups. In proteins, just as in the free amino acids, these groups exist in solution as charged species such as $—COO^-$, $—S^-$, and $—NH_3^+$. Because all proteins contain some of the acidic and basic amino acids, positive and negative charges are found throughout the molecule.

Each protein has a characteristic isoelectric pH at which the protein molecule as a whole is electrically neutral. It may contain many ionized groups, but the positively charged side chains are exactly balanced by negatively charged ones. The isoelectric pH depends on the number, kind, and arrangement of the acidic and basic groups within the molecule. Proteins that have a high proportion of basic amino acids usually have a relatively high isoelectric pH, and those with a preponderance of acidic amino acids have a relatively low isoelectric pH. Table 21.5 lists the isoelectric pH values of several proteins.

The differences in isoelectric pH can be used to separate and identify specific proteins in a mixture of proteins by electrophoresis (recall Figure 21.2). Proteins with ionized side chains behave as either cations or anions, depending upon the pH of the solution. Electrophoresis is a very powerful tool used to separate and identify specific proteins in a mixture by subjecting them to an electric field. In electrophoresis the protein mixture is applied on a solid support, such as paper or a gel, that is bathed in a buffer solution at a particular pH. A current is then applied, and the proteins migrate toward the electrodes. For proteins of comparable molar mass, those containing the greatest number of negative charges migrate most rapidly toward the positive electrode; those containing the greatest number of positive charges move most rapidly to the negative electrode. A dye (such as Coomassie Blue) is used to make the separated protein spots visible. This technique is used on blood samples in hospital laboratories to assess certain diseases by detecting the relative concentrations of the plasma proteins (Figure 21.18).

The solubility of proteins in water is greatly dependent on pH. The size of many proteins places them in the category of colloids (Section 8.10). At pH values other than the isoelectric pH, the molecules carry a net charge. These charges on the surface of the colloidal proteins repel the other colloidal particles and keep them from coalescing. Thus they form colloidal dispersions. At the isoelectric pH, however, the colloidal protein molecules are electrically neutral and no longer repel one another. Therefore they come together to form larger aggregates that eventually precipitate from solution. So, as a general rule, a protein is least soluble at its isoelectric pH.

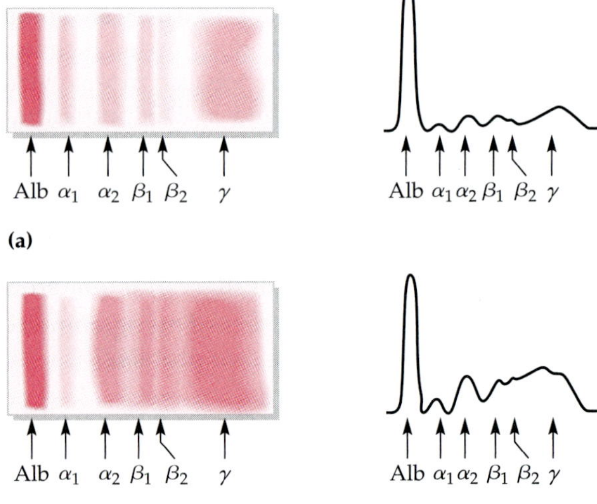

Figure 21.18 (a) An electrophoresis pattern of normal blood. (b) A pattern of abnormal blood that has elevated γ-globulin, indicating possible infection, collagen disorder, or liver disease.

Casein is the major protein component of milk, and it precipitates in the form of white curds at its isoelectric pH of 4.6. The souring of milk results from the production of lactic acid by bacteria. The lactic acid lowers the pH of milk from its normal value of about 6.6 to about 4.6. Casein is used in the manufacture of cheese. It can be obtained either by adding acid to milk or by bacterial action.

Review Questions

21.19 How do the properties of a protein at its isoelectric pH differ from its properties in solutions at other pH values?

21.20 What occurs when milk becomes sour?

21.11 Denaturation of Proteins

In many ways, proteins are remarkable compounds. Their highly organized structures are truly masterworks of chemical architecture. But highly organized structures tend to have a certain delicacy, and this is true of proteins. We define **denaturation** as the process in which there is a change in the three-dimensional structure of a protein that renders it incapable of performing its assigned function. If the protein's conformation is destroyed, we say it has been *denatured*, and it can't do its job. (Sometimes denaturation is equated with the precipitation or coagulation of a protein. Our definition is a bit broader.) The process is sometimes reversible, but often it is not. You have certainly observed the denaturation of egg albumin. The clear egg "white" turns to an opaque white when the egg is cooked. What you have observed is the denaturation and coagulation of the albumin. No one yet has reversed that process!

The primary structure of proteins is quite sturdy. In general, it takes fairly vigorous conditions to hydrolyze peptide bonds. At the secondary and tertiary levels, however, proteins are quite vulnerable to attack (Figure 21.19). A wide variety of reagents and conditions can cause protein denaturation. Some of them are outlined here.

Heat and Ultraviolet Radiation

Heat and ultraviolet radiation supply kinetic energy to protein molecules, causing their atoms to vibrate more rapidly and disrupting the relatively weak hydrogen bonds and dispersion forces. Most proteins are denatured when heated above 50 °C, and this results in coagulation of the protein. Heat and ultraviolet radiation are employed in sterilization techniques because they denature the enzymes in bacteria and, in so doing, destroy the bacteria. Denatured proteins are usually easier to chew and easier for enzymes to digest; hence we cook most of our protein-containing food.

Treatment with Organic Compounds

Ethyl alcohol, formaldehyde, urea, and rubbing alcohol are capable of forming intermolecular hydrogen bonds with protein molecules, thus disrupting the intramolecular hydrogen bonding within the molecule. A 70% isopropyl alcohol solution is used as a disinfectant to cleanse the skin before an injection. The alcohol denatures the protein (enzymes in particular) of any bacteria present in the area of the injection. A 70% alcohol solution effectively penetrates the bacterial cell wall, whereas 100% alcohol coagulates proteins at the surface, forming a crust that prevents the alcohol from entering into the cell (Figure 21.20).

Salts of Heavy Metal Ions

The heavy metal cations Hg^{2+}, Ag^{+}, and Pb^{2+} form very strong bonds with the carboxylate anions of the acidic amino acids and with the sulfhydryl groups of cysteine. Therefore they disrupt salt linkages and disulfide linkages and cause the

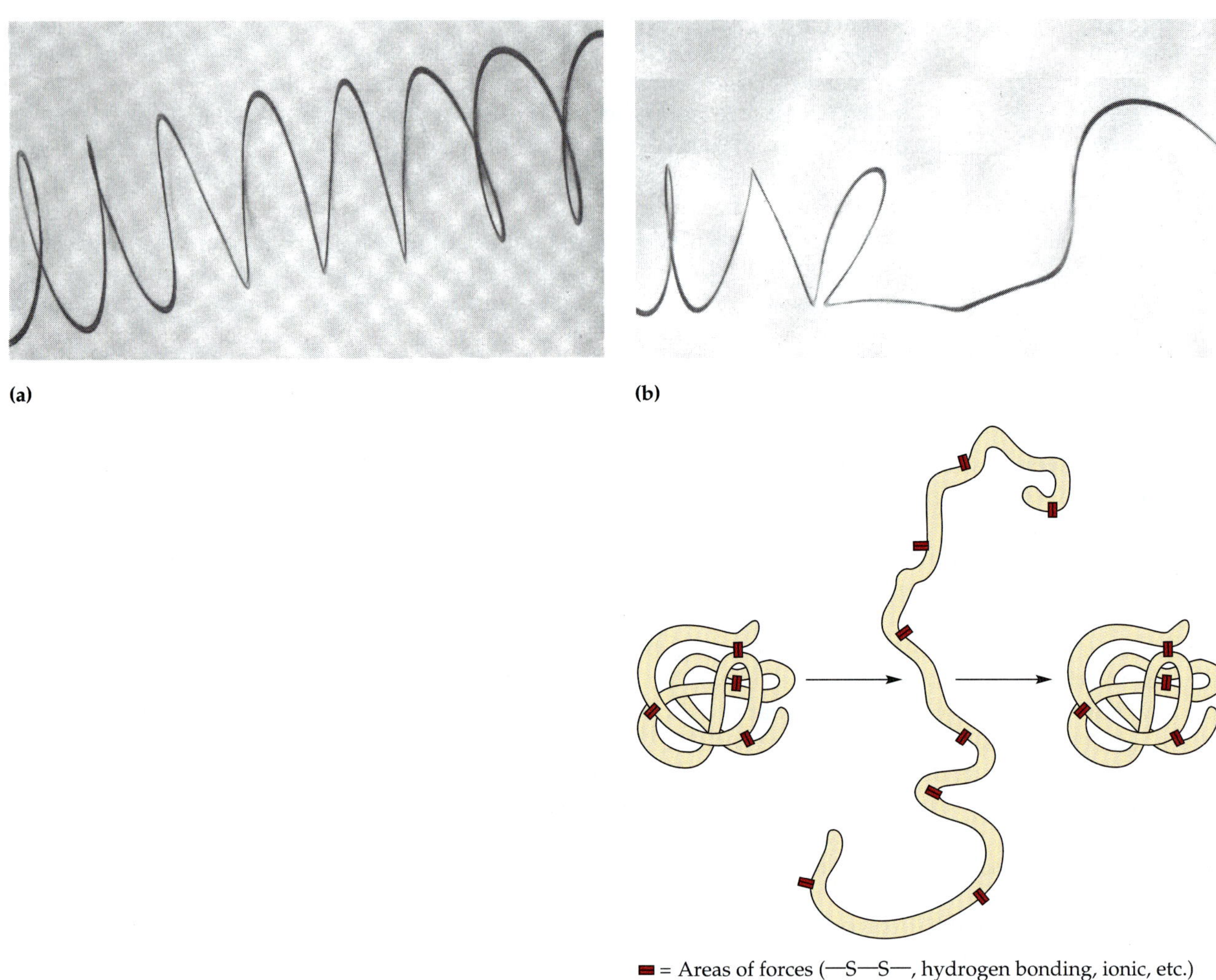

▲ **Figure 21.19**
Denaturation of a protein. (a) Irreversible denaturation. The coiled spring represents the helical structure of a protein when the elastic limit of the helix is exceeded, and (b) the shape is irreversibly altered. (c) The globular protein is folded into the tertiary conformation necessary for its functioning. The denatured protein can assume various random conformations. It is not active, but under proper conditions it may refold to the active conformation.

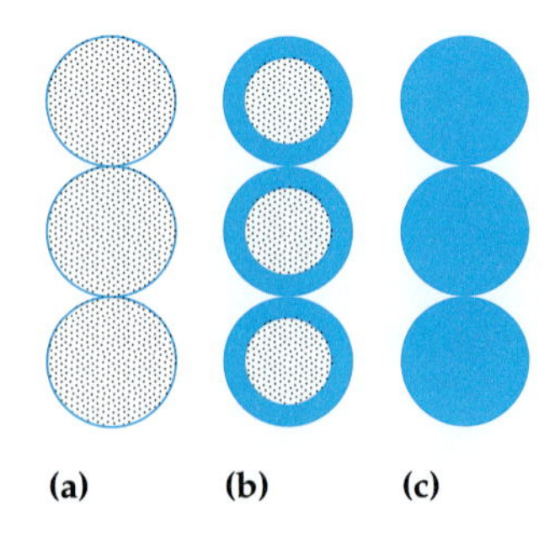

▲ **Figure 21.20**
Effect of isopropyl alcohol on bacteria. Dark areas represent coagulated protein. (a) Bacteria before application of alcohol. (b) After application of 100% alcohol. (c) After application of 70% alcohol, which is more effective than 100% alcohol.

protein to precipitate out of solution as insoluble metal–protein salts. This property makes some of the heavy metal salts suitable for use as antiseptics. For example, a 1% solution of silver nitrate (also called lunar caustic), which is used to prevent gonorrhea infections in the eyes of newborn infants, and mercuric chloride, another antiseptic, precipitate the proteins in infectious bacteria.

Most heavy metal salts are toxic when taken internally because they precipitate the proteins of all the cells with which they come into contact. Substances high in protein, such as egg whites and milk, are used as antidotes for heavy metal poisoning. If a person who has ingested mercury is fed raw eggs immediately, the mercury reacts with egg protein in the stomach rather than with other, more essential proteins. The stomach contents must then be pumped out or vomited to prevent the ultimate digestion of the egg protein and the consequent release of mercury ions within the body. Quite clearly, the technique works only for acute poisonings and not for the far more common chronic mercury poisoning.

Alkaloid Reagents

Picric acid and tannic acid are called alkaloid reagents because they were originally used to study the structures of the alkaloids (morphine, cocaine, quinine). They function in a manner analogous to the heavy metal cations, but the picrate and tannate anions combine with the positively charged amino groups in proteins to disrupt the salt linkages. In the manufacture of leather, tannic acid is used to precipitate the proteins in animal hides. This is the process called *tanning*.

There are many other ways of denaturing proteins that we have not discussed, such as the addition of detergents or radical changes in pH. The point should be clear, however. The very complexity that makes proteins so versatile also makes them vulnerable. There is a considerable range of vulnerability. The delicately folded globular proteins are much more readily denatured than are the tough, fibrous proteins of hair and skin.

On the other hand, a carefully unfolded protein, given the proper conditions and enough time, may refold and again exhibit biological activity. Such evidence suggests that, for these molecules, primary structure determines secondary and tertiary structure. A given sequence of amino acids seems to adopt its particular three-dimensional arrangement naturally if conditions are right.

Lead and Mercury Poisoning

Lead compounds are quite toxic. We can excrete about 2 mg of lead per day. Our intake from air, food, and water is generally less than that, so normally we do not accumulate toxic levels. If intake exceeds excretion, however, lead builds up in the body and chronic irreversible lead poisoning results.

Lead poisoning is a major problem with children, particularly those in areas containing old, run-down buildings. Some children develop a craving that causes them to eat unusual things. Children with this syndrome (called *pica*) eat chips of peeling, lead-based paints. These children can also pick up lead compounds from the streets, where they have been deposited by automobile exhausts, as well as other sources. Such poisoning often leads to mental retardation and neurological disorders through damage to the brain and nervous system. The U.S. Environmental Protection Agency estimates that lead poisoning contributes to 123,000 cases of hypertension and 680,000 miscarriages annually and retards the growth of 7000 children each year. These problems add $635 million to health-care costs annually. Because of these health concerns, lead salts are no longer used as pigments in paints, and most of the lead has been removed from gasoline.

Mercury compounds are also quite toxic. In August 1996 Karen Wetterhahn, a professor at Dartmouth College, spilled a tiny amount of dimethylmercury (an organic compound containing mercury) on the latex gloves she was wearing. The mercury-containing compound permeated the gloves and was absorbed by her skin. A few months later Wetterhahn became ill and died of mercury poisoning less than a year following the exposure. Follow-up work done at Dartmouth showed that dimethylmercury can penetrate disposable gloves in 15 seconds or less. Inorganic mercury salts are less volatile and less readily absorbed by the skin than organic compounds containing mercury, but they still present health and environmental hazards. Hatter's disease (which probably afflicted the Mad Hatter in *Alice's Adventures in Wonderland*) was a form of chronic mercury poisoning. Mercury compounds were used to convert fur to felt for felt hats. Mercury compounds still occur in water systems because large quantities of mercury and mercury salts have been dumped by industries into streams and lakes. The mercury is taken in by fish, and then humans eat the fish.

Permanent Waving

The chemistry of curly hair is interesting. Hair is protein (keratin), and adjacent protein chains are held together by disulfide linkages or bonds. To put a permanent wave in the hair, you use a lotion containing a reducing agent such as thioglycolic acid ($HSCH_2COOH$). This lotion reduces the disulfide linkages between cysteine side chains, breaking the covalent bonds and allowing the protein chains to be pulled apart as the hair is held in a curled position on rollers (Figure 21.21). The hair is then neutralized with a mild oxidizing agent such as hydrogen peroxide. Disulfide linkages then reform in new positions to give shape to the hair.

The same chemical process can be used to straighten naturally curly hair. The change in curliness depends only on how you arrange the hair after the disulfide bonds have been reduced and before new disulfide linkages can form. As with permanent dyes, permanent curls grow out as new hair is formed.

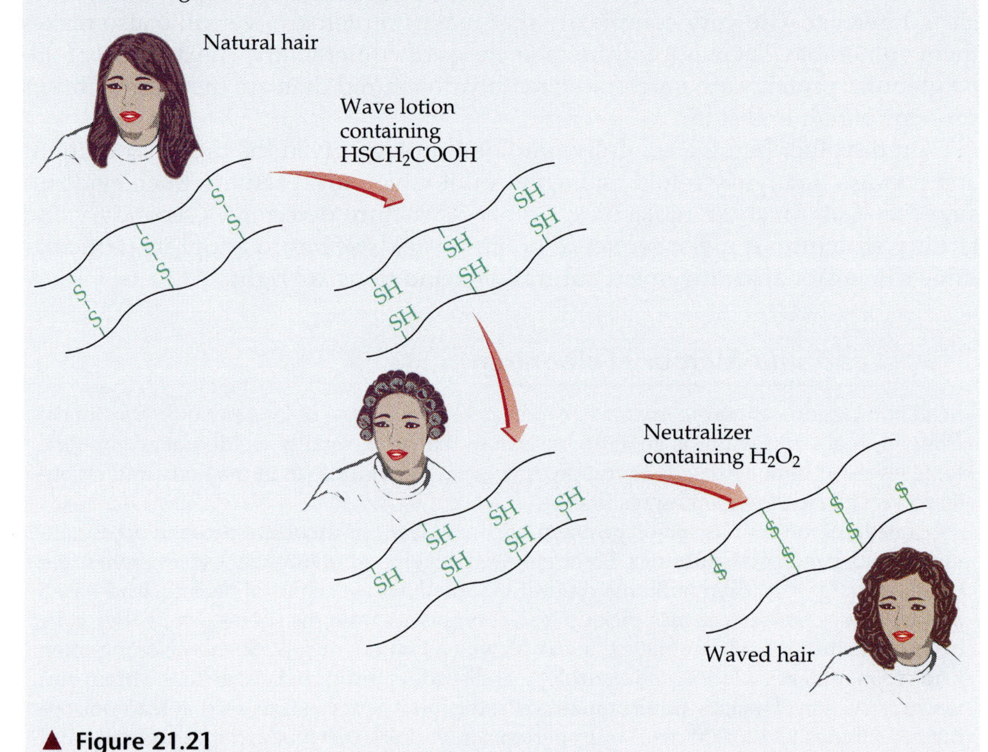

▲ **Figure 21.21**
Permanent waving of hair is accomplished by breaking disulfide linkages and then reforming them in new positions.

We have emphasized structure in this chapter. In other chapters we concentrate on the functions of several kinds of proteins, particularly the enzymes.

Review Questions

21.21 Describe some ways of denaturing a protein.

21.22 **(a)** What level(s) of structure is(are) ordinarily disrupted in denaturation? **(b)** Is the denaturation of proteins usually a reversible process?

Summary

Proteins are important biomolecules that serve a variety of roles in living organisms. A **protein** is a large polymer synthesized from **amino acids**—carboxylic acids containing an α-amino group. Nearly all the proteins in living organisms are made from the same 20 amino acids. In the solid state and in neutral solutions, amino acids exist as **zwitterions**, a form that is charged but is electrically neutral.

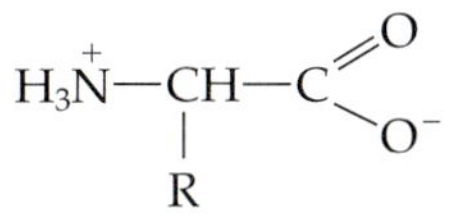

$$H_3\overset{+}{N}-\underset{R}{\underset{|}{CH}}-C(=O)O^-$$

Their existence in the zwitterion form makes amino acids behave much like inorganic salts. There are four classes of amino acids, based on the characteristics of the R group or amino acid side chain: nonpolar, polar but neutral, positively charged, and negatively charged.

Amino acids can act, depending on reaction conditions, as either acids or bases, with the result that proteins act as buffers:

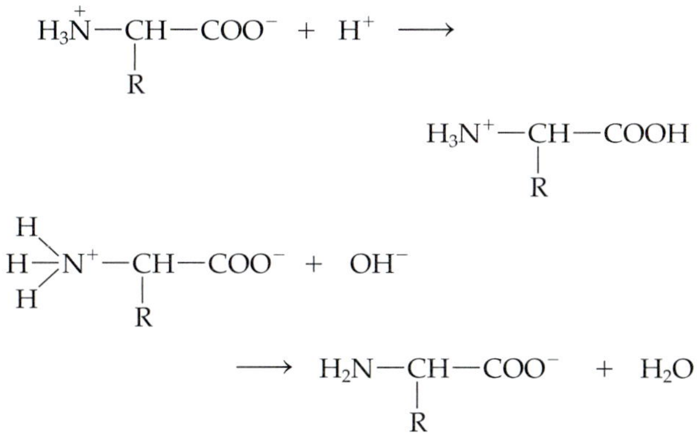

$$H_3\overset{+}{N}-\underset{R}{\underset{|}{CH}}-COO^- + H^+ \longrightarrow H_3N^+-\underset{R}{\underset{|}{CH}}-COOH$$

$$H_3N^+-\underset{R}{\underset{|}{CH}}-COO^- + OH^- \longrightarrow H_2N-\underset{R}{\underset{|}{CH}}-COO^- + H_2O$$

The pH at which an amino acid exists as the zwitterion is called the **isoelectric pH**. The isoelectric pH lies somewhere between the pH at which the amino acid acts as an acid and that at which it acts as a base. At its isoelectric pH, an amino acid behaves like the salt of a weak acid and a weak base.

The amino acids in a protein are linked together by **peptide bonds**:

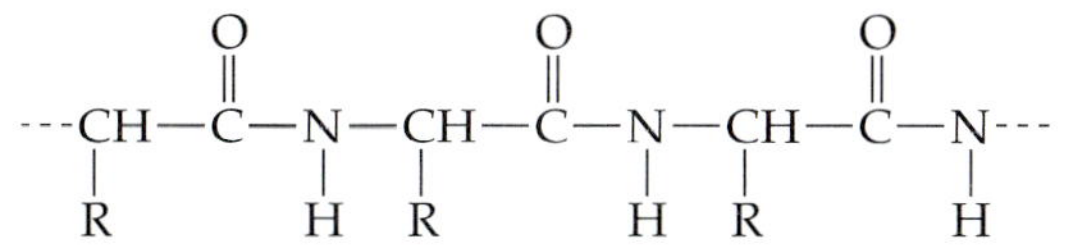

$$\cdots\underset{R}{\underset{|}{CH}}-\overset{O}{\overset{\|}{C}}-\underset{H}{\underset{|}{N}}-\underset{R}{\underset{|}{CH}}-\overset{O}{\overset{\|}{C}}-\underset{H}{\underset{|}{N}}-\underset{R}{\underset{|}{CH}}-\overset{O}{\overset{\|}{C}}-\underset{H}{\underset{|}{N}}\cdots$$

Protein chains containing ten or fewer amino acids are usually referred to as **peptides**, with a prefix—di-, tri-, and so forth, through deca—indicating the number of amino acids. Chains containing more than ten amino acid units are **polypeptides**, and when the molar mass of the chain exceeds about 10,000 daltons, the term *protein* is used. This naming scheme is used loosely, however, and in most contexts, the terms polypeptide and protein can be used interchangeably.

Proteins are classified as being globular or fibrous, depending on their structure and solubility in water. **Globular proteins** are nearly spherical and are soluble in water; **fibrous proteins** have elongated or fibrous structures and are insoluble in water.

Protein molecules can have as many as four levels of structure. The **primary structure** is the sequence of amino acids in the chain. The **secondary structure** is the conformation—helical or β-pleated sheet—of the chain. The **tertiary structure** is the overall three-dimensional shape of the molecule that results from the way the chain bends and folds in on itself. Proteins that consist of more than one chain also have **quaternary structure**, which refers to the way the multiple chains are packed together.

Four types of intramolecular and intermolecular forces contribute to secondary, tertiary, and quaternary structure: (1) *hydrogen bonding* between an oxygen or nitrogen atom and a hydrogen bound to an oxygen or nitrogen, either on the same chain or on a neighboring chain; (2) *ionic bonds* between one positively charged side chain and one negatively charged side chain; (3) *disulfide linkages* between cysteine units; and (4) *dispersion forces* between nonpolar side chains.

Proteins are amphoteric because they contain side chains that are positively and/or negatively charged. At the isoelectric pH of a given protein, these positive and negative charges occur in equal numbers. Proteins containing a preponderance of lysine, arginine, and histidine have a high isoelectric pH; those containing large numbers of aspartic acid or glutamic acid have a low isoelectric pH.

Because of their complexity, protein molecules are delicate and can be easily changed by a number of chemical agents. A *denatured* protein is one whose conformation has been changed (in a process called **denaturation)** so that it can no longer do its physiological job. Heat and ultraviolet radiation denature a protein by increasing the atomic vibrations in the molecule and thereby rupturing the hydrogen bonds and dispersion forces that hold the protein in its unique three-dimensional shape. Certain organic compounds, such as alcohols, denature a protein by forming *inter*molecular hydrogen bonds with the protein and thereby causing *intra*molecular hydrogen bonds to rupture. Heavy metal cations tend to bond with carboxylate anions and cysteine side chains and denature a protein by rupturing salt and disulfide linkages and forming metal–protein salts. Alkaloid reagents also denature a protein by breaking salt linkages, in this case as the anionic form of the reagent combines with the amino cation parts of the protein molecule.

Key Terms

amino acid (page 569)
dalton (page 569)
denaturation (21.11)
disulfide linkage (21.8)
diuretic (21.5)
electrophoresis (21.2)
fibrous protein (21.6)
globular protein (21.6)
heme (21.8)
isoelectric pH (21.2)
peptide (page 575)
peptide bond (21.3)
polypeptide (21.3)
primary structure (21.6)
protein (page 569)
quaternary structure (21.6)
secondary structure (21.6)
tertiary structure (21.6)
zwitterion (21.1)

Problems

Properties and Reactions of Amino Acids

1. Draw the side chains of:
   a. threonine b. lysine c. tyrosine
2. Draw the side chains of:
   a. glutamic acid b. cysteine c. histidine
3. Write the structural formulas for:
   a. glycine b. alanine c. valine
4. Write the structural formulas for:
   a. serine b. aspartic acid c. isoleucine
5. Identify the amino acid whose side chain (R group) fits each description:
   a. contains an amino group.
   b. contains a heterocyclic ring.
   c. contains an unsubstituted benzene ring.
   d. contains a carboxyl group.
6. Identify the amino acid whose side chain (R group) fits each description:
   a. contains a sulfhydryl group.
   b. contains a phenolic group.
   c. contains a branched chain.
   d. contains a hydroxyl group.
7. Draw the structure of alanine and determine the charge on the molecule in a(n): **(a)** acidic solution (pH = 1); **(b)** neutral solution (pH = 7); **(c)** basic solution (pH = 11).
8. Draw the structure of valine and determine the charge on the molecule in a(n): **(a)** acidic solution (pH = 1); **(b)** neutral solution (pH = 7); **(c)** basic solution (pH = 11).
9. Identify two amino acids that contain more than one chiral carbon atom.

The Peptide Bond and the Sequence of Amino Acids

10. Write structural formulas for:
    a. glycylalanine b. alanylglycine
11. Write the structural formula for phenylalanylglycylalanine.
12. Draw the structural formula for Cys-Val-Gly.
13. What is the amino acid sequence of this peptide (use the three-letter abbreviations for the amino acids)?

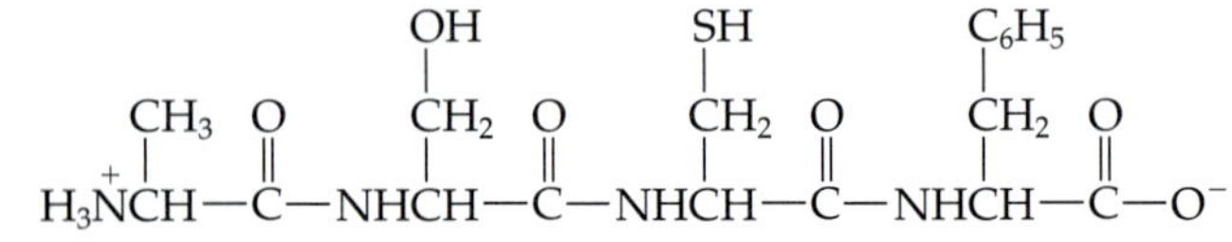

Peptide Hormones

14. a. How do the structures of oxytocin and vasopressin differ?
    b. What are the functions of these two peptide hormones?
15. a. What is the role of angiotensin II in the body?
    b. Captopril blocks the reaction in the body in which angiotensin I is converted to angiotensin II. Why is this drug effective in treating hypertension?

Classification of Proteins and Protein Structure

16. Describe the structure of silk and explain how its properties reflect its structure. What name is given to the predominant secondary structure of silk?
17. Describe the structure of wool and relate this structure to wool's elasticity. What name is given to the predominant secondary structure of wool protein?
18. Describe the structure of collagen.
19. The following sets of amino acids are involved in maintaining the tertiary structure of a peptide. In each case, identify the type of interaction involved.
    a. aspartic acid and lysine
    b. phenylalanine and alanine
    c. serine and lysine
    d. two cysteines
20. Classify these proteins as fibrous or globular:
    a. albumin b. myosin c. collagen
21. Classify these proteins as fibrous or globular:
    a. hemoglobin b. keratins c. myoglobin
22. For each of the following amino acids, state whether it is more likely to be on the inside or the outside of a globular protein. Explain your choice.
    a. phenylalanine b. aspartic acid c. serine
    d. lysine e. leucine f. glutamic acid

Electrochemical Properties of Proteins

23. Under what conditions does a protein have **(a)** a net positive charge, **(b)** a net negative charge, and **(c)** a net zero charge?
24. How do the properties of a protein at its isoelectric pH differ from its properties in solutions at other pH values?

Denaturation of Proteins

25. Which class of proteins is more easily denatured—fibrous or globular?
26. Why is a 70% alcohol solution more effective as a disinfectant than a 100% alcohol solution?

Additional Problems

27. A electric current was passed through a solution containing alanine, lysine, and aspartic acid at pH 6.0. One amino acid migrated to the cathode, one migrated to the anode, and one remained stationary. Match each behavior with the correct amino acid.
28. Give the structural formulas for the products of the acid hydrolysis of:

$$H_3\overset{+}{N}-\underset{\substack{|\\(CH_2)_4\\|\\+NH_3}}{\overset{\substack{H\\|}}{C}}-\overset{\substack{O\\\|}}{C}-\underset{\substack{|\\H}}{N}-\underset{\substack{|\\CH_3}}{\overset{\substack{H\\|}}{C}}-\overset{\substack{O\\\|}}{C}-\underset{\substack{|\\H}}{N}-\underset{\substack{|\\CH_2\\|\\C_6H_5}}{\overset{\substack{H\\|}}{C}}-\overset{\substack{O\\\|}}{C}-\underset{\substack{|\\H}}{N}-\underset{\substack{|\\(CH_2)_2\\|\\COO^-}}{\overset{\substack{H\\|}}{C}}-COO^-$$

29. Glutathione (γ-glutamylcysteinylglycine) is a tripeptide found in all cells of higher animals. It contains a glutamic acid joined in an unusual peptide linkage involving the carboxyl group of the R group (known as the γ-carboxyl group), rather than the usual carboxyl group (known as the α-carboxyl group). Draw the structure of glutathione.
30. What is the difference between the arrangement of hydrogen bonds in secondary structures of proteins and their arrangement in tertiary structures?
31. Proteins help to maintain the pH of an organism. How can they perform this function?
32. One of the neurotransmitters involved in the sensing of pain is a peptide called substance P that is composed of 11 amino acids. It is released by nerve terminals in response to pain. Its primary structure is arg-pro-lys-pro-gln-gln-phe-phe-gly-leu-met. Would you expect this peptide to be positively charged, negatively charged, or neutral at a pH of 6.0? Justify your answer.
33. Bacteria synthesize D-alanine and use it in the biosynthesis of cell walls. Draw the structure of D-alanine.
34. Draw the structure of the amino acid γ-aminobutyric acid (GABA). Is GABA found in proteins? What is its role in the body?
35. The isoelectric pH of silk fibroin is 2.2. Which amino acid is likely to be present in large amounts: **(a)** aspartic acid, **(b)** histidine, or **(c)** lysine? Justify your answer.
36. Write equations to show how alanine can act as a buffer.
37. Carbohydrates are incorporated into *glycoproteins*. How does the incorporation of sugar units affect the solubility of a protein?
38. Aspartame, L-aspartyl-L-phenylalanine methyl ester, is an artificial sweetener that is about 160 times sweeter than sucrose. It is sold commercially as NutraSweet®. Draw the structure of aspartame.
39. **a.** What are prions?
    b. Stanley Prusiner has hypothesized that diseases such as mad cow disease are transmitted by prions. How does he propose that these proteins transmit the disease?
40. What are some of the effects of lead poisoning in children?
41. Prior to the twentieth century, hat makers were often referred to as "mad hatters." Why?
42. **a.** When an individual's hair is given a permanent, a lotion containing a reducing compound such as thioglycolic acid is applied to the hair after it has been set on curlers. What is the purpose of the thioglycolic acid?
    b. After the lotion is washed off, a second solution containing a mild oxidizing agent such as hydrogen peroxide is applied to the hair. What is the purpose of the hydrogen peroxide?

CHAPTER 22 Enzymes

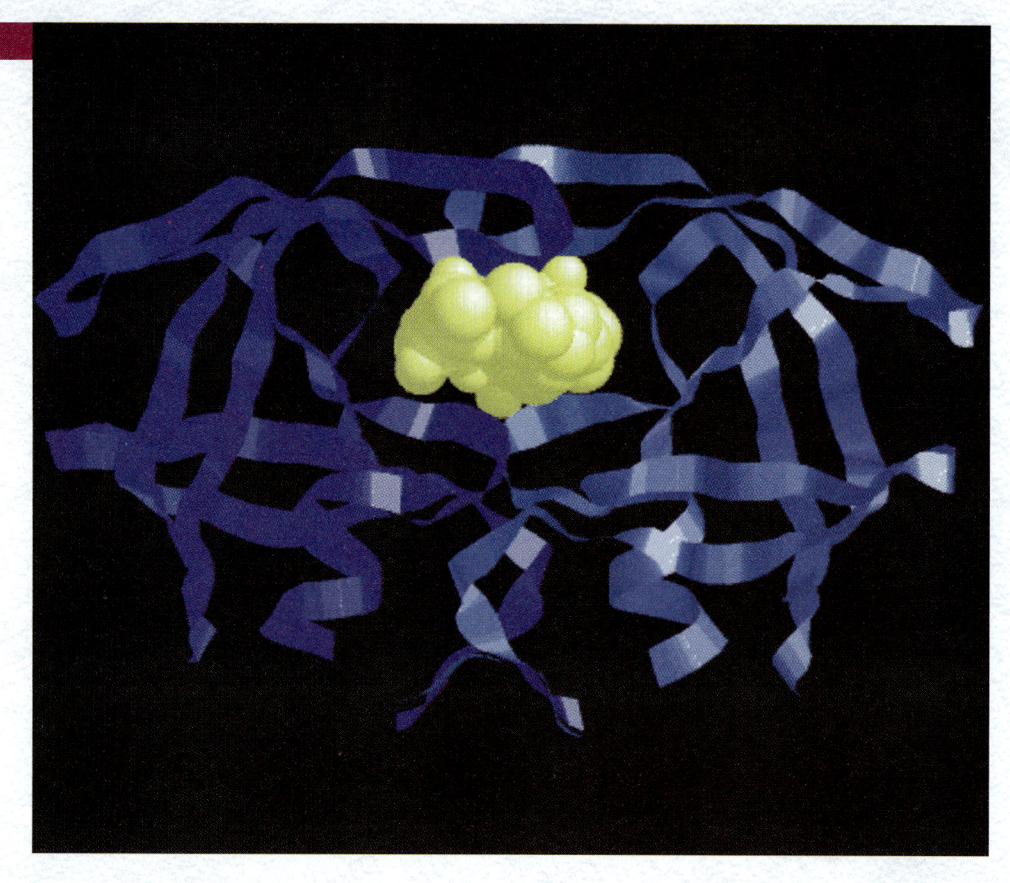

The crystal structure of HIV protease (blue) with an inhibitor bound (yellow). HIV protease is an enzyme produced by HIV, the virus that is the cause of AIDS. Several inhibitors of this enzyme are now being used by individuals with AIDS.

Learning Objectives/Study Questions

1. What is an enzyme?
2. How are enzymes classified and named?
3. What are some structural characteristics of enzymes that are important in understanding how enzymes work?
4. What is the active site of an enzyme? What types of interactions bind the substrate to the active site?
5. What is the difference between the lock-and-key theory and the induced-fit theory of enzyme action?
6. What is meant by enzyme specificity? How can enzyme specificity vary?
7. How do factors such as pH, temperature, and the concentration of enzyme and substrate influence enzyme activity?
8. What are enzyme inhibitors? What is the difference between a competitive inhibitor, a noncompetitive inhibitor, and an irreversible inhibitor?
9. What is chemotherapy?

Life would be impossible without enzymes because nearly all cell functions depend directly or indirectly on them. **Enzymes** are complex organic molecules, synthesized by living organisms, that act as catalysts. You'll recall that a *catalyst* is any substance that increases the *rate* of a chemical reaction without being consumed in the reaction (Section 4.9). Enzymes are truly amazing catalysts—reactions occur at a rate that is a million (10^6) or more times faster than the reaction in the absence of an enzyme. What is even more amazing is that they do this at body temperature (~37 °C) and in the physiological pH range (pH ~7), rather than at the typical conditions used to increase the rate of a reaction in the laboratory (high temperature and/or pressure; the use of strong oxidizing or reducing agents and/or strong acids or bases). In addition, enzymes are highly specific in their action; that is, an enzyme will catalyze the reaction of only one compound or group of structurally related compounds. The compound or compounds upon which an enzyme acts are known as its **substrate(s).**

The hydrolysis of sucrose provides a good example of how enzyme action differs from other catalysts. If we were to exclude bacteria and molds, a solution of sucrose in water could be kept indefinitely without undergoing hydrolysis to any appreciable extent. If we added hydrochloric acid, a catalyst, and heated the reaction mixture, hydrolysis would take place—producing glucose and fructose. If we added the enzyme sucrase as a catalyst instead of the acid, the reaction would take place more rapidly and the solution would not have to be heated at all. Furthermore, hydrochloric acid catalyzes the hydrolysis of lactose and maltose as well as sucrose, while sucrase is specific for sucrose alone. It will not catalyze the hydrolysis of any other disaccharide.

Hundreds of enzymes have been purified and studied in an effort to understand how they work so effectively and with such specificity. Understanding how enzymes work is also important in designing drugs that will inhibit or activate specific enzymes. An example is the intensive research to find a cure for AIDS (acquired immune deficiency syndrome). AIDS is caused by the human immunodeficiency virus (HIV), and researchers have studied the enzymes produced by this virus in an effort to design drugs that will block the action of the viral enzymes, but not interact with enzymes produced by the human body. Several of these drugs have now been approved for use by AIDS patients (Section G.3).

22.1 Classification and Naming of Enzymes

The first enzymes to be discovered were named according to their source or method of discovery. The enzyme pepsin, which aids in the hydrolysis of proteins, is found in the digestive juices of the stomach (Greek *pepsis*, digestion). Papain, an enzyme that also hydrolyzes protein and is used in meat tenderizers, is isolated from papayas. As more enzymes were discovered, it became apparent that a systematic nomenclature was needed to provide a unique name for each enzyme.

In 1956 an International Commission on Enzymes developed a scheme for classifying and naming enzymes. This scheme has been revised and is maintained by the Nomenclature Commission of the International Union of Biochemistry. Enzymes are arranged into groups, with subgroups and secondary subgroups that specify more precisely the reaction catalyzed. Each enzyme is assigned a four-digit number, preceded by the prefix EC. The first number indicates the general type of reaction catalyzed by the enzyme, which can be one of six types:

1. **Oxidoreductases** catalyze all the reactions in which one compound is oxidized and another is reduced. Included in this category are the *dehydrogenases,* which catalyze oxidation-reduction reactions involving hydrogen.

2. **Transferases** facilitate the transfer of groups such as methyl, amino, and acetyl from one molecule to another. For example, *transaminases* catalyze the transfer of an amino group. This reaction is involved in the removal of the amino group during the metabolism of amino acids (Chapter 27).
3. **Hydrolases** catalyze hydrolysis reactions. These include *lipases,* which catalyze the breakdown or hydrolysis of fats, and *proteases,* which catalyze the hydrolysis of proteins.
4. **Lyases** aid in the removal of certain groups without hydrolysis or the addition of groups to a double bond. The *decarboxylases,* which catalyze the removal of carboxyl groups, are an example of enzymes in this category.
5. **Isomerases,** as their name implies, catalyze the conversion of a compound to its isomer.
6. **Ligases** are involved in the formation of new bonds between carbon and nitrogen, oxygen, sulfur, or another carbon atom. These enzymes require energy, usually supplied by the hydrolysis of ATP (Section 16.10), to carry out their respective reactions.

The second of the four numbers indicates the subgroup and the third number the secondary subgroup; the fourth number is assigned sequentially as each enzyme in the secondary group is isolated and characterized. Table 22.1 shows the assignment of the four-digit classification number for alcohol dehydrogenase.

The Nomenclature Commission also gives each enzyme a name consisting of the root name of the substrate (or substrates) and the *-ase* suffix. Thus, urease is the enzyme that catalyzes the hydrolysis of urea. However, many substrates have complex names and the systematic name can become quite long, so each enzyme also has a trivial or recommended name.

Review Questions

22.1 What is the substrate for each of the following?
a. lactase; **b.** cellulase; **c.** peptidase; **d.** lipase.

22.2 Both hydrolases and lyases catalyze reactions in which bonds are broken. What is the difference between these two classes of enzymes?

Table 22.1 Assignment of an Enzyme Classification Number

Alcohol dehydrogenase EC 1.1.1.1
The first digit indicates that this enzyme is an oxidoreductase, that is, an enzyme that catalyzes an oxidation-reduction reaction.
The second digit indicates that this oxidoreductase catalyzes a reaction involving a primary or secondary alcohol, or a hemiacetal.
The third digit indicates that either the coenzyme NAD^+ or $NADP^+$ (see Figure F.4) is required for this reaction.
The fourth digit indicates that this was the first enzyme isolated, characterized, and named using this system of nomenclature.
The systematic name for this enzyme is *alcohol:NAD^+ oxidoreductase,* while the recommended or trivial name is alcohol dehydrogenase.
Reaction catalyzed:

$$RCH_2\text{—}OH + NAD^+ \rightleftharpoons R\text{—}\overset{\overset{\displaystyle O}{\|}}{C}\text{—}H + NADH + H^+$$

22.2 Characteristics of Enzymes

For many years it was thought that all enzymes were proteins. However, in the 1980s Thomas Cech and Sidney Altman identified some ribonucleic acids (RNA; see Chapter 23) that catalyze cellular reactions. The enzymes that we will study in this chapter are all globular proteins. Many enzymes, such as pepsin and lysozyme, are simple proteins—they consist entirely of one or more amino acid chains. Others contain a nonprotein component called a **cofactor** that is necessary for the proper functioning of the enzyme. There are two types of cofactors: inorganic ions (e.g., Zn^{2+}, Mn^{2+}) and organic molecules known as **coenzymes.** Many coenzymes are vitamins or are derived from vitamins and are discussed in Selected Topic F.

The polypeptide segment of any enzyme containing a cofactor is called an **apoenzyme.** Neither the cofactor nor the apoenzyme by itself can act as an effective catalyst. The catalytically active apoenzyme-cofactor complex is called the **holoenzyme.**

$$\underset{\substack{\text{Nonprotein}\\\text{(inactive)}}}{\text{Cofactor}} + \underset{\substack{\text{Protein}\\\text{(inactive)}}}{\text{Apoenzyme}} \rightleftharpoons \underset{\substack{\text{Complex}\\\text{(active)}}}{\text{Holoenzyme}}$$

Most enzymes operate within the cell that produces them and are thus termed *intracellular.* If the enzyme's usual site of catalytic activity is outside the cell that produces it (as in the case of the digestive enzymes), the enzyme is designated as *extracelluar.*

Some enzymes, such as the protein-digesting enzymes (Section 24.2) and the blood clotting factors (Section 28.4), are secreted in larger, inactive forms known as *zymogens* or **proenzymes.** These proenzymes can be quickly activated when needed; otherwise, they are kept inactive so that they do not digest the protein in the walls of the digestive tract or form unneeded (and potentially fatal) blood clots.

Review Questions

22.3 What is the difference between a cofactor and a coenzyme?

22.4 What is the difference between an apoenzyme and a holoenzyme?

22.3 Mode of Enzyme Action

In Section 4.9 we discussed the role of catalysts in increasing reaction rates by lowering the activation energy of a reaction. Enzymes reduce activation energies more effectively than other catalysts, thus enabling biochemical reactions to proceed at relatively low temperatures. Note that the overall amount of energy absorbed or released in the reaction is not altered by the enzyme (Figure 22.1).

Enzymatic reactions occur in at least two steps. In the first step, a molecule of the enzyme (E) and a molecule of the substrate (S) collide and react to form an intermediate compound called the *enzyme–substrate complex* (E–S). (This step is reversible because the complex can break apart, yielding the original substrate and the free enzyme.) The enzyme catalyzes the formation of a product, which is then released from the surface of the enzyme:

$$S + E \rightleftharpoons E\text{—}S$$

$$E\text{—}S \longrightarrow P + E$$

The existence of an enzyme-substrate complex has been verified by spectroscopic and kinetic experiments. The enzyme and substrate are held together by hydrogen bonds and electrostatic interactions between functional groups. In addition,

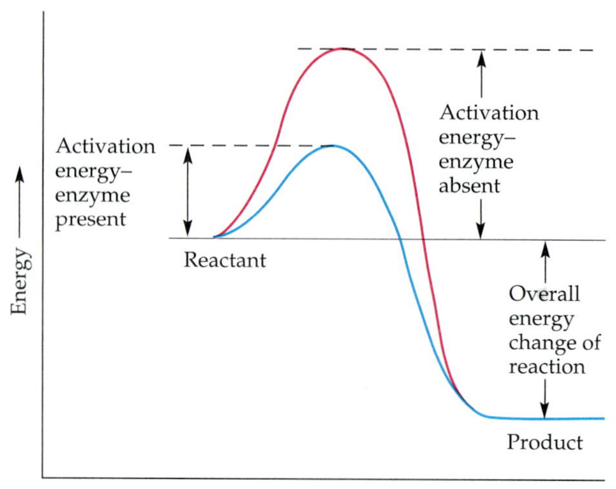

▶ **Figure 22.1**
Energy diagram for the progress of a chemical reaction, showing the effect of an enzyme.

it has been demonstrated that the structural features or functional groups essential to the formation of the enzyme-substrate complex occur at a specific location in a cleft or pocket of the enzyme surface. This section of the enzyme, which combines with the substrate and at which the substrate is transformed to product(s), is called the **active site** of the enzyme (Figure 22.2). The active site possesses a unique conformation (as well as correctly positioned bonding groups) that is complementary to the structure of the substrate. Thus the two molecules are able to fit together in much the same manner as a key fits into a tumbler lock. This **lock-and-key theory** of enzyme action is illustrated in Figure 22.3. It portrays an enzyme as conformationally rigid and able to bond only to substrates that are structurally suitable.

The use of X-ray crystallography in determining the precise three-dimensional structures of enzymes was a major advance in understanding their catalytic activity. It was observed that the binding of substrate could lead to a large conformational change in the enzyme. In 1963 D. E. Koshland, Jr., augmented the lock-and-key theory by suggesting that the binding site of an enzyme is not a rigid structure. Instead, enzymes can undergo a change in conformation when they bind

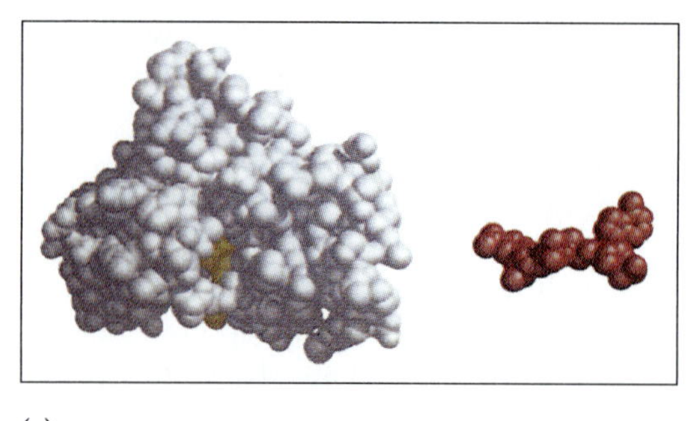

(a)

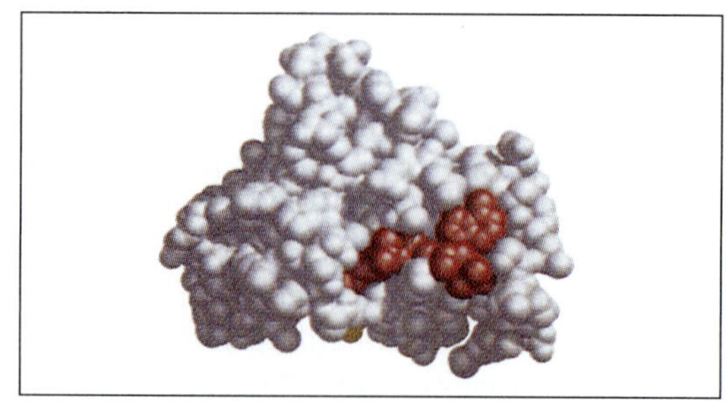

(b)

▲ **Figure 22.2**
An enzyme is a large molecule with the active site in a crevice. According to the lock-and-key model, only a substrate with a shape and structure complementary to those of the active site can fit the enzyme. (a) The enzyme dihydrofolate reductase is shown with one of its substrates, $NADP^+$ (red) unbound and (b) bound. The $NADP^+$ binds to a pocket that is complementary to it in shape and ionic properties.

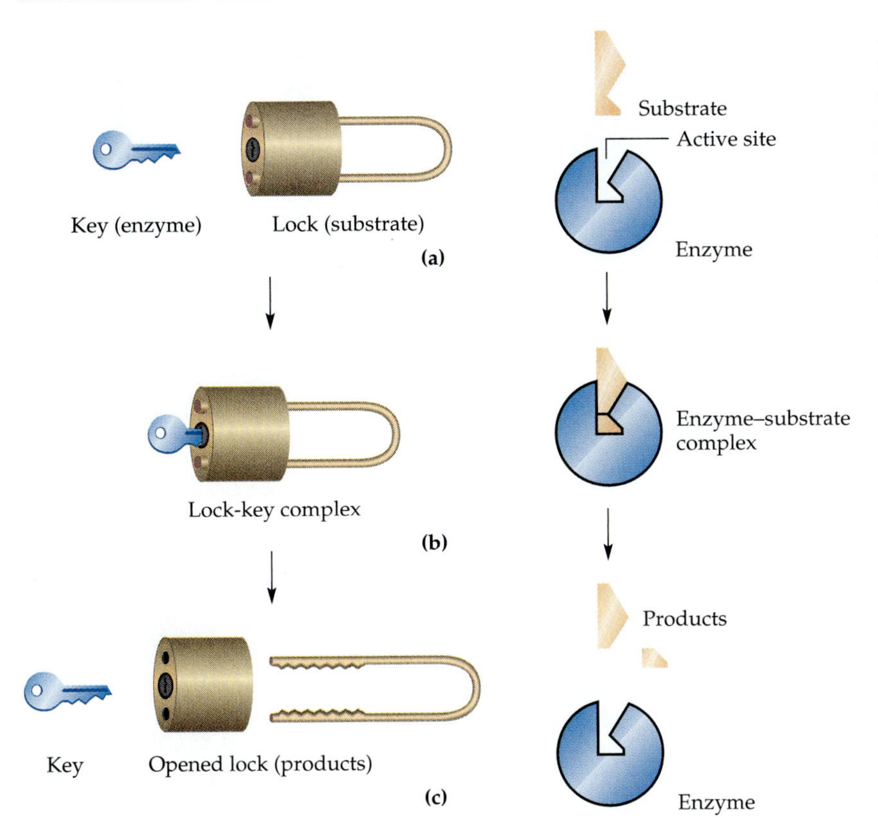

◀ **Figure 22.3**
The lock-and-key model of enzyme action. (a) The substrate and the active site of the enzyme have complementary structures and bonding groups; they fit together as a key fits a lock. (b) The catalytic reaction occurs as the two are bonded together in the enzyme-substrate complex. (c) The products of the reaction leave the surface of the enzyme, freeing it to combine with another substrate molecule.

substrate molecules. The active site has a shape complementary to that of the substrate only *after* the substrate is bound, as shown in Figure 22.4 for hexokinase. After catalysis, the enzyme resumes its original structure. This theory is known as the **induced-fit theory** because the active site is a flexible region that can be induced (or influenced) to fit several structurally similar compounds. However, only the proper substrate is capable of correct alignment with the catalytic groups (Figure 22.5).

This alignment of the substrate with specific groups in the active site is crucial. Amino acid side chains, such as the hydroxyl group of serine and the sulfhydryl group of cysteine, can act as acid or base catalysts, provide binding sites for functional groups being transferred from one substrate to another, or aid in the rearrangement of a substrate. These amino acids, which are usually widely separated in the primary sequence of the protein, are brought into proximity to each other in the active site as a result of the folding and bending of the polypeptide chain (or chains). When enzymes bind the reactants close to each other and in the proper alignment, they also increase the effective concentration of the reacting compounds.

Table 22.2 lists some important enzymes and indicates how rapidly some of them work. The **turnover number** is the number of substrate molecules converted to product in one second by a single enzyme active site.

✓ Review Questions

22.5 Distinguish between the lock-and-key theory and the induced-fit theory.

22.6 How is it possible that two amino acids that are relatively far apart in the primary structure of a protein chain can be in proximity at the active site?

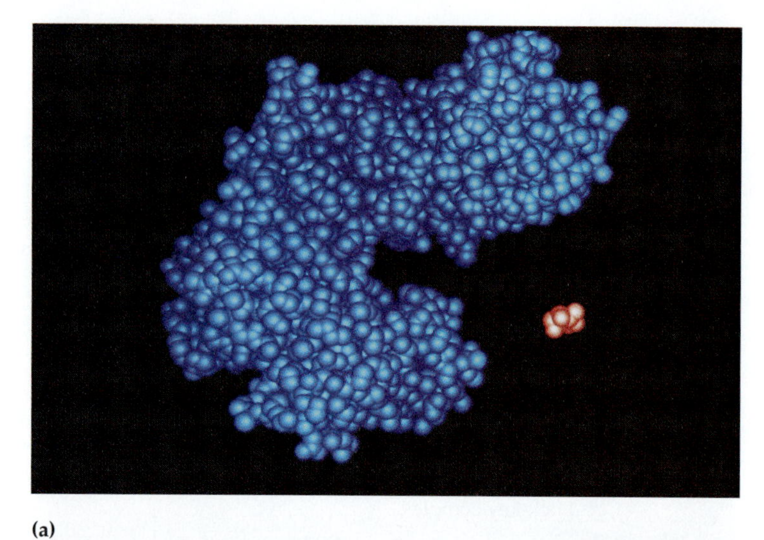

(a)

(b)

► **Figure 22.4**
Illustration of the induced-fit theory. (a) Hexokinase (blue) without substrate (glucose—shown in red) bound. (b) The enzyme conformation dramatically changes when glucose binds, resulting in additional interactions between the enzyme and glucose.

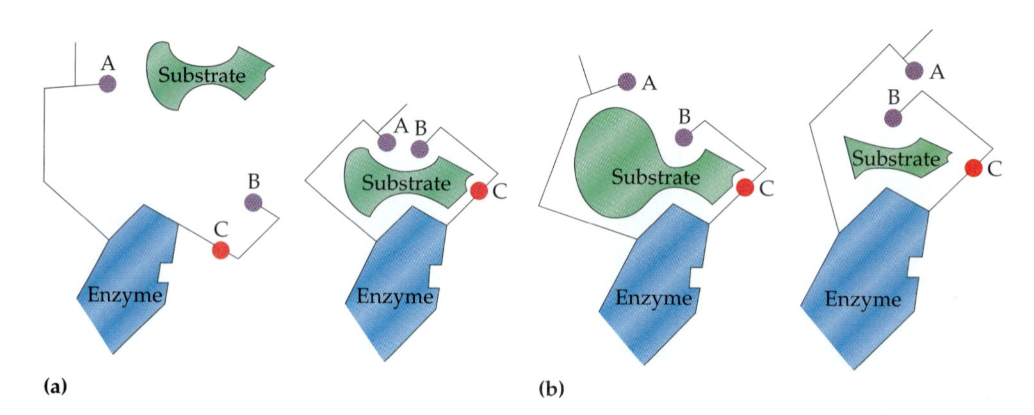

▲ **Figure 22.5**
Schematic representation of a flexible active site. (a) Substrate binding induces the proper alignment of catalytic groups A and B so that a reaction takes place. (b) Compounds that are too large or too small are bound but fail to induce proper alignment of catalytic groups; no reaction occurs.

Table 22.2 Examples of Enzyme Turnover Numbers

Enzyme	Reaction Catalyzed	Turnover Number (per second)
Catalase	$2\,H_2O_2 \rightleftharpoons 2\,H_2O + O_2$	40,000,000
Carbonic anhydrase	$H_2O + CO_2 \rightleftharpoons H_2CO_3$	1,000,000
Lactate dehydrogenase	Pyruvate $\rightleftharpoons$ Lactate	1,000
Fumarase	Fumarate $\rightleftharpoons$ Malate	800
Succinate dehydrogenase	Succinate $\rightleftharpoons$ Fumarate	20
DNA polymerase I	Addition of nucleotides to DNA chain	15
Lysozyme	Hydrolysis of specific polysaccharide bonds	0.5

22.4 Specificity of Enzymes

One characteristic that distinguishes an enzyme from all other types of catalysts is its *substrate specificity.* We have noted in previous chapters that acids catalyze the hydrolysis of disaccharides, polysaccharides, lipids, and proteins with complete impartiality, whereas a different enzyme is required for each compound. Enzyme specificity results from the uniqueness of the active site of each enzyme. This is a function of the chemical nature, electric charge, and spatial arrangements of the groups located there. Enzyme specificity is crucial in chemical reactions in the cell. It ensures that, for the most part, the proper reactions occur in the proper place at the proper time. A wide range of enzyme specificities exist, and they are arbitrarily grouped as follows.

Absolute Specificity

Enzymes that have *absolute specificity* catalyze a particular reaction for one particular substrate only and have no catalytic effect on substrates that are closely related. Urease, for example, catalyzes the hydrolysis of urea but not of methyl urea, thiourea, or biuret. Absolute specificity is rare among enzymes characterized to date.

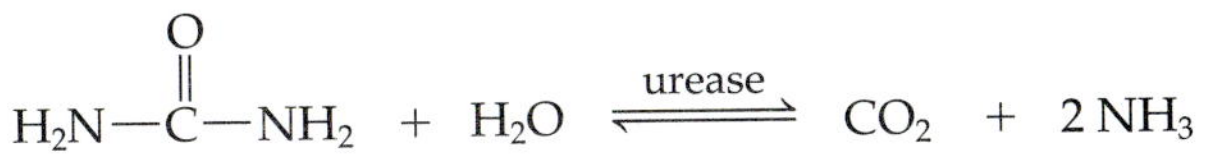

$$H_2N{-}\overset{\overset{\displaystyle O}{\|}}{C}{-}NH_2 + H_2O \xrightarrow{\text{urease}} CO_2 + 2\,NH_3$$

Urea

$$H_2N{-}\overset{\overset{\displaystyle O}{\|}}{C}{-}NH{-}CH_3 \qquad H_2N{-}\overset{\overset{\displaystyle S}{\|}}{C}{-}NH_2 \qquad H_2N{-}\overset{\overset{\displaystyle O}{\|}}{C}{-}NH{-}\overset{\overset{\displaystyle O}{\|}}{C}{-}NH_2$$

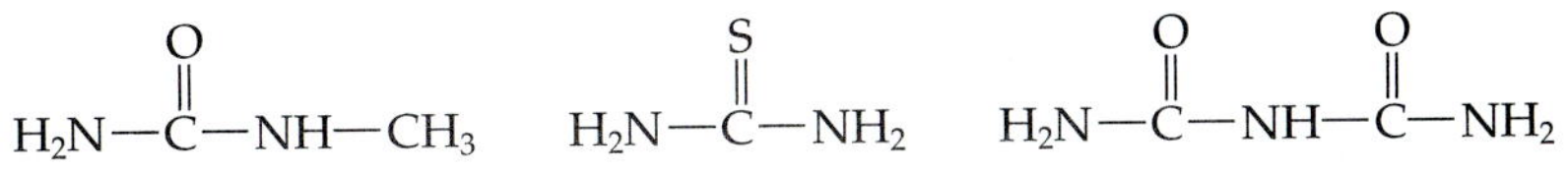

Methylurea Thiourea Biuret

Stereochemical Specificity

Because enzymes are chiral molecules, they show a high degree of *stereochemical specificity*—specificity for one stereoisomeric form of the substrate. This is analogous to the binding of monosodium glutamate to the taste buds (see Figure 18.5). L-Lactate dehydrogenase catalyzes the oxidation of L-lactate in muscle cells. D-Lactate, found in certain microorganisms, does not bind to the enzyme. Fumarase catalyzes the addition of water to fumarate but not to its *cis* isomer, maleate.

Maleate is the *cis* isomer of fumarate, while malate is the alcohol formed by the hydration of fumarate (Section 24.4).

$^{-}O \diagdown C \diagup\!\!= O$; $H—C—OH$; CH_3

D-Lactate

$^{-}O \diagdown C \diagup\!\!= O$; $HO—C—H$; CH_3

L-Lactate

Fumarate

Maleate

Group Specificity

Enzymes that have *group specificity* are less selective—they act upon structurally similar molecules that have the same functional groups. Many of the peptidases fall into this category. Chymotrypsin hydrolyzes peptide bonds involving the carboxyl groups of the aromatic amino acids. Carboxypeptidase attacks peptides from the carboxyl end of the chain, cleaving the amino acids one at a time.

Linkage Specificity

Enzymes with *linkage specificity* are the least specific of all because they attack a particular kind of chemical bond, irrespective of the structural features in the vicinity of the linkage. The lipases, which catalyze the hydrolysis of ester linkages in lipids, are an example of this type of enzyme.

✓ Review Questions

22.7 Distinguish between absolute specificity and group specificity.

22.8 Which enzyme is more specific—urease or carboxypeptidase? Explain your answer.

22.5 Factors That Influence Enzyme Activity

The single most important property of an enzyme is its catalytic activity. Because enzymes are proteins, they are affected by factors that disrupt protein structure and by factors that affect catalysts in general. The activity of an enzyme can be measured by monitoring the reaction it catalyzes. The reaction rate is determined by observing either the rate at which substrate disappears or the rate at which product forms.

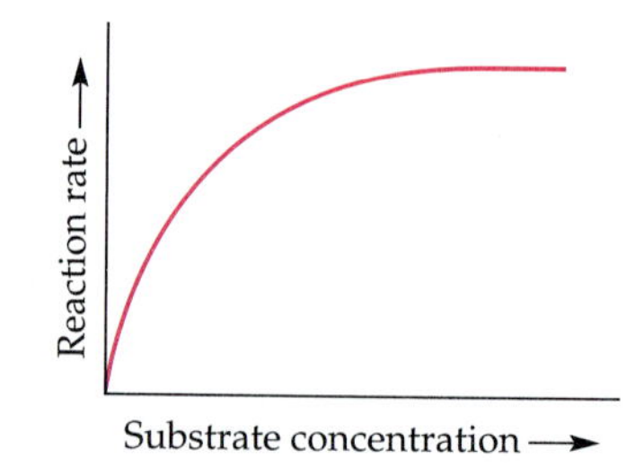

▲ **Figure 22.6**
Effect of substrate concentration on the rate of a reaction that is catalyzed by a fixed amount of enzyme.

Concentration of Substrate

The rate of an enzymatic reaction increases as the substrate concentration increases until a limiting rate is reached. At this point further increase in the substrate concentration produces no significant change in the reaction rate (Figure 22.6). At excess substrate concentrations, essentially all of the enzyme active sites have substrate bound to them. In other words, the enzyme molecules are saturated with substrate (Figure 22.7). For additional substrate molecules to react, the enzyme-substrate complex must dissociate to yield free enzyme and product(s) (or unreacted substrate).

Figure 22.7 Schematic representation of relative concentrations of enzyme (represented as crescents), substrate (shown as circles) and products (split circles). (a) Low substrate concentration. (b) Adequate substrate concentration. (c) Excess substrate concentration.

Let's consider an analogy. Ten taxis (enzyme molecules) are waiting at a taxi stand to take people (substrate) on a 10-minute trip to a concert hall, one passenger at a time. If only 5 people are present at the stand, the rate of their arrival at the concert hall is 5 people in 10 minutes. If the number of people at the stand is increased to 10, the rate increases to 10 arrivals in 10 minutes. With 20 people at the stand, the rate would still be 10 arrivals in 10 minutes. The taxis have been "saturated." If the taxis could carry 2 or 3 passengers each, the same principle would apply. The rate would simply be higher (20 or 30 people in 10 minutes) before it leveled off.

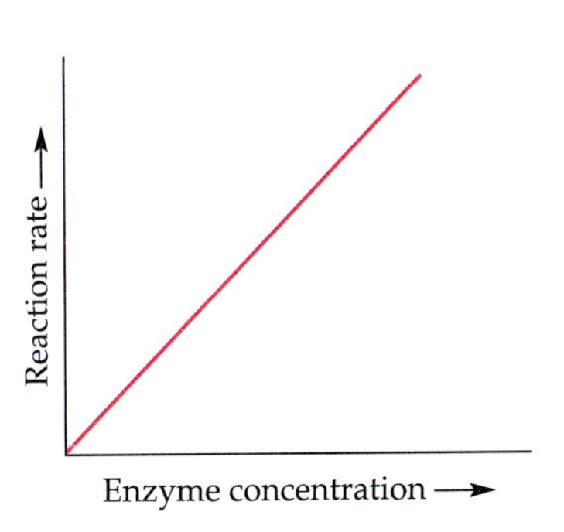

Figure 22.8 Effect of enzyme concentration on the reaction rate.

Concentration of Enzyme

When the concentration of enzyme is significantly lower than the concentration of substrate the rate of an enzyme-catalyzed reaction is directly dependent upon the enzyme concentration (Figure 22.8). This is true for any catalyst; the reaction rate increases as the concentration of the catalyst is increased. At any given time, the cellular concentration of enzyme is determined by its rate of synthesis and its rate of degradation. The cellular concentration can be increased (enzyme induction) or decreased (enzyme suppression) according to the needs of the organism.

Temperature

A rule of thumb for most chemical reactions is that a rise in temperature of 10 °C approximately doubles the reaction rate. (This is due to an increase in the number of molecules that possess sufficient kinetic energy to exceed the activation energy.) To some extent this rule holds for all enzymatic reactions. After a certain point, however, an increase in temperature causes a decrease in reaction rate due to denaturation of the protein structure (Figure 22.9a). It is difficult to determine at what temperature an enzyme has its maximum rate of reaction. At higher temperatures

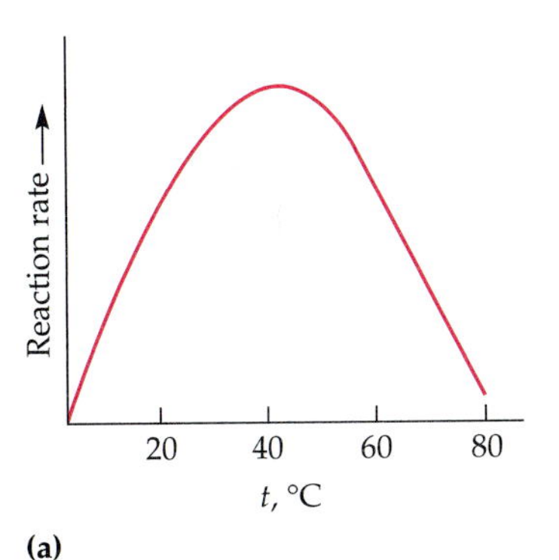

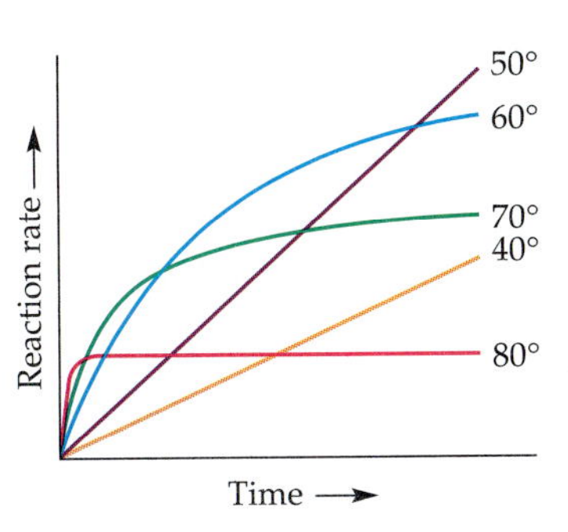

Figure 22.9 (a) The initial reaction rate of an enzyme-catalyzed reaction as a function of temperature. (b) Measurement of the reaction rate as a function of time at temperatures ranging from 40–80 °C. Note the slowing of the reaction rate over time at higher temperatures due to denaturation of the enzyme structure.

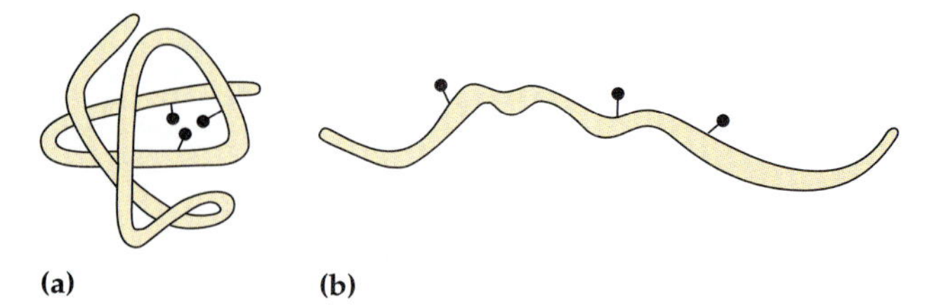

► **Figure 22.10**
(a) Representation of an active site in an enzyme. (b) Heating denatures the enzyme, and the groups of the active site are no longer close to one another.

the rate of reaction depends on how quickly the measurement is made and the rate of denaturation at that temperature (Figure 22.9b). Heating disrupts the secondary, tertiary, and quaternary structure of the enzymes, causing a disorientation of the active site. This means that the amino acid side chains that form the active site are no longer in the correct position to bind the substrate (Figure 22.10).

At 0 °C or 100 °C, the rate of enzyme-catalyzed reactions is nearly zero. This fact has several practical applications. We sterilize objects by placing them in boiling water; this denatures the enzymes of any bacteria that may be in or on the objects. We refrigerate and freeze our food to preserve it by slowing enzyme activity. When animals go into hibernation in winter, their body temperature drops, and, as a result, the rates of their metabolic processes decrease. The energy required to maintain this lowered metabolic rate is provided by fat reserves stored in their tissues (Section 26.1).

Hydrogen Ion Concentration

Being proteins, enzymes are sensitive to changes in pH. Extreme values of pH (whether high or low) can denature the protein. However, *any* change in pH, even a small one, alters the degree of ionization of acidic and basic groups on both the enzyme and on the substrate. Ionizable groups located at the active site must have a certain charge for the enzyme to bind its substrate. Thus, neutralization of even one of these charges alters an enzyme's catalytic activity.

An enzyme will exhibit maximum activity over a narrow pH range in which the molecule exists in its proper charged form. The median value of this pH range is known as the **optimum pH** of the enzyme (Figure 22.11). With the notable exception of gastric juice, most body fluids have pH values between 6 and 8. Not surprisingly, most enzymes exhibit optimal activity in this pH range. However, a few enzymes have optimum pH values outside this range. For example, the optimum pH for pepsin, an enzyme that is active in the stomach, is 2.0.

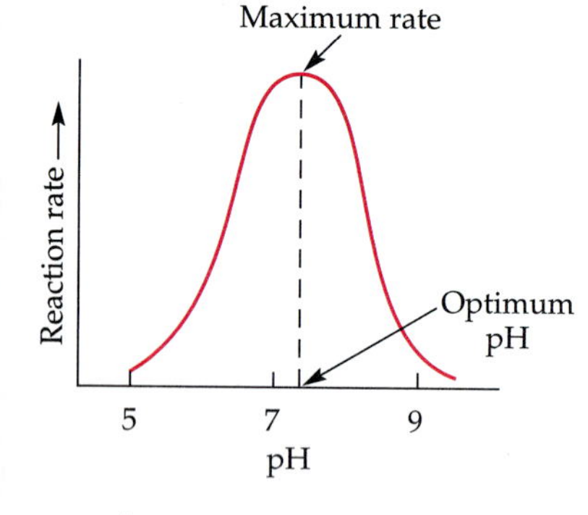

▲ **Figure 22.11**
Effect of pH on the rate of an enzymatic reaction.

✓ Review Questions

22.9 The concentration of substrate X is low. What happens to the rate of the enzyme-catalyzed reaction if the concentration of X is doubled?

22.10 How does an increase in the enzyme concentration affect the rate of a reaction?

22.6 Enzyme Inhibition

In the preceding section, we noted that enzymes are inactivated at high temperatures and by changes in pH. These are nonspecific effects that would inactivate any enzyme. The activity of enzymes can also be regulated by specific inhibitors. Many compounds are poisons because they bind covalently to specific enzymes and inactivate them (Table 22.3).

Irreversible Inhibition: Poisons

An **irreversible inhibitor** inactivates an enzyme by forming covalent bonds to a particular group at the active site. The inhibitor-enzyme bond is so strong that the inhibition cannot be reversed by addition of excess substrate. The nerve gases,

Table 22.3 Poisons as Enzyme Inhibitors

Poison	Formula	Example of Enzyme Inhibited	Action
Cyanide	CN^-	Cytochrome oxidase	Binds Fe^{3+} cofactor
Fluoride	F^-	Enolase	Binds Mg^{2+} cofactor
Sulfide	S^{2-}	Phenolase	Binds Cu^{2+} cofactor
Arsenate	AsO_4^{3-}	Glyceraldehyde 3-phosphate dehydrogenase	Substitutes for phosphate
Iodoacetate	ICH_2COO^-	Triose phosphate dehydrogenase	Binds to cysteine sulfhydryl group
Diisopropylfluorophosphate (Nerve poison)	$F{-}P(=O)(OCH(CH_3)_2){-}OCH(CH_3)_2$	Acetylcholinesterase	Binds to serine hydroxyl group

especially diisopropylfluorophosphate (DIFP), irreversibly inhibit biological systems by forming an enzyme-inhibitor complex with a specific hydroxyl group of serine situated at the active sites of certain enzymes. The peptidases trypsin and chymotrypsin contain serine groups at the active site and are inhibited by DIFP.

$$\text{Enzyme active site}{-}CH_2OH + F{-}\overset{\overset{O}{\|}}{P}(OCH(CH_3)_2){-}OCH(CH_3)_2 \longrightarrow {-}CH_2{-}O{-}\overset{\overset{O}{\|}}{P}(OCH(CH_3)_2){-}OCH(CH_3)_2 + HF$$

Metalloenzymes (enzymes that require a metal ion cofactor) are irreversibly inhibited by substances that form strong complexes with the metal. Traces of hydrogen cyanide inactivate iron-containing enzymes, such as catalase and cytochrome oxidase. Oxalate and citrate inhibit blood clotting by forming complexes with calcium ions, which are necessary for the activation of the enzyme thrombin (see Section 28.4).

Reversible Inhibition

A **reversible inhibitor** inactivates an enzyme through noncovalent interactions. Unlike an irreversible inhibitor, a reversible inhibitor can dissociate from the enzyme. We'll consider two types of reversible inhibitors—competitive inhibitors and noncompetitive inhibitors. A **competitive inhibitor** is any compound that bears a structural resemblance to a particular substrate and competes with that substrate for binding at the active site of the enzyme. The inhibitor is not acted upon by the enzyme, but it prevents the substrate from approaching the active site.

The degree of inhibition depends upon the relative concentrations of substrate and inhibitor. If the inhibitor is present in relatively large quantities, it will initially block most of the active sites. But, because the binding is reversible, some substrate molecules will eventually bind to the active site and be converted to product. Increased substrate concentration permits displacement of the inhibitor from the active site. Competitive inhibition can be completely reversed by adding substrate at a much higher concentration than inhibitor.

The reversible nature of competitive inhibition has provided much information about the enzyme-substrate complex and about the specific groups involved at the active sites of various enzymes. Pharmaceutical companies have synthesized drugs that can competitively inhibit metabolic processes in bacteria (Section 22.7) and in cancer cells (Selected Topic G).

Inhibition of Nerve Transmission

The organophosphorus compounds (Figure 22.12) are irreversible inhibitors that form covalent bonds to enzymes having a serine side chain in the active site. One of these enzymes is *acetylcholinesterase*, an enzyme that is an essential part of the process of nerve transmission.

Nerve cells (neurons) interact with other nerve cells, and with muscles and glands, at junctions called *synapses* (refer to Figure C.1). Nerve impulses are transported across synapses by small molecules known as *neurotransmitters.* Neurons use a number of different molecules as neurotransmitters, including acetylcholine.

Acetylcholine is synthesized from acetyl coenzyme A (Section 24.3) and choline and is stored in special vesicles at the axon ends of neurons. The arrival of a nerve impulse leads to the release of acetylcholine into the synapse. The acetylcholine molecules then diffuse across the synapse, where they combine with specific receptors embedded in the postsynaptic membrane of the adjacent neuron (or muscle, or gland). Binding of acetylcholine to the receptors causes a change in membrane permeability of the receiving neuron (or muscle, or gland). Sodium ions then move into the cell, potassium ions move out, and this ion flux causes the signal (the action potential) to be sent along the entire neuron until it reaches another synapse.

Once the impulse has been passed on, the acetylcholine must be immediately deactivated so that the receptor molecules can receive the next stimulus. The deactivation occurs by the hydrolysis of acetylcholine to choline and acetate, through the catalytic activity of the acetylcholinesterase (which is located in the synapse):

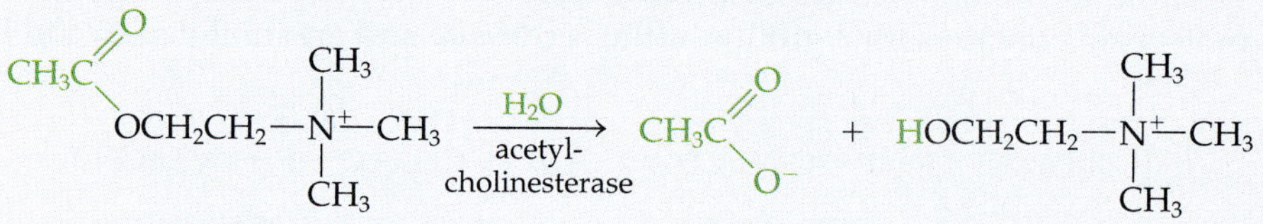

$$CH_3C(=O)OCH_2CH_2{-}\overset{+}{N}(CH_3)_3 \xrightarrow[\text{acetyl-cholinesterase}]{H_2O} CH_3C(=O)O^- + HOCH_2CH_2{-}\overset{+}{N}(CH_3)_3$$

Acetylchloline Acetate Choline

This enzyme is characterized by an extremely high turnover number. It is estimated that the enzyme-catalyzed hydrolysis reaction occurs in 40 μs (40×10^{-6} s). This speed is essential because neurons can transmit 1000 impulses per second as long as each postsynaptic membrane is continually available to receive new acetylcholine molecules.

After the breakdown of acetylcholine, the receptor neuron releases the hydrolysis products and is then ready to receive further impulses. Other enzymes convert the acetic acid and choline back to acetylcholine, completing the cycle.

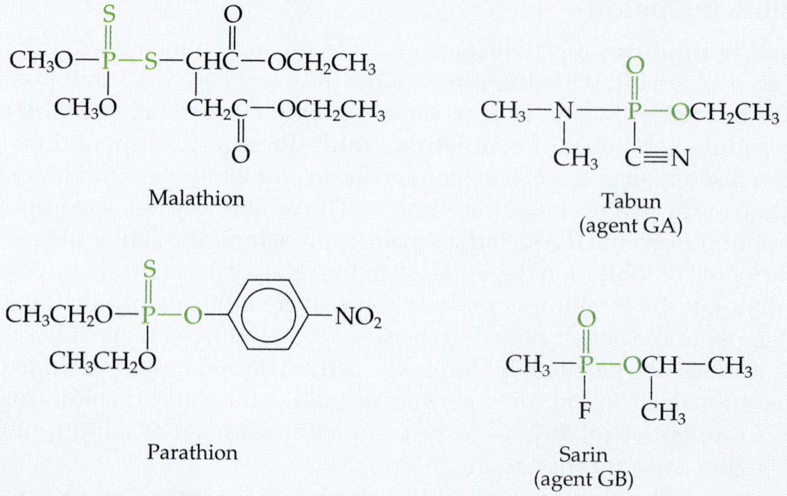

$$(CH_3O)_2P(=S){-}S{-}CH(CH_2C(=O){-}OCH_2CH_3){-}C(=O){-}OCH_2CH_3$$

Malathion

$$CH_3{-}N(CH_3){-}P(=O)(C{\equiv}N){-}OCH_2CH_3$$

Tabun
(agent GA)

$$(CH_3CH_2O)_2P(=S){-}O{-}C_6H_4{-}NO_2$$

Parathion

$$CH_3{-}P(=O)(F){-}OCH(CH_3){-}CH_3$$

Sarin
(agent GB)

▲ **Figure 22.12**
Some organophosphorus compounds. Malathion and parathion are insecticides. Tabun and sarin are nerve poisons designed for use in chemical warfare. Sarin was used in the 1995 terrorist attack on the Tokyo subway that killed 12 people and sickened thousands.

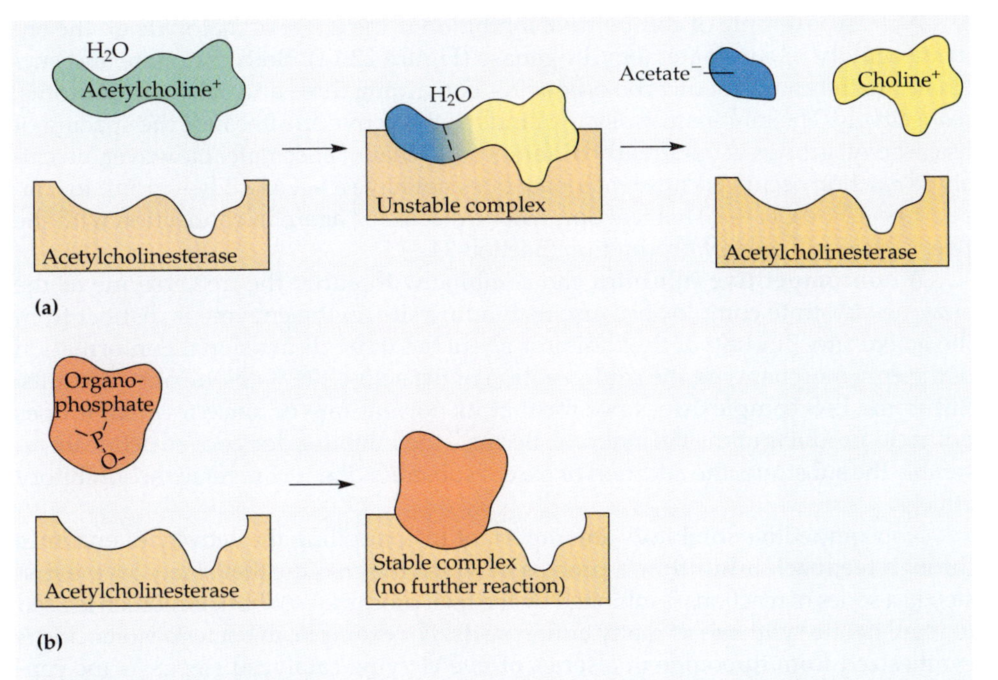

▲ **Figure 22.13**
(a) Acetylcholinesterase catalyzes the hydrolysis of acetylcholine to acetic acid and choline.
(b) An organophosphate ties up acetylcholinesterase, preventing it from breaking down acetylcholine.

The organophosphorus compounds inhibit acetylcholinesterase and other serine-containing enzymes. The polar phosphorus-oxygen bond attaches tightly to acetylcholinesterase, preventing the enzyme from performing its normal function (Figure 22.13). If the breakdown of acetylcholine is blocked, this messenger compound builds up, causing the receptor nerves to "fire" repeatedly, keeping it continuously "on." This overstimulates the muscles, glands, and organs. The heart beats wildly and irregularly. The victim goes into convulsions and dies quickly.

Botulism toxin, one of the most poisonous substances known, and certain snake venoms act by *inhibiting the release of acetylcholine* from the axon. Thus these toxic substances effectively block nerve transmissions that use acetylcholine as their neurotransmitter. Paralysis sets in and death occurs, usually by respiratory failure. Curare, used for centuries on the arrows of South American Indians, exerts its effects by competing with acetylcholine for the receptor sites on the postsynaptic muscle cell membranes. So, it *blocks the receptor sites* and prevents the message from being transmitted from nerve to muscle, leading to muscle paralysis. Nerve, muscle, and other cells sensitive to acetylcholine are said to be *cholinergic*. Pharmacologists have developed drugs that affect cholinergic nerves by either enhancing or blocking the action of acetylcholine on the receptor cells. Succinylcholine, an antagonist of acetylcholine, is used to produce muscular relaxation in surgical procedures. On the other hand, neostigmine, which inhibits acetylcholinesterase, is used to treat myasthenia gravis, a condition in which motor neurons secrete insufficient acetylcholine. A person with the disease suffers from muscular weakness and may have trouble contracting the muscles associated with breathing, chewing, eye movements, and speaking. If the enzyme is inhibited, enough acetylcholine may accumulate in the neuromuscular synapse to stimulate muscle contraction.

$$(CH_3)_3\overset{+}{N}CH_2CH_2O\overset{\overset{\displaystyle O}{\|}}{C}CH_2CH_2\overset{\overset{\displaystyle O}{\|}}{C}OCH_2CH_2\overset{+}{N}(CH_3)_3$$

Succinylcholine

$$(CH_3)_3\overset{+}{N}-C_6H_4-O-\overset{\overset{\displaystyle O}{\|}}{C}-N(CH_3)_2$$

Neostigmine

A classic example of competitive inhibition is the effect of malonate on the enzyme activity of succinate dehydrogenase (Figure 22.14). Both malonate and succinate are the anions of dicarboxylic acids, containing three and four carbon atoms, respectively. The malonate molecule binds to the active site because the spacing of its carboxyl groups is not greatly different from that of succinate. However, no catalytic reaction occurs because malonate does not have a —CH_2CH_2— group to convert to —CH=CH—. [This reaction will be discussed again in connection with the Krebs cycle and energy production (Section 24.4).]

A **noncompetitive inhibitor** can combine with either the free enzyme or the enzyme-substrate complex because its binding site on the enzyme is distinct from the active site. Binding of the inhibitor alters the three-dimensional conformation of the enzyme, changing the configuration of the active site with one of two results. Either the E–S complex does not form at its normal rate or, once formed, it does not yield products at the normal rate. Because the inhibitor does not structurally resemble the substrate, the addition of excess substrate does *not* reverse the inhibitory effects.

Noncompetitive inhibitors are important in controlling the activity of enzymes through **feedback inhibition,** a process in which the enzyme that catalyzes the first step in a series of reactions is inhibited by the final product. Feedback inhibition is used to regulate the synthesis of many amino acids. For example, in bacteria isoleucine is synthesized from threonine in a series of five enzyme-catalyzed steps. As the concentration of isoleucine increases, the organism needs a way to decrease the amount formed. This is accomplished by the binding of isoleucine, as a noncompetitive inhibitor, to the first enzyme of the series (threonine deaminase) (Figure 22.15).

Feedback inhibition is an example of *allosteric regulation.* In the example above, isoleucine binds at a site separate from the active site that is known as the *allosteric* site. Binding of isoleucine to this site leads to a change in the structure of threonine deaminase, which makes the enzyme less active. For other enzymes allosteric regulators may be either inhibitors or activators.

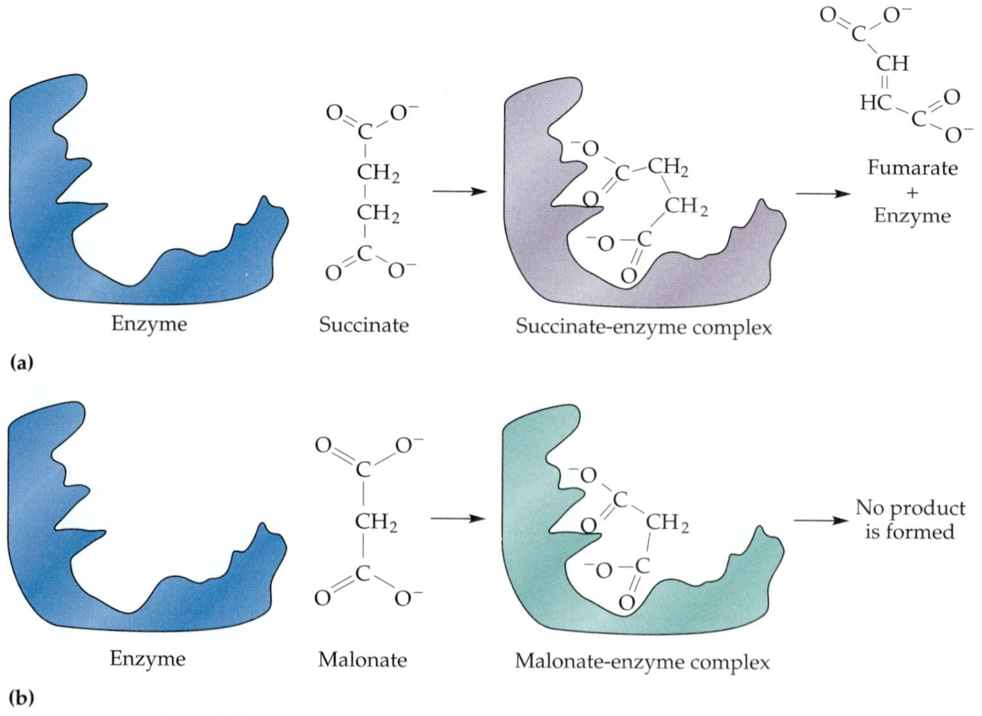

Figure 22.14
(a) Succinate binds to the enzyme succinate dehydrogenase. A dehydrogenation reaction occurs and the product, fumarate, is released from the enzyme. (b) Malonate also binds to the active site of succinate dehydrogenase. In this case, however, no subsequent reaction occurs while malonate remains bound to the enzyme.

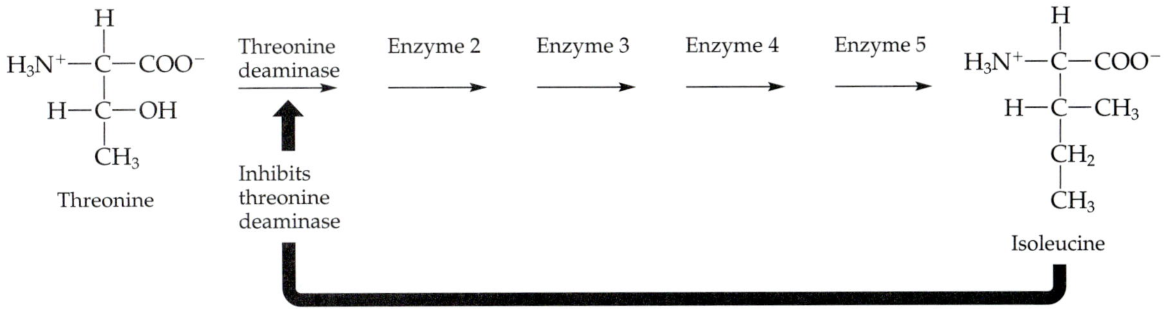

Threonine ($H_3N^+-CH(COO^-)-CH(OH)-CH_3$) —Threonine deaminase→ —Enzyme 2→ —Enzyme 3→ —Enzyme 4→ —Enzyme 5→ Isoleucine ($H_3N^+-CH(COO^-)-CH(CH_3)-CH_2-CH_3$)

Inhibits threonine deaminase

▲ **Figure 22.15**
Feedback inhibition of threonine deaminase by isoleucine, the final product of the reaction sequence.

✓ Review Questions

22.11 What are the characteristics of an irreversible inhibitor?

22.12 In what ways does a competitive inhibitor differ from a noncompetitive inhibitor?

22.7 Chemotherapy

Chemotherapy is the use of chemicals (drugs) to destroy infectious microorganisms or cancer cells while attempting to minimize damage to the cells of the host. From bacteria to humans, the metabolic pathways of all living organisms are quite similar, and so the discovery of safe and effective chemotherapeutic agents is a formidable task. It is now well established that many chemotherapeutic drugs function by inhibiting a critical enzyme in the cells of the invading organism.

In 1904 Paul Ehrlich, a German chemist, realized that certain chemicals could be used to control or cure infectious diseases. Ehrlich coined the term *chemotherapy,* a shorter version of the term "chemical therapy."

Chemotherapy is widely used in the treatment of cancer patients. **Antineoplastic drugs** (substances that inhibit the growth of cancer cells) include antimetabolites, such as 5-fluorouracil and 6-mercaptopurine, that interfere with the production of DNA and RNA in tumor cells by substituting for the pyrimidine and purine bases (see Section 23.1). An **antimetabolite** is a substance whose structure closely resembles that of the normal substrate (the *metabolite*) of an enzyme and *competitively* inhibits a significant metabolic reaction.

Antibacterial Agents

One of the earliest (1935) and best understood antimetabolites was the synthetic antibacterial agent sulfanilamide (Figure 22.16). Its effectiveness rests on its structural similarity to *p*-aminobenzoic acid,[1] a compound vital to the growth of many pathogenic bacteria. The vitamin folic acid (Section F.6) serves as a coenzyme for several important biochemical processes. Bacteria can synthesize the folic acid they need *only if* they have access to *p*-aminobenzoic acid. When bacteria encounter sulfanilamide, a bacterial enzyme readily incorporates the drug into psuedofolic acid. This compound cannot function as a proper coenzyme. In fact, it is a competitive inhibitor of the enzyme and is fatal to the bacteria because they cannot make crucial amino acids and nucleotides.

[1]It is interesting to note that *p*-aminobenzoic acid (PABA) has become widely known as an ingredient in suntan lotions. PABA acts as a sun filter by absorbing the short-wavelength ultraviolet rays that are responsible for causing sunburn. In most sunscreen products, the concentration of PABA is indicated by skin-protection factor (SPF) ratings. The larger the number, the higher the concentration of PABA and thus the greater the effectiveness against sunburn. The ethyl ester of PABA is benzocaine, a local anesthetic found in a wide variety of over-the-counter products, including first aid and sunburn sprays, foot powders, cough medicines, and appetite control products.

Figure 22.16 (a) Structure of *p*-aminobenzoic acid. (b) Some examples of sulfa drugs.

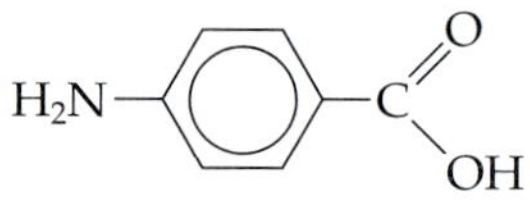

Sulfanilamide is not harmful to humans (or to other mammals) because we do not synthesize folic acid but obtain it from our diets. After the drug was recognized as an antibacterial agent, many other sulfanilamide derivatives (sulfa drugs) were synthesized and found to be even more effective. Many lives were saved during World War II as a result of these sulfa drugs. Soldiers carried packages of powdered sulfa drugs to sprinkle on open wounds to prevent infection.

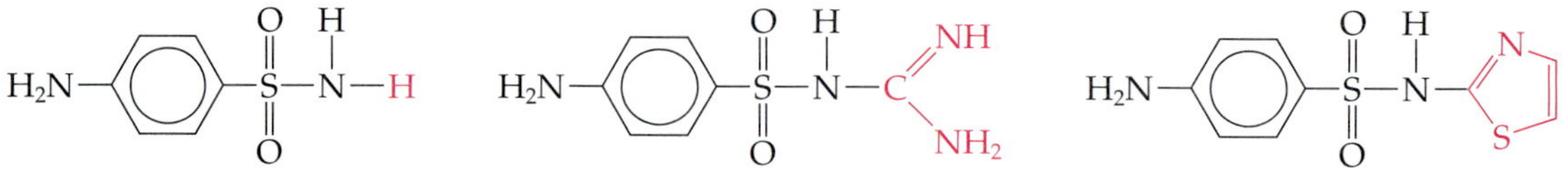

Unfortunately, prolonged use of sulfa drugs causes a number of side effects, particularly kidney damage, and so they have been largely replaced by the penicillins and other antibiotics. However, they are still prescribed for some specific infections against which they are highly effective, such as infections of the bladder and urinary tract. Some newer sulfa drugs are used in the treatment of tuberculosis and leprosy, and they are widely used in veterinary medicine.

Antibiotics

Many antibiotics are also effective against human cancer cells. In fact, several antibiotics are used *only* in cancer treatment because their activity against human cells is too great for them to be used in bacterial infections.

Although some antibiotics are believed to function as antimetabolites, the terms are not synonymous. An **antibiotic** is a compound that kills bacteria; it may come from a natural source such as molds, or be synthesized with a structure analogous to a naturally occurring antibacterial compound. Antibiotics constitute no well-defined class of chemically related substances. Instead, they possess the common property of effectively inhibiting a variety of enzymes essential to bacterial growth.

Penicillin, one of the most widely used antibiotics in the world, was fortuitously discovered by Alexander Fleming in 1928 from a mold growing on a bacterial culture plate, although he did not fully recognize its potential as a therapeutic drug. In 1938 Ernst Chain and Howard Florey began an intensive effort to isolate penicillin from the mold and to study its properties. Their need for large quantities of penicillin was aided by the development of a corn-based medium that the mold loved and the discovery of a higher-yielding strain of mold at a USDA research center near Peoria, Illinois. But it wasn't until 1944 that large quantities of penicillin were produced and made available for the treatment of bacterial infections.

Ironically, one of the highest-yielding strains came from a rotting cantaloupe found in a Peoria, Illinois, market!

Penicillin functions by interfering with the synthesis of cell walls of reproducing bacteria. Penicillin inhibits an enzyme (transpeptidase) that catalyzes the last step in bacterial cell wall biosynthesis. This step involves the joining of long polysaccharide chains by short peptide chains. The new cell walls are defective, and subsequently the bacterial cells burst. Because human cells have cell membranes and not cell walls, they are not affected.

Several naturally occurring penicillins have been isolated. All have the empirical formula $C_9H_{11}O_4SN_2R$ and contain a four-member ring (called a lactam ring) fused to a five-member ring (Figure 22.17). The various R-groups are obtained by the addition of the appropriate organic compounds to the culture medium.

Cysteine

Valine

When R is	The drug is
$-CH_2-C_6H_5$	Penicillin G
$-CH_2-S-CH_2-CH=CH_2$	Penicillin O
$-CH_2-O-C_6H_5$	Penicillin V
$-CH(NH_2)-C_6H_5$	Ampicillin
$-CH(NH_2)-C_6H_4-OH$	Amoxicillin
$-C_6H_3(OCH_3)_2$	Methicillin

◀ **Figure 22.17** The penicillins differ only in the identity of their R-groups. Note that the amino acids valine and cysteine are incorporated into the penicillin structure.

The penicillins are effective against gram-positive bacteria (bacteria that are stained by Gram's stain) and a few gram-negative bacteria (including *E. coli*). They have proved effective in the treatment of diphtheria, gonorrhea, pneumonia, syphilis, many pus infections, and certain types of boils. Penicillin G was the earliest penicillin to be used on a wide scale. However, it cannot be administered orally because it is quite unstable and the acidic pH of the stomach causes a rearrangement to an inactive derivative. The major oral penicillins—penicillin V, ampicillin, and amoxicillin—on the other hand, are acid-stable.

Some strains of bacteria become resistant to penicillin by a mutation that allows them to synthesize an enzyme, penicillinase, that breaks down the antibiotic (by cleavage of the amide linkage in the lactam ring). To combat these strains, scientists have been able to synthesize penicillin analogs (such as methicillin) that are not inactivated by penicillinase.

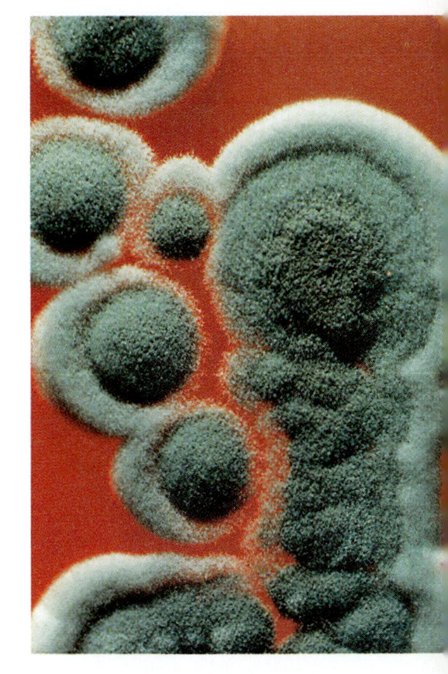

▲ Penicillin molds. These symmetrical colonies of mold are *Penicillium chrysogenum*, a mutant form of which now produces almost all of the world's commercial penicillin.

Some people (perhaps as many as 5% of the population) are allergic to penicillin and therefore must be treated with other antibiotics. The allergic reaction can be so severe that a fatal coma may occur if penicillin is inadvertently administered to a sensitive individual. Fortunately, a number of other antibiotics have been discovered (Figure 22.18). Most are the products of microbial synthesis (e.g., aureomycin, streptomycin). Others are made by chemical modifications of antibiotics (e.g., semisynthetic penicillins, tetracyclines), and some are manufactured entirely by chemical synthesis (e.g., chloramphenicol). They have proved to be as effective as penicillin in destroying infectious microorganisms. Many of these antibiotics exert their effects by blocking protein synthesis in microorganisms.

During their early history, antibiotics were considered miracle drugs, substantially reducing the number of deaths from blood poisoning, pneumonia, and other infectious diseases. Only six decades ago, a person with a major infection almost always died. Today, such deaths are rare. Six decades ago, pneumonia was a dread killer of people of all ages. Today, it kills only the very old or those ill from other causes. The antibiotics have indeed worked miracles in our time, but even miracle

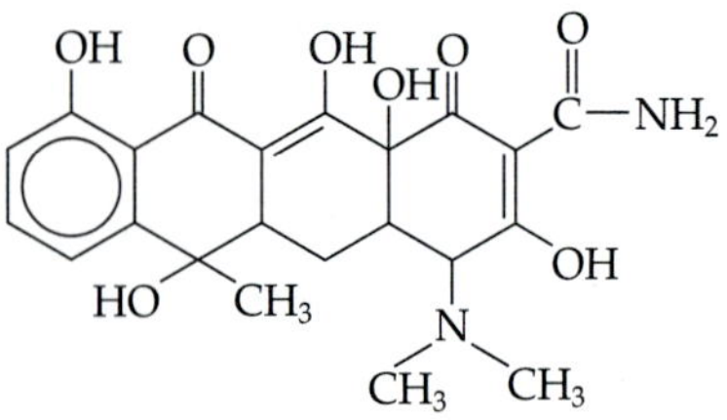

Tetracycline

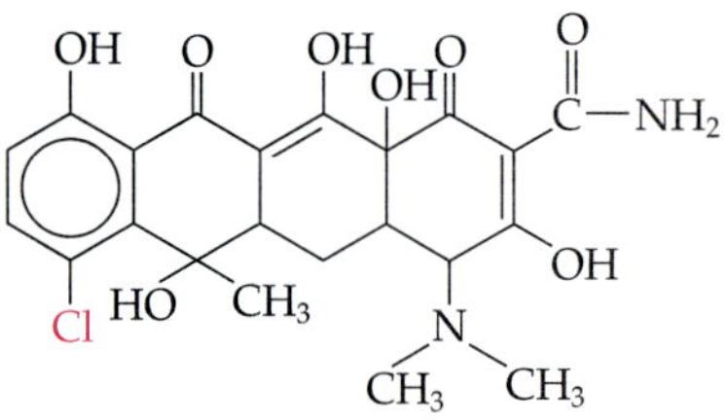

Aureomycin
(Chlorotetracycline)

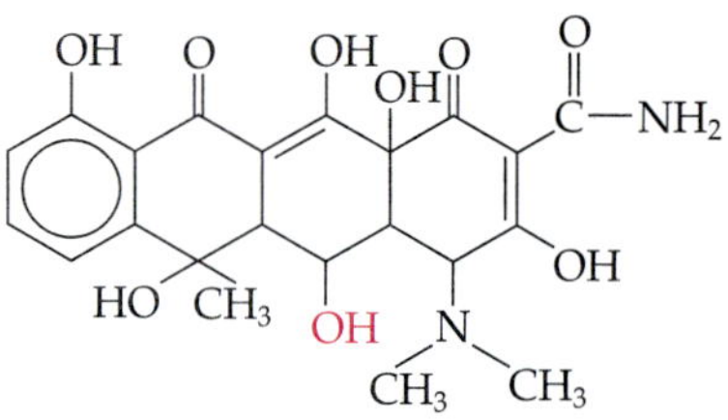

Terramycin
(Oxytetracycline)

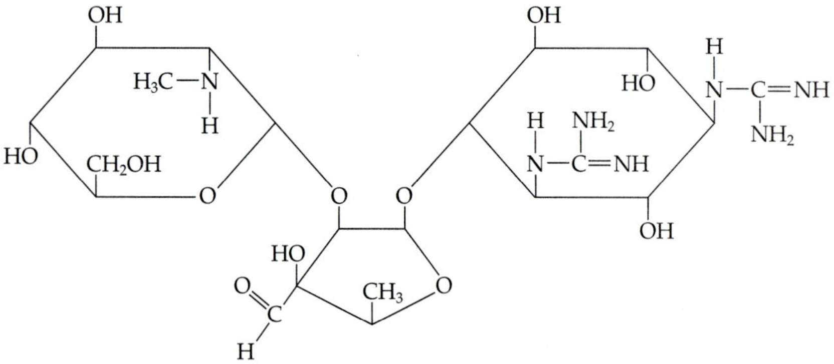

Streptomycin

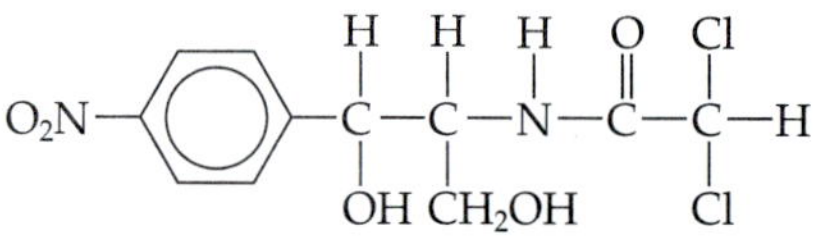

Chloramphenicol
(Chloromycetin)

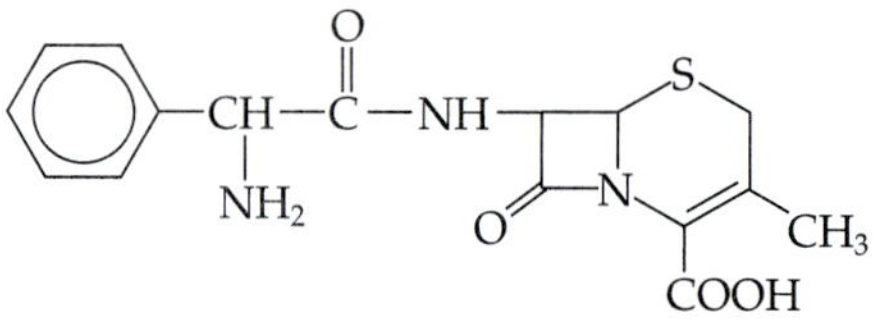

Cephalexin
(Keflex)

▲ **Figure 22.18**
Structures of some common antibiotics. Note the structural similarities among the tetracyclines. Streptomycin is a glycoside containing an amino derivative of glucose. Chloramphenicol bears a resemblance to epinephrine. Cephalexin is a member of the cephalosporins, antibiotics related to the penicillins.

drugs are not without problems. It wasn't long after the drugs were first used that disease organisms began to develop strains resistant to the drugs. Early on, all strains of staphylococci could be readily handled by erythromycin (an antibiotic obtained from *Streptomyces erythreus*). After extensive use, resistant strains began to appear and staph infections became a serious problem in hospitals. People who had gone to the hospital to be cured got serious bacterial infections instead, and some even died from these antibiotic-resistant staph infections.

In a race to stay ahead of resistant bacterial strains, scientists continue to seek new antibiotics. The penicillins now have been partially displaced by related compounds such as the cephalosporins and vancomycin. Unfortunately, some strains of bacteria have already shown resistance to these antibiotics, and drug companies have begun intensive research efforts aimed at identifying and marketing new antibiotics.

Drug-resistant tuberculosis now accounts for one out of every seven new cases.

✓ Review Questions

22.13 How do sulfa drugs kill bacteria?

22.14 **(a)** How does penicillin kill bacteria? **(b)** How do resistant bacteria deactivate penicillin?

Diagnostic Applications of Enzymes

The measurement of enzyme activity in such body fluids as plasma and serum has become a valuable tool in medical diagnosis. Certain enzymes that function in the plasma, such as those involved in blood clotting, are continually secreted into the blood by the liver. Most other enzymes, however, are normally present in plasma in very low concentrations. They are derived from the routine destruction of erythrocytes, leukocytes, and other cells. When cells die, their soluble enzymes leak out of the cells and enter the bloodstream. Because not all cells contain the same complement of enzymes, those that are specific to a particular organ can be important in aiding diagnosis.

An abnormally high level of a particular enzyme in the blood often indicates specific tissue damage, as in hepatitis and heart attack. For example, elevated blood levels of creatine kinase and glutamate-oxaloacetate transaminase (GOT) accompany some forms of severe heart disease. A blood analysis that shows high levels of creatine kinase (CK) may indicate that the heart muscle has suffered serious damage. On the other hand, many forms of strenuous (and healthful) physical activity also result in elevated CK levels. The enzyme mediates the reaction that serves as one source of energy for muscle contraction. Indeed, it is even possible for the CK level to rise simply because someone who hates needles has tensed up while waiting for the blood sample to be taken. Nonetheless, as Figure 22.19 suggests, analysis for specific enzymes is considered one valuable source of data on which to base a medical diagnosis. Table 22.4 lists the commonly assayed enzymes that are used in clinical diagnoses.

LAB NO. ______

☐ ROUTINE
☐ PRE-OP
☐ ASAP
☐ STAT

SPECIMEN:
☐ BLOOD
☐ OTHER
DATE ______
TIME ______
TECH ______

IF IMPRINT PLATE NOT USED PRINT HERE

PATIENT NO. ROOM NO.
PATIENT NAME AGE
ADDRESS
DOCTOR

✓	TEST	NORM	RESULT	✓	TEST	NORM	RESULT	✓	TEST	NORM	RESULT	✓	TEST	NORM	RESULT
	LDH				Amylase				Albumin				Calcium		
	LDH-1				Acid Phos.				Globulin				Phosphorus		
	CPK				Cholesterol				BUN				Magnesium		
	CPK-MB				Triglyceride				Creat				Ethanol		
	Alk. Phos.				HDL-Chol.				Total Bili				Hgb AIC		
	SGOT (AST)				Iron				Direct Bili				Ammonia		
	SGPT (ALT)				TIBC				Indirect Bili				Lithium		
	GGPT				Total Prot.				Uric Acid			TECH:		DATE:	

BRIGGS, DES MOINES, IA 50306 CHEMISTRY I PRINTED IN U.S.A.

▲ **Figure 22.19**
A hospital form for clinical analysis of a blood sample.

Table 22.4 Some Important Enzymes for Clinical Diagnoses

Enzyme Assayed	Organ or Tissue Affected
α-Amylase	Pancreas
Alkaline phosphatase	Bone, liver
Acid phosphatase	Prostate
Creatine kinase (CK)	Muscle, heart
Glutamate-oxaloacetate transaminase (GOT)	Heart, liver
Glutamate-pyruvate transaminase (GPT)	Liver
Lactate dehydrogenase (LDH)	Heart, liver
Alanine aminotransferase	Liver
Aspartate aminotransferase	Heart, liver

Modern medical practices have automated and computerized the assay procedures for most serum enzymes. It is important to note that the precise patterns of enzyme changes in certain tissue diseases are characteristic. For example, in a heart attack, the GOT-GPT ratio is usually high; the reverse is true in liver disease. A summary of changes in serum enzyme levels following a heart attack is illustrated in Figure 22.20.

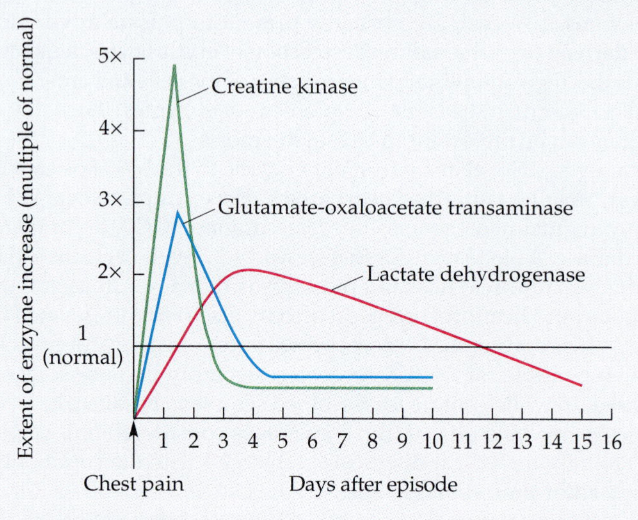

▲ **Figure 22.20**
Typical changes in serum enzyme levels following a heart attack.

Summary

An **enzyme** is an organic catalyst produced by a living cell. An enzyme works by lowering the activation energy required by a reaction. Because enzymes are such powerful catalysts, the reactions they promote occur rapidly at body temperature. Without the help of enzymes, these reactions would require high temperatures and long reaction times.

The molecule (or molecules) upon which an enzyme acts is the **substrate(s)** for that enzyme. The enzyme binds to its substrate to form an enzyme-substrate complex. The reaction occurs and the product molecule is released:

$$E + S \rightleftharpoons E\text{—}S \longrightarrow E + P$$

Most enzymes are globular proteins and are classified into six groups according to the reactions they catalyze: **oxidoreductases, transferases, hydrolases, lyases, isomerases,** and **ligases.** Simple enzymes contain only one or more amino acid chains. Complex ones comprise one or more amino acid chains joined to **cofactors** (an inorganic ion or an organic **coenzyme**). The polypeptide part of a complex enzyme is called an **apoenzyme.** The combined enzyme/cofactor entity is a **holoenzyme.**

Enzymes that would harm the cell in which they are synthesized are produced as an inactive form called a *zymogen* or a **proenzyme.**

The **lock-and-key theory** of enzyme and substrate binding pictures a rigid enzyme of unchanging configuration that binds to the appropriate substrate. The **induced-fit theory** holds that the enzyme active site is not rigid but changes its conformation upon binding to the substrate. Thus an enzyme may bind to *several substrates* (at different times) that are structurally similar.

The **turnover number** for an enzyme is the number of substrate molecules converted to product by a single enzyme active site per second.

Enzymes can vary widely in their substrate specificity. Enzymes with *absolute specificity* can catalyze a particular reaction for only one particular substrate, while enzymes with *stereochemical specificity* act on only one stereoisomeric form of a substrate. Enzymes that can act on structurally similar substrates with a common functional group are said to have *group specificity,* while enzymes that act on a particular type of linkage or bond in a wide variety of substrates display *linkage specificity.*

Several factors influence enzyme activity, among them substrate concentration, enzyme concentration, temperature, and pH. The rate of an enzyme-catalyzed reaction increases with an increase in substrate concentration until all the enzyme active sites are filled. As this point is reached, further increases in substrate concentration have no effect on the reaction rate because the binding sites are saturated with substrate. The reaction rate is directly dependent on the enzyme concentration—increasing the amount of enzyme present increases the reaction rate.

Reaction rate increases with temperature up to the point at which the heat denatures the enzyme, and then the reaction rate drops off sharply. Most enzymes have maximal activity in a narrow pH range centered around an **optimum pH.** In this pH range the enzyme is correctly folded and catalytic groups in the active site have the correct charge (positive, negative, or neutral). For most enzymes, the optimum pH lies between 7 and 8.

Substances that interfere with enzyme function are inhibitors. An **irreversible inhibitor** inactivates enzymes by forming covalent bonds to the enzyme, while a **reversible inhibitor** inactivates an enzyme by forming weaker noncovalent interactions, allowing it to dissociate. A **competitive inhibitor** is a reversible inhibitor that is structurally similar to the substrate and binds to the active site. When the inhibitor is bound, the substrate is blocked from the active site and no reaction occurs. Because the binding of inhibitor is reversible, a high substrate concentration will overcome the effect of the inhibitor because it is more likely that substrate will bind. A **noncompetitive inhibitor** binds reversibly at a site distinct from the active site. Thus it can bind either to the enzyme or the enzyme-substrate complex. The inhibitor changes the conformation of the active site so that the enzyme cannot function properly. Noncompetitive inhibitors are important in **feedback inhibition** in which the amount of product produced by a series of reactions is carefully controlled. The final product in the series of reactions acts as a noncompetitive inhibitor of the initial enzyme.

An **antimetabolite** is a chemotherapeutic substance that competitively inhibits some enzymatic reaction in a cancer cell or a microbial pathogen. Blockage of a necessary reaction leads to the death of the cell or pathogen. An **antibiotic** is a substance (or its synthetic analog) produced by one microorganism that is lethal to another type of organism.

Key Terms

active site (22.3)
antibiotic (22.7)
antimetabolite (22.7)
antineoplastic drug (22.7)
apoenzyme (22.2)
chemotherapy (22.7)
coenzyme (22.2)
cofactor (22.2)
competitive inhibitor (22.6)
enzyme (page 599)
feedback inhibition (22.6)
holoenzyme (22.2)
hydrolase (22.1)
induced-fit theory (22.3)
irreversible inhibitor (22.6)
isomerase (22.1)
ligase (22.1)
lock-and-key theory (22.3)
lyase (22.1)
noncompetitive inhibitor (22.6)
optimum pH (22.5)
oxidoreductase (22.1)
proenzyme (22.2)
reversible inhibitor (22.6)
substrate (page 599)
transferase (22.1)
turnover number (22.3)

Problems

Classification and Characteristics of Enzymes

1. To which of the six major types of enzymes does each of the following belong?
   a. decarboxylase **b.** peptidase **c.** transaminase
2. To which of the six major types of enzymes does each of the following belong?
   a. dehydrogenase **b.** kinase **c.** lipase
3. **a.** What enzyme is involved in the conversion of lactose to galactose and glucose?
   b. To what class of enzymes does it belong?
4. Commercially, glucose (in corn syrup) is converted to fructose (to form high-fructose corn syrup) by the use of an enzyme. What type of enzyme is involved? (Recall that both glucose and fructose have the formula $C_6H_{12}O_6$.)
5. Alcohol dehydrogenase catalyzes the conversion of ethanol to acetaldehyde. The active enzyme consists of a protein molecule and a zinc ion.
   a. Identify the substrate, the cofactor, and the apoenzyme.
   b. Can the zinc ion be a coenzyme? Explain.
6. Succinate dehydrogenase is active only in combination with a nonprotein organic molecule called flavin adenine dinucleotide (FAD). Is FAD a cofactor? Is it a coenzyme?
7. Why does the body synthesize trypsin, a peptidase, as the proenzyme trypsinogen rather than make the enzyme directly?

Mode of Enzyme Action and Specificity

8. **a.** In what ways are enzymes similar to ordinary chemical catalysts?
   b. How are they different?
   c. What is the effect of an enzyme on the overall energy change of a reaction?
9. How can a relatively few enzyme molecules catalyze the conversion of many substrate molecules to product?
10. Which type of interactions (i.e., ionic bonding, hydrogen bonding, dispersion forces, etc.) at the active site of an enzyme would bind each of the following groups on a substrate?
    a. —COOH **b.** $—COO^-$ **c.** $—NH_2$
    d. $—NH_3^+$ **e.** —OH **f.** —SH
    g. $—CH(CH_3)_2$ **h.** $—PO_4^{2-}$ **i.** $—C_6H_5$ (phenyl)
11. For each group shown in Problem 10, suggest an amino acid that could be in the active site of the enzyme to form the type of interaction needed.
12. Why should enzymes consist of 100 or more amino acid units when only a few amino acid side chains are involved in the active site?
13. What does the *turnover number* tell you about an enzyme?

Factors That Influence Enzyme Activity

14. In non-enzyme catalyzed reactions, the reaction rate increases as the concentration of reactant is increased. In an enzyme-catalyzed reaction, the reaction rates initially increase as the substrate concentration is increased but then begin to level off—that is, the increase in reaction rate becomes less and less as the substrate concentration increases. Explain this difference.

15. Why do enzymes become inactive at very low and at very high temperatures?

16. A patient has a fever of 40 °C. Would you expect the activity of enzymes in the body to increase or decrease as compared to their activity at normal body temperature (37 °C)?

17. An enzyme has an optimum pH of 7.4. What is most likely to happen to the activity of the enzyme if the pH drops to 6.3? Explain.

18. An enzyme has an optimum pH of 7.2. What is most likely to happen to the activity of the enzyme if the pH rises to 8.5? Explain.

Enzyme Inhibition and Chemotherapy

19. Experimentally, how could you distinguish a competitive inhibitor from a noncompetitive inhibitor?

20. Why is cyanide lethal at low concentrations?

21. Describe two ways to get around the problem of bacterial resistance to antibiotics like the penicillins.

22. Oxaloacetate ($^{-}OOCCH_2COCOO^{-}$) inhibits succinate dehydrogenase. Would you expect oxaloacetate to be a competitive or noncompetitive inhibitor? Explain.

Additional Problems

23. How would you classify the enzymes that catalyze each of the reactions shown below (oxidoreductase, transferase, etc.)?

a. The enzyme that catalyzes the conversion of ethylene glycol to oxalic acid?

b. The enzyme that catalyzes the breakdown of glucose-6-phosphate to glucose and inorganic phosphate ions?

c. The enzyme that catalyzes the removal of a carboxyl group from pyruvate to form acetate?

CH_3–C(=O)–C(=O)O^- Pyruvate

CH_3–C(=O)–O^- Acetate

24. Acetylcholinesterase has an aspartic acid and a histidine in the active site. At low pH, the enzyme is inactive, but activity increases as the pH rises. Explain.

25. The activity of a purified enzyme is measured at a substrate concentration of 1.0 μM and found to convert 49 μmol of substrate to product in one minute. The activity is measured at 2.0 μM substrate and found to convert 98 μmol of substrate to product/minute.

a. At a substrate concentration of 100 μM, how much substrate would you predict is converted to product in one minute? What if the substrate concentration were increased to 1000 μM (1.0 mM)?

b. The activities actually measured are 676 μmol product formed/minute at a substrate concentration of 100 μM and 698 μmol product formed/minute at 1000 μM (1.0 mM) substrate. Is there any discrepancy between these values and those you predicted in **(a)**? Explain.

26. Using your knowledge of factors that influence enzyme activity, what happens when milk is pasteurized?

27. Alcohol dehydrogenase is involved in the oxidation of both methanol and ethanol. How does ethanol work as an antidote for methanol poisoning?

28. Explain why antimetabolites and antibiotics can both be classified as antiseptics.

29. Acetylcholine is an important neurotransmitter. Describe the sequence of events following the release of acetylcholine into a synapse.

30. When the nerve gas sarin is inhaled there is an immediate increase in the levels of acetylcholine. Why?

31. Following a heart attack, the concentrations of what enzymes rapidly increase in the serum and indicate that the heart has been damaged?

SELECTED TOPIC 

Vitamins

Carbohydrates, fats, and proteins are the three major classes of foods. To remain healthy, we must take in relatively large amounts of these substances. They are not, however, the only nutrients we require. Some of our needs are satisfied only by vitamins and minerals. The minerals, inorganic ions of critical importance to our health, were discussed as a group in Section 11.8. We consider vitamins in this selected topic.

F.1 What Are Vitamins?

Like the hormones, **vitamins** are organic compounds that are essential in very small amounts for the maintenance of normal metabolism. Unlike hormones, vitamins generally cannot be synthesized at adequate levels by the body and must be obtained from the diet. The absence or shortage of a vitamin may result in a vitamin-deficiency disease. In the first half of the twentieth century a major focus of biochemistry was the identification, isolation, and characterization of vitamins.

Despite accumulating evidence that people needed more than just carbohydrate, fat, and protein in their diets for normal growth and health, it wasn't until the early 1900s that research established the need for trace nutrients in the diet. In 1897, Christiaan Eijkman, a Dutch physician stationed in Java, noted that chickens developed a nerve disorder that resembled beriberi, a disease common at that time in the Far East. (The symptoms of beriberi include damage to the cardiovascular and nervous systems.) He conducted feeding experiments and noted that symptoms of the nerve disorder appeared when the chickens were fed polished rice, but not when they were fed unpolished or whole-grain rice. He also noted that prisoners who were fed polished rice had a much higher incidence of beriberi than those who ate brown rice. He concluded that lack of "something" that was present in the hulls of whole-grain rice caused the disease beriberi.

Frederick G. Hopkins, a British scientist, and his colleagues fed mice a synthetic diet of carbohydrate, minerals, protein, and fat. The survival time for these mice was 28 days, suggesting that this diet was inadequate. Addition of a small amount of milk led to normal growth and survival times. In 1906 Hopkins stated that synthetic diets were inadequate because natural foods contained other substances vital for health. He indicated that rickets and scurvy were diseases that might be prevented or cured by proper diets.

By comparing the isolated proteins of polished rice and rice hulls, Casimir Funk, a Polish biochemist, demonstrated that beriberi was not due to an amino acid deficiency, as some researchers thought. He eventually isolated organic bases that were vital for good nutrition and, in 1912, he coined the word *vitamine* (from the Latin *vita,* life) for these missing factors, thinking that they all contained an amino group. The final *e* was dropped after it was found that not all the factors were amines, giving us the word *vitamin.*

Because organisms differ in their synthetic abilities, a substance that is a vitamin for one species may not be so for another. Over the years scientists have identified and isolated 13 vitamins required in the human diet.

Vitamins are divided into two broad categories, the **fat-soluble vitamins,** which include vitamins A, D, E, and K, and the **water-soluble vitamins,** made up of the B complex vitamins and vitamin C. All fat-soluble vitamins contain a high proportion of hydrocarbon structural elements. There are one or two oxygen atoms present, but the compounds as a whole are nonpolar. In contrast, a water-soluble vitamin contains a high proportion of the electronegative atoms oxygen and nitrogen, which can form hydrogen bonds to water. The vitamin content of food can deteriorate when the food is cooked in water and then drained because the water-soluble vitamins go down the drain with the water. Most water-soluble vitamins act as coenzymes or are required for the synthesis of coenzymes (see Section F.6). The fat-soluble vitamins are important for a variety of physiological functions.

Fat-soluble vitamins, if taken in high doses, can accumulate to toxic levels because they are stored in body fat. The body does not store appreciable amounts of the water-soluble vitamins, but excretes anything that cannot be immediately used. Thus water-soluble vitamins must be taken at frequent intervals, whereas a single dose of a fat-soluble vitamin can be used by the body over several weeks.

Minimum daily requirements (MDRs) of the vitamins have been set by examining the levels below which deficiency diseases occur. General warning signs of vitamin deficiency include slow healing of wounds, tiredness, and frequent illness. There is no agreement, however, on the optimum levels of dietary vitamins. Since 1941 the Food and Nutrition Board of the National Academy of Sciences has established recommended dietary allowances (RDAs) for each vitamin (Table F.1). This is an estimate of the daily amount needed by the average healthy person to maintain good nutrition and health.

The RDAs are now being replaced with **dietary reference intakes** (DRIs), a term that encompasses four categories of reference intakes of vitamins, minerals, and other food components important in diet and nutrition. These four categories are: (1) estimated average requirements (EARs)—the amount of each nutrient estimated to meet the requirement for that nutrient in 50% of a specified age and gender group; (2) recommended dietary allowances (RDAs)—the amount of each nutrient estimated to meet the nutrient requirements of nearly all individuals in a given group; (3) adequate intakes (AIs)—these are used in place of RDAs when insufficient research data or uncertainty in the data makes it difficult to establish an EAR; and (4) tolerable upper intake levels (ULs)—the maximum level of daily nutrient intake that is unlikely to pose adverse health risks in a given group. The dietary reference intakes for seven nutrient groups are being developed by committees of scientists from the U.S. and Canada. These nutrient groups are: (1) calcium, vitamin D, phosphorus, magnesium, and fluoride; (2) folate and other

Table F.1 Recommended Dietary Allowances (RDA)

AGE (YR)	Weight (kg)	Weight (lb)	Height (cm)	Height (inches)	(μg RE) VITAMIN A	(μg) VITAMIN D	(mg) VITAMIN E	(mg) Vitamin K	(mg) VITAMIN C	(mg) THIAMIN	(mg) RIBOFLAVIN	(mg) NIACIN	(mg) VITAMIN B_6	(μg) FOLATE	(μg) VITAMIN B_{12}
Infants															
0.0–0.5	6	13	60	24	375	7.5	3	5	30	0.3	0.4	5	0.3	25	0.3
0.5–1.0	9	20	71	28	375	10	4	10	35	0.4	0.5	6	0.6	35	0.5
Children															
1–3	13	29	90	35	400	10	6	15	40	0.7	0.8	9	1.0	50	0.7
4–6	20	44	112	44	500	10	7	20	45	0.9	1.1	12	1.1	75	1.0
7–10	28	62	132	52	700	10	7	30	45	1.0	1.2	13	1.4	100	1.4
Males															
11–14	45	99	157	62	1000	10	10	45	50	1.3	15.	17	1.7	150	2.0
15–18	66	145	176	69	1000	10	10	65	60	1.5	1.8	20	2.0	200	2.0
19–24	72	160	177	70	1000	10	10	70	60	1.5	1.7	19	2.0	200	2.0
25–50	79	174	176	70	1000	5	10	80	60	1.	1.7	19	2.0	200	2.0
51+	77	170	173	68	1000	5	10	80	60	1.2	1.4	15	2.0	200	2.0
Females															
11–14	46	101	157	62	800	10	8	45	50	1.1	1.3	15	1.4	150	2.0
15–18	55	120	163	64	800	10	8	55	60	1.1	1.3	15	1.5	180	2.0
19–24	58	128	164	65	800	10	8	60	60	1.1	1.3	15	1.6	180	2.0
25–50	63	138	163	64	800	5	8	65	60	1.1	1.3	15	1.6	180	2.0
51+	65	143	160	63	800	5	8	65	60	1.0	1.2	13	1.6	180	2.0
Pregnant					800	10	10	65	70	1.5	1.6	17	2.2	400	2.2
Lactating															
1st 6 mo.					1300	10	12	65	95	1.6	1.8	20	2.1	280	2.6
2nd 6 mo.					1200	10	11	65	90	1.6	1.7	20	2.1	260	2.6

Source: Reprinted with permission from *Recommended Dietary Allowances*, 10th edition © 1989 by the National Academy of Sciences. Courtesy of the National Academy Press, Washington, D.C.

(a)

Cleavage at this point can yield two molecules of vitamin A*

Beta-carotene, a precursor

(b)

(c)

*Sometimes cleavage occurs at other points as well, so that one molecule of beta-carotene may yield only one molecule of vitamin A. Futhermore, not all beta-carotene is converted to vitamin A, and absorption of beta-carotene is not as efficient as vitamin A. For these reasons, 6 μg of beta-carotene are equivalent to 1 μg of vitamin A. Conversion of other carotenoids to vitamin A is even less efficient.

Figure F.1
(a) 11-*trans*-retinol. (b) β-carotene is a precursor of vitamin A found in many plants. (c) A sampling of foods rich in vitamin A or β-carotene.

B vitamins; (3) antioxidants; (4) macronutrients (protein, fat, and carbohydrates); (5) trace elements (minerals); (6) electrolytes and water; and (7) other food components, such as fiber. For each nutrient group a book is to be published that presents what is known about the function of each nutrient in the human body, the methods used to determine how much is required, what factors, such as exercise, may affect how the nutrient works, and how it may be implicated in any chronic disease. Books for the first two nutrient groups have been published by the National Academy of Sciences, but work continues on establishing DRIs for the other groups.

Vitamins have been the subject of more fads and more misrepresentations than any other group of nutrients. Among the claims made are that vitamins cure cancer, arthritis, and mental illness; increase sexual potency; prevent colds; and overcome muscular weakness. It is small wonder that the public is baffled by such exaggerated claims.

F.2 Vitamin A

Several compounds exhibit vitamin A activity, but the parent compound is 11-*trans*-retinol (Figure F.1a). Vitamin A occurs *only* in the animal kingdom. It was first isolated from halibut oil and is present in high concentration in fish liver oils. Liver, eggs, butter, and cheese are also good sources. However, the plant pigment β-carotene is a precursor substance or **provitamin** that can be converted to vitamin A by animals, and thus, most green and yellow vegetables (carrots, lettuce, spinach, yams) are good sources of the vitamin (Figure F.1b). β-Carotene is not always utilized in the body as a provitamin. It also acts as an antioxidant, a function that will be discussed in Section F.8.

The recommended daily allowance of vitamin A for adults is 800–1000 retinol equivalents (RE).[1] For adults, ingesting vitamin A at >100 times the recommended level produces acute toxicity, early signs of which include nausea, vomiting, headache, and blurred vision. Abortions and some birth defects have also been linked to toxic levels of vitamin A intake. It is interesting to note that polar bear liver has so much vitamin A that it is toxic. These large animals eat seals that eat fish. Because vitamin A is fat-soluble, it is concentrated in each step of the food chain. Eskimos who kill a polar bear know better than to eat its liver. β-carotene, however, is not known to be toxic even when eaten in large quantities because as the level in the diet increases, the efficiency of its conversion to vitamin A decreases.

Adults are able to store enough vitamin A, primarily in the liver, to last for several months. On the other hand, infants and children do not store much of the vitamin and consequently are more likely to develop vitamin A deficiencies.

In Selected Topic D, we discussed the well-known role of vitamin A in vision. In addition to its direct role in the chemistry of vision, vitamin A also stimulates fluid secretion by the epithelial cells of the eye and helps maintain its mucous membranes. Thus, if the dietary supply of vitamin A is inadequate, the cornea of the eye becomes dried, or keratinized. Mucous membrane may harden, dry, and crack. The cornea then becomes extremely vulnerable, and even the slightest nick or scratch may cause it to perforate, which leads to blindness. One of the earliest manifestations of vitamin A deficiency is a loss of night vision. In cases of severe deprivation, victims may exhibit *xerophthalmia,* a condition characterized by inflammation of the eyes and eyelids, leading ultimately to infection and blindness. It is estimated that worldwide 500,000 preschool-age children become blind each year due to vitamin A deficiency.

Vitamin A promotes the differentiation (maturation) of epithelial cells that form the mucous membranes—the linings of the mouth, stomach, intestines, lungs, and other organs. These cells secrete mucus, which coats and protects the body from invasive microorganisms and other harmful substances. Vitamin A also supports growth by participating in the "remodeling" of smaller bones into larger bones as growth occurs. There is also increasing evidence that vitamin A plays a direct role in improving immunity to disease.

Some acne treatments utilize derivatives of vitamin A, such as the oral medication Accutane® (13-*cis*-retinoic acid) or the ointment RETIN-A® (all-*trans*-retinoic acid). Accutane® has been linked with birth defects in infants whose mothers used it during pregnancy. Thus women must use an effective form of birth control from at least one month before beginning the drug until one month after discontinuing its use. It is not clear what the effects are for the long term use of RETIN-A®.

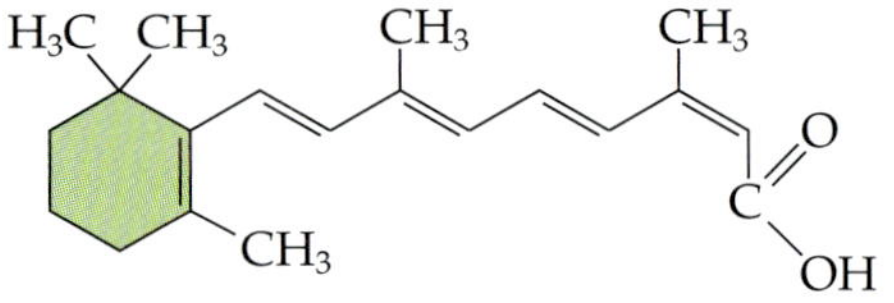

H_3C CH_3 CH_3 CH_3 CH_3 C O OH

Accutane®
(13-*cis*-retinoic acid)

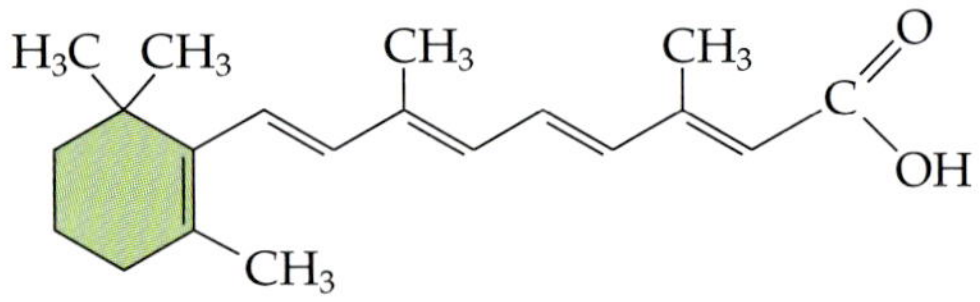

H_3C CH_3 CH_3 CH_3 CH_3 C O OH

RETIN-A®
(all-*trans*-retinoic acid)

F.3 Vitamin D

Several chemical compounds have vitamin D activity. Only two commonly occur in foods or are used in drugs and food supplements. Each is formed from a precursor by the action of ultraviolet light (Figure F.2). Vitamin D_2 (ergocalciferol) is synthesized by irradiation of ergosterol, a compound found in yeast and other molds. Vitamin D_3 (cholecalciferol) is formed in the skin of animals by the action of sunlight on 7-dehydrocholesterol. The two vitamins differ only in the structure of their side chains; D_2 contains an extra carbon and a double bond.[2]

Vitamin D increases the body's ability to absorb calcium and phosphorus. Individuals deficient in vitamin D may develop osteomalacia (softening of the bones), a condition in which the loss of calcium causes the bones to soften and become flexible. Deformities can result. When osteomalacia occurs in infants and growing children, it is known as rickets. Rickets is characterized by bowed legs, knobby bone growths where the ribs join the breastbone (called a "rachitic rosary"), pigeon breast (projection of the chest), and poor tooth development.

Vitamin D is the "sunshine vitamin." Individuals with a reasonable proportion of their skin exposed to sunlight rarely suffer from a vitamin D deficiency. Most foods contain little or no vitamin D. The best natural sources are fish liver oils and egg yolks. Irradiated ergosterol

[1]A retinol equivalent is defined as 1 μg retinol or 6 μg β-carotene.

[2]There is no vitamin D_1. The material that was originally given this designation proved to be a mixture of vitamins D_2 and D_3.

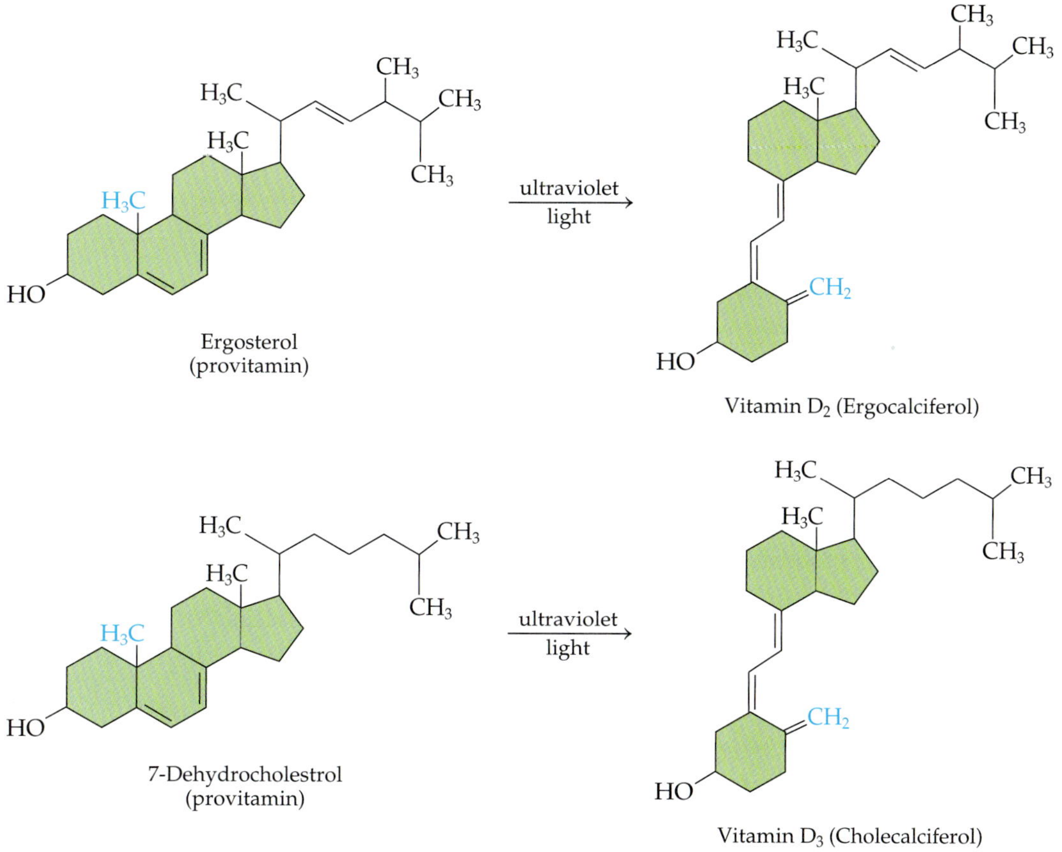

ultraviolet light

Ergosterol (provitamin)

Vitamin D_2 (Ergocalciferol)

ultraviolet light

7-Dehydrocholestrol (provitamin)

Vitamin D_3 (Cholecalciferol)

◀ Figure F.2 Formation of two forms of vitamin D by action of ultraviolet light on provitamins.

(from yeast) is added to milk (10 μg/quart) and margarine as a supplemental source of vitamin D. It is recommended that children receive about 10 μg of vitamin D in their daily diet.

Excess amounts of vitamin D, like vitamin A, are stored in body fat. Too much vitamin D can cause pain in the bones, nausea, diarrhea, kidney stones, and eventual death. Bonelike material may be deposited in kidney tubules, in blood vessels, in heart, stomach, and tissue, and in joints.

F.4 Vitamin E

As with vitamins A and D, there are several compounds that have vitamin E activity. The compounds are called tocopherols; the most potent of these is α-tocopherol:

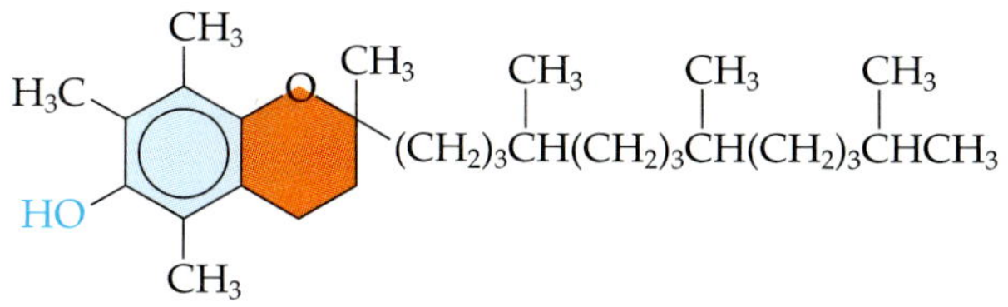

$(CH_2)_3CH(CH_3)(CH_2)_3CH(CH_3)(CH_2)_3CH(CH_3)CH_3$

α-Tocopherol

Vitamin E was first identified after studies indicated a component of vegetable oils was necessary for reproduction in rats. Because of this, vitamin E is sometimes called the antisterility vitamin.

Vitamin E is a potent fat-soluble antioxidant that can prevent oxidative damage of vulnerable lipids and other components of cell membranes. Recent evidence shows that an *oxidized* form of LDL cholesterol is deposited in arteries (Section 20.7). Vitamin E acts to prevent the oxidation of cholesterol; thus it seems to have considerable value in maintaining the cardiovascular system. It is generally believed that much of the physiological damage from aging is due to oxidative processes.

Vitamin E is available in wheat germ oil, green vegetables, vegetable oil, egg yolks, and meat. Vitamin E deficiency is rare and is usually found in individuals who have been placed on a no-fat diet or who have a disease in which fat is poorly absorbed, such as cystic fibrosis. Large doses of vitamin E do not seem to have harmful effects, although extremely high doses may interfere with the blood-clotting action of vitamin K.

F.5 Vitamin K

Vitamin K has a fused ring system related to the structure of naphthalene (Section 13.13). One of the rings contains two carbonyl groups, an arrangement that has the special name *quinone*. Attached to the quinone ring are alkyl groups. One of these is usually methyl, while the other contains 20 or more carbon atoms. Many compounds have vitamin K activity; the structure of one of them is:

Vitamin K_1 (Phylloquinone)

Vitamin K is necessary for the formation of prothrombin (Section 28.4), one of the proenzymes involved in blood clotting. (The vitamin got its name from the Danish word *koagulation*.) A vitamin K deficiency increases the time required for blood to clot. Symptoms are bleeding under the skin and in muscles, leading to ugly "bruises" from what would otherwise be minor blows. Infants lacking in vitamin K may die from hemorrhaging in the brain. Increased vitamin K intake by pregnant women has lowered the incidence of this syndrome in newborn infants. Good sources of vitamin K are spinach and other green leafy vegetables. Vitamin K deficiencies in humans are rare because intestinal bacteria usually synthesize as much as the body requires. Prolonged treatment with antibiotics has the adverse effect of killing these bacteria, and the body's supply of vitamin K is temporarily reduced.

F.6 The B Complex

In Chapter 22 we noted that some enzymes require coenzymes in order to function properly. Most coenzymes are synthesized from one of the B vitamins. There is no single vitamin B. What was once called vitamin B has since been recognized as a complicated mixture of factors. The term **B complex** is now used to designate a group of water-soluble vitamins found together in many food sources. There appears to be no toxicity connected with the B vitamins, with the possible exception of vitamin B_6, which apparently can cause neurological damage in some people if taken in extremely large daily doses. As a group, the B vitamins are important for maintaining the skin and the nervous system.

Thiamine (Vitamin B_1)

Thiamine is necessary for the normal metabolism of carbohydrates. In the body, a pyrophosphate group is attached to form the coenzyme thiamine pyrophosphate. This coenzyme is involved in the decarboxylation of pyruvate to acetyl-CoA (Section 25.2) and α-ketoglutarate to succinyl-CoA (Section 24.4). The coenzyme is also involved in the synthesis of ribose, a component of nucleotides and nucleic acids.

Thiamine (B_1)

Thiamine pyrophosphate (TPP)

As mentioned earlier, a deficiency of thiamine in the diet leads to *beriberi*, characterized by deterioration of the nervous system. This disease is a serious health problem in the Far East because rice, the major food in the region, has a relatively low thiamine content. A severe form of beriberi can also occur in infants of nursing mothers whose diets are deficient in thiamine. Alcoholism is the most common cause of thiamine deficiency (as well as deficiencies of other B vitamins) in the United States because alcohol, the major caloric contributor to an alcoholic's diet, is a poor source of vitamins.

Synthetic vitamin B_1 is added to enrich the vitamin content of bread and flour. Thiamine is not stored in the body to any significant degree; excesses are excreted

▲ These foods are generally good sources of the B complex vitamins.

in the urine. The vitamin is destroyed in foods that are cooked for prolonged periods at temperatures over 100 °C.

Riboflavin (Vitamin B_2)

Riboflavin, a yellow fluorescent solid at room temperature, is essential for the formation of the coenzymes flavin adenine dinucleotide (FAD) and flavin mononucleotide (FMN) (Figure F.3). These coenzymes function in a variety of oxidation-reduction reactions. No specific deficiency disease is associated with a lack of riboflavin, but well-defined symptoms of deficiency include dermatitis (skin inflammation), glossitis (tongue inflammation), burning and itching of the eyes, and anemia. Riboflavin is stable at ordinary cooking temperatures, but is destroyed by light. The addition of baking soda when cooking green vegetables to make them appear fresher will accelerate the degradation of riboflavin by light.

Niacin (Vitamin B_3)

Niacin is a general term for two compounds—nicotinic acid and its amide (Figure F.4a). Nicotinamide serves as a component of the coenzymes nicotinamide adenine dinucleotide (NAD^+) and nicotinamide adenine dinucleotide phosphate ($NADP^+$) (Figure F.4b). These coenzymes are utilized by a wide variety of enzymes that catalyze oxidation-reduction reactions.

Niacin is readily obtained from the diet, but the body can also synthesize it from tryptophan. The synthesis of 1 mg of niacin requires approximately 60 mg of tryptophan. The vitamin is best known for its ability to prevent *pellagra,* a disease characterized by the loss of appetite, followed by diarrhea, dermatitis, mental disorders, and possible death in severe cases. In the early 1900s pellagra was prevalent in the southern United States and directly associated with a diet that relied heavily on corn as a staple. The niacin in corn is tightly bound to macromolecules and not readily absorbed during digestion. In addition, corn proteins are low in tryptophan.

Clinical studies have shown that large doses of nicotinic acid (more than ten times the RDA) can lower total and LDL cholesterol levels, while increasing HDL cholesterol levels. However, administration of these doses of nicotinic acid must be closely monitored due to side effects that include fainting, dizziness, headache, upset stomach, and flushing of the skin. This latter effect is due to a dilation of the capillaries that also causes a sometimes painful tingling sensation. Nicotinamide has not been shown to be effective in lowering cholesterol levels; however, clinical trials are

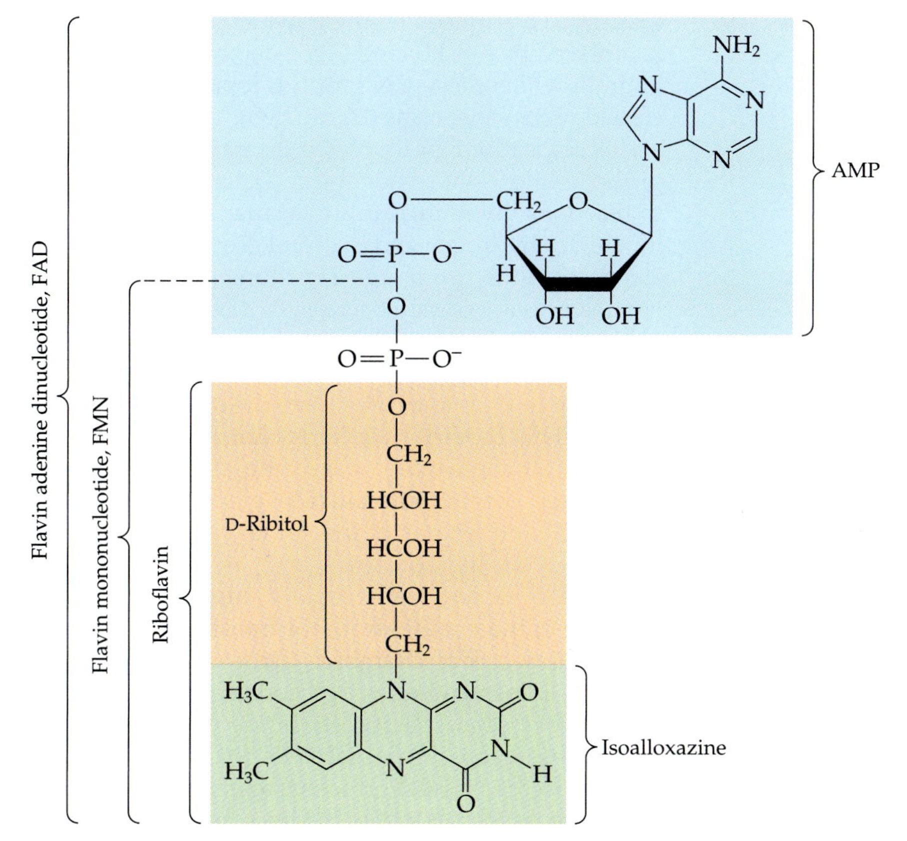

Figure F.3 The structures of riboflavin, flavin mononucleotide (FMN), and flavin adenine dinucleotide (FAD).

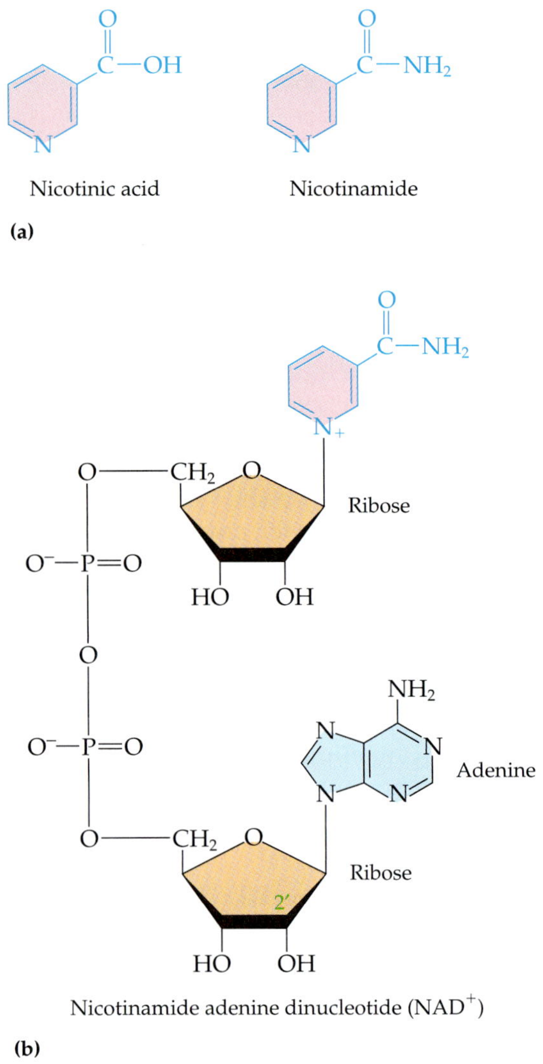

Nicotinic acid

Nicotinamide

(a)

Ribose

Adenine

Ribose

Nicotinamide adenine dinucleotide (NAD^+)

(b)

▲ Figure F.4
(a) Nicotinic acid and nicotinamide. (b) Nicotinamide adenine dinucleotide (NAD^+). When there is an additional phosphate group on the 2′-hydroxyl group of the ribose moiety, the coenzyme is named nicotinamide adenine dinucleotide phosphate ($NADP^+$).

currently studying its effects in the possible prevention and/or control of insulin-dependent diabetes.

Good sources of niacin include yeast, meats, cereals, legumes, seeds, milk, eggs, and nuts. The vitamin is not destroyed by cooking, but will dissolve in cooking water.

Vitamin B_6

Vitamin B_6, also known as pyridoxine, occurs naturally in three forms (Figure F.5a) and is found in a variety of foods including green and leafy vegetables, meats, legumes, fruits, and whole grains. The coenzyme pyridoxal phosphate (Figure F.5b) is required for a wide variety of metabolic reactions including the interconversion of amino acids and α-keto acids (Section 27.2). The vitamin B_6 content of food decreases when it is heated.

There is no deficiency disease specifically associated with a lack of vitamin B_6. Most vitamin B_6 deficiencies occur along with a deficiency of other nutrients, particularly other water-soluble vitamins. Some of the symptoms are similar to those for riboflavin deficiency and include dermatitis, glossitis, irritability, depression, and confusion. Until 1983 it was thought that, like other water-soluble vitamins, vitamin B_6 would not reach toxic levels in the body. Then a report described the cases of seven women who had been taking more than 2 g of vitamin B_6 daily (RDA for women is < 2 mg!) for at least two months and showed signs of irreversible nerve damage leading to numbness and muscle weakness. Thus, doses over 500 mg are not recommended!

A number of claims have promoted vitamin B_6 for premenstrual syndrome, carpal tunnel syndrome, Down's syndrome, and gestational diabetes. None of these claims have been supported by clearly defined clinical trials that included rigorous controls.

Cyanocobalamin (Vitamin B_{12})

Vitamin B_{12} is a complex cobalt-containing structure that has similarities to the heme group of hemoglobin (Figure F.6a). The vitamin is converted to two coenzymes: (1) methylcobalamin in which the —CN group is replaced by —CH_3; and (2) 5′-deoxyadenosylcobalamin in which the —CN group is replaced by the nucleoside deoxyadenosine (Figure F.6b). Methylcobalamin acts in some reactions involving the transfer of a methyl group, while 5′-deoxyadenosylcobalamin participates in reactions involving intramolecular rearrangements in which a hydrogen and a second group, bound to adjacent carbons, exchange positions. Vitamin B_{12} is formed only by certain bacteria that live in a symbiotic relationship with their hosts. It is stored in various tissues, particularly the liver.

Vitamin B_{12} is associated with the disease *pernicious anemia*. This dietary disease is characterized by the presence of abnormally large, immature, fragile red blood cells. It is accompanied by gastrointestinal disturbances and lesions of the spinal cord with loss of muscular coordination.

Pernicious anemia is usually caused by poor absorption of the vitamin from the intestinal tract rather than by any lack of vitamin B_{12}. Normally, cells of the stomach lining synthesize a glycoprotein, called the *intrinsic factor,* that specifically binds vitamin B_{12} and transports it into intestinal cells for its subsequent transfer to the blood. Pernicious anemia patients lack or have a deficiency of the intrinsic factor and cannot absorb the ingested vitamin B_{12}. Elderly people often

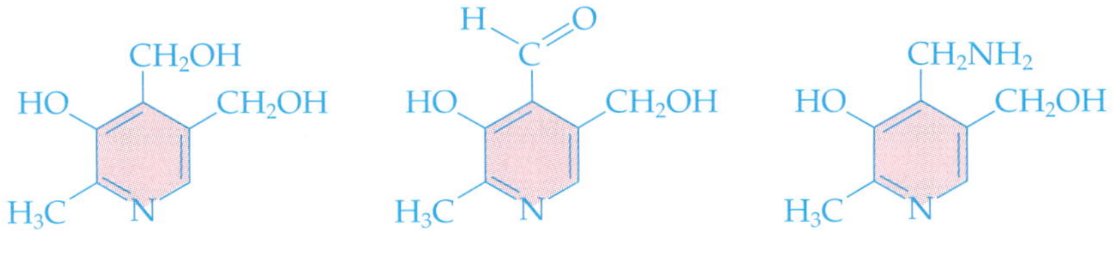

Pyridoxine

Pyridoxal

Pyridoxamine

(a)

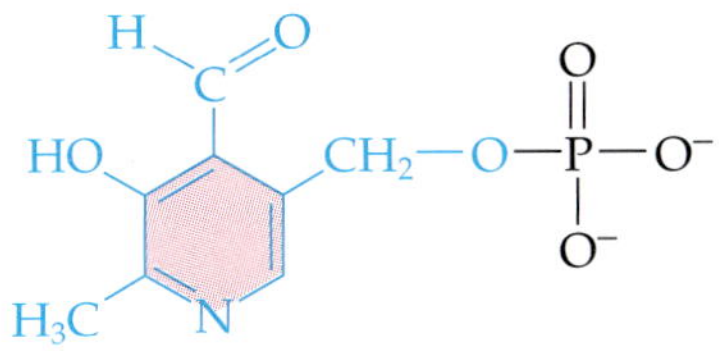

Pyridoxal phosphate

(b)

Figure F.5
(a) Vitamin B_6 exists in three forms: pyridoxine, pyridoxal, and pyridoxamine. (b) Pyridoxal phosphate.

have a decreased synthesis of intrinsic factor and must receive vitamin B_{12} by injection directly into the bloodstream in order to avoid anemia. Because plants do not contain vitamin B_{12}, pernicious anemia symptoms are sometimes observed among strict vegetarians. Vitamin B_{12} is stable during most cooking procedures.

Biotin

Biotin is widely distributed in mammalian tissues. It functions as a coenzyme in carboxylation reactions. It is a carbon carrier in both carbohydrate and lipid metabolism. Because biotin is synthesized by intestinal microorganisms in large quantities, biotin deficiency seldom occurs in humans. Deficiency can be produced, however, by antibiotics that inhibit the growth of intestinal bacteria. Also, raw egg white contains a protein, avidin, that binds biotin and prevents its absorption from the intestinal tract. An artificially produced deficiency of biotin in humans causes dermatitis, anorexia, nausea, muscle pains, and depression. Biotin is stable at normal cooking temperatures.

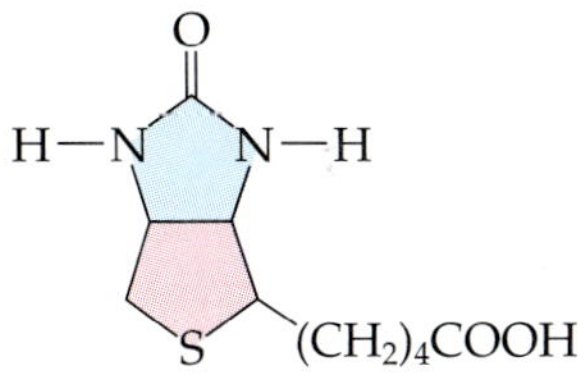

Biotin

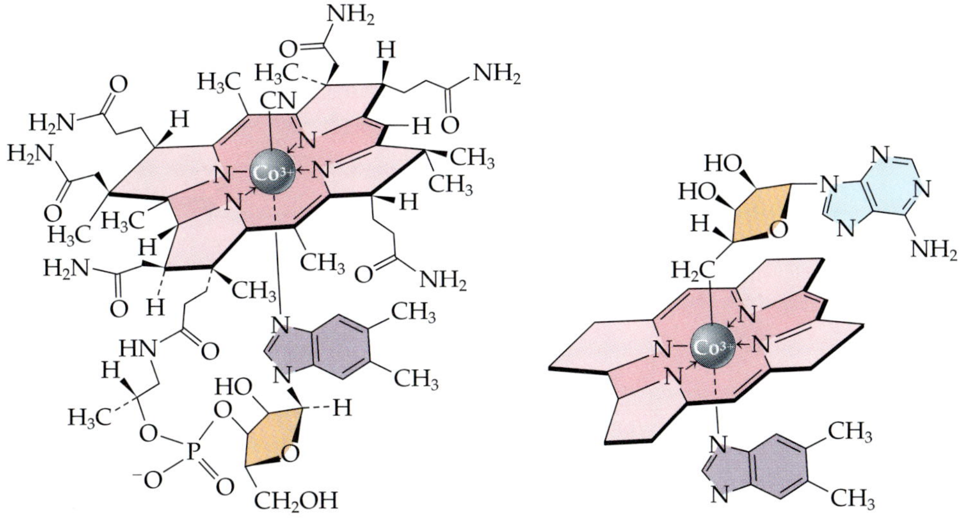

Figure F.6
(a) Cyanocobalamin. (b) Simplified structure of 5′-deoxyadenosylcobalamin.

(a) Cyanocobalamin

(b) 5′-Deoxyadenosylcobalamin

Folic Acid

Folic acid (Figure F.7a) was first discovered in green leafy vegetables, and its name is derived from the Latin word for leaf, *folium.* It is critically important in preventing both spina bifida and anencephaly (birth defects of the neural tube). Folic acid is reduced to its coenzyme, tetrahydrofolic acid (Figure F.7b), which acts as a carrier of one-carbon units (e.g., as formyl or methyl groups) in the formation of such compounds as choline, heme, and nucleic acids. Deficiency of folic acid affects purine biosynthesis, and clinical symptoms include anemia and gastrointestinal disturbances. Current research indicates that sufficient folic acid in the diet may be important in preventing cardiovascular disease.

Folic acid is synthesized by intestinal microorganisms, and it can be absorbed into the general circulation. It is readily destroyed by cooking. As we saw in Section 22.7, the sulfa drugs interfere with the bacterial synthesis of folic acid. The anticancer drug methotrexate inhibits the conversion of folic acid to tetrahydrofolic acid. Without the coenzyme, cells cannot grow because they cannot synthesize the deoxyribonucleotides needed for DNA replication.

Pantothenic Acid

Pantothenic acid (Figure F.8a) is needed for the biosynthesis of coenzyme A (Figure F.8b), which is important as a carrier of acyl groups. (The name coenzyme A resulted from its involvement in enzymatic acetylation reactions.) Pantothenic (from the Greek word meaning *from everywhere*) acid has a widespread distribution in foods, and deficiency in humans is practically unknown. Symptoms produced by experimental feeding of an antagonist include nausea, fatigue, and burning cramps in the limbs. This vitamin is stable at moderate cooking temperatures but is destroyed at high temperatures.

F.7 Vitamin C

Vitamin C is ascorbic acid, a white, crystalline solid quite soluble in water. As with many of the vitamins, the role of vitamin C in nutrition and disease prevention is still a subject of heated debate.

Scurvy, a disease in which collagen is weakened, was dreaded by seamen on long ocean voyages. In 1747, a Scottish navy surgeon, James Lind, showed that scurvy could be prevented by the inclusion of fresh fruits or vegetables in the diet. However, it was another fifty years before the British Admiralty required all

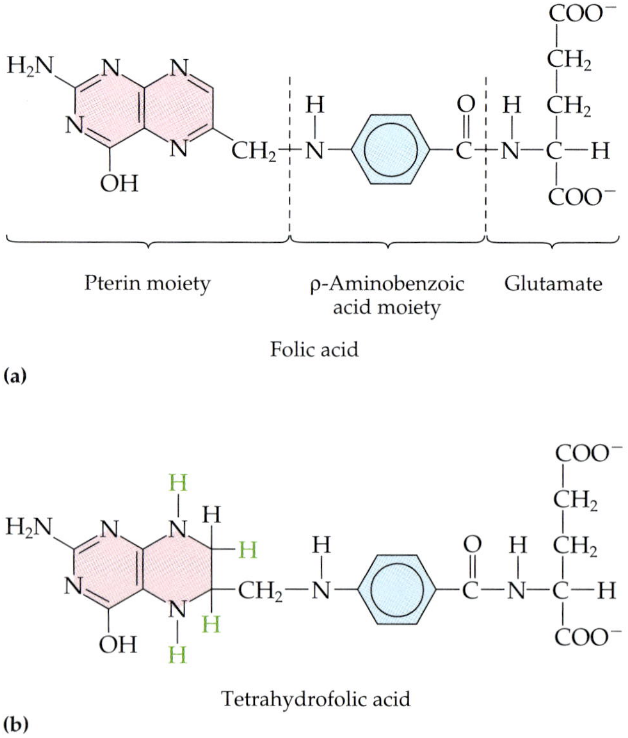

Pterin moiety

ρ-Aminobenzoic acid moiety

Glutamate

Folic acid

(a)

Tetrahydrofolic acid

(b)

▲ **Figure F.7**
(a) Folic acid. (b) Tetrahydrofolic acid.

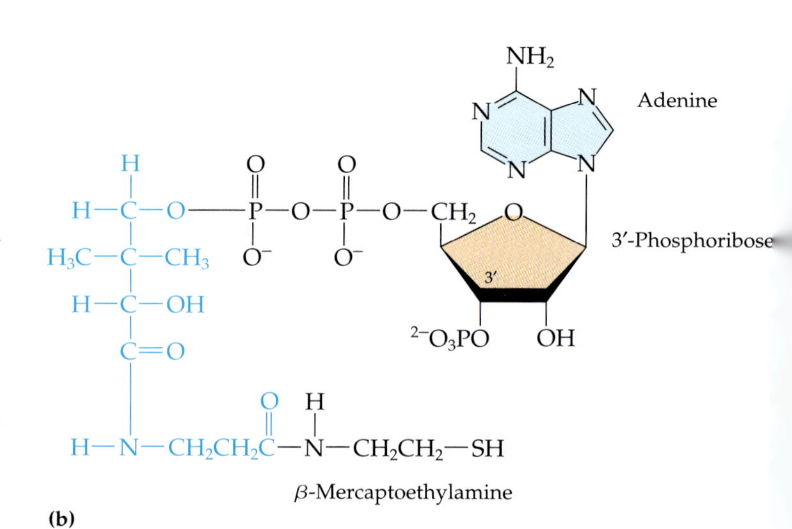

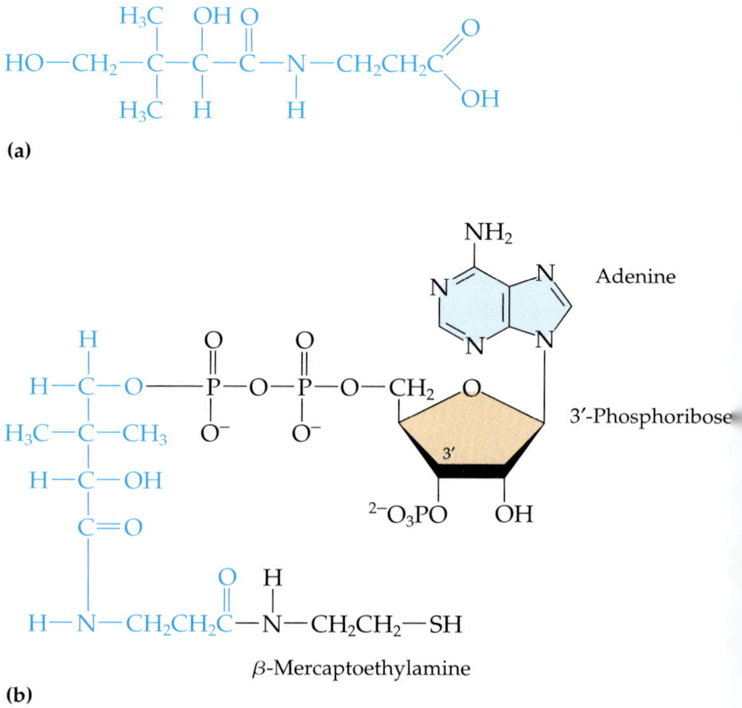

▲ **Figure F.8**
(a) Pantothenic acid. (b) Coenzyme A (CoASH).

British ships to carry barrels of limes or lime juice to prevent scurvy. British sailors came to be known as "lime eaters" or simply "limeys."

Although the cure for scurvy was known, it was not until 1932 that the actual substance responsible for preventing scurvy was isolated from citrus fruit and named ascorbic acid or vitamin C. It is widely distributed, but citrus fruits, members of the cabbage family, dark green vegetables, cantaloupe, strawberries, tomatoes, potatoes, and mangoes are excellent sources of this vitamin.

Vitamin C is a good reducing agent and thus acts as an antioxidant to help prevent damage to tissues. It reacts with oxygen and/or oxidizing agents to form dehydroascorbic acid:

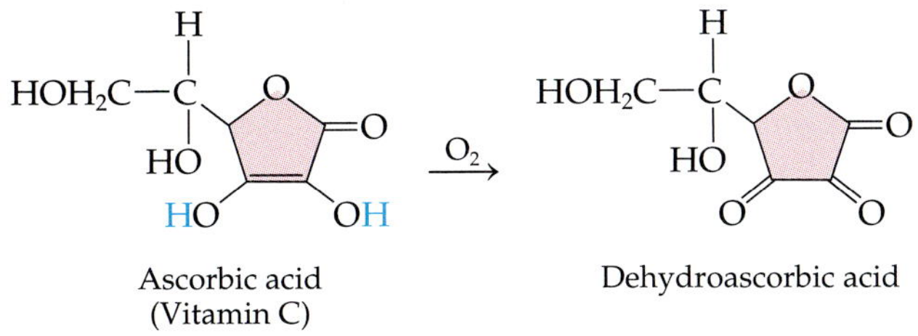

Ascorbic acid (Vitamin C) $\xrightarrow{O_2}$ Dehydroascorbic acid

Vitamin C enhances the absorption of iron from the small intestine by keeping it in a reduced form. It is also needed to help form collagen, the most important protein of connective tissues such as tendons and ligaments and the foundation or matrix on which bone and teeth are formed. Collagen contains a high proportion of proline amino acid units. After incorporation into the protein chain, some prolines are hydroxylated to hydroxyproline (Figure F.9) in a complex reaction that requires vitamin C. (Some lysines are also hydroxylated in an analogous reaction.)

F.8 Antioxidants and Disease Prevention

We have noted that β-carotene, vitamin E, and vitamin C can act as antioxidants in the body, but we haven't addressed the question of why antioxidants are needed. **Antioxidants** prevent damage from free radicals, highly reactive molecules that have unpaired electrons (Section 3.15). Free radicals are formed not only through metabolic reactions involving oxygen, but also through environmental factors such as radiation and pollution.

Free radicals are highly reactive, needing another electron to form a stable structure. They most commonly react with lipoproteins and unsaturated fatty acids in cell membranes, removing an electron and generating a new free radical. This initiates a chain reaction that leads to the oxidative degradation of these compounds. Antioxidants react with free radicals to stop these chain reactions by forming a more stable molecule or, in the case of α-tocopherol, by forming a free radical that is much less reactive. (α-Tocopherol is converted back to its original form through interaction with vitamin C.)

In addition to β-carotene and vitamins C and E, the mineral selenium is known to play an important role in the body as an antioxidant, and there is evidence that compounds known as phytochemicals also act as antioxidants. **Phytochemicals** are nonnutrient compounds found in plant-derived foods that have biological activity in the body. The exact role of phytochemicals in lowering the incidence of cancer is an area of intensive research. A summary of the food sources and actions of some selected phytochemicals is given in Table F.2.

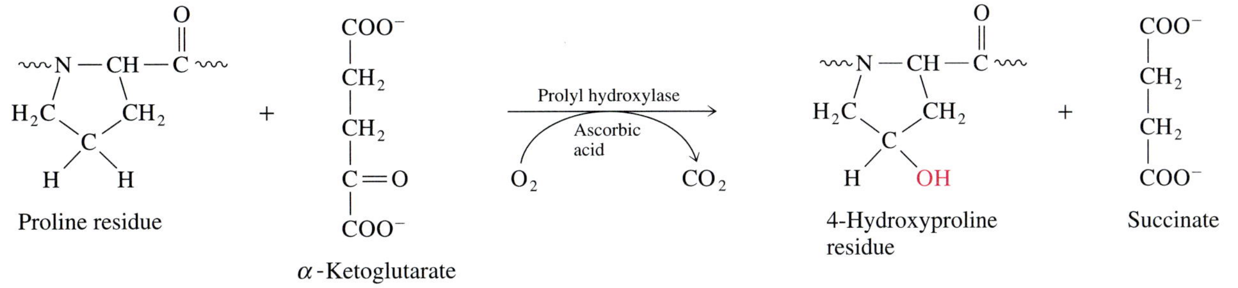

Proline residue + α-Ketoglutarate $\xrightarrow[\text{Ascorbic acid}]{\text{Prolyl hydroxylase}}$ 4-Hydroxyproline residue + Succinate (O_2 in, CO_2 out)

▲ Figure F.9
The hydroxylation of proline to form hydroxyproline is catalyzed by prolyl hydroxylase and requires molecular oxygen, α-ketoglutarate, and vitamin C.

Table F.2 Representative Phytochemicals and Their Food Sources and Actions

Name	Food Source	Action in the Body
Carotenoids (which includes β-carotene)	Darkly pigmented fruits and vegetables (carrots, tomatoes, etc.)	Act as antioxidants, reducing the risk of cancer
Limonene	Citrus fruits	Triggers enzyme production to facilitate carcinogen excretion.
Phenols	Citrus fruits	Inhibit lipid oxidation; block formation of nitrosamines in the body.
Allyl sulfides	Garlic/onions	Trigger enzyme production to facilitate carcinogen excretion
Dithiolthiones	Broccoli and other cruciferous vegetables	Trigger enzyme production to block carcinogen damage to cells' DNA
Indoles	Broccoli and other cruciferous vegetables	Trigger enzymes to inhibit estrogen action, reducing the risk of breast cancer
Phytosterols	Soy/legumes	Inhibit cell reproduction in G.I. tract, preventing colon cancer
Caffeic acid	Fruits (blueberries, prunes, grapes), oats, soybeans	Triggers enzyme production to make carcinogens water soluble, facilitating excretion

Supplements of the various antioxidants are highly touted, along with health claims regarding their probable role in the prevention of cancer and heart disease. But which are better: supplements or the natural foods or does it matter? Because research is still continuing on the effects of large doses, health-care professionals are hesitant to recommend the intake of large amounts of antioxidants. In addition, it is not always clear which chemicals have which specific effect.

This is illustrated by a study conducted in Finland that looked at the incidence of lung cancer in male smokers who received 50 mg of vitamin E (α-tocopherol), 20 mg β-carotene, both, or a placebo every day for five to eight years. This study was begun because of evidence that smokers with higher intakes of vitamin E or β-carotene in their diets seemed to have a lower incidence of lung cancer. The results of the study were unexpected. There was no reduction in the number of lung cancer cases diagnosed or in the total number of lung cancer deaths in men who took the vitamin E supplements. However, men who took β-carotene supplements had 18% more lung cancers and an 8% increase in mortality—hardly the expected resulted. But to further confuse the picture, men who had higher blood levels of β-carotene or vitamin E before the supplementation study began had fewer lung cancers regardless of which supplements they took during the study. This would indicate that there are other substances present in food that are important in disease prevention and that it is a wide variety of these antioxidants—vitamins, selenium, and phytochemicals—present in fruits, vegetables, legumes, and grains that makes a diet high in these compounds more likely to protect against the development of certain cancers.

Key Terms

antioxidant (F.8)
B complex (F.6)
dietary reference intakes (F.1)
fat-soluble vitamin (F.1)
phytochemical (F.8)
provitamin (F.2)
vitamin (F.1)
water-soluble vitamin (F.1)

Review Questions

1. Compare and contrast vitamins and minerals (Section 11.8) with regard to:
   a. inorganic or organic
   b. essentiality in the diet
   c. amounts needed by the body
2. Identify each vitamin as water-soluble or fat-soluble.
   a. vitamin A b. biotin c. vitamin K
   d. cholecalciferol e. niacin f. vitamin B_6
3. Which is more likely to be dangerous—an excess of a water-soluble vitamin or an excess of a fat-soluble vitamin? Why?
4. If boiled vegetables are served as part of a meal, would the water-soluble or fat-soluble vitamins originally present be lost? Explain your choice.
5. What is the structural difference between water-soluble and fat-soluble vitamins?
6. Why is vitamin D called the "sunshine vitamin"?
7. How are vitamins related to coenzymes?
8. What is meant by the term *B complex?*

9. Match each compound with its designation as a vitamin:

Compound	Designation
ascorbic acid	vitamin A
ergocalciferol	vitamin B_1
thiamine	vitamin C
retinol	vitamin D_2
tocopherol	vitamin E

10. Identify the vitamin deficiency associated with:
a. night blindness **b.** beriberi **c.** rickets

11. Identify the deficiency disease associated with a diet lacking in:
a. vitamin C **b.** niacin **c.** vitamin B_{12}

12. Match each vitamin with its coenzyme.

Vitamin	Coenzyme
folic acid	coenzyme A
nicotinamide	flavin adenine dinucleotide
pantothenic acid	nicotinamide adenine dinucleotide
pyridoxine	pyridoxal phosphate
riboflavin	tetrahydrofolic acid

13. Identify each vitamin or coenzyme that is described below.
a. These vitamins are converted into coenzymes that catalyze oxidation-reduction reactions.
b. This coenzyme is needed for reactions in which carboxyl groups are transferred from one molecule to another.
c. A pyrophosphate group is added to this vitamin to form its coenzyme.
d. In this coenzyme a methyl group has replaced the cyanide group found in the vitamin.
e. This vitamin is reduced to form its coenzyme.

14. The lack of vitamin B_{12} in the diet is usually *not* the reason why individuals develop pernicious anemia. What other factor leads to the development of pernicious anemia?

15. Each molecule of vitamin B_{12} has 63 carbon atoms. The molecular formula is $C_{63}H_{88}CoN_{14}O_{14}P$ and the molar mass is 1355 g/mol, and yet it is soluble in water. Explain.

16. Identify each vitamin as water-soluble or fat-soluble.

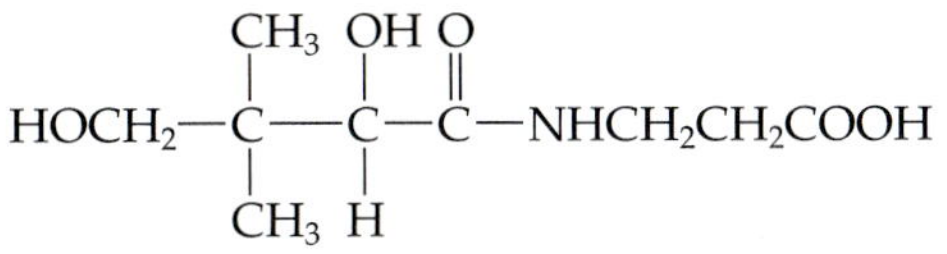

$$HOCH_2-C(CH_3)_2-CH(OH)-C(=O)-NHCH_2CH_2COOH$$

(a)

$$\text{(2,6,6-trimethylcyclohexenyl)}-CH{=}CHC(CH_3){=}CHCH{=}CHC(CH_3){=}CHCH_2OH$$

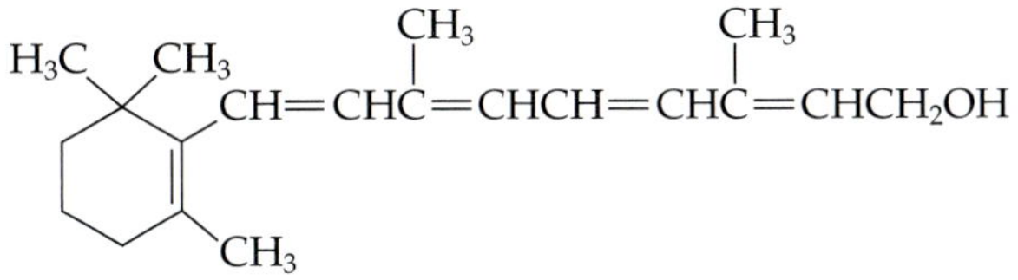

(b)

17. Is vitamin C a single chemical compound? Does synthetic vitamin C differ from natural vitamin C? In what way(s) could *tablets* labeled natural vitamin C differ from those made with synthetic vitamin C?

18. A salesperson is selling a supplement containing 1000 mg of α-tocopherol, telling you that this antioxidant has been shown to prevent oxidative damage to cells and prevent cancer. Would you purchase the supplement? Why or why not?

CHAPTER 23

Nucleic Acids and Protein Synthesis

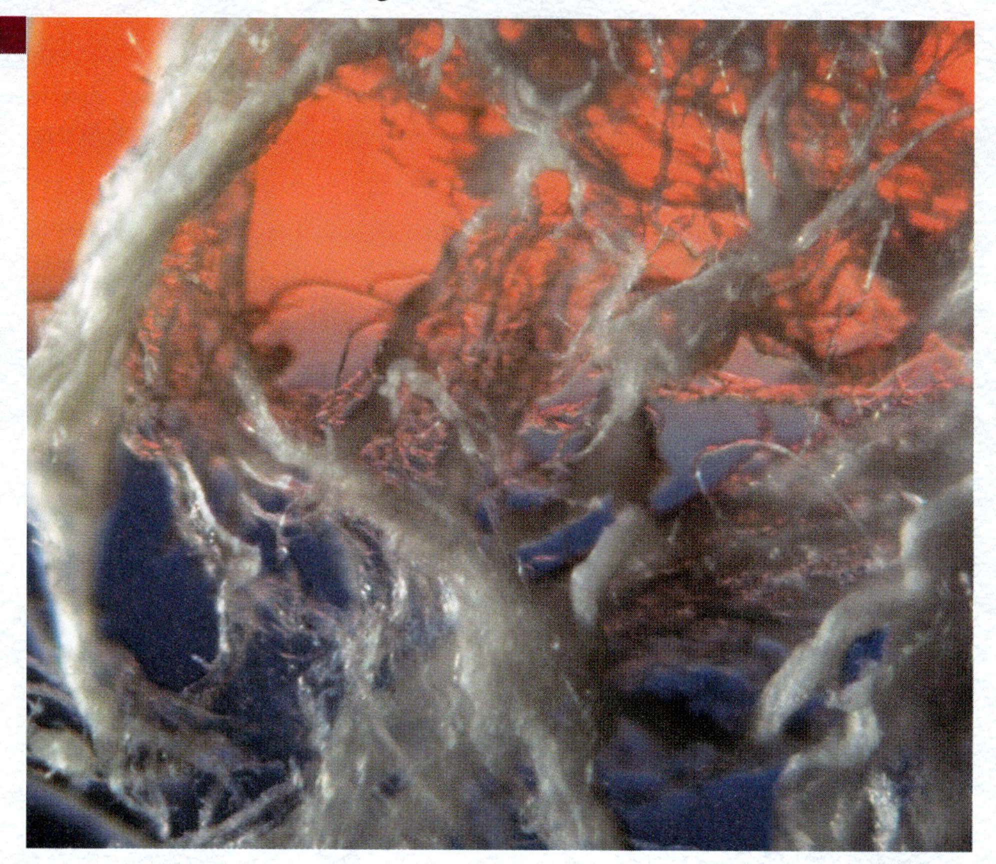

Artificial DNA synthesized from animal DNA. DNA is called the "blueprint of life" because it contains the information needed for the synthesis of all the body proteins.

Learning Objectives/Study Questions

1. What are the two types of nucleic acids, and what is the function of each? What molecules are needed to make nucleotides, the building blocks or monomer units of nucleic acids?
2. What are the different types of RNAs? What are the characteristics of each type of RNA?
3. What is meant by complementary base pairing?
4. How does the secondary structure of DNA differ from that of RNA?
5. How is a new copy of DNA synthesized?
6. How is mRNA synthesized from DNA?
7. How is a protein synthesized from mRNA?
8. What are the characteristics of the genetic code?
9. What are genetic mutations and how do they arise?
10. What are genetic diseases?
11. How is recombinant DNA technology used in biotechnology?

Cats have kittens that grow up to be cats. Bears have cubs that grow up to be bears. From viruses to humans, each species reproduces after its own kind. Within each multicellular organism, every tissue is composed of cells specific to that tissue. What accounts for this specificity at all levels of reproduction? How does a fertilized egg "know" that it should develop into a kangaroo and not a koala? Most higher organisms reproduce sexually—a sperm cell from the male unites with an egg cell from the female. The fertilized egg so formed must carry all the information needed to make the various cells, tissues, and organs necessary for the functioning of the new individual—legs, liver, lungs, heart, head, hair, and so forth. In addition, if the species is to survive, information must be set aside in germ cells—either sperm or eggs—for the production of new individuals.

The "blueprint" for reproduction and maintenance of each organism is found in the nuclei of all cells, concentrated in elongated, threadlike bodies called **chromosomes,** complex structures of DNA and proteins (Figure 23.1). The basic units of heredity, called **genes,** are arranged along the chromosomes in a linear fashion. The number of chromosomes varies with the species. Human body cells have 23 pairs of chromosomes containing a total of 80,000–100,000 different genes.

Human chromosomes are composed of about 25% DNA and 75% protein.

Sperm and egg cells contain only a single copy of each chromosome. Thus, in sexual reproduction, the entire complement of chromosomes is achieved only when the egg and sperm combine; a new individual receives half its hereditary material from each parent.

Calling the unit of heredity a gene merely gives it a name. What are genes? What are they made of? How is the information they contain expressed? One definition of genes is that they are distinct segments of DNA. *Each gene codes for a specific polypeptide.* If genes are segments of DNA, the first question that needs to be answered is, "What exactly is DNA?"

Nucleic Acid Structure

Nucleic acids are large polymers found in every cell. **Deoxyribonucleic acid (DNA)** occurs in the cell nucleus of cells and stores genetic information. It has been calculated that the total amount of DNA in a typical mammalian cell contains about 3×10^9 nucleotides. If all this DNA were stretched out end to end, it would extend more than two meters! **Ribonucleic acid (RNA)** is found in all parts of the cell and is responsible for transmitting or expressing the genetic information contained in the DNA into the thousands of proteins found in living organisms. As we will see, the two types of nucleic acids differ slightly in the composition of their repeating units or nucleotides.

Cellular RNAs are divided into three types. The distinctions among them are made primarily on the basis of biochemical function, which we'll discuss more fully in Section 23.5. Messenger RNA (mRNA) makes up only a few percent of the total amount of RNA within the cell. Each different mRNA contains the information needed for the synthesis of a unique protein (or proteins). Ribosomal RNA (rRNA) makes up 80% of the total cellular complement of ribonucleic acid and is needed for the formation of ribosomes. Transfer RNA (tRNA) is a relatively low-molar-mass nucleic acid. Different tRNAs bind specific amino acids, carrying them to the ribosome where the amino acid will be incorporated into a new protein.

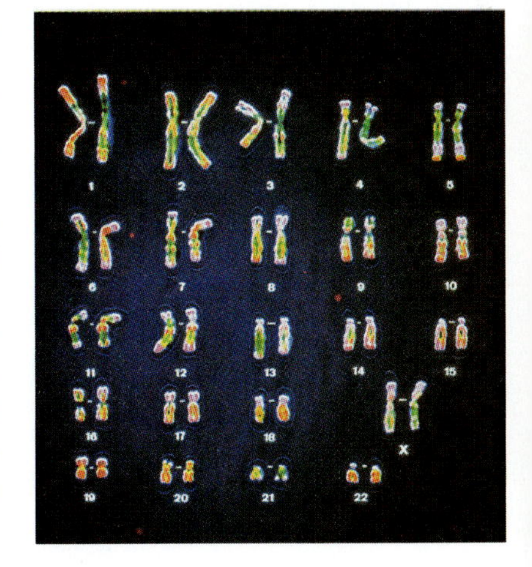

▲ **Figure 23.1**
Female chromosomes arranged in numbered homologous pairs. The male set differs from the female set only in the sex pair (fourth row, right); a male has an X and a Y instead of two Xs. The nucleus of each human body cell has 46 chromosomes, 23 from the mother and 23 from the father.

23.1 Nucleotides

Controlled hydrolysis of nucleic acids releases the repeating or monomer units known as **nucleotides,** which can be further hydrolyzed to phosphoric acid and compounds called **nucleosides.** In turn, a nucleoside can be hydrolyzed to a pentose sugar and a nitrogenous (nitrogen-containing) base.

$$\text{Nucleic acids} \xrightarrow{H_2O} \text{Nucleotides} \xrightarrow{H_2O} \left\{\begin{matrix}\text{Nucleosides} \\ + \\ H_3PO_4\end{matrix}\right. \xrightarrow{H_2O} \left\{\begin{matrix}\text{Nitrogenous bases} \\ + \\ \text{Pentose sugars}\end{matrix}\right.$$

If the pentose sugar is ribose, the nucleoside is more specifically referred to as a *ribonucleoside.* If 2-deoxyribose is the sugar involved, the nucleoside is a *deoxyribonucleoside.*

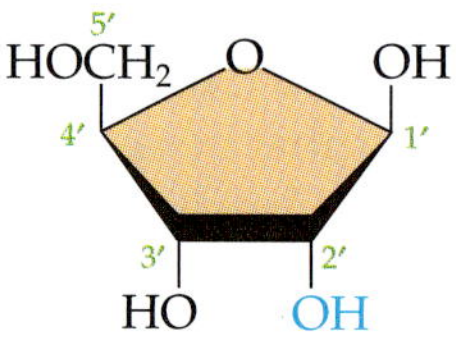

β-Ribose 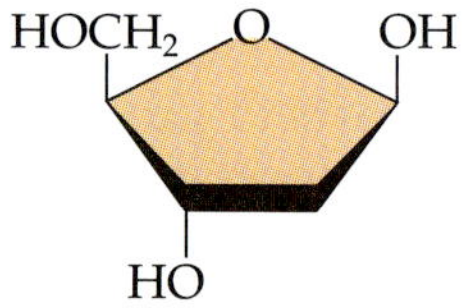 β-2-Deoxyribose

The nitrogenous bases found in nucleotides and nucleic acids are classified as **pyrimidines** or **purines,** based on whether they are substituted derivatives of the parent compounds, pyrimidine or purine. Pyrimidine is a heterocyclic six-member ring containing two nitrogen atoms in the ring. It does not occur free in nature, but its derivatives uracil, thymine, and cytosine are important components of nucleotides. Purine is a heterocyclic amine consisting of a pyrimidine ring fused to an imidazole ring. Adenine and guanine are the major purines found in nucleic acids (Figure 23.2).

The numbering convention is that atoms of the pentose ring are designated by primed numbers and atoms of the purine or pyrimidine ring are designated by unprimed numbers.

Several other pyrimidine or purine derivatives (called modified, or minor, bases) also are found in various nucleic acids. Methylation is the most common form of modification. Examples of modified bases are 5-methylcytosine, N^6-methyladenine, and 7-methylguanine.

An *N-glycosyl linkage,* always beta in naturally occurring nucleosides and nucleotides, joins the pentose to the nitrogen base. It is formed between C-1′ of the sugar and N-1 of the pyrimidine base or N-9 of the purine base, eliminating a molecule of water in the process.

Nucleotides are phosphate esters of the nucleosides and may be envisioned as resulting from the esterification of phosphoric acid with one of the free pentose hydroxyl groups. The equation on page 637 is given only to help you visualize the formation of nucleosides and nucleotides; these compounds are *not* synthesized in this fashion in the cell.

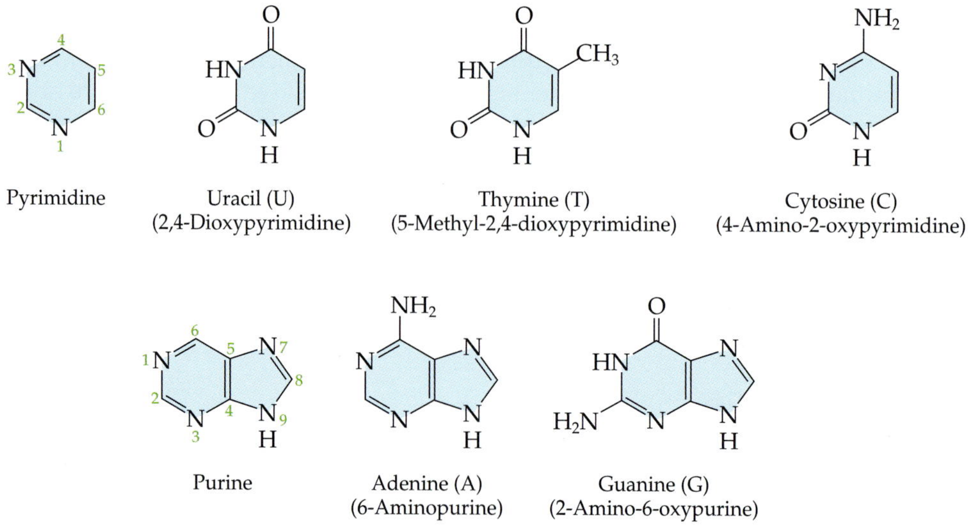

▶ Figure 23.2
Structures of nitrogenous bases most commonly incorporated into the nucleotides found in DNA and RNA.

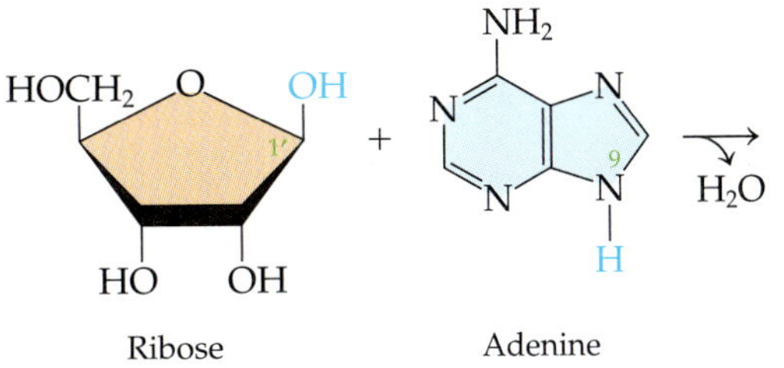

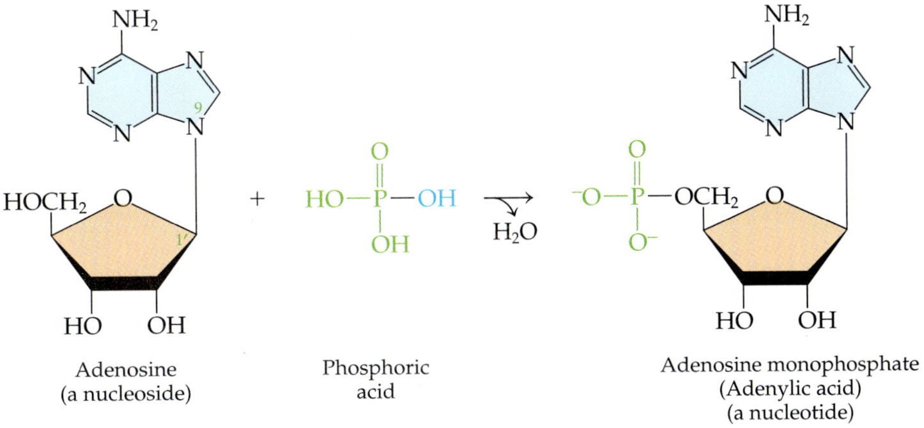

Ribose + Adenine → H_2O

Adenosine (a nucleoside) + Phosphoric acid → H_2O + Adenosine monophosphate (Adenylic acid) (a nucleotide)

The protons of the monophosphate ester are ionized at physiological pH. Thus, nucleotides exist as ions with a 2− charge in solution.

Table 23.1 indicates the similarities and differences in the composition of nucleotides from DNA and RNA.

The common names of the ribonucleosides are derived from the names of the nitrogenous bases. The suffix *-osine* denotes purine nucleosides, and the suffix *-idine* is used for pyrimidine nucleosides. The prefix *deoxy-* is used if the base is combined with deoxyribose—deoxyadenosine, deoxyguanosine, deoxycytidine, and deoxythymidine.

The nucleotides are named as acids or nucleoside phosphates. Thus the nucleoside uridine, upon esterification with phosphoric acid, becomes uridylic acid or uridine monophosphate (UMP). Similarly, guanosine becomes guanylic acid or guanosine monophosphate (GMP). The names and structures of the major ribonucleotides and one of the deoxyribonucleotides are given in Figure 23.3.

In addition to being the monomer units of DNA and RNA, the nucleotides and some of their derivatives perform a variety of other functions in the cell. Adenosine diphosphate (ADP) and adenosine triphosphate (ATP) are involved in many metabolic processes (Figure 23.4). We shall encounter them often in upcoming chapters. In addition, adenosine 3′,5′-monophosphate (cyclic AMP), a cyclic structure in

Table 23.1 Composition of Nucleotides from DNA and RNA

	DNA	RNA
Purine bases	Adenine Guanine	Adenine Guanine
Pyrimidine bases	Cytosine Thymine	Cytosine Uracil
Pentose sugar	2-Deoxyribose	Ribose
Inorganic acid	Phosphoric acid	Phosphoric acid

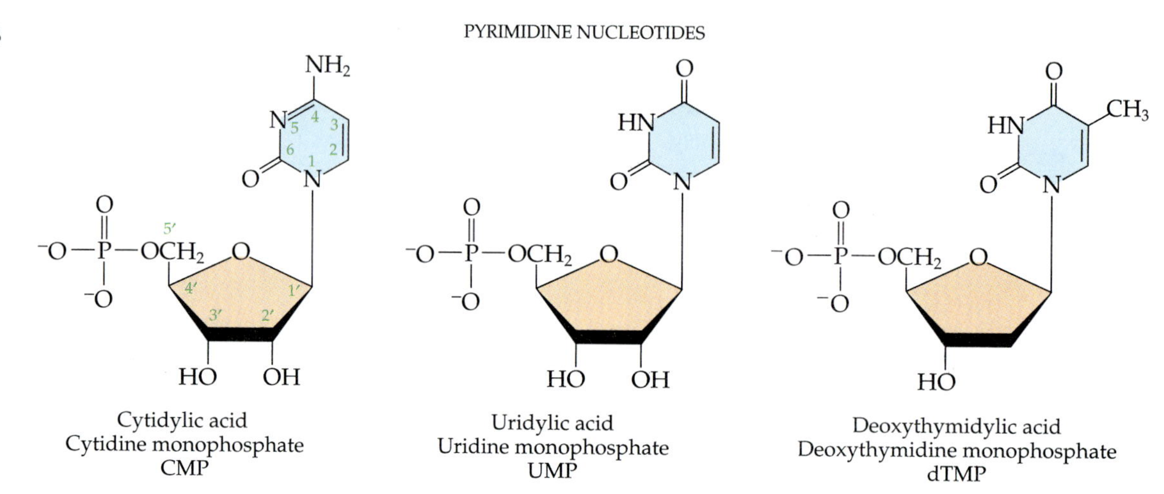

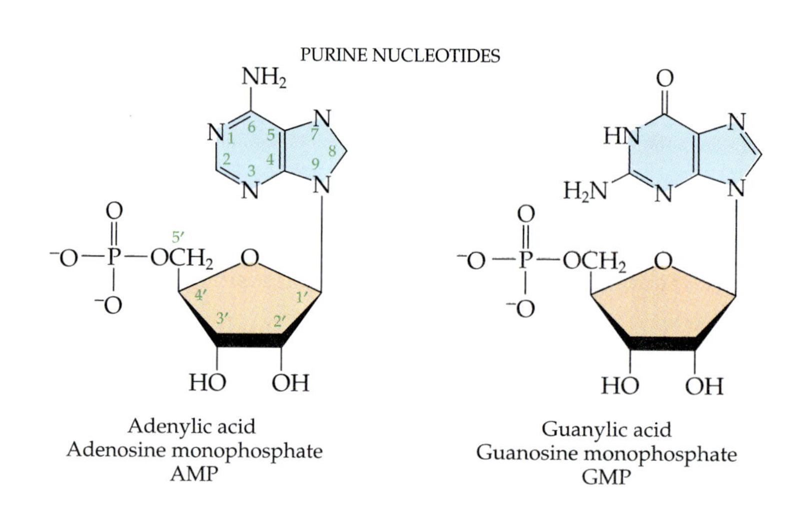

Figure 23.3 The pyrimidine and purine nucleotides.

which the phosphate group is bonded to both the 3′ and 5′ ribose carbons, plays a crucial role in metabolism (see Chapter 25). A number of coenzymes, including FAD, NAD^+, and coenzyme A, contain adenine nucleotides as structural components (Selected Topic F).

Review Questions

23.1 **(a)** Name the two kinds of nucleic acids. **(b)** Which is concentrated in the nucleus of the cell?

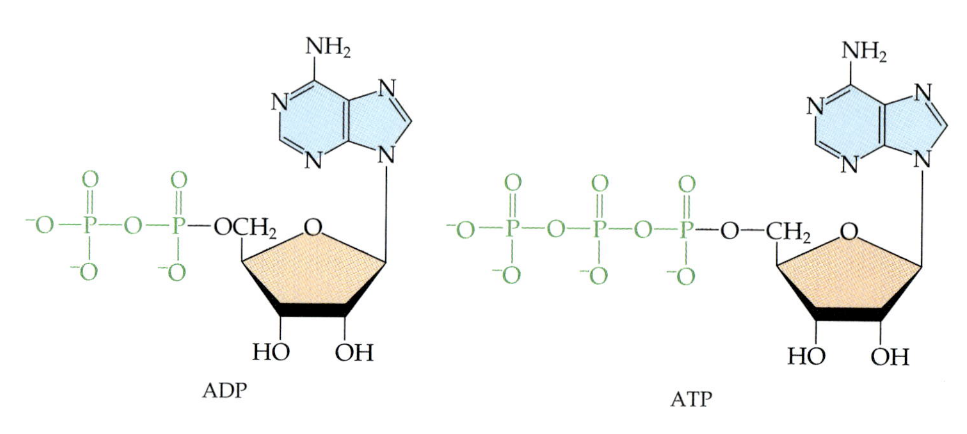

Figure 23.4 Structures of two important nucleotide derivatives.

23.2 Give the names of all the molecules that can be obtained from the complete hydrolysis of **(a)** DNA and **(b)** RNA.

23.2 The Primary Structure of Nucleic Acids

Nucleotides are joined to one another through the phosphate group to form nucleic acid chains. The phosphate unit on one nucleotide forms an ester linkage to the hydroxyl group on the third carbon atom of the sugar unit in a second nucleotide. This unit is in turn joined to another nucleotide, and the process is repeated to build up the long nucleic acid chain (Figure 23.5). The backbone of the chain consists of alternating phosphate and sugar units (2-deoxyribose in DNA and ribose in RNA). The purine bases (adenine and guanine) and pyrimidine bases (cytosine and thymine in DNA and cytosine and uracil in RNA) are branched off this backbone.

Note that each phosphate group has one acidic hydrogen that is ionized at physiological pH. That is what makes these compounds nucleic *acids.*

Nucleic acids resemble proteins in one respect. To completely specify the primary structure of a nucleic acid, one must specify the sequence of nucleotides. Unlike the proteins, which have 20 different amino acids, there are only four different

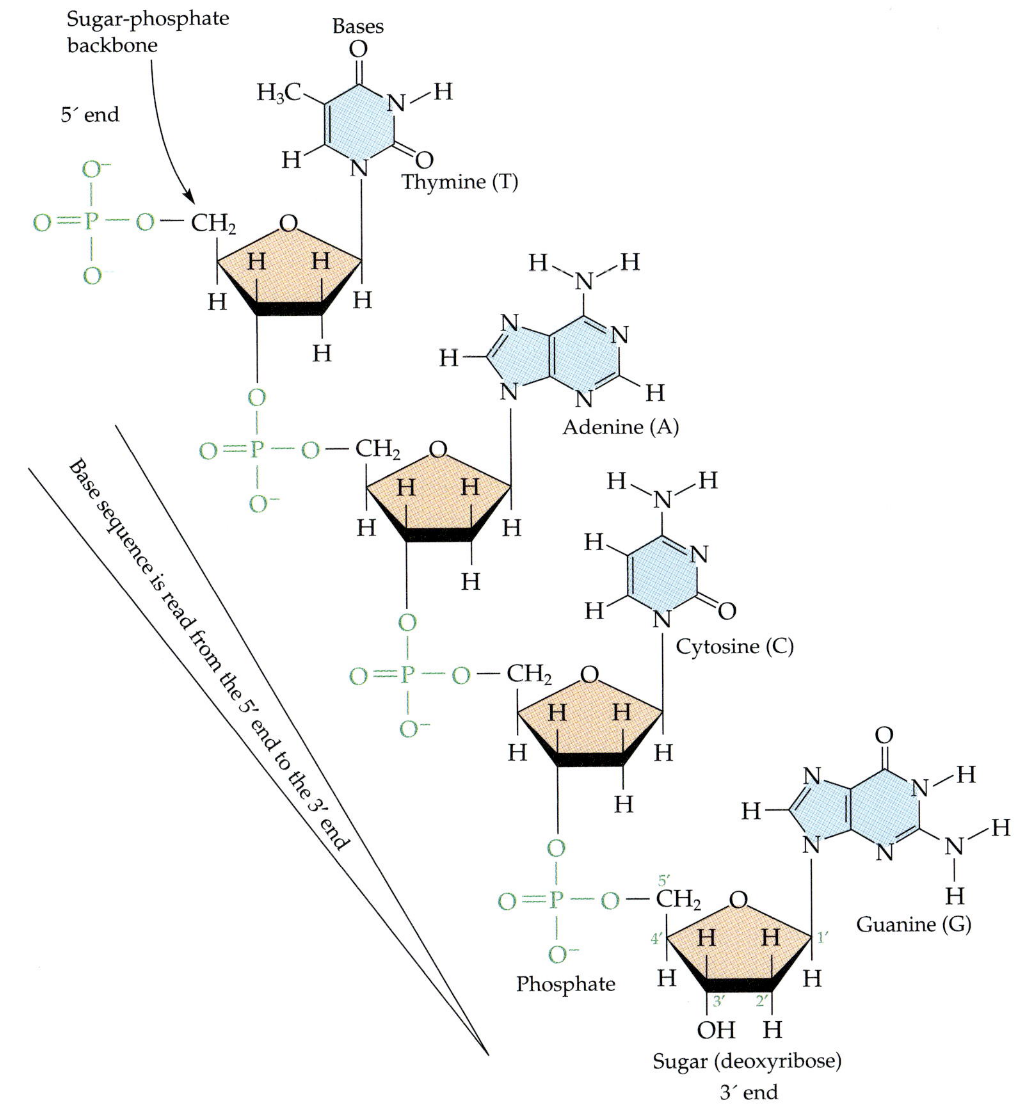

◀ Figure 23.5
Structure of a segment of DNA. A similar segment of RNA would have —OH groups on each C-2′ and uracil (U) would replace thymine (T).

nucleotides in a nucleic acid. For amino acid sequences in proteins the convention is to write the amino acids in order, starting with the N-terminal amino acid. In writing nucleotide sequences for nucleic acids, the convention is to start with the nucleotide having a free phosphate group (known as the 5′-end) and indicate the nucleotides in order (usually using the one-letter abbreviations for the bases). For DNA, a lower case *d* is written in front of the abbreviation to indicate that deoxyribonucleotides are present. The final nucleotide has a free —OH group on the 3′ carbon and is known as the 3′ end. Thus the sequence of nucleotides in the DNA segment shown in Figure 23.5 would be designated as 5′-dT-dA-dC-dG-3′, which can be further abbreviated to d-TACG.

✓ Review Questions

23.3 What constitutes the backbone of a DNA or RNA chain?

23.4 The primary structure of a protein is defined by the sequence of amino acids. What defines the primary structure of nucleic acids?

23.3 The Secondary Structure of Nucleic Acids

The determination of the three-dimensional structure of DNA was a topic of intensive research effort in the late 1940s to early 1950s. Initial work revealed that the polymer had a structure with a regular repeating pattern. In 1950 Erwin Chargaff of Columbia University showed that the molar amount of adenine (A) in DNA is always equal to that of thymine (T). Similarly, the molar amount of guanine (G) is the same as that of cytosine (C). Chargaff drew no conclusions from his work, but others soon did. At Cambridge University, in 1953, James D. Watson and Francis Crick announced that they had a model for the secondary structure of DNA. Using the information from Chargaff's experiments (as well as other experiments), data from the X-ray studies of Rosalind Franklin, which involved sophisticated chemistry, physics, and mathematics, and working with models not unlike a child's construction set, they determined that DNA is composed of two nucleic acid chains which are antiparallel—that is, the 5′ end of one nucleic acid chain is opposite the 3′ end of the other. The two chains form a **double helix**—a structure that can be thought of as a spiral staircase in which the phosphate and sugar groups (the backbone of the nucleic acid polymer) form the outside or banisters. The purine and pyrimidine bases are paired on the inside—with guanine always opposite cytosine and adenine always opposite thymine. These specific base pairs are referred to as **complementary bases.** In our staircase analogy, these base pairs are the stairsteps (Figure 23.6). In 1962 Watson and Crick received the Nobel Prize in physiology and medicine for discovering, as Crick put it, "the secret of life."

This structure can explain how cells are able to divide and go on functioning, how genetic data are passed on to new generations, and even how proteins are built to required specifications. It all depends on the base pairing. But there is still the unanswered question, "Why do the bases pair in that precise pattern, always A to T and T to A; always G to C and C to G?" The answer is hydrogen bonding and a truly elegant molecular design. Figure 23.7 shows the two sets of base pairs. Notice two things. First, a pyrimidine is paired with a purine in each case, and the long dimensions of both pairs are identical (1.08 nm). If two pyrimidines were paired or two purines were paired, the two pyrimidines would take up less space than a purine and a pyrimidine, and the two purines would take up more space, as illustrated in Figure 23.8. If this were the situation, the structure of DNA would be like a staircase made with stairs of different widths. In order for the two strands of the double helix to fit neatly, a pyrimidine must always be paired with a purine.

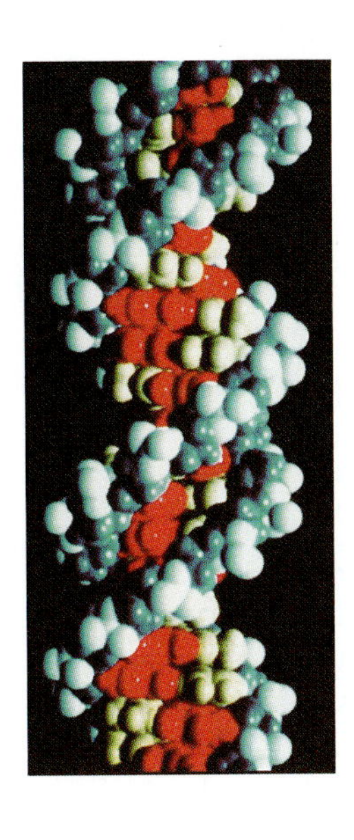

(a)

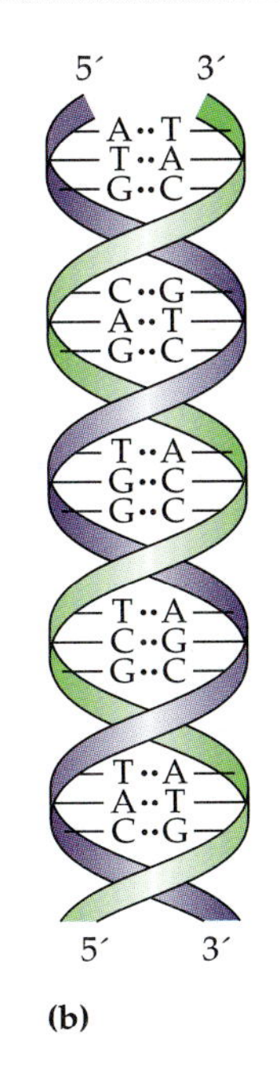

(b)

Figure 23.6
(a) A computer-generated model of the DNA double helix. The blue and white atoms represent the sugar-phosphate chains that wrap around the outside of the helix. On the inside are the bases, shown in red and yellow. (b) A schematic representation of the double helix, showing the complementary bases.

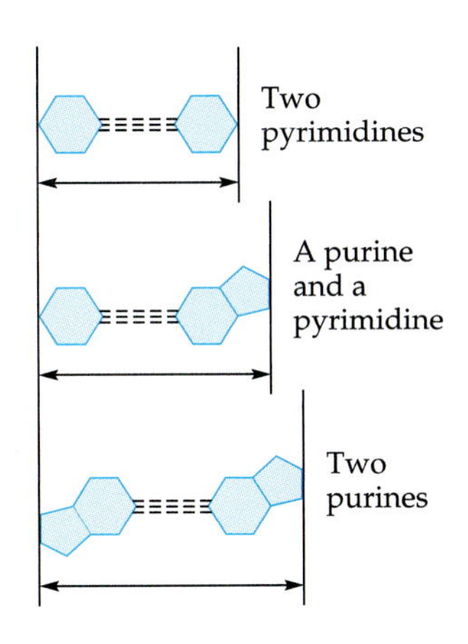

Figure 23.8
Difference in widths of possible base pairs.

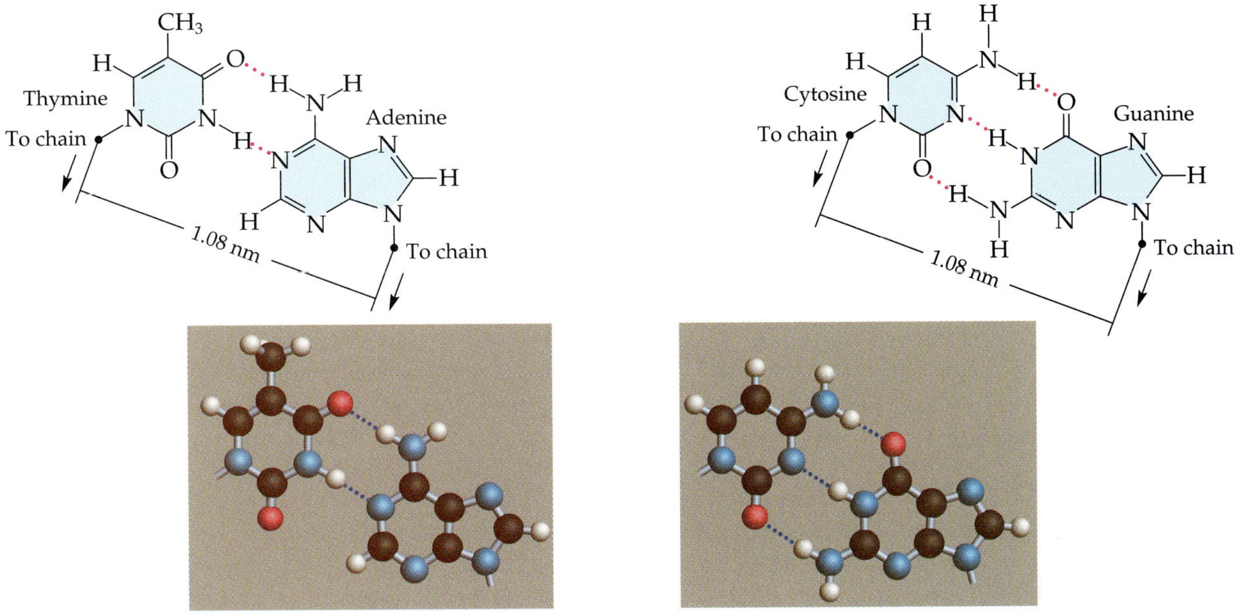

Thymine
Adenine
To chain
1.08 nm
To chain

(a)

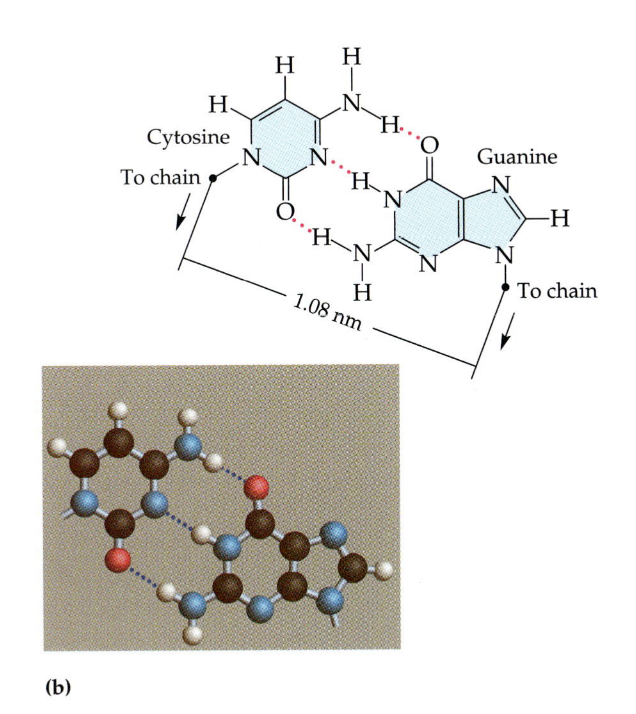

(b)

Figure 23.7
Pairing of the complementary bases (a) thymine and adenine and (b) cytosine and guanine. The pairing involves hydrogen bonding.

The other thing you should notice in Figure 23.7 is that this pairing enables formation of three hydrogen bonds between guanine and cytosine, and two between adenine and thymine. The additive contribution of all these hydrogen bonds imparts great stability to the DNA double helix.

While the secondary structure of DNA is described by the double helix model, a molecule of RNA consists of a single strand. Some internal (intramolecular) base pairing, however, may occur in sections where the molecule folds back on itself. Because of this, portions of the molecule may exist in a double-helix form, but overall, RNA is considered a single-stranded molecule.

Review Questions

23.5 Why is it structurally important that a purine base always pair with a pyrimidine base in the DNA double helix?

23.6 Explain what is meant by complementary bases.

23.7 What kind of intermolecular force is involved in base pairing?

Replication and Expression of Genetic Information

We have stated that DNA stores genetic information, while RNA is responsible for transmitting or expressing genetic information by directing the synthesis of the thousands of proteins found in living organisms. But how does this happen? Three processes are required: (1) *replication,* in which new copies of DNA are made (Section 23.4); (2) *transcription,* in which a segment of DNA is transcribed into RNA (Section 23.5); and (3) *translation,* in which the information in RNA is translated into a protein sequence (Section 23.6).

23.4 Replication of DNA

New cells are continuously forming in the body through the process of cell division. For this to happen, a copy of the DNA in one cell must be copied in a process known as **replication.** The complementary base pairing of the double helix provides a ready model for how genetic replication occurs. If the two chains of the double helix are pulled apart, disrupting the hydrogen bonds between base pairs, each chain can act as a *template,* or pattern for the synthesis of a new DNA chain.

In the nucleus are all the necessary enzymes and nucleotides needed for this synthesis (Figure 23.9). A short segment of DNA is "unzipped," separating the two strands. An enzyme, DNA polymerase, recognizes each base in the template strand and matches it to the complementary base in a free nucleotide. The enzyme then catalyzes the formation of an ester bond between the 5′ phosphate group of the nucleotide and the 3′—OH of the growing DNA chain. In this way, each strand of the original DNA molecule forms a duplicate of its former partner. Whatever information was encoded in the original DNA double helix is now contained in each of the replicates. When the cell divides, each daughter cell gets one of the DNA molecules and all of the information that was available to the parent cell.

Example 23.1

A segment of DNA was found to have the sequence of nucleotides shown below. What is the sequence of nucleotides in the opposite or complementary DNA chain?

5′ d-TCCATGAGTTGA 3′

Solution

Knowing that the two strands are antiparallel and that T base-pairs with A, and C base-pairs with G, the sequence of nucleotides is determined to be:

3′ d-AGGTACTCAACT 5′

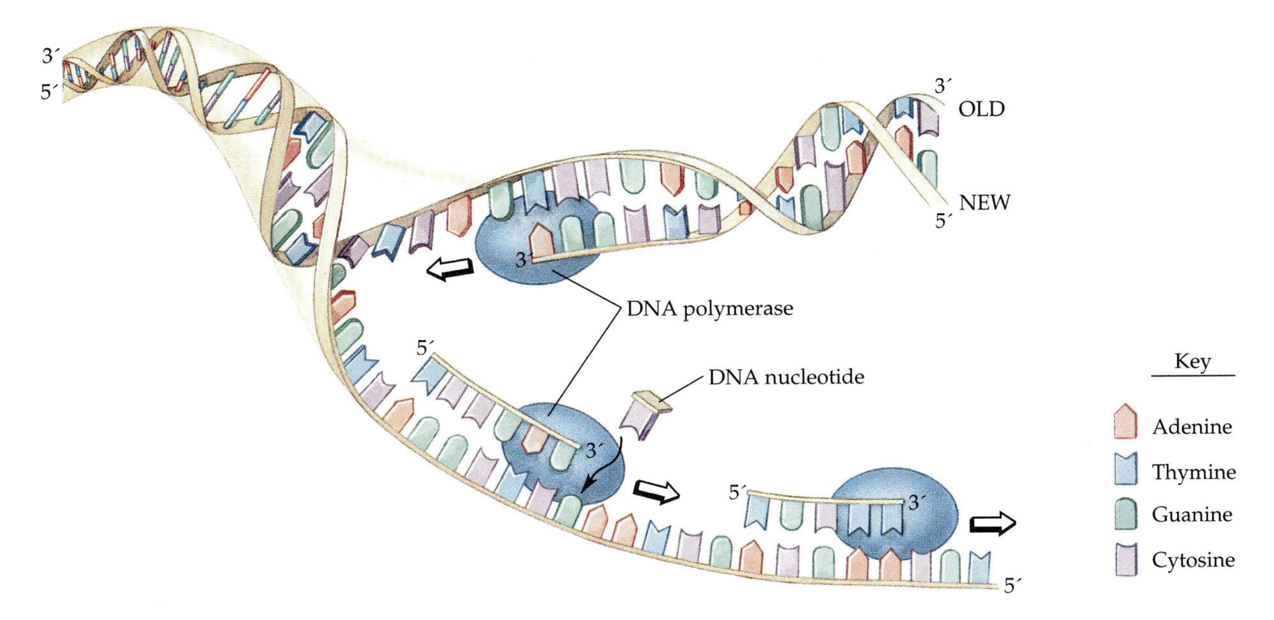

▲ **Figure 23.9**
A schematic diagram of DNA replication, which occurs by sequential "unzipping" of the double helix. The new nucleotides are brought into position by the enzyme DNA polymerase and phosphate bridges are formed, thus restoring the original double helix configuration. Each newly formed double helix consists of one old strand and one new strand, a process called semiconservative replication. (This representation is simplified; the nucleotides are actually triphosphate derivatives: dATP, dTTP, dGTP, and dCTP.)

Practice Exercise

A segment of DNA was found to have the sequence of nucleotides shown below. What is the sequence of nucleotides in the complementary DNA chain?

5′ d–CCAGTGAATTGCCTAT 3′

We keep saying that there is information encoded in the DNA molecule. DNA can be compared to a book containing directions for putting together a model airplane or for knitting a sweater. Knitting directions store information as words on paper. Letters of the alphabet are arranged in a certain way (e.g., "knit one, purl two"), and these words direct the knitter to perform a certain operation with needles and yarn. If all of the directions are correctly followed, the ball of yarn becomes a sweater.

How is information stored in DNA? It is the particular sequence of nucleotides along the DNA chain that encodes the directions for building an organism. Just as *saw* means one thing in English and *was* means another, the sequence of bases CGT means one thing, and TGC means something different. Although there are only four "letters"—the four nucleotides—in the genetic code of DNA, their sequence along the DNA strands can vary so widely that information storage is essentially unlimited. An *E. coli* bacterium only 2 μm long and 0.8 μm in diameter has 4.7 million base pairs, while the genetic material of a human cell consists of 3 billion base pairs. Thus each cell can carry all the information it needs to determine all its hereditary characteristics. We shall see how this information is expressed in the synthesis of proteins in Section 23.6.

Review Questions

23.8 Describe DNA replication.

23.9 In DNA replication, a parent DNA molecule produces two daughter molecules. What is the fate of each strand of the parent DNA double helix?

23.5 Transcription: Synthesis of RNA

All of the RNA found in a cell is synthesized from DNA by a template mechanism analogous to DNA replication in many respects. Because the RNA that is synthesized is a complementary copy of information contained in DNA, RNA synthesis is referred to as **transcription.** But there are key differences between replication and transcription. RNA molecules are much shorter than DNA molecules because only a portion of one DNA strand is copied or transcribed. The DNA sequence that is transcribed is the *template strand,* while the complementary sequence on the other DNA strand is the *coding* or *informational strand* (Figure 23.10).

To initiate RNA synthesis, the two DNA strands unwind at specific sites, called *promoters,* on the DNA template. Ribonucleotides are attracted to the uncoiling region of the DNA molecule, beginning at the 3′ end of the template strand. Nucleotides are bound according to the rules of base-pairing. Thymine in DNA calls for adenine in RNA, cytosine specifies guanine, guanine calls for cytosine, and adenine requires uracil. RNA polymerase is the enzyme needed to bind the correct

Polymerase Chain Reaction

The widely used technique known as the polymerase chain reaction (PCR), first described in the mid-1980s, allows a researcher to take a very small amount of DNA and make 10^6–10^9 copies of a specific DNA segment in two to three hours. The technique mimics the process of replication in the cell and requires that four components be present in a buffered solution: (1) the DNA sample that is to be copied; (2) sufficient quantities of the four deoxyribonucleotides; (3) heat-stable DNA polymerase; and (4) primers—short segments of ribonucleotides that have been synthesized to exactly match and border the DNA that is to be copied. The steps involved in PCR are shown at the left.

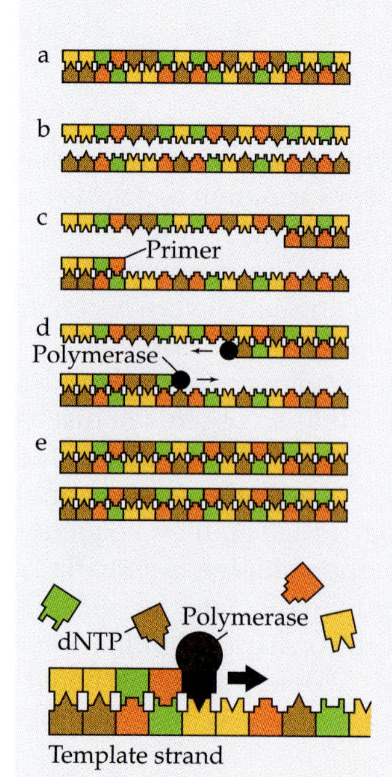

(a) The DNA to be copied must be double-stranded. (b) The sample is heated to 95 °C, which disrupts the hydrogen bonds holding the DNA strands together, leaving single-stranded DNA. (c) The sample is then cooled to 50–65 °C, and the DNA primers bind, using complementary base-pairing, to each strand at the region bordering the DNA segment to be copied. (d) The temperature is then raised to 72 °C, allowing DNA polymerase to attach nucleotides to the bound primers, using the original DNA strand as a template. (e) Two copies of DNA are formed, both identical to the original sample. This series of steps takes less than two minutes and can be repeated indefinitely, each time doubling the amount of that DNA segment (after 30 cycles there will be ~1 billion copies!).

The final illustration depicts DNA polymerase synthesizing the complementary DNA strand. It binds a free deoxyribonucleotide triphosphate (dNTP) that is complementary to the next unpaired nucleotide in the template strand of DNA. The enzyme then joins the dNTP to the end of the primer and moves on to the next nucleotide.

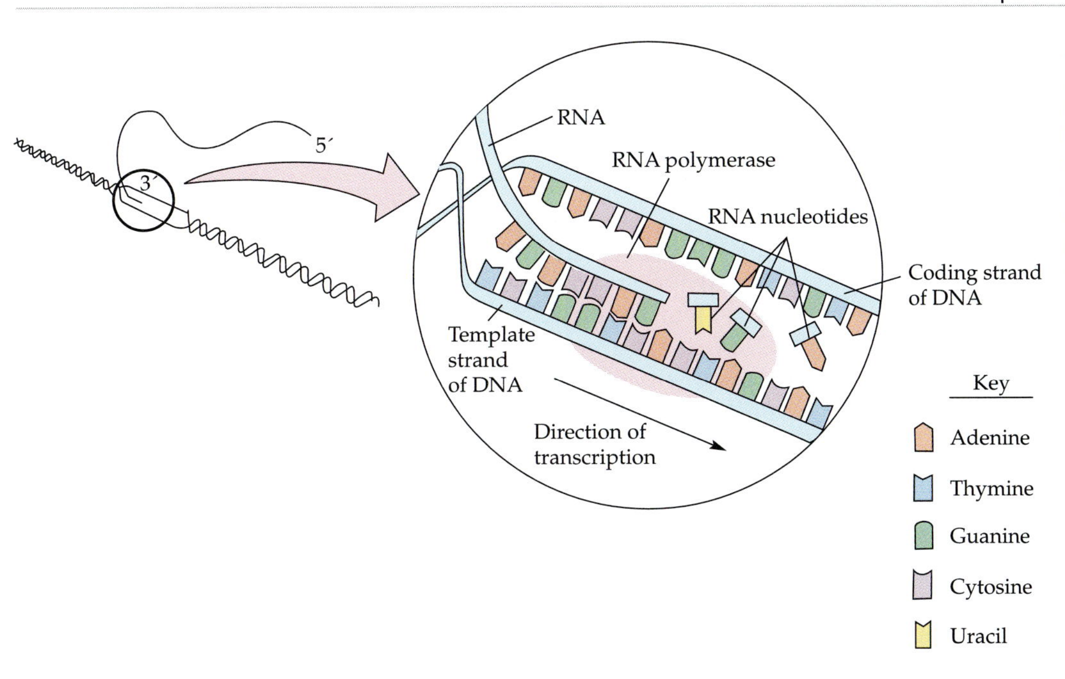

Figure 23.10 A schematic diagram of RNA transcription from a DNA template. RNA polymerase as shown is much smaller than the actual molecule, which encompasses about 50 nucleotides at a time.

nucleotide and catalyze the formation of the ester linkage between nucleotides, a reaction very similar to that of DNA polymerase. Synthesis of the RNA strand takes place in the 5′ to 3′ direction, antiparallel to the template strand. Only a short segment of the RNA molecule is hydrogen-bonded to the template strand at any one time during transcription. When transcription is completed, the RNA is released and the DNA helix re-forms. Note that the RNA strand formed during transcription is identical to the nucleotide sequence of the coding strand of the DNA except that U replaces T.

Example 23.2

A portion of the template strand of a gene has the sequence of nucleotides shown below. What is the sequence of nucleotides in the RNA that is formed from this template?

5′ d–TCCATGAGTTGA 3′

Solution

Four things must be remembered in answering this question: (1) the DNA strand and the RNA strand being synthesized are antiparallel; (2) mRNA is synthesized in a 5′→3′ direction, so transcription begins at the 3′ end of the template strand; (3) ribonucleotides are used in place of deoxyribonucleotides; (4) T base-pairs with A, A pairs with U (in RNA), and C base-pairs with G. The sequence of nucleotides is determined to be:

5′ UCAACUCAUGGA 3′

Practice Exercise

A portion of the template strand of a gene has the sequence of nucleotides shown below. What is the sequence of nucleotides in the RNA that is formed from this template?

5′ d–CCAGTGAATTGCCTAT 3′

Before considering what happens to the RNA and how it is used in the synthesis of proteins, we need to look more closely at the three types of RNA that are formed during transcription. The three types of RNA differ in their functions, size, and stability (Table 23.2).

Messenger RNA

Messenger RNA (mRNA) makes up only a few percent of the total amount of RNA within the cell, primarily because a molecule of mRNA exists for a relatively short time. Like proteins, it is continuously being degraded and resynthesized. The rate of mRNA degradation differs from species to species and also from one type of cell to another. In bacteria, one-half of the total mRNA is degraded every two minues, whereas in rat liver the half-life is several days.

The molecular dimensions of the mRNA molecule vary according to the amount of genetic information the molecule is meant to encode. It is known, however, that there is very little intramolecular hydrogen bonding in mRNA and that the molecule exists in a single stranded random coil. After transcription, which takes place in the nucleus, the mRNA passes into the cytoplasm, carrying the genetic message from DNA to the ribosomes, the sites of protein synthesis. In Section 23.6 we shall see how mRNA directly determines the sequence of amino acids during protein synthesis.

Ribosomal RNA

The **ribosome** is a cellular substructure that serves as the site for protein synthesis. Its composition is about 65% rRNA and 35% protein. The rRNA molecules and the proteins are held together by a large number of noncovalent forces, such as hydrogen bonds. Structurally, a ribosome is composed of two spherical particles of unequal size. The smaller of them has a distinct affinity for mRNA; the larger has an attraction for tRNA. Ribosomes are extremely small particles, visible only with the aid of an electron microscope. More often than not, they are seen as clusters known as *polyribosomes,* or *polysomes,* bound to the endoplasmic reticulum of animal and plant cells or to the cell membrane of microorganisms.

Transfer RNA

Each different tRNA molecule is attached to a specific amino acid in a reaction catalyzed by a unique enzyme. The tRNA carries that amino acid to the site of protein synthesis at the precise moment specified by a set of three nucleotides known as a codon (Section 23.6). Each of the 20 amino acids found in proteins has at least one corresponding tRNA, and most amino acids have more than one. For example, there are two different tRNAs specific for the transfer of lysine and six for serine.

The two-dimensional structure of a tRNA molecule has a distinctive "cloverleaf" structure (Figure 23.11). On one of the loops is a unique sequence of three bases that varies for each different tRNA. This triplet is called the **anticodon.** At the opposite end of the molecule is the acceptor stem where the amino acid is attached.

Table 23.2 Properties of Cellular RNA in *E. coli*

Type	Function	Approx. Number of Nucleotides	Stability	Percentage of Total Cell RNA
mRNA	Messenger	100–6,000	Unstable	~3
rRNA	Ribosome structure	120–2900	Stable	83
tRNA	Adapter	75–90	Stable	14

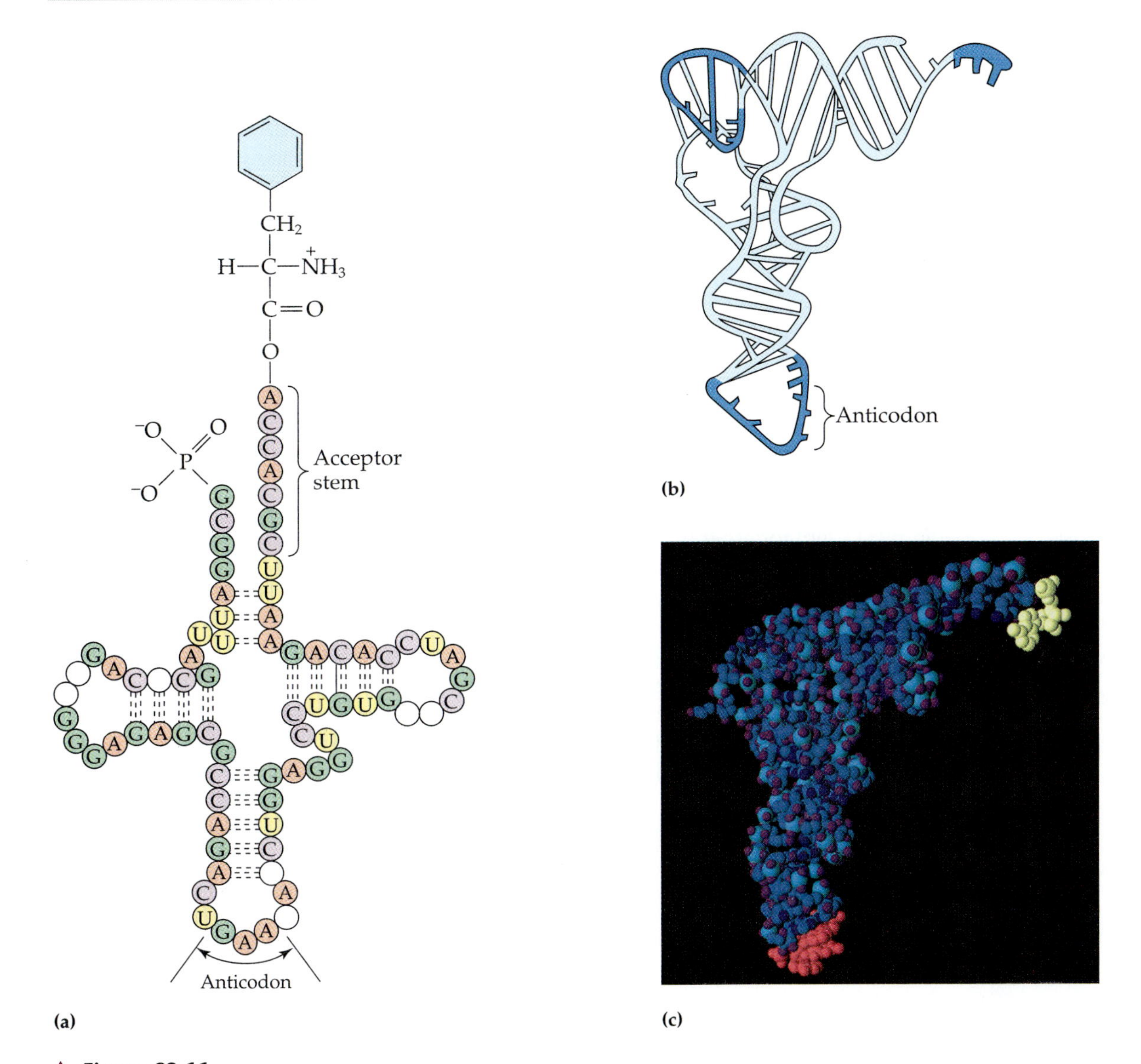

▲ **Figure 23.11**
(a) Two-dimensional structure of a yeast tRNA molecule for phenylalanine; the amino acid binds to the acceptor stem located at the 3′ end of the tRNA. (The nucleotides not specifically identified are slightly altered analogs of the four normal ribonucleotides.) (b) Three-dimensional structure of the yeast phenylalanine tRNA. Note that the anticodon loop is at the bottom and the acceptor stem is at the top right. (c) A space-filling model of the tRNA.

Review Questions

23.10 Which nucleic acid(s) is (are) involved in transcription?

23.11 Distinguish between the functions of mRNA, rRNA, and tRNA.

23.6 Protein Synthesis and the Genetic Code

In the previous section we noted that mRNA is a transcribed copy of the DNA sequence for a particular gene that codes for the synthesis of one polypeptide chain. (If a protein contains two or more different polypeptide chains, each chain is coded by a different gene.) How is a particular sequence of nucleotides in the RNA translated into an amino acid sequence? How can a molecule containing just four different nucleotides specify the sequence of the 20 amino acids that occur in proteins? If each nucleotide coded for one amino acid, then obviously the nucleic acids could code for only four amino acids. Suppose we consider the nucleotides in groups of two. There are 4^2, or 16, combinations of pairs of the four distinct nucleotides. Such a code is more extensive but still inadequate. If, however, the nucleotides are considered in groups of three, there are 4^3, or 64, combinations. Here we have a code that is extensive enough to direct the synthesis of the primary structure of a protein molecule. *Each group of three bases along the mRNA strand specifies a particular amino acid, and the sequence of these triplet groups dictates the sequence of the amino acids in the protein.* Because this **genetic code** involves three nucleotides per coding unit, it is referred to as a **triplet code.** The three-nucleotide coding unit is called a **codon.** How does this triplet code direct protein synthesis?

Protein synthesis is accomplished by orderly interactions between the ribonucleic acids and more than 100 enzymes. The mRNA formed during transcription is transported across the nuclear membrane into the cytoplasm (and hence to the ribosomes), carrying with it the genetic instructions. The cell faces the problem of lining up the amino acids according to the sequence specified by the mRNA and joining them together by peptide linkages. Because this process involves the transfer of the information encoded in the mRNA to the ultimate structure of the protein molecule, it is often referred to as **translation.**

Before an amino acid can be incorporated into a polypeptide chain, it must be attached to its unique tRNA. This crucial process requires an enzyme known as aminoacyl-tRNA synthetase and an ATP molecule (Figure 23.12). *Both the enzyme (aminoacyl-tRNA synthetase) and the tRNA are highly specific for a particular amino acid.* The high degree of specificity of the synthetase enzyme is vital to the correct incorporation of the amino acid into a protein. After the amino acid molecule has been activated and bound to its tRNA carriers, protein synthesis can take place. Figure 23.13 depicts a schematic stepwise representation of this all-important process.

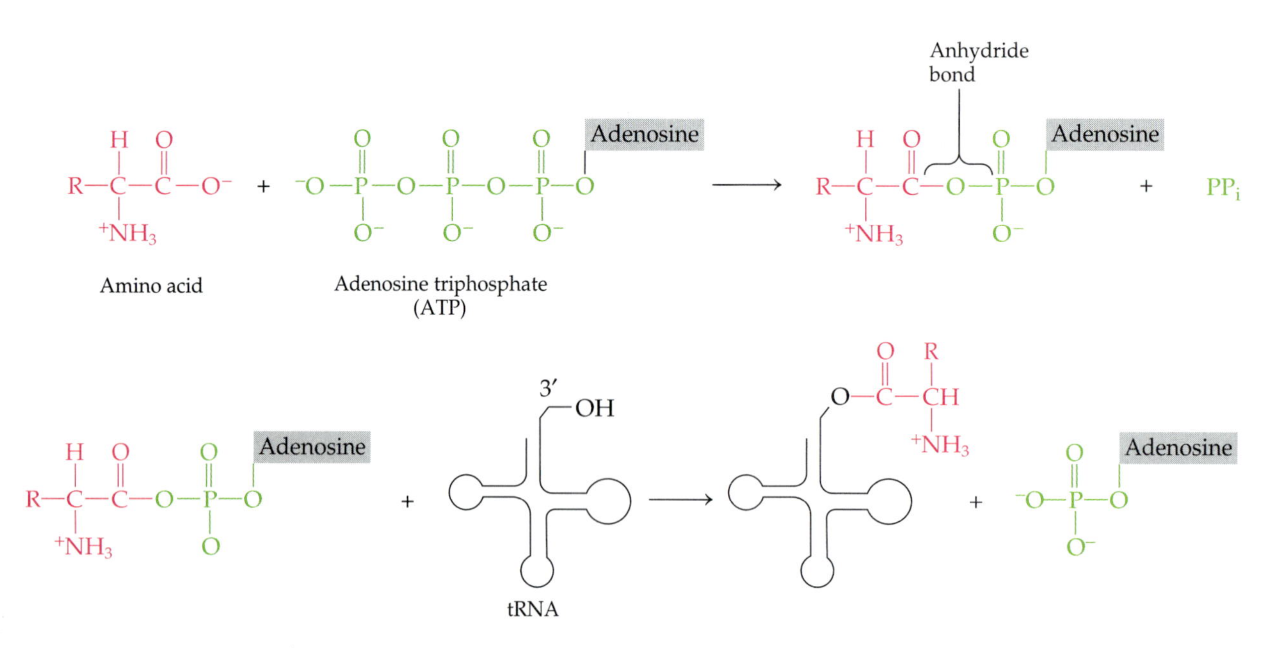

▲ Figure 23.12
Activation and binding of an amino acid to its tRNA.

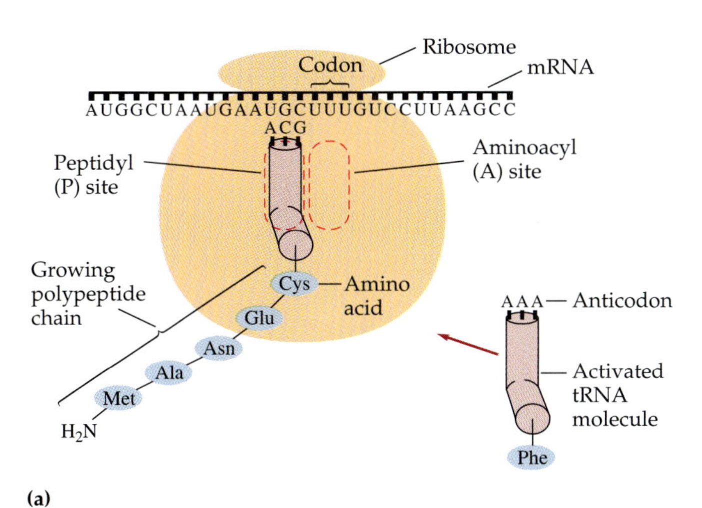

(a) Protein synthesis is already in progress at the ribosome. The growing polypeptide chain is bound to the peptidyl (P) site on the surface of the ribosome. At this point the aminoacyl (A) site is vacant. The codon UUU is lined up above the A site. An activated tRNA molecule, which has the anticodon AAA, moves up to the ribosome. (The tetracyclines—Figure 22.18—block the binding of the aminoacyl tRNA to the A site on bacterial ribosomes, inhibiting protein synthesis and stopping bacterial growth. The tetracyclines do not bind to mammalian ribosomes and thus do not hinder protein synthesis in host cells.)

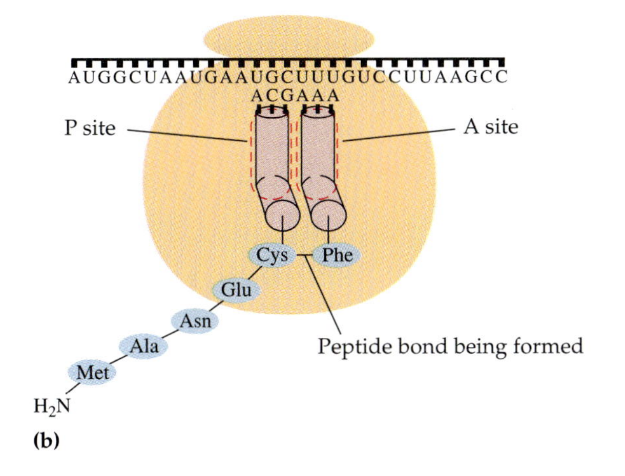

(b) The activated tRNA molecule is now bound to the ribosome at the A site. It is also bound to the mRNA molecule through base-pairing of the codon and anticodon. The amino acid Phe is being incorporated into the polypeptide chain by the formation of a peptide linkage between the carboxyl group of Cys and the amino group of Phe. This reaction is catalyzed by the enzyme peptidyl transferase, a component of the ribosome. (Chloramphenicol acts by blocking the action of peptidyl transferase.)

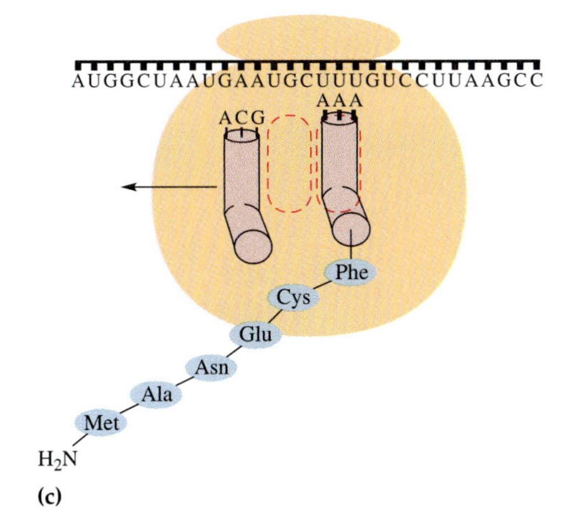

(c) The Cys–Phe linkage is now complete, and the growing polypeptide chain is attached to the A site. The tRNA molecule has detached from the P site. It can now move off the ribosome and into the cytoplasm and pick up another amino acid.

▲ **Figure 23.13**
The elongation steps in protein synthesis.

(continued on next page)

(d) The ribosome moves (translocates) to the right along the mRNA strand. The polypeptide chain and the tRNA molecule to which it is bound are shifted from the A site to the P site. This shift brings the next codon, GUC, into place over the A site. Notice that an activated tRNA molecule, containing the next amino acid to be attached to the chain is moving into position on the surface of the ribosome. Its anticodon is CAG. (Erythromycin blocks the translocation reaction in bacteria.)

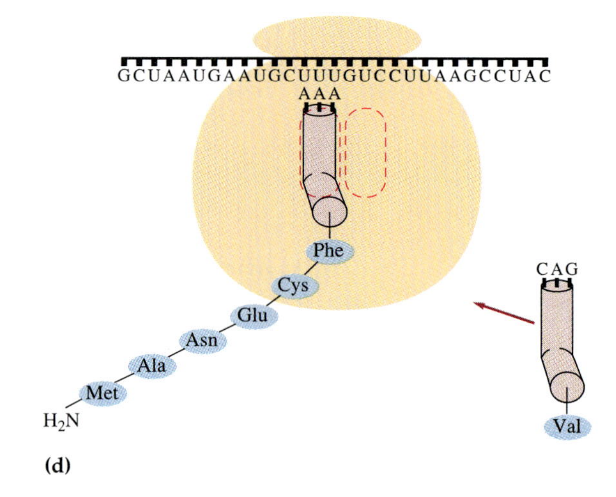

(e) The activated tRNA molecule, carrying the amino acid Val, is now in place on the ribosome. The peptide linkage between the carboxyl group of Phe and the amino group of Val is forming.

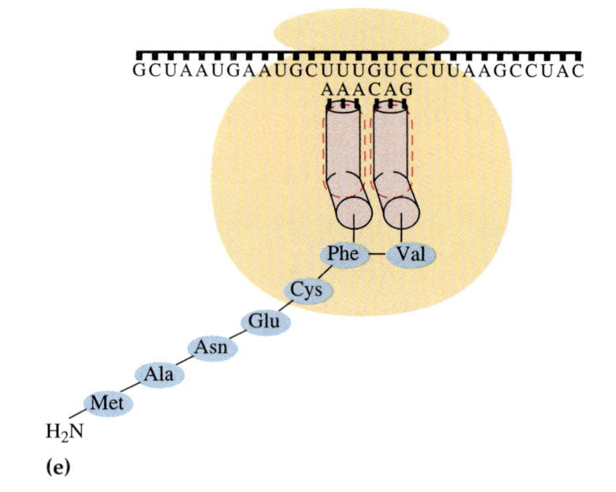

(f) The Phe–Val linkage is now complete, and the growing polypeptide chain is now attached, through a tRNA molecule, to the A site. The ribosome will translocate again, and the tRNA molecule plus the attached polypeptide chain will be in position at the P site. This process continues until the polypeptide chain is complete—that is, until one of the three termination codons appears at the A site. When the ribosome reaches the end of the message, both it and the polypeptide are released from the mRNA molecule.

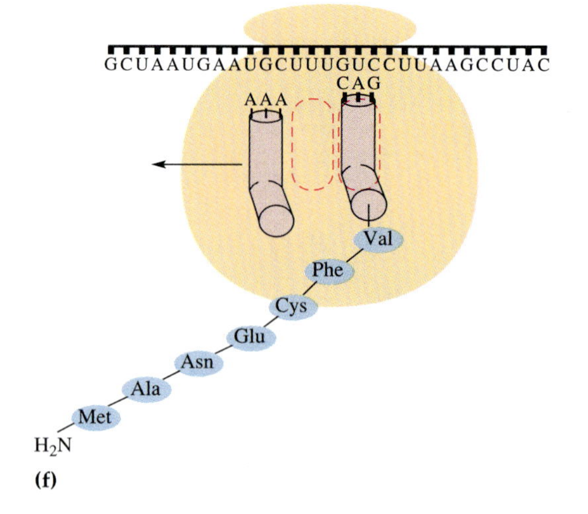

▲ **Figure 23.13 continued.**

After a portion of the mRNA strand has been "translated" by a given ribosome, another ribosome may attach itself to the strand and begin to read it as the first ribosome moves on to translate the remainder of the mRNA. Thus in cells active in protein synthesis we find clusters of ribosomes connected by a single strand of mRNA (Figure 23.14). On the average, five to eight ribosomes are simultaneously synthesizing the same polypeptide from the information in one mRNA strand (large proteins require long strands of mRNA, and as many as 100 individual ribosomes may be attached). The time required for the synthesis of an average-size polypeptide (~300 amino acids) is about 15 seconds in a bacterial cell and two or three minutes in a mammalian cell.

Figure 23.15 summarizes the steps involved in gene expression. The amount of any particular protein in a cell depends on the balance between the rate at which it is synthesized (which is largely controlled by the rate at which its mRNA is synthesized in the nucleus) and the rate at which it is degraded.

We have stated that the codon, a group of three adjacent nucleotides on the mRNA, directs the precise sequence of the amino acids for each protein. There are 64 possible triplet codons. Early experimenters were faced with the task of determining which codon (or perhaps codons) stood for each of the 20 amino acids. The cracking of the genetic code was the joint accomplishment of several well-known

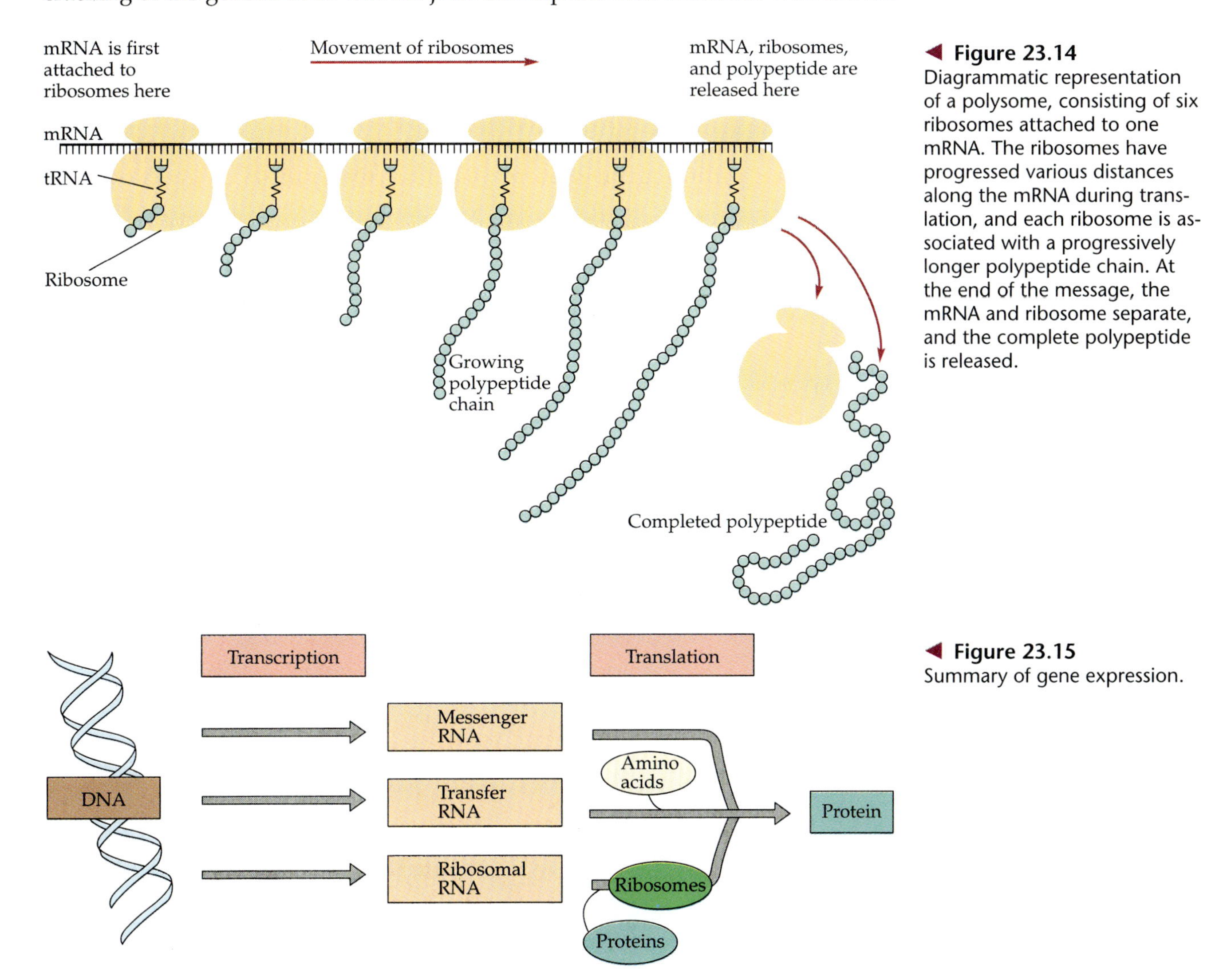

Figure 23.14 Diagrammatic representation of a polysome, consisting of six ribosomes attached to one mRNA. The ribosomes have progressed various distances along the mRNA during translation, and each ribosome is associated with a progressively longer polypeptide chain. At the end of the message, the mRNA and ribosome separate, and the complete polypeptide is released.

Figure 23.15 Summary of gene expression.

geneticists, notably H. Khorana, M. Nirenberg, P. Leder, and S. Ochoa, from 1961 to 1964. A genetic dictionary has been compiled and is given in Table 23.3. Of the 64 possible codons, 61 code for amino acids and three serve as signals for the termination of polypeptide synthesis (that is, as periods at the end of a sentence). Notice that only methionine (AUG) and tryptophan (UGG) have single codons. All other amino acids have two or more codons.

Example 23.3

A portion of a mRNA has the sequence of nucleotides shown below. What amino acid sequence does this code for?

5′ AUGCCACGAGUUGAC 3′

Solution

Use Table 23.3 to determine what amino acid each set of three nucleotides (codon) codes for. Remember that the sequence is read starting from the 5′ end and that the protein is synthesized starting with the N-terminal amino acid.

Met–Pro–Arg–Val–Asp

Practice Exercise

A portion of a mRNA has the sequence of nucleotides shown below. What amino acid sequence does this code for?

5′ AUGCUGAAUUGCGUAGGA 3′

Further experimentation threw much light on the nature of the genetic code. It now appears that:

Table 23.3 The Genetic Code

First Base	Second Base: U		Second Base: C		Second Base: A		Second Base: G		Third Base
U	UUU	Phe	UCU	Ser	UAU	Tyr	UGU	Cys	U
	UUC	Phe	UCC	Ser	UAC	Tyr	UGC	Cys	C
	UUA	Leu	UCA	Ser	UAA	Termination	UGA	Termination	A
	UUG	Leu	UCG	Ser	UAG	Termination	UGG	Trp	G
C	CUU	Leu	CCU	Pro	CAU	His	CGU	Arg	U
	CUC	Leu	CCC	Pro	CAC	His	CGC	Arg	C
	CUA	Leu	CCA	Pro	CAA	Gln	CGA	Arg	A
	CUG	Leu	CCG	Pro	CAG	Gln	CGG	Arg	G
A	AUU	Ile	ACU	Thr	AAU	Asn	AGU	Ser	U
	AUC	Ile	ACC	Thr	AAC	Asn	AGC	Ser	C
	AUA	Ile	ACA	Thr	AAA	Lys	AGA	Arg	A
	AUG	Met	ACG	Thr	AAG	Lys	AGG	Arg	G
G	GUU	Val	GCU	Ala	GAU	Asp	GGU	Gly	U
	GUC	Val	GCC	Ala	GAC	Asp	GGC	Gly	C
	GUA	Val	GCA	Ala	GAA	Glu	GGA	Gly	A
	GUG	Val	GCG	Ala	GAG	Glu	GGG	Gly	G

1. The code is essentially universal—animal, plant, and bacterial cells use the same codons to specify each amino acid (with a few exceptions).
2. The code is degenerate—in all but two cases (methionine and tryptophan), more than one triplet codes for a given amino acid.
3. The first two bases of each codon are most significant; the third base often varies. This suggests that a change in the third base by a mutation may still permit the correct incorporation of a given amino acid into a protein (see Section 23.7). The third base is sometimes called the "wobble" base.
4. In general, codons with C or U as the second base specify the nonpolar amino acids, whereas codons with A or G as the second base specify the polar amino acids (see Table 21.1).
5. The code is continuous and nonoverlapping—there are no special signals, and adjacent codons do not overlap (except in the case of a few viruses that do have overlapping genes).
6. There are three codons that do not code for any amino acid. These are the termination codons; they are read by special proteins (called release factors) and signal the end of the translation process.
7. The codon AUG codes for methionine and is also the initiation codon. Thus methionine is the first amino acid in each newly synthesized polypeptide. This first amino acid is usually removed enzymatically before the polypeptide chain is completed; the vast majority of polypeptides do not begin with methionine.

Review Questions

23.12 Explain the role of mRNA and tRNA in protein synthesis.

23.13 Which nucleic acid contains the codon? Which contains the anticodon?

Replication, Transcription, and Translation Expanded

It is essential at this point to note that the preceding discussions of replication, transcription, and protein synthesis (translation) are correct but oversimplified. Several molecular components (including several proteins and primers—short RNA sequences) are required to initiate DNA replication. Additional proteins are required for the elongation and termination processes. Furthermore, ultraviolet radiation and certain chemicals are known to damage DNA by disrupting the sugar-phosphate backbone and/or by altering the purine and pyrimidine bases. Several different DNA repair enzymes exist that can correct different types of damage to the DNA.

Often the initial RNA formed during transcription is modified or processed in some way. For example, the *E. coli* ribosomal RNA is transcribed as one long RNA molecule that contains the three ribosomal RNAs and one or more tRNA molecules. These must be cut apart and processed through the action of various nucleases. Also, most plant and animal genes occur in pieces, spread out along the DNA. Parts of the gene are expressed (*exons*); the intervening regions (*introns*) are not expressed. Therefore, the RNAs transcribed from these genes are spliced—that is, the entire length of DNA is copied in a longer transcript that is cut once or several times, and the exons are linked together to produce the mRNA that codes for the protein.

Protein synthesis is critically controlled at the level of transcription. We now know that there are structural genes, operator genes, promoter genes, regulatory genes, and repressor molecules (which are proteins). Modification of proteins occurs following their synthesis on the mRNA–ribosome complex and involves many types of modifications such as methylation or hydroxylation of amino acid side chains and breakage of peptide bonds to activate a proenzyme or a hormone.

23.7 Mutations and Genetic Diseases

We have seen that DNA directs the synthesis of proteins through the intermediary mRNA and that the sequence of nucleotides in the DNA is critical and specific for the proper sequence of amino acids in proteins. On rare occasions, however, the nucleotide sequence in DNA may be modified either spontaneously (about 1 in 10 billion) or by exposure to heat, radiation, or certain chemicals. Any chemical or physical change that alters the nucleotide sequence in the DNA molecule is termed a **mutation.** The most common types of mutations are *substitution* (a different nucleotide is substituted), *insertion* (addition of a new nucleotide), and *deletion* (loss of a nucleotide). These changes within the DNA are called **point mutations** because the change occurs at a single nucleotide position (Figure 23.16). Because an insertion or deletion results in a frame-shift that changes the entire amino acid sequence following the mutation, these are usually more harmful than a substitution in which only a single amino acid may be altered.

The chemical and/or physical agents that cause mutations are termed **mutagens.** Examples of physical mutagens are ultraviolet and gamma radiation. They exert their mutagenic effects either directly or via free radicals induced by the radiation. Radiation and free radicals are known to cause covalent modification (often cross-linkage) of bases already incorporated into DNA. For example, exposure to UV light can result in covalent linkage between two adjacent thymines on a DNA strand, producing a thymine dimer (Figure 23.17). If not repaired, the dimer prevents formation of the

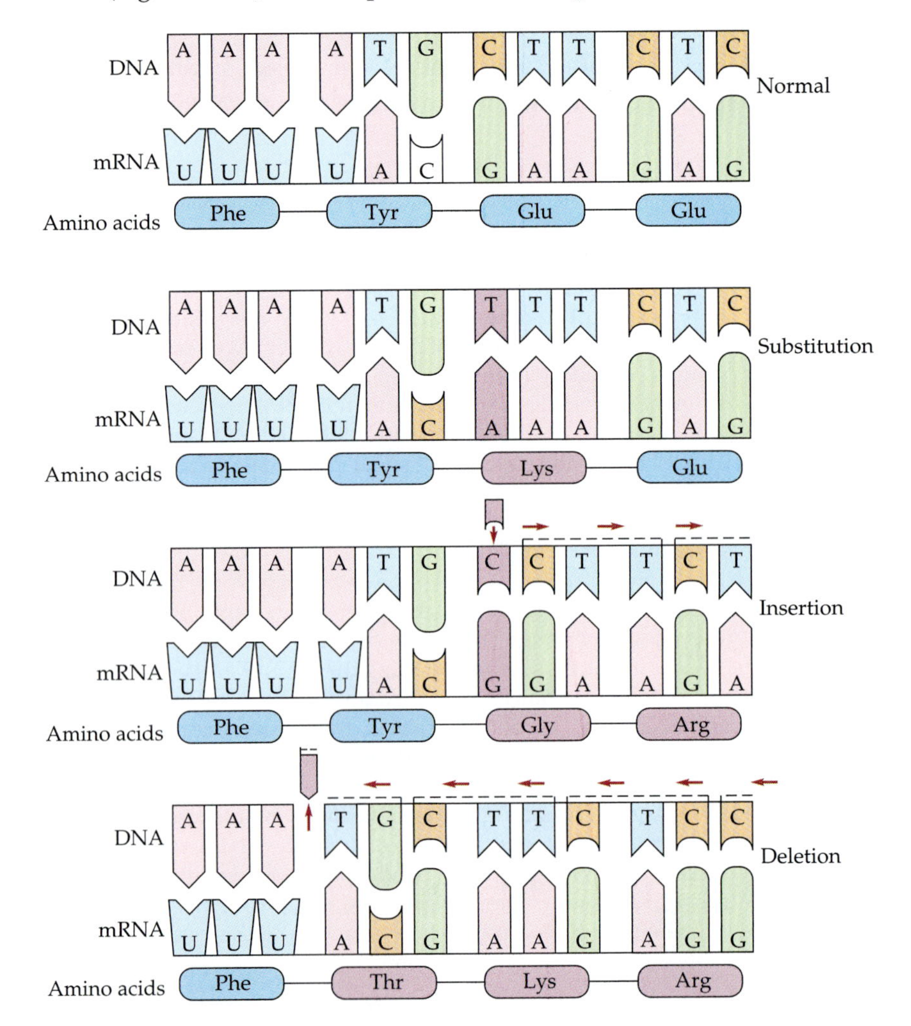

▶ Figure 23.16
Three types of point mutations.

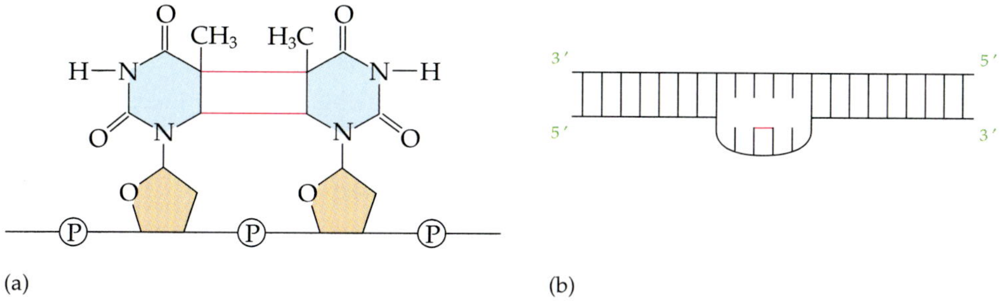

Figure 23.17
(a) Structure of a thymine dimer formed by the action of ultraviolet light. (b) Defect in the double strand produced by the thymine dimer. This temporarily stops DNA replication, but the dimer can be removed and the region repaired by an enzyme repair system.

double helix at the point at which it occurs. The genetic disease *xeroderma pigmentosum* is caused by a lack of the enzyme that cuts out the damaged thymine dimers in DNA. Individuals affected by this condition are abnormally sensitive to light and are more prone to skin cancer than normal individuals.

Among the chemical mutagens are two base analogs, 5-bromouracil and 2-aminopurine. They can be incorporated into the new DNA strand, but they exhibit faulty base-pairing. 5-Bromouracil is incorporated into DNA in place of thymine, but it can base-pair with guanine (instead of adenine). 2-Aminopurine substitutes for adenine, yet it sometimes base-pairs with cytosine (instead of thymine). Hydroxylamine and nitrous acid are other chemical mutagens. Hydroxylamine (NH_2OH) deaminates cytosine, yielding a product that pairs with adenine instead of guanine. Nitrous acid (HNO_2) can convert cytosine to uracil, which pairs with adenine instead of guanine. These compounds are also carcinogenic, indicating that mutations in DNA can lead to cancer (Selected Topic G).

When a mutation occurs in a germ cell, it will be inherited by all offspring. Although it is possible for a gene mutation to be beneficial, most mutations are detrimental. If a point mutation occurs at a crucial position, the defective protein will lack biological activity and may result in the death of the cell. In such cases the altered DNA sequence is lost and will not be copied into daughter cells. Nonlethal mutations often lead to metabolic abnormalities or to hereditary diseases. Such diseases are called *inborn errors of metabolism* or **genetic diseases.** A partial listing of genetic diseases is presented in Table 23.4, and a few specific conditions are discussed here. In most cases the defective gene results in the failure to synthesize a particular enzyme.

Phenylketonuria (PKU) results when the enzyme phenylalanine hydroxylase is absent. A person with PKU cannot convert phenylalanine to tyrosine, which is the precursor of the neurotransmitters dopamine and norepinephrine as well as the skin pigment melanin.

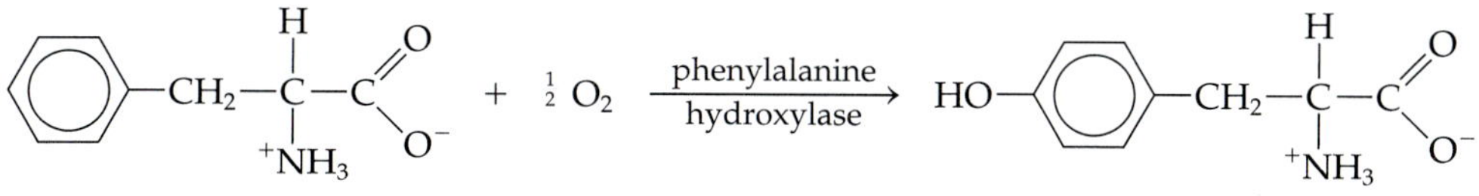

$$C_6H_5{-}CH_2{-}CH(NH_3^+){-}COO^- + \tfrac{1}{2}O_2 \xrightarrow[\text{hydroxylase}]{\text{phenylalanine}} HO{-}C_6H_4{-}CH_2{-}CH(NH_3^+){-}COO^-$$

Phenylalanine → Tyrosine

When this reaction cannot occur, phenylalanine accumulates, and the transamination (see Section 27.2) of phenylalanine to phenylpyruvate, normally a very minor process, becomes important. The disease acquired its name from the high levels of phenylpyruvate, a phenyl ketone, in the urine. Excessive amounts of phenylpyruvate impair normal brain development, causing severe mental retardation.

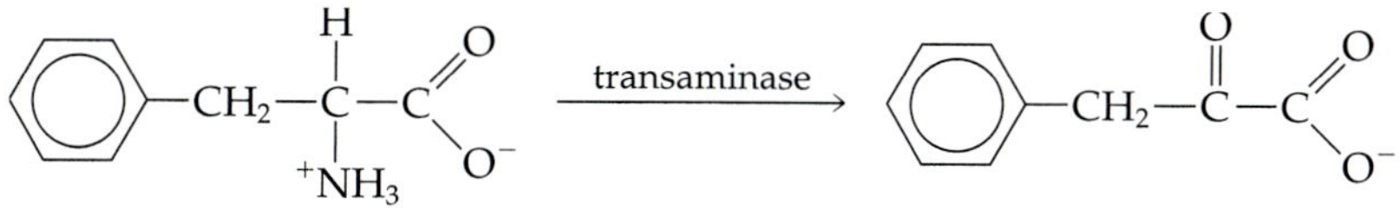

$$C_6H_5{-}CH_2{-}CH(NH_3^+){-}COO^- \xrightarrow{\text{transaminase}} C_6H_5{-}CH_2{-}C(=O){-}COO^-$$

Phenylalanine → Phenylpyruvate

Table 23.4 A Partial Listing of Genetic Diseases in Humans and the Protein or Enzyme Responsible

Disease	Responsible Protein or Enzyme
Acatalasia	Catalase (red blood cells)
Albinism	Tyrosinase
Alkaptonuria	Homogentistic acid oxidase
Cystathioninuria	Cystathionase
Fabry's disease	α-Galactosidase
Galactosemia	Galactose 1-phosphate uridyl transferase
Gaucher's disease	Glucocerebrosidase
Glycogen storage disease	Various types: α-Amylase Debranching enzyme Glucose 1-phosphatase Liver phosphorylase Muscle phosphofructokinase Muscle phosphorylase
Goiter	Iodotyrosine dehalogenase
Gout and Lesch–Nyhan syndrome	Hypoxanthine–guanine phosphoribosyl transferase
Hemolytic anemias	Various types: Glucose 6-phosphate dehydrogenase Glutathione reductase Phosphoglucose isomerase Pyruvate kinase Triose phosphate isomerase
Hemophilia	Antihemophilic factor (factor VIII)
Histidinemia	Histidase
Homocystinuria	Cystathionine synthetase
Hyperammonemia	Ornithine transcarbamylase
Hypophosphatasia	Alkaline phosphatase
Isovaleric acidemia	Isovaleryl-CoA dehydrogenase
Maple syrup urine disease	α-Keto acid decarboxylase
McArdle's syndrome	Muscle phosphorylase
Metachromatic leukodystrophy	Sphingolipid sulfatase
Methemoglobinemia	NADPH-methemoglobin reductase and NADH-methemoglobin reductase
Niemann-Pick disease	Sphingomyelinase
Phenylketonuria	Phenylalanine hydroxylase
Pulmonary emphysema	α-Globulin of blood
Sickle cell anemia	Hemoglobin
Tay–Sachs disease	Hexosaminidase A
Tyrosinemia	Hydroxyphenylpyruvate oxidase
Von Gierke's disease	Glucose-6-phosphatase
Wilson's disease	Ceruloplasmin (blood protein)

PKU may be diagnosed by assaying a sample of blood or urine for phenylalanine or one of its metabolites. It is recommended that blood be tested for phenylalanine within 24 hours to 3 weeks after birth. If the condition is detected, mental retardation can be prevented by immediately placing the infant on a diet containing little or no phenylalanine. Because phenylalanine is so prevalent in natural foods, the low-phenylalanine diet is composed of a synthetic protein substitute plus very small measured amounts of natural foods. Before dietary treatment was begun in the early 1960s, severe mental retardation was a common outcome in children with PKU. In 1953 it was reported that 85% of PKU patients had an intelligence quotient (IQ) less than 40, and 37% had IQ scores below 10. Since dietary treatments have been introduced, however, over 95% of children with PKU have developed normal or near-normal intelligence. In the past, strict adherence to a low phenylalanine diet was recommended for the first 4–8 years of life when the central nervous system was still developing, with a lessening of the restrictions after this time. Recent data, however, suggest that discontinuation of the diet may result in a deterioration of mental functioning, leading many to recommend that a low phenylalanine diet be continued into adulthood. The incidence of PKU in newborns is about 1 in 12,000 in North America.

Every state has mandated that screening for PKU be provided to all newborns, but participation is not required in four states: Delaware, Maryland, North Carolina, or Vermont.

Another genetic defect in the metabolism of tyrosine leads to *albinism.* Tyrosine serves as a precursor for the melanins, the pigments that color the skin, hair, and eyes. The absence of the enzyme tyrosinase prevents the occurrence of one of the reactions necessary for this conversion. The lack of pigmentation characteristic of albinism is the result.

Several genetic diseases are collectively categorized as *lipid-storage diseases.* Lipids are constantly being synthesized and broken down in the body. If the enzymes that catalyze lipid degradation are missing, the lipids tend to accumulate and cause a variety of medical problems. The enzymes that are responsible for lipid-storage diseases are known. Although the juncture at which the metabolic pathways go awry can be pinpointed, no cure for these diseases has yet been developed. At present, genetic counseling of prospective parents who carry the defective gene is the only approach to their control.

In *Niemann–Pick disease* sphingomyelins (Section 20.4) accumulate in the brain, liver, and spleen because the enzyme sphingomyelinase is lacking. This accumulation causes mental retardation and early death, usually between two and three years of age.

In *Gaucher disease,* a mutation has occurred in the gene that codes for the enzyme glucocerebrosidase, which cleaves glucocerebrosides (Section 20.4) into glucose and sphingosine.

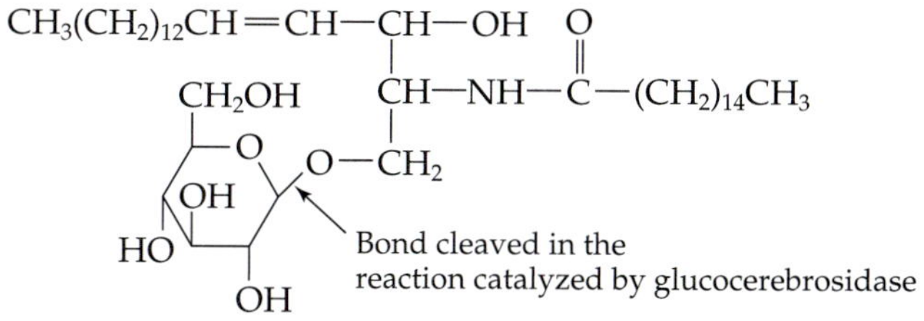

$CH_3(CH_2)_{12}CH{=}CH{-}CH{-}OH$

$CH{-}NH{-}C({=}O){-}(CH_2)_{14}CH_3$

$O{-}CH_2$

CH_2OH, O, OH, HO, OH

Bond cleaved in the reaction catalyzed by glucocerebrosidase

Thus, an individual with Gaucher disease does not produce the enzyme and cannot degrade glucocerebrosides. These accumulate primarily in the liver, spleen, and bone marrow, resulting in pain, fatigue, jaundice, bone damage, anemia, and even death. In 1991 the first effective treatment for Gaucher disease, enzyme replacement therapy, became available. A modified form of the missing enzyme is given intravenously. These treatments must be given regularly because the body eventually degrades the enzyme.

When a genetic mutation occurs in the gene for the enzyme hexosaminidase A, gangliosides (Section 20.4) cannot be degraded and they accumulate in brain tissue. The ganglion cells of the brain become greatly enlarged and nonfunctional in this genetic disease known as *Tay–Sachs disease.* This leads to a regression in development, dementia, paralysis, and blindness, with death usually occurring before the age of three. There is currently no treatment, but Tay–Sachs disease can be diagnosed during pregnancy by assaying the amniotic fluid (amniocentesis) for the amount of hexosaminidase A. Its absence indicates that the child has Tay–Sachs disease and permits a recommendation for a therapeutic abortion. Genetic screening can identify Tay–Sachs carriers because they produce only half the normal amount of hexosaminidase A (although they do not exhibit symptoms of the disease). Tay–Sachs is most common in persons of Eastern European Jewish ancestry.

✓ **Review Questions**

23.14 **(a)** What effect does UV radiation have on DNA? **(b)** Would UV radiation be an example of a physical mutagen or a chemical mutagen?

23.15 **(a)** How does the incorporation of 5-bromouracil into DNA lead to a genetic mutation? **(b)** Would 5-bromouracil be an example of physical mutagen or a chemical mutagen?

23.8 Genetic Engineering: Biotechnology

More than 3000 human diseases have a genetic component. Over the last decade or so researchers have linked mutations in specific genes to specific diseases. Now the ability to use this information to diagnose and cure genetic diseases appears to be within our grasp. By determining the location of genes on the DNA molecule, scientists have been able to identify and isolate those having specific functions. These genes can then be placed in another organism, such as a bacterium, which can easily be grown in culture. The techniques for doing this are known as **recombinant DNA technology** and are important techniques in the field of biotechnology.

Restriction enzymes have been isolated from a number of bacteria and are named after the bacterium of origin. *EcoRI* is a restriction enzyme obtained from the R strain of *E. coli.* The Roman numeral I indicates that it was the first restriction enzyme obtained from this strain of bacteria.

Isolating the specific gene (or genes) that causes a particular genetic disease is a monumental task. The first problem to be surmounted is dealing with the enormous amount of a cell's DNA, only a minute portion of which contains the gene sequence. Thus the first task is to obtain smaller pieces of DNA that can be more easily handled. The discovery of *restriction enzymes* (also known as restriction endonucleases) in 1970 significantly enhanced the ability of researchers to cut DNA at precisely known nucleotide sequences, yielding DNA fragments of defined length. For example, the restriction enzyme *EcoRI* recognizes the nucleotide sequence below and cuts both DNA strands at the sites indicated:

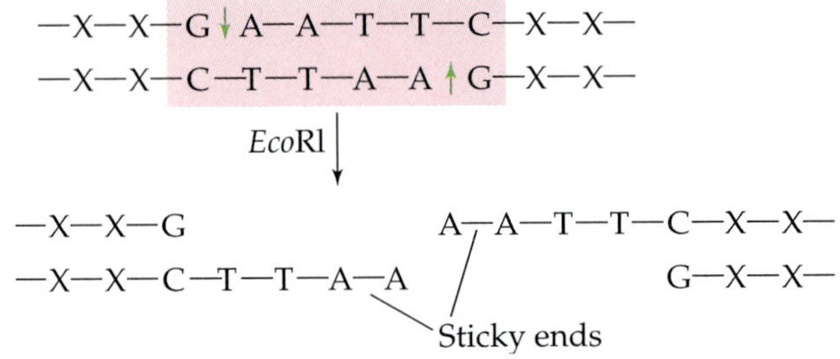

—X—X—G ↓ A—A—T—T—C—X—X—
—X—X—C—T—T—A—A ↑ G—X—X—

EcoRI ↓

—X—X—G   A—A—T—T—C—X—X—
—X—X—C—T—T—A—A   G—X—X—

Sticky ends

Once the DNA has been fragmented it must be **cloned;** that is, multiple identical copies of each unique DNA fragment must be made to insure sufficient amounts of material to detect and manipulate in the laboratory. The individual DNA fragments are inserted into phages (bacterial viruses) that can enter bacterial cells and

The Human Genome Project

The Human Genome Project, begun in 1990, is an international effort to determine the location of each of the ~80,000–100,000 human genes on the 23 pairs of chromosomes that comprise the human **genome** and to determine the complete nucleotide sequence of the nuclear DNA (the DNA found in the nucleus and assembled into chromosomes). In the United States this project is supported by the National Institutes of Health and the Department of Energy.

The coding regions or actual gene sequences comprise only about 2% of the total human genome (the function of the remaining 98% is unknown). Thus one of the initial efforts of the Human Genome Project has been to create a physical map of each chromosome. This work has led to the identification of about 100 disease-causing genes. We have discussed a few of the diseases that are the result of a mutation in a specific gene; for some of these the mutated gene has been identified, but for others it has not. Identifying the defective gene would give researchers the ability to detect and treat diseases sooner (before overt symptoms appear) and possibly design ways to prevent some of these diseases. Research also can concentrate on treatments that target the cause of the disease and not just the symptoms, through treatments such as gene therapy. In addition, it is possible to identify the protein that is not made, or made in an ineffective form, to better understand the molecular basis of a disease.

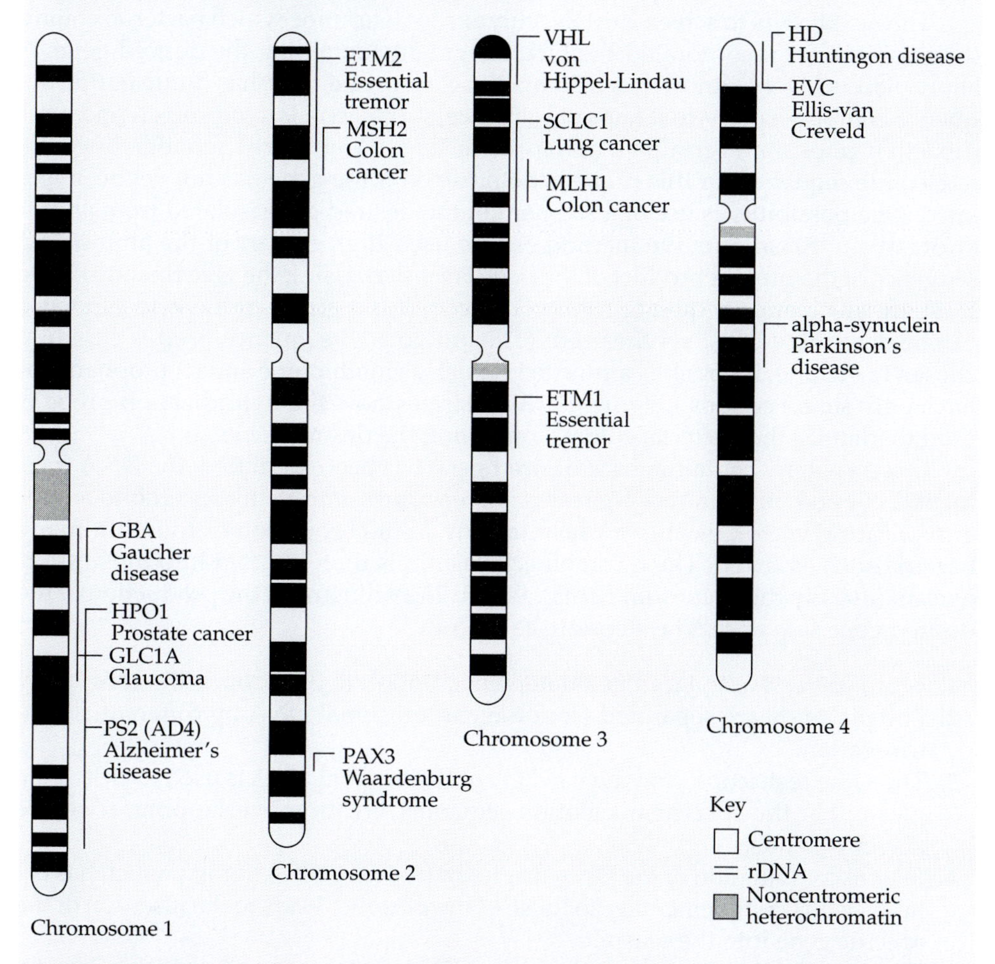

▲ The locations of several genes, mutations in which lead to specific genetic diseases, have been determined. For example, the defective gene that leads to Gauchers disease is found on chromosome 1.

The Human Genome Project has outlined eight major goals for the five-year period beginning in 1998. These goals are: (1) obtain a high-quality nucleotide sequence of the human genome that is publicly available; (2) improve techniques used to sequence DNA so that longer sequences can be done in a shorter period of time; (3) obtain information on the types, frequencies, and distribution of sequence variation in the human population; (4) develop techniques for determining the function of DNA sequences, including introns and other noncoding sequences; (5) compare genomes from a variety of organisms from bacteria to humans; (6) examine the ethical, legal, and social implications of this research, such as issues of privacy and fair use of genetic information; (7) develop databases and analytical tools that will allow researchers to readily access this tremendous amount of information in a useful way; and (8) train scientists in interdisciplinary areas involved with genomic research such as molecular biology, law, and the social sciences.

be replicated. When the bacterial cell infected by the modified phage is placed in the appropriate culture media, it forms a colony of cells, all containing copies of the original DNA fragment. Thus multiple copies of this fragment are produced. Many bacterial colonies are formed, each producing a unique DNA fragment. In this way, a *DNA library* is formed—a collection of bacterial colonies that together contain the entire genome of a particular organism.

The next task is to screen the DNA library to determine which bacterial colony (or colonies) has incorporated the DNA fragment containing the desired gene. A short piece of DNA, known as a **hybridization probe,** that has a nucleotide sequence complementary to a known sequence in the gene is synthesized and a radioactive phosphate group is added as a "tag." You might wonder how the nucleotide sequence for this probe is determined if the gene has not yet been isolated. One possibility is use of a segment of the desired gene isolated from another organism. An alternative method can be used if all or part of the amino acid sequence of the protein product of the gene is known. Using the genetic code (Table 23.3), the nucleotide sequence that would code for a given amino acid sequence can be determined and synthesized. (The amino acid sequence used is carefully chosen to include, if possible, amino acids such as methionine and tryptophan that have only single codons.) Figure 23.18 illustrates how the hybridization probe is used to identify the bacterial colony containing the desired gene.

There is a high degree of similarity among genes with the same functions from different organisms.

Once a colony containing the desired gene has been identified, the DNA fragment is clipped out, again using restriction enzymes, and then spliced into another replicating vector, usually a **plasmid,** tiny mini-chromosomes found in many bacteria such as *E. coli*. The recombined plasmid is then inserted into the host organism (usually the bacterium *E. coli*). Figure 23.19 illustrates the production of the desired gene sequence. The steps are as follows:

▲ Circular plasmids isolated from *E. coli.*

1. *E. coli* bacteria are placed in a detergent solution to break open the cells.
2. The plasmids are separated from the chromosomal DNA by differential centrifugation.
3. The same restriction enzyme used to cut the original DNA is used to cleave the plasmid at the specific nucleotide sequence, creating overlapping, cohesive ("sticky") ends.
4. *In vitro* combination of the DNA fragment from the DNA library, which has cohesive ends complementary to those of the plasmid, leads to the insertion of the specific gene into the plasmid.
5. The enzyme DNA ligase seals the foreign DNA segment into place in the plasmid.
6. The resealed plasmid is placed in a solution of calcium chloride containing *E. coli*. When the solution is heated, the bacterium cell membrane becomes permeable, allowing the plasmid to enter.

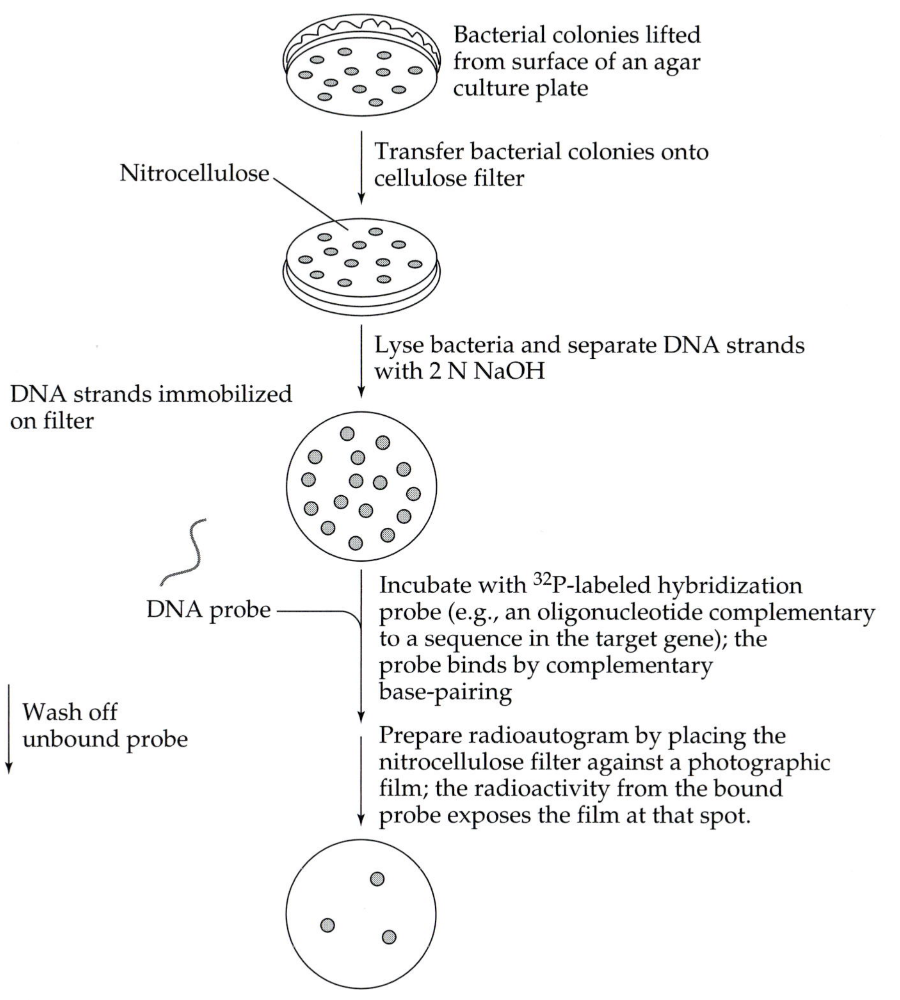

Bacterial colonies lifted from surface of an agar culture plate
Transfer bacterial colonies onto cellulose filter
Nitrocellulose
Lyse bacteria and separate DNA strands with 2 N NaOH
DNA strands immobilized on filter
DNA probe
Incubate with ³²P-labeled hybridization probe (e.g., an oligonucleotide complementary to a sequence in the target gene); the probe binds by complementary base-pairing
Wash off unbound probe
Prepare radioautogram by placing the nitrocellulose filter against a photographic film; the radioactivity from the bound probe exposes the film at that spot.
³²P-labeled probe hybridized with DNA from a particular colony makes that colony radioactive

Figure 23.18
A probe containing a radioactive phosphate group is used to identify bacteria containing DNA sequences complementary to the probe.

7. *Escherichia coli* reproduces by dividing (and thus doubling its population) at a rate of about once every 20 to 30 minutes, producing vats of material containing the modified plasmids. These plasmids replicate, the gene sequences, including the inserted gene sequence, are transcribed into mRNA, and the mRNA is then used to produce specific proteins.

Many valuable materials, difficult to obtain in any other way, are now made using recombinant DNA technology. People with diabetes formerly had to use insulin from pigs or cattle. Now human insulin is being produced in recombinant bacteria. All newly diagnosed insulin-dependent diabetics in the United States are now treated with human insulin produced through recombinant DNA technology. The hope for the future is that a functioning gene for insulin can be incorporated into the cells of insulin-dependent diabetics.

Human growth hormone, used to treat children who fail to grow properly, was formerly available only in tiny amounts obtained from cadavers. Now it is readily available through recombinant DNA technology. This technology also yields interferon, a promising anticancer agent. The gene for epidermal growth factor, which stimulates the growth of skin cells, has been cloned. It has been used to speed the healing of burns and other skin wounds. Scientists have even designed bacteria that "eat" the oil released in an oil spill, although success in actual spills has been minimal.

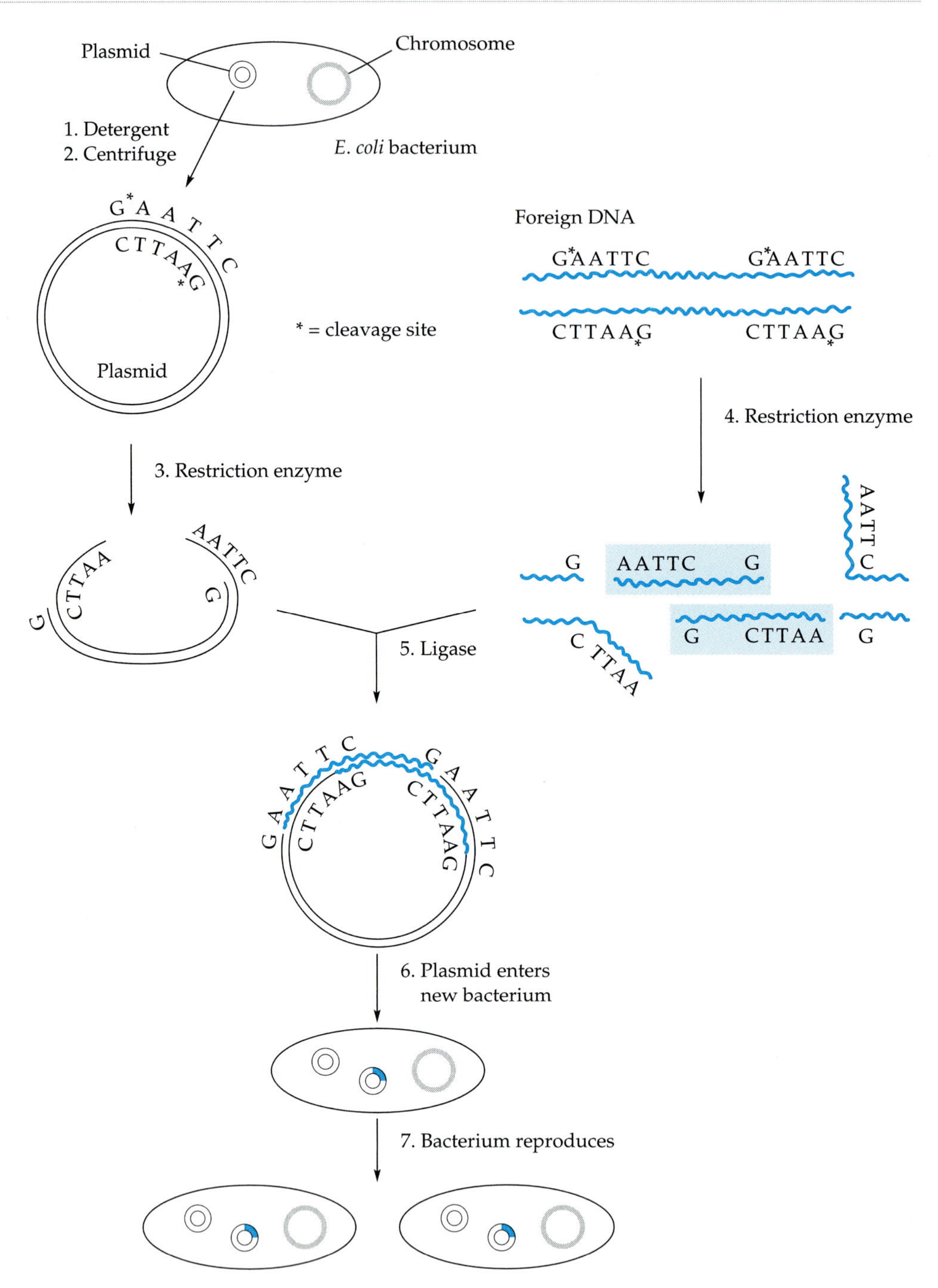

► **Figure 23.19**
Recombinant DNA in *E. coli.*

Proponents of recombinant DNA research are excited about its great potential benefits. Recombinant techniques are an enormous aid to scientists in mapping and sequencing genes and in determining the functions of different segments of an organism's DNA. The complete DNA sequences of more than 100 mammalian genes have been determined using recombinant DNA technology. An understanding of gene function and gene regulation is a primary goal of scientists working to cure cancer.

Human gene therapy has begun. Skin cancer patients have received injections of genetically altered white blood cells. The cells circulate in the blood seeking out malignant tissue. When one of these cells contacts a cancer cell, its foreign gene produces a toxic enzyme called tumor necrosis factor. Cancer cells get lethal doses, and the rest of the body is spared.

DNA Fingerprinting

In 1985 the British biologist Alec Jeffreys developed a technique that he termed "DNA fingerprinting." This technique relies on the fact that, like fingerprints, a person's DNA is unique to that individual (and an identical twin). DNA fingerprinting has been hailed as a major advance in criminal investigation. Several hundred criminal cases have been solved with this technology. Also, because children inherit half their DNA from each parent, DNA fingerprinting has been used to establish the parentage of a child of contested parentage.

Any cells—skin, blood, semen, saliva, and so forth—can supply the necessary DNA. The DNA is isolated from the biological sample and cut at specific sites using restriction enzymes. The fragmented DNA is then separated by size using gel electrophoresis (Section 21.2). The DNA fragments are visualized by staining or by imaging as an autoradiogram. In this final step, a "print" is represented as a series of horizontal bars resembling the bar codes imprinted on packaged goods sold in supermarkets. In a criminal case, the "print" made from the DNA of a suspect can be compared with the "print" of the DNA obtained from evidence at the crime scene.

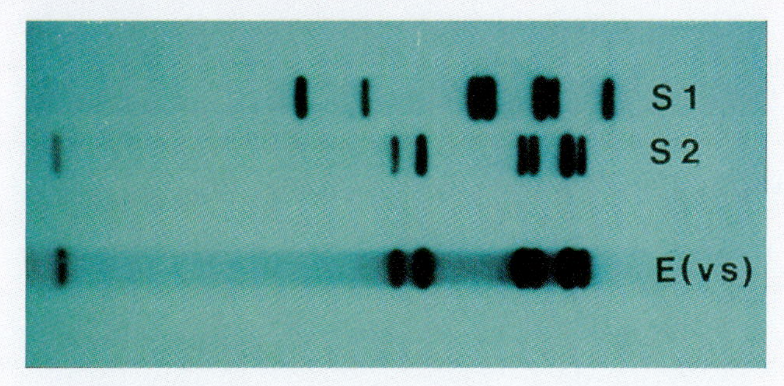

Rigorous procedures must be followed to obtain evidence that can be used in a court case. Internal controls must be run on the same gel to insure that two bands that appear to line up on the "print" in fact are the same size fragment of DNA. Questions still being debated include the statistical probability that a DNA sample is from a given suspect, based on the matching of bands.

It is conceivable that recombinant DNA could lead to cures for genetic diseases. When appropriate genes are successfully inserted into *E. coli,* the bacteria can become miniature pharmaceutical factories, producing great quantities of insulin, clotting factor for hemophiliacs, missing enzymes, hormones, vitamins, antibodies, vaccines, and so on. The production of DNA-recombinant molecules containing synthetic genes for the production of tissue plasminogen activator (TPA—a clot-dissolving enzyme that can rescue heart attack victims) has been accomplished in *E. coli.* Vaccines against hepatitis B (humans) and hoof-and-mouth disease (cattle) have been produced. In addition, it may be possible to breed human intestinal bacteria that can digest cellulose and to create new plants that can obtain their nitrogen directly from the air rather than from costly petroleum-based fertilizers.

Besides *E. coli,* scientists have used other bacteria as well as yeast and fungi in gene-splicing experiments. The enzyme rennin has been produced in fungi. Rennin is used commercially to coagulate milk into curds in the production of cheese. Plant molecular biologists use a bacterial plasmid as a vector to introduce genes for several foreign proteins (including animal proteins) into plants. The bacterium is *Agrobacterium tumefaciens,* which can cause tumors in many plants but whose tumor-causing ability has been eliminated. One practical application would be to enhance a plant's nutritional value by transferring into it the gene necessary for the synthesis of a deficient amino acid (e.g., the gene for methionine synthesis into pinto beans).

Concern over the potential for disaster in this type of research has lessened somewhat in recent years. Initially, scientists worried about the possibility of producing a deadly "artificial" organism. What if a gene that causes cancer were spliced into the DNA of a bacterium that normally inhabits our intestines? We would have no natural immunity against such an artificial organism. To protect against such a development, strict guidelines for recombinant DNA research have been instituted.

Another point of contention is that this research can be misused for political and social purposes. These techniques might be exploited for genetically engineered control of human behavior, even enhancing people's IQs. There are some people who believe that we should not attempt to create new forms of life different from any that exist on Earth.

The new molecular genetics has already resulted in some impressive achievements. Its possibilities are mind-boggling—elimination of genetic defects, a cure for cancer, a race of geniuses, and who knows what else? Knowledge gives power. It does not necessarily give wisdom. The greatest problem we are likely to face in our use of bioengineering is that of choosing who is to play God with the new "secret of life."

Review Questions

23.16 What is the basic process in recombinant DNA technology?

23.17 Discuss some applications of genetic engineering.

Summary

A cell's hereditary information is encoded in **chromosomes** in the cell nucleus. Each chromosome is composed of proteins and **deoxyribonucleic acid (DNA).** The DNA contains smaller hereditary units called **genes,** relatively short segments on the chromosome. This hereditary information is expressed or utilized through the synthesis of **ribonucleic acid (RNA).** The **nucleic acids,** DNA and RNA are polymers composed of smaller units known as **nucleotides,** which can be hydrolyzed into **nucleosides** and phosphoric acid. Nucleosides can be further hydrolyzed to a nitrogenous base and a five-carbon sugar.

The two types of *nitrogenous bases* most important in nucleic acids are **purines**—adenine (A) and guanine (G)—and **pyrimidines**—cytosine (C), thymine (T), and uracil (U). DNA contains the nitrogenous bases adenine, cytosine, guanine, and thymine, while the bases in RNA are adenine, cytosine, guanine, and uracil.

The five-carbon sugar in nucleotides obtained from RNA is ribose and in nucleotides obtained from DNA, it is 2-deoxyribose .

The sequence of nucleotides in any nucleic acid defines the primary structure of the molecule. RNA is a single-chain nucleic acid. DNA comprises two nucleic-acid chains intertwined in a secondary structure called a double helix. The sugar-phosphate backbone forms the outside of the double helix, with the purine and pyrimidine bases inside. The pairing of bases, one from each chain, is always A–T and C–G. Adenine and thymine are one pair of **complementary bases,** and cytosine and guanine are another pair.

Cell growth requires the synthesis of new strands of DNA. This involves the **replication** of the cell's DNA. The double helix unwinds, and the hydrogen bonds joining complementary bases break, so that there are two single strands of DNA, each a *template* for the synthesis of a new strand.

For protein synthesis, three types of RNA are needed: *messenger RNA, ribosomal RNA,* and *transfer RNA.* All are made from a DNA template by a process called **transcription.** The double helix uncoils just as in replication, and nucleotides base-pair to one DNA strand. Three of the pairings are the same as in DNA replication (C–G, G–C, T–A), but any template adenine calls for *uracil* on the RNA molecule (A-U). Once the RNA is formed, it dissociates from the template and leaves the nucleus, and the DNA double helix reforms.

The **ribosome** is the site of protein synthesis. It is located outside the nucleus and is made of rRNA and protein. Clusters of ribosomes banded together by a strand of mRNA are *polysomes.*

Protein synthesis depends on the **genetic** or **triplet code,** the 64 possible three-nucleotide combinations of the four nucleotides of DNA. Each three-nucleotide sequence on mRNA is a **codon.**

The sequence of nucleotides in an organism's DNA is the *genetic code* for that organism. The general term for any change in the code is **mutation,** and each specific change is a **point mutation.**

In **recombinant DNA technology,** a gene from one organism is inserted or "spliced" into the DNA of a host organism. The host then replicates large quantities of the altered DNA containing the foreign gene, and this DNA is harvested by humans and used in medical treatments and other applications.

Key Terms

anticodon (23.5)
chromosome (page 635)
clone (23.8)
codon (23.6)
complementary base (23.3)
deoxyribonucleic acid (DNA) (page 635)
double helix (23.3)
gene (page 635)
genetic code (23.6)
genetic disease (23.7)
genome (23.8)
hybridization probe (23.8)
mutagen (23.7)
mutation (23.7)
nucleic acid (page 635)
nucleoside (23.1)
nucleotide (23.1)
plasmid (23.8)
point mutation (23.7)
purine (23.1)
pyrimidine (23.1)
recombinant DNA technology (23.8)
replication (23.4)
ribonucleic acid (RNA) (page 635)
ribosome (23.5)
transcription (23.5)
translation (23.6)
triplet code (23.6)

Problems

Nucleotides

1. What is the sugar unit in RNA? in DNA?
2. What are the major bases present in DNA? in RNA?
3. For each of the following, indicate whether the compound is a nucleoside, a nucleotide, or neither.

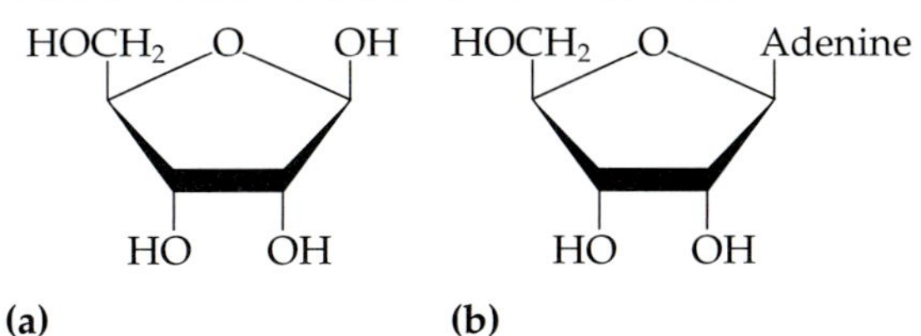

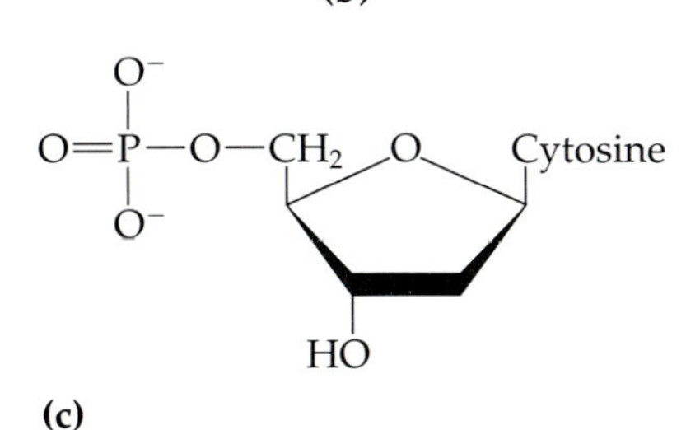

4. For each of the following, indicate whether the compound is a nucleoside, a nucleotide, or neither.

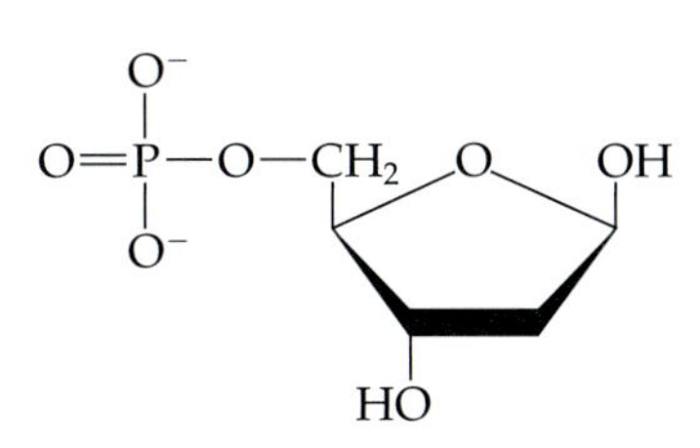

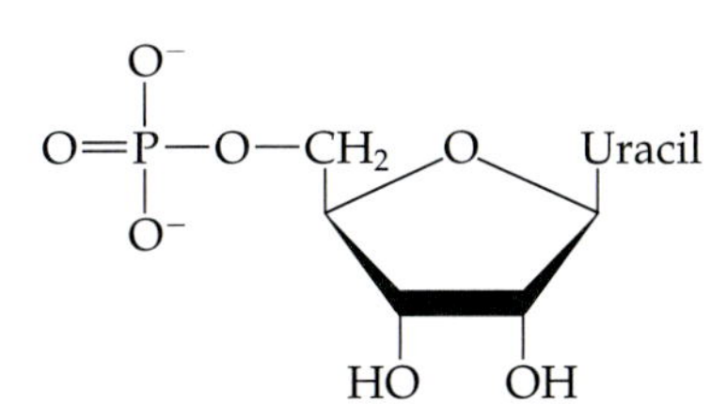

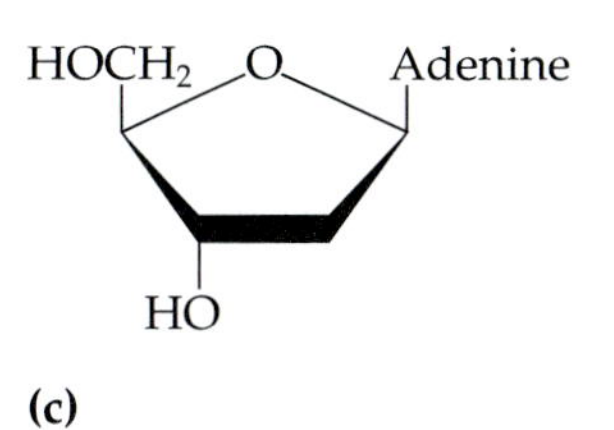

5. For each structure in Problem 3, indicate whether the sugar unit is ribose or deoxyribose.
6. For each structure in Problem 4, indicate whether the sugar unit is ribose or deoxyribose.
7. Draw structural formulas for the following minor pyrimidines and purines.
   a. N^6-methyladenine
   b. 5-methylcytosine
   c. 1-methylguanine
   d. 4-thiouracil
8. Draw structural formulas for the following nucleosides and nucleotides.
   a. deoxyadenosine
   b. 5-methylcytidine
   c. guanosine monosphosphate
   d. deoxythymidine triphosphate

Primary and Secondary Structure of Nucleic Acids

9. DNA and RNA are termed nucleic *acids*. What makes them acidic?
10. Using a schematic representation, show a length of nucleic acid chain (approximately 3 nucleotides in length), indicating the positions of the sugar, phosphoric acid, and base units.
11. With the same sort of schematic representation you used for Problem 10, show the overall design of the double helix. Be sure to clearly designate the 5′ end and 3′ end of each strand.
12. a. Draw the complete structure of the RNA trinucleotide CGU.
    b. Circle the atoms that comprise the backbone of the nucleic acid chain.
    c. Identify the 5′ end and the 3′ end of the molecule.

13. a. Draw the complete structure of the DNA trinucleotide dTAC.
    b. Circle the atoms that comprise the backbone of the nucleic acid chain.
    c. Identify the 5′ end and the 3′ end of the molecule.
14. In DNA, list the base that pairs with
    a. cytosine b. adenine
    c. guanine d. thymine
15. In RNA, list the base that pairs with
    a. cytosine b. adenine
    c. guanine d. uracil
16. The base sequence along one strand of DNA is 5′ d-ATTCG 3′. What is the sequence of the complementary strand of DNA?
17. How many total hydrogen bonds would exist between the two strands of DNA in Problem 16?
18. One of the key pieces of information that Watson and Crick used in determining the secondary structure of DNA was from experiments done by E. Chargaff in which he studied the nucleotide composition of DNA from many different species. Chargaff noted that the molar quantity of A was always approximately equal to the molar quantity of T, and the molar quantity of C was always approximately equal to the molar quantity of G. How were Chargaff's results explained by the structural model of DNA proposed by Watson and Crick?
19. Suppose Chargaff had used RNA instead of DNA (see Problem 18). Would his results have been the same; that is, would the molar quantity of A approximately equal the molar quantity of T? Explain your answer.

Replication and Transcription

20. In DNA replication, a parent DNA molecule produces two daughter molecules. What is the fate of each strand of the parent DNA double helix?
21. a. How are replication and transcription similar?
    b. How do these two processes differ?
22. a. A portion of the coding strand for a given gene has the sequence:

    5′ d–ATGAGCGACTTTGCGGGATTA 3′

    What is the sequence of the complementary strand?
    b. What is the sequence of the mRNA that would be produced during transcription from this segment of DNA?
23. a. A portion of the coding strand for a given gene has the sequence:

    5′ d–ATGGCAATCCTCAAACGCTGT 3′

    What is the sequence of the complementary strand?
    b. What is the sequence of the mRNA that would be produced during transcription from this segment of DNA?

Protein Synthesis and the Genetic Code

24. What anticodon on tRNA would pair with the mRNA codon:
    a. 5′ UUU 3′ b. 5′ CAU 3′
    c. 5′ AGC 3′ d. 5′ CCG 3′
25. What codon on mRNA would pair with the tRNA anticodon:
    a. 5′ UUG 3′ b. 5′ GAA 3′
    c. 5′ UCC 3′ d. 5′ CAC 3′
26. a. How many nucleotides are present in a codon?
    b. Why is this number of nucleotides needed? Why couldn't the number be one greater or one less?
27. The peptide hormone oxytocin contains nine amino acid units. What is the minimum number of nucleotides needed to code for this peptide?
28. Phosphoglucose isomerase, an enzyme needed to break down glucose for energy, has been isolated from pig muscle and is composed of two identical polypeptide chains, each chain containing 585 amino acid units. What is the minimum number of nucleotides that must be present in the mRNA that codes for this protein?
29. Using Table 23.3, identify the amino acids carried by the tRNA molecules in Problem 24.
30. Using Table 23.3, identify the amino acids carried by the tRNA molecules in problem 25.
31. Using Table 23.3, determine the amino acid sequence produced from the mRNA synthesized in Problem 22.
32. Using Table 23.3, determine the amino acid sequence produced from the mRNA synthesized in Problem 23.

Applications of Molecular Biology

33. A point mutation occurred in the portion of the gene sequence given in Problem 22 so that the sequence on the coding strand was:

    5′ d–ATGAGCGACCTTGCGGGATTA 3′

    a. Identify the mutation as substitution, insertion, or deletion.
    b. What effect would the mutation have on the amino acid sequence of the protein obtained from this mutated gene (use Table 23.3)?
34. For each genetic disease listed below, indicate which enzyme is lacking due to a genetic mutation and the characteristic symptoms of that disease.
    a. phenylketonuria
    b. Gauchers disease
    c. Tay-Sachs disease
35. Suppose 2-aminopurine is incorporated into a DNA strand being synthesized in your liver cell instead of adenine. This base pairs with cytosine when the DNA is replicated, introducing a point mutation into the DNA. Will this genetic mutation be passed on to any children that you have (assuming you have children)? Why or why not?
36. a. Why are restriction enzymes so important in recombinant DNA technology?
    b. Why are plasmids so important in recombinant DNA technology?

Additional Problems

37. What is the relationship among chromosomes, genes, and DNA?
38. a. Diagram the replication of the DNA segment shown below and indicate the final product of replication.

    5′ d–ATGTACCACGGTACG 3′ (coding strand)

    3′ d–TACATGGTGCCATGC 5′ (template strand)

    b. Write the mRNA sequence that would be obtained from transcription of the DNA segment in (a).
    c. What is the amino acid sequence of the peptide produced from this mRNA (see Table 23.3)?
39. What is the amino acid sequence of the peptide produced from this DNA segment (see Table 23.3)?

    5′ d–ATGCGGGTATTGCTAGCC 3′ (coding strand)

    3′ d–TACGCCCATAACGATCGG 5′ (template strand)

40. What are the two most important sites on tRNA molecules?
41. A hypothetical protein has a molecular weight of 18,700 daltons. Assume that the average molar mass of an amino acid is 120.
    a. How many amino acids are present in the protein?
    b. How many codons occur in the mRNA that codes for this protein?
    c. What is the minimum number of nucleotides needed to code for this protein?
42. The hormone somatostatin, produced in the pancreas, is composed of 14 amino acids. Somatostatin inhibits the release of a variety of hormones, including glucagon, insulin, and gastrin. It was the first peptide hormone to be synthesized using recombinant DNA technology.
    a. Given the following structure for somatostatin, postulate a base sequence in the mRNA that directs the synthesis of somatostatin. Include an initiation codon and a termination codon.

```
Ala-Gly-Cys-Lys-Asn-Phe-Phe
         |                 |
         S                Trp
         |                 |
         S                Lys
         |                 |
        Cys-Ser-Thr-Phe-Thr
```

    b. What is the nucleotide sequence of the DNA that codes for this mRNA?
43. We shall see in Section 28.9 that sickle cell hemoglobin differs from normal hemoglobin as a result of the substitution of valine for glutamic acid as the sixth amino acid from the N-terminal end of the polypeptide chain.
    a. What alteration in the nucleotide sequence of the DNA could have caused this substitution?
    b. What is the resulting nucleotide sequence in the mRNA?
44. The following table contains just a few of the 200 or so known hemoglobin variants. For each mutation, give a codon for the normal amino acid and indicate how a single substitution would give the codon for the amino acid that is found in the hemoglobin variant.

Type	Chain	Position	Normal	Mutant
J	α	5	Ala	Asp
I	α	16	Lys	Glu
M	α	58	His	Tyr
D	β	121	Glu	Gln
K	β	136	Gly	Asp

45. For the DNA coding segment

    5′ d–ACGTTAGCCCCAGCT 3′

    a. Write the sequence of nucleotides in the corresponding mRNA.
    b. What is the amino acid sequence formed during translation?
    c. What amino acid sequence results from each of the following?
        i. replacement of the red guanine by adenine
        ii. insertion of thymine immediately after the red guanine
        iii. deletion of the red guanine
46. Assume that a coding segment of a gene has the nucleotide sequence

    5′ d–TACGACGTAACAAGC 3′

    a. What effect results from a point mutation in which an adenine replaces the red guanine?
    b. What effect results from a point mutation in which a thymine replaces the red adenine?
47. Following are the results of two point mutations. Which is likely to be more serious and why?
    a. Valine is substituted for leucine.
    b. Glutamic acid is substituted for leucine.
48. What techniques discussed in this chapter require the use of restriction enzymes? Why are these enzymes important in these techniques?
49. a. What is the importance of the technique known as the polymerase chain reaction?
    b. Briefly describe how this technique works.
50. It has been found that most human gene sequences are composed of exons and introns, which are transcribed into an initial RNA molecule. Which sections—the exons or the introns—must be connected together to form the mature mRNA that will be used to synthesize the protein?
51. Outline the major goals of the Human Genome Project.
52. How could information obtained from the Human Genome Project be used to help find the cause of a genetic disease?
53. Outline how a "DNA fingerprint" is obtained.

Viruses

A discussion of viruses seems particularly appropriate here because viruses are composed almost entirely of proteins and nucleic acids. **Virology,** the study of viruses, is a rapidly expanding field. As Table G.1 indicates, a greater number of human diseases are of viral origin than of bacterial origin. Infectious diseases of viral origin (especially the common cold, influenza, and AIDS) are among the most significant health problems in our society.

G.1 The Nature of Viruses

Viruses are a unique group of infectious agents composed of a tightly packed central core of nucleic acids enclosed by a protective shell. The shell is composed of layers of one or more proteins and may also include lipid or carbohydrate molecules on the surface. Viruses are much smaller than bacteria and are visible only under the electron microscope. The tobacco mosaic virus, for example, is approximately 0.30 μm long by 0.018 μm wide and has a molar mass of about 40 million daltons. By comparison, the size of an average bacterial cell is 1.5 μm long by 0.75 μm wide.

Viruses are divided into two main classes on the basis of nucleic acid content. Viruses contain either DNA or RNA, *but never both.* (Recall that the cells of organisms, from bacteria to humans, contain both kinds of nucleic acids.) Viruses differ from one another in both size and shape. The influenza virus, for example, is about ten times bigger than the polio virus. Viruses are found in a variety of shapes ranging from spherical to rod-shaped.

Table G.1 Human Infectious Diseases

Diseases of Bacterial Origin	Diseases of Viral Origin	
Cholera	AIDS	Measles
Diphtheria	Chicken pox	Meningitis
Dysentery	Cold sores	Mumps
Gonorrhea	Common cold	Pneumonia
Plague	Encephalitis	Polio
Syphilis	Gastroenteritis	Rabies
Tetanus	Genital herpes	Shingles
Tuberculosis	German measles	Smallpox
Typhoid fever	Hepatitis	Warts
Whooping cough	Influenza	Yellow fever

A *DNA virus* enters a host cell, where the viral DNA is replicated and directs the host cell to produce viral proteins. The viral proteins and viral DNA assemble into new viruses, which are released by the host cell. These new viruses can then invade other cells and continue the process. Cell death and the production of new viruses account for the symptoms of viral infections.

Most *RNA viruses* use their nucleic acids in much the same way as do the DNA viruses. The virus penetrates a host cell, where the RNA strands are replicated and induce the synthesis of viral proteins. The new RNA strands and viral proteins are then assembled into new viruses. Some RNA viruses, called **retroviruses,** synthesize DNA in the host cell. This process is the opposite of the DNA-to-RNA transcription that normally occurs in cells. The synthesis of DNA from an RNA template is catalyzed by the enzyme reverse transcriptase. Figure G.1 depicts the life cycle of the human immunodeficiency virus (HIV) that causes AIDS (acquired immune deficiency syndrome), perhaps the best-known retrovirus.

G.2 Antiviral Drugs

The ideal drug for treating infectious diseases is one that kills the infectious agent (virus or bacterium), but is not toxic to the organism being infected (you). A wide variety of drugs have been developed to treat bacterial infections. Most of these drugs inhibit or block a particular

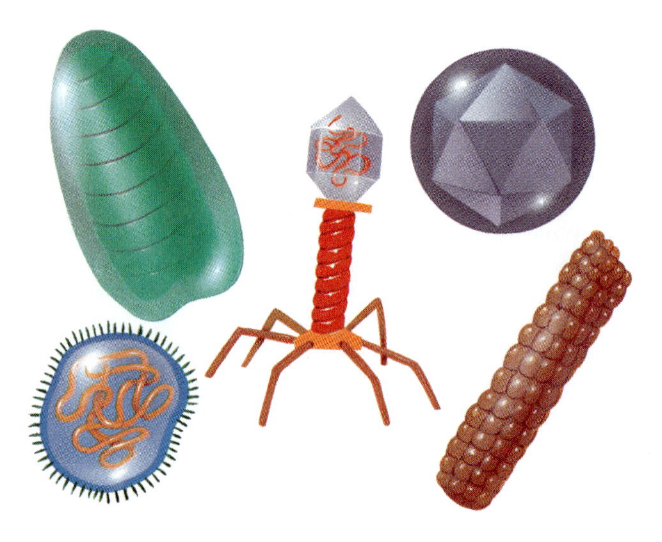

▲ Viruses come in a variety of shapes that are determined by their protein coats.

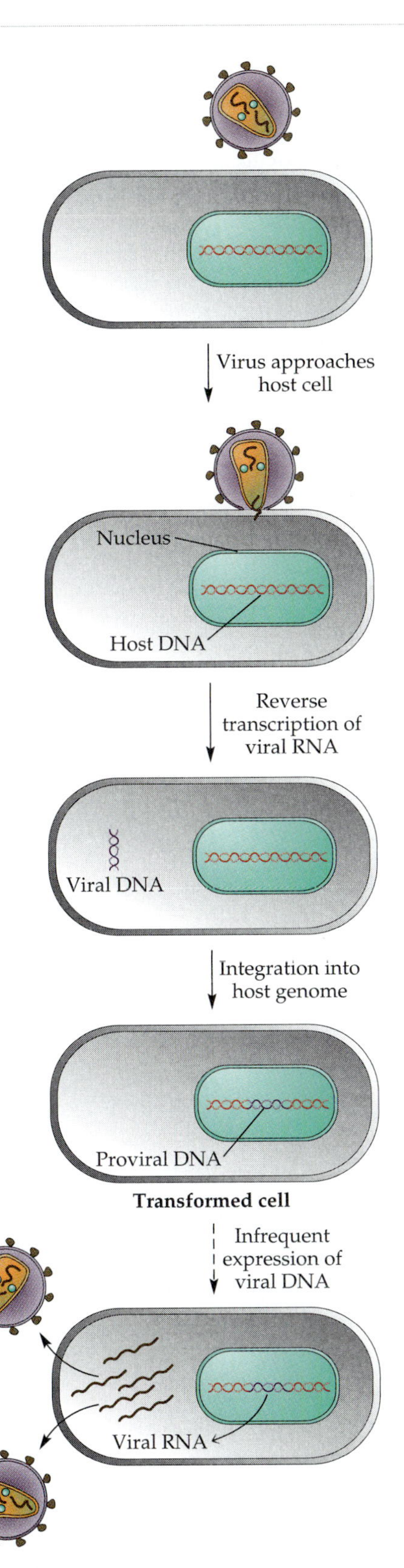

▲ **Figure G.1**
Life cycle of a retrovirus.

reaction uniquely required by the bacterium, but not the host (recall how sulfanilamide works; Section 22.7). Viruses are parasites; that is, they depend on the cells they infect—the host cells—to provide the enzymes and other molecules necessary to carry out replication, transcription, and translation. Thus there are few reactions that are unique to the virus. Because of the difficulty in developing drugs to treat viral infections, vaccines have been widely used to prevent them (Section 28.5).

One of the earliest antiviral drugs developed was acyclovir, a compound similar to deoxyguanosine in which the attached sugar is a short linear chain rather than deoxyribose:

O, HN, N, H_2N, N, N, $CH_2OCH_2CH_2OH$

Acyclovir

O, HN, N, H_2N, N, N, $HOCH_2$, O, HO

Deoxyguanosine

Acyclovir (Zovirax®) controls flare-ups of the herpes viruses that cause genital sores, chicken pox, shingles, mononucleosis, and cold sores. These viruses induce the synthesis of a viral thymidine kinase that is quite different from the host cell's thymidine kinase. Thymidine kinase catalyzes the reaction in which the nucleoside thymidine is converted to the nucleotide thymidine monophosphate:

O, CH_3, HN, O, N, $HOCH_2$, O, HO

Deoxythymidine

thymidine kinase →

O, CH_3, HN, O, N, ^-O—P(=O)(—O^-)—OCH_2, O, HO

Deoxythymidine monophosphate
dTMP

Acyclovir is a substrate of the viral thymidine kinase (but not the host's thymidine kinase) and is converted to acyclovir monophosphate. This in turn is converted to acyclovir triphosphate by cellular kinases.

Acyclovir monophosphate

Acyclovir triphosphate acts as a competitive inhibitor of the viral DNA polymerase. The DNA polymerase incorporates the acyclovir triphosphate into the growing DNA chain during replication. Because acyclovir lacks the 3′-hydroxyl group of a cyclic sugar, it blocks chain elongation, and thus replication (see Section 23.4); therefore, only short fragments of viral DNA are formed.

G.3 Human Immunodeficiency Virus (HIV) and AIDS

AIDS is now the leading cause of death for adults between the ages of 25 and 44.[1] HIV, the virus that causes AIDS, is shown in Figure G.2. The virus uses the glycoproteins on its outer surface to attach to specific receptors on the surface of T cells, a group of white blood cells that normally help protect the body from infections. The virus then enters the T cell where it replicates and eventually destroys the cell. With the T cells destroyed, the AIDS victim succumbs to pneumonia or other infectious diseases.

Azidothymidine (AZT, zidovudine or Retrovir®) was the first drug approved for the treatment of AIDS (1987). AZT binds to reverse transcriptase in place of deoxythymidine triphosphate. Like acyclovir, AZT does not have a 3′-OH group, and further replication is blocked. In the past ten years several other drugs have been approved that also act by inhibiting the viral reverse transcriptase (Figure G.3).

When HIV particles are formed in an infected cell, newly synthesized viral proteins must be cut by a specific viral-induced HIV protease to form shorter proteins. An intensive research effort was made to design drugs that specifically inhibited this proteolytic enzyme, but not other proteolytic enzymes (like trypsin) that are important for the host. In December, 1995 saquinavir (Invirase® and Fortovase®) was approved for the treatment of AIDS. This drug represented a new class of drugs known as protease inhibitors. Other protease inhibitors soon gained FDA approval: ritonavir (Norvir®) and indinavir (Crixivan®) [March, 1996] and nelfinavir (Viracept®) [March 1997].

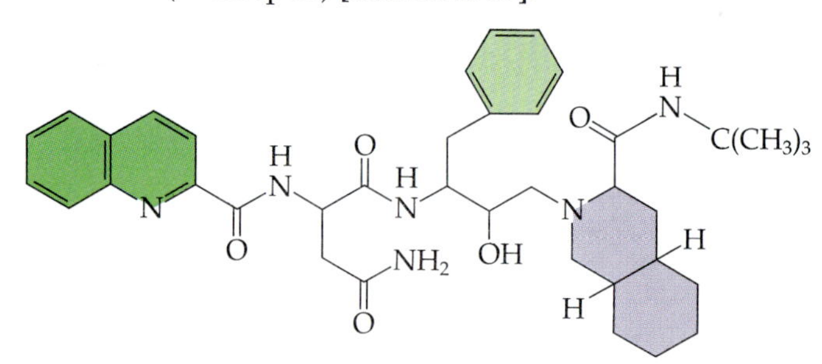

Saquinavir

The newest group of anti-AIDS drugs to be approved are the nonnucleoside reverse transcriptase inhibitors (NNRTIs), which lack the nucleoside structure found in drugs such as AZT and inhibit reverse transcriptase noncompetitively. These include nevirapine (Viramune®), efavirenz (Sustiva®), abacavir (Ziagen®), and delavirdine (Rescriptor®).

Nevirapine

A major problem in treating HIV infection is that the virus can become resistant to any of these drugs. One way to combat this has been to give a "cocktail" of drugs, typically a combination of two inhibitors of

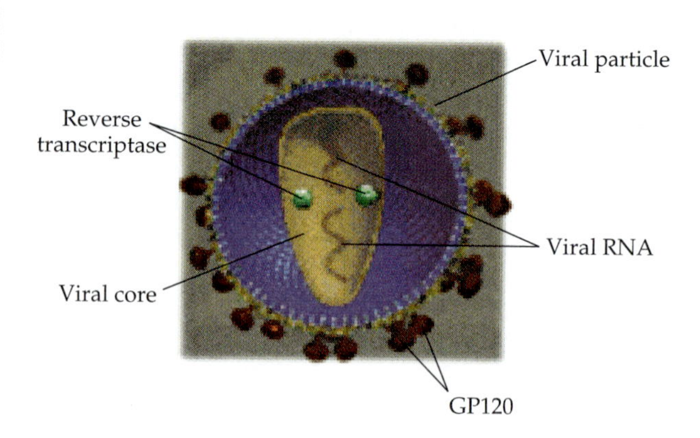

▲ Figure G.2
Cross section of the human immunodeficiency virus (HIV), which causes AIDS. The RNA on the inside is surrounded by a protein coat. GP120 is a glycoprotein (a protein with short carbohydrate chains attached) that is used by the virus to attach to the host cell.

[1]Worldwide, it is estimated that 33.4 million people are living with HIV/AIDS. An estimated 13.9 million people have died in this epidemic.

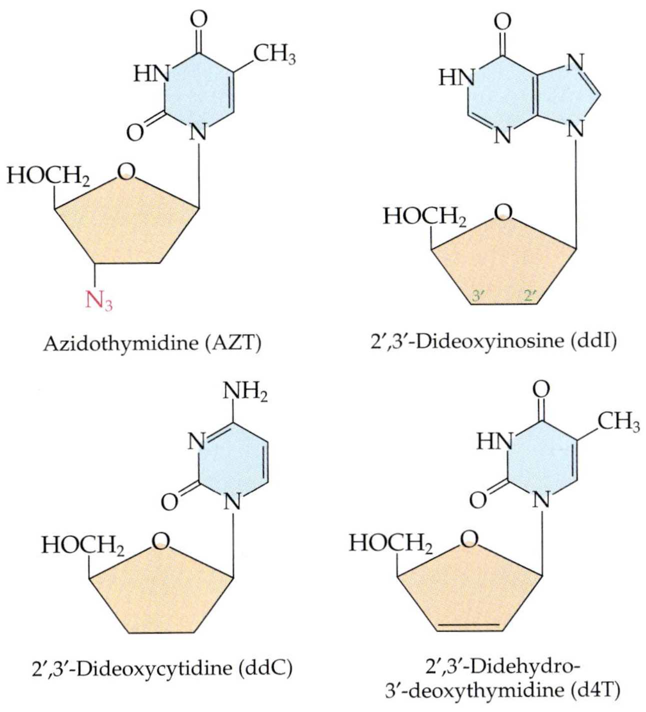

Azidothymidine (AZT)

2′,3′-Dideoxyinosine (ddI)

2′,3′-Dideoxycytidine (ddC)

2′,3′-Didehydro-3′-deoxythymidine (d4T)

▲ **Figure G.3**
Nucleoside analogs approved for use in the treatment of HIV infection.

reverse transcriptase along with a protease inhibitor. These treatments have been effective in drastically reducing the amount of HIV present in an infected person.

G.4 Cancer and Carcinogens

Viruses have been implicated as causes of some cancer, but to a lower degree than environmental factors. The World Health Organization estimates that 80–90% of cancer cases are caused by environmental factors, and 10–20% by genetic factors and viruses. Included most prominently among those "environmental" causes are cigarette smoking (~30%), lifestyle factors such as diet and lack of exercise (~30%), and occupational exposure (10%). Viruses that have been implicated as cancer-causing agents include the hepatitis viruses, which have been linked to liver cancer; papilloma viruses, associated with cervical cancer; Epstein–Barr virus, connected with certain lymphomas and nose and throat cancer; and herpes virus, associated with Kaposi's sarcoma.

The substances that cause the growth of malignant tumors are called carcinogens. A tumor, an abnormal growth of new tissue, may be either benign or malignant. **Benign tumors** are characterized by slow growth; they often regress spontaneously, and they do not invade neighboring tissues. **Malignant tumors** may grow slowly or rapidly, but their growth is generally irreversible. They often are called **cancers.** Malignant growths invade and destroy neighboring tissues. Cancers are classified by the organ in which they originate and the kind of cell involved. When considered in this way, there are about 100 distinct kinds of cancer.

If you read the newspapers or watch television, you may get the idea that everything causes cancer. How do we know that a chemical causes cancer? Obviously, we can't experiment on humans to see what happens. That leaves us with no way to prove beyond doubt that a chemical does or does not cause cancer in humans. There are three ways, however, to gain evidence against a compound: bacterial screening for mutagenesis, animal tests, and epidemiological studies.

The quickest and cheapest way to find out whether or not a substance may be carcinogenic is to use the screening test developed by Bruce N. Ames of the University of California at Berkeley. The Ames test is a simple laboratory procedure that can be carried out in a petri dish. It assumes that most carcinogens are also mutagens, altering the genes in some way.

The Ames test uses a special strain of salmonella bacteria that has been modified so that it requires histidine as an essential amino acid. The bacteria are placed in an agar medium containing all nutrients except histidine. Incubating the mixture in the presence of a mutagenic chemical causes the bacteria to mutate, so that they no longer require histidine and can grow like normal bacteria. Growth of bacterial colonies in the petri dish means that the chemical added was a mutagen, and probably also a carcinogen.

Chemicals suspected of being carcinogens can be tested on animals, although these studies cost about $1 million each and take at least two years. Tests are usually conducted on 30 or so rodents, with an equal number serving as controls. The control group is exposed to the same diet and environment as the experimental group, except that the control group does not get the suspected carcinogen. A higher incidence of cancer in the experimental animals than in the controls indicates that the compound is carcinogenic.

Animal tests are not conclusive. Humans usually are not exposed to comparable doses, and there may be a threshold below which a compound is not carcinogenic. Further, human metabolism is different from the metabolisms of the test animals. The carcinogen might be active in the rodent but not in humans (or vice versa!).[2]

[2]There is only a 70% correlation between the carcinogenesis of a chemical in rats and that in mice. The correlation between carcinogenesis in either rodent and that in humans probably is less.

The best evidence that a substance causes cancer in humans comes from epidemiological studies. A population that has a higher-than-normal rate for a particular kind of cancer is studied for common factors in their background. It was this sort of study, for example, that showed that cigarette smoking causes lung cancer, that vinyl chloride causes a rare form of liver cancer, and that asbestos causes cancer of the lining of the pleural cavity (the body cavity that contains the lungs). These studies require sophisticated mathematical analyses. There is always the chance that some other (unknown) factor is involved in the carcinogenesis.

Many carcinogens, particularly those that are ingested, are naturally occurring compounds, such as safrole in sassafras and the aflatoxins, produced by molds on foods. Plants produce natural pesticides to protect themselves from fungi, insects, and higher animals. Animal studies have been conducted on about 52 of these natural pesticides; twenty-seven were found to cause cancer in rodents. It is difficult to assess whether or not these compounds are carcinogens in humans, and if so, at what dose. In addition, many of these compounds are founds in herbs, fruits, and vegetables that also contain antioxidants (Section F.8), compounds that appear to protect against cancer.

G.5 Genetic Basis of Cancer

We seem to be surrounded by carcinogens, but not everyone develops some form of cancer. It must be that we have a way of protecting ourselves most of the time. For a cell to become cancerous, several mutations must occur in critical genes. In particular, two classes of genes have been implicated: oncogenes and tumor-suppressor genes.

Oncogenes are mutated versions of ordinary genes (proto-oncogenes) that promote cell growth and division. When mutated, the oncogenes are no longer dependent on specific signals, such as the binding of a hormone to its receptor, to became activated and lead to increased growth. **Tumor-suppressor genes** code for proteins that act as signals to inhibit cell growth. These are inactivated when mutated, thus removing the stop signals that prevent inappropriate cell growth. There is hope that someday suppressor genes can be produced, through genetic engineering, and used in therapy.

Cancers appear earlier in some individuals due to inheritance of mutations that have converted a proto-oncogene to an oncogene, inactivated a tumor-suppressor gene, or affected a gene that codes for an enzyme that is needed for the repair of damaged DNA.

G.6 Chemicals Against Cancer

Chemists have designed molecules to relieve headache, cure infectious diseases, and prevent conception. Why can't they do something about cancer? They have done a lot, but much remains to be done. We examine a few representative anticancer drugs here.

Many of the compounds used in cancer chemotherapy inhibit the synthesis of DNA (Section 22.7). Rapidly dividing cells, characteristic of cancer, require large quantities of DNA. Antimetabolites used in cancer chemotherapy block DNA synthesis and therefore block the increase of the number of cancer cells. Because cancer cells are undergoing rapid growth and cell division, they generally are affected to a greater extent than normal cells.[3] Two prominent antimetabolites are 5-fluorouracil and its deoxyribose nucleoside, 5-fluorodeoxyuridine:

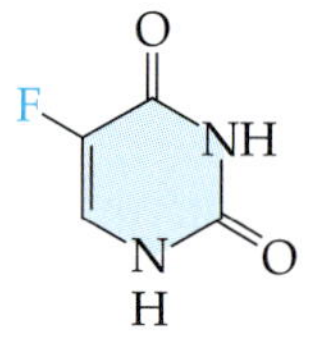

5-Fluorouracil 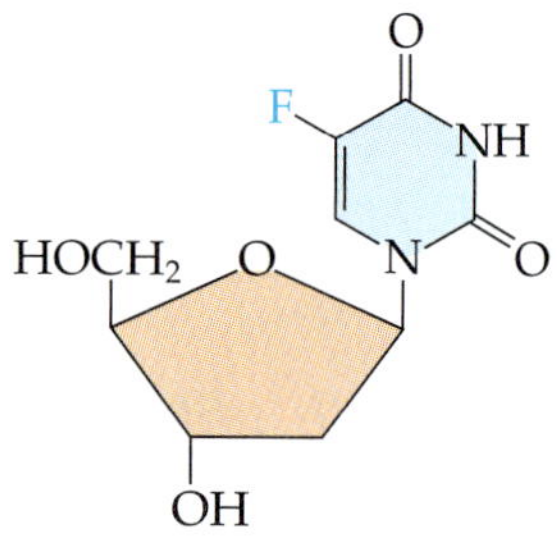 5-Fluorodeoxyuridine

In the body, both of these compounds can be incorporated into a nucleotide (Section 23.1). The fluorine-containing derivatives inhibit the formation of thymine-containing nucleotides required for DNA synthesis. Thus both compounds slow the division of cancer cells. These compounds have been employed against a variety of cancers, especially those of the breast and digestive tract.

Another common antimetabolite is 6-mercaptopurine, which substitutes for adenine in a nucleotide:

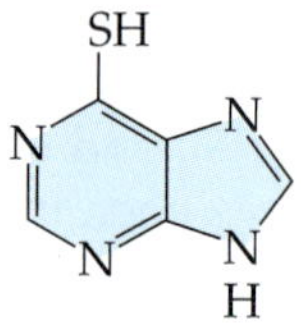

6-Mercaptopurine 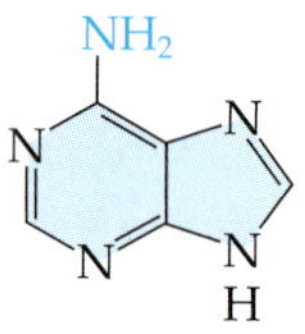 Adenine

The pseudonucleotide then inhibits the synthesis of nucleotides that incorporate adenine and guanine.

[3]Cancer chemotherapy also affects noncancerous body cells that undergo rapid replacements, including those that line the digestive tract and those that produce hair. Side effects of the therapy include nausea and loss of hair. Eventually the normal cells are affected to such a degree that treatment must be discontinued.

6-Mercaptopurine has been used in the treatment of leukemia.

Another antimetabolite, methotrexate, acts in a somewhat different manner. Note the similarity between its structure and that of folic acid:

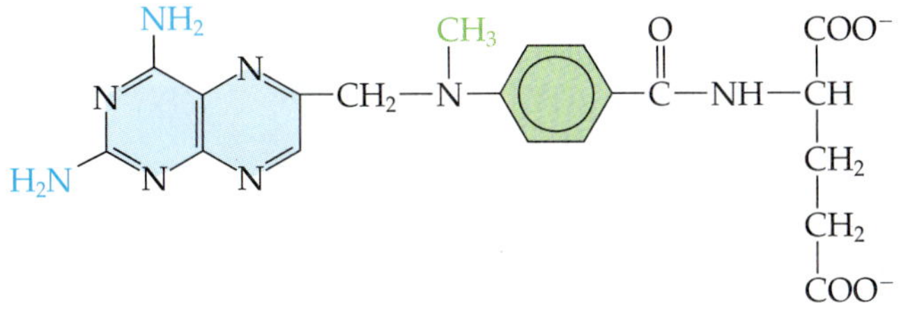

Methotrexate

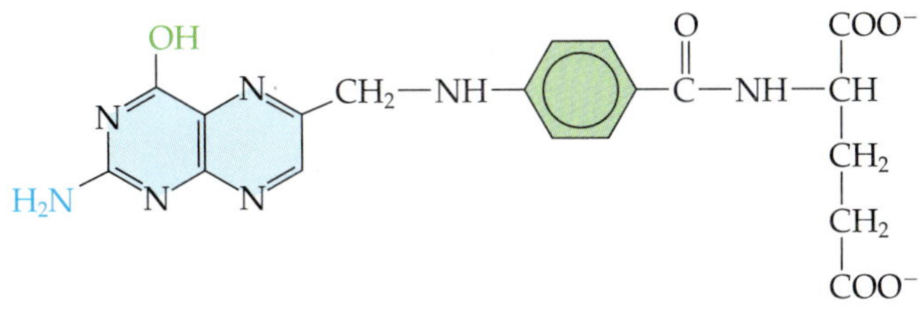

Folic acid

Methotrexate is a competitive inhibitor of the enzyme that converts folic acid to its coenzyme tetrahydrofolic acid (THF). THF is required for the synthesis of deoxythymidine monophosphate (dTMP) from deoxyuridine monophosphate (dUMP). A decrease in dTMP levels limits the amount of DNA that can be formed and cell division and cancer growth are slowed. Methotrexate is used frequently against leukemia.

One of the most widely used anticancer drugs is cisplatin, a platinum-containing compound. Cisplatin binds to DNA and blocks its replication. Transplatin, an isomer of cisplatin, is ineffective, indicating that the *shape* of the molecule is all-important.

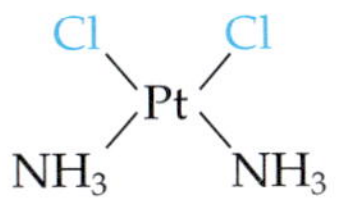

Cisplatin

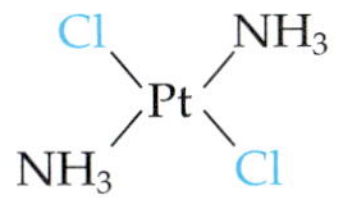

Transplatin

Sex hormones can be used against cancers of the reproductive system. For example, the female hormones estradiol (a natural hormone) and DES (a synthetic hormone) can be used against cancer of the prostate gland. Conversely, male hormones such as testosterone can be used against breast cancer. Such treatment often brings about a temporary cessation—or even a regression—in the growth of cancer cells.

Chemotherapy is only a part of the treatment of cancer. Surgical removal of tumors and radiation treatment (Section 12.9) remain major weapons in the war on cancer. Modern management of cancers can involve surgery, radiation, and one or more anticancer drugs. Indeed, a combination of drugs is often considerably more effective than any one alone. Steady progress is being made in the development of more effective drugs that have fewer side effects. Rates of cure should improve as research progresses, but emphasis must be placed on the prevention of cancer.

Key Terms

benign tumor (G.4)
cancer (G.4)
malignant tumor (G.4)
oncogene (G.5)
retrovirus (G.1)
tumor-suppressor gene (G.5)
virology (page 668)
virus (G.1)

Review Questions

1. a. What is the composition of viruses?
   b. How does a DNA virus differ from an RNA virus?
2. a. How does a DNA virus invade and destroy a cell?
   b. How does an RNA virus invade and destroy a cell?
3. Name five diseases of viral origin and five diseases of bacterial origin.
4. How does acyclovir work as an antiviral drug?
5. Why is HIV known as a retrovirus?
6. a. What two classes of drugs are currently approved for the treatment of AIDS?
   b. Briefly explain how each group works to reduce the amount of HIV in cells.
7. How are benign and malignant tumors different?
8. What environmental factors are the leading causes of cancers?
9. Name several natural carcinogens.
10. a. What is a mutagen?
    b. Why does the Ames test for mutagens provide evidence that a compound may be a carcinogen?
11. What are oncogenes? How are they involved in the development of cancer?
12. What are tumor suppressor genes? How are they involved in the development of cancer?
13. Why would a mutation in a gene that codes for a DNA repair enzyme increase the probability that a person will develop cancer?
14. How does each of the following antimetabolites act against cancer?
    a. 5-fluorouracil
    b. 6-mercaptopurine
    c. methotrexate
15. How does cisplatin work as an anticancer agent?

CHAPTER 24

Metabolism and Energy

Strenuous activity requires a rapid release of energy, but even while sleeping an input of energy is needed to keep our hearts beating and brains functioning. Emma George of Australia, the top female pole vaulter, used mainly anaerobic energy sources as she cleared a height of 15′0″ (4.57 m) on February 21, 1998.

Learning Objectives/Study Questions

1. What is the ultimate source of energy on Earth? How are living organisms interdependent on each other for energy?
2. Why is ATP (adenosine triphosphate) important for living organisms?
3. What is the purpose of each of the three stages in the breakdown of food (carbohydrates, fats, and proteins) to provide energy?
4. How are carbohydrates, fats, and proteins broken down during digestion (Stage I of catabolism)?
5. What is the role of the Krebs cycle in energy production? What products are obtained?
6. What is the role of the electron transport chain in energy metabolism? Why is oxygen required?
7. What is the role of oxidative phosphorylation in energy metabolism?
8. How do muscles obtain their energy?

Life requires energy. We need heat energy to maintain body temperature. We use mechanical energy to move. Chemical energy is required for the synthesis of compounds needed in each cell. Living cells remain organized and functioning properly only through a continual supply of energy. But living organisms can only use specific forms of energy. Supplying a plant with energy by holding it in a flame will do little to prolong its life. On the other hand, a green plant is able to absorb radiation from the sun, the most abundant source of energy on Earth. The plants use this energy first to form glucose and then to make other carbohydrates, lipids, and proteins.

The synthesis of glucose in plants from carbon dioxide, water, and solar energy is the process of **photosynthesis.** This process can be simplified to a general chemical equation:

$$6\,CO_2 + 6\,H_2O + \underset{\text{(From solar energy)}}{686\text{ kcal}} \longrightarrow \underset{\text{Glucose}}{C_6H_{12}O_6} + 6\,O_2$$

Photosynthesis is a distinguishing characteristic of green plants (as well as certain bacteria and algae). Animals cannot directly use the energy of sunlight. They must eat either plants or plant-eating animals in order to get carbohydrates, fats, and proteins and the chemical energy stored in them (Figure 24.1). Once digested and transported to the cell, a food molecule can be used in either of two ways. It can be used as a building block to make new cell parts or repair old ones, or it can be "burned" for energy.

The entire series of thousands of coordinated chemical reactions that keep cells alive is called **metabolism.** In general, metabolic reactions are divided into two classes. The breaking down of molecules to provide energy is **catabolism.** The process of building up the molecules of living systems is **anabolism.**

Any chemical compound involved in a metabolic reaction is a *metabolite.*

Nearly all the energy required by animals is generated from lipids and carbohydrates. These fuels must be burned or oxidized if energy is to be released. The oxidation process ultimately converts the lipid or carbohydrate to carbon dioxide and water.

$$C_6H_{12}O_6 + 6\,O_2 \longrightarrow 6\,CO_2 + 6\,H_2O + 686\text{ kcal} \quad \text{(carbohydrate)}$$

$$C_{16}H_{32}O_2 + 23\,O_2 \longrightarrow 16\,CO_2 + 16\,H_2O + 2340\text{ kcal} \quad \text{(lipid)}$$

These equations summarize the biological combustion of a carbohydrate and a lipid by the cell (respiration). The term **respiration** includes all metabolic processes in

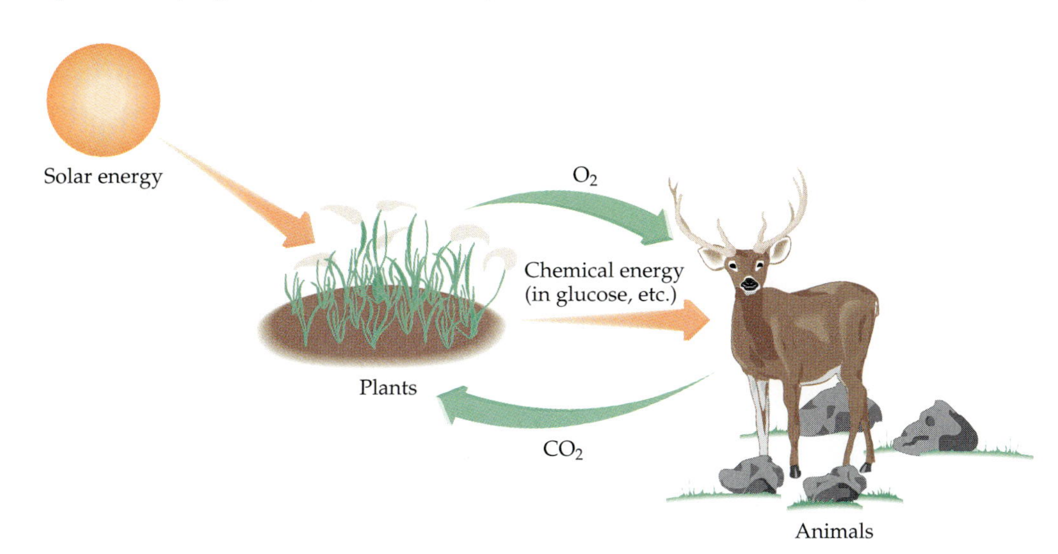

Figure 24.1 Some energy transformations in living systems. Plants and animals exist in a symbiotic cycle—each requires the products of the other.

which gaseous oxygen is used to oxidize organic matter to carbon dioxide, water, and energy.

Both respiration and the combustion of the common fuels (wood, coal, gasoline) use oxygen from the air to break down complex organic substances to carbon dioxide and water. The energy released in the burning of wood is manifested entirely in the form of heat, but excess heat energy is useless and even injurious to the living cell. Living organisms conserve much of the energy released by a series of stepwise reactions, some of which result in the direct synthesis of adenosine triphosphate (ATP) or other compounds that ultimately lead to the synthesis of ATP. The remainder of the energy is used to maintain body temperature. In the next section we'll look more closely at ATP and learn why this compound is considered the chemical energy carrier of the body.

✓ Review Questions

24.1 Define each of the following terms: **(a)** metabolism; **(b)** photosynthesis; **(c)** respiration.

24.2 Distinguish between anabolism and catabolism.

24.1 ATP: Universal Energy Currency

Probably the most important phosphate compound in metabolism is adenosine triphosphate. It was first isolated from skeletal muscle tissue and has since been shown to occur in all types of plant and animal cells. The concentration of ATP in the cell varies from 0.5 to 2.5 mg/mL of cell fluid. ATP is a nucleoside triphosphate composed of adenine, ribose, and three phosphate groups (Figure 24.2).

ATP is often called an *energy-rich compound.* Energy-rich compounds are substances having particular structural features that lead to a release of energy upon hydrolysis. For this reason, these compounds are able to supply energy for energy-requiring biochemical processes (Figure 24.3). The structural feature important in ATP is the phosphoric acid anhydride, or pyrophosphate, linkage:

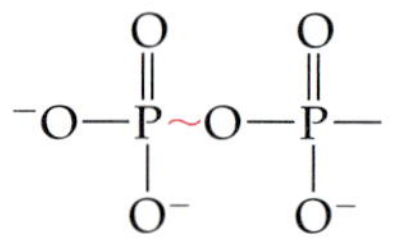

$$^{-}O-\overset{\displaystyle O}{\overset{\|}{\underset{\underset{\displaystyle O^{-}}{|}}{P}}}\sim O-\overset{\displaystyle O}{\overset{\|}{\underset{\underset{\displaystyle O^{-}}{|}}{P}}}-$$

The pyrophosphate bonds are referred to as high-energy bonds and are sometimes symbolized by a squiggle bond (~). There is nothing special about the bonds themselves. The symbol ~ is just a device to focus attention on a portion of the molecule that undergoes reaction. When this bond is broken in the hydrolysis reaction, a relatively large amount of energy (>7 kcal/mol) is released. One of the driving forces

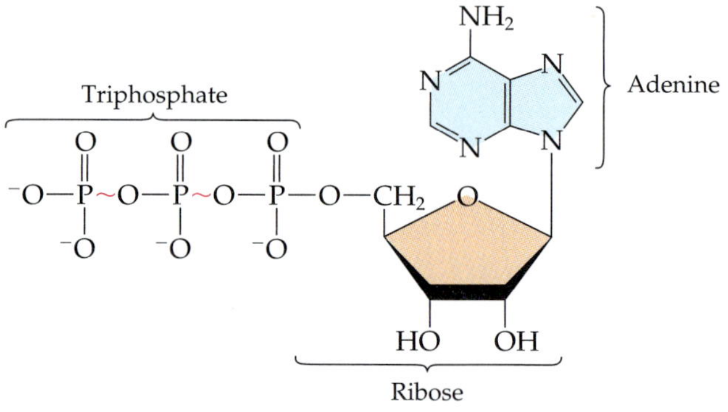

▶ Figure 24.2
Adenosine triphosphate.

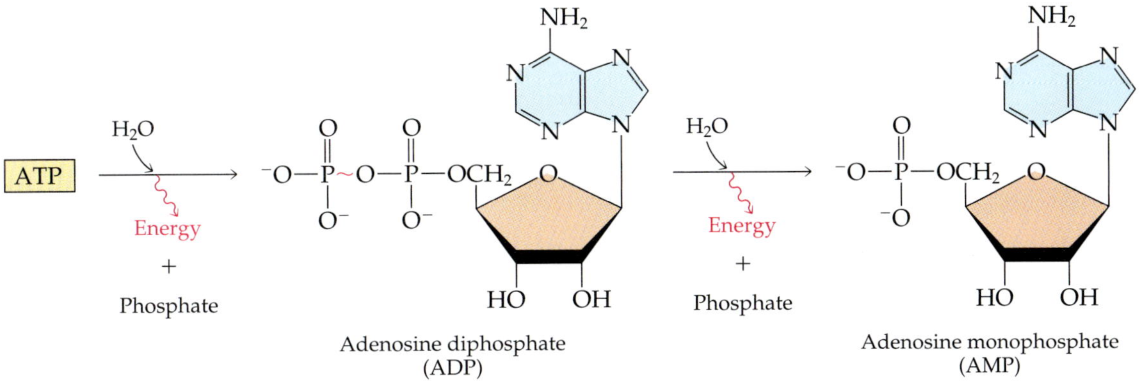

ATP $\xrightarrow{H_2O}$ Energy + Phosphate + Adenosine diphosphate (ADP) $\xrightarrow{H_2O}$ Energy + Phosphate + Adenosine monophosphate (AMP)

▲ **Figure 24.3**
The relationships among ATP, ADP, and AMP. Hydrolysis of each pyrophosphate bond releases energy. In metabolism this usually occurs in stepwise fashion, releasing energy when ATP is hydrolyzed to ADP, and when ADP is hydrolyzed to AMP.

for the reaction is to relieve the electron–electron repulsions associated with the negatively charged phosphate groups.

The general equation for ATP hydrolysis is

$$\text{ATP} \xrightarrow{H_2O} \text{ADP} + P_i + 7.5 \text{ kcal/mol}$$

The important feature of this biochemical reaction is its reversibility. The hydrolysis of ATP releases energy; its synthesis requires energy. Thus ATP is produced by those processes that supply energy to an organism (absorption of radiant energy of the sun in green plants and breakdown of food in animals), and it is hydrolyzed by those processes that require energy (syntheses of carbohydrates, lipids, proteins; transmission of nerve impulses; muscle contractions). These couplings of ATP synthesis to processes that release energy, and ATP breakdown to processes that require energy constitute one of the striking characteristics of living matter (Figure 24.4).

Because ATP is the principal medium of energy exchange in biological systems, it is called the energy currency of the cell. However, it is not the only high-energy compound. Several other phosphate esters provide energy for certain energy-requiring reactions. Table 24.1 lists a number of them. Notice that the free energy of hydrolysis of ATP is approximately midway between those of the high-energy and the low-energy phosphate compounds. This means that the hydrolysis of ATP can provide energy for the phosphorylation of the compounds below it in the table, for example, the phosphorylation of glucose to form glucose-1-phosphate. The hydrolysis of compounds, such as creatine phosphate, that appear above ATP in the table can provide the energy needed to resynthesize ATP from ADP.

P_i is the symbol for the inorganic phosphate anions $H_2PO_4^-$ and HPO_4^{2-} present in the intra- and extracellular fluids.

The values in the literature for the energy released when ATP is hydrolyzed to ADP vary. This is due in part to the fact that reaction conditions (concentrations of ATP, ADP, P_i, and Mg^{2+}, temperature, pH) have not always been the same in different laboratories and in part to the difficulties in obtaining exact values for the equilibrium constants. We shall use a value of -7.5 kcal/mol throughout this text. However, keep in mind that this is only an approximate value and varies depending on reaction conditions.

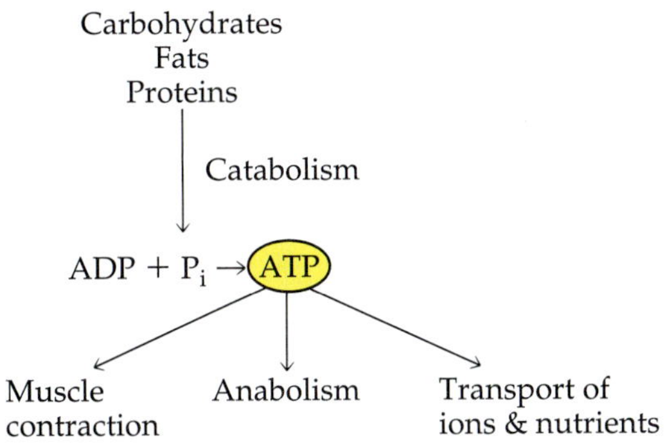

Carbohydrates
Fats
Proteins
↓ Catabolism
ADP + P_i → ATP
→ Muscle contraction
→ Anabolism
→ Transport of ions & nutrients

◀ **Figure 24.4**
ATP is called the energy currency of the cell. Energy stored by the formation of ATP from catabolic reactions, can be used for mechanical work such as muscle contraction, for chemical synthesis (anabolism), and for transport of nutrients.

Table 24.1 Standard Free Energies of Hydrolysis of Some Phosphate Compounds

Type	Example	ΔG° (kcal/mol)
Acyl phosphates	1,3-Bisphosphoglyceric acid	−11.8
$R-C(=O)-O-P(=O)(O^-)-O^-$	Acetyl phosphate	−11.3
Guanidine phosphates	Creatine phosphate	−10.3
$R-NH-C(=N-H)-NH-P(=O)(O^-)-O^-$	Arginine phosphate	−9.1
Pyrophosphates	$PP_i \rightarrow 2\ P_i$	−7.8
$R-O-P(=O)(O^-)-O-P(=O)(O^-)-O^-$	$ATP \rightarrow AMP + PP_i$	−7.7
	$ATP \rightarrow ADP + P_i$	−7.5
	$ADP \rightarrow AMP + P_i$	−7.5
Sugar phosphates	Glucose 1-phosphate	−5.0
$R-O-P(=O)(O^-)-O^-$	Fructose 6-phosphate	−3.8
	$AMP \rightarrow Adenosine + P_i$	−3.4
	Glucose 6-phosphate	−3.3
	Glycerol 3-phosphate	−2.2

✓ Review Question

24.3 Why is ATP referred to as the energy currency of the cell?

24.2 Digestion and Absorption of Major Nutrients

In the introduction we noted that animals require food to obtain the carbohydrates, fats, and proteins needed to provide chemical energy. The breakdown of these molecules to provide energy was defined as *catabolism.* But what is involved in catabolism and exactly how are food molecules broken down? We'll begin with a general overview of catabolism and then look more specifically at the chemical reactions involved. We can think of catabolism as occurring in three stages, as shown in Figure 24.5. In Stage I carbohydrates, lipids, and proteins are broken down into their individual monomer units: carbohydrates into simple sugars, lipids into fatty acids and glycerol, and proteins into amino acids. In Stage II these monomer units (or building blocks) are further broken down through specific reactions, some of which produce ATP, to form a common end product that can then be utilized in Stage III to produce even more ATP.

Digestion is a hydrolytic process that breaks down food molecules into simpler chemical units that can be absorbed by the body, and thus represents Stage I of catabolism. In humans digestion takes place in the digestive tract, and absorption

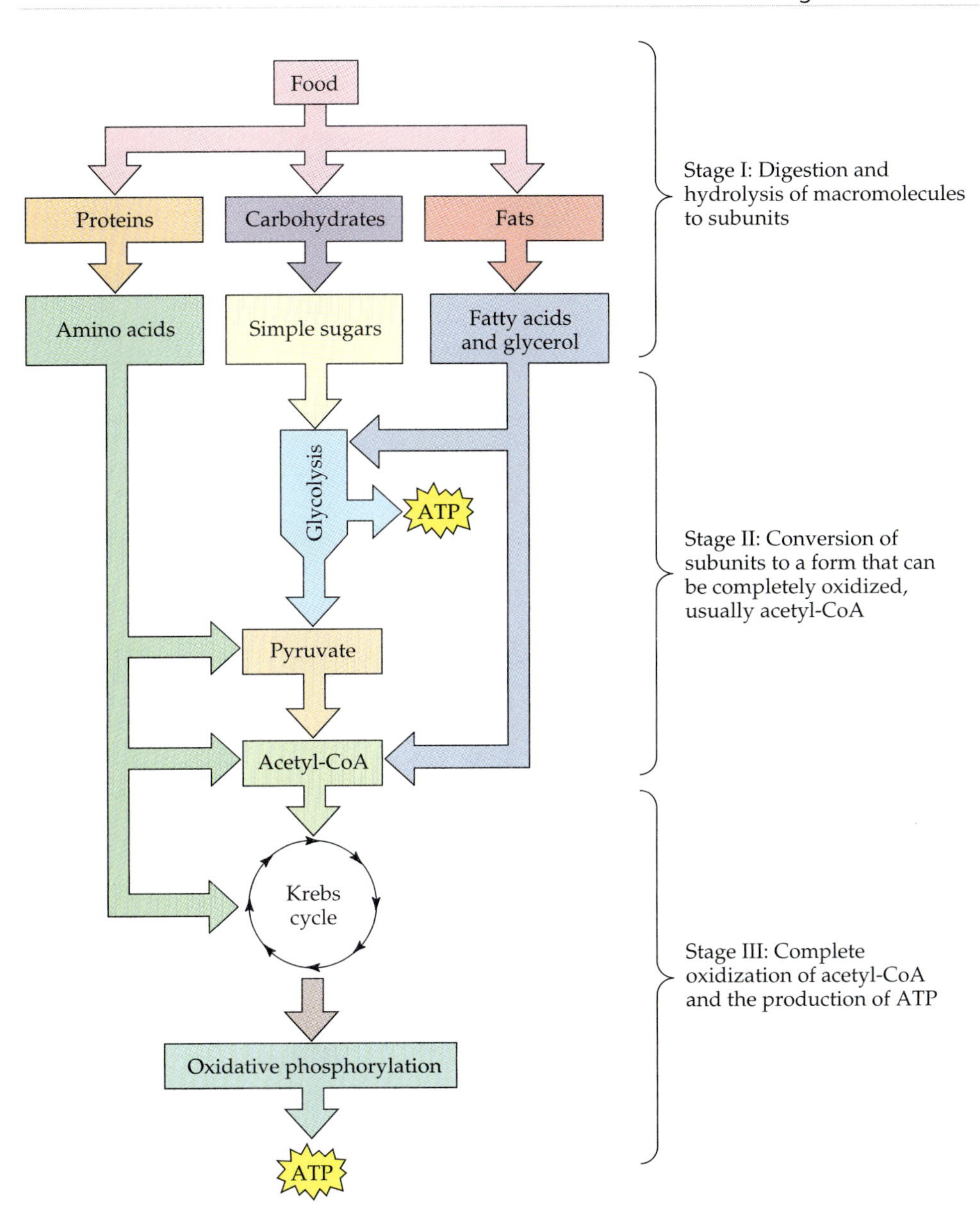

Figure 24.5 The conversion of food into cellular energy (as ATP) occurs in three stages.

occurs primarily in the small intestine. The digestive tract is a tunnel that runs *through* the body. Food in the digestive tract is in the tunnel, not in the body. The alternative name for the digestive tract, *alimentary canal*, perhaps conveys this image more clearly (Figure 24.6).

Mechanical Aspects of Digestion

The digestive tract acts upon food both mechanically and biochemically. Food is subdivided into smaller particles, mixed, and propelled forward through the tract. At each site enzymes act on specific nutrients to hydrolyze them into ever smaller molecules. Compounds that are changed into suitable forms during this journey are absorbed through the walls of the small intestine into the circulatory systems of the body. Materials that can't be absorbed make their way through the entire length of the canal and out again. Without the process of digestion, very little of the food we eat would nourish us.

In the mouth, food is chewed—that is, torn or crushed to a finer consistency. This reduction in the size of food particles facilitates digestion by providing greater

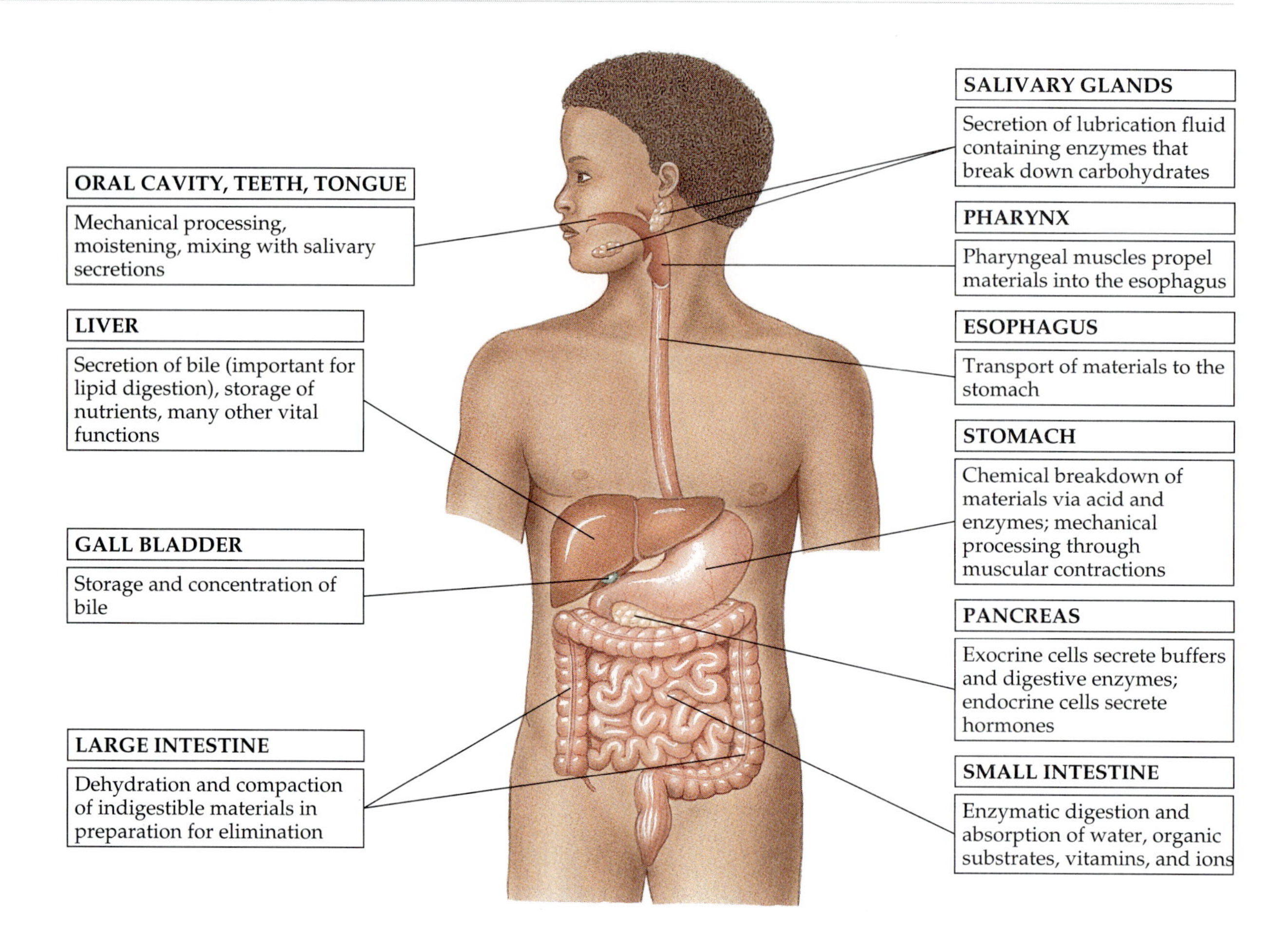
ORAL CAVITY, TEETH, TONGUE
Mechanical processing, moistening, mixing with salivary secretions
LIVER
Secretion of bile (important for lipid digestion), storage of nutrients, many other vital functions
GALL BLADDER
Storage and concentration of bile
LARGE INTESTINE
Dehydration and compaction of indigestible materials in preparation for elimination
SALIVARY GLANDS
Secretion of lubrication fluid containing enzymes that break down carbohydrates
PHARYNX
Pharyngeal muscles propel materials into the esophagus
ESOPHAGUS
Transport of materials to the stomach
STOMACH
Chemical breakdown of materials via acid and enzymes; mechanical processing through muscular contractions
PANCREAS
Exocrine cells secrete buffers and digestive enzymes; endocrine cells secrete hormones
SMALL INTESTINE
Enzymatic digestion and absorption of water, organic substrates, vitamins, and ions

▲ **Figure 24.6**
The human digestive system and the primary functions of key organs.

surface area for enzymes to work on. Chewing also coats the food particles with mucin, a glycoprotein constituent of saliva. The mucin lubricates the food and makes it easier to swallow. **Saliva,** a digestive fluid secreted by the salivary glands, also contains α-amylase, the principal digestive enzyme in the mouth. The secretion of saliva can be triggered by the sight, taste, smell, or even the thought of food.

An average person produces about 1.5 L of saliva a day.

From the esophagus to the large intestine, food is pushed forward by muscular movements of the digestive tract, called peristalsis. This forward flow is aided by a flow of digestive juices that also help liquefy the tract contents.

Food stays in the stomach for two to five hours. It is broken down by pepsin and by the mechanical churning of the stomach into a thin, watery liquid called *chyme.* This material then passes, in small portions, into the duodenum (the first 30 cm of the small intestine). (Much coiled, the small intestine is about seven meters long in adult humans.)

Digestion of Carbohydrates

Carbohydrate digestion begins in the mouth (Figure 24.7), where salivary α-amylase attacks the α-glycosidic linkages in starch, the main carbohydrate ingested by humans. The pH of saliva is about 6.8, the optimum pH for α-amylase. Cleavage of the glycosidic linkages produces a mixture of dextrins, maltose, and glucose (Figure 24.8). α-Amylase continues to function as food passes through the esophagus, but it is quickly inactivated when it comes into contact with the acidic environment of the stomach.

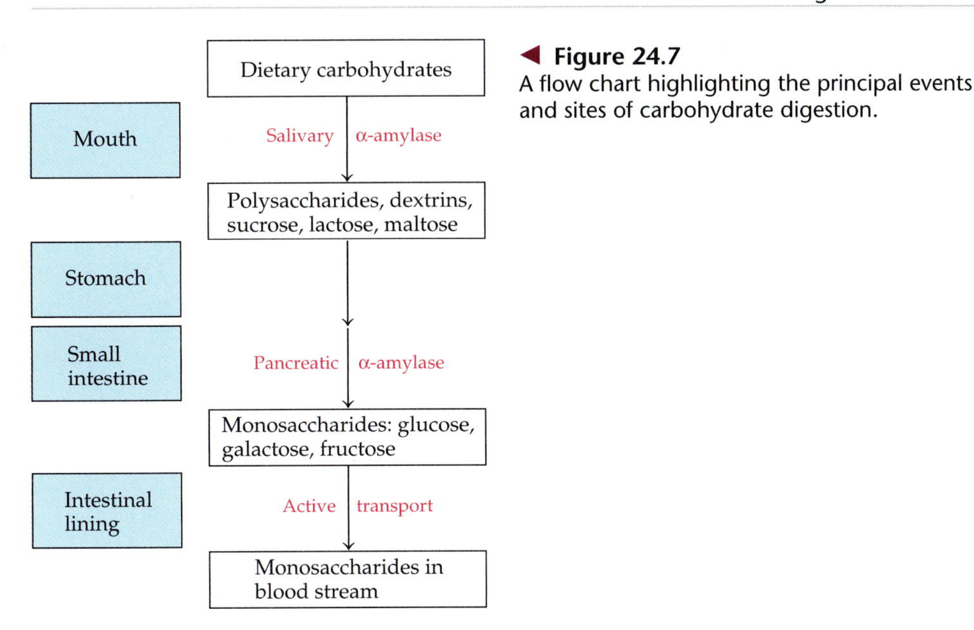

Figure 24.7
A flow chart highlighting the principal events and sites of carbohydrate digestion.

The primary site of carbohydrate digestion is the small intestine, where α-amylase converts the remaining starch molecules, along with the dextrins, to maltose. Maltose is then cleaved into two glucose molecules by the enzyme maltase. Disaccharides such as sucrose and lactose are not digested until they reach the small intestine, where they are acted upon by sucrase and lactase. These enzymes are located on the membranes of cells that line the inner surface of the small intestine. Ultimately, the complete hydrolysis of disaccharides and polysaccharides produces three monosaccharide units—glucose, fructose, and galactose. These monosaccharides are then absorbed through the wall of the small intestine into the bloodstream.

Digestion of Proteins

Protein digestion begins in the stomach (Figure 24.9), where the action of **gastric juice** hydrolyzes about 10% of the peptide bonds. Gastric juice is a mixture of substances secreted by the stomach. The chief components are water (more than 99%), mucin, inorganic ions, hydrochloric acid, and some enzymes. A protein hormone called *gastrin,* produced in the stomach, starts the flow of gastric juice. Flow can also be started by histamine. Indeed, it may well be that gastrin acts by releasing histamine (Section 27.5), which in turn stimulates the secretion of gastric juice.

Hydrochloric acid is secreted by certain glands in the stomach lining. The pH of freshly secreted gastric juice is about 1.0, but the contents of the stomach may react with it, raising the pH to between 1.5 and 2.5. (The pain of a gastric ulcer is at least partially due to the irritation of the ulcerated tissue by the acidic gastric juice.)

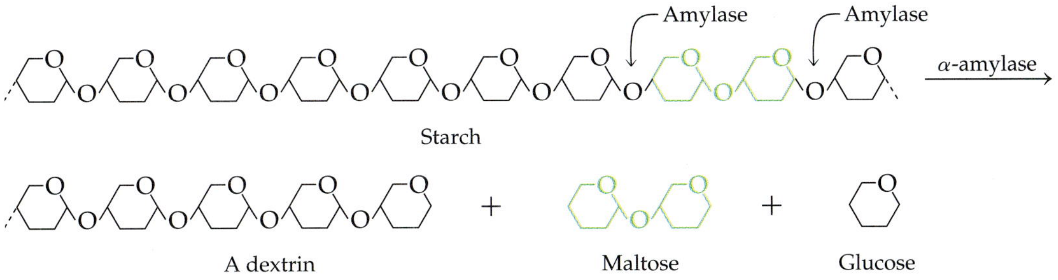

Amylase Amylase
α-amylase
Starch
A dextrin + Maltose + Glucose

Figure 24.8
A schematic representation of the hydrolysis of starch to dextrins, maltose, and glucose. See Chapter 19 for the complete structures of these compounds.

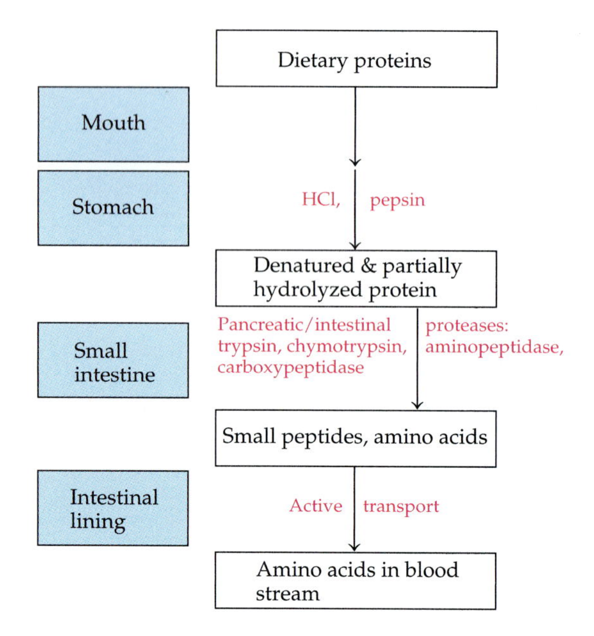

Figure 24.9
A flow chart highlighting the principal events and sites of protein digestion.

The gastric juice of infants contains *rennin,* an enzyme having a specificity very similar to that of pepsin.

The pancreatic juice is sufficiently alkaline pH (~7.5 to 8.5) to neutralize the acidic material passed on from the stomach.

Hydrochloric acid is involved in the denaturation of food protein—it opens up the folds in the protein molecule to expose the chains to more efficient enzyme action. The principal digestive component of gastric juice is *pepsinogen,* a zymogen produced in secretory cells located in the stomach wall. Pepsinogen is catalytically converted by hydrogen ions to its active form, *pepsin.* The newly formed pepsin then acts as an autocatalyst in the activation of the remaining pepsinogen. Pepsin is an endopeptidase that catalyzes the hydrolysis of peptide linkages within the protein molecule. It has a fairly broad specificity but acts preferentially on linkages involving the aromatic amino acids tryptophan, tyrosine, and phenylalanine as well as methionine and leucine.

Protein digestion is completed in the small intestine. Pancreatic juice, carried from the pancreas through the pancreatic duct, contains the proenzymes such as trypsinogen and chymotrypsinogen. The activation of these proenzymes is outlined in Figure 24.10. The intestinal mucosal cells secrete the proteolytic enzyme *enteropeptidase,* which converts trypsinogen to trypsin (Figure 24.11). Trypsin then activates chymotrypsinogen to chymotrypsin. Both of these active enzymes are endopeptidases. Chymotrypsin preferentially attacks peptide bonds involving the carboxyl groups of the aromatic amino acids (phenylalanine, tryptophan, and tyrosine). Trypsin attacks peptide bonds involving the carboxyl groups of the basic amino acids (lysine and arginine). Pancreatic juice also contains the proenzyme *procarboxypeptidase,* which is cleaved by trypsin to *carboxypeptidase.* The latter is an exopeptidase—it catalyzes the hydrolysis of the peptide linkages at the free carboxyl end of the peptide chain, resulting in the stepwise liberation of free amino acids from the carboxyl end of the polypeptide.

Two types of peptidases are secreted in intestinal juice: (1) an exopeptidase, *aminopeptidase,* which acts upon the peptide linkages of terminal amino acids

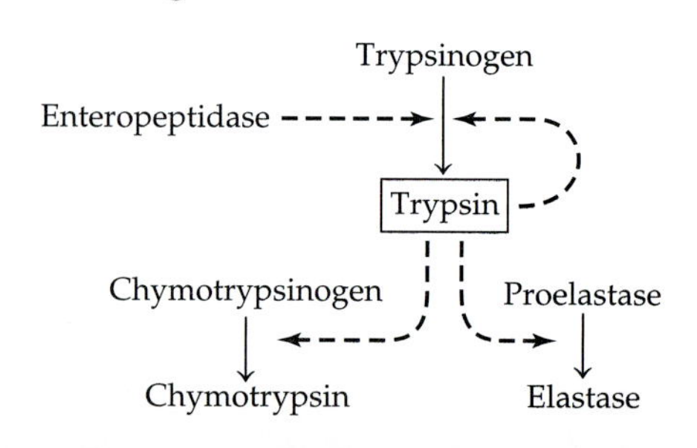

Figure 24.10
Activation of some pancreatic proenzymes. Elastase is an endopeptidase that preferentially cleaves peptide bonds whose carbonyl group is from amino acids having small uncharged side chains, such as alanine.

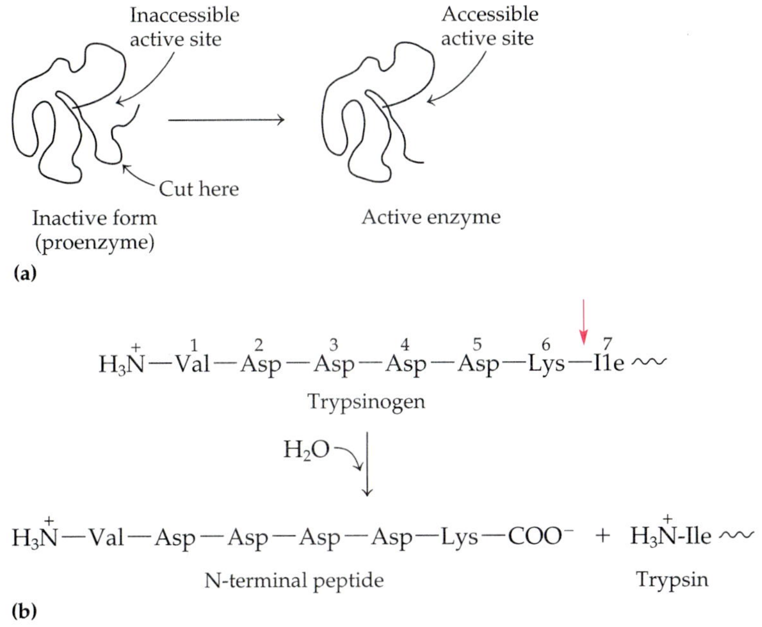

Inaccessible active site

Cut here

Inactive form (proenzyme)

Accessible active site

Active enzyme

(a)

$$\overset{+}{H_3N}-\overset{1}{Val}-\overset{2}{Asp}-\overset{3}{Asp}-\overset{4}{Asp}-\overset{5}{Asp}-\overset{6}{Lys}\downarrow-\overset{7}{Ile}\sim$$

Trypsinogen

$$\xrightarrow{H_2O}$$

$$\overset{+}{H_3N}-Val-Asp-Asp-Asp-Asp-Lys-COO^- + \overset{+}{H_3N}\text{-Ile}\sim$$

N-terminal peptide — Trypsin

(b)

◀ **Figure 24.11**
(a) Proenzymes are activated by the removal of a polypeptide segment that often is blocking the active site. (b) Trypsinogen is activated by the removal of a hexapeptide from the N-terminus. Hydrolysis of the peptide bond between Lys-6 and Ile-7 is catalyzed by either enteropeptidase or trypsin.

possessing a free amino group, and (2) *dipeptidase* and *tripeptidase,* which cleave dipeptides and tripeptides. Figure 24.12 illustrates the specificity of protein-digesting enzymes.

The combined action of the proteolytic enzymes of the gastrointestinal tract results in the hydrolysis not only of the dietary (exogenous) proteins but also of endogenous proteins (the digestive enzymes, other secreted proteins, and dead epithelial cells). The amino acids are then absorbed across the intestinal wall into the circulatory system.

Digestion of Lipids

Lipids are not digested until they reach the upper portion of the small intestine (Figure 24.13). A hormone secreted in this region stimulates the gallbladder to discharge bile into the duodenum. The principal constituents of bile are the bile salts (Section 20.6). These salts act as emulsifiers. This emulsification is essential to lipid digestion. Bile salts act much like soap molecules. They break down large, water-insoluble lipids into smaller globules (micelles) and keep the smaller globules suspended in the aqueous digestive medium. The greatly increased surface area of the lipid particles and the opportunity afforded for more intimate contact with the lipases result in rapid digestion of the fats. Another hormone then promotes the secretion of the pancreatic juice.

The lipases contained in the pancreatic juice catalyze the digestion of triglycerides first to diglycerides and then to 2-monoglycerides and fatty acids that can be absorbed into the bloodstream (see equation on page 684). Phospholipids and cholesteryl esters are also hydrolyzed into their component molecules, which are then absorbed.

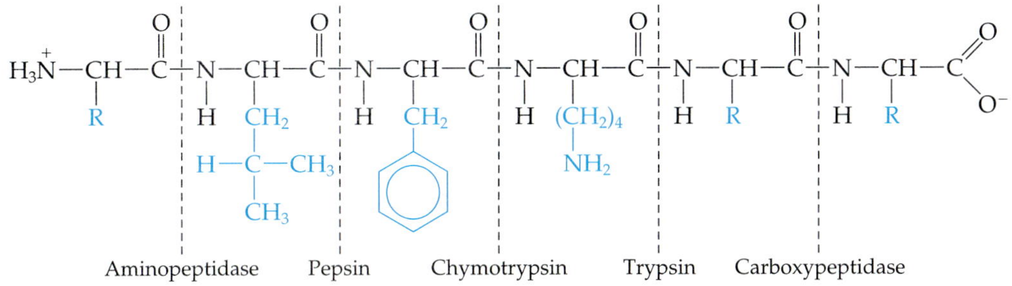

$$\overset{+}{H_3N}-\underset{R}{CH}-\overset{O}{\overset{\|}{C}} \,\vdots\, -\underset{H}{N}-\underset{CH_2-CH(CH_3)_2}{CH}-\overset{O}{\overset{\|}{C}} \,\vdots\, -\underset{H}{N}-\underset{CH_2-C_6H_5}{CH}-\overset{O}{\overset{\|}{C}} \,\vdots\, -\underset{H}{N}-\underset{(CH_2)_4-NH_2}{CH}-\overset{O}{\overset{\|}{C}} \,\vdots\, -\underset{H}{N}-\underset{R}{CH}-\overset{O}{\overset{\|}{C}} \,\vdots\, -\underset{H}{N}-\underset{R}{CH}-COO^-$$

Aminopeptidase — Pepsin — Chymotrypsin — Trypsin — Carboxypeptidase

◀ **Figure 24.12**
Specificity of peptidase hydrolysis.

▶ **Figure 24.13**
A flow chart highlighting the principal events and sites of lipid (primarily triglyceride) digestion.

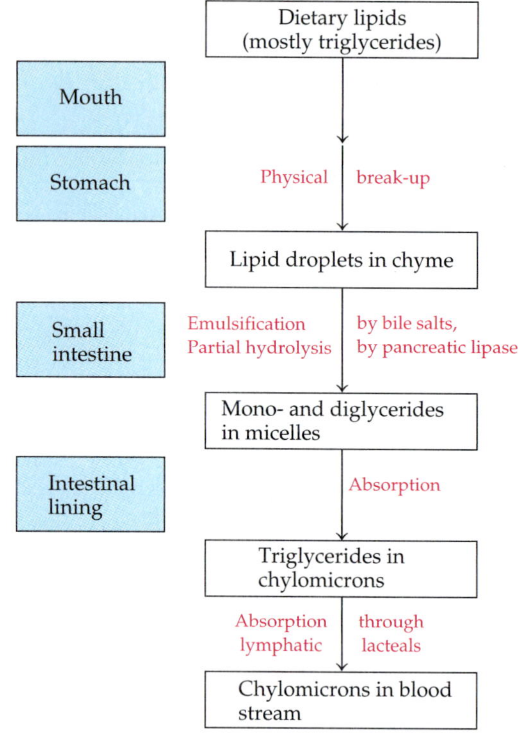

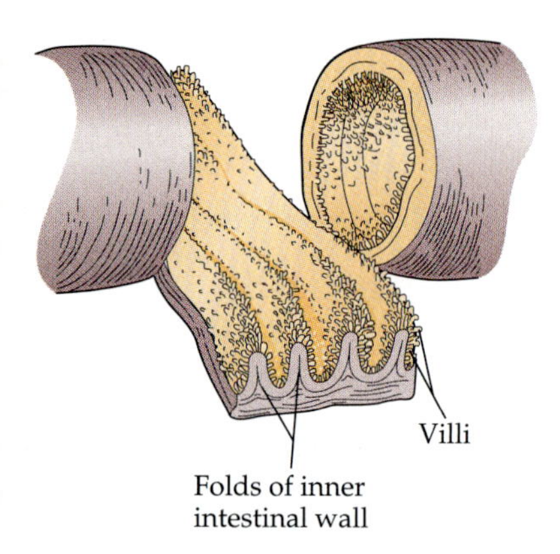

▲ **Figure 24.14**
A section of the small intestine opened to reveal the folds of the inner wall and the lining covered with villi.

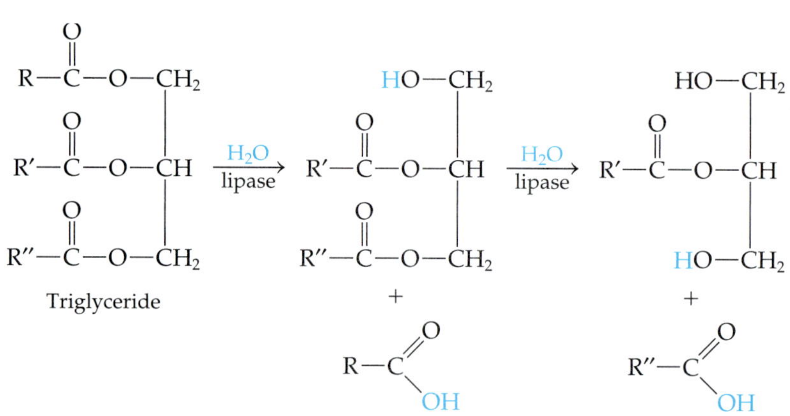

Any polysaccharides or disaccharides that escape hydrolysis by intestinal enzymes cannot be absorbed. Intestinal bacteria metabolize these carbohydrates to lactose, short-chain carboxylic acids, and gases (CO_2, CH_4, and H_2), a process that causes fluid secretion, increased intestinal mobility, and cramps.

Absorption of Digested Nutrients

Absorption of most digested food takes place in the small intestine through the fingerlike projections, called *villi,* that line the inner surface (Figure 24.14). Each villus is richly supplied with a fine network of blood vessels and a central lymphatic vessel (Figure 24.15). The fatty acids, monoglycerides, monosaccharides, and amino acids that were released during the digestive process pass through the semipermeable membrane of each villus and are absorbed into the blood capillaries. The absorption of fatty acids and monoglycerides occurs via a simple process of osmosis or diffusion through an inert membrane (i.e., *passive transport*). However, this is not true for monosaccharides and amino acids. All cell membranes act selectively, a fact that implies that they play an active role in absorption. The passage of these molecules across the intestinal wall requires energy. We therefore use the term *active transport* to describe this type of absorption.

Following absorption, food molecules are carried through the blood stream primarily to the liver where they are further metabolized or transported to other tissues where they are needed. The further metabolism of food molecules for energy occurs in Stages II and III of catabolism, which we will consider next.

Review Questions

24.4 Define each of the following terms: (a) peptidase; **(b)** bile salts; **(c)** lipases; **(d)** villi.

24.5 Distinguish between **(a)** pepsin and pepsinogen; **(b)** chymotrypsin and trypsin; **(c)** endopeptidase and exopeptidase.

24.6 What are the end products of: **(a)** carbohydrate digestion; **(b)** lipid digestion; **(c)** protein digestion?

24.7 In what section of the digestive tract is most carbohydrate digested? Lipid? Protein?

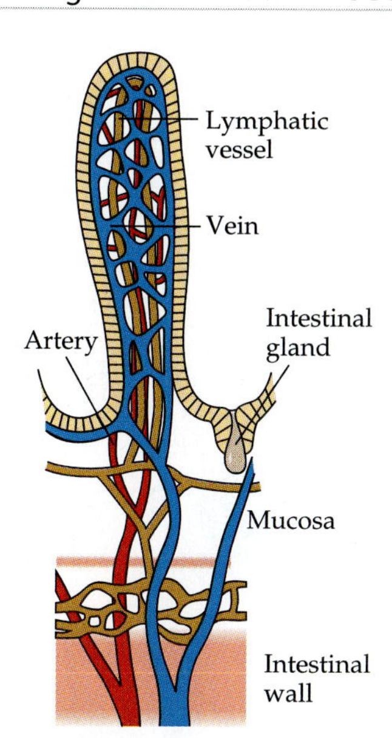

▲ Figure 24.15
Diagram of one of the intestinal villi. Each villus contains arteries, veins, and lymphatic vessels.

24.3 Overview of Stage II of Catabolism

To fully appreciate how food molecules are completely broken down to provide energy for the body, we must understand their metabolic pathways. A **metabolic pathway** is a series of biochemical reactions that enables us to explain how an organism converts a given reactant to a desired end product. As we shall discuss in the following three chapters, specific metabolic pathways, different for each compound, describe how simple sugars, fatty acids, and amino acids are broken down to produce a common end product, acetyl-coenzyme A (acetyl-CoA) in Stage II of catabolism.

Acetyl-CoA is shown in Figure 24.16. The two-carbon acetyl unit, derived from the breakdown of carbohydrates, lipids, and proteins, is attached to coenzyme A, which makes the acetyl group more reactive. The acetyl-CoA can be used in a myriad of biochemical pathways. It may be used as the starting material for biosynthesis of lipids (triglycerides, phospholipids, cholesterol, and other steroids). Acetyl-CoA is also used in the formation of ketone bodies (Section 26.4). Most importantly for energy generation, it may enter the Krebs cycle and be oxidized to produce energy if energy is needed and oxygen is available. The various fates or uses of acetyl-CoA are summarized in Figure 24.17.

Review Questions

24.8 What is a metabolic pathway?

24.9 What are the different ways in which acetyl-CoA is used in the body?

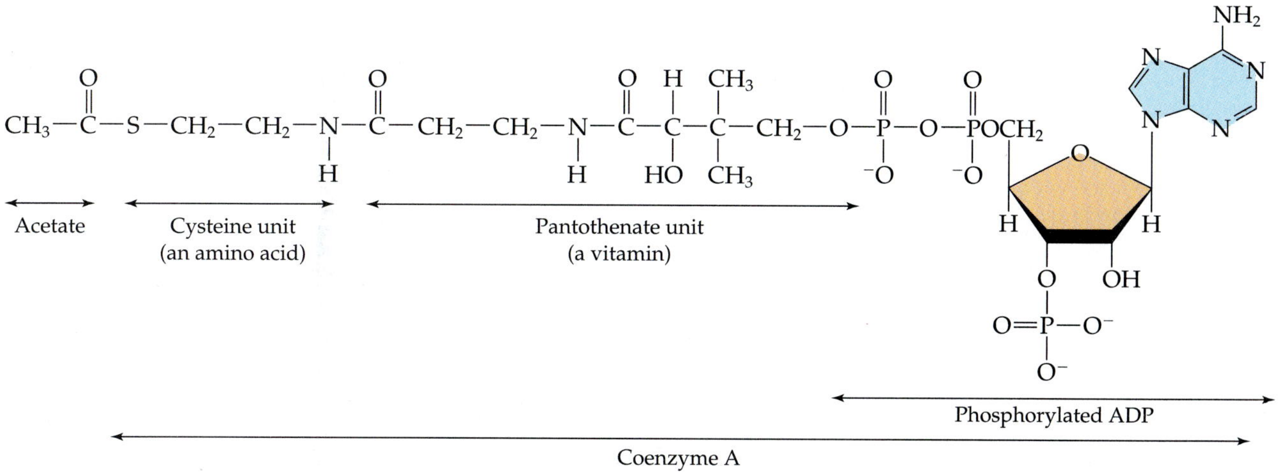

$$CH_3-\overset{O}{\overset{\|}{C}}-S-CH_2-CH_2-\underset{H}{\underset{|}{N}}-\overset{O}{\overset{\|}{C}}-CH_2-CH_2-\underset{H}{\underset{|}{N}}-\overset{O}{\overset{\|}{C}}-\overset{H}{\underset{HO}{C}}-\overset{CH_3}{\underset{CH_3}{C}}-CH_2-O-\overset{O}{\underset{^-O}{P}}-O-\overset{O}{\underset{^-O}{P}}OCH_2$$

Acetate | Cysteine unit (an amino acid) | Pantothenate unit (a vitamin) | Phosphorylated ADP | Coenzyme A

(Ribose ring bearing H, H, OH and $O-PO_3^{2-}$ ($O=P-O^-$, O^-); adenine ring with NH_2, N, N, N, N.)

▲ Figure 24.16
The structure of acetyl-coenzyme A (acetyl-CoA).

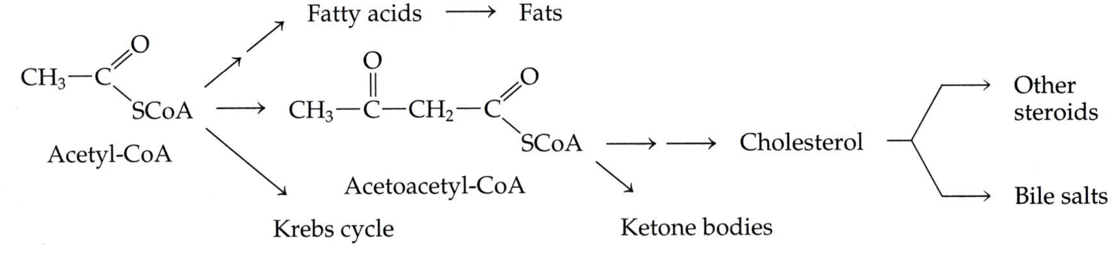

▶ **Figure 24.17** Acetyl-CoA plays a variety of roles in the chemistry of the cell.

▲ Hans Krebs (1900–1981) who determined the sequence of reactions now known as the Krebs cycle. He also discovered the urea cycle (Chapter 27) and another cycle, the glyoxylate cycle, that is not discussed in this text.

24.4 The Krebs Cycle

Acetyl-CoA's entrance into the Krebs cycle is the beginning of Stage III of catabolism. This scheme for the complex series of reactions that brings about the oxidation of a two-carbon unit to carbon dioxide and water was first proposed by Hans Krebs in 1937 (Nobel Prize in physiology or medicine, 1953). The two carbon unit is the acetyl group bound to coenzyme A (acetyl-CoA). The acetyl group enters a cyclic sequence of reactions known collectively as the **Krebs cycle** (also known as the citric acid cycle and the tricarboxylic acid [TCA] cycle). The Krebs cycle produces ATP, NADH, $FADH_2$, and metabolic intermediates for the synthesis of needed compounds.

Taken as a whole, the Krebs cycle appears rather complex (Figure 24.18). All the reactions, however, are familiar types in organic chemistry: condensation, dehydration,

Because succinate and fumarate are symmetrical molecules, the two-carbon group derived from acetyl-CoA can no longer be specifically labeled.

Oxaloacetate + $CH_3-C(=O)-CoA$ + H_2O → ① → Citrate + CoA
Citrate → ② → Isocitrate
Isocitrate + NAD^+ → ③ → α-Ketoglutarate + NADH + CO_2
α-Ketoglutarate + CoA + NAD^+ → ④ → Succinyl-CoA + NADH + CO_2
Succinyl-CoA + GDP + P_i → ⑤ → Succinate + GTP + CoA
Succinate + FAD → ⑥ → Fumarate + $FADH_2$
Fumarate + H_2O → ⑦ → Malate
Malate + NAD^+ → ⑧ → Oxaloacetate + NADH

▶ **Figure 24.18** Reactions of the Krebs cycle.

hydration, oxidation, decarboxylation, and hydrolysis. The difference here is one of *in vitro* versus *in vivo* conditions. In a living organism, these reactions take place at constant temperature (37 °C in humans) and constant pH. Each reaction is catalyzed by an enzyme.

Every enzyme and intermediate in the Krebs cycle has been identified, and the operation of this metabolic pathway *in vivo* has been completely verified by the use of isotopic tracers. Each reaction of the Krebs cycle (Figure 24.18) is numbered and the two acetyl carbons are highlighted in red. Each intermediate in the cycle is a carboxylic acid, existing as an anion at physiological pH. All of the reactions of the Krebs cycle are located within the mitochondria, small organelles within the cell. We'll look more closely at the structure of mitochondria in the next section.

Acetyl-CoA enters the Krebs cycle by condensing with oxaloacetate (step 1), yielding citrate. Note that this step regenerates coenzyme A. The Krebs cycle is partially regulated at this step. The reaction is catalyzed by *citrate synthase,* an enzyme that is inhibited by ATP and NADH (a coenzyme produced in the Krebs cycle and required for the production of ATP). When the cell has sufficient ATP for its immediate needs, ATP molecules interact with citrate synthase to reduce its affinity for acetyl-CoA. The acetyl-CoA is then shunted to the synthesis of fatty acids, and the Krebs cycle slows down.

In the second step, *aconitase* catalyzes the isomerization of citrate to isocitrate. In this reaction a tertiary alcohol, which cannot be oxidized, is converted to a secondary alcohol, which can be oxidized in the next step. In step 3 isocitrate undergoes a reaction known as oxidative decarboxylation because the alcohol is oxidized and the molecule shortened by one carbon with the release of CO_2 (decarboxylation). The reaction is catalyzed by *isocitrate dehydrogenase,* and the product of the reaction is α-ketoglutarate. An important reaction linked to this is the reduction of the coenzyme NAD^+ to NADH. NADH is ultimately reoxidized, and the energy released is used in the synthesis of ATP (Section 24.5).

The fourth step is another oxidative decarboxylation in which α-ketoglutarate is converted to succinyl-CoA and another molecule of NAD^+ is reduced to NADH. The reaction is catalyzed by the *α-ketoglutarate dehydrogenase complex.* This is the only irreversible reaction in the Krebs cycle. As such, it prevents the cycle from operating in the reverse direction.

So far, in steps 1–4 of the Krebs cycle, two carbons have entered the cycle as an acetyl group and two carbons have been released as molecules of CO_2. The remaining reactions of the Krebs cycle take the four carbons of the succinyl group and resynthesize a molecule of oxaloacetate, the compound needed to combine with the acetyl group and begin another round of the Krebs cycle.

In step 5 the energy released by the hydrolysis of the high-energy thioester bond of succinyl-CoA is used to form guanosine triphosphate (GTP) from guanosine diphosphate (GDP) and inorganic phosphate in the reaction catalyzed by *succinyl-CoA synthetase:*

$$^-OOC{-}CH_2{-}CH_2{-}C(=O)SCoA + H_2O \xrightleftharpoons[\text{Succinyl-CoA synthetase}]{P_i + GDP \;\rightarrow\; GTP} {}^-OOC{-}CH_2{-}CH_2{-}COO^- + CoASH$$

Succinyl-CoA Succinate

$$GTP + ADP \xrightleftharpoons{\text{kinase}} GDP + ATP$$

This step is the only reaction in the Krebs cycle that directly involves the formation of a high-energy phosphate compound. GTP can readily transfer its terminal phosphate group to ADP to generate ATP in the presence of *nucleoside diphosphokinase.*

In the next step (6), *succinate dehydrogenase* catalyzes the removal of two hydrogen atoms from succinate, forming fumarate. This oxidation-reduction reaction is the only one in the cycle that uses the coenzyme FAD rather than NAD^+. Succinate dehydrogenase is the only enzyme of the Krebs cycle located within the inner mitochondrial membrane. We will see the importance of this in the next section.

The addition of a molecule of water across the double bond of fumarate to form L-malate is catalyzed by *fumarase* (step 7). One revolution of the cycle is completed with the oxidation of L-malate to oxaloacetate, brought about by *malate dehydrogenase* (step 8). This is the third oxidation-reduction reaction that uses NAD^+ as the oxidizing agent. Oxaloacetate can accept an acetyl group from acetyl-CoA, and the cycle is ready for another spin.

Review Questions

24.10 What is the main function of the Krebs cycle?

24.11 What is GTP? How is GTP utilized to form ATP?

24.12 Two carbons are fed into the Krebs cycle as acetyl-CoA. In what form are two carbon atoms removed from the cycle?

24.5 Cellular Respiration

Earlier we defined respiration as the process by which cells oxidize organic molecules in the presence of gaseous oxygen to produce carbon dioxide, water, and energy in the form of ATP. We have seen that in the Krebs cycle two carbon atoms enter the cycle as acetyl-CoA (step 1), and two different carbon atoms exit the cycle as carbon dioxide (steps 3 and 4). Nowhere in our discussion of the Krebs cycle have we indicated how oxygen is utilized. However, we noted that four oxidation-reduction steps occur in the Krebs cycle in which the coenzyme NAD^+ or FAD is reduced to NADH or $FADH_2$, respectively. NAD^+ serves as the electron acceptor in steps 3, 4, and 8, while FAD is the electron acceptor in step 6. *Oxygen is needed to reoxidize these coenzymes.* Also, little ATP is directly obtained from the Krebs cycle. Oxygen participation and significant ATP production occur in two pathways that are closely linked—electron transport and oxidative phosphorylation.

All the enzymes and coenzymes for the Krebs cycle, the reoxidation of NADH and $FADH_2$, and the production of ATP are localized in the **mitochondria,** small organelles often referred to as the "power plants" of the cell (Figure 24.19). Mitochondria are oval, dual-membrane structures. They may be randomly distributed throughout the cytoplasm, or they may be organized in regular rows or clusters. A cell may contain 100 to 5000 mitochondria, depending on its function, and the mitochondria can reproduce themselves if the energy requirements of the cell increase.

A mitochondrion has two membranes—an *outer membrane* and an *inner membrane* that is extensively folded into a series of internal ridges called *cristae.* Thus there are two compartments in mitochondria—the *intermembrane space* and the *matrix,* which is surrounded by the inner membrane. The outer membrane is permeable, whereas the inner membrane is impermeable to most molecules and ions. (Water, oxygen, and carbon dioxide can freely penetrate both membranes.) The matrix contains all the enzymes of the Krebs cycle, with the exception of succinate dehydrogenase, which is embedded in the inner membrane. The enzymes that are needed for the reoxidation of NADH and $FADH_2$ and for ATP production are also contained within the inner membrane. They are positioned in geometrically specific

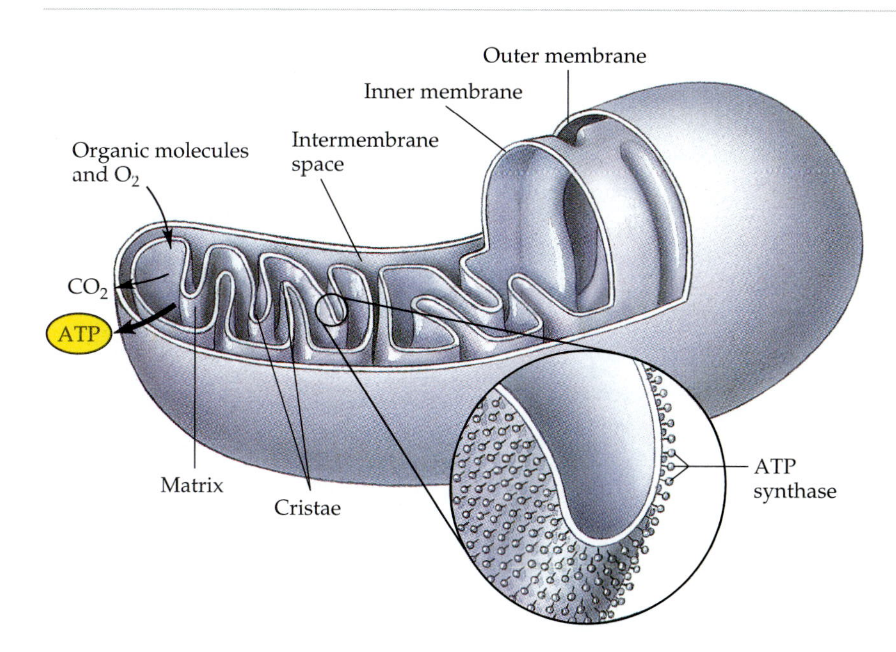

◀ **Figure 24.19**
Cellular respiration occurs in the mitochondria. The subdivisions in a mitochondrion reflect the compartmentalized reactions that take place there. The inner compartment, which contains the soluble enzymes of the matrix, is separated from the intermembrane space by the inner membrane.

arrays able to function in a manner analogous to a bucket brigade. This sequence of highly organized oxidation-reduction enzymes is known as the **electron transport chain** (or **respiratory chain**).

The Electron Transport Chain

Figure 24.20 illustrates the electron transport chain in the mitochondria. The components of the electron transport chain are organized into four complexes, designated as I, II, III, and IV. Each complex contains several enzymes, other proteins, and metal ions. The metal ions can be repeatedly reduced and then oxidized as electrons are passed from one component to the next. (Recall from Chapter 5 that a compound is reduced when it gains electrons or hydrogens and is oxidized when it loses electrons or hydrogens.) The sequence in which these electron carriers operate is determined by their respective reduction potentials. The **reduction potential** is a measure of the tendency of a substance to gain electrons compared with the standard hydrogen electrode (25 °C, 1 atm H_2, and 1 M H^+), which is arbitrarily assigned a value of 0.0 V. Because in biological systems we are usually interested in neutral solutions, a correction is made for changes in pH. At pH 7 the hydrogen electrode has a potential difference of −0.42 V when measured against the standard hydrogen electrode. The reduction potentials of some respiratory-chain intermediates are listed in Table 24.2.

Electrons can enter the electron transport chain through either Complex I or II. We'll look first at electrons entering at Complex I; these electrons come from NADH. NADH is formed in three reactions of the Krebs cycle; as an example, we'll use step 8, where L-malate is oxidized to oxaloacetate and NAD^+ is reduced to NADH. This reaction can be divided into two half-reactions:

Oxidation half-reaction:

$$^{-}OOC{-}CH_2{-}\underset{\underset{H}{|}}{\overset{\overset{OH}{|}}{C}}{-}COO^{-} \longrightarrow {}^{-}OOC{-}CH_2{-}\overset{\overset{O}{\|}}{C}{-}COO^{-} + 2\,H^+ + 2\,e^-$$

L-Malate   Oxaloacetate

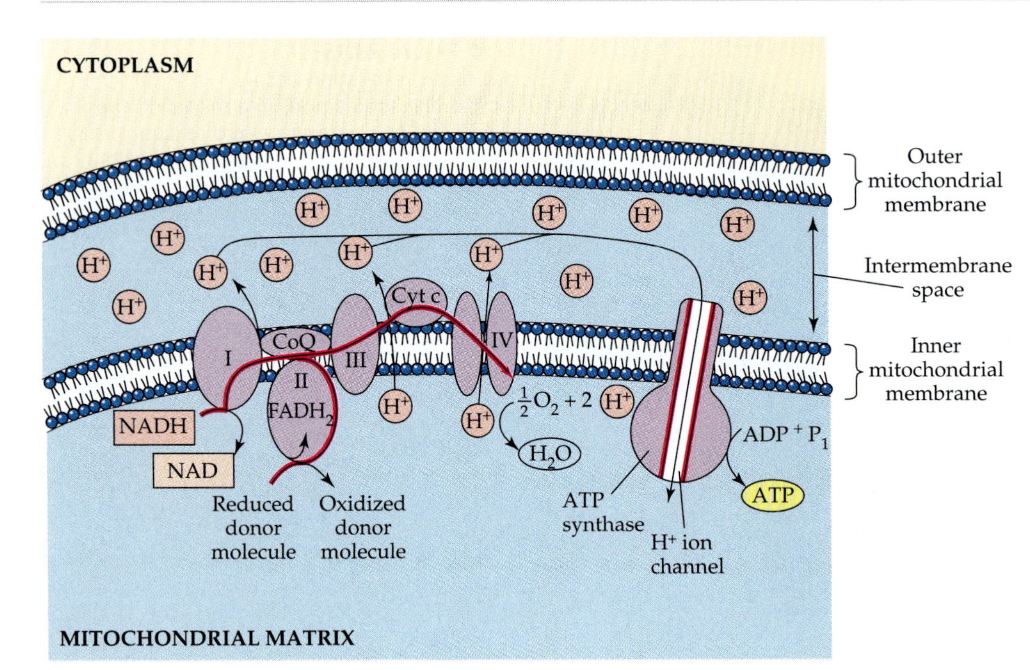

(a)

◀ **Figure 24.20**
(a) The mitochondrial electron transport chain and ATP synthase. The red line shows the path of electrons. Hydrogen ions are transported across the membrane at Complexes I, III, and IV of the electron transport chain, creating a higher concentration on the intermembrane side than on the matrix side. When the H^+ ions reenter the matrix through a channel in ATP synthase, the energy they release powers the synthesis of ATP. (b) A schematic diagram of the four complexes of the electron transport chain, showing the oxidized (magenta) and reduced (blue) forms of the carriers.

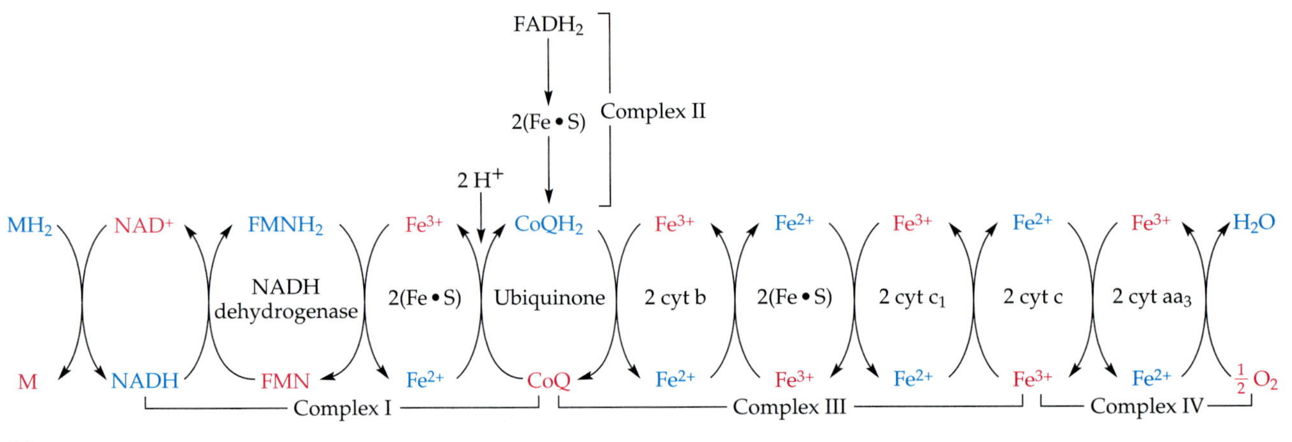

(b)

Table 24.2 Standard Reduction Potentials for Some Respiratory Chain Intermediates

System (Oxidant/Reductant)	$E°$ at pH 7 (V)
$NAD^+/NADH$	−0.32
$FMN/FMNH_2$	−0.22
$CoQ/CoQH_2$	+0.06
Cyt b-Fe(III)/cyt b-Fe(II)	+0.075
Cyt c_1-Fe(III)/cyt c_1-Fe(II)	+0.23
Cyt c-Fe(III)/cyt c-Fe(II)	+0.25
Cyt a-Fe(III)/cyt a-Fe(II)	+0.29
$\frac{1}{2}O_2/H_2O$	+0.82

Reduction half-reaction:

$$NAD^+ + 2\,H^+ + 2\,e^- \longrightarrow NADH + H^+$$

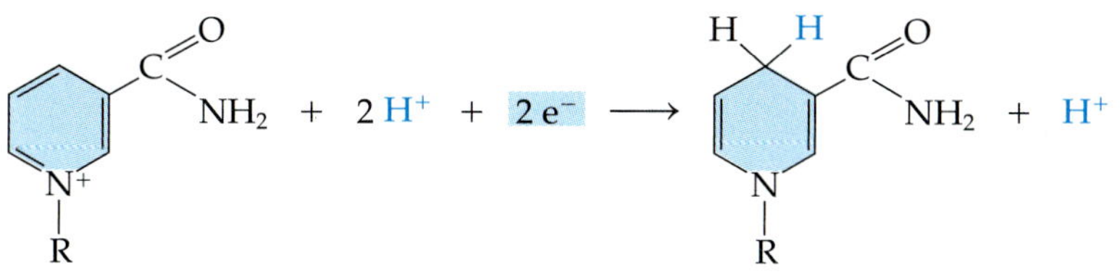

NAD+ NADH

Two hydrogen ions and two electrons are removed from the substrate. The NAD^+ molecule accepts both electrons and one hydrogen ion. The other hydrogen ion is transported from the matrix, across the inner mitochondrial membrane, and into the intermembrane space. The NADH diffuses through the matrix and is bound by a specific enzyme, *NADH dehydrogenase,* in Complex I of the electron transport chain. NADH dehydrogenase contains the coenzyme FMN, which accepts both electrons from NADH. By passing the electrons along, NADH is oxidized back to NAD^+ and FMN is reduced to $FMNH_2$. Again, this reaction can be illustrated by dividing it into its respective half-reactions.

Oxidation half-reaction:

$$NADH + H^+ \longrightarrow NAD^+ + 2\,H^+ + 2\,e^-$$

Reduction half-reaction:

$$FMN + 2\,H^+ + 2\,e^- \longrightarrow FMNH_2$$

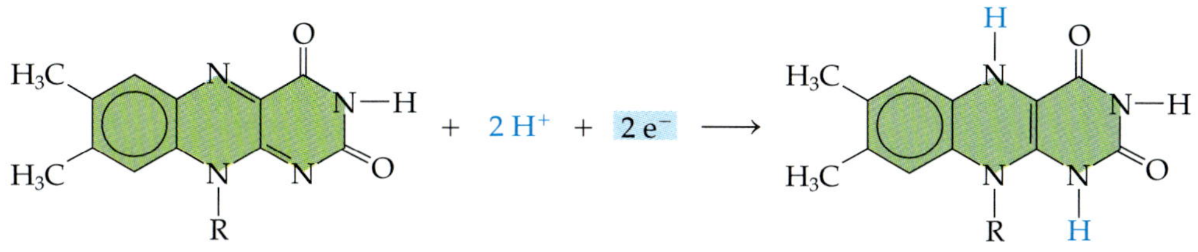

FMN FMNH$_2$

Complex I also contains several proteins that contain iron-sulfur centers (abbreviated Fe · S). The electrons that reduced FMN to $FMNH_2$ are now transferred to these proteins. The iron ions in the Fe · S centers are in the Fe(III) form; by accepting an electron, each ion is reduced to the Fe(II) form. Because each iron-sulfur center can transfer only one electron, two centers are needed to accept the two electrons that will regenerate FMN.

Oxidation half-reaction:

$$FMNH_2 \longrightarrow FMN + 2\,H^+ + 2\,e^-$$

Reduction half-reaction:

$$2\,Fe(III)\cdot S + 2\,e^- \longrightarrow 2\,Fe(II)\cdot S$$

Electrons from $FADH_2$ enter the electron transport chain through Complex II. Succinate dehydrogenase, the enzyme in the Krebs cycle that catalyzes the formation of $FADH_2$ from FAD (step 6), is part of Complex II. The electrons from $FADH_2$ are then transferred to an iron-sulfur protein.

Oxidation half-reaction:

$$FADH_2 \longrightarrow FAD + 2\,H^+ + 2\,e^-$$

Reduction half-reaction:

$$2\,Fe(III){\cdot}S + 2\,e^- \longrightarrow 2\,Fe(II){\cdot}S$$

Electrons from Complex I and Complex II are then transferred from the iron-sulfur protein to coenzyme Q (CoQ), a quinone derivative containing a long isoprenoid side chain. Coenzyme Q is a mobile electron carrier that acts as the electron shuttle between Complexes I and II and Complex III. The quinone ring of coenzyme Q is reduced to a hydroquinone.

Coenzyme Q is also called *ubiquinone* because it is ubiquitous in living systems.

Cyanide Poisoning

Hydrogen cyanide (HCN) is one of the strongest and most rapidly acting poisons. It is a gas at room temperature (bp 25.6–26.5 °C) and is readily absorbed by the body through the stomach, skin, or lungs. Small amounts of HCN are present in tobacco smoke and crude gas. During World War I it was tested for use as a chemical weapon. Cyanide ion (CN^-) is an irreversible inhibitor of cytochrome oxidase, the primary enzyme found in Complex IV of the electron transport chain. Cyanide binds to the Fe^{3+} ion in the cytochrome a_3 heme of cytochrome oxidase and completely inhibits its reduction to the Fe^{2+} state, thus blocking the transfer of electrons to molecular oxygen. This accounts for the extreme toxicity of cyanide to living organisms—just 60 mg of HCN constitutes a lethal dose for a human.

Cyanide occurs widely in nature, primarily as a component of amygdalin and other cyanogenic glycosides found, for example, in the pits of apricots and bitter almonds. When plant materials containing cyanogenic glycosides are crushed, they come in contact with enzymes that break down the glycosides to sugars and other organic compounds, as well

Amygdalin

▲ Amygdalin is a cyanogenic glycoside that is found in bitter almonds.

as hydrogen cyanide. Amygdalin has been marketed as a treatment for cancer under the name Laetrile® and/or Vitamin B_{17}. Extensive research on animals and clinical trials in cancer patients have failed to provide evidence that amygdalin is effective in shrinking tumors, increasing survival time, alleviating cancer symptoms, or enhancing well-being. Several reports have documented cases in which patients have received toxic doses of cyanide when taking large doses of amygdalin; some cases have resulted in death.

Cyanide poisoning can be treated if it is quickly diagnosed. The patient is injected with an oxidizing agent, such as a nitrite, which oxidizes the Fe^{2+} in hemoglobin to Fe^{3+}, forming methemoglobin. The Fe^{3+} of methemoglobin can effectively compete with cytochrome oxidase for cyanide, forming a methemoglobin—cyanide complex. Sodium thiosulfate (the "hypo" used in developing photographic film) is also injected (usually with the nitrite) and causes the cyanide to react with rhodanase, an enzyme present in the body, to form the nontoxic compound thiocyanate, which is excreted in the urine.

$HbO_2(Fe^{2+})$ —(NO_2^- → NO_3^-)→ $MetHbOH(Fe^{3+})$

$MetHbOH(Fe^{3+})$ + CN^- (from Cytochrome $a_3(Fe^{3+})—CN^-$ → Cytochrome $a_3(Fe^{3+})$) → $MetHbOH(Fe^{3+})—CN$

$$CN^- \xrightarrow[\text{Rhodanase}]{S_2O_3^{2-} \;\rightarrow\; SO_3^{2-}} SCN^-$$

▲ Reactions that occur in treating cyanide poisoning.

Oxidation half-reaction:

$$2\ Fe(II){\cdot}S \longrightarrow 2\ Fe(III){\cdot}S + 2\,e^-$$

Reduction half-reaction:

Coenzyme Q (2,3-dimethoxy-5-methyl-1,4-benzoquinone with 6-substituent $(CH_2—CH{=}C(CH_3)—CH_2)_{10}—H$) $+ 2\,H^+ + 2\,e^- \longrightarrow$ Coenzyme QH_2 (corresponding hydroquinone, OH groups in place of the two C=O)

Coenzyme Q Coenzyme QH_2

Complexes III and IV incorporate several heme-containing enzymes known as **cytochromes.** As in hemoglobin and myoglobin, their prosthetic groups are iron porphyrins (Figure 21.13). The various cytochromes differ with respect to: (1) their amino acid composition and sequence, (2) the manner in which the porphyrin is bound to the protein, and (3) the substituents on the periphery of the porphyrin ring. Such slight differences in structure bestow differences in reduction potential upon the different cytochromes. Like the iron in the Fe · S centers, the characteristic feature of the cytochromes is the ability of their iron atoms to exist as either Fe(II) or Fe(III). Thus each cytochrome in its oxidized form, Fe(III), can accept one electron and be reduced to the Fe(II) form. This change in oxidation state is reversible, and the reduced form can donate its electron to the next cytochrome, and so on. Complex III contains cytochromes b and c, as well as Fe · S proteins, with cytochrome c acting as the electron shuttle between Complex III and IV. Complex IV contains cytochromes a and a_3 in an enzyme known as *cytochrome oxidase.* This enzyme has the ability to transfer electrons to molecular oxygen, the ultimate electron acceptor. In this final step, water is formed.

Oxidation half-reaction:

$$2\ Cyt\ a_3—Fe(II) \longrightarrow 2\ Cyt\ a_3—Fe(III) + 2\,e^-$$

Reduction half-reaction:

$$\tfrac{1}{2}\,O_2 + 2\,H^+ + 2\,e^- \longrightarrow H_2O$$

Oxidative Phosphorylation

Each intermediate compound in the electron transport chain is reduced by the addition of electrons in one reaction and is subsequently restored to its original form when it delivers the electrons to the next compound. In addition, the successive electron transfers result in energy production. But how is this energy utilized for the synthesis of ATP? The process whereby ATP synthesis is linked to oxygen consumption in the electron transport chain is referred to as **oxidative phosphorylation.**

Electron transport is tightly coupled to oxidative phosphorylation. The reduced forms of the coenzymes NADH and $FADH_2$ are oxidized by the respiratory chain *only* if ADP is simultaneously phosphorylated to ATP. The currently accepted model that explains how these two processes are linked is known as the chemiosmotic hypothesis, proposed by Peter Mitchell (Nobel Prize in chemistry, 1978).

Look back at Figure 24.20a and note that as electrons are being transferred through the electron transport chain, hydrogen ions (H^+) are being transported across the inner mitochondrial membrane from the matrix to the intermembrane space. Because the concentration of H^+ is higher in the intermembrane space, energy is required to transport more H^+ into this region. This energy comes from the electron transfer reactions in the electron transport. But how does this difference in H^+ concentration lead to ATP synthesis? The buildup of hydrogen ions in the intermembrane space results in a proton gradient that is a large energy source (like water behind a dam). Current research indicates that the flow of H^+ down this concentration gradient through a fifth enzyme complex, known as ATP synthase, leads to a change in the structure of the enzyme and the synthesis and release of ATP.

Within energy-utilizing cells, the turnover of ATP is very high. These cells, then, contain high levels of ADP, and they must consume large quantities of oxygen to continuously obtain the energy required for the phosphorylation of ADP to form ATP. Resting skeletal muscles use about 30% of a resting adult's oxygen consumption, whereas during strenuous exercise these muscles account for almost 90% of the total oxygen consumption of the organism.

From the data in Table 24.2, we can calculate the maximum amount of energy (E) made available when a pair of electrons travels from NADH to oxygen along the chain:

$$\begin{array}{lcll} & & & E \\ \text{NADH} & \longrightarrow & NAD^+ + H^+ + 2\,e^- & +0.32\text{ V} \\ \tfrac{1}{2}O_2 + 2\,H^+ + 2\,e^- & \longrightarrow & H_2O & +0.82\text{ V} \\ \hline \text{NADH} + \tfrac{1}{2}O_2 + H^+ & \longrightarrow & NAD^+ + H_2O & +1.14\text{ V} \end{array}$$

The energy change for the reaction can be obtained from the equation

$$\text{Energy change} = -n\,\mathfrak{F}\,\Delta E$$

where n is the number of electrons transferred and $\mathfrak{F}$ is Faraday's constant (23.062 kcal/V).

$$\begin{aligned} \text{Energy change} &= -(2)(23.062\text{ kcal/V})(1.14\text{V}) \\ &= -52.6\text{ kcal} \end{aligned}$$

This value of 52.6 kcal represents a considerable amount of energy. If it were released all at once, much of it would be dissipated as heat and might damage the cell. Therefore, the electron transport chain delivers this energy in small increments to

Table 24.3 Yield of ATP From the Complete Oxidation of One Mole of Acetyl-CoA

Reaction	Comments	Yield of ATP (moles)
Krebs cycle		
isocitrate $\rightarrow$ α-ketoglutarate + CO_2	produces NADH	
α-ketoglutarate $\rightarrow$ succinyl-CoA + CO_2	produces NADH	
succinyl-CoA $\rightarrow$ succinate	produces GTP	+1
succinate $\rightarrow$ fumarate	produces $FADH_2$	
malate $\rightarrow$ oxaloacetate	produces NADH	
Electron transport chain and oxidative phosphorylation		
$FADH_2$ formed in the Krebs cycle	yields 1.5 ATP	+1.5
3 NADH formed in the Krebs cycle	each yields 2.5 ATP	+7.5
	Net yield of ATP	+10

be used to phosphorylate ADP. It has been experimentally observed that approximately 2.5 ATP molecules are formed for every molecule of NADH oxidized in the electron transport chain and approximately 1.5 ATP are formed for every molecule of $FADH_2$ oxidized. The net equation for the oxidation of NADH and formation of ATP is

$$\text{NADH} + 0.5\,\text{O}_2 + \text{H}^+ + 2.5\,\text{ADP} + 2.5\,\text{P}_\text{i} \rightarrow \text{NAD}^+ + \text{H}_2\text{O} + 2.5\,\text{ATP}$$

Recall that 7.5 kcal is required for the conversion of 1 mol of ADP to ATP. It can be determined that approximately one-third of the energy released in the electron transport chain is conserved in the formation of high-energy phosphate bonds:

$$\text{Energy conserved by respiration} = \frac{\text{energy conserved}}{\text{energy available}}(100\%)$$

$$= (2.5)(7.5\text{ kcal})(100\%)/(52.6\text{ kcal}) = 36\%$$

Table 24.3 provides a summary of the total amount of ATP produced by the complete oxidation of one mole of acetyl-CoA by the sequential action of the Krebs cycle, electron transport chain, and oxidative phosphorylation.

Review Questions

24.13 What are mitochondria, and what is their function in the cell?

24.14 What is oxidative phosphorylation?

24.15 How many molecules of ATP are formed by the oxidation of **(a)** NADH in respiration? **(b)** $FADH_2$ in respiration?

24.6 Muscle Power

Studies have shown that frequent exercise prolongs life and lowers the incidence of disease. One reason for this is that exercise can make muscles stronger, more flexible, and more efficient in their use of oxygen. Humans have about 600 muscles each. Strong muscles can do more work than weak ones; this includes the heart, which is an organ comprised mainly of muscle. With regular exercise, resting pulse and blood pressure usually decline. After several months of an effective exercise program, pulse and blood pressure remain lower even *during* exercise. The net result, called the **training effect,** is that a person who exercises regularly is able to do more physical work with less strain. People who expand their capacity to do more physical work under less strain often begin to think of doing more—faster and with more agility and accuracy. These people often become athletes. Exercise is an art, but it is increasingly also a science—a science in which chemistry plays a vital role.

The stimulation of muscle causes it to contract; that contraction is work and requires energy. The immediate source of energy for muscle contraction is ATP. The energy stored in this molecule powers the physical movement of muscle tissue. Two proteins, actin and myosin, play important roles in this process. Together they form a loose complex called **actomyosin,** the contractile protein of which muscles are made (Figure 24.21). When ATP is added to isolated actomyosin, the protein fibers contract. It seems likely that the same process occurs *in vivo*. Not only does myosin serve as part of the structural complex in muscles, it also acts as an enzyme for the removal of a phosphate group from ATP. Thus it is directly involved in liberating the energy required for the contraction.

In a resting person, muscle activity (including that of the heart muscle) accounts for only about 15 to 30% of the energy requirements of the body. Other activities, such as cell repair, transmission of nerve impulses, and even the maintenance of body temperature, account for the remaining energy needs. During intense physical activity, the energy requirements of muscle may be more than 200 times the resting level.

In contrast with anaerobic metabolism, aerobic metabolism cannot be switched on quickly. At least one to two minutes of hard exercise must pass until the increase in breathing and heart rate ensures delivery of oxygen to muscle cells.

Under usual conditions and during moderate exercise, the respiration of muscle cells is **aerobic** (in the presence of oxygen). During strenuous exercise, however, the energy demand on the muscles is enormous, and the respiratory and circulatory systems are unable to deliver sufficient oxygen to these cells. (This condition, called **oxygen debt,** occurs when not enough oxygen is available for cellular activities.) As a result, the muscle cells must obtain energy under **anaerobic** (in the absence of oxygen) conditions. Under anaerobic conditions, energy can only be obtained from carbohydrates through the breakdown of glycogen and anaerobic glycolysis, a topic we will discuss in Chapter 25.

Muscle tissue seems to have been designed to provide for both short, intense bursts of activity and sustained, moderate levels of activity. Muscle fibers are divided into two categories: *slow-twitch* (Type I) and *fast-twitch* (Type IIB). Table 24.4

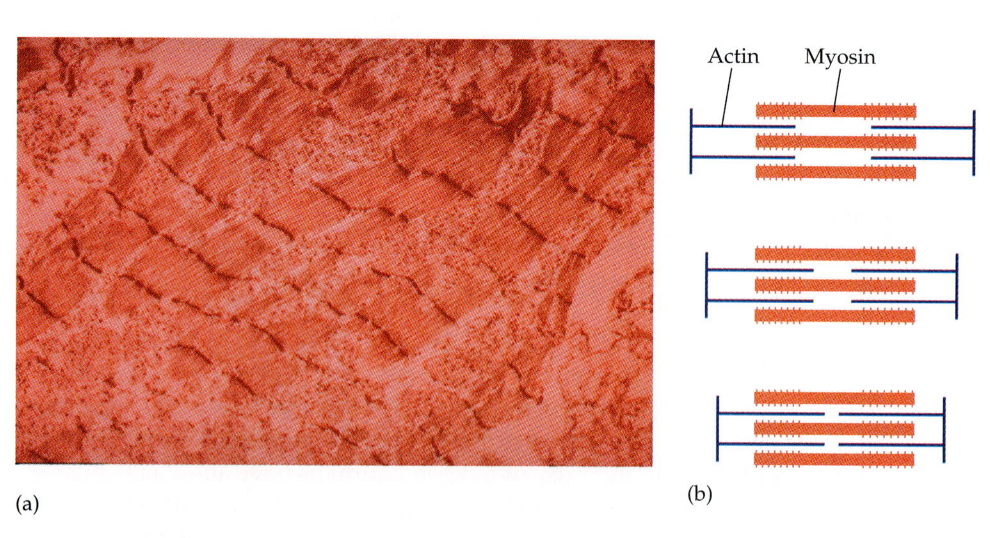

▲ **Figure 24.21**
(a) Skeletal muscle tissue has a banded or *striated* appearance, shown here in an electron micrograph of a striated insect flight muscle. (b) These schematic diagrams of the actomyosin complex in muscle show extended muscle fibers (top), resting fibers (middle), and partially contracted fibers (bottom).

Table 24.4 A Comparison of Types of Muscle Fibers

	Type I	Type IIB[a]
Category	Slow-twitch	Fast-twitch
Color	Red	White
Respiratory capacity	High	Low
Myoglobin level	High	Low
Catalytic activity of actomyosin	Low	High
Capacity for glycogen breakdown	Low	High
Number of mitochondria	High	Low

[a]There is a Type IIA fiber that resembles Type I in some respects and Type IIB in others. We discuss only the two types characterized in this table.

lists some characteristics of these different types of muscle fibers. Type I fibers are called on during light or moderate activity. Their respiratory capacity is high, which means they can provide much energy via aerobic pathways—they are geared to oxidative phosphorylation. Notice that for Type I fibers, myoglobin levels are also high. Myoglobin (Section 21.8) is the heme-containing protein in muscle that stores oxygen obtained from hemoglobin (Chapter 28). Aerobic oxidation requires oxygen, and this muscle tissue is geared to supply high levels of oxygen. It is not geared to anaerobic generation of energy. The number of mitochondria in Type I muscle tissue is high, as we would expect, because oxidative phosphorylation takes place in the mitochondria. The catalytic activity of its actomyosin complex is low. This means that the energy is parceled out more slowly. That is not good if you want to lift 200 kg, but it is perfect for a long, sustained run.

Type IIB fibers have characteristics just the opposite of those of Type I fibers. Low respiratory capacity, low myoglobin levels, and fewer mitochondria all argue against aerobic oxidation. A high capacity for glycogen breakdown and high catalytic activity of actomyosin allow this tissue to generate ATP rapidly and also to hydrolyze that ATP rapidly in intense muscle activity. Thus this type of muscle tissue gives you the capacity to do short bursts of vigorous work. We say *bursts* because Type IIB muscle fatigues relatively quickly and requires a period of recovery between brief periods of activity in order to clear lactate from the muscle (Section 25.2).

Fast-twitch fibers increase in size and strength with repeated anaerobic exercise. Weight training does not increase respiratory capacity.

The fields of sports medicine and exercise physiology have done much to increase our understanding of muscle action. Endurance training (e.g., running for long distances) increases both the size and number of mitochondria in skeletal muscles and the

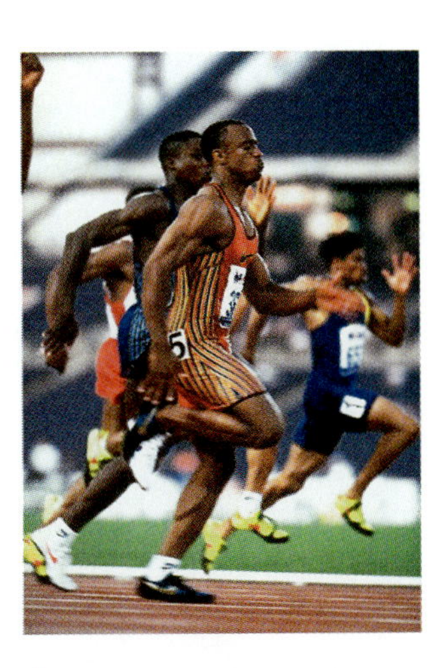

▲ The energy for running a 60-m race comes from anaerobic metabolism.

◀ The Ironman Triathalon is an extreme test of the respiratory capacity of muscle.

Creatine Phosphate

In 1927 a compound called *creatine phosphate* was isolated from mammalian muscle. It was subsequently demonstrated to be the storage form of energy in the muscles of vertebrates and in nerve tissue, where it serves as a reservoir of phosphate bond energy. At rest, mammalian muscle contains four to six times as much creatine phosphate as ATP. As ATP is utilized, creatine phosphate, in a reaction catalyzed by creatine kinase, reacts with ADP to produce more ATP and creatine:

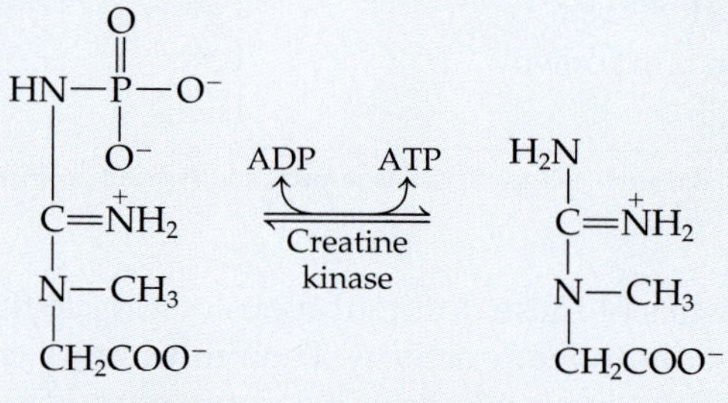

$$\mathrm{{}^{-}O{-}P(=O)(O^{-}){-}NH{-}\overset{+}{C}(=NH_2){-}N(CH_3){-}CH_2COO^-} \underset{\text{Creatine kinase}}{\xrightarrow{\text{ADP} \quad \text{ATP}}} \mathrm{H_2N{-}\overset{+}{C}(=NH_2){-}N(CH_3){-}CH_2COO^-}$$

Creatine phosphate (phosphocreatine) — Creatine

The reaction is readily reversible. When muscular activity is required, the reaction proceeds to the right. When there is abundant ATP (from the catabolism of food), the reaction proceeds to the left, and creatine phosphate is stored in muscle cells. However, the concentration of creatine phosphate in the muscle is limited, and this compound is quickly exhausted. It has been estimated that the creatine phosphate stored in muscles can provide energy for about 10–15 seconds of strenuous exercise (long enough for a short sprint).

Creatine is found in high amounts in meat and fish and is naturally produced in the body in a multistep synthesis from the amino acid arginine. Many athletes take creatine supplements. A study published in the *Journal of the American Dietetic Association* found that athletes who are participating in strength-training programs may benefit from creatine supplements. Male college athletes who used creatine monohydrate showed a significant increase in muscular performance and body mass during multiple sets of bench presses and jump squats. Consultation with a physician and registered dietician is strongly recommended before taking dietary supplements such as creatine.

level of enzymes required for the transport and oxidation of fatty acids (Chapter 26), for the Krebs cycle, and for oxidative phosphorylation. The increase in the mitochondrial enzymes is much greater for Type I fibers (used in prolonged, moderate-intensity activity, which is aerobic) than for Type IIB fibers (brief, intense activity, which is anaerobic). Endurance training also increases myoglobin levels in skeletal muscles, providing for faster oxygen transport, and stimulates additional capillaries to grow within the muscles, bringing additional oxygen-carrying blood. These changes can be observed after only one or two weeks of training. Muscle changes resulting from endurance training do not necessarily include a significant increase in muscle size,[1] in contrast to the effect of strength exercises such as weight lifting. Weight lifting develops fast-twitch muscles but does not result in the mitochondrial changes we have just described. The mitochondria of heart muscle, which is working constantly anyway, also undergo no change during endurance training.

Athletes usually emphasize one type of training (anaerobic or aerobic) over the other. For example, an athlete training for a 60-m dash will do mainly anaerobic work, but one planning to run a 13-km race will do mainly aerobic training. Muscle-fiber

[1]Conversely, a muscle that is not used to any great extent tends to decrease in size and strength. There is a reduction in the number of mitochondria within each muscle cell and a reduction in the number of capillaries within the muscle fibers. This condition is called atrophy and is particularly noticeable in an individual upon removal of a cast after several weeks of limb immobilization.

type seems to be inherited. Research shows that world-class marathon runners may possess up to 80 to 90% slow-twitch muscle fibers, compared with championship sprinters, who may have up to 70% fast-twitch fibers. Some exceptions have been noted, however.

Now that we have seen the central role ATP plays in many life-sustaining metabolic reactions and how some of the ways it is synthesized by the body, we shall look at the metabolic processes unique to carbohydrates, lipids, and proteins in the upcoming chapters.

Review Questions

24.16 What are the two functions of actomyosin?

24.17 Which type of metabolism—aerobic or anaerobic—is primarily responsible for providing energy for: **(a)** intense bursts of vigorous activity? **(b)** prolonged low levels of activity?

Summary

Metabolism is the general term for all chemical reactions in living organisms. The two types of metabolism are **catabolism,** those reactions in which complex molecules (carbohydrates, lipids, and proteins) are broken down to simpler ones with the concomitant release of energy, and **anabolism,** those reactions that consume energy in order to build complex molecules.

Oxidation of fuel molecules (primarily carbohydrates and lipids), a process called **respiration,** is the source of energy used by cells. Catabolic reactions release energy from food molecules and utilize some of that energy for the synthesis of *adenosine triphosphate* (ATP); anabolic reactions use the energy in ATP to create new compounds. Catabolism can be divided into three stages. In stage I carbohydrates, lipids, and proteins are broken down into their individual monomer units—simple sugars, fatty acids, and amino acids, respectively. In stage II, these monomer units are broken down by specific **metabolic pathways** to form a common end product—*acetyl-CoA*. In stage III, acetyl-CoA is completely oxidized to form carbon dioxide and water, and ATP is produced.

Digestion of carbohydrates begins in the mouth as saliva coats food particles and α-amylase breaks α-glycosidic linkages in the carbohydrate molecules. There is essentially no carbohydrate digestion in the stomach, and the saliva-coated particles pass through to the small intestine. Here α-amylase and other intestinal enzymes convert complex carbohydrate molecules (starches) to monosaccharides. The monosaccharides then pass through the lining of the small intestine and into the bloodstream for transport to all the body cells.

Protein digestion begins in the stomach as pepsinogen in gastric juice is converted to pepsin, the enzyme that hydrolyzes peptide bonds. The partially digested protein then passes to the duodenum, where the rest of protein digestion takes place through the action of several enzymes. The resulting amino acids cross the intestinal wall into the blood and are carried to the liver.

Lipid digestion begins in the duodenum. First bile salts emulsify the lipid molecules, and then lipases hydrolyze them to fatty acids and monoglycerides. The hydrolysis products pass through the intestine and are repackaged for transport in the bloodstream.

In cells that are operating aerobically, acetyl-CoA produced in stage II of catabolism is oxidized to carbon dioxide. The **Krebs cycle** describes this oxidation, which takes place with the formation of the reduced coenzymes NADH and $FADH_2$. The sequence of reactions needed to oxidize these coenzymes and transfer the resulting electrons to oxygen is called the **electron transport chain** (or **respiratory chain**). The reactions of the electron transport chain are a series of oxidation-reduction reactions involving **cytochromes,** iron-sulfur proteins, and other molecules that ultimately result in the reduction of molecular oxygen to water. Every time a 2-carbon compound is oxidized in the Krebs cycle, a respiratory chain compound accepts the electrons lost in the oxidation (and so is reduced) and then passes them on to the next metabolite in the chain. The energy released by the electron transport chain is used to transport H^+ from the mitochondrial matrix to the intermembrane space. The flow of H^+ back through ATP synthase leads to the synthesis and release of ATP from ADP and P_i in a process known as **oxidative phosphorylation.** Electron transport and oxidative phosphorylation are tightly coupled to each other. The enzymes and intermediates of the Krebs cycle, electron transport chain, and oxidative phosphorylation are found in organelles called **mitochondria.**

Much of the energy in ATP is used for muscle contraction. Muscle fibers are composed of actin and myosin, two proteins combined into a complex called **actomyosin.** Actomyosin acts as an enzyme for the conversion ATP $\rightarrow$ ADP, which releases the energy needed for a muscle to do work.

There are two main types of muscle fibers in humans. *Type I,* the *slow-twitch fibers,* are used to do light and moderate work. They operate under **aerobic** conditions. *Type IIB,* the *fast-twitch fibers,* are used in short bursts of intense activity. They operate under **anaerobic** conditions, when the body cannot inhale oxygen fast enough to meet the needs of the muscle cells involved in the activity.

Key Terms

actomyosin (24.6)
aerobic (24.6)
anabolism (page 675)
anaerobic (24.6)
catabolism (page 675)
cytochrome (24.5)
digestion (24.2)
electron transport chain (24.5)
gastric juice (24.2)
Krebs cycle (24.4)
metabolic pathway (24.3)
metabolism (page 675)
mitochondria (24.5)
oxidative phosphorylation (24.5)
oxygen debt (24.6)
photosynthesis (page 675)
reduction potential (24.5)
respiration (page 675)
respiratory chain (24.5)
saliva (24.2)
training effect (24.6)

Problems

ATP: Universal Energy Currency

1. What are the structural differences among ATP, ADP, and AMP?
2. Why is ATP referred to as the energy currency of the cell?
3. Referring to Table 24.1, indicate which compounds are high-energy phosphates:
   a. ATP b. glucose 6-phosphate
   c. creatine phosphate d. ADP
   e. AMP f. glucose 1-phosphate

Digestion and Absorption of Major Nutrients

4. What is the general type of reaction used in digestion?
5. What is mucin? What is its function in saliva?
6. Give the site of action and the function of:
   a. α-amylase b. lactase
   c. sucrase d. maltase
   e. chymotrypsin f. trypsin
   g. pepsin h. enteropeptidase
7. Describe the emulsifying action of bile salts. What function does emulsification serve?
8. Show, with equations, the chemical changes that triglycerides undergo during digestion.
9. Indicate the expected products from the enzymatic action of pepsin, chymotrypsin, and trypsin on:
   a. Gly-Ala-Phe-Tyr b. Ala-Ile-Tyr-Ser
   c. Val-Phe-Arg-Leu d. Leu-Thr-Glu-Lys
10. Indicate the cleavage sites of:
    a. an aminopeptidase b. a carboxypeptidase.

Krebs Cycle

11. Replace the question marks with the proper compounds:
    a. $? \xrightarrow{\text{aconitase}} \text{Isocitrate}$
    b. $? + ? \xrightarrow{\text{citrate synthase}} \text{Citrate}$
    c. $\text{Fumarate} \xrightarrow{\text{fumarase}} ?$
    d. $\text{Isocitrate} \xrightarrow{?} \alpha\text{-Ketoglutarate}$
12. Replace the question marks with the proper compounds:
    a. $\text{Malate} \xrightarrow{\text{malate dehydrogenase}} ?$
    b. $? + ? \xrightarrow{\text{nucleoside diphosphokinase}} \text{GDP} + \text{ATP}$
    c. $\text{Succinyl-CoA} \xrightarrow{\text{succinyl-CoA synthetase}} ?$
    d. $\text{Succinate} \xrightarrow{\text{succinate dehydrogenease}} ?$
13. Refer to Problems 11 and 12 and select the equations (by number and letter) in which the following processes occur:
    a. isomerization b. hydration
    c. dehydration
14. Refer to Problems 11 and 12 and select the equations (by number and letter) in which the following processes occur:
    a. oxidation b. decarboxylation
    c. phosphorylation
15. Write balanced half-reactions for steps 3, 4, and 6 in the Krebs cycle.

Cellular Respiration

16. What similar role do coenzyme Q and cytochrome c serve in the respiratory chain?
17. What is the electron acceptor at the end of the respiratory chain? To what product is this compound reduced?
18. What is the function of the cytochromes in the respiratory chain?
19. a. What is meant by the statement "Electron transport is tightly coupled to oxidative phosphorylation"?
    b. How are electron transport and oxidative phosphorylation coupled or linked?

Muscle Power

20. What is the role of myosin in muscle contraction?
21. Identify Type I and Type IIB muscle fibers as **(a)** fast-twitch or slow-twitch and **(b)** suited to aerobic or anaerobic metabolism.
22. Explain why high levels of myoglobin and mitochondria are appropriate for muscle tissue geared to aerobic oxidation.
23. Why does the high catalytic activity of actomyosin in Type IIB fibers suggest that these are the muscle fibers engaged in brief, intense physical activity?

Additional Problems

24. What does the following statement mean in terms of energy and fuel? *"Human existence on this planet is directly dependent upon the plant kingdom."*
25. What is the significance of the fact that ATP has an intermediate value of $\Delta G°$ of hydrolysis (its value isn't one of the highest nor the lowest)?
26. If a cracker, which is rich in starch, is chewed for a long time, it begins to develop a sweet, sugary taste. Why?
27. If the methyl carbon atom of acetyl-CoA is labeled, where does the label appear after the acetyl-CoA goes through one turn of the Krebs cycle?
28. The complete oxidation of 1 mol of acetic acid in a calorimeter yields about 200 kcal.
    a. How much energy is stored as ATP when 1 mol of acetic acid is converted to acetyl-CoA and metabolized via the Krebs cycle?
    b. What is the percent energy efficiency?
29. Which type of muscle fiber is more affected by endurance training? What changes occur in the muscle tissue?
30. Birds use large, well-developed breast muscles for flying. Pheasants can fly 80 km/h, but only for short distances. Great blue herons can fly only about 35 km/h, but they can cruise great distances. What kind of fibers would each bird have in its breast muscles?
31. **a.** Why is cyanide so toxic to humans?
    b. How is cyanide poisoning treated?
32. What is the role of creatine phosphate in muscles?

Carbohydrate Metabolism

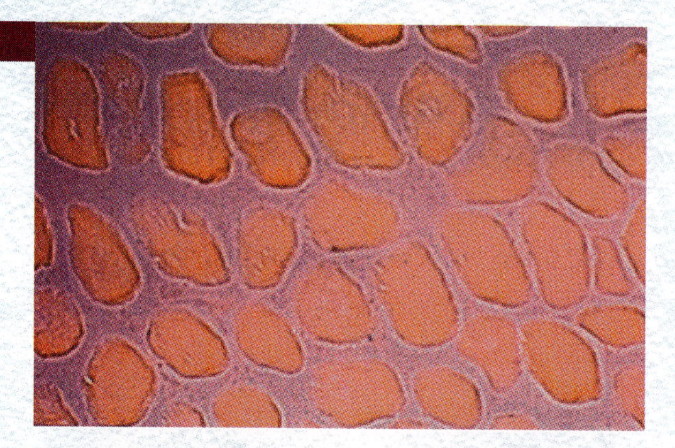

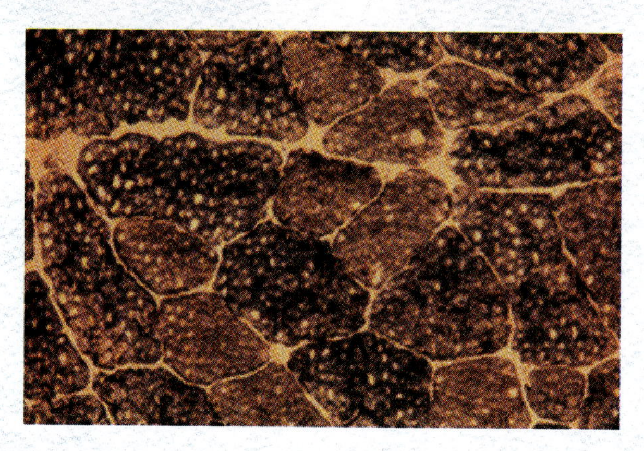

On the top are muscle cells taken from a patient with McArdle's disease. The muscle cells on the bottom contain glycogen phosphorylase, the enzyme missing in patients with McArdle's disease. Glycogen phosphorylase is an important enzyme in carbohydrate metabolism that is needed to degrade glycogen.

Learning Objectives/Study Questions

1. What is the function of glycolysis? What are the major products of glycolysis?
2. What happens to pyruvate? How does the presence or absence of oxygen determine the fate of pyruvate?
3. How much ATP is produced by the breakdown of glucose in the absence of oxygen? When oxygen is abundant?
4. What is the function of gluconeogenesis? What compounds can be used to synthesize glucose?
5. How is glycogen used to maintain the blood glucose level? How is glycogen synthesized and degraded?
6. What hormones regulate carbohydrate metabolism and the levels of glucose in the blood? What happens when blood sugar levels are too high or too low?

In 1951 a 30-year-old man visited his physician, Dr. Brian McArdle, complaining of muscle pain and stiffness following any exercise. Dr. McArdle, recognizing a new disease, described the man and his symptoms. It was another eight years before the cause of what became known as McArdle's disease was shown to be a lack of a muscle enzyme—glycogen phosphorylase. McArdle's disease is a rare genetic disease in which lack of glycogen phosphorylase prevents muscle cells from degrading glycogen when energy is needed. In this chapter we'll look at the metabolism of carbohydrates that occurs between digestion (Section 24.2) and the entrance of acetyl-CoA into the Krebs cycle (Section 24.4). We'll also look at glycogen metabolism, glucose synthesis, and the various control mechanisms that regulate these pathways in the body.

25.1 Glycolysis

The metabolic pathway known as **glycolysis** is a series of reactions that converts glucose (an aldohexose) into two molecules of pyruvate (a three-carbon compound) with the corresponding production of ATP. The individual steps in glycolysis were determined during the first part of the twentieth century. It was the first metabolic pathway to be elucidated, in part because the enzymes needed for glycolysis are found in a soluble form in the cell and are readily isolated and purified. The enzymes act in such a manner that the product of one enzyme-catalyzed reaction becomes the substrate of the next. The transfer of intermediates from one enzyme to the next occurs by diffusion.

Glycolysis is also known as the Embden–Meyerhof pathway in honor of Gustav Embden and Otto Meyerhof who, along with Otto Warburg, were instrumental in sorting out this sequence of reactions during the 1930s.

All of the intermediates in glycolysis have either six or three carbon atoms, and each is phosphorylated. The ten reactions of glycolysis from glucose to pyruvate are numbered and summarized in Figure 25.1. They are divided into two phases. The first five reactions comprise Phase I in which glucose is broken down to two molecules of glyceraldehyde-3-phosphate. In the last five reactions, Phase II, each glyceraldehyde-3-phosphate is converted into pyruvate and ATP is generated.

Phase I

When glucose enters a cell it is immediately phosphorylated to form glucose-6-phosphate. The phosphate donor in this reaction is ATP, and the enzyme, which requires Mg^{2+} ions for its activity, is *hexokinase.* In this reaction ATP is being used rather than synthesized. This reaction is unexpected in a catabolic pathway that is supposed to generate energy. However, it is necessary to activate the glucose molecule and to make this step essentially irreversible. Addition of the negatively charged phosphate group not only activates the glucose but also keeps the intermediates formed in glycolysis from diffusing through the cell membrane, as neutral molecules like glucose can do.

Recall from Section 24.2 that monosaccharides from digestion enter the bloodstream from the small intestine.

In step 2, *phosphoglucose isomerase* catalyzes the isomerization of glucose-6-phosphate to fructose-6-phosphate. This reaction is important because it creates a primary alcohol that can be phosphorylated more readily than a hemiacetal.

The phosphorylation reaction in step 3 to form fructose-1,6-bisphosphate is catalyzed by the enzyme *phosphofructokinase,* which requires Mg^{2+} ions for activity. ATP is again the phosphate donor that insures that this reaction is irreversible. In Section 25.6, we'll discuss phosphofructokinase and its role in controlling the rate of glycolysis.

When a molecule contains two phosphate groups on different carbon atoms, the convention is to use the prefix *bis.* When the two phosphate groups are bonded to each other on the same carbon atom (for example, ADP), the prefix is *di.*

Fructose-1,6-bisphophate is enzymatically cleaved by *aldolase* to form two triose phosphates—dihydroxyacetone phosphate and glyceraldehyde-3-phosphate. Isomerization of dihydroxyacetone phosphate into a second molecule of glyceraldehyde-3-phosphate is the final step in Phase I. The enzyme catalyzing this reaction

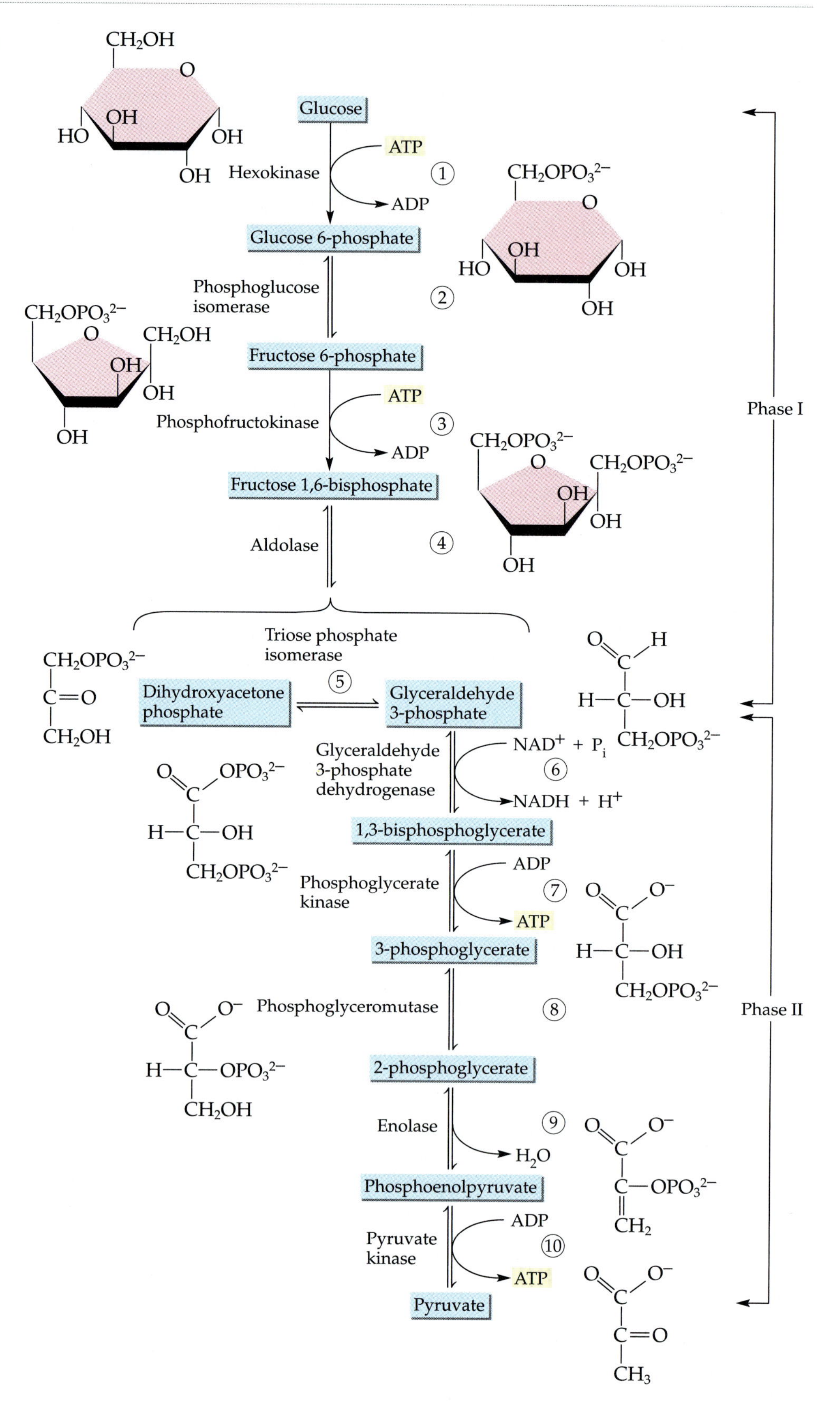

Figure 25.1
Glycolysis.

is *triose phosphate isomerase.* A summation of steps 4 and 5 indicates that aldolase and triose phosphate isomerase have effectively converted one molecule of fructose-1,6-bisphosphate into *two* molecules of glyceraldehyde-3-phosphate. Thus, Phase I requires energy in the form of two molecules of ATP and releases none of the energy stored in glucose.

Phase II

In step 6, glyceraldehyde-3-phosphate is both oxidized and phosphorylated in a reaction catalyzed by *glyceraldehyde-3-phosphate dehydrogenase,* an enzyme that requires the coenzyme NAD^+ as the oxidizing agent and inorganic phosphate as the phosphate donor. In the reaction, NAD^+ is reduced to NADH and 1,3-bisphosphoglycerate is formed.

1,3-Bisphosphoglycerate contains a high-energy acyl phosphate bond (see Table 24.1). This phosphate group on C-1 can be directly transferred to a molecule of ADP, thus forming ATP and 3-phosphoglycerate. The enzyme that catalyzes the reaction in step 7 is *phosphoglycerate kinase,* which like all other kinases, requires Mg^{2+} ions for activity. It is in this reaction that ATP is first produced in the pathway. Because the ATP is formed by a direct transfer of a phosphate group from a metabolite to ADP, the process is referred to as **substrate-level phosphorylation** to distinguish it from *oxidative phosphorylation* that was discussed in Section 24.5.

In step 8, the phosphate group is transferred from the hydroxyl group of C-3 to the hydroxyl group of C-2, forming 2-phosphoglycerate in a reaction catalyzed by *phosphoglyceromutase.* This is followed by a dehydration reaction (step 9), catalyzed by *enolase,* to form phosphoenolpyruvate (PEP), another compound with a high-energy phosphate group.

The final step is irreversible and is the second reaction in which substrate-level phosphorylation occurs. The phosphate group of PEP is transferred to ADP; one molecule of ATP is produced per molecule of PEP. The reaction is catalyzed by *pyruvate kinase,* which requires both Mg^{2+} and K^+ for activity. Thus in phase II, two molecules of glyceraldehyde-3-phosphate are converted to two molecules of pyruvate, along with the production of four molecules of ATP and 2 molecules of NADH.

✓ Review Questions

25.1 **(a)** Which step in glycolysis is an oxidation-reduction reaction? **(b)** What is the oxidizing agent in this reaction?

25.2 In glycolysis, how many molecules of pyruvate are produced from one molecule of glucose?

25.2 Metabolism of Pyruvate

The presence or absence of oxygen determines the fates of pyruvate and NADH, produced in glycolysis. When plenty of oxygen is available, pyruvate is completely oxidized to CO_2, with the release of much greater amounts of ATP through the combined actions of the Krebs cycle (Section 24.4), the electron transport chain, and oxidative phosphorylation (Section 24.5). In the absence of oxygen (anaerobic conditions), the fate of pyruvate depends on the organism. In vertebrates it is converted to lactate, while other organisms, such as yeast, convert pyruvate to ethanol and carbon dioxide. These possible fates of pyruvate are summarized in Figure 25.2. Each will be discussed more fully below.

Consider first the complete oxidation of pyruvate when oxygen supplies are plentiful. Pyruvate is enzymatically decarboxylated and oxidized (oxidative

▶ **Figure 25.2**
Metabolic fates of pyruvate.

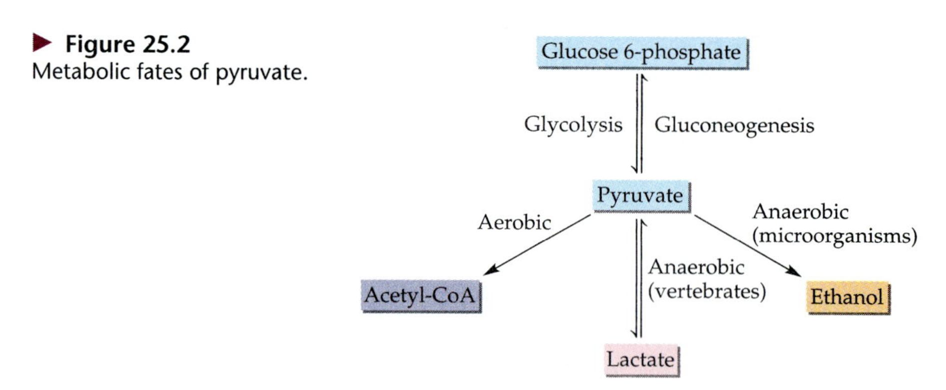

decarboxylation) to yield acetyl-CoA (Section 24.3), which enters the Krebs cycle. The formation of acetyl-CoA from pyruvate requires the sequential action of three enzymes and the participation of five coenzymes: coenzyme A, thiamine pyrophosphate (TPP), lipoic acid, NAD^+, and FAD.

$$\underset{\text{Pyruvate}}{CH_3-\overset{\overset{O}{\|}}{C}-COO^-} + \underset{\text{Coenzyme A}}{CoASH} + NAD^+ \xrightarrow[\text{TPP, FAD}]{\text{lipoic acid}} \underset{\text{Acetyl-CoA}}{CH_3-\overset{\overset{O}{\|}}{C}-SCoA} + NADH + CO_2 + H^+$$

The name given to this multienzyme system is the *pyruvate dehydrogenase complex*.

As we saw in Section 24.6, when strenuous exercise places enormous energy demands on muscles, the muscle cells must obtain energy under anaerobic conditions. When this happens, pyruvate is not oxidatively decarboxylated, but is converted to lactate. *Lactate dehydrogenase* catalyzes the reduction of pyruvate to lactate and the oxidation of NADH to NAD^+. This solves the problem of how to reoxidize NADH (needed in step 6 of glycolysis) in the absence of oxygen (you'll recall that the reoxidation of NADH through the electron transport chain requires oxygen as the ultimate acceptor of electrons).

Lactate is initially made as lactic acid, but at physiological pH, it ionizes, releasing H^+.

If lactate were allowed to accumulate in muscle cells (as well as the cells of other tissues[1]), it would make the cell more acidic, lowering the pH. The generation of energy via anaerobic glycolysis is self-limiting. Two processes act to maintain a proper level of lactate, and both require oxygen:

1. Most (70–80%) of the lactate diffuses out of the muscle and is transported to the liver. There it may be oxidized to pyruvate, which is completely oxidized (via the Krebs cycle) or converted back to glucose in a metabolic pathway known as gluconeogenesis (Section 25.5). The anaerobic catabolism of glucose to lactate in muscle cells, the transport of the lactate via the blood to the liver, and the reconversion of lactate to glucose comprise the **Cori cycle** (Figure 25.3).
2. When muscle cells receive an ample supply of oxygen, the 20 to 30% of the lactate that remains can be reoxidized to pyruvate and converted to acetyl-CoA, which enters the Krebs cycle.

The Cori cycle is named in honor of Gerty and Carl Cori, who first described it. The Coris won a Nobel Prize in 1947, the third husband/wife team to achieve this distinction (Marie and Pierre Curie were the first; Irène and Frédéric Joliot-Curie, the second).

We have seen that, when muscles use anaerobic pathways, they incur an oxygen debt. It is as if the body regards oxidation as the only proper source of energy

[1]The average individual, expending normal energy, produces about 120 g of lactate each day. (Vigorous exercise produces significantly more lactate.) It is estimated that one third of this daily lactate is produced by tissues that are strictly anaerobic (i.e., erythrocytes, retina, epidermis)—tissues that do not have mitochondria.

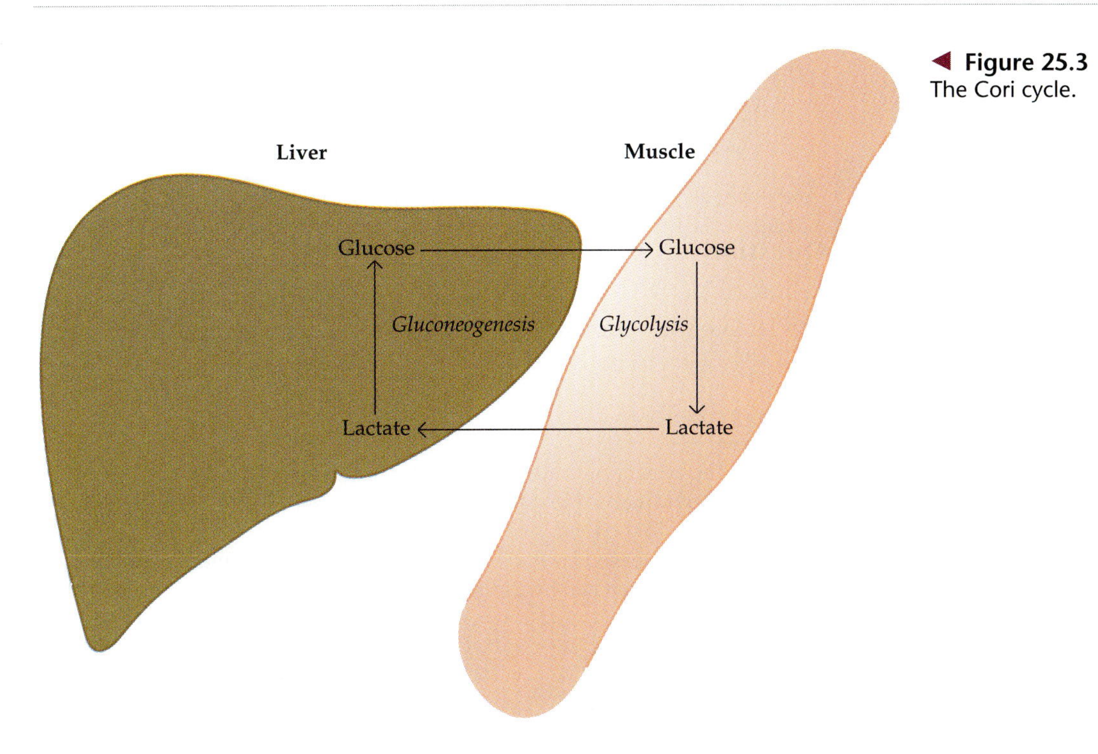

Figure 25.3
The Cori cycle.

for muscular activity and uses anaerobic metabolism as a temporary expedient. As soon as it can, the body oxidizes some of the resulting lactate back to pyruvate and ultimately to carbon dioxide and water. The energy released in the process is used to convert the rest of the lactate back to glucose and then to glycogen to be stored for future use.

When is this oxygen debt repaid? Just as soon as the very high level of muscular activity ceases. Sprinters running a 60-m dash breathe in oxygen but still obtain only a fraction of the energy required for their intense muscular activity through aerobic processes. When the race is over, the sprinters continue to take in great gulps of air. This air is used to repay the oxygen debt incurred during the race. We continue to breathe hard even after we stop vigorous activity because our body chemistry is still catching up.

Under anaerobic conditions, yeast and other microorganisms metabolize pyruvate to carbon dioxide and ethanol in two steps known as **alcoholic fermentation.** Pyruvate is first irreversibly decarboxylated to acetaldehyde in a reaction catalyzed by *pyruvate decarboxylase.* The enzyme requires the coenzyme thiamine pyrophosphate and Mg^{2+} ions:

$$\underset{\text{Pyruvate}}{CH_3COCOO^-} \xrightarrow[\text{TPP, } Mg^{2+}]{\text{decarboxylase}} \underset{\text{Acetaldehyde}}{CH_3CHO} + CO_2$$

In the next step, catalyzed by *alcohol dehydrogenase,* acetaldehyde is reduced to ethyl alcohol and NADH is oxidized to NAD^+, regenerating this coenzyme needed in step 6 of glycolysis. (When yeast is used in baking, the alcohol readily evaporates out during the baking process.)

$$CH_3CHO \text{ (Acetaldehyde)} + H^+ + NADH \xrightleftharpoons[\text{dehydrogenase, } Zn^{2+}]{} NAD^+ + CH_3CH_2OH \text{ (Ethanol)}$$

Review Questions

25.3 What is the fate of lactate formed by muscular activity?

25.4 Explain how lactate formation allows glycolysis to continue under anaerobic conditions.

Alcohol Metabolism

More than two-thirds of the adult population in the United States drinks alcoholic beverages at least occasionally. In humans, the principal route of metabolism of ingested alcohol is believed to be oxidation in the liver, first to acetaldehyde and subsequently to acetic acid. Most of the acetic acid is released by the liver and transported to other tissues, where it is converted to acetyl-CoA, which can be utilized in a variety of ways (Section 24.3).

$$CH_3CH_2OH \xrightarrow[\text{dehydrogenase}]{\text{alcohol}} CH_3CHO \xrightarrow[\text{dehydrogenase}]{\text{acetaldehyde}} CH_3COOH \xrightarrow[\text{synthetase}]{\text{acetyl-CoA}} CH_3COSCoA$$

In a pregnant woman, alcohol readily crosses the placental membrane and builds up in the fetus, because the liver of the fetus does not contain the enzymes needed to metabolize the alcohol. These children have a high risk of *fetal alcohol syndrome* (FAS). FAS consists of facial deformities, growth deficiency, and mental retardation. It is believed to be the third most common cause of mental deficiency (IQs average 35 to 40 points below normal). In the United States, more than 2000 FAS children are born each year. It has not been possible to establish the level of alcohol consumption that leads to FAS, so woman who are pregnant are advised not to drink any alcohol.

More than 200,000 people die because of alcohol addiction each year (mainly as a result of cirrhosis of the liver, but also of cardiovascular disease and cancers of the mouth and larynx), and alcohol is a factor in one of every ten deaths. People who drive while intoxicated are responsible for about half of U.S. traffic fatalities. Alcohol-impaired driving is the leading cause of death and injury among those under 25 years of age.

One method for the treatment of chronic alcoholism is to administer a drug that inhibits the second step of alcohol metabolism. The drug disulfiram (Antabuse) successfully competes with acetaldehyde for the active site on the enzyme acetaldehyde dehydrogenase. As a result, when alcohol is ingested by an individual previously treated with disulfiram, the blood acetaldehyde concentrations rise five to ten times higher than in an untreated individual. This leads to severe discomfort. Characteristic physiological responses are an increase of the heartbeat and a reduction in blood pressure. The individual experiences respiratory difficulties, nausea, sweating, vomiting, chest pains, and blurred vision. Once elicited, the effect lasts between 30 minutes and several hours. To avoid such discomfort, the patient must abstain from alcohol (the treatments are continued until the person can abstain voluntarily). Disulfiram should be administered only by a physician, and therapy is usually begun in a hospital.

$$(CH_3CH_2)_2N{-}C({=}S){-}S{-}S{-}C({=}S){-}N(CH_2CH_3)_2$$

Disulfiram

25.3 ATP Yield From Glycolysis

The net energy yield from anaerobic glucose metabolism can be readily determined as moles of ATP. One mole of ATP is expended in the initial phosphorylation of glucose (step 1) and again in the phosphorylation of fructose-6-phosphate (step 3). In step 7, two moles of 1,3-bisphosphoglycerate (recall that two moles of 1,3-BPG are formed for each mole of glucose) are converted to two moles of 3-phosphoglycerate and two moles of ATP are produced. In step 10, two moles of pyruvate and two moles of ATP are formed per mole of glucose.

For every mole of glucose degraded, 2 moles of ATP are initially consumed and 4 moles of ATP are ultimately produced. The net production of ATP is thus 2 moles per mole of glucose converted to lactate or ethanol. You'll recall that about 7.5 kcal of free energy is conserved per mole of ATP produced (Section 24.1). Recall also that the total amount of energy that can theoretically be obtained from the complete oxidation of 1 mole of glucose is 686 kcal (Chapter 24 Introduction). The energy conserved in the anaerobic catabolism of glucose to two molecules of lactate (or ethanol) would be:

$$[(2 \times 7.5 \text{ kcal})/686 \text{ kcal}] \times 100\% = 2.2\%$$

Thus anaerobic cells extract only a very small fraction of the total energy of the glucose molecule.

Contrast this with the amount of energy obtained when glucose is completely oxidized to carbon dioxide and water through glycolysis, the Krebs cycle, electron transport, and oxidative phosphorylation as summarized in Table 25.1. Note in the

Table 25.1 Yield of ATP From the Complete Oxidation of One Mole of Glucose

Reaction	Comments	Yield of ATP (moles)
Glycolysis		
glucose → glucose-6-phosphate (G6P)	consumes 1 mol ATP	−1
G-6-P → fructose-1,6-bisphosphate	consumes 1 mol ATP	−1
glyceraldehyde-3-phosphate → 1,3-bisphosphoglycerate	produces 2 mol of cytoplasmic NADH	
1,3-BPG → 3-phosphoglycerate	produces 2 mol ATP	+2
phosphoenolpyruvate → pyruvate	produces 2 mol ATP	+2
Oxidation of Pyruvate (mitochondrial reaction)		
pyruvate → acetyl CoA + CO_2	produces 2 mol NADH	
Krebs cycle (mitochondria)		
isocitrate → α-ketoglutarate + CO_2	produces 2 mol NADH	
α-ketoglutarate → succinyl-CoA + CO_2	produces 2 mol NADH	
succinyl-CoA → succinate	produces 2 mol GTP	+2
succinate → fumarate	produces 2 mol $FADH_2$	
malate → oxaloacetate	produces 2 mol NADH	
Electron transport chain and oxidative phosphorylation		
2 cytoplasmic NADH formed in glycolysis	each yields 1.5–2.5 mol ATP (depending on tissue)	+3 to +5
2 NADH formed in the oxidation of pyruvate	each yields 2.5 mol ATP	+5
2 $FADH_2$ formed in the Krebs cycle	each yields 1.5 mol ATP	+3
6 NADH formed in the Krebs cycle	each yields 2.5 mol ATP	+15
	Net yield of ATP	+30 to +32

table that a variable amount of ATP is synthesized from the NADH formed in the cytoplasm during glycolysis. This is because NADH is not transported into the inner mitochondrial membrane where the enzymes for the electron transport chain are located. Brain and muscle cells utilize a transport mechanism that passes electrons from the cytoplasmic NADH through the membrane to FAD molecules inside the mitochondria, forming $FADH_2$, which then feeds the electrons into the electron transport chain. This lowers the yield of ATP to 1.5 molecules of ATP, rather than the usual 2.5 molecules. A more efficient transport system is found in liver, heart, and kidney cells where one cytoplasmic NADH results in one mitochondrial NADH, leading to the formation of 2.5 molecules of ATP.

The total amount of energy conserved in aerobic catabolism of glucose, in the liver, would be:

$$[(32 \times 7.5 \text{ kcal})/686 \text{ kcal}] \times 100\% = 35\%$$

A conservation of 35% of the total energy released compares favorably with the efficiency of any machine, and it represents a remarkable achievement on the part of the living organism. In comparison, automobiles are only about 20–25% efficient in utilizing the energy released in the combustion of gasoline.

What happens to the 65% of the energy that is not conserved? It is released as heat to the surroundings—that is, to the cell. It is this heat that maintains body temperature. If we are exercising strenuously and our metabolism speeds up to provide the necessary energy for muscle contraction, more heat is also produced. We begin to sweat to dissipate some of that heat. As the sweat evaporates, the excess heat is carried away from the body by the departing water vapor.

✓ Review Questions

25.5 In anaerobic glycolysis, how many molecules of ATP are produced from one molecule of glucose?

25.6 When plenty of oxygen is available, how many molecules of ATP can be obtained from one molecule of glucose?

25.4 Gluconeogenesis

We have seen how glucose can be used for energy, but what happens when all of the glucose is gone? The brain must have a constant supply of energy to continue functioning, and it normally gets this energy through glycolysis. The body must have a way of synthesizing glucose from noncarbohydrate sources such as pyruvate and lactate when inadequate amounts of carbohydrates are obtained from the diet. This is done through a series of reactions known as **gluconeogenesis,** shown in Figure 25.4. If we compare this to Figure 25.1 we see that it looks very much like the reverse of glycolysis.

The irreversibility of metabolic pathways is a common phenomenon—it will be encountered again in the metabolism of lipids and proteins. Catabolic and anabolic processes may have many of the same intermediates and many of the same enzymes, and yet their pathways are not identical. The separation of degradative and synthetic pathways allows an organism to control one series of reactions without affecting the other.

The conversion of pyruvate into phosphoenolpyruvate occurs in two enzyme-catalyzed reactions, both requiring the hydrolysis of a high-energy phosphate compound. In the first reaction, oxaloacetate is formed from pyruvate in a carboxylation reaction catalyzed by *pyruvate carboxylase.* This enzyme requires biotin as a coenzyme. The energy needed to form the C—C bond is provided by the hydrolysis of ATP. In

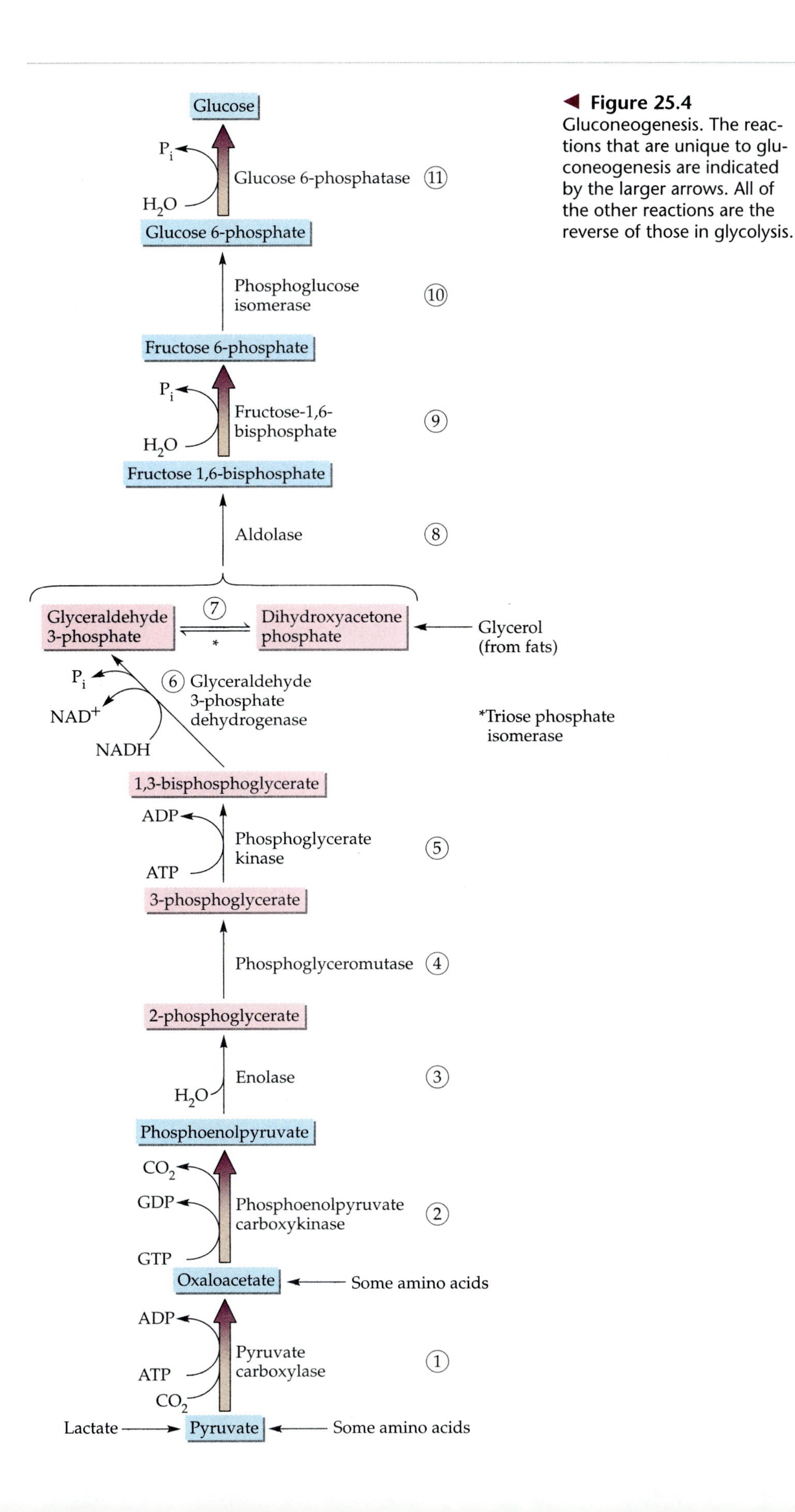

◀ Figure 25.4
Gluconeogenesis. The reactions that are unique to gluconeogenesis are indicated by the larger arrows. All of the other reactions are the reverse of those in glycolysis.

the second reaction, catalyzed by *phosphoenolpyruvate carboxykinase,* the carboxyl group is removed and a phosphate group is added as GTP is hydrolyzed to GDP.

$$\underset{\text{Pyruvate}}{\begin{array}{c}CO_2^-\\ |\\ C{=}O\\ |\\ CH_3\end{array}} + CO_2 + ATP + H_2O \xrightleftharpoons[]{\text{pyruvate carboxylase}} \underset{\text{Oxaloacetate}}{\begin{array}{c}CO_2^-\\ |\\ C{=}O\\ |\\ CH_2\\ |\\ CO_2^-\end{array}} + ADP + P_i$$

$$\underset{\text{Oxaloacetate}}{\begin{array}{c}CO_2^-\\ |\\ C{=}O\\ |\\ CH_2\\ |\\ CO_2^-\end{array}} + GTP \xrightleftharpoons[]{\text{phosphoenolpyruvate carboxykinase}} \underset{\text{Phosphoenolpyruvate}}{\begin{array}{c}CO_2^-\\ |\\ C{-}OPO_3^{2-}\\ \|\\ CH_2\end{array}} + CO_2 + GDP$$

Reactions 3–8 are the reverse of those in glycolysis. However, removal of a phosphate group from carbon 1 of fructose-1,6-bisphosphate (reaction 9) is catalyzed by an enzyme unique to gluconeogenesis—*fructose-1,6-bisphosphatase.* A second phosphatase, *glucose-6-phosphatase,* is needed for reaction 11 in which glucose-6-phosphate is hydrolyzed to form glucose. This last enzyme is found only in liver and kidney cells, and thus these are the only tissues capable of carrying out all of the reactions of gluconeogenesis.

The glucose that is formed in the liver and kidney can be transported out of these tissues and carried by the blood to other tissues (primarily brain and skeletal muscle). A genetic disease, von Gierke's disease, is caused by a lack of glucose-6-phosphatase in liver cells. The amount of glucose in the blood is abnormally low because glucose cannot be formed from glucose-6-phosphate.

To convert two molecule of pyruvate to one molecule of glucose requires the hydrolysis of six high-energy phosphate compounds—two from GTP and four from ATP.

✓ **Review Question**

25.7 What is the function or purpose of gluconeogenesis?

25.5 Glycogen Metabolism

Hers's disease is another genetic disease, like McArdle's disease, in which there is an inability to properly degrade glycogen. In Hers's disease there is a deficiency or absence of liver glycogen phosphorylase. Thus stored glycogen cannot be used to maintain the blood sugar level. Because glycogen cannot be degraded, the liver becomes enlarged.

In Chapter 19 we discussed glycogen—a branched polymer of glucose that animals use to store carbohydrate, particularly in the liver and in muscle. The glycogen stored in these two tissues serves different purposes in the body. The major function of muscle glycogen is to provide energy for short bursts of intense activity, while the major function of liver glycogen is to maintain a reserve of glucose to use when the blood sugar level is low.

The metabolic pathways for glycogen synthesis (**glycogenesis**) and degradation (**glycogenolysis**) are shown in Figure 25.5. In glycogenesis, the reaction catalyzed by glycogen synthase is a regulated step. For glycogenolysis the regulated step is catalyzed by glycogen phosphorylase. It is important that these two pathways are regulated in a coordinated manner so that when glycogen is being hydrolyzed to provide glucose, the same cell is not trying to synthesize glycogen from glucose.

In Figure 25.5, note that when glycogen is degraded in the reaction catalyzed by glycogen phosphorylase, glucose-1-phosphate is formed, rather than free glucose. The glucose-1-phosphate that is formed can be converted to glucose-6-phosphate,

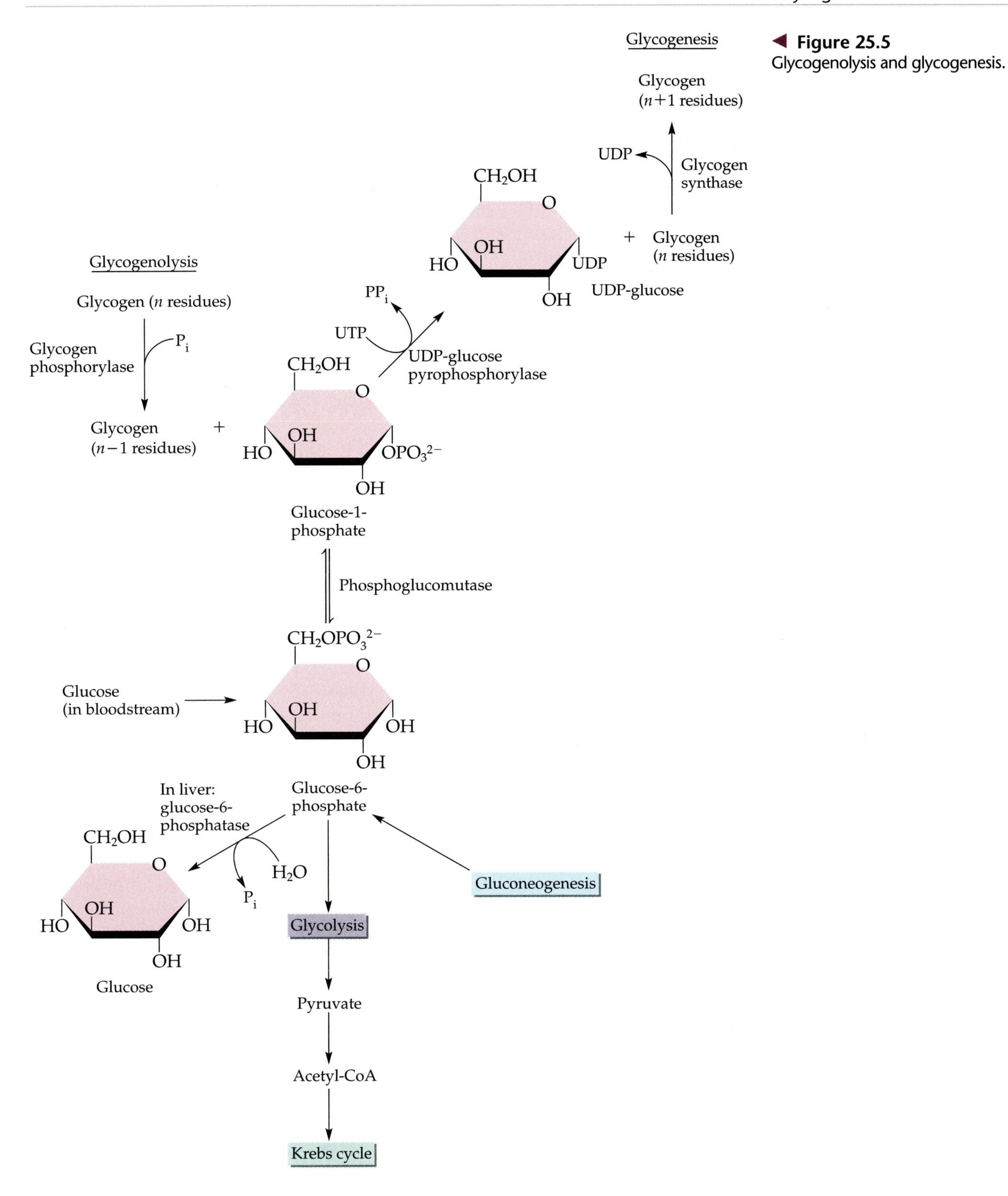

Figure 25.5
Glycogenolysis and glycogenesis.

in a reaction catalyzed by phosphoglucomutase, and enter glycolysis at step 2 (Figure 25.1), rather than at step 1. Thus when glycogen is the source of glucose, it is necessary to expend only one ATP in Phase I of glycolysis, rather than two ATP. This represents a 50% increase in efficiency compared with the utilization of free glucose, because there is a net production of three ATP, rather than two (Section 25.3).

The addition and removal of the branches in glycogen (joined with α-1,6-glycosidic bonds) requires the action of other enzymes known as the branching and debranching enzymes, respectively.

✓ Review Questions

25.8 **(a)** What is the storage form of carbohydrate in the body? **(b)** In what tissues or organs is this carbohydrate stored?

25.9 What is the enzyme that catalyzes the regulated step in: **(a)** glycogenolysis? **(b)** glycogenesis?

25.6 Regulation of Carbohydrate Metabolism

▲ The teaspoon of glucose represents the total amount that is circulating in the blood at any given time. The large pile of glucose is the amount of glucose used by the body each day.

Regulation of carbohydrate metabolism is critical in maintaining the blood sugar level (concentration of glucose in the blood), as well as providing sufficient energy to the brain, muscles, and other tissues. Under normal circumstances the blood sugar level remains remarkably constant at about 80 mg/dL of blood. However, because individuals differ in chemical makeup, a normal concentration of glucose may range from 70–100 mg/dL. (This amounts to a total of about 5–6 g of glucose, or 1 teaspoonful, circulating in the entire blood stream at any given time.) Soon after a meal, the blood sugar level may rise to 120 mg/dL, but it returns to the normal level within 2 hours.

The brain uses about 120 g of glucose per day. At rest, the total glucose requirement of all other tissues of the body (heart, liver, kidneys, muscles, etc.) is about 200 g/day. When we contrast the amount of glucose circulating in the blood at any given time with the total amount of glucose required by the brain and other tissues, we begin to glimpse the importance of the body's exquisite regulation of blood sugar levels.

When blood sugar levels are not controlled, there may be dire consequences. After severe starvation or vigorous exercise, the blood sugar concentration may fall below normal, leading to the condition called **hypoglycemia.** Extreme hypoglycemia can cause unconsciousness and lowered blood pressure and may result in death. Loss of consciousness is most likely due to the lack of glucose in the brain tissue.

Diabetes mellitus is a metabolic disease that has been recognized for thousands of years. It is named from its symptoms—*diabetes mellitus* means excessive, sweet urine.

The condition of high blood sugar is called **hyperglycemia.** The major cause of hyperglycemia is **diabetes mellitus** (see boxed essay). In an effort to lower the blood sugar level, the kidneys may excrete excess glucose. (The renal threshold value for glucose is fairly high, ranging from 150 to 170 mg of glucose per 100 mL of blood.)

In all metabolic pathways (not just carbohydrate metabolism), it is important that when one pathway is needed (for example, gluconeogenesis to synthesize glucose) the opposing pathway (in this example, glycolysis) be essentially inactive. In this section we'll consider how carbohydrate metabolism—glycolysis, gluconeogenesis, glycogenolysis, and glycogenesis—are regulated.

The major regulated step in glycolysis is the conversion of fructose-6-phosphate into fructose-1,6-bisphosphate, a reaction catalyzed by phosphofructokinase. This enzyme is inhibited by ATP, but activated by AMP. In gluconeogenesis the reverse reaction is the major regulatory step—the conversion of fructose-1,6-bisphosphate to fructose-6-phosphate, catalyzed by fructose-1,6-bisphosphatase. This enzyme is inhibited by AMP. Low ATP levels and high AMP levels indicate that

Diabetes

If the pancreas does not secrete enough insulin and/or if there are insufficient (or defective) insulin receptors on the cell membranes, diabetes mellitus develops. Although medical science has made significant progress against this disease, it continues to be a major health threat. Consider a few of the serious complications:

1. Diabetes is second only to trauma as the most frequent cause of lower limb amputations in the United States.
2. It is the leading cause of blindness in adults older than 20.
3. It is the leading cause of kidney failure.
4. It more than doubles the risk of having a heart attack or stroke.

Diabetes is characterized by abnormal metabolism of carbohydrates, proteins, and lipids. Because a diabetic is unable to use glucose properly, excessive quantities accumulate in the blood and urine. Characteristic symptoms of diabetes are constant hunger, weight loss, extreme thirst, and frequent urination, because the kidneys excrete large amounts of water in an attempt to remove excess sugar from the blood.

The **glucose tolerance test** is an important diagnostic test for diabetes mellitus. A patient's blood sugar level is determined after an overnight fast. Then a known amount (~75 g) of glucose is dissolved in about 400 mL of water and administered orally over a period of 5 to 10 minutes. Blood is drawn from the patient at 30-minute intervals after ingestion, and the blood sugar concentration is determined. After an initial rise, the blood sugar level falls rapidly in a normal individual. In a diabetic person, on the other hand, the increase in the blood sugar level is greater than normal and the level remains elevated for several hours (Figure 25.6).

There are two types of diabetes. In immune-mediated diabetes (formerly known as insulin-dependent or juvenile-onset diabetes) insufficient amounts of insulin are produced. This type of diabetes develops early in life and is also known as *Type 1 diabetes.* Symptoms are rapidly reversed by the administration of insulin, and Type 1 diabetics can lead active lives provided they receive insulin as needed. Because insulin is a protein that is readily digested in the small intestine, it cannot be taken orally, but must be injected at least once a day.

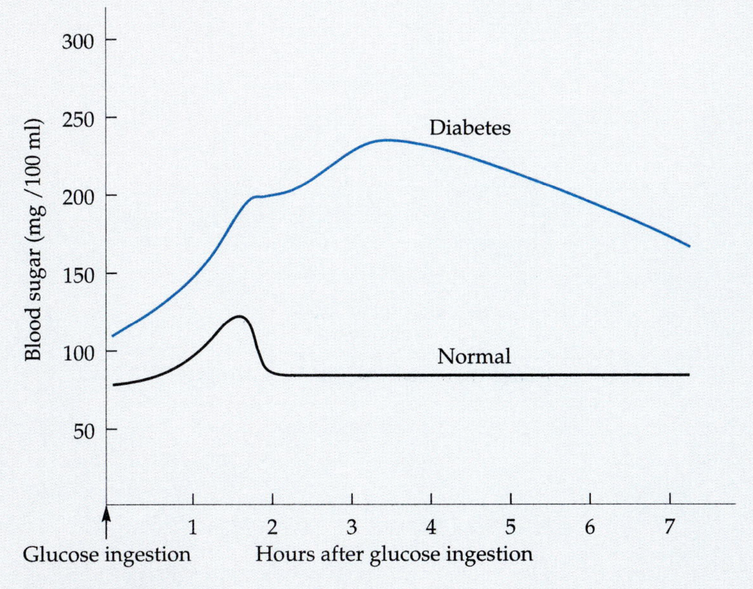

▲ **Figure 25.6**
Glucose tolerance test results for a normal person and for a diabetic person.

In Type 1 diabetes, insulin-producing cells of the pancreas are destroyed by the body's immune system. The reasons for this are areas of intensive research. Researchers have developed a simple blood test capable of predicting who will develop Type 1 diabetes several years before the disease becomes apparent. The blood test searches for antibodies that destroy the body's insulin-producing cells.

Type 2 diabetes (formerly known as noninsulin-dependent or adult-onset diabetes) is by far the more common (about 95% of diabetic cases—about 15 million Americans). Type 2 diabetics produce sufficient amounts of insulin, but either the insulin-producing cells in the pancreas do not secrete enough of it or it is not utilized properly (because there is a lack of insulin receptors on the target cells or the insulin receptors are defective). For many of these people, the disease can be controlled with a combination of diet and exercise alone. For some people who are overweight, losing weight is sufficient to bring their blood sugar level into the normal range. Medication is not required if they maintain an exercise plan and eat wisely. In the past, diabetics were told to avoid sucrose (or sugar) because it was thought that this would lead to a rapid and significant increase in the blood sugar level. This has been shown not to be true, but foods high in sugar are usually not high in nutrients, and should be used in small quantities for that reason.

For those individuals who require medication, oral antidiabetic drugs that stimulate the islet cells to secrete insulin can be used (see below). Second-generation antidiabetic drugs, such as glyburide, stimulate the release of insulin, as do the first-generation drugs, but they also increase the sensitivity of the cell receptors to insulin. All of these drugs carry warnings about the increased risk of cardiovascular disease, even though only one of these drugs, tolbutamide, was linked to an increased risk of heart disease in a single study, known as the UGDP study. This study has been criticized by many diabetes experts who believe that there were major flaws in the design of the study that negate the results. Thus there is not a well-established link between tolbutamide and an increased risk for heart disease.

Some individuals with Type 2 diabetes do not secrete sufficient insulin and do not respond to these oral medications; they must use insulin. In either Type 1 or Type 2 diabetes, the blood sugar level must be carefully monitored and adjustments made in diet or medication to keep the level as normal as possible (70–120 mg/dL).

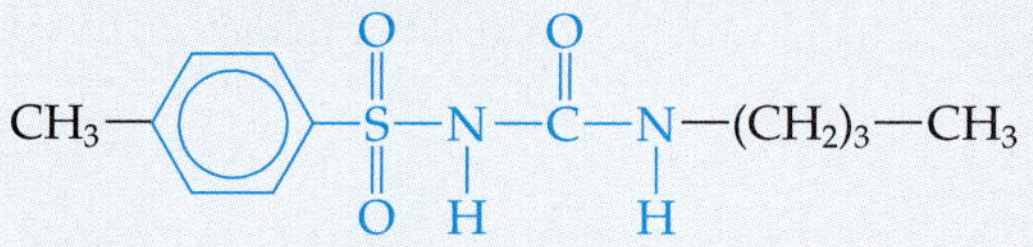

Tolbutamide
(Orinase)

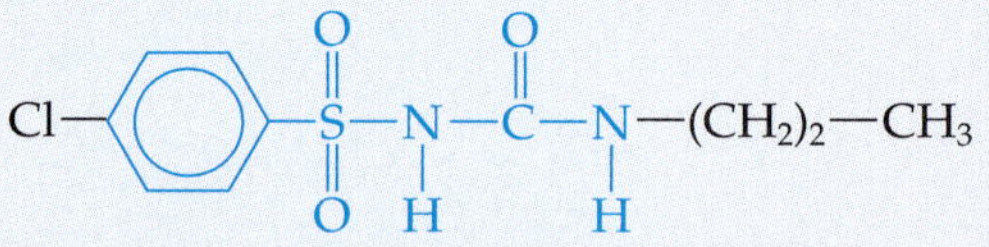

Chlorpropamide
(Diabinese)

First-Generation Antidiabetic Drugs

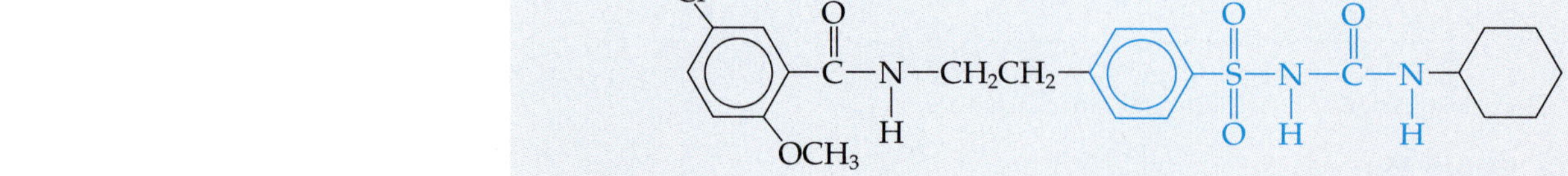

Glyburide
(Second-Generation Antidiabetic Drug)

the cell is degrading ATP for energy and glucose needs to be broken down to provide more ATP. Conversely, high ATP levels indicate that the cell does not need to break down glucose to provide more energy.

The most important process in the maintenance of a constant blood glucose concentration is the synthesis and breakdown of glycogen in the liver. The liver is responsible for removing glucose from the blood when the concentration is too high (after a meal) and releasing it to the blood when the blood sugar level is too low (e.g., between meals). The activity of the liver, in this regard, is controlled by several hormones, including insulin, epinephrine (adrenaline), and glucagon.

Hormonal Regulation

Insulin is produced by the beta cells of the islets of Langerhans in the pancreas. It is released in response to a high blood sugar level. In general, insulin promotes anabolic reactions and inhibits catabolic reactions in the liver, muscle, and adipose tissue. (It has no effect on carbohydrate metabolism in the brain or kidneys.) Specifically, insulin performs the following functions:

1. Enhances the formation of glycogen from glucose in both the liver and muscle.
2. Promotes the entry of glucose into muscle, liver, and adipose tissue. (Note, however, that red blood cells and cells in the brain, kidneys, and intestinal tract do not require insulin for glucose uptake.)
3. Accelerates the conversion of glucose to fatty acids and hence the synthesis and storage of triglycerides in adipose tissue.
4. Inhibits the breakdown of glycogen and stored fat.
5. Promotes the transport of amino acids into cells and stimulates protein synthesis; inhibits intracellular degradation of proteins.
6. Suppresses gluconeogenesis, which occurs primarily in the liver.

Hence, the principal role of insulin is to remove glucose rapidly from the blood, lowering the blood sugar level. The mechanism by which insulin accomplishes its prodigious tasks is slowly being determined. Insulin binds to specific receptor proteins in the membranes of target cells. It is known that the binding of insulin to its receptor activates an enzyme in the receptor that leads to the phosphorylation of specific tyrosine residues in other proteins. How this phosphorylation activity is coupled to the physiological responses observed upon insulin release is an area of active research.

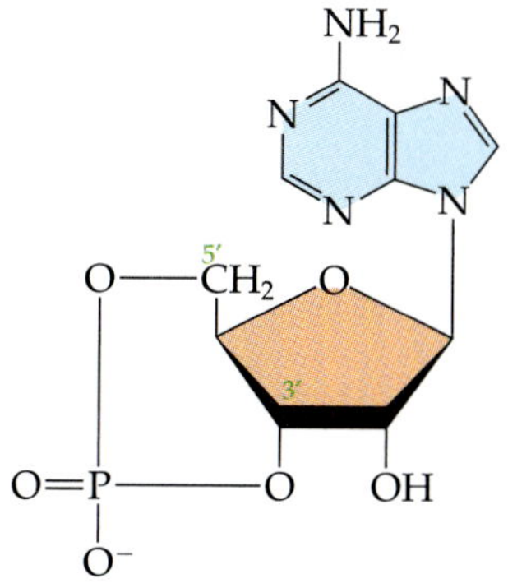

All other hormones that affect glucose metabolism act to raise the blood sugar level. Both epinephrine and glucagon exert their effects by binding to receptor proteins on the outside of the cell membrane (each binds to a different specific receptor). These receptor proteins are linked to the enzyme **adenylate cyclase,** which is bound to the inner membrane. When the outer receptor protein has no hormone bound to it, the enzyme is not active. The binding of epinephrine or glucagon to its respective receptor causes conformational changes in other membrane proteins, known as G proteins. These conformational changes activate adenylate cyclase, which catalyzes the conversion of ATP to adenosine 3′,5′-monophosphate (cyclic AMP or cAMP). Adenylate cyclase is extremely efficient, and many cAMP molecules can be synthesized by a single activated enzyme:

NH2 N N N N 5′ CH2 O O 3′ O=P O OH O−

Adenosine 3′,5′-monophosphate (cAMP)

$$\text{ATP} \xrightarrow[\text{adenylate cyclase}]{Mg^{2+}} \text{cAMP} + \text{PP}_i$$

cAMP is often referred to as a *second messenger* because it transmits messages (delivered via the blood by the extracellular hormones—the primary messengers) from the cell membrane to enzymes within the cell (Figure 25.7). cAMP binds to and activates certain inactive enzymes, thus beginning a cascade of cellular events that results in the stimulation of a wide range of catabolic processes and the inhibition of several anabolic reactions. It should be noted that cAMP serves as the second messenger for some neurotransmitters and several other hormones in addition to epinephrine and glucagon (Table 25.2).

Epinephrine (Section C.2) is secreted by the adrenal medulla during exercise or during periods of emotional stress, such as anger or fright, to provide the organism with additional energy. Epinephrine binds to its receptors (called *adrenergic receptors* after the hormone's other name, adrenaline), primarily on the membranes of muscle cells and to a lesser extent on the membranes of liver cells. It markedly stimulates the breakdown of glycogen to glucose. Epinephrine also promotes gluconeogenesis in the liver. Both of these processes increase the amount of glucose, making it available for energy production.

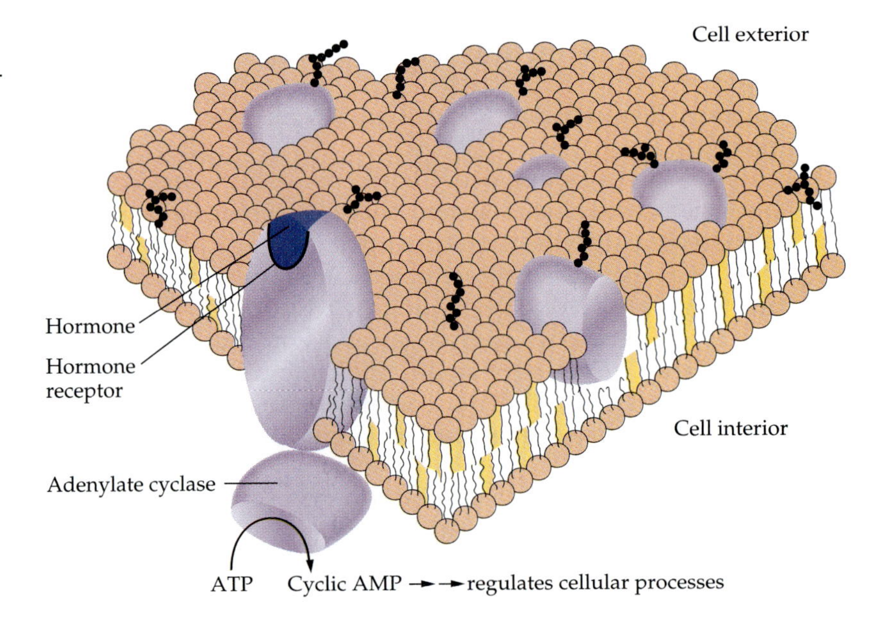

Figure 25.7
Binding of an extracellular hormone to a receptor activates adenylate cyclase inside the cell. This enzyme then catalyzes the synthesis of cAMP.

Table 25.2 Hormones That Increase cAMP Levels

Tissue	Hormone	Principal Response
Bone	Parathyroid hormone	Calcium resorption
Muscle	Epinephrine	Glycogenolysis
Adipose	Epinephrine	Lipolysis
	Adrenocorticotrophic hormone	Lipolysis
	Glucagon	Lipolysis
Brain	Norepinephrine	Discharge of Purkinje cells
Thyroid	Thyroid-stimulating hormone	Thyroxine secretion
Heart	Epinephrine	Increased contractility
Liver	Epinephrine	Glycogenolysis
Kidney	Parathyroid hormone	Phosphate excretion
	Vasopressin	Water reabsorption
Adrenal	Adrenocorticotrophic hormone	Hydrocortisone secretion
Ovary	Luteinizing hormone	Progesterone secretion

Glucagon, like insulin, is a peptide (29 amino acids) hormone produced by the pancreas (the alpha cells of the islets of Langerhans), but its actions are opposite to those of insulin. Glucagon is secreted in response to a low blood sugar level, and its task is to increase the glucose concentration. Glucagon acts primarily on the liver (and adipose tissue) and not on skeletal muscle because there are no receptors for glucagon on muscle cell membranes. Glucagon stimulates gluconeogenesis and glycogen breakdown, two processes that restore the blood glucose to its normal level. The concentration of glucagon in the blood of a diabetic is above normal, and this may be a significant contributor to the problems associated with diabetes.

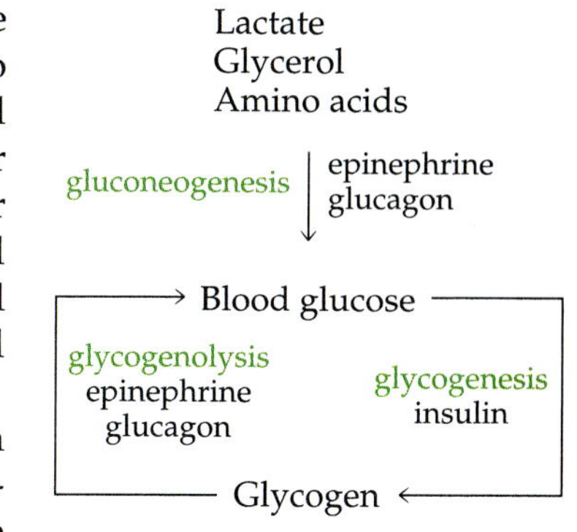

Glycogen metabolism is strongly influenced by the ratio of insulin to glucagon in the blood. Higher amounts of insulin lead to glycogen storage after a meal, whereas higher amounts of glucagon favor the breakdown of liver glycogen to add more glucose to the blood. It is the insulin/glucagon ratio that determines the outcome of carbohydrate metabolism. A high ratio leads to carbohydrate anabolism and storage, whereas a low ratio results in carbohydrate catabolism and utilization. The secretion of these hormones is directly governed by the blood-sugar level.

Review Questions

25.10 How are the activities of phosphofructokinase and fructose-1,6-bisphosphatase regulated?

25.11 What is the role of insulin in regulating the blood-sugar level?

Summary

Oxidation of carbohydrates is the source of over 50% of the energy used by cells. Glucose, obtained from the diet or the breakdown of glycogen, is oxidized to two molecules of pyruvate through a series of reactions known as **glycolysis.** Some of the energy released in these reactions is conserved by the formation of ATP from ADP. Glycolysis can be divided into two phases: Phase I consists of the first five reactions and requires energy to "prime" the glucose molecule for Phase II, the last five reactions in which ATP is produced through **substrate-level phosphorylation.**

The pyruvate produced by glycolysis has several possible fates, depending on the organism and whether or not oxygen is present. In animal cells, pyruvate can be further oxidized to acetyl-CoA and then to carbon dioxide (through the Krebs cycle) if oxygen supplies are sufficient. When oxygen supplies are insufficient, pyruvate is reduced to lactate. This provides a way for NADH, formed during glycolysis, to be reoxidized in the absence of oxygen (needed for the electron transport chain to function). Most of the lactate formed in muscle cells under anaerobic conditions is transported to the liver. There it can be reoxidized to pyruvate and converted back to glucose. The anaerobic catabolism of glucose to lactate in muscle cells, the transport of the lactate via the blood to the liver, and the reconversion of lactate to glucose is referred to as the **Cori cycle.**

In yeast and other microorganisms, pyruvate is not converted to lactate in the absence of oxygen, but instead is converted to ethanol and carbon dioxide in a process known as **alcoholic fermentation.**

The amount of ATP formed by the oxidation of glucose depends on whether or not oxygen is present. If oxygen is present, glucose is oxidized to CO_2 and 30–32 ATP molecules are produced for each glucose oxidized, using the combined pathways of glycolysis, Krebs cycle, electron transport chain, and oxidative phosphorylation. Thus approximately 35% of the energy released by the complete oxidation of glucose is conserved by the synthesis of ATP. In the absence of oxygen, only 2 molecules of ATP are formed for each molecule of glucose converted to lactate (2 molecules), and the amount of energy conserved is much less (2%).

When the amount of glucose in the blood drops, liver cells convert noncarbohydrate molecules such as lactate, oxaloacetate, and pyruvate to glucose through a series of reactions known as **gluconeogenesis.** Many of the reactions in gluconeogenesis are the reverse of reactions in glycolysis, but the two pathways differ in three steps.

The blood sugar level is the concentration of glucose in the blood. The condition in which this concentration is high is **hyperglycemia;** that in which this concentration is low is **hypoglycemia.** The **glucose tolerance test** is used to diagnose **diabetes mellitus** by determining the rate at which the blood sugar level drops following the intake of a large amount of glucose.

The body regulates the amount of glucose in the body by carefully controlling glycolysis, gluconeogenesis, **glycogenolysis** (glycogen breakdown), and **glycogenesis** (glycogen synthesis). When blood sugar levels are high, *insulin* secreted from the pancreas increases the uptake of glucose by liver and muscle cells. Glucose that is not needed for glycolysis is converted to glycogen. Insulin activates glycogen synthase, the primary enzyme needed for this process, and inactivates glycogen phosphorylase, the primary enzyme needed for glycogen breakdown.

When blood sugar levels are low, *glucagon* secreted from the pancreas stimulates liver cells to break down glycogen to glucose by activating glycogen phosphorylase and inactivating glycogen synthase. *Epinephrine* has an effect similar to glucagon, except that it is more active on muscle cells.

Key Terms

adenylate cyclase (25.6)
alcoholic fermentation (25.2)
Cori cycle (25.2)
diabetes mellitus (25.6)
gluconeogenesis (25.4)
glucose tolerance test (25.6)
glycogenesis (25.5)
glycogenolysis (25.5)
glycolysis (25.1)
hyperglycemia (25.6)
hypoglycemia (25.6)
substrate-level phosphorylation (25.1)

Problems

Glycolysis and Metabolism of Pyruvate

1. Replace the question marks with the proper compounds:

a. Fructose-1,6-bisphosphate $\xrightarrow{\text{aldolase}}$? + ?

b. ? $\xrightarrow{\text{pyruvate kinase}}$ pyruvate

c. Dihydroxyacetone phosphate $\xrightarrow{?}$ Glyceraldehyde–3–phosphate

d. Glucose $\xrightarrow{\text{hexokinase}}$?

2. Replace the question marks with the proper compounds:

a. Fructose-6-phosphate $\xrightarrow{?}$ Fructose-1,6-bisphospate

b. ? $\xrightarrow{\text{phosphoglucose isomerase}}$ Fructose-6-phosphate

c. Glyceraldehyde-3-phosphate $\xrightarrow{?}$ 1,3–Bisphosphoglycerate

d. 3-Phosphoglycerate $\xrightarrow{\text{phosphoglyceromutase}}$?

3. Refer to Problems 1 and 2, and select the equation(s) (by number and letter) in which each process occurs:
a. the hydrolysis of a high-energy phosphate compound
b. the formation of a high-energy phosphate compound
c. isomerization reaction
d. oxidation reaction

4. What critical role is played by both 1,3-bisphosphoglycerate and phosphoenolpyruvate (PEP) in glycolysis?

5. What coenzyme is needed as an oxidizing agent in glycolysis?

6. What type of enzyme is needed to catalyze each of the following reactions? Choose from dehydrogenase, isomerase, kinase, or mutase.
a. reaction of fructose-6-phosphate to form fructose-1,6-bisphosphate
b. reaction of 3-phosphoglycerate to form 2-phosphoglycerate
c. the reaction of glucose to form glucose-6-phosphate

7. List four phosphorylated and four nonphosphorylated metabolites of glycolysis and fermentation.

8. **a.** In muscle cells, what is the fate of pyruvate when oxygen supplies are abundant?
b. What happens to pyruvate when oxygen supplies in the muscle are limited (anaerobic conditions)?

9. **a.** In glycolysis, how many *total* molecules of ATP are produced for each molecule of glucose that is converted to pyruvate?
b. How many molecules of ATP are hydrolyzed in Phase I?
c. What is the *net* ATP production from glycolysis alone?

10. How is the NADH produced in glycolysis reoxidized when oxygen supplies are abundant?

11. How is the NADH produced in glycolysis reoxidized when oxygen supplies are limited: **(a)** in muscle cells; **(b)** in yeast?

12. **a.** Outline the steps in the Cori cycle.
b. Why is the Cori cycle important to muscle cells?

ATP Yield from Glycolysis/Gluconeogenesis

13. **a.** How many moles of ATP are produced by the aerobic oxidation of 1 mole of glucose in a typical liver cell?
b. Of this total, how many moles of ATP are produced by: (i) glycolysis alone; (ii) the Krebs cycle; (iii) the electron transport chain linked with oxidative phosphorylation?
c. What is the efficiency associated with the ATP production in the liver cell?

14. Explain why each molecule of NADH produced in muscles during glycolysis ultimately gives rise to just 1.5 molecules of ATP.

15. What noncarbohydrate compounds can be used to provide intermediates for gluconeogenesis?

16. **a.** Which reaction in glycolysis must be bypassed in gluconeogenesis?
b. How are these reactions carried out in gluconeogenesis?

Blood Glucose and Glycogen Metabolism

17. Distinguish between normal blood sugar level and the renal threshold value for glucose.

18. What is the major factor in maintaining a constant blood sugar level?
19. What is the activated form of glucose needed in glycogen synthesis?
20. How do muscle cells and liver cells differ in their metabolism of glucose-6-phosphate and glycogen?
21. Why is it more efficient for anaerobic cells to utilize glycogen as a source of energy than for them to use glucose?

Regulation of Carbohydrate Metabolism

22. What is the role of insulin in regulating blood sugar level?
23. How does glucagon act to raise the blood sugar level?
24. When a hormone causes an increase in cAMP levels in the cell, where is the binding site for the hormone located?
25. Explain how the binding of a hormone at its receptor site can release cAMP in a target cell.
26. What cellular or hormonal signals are used to regulate each of the following enzymes?
    a. phosphofructokinase
    b. glycogen synthase
    c. fructose-1,6-bisphosphatase
    d. glycogen phosphorylase
    e. hexokinase
27. For each enzyme in Problem 26, indicate in which metabolic pathway (glycolysis, gluconeogenesis, etc.) it is found.

Additional Problems

28. When aldolase catalyzes the cleavage of ketose mono- and bisphosphates, dihydroxyacetone phosphate is always one of the products. Complete the following catalytic reactions.

a.

CH_2O—Ⓟ
|
C=O
|
HO—C—H
|
H—C—OH
|
H—C—OH
|
H—C—OH
|
CH_2O—Ⓟ

Sedoheptulose 1,7-bisphosphate

$\xrightleftharpoons{\text{aldolase}}$

b. Fructose 1-phosphate $\xrightleftharpoons{\text{aldolase}}$

29. The average adult consumes about 65 g of fructose daily (either as the free sugar or as part of sucrose). In the liver, fructose is first phosphorylated to fructose-1-phosphate, which is then split into dihydroxyacetone phosphate and glyceraldehyde. The latter compound is phosphorylated to glyceraldehyde-3-phosphate. Write out equations (using structural formulas) for these three steps, and give the specific names of the enzymes. Indicate which steps utilize ATP.
30. Lactate dehydrogenase is not specific for pyruvate but will also catalyze the reduction of other keto acids (and their conjugate bases). Write an equation for the enzymatic reduction of phenylpyruvate.
31. Glycogen present in the liver can be used to maintain the blood glucose level, while muscle glycogen cannot. What makes the difference?
32. The alcohol we drink is detoxified by enzymes in the liver.
    a. Write out the sequence of reactions for the metabolism of alcohol.
    b. Which enzyme is inhibited by disulfiram?
    c. Does alcohol supply energy (calories) for the body? Explain.
33. Briefly describe the two types of diabetes mellitus.
34. **a.** Why can't insulin be taken orally?
    b. In structure and purpose, how do oral drugs such as Orinase differ from insulin?

CHAPTER 26

Lipid Metabolism

A camel's hump contains a high proportion of fat. Oxidation of this fat produces large quantities of water that allow the camel to go for extended periods without water.

Learning Objectives/Study Questions

1. Why do triglycerides store energy more efficiently than glycogen?
2. What role do the hormones insulin, epinephrine, and glucagon play in the synthesis and degradation of fatty acids and triglycerides?
3. What reactions are needed to completely oxidize a fatty acid?
4. How do you calculate the ATP yield from the complete oxidation of a fatty acid? From a triglyceride?
5. What are ketone bodies and how are they used by the body? What conditions cause the level of ketone bodies to increase?
6. What reactions are needed to synthesize fatty acids?
7. How are overweight and obesity defined? How are diet and exercise effective in the treatment of these conditions?

Triglycerides are the richest energy source in the human diet. The oxidation of 1 g of a typical fat or oil liberates about 9.5 kcal. (By comparison, the oxidation of an equal mass of carbohydrate liberates only about 4.2 kcal.) Lipid molecules are more highly reduced; that is, they contain a higher proportion of carbon–hydrogen bonds than do carbohydrate molecules. Therefore, lipids have a greater capacity to combine with oxygen and consequently have a higher heat content. Whereas carbohydrates provide a readily available source of energy, lipids function as the principal *energy reserve.* Fats may be thought of as analogous to combustible petroleum products (highly reduced), while carbohydrates are analogous to alcohols (more oxidized), which are not nearly as effective fuels.

The nutritional aspects of lipids are still not completely understood. A human adult can survive on an nearly fat-free diet if carbohydrates and proteins are supplied as sources of metabolic energy. Two unsaturated fatty acids—linoleic and linolenic acids—are required for normal growth and development and are termed **essential fatty acids.** They must be obtained from the diet because they are not synthesized by humans. Linoleic acid is used by the body to synthesize many of the other unsaturated fatty acids such as arachidonic acid, a precursor for the biosynthesis of prostaglandins (Selected Topic E). In addition, the essential fatty acids are incorporated into the structures of the membrane lipids (Section 20.4), and they are necessary for the efficient transport and metabolism of cholesterol. The average daily diet should contain about 4 to 6 grams of the essential fatty acids.

Infants lacking essential fatty acids in their diet lose weight and develop *eczema,* an inflammatory skin disease characterized by scaly and crusty skin.

Nuts, seeds, soybeans, and soybean oil are good sources of both linoleic and linolenic acids. Linoleic acid is also found in leafy vegetables.

Men and women differ in their capacity to store lipids. The average percent body fat is about 16% for an adult male and 25% for an adult female. Male athletes in superb condition will have less than 7% body fat and females, less than 12% body fat. Americans consume about 100 to 125 g of lipids each day. This represents 34% of the daily calorie requirements. The National Cancer Institute, the American Heart Association, and most other health authorities recommend that Americans reduce their lipid intake so that no more than 30% of total calories be provided by lipids. In this chapter, we look at how our bodies store and metabolize fats. We also consider some of the problems associated with lipid storage and metabolism. Lipids offer one of the best illustrations of an old saying: Too much of a good thing can be bad.

26.1 Storage and Mobilization of Fats

The fat reserves in the average person provide sufficient energy to survive starvation for 30–40 days (given sufficient water). In comparison, glycogen in the liver (about 100 g) is depleted within one day. From the standpoint of efficiency of fuel storage, fats are far superior to carbohydrates. Glycogen, because of its many hydroxyl groups, is extremely hydrated. (About 2 g of water is bound to every gram of stored glycogen.) Thus 1 g of body fat contains more than six times the stored energy content of 1 g of hydrated glycogen.

The ability to store greater amounts of the more energy-efficient lipids is especially important for migrating birds and some terrestrial animals. The camel's hump, for example, is almost all adipose tissue, and migratory birds rely on stored fats to supply the energy for long, sustained flight.

Fats are stored throughout the body. Principally, however, they are deposited in **fat cells** (or **adipocytes**) found in a special kind of connective tissue called **adipose tissue.** In these cells most of the cytoplasm is replaced by a large droplet of triglycerides (approximately 90% of the mass of the fat cell; Figure 26.1). Adults have approximately 30 to 40 billion fat cells that swell or shrink like a sponge depending on the amount of fat inside them. Adipose tissue is the only tissue in which free triglycerides occur in appreciable amounts. Elsewhere, in nonadipose cells or in the blood plasma, lipids are bound to proteins as lipoprotein complexes (Section 20.7).

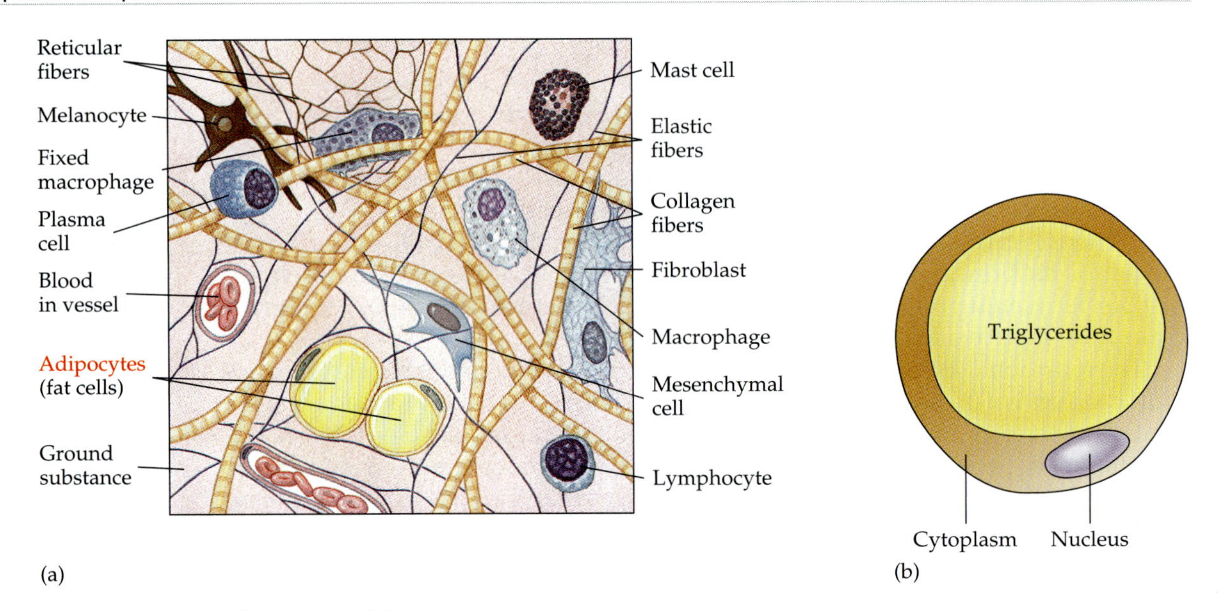

▲ **Figure 26.1**
(a) A schematic representation of adipose tissue. (b) A typical fat cell found in adipose tissue. Fat cells produce, store, and mobilize triglycerides; they are among the largest cells in the body.

Body locations containing large amounts of adipose tissue are called **fat depots.** Two such locations are just beneath the skin (subcutaneous fat depots) and in the abdominal area. Fat depots are also found around vital organs, such as the heart, liver, kidneys, and spleen. There the adipose tissue serves as a protective cushion, helping prevent injury to the organs. Subcutaneous fat depots help insulate against sudden temperature changes. The adipose tissue acts just like the insulation in the walls of a house, trapping body heat and preventing it from escaping to the surroundings. In some ways the tissue also acts like the furnace of a house. When the outside temperature drops, metabolic activity in the cells generates heat to compensate for the heat lost to the environment.

In order for the chemical energy in triglycerides to be released and so become available to the body, stored triglycerides must be cleaved to fatty acids and glycerol by the lipases in adipose tissue. Fat metabolism, like carbohydrate metabolism, is regulated by hormones, such as epinephrine and glucagon (Section 25.6). These two hormones, activated by low blood sugar levels, bind to receptor proteins in the cell membranes of adipose tissue. This binding activates the enzyme *adenylate cyclase* (Section 25.6) to form cAMP from ATP. The cAMP then stimulates the hydrolysis of triglycerides (by the lipases) and the release (*mobilization*) of fatty acids and glycerol from adipose tissue.

As we saw in Sections 20.7 and 24.2, lipids are digested and absorbed in the small intestine. The resulting triglycerides are transported in the blood as lipoproteins. After hydrolysis by lipases in the capillary walls, fatty acids and glycerol have the same fate as those mobilized from storage. We see the possible metabolic fates of lipids in Figure 26.2.

The fatty acids, bound to serum albumin, are transported by the blood to other tissues for oxidation (Section 26.2). The glycerol is transported to the liver, where it is readily converted in two steps to dihydroxyacetone phosphate. The first step is the phosphorylation of a primary hydroxyl group followed by the oxidation of a secondary hydroxyl group to a ketone:

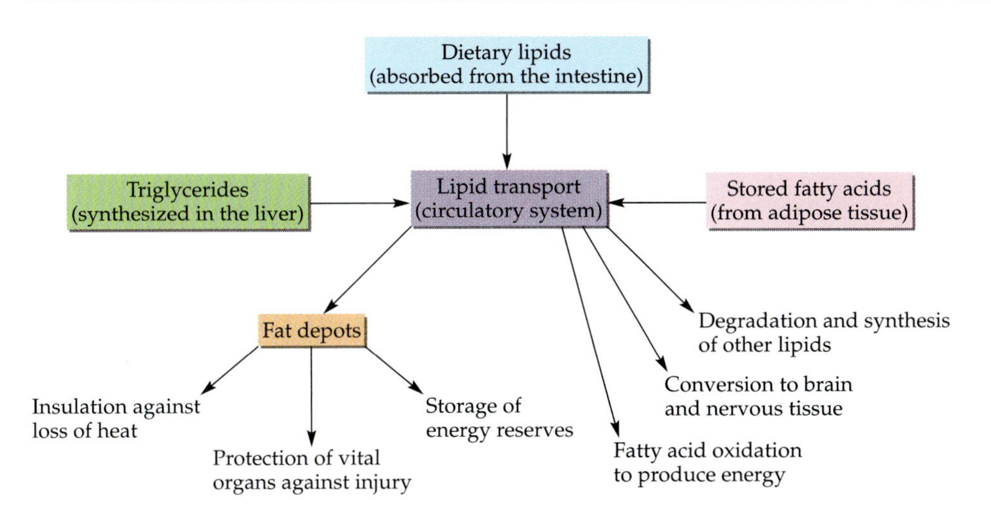

◀ **Figure 26.2**
Metabolic fates of lipids.

$$\underset{\text{Glycerol}}{HO\text{—}C(CH_2OH)_2\text{—}H} \xrightarrow[\text{ATP} \quad \text{ADP}]{\text{glycerol kinase}} \underset{\text{L-glycerol 3-phosphate}}{HO\text{—}C(CH_2OH)(CH_2OPO_3^{2-})\text{—}H} \xrightarrow[\text{NAD}^+ \quad \text{NADH} + \text{H}^+]{\text{glycerol phosphate dehydrogenase}} \underset{\text{Dihydroxyacetone phosphate}}{CH_2OH\text{—}C{=}O\text{—}CH_2OPO_3^{2-}}$$

Dihydroxyacetone phosphate can be used to provide energy by conversion to pyruvate through glycolysis (Section 25.1), or it can be transformed into glucose through gluconeogenesis (Section 25.4).

✓ **Review Questions**

26.1 What is the unique structural feature of the essential fatty acids?

26.2 What are the functions of adipose tissue?

26.2 Fatty Acid Oxidation

Fatty acids are broken down in a series of sequential reactions accompanied by the gradual release of utilizable energy. Some of these reactions are oxidative and require the coenzymes NAD^+ and FAD. The enzymes involved in fatty acid catabolism are localized in the mitochondria along with the enzymes of the Krebs cycle, electron transport chain, and oxidative phosphorylation (Chapter 24). This localization in the mitochondria is of the utmost importance because it provides for the efficient utilization of the energy stored in the fatty acid molecules.

At the beginning of the twentieth century Franz Knoop showed that the breakdown of fatty acids occurs in a stepwise fashion with the removal of two carbon atoms at a time. Subsequent investigations have resulted in the separation and purification of all of the enzymes involved in fatty acid oxidation.

Fatty acid oxidation is initiated on the outer mitochondrial membrane. Fatty acids, like carbohydrates and amino acids, are relatively inert and must first be activated by conversion to an energy-rich fatty acid derivative of coenzyme A (called *fatty acyl-CoA*). This activation is catalyzed by *acyl-CoA synthetase.* For each molecule of fatty acid activated, one molecule of coenzyme A and one molecule of ATP are used, with net utilization of the two high-energy bonds in one ATP molecule:

Brown Fat

In addition to the normal white adipose tissue, some organisms contain another type of fatty tissue called **brown fat.** Its brown color is due to the presence of large numbers of mitochondria (stocked with red-brown cytochromes). In these specialized mitochondria, the reactions of the electron transport chain continue to take place, but no ATP is synthesized—that is, electron transport is no longer linked to oxidative phosphorylation. The energy that would otherwise be used for ATP synthesis is released as heat, and the temperature of the organism rises. In essence, brown fat acts as a furnace to produce heat energy rather than chemical energy.

Brown fat is particularly abundant in hibernating animals and in the necks and upper backs of newborn infants (as well as other mammals that are born hairless). These cells revive hibernating animals and maintain the body temperature of young animals. Adult humans have few brown fat cells because metabolic reactions usually generate more than enough heat, allowing body temperature to be regulated by the dissipation of this heat.

$$\underset{\text{Fatty acid}}{R{-}\overset{\overset{\displaystyle O}{\|}}{C}{-}OH} + \underset{\text{Coenzyme A}}{HS{-}CoA} \xrightarrow[\text{ATP} \quad \text{AMP} + \text{PP}_i]{\text{synthetase}} \underset{\text{Fatty acyl-CoA}}{R{-}\overset{\overset{\displaystyle O}{\|}}{C}{-}S{-}CoA} + H_2O$$

At the inner mitochondrial membrane the fatty acyl-CoA combines with a carrier molecule known as carnitine in a reaction catalyzed by *carnitine acyltransferase.* The acyl-carnitine derivative is transported into the mitochondrial matrix (Figure 26.3) and converted back to the fatty acyl-CoA. Further oxidation of the fatty acyl-CoA occurs in the mitochondrial matrix via a sequence of four reactions known as **β-oxidation** (because the β carbon undergoes successive oxidations in the progressive removal of two-carbon units from the carboxyl end of the fatty acyl-CoA). (Figure 26.4)

The first step is an oxidation reaction catalyzed by *acyl-CoA dehydrogenase.* The coenzyme FAD accepts two hydrogen atoms, one from the α carbon and one from the β carbon. The enzyme is stereospecific in that only the *trans*-alkene is obtained. Each molecule of $FADH_2$ that is reoxidized back to FAD via the electron transport chain (Section 24.5) supplies energy to form 1.5 molecules of ATP.

In the second step, the *trans*-alkene is hydrated to form a secondary alcohol in a reaction catalyzed by *enoyl-CoA hydratase.* The enzyme is stereospecific and forms only the L-isomer.

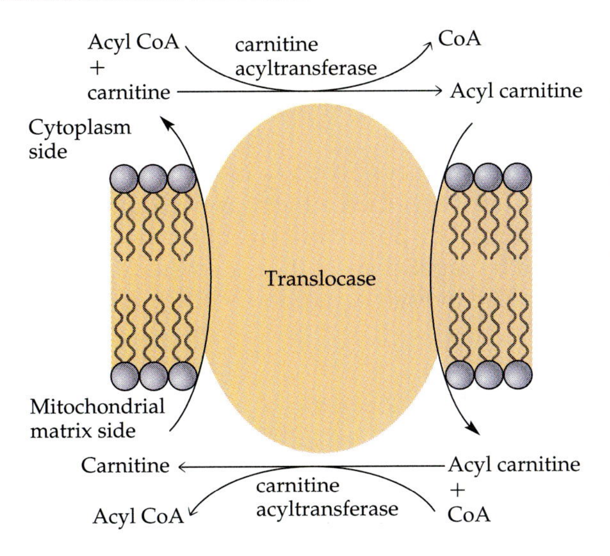

Figure 26.3
Formation and transport of the acyl-carnitine derivatives. Carnitine acyltransferase catalyzes the transfer of the fatty acyl group from coenzyme A to carnitine. The acyl-carnitine is transported into the mitochondrial matrix by an integral membrane protein known as translocase. Carnitine acyltransferase in the mitochondria catalyzes the transfer of the fatty acyl group back to coenzyme A and the free carnitine is transported by translocase back into the cytoplasm.

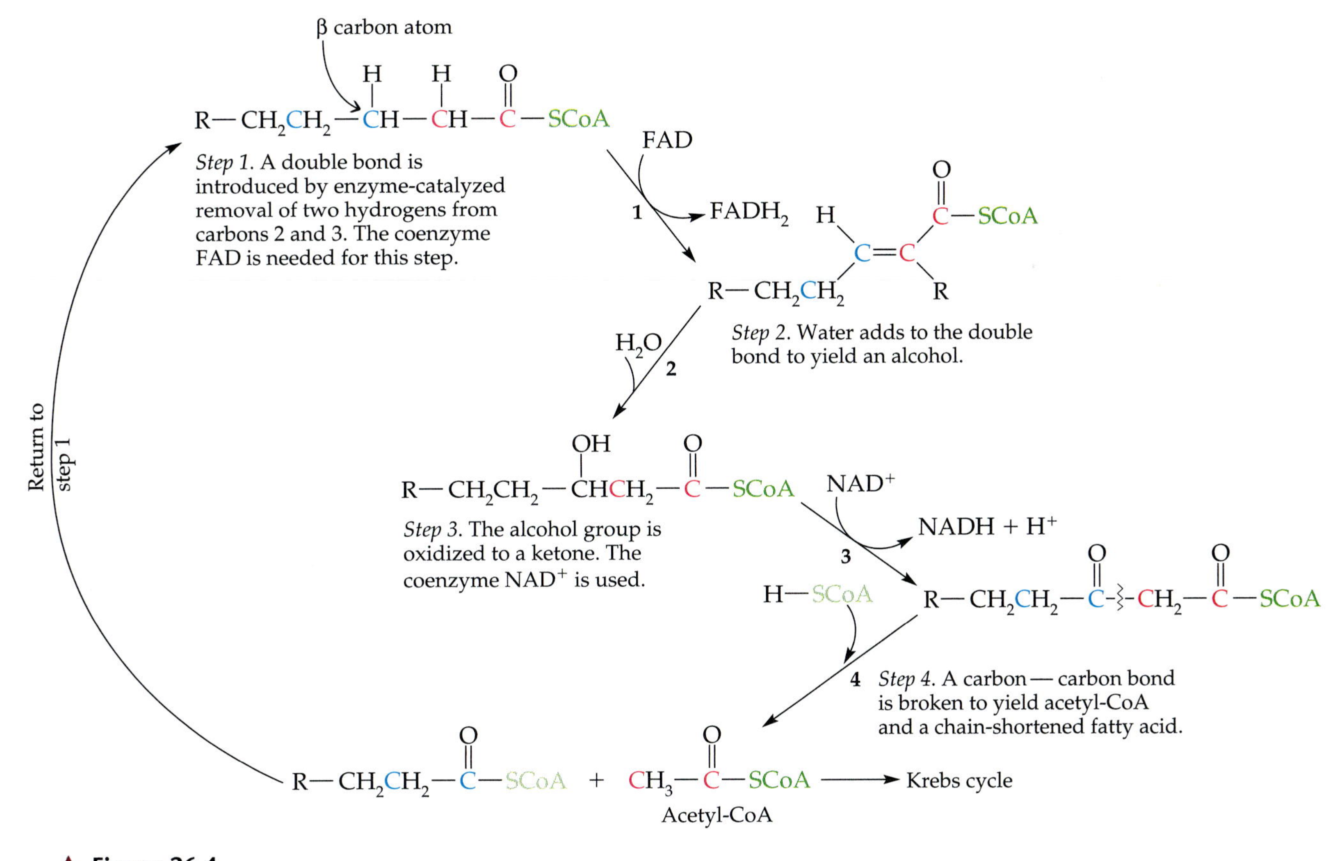

Figure 26.4
Fatty acid oxidation. The fatty acyl-CoA formed in the final step becomes the substrate for the first step in the next round of β-oxidation. β-Oxidation continues until two acetyl-CoA molecules are produced in the final step.

In the third step, the secondary alcohol is oxidized to a ketone by *β-hydroxyacyl-CoA dehydrogenase,* with NAD^+ acting as the oxidizing agent. The reoxidation of each molecule of NADH to NAD^+ by the electron transport chain furnishes 2.5 molecules of ATP.

Because each shortened fatty acyl-CoA cycles back to the beginning of the pathway, β-oxidation is sometimes referred to as the *fatty acid spiral.*

The final reaction is cleavage of the β-ketoacyl-CoA by a molecule of coenzyme A. The products are acetyl-CoA and a fatty acyl-CoA which has been shortened by two carbon atoms. The reaction is catalyzed by *β-ketothiolase* (or *thiolase*). The shortened fatty acyl-CoA is then degraded by repetitions of these four steps, each time releasing a molecule of acetyl-CoA. The overall equation for the β oxidation of palmitoyl-CoA (16 carbons) is:

$$CH_3(CH_2)_{14}C(=O)SCoA + 7\,FAD + 7\,NAD^+ + 7\,HSCoA + 7\,H_2O \longrightarrow$$

$$8\,CH_3C(=O)SCoA + 7\,FADH_2 + 7\,NADH + 7\,H^+$$

Unsaturated fatty acids, which are about 50% of the fatty acids found in humans are also degraded through β-oxidation; however, one or two ancillary enzymes are required to convert intermediates to substrates that can be used in these reactions.

The fate of acetyl-CoA obtained from fatty acid oxidation depends on the needs of the organism. It may enter the Krebs cycle and be oxidized to produce energy (Chapter 24), it may be used for the formation of ketone bodies (Section 26.4), or it may serve as the starting material for the synthesis of fatty acids (Section 26.5).

Review Questions

26.3 Why is the pathway for fatty acid oxidation known as β-oxidation?

26.4 How many molecules of each of the following are produced in the metabolism of one molecule of palmitic acid: **(a)** acetyl-CoA; **(b)** NADH; **(c)** $FADH_2$?

26.3 ATP Yield From Fatty Acid Oxidation

The amount of ATP obtained from fatty acid oxidation depends on the fatty acid being oxidized. We'll choose palmitic acid as a typical fatty acid of the human diet. Calculating its energy yield provides a guideline to use in determining the ATP yield of all other fatty acids.

The breakdown by an organism of 1 mol of palmitic acid requires 1 mol of ATP (for activation), and 8 moles of acetyl-CoA are formed. Recall from Table 24.3 that each mole of acetyl-CoA metabolized by the Krebs cycle yields 10 mol of ATP. The complete degradation of 1 mol of palmitic acid requires that the reactions comprising β-oxidation be repeated seven times. Thus, 7 mol of NADH and 7 mol of $FADH_2$ are produced (Section 26.2). Reoxidation of these compounds through respiration yields 2.5 and 1.5 mol of ATP, respectively. The energy calculations can be summarized as follows:

The number of times β-oxidation is repeated for a fatty acid containing *n* carbons is $(n/2) - 1$ because the final turn yields two acetyl-CoA molecules.

1 mol of ATP is split to AMP and 2 P_i		−2 ATP
8 mol of acetyl–CoA formed (8×10)		80 ATP
7 mol of $FADH_2$ formed (7×1.5)		10.5 ATP
7 mol of NADH formed (7×2.5)		17.5 ATP
	Total	106 ATP

We can calculate the percentage of available energy that is conserved by the cell in the form of ATP. The combustion of 1 mol of palmitic acid releases a considerable amount of energy:

$$C_{16}H_{32}O_2 + 23\ O_2 \longrightarrow 16\ CO_2 + 16\ H_2O + 2340\ \text{kcal}$$

$$\frac{\text{Energy conserved}}{\text{Total energy available}}(100\%) = \frac{(106\ \text{ATP})(7.5\ \text{kcal/ATP})}{(2340\ \text{kcal})}(100\%) = 34\%$$

The efficiency of fatty acid metabolism is comparable to that of carbohydrate metabolism (35%; see Section 25.3).

Note that the oxidation of fatty acids also produces large quantities of water, and it is this water that sustains migratory birds and animals (e.g., the camel) for long periods of time.

Let's consider the ATP yield for the complete oxidation of 1 mol of a triglyceride to CO_2 and H_2O. We'll use glyceryl palmitate, a simple triglyceride—that is, a triglyceride that has the same fatty acid (palmitic acid) attached to all three carbons of the glycerol backbone. Previously, we determined that 106 mol of ATP are produced for each palmitic acid that is completely oxidized to CO_2 and H_2O. Therefore, 3 mol of palmitic acid would yield $(3 \times 106) = 318$ mol of ATP. In addition to the palmitic acid, glycerol can also be oxidized. If the glycerol is converted to dihydroxyacetone phosphate and then oxidized through the combined actions of glycolysis, the Krebs cycle, electron transport chain, and oxidative phosphorylation, an additional 16.5–18.5 mol of ATP can be obtained, as outlined in Table 26.1. Thus the total yield of ATP would be 334.5–336.5 mol ATP.

Table 26.1 Yield of ATP From the Complete Oxidation of One Mole of Glycerol

Reaction	Comments	Yield of ATP (moles)
Conversion of glycerol to dihydroxyacetone phosphate		
glycerol → glycerol-3-phosphate (G3P)	consumes 1 mol ATP	−1
G3P → dihydroxyacetone phosphate	produces 1 mol of cytoplasmic NADH	
Glycolysis		
glyceraldehyde-3-phosphate → 1,3-bisphosphoglycerate	produces 1 mol of cytoplasmic NADH	
1,3-BPG → 3-phosphoglycerate	produces 1 mol ATP	+1
phosphoenolpyruvate → pyruvate	produces 1 mol ATP	+1
Oxidation of Pyruvate (mitochondria)		
pyruvate → acetyl-CoA + CO_2	produces 1 mol NADH	
Krebs cycle (mitochondria)		
isocitrate → α-ketoglutarate + CO_2	produces 1 mol NADH	
α-ketoglutarate → succinyl-CoA + CO_2	produces 1 mol NADH	
succinyl-CoA → succinate	produces 1 mol GTP	+1
succinate → fumarate	produces 1 mol $FADH_2$	
malate → oxaloacetate	produces 1 mol NADH	
Electron transport chain and oxidative phosphorylation		
2 cytoplasmic NADH	yields 1.5–2.5 mol ATP[a]	+3 to +5
1 NADH from the oxidation of pyruvate	yields 2.5 mol ATP	+2.5
1 $FADH_2$ formed in the Krebs cycle	yields 1.5 mol ATP	+1.5
3 NADH formed in the Krebs cycle	yields 2.5 mol ATP	+7.5
	Net yield of ATP	+16.5 to + 18.5

[a]Depending on the tissue (see Section 25.3).

Review Question

26.5 How many molecules of ATP are formed during the complete oxidation of one molecule of: **(a)** stearic acid? **(b)** glyceryl stearate?

26.4 Ketosis

In the liver, most of the acetyl-CoA obtained from fatty acid oxidation is oxidized by the Krebs cycle. However, some of the acetyl-CoA is used to synthesize a group of compounds known as ketone bodies. Two acetyl-CoA molecules combine, in a reversal of the final step of β-oxidation, to produce acetoacetyl-CoA. The acetoacetyl-CoA reacts with another molecule of acetyl-CoA and water to form β-hydroxy-β-methylglutaryl-CoA, which is then cleaved to acetoacetate and acetyl-CoA. Most of the acetoacetate is reduced to β-hydroxybutyrate, while a small amount is decarboxylated to carbon dioxide and acetone (Figure 26.5). Acetoacetate, β-hydroxybutyrate, and acetone are collectively referred to as **ketone bodies.** The acetoacetate and β-hydroxybutyrate synthesized by the liver are released into the blood for use as a metabolic fuel (to be converted back to acetyl-CoA) by other aerobic tissues, particularly the kidney and heart. (During prolonged starvation, ketone bodies provide about 70% of the energy requirements of the brain.) The kidneys excrete about 20 mg of ketone bodies each day. Normally, blood levels are maintained at about 1 mg of ketone bodies per 100 mL of blood.

Figure 26.5 The formation of ketone bodies.

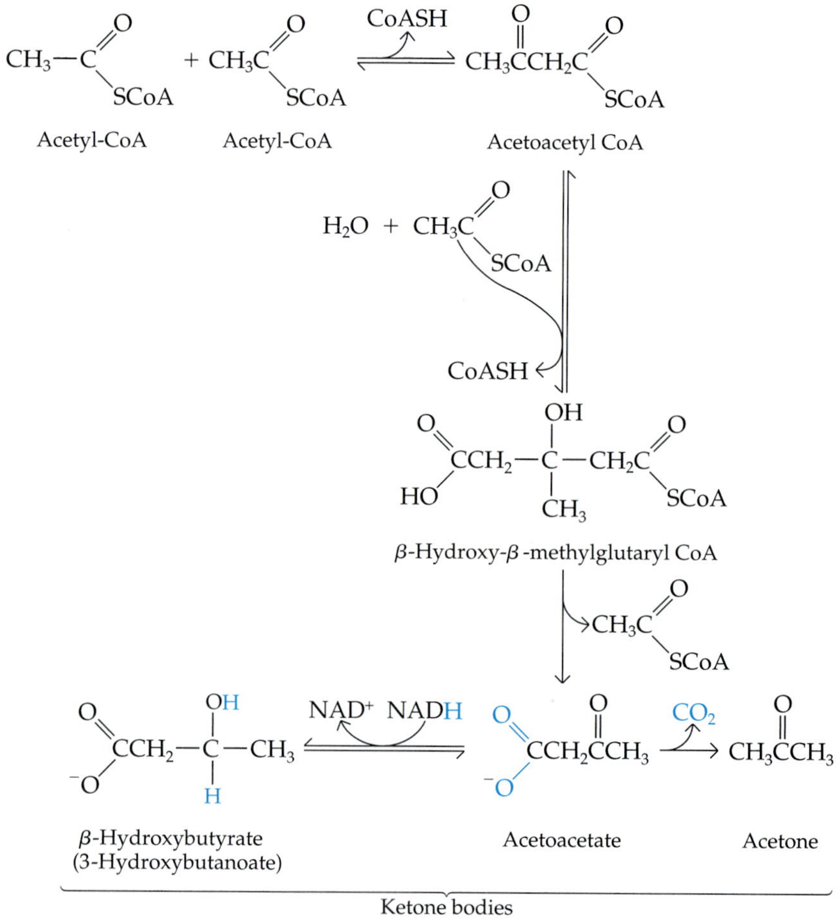

$$CH_3C(=O)SCoA + CH_3C(=O)SCoA \underset{}{\overset{CoASH}{\rightleftharpoons}} CH_3C(=O)CH_2C(=O)SCoA$$

Acetyl-CoA + Acetyl-CoA → Acetoacetyl CoA

$$CH_3C(=O)CH_2C(=O)SCoA + H_2O + CH_3C(=O)SCoA \rightleftharpoons HOC(=O)CH_2-C(OH)(CH_3)-CH_2C(=O)SCoA + CoASH$$

β-Hydroxy-β-methylglutaryl CoA

$$HOC(=O)CH_2-C(OH)(CH_3)-CH_2C(=O)SCoA \rightarrow {}^-OC(=O)CH_2C(=O)CH_3 + CH_3C(=O)SCoA$$

$${}^-OC(=O)CH_2-CH(OH)-CH_3 \underset{NAD^+ \quad NADH}{\rightleftharpoons} {}^-OC(=O)CH_2C(=O)CH_3 \xrightarrow{CO_2} CH_3C(=O)CH_3$$

β-Hydroxybutyrate (3-Hydroxybutanoate) — Acetoacetate — Acetone

Ketone bodies

In some physiological conditions, such as starvation or diabetes mellitus, cells don't receive sufficient amounts of carbohydrate and the rate of fatty acid oxidation increases to provide energy. This leads to an increase in the concentration of acetyl-CoA. The increased amounts of acetyl-CoA cannot be oxidized by the Krebs cycle because of a decrease in the concentration of oxaloacetate, needed for glucose synthesis (Section 25.4). Thus the rate of ketone body formation in the liver is increased to a level much higher than can be utilized by other tissues. The excess ketone bodies accumulate in the blood (*ketonemia*—concentrations greater than 3 mg/dL) and in the urine (*ketonuria*). These conditions together are referred to as **ketosis.**

The acetone produced in the metabolism of excess acetyl-CoA is transported to the kidneys and lungs. Being volatile, the acetone that reaches the lungs is expelled in the breath. The sweet smell of acetone, a characteristic of ketosis, is frequently noticed on the breath of severely diabetic patients.

Because two of the three ketone bodies are formed as weak acids, their presence in the blood in excessive amounts overwhelms the blood buffers and causes a marked decrease in blood pH (to 6.9 from a normal value of 7.4). This decrease in pH leads to a serious condition known as **acidosis** (see Section 10.8). Acidosis results in interference with the transport of oxygen by hemoglobin (see Section 28.2). In moderate to severe acidosis, breathing becomes labored and very painful. The body also loses fluids as the kidneys eliminate large quantities of water trying to get rid of the acids. The person becomes dehydrated. The short oxygen supply and dehydration lead to depression. Even mild acidosis leads to lethargy, loss of appetite, and a generally run-down feeling. Untreated patients may go into a coma. At that point prompt treatment is necessary if the patient's life is to be saved.

Acidosis resulting from the production of acidic metabolites is sometimes called metabolic acidosis to distinguish it from respiratory acidosis (see Section 28.3).

It is important to recall that the conversion of pyruvate to acetyl-CoA is irreversible (Section 25.2) in mammals. Therefore the acetyl-CoA molecules derived from fatty acids cannot be used to synthesize glucose. Mammals can convert carbohydrates to lipids (via acetyl-CoA). They cannot convert lipids to carbohydrates.

Review Questions

26.6 What three compounds are collectively known as ketone bodies?

26.7 Why does acidosis result from ketosis?

26.5 Fatty Acid Synthesis

When the body ingests more carbohydrate than it needs for energy and for glycogen synthesis, the excess is converted to fatty acids via the common intermediate, acetyl-CoA. As with opposing synthetic and degradative pathways in carbohydrate metabolism (e.g., glycolysis and gluconeogenesis), fatty acid synthesis is not simply the reverse of fatty acid breakdown (see Table 26.2).

In humans, the liver is the major site of fatty acid synthesis. Acetyl-CoA, obtained from the oxidation of carbohydrates or fatty acids, is first converted to a three-carbon acyl-CoA known as malonyl-CoA in a reaction catalyzed by *acetyl-CoA carboxylase.* Like pyruvate carboxylase in gluconeogenesis, acetyl-CoA carboxylase requires the coenzyme biotin for activity. This is an irreversible reaction and is the regulated step in fatty acid synthesis.

$$\underset{\text{Acetyl-CoA}}{CH_3\overset{\overset{O}{\|}}{C}\text{—SCoA}} + \underset{\text{Bicarbonate ion}}{HCO_3^-} \xrightarrow{\text{ATP} \;\; \text{ADP}} \underset{\text{Malonyl-CoA}}{{}^-O\text{—}\overset{\overset{O}{\|}}{C}\text{—}CH_2\text{—}\overset{\overset{O}{\|}}{C}\text{—SCoA}}$$

Acetyl-CoA carboxylase is activated by the release of insulin, which signals that food supplies in the body are abundant, and inhibited by glucagon and epinephrine, released when energy and blood sugar levels are low. As we've seen before, these hormones do not act directly on the enzyme, but initiate a series of reactions that lead to the activation or inhibition of acetyl-CoA carboxylase.

Table 26.2 Comparison of Fatty Acid Oxidation and Synthesis

Oxidation	Synthesis
Occurs in mitochondria	Occurs in cytosol
Carbon atoms removed two at a time	Carbon atoms added two at a time
Enzymes are separate proteins	Enzymes are components of a single multifunctional protein
Intermediates bound to coenzyme A	Intermediates bound to acyl carrier protein
Coenzymes: FAD, NAD^+	Coenzyme: NADPH

In plants and bacteria, the enzymes are separate proteins.

In mammals, all of the enzymes needed to synthesize palmitic acid from acetyl-CoA and malonyl-CoA are components of a single multifunctional protein known as **fatty acid synthase.** The fatty acid chain is synthesized two carbons at a time through a series of reactions, outlined in Figure 26.6, that are repeated, much like the reactions of fatty acid oxidation. To initiate fatty acid synthesis, one molecule of acetyl-CoA and one molecule of malonyl-CoA are each attached to separate molecules of acyl carrier protein (ACP) and then linked together, with the loss of CO_2, to form acetoacetyl-ACP.

The next three reactions are the reverse of the first three steps in β-oxidation. The ketone group is reduced to an alcohol in a reaction that requires NADPH as the reducing agent. The alcohol is dehydrated, which forms a double bond between

Preferred Fuels of Various Tissues

In a normal individual, certain organs use both glucose and fatty acids (in varying amounts) for their energy needs. The utilization rate of a specific fuel depends on the physiological state of the individual. For example, glucose (or glycogen) is the chief fuel for the immediate energy needs of active skeletal muscle (as in a sprint), whereas fatty acids are the major fuels for resting skeletal muscle. Fatty acids are the major fuel for cardiac muscle, although ketone bodies, glucose, and lactate are also used. During fasting and prolonged exercise, fatty acids are the preferred fuel for all types of muscles. The kidneys and adipose tissue use both glucose and fatty acids, whereas fatty acids are the preferred fuel of the liver. The brain prefers glucose, but will utilize ketone bodies to meet part of its energy needs during prolonged starvation. The brain does not use fatty acids because they are bound to proteins and cannot diffuse across the blood-brain barrier. Red blood cells can use only glucose because they lack mitochondria and hence cannot obtain energy via oxidative phosphorylation.

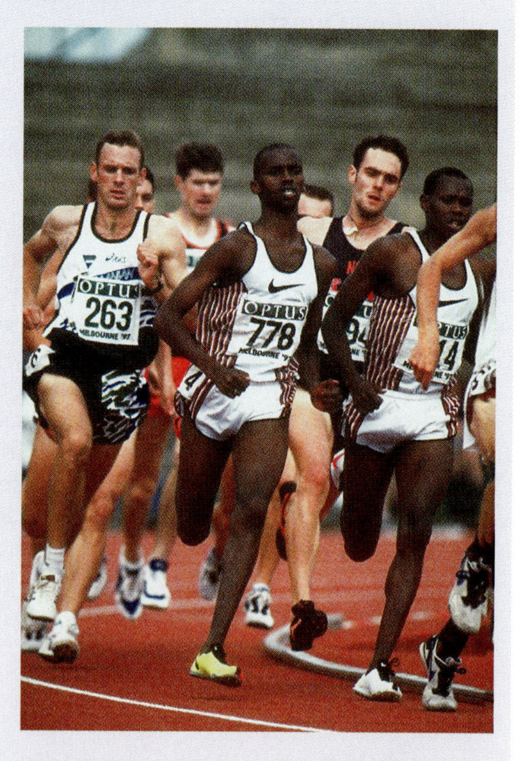

▲ When athletes run a race, their initial fuel is mainly glucose derived from muscle glycogen. In longer races, liver glycogen and triglycerides in adipose tissue are broken down for fuel.

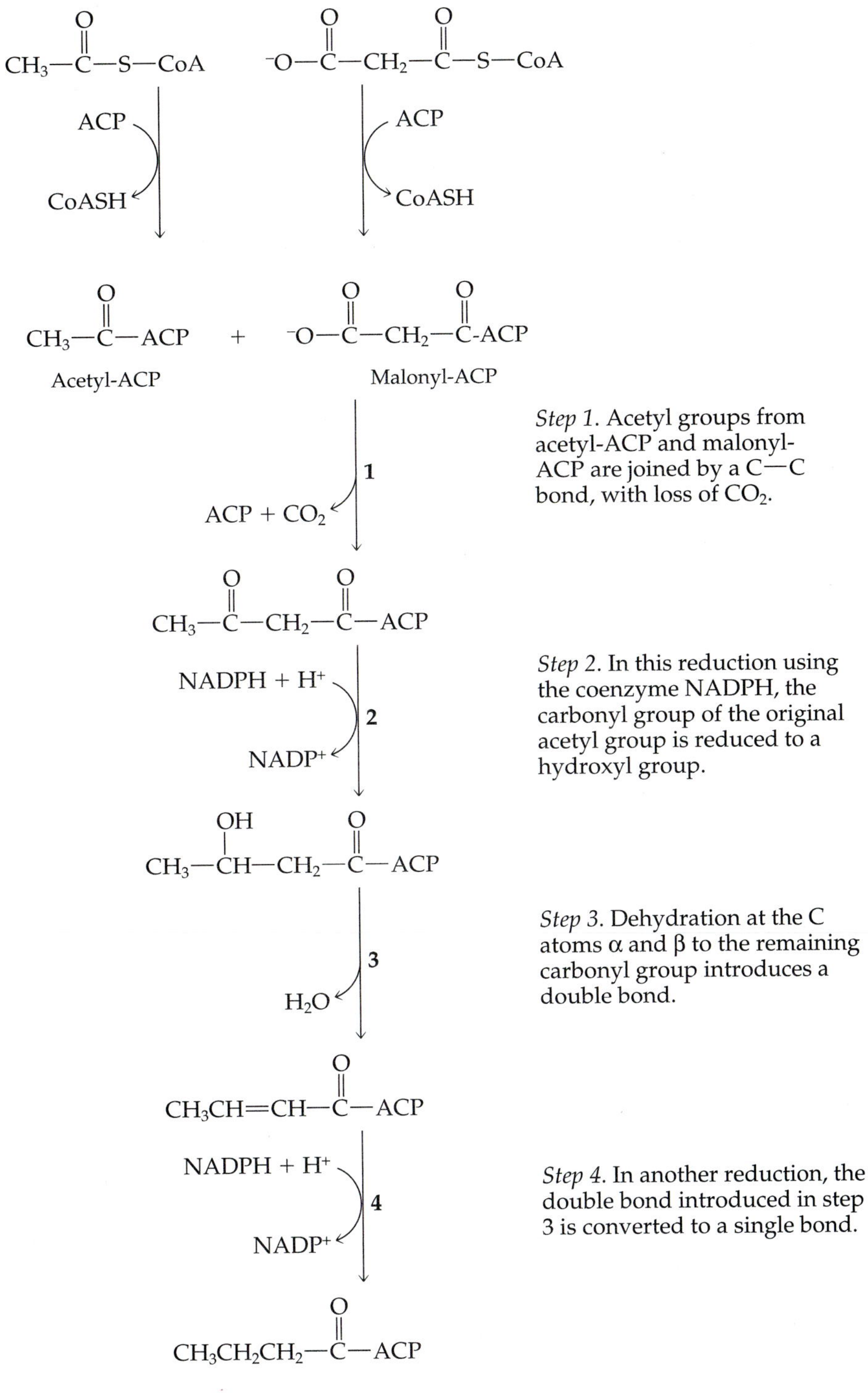

$$CH_3-\overset{O}{\overset{\|}{C}}-S-CoA \qquad {}^{-}O-\overset{O}{\overset{\|}{C}}-CH_2-\overset{O}{\overset{\|}{C}}-S-CoA$$

ACP → CoASH; ACP → CoASH

$$CH_3-\overset{O}{\overset{\|}{C}}-ACP \quad + \quad {}^{-}O-\overset{O}{\overset{\|}{C}}-CH_2-\overset{O}{\overset{\|}{C}}\text{-}ACP$$

Acetyl-ACP Malonyl-ACP

1 (ACP + CO_2)

Step 1. Acetyl groups from acetyl-ACP and malonyl-ACP are joined by a C—C bond, with loss of CO_2.

$$CH_3-\overset{O}{\overset{\|}{C}}-CH_2-\overset{O}{\overset{\|}{C}}-ACP$$

2 (NADPH + H^+ → $NADP^+$)

Step 2. In this reduction using the coenzyme NADPH, the carbonyl group of the original acetyl group is reduced to a hydroxyl group.

$$CH_3-\overset{OH}{\overset{|}{C}H}-CH_2-\overset{O}{\overset{\|}{C}}-ACP$$

3 (H_2O)

Step 3. Dehydration at the C atoms α and β to the remaining carbonyl group introduces a double bond.

$$CH_3CH{=}CH-\overset{O}{\overset{\|}{C}}-ACP$$

4 (NADPH + H^+ → $NADP^+$)

Step 4. In another reduction, the double bond introduced in step 3 is converted to a single bond.

$$CH_3CH_2CH_2-\overset{O}{\overset{\|}{C}}-ACP$$

◀ **Figure 26.6**
Fatty acid synthesis. Each new pair of carbon atoms is carried into the next cycle by another molecule of malonyl-ACP. The growing fatty acyl chain remains attached to ACP through the original acetyl-ACP.

carbons 2 and 3. The double bond is then reduced, again using NADPH as the coenzyme, to give butyryl-ACP. These reactions are now repeated with another molecule of malonyl-CoA binding to ACP, and then combining with the butyryl-ACP, with the loss of CO_2, to give a six-carbon intermediate that is reduced, dehydrated, and reduced. These reactions are repeated until palmitic acid (16 carbons) is synthesized. The overall reaction is:

Seven of the acetyl-CoA molecules are converted to malonyl-CoA.

$$8\ \text{Acetyl–CoA} + 7\ \text{ATP} + 14\ \text{NADPH} + 7\ \text{H}^+ \longrightarrow$$
$$\text{palmitic acid} + 14\ \text{NADP}^+ + 8\ \text{Co–A} + 7\ \text{ADP} + 7\ \text{P}_i$$

In subsequent reactions palmitic acid can be elongated or shortened, or double bonds can be formed, to obtain the diversity of fatty acids found in the body. (You'll recall that polyunsaturated fatty acids such as linoleic and linolenic acid are essential fatty acids and must be obtained from the diet [Introduction].) Synthesized fatty acids can be used to provide immediate energy or esterified with glycerol and stored as triglycerides in adipose tissue.

✓ Review Question

26.8 What compound, produced by glycolysis and fatty acid oxidation, is needed for the synthesis of all fatty acids?

26.6 Obesity, Exercise, and Diets

In the United States approximately 97 million adults (~55% of the population) are overweight or obese. This condition raises the risk of developing high blood pressure, type 2 diabetes, coronary heart disease, stroke, gallbladder disease, and breast, prostate, and colon cancers. Obesity is now the second leading cause of preventable death in the United States. But how does a person determine if they are overweight or obese? An index known as the **body mass index** (BMI), that has been shown to correlate well with total body fat content, is used. The BMI is defined as follows:

$$\text{BMI} = \frac{\text{Body weight (kg)}}{[\text{Height (m)}]^2} \quad \text{OR} \quad \frac{704.5 \times \text{Body weight (lb)}}{[\text{Height (in)}]^2}$$

For a person who is 5 ft 10 in. tall (70 in.) and weighs 170 lb, the body mass index is: (704.5 × 170 lb)/(70 in × 70 in) = 24. The classification of weight categories by BMI are shown in Table 26.3.

The overwhelming majority of obese people are that way because they eat more food than their bodies use as fuel. However, obesity is a chronic disease that develops because of the interaction of social, behavioral, cultural, physiological, metabolic, and genetic factors.

Because of the increased health risks and costs associated with being overweight or obese, many have attempted to lose weight. But often it is quite difficult to lose a significant amount of weight and more difficult to keep it off. Researchers have attempted to determine how the various signals that control appetite as well as the rates at which fats and carbohydrates are metabolized are combined to maintain a steady weight and why it is so difficult to move away from that maintained weight. Two theories have been proposed and are being tested: the set-point theory and the settling-point theory.

The signals that control appetite are poorly understood and include signals released by the digestive tract, as well as hormonal signals that include **leptin,** a hormone that is secreted by adipose tissues (see the boxed essay "Obesity Genes"). According to the **set-point theory,** weight is centrally controlled by the brain, probably the hypothalamus. Signals are sent that subconsciously adjust each individual's metabolic rate and behavior to maintain a particular set weight. Studies have shown that when a person loses a significant amount of weight (10–20%) the body burns fewer calories than expected. The reverse is true when a person gains a significant amount of weight (increase of ~10%); the body increases the rate at which it burns calories. However, a major problem with the set-point theory is

Table 26.3 Weight Classification by Body Mass Index

Classification	BMI
Underweight	<18.5
Normal	18.5–24.9
Overweight	25.0–29.9
Obesity	30.0–34.9
Extreme Obesity	≥40

that it doesn't explain why obesity is increasing at such an alarming rate in the United States and other countries where food (particularly energy-dense food) is abundant.

A newer theory, known as the **settling-point theory,** suggests that we maintain a particular weight when the various metabolic feedback loops are in equilibrium with our environment, but when this equilibrium is upset those individuals who have more genetic risk factors become obese. One of the prime factors affecting this equilibrium appears to be the increase in the amount of fat in our diets. A variety of genetic and biological factors, including exercise and changes in diet, can influence an individual's settling point and the rate at which fat is oxidized. Neither the set-point theory nor the settling point theory has been proven; active research continues in this area.

What is the best method for losing weight? A combined program of behavior therapy, a low-calorie diet, and increased physical activity has been the most effective. If positive results are not attained in six months, the use of appropriate drugs may be considered. In addition, any weight that is lost will be regained without implementation of an effective weight maintenance program that includes the same components involved in losing the weight.

Weight loss or gain is based on the law of conservation of energy. If we take in more calories than we use up, the excess calories are stored as fat. If we take in fewer calories than we need for our activities, our bodies burn some of the stored fat to make up for the deficit. One pound of fat is deposited in adipose tissue for every 3500 kcal in excess of the body's requirements. Therefore, if you reduce your intake by 100 kcal/day and keep your activity constant, you will burn off a pound of adipose tissue in 35 days. Unfortunately, people are seldom so patient, and they resort to more stringent diets. However many of these are ineffective and even hazardous, primarily because they are grossly unbalanced in terms of necessary nutrients. An unbalanced diet, especially over an extended period of time, causes a variety of nutritional deficiencies, a decrease in resistance to disease, and a decline in general health. The minerals, such as iron, calcium, potassium, also are deficient in many crash diets.

If your normal energy expenditure is 2400 kcal/day, the most fat you could lose by total fasting for a day would be 0.69 lb:

$$2400 \text{ kcal/day} \div 3500 \text{ kcal/lb adipose tissue} = 0.69 \text{ lb adipose tissue/day}$$

This assumes that your body would burn nothing but fat. It wouldn't. Recall that the brain runs on glucose, and if the glucose isn't supplied in the diet, it is obtained from protein (see Section 27.6). Any diet that restricts carbohydrate intake leads to loss of muscle mass as well as fat.

On a weight-loss diet (without exercise), about 62% of the loss is fat. About 11% is protein (muscle tissue). The rest is water and glycogen.

Example 26.1

If you ordinarily expend 2200 kcal/day and then go on a diet of 1500 kcal/day, how long will it take to lose 1 lb of fat?

Solution

You will use 700 kcal/day more than you consume. There are 3500 kcal in 1 lb of fat, so it will take:

$$3500 \text{ kcal} \times \frac{1 \text{ day}}{700 \text{ kcal}} = 5 \text{ days}$$

(Keep in mind, however, that your weight loss will not all be fat. You will probably lose more than 1 lb in the five days, but it will be mostly water with some protein and glycogen.)

Practice Exercise

A person who expends 1800 kcal/day goes on a diet of 1200 kcal/day without a change in activities. How much fat will the person lose if the diet is followed for 3 weeks?

Most quick weight-loss diets depend on factors other than fat metabolism to hook prospective customers. Many contain a *diuretic*, such as caffeine, to increase the output of urine. Much of the weight loss is water loss. Weight is regained when the body is rehydrated. Other diets deplete the body's stores of glycogen. When carbohydrates are eliminated from the diet, the body draws on its glycogen reserves, depleting them in about 24 hours. Recall that glycogen molecules have lots of hydroxyl groups that can form hydrogen bonds to water molecules. Each pound of glycogen carries with it about 2 lb of water held by hydrogen bonds. Depletion of about a pound of glycogen results in a weight loss of about 3 lb (1 lb glycogen + 2 lb water). No fat is lost, and the weight is quickly regained when the dieter resumes eating carbohydrates.

Current guidelines state that any weight-loss diet should include generous portions of complex carbohydrates (starches), along with fruits and vegetables, modest portions of protein, and very small amounts of fats, oils, and simple carbohydrates (sugars). (See the food guide pyramid in Figure 26.7 and the guidelines outlined in Table 26.4.) Diets with fewer than 1200 kcal/day are likely to be deficient in necessary nutrients, particularly in B vitamins and iron. The recommended weight-loss goal should be 1–2 lbs per week.

The calories listed on Nutrition Facts labels are actually kcal.

The following example problems show you how to determine the % calories from fat that are supplied by a particular food.

Example 26.2

One of McDonald's Big Mac sandwiches furnishes 541 kcal, of which 279 kcal comes from fat. What percentage of total calories is from fat?

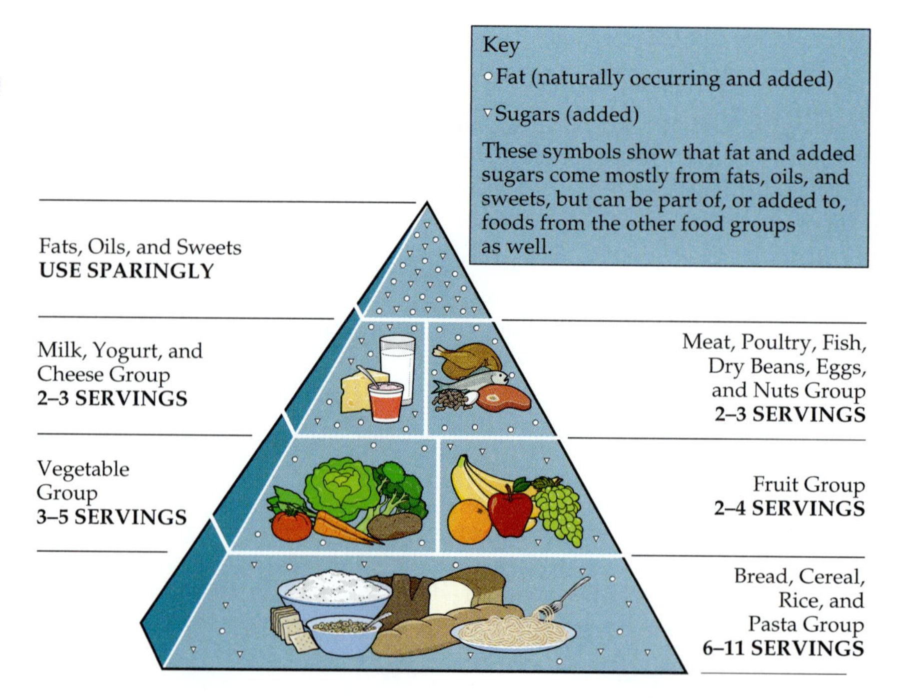

Figure 26.7 The food-guide pyramid. Start with bread, cereal, rice, and pasta, along with fruits and vegetables. Then add some servings from the milk and meat groups. Go easy on the fats, oils, and sweets.

Table 26.4 Recommended Low-Calorie Diet[a]

Nutrient	Recommended Intake
Calories	Approximately 500–1000 kcal/day reduction from usual intake
Total Fat	30% or less of total calories
Saturated Fatty Acids	8–10% of total calories
Monounsaturated Fatty Acids	up to 15% of total calories
Polyunsaturated Fatty Acids	up to 10% of total calories
Cholesterol	<300 mg/day
Protein	Approximately 15% of total calories
Carbohydrate[b]	55% or more of total calories
Sodium Chloride	no more than 100mmol/day (~6 g of sodium chloride)
Calcium	1000–1500 mg/day
Fiber	20–30 g

[a]*Clinical Guidelines on the Identification, Evaluation, and Treatment of Overweight and Obesity in Adults* (preprint June 1998). National Institutes of Health.

[b]Carbohydrates should be primarily complex carbohydrates, rather than simple carbohydrates or sugars.

Solution

Simply divide the calories from fat by the total calories. Then multiply by 100 to get percentage (parts per 100):

$$\% \text{ calories from fat} = \frac{279 \text{ kcal}}{541 \text{ kcal}} \times 100 = 51.6\%$$

Practice Exercise

One of Burger King's Whopper sandwiches furnishes 606 kcal, of which 288 kcal comes from fat. What percentage of total calories is from fat?

Example 26.3

The label on a macaroni-and-cheese dinner indicates that each $\frac{3}{4}$ cup serving furnishes 290 kcal, 9 g of protein, 34 g of carbohydrate, and 13 g of fat. Calculate the percentage of calories from fat.

Solution

First calculate the calories that are contributed by the fat:

$$13 \text{ g} \times 9 \text{ kcal/g} = 117 \text{ kcal}$$

Then calculate the percentage of calories from fat.

$$\% \text{ calories from fat} = \frac{117 \text{ kcal}}{290 \text{ kcal}} \times 100 = 40.3\%$$

Practice Exercise

The label on a can of cream-style corn indicates that each one-cup serving furnishes 90 kcal, 2 g of protein, 22 g of carbohydrate, and 1 g of fat. Calculate the percentage of calories from fat.

People who do not increase their food intake when they begin an exercise program will lose weight. Most of the weight loss from exercise is due to the increase in metabolic rate during the activity, but the increased metabolic rate continues for several hours after completion of the exercise. Exercise helps us maintain our fitness level and is a key factor in maintaining weight loss. Table 26.5 provides the number of calories expended in several forms of exercise. The principles of weight loss are met by decreasing caloric intake and increasing caloric expenditure.

Review Questions

26.9 How much energy, in calories, is stored in 1.0 lb of adipose tissue?

26.10 List two ways in which fad diets lead to a "quick weight loss." Why is this weight rapidly regained?

Obesity Genes

Numerous studies have been conducted with families or twins separated at birth in an attempt to determine the role that genetics plays in the development of obesity. These studies indicate that genetic factors are important and have spurred efforts to identify these genes and determine the function of each one.

In the mid-1990s at least five different genes were identified in rodents that have a role in controlling weight. *Obese* is the gene that codes for leptin, a hormone that has been implicated in regulating weight. A mutation in this gene will prevent leptin from being synthesized, or produce a mutant hormone that cannot function properly. There was a great deal of excitement when the first reports were published, in 1995, that injections of leptin led to weight loss in mice who carried a mutation in the *obese* gene. Unfortunately, humans are more complex than mice and leptin does not appear to be the magic drug for treating obesity in humans. But researchers are still very interested in understanding exactly what leptin does in the human body. Some evidence suggests that it may play a role in the proper functioning of the immune system, as well as in weight control.

The *diabetes* gene codes for the leptin receptor. Mutations in this gene lead to the formation of a receptor that cannot properly bind leptin. The gene products and functions of the other three genes (*fat, tubby,* and *agouti yellow*) have not yet been identified. Although geneticists have identified copies of these various genes in humans, they have not shown a link between obesity and a mutation in one or more of these genes.

▶ The mouse on the left has a mutation in the obese gene.

Table 26.5 Calories Expended in Exercise

Activity	Kcal per Hour[a]
Badminton, competitive singles	480
Basketball	360–660
Bicycling	
10 mph	420
11 mph	480
12 mph	600
13 mph	660
Calisthenics, heavy	600
Handball, competitive	660
Rope skipping, vigorous	800
Rowing machine	840
Running	
5 mph	600
6 mph	750
7 mph	870
8 mph	1020
9 mph	1130
10 mph	1285
Skating, ice or roller, rapid	700
Skiing, downhill, vigorous	600
Skiing, cross-country	
2.5 mph	560
4 mph	600
5 mph	700
8 mph	1020
Swimming, 25–50 yards per min	360–750
Walking	
Level road, 4 mph (fast)	420
Upstairs	600–1080
Uphill, 3.5 mph	480–900
Gardening, much lifting, stooping, digging	500
Mowing, pushing hand mower	450
Sawing hardwood	600
Shoveling, heavy	660
Wood chopping	560

[a]Caloric expenditure is based on a 150-lb person. There is a 10% increase in caloric expenditure for each 15 lb over this weight and a 10% decrease for each 15 lb under.

Adapted from E. L. Wynder, *The Book of Health: The American Health Foundation.* © 1981 Franklin Watts, Inc., New York. Used with permission.

Summary

Triglycerides are the richest energy source in the human diet because they are more highly reduced than carbohydrates or proteins. A human adult can maintain a nearly fat-free diet if two **essential fatty acids,** linoleic and linolenic acids, are present in the diet.

Body tissue that stores lipids is **adipose tissue,** which contains specialized cells known as **fat cells** (or **adipocytes**). Areas of the body containing large amounts of adipose tissue are **fat depots.** Lipases present in adipose tissue catalyze the hydrolysis of triglycerides to fatty acids and glycerol.

These lipases are activated following the release of the hormones epinephrine or glucagon, through the action of cAMP-dependent protein kinase. Glycerol is carried to the liver, where it enters glycolysis (or gluconeogenesis). The fatty acids are carried to all cells that can use them, converted to fatty acyl-CoAs, transported into the mitochondria, and oxidized through repeating a sequence of four reactions known as **β-oxidation.**

A fatty acid is activated for β-oxidation in a reaction that converts it to a fatty acyl-CoA molecule. The reaction is catalyzed by acyl-CoA synthetase. The equivalent of two high-energy phosphate bonds are consumed in the reaction in which the fatty acid combines with coenzyme A to form fatty acyl-CoA.

In each round of β-oxidation, the fatty acyl-CoA is shortened by two carbons as one molecule of acetyl-CoA is formed. When the chain has been shortened to four carbons, the final round of β-oxidation occurs because two molecules of acetyl-CoA are formed. Therefore, β-oxidation is repeated $(n/2) - 1$ times, where n is the number of carbons in the acid releasing a total of $n/2$ molecules of acetyl-CoA and $(n/2) - 1$ molecules of $FADH_2$ and NADH. Each round of β-oxidation produces 1 molecule of $FADH_2$ and 1 molecule of NADH, leading to the formation of 4 molecules of ATP when these are reoxidized through the combined action of the electron transport chain and oxidative phosphorylation. All of the acetyl-CoA formed through β-oxidation can be completely oxidized, yielding 10 molecules of ATP. The net ATP yield of fatty acid oxidation is therefore:

$(n/2) \times 10$ ATP from acetyl–CoA

$[(n/2) - 1] \times 4$ ATP from $FADH_2$ and NADH formed in each round of β–oxidation

MINUS the 2 high–energy phosphate bonds required to activate the fatty acid.

The efficiency of fatty acid oxidation in the human body is approximately 34%.

If carbohydrate consumption is inadequate to meet immediate energy needs, fatty acid oxidation will occur at an elevated rate, causing the acetyl-CoA concentration in the liver to increase. The acetyl-CoA produces acetoacetate, which is converted to β-hydroxybutyrate and acetone. These three compounds are collectively known as **ketone bodies.** They accumulate in the blood in a condition known as **ketosis.** High blood levels of the two acidic ketone bodies overwhelm the blood buffers, so that the blood becomes too acidic, a condition called **acidosis.**

When insulin is released, signaling abundant energy supplies, malonyl-CoA is synthesized from acetyl-CoA and bicarbonate ion (HCO_3^-) in a reaction catalyzed by acetyl-CoA carboxylase. The malonyl-CoA is then utilized to synthesize fatty acids in a series of reactions catalyzed by **fatty acid synthase.** In this way, the body stores excess energy that it does not immediately need. Many of the reactions catalyzed by fatty acid synthase are the reverse of the reactions in β-oxidation.

Approximately 55% of the U.S. population is either overweight or obese. One of the primary methods of classification is the use of the **body mass index (BMI).** The BMI is defined as:

$$\text{BMI} = \frac{\text{Body weight (kg)}}{[\text{Height (m)}]^2} \quad \text{or} \quad \frac{704.5 \times \text{Body weight (lb)}}{[\text{Height (in)}]^2}$$

Losing weight and maintaining the weight loss are extremely difficult. Two theories have been proposed to explain why this is true. According to the **set-point theory,** weight is centrally controlled by the brain, which sends out signals to maintain a particular set weight. The newer **settling-point theory** suggests that we maintain a particular weight when the various metabolic feedback loops are in equilibrium with our environment. This equilibrium can be upset, and a new settling point attained, by an increase in the amount of fat in our diet or by exercise. The best method for losing weight is a combined program of behavior therapy, a low calorie diet, and increased physical activity.

Key Terms

acidosis (26.4)
adipocyte (26.1)
adipose tissue (26.1)
body mass index [BMI] (26.6)
brown fat (26.1)
essential fatty acid (page 723)
fat cell (26.1)
fat depot (26.1)
fatty acid synthase (26.5)
ketone body (26.4)
ketosis (26.4)
leptin (26.6)
β-oxidation (26.2)
set-point theory (26.6)
settling-point theory (26.6)

Problems

Storage and Mobilization of Fats

1. Compare the energy released when 1.0 g of carbohydrate and 1.0 g of lipid are oxidized completely in the body.
2. In this chapter the statement is made that 1 g of fat has more than six times the energy content of 1 g of glycogen. Explain.
3. **a.** During a fast, which energy reserves are used first?
   b. Which energy reserves supply the major part of the body's needs during a fast?
4. **a.** Will the release of the hormone glucagon increase the rate of triglyceride synthesis or degradation?
   b. What is the effect when insulin is secreted?
5. Lipases catalyze the hydrolysis of triglycerides. Write out the reaction, using glyceryl palmitate as the triglyceride, showing the products that are formed.

6. What happens to fatty acids that are released when triglycerides are hydrolyzed?
7. Show, with equations, how the glycerol obtained from lipid hydrolysis is incorporated into the glycolytic pathway.

Fatty Acid Oxidation

8. How are fatty acids activated prior to being transported into the mitochondria and oxidized?
9. What types of reactions are involved in β-oxidation?
10. What are the end products of β-oxidation?
11. How many rounds of β-oxidation are necessary to metabolize the following fatty acids?
    a. myristic acid ($C_{13}H_{27}COOH$)
    b. cerotic acid ($C_{25}H_{51}COOH$)
12. When myristic acid is completely oxidized by β-oxidation, how many molecules of each of the following are formed?
    a. acetyl-CoA b. $FADH_2$ c. NADH
13. When stearic acid ($C_{17}H_{35}COOH$) is completely oxidized by β-oxidation, how many molecules of each of the following are formed?
    a. acetyl-CoA b. $FADH_2$ c. NADH
14. How and where is $FADH_2$ from β-oxidation converted back to FAD?
15. How many molecules of ATP are formed during the complete oxidation, in a liver cell, of one molecule of:
    a. myristic acid; b. glyceryl myristate;
    c. glyceryl lauropalmitostearate?

Ketosis

16. Why does a deficiency of carbohydrates in the diet lead to the formation of ketone bodies?
17. Why does starvation result in acidosis?
18. How does lack of insulin lead to ketosis?
19. a. What tissues normally use ketone bodies to supply a major part of their energy tissues?
    b. What organ will begin using ketone bodies for fuel during a prolonged fast?

Fatty Acid Synthesis

20. Why do most naturally occurring fat acids contain even numbers of carbon atoms?
21. Compare fatty acid oxidation and fatty acid synthesis in terms of:
    a. location in the cell where these two pathways are localized;
    b. coenzymes needed;
    c. form in which carbons are added or removed.
22. What fatty acid is formed by the action of fatty acid synthase?
23. How many moles of NADPH are required for the synthesis of palmitate?
24. Explain why excess glucose can be converted to fatty acids, but excess fatty acids cannot be used to synthesize glucose.
25. What is meant by the statement that the enzymes needed to synthesize palmitic acid are components of a single multifunctional protein?

Obesity, Exercise, and Diets

26. What is the principal cause of obesity?
27. Calculate BMI values for the following individuals and determine if they are underweight, normal, overweight, or obese.
    a. a person who is 5′3″ and weighs 125 lbs
    b. person who is 5′8″ and weighs 115 lbs
    c. a person who is 6′2″ and weighs 215 lbs
28. Calculate BMI values for the following individuals and determine if they are underweight, normal, overweight, or obese.
    a. a person who is 5′5″ and weighs 160 lbs
    b. a person who is 6′0″ and weighs 230 lbs
    c. a person who is 4′11″ and weighs 100 lbs
29. One snack cake furnishes 130 kcal, of which 55 kcal comes from fat. What percentage of total calories is from fat?
30. One serving of cream of mushroom soup furnishes 80 kcal, of which 10 come from fat. What percentage of total calories is from fat?
31. A large double-decker hamburger provides 600 kcal of energy. How long would you have to walk to burn off that energy if 1 h of walking uses about 300 kcal?
32. How long would you have to run to burn off the 110 kcal in one glass of beer if 1 h of running burns off 1100 kcal?
33. List some problems that result from low-calorie diets.
34. You want to lose 15 lbs. in 10 weeks. Currently, you eat approximately 2800 kcal/day. How many calories can you consume each day and achieve your goal (assuming all weight lost is due to loss of fat and you do not increase your amount of exercise)?

Additional Problems

35. Fat tissue has a density of about 0.90 g/mL, lean tissue a density of about 1.1 g/mL. Calculate the density of a person who has a body volume of 80 L and who weights 85 kg. Is the person fat or lean?
36. Compare and contrast glycogen and adipose tissue as reserve energy sources.
37. Use equations to show how acetone and β-hydroxybutyrate are formed from acetoacetate.
38. Use what you have learned in this chapter and chapter 24 to fill in the blanks in the diagram on page 742 (use words and/or structures).

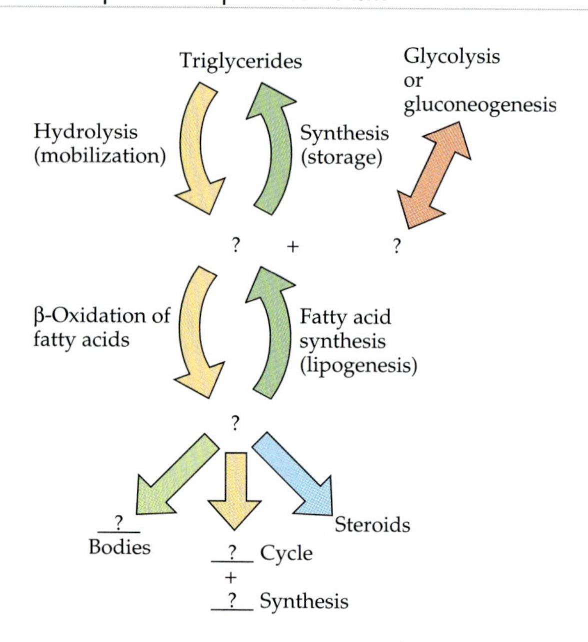

39. In severe, uncontrolled diabetes acetone can be smelled on the patient's breath. Explain how this occurs.
40. The ingestion of excess carbohydrates results in the deposition of fats in adipose tissue. Explain.
41. Why does a diet that restricts carbohydrate intake lead to loss of muscle mass as well as loss of fat?
42. What purpose does brown fat tissue serve in mammals? How does it accomplish its purpose?
43. List the chief fuel for each of the following:
    a. brain **b.** liver **c.** cardiac muscle
    d. kidneys **e.** resting skeletal muscle
    f. active skeletal muscle
    g. brain (prolonged starvation)
44. What is the product of each of the following genes identified in rodents?
    a. diabetes **b.** obese

Protein Metabolism

Meats, vegetables, and nuts supply the amino acids necessary to build body proteins.

Learning Objectives/Study Questions

1. How do mammals obtain the amino acids they need?
2. What metabolic processes require amino acids and which provide amino acids?
3. How are excess amino acids degraded?
4. How are small amounts of ammonia stored in the body? How is excess ammonia safely excreted?
5. How are nonessential amino acids synthesized?
6. What important compounds, besides proteins, are formed from amino acids?
7. How are hormones important in coordinating the regulation of metabolic pathways?
8. What strategies does the body use to maintain energy levels during periods of starvation?

Higher plants and most microorganisms are capable of synthesizing all of their amino acids from carbon dioxide, water, and inorganic salts. They obtain the required nitrogen either from soil nitrates or from atmospheric nitrogen (via nitrogen-fixing bacteria). Thus these organisms can grow on a medium that does not contain any preformed amino acids. From these amino acids they put together all the proteins they need. We are not quite so versatile. We can put together all the proteins we need, but *only* if we obtain the right proportions of specific amino acids in our diet.

The peptides and proteins in our bodies are in a state of dynamic equilibrium as "old" ones are hydrolyzed and "new" ones are synthesized. We take in proteins when we eat and digest them into their constituent amino acids (see Section 24.2). The liver and other body tissues incorporate these amino acids into new peptides and proteins (see Section 23.6). In addition, about 300 g of a normal adult's tissue proteins are degraded each day, and most of the amino acids released are reused to synthesize new proteins. However, some amino acids are catabolized, and the nitrogen is eliminated from the body.

The daily recommended intake of protein is 45–60 g. In the industrialized countries, the average daily intake of proteins is about 100 g.

Under normal conditions, an individual's intake of dietary nitrogen is equal to the amount of nitrogen lost in the feces, urine, and sweat. Such a condition is referred to as **nitrogen balance** or *nitrogen equilibrium.* Organisms are said to be in *positive* nitrogen balance (intake exceeds excretion) whenever tissue is being synthesized—for example, during periods of growth, pregnancy, and convalescence from disease. *Negative* nitrogen balance results from (1) an inadequate intake of protein (for example, fasting); (2) fever, infection, surgery, or a wasting disease; or (3) a diet that lacks, or is deficient in, any one of the essential amino acids. These factors accelerate breakdown of tissue protein (in an attempt to supply the missing amino acids), and nitrogen excretion exceeds intake. In this chapter we deal with the metabolism of the amino acids.

27.1 Overview of Amino Acid Metabolism

Carbohydrates are stored in the liver and muscles as glycogen. Fats are placed on reserve in the fat depots. For proteins, however, there are no comparable storage facilities. Proteins are digested in the small intestine, and the resulting amino acids are absorbed directly into the bloodstream. Here they join a circulating pool of amino acids. Members of the **amino acid pool** are there only on temporary assignment. The body can supply additional amino acids to this pool by synthesizing them from other compounds or by breaking down tissue protein. Amino acids are removed from the pool for the synthesis of new proteins or other nitrogen-containing compounds, such as heme. They can also be used as an energy source. The pathways in which amino acids are supplied to and removed from the amino acid pool are outlined in Figure 27.1.

The amount of protein degraded per unit time gives the **turnover rate,** which represents the average residence time for a protein molecule in a tissue. This rate varies for proteins in different body tissues and is usually expressed in terms of a half-life. As with radioactive substances (Section 12.5), the half-life of a protein is the time interval required for one-half of the protein molecules in a given tissue to be replaced. Liver proteins and those in the blood plasma have rapid turnovers, with half-lives of two to ten days. Muscle proteins are more stable, with half-lives of about six months, and some collagen molecules are not degraded for three years. The half-lives of enzymes vary widely from ten minutes to six hours depending on their metabolic importance and the cell in which they function. Hair has no half-life because only synthesis (and not degradation) of hair protein occurs within the ectodermal (skin) cells.

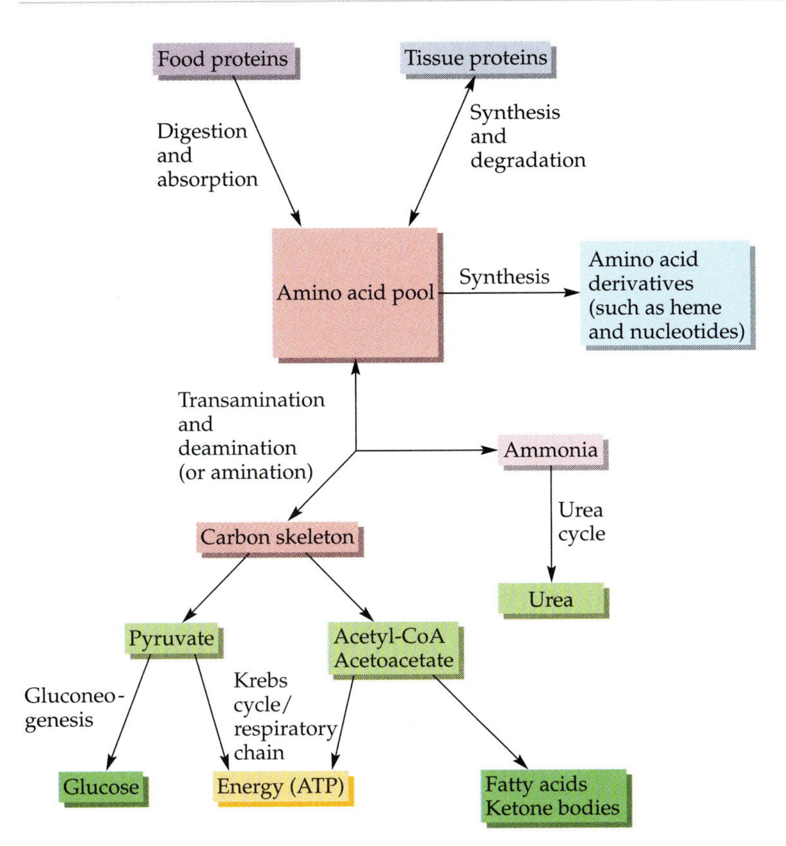

Figure 27.1 Schematic diagram of general paths of amino acid metabolism in humans.

Table 27.1 Essential and Nonessential Amino Acids for Humans

Essential	Nonessential
Lysine	Glycine
Leucine	Alanine
Isoleucine	Serine
Methionine	Tyrosine[a]
Threonine	Cysteine
Tryptophan	Aspartate
Valine	Asparagine
Phenylalanine	Glutamate
Histidine	Glutamine
Arginine[b]	Proline

[a]In the presence of adequate amounts of phenylalanine.
[b]Essential for growing children, not for adults.

Animals can synthesize only about half of the amino acids they need in order to function properly. The remainder must be supplied in the diet. An **essential amino acid** is one that cannot be synthesized by an organism at a rate rapid enough to supply the normal requirements of protein biosynthesis. A list of essential amino acids for humans is given in Table 27.1. Notice that, as a rule, essential amino acids contain carbon chains or aromatic rings that are not intermediates in carbohydrate or lipid metabolism. The inability to synthesize these amino acids results not from a lack of the necessary nitrogen but rather from the animal's inability to manufacture the correct carbon skeleton. Supplied with phenylpyruvate, for example, an animal can readily synthesize the amino acid analog, phenylalanine. (Lysine appears to be an exception because the entire preformed amino acid must be supplied.)

Phenylpyruvate

Phenylalanine

All of the amino acids required for the synthesis of a particular protein must be available to the cell at the time of protein synthesis. If just one amino acid is either absent or present in insufficient quantity, the protein is not synthesized. For example, if a given protein contains 4 phenylalanine and 40 are available, only 10 molecules of that protein can be made, even though there may be enough of all the other amino acids to make a hundred copies of that protein.

A **complete protein** source supplies all the essential amino acids in the quantities needed for growth and repair of body tissues. Casein, the protein from milk, is especially beneficial because it is well balanced in its amino acid distribution. Proteins that lack an adequate amount of one or more essential amino acids are termed **incomplete proteins.**

Most plant proteins are deficient in one or more essential amino acids. Zein, the protein in corn, is deficient in lysine and tryptophan. Protein from wheat and rice lacks lysine and threonine. Dried beans (legumes), with the exception of soybeans,

Amino Acid and Protein Supplements

Amino acids, either individually or in protein supplements, have been touted in articles and advertisements in newspapers, magazines, and health food stores for a variety of reasons: to increase muscle mass, to strengthen fingernails, to cure cold sores, depression, fatigue, insomnia, and pain. For example, there are claims that tryptophan and tyrosine cure depression and insomnia, and that leucine and phenylalanine are effective as pain relievers. However, the FDA states that it is dangerous for consumers to ingest large amounts of any one amino acid, and there is little evidence that they do any good. Animal studies have shown that excessive intake of specific individual amino acids can create amino acid imbalances.

Some extreme endurance athletes, such as triathletes and ultramarathoners, may need a bit more than the RDA for protein. Because nearly all Americans eat 50% more protein than they need, even these athletes seldom need protein supplements. Protein-rich foods are cheaper and more completely digested by the body.

Muscles are built through exercise, not through eating excess protein. When a muscle contracts against a resistance, creatine is released and stimulates the production of the protein myosin (Section 24.6), thus building more muscle tissue. If the exercise stops, the muscle begins to shrink after about two days. After about two months without exercise, muscle built through the exercise program is almost completely gone. (The muscle does *not* turn to fat, as some athletes believe. Former athletes often get fat, however, because they continue to take in the same number of calories and expend fewer.)

Table 27.2 Efficiencies of Protein Conversions

Food	Efficiency of Production (%)[a]
Beef or veal	4.7
Pork	12.1
Chicken or turkey	18.2
Milk	22.7
Eggs	23.3

[a]Calculated by dividing the mass of edible protein by the mass of the protein feed required to produce it, then multiplying the result by 100.

are lacking in the essential amino acids methionine and valine. Because most plant proteins are incomplete, vegetarians should eat a variety of foods that complement each other. Thus black beans and rice, when eaten together, provide all of the essential amino acids and are said to be *complementary proteins.*

Plants trap a small fraction of the solar energy that falls upon them. They use this energy to convert carbon dioxide, water, and mineral nutrients to protein. Cattle eat the plant protein, digest it, and convert a small portion of it to animal protein. People eat this animal protein, digest it, and reassemble some of the amino acids into human protein. Some of the energy originally trapped by the green plants is lost as heat at every step of the food chain (Table 27.2). If people ate the plant protein directly, one highly inefficient step would be skipped. A vegetarian diet can provide excellent nutrition, if a wide variety of foods are eaten.

✓ **Review Questions**

27.1 Compare the turnover rates of enzymes and muscle proteins.

27.2 **(a)** Name three processes that add amino acids to the amino acid pool. **(b)** For what purposes are amino acids removed from the amino acid pool?

27.2 Catabolism of Amino Acids

The liver is the principal organ for amino acid metabolism, but other tissues such as kidney, intestine, muscle, and adipose tissue are also involved. Generally, the first step in the breakdown of amino acids is the separation of the amino group from the carbon skeleton. The carbon skeletons resulting from the deaminated amino acids are used to form either glucose (via gluconeogenesis) or fats (via acetyl-CoA), or they are converted to a metabolic intermediate that can be oxidized by the Krebs cycle. Amino acid catabolism is particularly prevalent during hypoglycemia, fasting, and starvation (Section 27.6).

Transamination

Transamination is an exchange of functional groups between any amino acid (except lysine, proline, and threonine) and an α-keto compound. The amino group is usually transferred to the keto carbon of pyruvate, oxaloacetate, or α-ketoglutarate, converting it to alanine, aspartate, or glutamate, respectively. Transamination reactions are catalyzed by specific transaminases (also called aminotransferases), which require pyridoxal phosphate (Section F.6) as a coenzyme.

$$\underset{\text{Amino acid}}{R-CH(NH_3^+)-CO_2^-} + \underset{\alpha\text{-Keto acid}}{R'-C(=O)-CO_2^-} \xrightleftharpoons{\text{transaminase}} \underset{\text{New }\alpha\text{-keto acid}}{R-C(=O)-CO_2^-} + \underset{\text{New amino acid}}{R'-CH(NH_3^+)-CO_2^-}$$

The amino groups from alanine and aspartate are in turn transferred to α-ketoglutarate, forming glutamate, by a second transamination reaction (Figure 27.2). These two transamination reactions are of special clinical interest. Normally the blood contains a low concentration of transaminases. However, extensive tissue destruction is accompanied by rapid and striking increases in the blood transaminase levels. Glutamate-pyruvate transaminase (GPT) has a particularly high activity in the cytoplasm of the liver, and an elevated serum level of this enzyme is indicative of liver damage. Glutamate-oxaloacetate transaminase (GOT) is abundant in heart muscle (Section 22.7), and a sharp rise in the concentration of GOT in the blood is an indication of myocardial infarction.

Oxidative Deamination

In the catabolism of amino acids, the final acceptor of the α-amino group is α-ketoglutarate, forming glutamate. But there must be some way of removing the amino group completely. Glutamate can undergo **oxidative deamination** in which it loses its amino group as ammonium ion and is oxidized back to α-ketoglutarate:

$$\underset{\text{Glutamate}}{{}^-O_2C-CH(NH_3^+)-CH_2-CH_2-CO_2^-} + H_2O \xrightleftharpoons[\text{dehydrogenase}]{NAD^+ \;\; NADH + H^+} \underset{\alpha\text{-Ketoglutarate}}{{}^-O_2C-C(=O)-CH_2-CH_2-CO_2^-} + NH_4^+$$

This reaction primarily occurs in the liver mitochondria. Most of the ammonium ion formed from glutamate by oxidative deamination is converted to urea and excreted in the urine (Section 27.3).

The reverse reaction, synthesis of glutamate, is important and occurs primarily in the cytoplasm of the cell, where NADPH acts as the reducing agent. This reverse reaction is significant because it is one of the few reactions in animals that can incorporate inorganic nitrogen (NH_4^+) into an α-keto acid to form an amino acid. The amino group can be passed on through transamination reactions, producing other cellular amino acids from the appropriate α-keto acids.

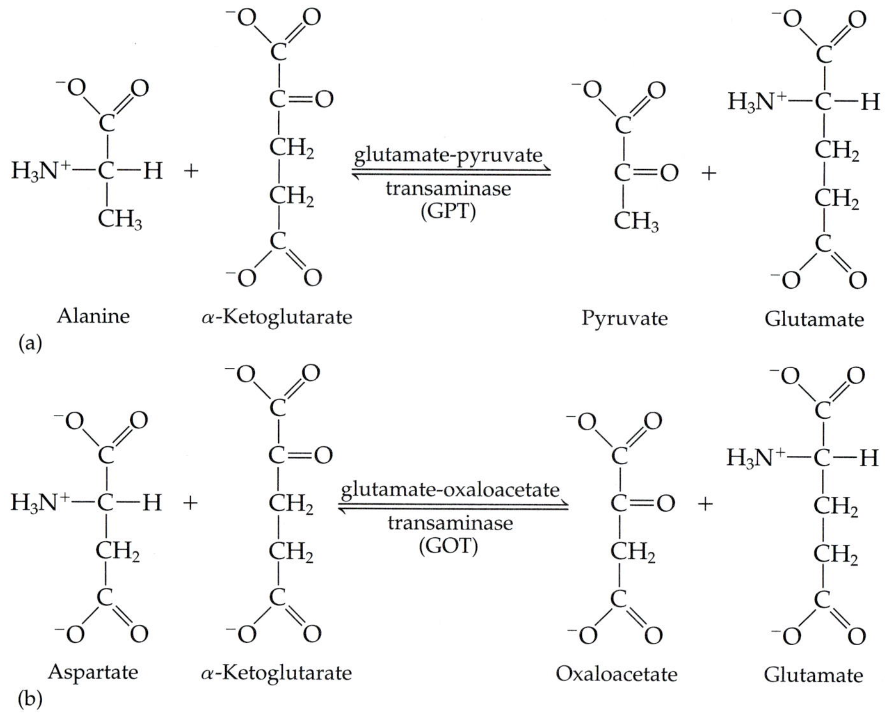

Figure 27.2
Transamination reactions catalyzed by (a) GPT and (b) GOT. In both reactions the final acceptor of the amino group is α-ketoglutarate and the final product is glutamate. The reactions are reversible.

(a) Alanine + α-Ketoglutarate ⇌ (glutamate-pyruvate transaminase (GPT)) Pyruvate + Glutamate

(b) Aspartate + α-Ketoglutarate ⇌ (glutamate-oxaloacetate transaminase (GOT)) Oxaloacetate + Glutamate

The Fate of the Carbon Skeleton

Any amino acid can be converted into an intermediate of the Krebs cycle. Once the amino group is removed from the amino acid, usually by transamination, the α-keto acid that remains is catabolized by a unique pathway of one or more reactions. For example, phenylalanine undergoes a series of six reactions before it splits into fumarate and acetoacetate. Fumarate is an intermediate in the Krebs cycle, while acetoacetate must be converted to acetoacetyl-CoA and then to acetyl-CoA before it enters the Krebs cycle.

Those amino acids that can form any of the intermediates of carbohydrate metabolism can be converted to glucose (via gluconeogenesis). They are referred to as **glucogenic amino acids.** Amino acids that give rise to acetoacetyl-CoA or acetyl-CoA (which can be used for the synthesis of ketone bodies, but not glucose) are called **ketogenic amino acids.** Certain amino acids fall into both categories. Leucine is the only amino acid that is exclusively ketogenic. Table 27.3 classifies the amino acids as glucogenic or ketogenic. Figure 27.3 summarizes the ultimate fates of the carbon skeletons of the twenty amino acids.

Table 27.3 Glucogenic and Ketogenic Amino Acids

Glucogenic		Ketogenic	Glucogenic and Ketogenic
Alanine	Glycine	Leucine	Isoleucine
Arginine	Histidine		Lysine
Asparagine	Methionine		Phenylalanine
Aspartate	Proline		Tyrosine
Cysteine	Serine		Tryptophan
Glutamate	Threonine		
Glutamine	Valine		

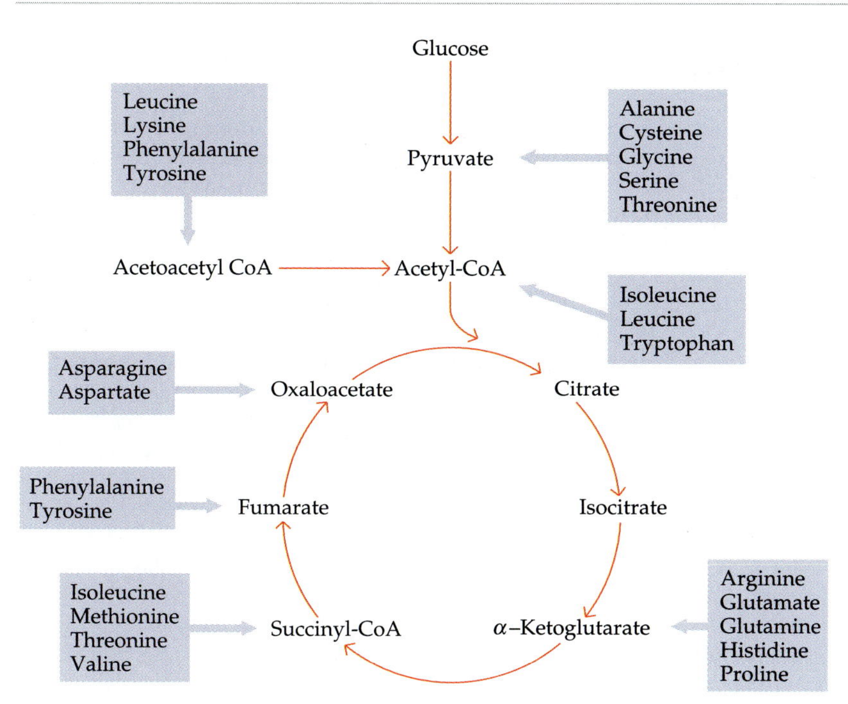

◀ **Figure 27.3**
Fates of the carbon skeletons of amino acids.

Review Questions

27.3 **(a)** Write the equation for the transamination reaction between alanine and oxaloacetate. **(b)** Name the two products that are formed.

27.4 What is oxidative deamination?

27.3 Storage and Excretion of Nitrogen

As mentioned earlier, excess protein is not stored to any appreciable extent in living organisms. The only significant way the body has to store small amounts of nitrogen is in the form of glutamine. This amide of glutamate is formed from glutamate and the ammonia produced by oxidative deamination reactions:

$$\underset{\text{Glutamate}}{{}^{-}OOC{-}CH(NH_3^+){-}CH_2{-}CH_2{-}COO^-} + NH_4^+ \xrightarrow[\text{synthetase}]{ATP \;\; ADP} \underset{\text{Glutamine}}{{}^{-}OOC{-}CH(NH_3^+){-}CH_2{-}CH_2{-}C(=O)NH_2} + H_2O$$

Glutamine is present in many tissues and in the blood and serves as a temporary storage and transport form of nitrogen. The formation of glutamine is the major method of disposing of ammonia from the brain. In the liver and kidneys, the amide group of glutamine can be donated to appropriate acceptor molecules in the biosynthesis of nitrogen-containing compounds, such as purines, pyrimidines, porphyrins (such as heme), and creatine (see Figure 27.1).

Genetic Diseases of Amino Acid Catabolism

In Section 23.7 we looked at several genetic diseases caused by mutations in specific genes coding for degradative enzymes in carbohydrate, lipid, or amino acid catabolism. For example, in *phenylketonuria* (PKU), the enzyme phenylalanine hydroxlase is absent and phenylalanine cannot be converted to tyrosine, the first step in the degradation of phenylalanine. There are many other genetic diseases in which the catabolism of a specific amino acid (or group of amino acids) is impaired because of a mutated gene.

In 1 out of every 216,000 live births a mutation occurs in one of the enzymes needed to degrade the branched-chain amino acids: leucine, isoleucine, and valine. These amino acids are transaminated to their respective α-keto acids, but cannot be degraded any further. The α-keto acids accumulate and are excreted in the urine imparting a characteristic odor similar to maple syrup, which gave this genetic disease its name—*maple syrup urine disease*. If untreated, severe mental retardation results. Dietary restrictions of the branched-chain amino acids is currently the only treatment available.

Alkaptonuria results from a defect in an enzyme needed for the complete degradation of tyrosine. A compound known as homogentisate accumulates in cartilage, the eyes, and kidneys, leading to a type of arthritis. Homogentisate is also excreted in the urine and is quickly oxidized to a black substance. One of the earliest records of a person with alkaptonuria was written in 1649. Treatments, such as cold baths and a watery diet, were prescribed for this man in an attempt to cool his body, because it was thought his body had too much heat, which was charring his urine. The only real treatment is to maintain a strict diet that includes minimal amounts of tyrosine and phenylalanine.

Nitrogen that is not needed for biosynthesis is converted to ammonia (or ammonium ion) and must be eliminated from the body. Organisms must have a mechanism for removing this ammonia from the cell because even low concentrations are poisonous. Levels of only 5 mg of ammonia per 100 mL of blood are toxic to humans. (The normal concentration of ammonia is about 1–3 μg/100 mL.) Organisms differ biochemically in the manner in which they excrete this ammonia as *nitrogenous wastes*. Most vertebrates and adult amphibia excrete them as urea in the urine. Birds, reptiles, and insects convert them to uric acid. All marine organisms, from unicellular organisms to fish, excrete free ammonia. The ammonia is very soluble in water and is rapidly diluted in the aqueous environment.

In humans, 95% of nitrogenous wastes results from amino acid catabolism; the remaining 5% comes from the catabolism of other nitrogen-containing compounds (e.g., pyrimidines, porphyrins).

In mammals, the liver is the principal organ concerned with the formation of urea. Urea is formed by a series of reactions, outlined in Figure 27.4, known as the **urea cycle.** This sequence of reactions was elucidated by Hans Krebs several years before he determined the sequence of reactions known as the Krebs cycle (Chapter 24). Extensive liver damage (e.g., cirrhosis) or a genetic defect in any urea cycle enzyme results in high levels of ammonia and causes tremor, slurred speech, and blurred vision. A continued rise in ammonia concentrations can lead to coma and death.

Ammonium ion—released from the oxidative deamination of glutamate—enters the urea cycle as carbamoyl phosphate, formed by reaction with bicarbonate ion. The enzyme for this reaction is *carbamoyl phosphate synthetase*. The carbamoyl group (—$CONH_2$) is transferred from carbamoyl phosphate onto ornithine to form citrulline in the first reaction of the urea cycle, catalyzed by *ornithine transcarbamoylase.* Citrulline combines with a molecule of aspartate to form argininosuccinate in a reaction that requires ATP and the enzyme *argininosuccinate synthetase.* This intermediate is cleaved by *argininosuccinase* in the third reaction to form arginine and fumarate, a by-product of the urea cycle (but an intermediate of the Krebs cycle). In the fourth reaction, catalyzed by *arginase,* arginine is split into two compounds, urea and ornithine. Ornithine is used to begin another round of the urea cycle, and

In the Krebs cycle, fumarate is converted to oxaloacetate. Oxaloacetate can be converted to aspartate in a transamination reaction, forming more aspartate for the second reaction of the urea cycle.

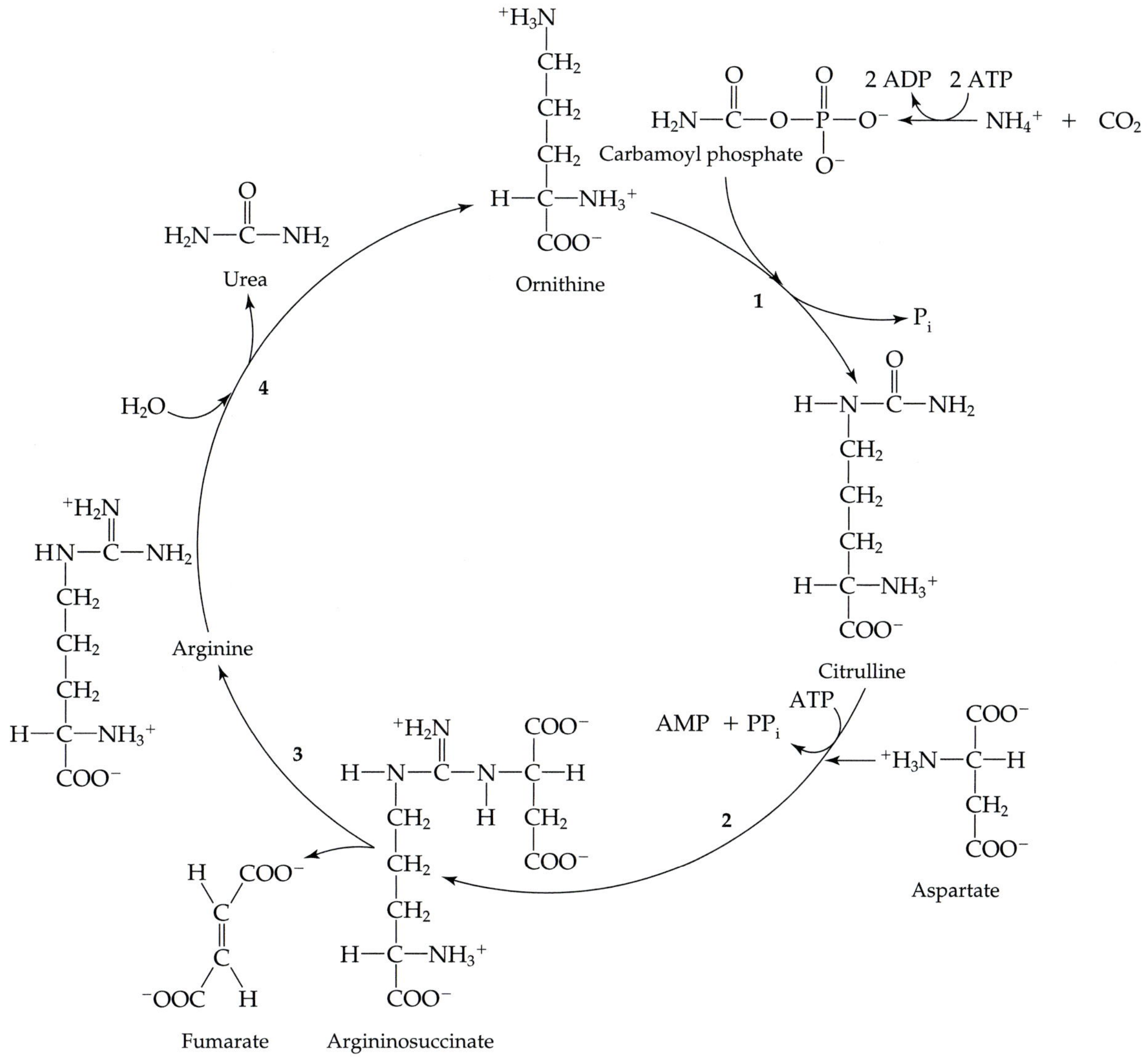

▲ **Figure 27.4**
The urea cycle. The formation of carbamoyl phosphate and reaction 1, in which citrulline is formed, occur in the mitochondrial matrix. Reactions 2–4 occur in the cytoplasm.

urea is transported by the bloodstream to the kidneys and eliminated in the urine. The overall reaction for the urea cycle is:

$$HCO_3^- + NH_4^+ + 3\ ATP + \text{aspartate} + 2\ H_2O \longrightarrow \text{urea} + 2\ ADP + 2\ P_i + AMP + PP_i + \text{fumarate}$$

The kidneys filter about 100 L of blood each day. The normal individual daily excretes about 1.5 L of water, containing approximately 30 g of urea. This value varies greatly from day to day and can rise dramatically (to about 100 g of urea) upon ingestion of a high-protein diet. A long-term decrease in urea excretion is indicative of liver and/or kidney disease.

✓ Review Questions

27.5 What organ is responsible for: **(a)** the synthesis of urea; **(b)** the excretion of urea?

27.6 What is the end product of purine metabolism (see page 752)?

Purine Metabolism

The breakdown of purines in the human body results in the production of uric acid, and very small concentrations (about 0.5 g/day) of this acid are found in the urine and in body fluids.

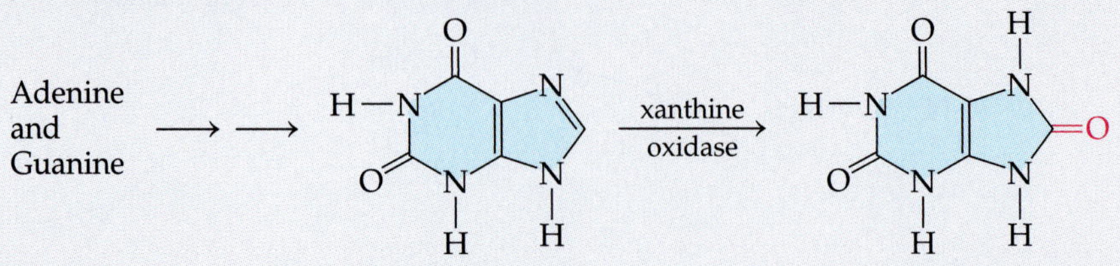

Adenine and Guanine ⟶ ⟶ Xanthine $\xrightarrow[\text{oxidase}]{\text{xanthine}}$ Uric acid

Large quantities of uric acid are produced when purine metabolism is impaired so that too many purines are produced or when uric acid is not properly excreted by the kidneys. In these conditions, the plasma concentration of uric acid becomes abnormally high (>7 mg/100 mL). At physiological pH, the excess uric acid is sparingly soluble and precipitates as the monosodium salt. When such deposits occur in the joints of the body's digital regions, the surrounding tissue can become inflamed, causing a painful arthritic condition known as **gout.** At concentrations of 12–13 mg/100 mL, crystals of monosodium urate may form kidney stones and impair renal function. Thus, kidney failure can be a serious medical consequence of gout. Only 5% of gout patients are women. The reason why men are more likely to develop gout is not known.

Another serious impairment of purine metabolism results from a genetic defect. In *Lesch-Nyhan syndrome,* an enzyme necessary for the normal reutilization of purines is absent. Affected children are mentally defective and exhibit a compulsive, aggressive behavior toward others (extreme hostility), as well as toward themselves (self-mutilation by chewing the tongue, lips, and fingers). The details of the relationship between the absence of the enzyme and the aberrant behavior are still unknown.

27.4 Synthesis of Nonessential Amino Acids

In Section 27.1 we defined essential amino acids as those that the body cannot synthesize in sufficient quantities. Humans are able to synthesize 11 of the 20 amino acids needed for protein synthesis. Three of the nonessential amino acids are synthesized in a single transamination reaction—alanine, aspartate, and glutamate (Section 27.2). Asparagine and glutamine are synthesized in an additional step from aspartate and glutamate, respectively. (The reaction for glutamine synthesis was given in Section 27.3.)

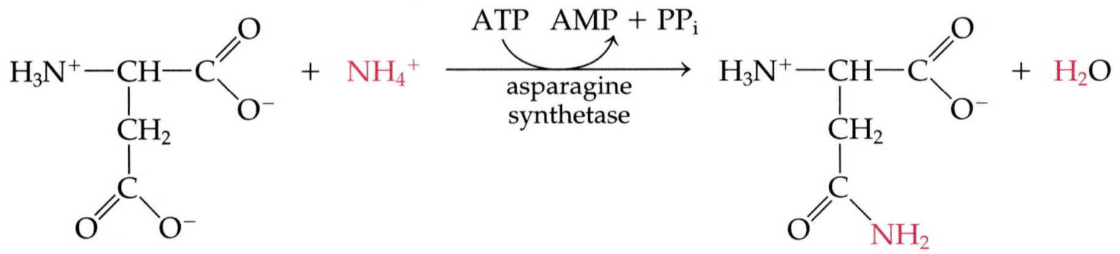

$$H_3N^+{-}CH(CH_2COO^-){-}COO^- + NH_4^+ \xrightarrow[\text{asparagine synthetase}]{\text{ATP} \quad \text{AMP} + \text{PP}_i} H_3N^+{-}CH(CH_2CONH_2){-}COO^- + H_2O$$

Proline is synthesized by a series of four enzyme-catalyzed reactions, starting with glutamate. Serine is synthesized from 3-phosphoglycerate, an intermediate found in glycolysis (Section 25.1), through a series of three reactions. In a subsequent reaction, serine can be converted to glycine.

Some of the nonessential amino acids can be synthesized only if sufficient amounts of an essential amino acid are supplied in the diet. Tyrosine can be synthesized from phenylalanine by a single reaction.

$$\text{C}_6\text{H}_5\text{—CH}_2\text{—}\overset{\text{H}}{\underset{^{+}\text{NH}_3}{\text{C}}}\text{—COO}^- + \tfrac{1}{2}\text{O}_2 \xrightarrow[\text{hydroxylase}]{\text{phenylalanine}} \text{HO—C}_6\text{H}_4\text{—CH}_2\text{—}\overset{\text{H}}{\underset{^{+}\text{NH}_3}{\text{C}}}\text{—COO}^-$$

Phenylalanine Tyrosine

Cysteine is a nonessential amino acid only if the diet contains adequate amounts of methionine. Methionine provides the sulfur for the cysteine side chain, while the remainder of the molecule is obtained from serine.

In the previous section we saw that arginine is synthesized in the urea cycle. Infants and children obtain insufficient quantities of arginine by these reactions and require additional arginine from the diet.

Review Question

27.7 Write out the transamination reaction in which aspartate is synthesized from oxaloacetate.

27.5 Formation of Amino Acid Derivatives

Amino acids act as precursors to many important biomolecules. Among the most complex is heme, an essential component of myoglobin, hemoglobin, and cytochromes (see Section 21.8 and Figure 21.13). Nearly all aerobic cells synthesize heme, using glycine and succinyl-CoA as precursors. The amino acids glycine, aspartate, and glutamine are precursors in the synthesis of the purines and pyrimidines. Primary amines are less complex, but no less important, amino acid derivatives.

Primary amines can be formed by the elimination of carbon dioxide from amino acids in reactions catalyzed by amino acid decarboxylases. Pyridoxal phosphate is the necessary coenzyme (Section F.6):

$$\text{R—}\overset{\text{H}}{\underset{^{+}\text{NH}_3}{\text{C}}}\text{—COO}^- \xrightarrow{\text{decarboxylase}} \text{R—}\overset{\text{H}}{\underset{^{+}\text{NH}_3}{\text{C}}}\text{—H} + \text{CO}_2$$

Some of the amines produced have important physiological effects. Many people are allergic to pollen, dust, insect stings, and so on. By a mechanism not completely understood, the body responds to these foreign substances by decarboxylating histidine to *histamine* (allergic individuals synthesize abnormally high amounts of histidine):

$$\text{Imidazolyl—CH}_2\text{—}\overset{\text{H}}{\underset{^{+}\text{NH}_3}{\text{C}}}\text{—COO}^- \xrightarrow[\text{decarboxylase}]{\text{histidine}} \text{Imidazolyl—CH}_2\text{—}\overset{\text{H}}{\underset{^{+}\text{NH}_3}{\text{C}}}\text{—H} + \text{CO}_2$$

Histidine Histamine

Histamine dilates blood vessels and thus initiates the inflammatory response (see Section E.2). The discomfort associated with hay fever and other allergies is due to the inflammation of the eyes, nose, and throat. Furthermore, the expansion of the blood capillaries causes a decrease in the blood pressure that, if severe enough, may induce shock. (Recall that prostaglandins are also released during an allergic reaction, and they increase the sensitivity to pain associated with inflammation; see Section E.4.)

Several neurotransmitters are synthesized by the decarboxylation of specific amino acids. (Neurotransmitters were discussed in Selected Topic C.) Tyrosine is converted to tyramine by bacterial action:

Antihistamines

An **antihistamine** is a compound structurally similar to histamine. It can occupy the receptor site normally occupied by histamine (e.g., on smooth muscles that surround capillaries), thereby preventing the physiological changes produced by histamine (i.e., they are antagonists). Figure 27.5 illustrates some of the common antihistamines found in nasal decongestants, combination pain relievers, and hay fever preparations. Antihistamines are not effective against colds, although they may temporarily relieve some cold symptoms such as congestion. The most common side effect of antihistamines is sedation, which accounts for their use as major ingredients in most over-the-counter sleeping pills. This effect can impair one's ability to operate machinery or drive a motor vehicle. Terfenadine (Seldane), an antihistamine approved for prescription use in 1985, does not cause drowsiness because the compound does not cross the blood-brain barrier.

▲ A variety of cold and allergy medications contain antihistamines.

There are two types of receptors for histamine, designated as H_1 and H_2. The H_1 receptors are found in the walls of capillaries and in the smooth muscle of the respiratory tract. They affect the vascular changes (dilation and increased permeability of capillaries) and muscular changes (bronchoconstriction) associated with hay fever and asthma. H_1 receptors are blocked by the classical antihistamines (such as Benadryl), which relieve the symptoms of allergies. The H_2 receptors occur mainly in the wall of the stomach, and their activation causes an increased secretion of hydrochloric acid. Cimetidine blocks H_2 receptors and thus, by reducing acid secretion, is an effective drug for people with ulcers.

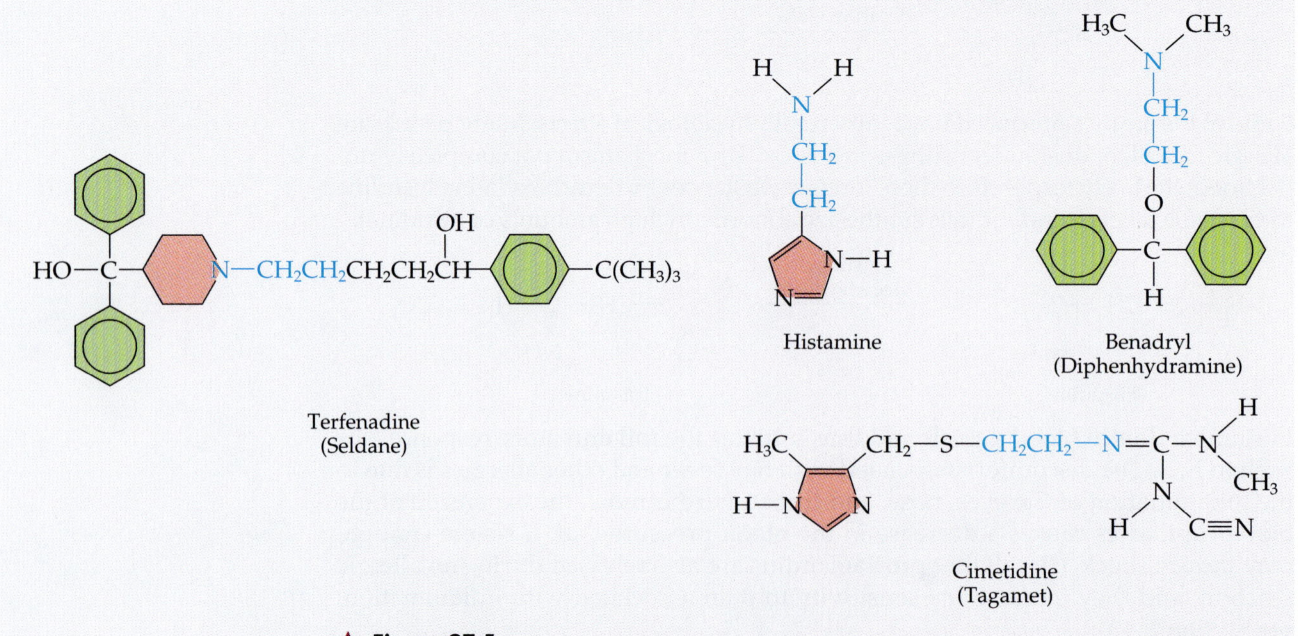

▲ **Figure 27.5**
The antihistamines act as competitive antagonists to histamine. Notice that the structural similarity to histamine is the substituted ethylamine moiety (blue).

$$\text{Tyrosine} \xrightarrow{\text{tyrosine decarboxylase}} \text{Tyramine} + CO_2$$

Tyrosine Tyramine

Tyramine has a physiological action similar to, but weaker than, that of norepinephrine, which it resembles structurally.

Serotonin (5-hydroxytryptamine) is formed by the action of a specific decarboxylase on 5-hydroxytryptophan:

$$\text{5-Hydroxytryptophan} \xrightarrow{\text{5-hydroxytryptophan decarboxylase}} \text{Serotonin} + CO_2$$

5-Hydroxytryptophan Serotonin (5-Hydroxytryptamine)

The cell bodies of serotonin-containing neurons are located almost exclusively in the upper brain stem, from which axons project to other areas of the central nervous system. Serotonin constricts blood vessels, stimulates smooth muscle, and has a potent inhibitory effect on its postsynaptic neurons.

An important inhibitory neurotransmitter, GABA (γ-aminobutyric acid), is formed in the brain and spinal cord from the decarboxylation of glutamate:

At physiological pH, GABA loses H^+ to form γ-aminobutyrate.

$$\text{Glutamate} \xrightarrow{\text{glutamate decarboxylate}} \gamma\text{-Aminobutyrate (GABA)} + CO_2$$

Glutamate γ-Aminobutyrate (GABA)

GABA is thought to inhibit dopamine neurons in particular, as well as other neurons throughout the central nervous system.

Another decarboxylase catalyzes the decarboxylation of 3,4-dihydroxyphenylalanine (L-dopa) to dopamine, which serves not only as a neurotransmitter itself but also as a precursor to norepinephrine, another important neurotransmitter:

$$\text{L-Dopa} \xrightarrow{\text{dopa decarboxylase}} \text{Dopamine} + CO_2$$

3,4-Dihydroxyphenylalanine (L-Dopa) 3,4-Dihydroxyphenylethylamine (Dopamine)

A deficiency of dopamine in the brain cells is a primary cause of Parkinson's disease, a disorder of the central nervous system that involves a progressive paralytic rigidity, tremors of the extremities, and unresponsiveness to external stimuli. Dopamine itself cannot be administered because it does not pass across the blood-brain barrier. A major breakthrough in the treatment of Parkinson's disease has been the use of L-dopa. (The L enantiomer is more effective and less toxic than the D form of the drug.) Large doses of L-dopa are administered orally; the drug is able to pass from the digestive system into the blood and then cross the blood-brain barrier. L-Dopa is decarboxylated to dopamine in the brain. (Incidentally, numerous studies have linked schizophrenia to an *overabundance* of dopamine in brain cells.)

Review Question

27.8 How are each of the following neurotransmitters formed in the body: **(a)** γ-aminobutyrate; **(b)** serotonin?

27.6 Relationships Among the Metabolic Pathways

A great variety of organic compounds can be derived from carbohydrates, lipids, and proteins. A brief summary of the interrelationships of the major metabolic pathways is given in Figure 27.6. All of these metabolic pathways must be coordinately regulated to keep the body supplied with needed energy, amino acids, and other compounds. We'll examine how these pathways are regulated under two sets of opposing conditions: immediately after a full meal (like Thanksgiving dinner) and after a period of total fasting.

After a meal blood glucose levels rise and insulin is secreted by the pancreas. The major biochemical actions of insulin are summarized in Table 27.4. These lead to a decrease in the blood glucose level and formation of glycogen and triglycerides.

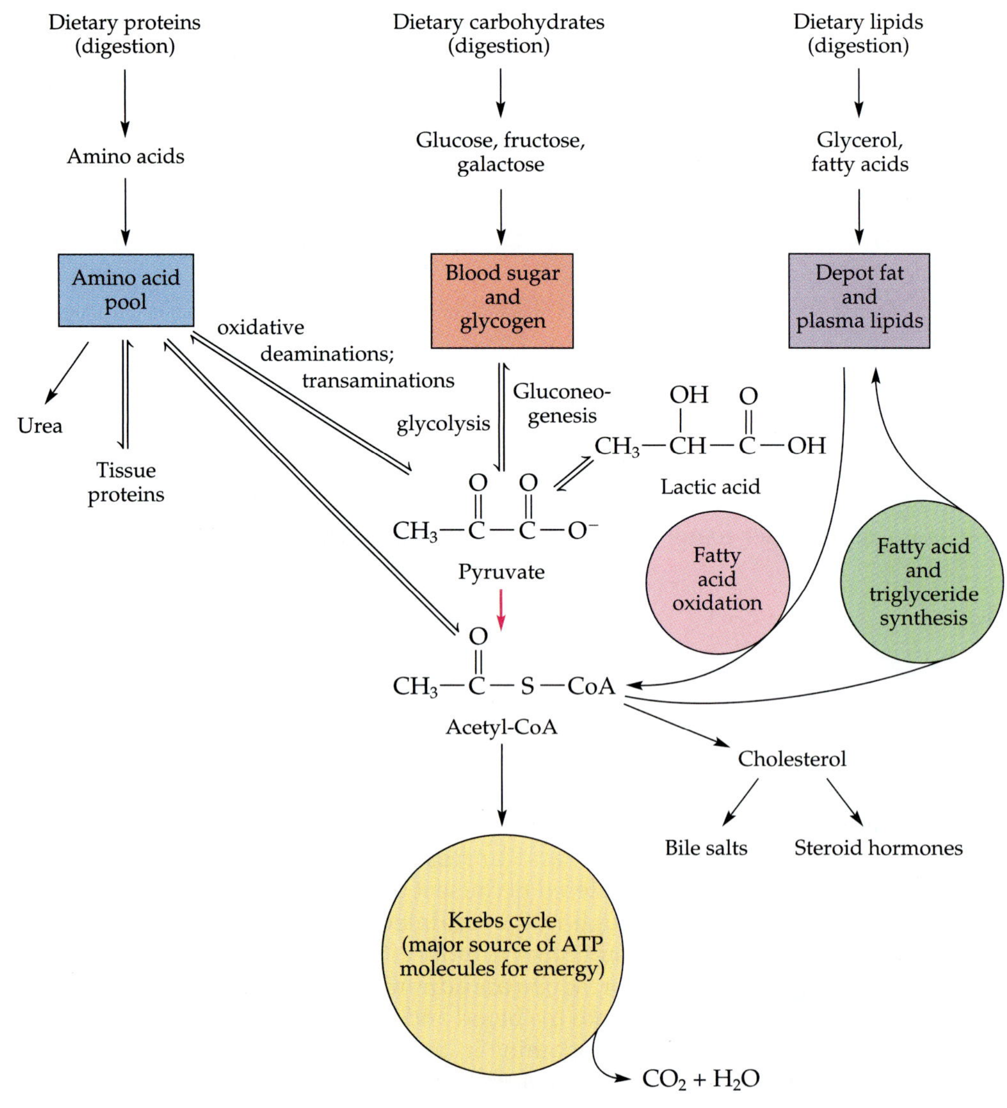

▶ Figure 27.6 Interrelationships of metabolic pathways.

Table 27.4 Major Hormones Controlling Fuel Metabolism in Mammals

Hormone	Biochemical Actions	Physiological Actions
Insulin	↑ Cell permeability to glucose (in muscle and adipose tissue) ↑ Glycolysis ↑ Glycogen synthesis ↑ Triglyceride synthesis ↓ Gluconeogenesis ↓ Triglyceride degradation ↓ Protein degradation ↑ Protxein, DNA, and RNA synthesis	Signals fed state ↓ Blood glucose level ↑ Fuel storage ↑ Cell growth and differentiation
Glucagon	↑ cAMP level in liver and adipose tissue ↑ Glycogenolysis ↑ Glycogen synthesis ↑ Triglyceride hydrolysis ↑ Gluconeogenesis ↑ Glycolysis	↑ Glucose release from liver ↑ Blood glucose level
Epinephrine	↑ cAMP level in muscle ↑ Triglyceride hydrolysis ↑ Glycogenolysis ↑ Glycogen synthesis	↑ Glucose release from liver ↓ Glucose use by muscle ↑ Blood glucose level

When the human body is totally deprived of food, whether voluntarily or involuntarily, the condition is known as **starvation.** During total fasting, the body's glycogen stores are depleted rapidly and the blood glucose level drops, leading to the release of glucagon. The biochemical actions that result are also summarized in Table 27.4.

The preferred energy source for brain cells is glucose. If none is available in the diet, cells make it from glucogenic amino acids obtained by degrading tissue protein (e.g., proteins from skeletal muscles and from membranes that line the digestive tract). A starving human may lose as much as 6% of her or his muscle mass per day. The loss of plasma proteins, especially albumins, occurs to an even greater extent. The nitrogenous metabolites from protein catabolism must be converted to urea and excreted through the kidneys, which requires large volumes of water.

After several days of starvation, the rate of protein breakdown slows considerably as the brain adjusts to using ketone bodies (Section 26.4) for some of its energy and other organs such as the heart and liver utilize the body's fat reserves for energy (either directly or from ketone bodies). When no more fat reserves remain, the body must again draw heavily on structural protein for its energy requirements. The emaciated appearance of a starving individual is due to depleted muscle protein.

Even low-carbohydrate diets high in complete proteins are hard on the body, which must rid itself of the nitrogen compounds—ammonia and urea—formed by the breakdown of proteins. This puts increased stress on the liver, where the waste products are formed (Section 27.3).

Involuntary starvation is a serious problem in much of the world. Even so, starvation is seldom the sole cause of death. The antibodies (Section 28.5) that are needed to remove foreign substances such as disease-causing bacteria or viruses are proteins. If the body is breaking down proteins for energy, it is not able to synthesize

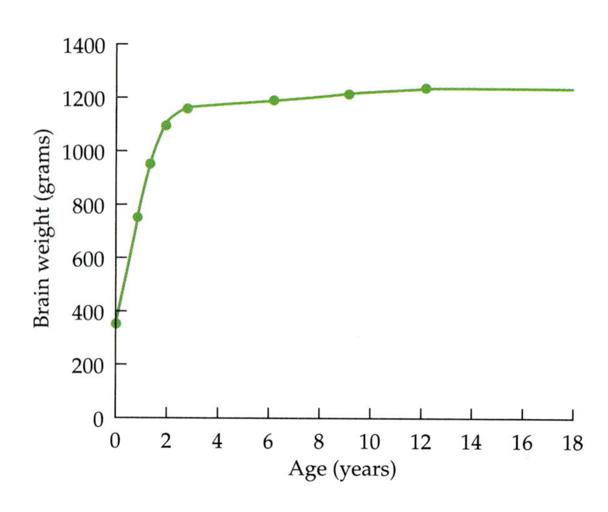

Figure 27.7
Growth of the human brain according to age.

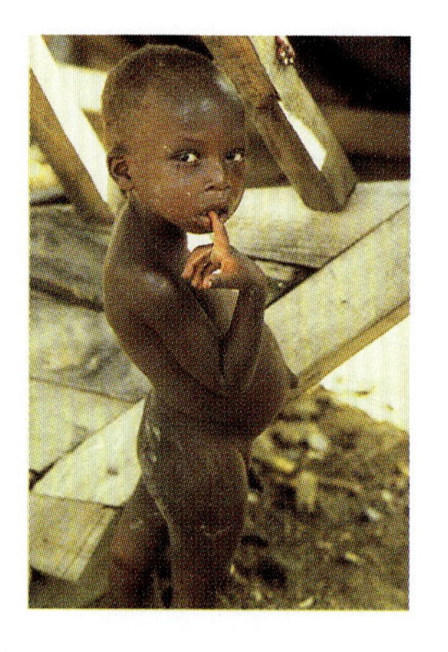

▲ An extreme lack of dietary protein causes a deficiency disease known as kwashiorkor.

these antibodies. Thus even usually minor disease, such as chicken pox and measles, become life-threatening disorders.

A similar situation occurs in the case of the protein deficiency disease kwashiorkor. Kwashiorkor results in extreme emaciation, bloated abdomen, mental apathy, diarrhea, lack of pigmentation of the skin and hair, and eventual death. This disease is prevalent in Latin America, Asia, and Africa, where corn or rice is the major food. It is said to be the most severe and widespread nutritional disorder among young children. Kwashiorkor can best be treated by the administration of adequate amounts of well-balanced protein. The problem, of course, is that complete proteins are a scarce commodity in many of these areas.

Nutrition is especially important in a child's early years. Protein deficiency leads to both physical and mental retardation. The effect on a human's mental capacity is readily apparent from a consideration of Figure 27.7, which shows that the human brain reaches nearly full size by the age of 2 years.

✓ Review Question

27.9 What is the difference between starvation and the disease known as kwashiorkor?

Summary

The body uses amino acids derived from food to make new tissues and other nitrogen-containing substances. At the same time, the body is breaking down old tissues and nitrogen-containing compounds. The relationship between these two processes, one of gain and one of loss, is called **nitrogen balance.** When nitrogen intake is greater than nitrogen loss and tissue is being built, the organism is in *positive nitrogen balance.* When nitrogen intake is less than nitrogen loss and body tissue is broken down in order to replenish the body's amino acid supply, the organism is in *negative nitrogen balance.*

Turnover rate is the amount of protein degraded per unit time and represents the average residence time for a protein molecule in a tissue.

Essential amino acids are those an organism needs but cannot synthesize for itself. These amino acids must be obtained from the diet. There are nine essential amino acids for human adults. A **complete protein** is one that contains all nine; an **incomplete protein** is one that lacks one or more.

Proteins are not stored in the body as carbohydrates and fats are. Instead, the amino acids obtained from food, the nonessential amino acids the body synthesizes, and those released as the body breaks down tissue proteins all circulate in the blood in the **amino acid pool.** The body takes from this pool the amino acids it needs to synthesize new proteins needed by the cells and new nonprotein molecules that contain nitrogen.

The amino acids from the pool can be catabolized to provide energy. The carbon skeleton is sent to the Krebs cycle and the amino group is ultimately excreted in the urine. Amino acids whose carbon skeletons are converted to intermediates that can be converted to glucose through gluconeogenesis are known as **glucogenic amino acids.** Those amino acids whose carbon skeletons are broken down to compounds used to form ketone bodies are known as **ketogenic amino acids.**

The first step in amino acid catabolism is separation of the amino group from the carbon skeleton. In a **transami-**

nation, the amino acid gives its —NH_2 to pyruvate, α-ketoglutarate, or oxaloacetate. The products of this reaction are a new amino acid and an α-keto acid containing the carbon skeleton of the original amino acid. Pyruvate is transaminated to alanine, α-ketoglutarate to glutamate, and oxaloacetate to aspartate. The amino groups used to form alanine and aspartate are ultimately transferred to α-ketoglutarate, forming glutamate. The glutamate then undergoes **oxidative deamination** to yield α-ketoglutarate and ammonia.

A small percentage of the ammonia formed in amino acid deamination reacts with glutamate to form glutamine. This substance can be stored in the body and so represents a way of storing excess nitrogen for a short time. Ammonia is toxic; thus, the amount of ammonia in the body must be carefully regulated. Humans incorporate excess ammonia into urea through a series of reactions known as the **urea cycle.**

Gout is a painful arthritis that develops when levels of uric acid, formed in the degradation of purines, become elevated due to an impairment of purine metabolism.

Some amino acids are *decarboxylated* to primary amines such as histamine and γ-aminobutyric acid, which serve a variety of functions in the body.

All of the metabolic pathways in the body must be coordinately regulated to keep the body supplied with energy, amino acids, and other compounds. When the body is totally deprived of food, whether voluntarily or involuntarily, the condition is known as **starvation.** For a short period of time, the body adapts to the absence of food, but long-term starvation leads to death, either directly or from diseases that become deadly due to an impaired immune system.

Key Terms

amino acid pool (27.1)
antihistamine (27.5)
complete protein (27.1)
essential amino acid (27.1)
glucogenic amino acid (27.2)
gout (27.3)
incomplete protein (27.1)
ketogenic amino acid (27.2)
nitrogen balance (page 744)
oxidative deamination (27.2)
starvation (27.6)
transamination (27.2)
turnover rate (27.1)
urea cycle (27.3)

Problems

Amino Acid Metabolism

1. a. What generalization can be made about the structures of the essential amino acids?
   b. What is the effect of a deficiency of one or more essential amino acids?
2. What are three functions of amino acids in the body?
3. a. Why is there a negative nitrogen balance during fasting?
   b. Why is there a positive nitrogen balance during pregnancy?
4. Why should a vegetarian obtain his or her protein from a wide variety of plants?

Catabolism of Amino Acids

5. Write the equation for transamination between phenylalanine and pyruvate.
6. The compound shown below can supply the body with an essential amino acid. Show, with an equation, how this is possible.

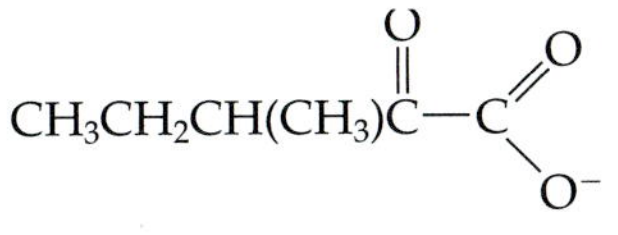

$$CH_3CH_2CH(CH_3)\overset{\overset{\displaystyle O}{\|}}{C}-C\begin{matrix}\nearrow O \\ \searrow O^-\end{matrix}$$

7. What reaction is catalyzed by the enzyme:
   a. GOT? b. GPT?
8. What product is formed by oxidative deamination of glutamate?
9. Identify each of the following amino acids as glucogenic, ketogenic, or both.
   a. phenylalanine b. leucine c. serine
10. Identify each of the following amino acids as glucogenic, ketogenic, or both.
   a. asparagine b. tyrosine c. valine

Storage and Excretion of Nitrogen

11. a. What compound serves as the temporary storage form of nitrogen in the body?
   b. Write the equation for its formation.
12. a. What toxic compound is formed in oxidative deamination?
   b. How does the body get rid of this compound?
13. The carbon and each of the two nitrogens found in urea are obtained from what compounds?
14. a. What amino acid is needed by the urea cycle?
   b. What amino acid is formed in the urea cycle?

Synthesis of Nonessential Amino Acids and Amino Acid Derivatives

15. a. Cysteine is a nonessential amino acid only if what essential amino acid is present in adequate amounts?
   b. Tyrosine is a nonessential amino acid only if what essential amino acid is present in adequate amounts?
16. What nonessential amino acids are synthesized from:
   a. glutamate? b. aspartate? c. serine?
17. Write equations for the formation of each of the following compounds from an amino acid:
   a. histamine b. tyramine c. dopamine
18. A primary cause of Parkinson's disease is a deficiency of dopamine. Why are patients treated with L-dopa, rather than dopamine?

Relationships Among the Metabolic Pathways

19. Which of the following pathways would you expect to be activated by the release of glucagon? By release of insulin?
    a. glycolysis b. triglyceride degradation
    c. glycogen synthesis
20. Which of the following pathways would you expect to be activated by the release of glucagon? By release of insulin?
    a. glycogenolysis b. fatty acid synthesis
    c. gluconeogenesis
21. In the first few days without food, the body rapidly breaks down body proteins. After a few days the rate of protein degradation slows. What events lead to a slowing of the rate of protein degradation?

Additional Problems

22. The RDA for protein is about 0.8 g per kg of body weight. How much protein is required each day by a 125-kg football player?
23. If 1 mol of alanine is converted to pyruvate in a muscle cell and the pyruvate is then metabolized via the Krebs cycle, how many moles of ATP are produced?
24. If the essential amino acid leucine (2-amino-4-methylpentanoic acid) is lacking in the diet, a keto acid can substitute for it. Give the structure of the keto acid and the reaction in which leucine is formed.
25. What is the significance of transamination in the metabolism of amino acids?
26. What compound in the urea cycle serves a role analogous to the role of oxaloacetate in the Krebs cycle?
27. Insulin is released when blood glucose levels are elevated, and brings about a lowering of the blood glucose level. When insulin is released, fatty acid synthesis is activated. How does this lead to a lowering of glucose levels?
28. Why is a person with alkaptonuria placed on a diet with minimal amounts of phenylalanine?
29. **a.** What type of compound is Benadryl?
    b. How does it exert its effect?
    c. What is the difference between an H_1 receptor and an H_2 receptor?

Body Fluids

CHAPTER 28

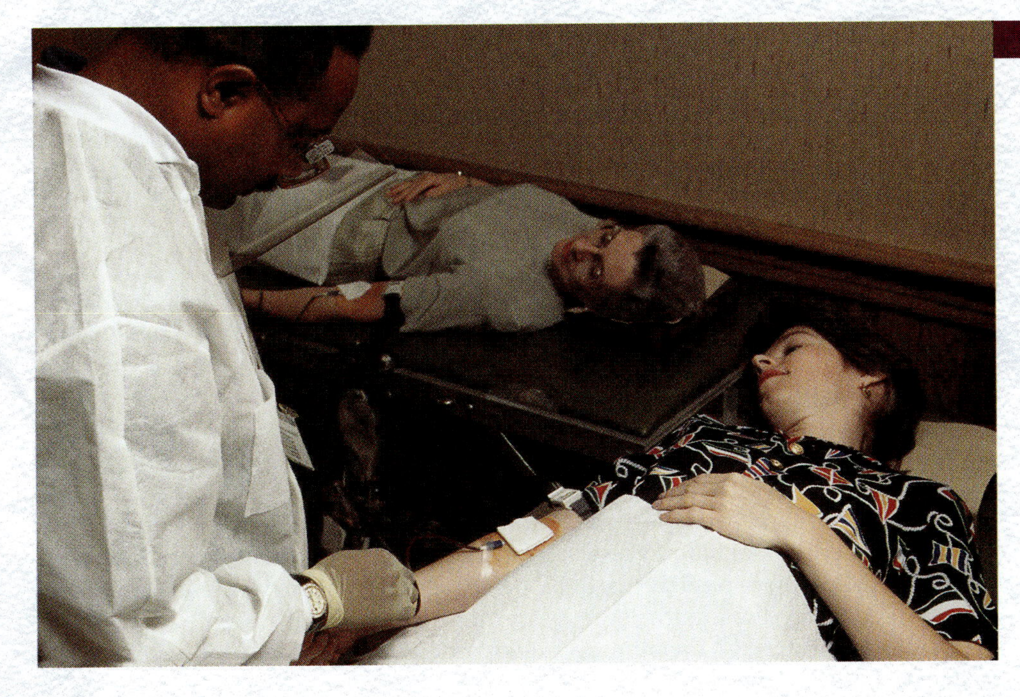

About 14 million units of blood are donated in the United States each year. On any given day, approximately 40,000 units of blood products are needed for people involved in accidents, undergoing surgery, or receiving treatment for a variety of blood disorders.

Learning Objectives/Study Questions

1. What are the major components of whole blood? What is the function of each component?
2. How are oxygen and carbon dioxide transported throughout the body?
3. How do buffers in the blood maintain a fairly constant pH?
4. How is a blood clot formed?
5. What are antibodies, and what is their role in the immune response?
6. What is blood pressure? What is hypertension?
7. What is the role of the kidneys in waste removal?
8. How does sweat help to regulate the temperature of the body?
9. Why is milk such an important food for infants?

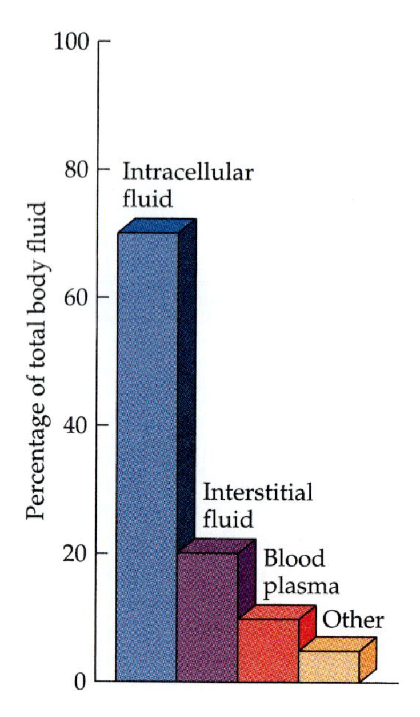

▲ **Figure 28.1**
Distribution of body fluids.

Most of the reactions that occur in the body take place in aqueous solutions. (The body is about 60% water, by mass.) The composition of the various body fluids are critically important. These fluids carry nutrients and chemical signals to cells and remove and dispose of waste products. They are categorized as **intracellular fluid**—fluid located inside cells, or **extracellular fluid**—fluid located outside cells. The two major extracellular fluids are interstitial fluid and blood plasma. **Interstitial fluid** is the fluid that surrounds most cells and fills any space between them. Other extracellular fluids, found in lesser amounts, are urine, lymph, digestive juices, and cerebrospinal fluid. The distribution of body fluids is shown in Figure 28.1.

The chemical compositions of interstitial fluid and blood plasma are quite similar, except plasma contains more protein than does interstitial fluid. The ion profiles of these two extracellular fluids differ markedly from that of intracellular fluid. Figure 28.2 depicts their chemical composition. Note the differences in the amounts of Na^+, K^+, Cl^-, and HPO_4^{2-} ions between intracellular and extracellular fluids.

The body fluids of which you are composed are not static—they constantly renew themselves, and their compositions depend on the state of your health. Analysis of body fluids represents one of the most powerful diagnostic techniques available to medical personnel.

28.1 Blood: Functions and Composition

Blood is the principal transport medium of the human body. It moves through a 100,000-km network of blood vessels. Some of these vessels are so small that blood cells have to line up to pass through. Blood carries (1) oxygen from the lungs to the tissues, (2) carbon dioxide from the tissues to the lungs, (3) nutrients from the intestines to the tissues, (4) metabolic wastes from the tissues to the excretory organs, (5) hormones from the endocrine glands to their target tissues, and (6) three major kinds of blood cells. In addition, blood helps maintain a fairly constant body temperature and the correct acid-base, electrolyte, and water balances. In short, blood is a rather remarkable substance. And we've only mentioned a few highlights!

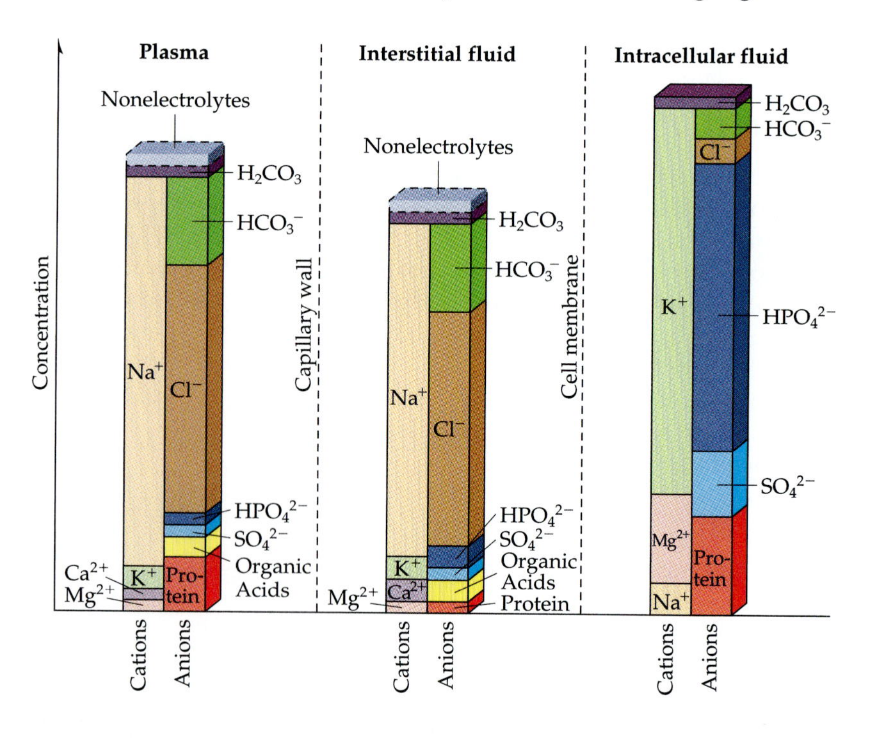

► **Figure 28.2**
Relative chemical composition of the major body fluids; plasma, interstitial fluid, and intracellular fluid.

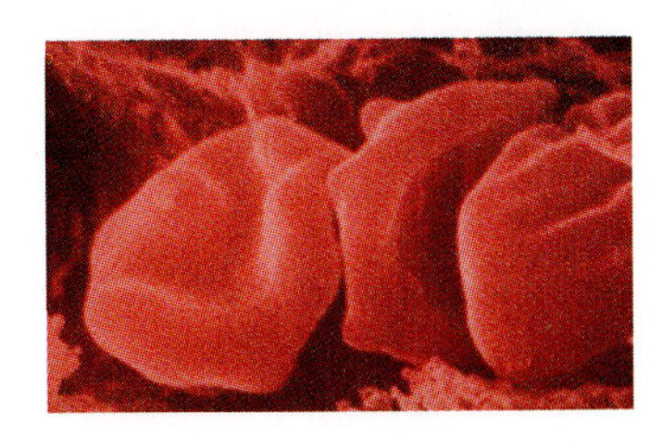

▲ Erythrocytes passing through a narrow capillary.

Blood accounts for about one-twelfth of the body weight of the average individual. The volume of blood varies with body size. A 150-lb (68-kg) human has about 5 L of blood. Moderate amounts lost through bleeding or blood donation are readily replaced. The medical condition called **shock** results from a much greater loss of fluid from the vascular system. This reduces the blood volume enough to cause a dramatic decrease in blood pressure and a consequent decrease in oxygen-transporting capability with potentially fatal results. Treatment of the condition involves bringing the blood volume back to normal levels.

The composition of a typical sample of whole blood is shown in Figure 28.3. Whole blood consists of a straw-colored liquid portion known as the **plasma** and the **formed elements: erythrocytes** (red blood cells), **leukocytes** (white blood cells), and **platelets**[1].

Plasma is an extremely complex solution, containing plasma proteins, organic nutrients (amino acids, carbohydrates, hormones, lipids, and vitamins), electrolytes (inorganic ions), and organic wastes (*i.e.,* urea and bilirubin). Although these compounds and ions are continuously entering and leaving the circulatory system, the overall composition of the plasma remains remarkably constant (a state of dynamic equilibrium exists).

Plasma Proteins

More than 100 proteins have been identified in the plasma; most of them are synthesized in the liver. These proteins remain in the circulatory system and ordinarily are not used as sources of energy. At times of protein deprivation, the plasma protein concentration is maintained at the expense of tissue protein. Plasma proteins are grouped into three main classes on the basis of their solubility properties and methods of isolation.

Albumins are the most abundant group of proteins. Their major function is to maintain osmotic pressure (see the boxed essay "Osmotic Pressure"). They are also important for their buffering capacity and for the transport of fatty acids, certain metal ions, and many drugs (particularly nonpolar ones). Normal concentrations of albumins range from 3.5 to 4.5 g per 100 mL of plasma.

Globulins have higher molar masses than the albumins (150,000 as compared to 70,000). Three subclasses of globulins are recognized. α-Globulins and β-globulins are synthesized in the liver. They form complexes (*i.e.,* VLDLs, LDLs, and HDLs) with the water-insoluble lipids and transport them through the aqueous media. The γ-globulins (also known as immunoglobulins or antibodies) are synthesized by leukocytes in the lymph nodes. They play a vital role in combating many infectious diseases (Section 28.5).

Fibrinogen is a large protein (MW 340,000) consisting of six polypeptide chains. It is synthesized in the liver and functions in blood coagulation (Section 28.4). The fibrinogen content of plasma increases when inflammatory or infectious conditions exist and during menstruation and pregnancy.

When fibrinogen and the formed elements are removed from the plasma (by centrifugation), the liquid that remains is called **serum.** The only distinction between blood plasma and blood serum is the presence of fibrinogen in plasma. *Because blood serum lacks fibrinogen, it is unable to clot.*

Erythrocytes

Mammalian erythrocytes are disk-shaped cells with a slight depression at the center (Figure 28.5). They are formed in the red bone marrow, and are the most numerous of the formed elements (Figure 28.3). Unlike most other cells, they contain neither

[1]Platelets in nonmammalian vertebrates are nucleated cells called thrombocytes. Because in humans they are cell fragments rather than individual cells, the term platelet is preferred.

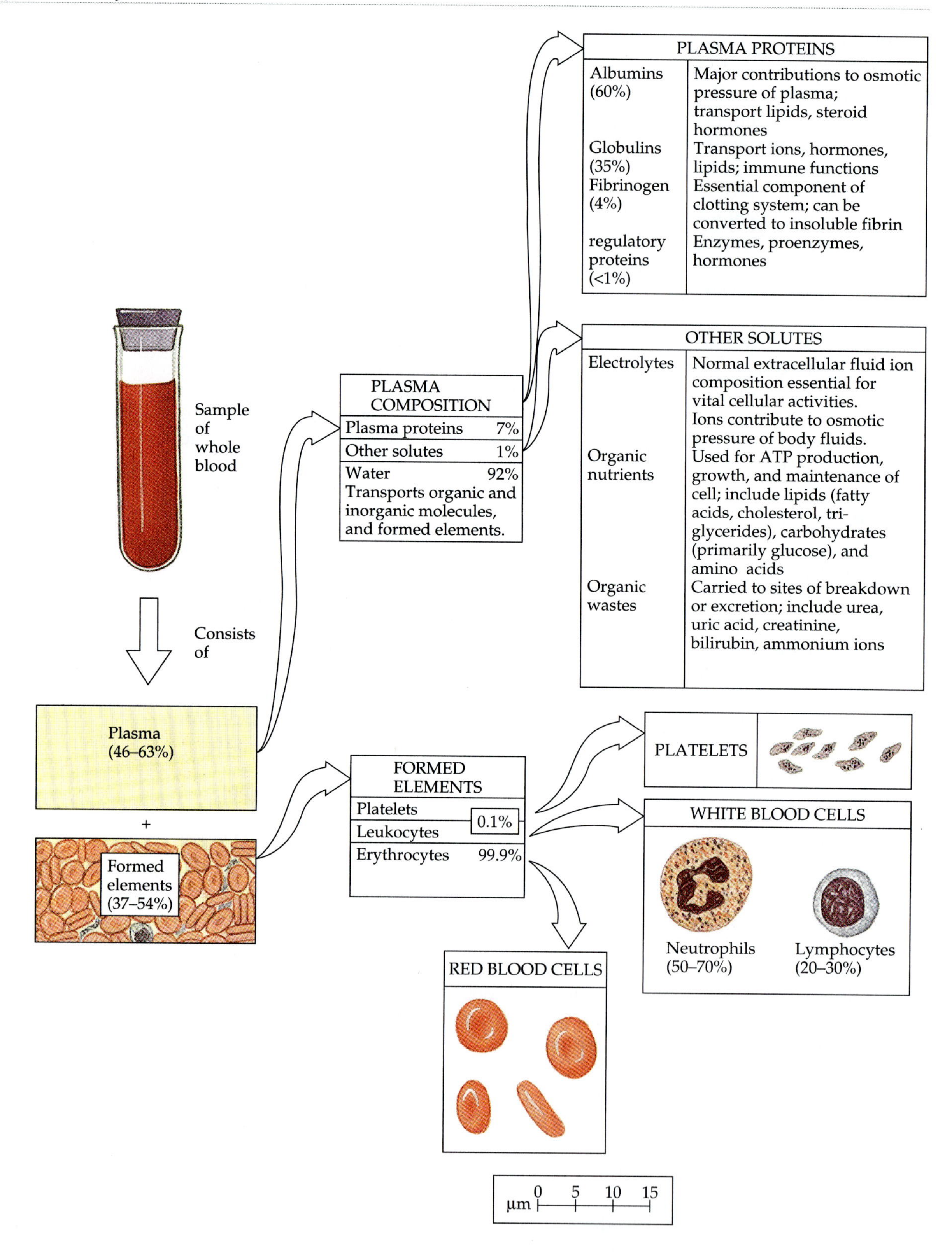

▲ **Figure 28.3**
The composition of a typical sample of whole blood.

Osmotic Pressure

How does material get from the blood to the cells? Water, electrolytes, glucose, amino acids, and other materials all diffuse rapidly through pores in the capillary walls. This diffusion occurs in both directions—from the blood into the interstitial fluid outside the capillary and from the interstitial fluid back into the blood. Ultimately material makes its way from the interstitial fluid through cell membranes and into (or out of) tissue cells.

The pressure that keeps blood moving around the circulatory system also has a tendency to push blood plasma out of the porous capillaries and into the interstitial space. In the absence of any counter effect, the rapid diffusion of material back and forth between blood and interstitial fluid would be accompanied by a slow but steady net loss of liquid from the blood to the interstitial space. Blood volume would drop, and tissues would swell with the extra fluid. There is, however, a counter effect—osmotic pressure (Section 8.9).

The concentration of protein in the plasma far exceeds the concentration of protein in the interstitial fluid outside the blood vessels (Figure 28.2). This concentration gradient results in an osmotic pressure of about 25 mmHg. Therefore, if no external force were applied, liquids would be expected to diffuse from the interstitial fluid into the bloodstream. However, the pumping action of the heart creates *blood pressure* (Section 28.6). This pressure is greater at the arterial end of a capillary (~32 mmHg) than at the venous end (~17 mmHg). Because the blood pressure at the arterial end is higher than the osmotic pressure, the natural tendency is reversed and there is a net flow *from* the capillary *into* the interstitial fluid. The fluid that leaves the capillary contains the dissolved nutrients, oxygen, hormones, and vitamins needed by the tissue cells. As the blood moves along the capillary branches, the blood pressure decreases until at the venous end the osmotic pressure is greater than the blood pressure and there is a net flow of fluid *from* the interstitial fluid *into* the capillary. The incoming fluid contains the metabolic waste products such as carbon dioxide and excess water (Figure 28.4). If we are in good health, everything balances nicely. Nutrients have diffused from blood to interstitial fluid and wastes have diffused from interstitial fluid to blood, but the volume of fluid in the two systems has remained constant.

Osmotic pressure, and hence the delicate balance of fluid exchange, is directly related to the concentration of albumins in the plasma. If the albumin level is low, as might be the case from (1) malnutrition (low protein intake), (2) abnormal protein synthesis (liver disease), or (3) the loss of protein in the urine as a result of kidney disease, the osmotic pressure decreases. This results in a net efflux of fluids from the capillaries into the interstitial and cellular regions. This abnormal accumulation of fluids within the interstitial space produces noticeable swelling, particularly in the lower extremities. This condition is known as *edema.*

You'll recall that children who suffer from the protein deficiency disease kwashiorkor (Section 27.6) characteristically have bloated abdomens. This swelling is caused by the accumulation of water, which leaves the blood because there are insufficient albumins to maintain the osmotic pressure of the blood.

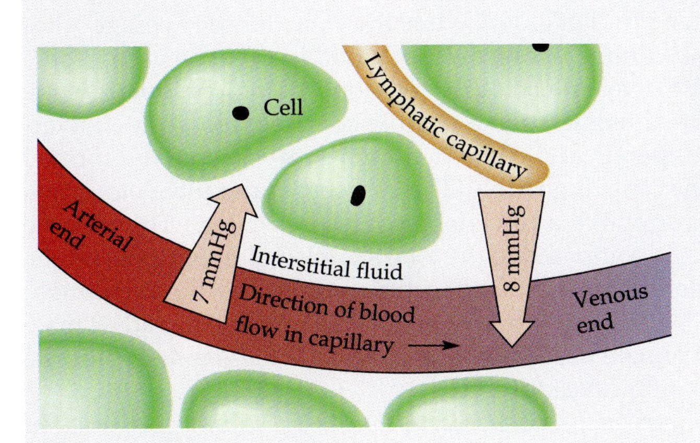

Figure 28.4 Oxygen and nutrients leave at the arterial ends of the capillaries. Carbon dioxide and other cellular waste products enter the venous ends of the capillaries.

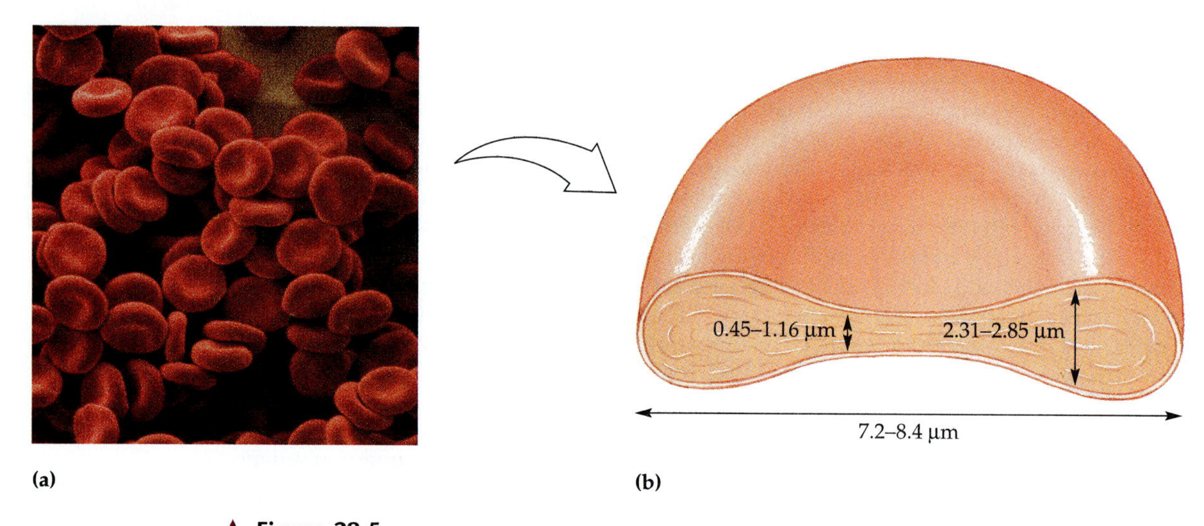

▲ **Figure 28.5**
(a) A scanning electron micrograph of erythrocytes shows their three-dimensional structure quite clearly (× 1195). (b) A sectional view of a mature erythrocyte showing average dimensions.

mitochondria nor a nucleus. (The nucleus is lost during the development and maturation of the erythrocyte.) Their most significant component is the protein hemoglobin (Section 21.9). Other important features of erythrocytes are specific short-chain polysaccharides bound to membrane proteins. These determine an individual's blood type (see the boxed essay "Blood Types.")

Because erythrocytes have no mitochondria, they must obtain all the energy they need from the anaerobic degradation of glucose through glycolysis and an alternate metabolic pathway called the pentose phosphate pathway (see an advanced biochemistry text). Because erythrocytes do not have nuclei, they cannot reproduce and are unable to synthesize new proteins. In humans, they have a life span of about four months. To maintain a constant level, new erythrocytes are formed in the bone marrow at the same rate that old ones are eliminated by special tissues in the liver and spleen. It has been estimated that of the approximately 30 trillion erythrocytes in an average adult male, about 3 million are destroyed each second. Assuming that there are 300 million hemoglobin molecules in each erythrocyte, 900 trillion molecules of hemoglobin must be synthesized every second (by cells in the bone marrow) in order to maintain a constant supply!

The average diet supplies about 12 to 15 mg of iron per day, only about 10% of which is absorbed. To maintain sufficient iron for hemoglobin synthesis, the body must retain the 20 to 25 mg of iron that is released each day by the destruction of erythrocytes. This is done by binding the iron to iron-storage proteins.

The blood of the average adult female contains about 5 million erythrocytes in every μL (or cubic millimeter, mm^3) of blood, while the value for the average adult male is 5.5 million/μL. (The volume of one drop of blood is about 50 μL, so there are about 250 million erythrocytes in each drop of blood.) The **hematocrit** is the volume (in percent) of formed elements in a sample of whole blood. It is determined by centrifuging a sample of whole blood at high speed to spin the formed elements to the bottom of the centrifuge tube. (The leukocytes and platelets form a very thin, light-colored layer above a thick layer of erythrocytes.) The volume occupied by the erythrocytes is compared to the total sample volume. For example, if 1.00 mL of whole blood contains 0.47 mL of erythrocytes (following centrifugation), the hematocrit is $(0.47/1.00) \times 100 = 47$. The normal range for an adult male is 40 –54, while for an adult female it is 37–47. Values outside these ranges indicate the existence of certain pathological conditions.

Anemia occurs when *the percentage of erythrocytes (and/or the percentage of hemoglobin) is abnormally low.* Anemia may result from (1) a decreased rate of erythrocyte production (aplastic anemia), (2) an increased rate of erythrocyte destruction (hemolytic anemia), or (3) an increased rate of erythrocyte loss (as in hemorrhaging). *Polycythemia* is a condition arising from the overproduction of erythrocytes.

Leukocytes

The composition of leukocytes resembles that of other tissue cells. They are nucleated, and contain glucose, lipids, proteins, and other soluble organic substances and inorganic salts. Leukocytes constitute the body's primary defenders against foreign organisms (*e.g.*, viruses, bacteria) (Section 28.5).

The different varieties of leukocytes have specialized functions. *Lymphocytes* are involved in the synthesis and storage of antibodies. *Phagocytes (macrophages)* are attracted to sites of inflammation by chemicals released from injured tissue. They contain lysosomal enzymes, and their function is to engulf and digest the invading organisms.

Billions of leukocytes are produced each day in the bone marrow to replace the ones that die. On the average, there are about 7000 leukocytes per μL of blood, but this value is subject to considerable variation. A higher-than-normal leukocyte count occurs during acute infections, such as *appendicitis* (16,000 to 20,000 per μL). High numbers of leukocytes may also appear during emotional disturbances and following vigorous exercise and/or excessive loss of body fluids. Viral diseases, such as chicken pox, influenza, measles, mumps, and polio, are accompanied by an abnormally low leukocyte count (<5000 per μL) because in fighting the viruses the leukocytes are killed faster than they can be produced. **Leukemia** is a cancer characterized by the uncontrolled production of leukocytes that fail to mature. Despite their numbers, these cells are unable to destroy invading pathogens, and the person has a lowered resistance to infections.

Platelets

There are about 350,000 platelets in every μL of blood. These small cell fragments contain proteins and relatively large amounts of phospholipids. They liberate proteins and other factors that are instrumental in blood clotting (Section 28.4). An abnormally low platelet count (<100,000 per μL) is related to a tendency to bleed.

✓ Review Questions

28.1 List five functions of the blood.

28.2 What is shock?

28.3 Name the three formed elements of the blood, and give a function for each.

28.2 Blood Gases

One of the main functions of blood is the transport of oxygen and carbon dioxide. The protein hemoglobin is a key component in this process. As discussed in Section 21.9, hemoglobin is composed of four polypeptide chains (two α subunits and two β subunits) each bound to a molecule of heme. Heme is a planar molecule containing a central Fe^{2+} to which a molecule of oxygen can be reversibly bound to form oxyhemoglobin. Because each hemoglobin molecule has four heme groups, it can transport a maximum of four molecules of oxygen. The binding of the first oxygen molecule to hemoglobin leads to a change in the tertiary structure of hemoglobin that facilitates the attachment of oxygen to the other subunits. This phenomenon is known as *positive cooperativity.*

Hemoglobin without bound oxygen is known as deoxyhemoglobin.

Hemoglobin makes up about 90% of the total protein of an erythrocyte. The characteristic red color of blood is due entirely to the presence of hemoglobin or, more precisely, to the presence of the heme groups, which absorb strongly in the blue region of the spectrum (~400 nm). Normally there is about 15 g of hemoglobin per 100 mL of blood. This amount of hemoglobin can combine with about 20 mL of gaseous oxygen (at STP). Without the hemoglobin, only 0.3 mL of gaseous oxygen could physically dissolve in 100 mL of plasma.

The human body requires an enormous amount of oxygen for cellular respiration—the linked processes of electron transport and oxidative phosphorylation (Section 24.5). The hemoglobin molecule is well suited to meet these demands because of its affinity for oxygen and because the attachment of oxygen to heme is readily reversible. In the alveoli of the lungs, hemoglobin comes into direct contact with a rich supply of oxygen (partial pressure of about 90–100 mmHg) and is converted to oxyhemoglobin. The oxyhemoglobin is carried by the arterial circulation to the cells in which there is a low oxygen concentration (25–40 mmHg) and a relatively high concentration of carbon dioxide (~60 mmHg). Oxygen is released to the

Sickle Cell Anemia

Many of the several hundred hemoglobin variants identified in humans lead to known diseases. Perhaps the most notorious variant is the one responsible for sickle cell anemia, an inherited disease that, if left untreated, may be fatal (due to infection, blood clots, or cardiac or renal failure). Sickle cell hemoglobin (HbS) differs from ordinary hemoglobin in only one amino acid—at the sixth position of the β chain—where HbS has valine rather than glutamic acid:

Normal Hemoglobin

Val-His-Leu-Thr-Pro-Glu-Glu-Lys . . .
1 2 3 4 5 6 7 8

Sickle Cell Hemoglobin

Val-His-Leu-Thr-Pro-Val-Glu-Lys . . .

The change from a polar amino acid (Glu) to a nonpolar one (Val) reduces the overall charge on the hemoglobin molecule. If the altered hemoglobin molecule is fully oxygenated, there is no problem and it remains in solution. However, if the level of oxygenation decreases (e.g., at high altitudes or during vigorous physical exercise), the less soluble deoxyhemoglobin molecules clump together, forming long, insoluble fibers, and force the erythrocytes to change from a round to a crescent, or sickle, shape (Figure 28.6). These abnormal cells become trapped in the capillaries and impair circulation. The resultant blockage of blood further decreases the oxygen supply to the affected areas of

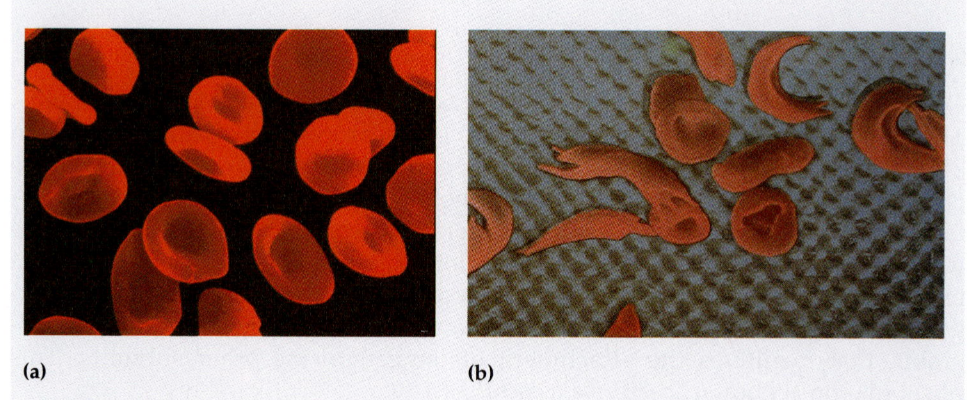

(a) (b)

▲ **Figure 28.6**
Scanning electron micrographs of (a) normal and (b) sickled erythrocytes.

the body and increases the sickling of additional erythrocytes. The abnormal cells are subsequently destroyed by the spleen, and this leads to anemia (and consequent tiredness). Pain can occur unpredictably in locations where the sickled blood cells block the supply of oxygen. Infections are also more common because of damage to the spleen from the sickled erythrocytes.

Persons who inherit one defective hemoglobin gene are heterozygous and carriers of the sickle cell trait. Generally they show no symptoms of the disease. A person who inherits a defective hemoglobin gene from each parent is homozygous and will have sickle cell anemia. In the United States about 2 million people carry the sickle cell trait, while approximately 72,000 have the disease.

Sickle cell anemia is found predominantly in people who originally lived in tropical areas of the world, such as Africa. Studies have shown that the occurrence of sickle cell anemia is highest in populations that came from regions with a high incidence of malaria. Individuals who carry the sickle cell trait (heterozygotes) have a higher resistance to malaria than individuals with normal hemoglobin. Thus they were more likely to survive and have children, who often inherited at least one variant hemoglobin gene.

Currently, the only known cure for sickle cell anemia is a bone marrow transplant. In addition to the surgical risks, the number of children that could benefit from this procedure is limited by the requirement for bone marrow from a healthy, tissue-compatible brother or sister. Blood transfusions are often given to increase the number of normal erythrocytes, which relieves the symptoms of anemia, such as tiredness. Many children are given antibiotics on a routine basis to prevent infections from bacteria that cause pneumonia, which can be deadly in children with sickle cell anemia.

New treatments for adults include the use of hydroxyurea. This drug appears to work by stimulating the production of fetal hemoglobin, a form of hemoglobin all humans produce before birth, but normally stop producing soon after birth. This fetal hemoglobin seems to interfere with the formation of the insoluble hemoglobin fibers that lead to sickling of the erythrocyte. Clinical trials are being conducted to determine if hyroxyurea is effective and safe for use in children.

Genetic screening tests can determine whether prospective parents carry the mutated gene. It is also possible to determine if an unborn baby has sickle cell anemia or sickle cell trait through tests on the amniotic fluid or tissue taken from the placenta.

cells and the hemoglobin carries some of the carbon dioxide back to the lungs to be expelled. In the lungs oxygen is again bound to the heme group, forming more oxyhemoglobin (see Figure 6.14).

Oxyhemoglobin does not transfer all of its oxygen to the tissue cells. Normally, every 100 mL of arterial blood combines with about 20 mL of oxygen. In the resting individual, the venous blood carries about 13 mL of oxygen per 100 mL of blood. Therefore, about 65% of the hemoglobin in venous blood is still combined with oxygen. When people are engaged in strenuous exercise, their oxygen demand is high and the percentage of oxyhemoglobin in the venous blood may fall to as low as 25%.

Arterial blood is crimson; venous blood is a darker red, but it is not purple or blue.

The greater release of oxygen to tissues that are actively working is triggered by the increased concentration of carbon dioxide and hydrogen ions formed during cellular respiration. Carbon dioxide (CO_2) is carried in the blood primarily as bicarbonate ion (HCO_3^-; 70%), although about 20% is covalently bound to hemoglobin. The binding of carbon dioxide to hemoglobin leads to a change in tertiary structure that causes oxygen to more readily dissociate. A similar effect is seen when the H^+ ion concentration increases. This is known as the **Bohr effect** after Christian Bohr (father of Niels Bohr) who first noted the relationship between the amount of oxygen released and the concentration of carbon dioxide and hydrogen ions (the pH) in blood. Figure 28.7 shows how changes in pH affect the amount of oxygen released in the tissues. The lower the pH, the greater the amount of oxygen released.

The carbon dioxide bound to hemoglobin does not bind to the heme group, but binds to the N-terminal amino groups of each polypeptide chain.

You'll recall from Section 10.3 that $pH = -\log[H^+]$.

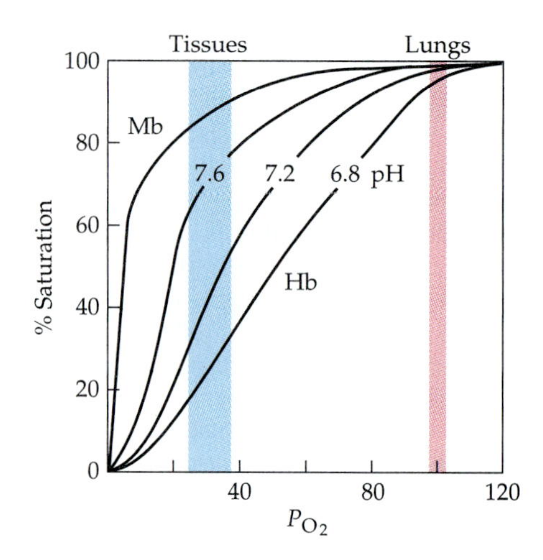

▶ Figure 28.7
The Bohr effect in hemoglobin. These oxygen-binding curves for hemoglobin were obtained at different pH values and show the saturation of hemoglobin with oxygen at varying partial pressures of oxygen. Note the greater release of oxygen in the tissues as the pH drops.

Bicarbonate ions are formed when CO_2 diffuses into the blood from the tissues and reacts with water:

$$CO_2 + H_2O \rightleftharpoons HCO_3^- + H^+$$

This reaction occurs primarily within the red blood cells, where it is catalyzed by carbonic anhydrase. Most of the HCO_3^- then diffuses back into the plasma. Carbonic anhydrase is also embedded in the walls of lung capillaries and catalyzes the reverse reaction, helping CO_2 diffuse out of the capillaries and into the alveoli (Figure 28.8), where it can be removed when a person exhales.

How does the body respond to the lower oxygen levels at high altitudes? In Leadville, Colorado, for example, where the altitude is 10,000 feet, the partial pressure of O_2 in the lungs is only about 68 mmHg (compared with 100 mmHg at sea level). Hemoglobin is only 90% saturated with O_2 at this pressure, meaning that less oxygen is available for delivery to tissues. People who climb suddenly from sea level to high altitude thus experience a feeling of oxygen deprivation, or *hypoxia*, as their bodies are unable to supply enough oxygen to the tissues. The body copes with this situation in two ways. Any condition that tends to lower the oxygen content of the blood causes an increase in the number of erythrocytes (and, thus, the amount of hemoglobin). Persons who live at high altitudes generally have higher erythrocyte counts than those who live at sea level. In addition, levels of a molecule known as 2,3-bisphosphoglycerate increase. This molecule binds to hemoglobin, causing it to release more of its oxygen to the tissue cells.

O O⁻ (C) — H—C—O—PO_3^{2-} — H—C—O—PO_3^{2-} — H

2,3-Bisphosphoglycerate

Various chemical substances act as poisons by interfering with the transport of oxygen by hemoglobin. Oxidizing agents such as potassium ferricyanide can oxidize the iron of hemoglobin to the Fe(III) state. The same result is achieved in vivo by the action of nitrites and certain organic compounds (e.g., acetanilide, nitrobenzene, the sulfa drugs). The resulting compound, which contains iron in the Fe(III) oxidation state, is called *methemoglobin* and cannot transport oxygen. Small amounts (about 0.3 g/100 mL of blood) of methemoglobin are normally present in the erythrocytes, but appreciable amounts of this substance result in the pathological condition *methemoglobinemia.*[2]

Methemoglobin is brown. During cooking, red meat turns brown because of the oxidation of hemoglobin to methemoglobin. Dried bloodstains turn brown for the same reason.

[2]Salami and other preserved meat products contain nitrite salts as preservatives. People who consume relatively large quantities of these foods have a tendency to develop methemoglobinemia. Also, nitrate ions are reduced to nitrite ions by microorganisms in the digestive tract. Concern exists over the high level of nitrates from fertilizer in the groundwater in some areas. Babies are particularly sensitive. Nitrites or nitrates cause methemoglobinemia and result in the blue baby syndrome.

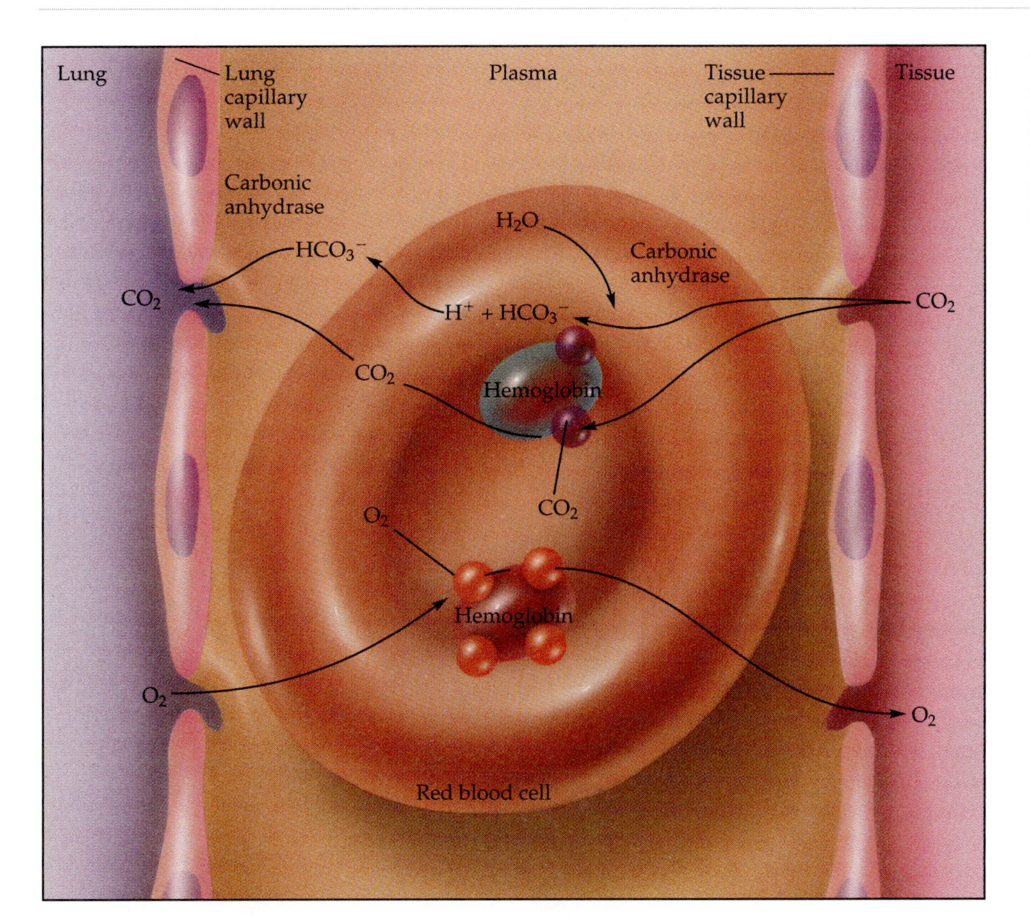

◀ **Figure 28.8**
A schematic diagram that shows the exchange of gases between capillary blood and the air in the alveoli (left) and between capillary blood and body tissues (right).

Carbon monoxide is poisonous because the Fe^{2+} ions in hemoglobin have a much greater affinity for carbon monoxide than they have for oxygen (by a factor of 200). Thus the heme groups preferentially combine with any carbon monoxide that is in the blood. When CO is bound to the Fe^{2+} ion, oxygen cannot bind. If sufficiently large numbers of hemoglobin molecules become saturated with carbon monoxide (about 60%), death occurs because the blood fails to supply the brain with oxygen. All except the most severe cases of CO poisoning are reversible if treated in time. The best antidote is the administration of pure oxygen. Artificial respiration may help if a tank of oxygen is not available. Because carbon monoxide poisoning impairs the blood's ability to transport oxygen, the heart has to work harder to supply oxygen to tissues. Chronic exposure, even to low levels of CO, as through cigarette smoking, puts an added strain on the heart and increases the chances of a heart attack.

✓ Review Questions

28.4 Contrast the contents of arterial blood and venous blood.

28.5 Why is carbon monoxide such a deadly poison? (b) How is carbon monoxide poisoning treated?

28.3 Blood Buffers

Recall that the maintenance of pH within narrow limits is vital to the well-being of an organism. Any slight change in hydrogen ion concentration (± 0.2 to 0.4 pH unit) affects the affinity of hemoglobin for oxygen (Section 28.2) and alters the rates

Metabolic Fate of Hemoglobin

When erythrocytes are destroyed, their hemoglobin molecules are completely catabolized. The porphyrin ring is first cleaved; then the globin and iron are removed. The iron is incorporated into specific iron-storage proteins to be reused in the synthesis of new heme molecules. The porphyrin skeleton is of no further use to the body. It undergoes a series of degradation reactions that lead to the production of the bile pigments, chiefly *biliverdin* (which has a green color) and *bilirubin* (which has an orange-yellow color). These colored substances give the bile its yellow color. The degraded pigments are stored in the gallbladder and released into the small intestine. As they travel down the intestinal tract, they undergo additional transformations that result in a darkening of their color, thus accounting for the characteristic colors of feces. The sometimes spectacular color changes observed in a bruise have a related cause. A bruise consists of blood released and trapped beneath the skin. As the blood is gradually broken down, degradation of the porphyrin skeleton from hemoglobin forms a series of colored products.

An excess of bilirubin in the blood is responsible for the yellow color of the skin in jaundice (French *jaune,* yellow). Jaundice can arise from (1) infectious hepatitis, a condition during which the liver malfunctions and cannot remove sufficient bilirubin; (2) the obstruction of bile ducts by gallstones; or (3) an acceleration of erythrocyte destruction in the spleen (hemolytic jaundice). Jaundice occurs in a large percentage of newborn infants because of insufficient synthesis of the liver enzymes that decompose bilirubin. A common treatment of neonatal jaundice is to shine a special fluorescent light onto the baby's skin. The energy of the light is able to decompose some of the bilirubin just beneath the surface of the skin.

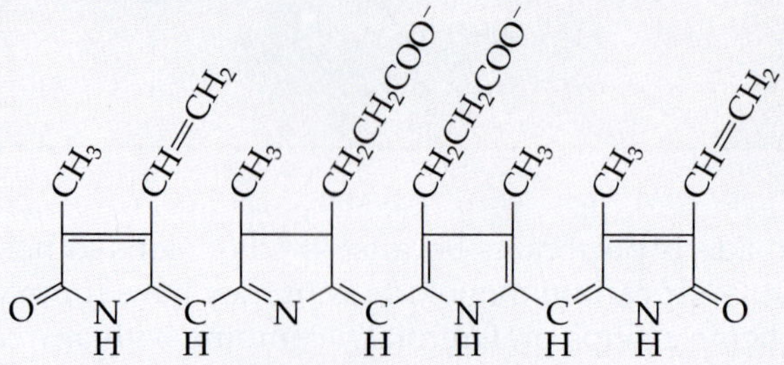

Biliverdin

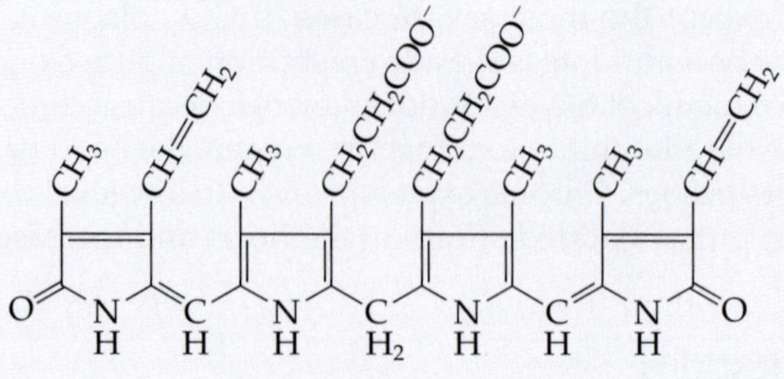

Bilirubin

of metabolic processes by (usually) decreasing the catalytic efficiency of enzymes (Section 22.5). Blood plasma and erythrocytes contain four buffering systems that maintain the pH of the blood between 7.35 and 7.45: (1) bicarbonate buffer, (2) phosphate buffer, (3) plasma proteins, and (4) hemoglobin. In Section 10.6 we mentioned the buffering actions of the bicarbonate buffer [H_2CO_3/HCO_3^-] and the phosphate buffer [$H_2PO_4^-/HPO_4^{2-}$]. In Section 21.2 we indicated that the amino acids, in their zwitterionic forms, can neutralize small concentrations of either acids or bases. Because the plasma proteins and hemoglobin contain both acidic and basic amino acids, they tend to minimize changes in pH by combining with, or liberating, hydrogen ions. Thus they serve as excellent buffering agents over a wide range of pH values.

The principal buffer in tissue cells is the phosphate buffer, whereas hemoglobin molecules are the most important buffer within the erythrocytes. The major buffer in the blood and interstitial fluid is the bicarbonate pair. Under normal conditions, the primary metabolic factor that tends to lower the pH is the continuous production of carbon dioxide and the acidic metabolites (acetoacetate, pyruvate, lactate, α-ketoglutarate, etc.). All of these compounds vary in concentration according to metabolic circumstances. When acids enter the blood, they are neutralized by bicarbonate ions, and the slightly dissociable carbonic acid is formed:

$$H^+ + HCO_3^- \rightleftharpoons H_2CO_3$$

It would seem that this reaction would alter the buffer ratio by decreasing the bicarbonate ion concentration and increasing the concentration of carbonic acid. The excess carbonic acid, however, is readily decomposed to water and carbon dioxide by carbonic anhydrase:

$$H_2CO_3 \xrightarrow[]{\text{carbonic anhydrase}} H_2O + CO_2$$

The respiration rate is increased, and the carbon dioxide is eliminated at the lungs, thus preserving the proper buffer ratio.

Respiration rate is another factor that influences blood pH. *Respiratory acidosis* results from **hypoventilation,** a condition that arises when the rate of breathing is too slow. Hypoventilation is brought on by an obstruction to respiration (e.g., asthma, pneumonia, or pulmonary emphysema), by coronary attack, or by drugs that depress the brain's respiratory center (e.g., morphine, barbiturates). When the respiration rate is very low, carbon dioxide is not expelled from the lungs fast enough, and the H_2CO_3/CO_2 equilibrium is shifted to the left. This increased carbonic acid concentration results in a higher-than-normal H_2CO_3/HCO_3^- ratio and a subsequent decrease in the blood pH.

Hyperventilation, when the rate of breathing is too rapid, causes *respiratory alkalosis.* Hyperventilation arises during strenuous exercise, anxiety, crying, and hysteria. At high altitudes, hyperventilation may occur in response to low oxygen pressure. An increased rate of respiration accelerates the removal of carbon dioxide from the lungs, and the H_2CO_3/CO_2 equilibrium shifts to the right. The concentration of carbonic acid in the blood decreases, and the H_2CO_3/HCO_3^- ratio becomes lower than normal, with a subsequent increase in blood pH.

✓ Review Questions

28.6 What is the normal pH range of the blood?

28.7 What is the principal buffer in: **(a)** blood; **(b)** tissue cells; **(c)** erythrocytes; **(d)** interstitial fluid?

28.4 Blood Clotting

Blood clotting, or coagulation, is of the utmost importance to the organism. If such a mechanism did not exist, life-threatening loss of blood would occur whenever a blood vessel was injured. The theory accounting for blood clotting has evolved from a simple two-step mechanism to the present model, which involves a multistep cascade in which many protein factors, most of them proteolytic enzymes, are activated sequentially (Table 28.1). A variety of blood disorders are due to mutations in the genes coding for these clotting factors.

The blood-clotting cascade becomes operative when a tissue is cut or injured and leads to the conversion of prothrombin to thrombin. Prothrombin is a proenzyme (Section 22.2), and its activation is analogous to the activation of the various digestive enzymes. *Thrombin* then converts fibrinogen to fibrin by hydrolysis of two peptide fragments from fibrinogen. The last two steps in the blood-clotting cascade can be summarized as:

$$\text{Prothrombin} \xrightarrow[\text{Ca}^{2+},\ \text{phospholipids}]{\text{blood-clotting factors}} \text{Thrombin}$$

$$\text{Fibrinogen} \xrightarrow{\text{thrombin}} \text{Fibrin}$$

$$\begin{array}{c} {}^{-}\text{OOC} \quad \text{COO}^{-} \\ \diagdown \;\; \diagup \\ \text{CH} \\ \gamma | \\ \text{CH}_2 \\ \beta | \\ \text{H}_3\text{N}^{+}\text{—C—COO}^{-} \\ \alpha | \\ \text{H} \end{array}$$

γ-Carboxyglutamate

Streptokinase, an enzyme isolated from a bacterium and now available through recombinant DNA techniques, is used to treat myocardial infarctions. It functions by converting plasminogen to plasmin.

When blood clots, the fibrin monomers undergo a polymerization reaction that results in the formation of insoluble needlelike threads. These threads enmesh the blood cells and effectively seal off the area where the blood vessel has been damaged (Figure 28.9).

Animal blood deficient in vitamin K (Section F.5) has a prolonged coagulation time because of a lack of thrombin in the plasma. Vitamin K (in its reduced form) is required in the oxygen-dependent carboxylation of the glutamic acid side chains of prothrombin to yield γ-carboxyglutamate residues. These residues must be present in prothrombin to enable it to bind calcium ions during its conversion to thrombin.

Normally blood takes five to eight minutes to form a clot. After the tissue is repaired by the body, the fibrin clot is eventually hydrolyzed into soluble components by the enzyme *plasmin* (which circulates in the blood as the proenzyme *plasminogen*).

A number of substances, known as *anticoagulants,* inhibit clotting by interfering with one or another of the reactions of the clotting cascade. Heparin is one of the principal anticoagulants. It is a polysaccharide rich in sulfate ester groups and is believed to block the catalytic activity of both thromboplastin and thrombin. Low concentrations of heparin are normally secreted into the circulatory system to prevent **thrombosis,** the formation of a clot within a blood vessel. Certain sodium salts are employed as anticoagulants when blood is collected for clinical purposes. The

Table 28.1 Blood Clotting Factors

Common Name	Roman Numeral Designation[a]	Associated Disorder
Fibrinogen	Factor I	Afibrinogenemia
Prothrombin[b]	Factor II	Prothrombin deficiency
Tissue factor	Factor III	...
Calcium ions	Factor IV	...
Proaccelerin	Factor V	Parahemophilia
Proconvertin[b]	Factor VII	ProSPCA deficiency
Antihemophilic factor	Factor VIII	Classic hemophilia (hemophilia A)
Christmas factor[b]	Factor IX	Christmas disease (hemophilia B)
Stuart factor[b]	Factor X	Stuart factor deficiency
Plasma thromboplastin antecedent (PTA)	Factor XI	PTA deficiency
Hageman factor	Factor XII	Hageman trait
Fibrin-stabilizing factor	Factor XIII	Factor XIII deficiency

[a]Originally, activated factor V was thought to be a new coagulation factor and was named factor VI. Following the withdrawal of factor VI as a separate coagulation factor, the Roman numeral VI has not been assigned to any other coagulation factor.

[b]Vitamin K–dependent protein.

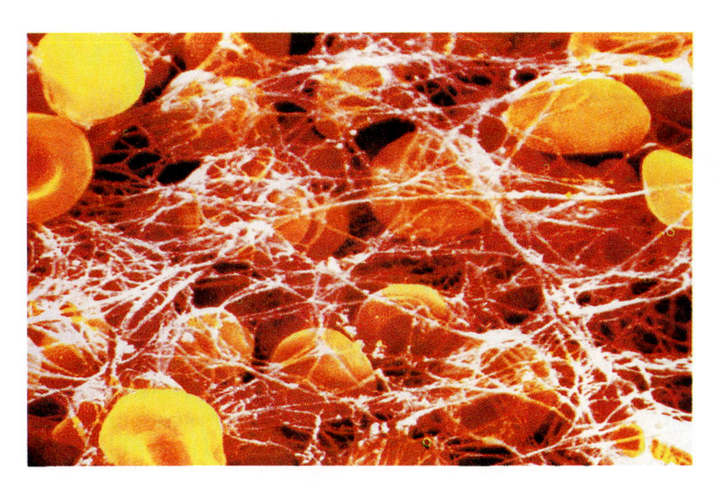

Figure 28.9 Scanning electron micrograph of erythrocytes enmeshed in fibrin fibrils—a part of a typical blood clot.

anions of these salts—citrate, oxalate, and fluoride—form strong complexes with calcium ions, thus preventing them from existing in the free ionic state. Without calcium ions in the plasma, blood will not clot.

If one wishes to store blood plasma, an anticoagulant must be added to the freshly drawn blood to prevent clotting.

Dicumarol, an anticoagulant originally isolated from decomposed clover, prevents blood clotting by inhibiting enzymes that keep vitamin K in its reduced form. A more potent anticoagulant, similar in structure and action to Dicumarol, is the sodium salt of warfarin (Coumadin).[3] Drugs such as Coumadin are usually administered before an operation or after heart attacks to minimize thrombosis.

OH OH CH_2 O O O O

Dicumarol

Na^+ O^- CH CH_2 C O CH_3 O

Sodium salt of Warfarin (Coumadin)

Hemophilia

Hemophilia is one of many inherited disorders characterized by inadequate production of clotting factors. Two types of hemophilia have been described. Approximately 85% of the hemophilia patients in the United States have hemophilia A in which little or no Factor VIII is produced. The other 15% have hemophilia B (or Christmas disease) in which little or no Factor IX is produced. Most hemophilia patients have less than 1% of the normal amount of the specific clotting factor, leading to severe hemophilia in which internal bleeding may occur at joints without apparent cause or injury. A patient with at least 5% of the normal level of clotting factor has a mild form of hemophilia in which bleeding is a problem only when undergoing surgery or following a serious injury.

The genes for both Factor VIII and IX are found on the X chromosome. Thus any male who inherits an X chromosome with a mutation in one of these genes will have hemophilia (males have only one X chromosome). A female who inherits one defective X chromosome is a carrier and may have mild bleeding problems due to lower levels of the clotting factor. She has a 50% chance of having a son with hemophilia. Hemophilia is much rarer in women because they must inherit a mutated X chromosome from both parents.

Transfusions of clotting factors can often reduce or control the symptoms of hemophilia, but plasma samples from many individuals must be combined to obtain adequate amounts. This increases the risk of blood-borne infections such as hepatitis and AIDS. Laboratory testing of donated blood and techniques to inactivate viruses in the blood have markedly improved the safety of blood products used to treat hemophilia. Gene-splicing techniques have been used to create recombinant clotting factors entirely free of blood products. Recombinant Factor VIII was approved by the FDA in 1993, while recombinant Factor IX was approved in 1997. These products cost more than the clotting factors obtained from blood donations, but are safer.

[3]Warfarin is also employed as a rat poison. It is safe for use as a rodenticide because regular ingestion of massive doses is fatal to rodents (producing internal hemorrhaging) whereas a single, accidental ingestion by a child or pet is usually harmless.

Aspirin and Thrombosis

Various medical protocols recommend that older people and patients with histories of heart attacks or strokes take aspirin regularly (one-quarter tablet to one tablet daily). Recall (Section E.4) that among its other physiological effects, aspirin inhibits the synthesis of prostaglandins and thromboxanes, thereby interfering with the aggregation of platelets and diminishing the rate at which they release the blood-clotting factors (thromboplastin). Hence, aspirin prevents heart attacks by inhibiting thrombosis.

The greatest danger from thrombosis is that the clot may become detached and travel through the blood to some vital organ, such as the heart or the brain. If the clot becomes lodged in and obstructs the blood vessels to these organs, their tissue cells are starved for oxygen and the cells die. If tissue death occurs in the brain, the condition is termed a *stroke*; if heart muscle tissue is destroyed, the condition is called a *coronary thrombosis* or a *myocardial infarction.* If the clot, or a fragment of it, breaks loose and is carried by the blood to lodge in a small blood vessel elsewhere (e.g., the lung), it is called an *embolism.*

✓ Review Question

28.8 What clotting factor is needed for the conversion of fibrin to fibrinogen?

28.5 The Immune Response

As mentioned in Section 28.1, antibodies are γ-globulins. An **antibody** (also called an *immunoglobulin*) is a specialized protein that plays a vital role in the sophisticated defense mechanism in the body termed the **immune response.** Whenever it is invaded by foreign substances, the body responds by increasing the synthesis of specific antibodies to eliminate the invader. Because it triggers the synthesis of antibodies, the invader is called an **antigen** (it causes *anti*body *gen*eration). Each distinctive antigen stimulates production of a unique set of antibodies.

Antibodies are produced by B cells, a specific type of leukocytes. Each B cell synthesizes a specific antibody. Each antibody contains four polypeptide chains: two heavy chains and two light chains that are linked by disulfide bonds (Figure 28.10). Each chain has a constant and a variable region. The constant regions for each antibody of a particular class (there are five classes) have the same amino acid sequence, while the variable regions have different amino acid sequences for each specific antibody. Each antibody has two regions where it can bind to its specific antigen.

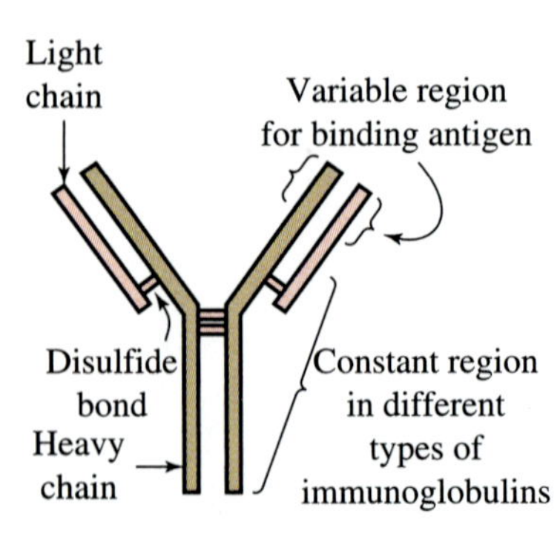

▲ **Figure 28.10**
Schematic diagram of an antibody.

The formation of an antigen-antibody complex is shown in Figure 28.11. This large complex can inactivate the antigen in several ways. It may simply prevent the antigen from binding to its target or, if the complex is large enough, it will precipitate and be degraded by phagocytes.

An important part of the immune response is the lymphatic system (Figure 28.12). This system, which returns components of the interstitial fluid to the bloodstream, is composed of veins and capillaries but no arteries. Interstitial fluid is absorbed into the lymph capillaries, in which it is called **lymph.** This lymph flows into larger and larger lymph veins. Eventually, two large lymph veins empty into veins of the blood circulatory system.

Lymph nodes serve as filters and as factories for the production of some forms of leukocytes. The leukocytes located there remove dead cells, bacteria, and other foreign elements from the lymph. Lymph nodes are also the sites of antibody synthesis. These nodes are lumpy enlargements in the lymph veins. Sometimes the

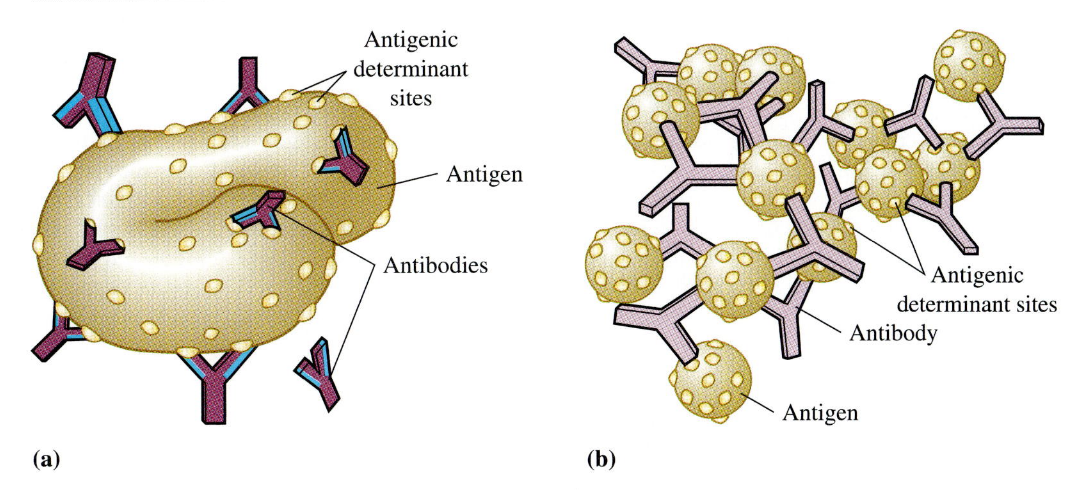

▲ **Figure 28.11**
(a) Antibodies bind to antigenic determinant sites on the surface of a bacterium. (b) Because each antibody has two binding sites, the interaction of many antigens and antibodies creates a large antigen-antibody complex that will precipitate.

nodes are so effective at filtering out bacteria from an infected area that they become swollen. Someone suffering from a sore throat may also exhibit swollen and tender lymph nodes in the neck area.

It is possible to prime the immune response. A **vaccine** containing a weakened form of the antigen (e.g., dead bacteria) will cause an individual to build up a certain level of antibodies for this particular antigen. If the same individual is then subjected to attack by a more virulent form of the antigen, the body responds much more rapidly and effectively to its presence. Having practiced on the dead invader, the system can easily handle the live one—it has gained an immunity to that particular organism.

Unfortunately, the body can't tell "good" antigens from "bad" ones. We want it to respond to a viral infection, but we wish it wouldn't attack purposely transplanted tissue. Nonetheless, the host body will respond to a donor skin graft, a kidney transplant, or a heart transplant as it does to a virus: by generating antibodies to attack it. When the body responds to foreign tissues in this way, the process is called *rejection.* This is a major problem associated with organ transplants. There are drugs that suppress the immune response, but when these are given to a transplant recipient to protect the transplanted organ, the patient also becomes more susceptible to infection. If you turn off the immune response, you turn off both the good and bad aspects of it.

A variety of diseases are associated with the immune system. An overactive or misguided immune system can attack its own body organs, causing an **autoimmune disease.** The system senses something as foreign even though it isn't. In multiple sclerosis, the body attacks and destroys the myelin sheath of nerves, while in arthritis it is the connective tissues in the joints that are destroyed. In type 1 diabetes, the insulin-producing cells of the pancreas are the target.

At the opposite end of the scale is *acquired immunodeficiency syndrome* (AIDS). In this deadly affliction, the immune system is destroyed by the HIV virus (Selected Topic G). The victims then succumb to a variety of infections or to a rare form of cancer.

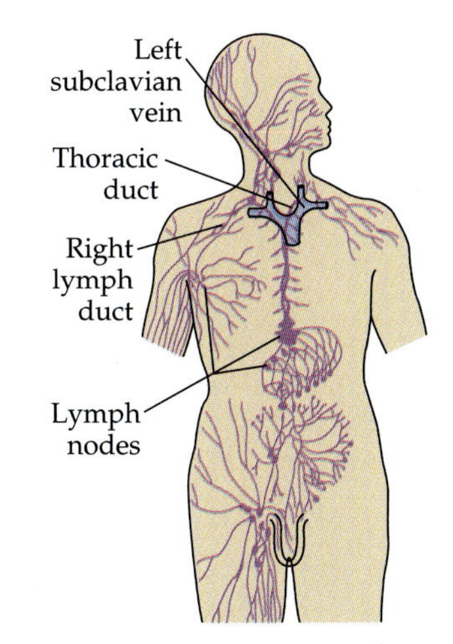

▲ **Figure 28.12**
The lymphatic system.

Review Question

28.9 Briefly describe the immune response.

Blood Types

You may know if your blood is type A, B, AB, or O, but do you know how they differ? Variations in blood type are due to slight structural differences in oligosacharides, known as *blood group antigens,* that are linked to erythrocyte membrane proteins. The only difference in these antigens is the presence or absence of a single monosaccharide: the A antigen has a terminal N-acetylgalactosamine, the B antigen has galactose, and the O antigen has no monosaccharide at this position (Figure 28.13). Individuals with type-A blood have the A antigen linked to erythrocyte membrane proteins, those with type-B blood have the B antigen, those with type-AB blood have a mixture of both, and type-O blood cells have the O antigen.

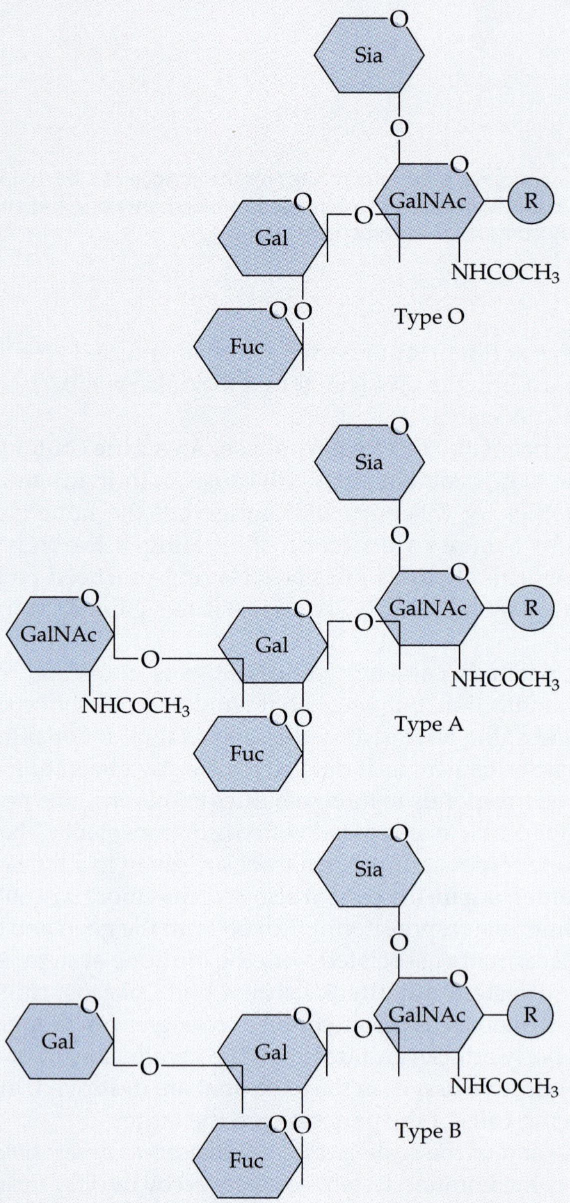

▲ **Figure 28.13**
The ABO blood group antigens. Fuc—fucose; Gal—galactose; GalNAc—N-acetylgalactosamine; Sia—sialic acid. (See an advanced biochemistry text for the structures of these monosaccharides.)

Humans can produce antibodies against the A and B blood group antigens, but not the O group antigens. However, a person does not produce antibodies against his or her own blood antigens. Thus, a person with type-A blood will have antibodies against the B antigen, but not the A antigen, and the reverse would be true of someone with type-B blood. A person with type-AB blood will not make antibodies against either antigen, while an individual with type-O blood has antibodies against both the A and B antigens. In theory, a person with type-AB blood can accept any type blood (although this is never done because of other possible interactions), while someone with type-O blood can only receive type-O blood. The opposite is true when donating blood. Because antibodies are not produced against the O antigen, individuals with type-O blood are known as "universal donors" and their blood can be safely donated to all blood types (Table 28.2).

Table 28.2 Transfusion Relationships Among ABO Blood Types

Person Has Blood Type	Makes Antibodies Against:	Can Safely Receive Blood from:	Can Safely Donate Blood to:
O	A, B	O	O, A, B, AB
A	B	O, A	A, AB
B	A	O, B	B, AB
AB	None	O, A, B, AB[a]	AB

[a]In principle, this relationship is true. However, ABs are never given donations from other types, because the donor's antibodies could react with the recipient's antigens.

28.6 Blood Pressure

Blood pressure is the force exerted by the blood against the inner walls of the arteries. The **systolic pressure** is the maximum pressure achieved during contraction of the heart ventricles. When the ventricles relax, the blood pressure drops, and the lowest pressure that remains in the arteries before the next ventricular contraction is called the **diastolic pressure.** The alternate expansion and contraction of the arterial walls can be felt as a *pulse* (where the artery is near the surface of the skin). Blood pressure measurements are reported as a *ratio of systolic pressure* (in mmHg) to *diastolic pressure* (in mmHg)—for example, 120/80.

The normal systolic and diastolic ranges for young adults are 100 to 120 and 60 to 80 mmHg, respectively. For older people, the corresponding normal ranges are 115 to 135 and 75 to 85 mmHg.

Blood pressure depends on several factors, including total blood volume, heart action, and action of the smooth muscles that surround the arteries. It is directly proportional to the volume of blood. Thus, if there is a loss of blood due to hemorrhaging, the blood pressure drops, but it returns to normal when the lost blood is replaced by a blood transfusion. Contraction of the smooth muscles in the walls of the arteries constricts the vessels (vasoconstriction), causing an increase in blood pressure. Conversely, dilation of the smooth muscles causes vasodilation and a decrease in blood pressure.

One of the major health problems throughout the world is high blood pressure, or **hypertension.** A person is diagnosed with hypertension when his or her systolic pressure is 140 mmHg or higher or the diastolic pressure is 90 mmHg or higher. Over 20% of the world's population has hypertension; in the United States it is estimated that it occurs in about 1 of every 4 adults. If hypertension is not brought under control, it can lead to serious medical problems such as arteriosclerosis, heart attack, enlarged heart, kidney damage, or stroke.

The lifestyle modifications listed in Table 28.3 can often prevent hypertension or be used to treat it. For some people, drugs such as diuretics, vasodilators, and those that block the action of the neurotransmitter epinephrine are often needed, as well.

Table 28.3 Lifestyle Modifications for Hypertension Prevention and Management

- Lose weight if overweight.
- Limit alcohol intake to no more than 1 oz (30 mL) ethanol (e.g., 24 oz [720 mL] beer, 10 oz [300 mL] wine, or 2 oz [60 mL] 100-proof whiskey) per day or 0.5 oz (15 mL) ethanol per day for women and lighter weight people.
- Increase aerobic physical activity (30 to 45 minutes most days of the week).
- Reduce sodium intake to no more than 100 mmol per day (2.4 g sodium or 6 g sodium chloride).
- Maintain adequate intake of dietary potassium (approximately 90 mmol per day).
- Maintain adequate intake of dietary calcium and magnesium for general health.
- Stop smoking and reduce intake of dietary saturated fat and cholesterol for overall cardiovascular health.

Review Questions

28.10 How are blood pressure measurements reported?

28.11 What lifestyle changes can prevent or manage hypertension?

28.7 Urine and the Kidneys

The kidneys remove metabolic waste products from the blood.[4] The functional units of the kidneys are called **nephrons** (Figure 28.14). Each nephron has a bulbous *Bowman's capsule,* which tails into a long, highly convoluted urinary tubule. The capsule surrounds a network of arterial capillaries called the *glomerulus.* Blood enters the glomerulus through a small artery (arteriole) and exits through another, which forms a network of capillaries (purple in Figure 28.14) that surround the tubule. Finally, these tubule-surrounding capillaries join again and form a small vein, which returns the filtered blood to the main circulation.

Most blood constituents, except formed elements and large protein molecules, filter through the capillary walls of the glomerulus and enter the Bowman's capsule. The glomerulus seems to act as a simple filter, with particle size being the main factor that decides what goes and what stays. The fluid that enters the Bowman's capsule and eventually collects in the tubule contains many valuable components as well as wastes. Consider water. About 170 L per day is filtered into the tubules. Most of this water—and valuable constituents such as glucose, amino acids, and salt—are reabsorbed by the blood through the walls of the tubules. Waste products such as urea, uric acid, and excess salts are passed on into collecting tubules and eventually excreted as **urine.** Knowing that healthy adults pass 1.1 to 1.5 L of urine each day, you can calculate that more than 99% of the water filtered through the glomerulus is reabsorbed through the tubules. We'd have quite a drinking problem if all that water weren't conserved!

The volume and composition of urine are quite variable. Its volume depends on liquid intake, amount of perspiration, presence of fever or diarrhea, and other factors. The composition varies with diet and state of health. Most substances have a renal threshold level. If the concentration of a substance in the blood exceeds this

[4]Waste removal is a secondary renal task. The principal role of the kidneys is to maintain stable concentrations of all the water and inorganic ions in the body.

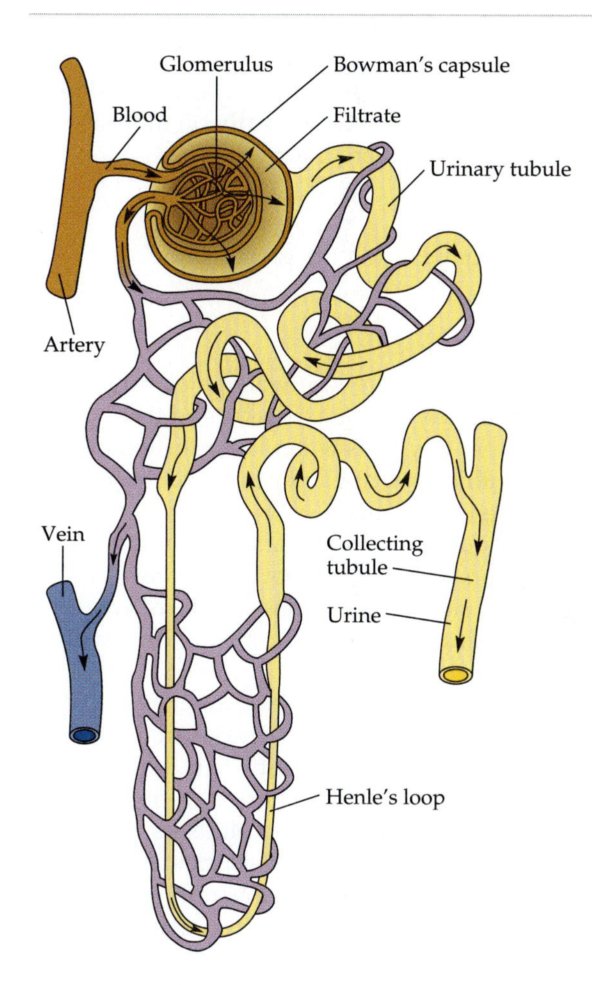

◄ **Figure 28.14**
Structure of a nephron. Nephrons are the functional units of the kidney, where metabolic wastes are removed from the blood. The arrows indicate the path of fluids through the nephron. Each human kidney has about a million nephrons.

threshold value, the excess is not reabsorbed through the tubule but appears in the urine. The threshold level for glucose (blood sugar), for example, is quite high (Chapter 25). Normally, nearly all glucose is reabsorbed. If the glucose level in the blood is too high, however (as happens in cases of uncontrolled diabetes), glucose shows up in the urine.

Some components of urine have relatively low solubility in water. When some condition (such as an increase in concentration or a change in pH) leads to the precipitation of these materials in the kidney, a stone is formed. *Kidney stones* (or renal calculi) usually consist of calcium phosphate [$Ca_3(PO_4)_2$], magnesium ammonium phosphate ($MgNH_4PO_4$), calcium carbonate ($CaCO_3$), calcium oxalate (CaC_2O_4), or a mixture of these compounds. Stone formation may accompany increased concentration of calcium ion caused by disease or increased ingestion of calcium ion. People who eat foods rich in oxalates (such as spinach) have a relatively high incidence of oxalate kidney stones. (What a great excuse for kids who don't want to eat their spinach!)

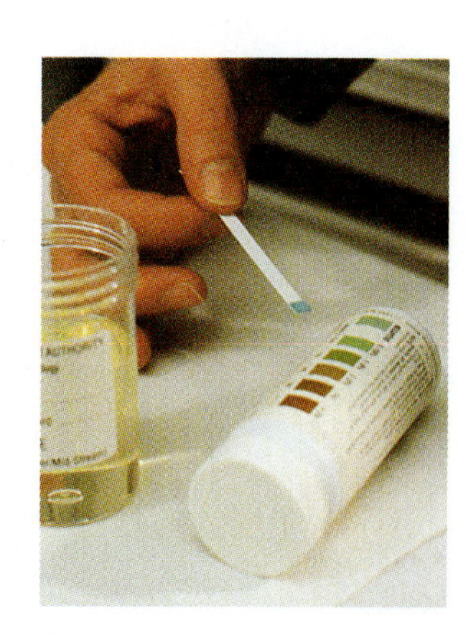

▲ Urinalysis test strips. The reagent strip is dipped into the urine sample and then compared with a color chart for each constituent.

The kidney does far more than just get rid of wastes. It plays a vital role, for example, in maintaining the balance of water, electrolytes, acids and bases, and other components of the body fluid. When the kidneys malfunction, the body is in trouble. Analysis of the urine can often give a good indication of the health of the individual.

Wilhelm Kolff invented the artificial kidney for dialysis at the Cleveland Clinic in the 1950s. Before that time, kidney failure meant death. Dialysis provides a way for waste products to be removed from the blood when the kidneys cannot do so. In the United States approximately 100,000 people are on dialysis due to chronic kidney failure, as well as many more who are temporarily on dialysis due to acute kidney failure resulting from surgery or severe injury.

A patient receiving dialysis treatment is shown on p. 219 (Chapter 8).

Review Question

28.12 What major blood component does not pass into the glomerulus?

28.8 Sweat and Tears

The skin is also an organ of excretion. Through it we lose water, electrolytes, nitrogenous wastes, and lipids. Water is lost directly through the skin and through the respiratory tract at a rate of about 700 mL per day. This water loss is called *insensible perspiration.* Perspiration from the 2.5 million sweat glands is called *sensible perspiration.* It is activated by a rise in blood temperature.

It is interesting to note that female mosquitoes, seeking their meal of blood, find us by following a warm stream of air laden with carbon dioxide and lactic acid. We could foil them by not sweating and not breathing!

Sweat is about 99% water. It contains sodium, chloride, and calcium ions, and smaller amounts of other minerals. A person working in a hot environment might lose 12 L of sweat per day, including 70 g of salt. Organic constituents of sweat include urea, lipids (body oils), amino acids, glucose, and lactic acid. Drugs such as morphine, nicotine, and alcohol also appear in sweat.

Replacing fluids lost through sweat is of critical importance. A water loss of 1–2% of body weight can reduce a person's ability to do muscular work. An endurance athlete can lose 1.5 L or more of fluid during each hour of activity. This fluid must be replaced during the event, and not just at the end. Even for shorter events, fluid replacement is critical. For most people water is the best choice. Minerals lost in the sweat are readily replaced by the next meal. However, commercial "thirst quenchers" or "sports drinks" may be needed in endurance events of more than three hours.

Carbohydrate replacement drinks may benefit athletes who exercise vigorously for more than one hour. These drinks delay the onset of exhaustion by a few minutes.

If the air around us is not too humid, the water in sweat evaporates. Perspiration carries off not only wastes but also heat. It helps us keep cool on hot days. Each gram of water that evaporates absorbs 0.540 kcal (2.26 kJ) of heat from the body (Section 7.7). **Heat stroke** occurs when the heat regulation system fails. Without prompt medical attention, a victim of heat stroke may die. To prevent heat stroke drink plenty of fluids, rest in the shade when tired, and wear lightweight clothing.

Tears, another important body fluid, are responsible for maintaining the health of your eyes by keeping them moist. Chemically, tears have about the same composition as other body fluids. They have approximately the same salt content as blood plasma. Tears contain *lysozyme,* an enzyme that ruptures bacterial cell walls. This bactericidal action helps prevent eye infections.

Tears are produced by the lacrimal (or tear) gland at the rate of ~ 1 mL/day (Figure 28.15). At regular intervals the eyelids sweep the secretions of the lacrimal glands over the surface of the eye where they mix with oils and other products. Tears are three-layered. There is an inner layer of mucus, then a layer of lacrimal secretions, and finally an outer layer of oily film that retards evaporation from the watery middle layer.

Following blinking, tears drain through the lacrimal canals into the lacrimal sac and eventually to the nasal cavity through the nasolacrimal duct.

A copious flow of tears may be caused by irritants, such as pepper, acid fumes, or a variety of chemical lachrymators. Tear gas, usually α-chloroacetophenone, is a specially designed eye irritant. The flow of tears can also be triggered by emotional upsets. Such crying is undoubtedly controlled by hormones. And if you feel like crying, go right ahead. Most psychologists say it's good for you. Indeed, many people have long believed that crying is beneficial. Richard Crashaw, an English poet of the seventeenth century, called tears "the ease of woe."

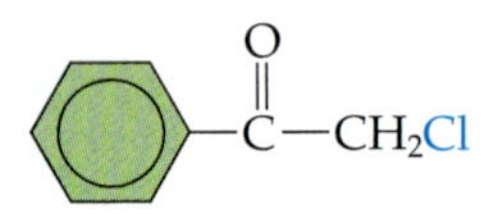

α-Chloroacetophenone

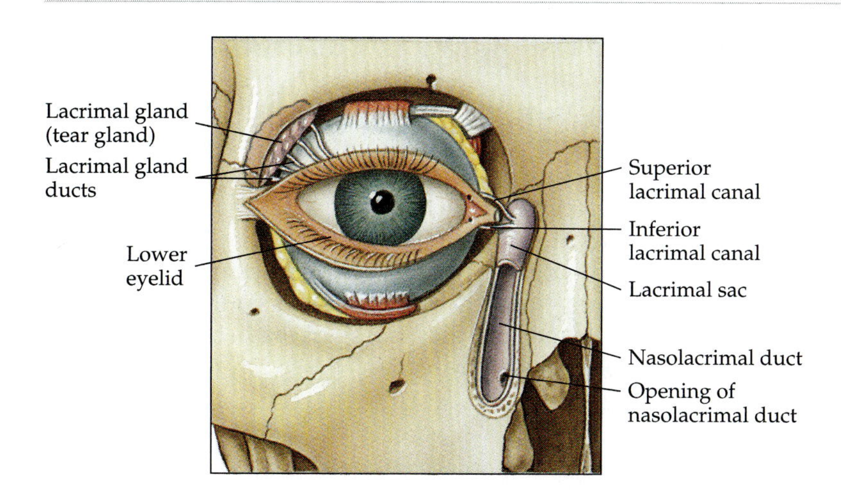

Figure 28.15
Organization of the lacrimal apparatus of the eye in which tears are produced and then carried away to the nasal cavity.

✓ Review Questions

28.13 What is the difference between sensible and insensible perspiration?

28.14 What are the three layers of tears?

28.9 The Chemistry of Mother's Milk

Newborn mammals are nourished by milk secreted from their mothers' mammary glands. Indeed, the presence of such glands in females characterizes the animal class *Mammalia.* The composition of milk varies from one species to another (Table 28.4). Amounts of fat and protein tend to be greater in marine mammals and those that live in cold climates.

Proteins and emulsified fats give milk its characteristic "milky" appearance. Casein is the principal protein in cow's milk. Humans beyond infancy often include some milk or milk products in their diet. This food is an excellent source of most nutrients. It includes all the essential amino acids. It also contains most vitamins and minerals. Cow's milk, however, is a poor source of iron, copper, and vitamin C. It is also short of vitamin D as it comes from the cow, so this vitamin is often added as a supplement to cow's milk intended for human consumption.

Cow's milk is sometimes given to infant humans as a substitute for their mothers' milk. Evidence now indicates that this is not a wise substitution. Human milk not only more closely matches the nutritional needs of human infants, but is also adds to the infant's immunological defenses against disease. Newborn infants are unable to make antibodies. They have *temporary passive immunity* because their blood contains antibodies made by the mother and passed across the placenta.

Table 28.4 Composition of Milks (g/100 g)

Mammal	Fat	Protein	Lactose
Cow	4.4	3.8	4.5
Human	3.8	1.6	7.5
Goat	4.1	3.7	4.2
Reindeer	22.5	10.3	2.5
Porpoise	49.0	11.0	1.3

Mother's milk produced within the first few days after childbirth is a rich source of γ-globulins.

And there you have it! Chemistry from atoms to mother's milk. Carbohydrate, lipid, and protein metabolism are all tied together. Indeed, in a living organism, everything is connected to everything else. Life is one huge, complicated set of chemical reactions—and, of course, a lot more. The whole of life is certainly much more than the sum of a set of chemical reactions.

We hope that we have enriched your life by helping you to learn something of the basis of chemistry and of the many ways chemistry touches your life every day. And we wish for you the proper reward for the hours you have spent studying chemistry: the joy of success in your chosen profession.

Review Question

28.15 What advantage does human milk possess over cow's milk as a food for a human infant?

Summary

The various fluids in the body may be either **intracellular** (inside cells) or **extracellular** (outside of cells). The two major extracellular fluids are interstitial fluid and blood plasma. **Interstitial fluid** is the fluid that surrounds most cells and fills any space between them. **Plasma** is a straw-colored liquid that represents 50–60% of the volume of a whole blood sample. The principal ions in plasma are sodium and chloride, with smaller amounts of potassium, calcium, magnesium, phosphate, bicarbonate, and sulfate. The principal proteins are **albumins, globulins,** and **fibrinogen.**

The remaining 40–50% of whole blood is comprised of the **formed elements: erythrocytes** (red cells), **leukocytes** (white cells), and **platelets.** The **hematocrit** is the volume (in percent) of formed elements in a sample of whole blood. **Anemia** occurs when the amount of erythrocytes (and/or hemoglobin) is abnormally low.

The major functions of erythrocytes are to carry inhaled oxygen from the lungs to tissue cells and to carry carbon dioxide from tissue cells to the lungs to be exhaled. Leukocytes defend the body when foreign substances enter. **Leukemia** is a cancer characterized by the uncontrolled production of immature leukocytes. Platelets have a role in blood clotting. Blood minus fibrinogen and formed elements is **serum.**

The medical condition known as **shock** results from a massive loss of blood, which leads to a dramatic decrease in blood pressure.

The most abundant protein in erythrocytes is *hemoglobin.* Hemoglobin is made up of four polypeptide chains (two α subunits and two β subunits) each surrounding a molecule of heme. Each heme group contains a central iron ion to which oxygen is reversibly bound in the lungs to form oxyhemoglobin. The oxygen is released in the tissues where it is needed. Increased concentrations of carbon dioxide and H^+ increase the amount of oxygen that is released in the tissues. This is known as the **Bohr effect.**

Buffers in the blood—mainly H_2CO_3/HCO_3^-, $H_2PO_4^-/HPO_4^{2-}$, and proteins—maintain the pH in the range of 7.35 to 7.45. Blood pH is also influenced by respiration rate. **Hypoventilation,** when the breathing rate is too slow, leads to a decrease in the blood pH, while **hyperventilation,** in which the breathing rate is too rapid, leads to an increase in pH.

Blood clotting occurs when a series of blood clotting factors are released following damage to a blood vessel. These factors lead to the conversion of *prothrombin* to *thrombin.* This latter enzyme catalyzes the conversion of fibrinogen to *fibrin,* which forms threads that seal off the damaged area. Individuals with **hemophilia** have very little or none of a specific blood-clotting factor. **Thrombosis** is the formation of a blood clot within a blood vessel, which is usually prevented by low concentrations of an anticoagulant *heparin* that is present in the circulatory system.

Antibodies (also called immunoglobulins) are specialized proteins that are synthesized in large amounts whenever a foreign substance enters the bloodstream. The invading substance is called an **antigen,** and its presence triggers a series of events, including the formation of specific antibodies, known as the **immune response. Lymph** is a type of interstitial fluid found in the lymphatic system, an important part of the immune response.

A **vaccine** is a weakened form or part of an antigen that causes an individual to build up antibodies towards that antigen. This primes the immune system so that it can respond more rapidly if a more virulent form of the antigen enters the body. An **autoimmune disease** occurs when the immune system attacks the body's own organs.

The heart pumps the blood through the circulatory system. **Blood pressure** is the force exerted by the blood against the inner walls of the arteries. **Systolic pressure** is the maximum pressure the blood exerts on a capillary wall; it occurs when the heart is contracting. **Diastolic pressure** is the minimum pressure the blood exerts on a capillary wall; it occurs when the heart is relaxed. A blood pressure measurement is the ratio of systolic pressure to diastolic pressure. **Hypertension** or high blood pressure is a systolic pressure of 140 mmHg or higher or a diastolic pressure of 90 mmHg or greater.

Waste products are removed from the blood as it flows through **nephrons,** the basic functional units that make up the kidneys. As the blood passes through the *glomerulus,* a clump of capillaries surrounded by the part of the nephron called *Bowman's capsule,* waste products leave the blood and enter the Bowman's capsule, then pass down the *urinary tubule* and out of the body in the **urine.** The kidneys regulate the water and electrolyte balance in the body.

Some water, electrolytes, nitrogenous wastes, and lipids pass through the skin in *sweat* and are eliminated from the body. The evaporation of sweat from the skin also serves as a mechanism for cooling the body. **Heat stroke** occurs when this heat regulation system fails.

Tears are an important body fluid that maintain the health of the eyes by keeping them moist.

Milk is secreted by the mammary glands of female mammals. It contains fat, protein, and lactose as well as other vitamins and minerals. It provides nourishment to young animals and protects them against disease because it contains antibodies produced by the mother.

Key Terms

albumins (28.1)
anemia (28.1)
antibody (28.5)
antigen (28.5)
autoimmune disease (28.5)
blood pressure (28.6)
Bohr effect (28.2)
diastolic pressure (28.6)
erythrocytes (28.1)
extracellular fluid (page 762)
fibrinogen (28.1)
formed elements (28.1)
globulins (28.1)
heat stroke (28.8)
hematocrit (28.1)
hemophilia (28.4)
hypertension (28.6)
hyperventilation (28.3)
hypoventilation (28.3)
immune response (28.5)
interstitial fluid (page 762)
intracellular fluid (page 762)
leukemia (28.1)
leukocytes (28.1)
lymph (28.5)
nephrons (28.7)
plasma (28.1)
platelets (28.1)
serum (28.1)
shock (28.1)
systolic pressure (28.6)
thrombosis (28.4)
urine (28.7)
vaccine (28.5)

Problems

Blood: Functions and Composition

1. List five inorganic ions present in blood.
2. What are the functions of: **(a)** albumins **(b)** α- and β-globulins, **(c)** γ-globulins, and **(d)** fibrinogen?
3. Where are the plasma proteins synthesized?
4. Compare erythrocytes to other cells.
   - **a.** How are they similar? How do they differ?
   - **b.** Where are erythrocytes formed?
   - **c.** Where are they broken down?
5. Identify the formed element being described in each of the following:
   - **a.** This formed element contains a nucleus and is the least numerous of the formed elements in whole blood.
   - **b.** This formed element has no nucleus and is needed to transport oxygen throughout the body.
   - **c.** This formed element has no nucleus and is needed for proper blood clotting.
6. What three things can lead to anemia?

Blood Gases

7. Describe the chemical structure of hemoglobin.
8. Distinguish between each of the following:
   - **a.** heme and hemoglobin
   - **b.** deoxyhemoglobin and oxyhemoglobin
   - **c.** oxyhemoglobin and CO-hemoglobin
   - **d.** oxyhemoglobin and methemoglobin
   - **e.** hemoglobin and sickle cell hemoglobin
9. What is the oxidation state of iron in **(a)** hemoglobin, **(b)** oxyhemoglobin, **(c)** methemoglobin, and **(d)** CO-hemoglobin?
10. How does each of the following affect the amount of oxygen released by hemoglobin in tissue cells?
    - **a.** increased concentration of carbon dioxide
    - **b.** increased concentration of H^+
    - **c.** increased concentration of 2,3-bisphosphoglycerate
    - **d.** increase in the pH of the blood

Blood Buffers

11. List the four buffering systems found in the blood.
12. Why is the bicarbonate buffer effective in spite of the fact that the ratio of $H_2CO_3:HCO_3^-$ is 1:10?
13. Illustrate, with equations, how the phosphate buffer ($H_2PO_4^-/HPO_4^{2-}$) keeps blood pH from changing when small amounts of acid or base are produced during metabolic reactions.
14. Use the following equilibrium to explain what causes **(a)** respiratory acidosis and **(b)** respiratory alkalosis:

$$H^+ + HCO_3^- \rightleftharpoons H_2O + CO_2$$

Blood Clotting

15. Why do blood-clotting factors circulate in the blood as proenzymes?
16. Describe how each of the following functions in blood clotting.
    a. prothrombin **b.** thrombin **c.** fibrinogen
    d. thromboplastin **e.** calcium ions **f.** fibrin
17. What is the role of vitamin K in blood coagulation?
18. How does each of the following act to prevent blood coagulation or thrombosis?
    a. heparin **b.** dicumarol
    c. anions such as oxalates, fluorides, and citrates

The Immune Response

19. Describe the structure of an antibody.
20. Antibodies are produced by what type of leukocytes?
21. What is an antigen?
22. List three autoimmune diseases and identify what part of the body is destroyed in each disease.

Blood Pressure

23. Distinguish between systolic pressure and diastolic pressure.
24. The following blood pressure measurements were made on four different patients. Which of the patients has hypertension? Explain your choices.
    a. 150/100 **b.** 145/70 **c.** 110/70 **d.** 135/95
25. Why are diuretics sometimes prescribed for a person with hypertension?
26. Why are individuals with hypertension advised to go on a low salt diet?

Urine and the Kidneys

27. How is urine formed in the kidneys?
28. List three organic components of urine.
29. List four factors that affect the volume of urine excreted.
30. What is dialysis and when is it used?

Sweat and Tears

31. List three inorganic and three organic constituents of sweat.
32. How does sweat help to regulate body temperature?
33. What is the function of lysozyme in tears?
34. Where are tears produced?

The Chemistry of Mother's Milk

35. **a.** What is the difference between milk from a cow and milk from a reindeer? Why are these differences important?
    b. What is the difference between milk from a goat and milk from a porpoise? Why are these differences important?
36. **a.** Is cow's milk a "perfect" food?
    b. Is it a source of complete protein?

Additional Problems

37. The blood accounts for about 8.0% of the total body weight. Calculate the volume of blood your body contains (density of blood = 1.06 g/mL).
38. How does acidosis affect the transport of oxygen by hemoglobin?
39. When pneumonia blocks respiratory passageways and limits the exhalation of carbon dioxide, blood CO_2 levels build up. In this situation, is the pH of the blood higher or lower than normal? Explain.
40. A common cause of death in young children is acidosis resulting from severe diarrhea. In severe diarrhea, large amounts of bicarbonate ion are excreted from the body. Why should this lead to acidosis?
41. **a.** Explain why an individual with a protein deficient diet might suffer from edema.
    b. Why does this person have a more difficult time fighting off an infection (their immune system is compromised)?
42. Explain the processes involved when fluid is exchanged between the plasma and the cells: **(a)** at the arterial end **(b)** at the venous end of a capillary.
43. If the amount of protein in the interstitial fluid matched the amount in the blood plasma, would more or less fluid be likely to filter from the capillaries to the interstitial space? Explain.
44. **a.** Why is the sickling of hemoglobin in sickle cell anemia such a problem?
    b. How is sickle cell anemia treated?
45. Why is sickle cell anemia most prevalent in people originally from tropical climates?
46. Name the two bile pigments formed from the porphyrin skeleton of hemoglobin.
47. What are the causes of jaundice?
48. What is the cause of: **(a)** hemophilia A and **(b)** hemophilia B?
49. A man who has hemophilia marries a woman who is not a carrier of the trait.
    a. What is the likelihood that any of their children will have hemophilia?
    b. What is the likelihood that any of their children will be carriers of the trait?
50. A woman who is a carrier of the hemophilia trait marries a man who does not have hemophilia.
    a. What is the likelihood that any of their children will have hemophilia?
    b. What is the likelihood that any of their children will be carriers of the trait?
51. Why do doctors prescribe one tablet of low-dose aspirin daily for older people and patients with histories of heart attacks or strokes?
52. **a.** Contrast the structural features of the A, B, and O antigens found on erythrocytes.
    b. Which of these antigens can trigger an immune response?

Some Mathematical Operations

I.1 Exponential Notation

Scientists often use numbers that are so large or small that they boggle the mind. For example, light travels at 300,000,000 m/s. There are 602,200,000,000,000,000,000,000 carbon atoms in 12.01 g of carbon. On the small side, the diameter of an atom is about 0.0000000001 m, and the diameter of an atomic nucleus is about 0.000000000000001 m. It is difficult to keep track of the zeros in such quantities. Scientists find it convenient to express such numbers in exponential notation.

A number is in *exponential notation*—sometimes called *scientific notation*—when it is written as the product of a coefficient—usually with a value between 1 and 10—and a power of 10. Two examples are

$$4.18 \times 10^3 \text{ and } 6.57 \times 10^4$$

Expressing numbers in exponential form generally serves two purposes: (1) We can write very large or very small numbers in a minimum of printed space and with a reduced chance of typographical error. (2) We can convey explicit information about the precision of measurements: The number of significant figures in a measured quantity is stated unambiguously.

In the expression 10^n, n is the exponent of 10, and the number 10 is said to be raised to the nth power. If n is a *positive quantity*, 10^n has a value *greater than 1*. If n is a *negative quantity*, 10^n has a value *less than 1*. We are particularly interested in cases where n is an integer. For example,

Positive powers of 10

$10^0 = 1$
$10^1 = 10$
$10^2 = 10 \times 10 = 100$
$10^3 = 10 \times 10 \times 10 = 1000$
and so on

The power of 10 determines the numbers of zeros that follow the digit 1.

Negative powers of 10

$10^0 = 1$
$10^{-1} = 1/10 = 0.1$
$10^{-2} = 1/(10 \times 10) = 0.01$
$10^{-3} = 1/(10 \times 10 \times 10) = 0.001$
and so on

The power of 10 determines the numbers of places to the right of the decimal point where the digit 1 appears.

To express 612,000 in exponential form,

$$612{,}000 = 6.12 \times 100{,}000 = 6.12 \times 10^5$$

To express 0.000505 in exponential form,

$$0.000505 = 5.05 \times 0.0001 = 5.05 \times 10^{-4}$$

We can use a more direct approach to converting numbers to the exponential form.

- Count the number of places a decimal point must be moved to produce a coefficient having a value between 1 and 10.
- The number of places counted then becomes the power of 10.
- The power of 10 is *positive* if the decimal point is moved to the *left*.

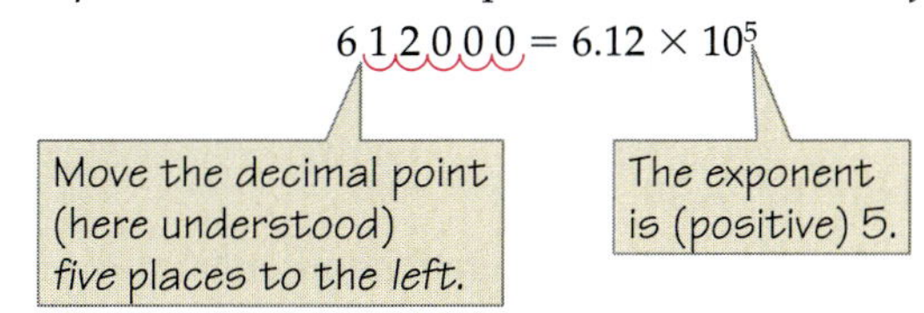

612000 = 6.12 × 10^5
Move the decimal point (here understood) five places to the left.
The exponent is (positive) 5.

The power of 10 is *negative* if the decimal point is moved to the *right*.

$$0.000505 = 5.05 \times 10^{-4}$$

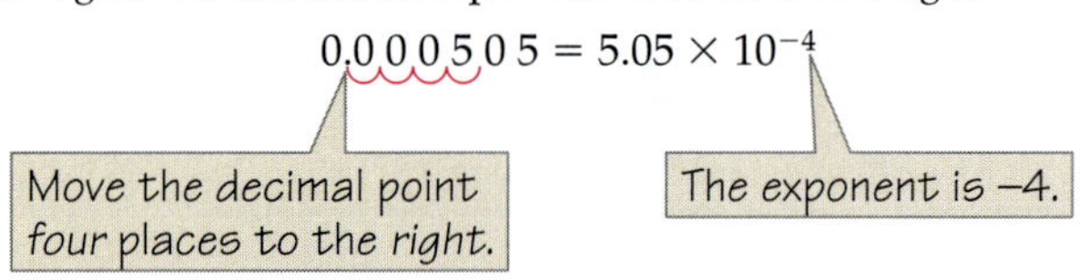
Move the decimal point
four places to the right.
The exponent is −4.

To convert a number from exponential form to the conventional form, move the decimal point in the opposite direction.

$$3.75 \times 10^{6} = 3{,}750000$$

The exponent is 6. Move the decimal point six places to the right.

$$7.91 \times 10^{-7} = 0.000000791$$

The exponent is −7. Move the decimal point seven places to the left.

It is easy to handle exponential numbers on most scientific calculators. A typical procedure is to enter the number, followed by the key [EXP]. The key strokes required for the number 2.85×10^{7} are [2] [.] [8] [5] [exp] [7], and the result displayed is 2.85^{07}.

For the number 1.67×10^{-5}, the key strokes are [1] [.] [6] [7] [exp] [5] [±], and the result displayed is 1.67^{-05}.

Many calculators can be set to convert all numbers and calculated results to the exponential form, regardless of the form in which the numbers are entered. Generally the calculator can also be set to display a fixed number of significant figures in results.

The keystrokes on your calculator may be different from those shown here. Check the instructions in the manual that came with the calculator.

Addition and Subtraction

To add or subtract numbers in exponential notation, it is necessary to express each quantity as the *same power of ten*. In calculations, this treats the power of ten in the same way as a unit—it is simply "carried along." In the following, each quantity is expressed with the power 10^{-3}.

$$(3.22 \times 10^{-3}) + (7.3 \times 10^{-4}) - (4.8 \times 10^{-4}) =$$

$$(3.22 \times 10^{-3}) + (0.73 \times 10^{-3}) - (0.48 \times 10^{-3})$$

$$= (3.22 + 0.73 - 0.48) \times 10^{-3}$$

$$= 3.47 \times 10^{-3}$$

Multiplication and Division

To multiply numbers expressed in exponential form, *multiply* all coefficients to obtain the coefficient of the result, and *add* all exponents to obtain the power of ten in the result.

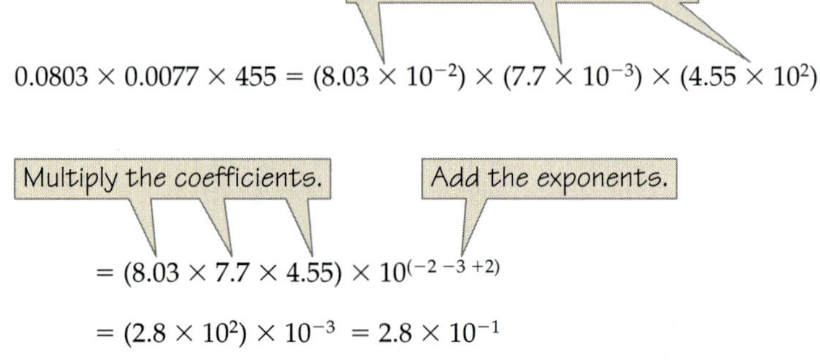
Rewrite in exponential form.
Multiply the coefficients.
Add the exponents.

$$0.0803 \times 0.0077 \times 455 = (8.03 \times 10^{-2}) \times (7.7 \times 10^{-3}) \times (4.55 \times 10^{2})$$

$$= (8.03 \times 7.7 \times 4.55) \times 10^{(-2\ -3\ +2)}$$

$$= (2.8 \times 10^{2}) \times 10^{-3} = 2.8 \times 10^{-1}$$

Generally, most calculators perform these operations automatically, and no intermediate results need be recorded.

To divide two numbers in exponential form, *divide* the coefficients to obtain the coefficient of the result, and *subtract* the exponent in the denominator from the exponent in the numerator to obtain the power of ten. In the example below, multiplication and division are combined. First, the rule for multiplication is applied to the numerator and to the denominator, and then the rule for division is used.

Rewrite in exponential form.

Apply the rule for multiplication to the numerator and denominator.

$$\frac{0.015 \times 0.0088 \times 822}{0.092 \times 0.48} = \frac{(1.5 \times 10^{-2})(8.8 \times 10^{-3})(8.22 \times 10^{2})}{(9.2 \times 10^{-2})(4.8 \times 10^{-1})}$$

Apply the rule for division.

$$= \frac{1.1 \times 10^{-1}}{4.4 \times 10^{-2}} = 0.25 \times 10^{-1-(-2)} = 0.25 \times 10^{1}$$

$$= 0.25 \times 10^{-1} \times 10^{1} = 2.5 \times 10^{0} = 2.5$$

Raising a Number to a Power and Extracting the Root of an Exponential Number

To raise an exponential number to a given power, raise the coefficient to that power, and multiply the exponent by that power. For example, we can cube a number (that is, raise it to the *third* power) in the following manner.

Rewrite in exponential form. Cube the coefficient. Multiply the exponent by 3.

$$(0.0066)^3 = (6.6 \times 10^{-3})^3 = (6.6)^3 \times 10^{3\times(-3)}$$

$$= (2.9 \times 10^{2}) \times 10^{-9} = 2.9 \times 10^{-7}$$

To extract the root of an exponential number, we raise the number to a fractional power—one-half power for a square root, one-third power for a cube root, and so on. Most calculators have keys designed for extracting square root and cube roots. Thus, to extract the square root of 1.57×10^{-5}, enter the number 1.57×10^{-5} into a calculator, and use the [$\sqrt{\ }$] key.

$$\sqrt{1.57 \times 10^{-5}} = 3.96 \times 10^{-3}$$

Some calculators allow you to extract roots by keying the root in as a fractional exponent. For example, we can take a cube root as follows.

$$(2.75 \times 10^{-9})^{1/3} = 1.40 \times 10^{-3}$$

Another approach for extracting the roots of numbers is to use logarithms, which we will discuss next.

I.2 Logarithms

The common logarithm (log) of a number (N) is the exponent (x) to which the base 10 must be raised to yield the number.

$\log N = x$ ← means that → $N = 10^x$ or that → $N = 10^{\log N}$

In the expressions below, the numbers N are printed in blue and their logarithms ($\log N$) are printed in red.

$$\log 1 = \log 10^0 = 0 \qquad \log 1 = \log 10^0 = 0$$

$$\log 10 = \log 10^1 = 1 \qquad \log 0.1 = \log 10^{-1} = -1$$

$$\log 100 = \log 10^2 = 2 \qquad \log 0.01 = \log 10^{-2} = -2$$

$$\log 1000 = \log 10^3 = 3 \qquad \log 0.001 = \log 10^{-3} = -3$$

Most of the numbers that we commonly encounter, of course, are not integral powers of ten, and their logarithms are not integral numbers. From the above pattern, though, we do have a general idea of what their logarithms might be. Consider, for example, the numbers 655 and 0.0078.

$$100 < 655 < 1000 \qquad 0.001 < 0.0078 < 0.01$$

$$2 < \log 655 < 3 \qquad -3 < \log 0.0078 < -2$$

We can see that log 655 is between 2 and 3 and that log 0.0078 is between −3 and −2. To get a more exact value, however, we must use a table of logarithms or the [LOG] key on a calculator.

$$\log 655 = 2.816 \quad \log 0.0078 = -2.11$$

In working with logarithms, we often need to find the number that has a certain value for its logarithm. The number is sometimes called an *antilogarithm,* and we can think of it in the following terms.

$$\text{If } \log N = 3.076, \text{ then } N = 10^{3.076} = 1.19 \times 10^3$$

$$\text{If } \log N = -4.57, \text{ then } N = 10^{-4.57} = 2.7 \times 10^{-5}$$

With a calculator, we simply enter the value of the logarithm (that is, 3.076 or −4.57) and then use the [10^x] key.

Problems

1. Express each of the following numbers in exponential form.
   a. 0.000017 **b.** 19,000,000 **c.** 0.0034 **d.** 96,500
2. Express each of the following numbers in exponential form.
   a. 4,500,000 **b.** 0.000108 **c.** 0.0341 **d.** 406,000
3. Carry out the following operations. Express the answers in exponential form.
   a. $(4.5 \times 10^{13})(1.9 \times 10^{-5})$ **b.** $\dfrac{9.3 \times 10^9}{3.7 \times 10^{27}}$
   c. $\dfrac{4.3 \times 10^{-7}}{7.6 \times 10^{22}}$ **d.** $\sqrt{1.7 \times 10^{-5}}$
4. Carry out the following operations. Express the answers in exponential form.
   a. $(6.2 \times 10^{-5})(4.1 \times 10^{-12})$ **b.** $(2.1 \times 10^{-6})^2$
   c. $\dfrac{2.1 \times 10^5}{9.8 \times 10^{-7}}$ **d.** $\dfrac{4.6 \times 10^{-12}}{2.1 \times 10^3}$
5. What is the logarithm of each of the following numbers?
   a. 937 **b.** 6.3×10^{-9} **c.** 2.6×10^{-5}
6. What is the logarithm of each of the following numbers?
   a. 0.00375 **b.** 2.6×10^{-10} **c.** 8.6×10^5
7. What is the antilogarithm of each of the following?
   a. 5.144 **b.** −1.19 **c.** 1.052
8. What is the antilogarithm of each of the following?
   a. −5.434 **b.** −2.414 **c.** −9.435

Glossary

Absolute alcohol is 100% ethanol.

The **accuracy** of a set of measurements refers to the closeness of the average of the set to the "correct" or most probable value.

An **acetal** is the product of a hemiacetal-alcohol reaction. It has two OR groups attached to the same carbon atom.

An **achiral** molecule is superimposable on its mirror image. It is *not* chiral.

An **acid** is (1) a compound that produces hydronium ions, H_3O^+, in water solution (Arrhenius theory), or (2) a proton donor (Brønsted–Lowry theory).

An **acidic anhydride** is a substance that reacts with water to produce an acid; a nonmetal oxide.

The **acid ionization constant** (K_a) is the equilibrium constant describing equilibrium in the reversible ionization of a weak acid.

Acidosis is the condition that results when the pH of the blood falls below 7.35. One important physiological effect is hindering oxygen transport.

Activation energy is the minimum energy necessary to initiate a reaction.

The **active site** of an enzyme is that portion into which the substrate fits.

The **activity series of metals** is a listing of metals in terms of their abilities to displace one another from solutions of their ions or to displace H^+ as H_2 from acidic solutions.

Actomyosin is the contractile protein of which muscles are made. It contains actin and myosin.

In **addition polymerization,** monomers add to one another to produce a polymeric product that contains all the atoms of the starting monomers. The monomers usually contain double bonds.

In an **addition reaction** substituent groups join to hydrocarbon (carbon-hydrogen compound) molecules at points of unsaturation—double or triple bonds. This type of reaction is typical of alkenes and alkynes.

Adenylate cyclase is the enzyme that catalyzes the conversion of ATP to adenosine 3′,5′-monophosphate (cyclic AMP or cAMP).

Adipocyte. *See* **fat cell.**

Adipose tissue is connective tissue where fat is stored.

An **aerobic** process is one that occurs in the presence of oxygen.

An **agonist** is a molecule that fits and activates a specific receptor.

Albumins are globular proteins that are abundant in blood plasma.

An **alcohol (ROH)** is an organic compound containing the —OH group on an aliphatic carbon atom.

Alcoholic fermentation is the process by which yeast produces alcohol.

An **aldehyde (RCHO)** is an organic compound with a *carbonyl* functional group that has a hydrogen atom attached. The other group on the carbonyl carbon atom may be a hydrocarbon group or a second hydrogen atom.

An **aldose** is a sugar with an aldehyde group.

An **aliphatic** compound is an open-chain compound or a ring compound that has no aromatic groups.

An **alkali metal** is an element in Group 1A of the periodic table.

An **alkaline earth metal** is an element in Group 2A of the periodic table.

An **alkaloid** is a nitrogen-containing organic compound obtained from plants that has physiological properties.

Alkalosis is a physiological condition in which the pH of the blood rises above 7.45.

An **alkane** is a hydrocarbon with only single bonds, a saturated hydrocarbon.

An **alkene** is a hydrocarbon containing one or more carbon–carbon double bonds.

An **alkyl group** is a hydrocarbon group derived from an alkane by removal of a hydrogen atom.

An **alkyl halide** is a compound resulting from the replacement of a hydrogen atom of an alkane with a halogen atom.

An **alkyne** is a hydrocarbon whose molecules contain one or more carbon–carbon triple bonds.

An **allergen** is a substance that triggers an allergic reaction.

In **allosteric regulation,** the activity of an enzyme is controlled by its interaction with a small molecule at a site other than the active site. This changes the shape of the enzyme and thus its activity.

Allotropes are two or more forms of an element that differ in their basic molecular structure. Diamond and graphite are allotropes of carbon.

An **alpha (α) particle** is identical to the helium nucleus and is emitted by the nuclei of certain radioactive atoms as they undergo decay.

An **amide** is an organic compound with a *carbonyl* functional group that has a nitrogen atom attached. The other group on the carbonyl carbon atom may be a hydrocarbon group or a second hydrogen atom.

An **amide linkage** is the bond between a carbonyl carbon atom and a nitrogen.

An **amine** is a compound derived from ammonia by replacement of one, two, or three of the hydrogens by alkyl or aryl groups.

An **amino acid** is a carboxylic acid with an amino substituent on the α-carbon.

The **amino acid pool** consists of the amino acids produced by hydrolysis of proteins in the digestive tract that are absorbed through the intestinal wall, as well as amino acids released by the hydrolysis of endogenous proteins.

The **amino group** is an NH_2 unit.

Amylase is an enzyme that catalyzes the hydrolysis of starches.

Amylopectin is a form of plant starch with branched chains of glucose units.

Amylose is a form of starch with the glucose units joined in an unbranched continuous chain.

An **anabolic steroid** is a synthetic androgen that stimulates protein synthesis, particularly in skeletal muscle cells.

Anabolism is the process of building up the molecules of living systems.

An **anaerobic process** is one that occurs in the absence of oxygen.

An **analgesic** is a pain reliever.

An **androgen** is a male sex hormone.

Anemia is a condition characterized by abnormally low levels of erythrocytes (red blood cells) or hemoglobin.

An **anesthetic** is a substance that produces insensitivity to pain. *See also* **general anesthetic** and **local anesthetic.**

Anhydrous means "without water."

An **anion** is a negatively charged ion.

The **anode** is the electrode at which oxidation occurs.

The **anomeric carbon** is the hemiacetal (latent carbonyl) carbon of a cyclic sugar.

Anomers are sugar isomers that differ in structure around the anomeric carbon atom.

An **antacid** is any basic substance used to neutralize stomach acid.

An **antagonist** is a drug that blocks the action of an agonist by blocking the receptor(s).

An **antibiotic** is a soluble substance produced by a mold or bacterium that inhibits growth of other microorganisms.

An **antibody** is a protein that binds to and destroys foreign substances.

An **anticoagulant** is a substance that inhibits the clotting of blood.

An **anticodon** is a sequence of three adjacent nucleotides in a tRNA molecule that is complementary to a codon on mRNA.

An **antidiuretic** is a water-conserving substance.

An **antigen** is a foreign substance that triggers the formation of antibodies.

An **antihistamine** is a compound that is structurally similar to histamine. It can bind to receptors normally occupied by histamine and prevent the physiological changes produced by histamine.

An **anti-inflammatory** substance inhibits inflammation.

An **antimetabolite** inhibits a significant metabolic reaction.

An **antineoplastic drug** inhibits the growth of cancer cells.

An **antioxidant** is a reducing agent that prevents the oxidation of molecules such as membrane lipids or oils in foods.

An **antipyretic** is a fever-reducing substance.

An **antiseptic** is a compound applied to living tissue to kill or prevent the growth of microorganisms.

An **apoenzyme** is the protein part of an enzyme.

An **aqueous** solution has water as the solvent.

An **aromatic hydrocarbon** is a hydrocarbon with a benzene-like structure. Resonance theory is used to describe the electronic structure of aromatic compounds.

An **artificial transmutation** is the conversion of one element into another by artificial means.

An **aryl group** is a group derived from an aromatic hydrocarbon by removal of a hydrogen atom.

An **atmosphere (atm)** of pressure is equal to 760 mmHg.

The **atmosphere** is the gaseous envelope surrounding the Earth (or other planet).

An **atom** is the smallest characteristic particle of an element.

The **atomic mass** of an element is the weighted average of the masses of the atoms of the naturally occurring isotopes of the element.

An **atomic mass unit (u)** is *exactly* one-twelfth the mass of an atom of carbon-12. The masses of the fundamental particles—electrons, protons, and neutrons—and of individual atoms are often expressed in these units.

The **atomic number** (Z) is the number of protons in the nucleus of an atom of an element.

The **atomic radius** is a measure of the size of an atom based on the measurement of internuclear distances.

The **aufbau principle** is a hypothetical process for building up an atom from the atom of preceding atomic number by adding a proton and the requisite number of neutrons to the nucleus and one electron to the appropriate atomic orbital.

In an **autoimmune disease,** the immune system attacks the body's own tissues.

Avogadro's hypothesis states that equal volumes of all gases at a fixed temperature and pressure contain an equal number of molecules.

Avogadro's law states that at a fixed temperature and pressure the volume of a gas is directly proportional to the number of molecules.

Avogadro's number is the number of particles of a substance in a mole of the substance: 6.02×10^{23}.

An **azeotrope** is a mixture of two or more liquids that boils at a constant temperature.

The **B complex** is the set of B vitamins. *See* Section F.6 for the list.

A **barbiturate** is a depressant anticonvulsant drug.

A **barometer** is an instrument to measure atmospheric pressure.

A **base** is (1) a compound that produces hydroxide ions, OH^-, in water solution (Arrhenius theory), or (2) a proton acceptor (Brønsted–Lowry theory).

The **base ionization constant** (K_b) is the equilibrium constant describing equilibrium in the reversible ionization of a weak base.

A **basic anhydride** is a substance that reacts with water to produce a basic solution; a metal oxide.

A **benign tumor** is an abnormal growth of new tissue that grows slowly and does not spread to other tissues.

Beta (β)-oxidation is the series of reactions in which fatty acids are degraded into acetyl-CoA by successive oxidations at the β-carbon.

A **beta (β^-) particle** is identical to an electron and is emitted by the nuclei of certain radioactive atoms as they undergo decay.

A **bilayer** is a layer two molecules thick. Phospholipids form bilayers with the polar "heads" sticking out in water and the nonpolar "tails" on the inside.

Bile is an alkaline fluid secreted by the liver that aids in the emulsification and digestion of fats.

Biochemistry is the study of all the substances and processes in living organisms.

Blood pressure is the force exerted by the blood against the inner walls of the arteries.

The **body-centered cubic (bcc)** crystal structure has as its unit cell a cube with a structural unit at each corner and one in the center of the cell. *See also* **unit cell.**

The **body mass index (BMI)** provides a measure of total body fat content using an individual's height and weight. *See* Section 26.6 for equation.

The **Bohr effect** is the relationship between the amount of oxygen released from hemoglobin and the concentration of carbon dioxide and hydrogen ion (H^+) in the blood.

The **boiling point** is the temperature at which the vapor pressure of a liquid becomes equal to the atmospheric pressure.

A **bonding pair (BP)** is a pair of electrons shared between two atoms in a molecule.

Boyle's law states that for a given mass of a gas at constant temperature, the volume varies inversely with the pressure.

Brown fat contains many mitochondria that synthesize no ATP but release energy as heat.

A **buffer solution** is a solution containing a weak acid and its salt, or a weak base and its salt. Small quantities of added acid are neutralized by one buffer component and small quantities of added base by the other. As a result, the solution pH is maintained nearly constant.

A **calorie (cal)** is the amount of energy needed to raise the temperature of 1 g of water by 1 °C (more precisely, from 14.5 to 15.5 °C). 1 cal = 4.184 J.

Cancer. *See* **malignant tumor.**

A **carbohydrate** is a polyhydroxy aldehyde or ketone or a compound that can be hydrolyzed to form a polyhydroxy aldehyde or ketone.

The **carbonyl group (C=O)** is a carbon atom doubly bonded to an oxygen atom.

The **carboxyl group (—COOH)** is the functional group of the organic acids.

A **carboxylic acid (RCOOH)** is an organic compound that contains the —COOH functional group.

A **carcinogen** is a substance or physical entity that causes cancer.

Catabolism is the process of breaking down molecules to provide energy in living systems.

A **catalyst** is a substance that increases the rate of a reaction without itself being consumed in the reaction.

Catalytic reforming is a process of converting aliphatic hydrocarbons with low octane numbers into aromatic compounds with higher octane numbers.

A **cathode** is the electrode at which reduction occurs.

A **cathode ray** is a beam of electrons that travels from the cathode to the anode when an electric discharge is passed through an evacuated tube.

A **cation** is a positively charged ion.

The **Celsius scale** of temperatures defines the freezing point of water as 0 °C and the boiling point of water as 100 °C.

A **cephalin** is a phosphatidate that contains ethanolamine as the alcohol.

A **cerebroside** is a glycosphingolipid found in nerve tissue. It has a sugar unit (usually galactose or glucose), a fatty acid unit, and a sphingosine unit.

Charles's law states that for a given mass of gas at constant pressure, the volume varies directly with the temperature (on the absolute scale).

A **chemical bond** is a force that holds atoms together in molecules or ions in crystals.

A **chemical equation** is a description of a chemical reaction that uses symbols and formulas to represent the elements and compounds involved in the reaction.

A **chemical property** describes how one substance reacts with other substances to produce new substances with altered compositions.

A **chemical symbol** is a representation of an element, made up of one or two letters derived from the English name of the element (or, sometimes, from the Latin name of the element or one of its compounds).

Chemistry is a study of the composition, structure, and properties of matter and of the changes that matter undergoes.

Chemotherapy is the use of chemicals to control or cure diseases.

Chirality means "handedness." A chiral molecule is *not* superimposable on its mirror image.

A **chiral center** is an atom in a molecule that is attached to four different groups.

Chlorination is addition or substitution of chlorine.

A **chlorofluorocarbon (CFC)** is a carbon compound that contains chlorine and fluorine.

Cholesterol is the most abundant steroid in the human body. It is a major component of all cell membranes and a precursor for the steroid hormones and bile salts.

A **chromosome** is a complex of protein and DNA found in the cell nucleus that contains the hereditary material.

The term **cis** is used to describe isomers in which two substituent groups are attached on the same side of a double bond or ring in an organic molecule, or along the same edge of a square planar or octahedral complex ion. *See also* **geometric isomerism.**

In **cloning** multiple identical copies are made of a unique DNA fragment.

A **codon** is a sequence of three adjacent nucleotides in mRNA that codes for one amino acid.

A **coenzyme** is an organic cofactor necessary for the function of an enzyme.

A **cofactor** is a substance necessary for the action of an enzyme. It may be a metal ion or a coenzyme.

A **colligative property** is a physical property of a solution that depends on the concentration of solute in the solution but not on the identity of the solute.

A **colloid** is a dispersion in which the dispersed matter has dimensions in the range from about 1 nm to 1000 nm.

The **combined gas law** is a combination of the three simple gas laws into one. The volume of n moles of a gas is inversely proportional to the pressure and directly proportional to the absolute temperature.

Combustion (burning) is the oxidation of a fuel by oxygen in an exothermic reaction.

The **common ion effect** refers to the ability of ions from a strong electrolyte to repress the ionization of a weak acid or weak base or the solubility of a slightly soluble substance.

A **competitive inhibitor** is one that blocks an enzymatic process by reversibly interacting with the enzyme.

Complementary bases are the base pairs adenine-thymine and guanine-cytosine.

A **complete protein** is one that supplies all the essential amino acids in the quantities needed for the growth and repair of body tissues.

A **compound** is a substance made up of atoms of two or more elements, with the different atoms joined in fixed proportions.

A **concentrated** solution contains a relatively large amount of solute in a given quantity of solvent.

Condensation is the conversion of a gas (vapor) to a liquid.

In **condensation polymerization** monomers with at least two functional groups link together by eliminating small-molecule byproducts.

A **condensed structural formula** is an organic chemical formula that shows the atoms of hydrogen (or other atoms or groups) right next to the carbon atoms to which they are attached.

Cones are the photoreceptor cells that provide us with color vision and function in bright light.

A **conjugate acid** is formed when a Brønsted–Lowry base accepts a proton. Every base has a conjugate acid.

A **conjugate base** is formed when a Brønsted–Lowry acid donates a proton. Every acid has a conjugate base.

A **continuous spectrum** is a spectrum in which there is a continuous variation from one color to another.

A **conversion factor** is a numerical factor by which we multiply a quantity that is expressed in a certain unit to express the quantity in another unit.

In the **Cori cycle,** glucose is oxidized to lactate in peripheral tissues, the lactate is transported to the liver where it is converted back to glucose, and the glucose is transported back to the peripheral tissues.

A **covalent bond** is a bond formed by a shared pair of electrons between atoms.

A **crystal** is a solid having plane surfaces, sharp edges, and a regular geometric shape.

In a **crystal lattice,** the fundamental units making up the crystal—atoms, ions, or molecules—are assembled in a regular, repeating manner, extending in three dimensions through the crystal.

The **curie (Ci)** is a measure of the rate of disintegration of a radioactive material.

A **cytochrome** is a heme-containing globular protein.

The **cytoplasm** is the portion of a plant or animal cell inside the cell membrane and external to the nucleus.

A **dalton** is a unit used by biochemists that is the same as the atomic mass unit.

Dalton's law of partial pressure states that in a mixture of gases each gas expands to fill the container and exerts its own pressure, called a partial pressure, and that the total pressure of the mixture is the sum of the partial pressures.

The ***d*-block** is the portion of the periodic table in which the *d* subshell of the next-to-outermost shell fills, in the aufbau process. The *d*-block comprises the B-group elements found in the main body of the periodic table.

A **deliquescent** substance is one that takes on water of hydration and then continues to absorb water until the hydrated solid dissolves.

Denaturation is any process that disrupts the naturally folded structure of a protein or nucleic acid. This is most often accomplished by heating, the addition of various chemicals, or change in pH.

Denatured alcohol is ethanol to which some toxic or noxious substance has been added to render it unsuitable for drinking.

The **density (*d*)** of a sample of matter is its mass per unit of volume, that is, the mass of the sample divided by its volume.

Deoxyribonucleic acid (DNA) is a polynucleotide containing the sugar D-2-deoxyribose that is found in the nuclei of cells and is the carrier of genetic information.

A **detergent** is a cleaning agent that has a water-soluble head and an oil-soluble tail.

A **dextrorotatory** substance rotates the plane of polarized light to the right.

Diabetes mellitus is a disease characterized by an abnormally high level of glucose in the blood.

Dialysis is a process that separates solvent and small molecules and ions from large ones by allowing the smaller particles to pass through a membrane that blocks the larger ones.

Diastereomers are stereoisomers that are not enantiomers.

Diastolic pressure is the minimum force exerted by the blood against the inner walls of the arteries between heartbeats.

Dietary Reference Intakes (DRI) is a term that encompasses four categories of reference intakes of vitamins, minerals, and other food components important in diet and nutrition. *See* Section F.1 for the specific categories.

Diffusion is the process by which one substance mixes with one or more other substances as a result of the movement of molecules.

Digestion is the hydrolytic process whereby food molecules are broken down into simpler chemical units.

A **dihydric alcohol** (glycol) is one with two OH functional groups.

A **dilute** solution is one that contains relatively little solute in a large quantity of solvent.

Dilution is a process of producing a more dilute solution from a more concentrated one, by the addition of an appropriate quantity of solvent.

A **dimer** is a molecule formed from two monomer units.

A **dipole** is a molecule that has a positive end and a negative end.

Dipole forces are the attractive forces that exist among polar covalent substances.

A **diprotic acid** is an acid that can donate two protons per molecule.

A **disaccharide** is a carbohydrate with molecules that can be hydrolyzed to two monosaccharide units.

A **dispersion force** is an attractive force between an instantaneous dipole and an induced dipole.

Dissociation is the separation of the ions of an ionic substance as it dissolves in water.

A **dissociative anesthetic** is one that causes unresponsiveness to the environment (dissociation) without sleep. It causes hallucinations in some patients.

Distillation is the separation of components of a solution by boiling off the most volatile compounds such as alcohol or water and then condensing their vapors. Solids and high-boiling compounds are left behind.

A **disulfide linkage** is a covalent linkage between two sulfur atoms.

A **diuretic** is a substance that increases the body's output of urine.

A **double bond** is a covalent linkage in which two atoms share *two* pairs of electrons between them.

A **double helix** is a two-stranded helical structure in which the two strands wrap around each other.

A **drug** is any substance that affects an individual in such a way as to bring about physiological, emotional, or behavioral change.

A **D sugar** has the OH group on the right at the chiral center farthest from the carbonyl group, when the structure is drawn in a Fischer projection.

Dynamic equilibrium occurs when two opposing processes occur at exactly the same rate, with the result that no net change occurs.

An **efflorescent** substance is a hydrate that gives up water to the atmosphere.

Electric current is the flow of electrons through a conductor or of ions through a solution or melt.

An **electrochemical cell** is a combination of two compartments in which metal electrodes are joined by a wire, and the solutions are brought into contact through a salt bridge or by other means.

An **electrode** is a metal strip or carbon rod dipped into a solution or molten compound to carry electricity to or from the liquid. *See also* **anode** and **cathode.**

Electrolysis is the decomposition of compounds by passing electricity through an ionic solution or a molten salt.

An **electrolyte** is a compound that, when melted or taken into solution, conducts an electric current.

The **electron** is a particle carrying the fundamental unit of negative electric charge. Electrons have a mass of 0.0005486 u and are found outside of the nuclei of atoms.

Electron affinity is the energy change that occurs when an electron is added to an atom in the gaseous state.

Electron capture (E.C.) is a type of radioactive decay in which a nucleus absorbs an electron from the first or second electronic shell.

The **electronegativity** of an element is a measure of the tendency of its atoms in molecules to attract electrons to themselves.

The **electronic configuration** of an atom describes the distribution of electrons among the atomic orbitals in the atom.

The **electron transport chain** is the sequence of highly organized oxidation-reduction enzymes found in the inner mitochondrial membrane that transfers electrons from $FADH_2$ and NADH to the ultimate electron acceptor O_2, forming H_2O.

Electrophoresis is the process of separating a mixture by application of a buffer solution and an electric current.

An **element** is a substance composed of a single type of atom. Elements are the fundamental substances from which all material things are made.

An **emulsifying agent** is a material that can stabilize an emulsion of two otherwise insoluble substances.

An **emulsion** is a colloid that has droplets of one liquid dispersed throughout another liquid; for example, droplets of oil dispersed in water.

Enantiomers are mirror-image isomers. An enantiomer is not superimposable on its mirror image.

In **endocrine communication,** chemical messengers, known as hormones, are released in one tissue and transported through the circulatory system to one or more other tissues.

An **endorphin** is a naturally occurring peptide that bonds to the same receptor site as an opiate drug.

Endothermic describes a process or reaction in which, in a nonisolated system, heat is absorbed from the surroundings.

Energy is the capacity for doing work.

An **energy level** is the state of an atom determined by the location of its electrons among the various principal shells and subshells.

Energy of activation. *See* **activation energy.**

An **enkephalin** is a morphine-like substance produced by the body that is composed of a peptide chain of five amino acid units.

The **enthalpy change (ΔH)** in a chemical reaction is equal to the heat of reaction at constant temperature and pressure.

Enzymes are biological catalysts.

Equilibrium is a condition that is reached when two opposing processes occur at equal rates. As a result, the concentrations of the reacting species remain constant with time.

The **equilibrium constant (K)** is the constant that relates the concentrations of the species in an equilibrium.

The **equilibrium constant expression** is a particular ratio of concentrations of products to reactants in a chemical reaction at equilibrium. The expression has a constant value at a given temperature.

The **equivalence point** of a titration is the point at which two reactants have been introduced into a reaction mixture in their stoichiometric proportions.

Erythrocytes are red blood cells. They contain the hemoglobin that transports oxygen to the tissues.

An **essential amino acid** is one not produced in adequate amounts in the body, so it must be included in the diet.

An **essential fatty acid** is one not produced in the body, so it must be included in the diet.

An **ester (R′COOR)** is a compound derived from a carboxylic acid and an alcohol. The —OH of the acid is replaced by an —OR group.

An **estrogen** is a female steroid sex hormone such as estradiol.

An **ether (R′OR)** is an organic compound that has an oxygen atom between two hydrocarbon groups.

An **excited state** of an atom is one in which one or more electrons has been promoted to a higher energy level than in the ground state. *See also* **ground state.**

Exothermic describes a process or reaction in which, in a nonisolated system, heat is given off to the surroundings.

The term **expanded valence shell** refers to a situation in which the central atom in a Lewis structure is able to accommodate more than the usual octet (eight) of electrons in its valence shell. Expanded valence shells are encountered in molecules and polyatomic ions in which the central atom is a nonmetal of the third period or beyond.

Extracellular fluid is fluid located outside cells.

A **face-centered cubic (fcc)** crystal structure has as its unit cell a cube with a structural unit at each of the corners and in the center of each face of the cube. *See also* **unit cell.**

The **Fahrenheit scale** is a temperature scale that defines the freezing point of water as 32 °F and the boiling point of water as 212 °F.

A **fat** is a triglyceride that is solid at room temperature.

A **fat cell** is found in adipose tissue. The cell contains a large droplet of triglycerides.

A **fat depot** is a location in the body that has large amounts of adipose tissue.

A **fat-soluble vitamin** is one that dissolves in the fatty tissue of the body and is stored for future use. They are vitamins A, D, E, and K.

Fatty acids are carboxylic acids found as structural components of fats and oils.

Fatty acid synthase is an enzyme complex that contains all of the enzyme activities needed for the synthesis of palmitic acid from acetyl-CoA and malonyl-CoA.

The ***f*-block** is the portion of the periodic table in which the *f* subshell of the second-from-outermost shell fills, in the aufbau process. The *f*-block consists of the lanthanides and actinides.

In **feedback inhibition,** the enzyme that catalyzes the first step in a series of reactions is inhibited by the final product.

Fibrinogen is the protein that is converted to fibrin by the enzyme thrombin in the clotting of blood.

A **fibrous protein** is a highly insoluble structural protein composed of peptides arranged in filaments or sheets.

A **Fischer projection** represents a molecule with one or more chiral centers, diagramming horizontal bonds projecting toward the viewer and vertical bonds projecting away from the viewer.

Fission is the splitting of an atomic nucleus into two large fragments.

Formed elements are cells and cell fragments in the blood: erythrocytes, leukocytes, and platelets.

Formula mass is the mass of a formula unit relative to that of a carbon-12 atom. It is the sum of the masses of the atoms or ions represented by the formula.

A **formula unit** is the simplest combination of atoms or ions consistent with the formula of a compound.

A **free radical** is a highly reactive atom or molecular fragment characterized by having one or more unpaired electron(s).

A **fuel cell** is a device in which chemical reactions are used to produce electricity directly from fuels and oxygen.

A **functional group** is an atom or grouping of atoms in or on an organic molecule that confers characteristic properties to the molecule as a whole.

Fusion is the combination of atomic nuclei to form a larger one.

Galactosemia is a genetic disease that results from the absence of an enzyme needed to convert galactose to glucose.

A **gamma (γ) ray** is a form of electromagnetic radiation emitted by the nuclei of certain radioactive atoms as they undergo decay. Gamma rays are similar to X-rays but have higher energy and are more penetrating.

A **ganglioside** is a glycosphingolipid with several monosaccharide units (one of which is N-acetylneuraminic acid), a sphingosine unit, and a fatty acid unit.

A **gas** is a state of matter in which the substance maintains neither shape nor volume.

Gastric juice is a mixture of substances secreted by the stomach that initiates the hydrolysis (digestion) of proteins.

A **gene** is a segment of a DNA molecule that codes for the biosynthesis of one polypeptide chain.

A **general anesthetic** is a substance that produces unconsciousness and insensitivity to pain.

The **genetic code** is the set of three-nucleotide codons that codes for the amino acids of proteins.

A **genetic disease** is caused by an inherited mutation that causes metabolic abnormalities.

The **genome** is all the genetic information present in a cell or virus.

Geometric isomers (cis, trans isomers) have different configurations because of the presence of a rigid structure such as a double bond or ring.

Global warming is the anticipated increase in Earth's average temperature resulting from the injection of CO_2 and other heat-absorbing gases into the atmosphere.

A **globular protein** is roughly spherical in shape and soluble in water as colloidal particles.

Globulins are proteins in the blood that are involved in fighting infectious diseases and that form complexes that transport lipids.

A **glucogenic amino acid** is an amino acid that can be converted to pyruvate or another intermediate that can be used to synthesize glucose.

Gluconeogenesis is the synthesis of glucose from noncarbohydrate precursors.

The **glucose tolerance test,** a test for diabetes, measures blood sugar levels after ingestion of glucose.

Glycogenesis is the formation of glycogen from glucose.

Glycogenolysis is the breakdown of glycogen to form glucose.

Glycol. *See* **dihydric alcohol.**

A **glycolipid** is any complex lipid with one or more carbohydrate units.

Glycolysis is the sequence of metabolic reactions in which glucose is converted to pyruvate.

A **glycosidic linkage** is the acetal linkage that joins the hemiacetal of a monosaccharide to another alcohol function by ether formation.

Gout is an affliction caused by abnormal purine metabolism that produces excess uric acid, which is deposited as urate salts in joints.

The **greenhouse effect** refers to the ability of $CO_2(g)$ and certain other gases to absorb and trap energy radiated by Earth's surface as infrared radiation.

The **ground state** of an atom is the state in which all electrons are in their lowest possible energy levels.

A **group** of the periodic table is a vertical column of elements having similar properties.

The **half-life** is the time in which one-half of the atoms of a radioisotope disintegrate.

A **half-reaction** is a portion of an oxidation-reduction reaction, representing either the oxidation process or the reduction process.

A **halogen** is an element in Group 7A of the periodic table.

A **halogenated hydrocarbon** is one in which one or more hydrogen atoms has been replaced by a halogen atom.

Heat is an energy transfer into or out of a system caused by a difference in temperature between a system and its surroundings.

The **heat index** is a measure of the discomfort caused by a combination of high temperature and high humidity.

The **heat of reaction** is the quantity of heat exchanged between a system and its surroundings when a chemical reaction occurs at a constant temperature and pressure.

A **heat stroke** occurs when the heat regulation system of the body fails, resulting in high fever and physical collapse.

The **hematocrit** is the volume (in percent) of formed elements in a sample of whole blood.

Heme is an organometallic complex containing an iron ion. It is found in cytochromes, hemoglobin, and myoglobin.

A **hemiacetal** has had an alcohol molecule added across the carbon–oxygen double bond of an aldehyde. It has an OH group and an OR group on the same carbon atom.

A **hemiketal** has had an alcohol molecule added across the carbon–oxygen double bond of a ketone. It has an OH group and an OR group on the same carbon atom.

Hemophilia is a set of genetic diseases characterized by the inability of the blood to clot properly, leading to excessive bleeding.

The **Henderson–Hasselbalch equation** relates the pH of a solution of a weak acid and a salt of the weak acid to the pK_a of the weak acid and to the molar concentrations of the weak acid and its salt.

Henry's law states that the solubility of a gas is directly proportional to the pressure maintained in the gas above the solvent.

A **heterocyclic compound** is a cyclic compound in which one or more atoms in the ring is an element other than carbon.

A **heterogeneous mixture** is a mixture in which the composition and properties vary from one region to another within the mixture.

A **holoenzyme** is a fully active enzyme consisting of an apoenzyme and a cofactor.

A **homogeneous mixture** is a mixture having the same composition and properties throughout the given mixture.

A **homologous series** of compounds differ by one carbon atom and two hydrogen atoms as one proceeds through the series. The properties of these compounds also vary in a regular manner.

A **hormone** is a chemical messenger that is secreted into the blood stream by an endocrine gland.

Hormone replacement therapy is the use by menopausal women of estrogen alone or in combination with progesterone (or its derivatives) to reduce the risk of osteoporosis and heart disease.

A **hybridization probe** is a short piece of DNA that has a nucleotide sequence complementary to a known sequence in a specific gene.

A **hydrate** is a solid compound that incorporates water molecules into its basic structure. A hydrate of an aldehyde or ketone has had water added across the carbon–oxygen double bond. It has two OH groups on the same carbon atom.

Hydration is the addition of water to a substance; in organic chemistry, the addition of water across the carbon–carbon double bond of an alkene or the carbon–oxygen double bond of an aldehyde or ketone.

A **hydrocarbon** is an organic compound that contains only carbon and hydrogen.

A **hydrogen bond** is a type of intermolecular force in which a hydrogen atom covalently bonded in one molecule is simultaneously attracted to a nonmetal atom in a neighboring molecule. Both the atom to which the hydrogen atom is bonded and the one to which it is attracted must be small atoms of high electronegativity—N, O, or F.

In a **hydrogenation** reaction, $H_2(g)$ reacts at a carbon–carbon double or triple bond or a carbon–oxygen double bond to add H atoms to C atoms.

A **hydrolase** is an enzyme that catalyzes a hydrolysis reaction.

Hydrolysis (literally, a splitting by water) is the reaction of a substance with water.

A **hydronium ion (H_3O^+)** is a water molecule to which a hydrogen ion (H^+) has been added; the characteristic ion of an aqueous acid.

A **hydrophilic** substance (literally, a water-loving substance) has an affinity for water.

A **hydrophobic** substance (water-hating substance) lacks an affinity for water.

The **hydrosphere** is the oceans, seas, rivers, and lakes of the Earth.

The **hydroxyl group** is the —OH group.

A **hygroscopic** substance is one that absorbs water vapor from the atmosphere to form a hydrate.

Hyperglycemia means the blood sugar concentration is above normal.

Hypertension is high blood pressure.

A **hypertonic solution** is a solution having an osmotic pressure greater than that of body fluids (blood, tears). A *hyper*tonic solution has a greater osmotic pressure than an *iso*tonic solution.

Hyperventilation is breathing too rapidly.

Hypoglycemia means the blood sugar concentration is below normal.

A **hypotonic solution** is a solution having an osmotic pressure less than that of body fluids (blood, tears). A *hypo*tonic solution has a lower osmotic pressure than an *iso*tonic solution.

Hypoventilation is breathing too little.

The **ideal gas equation** (or **ideal gas law**) states that the volume of a gas is directly proportional to the amount of a gas and its Kelvin temperature and inversely proportional to its pressure. Mathematically, it can be stated through the equation $PV = nRT$.

The **immune response** is the synthesis of antibodies in response to invading bacteria or viruses or other foreign matter.

An **incomplete protein** is one that does not supply all the essential amino acids in the quantities needed for the growth and repair of body tissues.

An **indicator dye** is a dye whose color depends on the acidity of the solution.

The **induced-fit theory** states that the active site on an enzyme changes its shape somewhat to fit the substrate, much as a glove changes shape to fit a hand.

Industrial smog is polluted air associated with industrial activities. The principal pollutants are oxides of sulfur and particulate matter.

Inflammation is a tissue response to injury or stress.

Inorganic chemistry is the chemistry of all the compounds except those containing carbon.

An **integral protein** is one that spans the lipid bilayer of the cell membrane.

Interstitial fluid is the fluid that surrounds most cells and fills any space between them.

Intermolecular forces are the forces of attraction between molecules.

The **International System of Units (SI)** is the measuring system used by scientists. It is based on seven base quantities and their multiples and submultiples.

Intracellular fluid is the fluid located inside cells.

Invert sugar is an equimolar mixture of glucose and fructose, formed by the hydrolysis of sucrose.

The **iodine number,** an indication of the degree of unsaturation, is the number of grams of iodine that will be consumed by 100 g of fat or oil.

An **ion** is an electrically charged particle comprised of one or more atoms.

Ionic bonds are attractive forces between positive and negative ions, holding them together in a solid crystal.

An **ionic compound** is a compound that consists of oppositely-charged ions held together by electrostatic attractions.

Ionization of an acid or base is the formation of ions by the reaction of a molecular acid or base with water.

Ionization energy is the energy required to remove the least tightly bound electron from a ground-state atom (or ion) in the gaseous state.

Ionizing radiation is radiation that causes the formation of ions from neutral particles.

The **ion product of water (K_W),** is the product of the concentration of hydronium ion, $[H_3O^+]$, and the concentration of hydroxide ion, $[OH^-]$, in pure water or a water solution. At 25 °C, its value is 1.0×10^{-14}.

An **irreversible inhibitor** is bound to an enzyme by a covalent bond and irreversibly destroys all activity.

The **isoelectric pH** is the pH value at which an amino acid or protein exists in the zwitterionic form.

An **isomerase** is an enzyme that catalyzes the interconversion of isomers.

Isomerization is the conversion of a compound into one of its isomers.

Isomers are compounds that have the same molecular formula but different structural formulas and properties.

An **isotonic solution** is one that has the same osmotic pressure as body fluids (blood, tears).

Isotopes are atoms that have the same number of protons—the same atomic number—but different numbers of neutrons—different mass numbers.

The **IUPAC system of nomenclature** is a systematic way of naming chemical substances so that each has a unique name.

A **joule** is the SI unit for energy.

A **kelvin (K)** is the unit of a temperature interval on the Kelvin temperature scale: One kelvin has the same magnitude as one degree on the Celsius scale.

The **Kelvin scale** is an absolute temperature scale with its zero at −273.15 °C.

A **ketal** is a product of a hemiketal-alcohol reaction. It has two OR groups attached to the same carbon atom.

A **ketogenic amino acid** is one that is degraded to a ketone body.

A **ketone** is an organic compound whose molecules have a *carbonyl* functional group between two hydrocarbon groups.

A **ketone body** is either acetoacetate, β-hydroxybutyrate, or acetone.

A **ketose** is a sugar with a ketone group.

Ketosis is a condition characterized by elevated levels of ketone bodies in the blood.

The **kilogram (kg)** is the SI base unit of mass.

Kinetic energy is the energy of motion.

The **kinetic-molecular theory** is a model that uses the motion of molecules to explain the behavior of gases.

The **Krebs cycle** is the complex series of reactions by which acetyl-CoA is oxidized to carbon dioxide and water.

Lactose intolerance is the inability to break down the sugar lactose, caused by a deficiency of lactase.

The **law of combining volumes** states that when gases measured at the same temperature and pressure are allowed to react, the volumes of gaseous reactants and products are in small whole-number ratios.

The $\mathbf{LD_{50}}$ of a substance is the dosage that is lethal to 50% of the population of test animals.

Le Châtelier's principle states that if a stress is applied to a system at equilibrium, the equilibrium shifts in the direction that will relieve the stress.

A **lecithin** is a phosphatidate that contains choline as the alcohol unit.

Leptin is a hormone secreted by adipose tissue that has been implicated in weight control. Leptin regulates body fat.

Leukemia is a cancer that is characterized by the uncontrolled production of leukocytes that fail to mature.

Leukocytes are white blood cells.

A **levorotatory** substance rotates the plane of polarized light to the left.

A **Lewis structure** is a representation of an element in which the chemical symbol stands for the core of the atom and dots are placed around the symbol for its valence electrons; a representation of covalent bonding through Lewis structures of atoms, shared electron pairs, and lone-pair electrons.

A **ligase** is an enzyme that catalyzes the union of two molecules.

The **limiting reagent** is the reactant that is used up first in a reaction, after which the reaction ceases no matter how much remains of other reactants.

Lipids are the components of biological systems that are relatively insoluble in water but soluble in nonpolar or slightly polar solvents such as hexane and diethyl ether.

A **lipoprotein** is a protein-lipid complex that transports lipids in the bloodstream.

A **liquid** is a state of matter in which the substance assumes the shape of its container, flows readily, and maintains a fairly constant volume.

A **liter (L)** is a metric unit of volume equal to one cubic decimeter or 1000 cubic centimeters: $1 \text{ L} = 1 \text{ dm}^3 = 1000 \text{ cm}^3$.

A **local anesthetic** is a substance that produces insensitivity to pain yet leaves the patient conscious.

The **lock-and-key theory** of enzyme action holds that an enzyme and its substrate fit together like a lock and its key.

Lone pairs (LP) are electron pairs assigned exclusively to one of the atoms in a Lewis structure. They are not shared, and hence not involved in the chemical bonding. (Also called *nonbonding pairs.*)

An **L sugar** has the OH group on the right at the chiral center farthest from the carbonyl group, when the structure is drawn in a Fischer projection.

A **lyase** is an enzyme, such as a decarboxylase, that catalyzes the nonhydrolytic cleavage of its substrate.

Lymph is interstitial fluid that has been absorbed into the lymph capillaries.

A **main-group** element is an element in which the subshell being filled in the aufbau process is either an *s* or *p* subshell of the principal shell of highest principal quantum number (the outermost shell). Main-group elements are located in the *s*- and *p*-blocks of the periodic table.

A **malignant tumor,** often called a cancer, grows and invades and destroys other tissues.

Marijuana is a psychoactive drug made from the leaves, flowers, seeds, and small stems of the *Cannabis* plant.

Markovnikov's rule states that when H—X adds across the double bond in an unsymmetrical alkene, the H goes on the carbon atom that already has the most hydrogen atoms attached. The X, which can be a halogen, OH, and so on, adds to the carbon atom at the other end of the double bond.

The **mass** of an object is a measure of the quantity of matter in the object.

The **mass number (A)** is the sum of the number of protons and neutrons in the nucleus of an atom.

Mass/volume percent is an expression of concentration in which the mass of the solute is divided by the volume of the solution and that quotient multiplied by 100%.

Matter is the stuff that makes up bodies detectable by the senses; it includes any entity that has mass when at rest.

A **mechanism** is a series of individual steps in a chemical reaction that gives the net overall change.

The **melting point** of a solid is the temperature at which it melts, that is, comes into equilibrium with the liquid phase.

A **meso compound** is a stereoisomer that contains at least two chiral centers but has an internal symmetry plane.

A **metabolic pathway** is a series of biochemical reactions that explains how an organism converts a given reactant to a desired end product.

Metabolism is the set of all chemical reactions in living systems that break down large molecules for energy and component parts and build large molecules from component parts.

A **metal** is an element having a distinctive set of properties: luster, good heat and electrical conductivity, malleability, and ductility. Metal atoms generally have small numbers of valence electrons and a tendency to form cations. Metals are found to the left of the stepped diagonal line in the periodic table. All *s*-block (except hydrogen and helium), *d*-block, and *f*-block elements are metals, as are a few in the *p*-block.

A **metalloid** is an element that has the physical appearance of a metal but some nonmetallic properties as well. Metalloids are located along the stepped diagonal line in the periodic table.

The **meter (m)** is the SI base unit of length.

A **micelle** is a cluster of molecules, such as phospholipids or soaps, that contain both polar and nonpolar groups. In water, the polar "heads" stick out into the water, and the nonpolar "tails" are turned to the inside.

A **millimeter of mercury (mmHg)** is a unit used to express gas pressure: 1 mmHg = 1/760 atm (exactly). Also called a *torr.*

Miscible substances can be mixed in all proportions.

Mitochondria are subcellular units (organelles) that contain the enzymes necessary for the Krebs cycle; they are the "power plants" of the cells.

A **mixture** is a type of matter that has a composition and properties that may vary from one sample to another.

Molar heat of fusion is the quantity of heat that must be absorbed to melt 1 mol of a solid at a constant temperature.

Molar heat of vaporization is the quantity of heat that must be absorbed to vaporize 1 mol of a given liquid at a constant temperature.

The **molar mass** of a substance is the mass of 1 mol of the substance. It is numerically equal to the atomic weight, molecular weight, or formula weight, and expressed as grams per mole.

The **molar volume of a gas** is the volume occupied by 1 mol of a gas at STP (22.4 L).

Molarity is an expression of the concentration of a solution in moles of solute per liter of solution.

A **mole (mol)** is the amount of substance that contains 6.02×10^{23} units of the substance.

The **molecular mass** is the average mass of a molecule of the substance relative to that of a carbon-12 atom. It is the sum of the masses of the atoms represented in the molecular formula.

A **molecule** is a discrete group of atoms held together by one or more shared pairs of electrons.

A **monohydric alcohol** is one with only one OH functional group.

A **monolayer** is a layer one molecule thick. Phospholipids in water form monolayers with the polar "head" in the water and the nonpolar "tails" sticking up into the air.

Monomers are small molecules that can be combined to make polymers.

A **monoprotic acid** is an acid that can donate only one proton per molecule.

A **monosaccharide** is a carbohydrate that cannot be hydrolyzed to simpler sugars.

A **monounsaturated fatty acid** contains one carbon–carbon double bond.

A **mutagen** is any chemical or physical agent that causes mutations.

Mutarotation is the change in observed optical rotation of plane polarized light as + and/or − forms of sugars move toward an equilibrium value.

A **mutation** is any chemical or physical change that alters the sequence of bases in DNA.

A **narcotic** is a drug that produces both narcosis (a profound stupor) and relief of pain.

Natural gas is a fuel composed of low molar mass hydrocarbons, mainly methane, with some ethane, propane, butane, and traces of higher alkanes. It is formed beneath the surface of the ground by the decay of plant and animal matter.

A **nephron** is the basic functional unit of the kidney.

A **net ionic equation** is an equation that represents the actual molecules or ions that participate in a chemical reaction, eliminating all nonparticipating species ("spectator" ions).

A **neuron** is a nerve cell.

A **neurotransmitter** is a chemical that carries an impulse across the synapse from a nerve cell to a receptor on a receiving cell.

Neutralization is the reaction of an acid and a base to produce a salt and water.

A **neutron** is a fundamental particle, found in the nucleus of atoms, that has a mass of 1.0087 u and no electric charge.

Nitrogen balance is the state in which an individual's intake of dietary nitrogen is equal to the amount of nitrogen excreted.

The **noble gases** are the elements in Group 8A of the periodic table. They have the valence shell electron configuration ns^2np^6 (except helium, $1s^2$).

A **noncompetitive inhibitor** binds to an enzyme at a different site than does the substrate. By changing the shape of the enzyme molecule, the noncompetitive inhibitor blocks enzyme action.

A **nonelectrolyte** is a substance that exists exclusively or almost exclusively in molecular form, whether in the pure state or in solution.

Nonmetals are elements that lack metallic properties. They are generally poor conductors of heat and electricity and brittle when in the solid state. They usually have larger numbers of valence electrons than do metals.

A **nonpolar covalent bond** is a covalent bond in which electrons are shared equally.

Nonpolarized light is ordinary light, light that is not polarized.

The **normal boiling point** of a liquid is the temperature at which the liquid boils when the prevailing atmospheric pressure is 1 atm.

Nuclear fission is the splitting of a large unstable nucleus into two lighter fragments and two or more neutrons. Mass destroyed in this process is converted to an equivalent quantity of energy, which may be used for power generation.

Nuclear fusion is the joining together or fusing of lighter nuclei into a heavier one. In the process some matter is converted to energy, which is released in an uncontrolled manner. Scientists hope to use controlled nuclear fusion for power generation.

A **nucleic acid** is a polynucleotide.

A **nucleoside** consists of a five-carbon sugar and a purine or pyrimidine base. It is a nucleotide without the phosphoric acid unit.

A **nucleotide,** the monomer unit of nucleic acids, is a molecule containing a heterocyclic amine, a pentose sugar, and phosphoric acid.

The **nucleus** is (1) the concentrated, positively charged matter at the center of an atom, composed of protons and neutrons; or (2) a membrane-enclosed structure within a plant or animal cell that contains the genetic material (also referred to as the *cell nucleus*).

The **octane rating** measures the antiknock properties of a sample of gasoline on a scale that has isooctane at 100 and heptane at 0.

The **octet rule** states that most covalently bonded atoms represented in a Lewis structure have eight electrons in their valence shells.

An **odorant** is a chemical capable of stimulating an olfactory receptor.

A food **oil** is a triglyceride that is a liquid at room temperature.

An **oncogene** is a gene that, when mutated, contributes to the development of a cancer.

An **optically active** substance is one that rotates a beam of polarized light.

The **optimum pH** for an enzyme is the pH at which it exhibits maximum activity.

An **orbital** is a wave function that describes the space occupied by electrons with specific values for the main energy level, sublevel, and directional qualities.

Organic chemistry is the chemistry of compounds of carbon.

An **osmole (osmol)** is the number of moles of a substance multiplied by the number of particles formed by each formula unit of solute.

Osmosis is the net flow of a solvent through a semipermeable membrane, from pure solvent into a solution or from a solution of a lower concentration into one of a higher concentration.

The **osmotic pressure** of a solution is the pressure that must be applied to the solution to prevent the flow of solvent molecules into the solution when the solution and pure solvent are separated by a semipermeable membrane.

Oxidation is a process in which the oxidation state of an element increases; that is, in which electrons are "lost."

The **oxidation number** of an element is a means of designating the number of electrons that its atoms lose, gain, or share in forming compounds.

Oxidative deamination is a reaction that removes the amino group of an amino acid as NH_4^+. The carbon atom that bore the amino group is oxidized to a carbonyl group.

Oxidative phosphorylation is the process that links ATP synthesis to oxygen consumption in the electron transport chain.

An **oxidizing agent** is a substance that causes oxidation and is itself reduced.

An **oxidoreductase** enzyme catalyzes a reaction in which one substance is oxidized and another is reduced.

An **oxygen debt** is an oxygen deficit resulting from anaerobic activity.

In **paracrine communication,** chemical messengers, known as paracrine factors, move from one cell to another within a single tissue, such as the liver.

Paracrine factors are chemical messengers that transfer information from cell to cell within a single tissue. Examples are the prostaglandins.

A **pascal (Pa)** is the basic unit of pressure in SI. It is a pressure of 1 newton per square meter, $1\ N/m^2$.

The ***p*-block** is the portion of the periodic table in which the *p* subshell of the outer shell fills, in the aufbau process. The *p*-block elements are all main-group elements.

A **peptide** is a compound that has two or more amino acids joined through peptide bonds.

A **peptide bond** is the amide linkage between amino acids in chains of peptides, polypeptides, and proteins.

Percent by mass is an expression of solution concentration in which the mass of the solute is divided by the mass of the solution and that quotient multiplied by 100%.

Percent by volume is an expression of solution concentration in which the volume of the solute is divided by the volume of the solution and that quotient multiplied by 100%.

The **periodic law** states that certain sets of physical and chemical properties recur at regular intervals (periodically) when the elements are arranged according to increasing atomic number.

The **periodic table of the elements** is a tabular arrangement, organized according to increasing atomic number, that places elements having similar properties into the same vertical columns. (Mendeleev's original periodic table was arranged according to atomic weights, not atomic numbers.)

Periods are the horizontal rows of elements in a periodic table. In a modern table, the periods range in width from two members (first period) to 32 members (sixth and seventh periods).

A **peripheral protein** is loosely bound to one side of a cell membrane through hydrogen bonds or ionic interactions.

A **petrochemical** is a synthetic substance made from petroleum or natural gas.

Petroleum is a naturally occurring liquid mixture, consisting mainly of hydrocarbons.

The **pH** is the negative of the logarithm of the hydronium ion concentration in a solution: $pH = -\log [H_3O^+]$.

The **pH scale** is an exponential scale of acidity; a pH below 7 is acidic; exactly 7, neutral; above 7, basic.

A **phenol (ArOH)** is an aromatic compound with a hydroxyl group attached directly to a benzene ring.

Pheromones are chemicals that are used for communication between members of the same species of insects.

The **phosphatidates** are esters of glycerol in which there are two fatty acid groups and a phosphoric acid unit esterified with another alcohol molecule, such as ethanolamine.

A **phospholipid** is a phosphate-containing polar lipid.

Photochemical isomerization is the conversion of a compound into one of its isomers by the action of light.

Photochemical smog is air that is polluted with oxides of nitrogen and unburned hydrocarbons, together with ozone and several other components produced by the action of sunlight.

Photosynthesis is the process by which green plants use solar energy to form glucose from carbon dioxide and water.

A **physical property** is a property that can be observed and specified without reference to any other substance and that does not produce changes in composition.

Phytochemicals are nonnutrient compounds found in plant-derived foods that have biological activity in the body.

Plasma is the liquid, noncellular part of blood.

A **plasmid** is a circular piece of DNA found in a bacterium that replicates independently of the bacterial chromosome.

Platelets are small cell fragments that are involved in blood clotting.

The **pOH** is the negative of the logarithm of the hydroxide ion concentration in a solution: $pOH = -\log [OH^-]$.

A **point mutation** is a mutation that occurs at a single nucleotide position.

In a **polar covalent bond** between two atoms, electrons are drawn closer to the more electronegative atom, creating a separation of charge. One end of the bond has a small negative charge, $\delta-$, and the other end, a small positive charge, $\delta+$.

A **polarimeter** is an instrument that is used to measure the optical activity of compounds.

Polarized light is light that vibrates in a single plane.

A **polyamide** is a condensation polymer in which the monomer units are joined by an amide linkage.

A **polyatomic ion** is an ion consisting of two or more atoms bonded together.

A **polyester** is a condensation polymer in which the monomer units are joined by an ester linkage.

A **polymer** is a giant molecule formed by the combination of smaller molecules (monomers) in a repeating manner.

Polymerization is a type of reaction in which small repeating units (monomers) combine to form giant molecules (polymers).

A **polypeptide** is a polymer of amino acids, usually of lower molar mass than a protein.

A **polyprotic acid** is capable of donating more than one proton to an appropriate base.

A **polysaccharide** is a polymeric carbohydrate that can be hydrolyzed into many monosaccharide units.

A **polyunsaturated fatty acid** contains two or more carbon–carbon double bonds.

A **positron (β^+)** is a positively charged particle having the same mass as β^- particles. Sometimes called "positive" electrons, positrons are emitted by certain radioactive nuclei.

Potential energy is energy due to position or arrangement. It is the energy associated with forces of attraction and repulsion between objects.

A **precipitate** is an insoluble substance formed in a chemical reaction between ions in a solution.

The **precision** of a set of measurements refers to how closely members of a set of measurements agree with one another. It reflects the degree of reproducibility of the measurements.

Pressure *(P)* is a force per unit area, that is, $P = F/A$.

A **primary (1°) alcohol** is one that bears the OH group on a carbon atom that is attached to only one other carbon atom.

A **primary (1°) amine** is one that has only one alkyl or aryl group on the nitrogen atom.

The **primary structure** of a protein is its sequence of amino acids.

The **products** are the substances that are produced in a chemical reaction. Their formulas appear on the right side of a chemical equation.

A **proenzyme** is an inactive protein from which an enzyme is formed.

A **progestin** is a compound that mimics the action of progesterone.

Prostaglandins are hormonelike compounds derived from arachidonic acid that are involved in increased blood pressure, the contraction of smooth muscle, or other physiological processes.

A **protein** is a polymer of amino acids.

A **proton** is a subatomic particle carrying the fundamental unit of positive charge. Protons have a mass of 1.0073 u and are found in the nuclei of atoms.

A **proton acceptor,** a base, is a substance that accepts H^+ (a proton).

A **proton donor,** an acid, is a substance that gives up H^+ (a proton).

A **provitamin** is a substance that the body can convert into a vitamin.

A **purine** is a nitrogenous base with two fused rings. Adenine and guanine are examples.

A **pyrimidine** is a nitrogenous base with a single ring. Examples are thymine, cytosine, and uracil.

A **pyrogen** is a substance that causes a fever.

A **quaternary (4°) ammonium salt** has four alkyl or aryl groups on a nitrogen atom.

The **quaternary structure** of a protein is the arrangement of two or more polypeptide chains into a specific arrangement of subunits.

A **racemic mixture** is a mixture of enantiomers that is optically inactive because it contains equimolar amounts of molecules with opposite rotatory power.

The **rad** is a unit of absorbed radiation equal to 0.01 J/kg.

Radioactive decay is a nuclear process involving the emission of particles or energy by which a nucleus is eventually converted to a stable nonradioactive nucleus.

Radioactivity is the spontaneous emission of ionizing radiation by the atomic nuclei of certain isotopes.

A **radioisotope** is a radioactive isotope.

The **reactants** are the starting materials or substances consumed in a chemical reaction. Their formulas appear on the left side of a chemical equation.

Recombinant DNA technology is a set of techniques that incorporates genetic material from one organism into the DNA of another organism.

A **reducing agent** is a substance that causes reduction and is itself oxidized.

A **reducing sugar** is any carbohydrate that is capable of reducing a mild oxidizing agent such as Tollens's reagent.

Reduction is a process in which the oxidation state of an element decreases, that is, in which electrons are "gained."

The **reduction potential** is a measure of the tendency of a substance to gain electrons compared with the standard hydrogen electrode.

The **relative humidity** expresses the water vapor content of air as a percent of the maximum water vapor content possible.

The **rem** is a unit of ionizing radiation that produces the same damage to humans as 1 roentgen of high voltage X-rays.

The **renal threshold** is the blood level of a substance, such as glucose, above which the substance appears in the urine.

Replication is the process in which all of the DNA in a cell is copied prior to cell division.

Resonance is a term used to describe a situation in which two or more plausible Lewis structures can be written to represent a species but in which the true structure cannot be written. The plausible structures are called contributing structures.

A **resonance hybrid** is a composite of the contributing resonance structures and is the true structure of a molecule that exhibits resonance.

Respiration includes all metabolic processes in which gaseous oxygen is used to oxidize organic matter to carbon dioxide, water, and energy.

Respiratory chain. *See* **electron transport chain.**

A **retrovirus** has RNA as its genetic material. It synthesizes DNA in the host cell.

A **reversible inhibitor** binds to an enzyme through noncovalent interactions and lowers the activity of the enzyme.

A **reversible reaction** can proceed in either the forward or the reverse direction, depending on conditions.

Reye's syndrome is a liver disorder occurring primarily in children and associated with their use of aspirin.

Ribonucleic acid (RNA), a polynucleotide containing the sugar D-ribose, is the form of nucleic acid responsible for translating the information in DNA into protein synthesis. It is found mainly in the cytoplasm but is present in all parts of the cell.

A **ribosome** is a cellular substructure that serves as the site for protein synthesis.

Rods are photoreceptor cells that function in dim light, but do not discriminate among different colors of light.

The **roentgen** is an exposure dose of X- or γ-radiation that produces ions with charges of 2.58×10^{-4} coulomb per kilogram of air.

Saliva is a secretion of the oral glands that aids in the swallowing and digestion of foods.

A **salt** is an ionic compound in which hydrogen atoms of an acid are replaced by metal ions.

Saponification (literally, soap making) is the alkaline hydrolysis of fat.

A **saturated fatty acid** contains no carbon–carbon double bonds.

A **saturated hydrocarbon** (alkane) has molecules that contain the maximum number of hydrogen atoms for the carbon atoms present. All bonds in the molecules are single covalent bonds.

A **saturated solution** contains the maximum amount of solute that can be dissolved in a particular quantity of solvent at equilibrium at a given temperature.

The ***s*-block** is the portion of the periodic table in which the *s* subshell of the outer shell fills, in the aufbau process.

A **scientific law** is a brief statement, sometimes in mathematical terms, used to summarize and describe patterns in large collections of scientific data.

A **secondary (2°) alcohol** is one that bears the OH group on a carbon atom that is attached to two other carbon atoms.

A **secondary (2°) amine** is one that has two alkyl or aryl groups on the nitrogen atom.

The **secondary structure** of a protein refers to regularities in local arrangements maintained by hydrogen bonds. Two common examples are the alpha helix and pleated sheets.

A **semipermeable** membrane is permeable to some solutes but not to others.

Serum is blood plasma from which fibrinogen has been removed.

The **set-point theory** holds that the hypothalamus monitors the level of circulating fatty acids in the blood; when the level is too low, the person is hungry. Each person has a unique set point.

The **settling-point theory** suggests that we maintain a particular weight when the various metabolic feedback loops are in equilibrium with our environment.

Shock is a condition characterized by the loss of fluid from the vascular system.

The **significant figures** in a measured quantity are all the digits known with certainty plus the first uncertain digit.

A **simple cubic** crystal structure has as its unit cell a cube with a structural unit at each of the corners of the cube. *See also* **unit cell.**

A **single bond** is a covalent linkage in which two atoms share one pair of electrons.

Smog (a contraction of the words smoke and fog) is air that is visibly polluted.

A **solid** is a state of matter in which the substance maintains its shape and volume.

The **solubility product constant (K_{sp})** describes the equilibrium that exists between a slightly soluble solute and its ions in a saturated solution.

A **solute** is a solution component that is dissolved in a solvent. A solution may have several solutes, which are generally present in lesser amounts than is the solvent.

A **solution** is a homogeneous mixture of two or more substances. The composition and properties are uniform throughout a solution.

The **solvent** is the solution component (usually present in greatest amount) in which one or more solutes are dissolved to form the solution.

The **specific gravity** of a substance is the ratio of the mass of a given volume of a substance to that of an equal volume of water.

The **specific heat** of a substance is the quantity of heat required to raise the temperature of one gram of substance by 1 °C (or 1 K).

The **specific rotation** of a substance is the amount it rotates plane-polarized light under standard conditions. This quantitative measure of optical activity is a physical property characteristic of the substance.

A **sphingolipid** is a lipid that contains sphingosine rather than glycerol.

Sphingomyelin is a sphingolipid with units derived from sphingosine, phosphoric acid, choline, and a fatty acid.

The **standard temperature and pressure (STP)** for a gas are 273.15 K (0 °C) and 1 atm (760 mmHg). (The formal definition of standard pressure is 1 bar = 0.98692 atm. Our use of the more familiar 1 atm introduces little error.)

Starvation is the withholding of nutrition from the body whether voluntary or involuntary.

Stereoisomers are isomers that have the same molecular formulas but differ in the arrangement of atoms in three-dimensional space.

A **steroid** is a lipid characterized by a structure with a particular arrangement of four fused rings.

A **stimulant** is a drug that increases alertness, speeds up mental processes, and generally elevates the mood.

STP. *See* **standard temperature and pressure.**

A **stoichiometric factor** is a conversion factor relating molar amounts of two species involved in a chemical reaction (i.e., a reactant to a product, one reactant to another, etc.).

Straight-run gasoline is gasoline as it comes off the distilling column of a petroleum refinery.

A **strong acid** is an acid that is essentially completely ionized in solution. *See also* **acid.**

A **strong base** is a base that is essentially completely ionized in solution. *See also* **base.**

A **strong electrolyte** is a substance that exists exclusively or almost exclusively in ionic form in solution.

A **structural formula** is a chemical formula that shows how the atoms of a molecule are attached to one another.

Structural isomers have the same molecular formula, but they differ in the order of attachment of atoms and groups.

Sublimation is the direct passage of molecules from the solid state to the vapor state.

A **substance** is a type of matter having a definite, or fixed, composition and properties that do not vary from one sample to another.

A **substrate** is the substance acted upon by an enzyme.

Substrate-level phosphorylation is the formation of ATP (adenosine triphosphate) by direct transfer of a phosphate unit from a metabolite to ADP (adenosine diphosphate).

A **supersaturated solution** contains more solute than is present in a saturated solution, with the excess solute remaining in solution.

Surface tension is the amount of work required to extend a liquid surface.

A **synapse** is the gap between the axon terminal of a neuron and the receptor(s) on the receiving cell.

Synergism is the interaction of two or more substances to give an effect that is usually greater than the sum of their separate effects.

Systolic pressure is the maximum force exerted by the blood against the inner walls of the arteries during a heartbeat.

A **tertiary (3°) alcohol** is one that bears the OH group on a carbon atom that is attached to three other carbon atoms.

A **tertiary (3°) amine** is one that has three alkyl or aryl groups on the nitrogen atom.

The **tertiary structure** of a protein is the folding of polypeptide chains into compact shapes such as globules.

A **theory** provides explanations of observed phenomena by predictions that can be tested experimentally. It is the intellectual framework for explaining scientific data and scientific laws.

Thrombosis is the formation of a clot within a blood vessel.

Titration is a laboratory procedure in which one reactant in solution is added quantitatively to another until the reaction is stoichiometrically complete. It is used to determine the concentration of a solution, such as an acid (reacted with a base).

The **training effect** of regular exercise gives a lower pulse rate and lower blood pressure; it enables a person to do more physical work with less strain.

The term **trans** is used in organic chemistry to indicate geometric isomers in which two groups are attached to opposite sides of a double bond or ring in a molecule.

Transamination is a reaction that transfers the amino group from one amino acid to an α-keto acid. This leads to the formation of a new amino acid and α-keto acid.

Transcription is the process by which DNA directs the synthesis of RNA molecules during protein synthesis.

A **transferase** enzyme catalyzes the transfer of a group from one molecule to another.

Transition elements are elements in which the subshell being filled in the aufbau process is in a principal shell of less than the highest quantum number (an inner shell).

Translation is the process by which the information contained in an mRNA molecule is converted to a protein structure.

A **triglyceride** (triacylglycerol) is an ester of glycerol with three fatty acids.

A **trihydric alcohol** is one with three OH functional groups.

A **triple bond** is a covalent linkage in which two atoms share *three* pairs of electrons.

Triplet code. *See* **genetic code.**

A **triprotic acid** is an acid that can donate three protons per molecule.

Tumor suppressor genes code for proteins that act as signals to inhibit cell growth.

The **turnover number** is the number of substrate molecules converted to product in 1 s by a single enzyme-active site.

The **turnover rate** is the amount of protein degraded per unit time in the body.

The **Tyndall effect** is the scattering of a beam of light as it passes through a colloid. This makes a colloidal dispersion distinguishable from a true solution.

The **unit cell** of a crystal structure is the simplest parallelepiped that can be used to generate the entire crystalline lattice through straight-line displacements in all three dimensions.

In the **unit conversion method** of problem solving, the units are carried along in the calculation. The problem is set up in a manner that allows all the units to be canceled except the desired unit(s) for the answer.

The **universal gas constant (R)** is the numerical constant required to relate pressure, volume, amount, and temperature of a gas in the ideal gas equation, $PV = nRT$. Its numerical value is 0.082057 L atm mol^{-1} K^{-1}.

An **unsaturated hydrocarbon** (an alkene or alkyne) is a carbon-hydrogen compound having one or more multiple (double or triple) bonds between carbon atoms.

An **unsaturated solution** contains less of a solute in a given quantity of solution than is present in a saturated solution. It is a solution having a concentration less than the solubility limit.

The **urea cycle** is the metabolic pathway by which excess nitrogen is converted to urea.

Urine is the fluid formed in the kidneys and excreted through the urinary tract.

A vaccine is a preparation of a weakened antigen administered to cause the body to build up its immune system against an infectious disease organism.

Valence electrons are electrons with the highest principal quantum number. They are found in the outermost electronic shells of atoms.

The **valence shell** is the outermost shell of electrons of an atom.

The **valence-shell electron-pair repulsion (VSEPR) theory** of chemical bonding describes the geometrical shape of a molecule or polyatomic ion based on the mutual repulsions among electron groups surrounding the central atom(s) in the structure.

The **van Slyke method** is used to determine the primary amine functions in amino acids or proteins.

Van't Hoff's rule states that the maximum number of stereoisomers for a given structural formula is 2^n, where n is the number of chiral centers in the molecule.

Vaporization or **evaporation** refers to the conversion of a liquid to a gas (vapor).

The **vapor pressure** of a liquid is the pressure exerted by the vapor in dynamic equilibrium with the liquid at a constant temperature.

Virology is the study of viruses.

A **virus** is a subcellular infectious agent that has a core of nucleic acid surrounded by proteins. It uses the host cell equipment to replicate.

Viscosity is a resistance to flow in a fluid (gas or liquid) produced by intermolecular forces. The stronger the intermolecular forces, the more viscous is the fluid.

Visual pigments are organic compounds called rhodopsins that contain a protein opsin and an aldehyde, 11-*cis*-retinal.

The **vital capacity** is the maximum amount of air that can be forced from the lungs.

A **vitamin** is an organic compound required in the diet of some animals in small amounts to maintain good health. Many vitamins are precursors for coenzymes.

A **water-soluble vitamin** is one that is soluble in water: the B vitamins and vitamin C.

A **weak acid** is an acid that exists partly in ionic form and partly in molecular form in solution. *See also* **acid.**

A **weak base** is a base that exists partly in ionic form and partly in molecular form in solution. *See also* **base.**

A **weak electrolyte** is a substance that is present partly in molecular form and partly in ionic form in its solutions.

Weight measures the force of attraction between two objects and is related to the masses of the objects. It is a measure of the force of attraction of the Earth for an object.

Work is the result of a force acting through a distance, or an energy transfer into or out of a system, that can be expressed as the product of a force and a distance.

An **X-ray** is a type of electromagnetic radiation produced by the impact of cathode rays (electrons) on a solid, such as on a dense metal anode (a target) in a cathode-ray tube.

A **zwitterion** is a molecule that has a positive charge on one atom and a negative charge on another. It is a neutral dipolar ion.

Answers to Selected Review Questions, Practice Exercises, and Odd-Numbered Problems

Chapter 1

Selected Review Questions

1.2 Science is testable, reproducible, explanatory, predictive, and tentative. **1.12** Like charges repel; unlike charges attract. **1.13** meter (m); kilogram (kg); kelvin (K) **1.14** meter (m); square meter (m^2); cubic meter (m^3) **1.16** At 5 ft, 3in. tall and 207 lb, he is a bit chubby. **1.20** 1 food Calorie = 1000 (scientific) cal = 1 kcal

Practice Exercises

1.1 a. 1.00 kg **b.** 475 lb **1.2** Chemical: b; Physical: a, c **1.3** Potential: a, c; Kinetic: b **1.5 a.** 7.24 mg **b.** 5.14 μm **c.** 1.91 ns **d.** 5.58 km **1.6 a.** 7.45×10^{-7} m **b.** 5.25×10^{-7} s **c.** 1.415×10^6 m **d.** 2.06×10^3 kg **1.7** 0.12 m^3 **1.8** 56.8 g **1.9 a.** 100.5 m **b.** 150 g **c.** 6.3 L **d.** 1.80×10^3 m^2 **e.** 2.33 g/mL **f.** 0.63 g/cm^3 **1.10 a.** 0.0163 g **b.** 1.53 lb **c.** 3.70×10^2 mL **1.11 a.** 0.0903 m **b.** 0.2224 km **c.** 1.50×10^2 fl oz **1.12 a.** 24.4 m/s **b.** 1.34 km/h **c.** 0.136 oz/qt **1.13 a.** 0.256 m^3 **b.** 1.03×10^4 kg/m^2 **1.14** 9.30 g/cm^3 **1.15** 63.2 g **1.16** 3.70 mL **1.17** 1.39 g/mL **1.18** 195 K **1.19 a.** 185 °F **b.** 10.0 °F **c.** 179 °C **d.** −29.3 °C **1.20** 8.00×10^4 cal; 80.0 kcal; 335 kJ **1.21** 1.23 °C

Odd-Numbered Problems

1. a, c **3.** Yes. Because the gravitational pull on each object is equal, the mass will be proportional to the weight. **5.** Physical change: a; Chemical change: b, c **7.** Physical property: a, c, d; Chemical property: b **9. a** The sprinter (moving faster) **b.** The automobile (greater mass and moving faster). **11.** The diver on the 10 m platform. **13.** Elements: a and c. Their composition is made up of only one type of fundamental substance. **15.** Substance: a Mixture: b, c **17.** Homogeneous: a; Heterogeneous: b; (tea solution; chunks of ice) **19. a.** helium **b.** nitrogen **c.** fluorine **d.** potassium **e.** iron **f.** copper **21. a.** H **b.** C **c.** O **d.** Zn **e.** I **f.** Hg **23. a.** 4.54 mg **b.** 3.76 cm **c.** 6.34 μg **25. a.** cm **b.** kg **c.** dL **d.** lb **27. a.** 4 **b.** 3 **c.** 5 **d.** 4 **e.** 4 **f.** 2 **29. a.** 100.5 m **b.** 153 g **c.** 54.4 cm **d.** 436 g **e.** 111 mL **f.** 2.4 cm **31. a.** 5.00×10^4 m **b.** 0.546 m **c.** 9.75×10^4 g **d.** 0.0479 L **e.** 0.577 mg **f.** 23.7 cm **33. a.** 11.5 yd **b.** 5.39 lb **c.** 2.00 qt **d.** 8.59 mi/h **35. a.** 41.7 cm **b.** 3.94 L **c.** 3.55 lb **d.** 0.329 oz **37.** 25.0 m/s **39. a.** 1.17 g/mL **b.** 1.26 g/mL **41. a.** 1.20×10^2 g **b.** 21.8 g **43. a.** 53.1 cm^3 **b.** 18.7 mL **45.** 5.23 g/cm^3 **47.** 1.02 **49. a.** °C **b.** Cal **51. a.** 98.6 °F **b.** −14.7 °C **c.** 523 °F **53. a.** 2750 cal **b.** 3.10×10^3 J **c.** 2060 cal **55.** 1.50×10^3 cal **57.** 586 kJ **59.** 388 °C

Chapter 2

Selected Review Questions

2.2 "Atomic:" a, b, e; continuous: c, d, f **2.7 a.** 11 electrons **b.** 12 neutrons **2.10** 18; 12 **2.11 a.** 2, 8, 4 **b.** 2, 5 **c.** 2, 8, 6 **2.12 a.** 6 **b.** dumbbell **c.** 3

Practice Exercises

2.1 A. 20.18 u **B.** 69.16% copper-63; 30.84% copper-65 **2.2 A.** $^{90}_{42}Mo$ **B.** $^{45}_{20}Ca$ **2.3 A.** 38 protons; 52 neutrons; 38 electrons **B.** $^{40}_{18}Ar$ **2.4 A.** two pairs: $^{90}_{37}E$ and $^{88}_{37}E$; $^{90}_{38}E$ and $^{93}_{38}E$ **B.** $^{238}_{92}U$; $^{15}_{7}N$; $^{24}_{11}Na$ **2.5 A.** 32 **B.** 98 **2.6 a.** $1s^22s^22p^63s^23p^3$; $[Ne]3s^23p^3$ **b.** $1s^22s^22p^63s^23p^5$; $[Ne]3s^23p^5$ **2.7** Ga: $4s^24p^1$; Te: $5s^25p^4$ **2.8 a.** O **b.** S **c.** F **2.9 a.** F < N < Be **b.** Be < Ca < Ba **c.** F < Cl < S **d.** Mg < Ca < K **2.10 a.** Be < N < F **b.** Ba < Ca < Be

Odd-Numbered Problems

1. Radioactivity indicated that atoms are not indivisible. **3.** No. Dalton thought atoms of different elements had different weights. **5.** Yes. Dalton thought atoms were indivisible. **7.** Attract. **9. a.** 20 each **b.** 11 each **c.** 9 each **d.** 18 each **11. a.** 30 protons; 32 neutrons **b.** 94 p; 147 n **c.** 43 p; 56 n **d.** 42 p; 57 n **13.** Isotopes: b, c **15.** Lithium-7 **17.** ^{109}Ag; the atomic mass of Ag is 107.9 u, about half way between ^{107}Ag and ^{109}Ag **19. a.** $1s^22s^22p^63s^23p^2$ **b.** $1s^22s^22p^3$ **c.** $1s^22s^22p^63s^23p^4$ **21. a.** Be **b.** N **c.** Al **23.** Ar **25.** 7 **27.** A ground-state atom has all its electrons in the lowest possible energy levels. **29.** An electron has dropped from a higher to a lower energy level. **31.** Fluorine has one more electron in its $2p$ subshell than does oxygen. **33.** Metals have a characteristic luster, are generally good conductors of heat and electricity, and are malleable and ductile; they are found to the left and toward the bottom of the periodic table. **35. a.** Group 4A, Period 2, nonmetal **b.** Group 2A, Period 4, metal **c.** Group 2B, Period 5, metal **d.** Group 7A, Period 3, nonmetal **e.** Group 3A, Period 2, nonmetal **f.** Group 2A, Period 6, metal; **g.** Group 5A, Period 6, metal **h.** Group 7A, Period 4, nonmetal **37. a.** Ga **b.** Cu **c.** I **39.** b, e **41.** b, d, e **43. a.** S; atomic radii decrease from left to right in a period (higher effective nuclear charge) **b.** Mg **45. a.** Al < Mg < Na; atomic radii decrease from left to right in a period. **b.** Mg < Ca < Sr; atomic radii increase from top to bottom within a group (electrons enter into the next higher principal shell) **47. a.** Ba < Ca < Mg; ionization energies decrease from top to bottom within a group (electrons are more readily removed from the higher principal shells of the larger atoms). **b.** Al < P < Cl; ionization energies generally increase from left to right in a period (electrons are less readily removed from the smaller atoms with higher effective nuclear charges). **49.** 7A; electron affinity increases from left to right.

Chapter 3

Selected Review Questions

3.1 Group 8A (the noble gases) **3.2** An ion is an atom (or a group of conjoined atoms) bearing an electrical charge. An ion is formed from an atom by addition or removal of one or more electrons. **3.3** A sodium ion has only 10 electrons whereas a sodium atom has 11. A sodium ion and a neon atom both have 10 electrons, but sodium has one more proton than neon. **3.5** :I:I: (with lone pairs above and below each I) — : represents bonding pair; : represents lone pair **3.7 a.** 1 **b.** 4 **c.** 2 **d.** 1 **e.** 3 **f.** 1

Practice Exercises

3.1 A. 2− **B.** 2+, 3+ **3.2 A a.** :Ar: **b.** ·Sr· **c.** :F· **d.** ·N· **e.** K· **f.** ·S: **B. a.** Rb· **b.** :Xe: **c.** :I· **d.** ·As· **e.** ·Ra· **f.** ·Se:

3.3 A. Li· + :F· ⟶ Li^+ + :F:$^-$

B. K· + ·I: ⟶ K^+ + :I:$^-$

3.4 A. ·Al· :O· Al^{3+} :O:$^{2-}$

·Al· + :O· ⟶ Al^{3+} + :O:$^{2-}$

:O· :O:$^{2-}$

B. 3·Ca· + 2·P· ⟶ 3 Ca^{2+} + 2:P:$^{3-}$ **3.5 A.** $MgBr_2$ **B.** Na_3N **3.6 A.** Ca_3N_2 **B.** AlP **3.7 A.** lithium oxide **B.** magnesium phosphide **3.8 A.** copper(II) bromide **B.** iron(III) oxide **3.9** bromine trifluoride; bromine pentafluoride **3.10 A.** N_2O_5 **B.** P_4Se_3 **3.11 A. a.** :Br· + ·Br: ⟶ :Br:Br: **b.** H· + ·Br: ⟶ H:Br: **B.** :I· + ·Cl: ⟶ :I:Cl:

3.12 A. :Cl—C(H)(H)—C(H)(H)—H **B.** ·O· bonded to two ·F· atoms (:F—O—F:)

3.13 A. [H—P(H)(H)—H]$^+$ **B.** :F—N(—O:)(=O:) **3.14 A.** linear **B.** pyramidal **3.15 A.** K_3PO_4 **B.** $Ca(CH_3CO_2)_2$ or $Ca(C_2H_3O_2)_2$ **3.16 A.** Calcium carbonate **B.** Potassium dichromate

Odd-Numbered Problems

1. a. Na· **b.** ·O: **c.** ·F: **d.** ·Al·

3. a. ·Ba· ⟶ Ba^{2+} + 2 e^- **b.** ·Br: + e^- ⟶ :Br:

5. a. 2+ **b.** 1−

7. a. Ca: + :Br· + :Br· ⟶ Ca^{2+} + :Br:$^-$ + :Br:$^-$

b. Mg: + :S ⟶ Mg^{2+} + :S:$^{2-}$ **9. a.** Ca: and Ca^{2+}

b. ·S: and :S:$^{2-}$ **c.** Rb· and Rb^+ **d.** ·P· and :P:$^{3-}$

11. a. Na^+:F:$^-$ **b.** K^+:Cl:$^-$ **c.** 2 Na^+:O:$^{2-}$

d. Ca^{2+} 2:Cl:$^-$ **e.** Mg^{2+} 2:Cl:$^-$ **13. a.** sodium ion **b.** magnesium ion **c.** aluminum ion **d.** chloride ion **e.** oxide ion **f.** nitride ion **15. a.** iron(III) ion **b.** copper(II) ion **c.** silver ion **17. a.** Br^- **b.** Ca^{2+} **c.** K^+ **d.** Fe^{2+} **19. a.** sodium bromide **b.** calcium chloride **c.** iron(II) chloride **d.** lithium iodide **e.** potassium sulfide **d.** copper(I) bromide **21. a.** carbonate ion **b.** hydrogen phosphate ion **c.** permanganate ion **d.** hydroxide ion **23. a.** NH_4^+ **b.** HSO_4^- **c.** CN^- **d.** NO_2^- **25.** $MgSO_4$ **b.** $NaHCO_3$ **c.** KNO_3 **d.** $CaHPO_4$ **27. a.** $Fe_3(PO_4)_2$ **b.** $K_2Cr_2O_7$ **c.** CuI **d.** NH_4NO_2 **29. a.** H:P:H (with H below P) **b.** :F:C:F: (C with :F: above and below) **31. a.** N_2O **b.** P_4S_3 **c.** PCl_5 **d.** SF_6 **33. a.** carbon disulfide **b.** dinitrogen tetrasulfide **c.** phosphorus pentafluoride **d.** disulfur decafluoride **35. a.** N **b.** Cl **c.** F **37. a.** B < N < F **b.** Ca < As < Br **39. a.** ionic **b.** polar covalent **c.** ionic **d.** nonpolar covalent

41. a. H—C(H)(H)—O—H **b.** H—C(H)=O

c. H—N(H)—O—H **d.** H—N(H)—N(H)—H **e.** :F—C(=O:)—F:

f. :Cl—P(—Cl:)—Cl: **43. a.** :N=O: **b.** :I—Be—I:

c. PCl₅-type structure: :Cl—P—Cl: with two Cl above and one :Cl: below **42. a.** :Cl—O:$^-$ **b.** [:O—Cl—O:]$^-$

c. [H—O—P(=O:)(—O:)—O:]$^{2-}$ **d.** [:O—Br(—O:)—O:]$^-$

47. a. potassium nitrite **b.** lithium cyanide **c.** ammonium iodide **d.** sodium nitrate **e.** potassium permanganate **f.** calcium sulfate **49. a.** sodium monohydrogen phosphate **b.** ammonium phosphate **c.** aluminum nitrate **d.** ammonium nitrate **51. a.** linear **b.** triangular (trigonal planar) **c.** bent **d.** tetrahedral **53. a.** pyramidal **b.** bent **55. a.** polar **b.** polar **c.** nonpolar **57. a.** $\overset{\delta+}{H}$—$\overset{\delta-}{O}$ **b.** $\overset{\delta+}{N}$—$\overset{\delta-}{Cl}$ **59.** Nonpolar; the two polar bonds point in opposite directions, canceling one another. **61. a.** F—F < Cl—F < H—F **b.** H—H < H—Br < H—F **63. a.** $\overset{\delta+}{Cl}$—$\overset{\delta-}{F}$ $\overset{\delta+}{H}$—$\overset{\delta-}{F}$ **b.** $\overset{\delta+}{H}$—$\overset{\delta-}{Br}$ $\overset{\delta+}{H}$—$\overset{\delta-}{F}$

Chapter 4

Selected Review Questions

4.2 Hydrogen gas reacts with oxygen gas to form gaseous water. **4.3 a.** balanced **b.** not balanced **c.** not balanced **4.6** The atomic mass of N is 14.0067 u, read directly from the periodic table. The molecular mass is that of the N_2 molecule; twice the atomic mass. **4.8** The molecular mass of carbon dioxide is 44.011u; the molar mass, 44.011g. Each is determined by summing the atomic masses of 2 C and 1 O; then using the appropriate unit. **4.9** 6.02×10^{23} O^2 molecules; 12.04×10^{23} O atoms **4.10** 6.02×10^{23} Ca^{2+} ions; 12.04×10^{23} Cl^- ions **4.11** Molecular: 1 molecule of $CH_4(g)$ reacts with 2 molecules of $O_2(g)$ to form 1 molecule of $CO_2(g)$ and 2 molecules of $H_2O(g)$. Molar: 1 mole of CH_4 reacts with 2 moles of O_2 to form 1 mole of CO_2 and 2 moles of H_2O. Mass: 16.0 g of CH_4 react with 64.0 g of O_2 to form 44.0 g of CO_2 and 36.0 g of H_2O. **4.12 a.** Two moles of solid potassium chlorate yield two moles of solid potassium chloride and three moles of oxygen gas. **b.** Two moles of solid aluminum react with six moles of aqueous hydrochloric acid to form two moles of aqueous aluminum chloride and three moles of hydrogen gas. **4.14 a.** C **b.** B **c.** A **d.** exothermic; there is a net release of energy **4.15** endothermic; there would have to be a net input of energy **4.17 a.** increase **b.** increase **c.** increase

Practice Exercises

4.1 $P_4 + 6\,H_2 \longrightarrow 4\,PH_3$ **4.2 a.** $3\,Mg + B_2O_3 \longrightarrow 2\,B + 3\,MgO$ **b.** $3\,NO_2 + H_2O \longrightarrow 2\,HNO_3 + NO$ **c.** $3\,H_2 + Fe_2O_3 \longrightarrow 2\,Fe + 3\,H_2O$ **4.3 a.** $2\,H_3PO_4 + 3\,Ca(OH)_2 \longrightarrow Ca_3(PO_4)_2 + 6\,H_2O$ **b.** $6\,CaO + P_4O_{10} \longrightarrow 2\,Ca_3(PO_4)_2$ **c.** $2\,Al(OH)_3 + 3\,H_2SO_4 \longrightarrow Al_2(SO_4)_3 + 6\,H_2O$ **4.4 A.** 1.48 L CO_2 **B.** 21.7 L CO_2 **4.5 a.** 147.004 u **b.** 98.96 u **c.** 97.9951 u **4.6 a.** 138.206 u **b.** 294.184 u **c.** 342.22 u **4.7 a.** 1000 g H_2O **b.** 0.756 g $C_4H_{10}O$; **c.** 73.7 g C_2H_6 **4.8 a.** 0.0664 mol Fe **b.** 0.776 mol H_3PO_4 **c.** 2.84 mol C_4H_{10} **4.9** Molecular: 2 molecules of H_2S react with 3 molecules of O_2 to form 2 molecules of SO_2 and 2 molecules of H_2O. Molar: 2 moles of H_2S react with 3 moles of O_2 to form 2 moles of SO_2 and 2 moles of H_2O. Mass: 68.2 g of H_2S react with 96.0 g of O_2 to form 128.2 g of SO_2 and 36.0 g of H_2O. **4.10 a.** 1.59 mol CO_2 **b.** 305 mol H_2O **c.** 0.6060 mol CO_2 **4.11 A.** 0.763 g O_2 **B. a.** 2130 g CO_2 **b.** 2350 g CO_2 **4.12 A.** 39.0 g NH_3 **B.** 0.967 g O_2 **4.13 A.** The reaction will shift to the right, decreasing the concentration of CO. **B.** $N_2O_4(g) + \text{heat} \rightleftharpoons 2\,NO_2(g)$. Addition of NO_2 will shift the reaction to the left, increasing the equilibrium concentration of N_2O_4.

Odd-Numbered Problems

1. balanced: a, b; not balanced: c, d **3. a.** $Cl_2O_5 + H_2O \longrightarrow 2\,HClO_3$ **b.** $V_2O_5 + 2\,H_2 \longrightarrow V_2O_3 + 2\,H_2O$ **c.** $4\,Al + 3\,O_2 \longrightarrow 2\,Al_2O_3$ **d.** $Sn + 2\,NaOH \longrightarrow Na_2SnO_2 + H_2$ **e.** $PCl_5 + 4\,H_2O \longrightarrow H_3PO_4 + 5\,HCl$ **f.** $Na_3P + 3\,H_2O \longrightarrow 3\,NaOH + PH_3$ **g.** $Cl_2O + H_2O \longrightarrow 2\,HClO$ **h.** $2\,CH_3OH + 3\,O_2 \longrightarrow 2\,CO_2 + 4\,H_2O$ **i.** $3\,Zn(OH)_2 + 2\,H_3PO_4 \longrightarrow Zn_3(PO_4)_2 + 6\,H_2O$ **j.** $C_3H_8 + 5\,O_2 \longrightarrow 3\,CO_2 + 4\,H_2O$ **5.** 5.00 L $CH_4(g)$ **7. a.** 2.65 L $H_2O(g)$ **b.** 105 L $O_2(g)$. **9. a.** 4 **b.** 4 **c.** 8 **d.** 6 **11. a.** 12 **b.** 8 **13. a.** 157.0 u **b.** 98.0 u **c.** 294.2 u **15. a.** 0.435 g **b.** 47.2 g **c.** 45.8 g **17. a.** 1.56 mol **b.** 0.0285 mol **c.** 0.0600 mol **d.** 0.0356 mol **19. a.** 16.7 mol CO_2 **b.** 55.9 mol O_2 **21.** 2500 g NH_3 **23.** 11.3 g O_2 **25.** 2040 g TNT **27.** 3590 g HNO_3 (balanced equation: $NH_3 + 2\,O_2 \longrightarrow HNO_3 + H_2O$) **29.** endothermic **31.** The mechanism of a chemical reaction is the step-by-step process by which the reaction takes place. **33. a.** Equilibrium shifts to the left, producing more H_2 and Cl_2 and using up HCl. **b.** Equilibrium shifts to the right, producing more CO and O_2 and consuming CO_2. **c.** Equilibrium shifts to the right, producing more O_3 and consuming O_2.

Chapter 5

Selected Review Questions

5.1 Oxygen occurs as O_2 in the atmosphere and in compounds in water and Earth's solid crust. **5.2** by distillation of liquefied air **5.3** Example reactions: **a.** $2\,Ca(s) + O_2(g) \longrightarrow 2\,CaO(s)$ [a basic oxide] **b.** $P_4(s) + O_2(g) \longrightarrow P_4O_{10}(s)$ (an acidic oxide) **5.4 a.** $C_3H_8(g) + 5\,O_2(g) \longrightarrow 3\,CO_2(g) + 4\,H_2O(g)$ (both products are oxides) **b.** $2\,H_2S(g) + 3\,O_2(g) \longrightarrow 2\,SO_2(g) + 2\,H_2O(g)$ (both products are oxides) **5.9** Oxidation is an increase in oxidation number; reduction, a decrease.

Practice Exercises

5.1 A. $2\,Zn + O_2 \longrightarrow 2\,ZnO$ **B.** $4\,Al + 3\,O_2 \longrightarrow 2\,Al_2O_3$ **5.2 A.** $SiH_4 + 2\,O_2 \longrightarrow SiO_2 + 2\,H_2O$ **B.** $2\,PbS + 3\,O_2 \longrightarrow 2\,PbO + 2\,SO_2$ **5.3** Oxidation: a, b, c, d; reduction: none **5.4** Reduction: a; oxidation: b **5.5 a.** Al, +3; O, −2 **b.** P, 0 **c.** Na, +1; Mn, +7; O, −2 **d.** H, +1; O, −1 **e.** C, −2; H, +1; F, −1 **f.** C, +2; H, +1; Cl: −1 **5.6** Oxidation: b, d; reduction: a; neither: c **5.7** Yes; iodine is oxidized from 0 to + 5, and Cl is reduced from 0 to −1. **5.8 A. a.** Oxidizing agent: O_2; reducing agent: Se **b.** Oxidizing agent: Br_2; reducing agent: K **B. a.** Oxidizing agent: V_2O_5; reducing agent: H_2 **b.** Oxidizing agent: $CH_3C{\equiv}N$; reducing agent: H_2

Odd-Numbered Problems

1. a. $C + O_2 \longrightarrow CO_2$ **b.** $2\,C_2H_6 + 7\,O_2 \longrightarrow 4\,CO_2 + 6\,H_2O$ **c.** $N_2 + O_2 \longrightarrow 2\,NO$ **d.** $C_3H_8 + 5\,O_2 \longrightarrow 3\,CO_2 + 4\,H_2O$ **3. a.** 0 **b.** +4 **c.** −2 **d.** +6 **5.** H, +1 (exception: −1 in metal hydrides); O, −2 (exception: −1 in peroxides **7. a.** oxidation (Cl, +4 to +5) **b.** oxidation (Mn, +2 to +4) **c.** reduction (Br, +1 to 0) **d.** oxidation (Sb: −3 to 0) **9.** Some of the S atoms in the H_2SO_4 are reduced to form SO_2 (+6 to +4). **11. a.** Oxidizing agent, O_2; reducing agent, Al **b.** Oxidizing agent, O_2; reducing agent, SO_2 **c.** Oxidizing agent, HCl (actually, H^{+}); reducing agent, Fe **d.** Oxidizing agent, O_2; reducing agent, CS_2 **13. a.** SO_2 is oxidized; HNO_3 is reduced. **b.** HI is oxidized; CrO_3 is reduced. **15.** I^- is oxidized; Cl_2 is reduced. **17.** Reduced. **19.** Acetylene is reduced; it gains hydrogen atoms. (H_2 is oxidized to the +1 state.) **21.** NO_2^- is reduced; ascorbic acid is a reducing agent.

Chapter 6

Selected Review Questions

6.1 the troposphere; the stratosphere **6.2** N_2, 78%; O_2, 21%; Ar, 1% **6.6** A mercury barometer measures atmospheric pressure. Mercury is quite dense; a much shorter barometer can be made with mercury than with water. **6.9** The gases can be contained in a much smaller volume. **6.15 a.** volume decreases **b.** volume decreases **c.** volume increases **6.16 a.** pressure increases **b.** pressure increases **c.** pressure increases **6.19** No. Heating would remove vital dissolved oxygen. **6.22** The air will be near body temperature (37 °C), at which the vapor pressure of water is about 50 mmHg (estimate from Table 6.2).

Practice Exercises

6.1 400 mmHg **6.2** 1800 mL (1.80 L) **6.3** 1.33 atm **6.4 A.** 2.98 L **B.** −167 °C **6.5 A.** 1.72 L **B.** 9.82 g CO_2 **6.6 A.** 40.9 mL **B.** 2.71 atm **6.7 A.** 0.0900 g/L **B.** 1.34 g/L **6.8 A. (a)** 0.0471 atm **(b)** 5.00 L **B.** 97.4 K **6.9** 0.0195 g H_2

Odd-Numbered Problems

1. (a) The temperature is decreasing **(b)** the pressure is decreasing **3. a.** Container A **b.** The density is the same in each. **c.** B **5. A.** 749 mmHg **b.** 1.12 atm **c.** 0.949 atm **7.** 213 mmHg (8.39 in) **9. (a)** 1090 mL **(b)** 2600 mmHg **11. a.** 9120 L **b.** 19 h **13.** 117 mL **15.** 160 °C **17. a.** 2.5 mol H_2 **b.** 2.23 mol SF_6 **c.** 1.66 mol CO_2; 5.0 g H_2 has the greatest number of molecules. **19.** 143 mL **21.** 2490 mmHg **23.** 0.489 m^3 **25.** 4.52 L **27.** 22.3 L **29.** 29.1 atm **31.** 0.0781 mol Kr **33. a.** 1.25 g/L **b.** 3.48 g/L **c.** 1.78 g/L **d.** 1.25 g/L **35.** 710 mmHg **37.** 250 mmHg **39.** 95%

Chapter 7

Selected Review Questions

7.2 Liquids and solids are similar in that the constituent particles are close together; they differ in that the particles of a liquid are much freer to move than those of a solid. **7.4** Generally, the higher the charges on the ions and the smaller the ions; the higher the melting point of the solid. **7.5** Polar liquids have intermolecular attractions between dipoles. (Both polar and nonpolar liquids have dispersion forces.) **7.6** Water molecules are associated through hydrogen bonding; methane molecules are not. **7.7** Methanol molecules are associated through hydrogen bonding; ethane molecules are not. **7.8 a.** Xe; xenon atoms are larger than Ne atoms and thus have greater dispersion forces. **b.** Ethane; C_2H_6 molecules are larger than CH_4 molecules and thus have greater dispersion forces. **7.10** O_2 molecules attract one another through dispersion forces. **7.14** The stronger the intermolecular forces in a liquid, the higher its heat of vaporization.

Practice Exercises

7.1 A. KCl; the chloride ion is smaller than the iodide ion. This allows the + and − charges to be closer together than those in KI, and thus the attractive forces in KCl are greater than those in KI. **B.** HgS; the doubly charged ions in HgS attract one another more strongly than do the singly charged ions in AgCl. **7.2 A.** HCl, with polar molecules, has a higher boiling point than F_2, with nonpolar molecules **B.** H_2S, with polar molecules, has a higher boiling point than SiH_4, with nonpolar molecules **7.3** NH_3: yes; CH_4: no; C_6H_5OH: yes; H_2S: no; H_2O_2: yes **7.4** 59.1 cal/g **7.5** 25.1 kcal **7.7 A.** 54.5 kcal **B.** 29.6 kJ **7.8 A.** 1180 cal **B.** 10.0 kcal

Odd-Numbered Problems

1. As compared to gases, liquids and solids are only slightly compressible, have molecules very close together, and have stronger intermolecular forces. **3.** At the lower pressure at high altitude, the boiling point of water is less than 100 °C and it takes longer to cook the egg at the lower temperature. It would not take longer to fry an egg because heat is transferred directly to the egg from the skillet. **5. a.** melting **b.** vaporization **7.** CS_2; neither is polar, but CS_2 is smaller and has smaller dispersion forces than CCl_4. **9.** C_2H_5OH; the alcohol molecules have the ability to hydrogen bond to one another; C_3H_8 molecules do not. Hydrogen bonding creates a network of molecules held by intermolecular forces that would tend to stay in the liquid phase at a higher temperature. **11.** H_2S (lowest), H_2Se, H_2Te (highest); they are all slightly polar molecules, but the dispersion forces increase with molar mass. **13.** 7.2 kcal/mol **15.** 0.0968 kcal **17.** 1.05 kcal **19.** 0.344 kcal **21.** 28.4 kcal

Chapter 8

Selected Review Questions

8.3 Na^+ and Cl^- ions must be separated (ionic bonds broken); water molecules must be separated (hydrogen bonds broken); ion-dipole bonds are formed. **8.4 (a)** No. Benzene is a nonpolar solvent; no ion-dipole forces can be formed to compensate for breaking ionic bonds in the NaCl. **(b)** No. Motor oil is nonpolar; no hydrogen bonds can be formed between the hydrocarbon molecules that make up motor oil and the water molecules; there is no compensation for the hydrogen bonds that would have to be broken in the water. **(c)** Yes. Both benzene and motor oil are nonpolar; they would mix by diffusion. **8.5** The OH group of ethyl alcohol can form hydrogen bonds to water; the Cl of ethyl chloride cannot. **8.8** The solution is saturated; the solute is in equilibrium with the precipitated solid. **8.9** The pressure on the surface of the soda pop is decreased and the solubility of the carbon dioxide gas decreases. Some of the CO_2 escapes from the solution. **8.10** Fish require oxygen; cold water holds more dissolved oxygen than does warm water.

Practice Exercises

8.1 A. 0.00870 M **B.** 0.968 M **8.2 a.** 9.00 M **b.** 1.26 M **c.** 0.274 M **d.** 0.0242 M **e.** 0.123 M **f.** 9.23 M **8.3 a.** 673 g KOH **b.** 5.61 g KOH **c.** 0.0561 g KOH **d.** 4.63 g KOH **8.4 A.** 0.0297 L **B.** 16.3 mL **8.5** 46.8% **8.6** Dilute 22.3 mL of acetic acid with enough water to make 67.5 mL of solution. **8.7** 2.81% **8.8** Add 15.1 g of glucose to 260 g (260 mL) water. **8.9 a.** 1.5 osmol/L; **b.** 0.30 osmol/L; **c.** 1.32 osmol/L

Odd-Numbered Problems

1. Soluble: a, b **3.** Dissolving of the solid and precipitation of the solute **5. (a)** unsaturated **(b)** 38 °C **7.** Soluble: a, b; each has an oxygen atom which can form hydrogen bonds to water. **9. a.** 2.40 M **b.** 0.700 M **11. a.** 0.907 M **b.** 1.95 M **13. a.** 80.0 g NaOH **b.** 7.65 g $C_6H_{12}O_6$ **15.** 0.208 L **17.** 2.05 L **19. a.** 4.83% **b.** 5.09% **21. a.** 4.12% **b.** 7.71% **23.** Add 77.5 g NaCl to 697.5 g of water. **25.** Add 0.0400 L (40.0 mL) of acetic acid to 1.96 L of water. **27.** Dissolve 3.75 g of $MgSO_4$ in enough water to make 250 mL of solution **29.** 100 mg/dL **31.** Solute Particles: **a.** two **b.** one **c.** three. Osmol per one mole: **a.** two **b.** one **c.** three. **33. a.** 0.1 M $NaHCO_3$ **b.** 1 M NaCl **35. a.** 0.5 mol **b.** 1.0 mol **c.** 0.33 mol **37. a.** same **b.** 2 osmol/L glucose ($C_6H_{12}O_6$) **39.** A suspension has larger particles and settles on standing. A colloid has smaller particles and does not settle.

Chapter 9

Selected Review Questions

9.1 Acidic solutions taste sour, produce a prickling sensation on the skin, turn litmus from blue to red, react with many metals to produce ionic compounds and hydrogen gas, and neutralize bases. Basic solutions taste bitter, feel slippery on the skin, turn litmus blue, and neutralize acids. None of these properties remain after neutralization. **9.2** Acidic solutions: H_3O^+; Basic solutions: OH^- **9.4 (a)** H_2O **(b)** NH_3 **(c)** H_3O^+ **(d)** $HCOO^-$ **9.5 a.** acid (COOH group) **b.** acid (ionizable H atom written first) **c.** base (conjugate base of HCN) **d.** acid (conjugate acid of an amine) **e.** acid (COOH group) **f.** base (an amine) **9.7** A polyprotic acid has more than one ionizable H atom. No; CH_4 has no ionizable H. **9.9** An acidic anhydride reacts with water to form an acid; a basic anhydride reacts with water to form a base. **9.10** $H_2CO_3 \longrightarrow CO_2 + H_2O$; CO_2 **9.11** sulfuric acid **9.14** Calcium carbonate; magnesium hydroxide; aluminum hydroxide **9.16** The Mg^{2+} ion draws water into the large intestine. Yes.

Practice Exercises

9.1 H_3AsO_4 **9.2 A.** HIO; $NaIO_3$; $NaIO_2$; NaIO **B.** $HMnO_4$ **9.3 A.** HNO_3; nitric acid **B.** KOH; potassium hydroxide **9.4 A. (a)** $Ca(OH)_2 + 2\,HCl \longrightarrow CaCl_2 + 2\,H_2O$ **(b)** $Ca^{2+} + 2\,OH^- + 2\,H_3O^+ + 2\,Cl^- \longrightarrow Ca^{2+} + 2\,Cl^- + 4\,H_2O$ **(c)** $2\,OH^- + 2\,H_3O^+ \longrightarrow 4\,H_2O$ or $OH^- + H_3O^+ \longrightarrow 2\,H_2O$ **B. (a)** $3\,NH_3 + H_3PO_4 \longrightarrow (NH_4)_3PO_4$ **(b)** $3\,NH_3 + H_3PO_4 \longrightarrow 3\,NH_4^+ + PO_4^{3-}$ **(c)** Same as (b). (Because ammonia is a weak base and phosphoric acid a weak acid, neither is written in ionic form.)

Odd-Numbered Problems

1. a. $HClO_2$ [Acid (1)] + H_2O [Base (2)] $\rightleftharpoons$ H_3O^+ [Acid (2)] + ClO_2^- [Base (1)] **b.** HSO_4^- [Acid (1)] + NH_3 [Base (2)] $\rightleftharpoons$ NH_4^+ [Acid (2)] + SO_4^{2-} [Base (1)] **c.** HCO_3^- [Acid (1)] + OH^- [Base (2)] $\rightleftharpoons$ CO_3^{2-} [Base (1)] + H_2O [Acid (2)] **3.** Strong: b; Weak: c, d; Neither: a **5. a.** salt **b.** strong base **c.** salt **d.** weak acid **7. a.** strong base **b.** strong acid **9.** an acid, weak **11. a.** HCl **b.** H_2SO_4 **c.** H_2CO_3 **d.** LiOH **e.** $Mg(OH)_2$ **f.** KOH **13. a.** sodium hydroxide **b.** phosphoric acid **c.** nitric acid **d.** sulfurous acid **e.** calcium hydroxide **f.** hydrosulfuric acid **15. a.** bromide ion; hydrobromic acid **b.** nitrite ion; nitrous acid **17.** hydroselenic acid

19. a. $HI(aq) \longrightarrow H^+(aq) + I^-(aq)$ **b.** $CH_3CH_2COOH(aq) \longrightarrow CH_3CH_2COO^-(aq) + H^+(aq)$ **c.** $HNO_2(aq) \longrightarrow NO_2^-(aq) + H^+$ **d.** $H_2PO_4^-(aq) \longrightarrow HPO_4^{2-}(aq) + H^+(aq)$ **21. a.** $HNO_3(aq) \longrightarrow NO_3^-(aq) + H^+(aq)$ **b.** $KOH(aq) \longrightarrow K^+(aq) + OH^-(aq)$ **c.** $HCOOH(aq) \longrightarrow HCOO^-(aq) + H^+(aq)$ **d.** $CH_3NH_2(aq) + H_2O \longrightarrow CH_3NH_3^+(aq) + OH^-(aq)$ **23. a.** base **b.** acid **c.** base **25. (a)** H_2SO_4, an acid **(b)** KOH; a base **27.** SeO_3 **29.** $NaOH + HCl \longrightarrow NaCl + H_2O; OH^- + H^+ \longrightarrow H_2O$ **31.** $Ca(OH)_2 + 2\,HCl \longrightarrow CaCl_2 + 2\,H_2O; OH^- + H^+ \longrightarrow H_2O$ **33.** $HCO_3^- + H^+ \longrightarrow H_2O + CO_2$ **35.** $NaHCO_3(s) + CH_3COOH(aq) \longrightarrow CH_3COONa(aq) + H_2O + CO_2(g)$; $NaHCO_3(s) + CH_3COOH(aq) \longrightarrow CH_3COO^-(aq) + Na^+(aq) + H_2O + CO_2(g)$

Chapter 10

Selected Review Questions

10.1 Generally, a concentrated acid or base is the strongest concentration commercially available. **10.2** Yes, the concentration is changed by dilution, but the number of moles of solute is not. **10.5 (a)** No (only 1/3 mole.) **(b)** Yes. **10.6 (a)** No. **(b)** Yes. **10.8** Basic: a; acidic, b, d; neutral: c **10.9** Greater than 7: b, d; lower than 7: a, c **10.10** a (Neutral; all the others are basic.) **10.11** a (lowest pH) < c < d < b (highest pH) **10.14** decrease **10.15 a.** $HBO_2 \rightleftharpoons H^+ + BO_2^-$ **b.** $HClO_2 \rightleftharpoons H^+ + ClO_2^-$ **c.** $HC_9H_7O_4 \rightleftharpoons H^+ + C_9H_7O_4^-$ **d.** $H_2Se \rightleftharpoons H^+ + HSe^-$ **10.18** c **10.20** COO^- reacts with added acid; NH_3^+ reacts with added base. **10.22** too high

Practice Exercises

10.1 0.208 L **10.2** 0.758 M **10.3** 0.0878 M **10.4 A.** 401.6 mL **B.** 0.1724 M **10.5 A.** $[H^+] = 2.2 \times 10^{-3}$ M; $[OH^-] = 4.5 \times 10^{-10}$ M **B.** $[H^+] = 4.0 \times 10^{-13}$ M; $[OH^-] = 2.5 \times 10^{-2}$ M **10.6** 9.00 **10.7** 8.57 **10.8** $[H^+] = 1.6 \times 10^{-11}$ M; $[OH^-] = 6.2 \times 10^{-4}$ M

10.9 a. $K = \dfrac{[H_2] \times [I_2]}{[HI]^2}$ **b.** $K = \dfrac{[O_3]^2}{[O_2]^3}$ **c.** $K = \dfrac{[XeF_4]}{[Xe] \times [F_2]^2}$

10.10 A. 9.1×10^{-6} M **B.** 2.8×10^{-7} M **10.11** 3.7×10^{-5} M **10.12 a.** neutral **b.** basic **c.** (can't determine) **d.** acidic **10.13** 2.5×10^{-9} M **10.14** 9.1×10^{-10} M **10.15** 3.18 **10.16** 2.94

Odd-Numbered Problems

1. (a) 0.167 L **(b)** 0.481 L **3. a.** 0.417 L **b.** 3.19 mL **5.** 0.249 M HCl **7.** 0.0230 M $Ca(OH)_2$ **9.** 0.12 M HCl **11.** 20.8 mL H_2SO_4 **13. (a)** 46.8 mL HCl **(b)** 11.9 mL HCl **(c)** 23.1 mL HCl **15. a.** 2.00 **b.** 4.00 **17. a.** 2.00 **b.** 3.00 **19. a.** 12.00 **b.** 10.00 **21. a.** 2.48 **b.** 4.24 **c.** 3.09 **23.** 7.34 **25.** 11.70 **27.** 7.9×10^{-6} M **29.** $CH_3COO^- + H_2O \rightleftharpoons CH_3COOH + OH^-$ **31. a.** neutral **b.** basic **c.** (can't determine) **33.** Basic. The product is the salt of a weak acid and a strong base.

35. a. $K = \dfrac{[H^+][OCl^-]}{[HOCl]}$ **b.** $K = \dfrac{[H^+][C_6H_7O_6^-]}{[HC_6H_7O_6]}$

c. $K = \dfrac{[H^+][HCO_2^-]}{[HCO_2H]}$

37. a. $[H^+] = 4.2 \times 10^{-4}$ M; $[OH^-] = 2.4 \times 10^{-11}$ M **b.** $[H^+] = 3.5 \times 10^{-3}$ M; $[OH^-] = 2.9 \times 10^{-12}$ M **c.** $[H^+] = 1.8 \times 10^{-5}$ M; $[OH^-] = 5.7 \times 10^{-10}$ M **39. a.** $[OH^-] = 6.7 \times 10^{-4}$ M; $[H^+] = 1.5 \times 10^{-11}$ M **b.** $[OH^-] = 6.5 \times 10^{-3}$ M; $[H^+] = 1.5 \times 10^{-12}$ M **c.** $[H^+] = 1.5 \times 10^{-9}$ M; $[OH^-] = 6.5 \times 10^{-6}$ M **41. a.** 6.2×10^{-10} M **b.** 6.6×10^{-4} M **c.** 4.6×10^{-5} M **43.** $[OH^-] = 1.8 \times 10^{-4}$ M; $[H^+] = 5.6 \times 10^{-11}$ M **45.** 9.21 **47.** 4.20 **49.** 4.34 **51.** 4.24 **53.** 10.00

Chapter 11

Selected Review Questions

11.3 In solid NaCl the ions are fixed in place; in molten NaCl they are free to move and conduct electric charge. **11.4** covalent **11.5** Ionization describes the formation of ions from a covalent substance; dissociation, the separation in solution of ions of an ionic solid. **11.7** Hydrogen chloride molecules ionize in water. **11.8** Cathode. Ions from the solution pick up electrons (are reduced to metal atoms) at the cathode.

Practice Exercises

11.1 a. Strong (a salt) **b.** Weak (a weak acid) **c.** Nonelectrolyte (covalent compound) **11.2 A.** $K_{sp} = [Cu^{2+}][S^{2-}]$ **B.** $K_{sp} = [Mg^{2+}][F^-]^2$ **11.3 A.** Yes. **B.** Yes.

Odd-Numbered Problems

1. Strong: a, b, d, e; Weak: c **3.** Lead chromate is essentially insoluble in water; there are too few ions in a saturated solution to conduct significant electric current. **5.** NaCl furnishes two ions per formula unit; sugar, one molecule per formula unit. The freezing point depression isn't quite twice that of the sugar because some of the ions are associated as ion pairs. **7. a.** $Ca(s) + 2\,HCl(aq) \longrightarrow H_2(g) + CaCl_2(aq)$ **b.** $Ni(s) + 2\,HCl(aq) \longrightarrow H_2(g) + NiCl_2(aq)$ **c.** $Mg(s) + 2\,HNO_3(aq) \longrightarrow H_2 + Mg(NO_3)_2(aq)$ **9. a.** $2\,Na(s) + 2\,H_2O \longrightarrow H_2(g) + 2\,NaOH(aq)$ **b.** $Ba(s) + 2\,H_2O \longrightarrow H_2(g) + Ba(OH)_2(aq)$ **11. a.** $Mg(s) + Cu^{2+}(aq) \longrightarrow Mg^{2+}(aq) + Cu(s)$ **b.** $Ag(s) + Pb^{2+}(aq) \longrightarrow$ no reaction **c.** $Fe(s) + Zn^{2+}(aq) \longrightarrow$ no reaction **d.** $2\,Al(s) + 3\,Ni^{2+}(aq) \longrightarrow 2\,Al^{3+}(aq) + 3\,Ni(s)$ **13.** Yes. **15.** Yes.

Selected Topic A

Selected Review Questions

2. Na^+ and K^+ **3.** Chlorophyll **5.** Water containing Ca^{2+}, Mg^{2+}, and/or Fe^{2+}. These ions precipitate soaps. **6.** Lithium carbonate **8.** aluminum **9.** Because it is lighter and less subject to corrosion than is iron **10.** Low oxygen (O_2) availability **11.** By binding tightly to hemoglobin and thus hindering oxygen transport **13.** Allotropes are two or more forms of an element that differ in their basic molecular structure. Diamond and graphite are allotropes of carbon. **20.** A lack of reactivity; filled valence subshells **21.** Helium is unreactive while hydrogen is combustible.

Odd-Numbered Problems

31. a. *s* **b.** *p* **33.** F and Cl both have ns^2np^5 valence-shell configurations. F has electrons in only two shells; Cl, in three. Oxygen has one fewer valence electrons than does F. **35.** All are soft, low-melting, highly reactive metals. They have ns^1 valence-shell configurations. **37.** Main group elements have their valence electrons in *s* and *p* subshells. Transition elements have a subshell of an inner shell being filled in the aufbau process. **39. a.** $4\,Li(s) + O_2(g) \longrightarrow 2\,Li_2O(s)$ **b.** $2\,Mg(s) + O_2(g) \longrightarrow 2\,MgO(s)$ **c.** $S(s) + O_2(g) \longrightarrow SO_2(g)$

Chapter 12

Selected Review Questions

12.1 Both X-rays and visible light are forms of electromagnetic radiation. X-rays have a much higher energy than visible light. **12.2 a.** beta particle **b.** alpha particle **12.3** positron emission and electron capture **12.4** None **12.5** Move away from the source or use shielding. **12.6** alpha particles **12.7** gamma rays **12.12** neutrons **12.14** to force like-charged particles together **12.16** iodine-131 **12.19** CT: X-rays; PET: gamma rays

Practice Exercises

12.1 a. ${}^{250}_{100}\text{Fm} \longrightarrow {}^{4}_{2}\text{He} + {}^{246}_{98}\text{Cf}$ (californium) **b.** ${}^{85}_{34}\text{Se} \longrightarrow {}^{0}_{-1}\text{e} + {}^{85}_{35}\text{Br}$ (bromine) **c.** ${}^{188}_{79}\text{Au} \longrightarrow {}^{0}_{+1}\text{e} + {}^{188}_{78}\text{Pt}$ (platinum) **d.** ${}^{192}_{77}\text{Ir} + {}^{0}_{-1}\text{e} \longrightarrow {}^{192}_{76}\text{Os}$ (osmium) **12.2 A.** 0.038 mg **B.** 0.31 mg **12.3** 11,460 y old **12.4** ${}^{96}_{42}\text{Mo} + {}^{2}_{1}\text{H} \longrightarrow {}^{97}_{43}\text{Tc} + {}^{1}_{0}\text{n}$

Odd-Numbered Problems

1. a. alpha particle **b.** beta particle **c.** neutron **d.** deuteron **3. a.** ${}^{209}_{82}\text{Pb} \longrightarrow {}^{0}_{-1}\text{e} + {}^{209}_{83}\text{Bi}$ **b.** ${}^{225}_{90}\text{Th} \longrightarrow {}^{4}_{2}\text{He} + {}^{221}_{88}\text{Ra}$ **5.** ${}^{31}_{16}\text{S} \longrightarrow {}^{0}_{+1}\text{e} + {}^{31}_{15}\text{P}$ **7.** ${}^{87}_{35}\text{Br} \longrightarrow {}^{1}_{0}\text{n} + {}^{86}_{35}\text{Br}$ **9.** ${}^{24}_{18}\text{Mg} + {}^{1}_{0}\text{n} \longrightarrow {}^{1}_{1}\text{H} + {}^{24}_{11}\text{Na}$ (sodium-24) **11. a.** ${}^{10}_{5}\text{B} + {}^{1}_{0}\text{n} \longrightarrow {}^{10}_{4}\text{Be} + {}^{1}_{1}\text{H}$ **b.** ${}^{121}_{51}\text{Sb} + {}^{1}_{1}\text{H} \longrightarrow {}^{121}_{52}\text{Te} + {}^{1}_{0}\text{n}$ **c.** ${}^{59}_{27}\text{Co} + {}^{1}_{0}\text{n} \longrightarrow {}^{56}_{25}\text{Mn} + {}^{4}_{2}\text{He}$ **13. a.** $2\,{}^{2}_{1}\text{H} \longrightarrow {}^{3}_{2}\text{He} + {}^{1}_{0}\text{n}$ **b.** ${}^{241}_{95}\text{Am} + {}^{4}_{2}\text{He} \longrightarrow {}^{243}_{97}\text{Bk} + 2\,{}^{1}_{0}\text{n}$ **c.** ${}^{121}_{51}\text{Sb} + {}^{4}_{2}\text{He} \longrightarrow {}^{124}_{53}\text{I} + {}^{1}_{0}\text{n}$
15. 24.1 days **17.** 335 hr

Chapter 13

Selected Review Questions

13.2 methane and ethane **13.3 a.** 2 **b.** 7 **c.** 4 **d.** 9 **13.6** Natural gas is mostly methane; bottled gas is propane and/or butanes. **13.8** The lighter liquid alkanes dissolve and wash away natural body oils, causing dermatitis. Heavier liquid alkanes act as emollients (skin softeners). **13.12** Both act as anesthetics. **13.15** The monomer molecules have a carbon-to-carbon double bond. **13.16** Alkenes have carbon-to-carbon double bonds; alkynes, triple bonds. Both are unsaturated hydrocarbons and undergo addition reactions. **13.18 a.** para **b.** ortho **c.** meta **13.19** aromatic: a, d; aliphatic: b, c **13.20** Many polycyclic aromatic hydrocarbons are carcinogens.

Practice Exercises

13.1 a. 3-methylhexane **b.** 2,4-dimethylpentane **c.** 3-ethylhexane **d.** 4-isopropylheptane

13.2 a. $CH_3CH_2CH_2CHCH_2CH_2CH_3$ (with $CH_2CH_2CH_3$ on the CH) **b.** $CH_3CHCHCH_2CH_3$ (with CH_3 on C-2 and CH_2CH_3 on C-3)

c. $CH_3CH_2CCH_2CH_2CH_2CH_2CH_3$ (with CH_3 and $HC(CH_3)_2$ on C-3) **13.3 a.** methyl iodide

b. propyl fluoride **13.4 a.** 2-chloro-3-methylbutane

b. 1-bromo-3-chloro-4-methylpentane **13.5 a.** 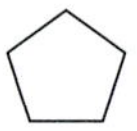 (cyclopentane)

b. 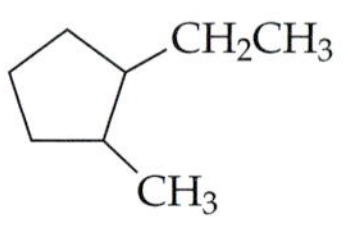 (cyclopentane with CH_2CH_3 and CH_3) **c.** 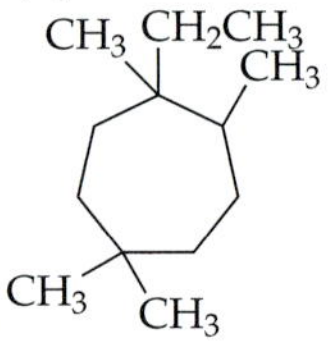 (cycloheptane with CH_3, CH_2CH_3, CH_3, CH_3, CH_3)

13.6 a. 4-ethyl-2-methyl-2-hexene **b.** 1-methylcyclohexene

c. 2,4,4-trimethyl-2-pentene **13.7 a.** $CH_2{=}CCHCH_2CH_2CH_3$ (with CH_2CH_3 on C-2 and CH_3 on C-3)

b. 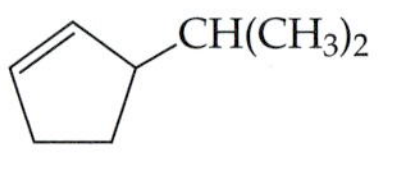 (cyclopentene with $CH(CH_3)_2$) **13.8 a.** 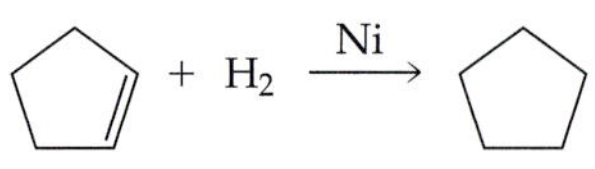 cyclopentene $+ H_2 \xrightarrow{\text{Ni}}$ cyclopentane

b. 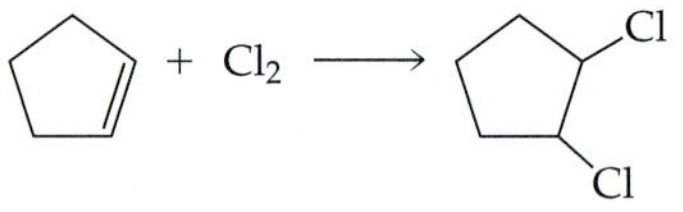 cyclopentene $+ Cl_2 \longrightarrow$ 1,2-dichlorocyclopentane (Cl, Cl)

c. 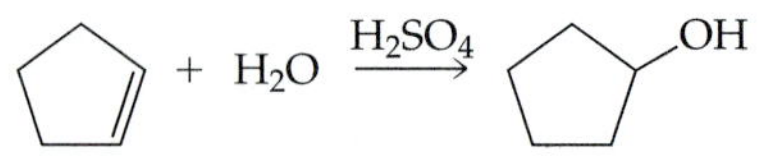 cyclopentene $+ H_2O \xrightarrow{H_2SO_4}$ cyclopentanol (OH)

Odd-Numbered Problems

1. organic: a, c; inorganic: b **3. a.** NaOH **b.** KCl

5. a. $CH_3CH_2CH_2CH_2CH_2CH_2CH_3$ **b.** $CH_3CH_2CHCH_2CH_3$ (with CH_3 on C-3)

c. $CH_3CCH_2CH_2CHCH_3$ (with CH_3, CH_3 on C-2 and CH_3 on C-5) **d.** $CH_3CH_2CHCHCH_2CH_2CH_2CH_3$ (with CH_3 on C-3 and CH_2CH_3 on C-4)

7. a. 3-methylpentane; **b.** 2,3-dimethylbutane

9. a. $CH_3CH_2{-}$ **b.** $CH_3CH{-}$ (with CH_3)

11. 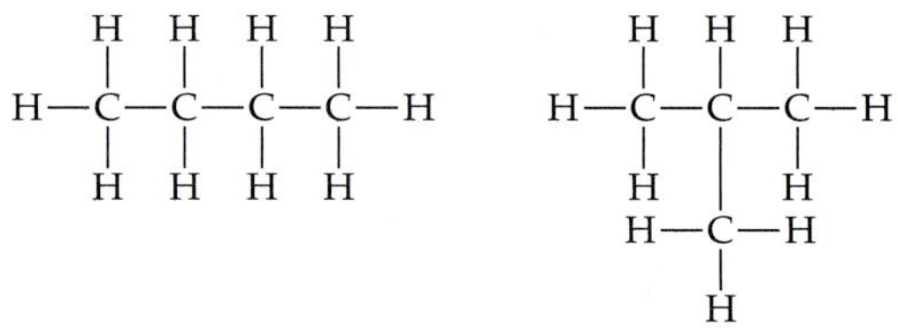 Butane: $H_3C{-}CH_2{-}CH_2{-}CH_3$ (all H drawn) Isobutane or methylpropane: $(CH_3)_3CH$ (all H drawn)

Butane

Isobutane or methylpropane

13. a. pentane **b.** $CH_3(CH_2)_4CH_3$ **c.** cyclohexane **d.** $CH_3(CH_2)_7CH_3$ **15. a.** methylcyclopropane **b.** 1,2-diethyl-4-methylcyclopentane **c.** cyclobutene **d.** 3-ethylcyclohexene

17. a. 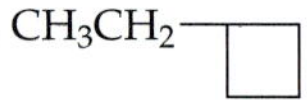 CH_3CH_2–(cyclobutane) **b.** 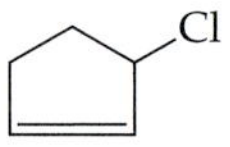 (cyclopentene with Cl) **19. a.** CH_3Cl

b. $CHCl_3$ **21.** CH_3CHCH_3 (with Br on C-2) $BrCH_2CH_2CH_3$

Isopropyl bromide (2-bromopropane) Propyl bromide (1-bromopropane)

23. Saturated: c; unsaturated: a, b **25. a.** $HC{\equiv}CH$ **b.**  (cyclohexene)

c. $HC{\equiv}CCHCH_2CH_2CH_3$ (with $CH{-}CH_3$ / CH_3 on C-3) **d.** $CH_3C{=}CCH_3$ (with CH_3, CH_3)

27. a. $CH_3C{=}CHCH_2CH_3$ (with CH_3 on C-2) **b.** $CH_2{=}CHCH_2CH_2CHCH_3$ (with CH_3 on C-5)

29. a. 2-methyl-1-pentene **b.** 2-methyl-2-pentene **c.** 2,5-dimethyl-2-hexene **31. a.** same **b.** same **c.** isomers **33. a.** isomers **b.** same **c.** isomers **35. a.** $(CH_3)_2CBrCH_2Br$

b. $CH_3CH(CH_3)CH_2CH_3$ **c.** 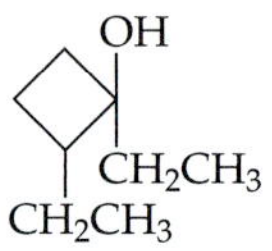 (cyclobutane with OH, CH_2CH_3, CH_2CH_3) **37. a.** H_2, Ni

b. H_2O, H^+ **39. a.**  (cyclohexene) **b.** 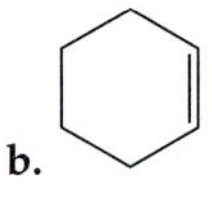 (cyclohexene) **41. a.** 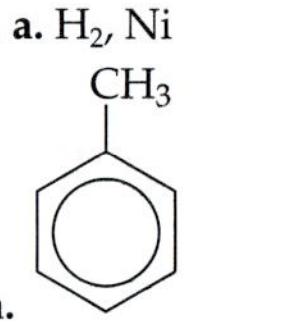 (toluene, CH_3)

b. 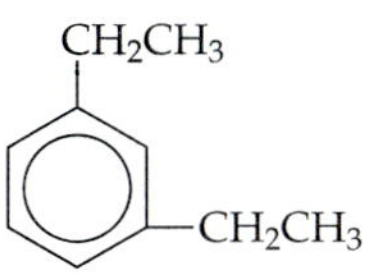 (1,3-diethylbenzene: benzene ring with CH_2CH_3 and CH_2CH_3) c. 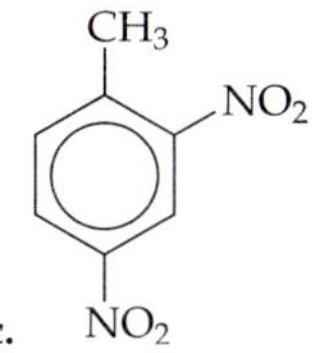 (benzene ring with CH_3, NO_2, NO_2) **43. a.** ethylbenzene **b.** isopropylbenzene **c.** 2-nitrotoluene **d.** 3,5-dichlorotoluene

Chapter 14

Selected Review Questions

14.1 A functional group is a group of atoms in an organic molecule that confers characteristic properties to the molecule. **(a)** C=C; carbon-to-carbon double bond **(b)** —OH; hydroxyl group **(c)** —O—; alkoxy group **14.2** Both have an —OH functional group. In an alcohol, the OH is attached to an aliphatic group; in phenols, to an aryl group. **14.3** Methanol has a smaller nonpolar group than does 1-hexanol. **14.4** tert-Butyl alcohol is more compact and thus a smaller surface area than does 1-butanol. Its intermolecular forces are therefore weaker than those between 1-butanol molecules. **14.11** Ethanol; by blocking the enzyme that oxidizes methanol to formaldehyde. **14.12** dehydration, oxidation, esterification **14.14** Higher temperatures would produce more of ethylene as a by-product. **14.17** Antiseptics kill microorganisms on living tissue; disinfectants kill microorganisms on inanimate objects. **14.18** Diethyl ether is highly flammable and forms explosive peroxides on standing exposed to air.

Practice Exercises

14.1 a. 1,2-propanediol **b.** 1-methylcyclopentanol

14.2 a. $CH_3C(CH_3){=}CHCH_2CH_3 + H_2O \xrightarrow{H^+}$ $CH_3C(OH)(CH_3)CH_2CH_2CH_3$ (2-methyl-2-pentanol)

b. 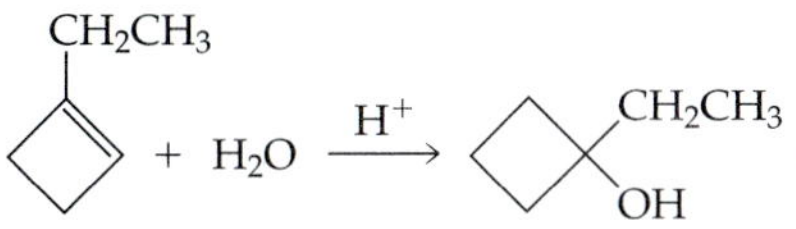 (1-ethylcyclobutene, ring with CH_2CH_3) + $H_2O \xrightarrow{H^+}$ (cyclobutane with CH_2CH_3 and OH) (1-ethylcyclobutanol)

14.3 a. $CH_3CH_2CH_2CH_2CH_2OH \xrightarrow[H_2SO_4]{K_2Cr_2O_7}$ $CH_3CH_2CH_2CH_2C(=O){-}OH$

b. $CH_3CH_2CH_2CH(OH)CH_3 \xrightarrow[H_2SO_4]{K_2Cr_2O_7}$ $CH_3CH_2CH_2C(=O)CH_3$

c. $CH_3CH_2C(OH)(CH_3)_2 \xrightarrow[H_2SO_4]{K_2Cr_2O_7}$ no reaction

Odd-Numbered Problems

1. a. 1-hexanol **b.** 2-hexanol **3. a.** 4,4-dichloro-2-butanol **b.** 3,3-dibromo-2-methyl-2-butanol **5. a.** $CH_3CH_2CHOHCH_2CH_2CH_3$ **b.** $CH_3CHOHC(CH_3)_2CH_3$ **7. a.** $CH_3CH_2CHOHCH(CH_3)CH(CH_3)CH_2CH_3$ **b.** 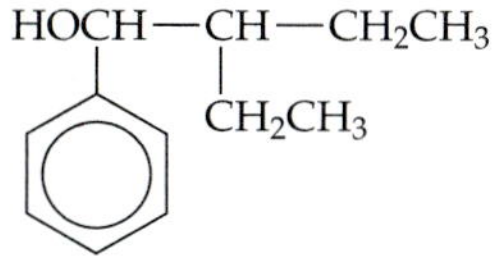 (benzene ring bearing $HOCH{-}CH(CH_2CH_3){-}CH_2CH_3$) **9.** methanol (lowest) < ethanol < 1-propanol (highest) **11.** 1-octanol (least) < 1-butanol < methanol (most) **13. a.** $CH_2{=}CHCH_3$ **b.** (1-methylcyclopentene, ring with CH_3) **c.** $CH_3C(CH_3){=}CH_2$ **15. a.** oxidation **b.** dehydration **c.** hydration

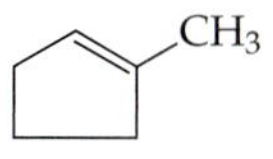

17. $CH_3CH_2CH_2CH_2OH \longrightarrow CH_3CH_2CH_2C(=O)OH$

$CH_3CH_2CHOHCH_3 \longrightarrow CH_3CH_2C(=O)CH_3$

$(CH_3)_2CHCH_2OH \longrightarrow (CH_3)_2CHC(=O)OH$

$(CH_3)_3COH \longrightarrow$ no reaction

19. 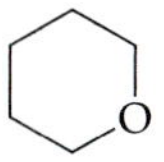 (tetrahydropyran ring, O in ring) **21. a.** $CH_3CHOHCH_2CH_3$ **b.** 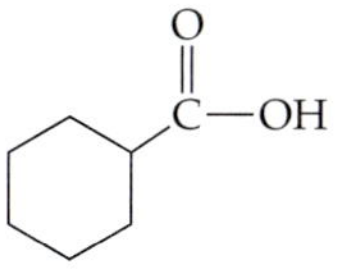 (cyclohexane with $C(=O){-}OH$)

23. a. H^+, H_2O **b.** $K_2Cr_2O_7$, H^+ **c.** conc. H_2SO_4, 140 °C, excess alcohol **25. a.** $CH_3CHOHCH_2CH_2CH_2OH$ **b.** $CH_3CHOHCH_2OH$ **27. a.** catechol (1,2-dihydroxybenzene) **b.** 4-bromophenol (*p*-bromophenol) **29. a.** 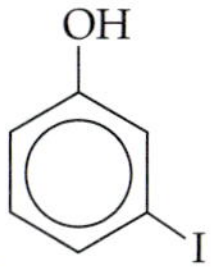 (benzene ring with OH and I, meta) **b.** 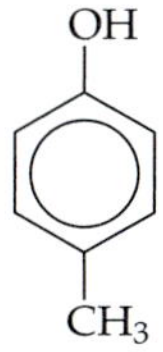 (benzene ring with OH and CH_3, para) **31. a.** 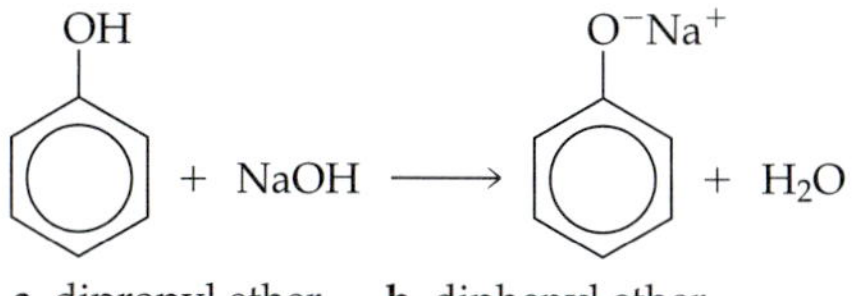 (phenol with OH) + NaOH ⟶ (benzene ring with O^-Na^+) + H_2O **b.** No reaction. **33. a.** dipropyl ether **b.** diphenyl ether **35. a.** $CH_3CH_2OCH_3$ **b.** 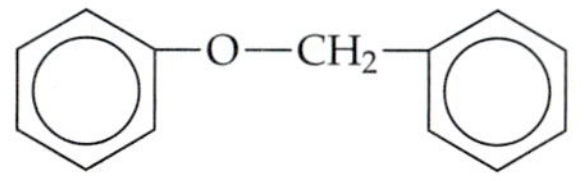 (benzene)—O—CH_2—(benzene)

Chapter 15

Selected Review Questions

15.1 Both aldehydes and ketones have a carbonyl functional group, but aldehydes have at least one hydrogen atom attached to the carbonyl carbon atom whereas ketones have the carbonyl carbon bonded to two other carbon atoms. **15.2** 1-Butanol molecules are associated through hydrogen bonding, butanal molecules have fairly strong dipole interactions, and diethyl ether is only slightly polar. **15.3** Some of the aldehyde formed from the 1° alcohol is oxidized further to a carboxylic acid. The ketone formed from the 2° alcohol is resistant to further oxidation. **15.4** Reaction with water (hydration); aldehyde ⟶ hydrate and ketone ⟶ hydrate. Reaction with alcohols; aldehyde ⟶ hemiacetal and ketone $\longrightarrow$ hemiketal. Reduction: aldehyde $\longrightarrow$ 1° alcohol; ketone $\longrightarrow$ 2° alcohol. **15.7** Acetal formation requires an anhydrous acid; acetal conversion to an aldehyde requires an aqueous acid. **15.8** benzaldehyde, cinnamaldehyde.

Practice Exercises

15.1 A. 3,3-dimethylbutanal

B. 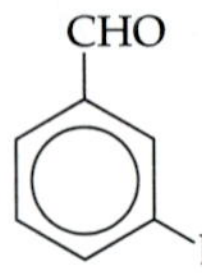 (benzene ring with CHO and Br) Br (3-bromobenzaldehyde)

15.2 A. $CH_3CH_2CH(Br)CH_2CH(I)CH_2C(=O)—H$

B. 2,3-dihydroxypropanal **15.3** $CH_3CH_2CH(Br)CH(CH_2CH_3)CH_2C(=O)CH_2Br$

15.4 A. isopropyl propyl ketone, 2-methyl-3-hexanone
B. 5-methyl-3-hexanone, ethyl isobutyl ketone

15.5 A. a. $CH_3CH(CH_3)—CH(OCH_3)—OCH_3$ **b.** 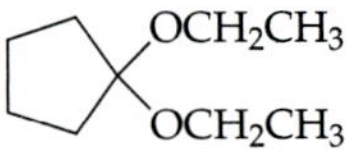 cyclopentane with OCH_2CH_3 and OCH_2CH_3

B. 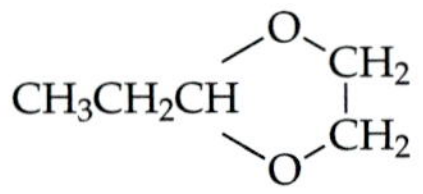 CH_3CH_2CH (ring: O—CH_2—CH_2—O)

Odd-Numbered Problems

1. a. benzaldehyde **b.** 3-hydroxypropanal **c.** 4,4-dimethylpentanal **d.** 2-chlorobenzaldehyde **3. a.** 5-methyl-3-hexanone **b.** cyclopentanone **c.** 2-pentanone **d.** 4-bromo-2,2-dimethyl-3-pentanone **5. a.** $CH_3CH_2CH_2C(=O)H$

b. $CH_3CH_2CH_2CH_2CH(CH_3)CH_2C(=O)H$ **c.** 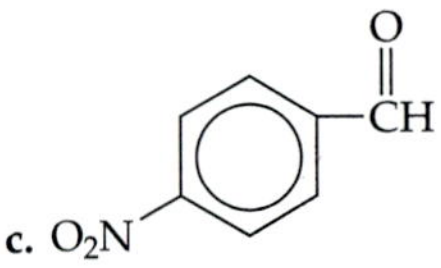 O_2N—(benzene ring)—$C(=O)H$

7. a. $CH_3C(=O)CH_2CH_2CH_2CH_3$ **b.** $CH_3C(=O)CHBrCH_2CH_2CH_2CH_3$

c. 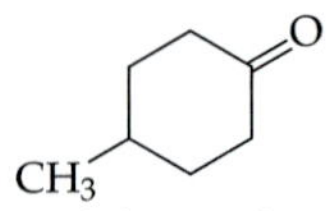 CH_3—(cyclohexane ring)=O **9.** 2-Propanol has a higher boiling point because it is associated through hydrogen bonding; acetone is polar but has no hydrogen bonding **11.** Acetaldehyde has a higher boiling point because it has strong dipoles; dimethyl ether has only weak dipoles. **13. a.** 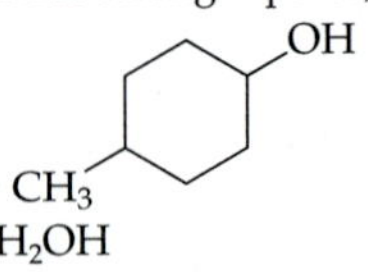 CH_3—(cyclohexane ring)—OH **b.** $(CH_3)_3CCH_2OH$
c. $CH_3CH_2CHBrCH_2CH_2OH$

15. a. $CH_3C(=O)H + CH_3OH \longrightarrow CH_3C(OH)(H)—O—CH_3$

b. $CH_3C(=O)H + 2CH_3OH \xrightarrow{H^+} CH_3—C(O—CH_3)(H)—O—CH_3$

c. 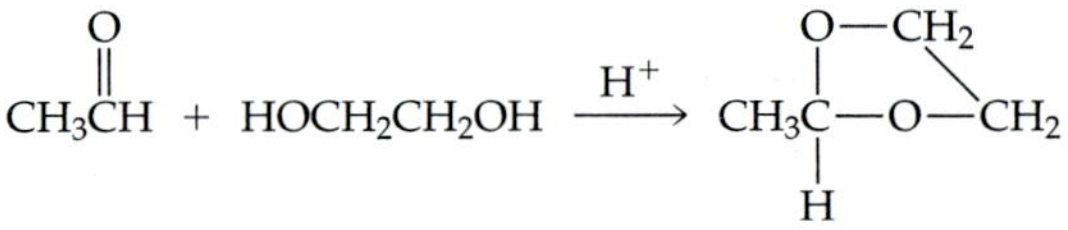 $CH_3C(=O)H + HOCH_2CH_2OH \xrightarrow{H^+} CH_3C(H)$ (ring: —O—CH_2—CH_2—O—)

17. a. $CH_3C(=O)CH_3 + CH_3OH \longrightarrow CH_3C(O—CH_3)(OH)CH_3$

b. $CH_3C(=O)CH_3 + 2CH_3OH \xrightarrow{H^+} CH_3C(O—CH_3)(O—CH_3)—CH_3$

c. $CH_3C(=O)CH_3 + HOCH_2CH_2OH \longrightarrow$ $(CH_3)_2C$ (ring: —O—CH_2—CH_2—O—)

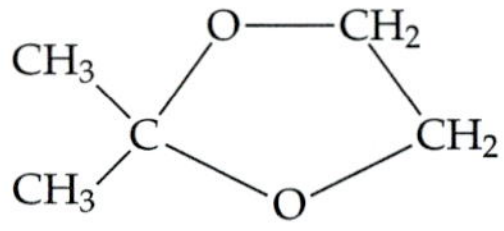

19. a. Yes; only pentanal would test positive **b.** No; both would test negative **c.** Yes, only pentanal would test positive **d.** Yes, only pentanal would test positive **e.** No; both would test negative **21.** $K_2Cr_2O_7$ and H^+ will oxidize 2-pentanol; a greenish precipitate of chromium(III) compounds would form. **23. a.** Ag^+, NH_3 **b.** CrO_3, HCl, pyridine, CH_2Cl_2 **c.** 2 CH_3OH, dry HCl **25.** Hemiacetals: a, d
27. Hydrates: c, e

Chapter 16

Selected Review Questions

16.2 a. formic acid, methanoic acid **b.** acetic acid, ethanoic acid **c.** propionic acid, propanoic acid **d.** butyric acid, butanoic acid **16.3 a.** 1-pentanol **b.** pentanal **16.4** More soluble; both have an OH group, but butyric acid has two oxygen atoms. **16.7** Unpleasant: carboxylic acids; Pleasant: esters **16.9** Esterification is the conversion of a carboxylic acid to an ester. Neutralization forms salts—ionic compounds; esterification forms covalent compounds. **16.10** Acidic hydrolysis produces a carboxylic acid and an alcohol; it yields an equilibrium mixture. Basic hydrolysis produces a carboxylate salt and an alcohol; it goes to completion. **16.11** basic hydrolysis **16.13** Lower; it has no intermolecular hydrogen bonding whereas acetamide does. **16.14** decanedioic acid and 1,6-hexanediamine **16.15** Acidic hydrolysis produces a carboxylic acid and an ammonium salt. Basic hydrolysis produces a carboxylate salt and ammonia or an amine.

Practice Exercises

16.1 A. 6-chlorohexanoic acid **B.** 4-methylpentanoic acid

16.2 A. $CH_3CH(CH_3)CH(Br)CH_2CH_2C(=O)—OH$ **B.** $H_2NCH_2CH_2CH_2COOH$

16.3 A. a. 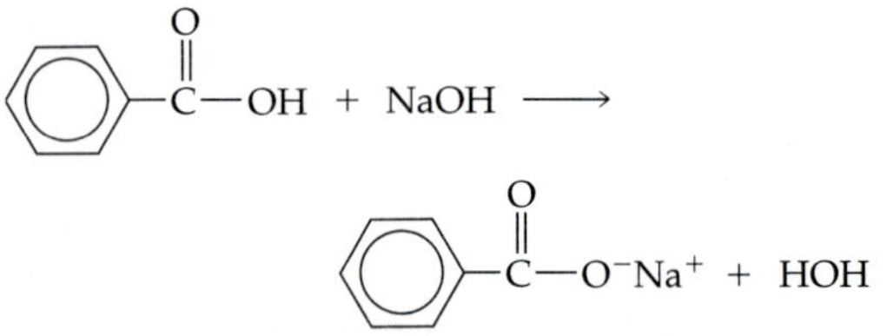 (benzene ring)—$C(=O)—OH + NaOH \longrightarrow$ (benzene ring)—$C(=O)—O^-Na^+ + HOH$

b. $C_6H_5-\overset{O}{\overset{\|}{C}}-OH + NaHCO_3 \longrightarrow$

$C_6H_5-\overset{O}{\overset{\|}{C}}-O^-Na^+ + HOH + CO_2$

B. a. $CH_3CH_2CH_2CH_2COOH + KOH \longrightarrow CH_3CH_2CH_2CH_2COO^-K^+ + H_2O$ **b.** $2\ CH_3CH_2CH_2CH_2COOH + Na_2CO_3 \longrightarrow 2\ CH_3CH_2CH_2CH_2COO^-Na^+ + CO_2 + H_2O$ **c.** $2\ CH_3CH_2CH_2CH_2COOH + Ca(OH)_2 \longrightarrow (CH_3CH_2CH_2CH_2COO^-)_2Ca^{2+} + 2\ H_2O$ **16.4 A. a.** ethyl butyrate, ethyl butanoate **b.** butyl propionate, butyl propanoate **B. a.** propyl hexanoate **b.** methyl octanoate

16.5 (a) $CH_3CH_2CH_2\overset{O}{\overset{\|}{C}}-O-C_6H_5$

(b) $CH_3CH_2CH_2CH_2CH_2COOCH(CH_3)_2$

16.6 $CH_3CH_2\overset{O}{\overset{\|}{C}}-OCH(CH_3)_2 + HOH \xrightleftharpoons{H^+}$

$CH_3CH_2\overset{O}{\overset{\|}{C}}-OH + CH_3\overset{OH}{\overset{|}{C}}HCH_3$

16.7 $H-\overset{O}{\overset{\|}{C}}-O-C_6H_5 + NaOH \longrightarrow$

$H-\overset{O}{\overset{\|}{C}}-O^-Na^+ + C_6H_5-OH$

16.8 A. a. butanamide, butyramide **b.** heptanamide

B. $CH_3CH_2CH_2CH_2\overset{O}{\overset{\|}{C}}-NH_2$ **16.9 A.** N,N-dimethylacetamide

B. $CH_3CH_2\overset{O}{\overset{\|}{C}}-N(CH_3)_2$

16.10 (a) $C_6H_5-\overset{O}{\overset{\|}{C}}-NH_2 + HCl + HOH \longrightarrow$

$C_6H_5-\overset{O}{\overset{\|}{C}}-OH + NH_4Cl$

(b) $CH_3CONH_2 + KOH(aq) \longrightarrow CH_3COOK(aq) + NH_3(g)$

Odd-Numbered Problems

1. a. $CH_3CH_2CH_2CH_2CH_2CH_2\overset{O}{\overset{\|}{C}}OH$ **b.** $(CH_3)_2CHCH_2\overset{O}{\overset{\|}{C}}OH$

c. 2,3-dibromobenzoic acid: $C_6H_3Br_2-\overset{O}{\overset{\|}{C}}-OH$ (Br, Br) **d.** $C_6H_4(CH(CH_3)_2)-\overset{O}{\overset{\|}{C}}-OH$ (meta $CH(CH_3)_2$)

3. a. $HO\overset{O}{\overset{\|}{C}}-\overset{O}{\overset{\|}{C}}OH$ **b.** $CH_3\overset{OH}{\overset{|}{C}}HCH_2\overset{O}{\overset{\|}{C}}OH$

5. a. 3-methylbutanoic acid **b.** 3,4,4-trimethylpentanoic acid **c.** 4-hydroxybutanoic acid **d.** 2,4-dimethylpentanoic acid

7. a. $CH_3\overset{O}{\overset{\|}{C}}O^-K^+$ **b.** $(CH_3CH_2\overset{O}{\overset{\|}{C}}O^-)_2Ca^{2+}$

9. a. $CH_3\overset{O}{\overset{\|}{C}}-O-CH_3$ **b.** $CH_3\overset{O}{\overset{\|}{C}}-O-C_6H_5$

11. a. $C_6H_5-\overset{O}{\overset{\|}{C}}-O-CH_2CH_3$ **b.** $C_6H_5-\overset{O}{\overset{\|}{C}}-O-C_6H_5$

13. a. methyl benzoate **b.** methyl formate (or methyl methanoate) **c.** ethyl propionate (or ethyl propanoate)

15. a. $CH_3CH_2CH_2\overset{O}{\overset{\|}{C}}-NH_2$ **b.** $CH_3CH_2CH_2CH_2CH_2\overset{O}{\overset{\|}{C}}-NH_2$

c. $CH_3\overset{O}{\overset{\|}{C}}-NH-CH_3$ **17. a.** benzamide **b.** 2-methylbutanamide **c.** acetamide **19.** II; Butanoic acid can form dimers which are hydrogen bonded to one another; the ether can't hydrogen bond. **21.** I; The amide can hydrogen bond while the ester can't. **23.** I; The acid can both hydrogen bond with water and ionize slightly in water. This helps its solubility. The alkane is completely insoluble in water due to its lack of polarity. **25.** I; The longer hydrocarbon chain of II will make it less able to mix with the polar water.

27. a. $CH_3CH_2CH_2\overset{O}{\overset{\|}{C}}OH + NaOH \longrightarrow$

$CH_3CH_2CH_2\overset{O}{\overset{\|}{C}}-O^-Na^+ + H_2O$

b. $CH_3CH_2CH_2\overset{O}{\overset{\|}{C}}OH + NaHCO_3 \longrightarrow$

$CH_3CH_2CH_2\overset{O}{\overset{\|}{C}}-O^-Na^+ + CO_2 + H_2O$

29. $CH_3\overset{O}{\overset{\|}{C}}O-CH_2CH_3 + H_2O \xrightarrow{H^+}$

$CH_3\overset{O}{\overset{\|}{C}}OH + HOCH_2CH_3$

31. $C_6H_5-\overset{O}{\overset{\|}{C}}-NH_2 + H_2O \xrightarrow{H^+}$

$C_6H_5-\overset{O}{\overset{\|}{C}}OH + NH_4^+$

33. a. $CH_3CH_2COO^-Na^+$

b. $C_6H_4(COO^-Na^+)_2 + 2\ H_2O + 2\ CO_2$

35. a. $C_6H_5-\overset{O}{\overset{\|}{C}}-O^-Na^+ + HOCH_2CH_2CH_3$

b. $C_6H_{11}-O-\overset{O}{\overset{\|}{C}}CH_3 + H_2O$

37. a. $CH_3C(=O)—O—CH_2CH_2CH_3 + H_2O$

b. $CH_3—O—C(=O)CH_2C(=O)—O—CH_3 + 2\,H_2O$

39. a. $CH_3C(=O)—OH + NH_4Cl$

b. $C_6H_5—C(=O)—O^-Na^+ + HN(CH_3)_2$

41. a. $K_2Cr_2O_7, H^+$ **b.** $K_2Cr_2O_7$, H+ **c.** NaOH **43. a.** $CH_3C(=O)OH, H^+$ **b.** LiOH

45. a. $CH_3CH_2O—P(=O)(OH)—OCH_2CH_3$ **b.** $CH_3O—P(=O)(OH)—OH$

c. $HO—P(=O)(OH)—O—P(=O)(OH)—O—P(=O)(OH)—OH$

Selected Topic B

Selected Review Questions

2. $C_6H_4(COOH)(O—C(=O)CH_3)$ acetylsalicylic acid; different doses; different additives (such as buffering agents, acetaminophen, caffeine). All must contain acetylsalicylic acid. **12.** Tetrahydrocannabinol interacts weakly with estrogen receptors. **13.** THC is fat-soluble. **14.** 3

Odd-Numbered Problems

15. Carboxylic acid **17. a.** $HOCH_3, H^+$ **b.** NaOH

Chapter 17

Selected Review Questions

17.5 Amine molecules have a lone pair of electrons on a nitrogen atom; they can act as proton acceptors. **17.6** Neutralization. The volatile, smelly amine is converted to a less volatile salt.

Practice Exercises

17.1 A. a. isopropylmethylamine, secondary **b.** diethylmethylamine, tertiary **B. a.** butylamine, primary **b.** *tert*-butylamine, primary **17.2 (a)** $CH_3CH_2NHCH(CH_3)_2$ **(b)** $CH_3CH_2CH_2N(CH_2CH_3)_2$ **17.3** 4-propylaniline (*p*-propylaniline)

17.4 $C_6H_5—N(H)—C_6H_5$ and $(C_6H_5)_3N$

17.5 $C_6H_5—CH_2CH(NH_2)CH(CH_2CH_3)CH_2CH_2CH_2CH_3$

17.6 a. ethylammonium ion **b.** triethylammoniumion **c.** dipropylammonium ion **d.** tetrabutylammonium ion
17.7 tetrapropylammonium iodide

Odd-Numbered Problems

1. a. amide **b.** neither **c.** both **3. a.** alcohol (1°) **b.** amine (1°) **c.** alcohol (2°) **d.** amine (1°) **e.** ether **f.** phenol

5. a. CH_3NHCH_3 **b.** $CH_3CH_2N(CH_3)—CH_2CH_3$ **c.** cyclobutane with OH and NH_2 on adjacent carbons

d. $HOCH_2CH_2NH_2$ **7. a.** $C_6H_5NH_2$ **b.** Br—C_6H_4—NH_2 (meta)

c. pyrimidine **d.** $C_6H_5—NHCH_2CH_3$ **9. a.** propylamine **b.** isopropylmethylamine **c.** triethylamine **d.** 2-aminopentane

11. a. $C_6H_5NH_3^+Br^-$ **b.** $[(CH_3)_4N]^+\ Cl^-$

13. a. diethylammonium bromide **b.** tetraethylammonium iodide **15.** Butylamine; it can hydrogen bond (while pentane cannot). **17.** Propylamine (1°). Trimethylamine (a 3° amine) has no hydrogen bonded to the nitrogen; therefore cannot undergo hydrogen bonding. **19.** $CH_3CH_2NH_2$. It forms hydrogen bonds with water (while propane cannot). **21.** $CH_2(NH_2)CH_2CH(NH_2)CH_2CH(NH_2)CH_3$ The more hydrogen bonding groups on a compound, the greater its solubility in water. **23. a.** $CH_3NH_3^+Br^-$ **b.** $[(CH_3)_3NH^+]_2SO_4^{-2}$

25. (a) $CH_3(CH_2)_4C(=O)—NH(CH_2)_3CH_3$ **(b)** $C_6H_5—C(=O)NH—C_6H_5$

27. a. $CH_3CH_2NHCH_3$ and CH_3CH_2COOH

b. $C_6H_5—NH_2 + HOC(=O)CH_2CH_3$ **29.** $C_6H_5—C(=O)—NH_2$

31. a. HCl **b.** HNO_3

Selected Topic C

Selected Review Questions

3. Tryptophan is the precursor for serotonin; tyrosine, for dopamine and norepinephrine. **9.** Amphetamines are stimulants; barbiturates, depressants.

Odd-Numbered Problems

13. Barbiturates all contain a six-membered barbituric acid ring structure. Groups are added to a carbon located between a pair of carbonyl groups to alter the drug's properties such as effectiveness and length of duration. **15.** Amide, ketone, ether, alcohol, amine **17.** Cocaine

Chapter 18

Selected Review Questions

18.1 Polarized light is light that vibrates in a single plane.
18.2 (a) An optically active substance rotates a beam of polarized light. **(b)** A dextrorotatory substance rotates the plane of polarized

light to the right, and **(c)** a levorotatory substance rotates the plane to the left. **18.5** No. The molecule and its mirror image may be identical (superimposable). **18.8 a.** same **b.** same **c.** same magnitude; different sign **d.** same **e.** same **f.** different rate, extent, and product. **18.9** The melting point, boiling point, and density are the same. The specific rotation is +50°. **18.10** (a) Yes. **(b)** No. **(c)** No.

Practice Exercises

18.1

CH_3CH_2 / HO—C—H / CH_3 and CH_3CH_2 / H—C—OH / CH_3

18.2

CHO / H—C—Br / H—C—Br / CH_3; CHO / Br—C—H / Br—C—H / CH_3; CHO / Br—C—H / H—C—Br / CH_3; CHO / H—C—Br / Br—C—H / CH_3

18.3

(H)(Br)C=C(CH_3)(Br) and (H)(Br)C=C(Br)(CH_3)

cis-1,2-dibromopropane; *trans*-1,2-dibromopropane

(H)(Br)C=C(H)(CH_2Br) and (H)(Br)C=C(CH_2Br)(H)

cis-1,3-dibromopropane; *trans*-1,3-dibromopropane

(Br)(Br)C=C(CH_3)(H) and (H)(H)C=C(CH_2Br)(Br)

1,1-dibromopropane (no isomers); 2,3-dibromopropane (no isomers)

and (H)(H)C=C($CHBr_2$)(H)

3,3-dibromopropane (no isomers)

Odd-Numbered Problems

1. Yes, no **3. a.** C_6H_5—$CH_2CH(CH_3)$—NH_2

b. $H_2NCOCH_2C(CH_3)(CH_2CH_2CH_3)CH_2OCNH_2$ (with C=O on each carbamate carbon) **5. a.** CH_3(CH)CH_2OH with OH on the circled CH

b. CH_3(CH)COOH with NH_2 on the circled CH **c.** $C_6H_5CH_2$(CH)CH_3 with NH_2 on the circled CH **d.** CH_3(CH)CH_2CH_3 with Br on the circled CH

e. CH_3(CH)CHO with OH on the circled CH **f.** CH_3(CH)—(CH)CH_3 with OH on each circled CH **7. a.** Yes **b.** No

c. Yes. **d.** No. **9. a.** CH_3 / H—C—OH / CH_2CH_3 and CH_3 / HO—C—H / CH_2CH_3

b. COOH / H—C—CH_3 / CH_3 and COOH / CH_3—C—H / CH_3

11. CHO / H—C—Cl / Cl—C—H / CH_3 and CHO / Cl—C—H / H—C—Cl / CH_3

Enantiomers

CHO / H—C—Cl / H—C—Cl / CH_3 and CHO / Cl—C—H / Cl—C—H / CH_3

Enantiomers

13. a, d **15. a.** 4 **b.** 8

17. a. (Br)(CH_3)C=C(H)(CH_2CH_3) cis; (CH_3)(Br)C=C(H)(CH_2CH_3) trans

b. (H)(CH_3CH_2)C=C(H)(CH_2CH_3) cis; (CH_3CH_2)(H)C=C(H)(CH_2CH_3) trans

c. none; **d.** none

e. (H)(CH_3)C=C(H)(COOH) cis; (CH_3)(H)C=C(H)(COOH) trans

f. (H)(CH_3)C=C(H)($CH(CH_3)_2$) cis; (CH_3)(H)C=C(H)($CH(CH_3)_2$) trans

Selected Topic D

Selected Review Questions

1. Retinal is converted from the *cis* isomer to the *trans* isomer. **2.** Vitamin A is the reduced form of retinal, the molecule involved in the visual process. **5.** (+)-Carvone and (−)-carvone are enantiomers (mirror images) of each other.

Chapter 19

Selected Review Questions

19.4 The D and L prefixes indicate the absolute configuration of a molecule. A D sugar is one that has the same configuration as D-glyceraldehyde about the chiral carbon farthest from the aldehyde or ketone group; an L sugar, as L-glyceraldehyde. **19.6 a.** aldose **b.** aldose **c.** aldose **d.** ketose **19.8** If one measures the specific rotation of the solution with a polarimeter, one will find that the value will change from +112° to +52.7° (when equilibrium is reached). **19.12** beta; alpha **19.13** Starch serves as fuel for the plant; cellulose is the structural support of the plant.

Odd-Numbered Problems

1. a. aldose; hexose **b.** aldose; pentose **c.** ketose; hexose **d.** aldose; hexose **3. a.** D sugar; **b.** L sugar **5.** The D,L designations are not related to the rotation of plane polarized light, but indicate the relationship between the stereochemistry at the chiral center farthest from the carbonyl group and that of D- and L-glyceraldehyde.

7.

CHO | H—C—OH | HO—C—H | H—C—OH | H—C—OH | CH_2OH D-Glucose

CHO | HO—C—H | HO—C—H | H—C—OH | H—C—OH | CH_2OH D-Mannose

9. CH_2OH, O, OH, OH, OH, OH

11. CH_2OH, O, OH, OH, OH, HO

13. a, b, and c will all give a positive test.

15.

O=C—OH | H—C—OH | HO—C—H | H—C—OH | H—C—OH | CH_2OH D-gluconic acid

O=C—H | H—C—OH | HO—C—H | H—C—OH | H—C—OH | C(=O)OH D-glucuronic acid

O=C—OH | H—C—OH | HO—C—H | H—C—OH | H—C—OH | C(=O)OH D-glucaric acid

17. a. beta **b.** alpha **19. a.** for **(a)** it is alpha; there is no hemiacetal carbon in **(b)** **b. (b)** is not a reducing sugar

21. CH_2OH, O, OH, HO, OH, O, CH_2, O, OH, HO, OH, OH

23. H_2O/H^+ or maltase **25.** Amylopectin contains an occasional α-1,6-glycosidic linkage while amylose does not. They both have α-1,4-glycosidic linkages between glucose units as their primary polymeric linkage. **27.** Both contain α-1,4- and α-1,6-glycosidic linkages, but glycogen has more frequent α-1,6 linkages than amylopectin.

Chapter 20

Selected Review Questions

20.5 Fats are the body's primary energy reserve. They also help maintain body temperature and insulate organs & tissues.

20.6
$CH_2—O—C(=O)(CH_2)_7CH=CH(CH_2)_7CH_3$
$CH—O—C(=O)(CH_2)_7CH=CH(CH_2)_7CH_3$ $+ 3\,H_2 \xrightarrow{Ni}$
$CH_2—O—C(=O)(CH_2)_7CH=CH(CH_2)_7CH_3$

$CH_2—O—C(=O)(CH_2)_{16}CH_3$
$CH—O—C(=O)(CH_2)_{16}CH_3$
$CH_2—O—C(=O)(CH_2)_{16}CH_3$

20.9
$CH_2—O—C(=O)(CH_2)_{14}CH_3$
$CH—O—C(=O)(CH_2)_7CH=CH(CH_2)_7CH_3$
$CH_2—O—P(=O)(O^-)—OCH_2CH_2—\overset{+}{N}H_3$

20.14 A saponifiable lipid can be hydrolyzed under alkaline conditions, whereas a nonsaponifiable lipid cannot undergo hydrolysis because there are no ester linkages in the molecule.

Odd-Numbered Problems

1. a. saturated, 6 carbons **b.** unsaturated; 18 carbons **c.** saturated; 18 carbons **3. a.** $CH_3(CH_2)_{10}COOH$ **b.** $CH_3(CH_2)_5CH=CH(CH_2)_7COOH$ **c.** $CH_3(CH_2)_{14}COO^-K^+$ **5.** The *cis* configuration of the double bonds in unsaturated fatty acids causes severe kinks or bends in the hydrocarbon chain. This prevents molecules from packing tightly together, weakening the attractions between adjacent fatty acids.

7. a.
$CH_2—O—C(=O)(CH_2)_{16}CH_3$
$CH—O—C(=O)(CH_2)_{16}CH_3$
$CH_2—O—C(=O)(CH_2)_{16}CH_3$

b.
$CH_2—O—C(=O)(CH_2)_{14}CH_3$
$CH—O—C(=O)(CH_2)_7CH=CH(CH_2)_7CH_3$
$CH_2—O—C(=O)(CH_2)_{16}CH_3$

(other answers are possible)

9.
$$\begin{array}{l} CH_2{-}O{-}\overset{O}{\overset{\|}{C}}(CH_2)_{14}CH_3 \\ | \\ CH{-}O{-}\overset{O}{\overset{\|}{C}}(CH_2)_{14}CH_3 \quad + 3\,KOH \xrightarrow{\Delta} \\ | \\ CH_2{-}O{-}\overset{O}{\overset{\|}{C}}(CH_2)_{14}CH_3 \end{array}$$

$$\begin{array}{l} CH_2OH \\ | \\ CHOH + 3\,CH_3(CH_2)_{14}C(=O)O^-K^+ \\ | \\ CH_2OH \end{array}$$

11. corn oil; An oil has a lower melting point, which indicates a higher number of double bonds and thus, a higher iodine number.

13.
$$\begin{array}{l} CH_2{-}O{-}\overset{O}{\overset{\|}{C}}(CH_2)_{10}CH_3 \\ | \\ CH{-}O{-}\overset{O}{\overset{\|}{C}}(CH_2)_{10}CH_3 \quad + 3\,NaOH \longrightarrow \\ | \\ CH_2{-}O{-}\overset{O}{\overset{\|}{C}}(CH_2)_{10}CH_3 \end{array}$$

$$\begin{array}{l} CH_2OH \\ | \\ CHOH + 3\,CH_3(CH_2)_{10}C(=O)O^-Na^+ \\ | \\ CH_2OH \end{array}$$

15. **a.** phospholipid **b.** glycolipid and sphingolipid **c.** phospholipid and sphingolipid

17. $CH_3(CH_2)_{12}CH{=}CH{-}CH{-}OH$; $CH{-}NH{-}\overset{O}{\overset{\|}{C}}{-}(CH_2)_{14}CH_3$; CH_2–O–(sugar ring with CH_2OH, OH, HO, OH)

19. b 21. (steroid ring system with rings A, B, C, D) 23. VLDLs contain a much higher proportion of triglycerides than do LDLs or HDLs, while LDLs have the highest percentage of cholesterol, and HDLs have the highest percentage of protein. VLDLs transport triglycerides, while LDLs transport cholesterol from the liver to cells that need it. One role of HDLs is to transport excess cholesterol from various tissues to the liver.

Selected Topic E

Selected Review Questions

1. Paracrine factors move from one cell to another within a single tissue, whereas hormones are transported from one tissue to another. 8. progestins and estrogens 10. the $-C{\equiv}CH$ group 11. arachidonic acid

Chapter 21

Selected Review Questions

21.2 **a.** asparagine **b.** glycine **c.** glycine **d.** arginine

21.4 **(a)** $H_2N{-}CH_2{-}COO^-$ **(b)** $H_3N^+{-}CH_2{-}COOH$

21.7 **a.** $H_3\overset{+}{N}{-}\underset{CH(CH_3)_2}{CH}{-}\overset{O}{\overset{\|}{C}}{-}NH{-}\underset{CH(OH)CH_3}{CH}{-}COO^-$

b. $H_3\overset{+}{N}{-}\underset{CH_2C_6H_4OH}{CH}{-}\overset{O}{\overset{\|}{C}}{-}NH{-}\underset{CH_2CH_2C(=O)NH_2}{CH}{-}COO^-$

21.13 hydrogen bonding

21.16 $H_3\overset{+}{N}{-}\underset{COO^-}{CH}{-}CH_2{-}S{-}S{-}CH_2\underset{COO^-}{CH}{-}\overset{+}{N}H_3$

21.22 **(a)** secondary, tertiary, and quaternary **(b)** most often it is irreversible

Odd-Numbered Problems

1. **a.** $-CHOHCH_3$ **b.** $-(CH_2)_4NH_3^+$ **c.** $-CH_2-C_6H_4-OH$

3. **a.** $\overset{+}{N}H_3{-}CH_2{-}COO^-$ **b.** $\overset{+}{N}H_3{-}\underset{CH_3}{CH}{-}COO^-$ **c.** $\overset{+}{N}H_3{-}\underset{CH(CH_3)_2}{CH}{-}COO^-$

5. **a.** lysine **b.** tryptophan, proline, or histidine **c.** phenylalanine **d.** aspartic acid or glutamic acid

7. $^+NH_3{-}CH(CH_3){-}COO^-$ **a.** +1 **b.** 0 **c.** −1

9. threonine and isoleucine

11. $H_3\overset{+}{N}{-}\underset{CH_2C_6H_5}{CH}{-}\overset{O}{\overset{\|}{C}}{-}NH{-}CH_2{-}\overset{O}{\overset{\|}{C}}{-}NH{-}\underset{CH_3}{CH}{-}C(=O)O^-$

13. ala-ser-cys-phe 15. **a.** It acts as a vasoconstrictor. b. Less angiotensin II leads to a lowered constriction of blood vessels and lowers blood pressure. 17. Wool has an arrangement of polypeptide chains where the polypeptide chain coils in a helical manner. Hydrogen bonds form between adjacent groups in the helix. Wool is elastic because these hydrogen bonds can be stretched and compressed without disrupting the structure. The predominant secondary structure is known as the α-helix.

19. **a.** ionic bond **b.** dispersion forces **c.** hydrogen bond **d.** disulfide linkage 21. **a.** globular **b.** fibrous **c.** globular 23. **a.** low pH **b.** high pH **c.** isoelectric pH 25. globular

Chapter 22

Selected Review Questions

22.1 a. lactose **b.** cellulose **c.** peptides (and proteins) **d.** lipids **22.2** Hydrolases require water to carry out the hydrolysis reaction; lyases remove a group without water participating as a reactant. **22.8** Urease is more specific because it catalyzes the hydrolysis of one compound (urea), while carboxypeptidase will catalyze the removal of the C-terminal amino acid from a wide variety of peptides and proteins. **22.14 (a)** Penicillin inhibits an enzyme that catalyzes the final step in synthesis of cell walls of reproducing bacteria. **(b)** Resistant strains of bacterial synthesize penicillinase, an enzyme that breaks down penicillin by cleaving the amide linkage in the lactam ring.

Odd-Numbered Problems

1. a. lyase **b.** hydrolase **c.** transferases **3. a.** lactase **b.** hydrolase **5. a.** Ethanol is the substrate, zinc is the cofactor, and alcohol dehydrogenase without the zinc ion is the apoenzyme. **b.** No, a coenzyme is an organic molecule and Zn^{2+} is inorganic. **7.** So that trypsin will not be active in the pancreas (where it's synthesized) and, thus, degrade important proteins in that tissue. **9.** The substrate binds to the active site, where the reaction occurs that converts the substrate to a product that is then released. The enzyme is unchanged and can bind another substrate molecule and catalyze its conversion to product. This cycle is continuously repeated. **11. a.** asp or glu **b.** asp or glu **c.** lys **d.** lys **e.** ser, thr, or tyr **f.** cys **g.** val **h.** none **i.** phe **13.** It tells you how rapidly an enzyme catalyzes a particular reaction. **15.** At high temperatures the enzyme is denatured and cannot properly bind the substrate and catalyze its conversion to product. At low temperatures chemical reactions occur much more slowly because of a decrease in the kinetic energy of the molecules. Slow-moving molecules collide less frequently and therefore the probability of product formation decreases. **17.** The enzyme will become less active due to the change in ionization of a key group in the active site. **19.** If the inhibitor is a competitive inhibitor, addition of increasing amounts of substrate will reverse the inhibition. This is not the case for noncompetitive inhibition. **21.** Penicillin-like compounds can be synthesized which resist cleavage by penicillinase or a penicillinase inhibitor (such as clauvulinic acid) can be combined with the penicillin to prevent penicillin degradation.

Selected Topic F

Selected Review Questions

2. a. fat-soluble **b.** water-soluble **c.** fat-soluble **d.** fat-soluble **e.** water-soluble **f.** water-soluble **5.** Water-soluble vitamins have a high proportion of polar groups attached which make them soluble while fat-soluble vitamins are comprised primarily of carbons and hydrogens. **9.** Ascorbic acid (vitamin C); ergocalciferol (vitamin D_2); thiamine (vitamin B_1); retinol (vitamin A); tocopherol (vitamin E) **10. a.** vitamin A **b.** vitamin B_1 **c.** vitamin D **13. a.** riboflavin (vitamin B_2), niacin (vitamin B_3) **b.** biotin **c.** thiamine (vitamin B_1) **d.** methylcobalamin **e.** folic acid **15.** Despite its high molar mass, vitamin B_{12} has a high proportion of electronegative nitrogens and oxygens (primarily as amide groups) that can hydrogen bond with water, making the vitamin water-soluble.

Chapter 23

Practice Exercises

23.1 3′ d-GGTCACTTAACGGATA 5′ **23.2** 5′ AUAGGCAAUUCACUGG 3′ **23.3** met-leu-asn-cys-val-gly

Selected Review Questions

23.2 (a) 2-deoxyribose, phosphoric acid, adenine, guanine, thymine, cytosine **(b)** ribose, phosphoric acid, adenine, guanine, uracil, cytosine **23.9** Each is paired with a daughter strand. **23.10** DNA and RNA **23.12** mRNA carries to the ribosome the instructions for the sequence of amino acids in a protein. tRNA molecules transport amino acids to the ribosome in an order dictated by mRNA, where they are incorporated into a growing protein chain. **23.13** mRNA; tRNA **23.15 (a)** 5-Bromouracil can be incorporated into a DNA strand in place of thymine, but it base-pairs with guanine rather than adenine, as thymine would do. **(b)** chemical mutagen

Odd-Numbered Problems

1. ribose; 2-deoxyribose **3. a.** neither **b.** nucleoside **c.** nucleotide **5. a.**ribose **b.** ribose **c.** deoxyribose

7. a.

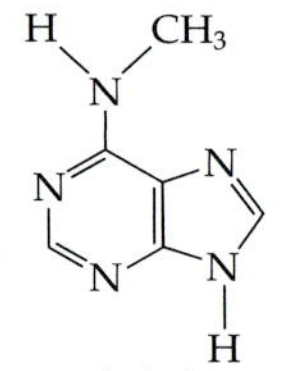

6-methyladenine

b.

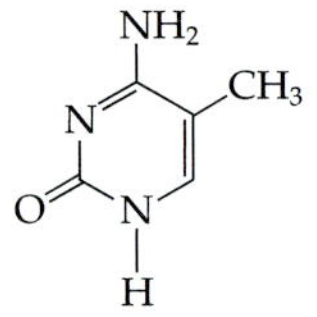

5-methylcytosine

c.

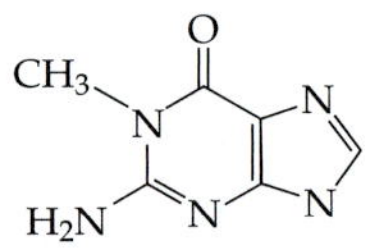

d.

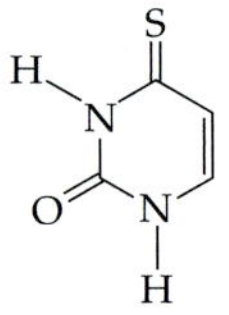

9. Phosphoric acid is a structural component of the backbone of nucleic acid molecules. At physiological pH, the hydrogens of the acid are dissociated, and the phosphate group is negatively charged.

11.

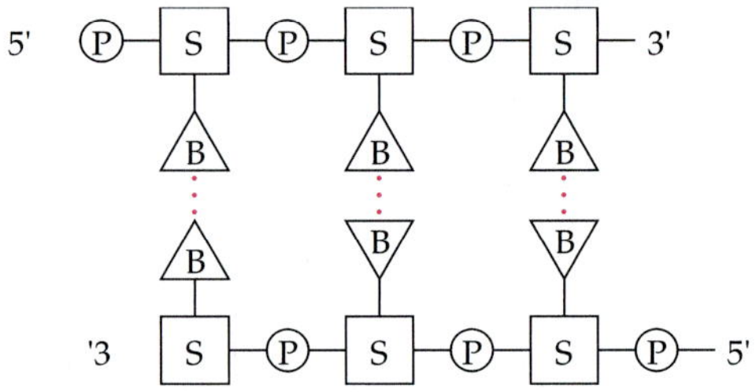

13.

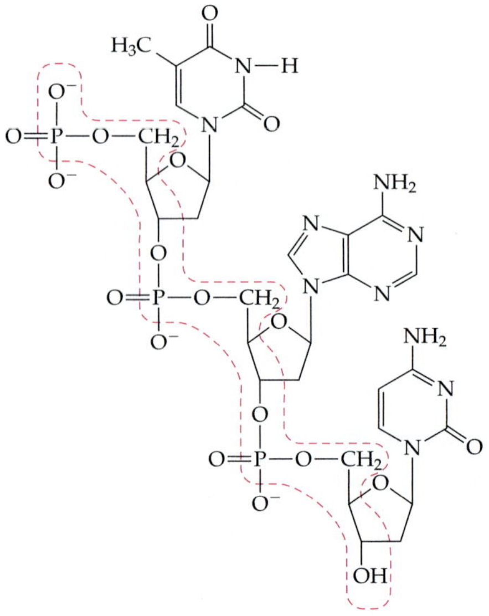

15. a. guanine **b.** uracil **c.** cytosine **d.** adenine **17.** 12 **19.** No; RNA is mostly single-stranded and does not have the extent of base-pairing observed in DNA. **21. a.** Both replication and transcription utilize one DNA strand as a template and synthesize the new strand in a 5′ to 3′ direction, using specific base-pairing to determine the sequence of nucleotides. **b.** In replication the entire DNA template strand is copied and the two strands remain bound together, while in transcription only a segment of the strand is transcribed and the RNA strand dissociates from the template. In replication deoxyribonucleotides are used, while ribonucleotides are used in transcription. **23. a.** 3′ d-TACCGTTAGGAGTTTGCGACA 5′ **b.** 5′ AUGGCAAUCCUCAAACGCUGU 3′ **25. a.** 3′ AAC 5′ **b.** 3′ CUU 5′ **c.** 3′ AGG 5′ **d.** 3′ GUG 5′ **27.** 27 **29. a.** phe **b.** his **c.** ser **d.** pro **31.** met-ser-asp-phe-ala-gly-leu **33. a.** substitution **b.** the fourth amino acid would be leu, rather than phe **35.** No; the mutation must occur in the DNA of a sperm or egg cell for it to be passed on to any children.

Selected Topic G

Selected Review Questions

4. In the body acyclovir is converted to acyclovir triphosphate which is a competitive inhibitor of the viral DNA polymerase. When acyclovir triphosphate is incorporated into the viral DNA that is being formed, it blocks further elongation of the DNA chain because it lacks the 3′-hydroxyl group. **6. a.** reverse transcriptase inhibitors; HIV protease inhibitors **b.** Reverse transcriptase inhibitors bind to reverse transcriptase and block the formation of the DNA copy of the viral RNA. HIV protease inhibitors bind to the HIV protease and prevent the enzyme from cutting the newly synthesized viral proteins into shorter, mature proteins. **13.** Cancer is caused by mutations in specific genes. If a DNA repair enzyme cannot function properly, this increases the number of mutations in the DNA and increases the likelihood of a mutation occurring in key genes that lead to the development of cancer.

Chapter 24

Selected Review Questions

24.3 The energy of hydrolysis of ATP is used by many enzymes and other biomolecules to drive unfavorable reactions. **24.6 a.** glucose, fructose, and galactose **b.** fatty acids, monoglycerides, glycerol **c.** amino acids **24.12** as two molecules of CO_2 **24.15 (a)** 2.5 ATP **(b)** 1.5 ATP **24.16** Actomyosin is the structural protein of muscle. It catalyzes the hydrolysis of ATP and couples this energy to drive muscle contraction.

Odd-Numbered Problems

1. AMP contains a single phosphate group attached to the ribose sugar, while ADP has two linked phosphate groups attached to the ribose, and ATP has three. **3.** a,c,d **5.** Mucin is a glycoprotein which attaches onto food particles. This lubricates them for easier swallowing. **7.** Bile salts act like soap molecules by breaking down large, water-insoluble lipids into smaller micelles, with the bile salts surrounding the lipids. The lipases can act on these smaller micelles more efficiently than the larger lipid particles. **9. a.** pepsin: gly-ala-phe and tyr; chymotrypsin: gly-ala-phe and tyr; trypsin: gly-ala-phe-tyr **b.** pepsin: ala-ile-tyr and ser; chymotrypsin: ala-ile-tyr and ser; trypsin: ala-ile-tyr-ser **c.** pepsin: val-phe and arg-leu; chymotrypsin: val-phe and arg-leu; trypsin: val-phe-arg and leu **d.** pepsin: leu and thr-glu-lys; chymotrypsin: leu-thr-glu-lys; trypsin: leu-thr-glu-lys **11. a.** citrate **b.** oxaloacetate and acetyl-CoA **c.** malate **d.** isocitrate dehydrogenase **13. a.** 11a **b.** 11c **c.** 11a

15. *Oxidation half-reactions:*

Step 3: $^{-}OOCCH_2CH(COO^-)CHOHCOO^- + H^+ \longrightarrow$

$$^{-}OOCCH_2CH_2\overset{O}{\overset{\|}{C}}COO^- + CO_2 + 2\,H^+ + 2\,e^-$$

Step 4: $^{-}OOCCH_2CH_2\overset{O}{\overset{\|}{C}}COO^- + H^+ + CoASH \longrightarrow$

$$^{-}OOCCH_2CH_2\overset{O}{\overset{\|}{C}}SCoA + CO_2 + 2\,H^+ + 2\,e^-$$

Step 6: $^{-}OOCCH_2CH_2COO^- \longrightarrow$

$$^{-}OOCCH{=}CHCOO^- + 2\,H^+ + 2\,e^-$$

Reduction half-reactions:

Steps 3&4: $NAD + 2\,H^+ + 2\,e^- \longrightarrow NADH + H^+$

Step 6: $FAD + 2\,H^+ + 2\,e^- \longrightarrow FADH_2$

17. O_2; H_2O **19. a.** The reduced coenzymes NADH and $FADH_2$ are oxidized by the electron transport chain only if ADP is simultaneously phosphorylated to ATP. **b.** The transport of H^+ out of the mitochondrial matrix forms a difference in H^+ concentration across the inner mitochondrial matrix that provides the energy for ATP synthesis. **21.** Type I: Slow twitch, suited to aerobic oxidation; Type IIB: Fast twitch, suited to anaerobic glycolysis **23.** The high catalytic activity of actomyosin in Type IIB fibers suggests that the tissue can hydrolyze ATP at high rates and, thus, is important for bursts of vigorous physical activity.

Chapter 25

Selected Review Questions

25.1 (a) Step 6 in which glyceraldehyde-3-phosphate is oxidized (and phosphorylated) to form 1,3-bisphosphoglycerate. **(b)** NAD^+ **25.4** In the formation of lactate, NADH is reoxidized to NAD^+. This solves the problem of how to reoxidize NADH (needed in step 6) in the absence of oxygen. **25.5** two **25.11** Insulin can bind to liver, muscle, and adipose cells and activate them to take up glucose from the blood, lowering blood-sugar levels.

Odd-Numbered Problems

1. a. glyceraldehyde-3-phosphate and dihydroxyacetone phosphate **b.** phosphenolpyruvate **c.** triose phosphate isomerase **d.** glucose-6-phosphate **3. a.** 1b **b.** 2c **c.** 1c and 2b **d.** 2c **5.** NAD^+ **7.** Lactate, pyruvate, acetaldehyde, and ethanol are the only nonphosphorylated metabolites of glycolysis and fermentation. All the other metabolites are phosphorylated. **9. a.** 4 **b.** 2 **c.** 2 **11. a.** NADH is reoxidized in muscle cells in the reaction in which pyruvate is reduced to lactate. **b.** In yeast cells, NADH is reoxidized in the reaction in which acetaldehyde is reduced to ethanol. **13. a.** 32 **b.** (i) 2; (ii) 2; (iii) 28 **c.** 35% **15.** pyruvate, lactate, oxaloacetate, and some amino acids

17. The normal blood sugar level may range from 70–100 mg/dL, while the renal threshold value is higher—150 to 170 mg/dL. **19.** UDP-glucose. **21.** Only 2 ATP/glucose are obtained when glucose is the starting material; however, 3 ATP/glucose are obtained from the degradation of glycogen because each glucose obtained from glycogen degradation is already phosphorylated (glucose-1-phosphate which is readily converted to glucose-6-phosphate), thus eliminating the need to hydrolyze an ATP to form glucose-6-phosphate from glucose. **23.** Glucagon binds to a specific receptor, which leads to the formation of cAMP. Release of cAMP leads to the activation of phosphorylase (and thus, glycogen degradation) and the inhibition of glycogen synthase (and thus, glycogen synthesis), which leads to an increase in the amount of glucose. **25.** Binding of the hormone to its receptor activates adenylate cyclase, which is bound to the inner cell membrane. This enzyme catalyzes the conversion of ATP to cAMP within the cell. **27. a.** glycolysis **b.** glycogen synthesis (glycogenesis) **c.** gluconeogenesis **d.** glycogen degradation (glycogenolysis) **e.** glycolysis

Chapter 26

Practice Exercises

26.1 3.6 lb **26.2** 47.5% **26.3** 10%

Selected Review Questions

26.1 All of the essential fatty acids contain two or more double bonds. **26.3** The carbon atom beta to the carboxyl group of the fatty acid undergoes successive oxidations. **26.4 (a)** 8 **(b)** 7 **(c)** 7 **26.5 (a)** 120 **(b)** 376.5–378.5 **26.8** acetyl-CoA

Odd-Numbered Problems

1. The oxidation of 1.0 g of carbohydrate liberates about 4.2 kcal. The oxidation of 1.0 g of lipid liberates about 9.5 kcal. **3. a.** glycogen in the liver **b.** fat reserves in adipose tissue

5. $CH_2{-}O{-}C(=O)(CH_2)_{14}CH_3$ / $CH{-}O{-}C(=O)(CH_2)_{14}CH_3$ / $CH_2{-}O{-}C(=O)(CH_2)_{14}CH_3$ $\xrightarrow{\text{lipase}}$ $CH_2{-}OH$ / $CH{-}O{-}C(=O)(CH_2)_{14}CH_3$ / $CH_2{-}O{-}C(=O)(CH_2)_{14}CH_3$ + $CH_3(CH_2)_{14}COOH$ $\xrightarrow{\text{lipase}}$

$CH_2{-}OH$ / $CH{-}O{-}C(=O)(CH_2)_{14}CH_3$ / $CH_2{-}OH$ + $CH_3(CH_2)_{14}COOH$ $\xrightarrow{\text{lipase}}$ $CH_3(CH_2)_{14}COOH$ + $CH_2{-}OH$ / $CH{-}OH$ / $CH_2{-}OH$

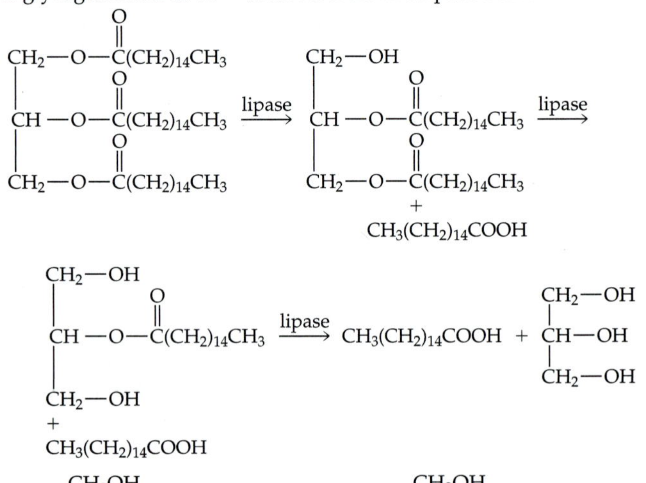

7. $HO{-}C(H)(CH_2OH)_2$ (glycerol) $\xrightarrow[\text{ATP} \to \text{ADP}]{\text{glycerol kinase}}$ $HO{-}C(H)(CH_2OH)(CH_2OPO_3^{2-})$ (L-glycerol 3-phosphate)

$\xrightarrow[\text{NAD}^+ \to \text{NADH} + \text{H}^+]{\text{glycerol phosphate dehydrogenase}}$ $CH_2OH{-}C(=O){-}CH_2OPO_3^{2-}$ (dihydroxyacetone phosphate)

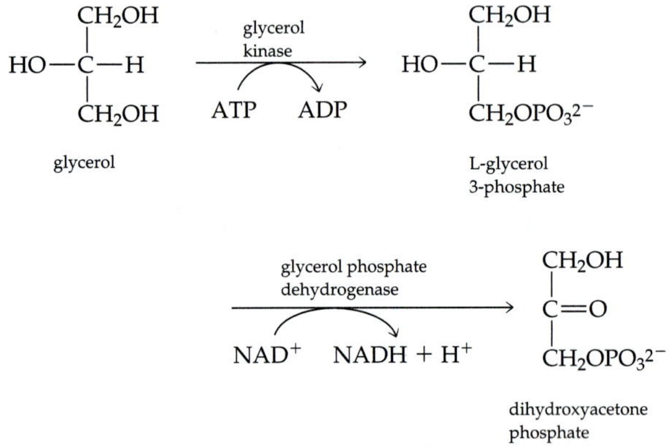

9. oxidation-reduction; hydration; lysis (cleavage) **11. a.** 6 **b.** 12 **13. a.** 9 **b.** 8 **c.** 8 **15. a.** 92 **b.** 294.5 **c.** 324.5 **17.** In starvation cells rely on fatty acid oxidation to supply energy needs. This results in higher amounts of acetyl-CoA that cannot be oxidized by the Krebs cycle due to a decrease in the concentration of oxaloacetate, needed for glucose synthesis. The acetyl-CoA is then used to synthesize ketone bodies, two of which are weak acids. **19. a.** kidney and heart **b.** brain **21. a.** Fatty acid oxidation occurs in the mitochondria, while fatty acid synthesis occurs in the cytoplasm. **b.** Fatty acid oxidation utilizes NAD^+ and FAD, while fatty acid synthesis utilizes $NADP^+$. **c.** In fatty acid oxidation, carbons are removed in units of two, while in fatty acid synthesis they are added in units of two. **23.** 14 **25.** A single protein has several active sites, each of which catalyzes one of the reactions in fatty acid synthesis. **27. a.** 22, normal **b.** 18, underweight **c.** 28, overweight **29.** 42% **31.** 2 hours **33.** These diets may be deficient in necessary nutrients, particularly in B vitamins and iron.

Chapter 27

Selected Review Questions

27.3 $^-OOC{-}CH(\overset{+}{N}H_3){-}CH_3$ (Alanine) + $^-OOC{-}C(=O){-}CH_2{-}COO^-$ (Oxaloacetate) $\xrightleftharpoons{\text{transaminase}}$ $^-OOC{-}C(=O){-}CH_3$ (Pyruvate) + $^-OOC{-}CH(\overset{+}{N}H_3){-}CH_2{-}COO^-$ (Aspartate)

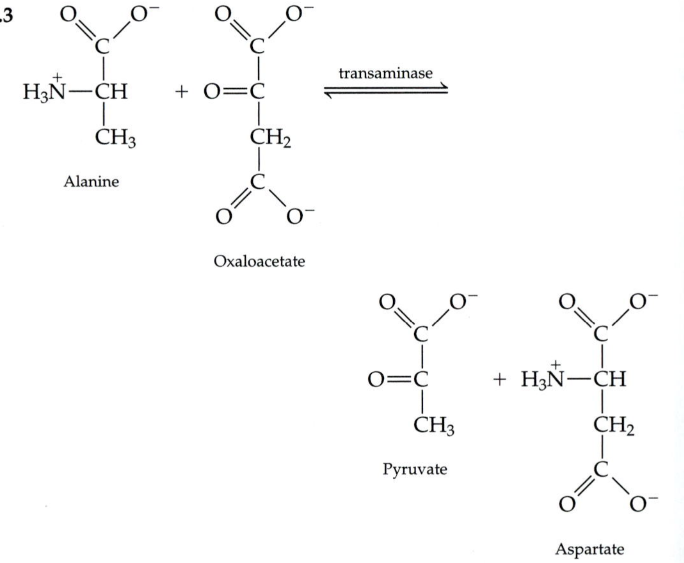

27.6 uric acid **27.9** Starvation is a condition in which the body is totally deprived of food, while kwashiorkor is a disease that is the result of insufficient protein in the diet.

Odd-Numbered Problems

1. a. Essential amino acids contain carbon chains or aromatic rings that are not intermediates in carbohydrate or lipid metabolism. **b.** Essential proteins cannot be synthesized. **3. a.** There is an inadequate intake of protein, while the body is degrading protein to provide precursors for glucose synthesis. **b.** Large amounts of protein are being synthesized during pregnancy, so very little is being excreted.

5. $^-OOC{-}CH(^+NH_3){-}CH_2{-}C_6H_5$ + $^-OOC{-}C(=O){-}CH_3$ $\xrightarrow{\text{transaminase}}$ $^-OOC{-}C(=O){-}CH_2{-}C_6H_5$ + $^-OOC{-}CH(^+NH_3){-}CH_3$

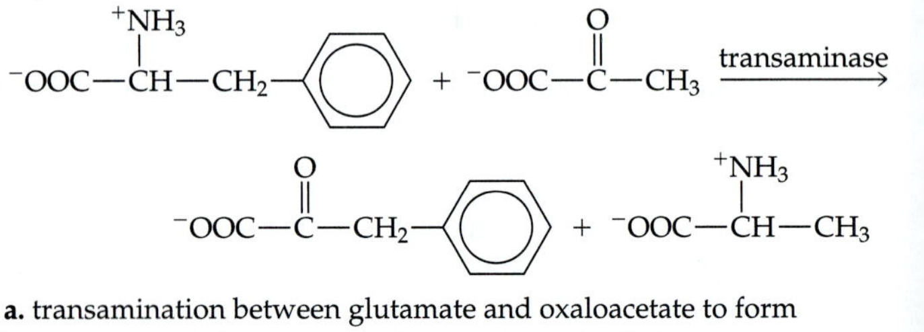

7. a. transamination between glutamate and oxaloacetate to form α-ketoglutarate and aspartate **b.** transamination between

glutamate and pyruvate to form α-ketoglutarate and alanine
9. a. both **b.** ketogenic **c.** glucogenic **11. a.** glutamine

b.

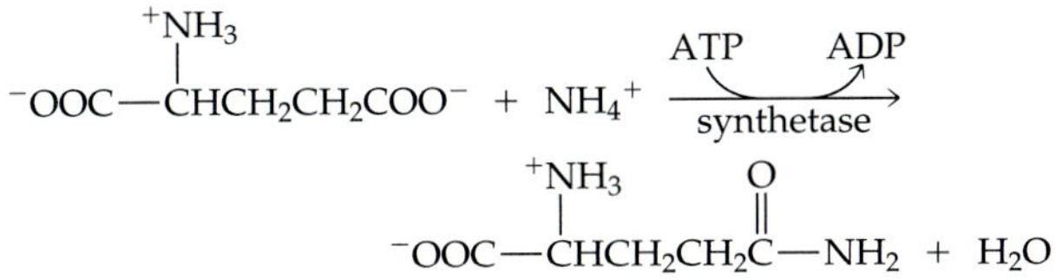

$$^{-}OOC-\underset{^{+}NH_3}{CHCH_2CH_2COO^-} + NH_4^+ \xrightarrow[\text{synthetase}]{\text{ATP} \quad \text{ADP}} {}^{-}OOC-\underset{^{+}NH_3}{CHCH_2CH_2}\overset{O}{\overset{\|}{C}}-NH_2 + H_2O$$

13. nitrogens from ammonium ion and aspartate; carbon from bicarbonate ion **15. a.** methionine **b.** phenylalanine

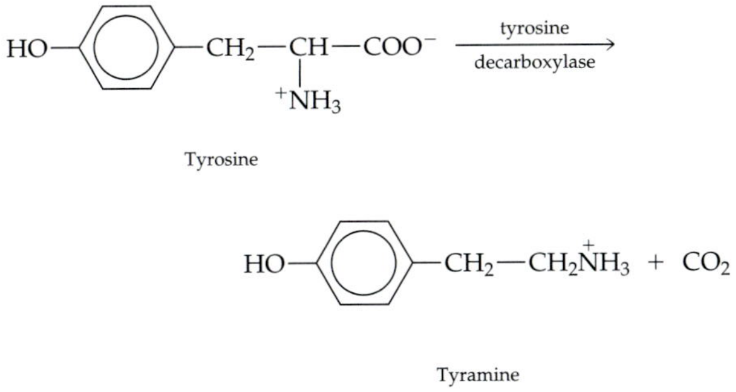

17. a. Histidine (imidazole ring)$-CH_2\underset{^{+}NH_3}{CH}COO^-$ $\xrightarrow{\text{histidine decarboxylase}}$ Histamine (imidazole ring)$-CH_2CH_2(^{+}NH_3) + CO_2$

Histidine

Histamine

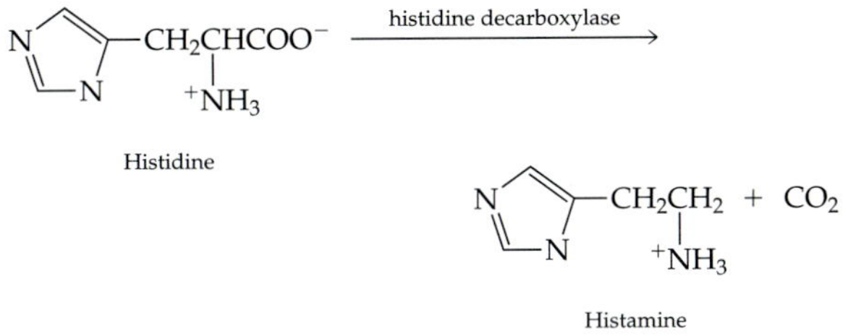

b. $HO-C_6H_4-CH_2-\underset{^{+}NH_3}{CH}-COO^-$ $\xrightarrow[\text{decarboxylase}]{\text{tyrosine}}$ $HO-C_6H_4-CH_2-CH_2\overset{+}{N}H_3 + CO_2$

Tyrosine

Tyramine

c. $(HO)_2C_6H_3-CH_2-\underset{^{+}NH_3}{CH}-COO^-$ $\xrightarrow[\text{decarboxylase}]{\text{dopa}}$ $(HO)_2C_6H_3-CH_2CH_2\overset{+}{N}H_3 + CO_2$

3,4-Dihydroxyphenylalanine

Dopamine

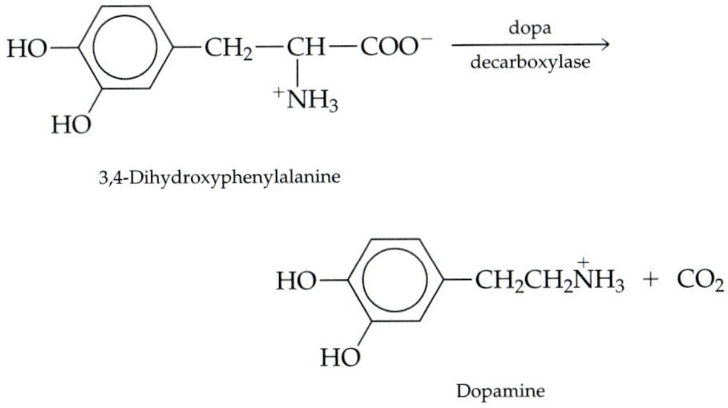

19. a. insulin **b.** glucagon **c.** insulin **21.** The brain requires less glucose, because it is using ketone bodies for some of its energy needs.

Chapter 28

Selected Review Questions

28.1 To help maintain osmotic relationships between the tissues; to transport oxygen and nutritive materials to the cells and waste products to the excretory organs; to regulate body temperature; to control the pH of the body; and to protect the organism against infection. **28.4** Arterial blood contains dissolved nutrients, oxygen, hormones, and vitamins. Venous blood contains metabolic waste products and has a higher carbon dioxide content and a lower oxygen content. **28.7 a.** bicarbonate buffer **b.** phosphate buffer **c.** hemoglobin **d.** bicarbonate buffer **28.12** formed elements

Odd-Numbered Problems

1. Na^+, K^+, Cl^-, HPO_4^{2-}, HCO_3^- **3.** Most are synthesized in the liver. **5. a.** leukocyte **b.** erythrocyte **c.** platelet **7.** Hemoglobin is composed of four polypeptide chains, each bound to a molecule of heme. **9. a.** 2+ **b.** 2+ **c.** 3+ **d.** 2+ **11.** bicarbonate buffer; phosphate buffer; plasma proteins; hemoglobin

13. $H_2PO_4^- \rightarrow HPO_4^{2-}$ with OH^- (from added base) → H_2O; $HPO_4^{2-} \rightarrow H_2PO_4^-$ with H^+ (from added acid)

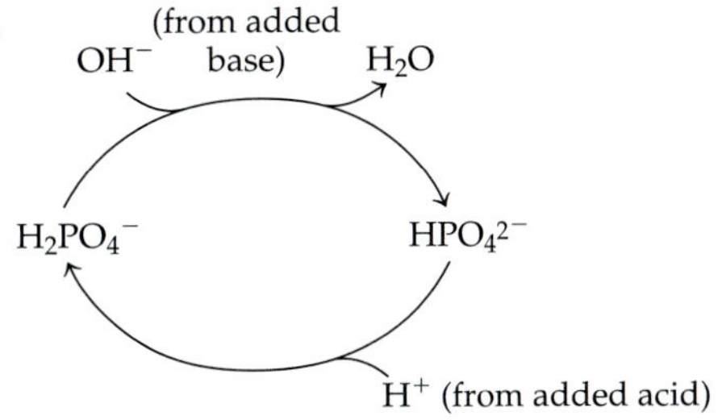

15. to prevent blood clotting when it is not needed **17.** Several blood-clotting factors require vitamin K for the reaction in which glutamate side chains are carboxylated to form γ-carboxyglutamate. **19.** An antibody is composed of four polypeptide chains: two heavy chains and two light chains that are linked by disulfide bonds. **21.** a substance that triggers the synthesis of antibodies **23.** The systolic pressure is the maximum pressure achieved during contraction of the heart ventricles, while diastolic pressure is the lowest pressure in the arteries before the next ventricular contraction. **25.** Diuretics reduce the blood volume, and thus the pressure exerted against the walls of the arteries. **27.** In the kidneys, the glomerulus filters most components (except proteins and formed elements) into the kidney tubules. As this fluid passes through the tubules there is a selective re-uptake into the blood of important constituents that the body wishes to save. Waste products not desirable to the body are not reabsorbed and are excreted as urine. **29.** liquid uptake; amount of perspiration; presence of fever; presence of diarrhea **31.** inorganic: Na^+, Cl^-, Ca^{2+}; organic: urea, amino acids, lipids **33.** It acts as an anti-bacterial agent. **35. a.** Reindeer milk contains much higher amounts of fat and protein. This is important because of the colder climate in which the reindeer live. **b.** Porpoise milk contains much higher amounts of fat and protein. These provide more energy and insulation from colder temperatures.

Credits

Chapter 1

Page 1: Blair Seitz/Photo Researchers, Inc. Page 3 (L): Giovanni Stradano, *The Alchemist.* Studiolo, Palazzo Vecchio, Florence, Italy. Scala/Art Resource, N.Y. Page 3 (R): National Library of Medicine. Page 4 (a): Chad Ehlers/Tony Stone Images. Page 4 (b): Charles Krebs/Tony Stone Images. Figure 1.1: NASA/Photo Researchers, Inc. Figure 1.2: Carey Van Loon. Page 9: Richard Megna/Fundamental Photographs. Figure 1.4: Rachel Epstein/Stuart Kenter Associates. Page 13: Mark Burnett/Stock Boston. Figure 1-6: Carey Van Loon. Page 18 (a): Science VU/Visuals Unlimited. Page 18 (b): Jan Robert Factor/Science Source/Photo Researchers, Inc. Page 18 (c): David M. Phillips/Visuals Unlimited. Page 18 (d): Rachel Epstein/Stuart Kenter Associates. Page 18 (e): The Image Works. Page 18 (f): European Space Agency/Science Photo Library/Photo Researchers, Inc. Page 23 (T): Leonard Lessin/Peter Arnold, Inc. Page 23 (B): Bob Daemmrich/Stock Boston. Page 26: Richard Megna/Fundamental Photographs. Page 30: Joyce Photographics/Photo Researchers, Inc. Page 37 (a): Norm Perdue/NBA Entertainment, Inc. Page 37 (b): William R. Sallaz/Duomo Photography Incorporated. Page 37 (c): Scott Quintard/UCLA.

Chapter 2

Page 38: Philippe Plaily/Science Photo Library. Page 39: Corbis. Figure 2.1: Carey Van Loon. Page 42: Corbis. Page 43: Bob Daemmrich/Stock Boston. Page 48: Ed Pritchard/Tony Stone Images. Figure 2.7: Richard Megna/Fundamental Photographs. Page 55: Nocosti/Science Photo Library/Photo Researchers, Inc. Page 55: Stamp from the private collection of Professor C. M. Lang, photography by Gary J. Shulfer, University of Wisconsin, Stevens Point. "1957, Russia (Scott #1906) and 1969, Russia (Scott #3607)"; Scott Standard Postage Stamp Catalogue, Scott Pub. Co., Sidney, Ohio. Page 57: Richard Megna/Fundamental Photographs.

Chapter 3

Page 67: Dr. E. R. Degginger/Bruce Coleman Inc. Page 70: University of California, Berkeley. Page 71: Richard Megna/Fundamental Photographs. Figure 3.1: Carey Van Loon. Figure 3.2: Albert Copley/Visuals Unlimited. Page 73: Richard Megna/Fundamental Photographs. Page 78: Richard Megna/Fundamental Photographs.

Chapter 4

Page 97: Richard Megna/Fundamental Photographs. Page 102: Stamp from the private collection of Professor C. M. Lang, photography by Gary J. Shulfer, University of Wisconsin, Stevens Point. "Italy #714 (1956)"; Scott Standard Postage Stamp Catalogue, Scott Pub. Co., Sidney, Ohio. Figure 4.5 (L): Tom Pantages. Figure 4.5 (R): Cosmo Condina/Toney Stone Images. Figure 4.6: Richard Megna/Fundamental Photographs. Page 112 (a): Carey Van Loon. Page 112 (b): Carey Van Loon/Ralph H. Petrucci. Page 112 (c): Carey Van Loon. Page 113: Barry L. Runk/Grant Heilman Photography, Inc. Figure 4.9: Peter Pearson/Tony Stone Images. Page 117: James King-Holmes/Imperial Cancer Research Fund/Science Photo Library/Photo Researchers, Inc. Figure 4.14: Novosti/Science Photo Library/Photo Researchers, Inc. Figure 4.16: Peter Pearson/Tony Stone Images. Figure 4.17: Michal Heron/Pearson Education/PH College. Page 122: Stamp from the private collection of Professor C. M. Lang, photography by Gary J. Shulfer, University of Wisconsin, Stevens Point. Scott Standard Postage Stamp Catalogue, Scott Pub. Co., Sidney, Ohio.

Chapter 5

Page 128: Ray Ellis/Photo Researchers, Inc. Page 129 (a): Romilly Lockyer/The Image Bank. Page 129 (b): John Kaprielian/Science Source/Photo Researchers, Inc. Page 129 (c): Chris Jones/The Stock Market. Page 129 (R): Richard Megna/Fundamental Photographs. Figure 5.2 (L): Dan and Coco McCoy/Rainbow. Figure 5.2 (C): David Madison/Bruce Coleman, Inc. Figure 5.2 (R): Jose L. Pelaez/The Stock Market. Page 131: Richard Megna/Fundamental Photographs. Page 133: Royal Observatory, Edinburgh, Scotland/Anglo-Australian Telescope Board/Science Photo Library/Photo Researchers, Inc. Figure 5.4 (L): SuperStock, Inc. Figure 5.4 (R): Alan Pitcairn/Grant Heilman Photography, Inc. Figure 5.5: Donald Clegg/Pearson Education/PH College. Figure 5.7: Richard Megna/Fundamental Photographs. Figure 5.8: Richard Megna/Fundamental Photographs. Page 142: Richard Megna/Fundamental Photographs. Figure 5.9: Carey Van Loon. Page 144: Carey Van Loon. Figure 5.10: Stuart Kenter Associates. Figure 5.11: Gary Holscher/Tony Stone Images. Figure 5.12 (L): Craig Turtle/The Stock Market. Figure 5.12 (R): Joseph Nettis/Photo Researchers, Inc.

Chapter 6

Page 149:Rachel Epstein/Stuart Kenter Associates. Page 153: Mary Evans Picture Library/Photo Researchers, Inc. Figure 6.7 (a): SuperStock, Inc. Figure 6.7 (b): Science Photo Library/Custom Medical Stock Photo, Inc. Figure 6.9: Richard Megna/Fundamental Photographs. Figure 6.10: Carey Van Loon. Page 166: Norbert Wu/Peter Arnold, Inc. Page 167: Kristen Brochmann/Fundamental Photographs.

Chapter 7

Page 174: Ralph A. Clevenger/Corbis. Page 182 (T): Kristen Brochmann/Fundamental Photographs. Page 182 (B): Herman Eisenbeiss/Photo Researchers, Inc. Figure 7.11: Carey Van Loon. Figure 7.12: Richard Megna/Fundamental Photographs. Figure 7.14: Carey Van Loon. Page 192: Kim Heacox Photography/DRK Photo. Figure 7.15: Randy Brandon/Peter Arnold, Inc. Figure 7.16: Jim Zuckerman/Corbis.

Chapter 8

Page 197: David Hall/The Image Works. Page 198: Rachel Epstein/Stuart Kenter Associates. Page 202: Richard Megna/Fundamental Photographs. Figure 8.6: Richard Megna/Fundamental Photographs. Page 205: Rachel Epstein/Stuart Kenter Associates. Figure 8.9: Richard Megna/Fundamental Photographs. Page 211: Richard Hutchings/PhotoEdit. Figure 8.12: David Phillips/Science Source/Photo Researchers, Inc. Page 214: Pearson Education/PH College. Figure 8.14: Carey Van Loon. Figure 8.15 (a): Stephen Frisch/Stock Boston. Figure 8.15 (b): John Lund/Tony Stone Images. Page 219: Jeff Greenberg/Visuals Unlimited.

Chapter 9

Page 223: John Curtis/The Stock Market. Figure 9.1: Richard Megna/Fundamental Photographs. Page 225: Richard Megna/Fundamental Photographs. Page 227: Richard Megna/Fundamental Photographs. Page 228: Carey Van Loon. Page 235: Carey Van Loon. Page 236: Michael P. Gadomski/Photo Researchers, Inc. Page 237: Lionel Delevingue/Phototake NYC. Figure 9.5: Ray Pfortner/Peter Arnold, Inc. Page 238: Robert Mathena/Fundamental Photographs. Page 239: Carey Van Loon.

Chapter 10

Page 243: L. S. Stepanowicz/Bruce Coleman Inc. Page 246: Rachel Epstein/Stuart Kenter Associates. Figure 10.1: Carey Van Loon. Figure 10.3: Yoav Levy/Phototake NYC. Figure 10.4: Richard Megna/Fundamental Photographs. Page 255: Richard Megna/Fundamental Photographs. Figure 10.5: Carey Van Loon. Page 258: Richard Megna/Fundamental Photographs. Page 259: Richard Megna/Fundamental Photographs. Figure 10.7: Donald Clegg/Pearson Education/PH College. Page 262: Carey Van Loon. Page 265: Dan McCoy/Rainbow.

Chapter 11

Page 271: Jim Olive/Peter Arnold, Inc. Page 273: Photofest. Page 274: Richard Megna/Fundamental Photographs. Figure 11.5: Carey Van Loon. Figure 11.11: Carey Van Loon. Page 285: Porter Gifford/Liaison Agency, Inc. Page 288: Carey Van Loon.

Selected Topic A

Page 295: Richard Megna/Fundamental Photographs. Page 296: Felicia Martinez/PhotoEdit. Page 297 (L): Rachel Epstein/Stuart Kenter Associates. Page 297 (R): General Electric Corporate Research & Development Center. Page 299: Geoff Tompkinson/Science Photo Library/Photo Researchers, Inc. Page 300: Richard Megna/Fundamental Photographs. Page 301: Marty Corando/DRK Photo. Page 302: Day Williams/Photo Researchers, Inc. Page 303: Donald Clegg and Roxy Wilson/Pearson Education/PH College.

Chapter 12

Page 305: Matt Meadows/Peter Arnold, Inc. Page 306: Science Photo Library/Photo Researchers, Inc. Page 307: Courtesy of the Othmer Library, Chemical Heritage Foundation. Figure 12.5: Yoav Levy/Phototake NYC. Figure 12.6: Wedgworth/Custom Medical Stock Photo, Inc. Figure 12.8: Gianni Tortoli/Science Source/Photo Researchers, Inc. Figure 12.9: Ron Sherman/Stock Boston. Page 317: Corbis. Page 318: Atomic Energy Commission/FPG International LLC. Page 12.12: Martin Dohrn/Science Photo Library/Photo Researchers, Inc. Figure 12.13: DuPont Merck Pharmaceutical Company. Figure 12.14: Science Source/Photo Researchers, Inc. Figure 12.15: Science Photo Library/Custom Medical Stock Photo, Inc. Figure 12.16 (c): Drs. M. J. Fulham and Giovanni Di Chiro, The Neuroimaging Section, NINDS, National Institutes of Health, Bethesda, Maryland. Figure 12.17: Alexander Tsiaras/Stock Boston. Page 324: Simon Fraser/Science Photo Library/Photo Researchers, Inc. Figure 12.18: Peticolas/Megna/Fundamental Photographs.

Chapter 13

Page 328: Ken Graham/Bruce Coleman Inc. Page 329: Charles D. Winters/Photo Researchers, Inc. Page 338: Ken Graham/Bruce Coleman Inc. Page 343: NASA/Science Photo Library/Photo Researchers, Inc. Page 344: Richard Megna/Fundamental Photographs. Page 350: PhotoDisc, Inc. Figure 13.9: Ed Degginger/Color-Pic, Inc. Page 352: Bob Masini/Phototake NYC. Page 354: Richard Megna/Fundamental Photographs. Figure 13.15 (b): Ed Degginger/Color-Pic, Inc.

Chapter 14

Page 367: Bill Banaszewski/Visuals Unlimited. Page 378: Richard Megna/Fundamental Photographs. Page 387: Corbis. Page 390: SIU/Peter Arnold, Inc.

Chapter 15

Page 394: Arthur Beck/Photo Researchers, Inc. Page 403: Tom Pantages. Page 408: Phil Degginger/Color-Pic, Inc. Page 409: Michael Krasowitz/FPG International LLC.

Chapter 16

Page 413: A. G. E. FotoStock/Corbis. Figure 16.1: Richard Megna/Fundamental Photographs. Page 416: Richard Megna/Fundamental Photographs. Page 421: Bruce Coleman Inc. Figure 16.2: Amoco Fabrics & Fibers Co. Page 432: Ed Degginger/Color-Pic, Inc.

Selected Topic B

Page 441 (L): Ed Degginger/Color-Pic, Inc. Page 441 (R): VU/Cabisco/Visuals Unlimited. Page 443: Drug Enforcement Administration. Page 444: Scott Camazine/Photo Researchers, Inc.

Chapter 17

Page 447: Alain Dex/Publiphoto/Photo Researchers, Inc. Page 455: Donald Clegg and Roxy Wilson/Pearson Education/PH College. Page 457: Rachel Epstein/Stuart Kenter Associates. Page 459: The Metropolitan Museum of Art, Catharine Lorillard Wolfe Collection, Wolfe Fund, 1931.

Chapter 18

Page 478: Richard Megna/Fundamental Photographs. Figure 18.5: Bellingham + Stanley Ltd. Page 486: Stuart J. Baum. Page 488: Science Photo Library/Photo Researchers, Inc.

Selected Topic D

Figures D.1, D.2, D.4, and D.5: © 1998 Prentice Hall from F. H. Martini, *Fundamentals of Anatomy and Physiology,* 4th edition.

Chapter 19

Page 505: Rachel Epstein/Stuart Kenter Associates. Page 506: Rachel Epstein/Stuart Kenter Associates. Figure 19.2 (L): Foodpix The All Food Stock Picture Agency. Figure 19.2 (LC): Rachel Epstein/Stuart Kenter Associates. Figure 19.2 (RC): Foodpix The All Food Stock Picture Agency. Figure 19.2 (R): Foodpix The All Food Stock Picture Agency. Page 510: Denise Van Patton. Page 511: Tony Freeman/PhotoEdit. Page 514: Saturn Stills/Science Photo Library/Photo Researchers, Inc. Page 517: Rachel Epstein/Stuart Kenter Associates. Page 518: Gregory K. Scott/Photo Researchers, Inc. Page 519: Frank LaBua/Pearson Education/PH College. Figure 19.7: Andrew Syred/Science Photo Library/Photo Researchers, Inc. Page 520: Tom Pantages. Figure 19.10: Mary Ginsburg. Figure 19.11: Biophoto Associates/Photo Researchers, Inc. Page 523: Foodpix The All Food Stock Picture Agency.

Chapter 20

Page 528: Inga Spence/Visuals Unlimited. Page 529: Rachel Epstein/Stuart Kenter Associates. Page 534: Richard Megna/Fundamental Photographs. Page 536: Tom Hubbard/Black Star. Page 539: Francois Gohier/Photo Researchers, Inc. Page 542:

Tom Pantages. Page 547: Southern Illinois University/Science Source/Photo Researchers, Inc. Figure 20.21 (a): Cabisco/Visuals Unlimited. Figure 20.21 (b): Biophoto Associates/Photo Researchers, Inc.

Selected Topic E

Table E.1 and Figure E.3: © 1998 Prentice Hall from F. H. Martini, *Fundamentals of Anatomy and Physiology,* 4th edition. Page 564: Scott Camazine/Sue Trainor/Photo Researchers, Inc.

Chapter 21

Page 568: Photo Researchers, Inc. Figure 21.5 (a): Chemical Design LTD/Science Photo Library/Photo Researchers, Inc. Figure 21.5 (b): Ken Eward/Science Source/Photo Researchers, Inc. Figure 21.10: CNRI/Science Photo Library/ Photo Researchers, Inc. Figure 21.11 (a): Tony Freeman/PhotoEdit. Figure 21.11 (b): Ken Eward/Science Source/Photo Researchers, Inc. Figure 21.12 (a): Ken Eward/BioGrafx/Science Source/Photo Researchers, Inc. Figure 21.15 (c): Scott Camazine/Photo Researchers, Inc. Page 588: Fred E. Cohen. Figure 21.15 (a): Peter Arnold, Inc. Figure 21.15 (b): Ken Eward/BioGrafx/Photo Researchers, Inc. Figure 21.19 (a): Stuart J. Baum. Figure 21.19 (b): Stuart J. Baum.

Chapter 22

Page 598: Dr. Irene Weber, Thomas Jefferson University. Figure 22.4: Thomas A. Steitz. Page 615: Dr. Jeremy Burgess/Science Photo Library/Photo Researchers, Inc.

Selected Topic F

Page 623: Rachel Epstein/Stuart Kenter Associates. Page 626: Rachel Epstein/Stuart Kenter Associates.

Chapter 23

Page 634: Phil A. Harrington/Peter Arnold, Inc. Figure 23.1: CNRI/Science Photo Library/Photo Researchers, Inc. Figure 23.6 (a): NIH/Photo Researchers, Inc. Figure 23.11: Ken Eward/Science Source/Photo Researchers, Inc. Page 660: K. G. Murti/Visuals Unlimited. Page 663: Leonard Lessin/Peter Arnold, Inc.

Chapter 24

Page 674: The Sporting Image, Inc. Figure 24.6: © 1998 Prentice Hall from F. H. Martini, *Fundamentals of Anatomy and Physiology,* 4th edition. Page 686: Science Photo Library/Photo Researchers, Inc. Page 692: Inga Spence/Visuals Unlimited. Figure 24.12 (a): Michael Webb/Visuals Unlimited. Page 697 (L): Tony Svensson/Trimarket Company. Page 697 (R): David Madison/Duomo Photography Incorporated.

Chapter 25

Page 702: Dr. N. Larry Edwards. Page 714: Tom Pantages.

Chapter 26

Page 722: Tomas D. W. Friedmann/Photo Researchers, Inc. Figure 26.1: © 1998 Prentice Hall from F. H. Martini, *Fundamentals of Anatomy and Physiology,* 4th edition. Page 726: Ken Highfill/Photo Researchers, Inc. Page 732: SportsChrome-USA. Page 738: Jackson/Visuals Unlimited.

Chapter 27

Page 743: Rachel Epstein/Stuart Kenter Associates. Page 754: Erich Schrempp/Photo Researchers, Inc. Page 758: Dagmar Fabricius/Stock Boston.

Chapter 28

Page 761: Will & Deni McIntyre/Photo Researchers, Inc. Page 763: Fred Hossler/Visuals Unlimited. Figure 28.3: © 1998 Prentice Hall from F. H. Martini, *Fundamentals of Anatomy and Physiology,* 4th edition. Figure 28.5 (a): Hulton Getty. Figure 28.5 (b): © 1998 Prentice Hall from F. H. Martini, *Fundamentals of Anatomy and Physiology,* 4th edition. Figure 28.6 (a): Dr. Gopal Murti/Science Photo Library/Custom Medical Stock Photo, Inc. Figure 28.6 (b): Science Source/Photo Researchers, Inc. Figure 28.9: CNRI/Science Photo Library/Photo Researchers, Inc. Page 781: Damien Lovegrove/Science Photo Library/Photo Researchers, Inc. Figure 28.15: © 1998 Prentice Hall from F. H. Martini, *Fundamentals of Anatomy and Physiology,* 4th edition.

Index

f = figure; n = note; t = table

B

C

D

E

F

G

H

I

M

N

O

Q

R

S

T

U

V